KALANCHOE ROTUNDIFOLIA

FLORA ZAMBESIACA

MOZAMBIQUE

MALAWI, ZAMBIA, RHODESIA

BOTSWANA

VOLUME FOUR

Edited by

E. LAUNERT

on behalf of the Editorial Board

J. P. M. BRENAN

Royal Botanic Gardens, Kew

A. FERNANDES

Coimbra

E. LAUNERT

British Museum (Natural History)

H. WILD

University of Rhodesia

Published by the Managing Committee on behalf of
the contributors to Flora Zambesiaca
1978

Printed by Robert MacLehose & Company Limited, Glasgow
Printers to the University of Glasgow

ISBN 0.85592–044.0

CONTENTS

EDITOR'S PREFACE

RECENT political changes in both Portugal and southern tropical Africa did not leave Flora Zambesiaca unaffected. The production of Volume IV was fraught with great managerial and financial difficulties, and its completion delayed by more than a year. It is the first volume which is published in its entirety, instead of, as previous ones, in two or more parts. Unfortunately family no. 67 (Crassulaceae) could not be included without delaying publication unduly. This is because of the death of the world authority on this family, Dr. Raymond-Hamet. It will now be written by Dr. Rosette Fernandes of Coimbra University, and published out of sequence in the next volume or part to appear. It is regretted that it was impossible at this stage to take account of many changes in locality names, especially with regard to Mozambique. The new names will be used in future volumes. Special recognition must be given to the editorial assistance of Dr. Frances Kupicha, B. A. Krukoff Curator of African Botany.

E.L.

LIST OF NEW NAMES AND TAXA
PUBLISHED IN THIS WORK

LIST OF FAMILIES INCLUDED IN
VOLUME IV

ANGIOSPERMAE

62. ROSACEAE

By E. J. Mendes

(*Rubus* by F. K. Kupicha)

Trees, shrubs, scramblers, brambles or herbs, rarely annuals in the F.Z. area. Leaves alternate, rarely opposite or borne in groups of 2 or 3 at each node, simple or compound; stipules usually present and paired. Inflorescence corymbose, racemose or paniculate, or flowers solitary. Flowers actinomorphic or almost so, mostly bisexual, rarely dioecious or polygamo-dioecious. Calyx-tube short or elongated, free or adnate to the gynoecium; sometimes with an "epicalyx" of bracteoles; calyx-lobes usually 5, imbricate; disk usually present, lining the mouth of the calyx-tube, usually entire-margined. Petals inserted below the margin of the disk, as many as the calyx-lobes, imbricate or convolute, usually free, often deciduous, or absent. Stamens usually numerous, usually in a complete ring at the margin of or above the disk; filaments usually free and filiform; anthers mostly small, dehiscing longitudinally. Carpels 1–∞, free or connate and then usually ± adnate to the calyx-tube; styles terminal or basal; ovules often 2, rarely 1 or several. Fruits various, inferior to superior, naked or enclosed by the persistent calyx or calyx-tube, drupaceous, pomaceous, follicular, or composed of an indefinite number of achenes or drupelets, rarely capsular. Seeds erect or pendulous; endosperm absent or very rarely scantily present; cotyledons usually fleshy.

A cosmopolitan family, most abundant in temperate regions of the northern hemisphere. *Hagenia* is confined to NE. tropical Africa, *Leucosidea* to SE. Africa, most species of *Cliffortia* and *Grielum* to S. Africa, and *Neuradopsis* to Botswana and SW. Africa.

Various species and cultivars of the genera *Cotoneaster, Crataegus, Cydonia, Eriobotrya, Fragaria, Prunus, Pyracantha, Pyrus, Rhaphiolepis, Rosa, Rubus, Sorbaria* and *Spiraea* have been cultivated either for ornamental purposes in nurseries, parks and gardens, or as fruit trees in orchards. Some of them may be found as escapes on road verges, railway embankments, on forest margins and similar localities. F. White (F.F.N.R.: 67–68, 1968) has provided a key to some of these genera.

Prickly shrubs or scramblers (brambles); leaves trifoliolate or imparipinnate; achenes
    indefinite, drupaceous, borne on a convex receptacle surrounded by the 5-lobed
    persistent calyx - - - - - - - - - **3. Rubus**
Unarmed trees, shrubs or herbs:
    Leaves trifoliolate or rarely unifoliolate; ericoid shrubs; flowers solitary; stigma
        lacerate-multifid; fruit a solitary achene included in the calyx-tube
                                                 6. Cliffortia
    Leaves and stigma not as above:
        Trees or shrubs:
            Leaves imparipinnate:
                Flowers in much-branched drooping panicles up to 60 cm. long, dioecious or
                    polygamo-dioecious tree; main leaflet-pairs 5–7(8); stipules 6–9 cm. long,
                    adnate to the petiole, deciduous with the leaves, thus branches ringed by the
                    scars of the fallen leaf-bases - - - - - - **4. Hagenia**
                Flowers in erect racemes up to 10 cm. long, monoecious shrub or small tree;
                    main leaflet-pairs 2–3, stipules c. 10 mm. long, sheathing the stem, persistent
                    for many years after the leaves have fallen - - - **5. Leucosidea**
            Leaves simple, fruit a coriaceous drupe - - - - - **2. Prunus**
        Herbs or weakly suffruticose plants:
            Flowers 4-merous, apetalous; leaves palmately nerved - - **8. Alchemilla**
            Flowers 5-merous, with yellow petals; leaves penninerved:
                Herbs tall, erect; leaves imparipinnate; flowers in spicate racemes; carpels 2
                                          7. Agrimonia
                Herbs low, with prostrate or decumbent branches; leaves simple; flowers
                    solitary, axillary; carpels 10 - - - - - **1. Neuradopsis**

62. ROSACEAE

1. NEURADOPSIS Bremek. & Oberm.

Neuradopsis Bremek. & Oberm. in Ann. Transv. Mus. **16**: 415 (1935).

Low-growing annual herbs with somewhat woody stems, diffusely to densely branched, woolly-tomentose. Leaves simple, in groups of 3 at each node, these groups alternate on the stem, their members unequal in size, with petioles connate at the very base; laminae crenate or serrate to pinnately lobed; stipules absent. Flowers solitary, axillary, showy. Calyx-tube cyathiform, concrescent with the carpels in fruit, surrounded by 2–3 series of inconspicuous spinules, ebracteolate; calyx-lobes 5, narrowly oblong-acute. Petals 5, large, with convolute aestivation, inserted at the mouth of the calyx-tube, spreading. Stamens 10, inserted with the petals; filaments filiform. Carpels 10, verticillate; styles 10, free, persistent in fruit; stigmas capitate; ovule solitary, pendulous. Fruit dry, ± depressed-discoid, 10-locular, indehiscent, invested by the enlarged spiny tube of the calyx which develops a circular wing, and crowned by the marcescent calyx-lobes. Seeds solitary in each locule.

A xerophytic genus of S. tropical Africa whose 2 species are hardly distinguishable except by the fruits.

Fruit discoid, depressed, conspicuously winged, c. 15 mm. in diameter (including the wing); wing up to 6 mm. broad, usually ± concave, with its enlarged nerves protruding 1–2 mm. as small spines; fruit crowned by 15 prickles, apparently uniseriate and often curved or hooked, up to 6 mm. long, surrounding the marcescent calyx-lobes - - - - - - - - - - - 1. *austroafricana*
Fruit discoid, slightly depressed, narrowly winged, c. 10 mm. in diameter (including the wing); wing up to 2 mm. broad with its enlarged nerves protruding 2–3 mm. as ± stout spines disposed in 10 groups each of 3 elements (the lateral ones shorter than the central); fruit crowned by 2 series of usually straight prickles; the inner series of 10 prickles 6–8 mm. long, ± erect; the outer series of 5 prickles c. 5 mm. long; both series surrounding the marcescent calyx-lobes - - - - 2. *bechuanensis*

1. **Neuradopsis austroafricana** (Schinz) Bremek. & Oberm. in Ann. Transv. Mus. **16**: 415 (1935) " austro-africana ".—Merxm. & Roessler in Prodr. Fl. SW. Afr. **56**: 2 (1968). TAB. **1** fig. B. Type from SW. Africa.
    Neurada austroafricana Schinz in Bull. Herb. Boiss., Sér. 2, **1**: 874 (1901).—Engl., Pflanzenw. Afr. **3**, 1: 301 (1915). Type as above.

Annual herb with numerous prostrate or decumbent branches which are densely branched, up to 40 cm. long, usually becoming somewhat woody. Leaves up to 3·5 × 1·0 cm.; lamina ovate in outline with broadly cuneate base or rounded to elliptic with narrowly cuneate base, grey-cinnamomeous lanate-tomentose; margins irregularly crenate or serrate to pinnately lobed; petiole up to 10 mm. long, lanate. Flowers showy; peduncles 3–6(18) mm. long. Calyx c. 12 mm. long, cinnamomeous-lanate; calyx-tube cyathiform, c. 15 × 25 mm., prolonged into a terete false tube c. 4 mm. long formed by the basal part of the calyx-lobes which cohere by the interlocking of their indumentum; calyx-lobes narrowly oblong, ending in a triangular point c. 6 × 2 mm. Petals c. 20 × 10 mm., obovate, unguiculate, lemon-yellow, spreading. Stamens dimorphous; the 5 oppositipetalous ones c. 4 mm. long, the alternipetalous ones up to 8 mm.; filaments glabrous. Styles hairy, of 2 sizes in each flower; c. 5 or c. 8 mm. long. Fruit c. 15 mm. in diameter, depressed-discoid, crowned by the persistent styles and marcescent calyx-segments; upper part bearing a ring of 15 straight, curved or hooked prickles up to 6 mm. long; lower part produced laterally into a circular wing up to 6 mm. wide; wing somewhat concave, its margin bearing spines 1(2) mm. long which are extensions of the nerves. The seedling (? always) perforates the fruit, which may be found, ± withered, at the base of some adult plants.

Botswana. SW: 24 km. NE. of Twee Rivieren 4·8 km. E. of Nossob R., immat. fr. 8.iv.1961, *Leistner* 2241 (SRGH).
Also known from SW. Africa. On loose red sands on slopes of dunes, c. 1000 m.

2. **Neuradopsis bechuanensis** Bremek. & Oberm. in Ann. Transv. Mus. **16**: 416 (1935). TAB. **1** fig. A. Type: Botswana, between Kuke and Gomodimo, *Van Son* in Transv. Mus. 28848 (BM; K; PRE, lectotype).
    Neuradopsis grieloidea Bremek. & Oberm., tom. cit.: 415 (1935). Type: Botswana, Damara Pan, *van Son* in Transv. Mus. 28845 (BM; PRE, lectotype).

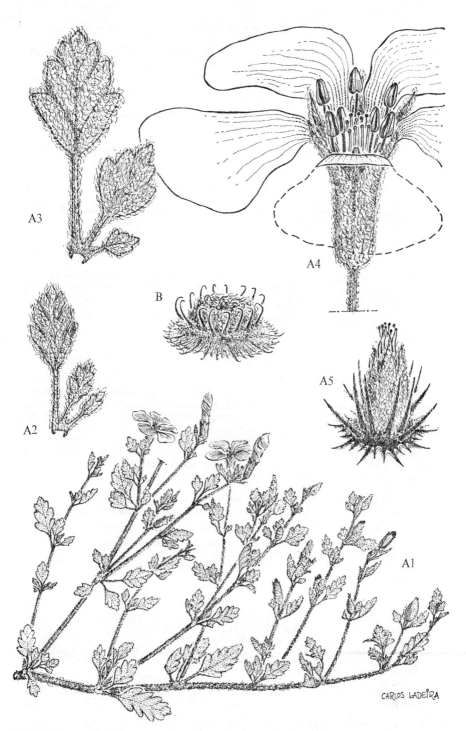

Tab. 1. A.—NEURADOPSIS BECHUANENSIS. A1, flowering branch (×⅔) *Wild* 4999; A2, group of three leaves from one node (×2) *van Son* 28848; A3, group of three leaves from one node (×2) *van Son* 28845; A4, flower, front petal partially removed (×4) *van Son* 28848; A5, fruit (×2) *Wild* 5027. B—N. AUSTROAFRICANA, fruit (×2 *Volk* 11543.

Annual herb with numerous prostrate or decumbent branches which are densely branched, up to 40 cm. long, usually becoming somewhat woody. Leaves (0·8)3·5 × (0·5)1·5 cm.; lamina narrowly obovate or ovate in outline with cuneate base, to broadly ovate with rounded base; margins irregularly crenate or serrate to pinnately lobed; petiole 0·6–1·0(2·0) cm. long, lanate. Flowers showy; peduncles (3)5–8(14) mm. long. Calyx c. 10 mm. long, cinnamomeous-lanate; calyx-tube cyathiform, c. 2 × 3·5–4 mm., prolonged into a terete false tube 3–5 mm. long as in *N. austroafricana*; calyx-lobes narrowly oblong, ending in a triangular point c. 4 × 1·3 mm. Petals c. 21 × 11 mm., obovate, unguiculate, lemon-yellow. Stamens dimorphous; the 5 oppositepetalous ones c. 5 mm. long, the alternipetalous ones up to 12 mm.; filaments glabrous. Styles hairy, of 2 sizes in each flower, c. 5 or c. 7 mm. long. Fruit c. 10 mm. in diameter, discoid, slightly depressed, crowned by the persistent styles and marcescent calyx-segments; upper part bearing 2 series of stout, usually straight prickles; inner series comprising 10 erect prickles 6–8 mm. long; outer series comprising 5 prickles c. 5 mm. long; lower part of fruit produced laterally into a circular wing 1–2 mm. wide, the margin bearing spines (in 10 groups of 3) 2–3 mm. long, which are extensions of the nerves of the wing.

Botswana. SW: Ghansi, 24 km. N. of Kang, fl. & immat. fr. 18.ii.1960, *Wild* 5027 (K; SRGH); 56 km. NE. of Kang, fl. 18.ii.1960, *Yalala* 60 (K; SRGH). SE: Mahalapye, western Lephephe, fl. 13.i.1958, *De Beer* 558 (K; LISC; SRGH); 6·4 km. N. of Murumush, fl. 21.ii.1960, *Wild* 4999 (BM; K; SRGH).

Only known from Botswana. *Terminalia*-bush or *Acacia*-tree grasslands on Kalahari sands, c. 1000 m.

2. PRUNUS L.

Prunus L., Sp. Pl. **1**: 473 (1753); Gen. Pl. ed. 5: 213 (1754).

Trees or shrubs, sometimes spiny. Leaves alternate, deciduous or persistent, simple, entire or incised; stipules small, caducous. Flowers usually bisexual, solitary or fasciculate-corymbose or in racemes, sometimes precocious. Perianth usually biseriate and 5-merous, sometimes irregularly so. Calyx-tube obconic, campanulate, cyathiform or tubular; calyx-lobes imbricate. Petals inserted at the mouth of the calyx-tube, or absent. Stamens 10–∞, inserted with the petals; filaments filiform, free. Carpel 1, style terminal, stigma peltate, capitate or truncate; ovules 2, collateral. Fruit drupaceous, fleshy or dry, indehiscent, 1- or rarely 2-seeded.

Over 200 species mainly in the N. temperate zone, also in tropical and southern Africa, S. Asia, Ceylon, Malay Archipelago and New Guinea. Only 2 species indigenous in Africa south of the Sahara.

Prunus africana (Hook. f.) Kalkm. in Blumea **13**: 26, 33, fig. 1e, 6c, 7a, 11, 13 (1965).— Mendes in C.F.A. **4**: 1 (1970).—Palmer & Pitman, Trees of S. Afr. **1**: 679, cum 2 fig. (1972). TAB. 2. Lectotype from Cameroon.
    Pygeum africanum Hook. f. in Journ. Linn. Soc., Bot. **7**: 191 (1864).—Oliv. in F.T.A. **2**: 373 (1871).—Sim, For. Fl. Col. Cape Good Hope: 215, t. 63 (1907).— Engl., Pflanzenw. Afr. **3**, 1: 302, fig. 192 (1915).—Burtt Davy, F.P.F.T.: 312, fig. 48 (1936).—Gardner, Trees & Shrubs of Kenya Col.: T.57 (1936).—Robyns & Tournay, Fl. Parc Nat. Alb. **1**: 256 (1948).—Brenan, T.T.C.L.: 477 (1949).— Hauman in F.C.B. **3**: 32 (1952).—Eggeling & Dale, Indig. Trees Uganda ed. 2: 335 (1952).—Brenan in Mem. N.Y. Bot. Gard. **8**: 431 (1954).—Keay in F.W.T.A. ed. 2, **1**: 426 (1958).—R. A. Grah. in F.T.E.A., Rosaceae: 45, fig. 6 (1960).— F. White, F.F.N.R.: 70 (1962).—Boughey in Journ. S. Afr. Bot. **30**: 157 (1964). Type as above.
    Pygeum crassifolium Hauman in Bull. Jard. Bot. Brux. **22**: 93 (1952); in F.C.B. **3**: 33 (1952) saltem pro parte quoad specim. *Stolz* 2279 (BM; BR; K). Lectotype from Zaire, cf. Kalkman, op. cit.: 34 (1965); R. A. Grah., op. cit.: 47 (1960).

A much-branched evergreen tree c. 10 (up to 25) m. high, or rarely a shrub 3–5 m. tall, entirely glabrous except for the flowers. Bark dark brown to grey, rugged. Leaves petiolate; lamina (4)6–11(15) × 2–4(5·5) cm., elliptic, lanceolate-elliptic or oblong-lanceolate, glabrous, coriaceous or subcoriaceous, the apex obtuse to subacuminate, the base broadly cuneate to rarely rounded, margins coarsely crenate-serrate to subentire with dark glandular dots in the incisions (the most

Tab. 2. PRUNUS AFRICANA. 1, leaf (×1) *Parry* 70; 2, apex of flowering branch (×1) *Moon* 765; 3, flower (×4); 4, flower from beneath showing sepals, petals and lower part of filaments (×4); 5, petal (×8); 6, part of flower in longitudinal section to show pistil and ovule (×4), 3–6 from *Milne-Redhead & Taylor* 10877; 7, fruiting raceme (×1); 8, fruit (×2), 7–8 from *Bally* 4356.   From F.T.E.A.

proximal gland on one or both margins of the leaf sometimes conspicuous); petiole
1–2 cm. long, channelled, often reddish; stipules ± linear, 1·5–2·0 cm. long,
caducous. Racemes usually solitary, 2–5(8) cm. long, 7–15-flowered, arising from
the axils of scales at the base of lateral shoots which may also produce leaves in their
upper part; peduncle usually 5–10 mm. long; pedicels (3)5–7(10) mm. long;
bracts triangular, small, soon caducous. Perianth (4)5(6)-merous, sometimes
irregularly so. Calyx-tube ± cyathiform, (3·0)3·5–4·0(5·0) mm. in diameter at the
mouth, rather fleshy, glabrous outside, usually hairy inside; calyx-lobes 1–1·5 mm.
long, triangular, glabrous except for the ciliate apex. Petals up to 2 mm. long,
creamy-white, elliptic to oblong, reflexed, hairy abaxially especially towards their
margins and sometimes adaxially also. Stamens 25–35; filaments c. 1·5 mm. long,
glabrous; anthers didymous, 0·5–1·0 mm. long. Ovary ± ovoid, sparsely long-
haired; style c. 1·5 mm. long, sparsely haired; stigma peltate, slightly 2- or 3-lobed.
Fruit 5–8 × 8–12 mm., transversely ellipsoid, broader than long, slightly didymous
and thus appearing as if 2-locular, dry, usually glabrous, red to purplish-brown.

Zambia. N: N. side of L. Young, fr. 20.ix.1938, *Greenway & Trapnell* 5728 (EA;
FHO; K); Mbala (Abercorn), Ndundu, fl. vi.1961, *Procter* 1883 (EA; K). W: Chingola,
fl. & immat. fr. 27.viii.1954, *Fanshawe* 1505 (FHO; K; SRGH); Solwezi, fr. 23.vii.1964,
Fanshawe 8832 (FHO; K). Rhodesia. C: Rusape Distr., fl. 2.xi.1952, *Dehn* R75/53
(SRGH); Makoni Distr., Silverbow road 19 km. N. of Rusape, st. iv.1961, *Guy* 55/61
(SRGH). E: Umtali Distr., road to Honde Valley, fl. buds & immat. fr. 5.xi.1948, *Chase*
1325 (BM; COI; LISC; SRGH); Umtali Commonage, Inyamatshira Mt. range, S. side
of the Gomo, fl. 9.ix.1951, *Chase* 3946 (BM; COI; LISC; PRE; SRGH). Malawi.
N: Matipa Forest Reserve, fr., *Chapman* 209 (FHO); Nyika Plateau near L. Kaulime,
fl. 3.ix.1962, *Tyrer* 792 (BM; SRGH). C: Dedza Distr., Chongoni, Ciwau, fr. 9.x.1960,
Chapman 983 (SRGH); on top of Kalichero, fr. 12.vi.1961, *Chapman* 1369 (FHO;
SRGH). S: Palombe R., fl. 29.iv.1958, *Chapman* 563 (FHO; K); Blantyre, fl., *Buchanan*
in J. M. Wood Herb. 7022 (PRE; SRGH). Mozambique. N: Amaramba, foot of
Tchirassulo ("Tshiradzuru") Mt., fr. 3.x.1858, *Kirk* (K). Z: Gúruè, Lioma, Tetete,
source of R. Lucasse, fr., *Serrano* 31 (LMA). MS: Manica Distr., Chimoio, Garuso,
st. iv.1935, *Gilliland* 1868 (FHO); Zuira Mts., Tsetsera, fl. 3.iv.1966, *Torre & Correia*
15659 (COI; EA; K; LISC; LMU).

Also known from Ghana to Ethiopia, Fernando Po, S. Tomé, Zaire, Uganda to
Tanzania, S. Africa, Madagascar and Grand Comore. In upland rain forest, riverine
forest, or rarely solitary in plateau grasslands, 1000–2100 m.

3. RUBUS L.

Rubus L., Sp. Pl. 1: 492 (1753); Gen. Pl. ed. 5: 218 (1754).

Prickly shrubs, sometimes glandular (but rarely so in the F.Z. area); stems
scrambling, trailing or occasionally prostrate, often becoming thick and woody and
hollow with age. Leaves alternate, petiolate, stipulate, trifoliolate, imparipinnate
with 2–3(4) pairs of lateral leaflets, or the uppermost leaves sometimes simple.
Leaflets with serrated margins, glabrous to densely hairy. Stipules filiform, linear
or obovate, free or adnate to the base of the petiole. Flowers sometimes solitary,
more usually in dense or open many-flowered inflorescences borne at and near the
ends of branches. Flowers ⚥, actinomorphic, slightly perigynous. Perigynous
zone (= calyx-tube) shallowly cup-shaped; calyx segments 5, equal, longer than
calyx-tube; petals 5, alternating with sepals, inserted on margin of perigynous
cup, conspicuous or inconspicuous, or sometimes absent. Stamens ∞, inserted on
rim of cup. Carpels many (c. 15–100 +), borne on a spongy elongate receptacle,
developing into drupelets, the fruit usually sweet and edible. Ovules 2, one of
them aborting.

A world-wide genus, most abundant in temperate regions. Some of its sections
are notoriously difficult to classify, due to apomictic and part-sexual methods of
reproduction, but the African members are not involved in this problem. This
account shows that the ranges of *R. iringanus*, *R. kirungensis* and *R. scheffleri* are
wider than was previously thought, and that a newly discovered species,
R. chapmanianus, is also present.

Most members of *Rubus* in the F.Z. area have edible fruits similar to the
European blackberry.

Mature fruit up to 2 cm. long, with carpels " in hundreds "; plants usually covered with
    glistening sessile greenish-yellow glands   -   -   -   -   -   1. *rosifolius*

Mature fruit up to c. 1 cm. long; carpels fewer than above; plants eglandular or glands long-stipitate, reddish or brownish:
Stems at end of flowering branches covered with stiff reddish bristles up to 5 mm. long like very slender prickles; leaflets obtuse - - - - - 2. *ellipticus*
Stems at end of flowering branches glabrous or hairy but never as above; leaflets acute to acuminate, very rarely obtuse:
Leaves 2–3-jugate, superior surface glabrous or sparsely appressed-pilose, inferior surface white-felted; stems white-pruinose at first, becoming dark red or purplish and glossy - - - - - - - - 3. *niveus*
Plants not as above: if leaves white-felted below then stems villous or leaves mostly ternate; if stems white-pruinose then inferior leaf-surface ± glabrous:
Stems and leaves densely greyish-green villous; stipules 6–10 mm. broad, obovate-acute - - - - - - - - - 9. *chapmanianus*
Stems and leaves glabrous or hairy; stipules filiform to narrowly linear or rarely narrowly obovate-acute, never more than 5 mm. broad:
Leaves mostly ternate; petals slightly shorter than calyx, broadly spathulate to suborbicular or rarely absent:
Leaflets ovate-acute to broadly elliptic or obovate; the lower leaf-surface pilose on nerves and glabrous between, the lateral nerves sparsely prickly; inflorescence usually shorter than leaves; petals white - 4. *iringanus*
Leaflets ovate-acute; lower leaf-surface greyish-tomentose or velutinous (very rarely pilose on nerves and glabrous between), the lateral nerves not prickly; inflorescence usually much longer than leaves; petals pink or purple - - - - - - - - - 5. *rigidus*
Leaves mostly pinnate (2–4-jugate), the uppermost sometimes ternate or simple; petals either absent or very small and narrowly spathulate, or slightly exceeding to 2 × calyx length, with subcircular limb:
Leaves glabrous to densely villous; petals absent or when present shorter than calyx, narrowly spathulate:
Stems ± glabrous; leaves glabrous above, entirely glabrous below or villous on main veins; petals usually present - - - 6. *pinnatus*
Stems sparsely to densely villous; leaves thinly appressed-pilose above, villous, velutinous or tomentose below; petals usually absent:
Stem densely fulvous-villous; inferior leaf-surface fulvous or pale yellowish-green, villous on main veins and tomentose between; inflorescence shortly cylindrical with short thick branches 7. *exsuccus*
Stem sparsely brownish-villous; inferior leaf-surface pale brownish-green, villous on main veins and pilose between, or whitish-velutinous; inflorescence broadly pyramidal to long-cylindrical, with spreading branches - - - - - - 8. *apetalus*
Leaves ± glabrous; petals slightly exceeding to 2 × calyx length, with subcircular limb and narrow claw:
Leaves rather stiff and leathery in texture; petiole and rhachis glabrous or sparsely pilose on adaxial side - - - - - 10. *kirungensis*
Leaves thin, not leathery, in texture; petiole and rhachis glaucous-pubescent or glabrous except for a densely fulvous-villous line along the upper surface - - - - - - - 11. *sheffleic*

1. **Rubus rosifolius** J. E. Sm., Pl. Ic. Ined.: t. 60 (1791).—Harv. in Harv. & Sond., F.C. 2: 286 (1862).—Focke in Bibl. Bot. 72, 2: 153, t. 65 (1911).—C. E. Gust. in Arkiv Bot. 26A, 7: 11 (1934).—R. A. Grah. in F.T.E.A., Rosaceae: 25 (1960). Type from Mauritius.

Low straggling perennial herb, the stems, leaves and carpels usually dotted with glistening sessile greenish-yellow glands. Flowering branches villous, reddish-brown, sparsely prickly; prickles up to 3 mm. long, straight or weakly decurved. Leaves 7–15 × 3·5–10 cm., 2–3-jugate or the uppermost ternate or simple; petiole and rhachis villous, with a few weakly hooked prickles 1–2 mm. long; lateral leaflets subsessile, the terminal one with petiolule 0·3–1·0 cm. long. Leaflets narrowly ovate-acute to -acuminate, the terminal one larger than the laterals and more gradually tapered at the apex; margins jaggedly doubly serrated, the primary serratures (2)6 mm. deep; superior surface green, sparsely pilose; inferior surface paler green, pilose on veins, the midrib not or very sparsely prickly. Stipules delicate, filiform. Flowers solitary, terminal and axillary; pedicels 0·5–2·0 cm. long. Calyx tomentose and weakly villous, deeply divided; lobes c. 13 mm. long, ovate-lanceolate with caudate apex. Petals a little shorter than to slightly exceeding sepals, broadly obovate, white. Receptacle elongating when mature. Fruit ellipsoid, up to 2 cm. long, yellow to scarlet and edible when ripe; carpels borne " in hundreds ", c. 1 mm. long, glabrous.

Zambia. N: side of stream coming from Kasulo Dam, 1520 m., 11.i.1955, *Richards* 4023a (K). **Rhodesia.** C: Salisbury, Avondale, cultivated, vii.1958, *Martin* in SRGH 91012 (SRGH). **Malawi.** C: Ntchisi Mt., 14.iii.1967, *Salubeni* 605 (LISC; SRGH).
Scattered in E. Africa. Introduced from E. Asia, and becoming naturalized. *Brachystegia* woodland, by streams, and as an escape from cultivation.

2. **Rubus ellipticus** J. E. Sm. in Rees, Cycl. **30**, no. 16 (1815).—Focke in Bibl. Bot. **72**, 2: 198 (1911).—R. A. Grah. in F.T.E.A., Rosaceae: 25 (1960). Type from Nepal.

Robust scrambling shrub up to 6 m. tall, with very stout canes. Flowering branches straight or zig-zag, reddish-brown, somewhat villous, densely covered with stiff, straight, reddish bristles up to 5 mm. long; prickles 3–5 mm. long, straight or weakly decurved, glossy. Leaves 10–19 × 7–13 cm., ternate; petiole and rhachis bristly like the stem, bearing smaller but more strongly hooked prickles; lateral leaflets subsessile, terminal one with petiolule 2·5–4 cm. long. Leaflets broadly elliptic to obovate, markedly obtuse, sometimes apiculate, the terminal one larger than the laterals; margins minutely serrate; superior surface dark greyish-green, glabrescent; inferior surface pale greenish-grey, very shortly tomentose, with prominent veins, the midrib bristly and prickly. Stipules stiffly filiform. Inflorescences terminal and axillary, shorter than leaves; peduncles and pedicels villous and bristly, longer than bracts and bracteoles. Calyx rather leathery, dorsally bristly near the base, divided into broadly ovate lobes 5–7 mm. long. Petals white or yellow, c. twice as long as sepals, with broad limb contracting abruptly into narrow claw. Fruit enclosed within persistent calyx, yellow when ripe; carpels subglabrous.

Malawi. N: Rumpi Distr., Kaziwizwi R., 1350 m., 8.i.1959, *Richards* 10543 (K). S: Zomba Plateau, 3.viii.1960, *Leach* 10402 (LISC; PRE; SRGH). **Mozambique.** N: Metonia, near Massangulo Catholic Mission, 13.x.1942, *Mendonça* 801 (LISC; LMU).
A native of India and Ceylon, found naturalized in scattered localities throughout Africa. Regenerated forest, scrub, grassy clearings.

3. **Rubus niveus** Thunb., Diss. Bot. Med. Rubo: 9, t. 1 fig. 3 (1813).—Focke in Bibl. Bot. **72**, 2: 182 (1911).—R. A. Grah. in F.T.E.A., Rosaceae: 40 (1960). Type unknown.

Shrub to c. 2 m. high. Flowering branches glabrous except in the inflorescence, white-pruinose at first, becoming glossy and dark reddish or purplish; prickles 4–7 mm. long, straight to weakly decurved, at first pruinose and then glossy like the stems. Leaves 10–17 × 4–8 cm., 2–3-jugate; petiole and rhachis tomentose, glabrescent, bearing hooked prickles; lateral leaflets subsessile, the terminal one with petiolule 1·5–2 cm. long. Leaflets elliptic to ovate-acute, the terminal one broader than the laterals and often somewhat cordate at the base and cuspidate at the apex; margins shallowly singly or doubly serrate, the veins excurrent; superior surface dark greyish-green, glabrous; inferior surface white-felted, with prominent veins, the midrib not prickly; leaflets somewhat plicate along the lines of lateral veins. Stipules lanceolate. Inflorescences terminal and axillary, shorter than the leaves, often few-flowered; peduncles and pedicels pale greyish-tomentose, longer than bracts and bracteoles. Calyx whitish-tomentose, deeply divided into narrowly ovate, shortly acuminate lobes 4–5 mm. long. Petals shorter than sepals, white to mauvish-pink or red, with broad subcircular limb and narrow claw, caducous. Fruit 5–7 mm. long, 8–10 mm. broad, cordate in longitudinal section, purplish when ripe; carpels pubescent.

Zambia. N: Mbala (Abercorn), 10.vii.1964, *Mutimushi* 884 (FHO; K; SRGH). **Rhodesia.** N: 15 km. N. of Umvukwes village, 16.iv.1972, *Biegel & Pope* 3937 (LISC; PRE; SRGH). C: Salisbury, Alexandra Park on rocky outcrop, 4.vii.1961, *Wild* 5488 (K; LISC; SRGH). E: Melsetter Orange Grove, 16.vii.1964, *Wild* 6570 (K; LISC; SRGH). **Malawi.** N: Rumpi Distr., Nyika Plateau, Chelinda Camp by Chalet 3, 2285 m., 19.v.1970, *Brummitt* 10906 (K; LISC; MAL; PRE; SRGH).
Scattered throughout E. and southern Africa; an introduced bramble native to India and Malaya. *Hyparrhenia* grassland and as an escape from cultivation.

4. **Rubus iringanus** C. E. Gust. in Kew Bull. **1938**: 184 (1938).—R. A. Grah. in F.T.E.A., Rosaceae: 34 (1960).—Brummitt in Wye Coll. Malawi Proj. Rep.: 73 (1973); in Kew Bull. **31**: 168 (1976). Type from Tanzania.

Low-growing shrub with creeping stems. Flowering branches pale green to purplish, covered with short stiff setose hairs (1 mm. long or less) especially distally, and prickly; prickles 1–3 mm. long, ochraceous, slender, straight or deflexed. Leaves 4–7·5 × 2·5–7·5 cm., ternate or the uppermost ones simple, rather stiff and leathery in texture; lateral leaflets sessile, terminal ones with petiolule 0·5–1·5 cm. long; petiole and rhachis densely setose and prickly like the stems. Leaflets ovate-acute to broadly elliptic or obovate; superior surface green, appressed-pilose especially between the nerves; inferior surface paler, pilose on nerves but glabrous between, with midrib and lateral nerves minutely prickly. Stipules 10–20 × 1–2·5 mm., linear-acuminate or lanceolate, often quite leaf-like in colour and texture. Leaf margins finely to coarsely serrate. Inflorescences terminal, leafy, 3–10 cm. long, ± cylindrical, few-flowered; peduncles and pedicels short, densely villous-setose; bracts often exceeding pedicels. Calyx greenish to brownish, villous and prickly, divided into triangular-acuminate lobes c. 8 mm. long. Petals a little shorter than sepals, broadly spathulate, white. Fruits 0·8–1·0 cm. long, globose, red or orange when ripe; carpels 3–3·5 mm. long, relatively few in number, glabrous.

Malawi. N: Chitipa Distr., Nyika Plateau, NW. slopes of Nganda, fl. & fr. 27.vii.1972, *Brummitt & Synge* WC 45 (K; MAL).
Also from Tanzania (Iringa Distr.) and Kenya (Nanyuki Distr.). Among bracken and grass.

5. **Rubus rigidus** J. E. Sm. in Rees, Cycl. **30**, no. 5 (1815).—Harv. in Harv. & Sond., F.C. **2**: 287 (1862).—Oliv. in F.T.A. **2**: 375 (1871).—Focke in Bibl. Bot. **72**, 2: 174 (1911).—C. E. Gust. in Archiv Bot. **26A**, 7: 58 (1934); in Kew Bull. **1938**: 186 (1938).—R. A. Grah. in F.T.E.A., Rosaceae: 33 (1960).—Mendes in C.F.A. **4**: 3 (1970). TAB. **3** fig. B. Type from S. Africa.
Rubus chrysocarpus Mundt in Linnaea **2**: 17 (1827). Type from S. Africa (Cape Prov.).
Rubus mundtii Chamisso & Schlechtendal in Linnaea **2**: 18 (1827). Type from S. Africa (Cape Prov.).
Rubus inedulis Rolfe in Journ. Linn. Soc., Bot. **37**: 514 (1906).—C. E. Gust, tom. cit.: 62 (1934); loc. cit. Syntypes from Uganda.
Rubus rigidus var. *buchananii* Focke, loc. cit.—C. E. Gust., tom. cit.: 61 (1934). Type from Malawi (further details not given).
Rubus rigidus var. *chrysocarpus* (Mundt) Focke, loc. cit. Type as for *R. chrysocarpus*.
Rubus rigidus var. *mundtii* (Chamisso & Schlechtendal) Focke, loc. cit.—C. E. Gust., tom. cit.: 59 (1934). Type as for *R. mundtii*.
Rubus rigidus var. *incisus* C. E. Gust. in Bot. Notis. **1932**: 18 (1932); tom. cit.: 60 (1934); loc. cit. Syntypes: Rhodesia, Inyanga, Inyangani, *Fries, Norlindh & Weimarck* 3475 (LD); Chipinga, *Michelmore* 249 (K; SRGH); Melsetter, Jansen's Hill to Lemonkop, *Michelmore* 227 (K; SRGH).
Rubus rigidus forma *lachnocarpus* C. E. Gust., loc. cit.; tom. cit.: 58 (1934); loc. cit. Syntypes: Rhodesia, Distr. Manica, Div. Umtali, Odzani R. valley, *Teague* 195 (K); Chipinga, *Michelmore* 262 (K; SRGH).
Rubus atrocoeruleus C. E. Gust., tom. cit.: 54, t. 11 (1934); tom. cit.: 185 (1938). Type from Kenya.
Rubus intercurrens var. *confluens* C. E. Gust., loc. cit. (1934); loc. cit. (1938) pro parte excl. specim. *Buchan.* Malawi 1891. Type from Cameroon.

A very variable scrambling shrub 1–3·5 m. high. Flowering branches pale greyish-green-tomentose, partially glabrescent, sparsely to moderately prickly; prickles 3–5 mm. long, straight to moderately decurved. Leaves (5)7- 14 × (5)6–11 cm., predominantly ternate but sometimes pinnate (2-jugate) below, or the upper-most simple; petiole and rhachis tomentose and aculeate like the stems; lateral leaflets subsessile, the terminal one with petiolule 1·5–4 cm. long. Leaflets broadly ovate-acute or ovate-acute, with evenly once-serrate margins; superior surface dark green, subglabrous to sparsely appressed-pilose; inferior surface densely greyish-tomentose or -velutinous, the midrib sparsely and minutely prickly. Stipules narrowly linear. Inflorescences terminal and axillary, 5–26 cm. long, many-flowered, cylindrical or narrowly pyramidal; peduncles and pedicels with indumentum like stems. Calyx densely greenish-tomentose, deeply divided; lobes broadly ovate-acute to ovate with long caudate apex, 5–13 mm. long, spreading to

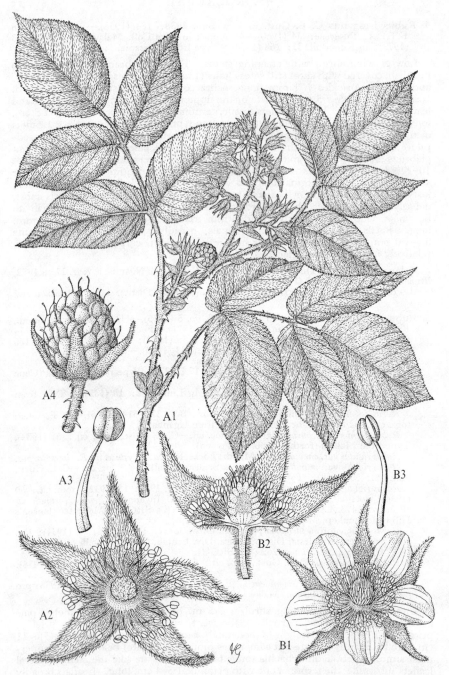

Tab. 3. A.—RUBUS CHAPMANIANUS. A1, habit (×⅔); A2, flower (×4); A3, stamen (×12); A4, fruit (×3), all from *Chapman* 489. B.—RUBUS RIGIDUS. B1, flower (×3); B2, longitudinal section of flower, with petals removed (×4); B3, stamen (×12), all from *Leach* 10401.

somewhat reflexed in fruit. Petals pale pink to purple, slightly shorter than the sepals, broadly spathulate or suborbicular with narrow claw, very occasionally absent. Fruits globose, c. 0·8 cm. long, but very often galled and becoming much larger, deep orange when ripe; carpels glabrous or pubescent.

Zambia. B: Shangombo, c. 1040 m., 15.viii.1952, *Codd* 7557 (BM; COI; EA; K; PRE; SRGH). N: Mbala (Abercorn), Kawimbe, 1600 m., 14.ix.1950, *Bullock* 3314 (EA; K; SRGH). W: Chichele, 10.iv.1951, *Holmes* 692 (FHO). C: Kabwe (Broken Hill), xi.1928, *Van Hoepen* 1342 (PRE). E: Nyika Plateau, c. 4 km. SW. of Rest House, 2100 m., 22.x.1958, *Robson & Angus* 266 (LISC). **Rhodesia.** N: Mazoe, 1370 m., ix.1906, *Eyles* 472 (BM; SRGH). C: Salisbury, Chacoma, 15.vii.1947, *Wild* in SRGH 16856 (SRGH). E: Tarka Forest Reserve, 1190 m., vi.1956, *Barrett* 98/56 (LISC; LMA; SRGH). S: Bikita, Old Bikita, 1310 m., 16.xii.1953, *Wild* 4404 (SRGH). **Malawi.** N: Rumpi Distr., Nyika Plateau road to Chelinda, 2100 m., 5.i.1959, *Richards* 10516 (K). S: Dedza, near main ? Mputa Milonde road, 10.ix.1960, *Chapman* 919 (SRGH). S: Zomba (Mt., 20.i.1960, *Banda* 388 (BM; K; SRGH). **Mozambique.** MS: Vila Pery Distr., Mossurize near the Missão Catolica de Espungabera, 10.vi.1942, *Torre* 4289 (COI; LISC; LMU; SRGH). SS: Gaza Distr., Chipenhe, Régulo Chiconela, Chirindzeni forest, R. Machacane, 13.x.1957, *Barbosa & Lemos* 8063 (COI; LISC). LM: Maputo, Marracuene, Bobole, Reserva Botanica, 3.x.1957, *Barbosa & Lemos* 7938 (COI; LISC).

Cameroon, Zaire, Ethiopia, Uganda and most of Africa to the south. Swamp forest margins, stream banks and roadsides, 1040–2100 m.

R. rigidus is apparently the only member of the genus in the F.Z. area with galled fruits, and as this condition is quite common it provides a useful additional character for specific recognition.

6. **Rubus pinnatus** Willd., Sp. Pl. 2: 1081 (1799).—Harv. in Harv. & Sond., F.C. 2: 287 (1862).—Oliv. in F.T.A. 2: 374 (1871).—Focke in Bibl. Bot. 72, 2: 177 (1911).— C. E. Gust. in Arkiv Bot. 26A, 7: 45 (1934).—R. A. Grah. in F.T.E.A., Rosaceae: 37 (1960). Type from St. Helena.
   Rubus kingaensis Engl., Bot. Jahrb. 30: 313 (1901).—Focke, tom. cit.: 173 (1911).—C. E. Gust., tom. cit.: 47 (1934). Type from Tanzania.
   Rubus pinnatus subsp. *afrotropicus* Engl., Pflanzenw. Afr. 3, 1: 294 (1915).— Mendes in C.F.A. 4: 2 (1970). Syntypes from Cameroon, Tanzania etc.
   Rubus pinnatus var. *afrotropicus* (Engl.) C. E. Gust., tom. cit.: 46 (1934).— R. A. Grah., tom. cit.: 38 (1960). Syntypes as above.
   Rubus pinnatus forma *subglandulosus* C. E. Gust. in Kew Bull. 1938: 183 (1938). Type from Kenya.
   Rubus pinnatus var. *subglandulosus* (C. E. Gust.) R. A. Grah. in Kew Bull. 12: 406 (1958). Type as above.

Straggling bush or scrambler up to c. 3·5 m. high. Flowering branches glabrous or occasionally somewhat tomentose, dark reddish-brown, rarely whitish-pruinose below, with glabrous yellowish or brownish decurved prickles 2–5 mm. long. Leaves 10–19 × 8–12 cm., 2–4-jugate or the uppermost ones ternate, rather thin and membranous in texture; petiole and rhachis aculeate like the stems; lateral leaflets shortly petiolulate, the terminal one with petiolule 0·5–2 cm. long. Leaflets ovate-acute to oblong-acuminate, with finely serrate margin; superior surface dark green, glabrous; inferior surface pale green, entirely glabrous or villous on main veins, the midrib not or finely aculeate, the lateral veins without prickles. Stipules narrowly linear, very occasionally with a few stipitate glands. Inflorescences terminal and axillary, (3)8–36 cm. long, narrowly pyramidal or pyramidal, usually much longer than the leaves, the branches spreading, greyish-tomentose. Calyx greyish-tomentose, very occasionally glandular-hairy, divided into lanceolate-acuminate lobes 5–8 mm. long. Petals usually present although inconspicuous, pink, narrow, shorter than the sepals. Fruits globose, c. 0·6 mm. long, red when ripe; carpels relatively few (10–35), glabrous.

Zambia. N: Mpongwe, 17.ix.1963, *Fanshawe* 7990 (K). C: Serenje, 22.ix.1961, *Fanshawe* 6704 (FHO). E: Nyika Plateau, Chowo Forest, 2230 m., 14.xi.1967, *Richards* 22543 (K; LISC). **Rhodesia.** E: Melsetter, Kasipiti, c. 1295 m., 20.ix.1964, *Loveridge* 1164 (LISC; SRGH). **Mozambique.** N: Maniamba, 20.iv.1934, *Torre* 846 (COI; LISC). Z: Morrumbala, 7.viii.1942, *Torre* 4542 (COI; LISC; LMU). MS: Manica Distr., Dombe, Maronga, 7.viii.1945, *Simão* 435 (LISC; LM).

Widespread in tropical and southern Africa; also in St. Helena and Ascension Is. Clearings in forests by streams, swamp forest margins, 1100–2230 m.

18 62. ROSACEAE

7. **Rubus exsuccus** Steud. ex A. Rich., Tent. Fl. Abyss. **1**: 256 (1847).—Focke in Bibl. Bot. **72**, 2: 176 (1911) in syn. sub *R. apetalus*. Type from Ethiopia.
Rubus adolphi-friedericii Engl. in Wiss. Ergebn. Deutsch Zentr.-Afr.-Exped. 1907–1908, **2**: 223 (1911).—C. E. Gust. in Arkiv Bot. **26A**, 7: 35 (1934).—R. A. Grah. in F.T.E.A., Rosaceae: 39 (1960).—Mendes in C.F.A. **4**: 3 (1970). Type from Ruanda-Urundi.

Scrambling shrub 2–5 m. high. Flowering branches densely fulvous-villous, with decurved, glabrous, brownish-yellow prickles 3–5 mm. long. Leaves 9–20 × 8–14 cm., 2-jugate or the uppermost ternate; petiole and rhachis villous and aculeate like the stems; lateral leaflets subsessile, the terminal one with petiolule 1–2 cm. long. Leaflets ovate-acute or -acuminate, with finely serrate margin; superior surface yellowish-green, thinly appressed-pilose; inferior surface fulvous or pale yellowish-green, villous on the main veins and densely and softly long-tomentose between, the midrib very sparsely and minutely prickly, the lateral nerves without prickles. Stipules narrowly linear. Inflorescences terminal and axillary, 2–15 cm. long, cylindrical, not much exceeding leafy part of stem, the branches short and thick, densely villous like the stem. Calyx densely greenish-grey-villous or -tomentose, divided into narrowly triangular segments c. 5 mm. long, clasping or spreading in fruit. Petals almost always absent, when present narrow, shorter than sepals, fugacious. Fruits globose, 0·5–1·0 cm. long, orange-red to black and edible when ripe; carpels few to many (13–64), glabrous or densely tomentose.

Zambia. N: Mansa (Ft. Rosebery) near Samfya Mission, Lake Bangweulu, 30.viii.1952, *White* 3172 (BM; FHO; K). **Rhodesia. C**: Makoni, 1490 m., vii.1917, *Eyles* 752 (BM; SRGH). **E**: Chipinga, 910–1130 m., 27.vi.1934, *Michelmore* 278 (K; SRGH). **S**: Fort Victoria, R. Rungwe, viii.1932, *Cuthbertson* SRGH 6070 (SRGH). **Malawi. N**: Nkhata Bay from beside the *Hagenia* Plot on Mukukwa road, 13.vi.1960, *Chapman* 732 (BM; FHO; SRGH). **C**: Lilongwe, Kasitu, 4.xi.1962, *Chapman* 1730 (SRGH). **S**: Cholo Mt., 1200 m., 19.ix.1946, *Brass* 17653 (K; SRGH). **Mozambique. N**: Metónia, near the Missão Catolica de Massangulo, 13.x.1942, *Mendonça* 800 (LISC). **Z**: Serra do Gúruè, 18.x.1949, *Barbosa & Carvalho* 4479 (LISC; LM). **MS**: Vila Pery Distr., Serra da Gorongosa on road to Pico Gogôgo, 10.x.1944, *Mendonça* 2443 (COI; LISC; LMU; SRGH).
Zaire, Rwanda, Burundi, Sudan, Ethiopia, Angola, Uganda, Kenya, Tanzania and S. Africa (Transvaal). Swamp forest, rain forest regrowths, and margins of relict montane evergreen forest, 910–1200 m.

8. **Rubus apetalus** Poir., Encycl. Méth. Bot. **6**: 242 (1804).—Oliver in F.T.A. **2**: 374 (1871) excl. syn.—Focke in Bibl. Bot. **72**, 2: 176, t. 72 (1911).—C. E. Gust. in Archiv Bot. **26A**, 7: 53 (1934).—R. A. Grah. in F.T.E.A., Rosaceae: 39 (1960) excl. syn. Type from Réunion.
Rubus pinnatiformis C. E. Gust. in Bot. Notis. **1932**: 17 (1932); tom. cit.: 48, t. 8 (1934). Syntypes: Rhodesia, Inyanga near Pungwe R., c. 1800 m., fr. 6.xi.1930, *Fries, Norlindh & Weimarck* 2726 (BM; LD); same locality, c. 1700 m., 17.xii.1930, *Fries, Norlindh & Weimarck* 3866 (BM; LD).

Scrambling shrub to c. 1·5 m. high. Flowering branches sparsely villous, brownish, with few to many decurved brownish-yellow prickles 2–5 mm. long. Leaves 7–16 × 5·5–13 cm., 2–3-jugate or the uppermost ternate; petiole and rhachis villous and aculeate like the stems; lateral leaflets subsessile, terminal one with petiolule 1–2 cm. long. Leaflets ovate-acute or elliptic-acute, with finely to coarsely serrate margin; superior surface brownish-green, thinly appressed-pilose; inferior surface pale brownish-green, villous on main veins and pilose between, or whitish-velutinous, the midrib very sparsely and minutely prickly, the lateral veins without prickles. Stipules narrowly linear. Inflorescences terminal and axillary, 3–19 cm. long, broadly pyramidal to cylindrical, usually much longer than leaves, the branches spreading, villous and aculeate like the stem. Calyx greyish-tomentose, divided into narrowly triangular segments 4–6 mm. long, ± clasping in fruit. Petals almost always absent, when present narrow, shorter than sepals, fugacious. Fruits globose, 0·5–0·8 cm. long, purplish-black and edible when ripe; carpels relatively few (15–36), glabrous or rarely tomentose at the apex.

Zambia. N: Chishinga Ranch near Luwingu, 1580 m., 2.v.1961, *Astle* 618 (K; SRGH). **W**: Ndola, 5.ix.1967, *Mutimushi* 2048 (K; SRGH). **C**: Chakwenga Head-waters, 100–129 km. E. of Lusaka, 8.ix.1963, *Robinson* 5652 (K; SRGH). **E**: Nyika

Plateau, 1·6 km. N. of Rest House, 2130 m., 27.xi.1955, *Lees* 92 (K). **Rhodesia.** E: Pungwe R., Honde Valley, c. 760 m., 16.ix.1964, *Loveridge* 1118 (COI; FHO; K; LISC; SRGH). **Malawi.** N: Nyika Plateau below Rest House on path to N. Rukuru waterfall, 2150 m., 27.x.1958, *Robson* 408 (K; LISC). S: Limbe, Bangwe Hill, 4.x.1960, *Chapman* 932 (FHO; K; SRGH). **Mozambique.** MS: Manica Distr., Báruè, serra de Choa, 4.vii.1941, *Torre* 3019 (COI; LISC; LMU; SRGH).

Gabon, Rwanda, Burundi, Ethiopia, Uganda, Kenya, Tanzania and the Mascarenes. Thickets, shrubby zone at forest margins, riparian fringes, 760–2130 m.

R. apetalus is intermediate between *R. pinnatus* and *R. exsuccus* in most morphological characteristics, and the three form a taxonomically difficult complex within which there has been much confusion, both in delimitation of taxa and application of names. Typical *R. pinnatus* and *R. exsuccus* (easily distinguished from one another) are common in Mozambique, which has, however, but few representatives of *R. apetalus*, suggesting that although the three taxa are difficult to define, they may be good species. This conclusion is also supported by the slightly different geographical ranges of the species outside our area.

9. **Rubus chapmanianus** Kupicha in Bull. Soc. Brot., Sér. 2, **49**: 5, t. 1 (1976). TAB. 3 fig. A. Type: Malawi, Mt. Mlanje, Lukulezi valley, *J. D. Chapman* 489 (BM, holotype; FHO; K; PRE).

Scrambling shrub climbing to c. 3·5 m. high. Flowering branches shortly and densely greyish-green-villous, with hooked brownish-yellow prickles up to 2·5 mm. long. Leaves 16–20 × 9–18 cm., 2-jugate or the uppermost ternate; petiole and rhachis villous and aculeate like the stems; lateral leaflets subsessile or with petiolule up to 5 mm. long; terminal one with petiolule 1–2·5 cm. long. Leaflets elliptic-acute or broadly ovate-acute, the terminal one largest or sometimes exceeded by those of the basal pair; leaf margins minutely and ± evenly serrate, the nerves excurrent; superior surface dark yellowish-green, appressed-villous, the indumentum glistening; inferior surface paler, greyish-green, densely villous especially on veins, the midrib minutely prickly. Stipules 1·0–1·8 × 0·6–1·0 cm., obovate-acute, densely appressed-villous. Inflorescence 4–14 cm. long, terminal and axillary, many-flowered, cylindrical or narrowly pyramidal; peduncles and pedicels densely greyish-green-villous and very prickly. Calyx greyish-green-villous, deeply divided into lanceolate segments 5–8 mm. long, lobes not reflexed in fruit. Petals absent, or when present broadly spathulate, shorter than calyx segments, white, caducous. Fruit globose, c. 1 cm. in diameter, red or black and edible when ripe; carpels up to 3 mm. long, glabrous or villous.

Zambia. E: Nyika, 31.xii.1962, *Fanshawe* 7364 (K). **Malawi.** N: near Nganda Hill, Nyika Plateau, 2290 m., 5.ix.1962, *Tyrer* 800 (BM; SRGH). S: Mt. Mlanje, 1891, *Whyte* s.n. (K).

Also from Tanzania (Mbeya Distr.) and Uganda (Kigezi Distr.). Dense deciduous woodland, near streams in moist valleys, path sides in Cedar forest.

R. chapmanianus is closely related to *R. pinnatus*, *R. apetalus* and *R. exsuccus*, especially the latter, which it resembles in being densely villous. Like the members of this complex, *R. chapmanianus* has flowers in which the petals are inconspicuous or absent. It is, however, easily distinguished from all other *Rubus* species in the F.Z. area by its very large, broad stipules. Similar stipules are found in *R. keniensis* Standl. (endemic to Kenya), but this species has ternate leaves and large petals; they are also present in some specimens of *R. runssorensis* Engl. (Uganda and Zaire) which again differs from *R. chapmanianus* in its much larger, more showy flowers.

10. **Rubus kirungensis** Engl. in Goetzen, Durch. Afr.: 378 (1895); in Mildbr., Wiss. Ergebn. Deutsch. Zentr.-Afr. Exped. 1907–1908, **2**: 224, t. 20 (1911).—R. A. Grah. in F.T.E.A., Rosaceae: 36 (1960). Type from Zaire.

Scrambling shrub up to 2 m. tall. Flowering branches ± glabrous or sparsely weakly pilose, reddish, sometimes whitish-pruinose below, moderately to very prickly; prickles 1–3 mm. long, straight, deflexed. Leaves 4–23 × 2·5–15 cm., 2–3-jugate or the uppermost ternate or simple, tending to be rather stiff and leathery in texture; petiole and rhachis ± glabrous and aculeate like the stems, or sparsely pilose on the adaxial side, occasionally with scattered stipitate glandular hairs; distal leaflets subsessile, proximal ones with petiolules c. 2–3 mm. long; terminal leaflet with petiolule 0·5–1·5 cm. long. Leaflets ovate, ovate-acute or more rarely narrowly ovate-acute to elliptic-acute, shallowly and irregularly serrate;

superior surface dark yellowish-green, glabrous; inferior surface paler green,
glabrous or with nerves sparsely appressed-pilose, the midrib and quite often lateral
nerves prickly. Stipules narrowly elliptic to narrowly obovate-acute. Flowers
solitary or in axillary clusters, usually hidden among leaves (see note at end);
pedicels very prickly. Calyx tomentose, glabrescent, sometimes tomentose inside
and glabrous outside, sometimes with glandular hairs, deeply divided into
triangular-acuminate lobes 7–10 mm. long. Petals white to pink, slightly exceeding
to 2 × calyx length, with subcircular limb and narrow claw. Fruit globose, c.
1·2 mm. long, black and edible when ripe; carpels few, large, glabrous or apically
pubescent.

Malawi. N: Nyika, fl. 27.xii.1962, *Fanshawe* 7268 (FHO; K). C: Nkhota Kota
Distr., Ntchisi Mt., 1650 m., st. 3.ix.1929, *Burtt Davy* 21124 (FHO).
From Uganda and Tanzania. Secondary montane forest by streams.

The specimens cited are the first records of *R. kirungensis* from the F.Z. area. *Burtt
Davy* 21124 cannot be identified with complete certainty as it is sterile, but its leaves are
typical of this species. *Fanshawe* 7268, on the other hand, is unusual within *R. kirungensis*
in having an extended terminal leafless inflorescence. It closely resembles a Tanzanian
specimen cited under *R. kirungensis* by Graham (op. cit. p. 37: Rungwe Distr., Upper
Kiwira R., v.1938, *McInnes* 408), who questioned whether another species might be
involved. In their leaf characters *Fanshawe* 7268 and *McInnes* 408 agree with *R.
kirungensis*, while their inflorescences suggest affinity with *R. scheffleri*. These species
belong to a complex including also *R. runssorensis* (Uganda) and *R. pinnatus*; more
research is needed in this area before all taxa can be delimited and their relationships
properly understood.

11. **Rubus scheffleri** Engl., Bot. Jahrb. **46**: 125 (1911).—R. A. Grah. in F.T.E.A.,
      Rosaceae: 34 (1960). Type from Kenya.

Straggling scrambling shrub. Flowering branches glaucous-pubescent, sometimes glabrescent, often whitish-pruinose below, moderately prickly; prickles
1–3·5 mm. long, almost straight or decurved. Leaves 7–18 × 4–16 cm., 2–3-jugate
or the uppermost ternate, thin, not leathery; petiole and rhachis pubescent and
aculeate like the stems or sometimes glabrous except for a densely fulvous-villous
line along the adaxial surface; distal leaflets sessile; proximal ones with petiolules
c. 2 mm. long; terminal leaflet with petiolule 0·5–2(4) cm. long. Leaflets ovate-
acute to oblong-acuminate, with shallowly, rather irregularly serrated margin;
superior surface dark green, glabrous to sparsely appressed-pilose; inferior surface
paler green with pilose veins and glabrous lamina, the midrib prickly, prickles very
occasionally appearing also on lateral veins. Stipules filiform to linear. Inflorescences few-flowered to many-flowered, axillary and terminal, 11–28 × 3–7 cm.,
exceeding leafy part of shoot; peduncles and pedicels with indumentum like stems,
almost without prickles to very prickly; pedicels and calyx sometimes with sparse
long-stipitate glands. Calyx densely greyish-tomentose, deeply divided into ovate
to triangular lobes, acute, acuminate to long-cuspidate, 6–9 mm. long, spreading in
fruit. Petals white, pale pink to purple or pale yellow, slightly exceeding to
2 × length of sepals, with suborbicular limb and narrow claw. Fruits globose,
c. 0·8–1·0 cm. long, black and edible when ripe; carpels few, pubescent.

Malawi. S: Mt. Mlanje, Nayawani Forest, 1950 m., fl. 23.viii.1956, *Newman &
Whitmore* 545 (BM; SRGH).
From Uganda, Kenya and Tanzania. Forest undergrowth by streams, c. 1950 m.

This is the first record of *R. scheffleri* from the F.Z. area.

4. HAGENIA J. F. Gmel.

Hagenia J. F. Gmel., Syst. Nat. ed. 13, **2**: 600, 613 (1791).
Banksia Bruce, Trav. **5** App.: 73 " *Banksia* ", tab. 22 & 23 (1790) *nom. illegit.*,
non J. R. & G. Forst. (1776) *nom. rej. Thymeleacearum*, nec L. f. (1781) *nom.
conserv. Proteacearum.*

Dioecious or polygamo-dioecious tree; branchlets ringed by the scars of the
fallen leaf-bases. Leaves imparipinnate, crowded; stipules large, adnate to the
base of the petiole, sheathing the branchlets. Panicles large, terminal, drooping.
Flowers with 2–3 bracts embracing the base of the calyx-tube; calyx-tube with

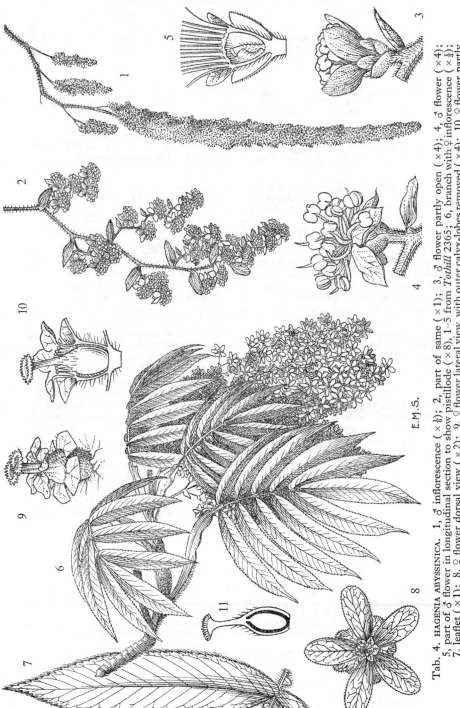

Tab. 4. HAGENIA ABYSSINICA. 1, ♂ inflorescence (×⅓); 2, part of same (×1); 3, ♂ flower partly open (×4); 4, ♂ flower (×4); 5, part of ♂ flower in longitudinal section to show pistillode (×8), 1–5 from *Tothill* 2365; 6, branch with ♀ inflorescence (×⅓); 7, leaflet (×1); 8, ♀ flower dorsal view (×2); 9, ♀ flower lateral view, with outer calyx-lobes removed (×4); 10, ♀ flower partly in longitudinal section to show staminodes and pistil (×4); 11, pistil (×4), 6–11 from *Richards* 6666. From F.T.E.A.

E.M.S.

(4)5 " epicalyx "-lobes alternating with and inserted below the (4)5 calyx-lobes; calyx-lobes in ♂ flowers larger than the epicalyx, in ♀ and ⚥ flowers smaller than the epicalyx; petals (4)5, alternating with the calyx-lobes, or 0; stamens up to 20, inserted on the annular hairy disk lining the mouth of the calyx-tube; carpels (1)2, free within the receptacle; styles (1)2, terminal; stigmas asymmetrically discoid; ovule single, pendulous. Achenes with fragile pericarp.

A monospecific upland genus from NE. Africa.

Hagenia abyssinica (Bruce) J. F. Gmel., Syst. Nat. ed. 13, **2**: 598, 613 (1791).—Engl., Pflanzenw. Afr. **1**: 103, fig. 85 (1910); op. cit. **3**, 1: 299 (1915).—Robyns & Tournay, Fl. Parc Nat. Alb. **1**: 254, fig. 13 (1948).—Brenan, T.T.C.L.: 475 (1949).—Hauman in F.C.B. **3**: 16 (1952).—Eggeling & Dale, Indig. Trees Uganda ed. 2: 332, photo 53 (1952).—Brenan in Mem. N.Y. Bot. Gard. **8**: 432 (1954).—R. A. Grah. in F.T.E.A., Rosaceae: 43, fig. 5 (1960).—Dale & Greenway, Kenya Trees & Shrubs: 401, fig. 80 (1961).—F. White, F.F.N.R.: 68 (1962). TAB. **4**. Type from Ethiopia.

   Banksia abyssinica Bruce, Trav. **5** App.: 73 " *Bankesia* ", tab. 22 & 23 (1790). Type as above.

   Brayera anthelmintica Kunth ex Brayer, Not. Nouv. Pl. Fam. Rosaceae, cum tab. (1822).—A. Rich. in Bull. Sci. Soc. Philom. Paris **1822**: 156 (1822).—DC., Prodr. **2**: 588 (1825).—Oliv., F.T.A. **2**: 380 (1871). Type uncertain.

   Hagenia abyssinica var. *viridifolia* Hauman in Bull. Jard. Bot. Brux. **22**: 90 (1952); in F.C.B. **3**: 17 (1952). Syntypes from Zaire, Rwanda, Burundi, Uganda and Kenya.

Tree up to 18 m. high, with globose or umbrella-shaped open crown; bark ridged and flaky, red-brown to brown. Branchlets densely sericeous-villous with golden antrorse hairs 3–4 mm. long, ringed by the scars of the fallen sheathing leaf-bases; scars at first villous, later glabrescent. Leaves imparipinnate, up to 40 cm. long, viscid; leaflets 5–7(8) on each side, usually 12–15 × 3·5–5·5 cm., subopposite, sessile or almost so, narrowly oblong, acuminate at apex, obliquely rounded to subcordate at the base, ± glabrescent above, densely silvery sericeous-villous to glabrescent below; margins finely dentate; main leaflet pairs alternating with 1(3) reduced subcircular ones, the latter with up to 5 teeth or minute and entire; rhachis and petiole with soft patent hairs up to 4 mm. long; petiole up to 15 cm. long; stipules (3)6–9 cm. long, membranous, reddish at first but becoming brown, adnate to petiole throughout almost their whole length. Panicles many-flowered, much-branched, terminal, drooping, up to 60 cm. long; ♀ panicles bulkier than the ♂ ones; flowers subtended by 2(3) broadly rounded bracts. ♂ flowers orange-buff to white, c. 8 mm. in diameter; calyx-tube obconical, c. 1·5 mm. long, densely hairy outside; epicalyx-lobes c. 1·5 × 0·5 mm., oblong; calyx-lobes 4–5 × 2·5–3 mm., oblong to obovate, recurved and with the abaxial surface concave, clearly veined; petals 1·5 mm. long, linear, fugacious (? or absent); stamens (8)10–15(20), anthers sparsely hairy; carpels vestigial, enclosed by the receptacle. ♀ and ⚥ flowers red, viscid, up to 1·5 cm. in diameter; calyx-tube c. 1 mm. long, obconical, densely hairy outside; epicalyx-lobes c. 1 × 0·4 cm., unequal, oblong-elliptic to obovate with apex obtuse, clearly veined, accrescent in fruit; calyx-lobes c. 3 × 2 mm., broadly ovate-acute with the apex acute, clearly veined; petals not seen (? absent or fugacious); stamens 0, or if present then fewer and smaller than in ♂ flowers; carpels (1)2, apically villous, with styles c. 1·5 mm. long; stigmas 0·75 mm. wide, hairy. Achenes enclosed within the calyx-tube with its accrescent epicalyx and persistent calyx-lobes; often only one of the achenes developing; pericarp thin, fragile, brown, reticulately rugose.

Zambia. E: Nyika Plateau, ♀ fl. 27.xi.1965, *Lees 96* (K), ♀ fl. 25.xii.1958, *Lawton 524* (FHO; K). **Malawi. N:** Nyika Plateau, ♀ fl. 17.viii.1946, *Brass 17298* (BM; K; SRGH); Nyika Plateau near L. Kaulime, ♂ fl. 12.ix.1962, *Tyrer 885* (BM).

Also from Ethiopia, Sudan, Uganda, Kenya, Tanzania, Zaire, Rwanda and Burundi. Fringe of upland rain-forest, deciduous woodland and evergreen bushland, 1950–2350 m.

Note: The dried ♀ inflorescence has highly effective anthelmintic properties.

5. LEUCOSIDEA Eckl. & Zeyh.

Leucosidea Eckl. & Zeyh., Enum. Pl. Afr. Austr. Extratrop.: 265 (1836).

Much-branched, densely leafy, monoecious shrub or small tree. Leaves alternate, imparipinnate; leaflets inciso-dentate, the main pairs sometimes

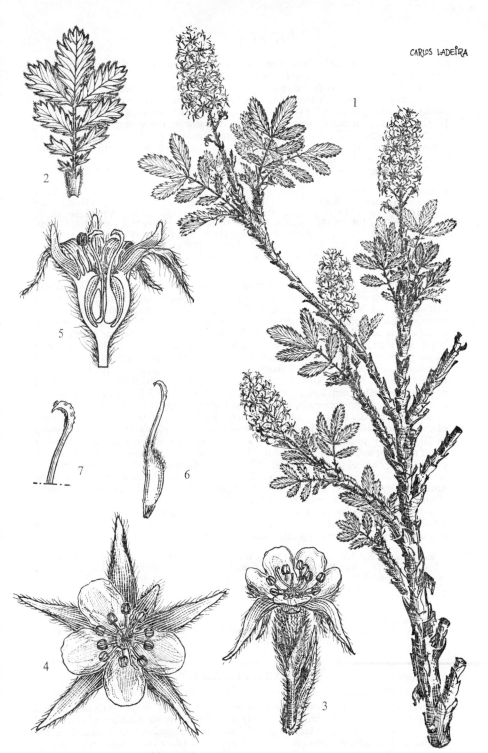

CARLOS LADEIRA

Tab. 5. LEUCOSIDEA SERICEA. 1, fruiting branch (× ⅔); 2, leaf (× ⅔), 1–2 from *Torre & Pereira* 12857; 3, flower, lateral view, showing bracts (× 4); 4, flower, top view, with one petal removed to show the underlying epicalyx-lobe (× 6); 5, flower, longitudinal section (× 6); 6, carpel (× 8); 7, apical part of style, and stigma (× 16), 3–7 from *Torre & Pereira* 12716.

alternating with reduced minute ones; stipules adnate to the petiole, sheathing the
stem, long-persistent.   Flowers bisexual, borne in dense terminal racemose
inflorescences.   Calyx-tube obconical, the mouth almost closed by the annular
disk, bearing 2 series, each consisting of 5(6) free elements: outer ones ("epicalyx"-
lobes) alternating with inner (calyx-lobes).   Petals 5(6), soon deciduous.   Stamens
10(12), inserted on the margin of the disk, dimorphic, the shorter ones oppositipetalous.   Carpels 2(3); styles terminal, filiform, exserted; stigmas hooked; ovules
solitary, pendulous.   Achenes membranous, enclosed within the persistent
indurated calyx-tube, often only one maturing.
    Monospecific genus, confined to SE. Africa.

Leucosidea sericea Eckl. & Zeyh., Enum. Pl. Afr. Austr. Extratrop.: 265 (1836).—
    Harv. in Harv. & Sond., F.C. **2**: 289 (1862).—Sim, For. Fl. Col. Cape Good Hope:
    216, t. 64 fig. 2 (1907).—Engl., Pflanzenw. Afr. **1**, **1**: 455, fig. 383 (1910); op. cit. **3**,
    1: 299, fig. 191 (1915).—Weim. in Bot. Notis. **1932**: 20 (1932).—Henkel, Woody
    Pl. Natal & Zulul.: 215 (1934).—Burtt Davy, F.P.F.T.: 317 (1936).—Pardy in
    Rhod. Agric. Journ. **53**: 961 cum 2 photogr. (1956).—Killick in Fl. Pl. Afr. **40**:
    t. 1566 (1969).—Palmer & Pitman, Trees of S. Afr. **1**: 675 (1972). TAB. **5**. Type
    from S. Africa.

    Much-branched shrub c. 3 m. tall or small tree up to 5(?10) m., usually appearing
shaggy shining white at a distance; young twigs densely leafy; branches flexuous,
exfoliating, reddish-brown.   Leaves (2)5–7(10) cm. in outline; upper surface dark
green, turning dark brown when dried, shortly tomentose, lower surface shining
white sericeous-villous; lateral leaflets 2–3-paired, oblong to obovate with cuneate
base; margins regularly inciso-dentate with c. 15 uninerved teeth 2–4 mm. long;
distal pair of leaflets up to 3·5 × 1·5 cm., the more proximal ones smaller and
sometimes interposed with 1–2 pairs of reduced minute leaflets with only 1–3(5)
teeth; rhachis, midribs and lateral main nerves densely white-sericeous-villous;
petiole 0·5–0·8(1·2) cm. long, articulated; stipules c. 10 × 3·5 mm., entire, brown,
membranous, white-sericeous, adnate to and exceeding petiole, sheathing the stem,
persistent for years after the leaves are shed, usually ± imbricate towards the base
of the branches.   Flowers borne in many-flowered ± cylindrical erect racemes up
to 10 × 3 cm.; pedicels up to 6 mm. long, ± hidden within a basal cymbiform
bract and also subtended by 2 smaller (c. 2·5 mm. long) ovate-acute bracts which
embrace the base of the calyx-tube.   Calyx-tube obconical, 3–4 mm. long, dorsally
sericeous-villous; epicalyx-lobes c. 2 × 2 mm., ovate-rhombic, with the apex
obtuse; calyx-lobes c. 4 × 1·5 mm., triangular.   Petals c. 3 × 1·6 mm., broadly
elliptic-rhombic, lemon-yellow.   Disk annular, hairy.   Filaments of longer stamens
c. 1·1 mm. long, of shorter ones c. 0·7 mm.   Ovary sericeous in upper ⅓; styles
1·5–2·0 mm. long, glabrous.   Achenes 1(2–?3), enclosed within the indurated
calyx-tube with persistent epicalyx and calyx-lobes.

    Rhodesia. E: Umtali Distr., Stapleford, fl. 29.ix.1948, *Chase* 1276 (BM; COI;
LISC; SRGH); Nuza Plateau, Wenya, fl. x.1934, *Gilliland* 883 (BM; FHO; K; PRE).
Mozambique. MS: Manica Distr., Manica, Zuira Mts., Tsetsera Plateau, fl. 5.xi.1965,
Torre & Pereira 12716 (COI; EA; K; LISC; LMU; SRGH); Tsetsera Plateau, fr.
11.xii.1965, *Torre & Pereira* 12857 (COI; LISC; LMU; PRE).
    Only known from SE. Africa, from Rhodesia and Mozambique to the Cape Prov.   On
mountain grassland slopes, especially along streams and river banks, locally common or
isolated; in F.Z. area at altitudes ranging from 1350 to 2200 m. (in S. Africa from above
1067 m.).

6. CLIFFORTIA L.

Cliffortia L., Sp. Pl. **2**: 1038 (1753); Gen. Pl. ed. 5: 460 (1754).

    Erect or procumbent, dioecious or monoecious shrubs or rarely trees.   Leaves
of species in F.Z. area trifoliolate (rarely unifoliolate by suppression of the lateral
leaflets); leaflets oblanceolate to linear; stipules triangular, acute, adnate to the
petiole and sheathing the stem.   Flowers unisexual (very rarely bisexual), axillary,
usually solitary, petals 0.   ♂ flowers with (3)4 petaloid, free or ± connate calyx-
lobes and (3)4 or (6)8 or up to 50 stamens.   ♀ flowers with calyx segments connate
below into a ribbed, winged or smooth tube enveloping the receptacle; free lobes
of the calyx (3)4, deciduous or persistent in fruit; styles 1 or 2, long and filiform

or very short; stigma lacerate-multifid.   Achenes 1 or 2, enveloped by the ±
inflated, persistent calyx-tube.
An African genus of c. 80 species, most of them confined to S. Africa.

Leaflets linear, with rounded apex; venation apparently consisting of a single stout
   midrib; margins thickened but not revolute; young branches antrorsely villous
   stipules usually densely villous on abaxial surface -   -   -   - 1. *linearifolia*
Leaflets oblanceolate to rhombic, with acute apex; venation pinnate; margins not too
   strongly revolute; young branches antrorsely or retrorsely hairy; stipules glabrous or
   hairy on abaxial surface:
   Young branches antrorsely villous; leaflets with strongly revolute margins which almost
     conceal the midrib, thus appearing linear; stipules usually glabrous on abaxial
     surface except for hairs along margins and at apex   -   -   -   2. *nitidula*
   Young branches retrorsely sericeous; leaflets with margins not or slightly revolute, the
     midrib thus not concealed and leaflets evidently rhombic or oblanceolate; stipules
     villous on abaxial surface -   -   -   -   -   -   - 3. *serpyllifolia*

1. **Cliffortia linearifolia** Eckl. & Zeyh., Enum. Pl. Afr. Austr. Extratrop.: 270 (1836).—
   Harv. in Harv. & Sond., F.C. **2**: 301 (1862) pro parte.—Sim, For. Fl. Col. Cape Good
   Hope: t. 137 fig. V (1907).—Burtt Davy, F.P.F.T.: 316 (1932) pro parte min.—
   Weim., Monogr. Cliffortia: 53, fig. 12 F–M (1934) pro parte max.   TAB. **6** fig. C.
   Type from S. Africa.

Ericoid shrub up to 2 m. high, erect, virgate, dioecious.   Branches rigid, erect,
densely antrorsely villous when young, the outer bark tending to flake off.   Leaves
trifoliolate (rarely unifoliolate by suppression of the lateral leaflets), usually densely
arranged on short shoots; leaflets up to 9 × 0·6 mm., linear, obtuse at apex, rigid,
glabrous, rather thick with thickened margins, apparently very stoutly 1-nerved;
stipules triangular, scarious, usually densely villous on the abaxial surface.   Stamens
always 4.   Stigma 1, almost sessile.

   Rhodesia.   E: Umtali, Inodzi, Umtali R., fl. 26.viii.1949, *Chase* 1734 (BM;   COI;   K;
LISC;   SRGH);   Inyanga, Niarerua R., fr. 29.x.1930, *Fries, Norlindh & Weimarck* 2406
(BM;   LD;   SRGH).   **Mozambique.**   MS: Manica Distr., Zuira Mts., Tsetsera Plateau,
road to Manica, fl. 4.xi.1965, *Torre & Correia* 12672 (COI   K   LISC;   LMU;   SRGH).
   From S. Africa to Rhodesia and the western border of Mozambique.   Along mountain
river banks, above 1550 m.

   When leaflets are macerated in heated lactic acid the venation is seen to be densely and
irregularly reticulate.   The leaves of plants growing in relatively sunny and/or dry habitats
have much more prominent venation than those of plants from shadier or more humid
environments.

2. **Cliffortia nitidula** (Engl.) R. E. & T. C. E. Fr.* in Notizbl. Bot. Gart. Berl. **8**: 649
   (1923).—Weim., Monogr. Cliffortia: 47, fig. 10 A–E (1934) pro parte max.—Brenan,
   T.T.C.L.: 475 (1949); in Mem. N.Y. Bot. Gard. **8**: 432 (1954) excl. specim. *Fries,
   Norlindh & Weimarck* 3685.—Van der Veken in Bull. Soc. Roy. Bot. Belg. **91**: 100
   (1958).—R. A. Grah. in F.T.E.A., Rosaceae: 42, fig. 4 (1960).—White, F.F.N.R.:
   68 (1968).—Mendes in C.F.A. **4**: 4 (1970).   TAB. **6** fig. B.   Lectotype from Tanzania.
   Cliffortia linearifolia sensu Bak. f. in Trans. Linn. Soc., Ser. 2, Bot. **4**: 13 (1894).
   Cliffortia linearifolia var. *nitidula** Engl. in Bot. Jahrb. **26**: 376 (1899) pro parte
   quoad specim. Tanzanianum.   Syntypes from Tanzania and Angola.

Ericoid shrub c. 2(?5) m. high, much-branched, erect, dioecious or monoecious.
Branches rigid, erect, densely antrorsely villous when young, the outer bark tending
to flake off.   Leaves trifoliolate, usually densely arranged on short shoots; leaflets
up to 9 × 1 mm., rigid, glabrous, shining, oblanceolate with acute, red-tipped apex
but appearing linear as the margins are usually strongly revolute so as almost to
conceal the midrib; venation weakly pinnate; stipules triangular, scarious,
glabrescent except at margins and apex or (in some Inyanga specimens) sparsely
hairy on abaxial surface.   Stamens usually (3)4(5) or, in Malawian specimens, (6)8.
Stigma 1, almost sessile.

   * Graham (loc. cit.) points out, in using the epithet " *nitidula* ", R. E. & T. C. E. Fries
evidently intended their species to be based on Engler's variety of *C. linearifolia*; however,
as they chose partly different types, Engler's name is omitted from his citation.   This is
indefensible: Engler based his var. on *Ramalho* s.n. from Angola and *Stuhlmann* 9160
from Tanzania.   The Fries's cited *Stuhlmann* 9160 and *Goetze* 257, both from Tanzania.
Their use of the name must be seen as a new combination, and implicit lectotypification
by the *Stuhlmann* collection.

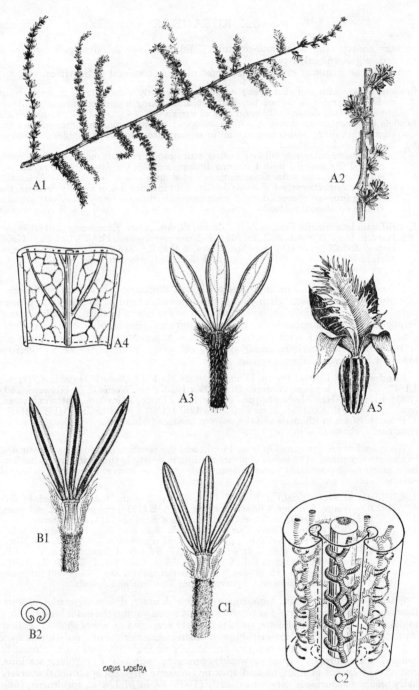

CARLOS LADEIRA

Tab. 6. A.—CLIFFORTIA SERPYLLIFOLIA. A1, adult branch (×½); A2, section of stem
showing bark flaking off (× ⅔); A3, leaf, abaxial view, and section of young stem
(×8); A4, semi-diagrammatic section of leaflet showing weak pinnate venation
(c. ×20), A1–A4 from *Torre & Correia* 15570; A5, ♀ flower (×25) from *E. M. & W.*
84. B.—CLIFFORTIA NITIDULA. B1, leaf, abaxial view, and section of young stem
(×8); B2, diagrammatic cross-section of leaflet (×16), B1–B2 from *Brass* 16457.
C.—CLIFFORTIA LINEARIFOLIA. C1, leaf, abaxial view, and section of young stem
(×8); C2, semi-diagrammatic section of leaflet showing reticulate venation from the
stout midrib (c. ×35), C1–C2 from *Rattray* 910.

Zambia. W: Mwinilunga Distr., Luakera R. near 16 mile post on Kalene hill–Mwinilunga road, fl. 7.xi.1952, *Holmes* 972 (FHO). E: Nyika plateau, fl. 30.xii.1962, *Fanshawe* 7321 (FHO; K; SRGH). **Rhodesia.** E: Inyanga Distr., Mare R., fl. 21.x.1946, *Wild* 1442 (K; LD; SRGH); Inyanga, fl. 27.x.1942, *Hopkins* SRGH 9383 (K SRGH). **Malawi.** N: Vipya Plateau, Rumpi stream, st. 17.ii.1962, *Chapman* 1598 (FHO; SRGH). S: Mlanje Distr., Luchenya Plateau, fl. 27.vi.1946, *Brass* 16457 (BM; EA; FHO; K; SRGH); Luchenya Plateau, fl. 24.xii.1956, *Chapman* 322 (BM; FHO; K; SRGH).

Also in Tanzania, Zaire and Angola. Along stream-banks in upland grassland and on rocky slopes in *Widdringtonia* forest and in mushitu; above 1750 m.

A few specimens of *C. nitidula* from Inyanga have a stout reticulation between the secondary veins of the leaflet, and stipules sparsely hairy on their abaxial surface; an example is *Hopkins* SRGH 9383 (K; SRGH) cited above. The characters of these plants suggests that some introgression may have occurred between *C. nitidula* and *C. linearifolia*, the only other member of *Cliffortia* known from this area. A specimen from Mozambique, MS: Tsetsera Mt., st. 6.xi.1946, *Simão* 1154 (LISC; LMA) should also be considered as a putative hybrid between these species. *C. linearifolia* has been recorded from this locality, and the area is ecologically compatible with *C. nitidula*.

3. **Cliffortia serpyllifolia** Cham. & Schlechtend. in Linnaea 2: 34 (1827).—Harv. in Harv. & Sond., F.C. 2: 301 (1862) pro parte excl. var. *α penninervis* et var. *β chamissonis*.—Weim., Monogr. Cliffortia: 51, fig. 12 A–E (1934). TAB. 6 fig. A. Type from S. Africa.

? Cliffortia polyphylla Eckl. & Zeyh., Enum. Pl. Afr. Austr. Extratrop.: 268 (1836). Type from S. Africa.

Cliffortia serpyllifolia var. *γ foliis angustioribus* Eckl. & Zeyh., loc. cit. Type from S. Africa.

Cliffortia serpyllifolia var. *γ polyphylla* (Eckl. & Zeyh.) Harv. in Harv. & Sond., loc. cit. Type from S. Africa.

Cliffortia linearifolia sensu Bak. f. in Journ. Linn. Soc., Bot. 40: 67 (1911).

Cliffortia tychonis Weim. in Bot. Notis. 1932: 20 (1932). Type: Rhodesia, Inyanga, *Norlindh & Weimarck* 4276 (LD, holotype).

Cliffortia nitidula sensu Weim., op. cit.: 47–48 (1934) pro parte quoad specim. *Cooper* s.n. in *Eyles* 2753; *Fries, Norlindh & Weimarck* 2725 & 3685a; *Norlindh & Weimarck* 4276; et *Galpin* 9265 in *Eyles* 5302.

Cliffortia nitidula sensu Brenan in Mem. N.Y. Bot. Gard. 8: 432–433 (1954) pro parte quoad specim. *Fries, Norlindh & Weimarck* 3685.

Shrub up to 1·8 m. tall, usually erect, much-branched, densely bushy. Branches ± diffuse, flexible, widely spreading, densely retrorsely sericeous when young. Leaves trifoliolate, ± closely arranged on short shoots; leaflets up to 8(10) × 1·7 mm., oblanceolate to rhombic with acute apex, glabrous, flat (in shade forms) or with margins slightly revolute but never concealing the midrib as in *C. nitidula*; venation weakly pinnate; stipules triangular, usually hairy on abaxial surface. Stamens 4(5). Stigma 1, almost sessile.

Rhodesia. E: Umtali Distr., Vumba Mts., st. 28.x.1948, *Chase* 1182 (BM; K; SRGH); Inyanga Distr., Mt. Inyangani, fl. 8.xii.1930, *Fries, Norlindh & Weimarck* 3685a (BM; LD). **Mozambique.** MS: Sofala Distr., Gorongosa Mts., Gogogo Peak, fl. 9.x.1944, *Mendonça* 2413 (K; LISC; LMU); Manica Distr., Zuira Mts., Tsetsera Plateau, road to Manica, fl. 2.iv.1966, *Torre & Correia* 15570 (COI; EA; LISC; LMA; PRE).

From S. Africa to Rhodesia and Mozambique. By streams, frequently in shade of forest trees; above 1700 m.

I have not seen the type collection of *C. serpyllifolia* Cham. & Schlechtend. (Cape of Good Hope, Congo R., v.1819, *Mundt & Maire* s.n.); it was probably destroyed in Berlin. In the protologue of *C. polyphylla*, Ecklon & Zeyher state that this species is near to *C. serpyllifolia*, and I follow H. Weimarck (op. cit.: 51–53) in accepting that the two are probably synonymous.

7. AGRIMONIA L.

Agrimonia L., Sp. Pl. **1**: 446 (1753); Gen. Pl. ed. 5: 206 (1754).

Tall herbs with imparipinnate, alternate leaves; leaflets dentate; stipules adnate to the base of the petiole. Flowers usually in terminal, spicate racemes; pedicels bracteate at the base, 2-bracteolate in the middle; calyx persistent, tube turbinate with ∞ hooked spinules or 5 teeth below the calyx-lobes; calyx-lobes 5, triangular,

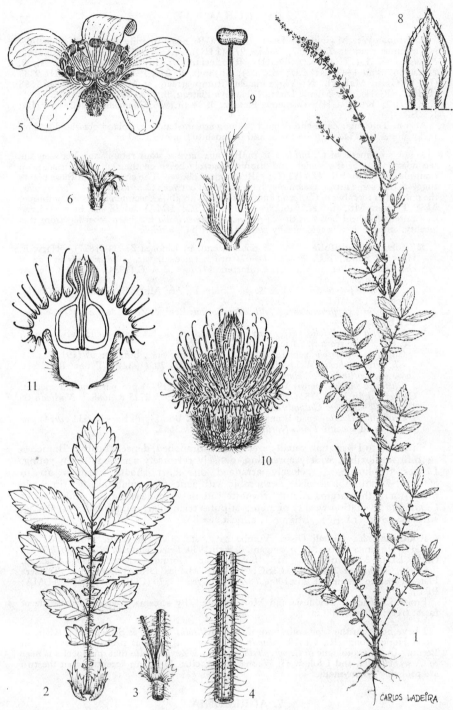

Tab. 7. AGRIMONIA BRACTEATA. 1, habit ($\times \frac{1}{6}$); 2, leaf ($\times \frac{1}{2}$); 3, node embraced by stipules ($\times \frac{1}{2}$); 4, section of stem ($\times 1\frac{1}{2}$); 5, flower, with pedicel, bract and bracteoles removed ($\times 4$); 6, pedicel and bracteoles, with bract removed ($\times 4$); 7, lower bract ($\times 4$); 8, calyx-lobe ($\times 10$); 9, stamen ($\times 10$); 10, fruit, lateral view ($\times 4$); 11, longitudinal section of fruit, showing achenes ($\times 4$), all from *Norlindh & Weimarck* 4658.

connivent in fruit; petals 5, yellow, circular to oblong, larger than the calyx-lobes, deciduous; disk annular, glandular, lining the mouth of the calyx-tube; stamens 5–20, uniseriate, inserted on the disk; carpels 2, styles filiform, exserted, with the stigma dilated. Achenes (1)2, included in the hardened calyx-tube which is beset near the apex with a crown of hooked prickles.

A genus of c. 20 species from N. temperate regions and tropical mountains. Only 1 species known from tropical and southern Africa.

Agrimonia bracteata E. Mey. ex C. A. Mey. in Bull. Acad. Sci. St. Petersb. **10**: 349 (1842); in Ann. Sci. Nat. Bot., Sér. 2, **18**: 380 (1842).—Skalický in Kew Bull. **17**: 87 (1963). TAB. **7**. Lectotype from S. Africa (Cape Prov.).

Agrimonia repens sensu auctt. et collectt. afric. pluribus, e.g. Cham. & Schlechtend. in Linnaea **2**: 28 (1827).—Eckl. & Zeyh., Enum. Pl. Afr. Austr. Extratrop.: 264 (1836) non L. (1759).

Agrimonia eupatoria sensu auctt. et collectt. afric. pluribus, e.g. Eckl. & Zeyh. loc. cit.—Engl., Pflanzenw. Afr. **3**, 1: 299 (1915).—Phillips in Ann. S. Afr. Mus. **16**: 90 (1917) non L. (1753).

Agrimonia caffra E. Mey. ex C. A. Mey. locis cit. Type from S. Africa.

Agrimonia bracteosa E. Mey. ex Drège, Zwei Pflanzengeogr. Doc.: 51, 142, 162 (1843) *nom. nud.*

Agrimonia eupatoria L. var. *capensis* Harv. in Harv. & Sond., F.C. **2**: 290 (1862).— Burtt Davy, F.P.F.T.: 318 (1936). Syntypes from S. Africa.

Agrimonia odorata sensu auctt. et collectt. afric. pluribus, e.g. Weim. in Bot. Notis. **1932**: 20 (1932) non Mill. (1768).

Rhizomatous perennial herb with an ephemeral basal rosette of leaves. Flowering stems up to 1 m. high, sparsely covered with both short glandular hairs and long, spreading hairs often of unequal length. Leaves up to 25 cm. long, alternate, imparipinnate; main leaflet pairs 3–4, alternating with 1–2(3) pairs of minute leaflets reduced to 1–3(5) teeth; leaflets of main pairs up to 6 × 3 cm., ovate-oblong to obovate, sessile, coarsely and acutely serrate almost to the base; superior surface sparsely pilose and beset with sparse sessile glands, inferior surface tomentose-pubescent and densely glandular; stipules foliaceous, broadly ensiform, dentate, shortly adnate to the base of the petiole. Inflorescences spicate, terminal, ± lax, mostly simple racemes up to 40 cm. long; pedicels short, bracteate at the base and bibracteolate at the articulation; bract and bracteoles usually 3-fid or rarely bract triangular and entire and then bracteoles subtrilobate to 3-fid. Calyx-tube turbinate, densely hirsute outside and armed with ∞ hooked spinules below the calyx-lobes; calyx-lobes triangular, c. 2 × 1 mm. long, conspicuously 3-nerved. Petals c. 6 × 4 mm., oblong to obovate, yellow, deciduous. Stamens 10–15, arising from the margin of the annular thickened disk which almost closes the mouth of the calyx-tube. Carpels 2; styles c. 2·5 mm. long, filiform, exserted; stigma dilated. Achenes (1)2, enclosed in the hardened calyx-tube, the latter turbinate-campanulate, hairy, grooved along almost its whole length, crowned with the accrescent hooked bristles (the outer bristles 1·5–2·5 mm. long, and deflexed, innermost ones 3·5–5 mm. long, erect); calyx-lobes connivent in fruit.

Rhodesia. E: Inyanga Distr., Inyanga, fl. 29.i.1931, *Norlindh & Weimarck* 4658 (BM; PRE; SRGH); Stapleford Distr., fl. & immat. fr. 22.ii.1946, *Wild* 842 (COI; K; SRGH).

Also known from mountains in S. Africa (Cape Prov. and Natal). Upland grasslands and underwoods, often riverine, 1700–1900 m.

8. ALCHEMILLA L.

Alchemilla L., Sp. Pl. **1**: 123 (1753); Gen. Pl. ed. 5: 58 (1754).

Perennial herbs, or (outside the F.Z. area) low shrubs or annual herbs. Stems erect or decumbent, often arising from a central rosette developing rooting or non-rooting stolons. Leaves circular or reniform in outline, ± deeply lobed, rarely entire, palmately veined; stipules membranous or foliaceous, adnate to the petiole and ± sheathing the stem. Inflorescences axillary, simple or branched, few- to many-flowered, exceeding or ± hidden by the subtending leaf, or reduced to 1–2 flowers concealed by the stipules. Flowers small, bisexual, usually tetramerous, pedicellate to sessile. Calyx-tube urceolate, ± membranous, persistent, bearing around the constricted mouth 2 series each of 4 free elements: the outer (" epicalyx-lobes ")

usually smaller but sometimes larger, alternating with the inner (calyx-lobes).
Disk ± fleshy, closing the mouth of the calyx-tube. Petals 0. Stamens usually 4,
sometimes less, small, inserted on the rim of the disk, alternating with the calyx-
lobes. Carpels 1–8(12), sessile or stalked, inserted at the base of the receptacle;
styles basal or central; stigma capitellate. Achenes 1–8(12), included within a ±
inflated, membranous urceolate calyx.

A large genus of c. 100 species of temperate regions; in the tropics confined to
mountainous areas.

Calyx glabrous; epicalyx definitely shorter than to almost equalling calyx-lobes; leaves
    with basal sinus wide, sometimes almost 180°; basal rosette absent
                                                         1. *ellenbeckii* subsp. *nyikensis*
Calyx hairy or rarely glabrous; epicalyx exceeding calyx-lobes; leaves with basal sinus
    narrower than above, usually c. 90°; basal rosette present or absent:
Basal rosette absent or fugacious, when present basal leaves 5(7)-lobed; leaf-lobes
    broader than long, rounded or broadly obovate, usually not deeper than ⅓ radius of
    lamina; superior leaf-surface usually glabrous or subglabrous, inferior surface
    usually ± densely hairy especially along the main nerves; inflorescence usually
    reduced to 1–2 flowers concealed by stipules, less often in short cymes or panicles
                                                                          2. *cryptantha*
Basal rosette present, usually persistent, basal leaves (5)7(9)-lobed; leaf-lobes longer
    than broad, oblong-elliptic to obovate; leaves usually divided to more than ½ radius
    of lamina; superior leaf surface usually sparsely appressed-sericeous (the hairs
    pointing towards the lobe-margins), inferior surface ± hairy; inflorescence usually
    a shortly branched panicle of cymes, rarely reduced to 1–2 flowers concealed by
    stipules -   -   -   -   -   -   -   -   -   -   3. *kivuensis*

1. **Alchemilla ellenbeckii** Engl., Bot. Jahrb. **46**: 135, fig. 1A (1911). Type from
Ethiopia.

Subsp. **nyikensis** (De Wild.) R. A. Grah. in Kew Bull. **12**: 406 (1958).—Piovano in Boll.
Soc. Bot. Ital., Sér. 2, **72**: 664 (1967). TAB. **8**. Type: Malawi, Nyika Plateau,
xi.1903, *Henderson* s.n. (BM, holotype).
    Alchemilla nyikensis De Wild. in Bull. Jard. Bot. Brux. **7**: 376 (1921).—Hauman &
Balle in Rev. Zool. Bot. Afr. **24**: 340, 346, fig. 14 (1934).—Brenan in Mem. N.Y.
Bot. Gard. **8**: 431 (1954). Type as above.

Creeping and scrambling perennial herb without basal rosette. Stems slender,
green, becoming brownish-green to red, densely pilose to glabrous. Leaves
petiolate, 1–1·5(2·0) × 1·2–2·5(3·0) cm., reniform to circular-reniform in outline,
(3)5-lobed, densely hairy to glabrous, paler on inferior surface; lobes quite shallow,
usually not as deep as ½ radius of lamina; central lobe with 9–11 shallow teeth, the
latter usually with revolute margins; basal sinus wide, sometimes almost 180°;
petioles 0·5–1·2(1·8) cm. Inflorescences 4–10 cm. long, slender, little-branched,
hairy; flowers 1–3 together in cymose clusters. Calyx glabrous; epicalyx-lobes
0·5–0·7 mm. long; calyx-lobes 0·7–0·8(1·0) mm. long, exceeding epicalyx-lobes.
Achenes usually solitary.

Zambia. E: Lundazi Distr., Nyika Plateau, fl. 24.xii.1962, *Fanshawe* 7218 (K;
SRGH); Nyika Plateau, fl. 3.i.1959, *Richards* 10414 (EA; K; LISC). **Malawi.** N:
Nyika Plateau, fl. 13.viii.1946, *Brass* 17201 (K; SRGH); Vipya Mts., Rumpi stream,
Nkalapia, fl. v.1954, *Chapman* 285 (BM).
    Also in Tanzania. Marshy upland grassland, bogs and *Widdringtonia* forest; alt.
1950–2300 m.

Note: A single gathering from Zambia, E: Nyika Plateau, *Fanshawe* 7234 (K; SRGH)
is too poor to be properly described, but it may represent a hybrid between *A. kivuensis*
and *A. ellenbeckii* subsp. *nyikensis*: it has the general habit and peculiar stem-colour of
the latter, while both the flower type and superior leaf-surface indumentum are
characteristic of the former. Both these taxa occur on the Nyika Plateau.

2. **Alchemilla cryptantha** Steud. ex A. Rich., Tent. Fl. Abyss. **1**: 259 (1847).—Oliv. in
F.T.A. **2**: 377 (1871).—Hauman & Balle in Bull. Jard. Bot. Brux. **14**: 8, fig. 2
(1936).—Robyns & Tournay, Fl. Parc Nat. Alb. **1**: 246 (1948).—Suesseng. & Merxm.
in Trans. Rhod. Sci. Ass. **43**: 15 (1951).—Hauman in F.C.B. **3**: 7 (1952).—Keay in
F.W.T.A. ed. 2, **1**: 424 (1958).—R. A. Grah. in F.T.E.A., Rosaceae: 15 (1960).—
Piovano in Boll. Soc. Bot. Ital., Sér. 2, **72**: 664 (1967). Type from Ethiopia.
    Alchemilla subreniformis De Wild. in Bull. Jard. Bot. Brux. **7**: 383 (1921).—

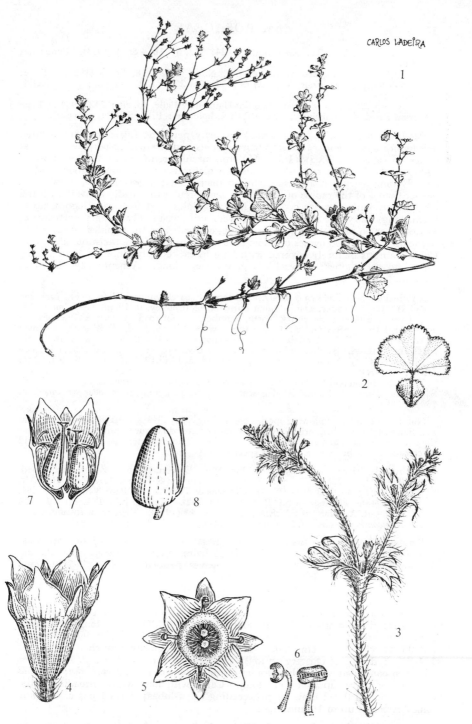

CARLOS LADEIRA

Tab. 8. ALCHEMILLA ELLENBECKII subsp. NYIKENSIS. 1, flowering plant, habit (×½)
Fanshawe 7258; 2, leaf (×1); 3, branch of inflorescence (×4); 4, flower, lateral
view, part of pedicel cut off (×16); 5, flower, top view (×16); 6, stamens, side and
adaxial view (×35). 7, longitudinal section of fruiting calyx showing immature
achenes (×16); 8, mature achene (×16), 2–8 from *Richards* 10414.

Hauman & Balle in Rev. Zool. Bot. Afr. **24**: 312, fig. 1b, 4 & 14 (1934). Type from Zaire.
Alchemilla inyangensis Weim. in Bot. Notis. **1932**: 18, fig. 6 (1932). Type: Rhodesia, Inyanga, *Fries, Norlindh & Weimarck* 2638 (LD, holotype; BM, SRGH, isotypes).
Alchemilla kiwuensis subsp. *rhodesica* Hauman & Balle, op. cit.: 342 (1954). Type from Rhodesia: Salisbury, *Eyles* 1911 (K, holotype; PRE, SRGH, isotypes).

Perennial herb with basal rosette of shortly petiolate, fugacious leaves, often developing long, slender, densely hairy stolons usually rooting here and there at nodes. Leaves 1–1·5(3·5) × 1·5–3(4·5) cm., circular-reniform in outline, those of rosette 5(7)-lobed, the rest 5-lobed, usually glabrous or subglabrous above and ± densely long-hairy beneath especially on main veins; lobes shallow, usually not deeper than ⅓ of the radius of lamina, ± semicircular to broadly obovate; central lobe usually with 15 teeth c. 1 mm. long; petioles of stem leaves slender, up to 5 cm. long, hairy like the stems. Flowers axillary, usually solitary or twinned, subsessile or on pedicels up to 8 mm. long, concealed by the stipules, or less often borne in ± elongated cymes or panicles usually not exceeding the subtending leaf. Calyx glabrous or rarely covered with long hairs; epicalyx-lobes c. 1 mm. long, exceeding calyx-lobes; calyx-lobes c. 0·7 mm. long. Achenes (2)4–8, up to 1·5 mm. long.

Rhodesia. N: Trelawney Distr., Tobacco Research Station, fl. xi.1944, *Jack* 265 (SRGH). C: Salisbury, Danga Lima Farm N. of Bromley, fl. 10.xii.1964, *Chase* 8200 (K; LISC; SRGH); Salisbury North, Sumber, Gwebi road, fl. early xi.1959, *Phipps* 2494 (K; LISC; SRGH). E: Inyanga, fl. 5.xi.1930, *Fries, Norlindh & Weimarck* 2638 (BM; LD; SRGH); Umtali Distr., Penhalonga, fl. 31.x.1956, *Robinson* 1829 (K; SRGH). **Malawi.** S: Ncheu Distr., road from Dedza to Ncheu, fl. 2.xi.1960, *Chapman* 1023 (K; SRGH); Lower Kirk Range, Chipusiri, fl. 17.iii.1955, *E. M. & W.* 970 (BM; LISC; SRGH).
From the Sudan and Ethiopia westwards to Cameroon and southwards to Natal, also in São Tomé, Fernando Po and Madagascar. Upland grasslands, often near streams; alt. 1300–1700 m.

Note: This and the following species are usually easily distinguishable in F.T.E.A. and F.C.B. areas; in the F.Z. area, however, intermediates do occur (? putative hybrids) and the two species seem to merge into a wide complex. The characters used in the key should be considered as the average for each species.

3. **Alchemilla kiwuensis** Engl. in Mildbr., Wiss. Ergebn. Deutsch. Zentr.-Afr. Exped. 1907–1908, **2**: 225, t. 21 fig. F–G (1911).—Hauman & Balle in Rev. Zool. Bot. Afr. **24**: 308, fig. 1a, 2, 3 & 14 (1934).—Robyns & Tournay, Fl. Parc Nat. Alb. **1**: 244, t. 23 (1948).—Hauman in F.C.B. **3**: 6, t. 1 (1952).—Keay in F.W.T.A. ed. 2, **1**: 424 (1958).—R. A. Grah. in F.T.E.A., Rosaceae: 14 (1960). Type from Rwanda.
Alchemilla sp.—Bak. f. in Trans. Linn. Soc., Ser. 2, Bot. **4**: 13 (1894).

Perennial herb with usually persistent basal rosette, sometimes stoloniferous; stolons slender with long spreading white hairs, often rooting at intervals and developing secondary rosettes. Rosette-leaves petiolate, 2–4(6) × 4–7(11) cm., orbicular to reniform in outline, (5)7(9)-lobed, white-hairy on both surfaces; indumentum on superior surface scattered, with hairs pointing towards lobe-margins; lobes deep, longer than ½ the radius of lamina, oblong-elliptic to obovate, marginally dentate almost to the base, the central lobe with c. 19 mucronulate ogive-shaped teeth up to 2 mm. long; petiole 6–12(18) cm. long, with long spreading hairs. Stem-leaves smaller, c. 1·5 × 1·5–2·5 cm., usually 5-lobed; petiole hairy, 1·5–2·5 cm. long. Inflorescence a shortly branched panicle of 1–4-flowered cymose clusters up to 25 cm. long, or rarely reduced to 1–2 flowers and then concealed within the stipules of the stolons as in *A. cryptantha*; pedicels glabrous or hairy, up to 1 mm. long. Calyx with long white spreading hairs; epicalyx-lobes 1·0–1·2 mm. long, exceeding calyx-lobes; calyx-lobes c. 0·7 mm. long. Achenes up to 1·2 mm. long.

Zambia. E: Nyika Plateau, fl. 27.xi.1962, *Fanshawe* 7265 (K). **Rhodesia.** E: Umtali Distr., Vumba Mts., near the Leopard Rock Hotel, fl. 9.xii.1948, *Chase* 1542 (BM; K; SRGH); Melsetter Distr., Chirinda, fl. iii.1962, *Goldsmith* 105/62 (K; LISC; SRGH). **Malawi.** N: Rumpi Distr., Nyika Plateau, *Richards* 10534 (EA; K). S: Kirk Range near Goche, fl. 14.xi.1950, *Jackson* 274 (EA; K); Mt. Mlanje, st. 1891, *Whyte* 112 (BM). **Mozambique.** MS: Manica Distr., Báruè, Choa Mts., 21 km. from Vila

Gouveia to the Rhodesian border, fl. 28.iii.1966, *Torre & Correia* 15460 (COI; EA; K; LISC; LMU; PRE; SRGH); Sofala Distr., Gorongosa Mts., near the Morombosi Falls, 23.x.1945, *Pedro* 437 (LMA; PRE).
Also in Cameroon, Zaire, Rwanda, Burundi, Uganda, Kenya and Tanzania. Upland wet grassland and edges of mountain forests, usually near streams; alt. 1200–2000 m.

Note: See notes under *A. ellenbeckii* subsp. *nyikensis* (p. 30) and under *A. cryptantha* (p. 32).

63. CHRYSOBALANACEAE

By F. White

Trees, shrubs or rhizomatous, geoxylic suffrutices. Wood always with abundant silica inclusions. Leaves simple, entire, alternate, often coriaceous, usually with two glands at base of lamina or near apex of petiole. Stipules small and caducous to large and persistent. Inflorescence a cyme, panicle or raceme. Flowers mostly (in our area always) bisexual, actinomorphic to zygomorphic, strongly perigynous. Receptacle-tube short to elongate, straight or curved, often gibbous at the base, always lined with nectariferous tissue which is extended at the throat as a short annular disk; throat at least partly blocked by long hairs. Sepals 5, free, imbricate, often unequal, ascending or reflexed. Petals 5, rarely absent (not in our area), sometimes unequal, imbricate, often caducous. Stamens 2–100 or more, included or exserted, inserted in 1 or 2 rows on the margin of the disk or adnate to its abaxial surface, either all fertile and forming a complete circle or partly staminodial; filaments free or appearing connate at the base or (not in our area) ligulately connate; anthers small, 2-thecous, dehiscing longitudinally. Ovary superior, basically of 3 carpels and gynobasic but usually with only 1 carpel fully developed, attached to base, middle or mouth of receptacle-tube, sessile or on a short gynophore, always hairy, each carpel 1-locular with 2 ovules, or 2-locular, owing to a false septum, with 1 ovule in each compartment; style filiform; stigma distinctly or indistinctly 3-lobed. Fruit a dry or fleshy drupe; endocarp thick or thin, fibrous, granular or bony, often with a special mechanism for seedling escape, often densely hairy inside. Seed erect, exalbuminous; cotyledons plano-convex, fleshy, sometimes ruminate. Germination hypogeal or epigeal; first leaves of seedling opposite or alternate.
A medium-sized family of 16 genera and 450 species distributed throughout the tropical regions of both hemispheres, but with the greatest concentration of genera in Africa and Madagascar, and of species in tropical America.

The Chrysobalanaceae was first recognized as a family by Robert Brown in 1818, and has been maintained as such by all subsequent workers with detailed knowledge of the group. In most general systems of classification, however, it appears as a tribe or subfamily of Rosaceae. Convincing evidence for its claim to family rank is summarized by Prance (Fl. Neotrop. 9, Chrysobalanaceae).
Parinari, in its usual circumscription, is a highly artificial assemblage—a dumping ground for all species of *Chrysobalanaceae* with a 2-locular ovary. Prance has convincingly shown that it should be split into several, more natural, units. Of these, only *Maranthes* occurs in our area. The differences between *Maranthes* and *Parinari* sens. strict. are as great as those between any other pair of genera in the family.
A striking feature of the African flora is the frequent occurrence of tree species, the distributions of which transgress important chorological boundaries and which also occur in markedly different vegetation types. No less remarkable is the frequent occurrence of pairs, or larger groups, of closely related species, the individual members of which are very different in their ecology and often also in their habit, but in other structural features are almost indistinguishable. These are the *séries écophylétiques* of Aubréville (Contribution à la paléohistoire des forêts de l'Afrique tropicale, 1949). Examples of both kinds of relationship are well represented in the Chrysobalanaceae. *Chrysobalanus icaco* and *Magnistipula butayei* are ecological and chorological transgressors which show subspecific differentiation.

B

Parinari excelsa is an ecological and chorological transgressor, which, in my opinion, does not show subspecific differentiation. The species of *Hirtella, Maranthes* and *Parinari* occurring in our area are all members of ecophyletic series. The structure of the leaves of an ecophyletic series in *Parinari* has been described by Homès, Duvigneaud, Balasse & Dewit (in Bull. Soc. Roy. Bot. Belg. **84**: 83, 1951). Within F.Z. area those members of ecophyletic series that have sympatric distributions appear to behave as perfectly good biological species. Intermediates that may be of hybrid origin have either not been detected or are very localized and require confirmation (see p. 41).

Since this account was written a paper by White on " The taxonomy, ecology and chorology of African Chrysobalanaceae (excluding *Acioa*) " has been published (Bull. Jard. Bot. Nat. Belg. 46: 265–350, 1976), as a companion to a series of distribution maps (Distr. Pl. Afr. **10**: 281–334, 1976) of the African taxa, including all those mentioned in the *Flora Zambesiaca* account.

1. Chrysobalanus

Ovary inserted at or near the base of the receptacle; endocarp with 4–8 longitudinal ridges:
Ovary inserted laterally in upper half or at the mouth of the receptacle-tube:
  Epicarp closely verrucose; endocarp with 2 small basal " plugs " or " stoppers " which
    allow the seedlings to escape; bracts and bracteoles completely concealing flower-
    buds up to anthesis, both individually and in small groups; leaf-undersurface with
    small stomatal cavities filled with hairs - - - - - **2. Parinari**
  Epicarp not verrucose; endocarp without basal plugs but sometimes dehiscing at
    germination by means of 2 lateral plates; bracts and bracteoles not enclosing
    flowers in groups; leaf-undersurface without hair-filled stomatal cavities:
    Endocarp with 2 lateral plates; carpels 2-locular; receptacle not ventricose; sepals
      suborbicular, concave; stamens 20–60 - - - - **3. Maranthes**
    Endocarp without lateral plates; carpels 1-locular; receptacle often ventricose;
      sepals acute, not concave; stamens 6–10:
      Flowers slightly zygomorphic; receptacle-tube not oblique at throat; sepals
        subequal; filaments far exserted; staminodes almost free; endocarp with 4–7
        shallow longitudinal channels - - - - - **4. Hirtella**
      Flowers strongly zygomorphic; receptacle-tube usually markedly oblique at throat;
        sepals unequal; filaments included; staminodes largely or completely united;
        endocarp without longitudinal channels - - - **5. Magnistipula**

1. CHRYSOBALANUS L.

Chrysobalanus L., Sp. Pl. **1**: 513 (1753); Gen. Pl. ed. 5: 229 (1754).

Small or medium-sized trees or shrubs. Lower leaf-surface glabrous or with a few stiff appressed hairs. Inflorescence few-flowered, a short raceme of cymules or cymose throughout, or a false raceme or a subsessile fascicle. Bracts small, eglandular. Flowers actinomorphic. Receptacle-tube cupuliform, interior and exterior puberulous. Sepals 5, acute, subequal. Petals 5, longer than the sepals. Stamens 12–26, forming a complete circle; filaments hairy, c. twice as long as sepals, appearing slightly united at the base. Ovary monocarpellary, 1-locular, inserted at base of receptacle, covered with a dense mass of hairs. Styles puberulous; stigma slightly expanded, shallowly 3-lobed. Drupe small, glabrous; endocarp thin, hard, interior glabrous, exterior smooth, with 4–8 prominent longitudinal ridges corresponding to the lines of fracture that allow the seedling to escape.

A small genus with 2 species in tropical America, one of which is also widespread in Africa. Except for *C. icaco* subsp. *atacorensis*, the distribution is mainly coastal.

Chrysobalanus icaco L., Sp. Pl. **1**: 514 (1753).—Prance, Fl. Neotrop. **9**, Chrysobal.: 14, t. 2 (1972). Type from Jamaica.

Evergreen shrub or tree up to 30 m. tall. Leaf-lamina suborbicular to lanceolate-elliptic, apex emarginate to acuminate, glabrous except for a few appressed hairs on both surfaces; petiole 0·2–0·4 cm. long; stipules intrapetiolar, 0·3 cm. long, boat-shaped, 2-fid, caducous. Inflorescence axillary, up to 3 cm. long, but usually much less, a raceme of cymules, or a congested complex cyme, or flowers in subsessile fascicles. Receptacle-tube c. 0·15 cm. long. Sepals c. 0·15 cm. long. Petals c. 0·25 cm. long. Stamens c. 0·3 cm. long. Drupe variable in shape and size, ovoid, ellipsoid or obovoid, up to 5 cm. long, usually smaller.

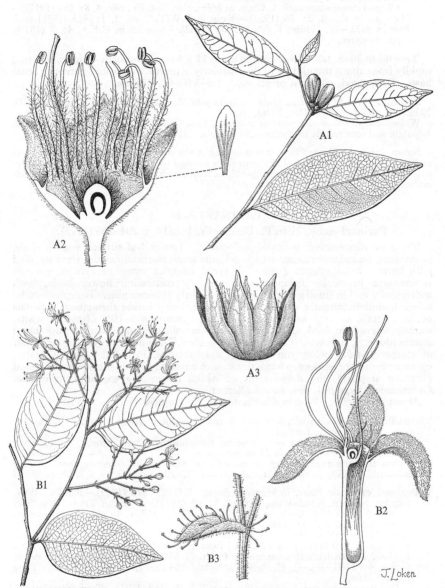

Tab. 9. A.—CHRYSOBALANUS ICACO subsp. ATACORENSIS. A1, fruiting branchlet (× ¾)
White 3327; A2, vertical section of flower and a single petal (× 10) *Vigne* 1966; A3,
endocarp after escape of seedling (× 1¼) *Prance & Silva* 58762. B.—HIRTELLA
ZANZIBARICA. B1, flowering branchlet (× ¾) *Andrade* 1976; B2, vertical section of
flower (× 5) *Gomes e Sousa* 4251; B3, bract showing stalked glands (× 5) *Gomes e
Sousa* 750.

Subsp. **atacorensis** (A. Chev.) F. White in Bull. Jard. Bot. Nat. Belg. **46**: 275 (1976).
TAB. **9** fig. A. Type from Dahomey.
  Chrysobalanus atacorensis A. Chev. in Mém. Soc. Bot. Fr., Sér. 4, **8**: 169 (1912).—
Hauman in F.C.B. **3**: 36 (1952).—Keay in F.W.T.A. ed. 2, **1**: 426 (1958) excl.
Brenan 8935.—F. White, F.F.N.R.: 68 (1962).—Mendes in C.F.A. **4**: 7 (1970).
Type as above.

Tree up to 30 m. tall. Leaf-lamina up to 11 × 4 cm., lanceolate-elliptic, tapering
rapidly from about the middle to the distinctly acuminate apex and narrowly acute
base. Drupe red, ellipsoid or obovoid, 1·8–3·0 × 0·5–1·6 cm.

Zambia. W: Mwinilunga Distr., tributary of Zambezi R., W. of Kalene Hill, fr.
22.ix.1952, *White* 3327 (COI; FHO; K).
Widespread in the wetter parts of tropical Africa from Liberia to the Central African
Republic and southwards to Zambia and Angola. Always in fringing forest.

Subsp. *icaco*, the Cocoa Plum, is widespread in the coastal regions of tropical America
and along the Atlantic seaboard of Africa from Sénégal to Angola. It differs from subsp.
atacorensis in its suborbicular to oblong-elliptic leaves with emarginate to shortly cuspidate
apices.

2. PARINARI Aubl.

Parinari Aubl., Hist. Pl. Guiane Fr. **1**: 514, t. 204–206 (1775).

Trees or rhizomatous geoxylic suffrutices. Lower leaf-surface with a close,
prominent, reticulate venation which delimits small stomatal cavities (crypts) filled
with hairs. Inflorescence a many-flowered complex cyme or cymose panicle.
Bracts and bracteoles eglandular, completely concealing flower buds, both
individually and in small groups. Flowers slightly zygomorphic. Receptacle-tube
subcampanulate, slightly swollen on one side, hairy inside throughout. Sepals
acute. Petals as long as sepals, caducous. Stamens 6–10; filaments white,
slightly curved in bud and during anthesis, slightly expanded at the base;
staminodes c. 6, subulate. Ovary monocarpellary, 2-locular, inserted in upper half
of receptacle-tube below the mouth; style arcuate, included. Drupe fleshy;
epicarp verrucose; endocarp hard, thick, with a rough, fibrous surface, with 2 basal
plugs or stoppers, the detachment of which allows the seedlings to escape.
Germination hypogeal, first leaves alternate.
About 40 species. Widely distributed in the tropics.

Rhizomatous suffrutex, usually less than 30 cm. in height, rarely up to 2 m.   1. *capensis*
Trees more than 4 m. in height:
  Forest tree; leaves always acuminate or subacuminate, base usually acute, never
    subcordate; lateral nerves in 22 or more pairs   -   -   -   3. *excelsa*
  Savanna tree; leaves rounded or emarginate, very rarely subacute, base usually
    subcordate, rarely subacute; lateral nerves in 12–20 pairs   -   - 2. *curatellifolia*

1. **Parinari capensis** Harv. in Harv. & Sond., F.C. **2**: 596 (1862).—R.E.Fr., Wiss.
Ergebn. Schwed. Rhod.-Kongo-Exped. **1**: 61 (1914).—Eyles in Trans. Roy. Soc.
S. Afr. **5**: 273 (1916).—Burtt Davy, F.P.F.T. **2**: 315, t. 49 (1932).—Steedman,
Trees etc. S. Rhod.: 11 (1933).—O. B. Mill., B.C.L.: 15 (1948); in Journ. S. Afr.
Bot. **17**: 18 (1952).—Wild, S. Rhod. Bot. Dict.: 110 (1953).—R. A. Grah. in Kew
Bull. **1957**: 230 (1957); in F.T.E.A., Rosaceae: 49 (1960).—F. White, F.F.N.R.:
69 (1962).—Friedrich-Holzhammer in Prodr. Fl. SW. Afr. **57**: 1 (1968).—Mendes in
C.F.A. **4**: 12 (1970).—J. H. Ross, Fl. Natal: 184 (1972). TAB. **10** fig. C. Type
from S. Africa.
  Parinari capensis var. *latifolia* Oliv. in F.T.A. **2**: 369 (1871). Type from Angola.
  Parinari curatellifolia Planch. ex Benth. var. *fruticulosa* R.E.Fr., tom. cit.: 60
(1914). Type: Zambia, Lake Bangweulu, Kamindas, fl. 9.x.1911, *R. E. Fries* 649a &
659 (UPS, syntypes).
  Parinari pumila Mildbr., Wiss. Ergebn. Zweit. Deutsch. Zentr.-Afr. Exped. 1910–
1911, **2**: 4–5 (1922).—Hauman in F.C.B. **3**: 67 (1952).—Mendes, loc. cit. Type as
for *P. capensis* var. *latifolia*.
  Parinari latifolia (Oliv.) Exell in Journ. Bot., Lond. **66**, Suppl.: 160 (1928). Type
as above.
  Parinari curatellifolia sensu Burtt Davy, F.P.F.T. **2**: 316 (1932) quoad syn.
tantum; sensu J. H. Ross, Fl. Natal: 184 (1972).
  Parinari capensis subsp. *latifolia* (Oliv.) R. A. Grah. in Kew Bull. **1957**: 230 (1957);
in F.T.E.A., Rosaceae: 49 (1960). Type as above.

Extensively rhizomatous, geoxylic suffrutex; stems nearly always less than 30 cm. tall, exceptionally (only in S. Mozambique and Natal) up to 1–2(?5) m. tall. Leaf-lamina from 9 × 1·2 to 10·5 × 4 cm., more or less elliptic or oblanceolate-elliptic, usually gradually tapering to the acute (very rarely subacuminate) or subacute apex and base, lower surface whitish; petiole 0·2–0·5 cm. long; stipules c. 1·5 × 0·4 cm., papery, caducous. Inflorescence terminal and axillary, up to 6·5 × 4 cm.; inflorescence-axes and outside of flowers whitish-tomentose. Receptacle-tube 0·35 cm. long. Sepals, petals and stamens 0·2 cm. long; staminodes 0·05 cm. long. Style 0·3 cm. long. Drupe ellipsoid, up to 3 × 2 cm.

Botswana. N: border with SW. Africa, 29 km. S. of Khardoum Valley, fr. 14.iii.1965, *Wild & Drummond* 7027 (K; SRGH). **Zambia.** B: Sesheke Distr. nr Masese, fl. 11.viii.1947, *Brenan & Keay* 7671 (FHO; K). N: Mporokoso Distr., Itabu dambo, fl. 14.xi.1961, *Lawton* 790 (FHO). W: Solwezi to Kasempa 34 km., fl. 30.ix.1947, *Brenan & Greenway* 7996 (FHO; K). C: 19 km. NNE. of New Serenje, fl. 8.x.1949, *Hoyle* 1267 (FHO). S: Livingstone Distr., Dambwa Forest Reserve, fr. 10.i.1952, *White* 1887 (FHO; K). **Rhodesia.** N: Lengwe R., fl. ix.1910, *Nobbs* 948 (SRGH) W: Wankie Distr., Gwaai Forest, fl. 4.ix.1955, *Chase* in SRGH 57714 (BM; LISC). C: Salisbury, Ruwa Park, fl. 3.x.1944, *Hopkins* in SRGH 12794 (SRGH). E: Odzani Valley, fl. 1934, *Gilliland* 1181 (BM; FHO; K). **Malawi.** N: Vipya Plateau, near Luwawa Dam, immature fr. 27.xii.1955, *Chapman* 261 (BM; FHO; K). **Mozambique.** MS: Manica Distr., Rotanda (Mavita), Monte Xiroco, fl. 26.x.1944, *Mendonça* 2661 (LISC). SS: Gaza, João Belo, Chongoene, fl. 11.x.1968, *Balsinhas* 1369 (LISC). LM: Maputo, Ponta do Ouro, fl. 18.xi.1944, *Mendonça* 2936 (LISC).

Widespread in S. tropical Africa from Congo (Brazzaville), Zaire and Tanzania to northern SW. Africa, Botswana, Transvaal and northern Natal. Often on sandy, especially seasonally waterlogged, soils at the edges of dambos and on Kalahari sand, where it also occurs in secondary grassland following the destruction of woodland by fire and cultivation.

The fruit is edible and is used to make beer.

Except in habit, *P. capensis* is very similar to *P. curatellifolia* and has sometimes been regarded as a variety of it. It does, however, seem to behave as a perfectly good biological species, albeit a sibling species. Over most of its range *P. capensis* is sympatric with *P. curatellifolia*. Genuine intermediates seem not to occur. The stems of *P. capensis* are normally burnt back to ground-level by fire each year, and flowering takes place at the ends of the current year's shoots or from the base of the remains of the previous year's shoots. Over the greater part of its range there is no evidence that in the absence of fire *P. capensis* would grow taller than c. 30 cm. It seems that unburnt shoots normally die back and further growth is from the axils of fallen leaves towards the base of the previous year's shoot. From the whole of its range, except for the south-eastern extremity, I have only seen two specimens which had produced second-year or older stems more than 30 cm. high. *White* 1887 (from Livingstone Distr., Zambia) consists of non-flowering shoots 45 cm. tall collected from a population of otherwise normal *P. capensis*. *Fanshawe* 6680 (from a dambo margin in Kasempa Distr., Zambia) was collected from a " shrub 0·6–1 m. high with woody rhizome ". It looks more like flowering coppice of *P. curatellifolia* than typical *P. capensis*. In the extreme S. of Mozambique and adjacent parts of Natal, however, *P. capensis* behaves differently. When subjected to fire it flowers on the current year's shoots. But most specimens from this area have been described as shrubs from 1–2 m. high. When protected from fire, in this part of its range, *P. capensis* is capable of a limited amount of upward growth and occurs as rather laxly branched individuals. The older stems, however, are not very floriferous and extensive die-back occurs (White, field observations). There is no evidence that even with complete fire-protection *P. capensis* is capable of more than limited upward growth. Since this part of its range lies entirely outside that of *P. curatellifolia*, hybridization is unlikely to be responsible for its unusual behaviour there. Subsequent to the completion of this account the populations from southern Mozambique and the coastal parts of Natal have been referred to a new subspecies, subsp. *incohata* F. White (in Bull. Jard. Bot. Nat. Belg. **46**: 320, 1976).

In leaf-shape most specimens of *P. capensis* differ from *P. curatellifolia* but there is a small amount of overlap. The leaves of *P. capensis* are narrower with a length/breadth ratio of 2·4–7·0. That of *P. curatellifolia* is from 1·8–2·8. On this character alone there is about 30% overlap, but the leaves of *P. capensis* are more tapered towards the ends. When the precise outline of the leaf is also taken into account, only about 15% of the individuals are found to be intermediate.

The leaves of *P. capensis* vary from very narrow to very broad. Broad-leaved individuals have previously been regarded as belonging to a distinct variety or subspecies or even a distinct species (*P. latifolia*, *P. pumila*). Although the distribution of extreme variants shows some correlation with geography, the differences between different populations are perhaps too small to justify recognition on taxonomic grounds alone. " *latifolia* " and " *capensis* " have a checkerboard distribution. The former occurs both in the extreme northern and the extreme southern parts of the range. The latter is the only variant occurring on the Kalahari Sands of southern Barotseland and adjacent regions. Elsewhere populations are somewhat mixed.

2. **Parinari curatellifolia** Planch. ex Benth. in Hook., Niger Fl.: 333 (1849).—Bak. f. in Journ. Linn. Soc., Bot. **40**: 66 (1911).—R.E.Fr., Wiss. Ergebn. Schwed. Rhod.-Kongo-Exped. **1**: 60 (1914).—Eyles in Trans. Roy. Soc. S. Afr. **5**: 273 (1916).—Burtt Davy, F.P.F.T. **2**: 316 excl. synon. (1932).—Hauman in F.C.B. **3**: 66 (1952).—Wild, S. Rhod. Bot. Dict.: 110 (1953).—Brenan in Mem. N.Y. Bot. Gard. **8**: 430 (1954).—Williamson, Useful Pl. Nyasal.: 90 (1955).—R. A. Grah. in Kew Bull. **1957**: 229 (1957); in F.T.E.A., Rosaceae: 50 (1960).—Keay in F.W.T.A. ed. 2, **1**: 429 (1958).—Gomes e Sousa, Dendrol. Moçamb. **5**: 183 cum tab. (1960); Dendrol. Moçamb. Estudo Geral **1**: 216 (1966).—Dale & Greenway, Kenya Trees & Shrubs: 403, t. 81 (1961).—F. White, F.F.N.R.: 69 (1962).—Friedrich-Holzhammer in Prodr. Fl. SW. Afr. **57**: 2 (1968).—Mendes in C.F.A. **4**: 9 (1970).—Palmer & Pitman, Trees of S. Afr. **1**: 681 cum tab. & photogr. (1972). TAB. **10** fig. A. Types from W. Africa.

Parinari mobola Oliv. in F.T.A. **2**: 368 (1871).—Sim, For. Fl. Port. E. Afr.: 61 (1909).—Eyles, loc. cit.—Burtt Davy, tom. cit.: 316, t. 50 (1932).—Steedman, Trees etc. S. Rhod.: 11 (1933).—O. B. Mill., B.C.L.: 15 (1948); in Journ. S. Afr. Bot. **18**: 17 (1952).—Codd, Trees & Shrubs Kruger Nat. Park: 33 cum photogr. (1951).—Pardy in Rhod. Agric. Journ. **48**: 264 cum photogr. (1951).—Hauman, tom. cit.: 64 (1952).—Williamson, tom. cit.: 91 (1955).—Palgrave, Trees of Centr. Afr.: 375 cum tab. & photogr. (1957). Types: Lectotype from Angola; Zambia, Batoka Highlands, fl.-buds vii–x 1860, *Kirk* s.n. (K, paratype); Malawi. Lower Zambezi, Muata Manga R., fl.-buds, fr., Kirk s.n. (K, paratype).

Parinari curatellifolia subsp. *mobola* (Oliv.) R. A. Grah., loc. cit.; tom. cit.: 51 (1960).—B. & M. de Winter & Killick, Sixty-six Transvaal Trees: 40 cum photogr (1966).—Mendes, loc. cit. Types as above.

Small or medium-sized evergreen tree 3–20 m. tall. Bark grey-black, rough, deeply and closely rectangularly fissured. Crown dense, rounded, umbrella-shaped, casting heavy shade. Leaf-lamina up to 8·5 × 5·5 or 10 × 4 cm., narrowly or broadly oblong-elliptic, apex rounded or emarginate, very rarely subacute, base usually subcordate, rarely subacute, lower surface usually yellowish; petiole 0·5–1 cm. long; stipules up to 2 × 0·5 cm., papery, caducous. Inflorescence terminal and axillary, up to 15 × 15 cm.; inflorescence-axes and outside of flower usually (in our area) fulvous- or ferruginous-tomentose, sometimes whitish-tomentose. Flowers white tinged with pink, fragrant. Receptacle-tube 0·45 cm. long. Sepals and petals 0·25 cm. long. Stamens 8; filaments 0·2–0·3 cm. long; staminodes 0·05 cm. long. Style 0·4 cm. long. Drupe broadly ellipsoid, up to 4 × 3 cm.

Zambia. B: Senanga Distr., Mashi R., Kaunga, fl. 2.x.1961, *Mubita* B44 (LISC; SRGH). N: Abercorn Distr., Chilongowelo, fl. 30.viii.1960, *Richards* 13169 (K; LISC). W: Ndola Distr., Kitwe, fl. 15.viii.1952, *Holmes* 940 (FHO; K). C: 33 km. E. of Lusaka, fl., fr. 13.vi.1957, *Angus* 1622 (FHO; K). E: Fort Jameson, fr. 29.viii.1929, *Burtt Davy* 21010 (FHO; K). S: Choma to Namwala 96 km., fr. 23.vii.1952, *White* 2967 (FHO; K). **Rhodesia.** N: Mafungabusi Plateau, escarpment W. of Chikombera source, fr. 25.vi.1947, *Keay* in FHI 21366 (FHO). W: Matopos, Boomerang Farm, fl. 7.ix.1952, *Plowes* 1466 (FHO). C: Salisbury Distr., Hatfield, fl. 26.ix.1965, *Simon* 475 (K; LISC; SRGH). E: Chipinga Distr., Chirinda Forest margin, fl. x.1967, *Goldsmith* 133/67 (FHO; K). S: Ndanga to Bikita 40 km., fl. 21.x.1930, *Fries, Norlindh & Weimarck* 2163 (BM; LISC). **Malawi.** N: Nyika Plateau, 7 km. N. of Rest House, 1600 m., fl. 29.x.1958, *Robson* 464 (BM; FHO; K). C: Kota-Kota Distr., Chia area, 480 m., fl. i.ix.1946, *Brass* 17467 (K; LISC; SRGH). S: Mt. Mlanje, base of Chambe Peak, 900 m., fr. 4.v.1957, *Chapman* 370 (BM; FHO; K). **Mozambique.** N: Quiterajo to Mocímboa da Praia 12 km., fl. 11.xi.1953, *Balsinhas* 77 (LISC). Z: between Mopeia and Marral, fl. 28.vii.1942, *Torre* 4434 (LISC). T: Macanga Distr. near Furancungo, fl. 13.viii.1941, *Torre* 3306 (LISC). MS: Chimoio Distr., Garuso, fl., *Simão* 557 (LISC). SS: Inhambane Distr., between Vilanculos and Macorane, fl. 1.ix.1942, *Mendonça* 87 (LISC).

Widespread in tropical Africa from Senegal to Kenya and southwards to northern SW. Africa, Botswana and the Transvaal. Widely scattered, though normally not very common, in *Brachystegia, Julbernardia, Isoberlinia* woodland. It is one of the few species left standing when woodland is cleared for cultivation. Because of its resistance to fire it also features prominently in fire-maintained wooded grassland, especially in upland

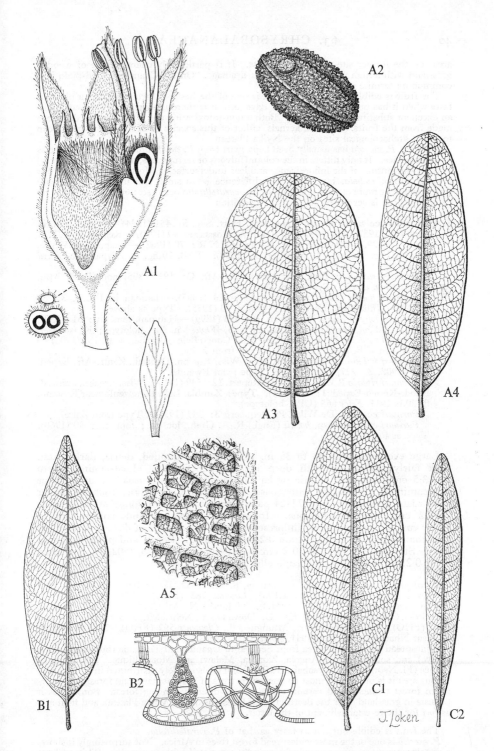

Tab. 10. A.—PARINARI CURATELLIFOLIA. A1, vertical section of flower, petal and t.s. ovary (×10) *Richards* 13169; A2, endocarp showing basal "stopper" (×1) *Angus* 1622; A3, leaf (× ¾) *Holmes* 940; A4, leaf (× ¾) *Angus* 1962; A5, leaf undersurface (×50) *White* 2967. B.—P. EXCELSA. B1, leaf (× ¾) *Fries* 665; B2, t.s. leaf showing stomatal crypt (×25) after *Homès et al.* C.—P. CAPENSIS. C1, leaf (× ¾) *Mendonça* 2936; C2, leaf (× ¾) *Brenan & Keay* 7671.

areas on the former site of montane forest. It is particularly characteristic of wooded
grassland with a high water-table and poor drainage. On the Chambezi flood-plain it is
common on termite mounds.

The fruit is edible but astringent and is one of the best wild fruits; it has a pleasant
taste when it has been stored for a few days until it is thoroughly ripe. The seeds make
an excellent substitute for almonds. Both a non-intoxicating and a very potent drink are
made from the fruits. Charred kernels, either of this species or of *P. excelsa*, have been
found at archaeological sites on the Nyika Plateau.

Hitherto, *P. mobola* has usually been kept apart from *P. curatellifolia*, either as a species
or a subspecies. It only differs in the colour (fulvous or ferruginous, not whitish or silvery)
of the indumentum of the inflorescence and leaf-undersurface. Most specimens from our
area fit subsp. *mobola*, but, since the difference is so small and individuals that are
intermediate or more closely resemble subsp. *curatellifolia* occur scattered throughout, no
useful purpose is served by its formal recognition.

3. **Parinari excelsa** Sabine in Trans. Roy. Hort. Soc. **5**: 451 (1824).—R. A. Grah. in
Kew Bull. **1957**: 229 (1957); in F.T.E.A., Rosaceae: 49 (1960).—Keay in F.W.T.A.
ed. 2, **1**: 429, t. 141 (1958).—F. White, F.F.N.R.: 70 (1962).—Chapman & White,
Evergr. For. Malawi: 40, photogr. 38 (1970). TAB. **10** fig. B. Type from Sierra
Leone.
    Parinari salicifolia Engl., Pflanzenw. Ost-Afr. **C**: 191 (1895) non (Presl) Miq.
(1855), *nom. illegit.* Type from Tanzania.
    Parinari holstii Engl., tom. cit.: 191, 423 (1895).—Hauman in F.C.B. **3**: 59
(1952).—Williamson, Useful Pl. Nyasal.: 90 (1955). Type as above.
    Parinari whytei Engl., Bot. Jahrb. **26**: 378 (1899).—Hauman, tom. cit.: 61 (1952).
Type: Malawi, Mt. Malosa, fl. xi–xii.1896, *Whyte* s.n. (B†, holotype; K, isotype).
    Parinari verdickii De Wild. in Ann. Mus. Congo Belge, Bot. Sér. 4, **1**: 182 (1903).—
Mendes in C.F.A. **4**: 11 (1970). Type from Zaire.
    Parinari mildbraedii Engl. in Mildbr., Wiss. Ergebn. Deutsch. Zentr.-Afr. Exped.
1907–1908, **2**: 227, t. 23 (1911). Type from Rwanda.
    Parinari riparia R.E.Fr. in Fedde, Repert. **12**: 539 (1913); Wiss. Ergebn. Schwed.
Rhod.-Kongo-Exped. **1**: 61 (1914). Type: Zambia, Lake Bangweulu near Kasomo,
fl. 20.ix.1911, *Fries* 665 (UPS, holotype).
    Parinari nalaensis De Wild., Pl. Bequaert. **3**: 289 (1931). Type from Zaire.
    Parinari excelsa subsp. *holstii* (Engl.) R. A. Grah., loc. cit.; tom. cit.: 50 (1960).
Type as for *P. holstii*.

Large evergreen tree up to 35 m. tall. Crown rounded, dense, dark green.
Bark fairly smooth or with deep longitudinal fissures. Leaf-lamina up to
10 × 3·5 cm., narrowly elliptic or lanceolate-elliptic, apex usually acuminate or
subacuminate, very rarely acute, base usually acute, sometimes rounded, never
subcordate, lateral nerves in 22–24 pairs; petiole 0·5–1 cm. long; stipules up to
2 × 0·2 cm., papery, caducous. Inflorescence terminal and axillary, up to
12 × 9 cm., lax or congested; inflorescence-axes and outside of flowers fulvous- or
grey-tomentose. Receptacle-tube 0·2–0·3 cm. long. Sepals and petals 0·25 cm.
long. Stamens 8; filaments 0·2 cm. long; staminodes 7–8, 0·02–0·1 cm. long.
Style 0·25–0·35 cm. long. Drupe ellipsoid c. 4 × 2·5 cm.

Zambia. B: Balovale Distr., Chavuma, fl. 13.x.1952, *White* 3503 (FHO; K). N:
Abercorn Distr., Chizisi, fl. 19.ix.1961, *Lawton* 783 (FHO). W: 112 km. S. of
Mwinilunga, fr. 8.xi.1932, *Duff* 37 (FHO). **Malawi.** N: Misuku Hills, Mugesse Forest,
fr. ix.1953, *Chapman* 137 (FHO; K). C: Dowa Distr., Nchisi Mt., st. 3.v.1961, *Chapman*
1260 (FHO). S: Zomba Distr., Mlungusi, fl., *Clements* 562 (FHO). **Mozambique.**
Z: near Nhamarroe, fl. 25.ix.1941, *Torre* 3508 (LISC).

Widespread in tropical Africa from Sénégal to Uganda and Tanzania (but absent from
Kenya) and southwards to Angola, Zambia, Malawi and Mozambique. According to
Prance (Fl. Neotrop. 9, Chrysobal.: 185, 1972) it is also widely distributed in S. America.
In our area it is one of the most characteristic trees of Afromontane rain forest, dry evergreen forest (*mateshi*) and certain types of well-drained fringing forest. Sometimes it
persists in grassland after the destruction of forest, as on the Vipya Plateau, and then acts
as *foci* for the re-establishment of the latter.

The fruit is edible but not as tasty as that of *P. curatellifolia*.

P. excelsa is one of the most widespread forest trees in Africa. Not surprisingly it shows
a certain amount of variation which is partly correlated with geography. This, at least
in part, accounts for its extensive synonomy. Plants from lowland rainforest in W. Africa
tend to have silvery indumentum, a lax inflorescence and smaller flowers. Those from
upland rainforest and fringing forest in E. and S. tropical Africa tend to have fulvous
indumentum, a more congested inflorescence and larger flowers (not smaller as stated by

Graham (in F.T.E.A., Rosaceae: 50, 1960)). They belong to subsp. *holstii* of Graham. The overall pattern, however, is too diffuse to justify the formal recognition of subspecies. Several specimens from W. Africa typologically are subsp. *holstii* and several from E. Africa are inseparable from subsp. *excelsa* or are intermediate. This would not matter if the anomalous specimens were few in number and if the differences between the subspecies were greater. Not only are the subspecies weakly defined, but most of the specimens from the Congo basin are intermediate in character.

Within our area, the range of *P. excelsa* lies entirely within that of *P. curatellifolia*, but the ecology of the two species is very different. *P. excelsa* is a species of various types of evergreen forest, in which it is often dominant or co-dominant. *P. curatellifolia* is a plant of woodland and tree savanna. In some situations, especially where *P. curatellifolia* has invaded secondary grassland on sites formerly occupied by forest, the two species may occur in close proximity. Notwithstanding this, and their great similarity in virtually all structural features other than habit and leaf shape, I have seen no intermediates other than the two specimens mentioned below. This suggests that the species are either reproductively isolated or that no niche is available for any hybrids they may produce.

The intermediates are both from Mozambique, where *P. excelsa* is only known from a small area centred on the Gúruè Mts., and were collected at localities quite close to or a short distance S. of the area where *P. excelsa* occurs. It is possible that these intermediates are of hybrid origin, and that it is only at the southern limit of the range of *P. excelsa* that habitats suitable for colonization by the hybrids are found. Careful observations in the field are needed to test this hypothesis. The specimens are: *Gomes e Sousa* 865 (K) from the Ribáuè Mts. (14° 50′ S, 38° 20′ E) at 900 m., where the plant is said to be common in xerophilous forest on granite, and *Gomes e Sousa* 4256 (K) from Inhaminga.

3. MARANTHES Blume

Maranthes Blume, Bijdr. Fl. Nederl. Ind.: 89 (1825).

Trees. Lower leaf-surface glabrous or with dense arachnoid indumentum. Inflorescence a many-flowered corymb or panicle. Bracts and bracteoles eglandular, caducous, not concealing young flowers. Flowers slightly zygomorphic. Receptacle-tube often slightly curved, obconoidal or subcampanulate, always gradually narrowed into the pedicel, nearly always almost solid and almost completely filled with nectariferous tissue, glabrous inside on one side, hairy on the other, or completely glabrous. Sepals suborbicular, deeply concave. Petals suborbicular to broadly lingulate. Stamens 20–60; filaments white, inserted in two rows on the free margin of the disk; tightly undulate in bud with 2 or more undulations, much longer than the sepals, usually occurring in a tangled mass; staminodes few and vestigial or absent. Ovary of 1 or sometimes 2–3 bilocular carpels, inserted at mouth of receptacle-tube; style curved upwards, much longer than the sepals, glabrous except at base. Drupe fleshy; endocarp very hard, fibrous with a rough exterior; glabrous inside, with 2 lateral plates which break away on germination and allow the seedlings to escape. Germination epigeal; first leaves opposite.

About 12 species. Confined to Africa except for 1 species which is widespread in Malaysia, New Guinea and Australia and a second which is only known from Panama.

Leaves rounded or subacute at apex, hairy; inflorescence-axes coarsely fulvous-tomentose or glabrous; pedicel plus receptacle-tube 1·3–2·2 cm. long; petals scarcely longer than the sepals; fruit up to 3 ×2 cm., hairy - - - - - 1. *floribunda*
Leaves acutely acuminate, glabrous; inflorescence-axes grey-tomentellous; pedicel plus receptacle-tube 0·8–1·0 cm. long; petals nearly twice as long as sepals; fruit c. 4·5 ×2·5 cm., glabrous - - - - - - - - 2. *goetzeniana*

1. **Maranthes floribunda** (Bak.) F. White in Bull. Jard. Bot. Nat. Belg. **46**: 297 (1976). TAB. 11. Type: Malawi, Karonga Distr., Fort Hill, fl. vii.1896, *Whyte* s.n. (K, holotype).
  Parinari floribunda Bak. in Kew Bull. **1897**: 265 (1897).—R.E.Fr., Wiss. Ergneb. Schwed. Rhod.-Kongo-Exped. **1**: 62 (1914).—Hauman in F.C.B. **3**: 57 (1952). Type as above.
  Parinari bequaertii De Wild. in Fedde, Repert. **13**: 108 (1914).—R.E.Fr., loc. cit.—Hauman, tom. cit.: 57 (1952). Type from Zaire.
  Parinari bequaertii var. *longistaminea* Hauman in Bull. Jard. Bot. Brux. **21**: 188 (1951). Type: Zambia, Mufulira, fl. 29.v.1934, *Eyles* 8201 (BM; K, holotype; SRGH).
  Parinari polyandra Benth. subsp. *floribunda* (Bak.) R. A. Graham in Kew Bull.

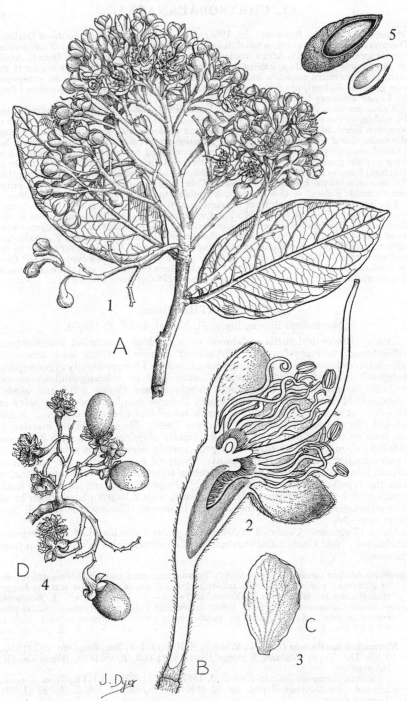

Tab. 11. MARANTHES FLORIBUNDA. 1, flowering branchlet (×½) *Astle* 753; 2, vertical section of flower (×4); 3, petal (×4) 2–3 from *Whellan* 1553; 4, flowers and developing fruit (×½); 5, endocarp showing " lateral plate " and seedling escape mechanism (×1) 4–5 from *White* 3484.

63. CHRYSOBALANACEAE

43

1957: 230 (1957); in F.T.E.A., Rosaceae: 53, t. 7 fig. 3–7 (1960).—F. White, F.F.N.R.: 70 (1962). Type as for *Maranthes floribunda*.
Maranthes polyandra (Benth.) Prance subsp. *floribunda* (Bak.) Prance in Bol. Soc. Brot., Sér. 2, **40**: 184 (1966).—Mendes in C.F.A. **4**: 16 (1970). Type as above.

Small evergreen tree (4)6–10 m. tall. Crown dense, dark green, rounded. Bark grey, finely reticulate or almost smooth. Leaf-lamina up to 15 × 8 cm., mostly ovate or broadly or narrowly elliptic-oblong, apex usually rounded, sometimes subacute, base rounded to subcordate, rarely acute, lower surface densely arachnoid-tomentose at first, glabrescent, though some hairs always persisting, venation closely reticulate, prominent above, less so beneath; petiole 0·3–0·6 cm. long; stipules up to 0·7 × 0·2 cm., caducous. Inflorescence a terminal corymbose panicle up to 15 × 20 cm.; inflorescence-axes and outside of flowers usually coarsely fulvous-tomentose, rarely completely glabrous; bracts c. 0·5 cm. long, boat-shaped, keeled; pedicel gradually expanded into receptacle-tube, the two together 1·3–2·2 cm. long. Receptacle-tube obconoidal, slightly curved, 0·5–0·6 cm. long. Sepals 0·7–0·8 cm. long. Petals up to 0·8 × 0·5 cm., white. Stamens c. 60; filaments 1·5–2·2 cm. long, white. Ovary densely tomentose; style c. 2·0 cm. long. Drupe up to 3 × 2 cm., ellipsoid, reddish-purple, tomentose when young, pubescent when mature.

Zambia. B: Balovale Distr., Chavuma, fr. 12.x.1952, *White* 3484 (FHO; K). N: Shiwa Ngandu–Chambezi road, 1525 m., fl. 27.vii.1938, *Greenway & Trapnell* 5530 (FHO; K). W: Solwezi to Nchanga 30 km., fr. 1.x.1947, *Brenan & Greenway* 7998 (BM; FHO; K). C: Serenje Distr., Chitambo Mission, fl. 30.vi.1929, *Stevenson* 65 (FHO). E: Lundazi to Mzimba 3 km., 1200 m., fr. 18.x.1958, *Robson & Angus* 146 (BM; FHO; K). **Malawi.** N: Fort Hill, fl. vii.1896, *Whyte* s.n. (K).
Also in Angola, Zaire and Tanzania. In higher-rainfall *Brachystegia, Julbernardia, Isoberlinia* woodland. Sometimes left standing in villages as a shade tree, 1000–1525 m.

Maranthes is exclusively a forest genus except for *M. floribunda* and *M. polyandra* (Benth.) Prance, which occur in woodland in the Zambezian and Sudanian domains respectively. They are separated by an interval of 1000 km. Although somewhat similar in their leaves and habit, they are so different in the size, proportions and indumentum of their flowers that specific rather than subspecific rank seems appropriate.
Most specimens of *M. floribunda* have densely hairy inflorescences. A few, however, from widely separated localities, including the type, are completely glabrous.

2. **Maranthes goetzeniana** (Engl.) Prance in Bull. Jard. Bot. Nat. Belg. **46**: 303 (1976). Type from Tanzania.
Parinari goetzeniana Engl., Bot. Jahrb. **34**: 153 (1904).—Brenan, T.T.C.L.: 476 (1949).—R. A. Grah. in F.T.E.A., Rosaceae: 51 (1960). Type as above.
Parinari gilletii sensu Bak. f. in Journ. Linn. Soc., Bot. **40**: 66 (1911).—Eyles, Trans. Roy. Soc. S. Afr. **5**: 273 (1916).—Steedman, Trees etc. S. Rhod.: 11 (1933).
Parinari polyandra subsp. *floribunda* sensu R. A. Grah., tom. cit.: 53 quoad cit. Rhodesia tantum (1960).—Boughey in Journ. S. Afr. Bot. **30**: 157 (1964).
Maranthes polyandra subsp. *floribunda* sensu Wild & Barbosa, Fl. Zamb. Suppl.: 4, 5 (1967).

Medium sized to large evergreen tree (12)20–35 m. tall, with a wide, rounded crown. Bark grey or grey-black, more or less smooth. Leaf-lamina up to 15 × 8 cm., mostly elliptic or oblong-elliptic, apex suddenly and acutely acuminate, acumen 1–1·5 cm. long, base rounded or acute, lower and upper surface completely glabrous, venation reticulate, not prominent; petiole (0·4)0·6–0·9 cm. long; stipules 0·4 × 0·25 cm., caducous. Inflorescence a terminal corymbose panicle up to 15 × 25 cm.; inflorescence-axes and outside of flower grey-tomentellous; bracts c. 0·5 cm. long, boat-shaped, keeled; pedicel gradually expanded into receptacle-tube the two together 0·8–1·0 cm. long. Receptacle-tube slightly curved, obconoidal, 0·5–0·6 cm. long. Sepals 0·3–0·4 cm. long. Petals up to 0·8 cm. long, white. Stamens c. 50; filaments c. 1·5 cm. long, white. Ovary densely tomentose; style c. 1·5 cm. long. Drupe up to 4·5 × 2·5 cm., obovoid, glabrous.

Rhodesia. E: eastern Vumba Mts., fl. 18.vii.1957, *Chase* 6639 (BM; COI; K; LISC; SRGH). **Mozambique.** Z: Serra do Gúruè, Mt. Murrece, 1100 m., fl. 4.xi.1967, *Torre & Correia* 15901 (LISC). MS: Báruè Distr., Serra de Choa, 10 km. from Vila Gouveia towards the frontier, fr. 15.xii.1965, *Torre & Correia* 13670 (LISC).
Also in Tanzania. One of the most characteristic species of moist evergreen forest at low and medium elevations in Rhodesia and Mozambique. Often associated with *Khaya*

nyasica and *Newtonia buchananii*. Also in fringing forest with *Adina microcephala*, *Erythrophleum suaveolens* and *Treculia africana*, 300–1500 m.

4. HIRTELLA L.

Hirtella L., Sp. Pl. **1**: 34 (1753); Gen. Pl. ed. 5: 20 (1754).

Trees or shrubs. Lower leaf surface glabrous or with a few strigose or strigulose hairs. Inflorescence many-flowered, usually a lax raceme or an elongate thyrse with patent flowers or lateral branches; the latter usually bear several sterile bracts and end in a single flower or a few cymosely arranged flowers. Flowers slightly zygomorphic. Bracts and bracteoles often with stalked or sessile glands. Receptacle-tube subcampanulate to narrowly cylindric, slightly gibbous, usually shorter than the sepals, usually glabrous inside except near the throat, and hairy outside. Sepals 5, subequal, usually spreading or reflexed. Petals 5, shorter than the sepals. Stamens 3–9; filaments laxly undulate in bud with a single undulation, inserted on abaxial surface of disk (at least in our area), far-exserted, usually much longer than the combined length of calyx and receptacle-tube. Staminodes short, free. Ovary monocarpellary, 1-locular, usually inserted at mouth of receptacle-tube, style filiform, far-exserted. Drupe with exiguous mesocarp and smooth, thin, hard, non-granular endocarp with 4–7 longitudinal shallow channels which represent lines of weakness that permit the seedling to escape. Germination hypogeal, first leaves alternate.

90 species. Of these 88 occur in Central and S. America and 1 in E. Africa, and Madagascar.

Hirtella zanzibarica Oliv. in Hook., Ic. Pl. **12**: 81, t. 1193 (1876).—Brenan in Trop. Woods **86**: 5 (1946); in Mem. N.Y. Bot. Gard. **8**: 431 (1954).—R. A. Grah. in F.T.E.A., Rosaceae: 54, t. 8 fig. 5–6 (1957).—Dale & Greenway, Kenya Trees & Shrubs: 403 (1961).—Gomes e Sousa, Dendrol. Moçamb. Estudo Geral **1**: 215 (1966).—Wild & Barbosa, Fl. Zamb. Suppl.: 12, 13, 26 (1967).—Chapman & White, Evergr. For. Malawi: 40 (1970). TAB. **9** fig. B. Type from Mafia I., East Africa.
   Hirtella thouarsiana Baill. ex Laness., Pl. Ut. Col. Fr.: 874 (1886). Type from Madagascar.
   Acioa goetzeana Engl., Bot. Jahrb. **30**: 315, t. 12 (1901). Type from Tanzania.
   Hirtella zanzibarica var. *cryptadenia* Brenan, tom. cit.: 11 (1946). Type from Zanzibar.

Small or medium-sized evergreen tree up to 20 m. tall, but sometimes flowering as a shrub 1–5 m. tall. Leaf-lamina up to 12 × 5 cm., lanceolate or lanceolate-elliptic, apex acutely subacuminate, base rounded, slightly asymmetric, glabrous on both surfaces except for a few strigulose hairs, venation raised and closely reticulate on upper surface; petiole 0·3–0·4 cm. long; stipules up to 0·5 cm. long, subulate, persistent. Inflorescence many-flowered, somewhat congested, a terminal or axillary thyrse; lateral branches 1–3-flowered and also bearing several sterile bracts; bracts c. 0·3 cm. long, margin with many sessile to long-stipitate glands; pedicels 0·1–0·15 cm. long. Receptacle-tube c. 0·8 cm. long, straight, narrowly cylindric, ending in a free annular disk at the throat, densely tomentellous outside, glabrous in lower ½ inside. Sepals 0·6–0·7 × 0·35 cm., spreading, usually with a few shortly stipitate glands on the recurved margins, both surfaces tomentellous, the inner much paler than the outer. Petals 0·3–0·4 cm. long, oblong-elliptic, slightly asymmetric, very shortly unguiculate. Stamens c. 9; filaments c. 1·0 cm. long, inserted on abaxial surface of the disk; staminodes c. 3, subulate, 0·05 cm. long. Style 1·0 cm. long, hirsute for greater part of length. Drupe up to 2·8 × 1·7 cm., ellipsoid or obovoid, suddenly contracted at the base, sparsely pubescent.

Malawi. N: eastern slopes of Vipya Plateau, occurring on banks of Lonjoswa R. for several km. upstream from Luweya R., 915–1065 m., st. 29.x.1964, *Chapman* 2275 (FHO). C: Kota-kota Distr., Chia area, 480 m., fl. fr. 3.ix.1946, *Brass* 17514 (EA; K; LISC). **Mozambique.** N: Ribáuè, 900 m., fr. ix.1931, *Gomes e Sousa* 750 (K; LISC; LM). Z: serra do Gurué, waterfall of Lucungo R., 1300 m., fr. 9.xi.1967, *Torre & Correia* 16032 (LISC). MS: Beira Distr., nr. Dondo, fl. 5.ix.1944, *Mendonça* 2009 (LISC).
   Also in Kenya, Tanzania, Mafia I., Zanzibar and Madagascar. Widespread in fringing forest and in various types of coastal forest. In Mozambique it is one of the most characteristic and abundant trees in dense semi-deciduous forest at low altitudes

dominated by *Pteleopsis myrtifolia* and *Erythrophleum suaveolens*, especially in wetter areas. 50–1350 m.

5. MAGNISTIPULA Engl.

Magnistipula Engl., Bot. Jahrb. **36**: 226 (1905).

Trees, shrubs or geoxylic suffrutices. Lower leaf surface glabrous or with a few strigose hairs. Inflorescence a many-flowered panicle, usually thyrsoid, sometimes extremely contracted and racemoid. Flowers markedly zygomorphic. Receptacle-tube always curved, obliquely campanulate, slightly to strongly gibbous, nearly always markedly oblique at the mouth, always longer than the sepals, glabrous inside only at the base. Sepals 5, acute, usually very unequal in size. Petals 5, longer than the sepals, persistent. Stamens 5–9; filaments white, coiled in bud, arcuate at anthesis, scarcely longer than the sepals, flattened at the base and appearing united for at least ⅓ of length; staminodes appearing partly or completely united to form a short comb- or tongue-like structure. Ovary monocarpellary, 1-locular (2-locular in only 1 species), inserted at mouth of receptacle-tube. Style arcuate, scarcely longer than the sepals, glabrous or hairy only at the base. Drupe fleshy; endocarp hairy inside, on germination breaking up in an irregular manner. Germination hypogeal; first leaves opposite.

About 10 species in tropical Africa and 2 in Madagascar.

Tree; inflorescence a branched raceme of sessile cymules; lateral branches patent; bracts usually subulate, and less than 0·2 cm. long; flowers up to 0·9 cm. long   1. *butayei*
Rhizomatous suffrutex; inflorescence an elongate very contracted thyrse with very few, markedly ascending lateral branches; bracts narrowly deltate, c. 0·3 cm. long; flowers 1·0–1·2 cm. long - - - - - - - - - -   2. *sapinii*

1. **Magnistipula butayei** De Wild. in Ann. Mus. Congo Belge, Bot. Sér. 5, **2**: 255 (1908). Type from Zaire.
    Hirtella butayei (De Wild.) Brenan in Trop. Woods **86**: 4 (1946).—Hauman in Bull. Jard. Bot. Brux. **21**: 178 (1951); in F.C.B. **3**: 41, t. 2 (1952).—Mendes in C.F.A. **4**: 18 (1970). Type as above.

Small or medium-sized evergreen tree with a wide-spreading, dense, rounded crown. Bark rough, dark grey. Leaf-lamina up to 17 × 7 cm., more or less oblong, obovate- or oblanceolate-oblong, or oblanceolate, apex broadly rounded to acuminate, base often subcordate, sometimes rounded or acute, sparsely strigulose on both surfaces, especially on the nerves, venation closely reticulate, scarcely prominent; petiole c. 0·2 cm. long; stipules 0·3–0·5 cm. long, subulate, caducous or persistent. Inflorescence up to 12 × 10 cm., terminal and from upper leaf-axils, a branched raceme of sessile cymules; inflorescence-axes, bracts and outside of flowers densely fulvous-tomentose; bracts persistent, up to 0·2 cm. long, subulate, with a pair of sessile glands at base partly concealed by indumentum; pedicels 0·2–0·3 cm. long. Receptacle-tube slightly to strongly curved, weakly to strongly gibbous, adaxial length 0·3 cm., abaxial length 0·4–0·5 cm. Sepals 0·3–0·4 cm. long. Petals 0·5–0·6 cm. long, lingulate, very shortly unguiculate. Fertile stamens 7; filaments 0·3–0·4 cm. long; staminodes 8, of variable length, 0·1–0·2 cm. long, deltate, united in lower ½. Ovary very sparsely pilose; style 0·45 cm. long, glabrous except at base. Drupe up to 5 × 3 cm., ellipsoid or obovoid, fulvous-tomentose.

The typical subspecies is widespread in the rain forests of the Guineo-Congolian region where its most characteristic habitat is the banks of rivers. In contrast, subsp. *bangweolensis* occurs in *Brachystegia*, *Julbernardia*, *Isoberlinia* woodland in a small part of the Zambezian region. Subsp. *butayei* differs from subsp. *bangweolensis* in its acuminate leaves, persistent stipules, less coarse indumentum, and smaller, less strongly curved, less gibbous flowers. Were it not for the existence of intermediates these two taxa would certainly deserve specific rank. The intermediates, however, occupy an area larger than that of subsp. *bangweolensis*, and are treated here as a third subspecies, subsp. *transitoria*.

Subsp. **bangweolensis** (R. E. Fr.) F. White in Bull. Jard. Bot. Nat. Belg. **46**: 283 (1976). TAB. **12** fig. B. Types: Zambia, Lake Bangweulu, Mano R., fl. ix.1911, *Fries 732* (K; UPS, lectotype of R. Graham); Kawendimusi, fl. ix.1911, *Fries 780* (UPS, paratype); Kamindas, *Fries 780a* (UPS, paratype, not seen).

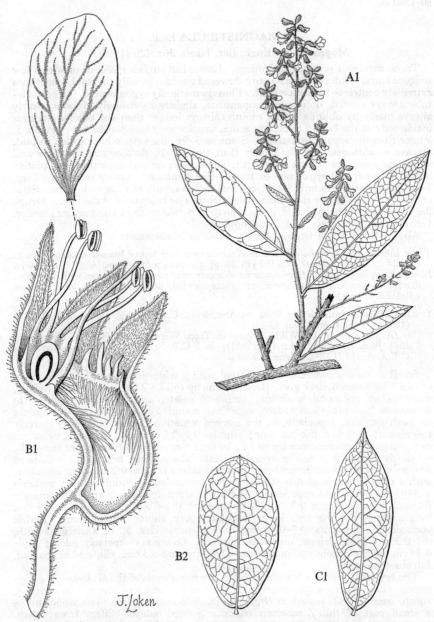

J.Loken

Tab. 12. A.—MAGNISTIPULA SAPINII. A1, habit (× ½) *Robinson* 5503. B.—M. BUTAYEI subsp. BANGWEOLENSIS. B1, vertical section of flower and a single petal (× 10) *Hoyle* 1274; B2, leaf (× ½) *Bredo* 6203. C.—M. BUTAYEI subsp. TRANSITORIA. C1, leaf (× ½) *Fanshawe* 3548.

Parinari bangweolensis R.E.Fr. in Fedde, Repert. **12**: 540 (1913); Wiss. Ergebn. Schwed. Rhod.-Kongo-Exped. **1**: 61, t. 2 fig. 2 & t. 7 (1914). Types as above.
Hirtella bangweolensis (R.E.Fr.) Greenway in Kew Bull. **1928**: 199 (1928).—Brenan in Mem. N.Y. Bot. Gard. **8**: 431 (1954).—F. White, F.F.N.R.: 69 (1962). Types as above.
Magnistipula bangweolensis (R. E. Fr.) R. A. Grah. in Kew Bull. **1957**: 230 (1957); in F.T.E.A., Rosaceae: 56, t. 8 fig. 1–4 (1960). Types as above.

Leaves oblong, obovate- or oblanceolate-oblong, apex broadly rounded or very shortly, suddenly and bluntly cuspidate, base usually subcordate; flowers c. 0·9 cm. long, strongly curved and strongly gibbous.

Zambia. N: Kawambwa Distr., Kawambwa to Fort Rosebery 64 km., fr. 29.x.1952, *White* 3529 (FHO; K). W: Ndola Distr., Chinama, 1370 m., fl.-buds 10.viii.1938, *Greenway & Trapnell* 5587 (EA; K). C: 17 km. NNE. of Serenje, fl. 11.x.1949, *Hoyle* 1274 (FHO). **Malawi.** N: Mugesse F.R., fl. 13.xi.1952, *Chapman* 40 (FHO; K). C: Kota-kota Distr., Chia area, 480 m., fl. 1.ix.1946, *Brass* 17463 (BM; K; SRGH).
Also in Tanzania. In wetter *Brachystegia, Julbernardia, Isoberlinia* woodlands, especially on damp sandy soil. Sometimes on termite mounds on floodplain grassland. 480–1500 m.

Subsp. **transitoria** F. White in Bull. Jard. Bot. Nat. Belg. **46**: 287 (1976). TAB. 12 fig. C. Type: Zambia, Kawambwa, in fringing forest, fl. 23.viii.1957, *Fanshawe* 3548 (K, holotype; FHO, isotype).
Hirtella katangensis Hauman in Bull. Jard. Bot. Brux. **21**: 179 (1951); in F.C.B. **3**: 40 (1952). Type from Zaire.
Hirtella bangweolensis (R.E.Fr.) Greenway var. *glabriuscula* Hauman, tom. cit.: 183 (1951); tom. cit.: 39 (1952). Type from Zaire.
Hirtella butayei var. *greenwayi* (Brenan) Hauman, loc. cit.; tom. cit.: 42 (1952) quoad, specim. Zaire tantum.
Magnistipula katangensis (Hauman) Mendes in Mem. Junta Invest. Ultramar, Sér. 2, **19**: 11 (1960); in C.F.A. **4**: 18 (1970). Type as for *H. katangensis*.
Hirtella sp. 1; F. White, F.F.N.R.: 69 (1962).

Leaves usually oblanceolate and acuminate or subacuminate at apex and acute at base, but sometimes indistinguishable from subsp. *bangweolensis*; flowers 0·7–0·8 cm. long, usually less strongly curved and less gibbous, but sometimes the same shape as in subsp. *bangweolensis*.

Zambia. N: Kawambwa Distr., Chishinga Ranch, fl. 6.ix.1966, *Lawton* 1418 (FHO; K). W: Mwinilunga Distr., Zambezi R., 7 km. N. of Kalene Hill, fr. 26.ix.1952, *White* 3379 (BM; COI; EA; FHO; K).
Also in Tanzania and Zaire. Inside and at the edges of fringing forest.

2. **Magnistipula sapinii** De Wild. in Bull. Jard. Bot. Brux. **3**: 262 (1911); **4**: 81 (1914). TAB. 12 fig. A. Type from Zaire.
Hirtella eglandulosa Greenway in Kew Bull. **1928**: 199 (1928).—Hauman in F.C.B. **3**: 39 (1952).—F. White, F.F.N.R.: 69 (1962). Type: Zambia, Barotseland, Mongu Distr., Sefula, 915 m., fl. viii.1921, *Borle* 254 (FHO; K, holotype; PRE).
Hirtella sapinii (De Wild.) A. Chev. in Bull. Mus. Hist. Nat. Paris, Sér. 2, **3**: 195 (1931).—Hauman, tom. cit.: 38 (1952). Type as for *Manistipula sapinii*.
Hirtella fruticulosa Hauman in Bull. Jard. Bot. Brux. **21**: 182 (1951). Types from Angola and Zaire.
Magnistipula eglandulosa (Greenway) R. A. Grah. in Kew Bull. **1957**: 407 (1957).—Mendes in C.F.A. **4**: 19 (1970). Type as for *Hirtella eglandulosa*.

Rhizomatous geoxylic suffrutex usually less than 20 cm. tall; young branchlets, petioles and inflorescence-axes fulvous-tomentellous. Leaf-lamina up to 16 × 4·5 cm., oblanceolate, apex usually acutely subacuminate, rarely cuspidate or rounded, base usually obtuse, rarely acute, sparsely strigulose on both surfaces, especially on the nerves, venation closely reticulate, slightly prominent beneath, more definitely so above; petiole c. 0·2 cm. long; stipules c. 0·6 cm. long, subulate, persistent. Inflorescence up to 17 × 4 cm., terminal, an elongate, very contracted thyrse with very few, markedly ascending lateral branches, most flowers being borne in subsessile cymules on the main axis; bracts c. 0·3 cm. long, narrowly deltate, with a pair of large sessile glands at base; pedicels 0·2–0·5 cm. long. Receptacle-tube slightly curved, slightly gibbous, adaxial length 0·4 cm., abaxial length 0·7 cm., sepals 0·35–0·5 cm. long. Petals 0·5–0·6 cm. long, lingulate, very shortly unguiculate. Fertile stamens 7; filaments 0·35–0·4 cm. long; staminodes 8,

0·2 cm. long, narrowly deltate, fused in lower ½ to form a " comb ". Style 0·4 cm·
long, glabrous. Drupe c. 3·5 × 2·0 cm., ellipsoid, fulvous-tomentose.

Zambia. B: Mongu Distr., Mombo, fl. 17.xi.1962, *Robinson* 5503 (K; SRGH).
W: Mwinilunga Distr., N. of Kalene Hill, fl. 24.ix.1952, *Angus* 535 (BM; COI; FHO;
K).
   Also in Angola and Zaire. Apparently confined to Kalahari Sands, where it is
characteristic of seasonally waterlogged sites such as the edges of dambos.

   The plant from which the type of *H. eglandulosa* was obtained was described by its
collector as a " small tree ". No subsequent observer in the field (including myself) has
been able to confirm this. Although the type specimen consists of flowering shoots arising
from the previous year's growth and is not attached to a subterranean rootstock, there can
be little doubt that it was collected from a geoxylic suffrutex which had escaped burning
the year before. It can easily be matched with similar specimens which are demonstrably
suffrutices or have been described by their collectors as such. *Magnistipula sapinii* was
also described by its author as a small tree, but this again has not been subsequently
confirmed. Some of the material cited by Hauman (e.g. *Overlaet* 941) was undoubtedly
collected from suffruticose plants. Material cited by Hauman as *H. eglandulosa* and
H. sapinii is, to my eye, indistinguishable.

64. VAHLIACEAE

By D. M. Bridson

   Annual or perennial herbs or subshrubs. Leaves opposite, simple, entire,
exstipulate. Flowers in axillary pairs, hermaphrodite, regular. Calyx-tube
campanulate, adnate to ovary; lobes 5, valvate, persistent. Petals 5, valvate,
alternate with calyx-lobes. Stamens 5, alternate with petals, inserted on margin of
epigynous disk; filaments free, subulate; anthers 2-thecous, dehiscing longitudinally, introrse, dorsifixed. Ovary inferior, 1-locular; styles 2 (or 3?), stigmata
capitellate; ovules numerous, attached to 2 thick, pendulous, flattened, rounded
placentas hanging from the apex of the loculus. Fruit a capsule, dehiscing by
sutures between the style bases into 2(3?) valves. Seeds numerous, oblong,
minute.

VAHLIA Thunb.

Vahlia Thunb., Nov. Gen. Pl. **2**: 36 (1782).—Hook. f. in Benth. & Hook. f., Gen.
   Pl. **1**: 637 (1865).—Engl. in Engl. & Prantl, Pflanzenfam. **III**, 2A: 65 (1890);
   ed. 2, **18A**: 166 (1930).—Maheshwari in Taxon **15**: 333 (1966) *nom. conserv.*
Bistella Adans., Fam. Pl. **2**: 226 (1763).—Bullock in Acta Bot. Neerl. **15**: 84–85
   (1966); in Meded. Bot. Mus. Herb. Rijksuniv. Utrecht **229**: 84–85 (1966).
      Russelia L. f., Suppl. Pl.: 24 (1781) non Jacq. (1760).

   Annual herbs or suffrutices, branched and erect, glabrous or pubescent with
multicellular, often gland-tipped hairs. Leaves sessile to subsessile, ovate to linear,
midrib apparent but lateral veins obscure. Inflorescences erect, sessile to subsessile
or pedunculate; flowers sessile to subsessile or pedicellate. Calyx-tube
campanulate to subglobose, often distinctly 5-veined. Petals shorter than or just
exceeding the calyx-lobes, lanceolate to orbicular; margin entire to crenate or
irregularly dentate. Filaments with or without a widened membranous area or
scale-like appendage at the base. Style-arms spreading, shorter than or exceeding
the calyx-lobes; style-bases separate to the disk or shortly above it. Capsule
subglobose or obovoid, all floral parts (except anthers) persistent.
   A genus of 5 species occurring throughout Africa south of the Sahara with the
exception of the Congo Basin; also found in Egypt, Iraq, Iran, Pakistan, India,
Madagascar and S. Vietnam.

Peduncles and pedicels absent or not exceeding 1 mm.; leaves ovate to lanceolate;
   gynoecium with style-bases dividing 0·15–0·3 mm. above the disk -   3. *digyna*
Peduncles and pedicels clearly present; leaves lanceolate to linear; gynoecium with style-
   bases free to disk:

Calyx-lobes seldom equalling the tube; styles shorter than calyx-lobes, up to 2 mm.
long - - - - - - - - - - 1. *dichotoma*
Calyx-lobes usually exceeding the tube; styles exceeding or at least equalling the
calyx-lobes, over 4 mm. long - - - - - - - 2. *capensis*

1. **Vahlia dichotoma** (Murray) Kuntze, Rev. Gen. Pl. **1**: 227 (1891).—Keay in
F.W.T.A. ed. 2, **1**: 120, fig. 41 (1954).—Bridson in Kew Bull. **30**, 1: 164, fig. 1 U–Z
(1975). TAB. **13** fig. B. Type cultivated at Göttingen from seeds of unknown
origin; present locality unknown.
Heuchera dichotoma Murray in Nov. Comm. Soc. Reg. Sci. Gött. **3**: 64, t. 1 (1773).
Type as above.
Oldenlandia pentandra Retz., Obs. Bot. **4**: 22 (1786). Type from India.
Oldenlandia dichotoma (Murray) Spreng., Pl. Min. Cog. Pug. Prim. **2**: 36 (1815).
Type as for *Vahlia dichotoma*.
Vahlia oldenlandiae DC., Prodr. **4**: 54 (1830) nom. superfl.
Vahlia silenoides DC., loc. cit.—Hook., Niger Fl.: 374 (1849).—A. Chev., Fl. Afr.
Occ. Fr. **1**: 284 (1938). Type from Senegal.*
Vahlia tomentosa DC., loc. cit.—Hook., loc. cit. Type from Senegal.
Vahlia oldenlandioides Roxb., Fl. Ind. **2**: 89 (1832).—Wight, Ic. Pl. Ind. Or. **2**, 3:
5, t. 562 (1843).—Hook. f. & Thomson in Journ. Linn. Soc., Bot. **2**: 74 (1857).—
Oliv. in F.T.A. **2**: 384 (1871).—Engl., Pflanzenw. Ost-Afr. **C**: 189 (1895).—Schinz
in Denkschr. Akad. Wiss. Wien, Math.-Nat. Kl. **78**: 415 (1905).—Engl., Pflanzenw.
Afr. **3**, 1: 286 (1915).—Eyles in Trans. Roy. Soc. S. Afr. **5**, 4: 359 (1916).—Hutch. &
Dalz. in F.W.T.A. ed. 1, **1**: 106, fig. 36 (1927).—Engl. in Engl. & Prantl, Pflanzenfam.
ed. 2, **18a**: 166, fig. 93 e–g (1930).—Gomes e Sousa in Bol. Soc. Estud. Col. Moçamb.
32: 304 (1936).—A. Chev., Fl. Afr. Occ. Fr. **1**: 283 (1938) nom illegit. Type from
India.
Vahlia cordofana Hochst. in Flora, Jena **24**, Intell. 1, 3: 43 (1841).—Schweinf.,
Beitr. Fl. Aethiop.: 271 (1867) nom. nud.
Vahlia pentandra (Retz.) C. E. C. Fischer in Kew Bull. **1932**: 56 (1932). Type as
for *Oldenlandia pentandra*.
Bistella dichotoma (Murray) Bullock in Acta Bot. Neerl. **15**: 85 (1966); in Meded.
Bot. Mus. Herb. Rijksuniv. Utrecht **229**: 85 (1966). Type as for *Vahlia dichotoma*.

Erect branching annual (4·5)12–51(66) cm. high, rarely with a slightly woody
base, sparsely covered with crisped pubescence, sometimes glandular. Leaves
0·6–7 × 0·1–0·6 cm., lanceolate to linear, acute at apex, narrowed towards base.
Peduncles 0·1–2·2 cm. long, (1)2-flowered; pedicels 0–0·5(0·8) cm. long. Calyx-
tube 1·3–2·5 mm. long, sparsely to ± densely pubescent; lobes ovate-lanceolate to
triangular, 1–2·8 × 0·7–1·1 mm., acute, glabrous to pubescent outside. Petals white
to yellow, orbicular or obovate-oblong, 0·9–2·6 × 0·6–1·7 mm., rounded to tapering
or shortly acuminate at apex, entire or sometimes becoming crenate towards apex;
midvein and usually at least 1 pair of lateral veins apparent. Filaments 0·5–1·7 mm.
long, usually glabrous but occasionally with a few hairs, not appendaged at base;
anthers 0·3–1 mm. long. Gynoecium with style-bases distinctly thickened, free to
the disk; styles 0·6–2 mm. long, usually glabrous but occasionally with a few hairs.
Capsule sparsely hairy, 2–3·3 mm. long. Seeds straw-coloured, ovoid to cylindrical,
0·2–0·3 mm. long, with low longitudinal ridges.

Zambia. N: Luangwa Forest Game Reserve South, along the Katete R., fl. & fr.
18.viii.1966, *Astle* 4318 (K; SRGH). E: Luangwa R., Nsefu Game Camp, fl. & fr.
15.x.1958, *Robson & Angus* 131 (K; LISC; SRGH). **Malawi.** C: near Salima, Nyasa
Hotel, fl. & fr. 15.ii.1959, *Robson & Steele* 1610 (BM; K; LISC; SRGH).
Mozambique. N: Nametil, fl. & fr. 15.v.1937, *Torre* 1415 (COI; LISC). Z:
Namacurra on R. Licungo, fl. 28.viii.1949, *Barbosa & Carvalho* 3868 (K; LISC; LMA).
T: Baroma Prov., Sisitso, R. Zambese., fl. & fr. 18.vii.1950, *Chase* 2764 (BM; SRGH).
MS: Expedition I., fl. & fr. vii.1858, *Kirk* (K).
Widespread from Senegal to Sudan (reaching as far north as Tibesti) and in E. tropical
Africa, Madagascar, India and Vietnam (Phu Quoc I.). Sandy river banks.

2. **Vahlia capensis** (L. f.) Thunb., Nov. Gen. Pl. **2**: 36 (1782).—Lam. Illustr. Gen. **2**: 328
(1819) et tab. 183 (1792).—DC., Prodr. **4**: 53 (1830).—Harv. in Harv. & Sond.,
F.C. **2**: 306 (1862).—Oliv. in F.T.A. **2**: 384 (1871).—Baill., Hist. Pl. **3**: 335,
fig. 374, 375 (1872).—Hiern, Cat. Afr. Pl. Welw. **1**: 324 (1896).—Engl. & Gilg in

* Erroneously placed in synonymy under *V. digyna* (Retz.) Kuntze by Keay in
F.W.T.A., ed. 2, **1**: 120 (1954) and by Bullock in Acta Bot. Neerl. **15**: 85 (1966) and in
Meded. Bot. Mus. Herb. Rijksuniv. Utrecht 229: 85 (1966).

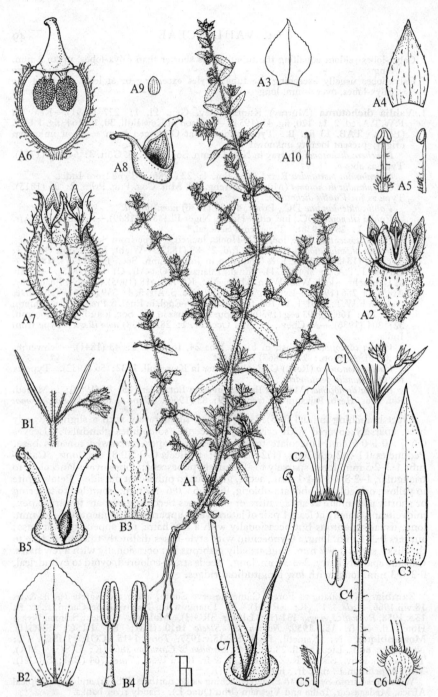

Tab. 13. A.—VAHLIA DIGYNA. A1, habit (× ⅔); A2, flower with 2 calyx-lobes and petals removed (×8); A3, superior surface of petal (×14); A4, calyx-lobe (×14); A5, stamen in frontal & lateral view (×14); A6, section through gynoecium (×14); A7, fruit (×10); A8, valve (style base in fruit) (×14); A9, seed (×20); A10, multicellular hair (×20), all from *Robinson* 1309. B.—VAHLIA DICHOTOMA. B1, flowering node (×1); B2, petal (×14); B3, calyx-lobe (×14); B4, stamens (×14); B5, styles (×14), all from *Astle* 4937. C.—VAHLIA CAPENSIS. C1, flowering node (×1); C2, petal (×10); C3, calyx-lobe (×10); C4, stamens in frontal view & dorsal view of anther (×8); C5, filament base in lateral view (×10); C6, filament base in frontal view (×10); C7, styles (×8). C1–C4 & C7 are subsp. VULGARIS var. LINEARIS, from *Harbor* 6494, C5–C6 are subsp. MACRANTHA, from *Barbosa & Carvalho* 3177

Warb., Kunene-Samb.-Exped. Baum: 242 (1903).—Schinz in Denkschr. Akad. Wiss. Wien, Math.-Nat. Kl. **78**: 415 (1905).—Dinter, Deutsch SW.-Afr. Fl. For. Landt. Frag.: 71 (1909).—Engl., Pflanzenw. Afr. **3**, 1: 286 (1915).—Eyles in Trans. Roy. Soc. S. Afr. **5**, 4: 358 (1916).—Engl. in Engl. & Prantl, Pflanzenfam. ed. 2, **18a**: 166 (1930).—Gomes e Sousa in Bol. Soc. Estud. Col. Moçamb. **32**: 304 (1936).—Friedrich-Holzhammer in Prodr. Fl. SW. Afr. **53** (1968).—Mendes in C.F.A. **4**: 22 (1970).—Bridson in Kew Bull. **30**, 1: 168, fig. 2 A–M, map 2 (1975). TAB. **13** fig. C. Type from S. Africa.

Russelia capensis L. f., Suppl. Pl.: 175 (1781). Type as above.

Vahlia glandulosa Schlechter ex Engl. in Engl. & Prantl, loc. cit. Type from S. Africa.

Bistella capensis (L. f.) Bullock in Acta Bot. Neerl. **15**: 85 (1966); in Meded. Bot. Mus. Herb. Rijksuniv. Utrecht **229**: 85 (1966). Type as for *V. capensis*.

An erect to straggling or diffusely branched, cushion-like herb or suffrutex, annual or biennial or perennial, 3·5–55 cm. tall, often with a woody tap-root or rhizome, entirely glabrous, pubescent or densely glandular-hairy. Leaves linear-lanceolate, 0·7–5·9 × 0·05–0·75 cm., linear or narrowly linear (or elliptic in 1 S. African subsp.), acute at apex, narrowed to base. Peduncles 0·2–1·5 cm. long; pedicels 0–4 mm. long (sessile or subsessile in Angolan and SW. African varieties). Calyx-tube 1·1–2(3) mm. long, glabrous to densely glandular hairy; calyx-lobes lanceolate or linear-lanceolate, (2·2)3–5(5·4) × 0·9–1·7(2) mm., frequently exceeding the tube in length, acute to acuminate at apex, the indumentum various but generally less dense than that of the tube, erect (or reflexing in subsp. *capensis* which is restricted to SW. Cape Prov.). Petals pale to bright yellow, sometimes drying with a maroon area at base, varying from strap-like to orbicular, 2·3–7 × 0·8–5 mm., rounded, acute or occasionally acuminate at apex; margin ± entire to very irregularly dentate; mid and lateral veins usually apparent, usually hairy at the base inside. Filaments 2–5·4 mm. long, glabrous or sparsely hairy, base with a membranous, hairy scale (variable in shape and size) or only with hairs; anthers 1–2·5 mm. long. Style-bases gradually and not markedly thickened, free to the disk, glabrous or rarely with short patent hairs; styles 4–6·6(8) mm. long, glabrous. Capsule 1·5–2·9 mm. long, glabrous to densely pubescent. Seeds straw-coloured to pale brown, ± ovoid or spindle-shaped, 0·3–0·55 mm. long, with distinct longitudinal ridges.

Plants robust, erect, over 27 cm. tall; filaments with a large basal appendage
          subsp. *macrantha*
Plants short or tall but never robust; filaments with basal appendage present or absent, infrequently large: subsp. *vulgaris*

Subsp. **macrantha** (Klotzsch) Bridson in Kew Bull. **30**, 1: 171, fig. 2 A (1975). Type: Mozambique, R. Zambezi, *Peters* (B†). TAB. **C5–C6**.

Vahlia macrantha Klotzsch in Peters, Reise Mossamb. Bot.: 175 (1861).—Walp., Ann. Bot. Syst. **7**: 900 (1871). Type as above.*

Robust erect herbs, 27–34 cm. tall; stems sparsely hairy. Leaves lanceolate or very narrowly elliptic, 3–5·6 × 0·3–0·6 cm., glabrous, scabrid or sparsely pubescent. Peduncles 3–7 mm. long; pedicels 1·5–4 mm. long. Flowers large; petals ovate to lanceolate, 3·5–6 × 1–2·6 mm.; filaments with a large basal appendage 0·5–0·8 × 0·6–1 mm.

Mozambique. MS: Expedition I., fl. & fr. viii.1858, *Kirk* (K). Known only from Mozambique along the Zambezi. On sand banks.

Subsp. **vulgaris** Bridson in Kew Bull. **30**, 1: 173, fig. 2 F, H, K (1975). Type: Rhodesia, Sebungwe near Kariyangwe R., *Davies* 2029 (K, holotype; LISC; PRE; SRGH, isotypes).

Cushion-like or tall, straggling to erect, but never robust plants, 3·5–53 cm. tall, entirely glabrous to densely glandular-hairy. Leaves narrowly linear, linear-lanceolate or oblanceolate. Filaments with or without a basal appendage, only occasionally large.

* Placed in synonymy with *V. dichotoma* (Murray) Kuntze by Bullock (following Index Kewensis) in Acta Bot. Neerl. **15**: 85 (1966) and in Meded. Bot. Mus. Herb. Rijksuniv. Utrecht **229**: 85 (1966), but the description states that the styles and anthers are exserted, thus placing it in *V. capensis* (L. f.) Thunb. The measurements given moreover conform with those of specimens of *V. capensis* subsp. *macrantha*.

Leaves (linear) linear-lanceolate or oblanceolate, sparsely to densely glandular-hairy or
    less often glabrous - - - - - - - - - - var. *vulgaris*
Leaves narrowly linear to linear, always glabrous:
    Small many-stemmed plants 3·5–15(22) cm. tall; stems not or only slightly woody at
      the base - - - - - - - - - - - var. *linearis*
    Taller few-stemmed plants, 13–40 cm. tall; stems usually becoming woody at the
      base, with corky bark - - - - - - - - var. *longifolia*

Var. **vulgaris**.

Few- or several-stemmed plants, 7–38 cm. tall, stems and leaves sparsely to
densely glandular-hairy or occasionally glabrous. Leaves rarely linear, more often
linear-lanceolate or occasionally oblanceolate, (1)1·5–4(7) mm. wide. Petals up to
4·5 × 2·5 mm. Filaments with basal appendage absent or variously shaped but not
exceeding 0·9(1) × 0·6 mm. or 0·8 × 0·7 mm.

    Caprivi Strip. 36 km. NW. of Ngoma Ferry, fl. 16.vii.1952, *Codd* 7082 (BM; K;
PRE; SRGH). **Botswana.** N: Gweta crossing on main Maun-Francistown road, fl. &
fr. 25.iv.1967, *Lambrecht* 170 (K; LISC; SRGH). SW: Ghanzi, Farm 56, fl. & fr.
29.iv.1969, *De Hoogh* 259 (K; SRGH). **Zambia.** B: c. 22 km. SE. of Kalabo on
Sandaula Pontoon road, fl. & fr. 17.xi.1959, *Drummond & Cookson* 6581 (K; LISC;
PRE; SRGH). C: 9·6 km. E. of Lusaka, fl. & fr. 20.xii.1957, *King* 396 (K). S:
Namwala, fl. & fr. 14.xii.1962, *van Rensburg* 1087 (K; SRGH). **Rhodesia.** N:
Urungwe, Mensa Pan, 17 km. ESE. of Chirundu Bridge, fl. & fr. 30.i.1958, *Drummond*
5381 (K; PRE; SRGH). W: Main Camp along Dopi Pan road, Wankie National Park,
fl. 30.xi.1968, *Rushworth* 1307 (K; LISC; PRE; SRGH). C: Poole Farm, Hartley,
fr. 2.ii.1964, *Hornby* 3437 (K; SRGH). E: Inyanga, Cheshire, fl. & fr. 5.ii.1931,
Norlindh & Weimarck 4880 (K). S: Rhino Hotel, Lundi R., Nuanetsi, fl. xii.1955,
Davies 1744 (K; SRGH). **Mozambique.** T: Mudzi R., 16 km. from Rhodesian
border, fl. & fr., *Wild* 2631 (K; PRE; SRGH). SS: Pafuri, fl. & fr. 18.ii.1954, *Schijff*
3578 (K; PRE).
    Also in SW. Africa, Cape Prov. and northern Transvaal. Usually on river banks etc.
and sandy soil or broken ground but also recorded from grasslands and vleis, 450–1300 m.

    This variety is very variable and broadly defined and many intermediates exist between
it and the other more strictly defined varieties of subsp. *vulgaris*. Several dissimilar plants
have been found from neighbouring localities and even from within the same gathering.

Var. **linearis** E. Mey. ex Bridson in Kew Bull. **30**, 1: 175, fig. 2B, J (1975). Type from
    S. Africa (Transvaal). TAB. **C1–C4** & **C7**.
    Vahlia cynodontetii Dinter, Deutsch-SW.-Afr. Fl. For. Landt. Frag.: 71 (1909).—
    Engl. in Engl. & Prantl, Pflanzenfam. ed. 2, **18a**: 166 (1930). Type from SW. Africa.

    Small several-stemmed, often compact and cushion-like plants, usually under
15 cm. tall, but can reach 22(30) cm.; stems glabrous or sparsely hairy. Leaves
narrowly linear or linear, not exceeding 2 mm. wide, often with revolute margins,
always glabrous. Filaments with basal appendage absent.

    Botswana. N: Kwebe Hills, fl. & fr. 2.ii.1898, *Lugard* 145 (K). SE: 3·2 km. S. of
Lobatsi, E. of railway, fl. 17.i.1960, *Leach & Noel* 155 (K; PRE; SRGH). **Rhodesia.**
W: Bulawayo, fl. 15.iv.1905, *Gardner* 83 (K).
    The main concentration occurs in the Transvaal, but also in SW. Africa, Cape Prov.,
Orange Free State and Lesotho. Swampy places and vleis, ± 1000 m.

    The following specimens are intermediate between var. *vulgaris* and var. *linearis*.
Botswana. SW: Ghanzi, Okwa, fl. & fr. 11.ix.1969, *Brown* 6058 (K; PRE; SRGH).
Zambia. C: Lusaka, fl. & fr. 30.xi.1964, *Robinson* 6262 (K; SRGH*). Rhodesia. W:
Gwaai, fl. & fr. ii.1949, *Davies* 224 (K; SRGH).

Var. **longifolia** (Gandog.) Bridson in Kew Bull. **30**, 1: 176 (1975). Type: Mozambique,
    Lourenço Marques, Matola, *Quintas* 132 (?LY, holotype; COI, isotype).
    Vahlia longifolia Gandog. in Bull. Soc. Bot. Fr. **65**: 29 (1918). Type as above.
    Vahlia capensis sensu J. Ross, Fl. Natal: 180 (1973).

    Erect few-stemmed plants, 13–40 cm. tall with unbranched or occasionally
branched stems usually becoming woody at the base, often with corky bark. Plants
entirely glabrous or less often with sparse hairs on the stems and rarely on the
leaves. Leaves linear, often a few much longer (up to 5·5 cm.) dispersed along the

    * The SRGH sheet is somewhat closer to var. *linearis* than the K sheet.

stem. Base of filaments gradually widened or with an inconspicuous appendage (not exceeding 0·5 × 0·4 mm.).

Mozambique. MS: Sofala, Mucarangue, fl. & fr. 19.x.1935, *Lea* 81 (PRE). SS: Gaza, between Maniquenique and Xai-Xai, fl. & fr. 6.vi.1957, *Barbosa & Lemos* 7610 (COI; LISC; SRGH). LM: Lourenço Marques, fl. & fr., *Borle* 220 (K; PRE). Also in northern Natal. In sandy soil, often near the coast, on river banks or in grassland, 0–45 m.

Grosvenor 615 (PRE; SRGH) from Rhodesia (S) is a gathering of 10 sheets, most of which are intermediate between var. *vulgaris* and var. *longifolia*; one, however, is true var. *vulgaris*. Some intermediate specimens also occur in E. Transvaal.

3. **Vahlia digyna** (Retz.) Kuntze, Rev. Gen. Pl. **1**: 227 (1891).—C. E. C. Fischer in Kew Bull. **1932**: 56 (1932) pro parte.—Keay in F.W.T.A. ed. 2, **1**: 120 (1954) pro parte.—Rampi in Adumb. Fl. Aethiop., *Saxifragaceae*: 25, Webbia **28**, 2: 528–530, fig. 2A, 3A, 3D, 4 (1973).—Bridson in Kew Bull. **30**, 1: 177, fig. 1 A–H (1975). TAB. **13** fig. A. Type from India.

Oldenlandia digyna Retz., Obs. Bot. **4**: 23 (1786). Type as above.
Oldenlandia decumbens Spreng., Pl. Min. Cog. Pug. Prim. **2**: 36 (1815) *nom. superfl.*
Vahlia ramosissima DC., Prodr. **4**: 54 (1830).—Hook., Niger Fl.: 374 (1849).—Oliv. in F.T.A. **2**: 383 (1871). Type from Senegal.*
Vahlia sessiliflora DC., loc. cit. *nom. superfl.*
Vahlia viscosa Roxb., Fl. Ind. **2**: 89 (1832).—Wight, Ic. Pl. Ind. Or. **2**: t. 563 (1843).—Hook. f. & Thomson in Journ. Linn. Soc., Bot. **2**: 74 (1857) pro parte.—Oliv., loc. cit. pro parte.—Engl., Pflanzenw. Ost-Afr. **C**: 189 (1895).—Engl., Pflanzenw. Afr. **3**, 1: 286 (1915) pro parte.—Hutch. & Dalz. in F.W.T.A. ed. 1, **1**: 106 (1927) pro parte.—Engl. in Engl. & Prantl, Pflanzenfam. ed. 2, **18a**: 166 (1930) pro parte.—A. Chev., Fl. Afr. Occ. Fr. **1**: 283–284 (1938) pro parte.
Vahlia menyharthii Schinz in Bull. Herb. Boiss., Sér. 2, **2**: 944 (1902). Type: Mozambique, Boruma, *Menyhart* 1069 (Z, holotype; K, isotype).
Haloragis jerosioides Perrier de Bâthie in Not. Syst. **14**: 305 (1966). Type from Madagascar.
Bistella digyna (Retz.) Bullock in Acta Bot. Neerl. **15**: 85 (1966); in Meded. Bot. Mus. Herb. Rijksuniv. Utrecht **229**: 85 (1966) pro parte. Type as for *Vahlia digyna*.

An erect, sparsely to much-branched annual herb 5–33·5 cm. tall; stems covered with patent or crisped, often glandular hairs. Leaves ovate to ovate-lanceolate, 0·5–2·5(4·8) × 0·2–0·6(0·9) cm., acute at apex, tapering or rounded at base, usually with only a few hairs but occasionally pubescent. Flowers sessile or subsessile, peduncles and pedicels up to 1 mm. long. Calyx-tube 1·1–2 mm. long, sparsely covered in patent hairs or shorter crisped hairs; calyx-lobes ovate, 1·3–2 × 0·6–1·2 mm., acute at apex, glabrous or sparsely pubescent outside. Petals yellow fading to white, always shorter than sepals, orbicular to ovate, 1–1·3(1·6) × 0·55–1 mm., usually distinctly apiculate at apex, narrowed to base; margin entire to finely or irregularly crenate towards apex; only mid-vein distinct. Filaments 0·3–1·1 mm. long, translucent with vein apparent, small, membranous, hairy, basal scale-like appendage present; anthers 0·15–0·4 mm. long. Gynoecium with style-bases distinctly thickened, not separating for 0·15–0·3 mm. above the disk; styles 0·7–1·3 mm. long, glabrous. Capsule 1·4–2·2 mm. long, sparsely pubescent. Seeds straw-coloured, ovoid-cylindric, 0·15–0·27 mm. long, with indistinct longitudinal ridges.

Botswana. SE: 8 km. E. of Lothlekane (Lotlakani), fl. & fr. 24.iii.1965, *Wild & Drummond* 7264 (K; LISC; PRE; SRGH). **Zambia.** N: Mpika Distr., Luangwa Valley, Luangwa R., W. bank between Kapamba R. & Mfuwe, fl. & fr. xi.1970, *Abel* 353 (SRGH). prob. C: Luangwa Valley National Park, fl. & fr. 1963, *Ansell* SRGH 200963 (K; SRGH specimen is 200969). S: Kafue Flats, Mazabuka, fl. & fr. 18.vi.1955, *Robinson* 1309 (K; SRGH). **Rhodesia.** N: Urungwe Distr. near Katupatupa Hotspring, Cheware Wilderness area, fl. & fr. v.1971, *Guy* 1759 (SRGH). **Mozambique.** T: Boruma (Boroma), fl. & fr. v.1891, *Menyhart* 1069 (K; Z).
Also throughout tropical Africa S. of the Sahara, Egypt, Madagascar, India and Pakistan. Usually along river banks and dambos but occasionally in drier situations.

Bistella geminiflora Del. (*Vahlia geminiflora* (Del.) Bridson in Kew Bull. **30**, 1: 179, fig. 1 J, K (1975)) has frequently been included in synonymy under *Vahlia digyna*, but it is actually a distinct species not occurring south of 4° N. In the literature *V. geminiflora* is usually referred to as *V. weldenii* Reichenb.

* Erroneously placed in synonymy under *V. dichotoma* by Bullock in Acta Bot. Neerl. **15**: 85 (1966); in Meded. Bot. Mus. Herb. Rijksuniv. Utrecht **229**: 85 (1966).

65. MONTINIACEAE

By E. J. Mendes

Small trees or shrubs, dioecious. Leaves alternate, opposite or subopposite, simple, entire, penninerved, usually petiolate, deciduous; stipules absent. Flowers in terminal or axillary inflorescences, the ♂ flowers in few-flowered panicles, the ♀ flowers solitary or in groups of 2(3)-nate. ♂ flowers 3–4-merous; calyx-tube expanded or cupular; petals imbricate, slightly fleshy, deciduous, inserted below the margin of the fleshy disk; stamens alternipetalous and inserted with the petals; anther-thecae dehiscing by longitudinal slits; pollen grains 3-colporate, spheroidal; rudimentary ovary absent or minute. ♀ flowers 4(?5)-merous; calyx-tube ± ellipsoid; petals imbricate, slightly fleshy, deciduous; disk fleshy, epigynous; staminodes alternipetalous, inserted with the petals below the margin of the disk; ovary inferior, 2-locular; style short, columnar or 2-branched; stigmas 2, large; ovules few, 2-seriate on axile placentas. Fruit a loculicidally dehiscent capsule, or indehiscent becoming 1-locular. Seeds ± compressed, winged or not; endosperm absent or abundant; embryo straight; cotyledons compressed.

A small African family comprising 2 distantly related genera.

Leaves alternate; ♂ flowers usually 4-merous, bracteate; capsule ellipsoid, smooth,
    loculicidal; seeds with a broad papery wing -   -   -   -   -   **1. Montinia**
Leaves opposite or subopposite; ♂ flowers usually 3-merous, ebracteate; capsule flask-
    shaped, tuberculate and spiny, indehiscent; seeds wingless -   -   **2. Grevea**

1. MONTINIA Thunb.

Montinia Thunb. in Phys. Saellsk. Handl. **1**: 108 (1777).

Dioecious shrub. Leaves alternate, simple, usually petiolate. ♂ flowers in terminal and subterminal corymbose inflorescences; calyx-tube expanded, usually 4-lobed; petals usually 4, imbricate; disk fleshy, lining the calyx-tube; stamens usually 4, alternipetalous, inserted below the margin of the disk; anthers extrorse. ♀ flowers terminal or subterminal, solitary or in groups of 2(3); calyx-tube fusiform, calyx-lobes usually 4; petals as in the ♂; disk epigynous; staminodes usually 4; gynoecium bicarpellary, 2-locular; style short, 2-branched; stigmas large; ovules numerous, 2-seriate. Capsule ellipsoid, 2-valved, loculicidal; seeds compressed, winged; cotyledons broadly oblong, apex obtuse.

Monospecific African genus with a southern and south-western distribution from the Cape to Angola.

Montinia caryophyllacea Thunb. in Phys. Saellsk. Handl. **1**: 109 (1777).—Marloth, Fl. S. Afr. **2**, 1: 25, t. 9 fig. A (1925).—Engl. & Prantl, Pflanzenfam. ed. 2, **18a**: 233, fig. 130 (1930).—Milne-Redh. in Hook., Ic. Pl. **36**: t. 3541A, B fig. 1–4 & 3542; text pp. 5 & 15, with map (1955).—Friedrich-Holzhammer in Prodr. Fl. SW. Afr. **54**: 1 (1968).—Mendes in C.F.A. **4**: 24 (1970). TAB. **14**. Type from S. Africa.
    Montinia acris L. f., Suppl. Pl.: 427 (1781) *nom. illegit.*—Harv. in Harv. & Sond., F.C. **2**: 307 (1862); Thes. Cap. **2**: 14, t. 122 (1863).—Engl., Pflanzenw. Afr. **1**, 2: 493, fig. 414 (1910); op. cit. **3**, 1: 286, fig. 186 (1915). Type from S. Africa.

A twiggy shrub, 1–2(?4) m. high, almost glabrous (except for the leaf-axils and the base of the midribs). Leaves entire, $1 \cdot 5$–$7 \cdot 0 \times (0 \cdot 5)1$–$2 \cdot 5$ cm., linear to ovate or elliptic, apex obtuse, base cuneate narrowing into a petiole up to 1 cm. long which is dilated at the very base into an accrescent cushion which persists long after the fall of the leaf, becoming ivory-coloured. Inflorescences terminal or subterminal. ♂ flowers ± numerous in corymbs 4–7 cm. long, erect; pedicels up to 1 cm. long, bracteate; bracts subulate; calyx-tube expanded, c. 3 mm. in diameter with (?3)4(?5) triangular, very short, calyx-lobes; petals (?3)4(?5), 3–5 mm. long, broadly ovate, with a broad crescent-shaped base, white; disk flat, subcircular, c. $2 \cdot 5$ mm. in diameter; stamens (?3)4(?5), with anthers broadly oblong c. $1 \cdot 2$ mm. long and filaments c. $1 \cdot 5$ mm. long. ♀ flowers solitary or in groups of 2(3), erect; calyx-tube c. 7 mm. long, fusiform; calyx-lobes, petals and disk as in the ♂; staminodes fleshy, very short; placentas 2, axile, elliptic, compressed and fused with the septum which in fruit becomes free from the walls of the capsule, hanging from its top; each placenta with up to 12 ovules in 2 longitudinal series; styles

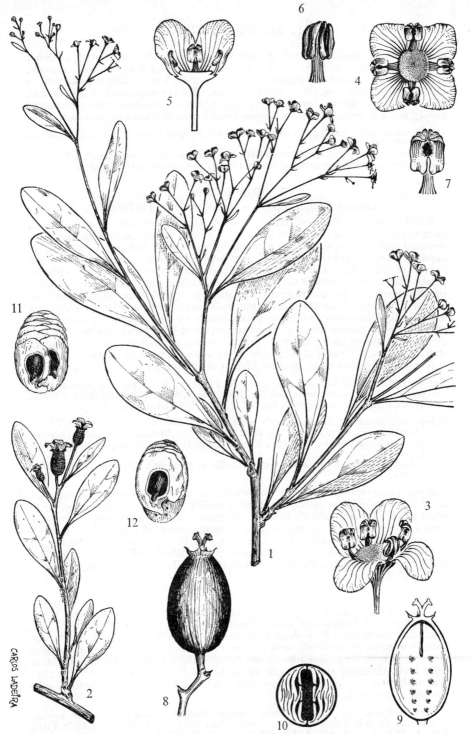

Tab. 14. MONTINIA CARYOPHYLLACEAE. 1, branchlets with ♂ inflorescences (×1) *Torre* 8460; 2, branchlet with ♀ flowers (×1) *Santos* 866; 3, ♂ flower, side view (×4); 4, ♂ flower, top view (×4); 5, ♂ flower, longitudinal section (×4); 6, stamen, abaxial view (×8); 7, stamen, adaxial view (×8), 3–7 from *Torre* 8460; 8, capsule (×2); 9, longitudinal section of capsule, seeds removed to show placenta (×2); 10, semi-diagrammatic transverse section of capsule (×2); 11, seeds, as disposed along one placenta (×2); 12, seeds as disposed along one placenta, four seeds removed (×2), 8–12 from *Torre* 8474.

c. 1 mm. long, thick, 2-branched from the middle, each branch with a broad,
subreniform stigma c. 1 mm. wide.   Capsule 1·5–2·0 × 1 cm., ellipsoid, smooth,
crowned by the persistent calyx-lobes, disk and style, loculicidally dehiscent from
the apex; 2-valved, the valves cartilagineous.   Seeds compressed, c. 10 × 8 mm.
including the ± elliptic, broad papery wing which is auriculate at the attachment
point; seeds disposed on each placenta along 2 rows like the tiles of a roof.

Botswana. SW: between Ghanzi and Gobabis, Mamuno Gate, ♂ fl. 13.ii.1970, *Brown*
34 (K; SRGH).
Known also from the Cape Prov. to SW. Africa and south-western Angola.   Tree and
shrub steppe with *Albizia anthelmintica* and *Combretum apiculatum*, c. 1000 m.

2. GREVEA Baill.
Grevea Baill. in Bull. Soc. Linn. Paris **1**: 420 (1884).

Small dioecious trees or shrubs.   Leaves simple, opposite or subopposite,
penninerved, petiolate; stipules absent.   ♂ flowers in terminal or axillary few-
flowered inflorescences; bracts very small, obsolete or absent; calyx-tube ±
cupular, truncate, subentire or shallowly 3-lobed; petals 3, inserted below the
margin of the disk, imbricate; stamens (?2)3, alternipetalous, inserted with the
petals, with introrse anthers dehiscing by longitudinal slits; disk fleshy, cupular,
usually elevated in the centre into a vestigial ovary.   ♀ flowers solitary, terminal;
calyx-tube elongated, smooth and longitudinally striate or tuberculate to echinulate,
with usually 4 inconspicuous lobes, petals usually 4; disk epigynous; staminodes
present; ovary bicarpellary; style columnar, short and thick; stigmas 2-lobed;
ovules 2-seriate on each placenta.   Fruit indehiscent, crowned by the persistent
style.   Seeds subellipsoid to subglobose; endosperm abundant, horny.
A small genus with an E. tropical African distribution, also in Madagascar.

Grevea eggelingii Milne-Redh. in Hook., Ic. Pl. **36**: t. 3541B fig. 6–7, 3543 & 3544;
text p. 7 (1955).—Mendes in Bol. Soc. Brot., Sér. 2, **44**: 299, fig. 1C (1970).—Verdc.
in F.T.E.A., Montiniaceae: 3, fig. 1–2 (1973).   Type from Tanzania.

Subsp. **echinocarpa** (Mendes) Verdc. in Kew Bull. **28**: 149 (1973); in F.T.E.A.,
Montiniaceae: 6 (1973).   TAB. **15** & **16**.   Type: Mozambique, Cabo Delgado,
between Montepuez and Nantulo, *Torre & Paiva* 11708 (COI; K; LISC, holotype;
P).
 Grevea eggelingii var. *echinocarpa* Mendes in tom. cit.: 299, fig. 1A–B (1970).
Type as above.

A deciduous tree up to 7·5 m. high, or a shrub.   Leaves opposite or subopposite,
usually 2 pairs only on each twig; lamina usually 10–24 × 4–12 cm., often much
smaller (3·5 × 1·0 cm.), elliptic to ovate, shortly and acutely acuminate at the apex,
broadly to narrowly cuneate at the base, margins entire, glabrous; lateral nerves
4–5-paired, arcuate; petiole (0·8)3–5 cm. long.   ♂ inflorescence ± corymbose, up
to 8-flowered, in the axils of juvenile leaves; peduncle (2)3–5 cm. long; pedicels
0·7–1·8 cm. long; bracts absent.   ♂ flowers: calyx-tube ± cupuliform, c. 1·5 mm.
deep and 2·5 mm. in diameter at the shallowly 3-lobed mouth; petals 3, c.
4 × 2 mm., oblong, with apex obtuse and slightly cucullate and base truncate and
1·5 mm. wide, strongly reflexed; stamens 3, with filaments c. 3 mm. long and
anthers c. 2 mm. long, oblong and ± horizontally disposed; rudimentary ovary
1 mm. long.   ♀ flowers not seen.   Fruit shortly stipitate, up to 3·5 × 2·3 cm.
(excluding the spines), flask-shaped, the base ± orbicular and densely beset with
both ± unequal spines 5–6(8) mm. long and smaller rounded tubercules, the neck
terete, c. 8 mm. long, also tuberculate but spineless, crowned by the persistent
calyx-limb, disk and style.   Seeds up to 12 per fruit, mostly 6 × 4·5 × 3 mm.,
reddish-brown, ± ellipsoid, asymmetric and compressed; cotyledons subrhombic, apex acute.

Mozambique. N: Cabo Delgado, between Montepuez and Nantulo, fr. immat.
27.xii.1963, *Torre & Paiva* 9740 (BM; BR; COI; EA; FHO; LISC; LMU; PRE;
SRGH; WAG); fr. 7.iv.1964, *Torre & Paiva* 11708 (COI; K; LISC; P).
Also known from Tanzania.   Riverine forest of the Rovuma basin, up to 500 m.

Only future collections will allow the proper rank of this taxon to be established.   It was
first described as a variety, is now accepted as a subspecies and may yet prove to be a good
species (see also Verdcourt, locis cit.).

Tab. 15. GREVEA EGGELINGII subsp. ECHINOCARPA, ♂. 1, flowering branchlets (×½);
2, leaf (×½); 3, inflorescence, with most flowers removed (×4); 4, longitudinal
section of flower (×8); 5, petal (×8); 6, stamen (×8), all from *Mason* 1.

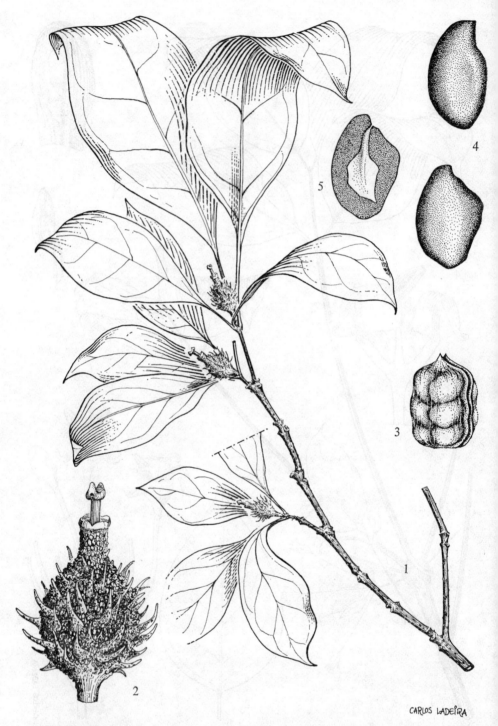

CARLOS LADEIRA

Tab. 16. GREVEA EGGELINGII subsp. ECHINOCARPA, ♀. 1, fruiting branchlets (× ⅔) *Torre &
Paiva* 9740; 2, fruit (×1⅔); 3, seeds, covered with endocarp, adhering to septum
(×1⅔); 4, seeds (×4); 5, seed, split open to show embryo (×4), 2–5 from *Torre &
Paiva* 11708.

66. BREXIACEAE
By N. K. B. Robson

Trees or shrubs, rarely climbing (*Roussea*), glabrous or rarely with simple hairs (*Roussea*). Leaves simple, alternate to subopposite or subverticillate, entire or with crenate to glandular-serrate or spinose-dentate margins, penninerved, rarely glandular beneath (*Roussea*); stipules present or absent (*Ixerba*), sometimes united by a transverse ridge. Flowers bisexual, actinomorphic, axillary, solitary or in cymes or false umbels or fascicles. Sepals 4–5(6), free or adnate to the ovary base, imbricate or valvate (*Roussea*), persistent (*Roussea*) or deciduous. Petals 4–5(6), free or united at the base (*Roussea*), imbricate or valvate (*Roussea*), deciduous or persistent (*Roussea*). Stamens 4–5(6?), antisepalous, free or united with the disk, inserted outside or inside (*Roussea*) the disk; anthers dithecous, introrse, versatile or dorsifixed, deciduous. Disk 4–5(6?)-lobed, with the lobes entire or bearing filaments (staminodes?), fleshy, not nectariferous. Ovary superior, free, sessile, syncarpous, completely or incompletely 4–5(7)-locular, with ∞ or 2 (*Ixerba*), collateral ovules in each loculus; styles as many as the loculi, completely united; stigmas capitate, lobed to punctiform. Fruit loculicidal, capsular (*Ixerba*) or baccate (*Roussea*) or drupaceous (*Brexia*). Seeds exarillate, not winged, with thin layer of endosperm; embryo erect, with cotyledons free, flat.

A family of 3 distantly related genera, *Brexia* Noronha ex Thouars (E. Africa, Madagascar, Mascarene Is.), *Roussea* Smith (Mauritius, 1 species) and *Ixerba* A. Cunningham (New Zealand, 1 species). It is closely related to Escalloniaceae (in which it is often included) and Celastraceae, but is in some ways more primitive than either family (e.g. in the disk in *Brexia*, which appears to be a whorl of sterile stamen fascicles). *Brexia* is near Celastraceae, but differs from it in disk structure and fruit. The other genera are nearer Escalloniaceae.

BREXIA Noronha ex Thouars

Brexia Noronha ex Thouars, Gen. Nov. Madag.: 20 (1806) *nom. conserv.*
  Venana Lam., Tabl. Encycl. Méth. Bot. 2: t. 131 (1792).
  Thomassetia Hemsl. in Hook., Ic. Pl. 28: t. 2736 (1902).

Trees or shrubs, glabrous. Leaves alternate, entire or dentate to spinose-dentate, petiolate; stipules minute, caducous. Flowers in condensed pedunculate cymes or false umbels or fascicles, sometimes cauliflorous. Sepals (4)5, free, imbricate, deciduous. Petals (4)5, free, imbricate, spreading, deciduous. Stamens (4)5, slightly perigynous, inserted between and united with the disk lobes, with filaments dorsally compressed, slightly dilated near the base; anthers oblong-sagittate, versatile. Disk (4)5-lobed, each lobe bearing 3–4(–6) stiff filaments (staminodes?). Ovary narrowly ovoid-pyramidal, 5–10-angled, completely or incompletely 5(6–7)-locular, with ∞ horizontal ovules in 2 rows; style thick, simple; stigma shallowly 5(6–7)-lobed. Fruit drupaceous, woody, 5(6–7)-locular or eventually 1-locular, each loculus ∞-seeded. Seeds obovoid-oblong, angular, with black testa; embryo as long as seed.

A small genus, usually considered to comprise 1 rather variable species; but Perrier de la Bâthie (see below) recognised 6 species of which 5 are said to be endemic to Madagascar and 1 occurs also in the Seychelles, the Comoro Is. and the E. African mainland. Verdcourt (see below) suggests that the size and ribbing of the fruit may provide useful taxonomic characters.

Brexia madagascariensis (Lam.) Ker-Gawl., Bot. Reg. 9: t. 730 (1823).—Harv. in Harv. & Sond., F.C. 2: 598 (1862).—Baker, Fl. Maurit. Seychelles: 97 (1877).— Thonn., Blütenfl. Afr.: t. 61 (1908).—Sim, For. Fl. Port. E.A.: 61 (1909).—Engl., Pflanzenw. Afr. 1, 1: 244, t. 211 (1910).—Thonn., Fl. Pl. Afr.: t. 60 (1915).— Engl., Pflanzenw. Afr. 3, 1: 286 (1915); in Engl. & Prantl, Pflanzenf., ed. 2, 18a: 186, t. 104 fig. a–f (1930).—Perr. in Bull. Soc. Bot. Fr. 80: 202 (1933).—Brenan & Greenway, T.T.C.L. 2: 195 (1949).—Williams, Useful & Ornament. Pl. Zanz. & Pemba: 154 (1949).—Gomes e Sousa, Dendr. Moçamb. Estudo Geral. 1: 211, t. 23

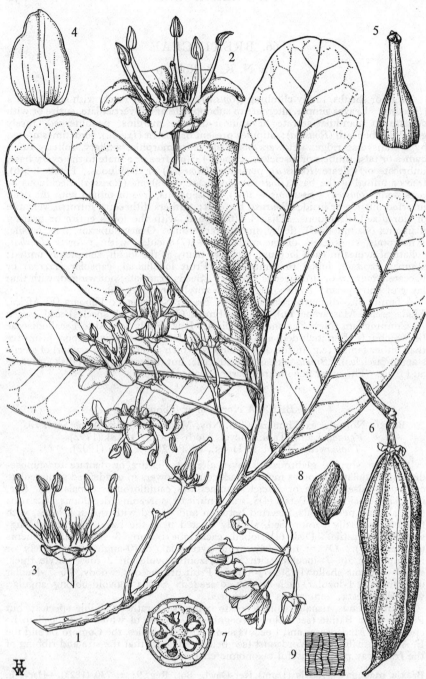

Tab. 17. BREXIA MADAGASCARIENSIS. 1, flowering branch (×⅔); 2, flower (×1); 3, flower with petals removed (×1); 4, petal (×2); 5, gynoecium (×3), 1–5 from *Faulkner* 2669; 6, fruit (×⅔) *Faulkner* 2749; 7, fruit in cross-section (×⅔) *Gomes e Sousa* 4426; 8, seed (×3); 9, surface of seed (×50), both from *Faulkner* 2749. From F.T.E.A.

(1967).—Verdcourt in F.T.E.A., Brexiaceae: 1, t. 1 (1968). TAB. **17.** Type from Madagascar.
Venana madagascariensis Lam., Tabl. Encycl. Méth. Bot. **2:** t. 131 (1792); 99 (1797).—Poir., Encycl. Méth. Bot. **8:** 450 (1808). Type as above.
Brexia spinosa Lindl., Bot. Reg. **11:** t. 872 (1825).—Colla in Mem. Acad. Torino **35:** 151, t. 3 (1831).—Type cultivated ex Madagascar.
Brexia chrysophylla Sweet, Hort. Brit.: 492 (1827) *nom. nud.*
Brexia serrata C. B. Presl in Walp. Repert. **1:** 190 (1834). Type cultivated ex Madagascar.
Brexia heterophylla Bojer, Hort. Maurit.: 52 (1837).—Tul. in Ann. Sci. Nat. Bot., Sér. 4, **8:** 159 (1857) *nom. illegit.*
Brexia acanthifolia Noronha ex Tul., loc. cit. *nom. nud.* in adnot.
Brexia amplifolia Noronha ex Tul., loc. cit. *nom. nud.* in adnot.
Brexia digyna Noronha ex Tul., loc. cit. *nom. nud.* in adnot.
Brexia ovatifolia Noronha ex Tul., loc. cit. *nom. synon.*
? *Brexia microcarpa* Tul., tom. cit.: 160 (1857).—Perrier, tom. cit.: 199 (1933). Type from the Seychelles.
Brexia madagascariensis var. *mossambicensis* Oliv. in F.T.A. **2:** 385 (1871). Type: Mozambique, mouth of Zambesi, *Kirk* (K).
Thomassetia seychellarum Hemsl. in Hook., Ic. Pl. **28:** t. 2736 (1902). Type from the Seychelles.

Shrub or small tree, (2)3–7(10) m. high, much branched; stems smooth, angular when young, later terete. Leaves evergreen; petiole 1–2 cm. long; lamina 3·5–35 × 2–7·6 cm., very variable in shape, those on young shoots narrowly oblong to oblong-linear with margin spinose-dentate, those on mature shoots narrowly to broadly obovate with margin crenate to entire, all rounded to retuse at the apex, cuneate to rounded at the base, coriaceous; stipules subulate. Inflorescence (1)3–12(17)-flowered, pedunculate, in leaf axils; peduncle 1–9 cm. long, usually ± flattened, 1·5–5 mm. wide, sometimes bearing at the apex 1–2 leaf-like bracteoles c. 1 × 1 cm.; bracteoles otherwise small, scale-like, deciduous; pedicels 0·6–1·8 mm. long. Sepals c. 2·5 × 3·5–4 mm., united in the lower ½, with lobes triangular-oblate, apex rounded, margin entire. Petals greenish- or yellowish-white, 1·2–1·7 × 0·9–1·2 cm., broadly oblong-ovoid, obtuse, ± fleshy. Stamens with filaments c. 1·2 mm. thick; anthers c. 5 × 2·5 mm. Disk-lobes with filaments 4–6, linear, unequal. Ovary 5-angled, c. 8–10 mm. long including style. Drupe 4–10 × 1·9–3 cm., ovoid to oblong-fusiform or cylindric, prominently 5-ribbed (in our area), with mesocarp woody at first but said to become eventually pulpy and edible. Seeds 4·5–7·5 × 3–3·5 mm., brown or blackish, irregularly compressed-ellipsoid, carinate, minutely rugulose in ridges.

Mozambique. Z: 32 km. N. of Quelimane, coast, fl. 10.vii.1962, *Wild* 5873 (COI; K; LISC; SRGH). MS: Beira Distr., R. Chiniziua near sawmills, fr. 28.x.1957, *Gomes e Sousa* 4426 (COI; K; LMJ). SS: Gaza, praia da Vila João Belo, fl. 26.xi.1941, *Torre* 3880 (BM; LISC). LM: Inhaca I., west coast, fl. 31.viii.1959, *Watmough* 329 (K; LISC).
Also in Tanzania (Tanganyika and Zanzibar), Comoro Is., Madagascar and the Seychelles. In sclerophyllous coastal scrub, mangrove swamps and fringing forest near the sea, 0–40 m.

The variation in *Brexia madagascariensis* is greatest in Madagascar. No specimens with the long narrow spinous leaves have been seen from Africa or Zanzibar. Plants from these areas have relatively broad obovate entire or subentire leaves which, however, can be matched with Madagascar material. If the lack of spinous leaves proves to be constant on mainland plants, then some infraspecific taxonomic recognition of these populations might be desirable. Likewise, the small-fruited plants from the Seychelles may prove to belong to a distinct species.

67. CRASSULACEAE

See p. 5.

68. DROSERACEAE

By J. R. Laundon

Annual or perennial insectivorous herbs. Leaves in whorls or alternate, frequently in basal rosettes; lamina with glandular excrescences; stipules usually present. Flowers in racemes or cymes or occasionally solitary, actinomorphic, hypogynous, bisexual. Sepals 4–8, basally connate, imbricate. Petals 4–8, free, imbricate. Stamens 5–20 in 1 or more whorls; filaments free or united at the base; anthers 2-locular, dehiscing by longitudinal slits. Ovary superior, syncarpous, 3–5-carpellary, 1-locular; styles 3–5, free or more or less united, simple or branched; ovules numerous, on 3–5 parietal placentas or a free basal placenta. Fruit a loculicidal capsule. Seeds small, with endosperm.

A family of 4 genera, 3 of which are monotypic. Famous for its insectivorous characteristics.

Plant submerged in water, floating, without roots; leaves in whorls; lamina articulate
                                                                    1. Aldrovanda
Plant terrestrial, with pseudo-roots; leaves in a basal rosette or alternate on stems; lamina
simple   -   -   -   -   -   -   -   -   -   -   -   -   **2. Drosera**

1. ALDROVANDA L.

Aldrovanda L., Sp. Pl. **1**: 281 (1753); Gen. Pl. ed. 5: 136 (1754).

Submerged floating aquatic plant, without roots. Stems simple or branched. Leaves in whorls, the lamina articulate, reduced in the flowering whorl. Flowers solitary, axillary, emergent. Sepals and petals 5. Stamens 5. Ovary 5-carpellary; styles 5, free, digitately branched at the apex; ovules on 5 parietal placentas. Capsules with 5 valves; seeds 6–8, ovoid, black.

Monotypic.

Aldrovanda vesiculosa L., Sp. Pl. **1**: 281 (1753).—Diels in Engl., Pflanzenr. **IV**, 112: 59 (1906).—F. W. Andr., Fl. Pl. Anglo-Egypt. Sudan **1**: 82 (1950).—Phillips, Gen. S. Afr. Fl. Pl. ed. 2: 359 (1951).—Oberm. in Fl. Southern Afr. **13**: 189 (1970). TAB. **18**. Type from Italy.

Leaves 6–9 in a whorl; petioles 3–9 mm. long, connate at the base, with c. 4–6 subulate dentate apical segments 0·4–1·2 cm. long; lamina 3–7 × 4–10 mm., reniform, articulate at the midrib, with bristles inside near the midrib which cause the 2 lobes to close when irritated. Inflorescence 1-flowered, axillary, pedicel 5–15 mm. long. Sepals 3–4 × 1·5 mm. Petals 4–5 × 2·5 mm. Stamens 3–4 mm. long. Ovary 2–2·5 mm. in diameter, subglobose; styles c. 2 mm. long. Seeds normally 6–8, 1·5 × 1 mm.

Botswana. N: Linyati R. on N. boundary, c. 910 m., vii.1930, *Stephens* (BM; K). **Zambia.** N: Lake Bangweulu, fl. 29.iv.1958, *Bands* B/SW 75 (SRGH). C: Lake Lusiwasi, E. of Kanona, 1580 m., *Van Zinderen Bakker* 909 (BLFU; K).

In Africa in widely scattered localities in Chad (Lake Chad), Sudan (Sudd Distr.), Rwanda, Burundi, Tanzania; also in central Europe, India, Manchuria, Japan, Timor and Australia. Shallow stagnant water at low altitudes.

This species is also reported from Mozambique, where it was collected by Stephens in 1928, according to the information on Stephens' Botswana specimen in K.; the material has not been traced.

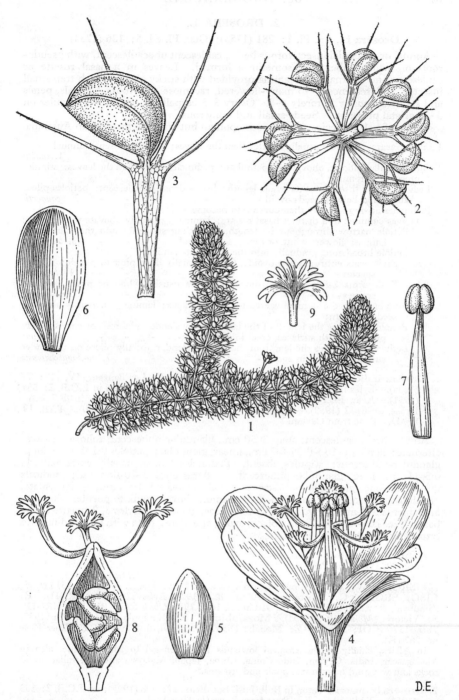

Tab. 18. ALDROVANDA VESICULOSA. 1, plant in flower (×⅔); 2, whorl of leaves (×2); 3, leaf (×6); 4, flower (×8); 5, sepal (×8); 6, petal (×8); 7, stamen (×8); 8, pistil dissected (×8); 9, stigma (×16), all from *Stephens* s.n.

2. DROSERA L.

Drosera L., Sp. Pl. **1**: 281 (1753); Gen. Pl. ed. 5: 136 (1754).

Annual or perennial insectivorous herbs, caulescent or acaulescent, with pseudo-roots on which tubers are sometimes formed. Leaves in a basal rosette or caulescent and alternate, the lamina provided with sticky glands which trap small insects. Inflorescence 1- to many-flowered, racemose or cymose. Sepals, petals and stamens usually 5, rarely 4–8. Ovary 3–5-carpellary; styles 3–5; ovules on 3–5 parietal placentas. Seeds small and numerous.

Genus of c. 100 species, almost cosmopolitan but with the majority in Australia.

Lamina linear; peduncle usually shorter than the leaves; stipules absent; annual
1. *indica*
Lamina circular, elliptic, obovate or spathulate; peduncle longer than the leaves; stipules present; perennial:
Peduncle erect throughout its whole length; leaves densely caulescent; petioles pilose above and below; seeds ovoid - - - - - - - 2. *bequaertii*
Peduncles arising laterally then curving to become erect:
Leaves confined to a distinct basal rosette; lamina ± broadly obovate to orbicular:
Petiole narrow throughout its length, expanding suddenly into the subcircular lamina; flowers white or pink; seeds ovoid - - - 3. *burkeana*
Petiole broadening gradually into the spathulate lamina:
Style arms entire; seeds ovoid; flowers usually deep pink to purple; upland species - - - - - - - - - 4. *dielsiana*
Style arms 2-fid; seeds fusiform; flowers usually white or pink; lowland species - - - - - - - - - 5. *natalensis*
Leaves borne on the stem, clustered towards the apex; lamina ± narrowly obovate; seeds fusiform:
Petiole 2–5 times the length of the lamina, very slender, glabrous or very sparsely pilose on both surfaces, erect in age - - - - 6. *affinis*
Petiole 1–2(3) times the length of the lamina, slender, usually pilose on the under surface, reflexed in age - - - - - - 7. *madagascariensis*

1. **Drosera indica** L., Sp. Pl. **1**: 282 (1753).—Diels in Engl., Pflanzenr. **IV**, 112: 77 (1906).—Eyles in Trans. Roy. Soc. S. Afr. **5**: 357 (1916).—Taton in F.C.B. **2**: 551 (1951).—Keay in F.W.T.A. ed. 2, **1**, 1: 122 (1954).—Laundon in F.T.E.A., Droseraceae: 2 (1959).—Oberm. in Fl. Southern Afr. **13**: 201 (1970). TAB. **19** fig. D. Type from Ceylon.

Annual herb, caulescent, stem 3–50 cm., glandular-pubescent, foliose. Leaves alternate; lamina 1–10 × 0·05–0·3 cm., linear, glandular; petioles 0·1–1·5 cm. long, glandular-pubescent; stipules absent. Peduncles lateral, usually extra-axillary, 0·5–15 cm. long, glandular-pubescent. Inflorescence 3–20-flowered; pedicels 0·3–2 cm. long, glandular-pubescent. Sepals 5, 3–5 × 1–1·5 mm., ± lanceolate, glandular-pubescent. Petals 5, 6–8 × 3–6 mm., obovate, pink to purple. Stamens 5, filaments 3–4 mm. long. Ovary subglobose, glabrous; styles 3, 2-partite to the base. Seeds 0·4–0·5 × 0·3 mm., black, ovoid, apiculate; testa with longitudinal and transverse ridges.

Zambia. N: Mbala (Abercorn), Kambole Escarpment, 1500 m., fl. & fr. 4.vi.1957 *Richards* 9992 (K). S: Pangama, Mapanza, Choma, 1070 m., fl. & fr. 27.iv.1958, *Robinson* 2853 (K; SRGH). **Rhodesia.** N: Shawanoe R., Mrewa Distr., fl. 14.iii.1954, *Whellan* 784 (SRGH). W: Matopos, near World's View, fl. 14.iv.1955, *E. M. & W.* 1513 (BM; LISC; SRGH). C: Salisbury, 1550 m., fl. vi.1920, *Eyles* 2267 (K; SRGH). E: Zimunya's Reserve, Umtali Distr., 910 m., fl. 16.iii.1958, *Chase* 6855 (BM; K; SRGH). S: Victoria, *Monro* 1032 (BM). **Mozambique.** N: Nampula, fl. & fr. 11.iv.1937, *Torre* 1376 (COI; LISC). Z: Mocuba Distr., Namagoa, fl. 18.v.1948, *Faulkner* Kew no. 267 (K).

In Africa, chiefly in the tropical lowlands from Senegal to Mozambique, also in Madagascar, India, Ceylon, Indo-China, China, Japan, Malaysia and Australia. Acid rocks and wet mud in swamps, pools and streams.

2. **Drosera bequaertii** Taton in Bull. Jard. Bot. Brux. **17**: 308 (1945); in F.C.B. **2**: 553 (1951). TAB. **19** fig. E. Type from Zaire (Katanga).
Drosera compacta Exell & Laundon in Bol. Soc. Brot., Sér. 2, **29**: 44, t. 7 (1955). Type from Angola.

Perennial herb, caulescent, stem ± 3–4 cm. long, densely pilose. Leaves alternately erect and dense at the top of the stem, horizontal or deflexed and less closely arranged on the lower part of the stem; lamina 4–8 × 3–4 mm., obovate to spathulate, glandular towards the margin above, pilose below; stipules 2–4 mm. long, membraneous, lacerated at the apex; petioles 0·5–1·2 × 0·05–0·1 cm., pilose above and densely pilose below. Peduncles 1–2, 10–17 cm. long, erect throughout their length, glandular and pilose below, glandular above. Inflorescence 2–8-flowered; pedicels 2–7 mm. long, glandular; bracts 2–3 mm. long, almost linear, pilose. Sepals 5, 3–4·5 × 1·5–2 mm., connate at the base, ovate to elliptic, glandular. Petals 5, 7 × 4 mm., bright pink, obovate. Stamens 5, filaments 4 mm. long. Ovary 2–3 mm. long, subglobose, glabrous; styles 3, 2 mm. long, 2-partite to the base. Seeds numerous, 0·3 × 0·2 mm., black and shining, ovoid, testa smooth.

Zambia. B: 21 km. W. of Balovale pontoon, fl. & fr. 26.v.1960, *Angus* 2274 (BM; K; SRGH). N: Chipili, 1220 m., fl. & fr. 15.vi.1957, *Robinson* 2256 (BM; K; SRGH). C: Walamba, fl. 23.v.1954, *Fanshawe* 1242 (K; SRGH).
Also in Angola and Zaire (Katanga). Bogs.

3. **Drosera burkeana** Planch. in Ann. Sci. Nat. Bot., Sér. 3, **9**: 192 (1848).—Diels in Engl., Pflanzenr. **IV**, 112: 88 (1906).—Eyles in Trans. Roy. Soc. S. Afr. **5**: 357 (1916).—Suesseng. in Proc. Trans. Rhod. Sci. Ass. **43**: 87 (1951).—Taton in F.C.B. **2**: 552 (1951).—Wild, Fl. Vict. Falls: 143 (1952); in S. Rhod. Bot. Dict.: 36, 77 (1953).—Brenan in Mem. N.Y. Bot. Gard. **8**: 436 (1954).—Keay in F.W.T.A. ed. 2, **1**, 1: 121 (1954).—Exell & Laundon in Bol. Soc. Brot., Sér. 2, **30**: 217–219, t. 2 (1956).—Laundon in F.T.E.A., Droseraceae: 2 (1959).—Oberm. in Fl. Southern Afr. **13**: 194 (1970). TAB. **19** fig. C. Type from S. Africa (Transvaal).
Drosera sp.—Eyles in Trans. Roy. Soc. S. Afr. **5**: 357 (1916).

Perennial herb, acaulescent. Leaves in a basal rosette; lamina 2–10 × 2–9 mm., suborbicular, glandular on and around the margins of the upper side, glabrous below; stipules 3 mm. long, connate at the base, apices lacerated; petiole 0·2–2 cm. long, narrow throughout its length and broadening abruptly into the lamina, glabrous or pilose. Peduncles 1–4, 4–30 cm. long, arising laterally from the rosette then curving to become erect, canaliculate, glandular. Inflorescence racemose, often secund, 2–12-flowered; pedicels 0·2–1·2 cm. long, glandular; bracts 1–2 mm. long, narrowly obovate, glandular or glabrous abaxially, glabrous adaxially, caducous. Sepals 5, 4–5 × 2 mm., connate at the base, ± elliptic, acute or obtuse, irregularly serrulate at the apex, glandular. Petals 5, 5–7 × 3–4 mm., white or pink. Stamens 5, filaments 4 mm. long. Ovary subglobose, glabrous; styles 3, 2-partite to the base with spathulate apices. Seeds 0·3–0·4 × 0·15–0·2 mm., black, ovoid, testa smooth.

Zambia. B: 16 km. N. of Senanga, 1040 m., fl. 3.viii.1952, *Codd* 7362 (BM; K; PRE; SRGH). N: Mbala (Abercorn) Distr., Lungwa R., 1220 m., 28.iii.1955, *E. M. & W.* 1250 (BM; SRGH). W: Mufulira, 1222 m., fl. 14.vi.1934, *Eyles* 8168 (BM; K; SRGH). C: Rufunsa-Lusaka, fl. & fr. 7.ix.1947, *Greenway & Brenan* 8181 (EA). E: 48 km. W. of Fort Jameson by Mpangwe Hills, 910 m., fr. 18.vi.1961, *Wright* 304 (K). S: 34 km. N. of Choma, 1190 m., fl. 17.v.1954, *Robinson* 763 (K; SRGH). **Rhodesia.** N: Umvukwes, fl. ix.1945, *Martineau* 759 (SRGH). W: Matopos, 1370 m., fl. 10.xi.1902, *Eyles* 1099 (K; SRGH). C: Makoni, 1450 m., fl. & fr. 26.x.1930, *F. N. & W.* 2277 (BM; SRGH). E: Inyanga road, fl. & fr. x.1934, *Gilliland* 1021 (BM; K; SRGH). S: Bikita, Ruwara, 3 km. below Cherene School on W. bank of Turgwe R., 940 m., fl. 5.v.1969, *Pope* 92 (BM; K; SRGH). **Malawi.** N: Mzimba, stream on Vipya Mt., 12.vi.1947, *Benson* 1290a pro parte (BM). C: Nchisi Mt., Kota-Kota Distr., 1400 m., fl. & fr. 29.vii.1946, *Shortridge* 17013 (K; NY; SRGH). **Mozambique.** N: Miandica-Maniamba, fl. 16.ix.1934, *Torre* 223 (COI; LISC). MS: Mossurize, near Catholic Mission, fl. 14.vi.1942, *Torre* 4332 (LISC).
Uganda, Tanzania, Zaire (Katanga), Angola, Swaziland, S. Africa (Transvaal) and Madagascar. In marshes, bogs and swamps.

4. **Drosera dielsiana** Exell & Laundon in Bol. Soc. Brot., Sér. 2, **30**: 214 (1956).—Oberm. in Fl. Southern Afr. **13**: 193 (1970). TAB. **19** fig. F. Type from S. Africa (Transvaal).
Drosera sp. Brenan in Mem. N.Y. Bot. Gard. **8**: 436 (1954).

Perennial herb, acaulescent. Leaves in basal rosettes; lamina 0·3–1·5 × 0·2–0·7 cm., obovate-spathulate, glandular; stipules 2–3 mm. long, connate at the base, apex lacerated; petioles 0·3–3·5 cm. long, broadening into the lamina, glabrous or sparsely pilose. Peduncles 1–2(6), 2–17 cm. long, arising laterally from the rosette

c

then curving to become erect, canaliculate, glabrous or sparsely pilose in the lower part, glandular in the upper part. Inflorescence racemose, 3–12-flowered; pedicels 1–5 mm. long, glandular; bracts 1–3 mm. long, linear-elliptic, caducous. Sepals 5, 3–6 × 2–2·5 mm., connate at the base, ± elliptic, obtuse, slightly irregularly serrulate at the apex, glandular. Petals 5, 7–8 mm. long, pink or crimson to reddish-purple. Stamens 5, filaments 5 mm. long. Ovary subglobose, glabrous; styles 3, 2 mm. long, 2-partite to the base with spathulate apices. Seeds 0·3–0·6 × 0·2 mm., black, ovoid; testa usually foveolate.

Rhodesia. E: Nuza, 1980 m., fl. i.1935, *Gilliland* 1399 (BM; K). **Malawi.** S: Mt. Mlanje, Tuchila Plateau, 1950 m., 21.vii.1956, *Newman & Whitmore* 130 (BM; SRGH). **Mozambique.** MS: Musapa Gap, Chimanimani Mts., 910 m., fl. & fr. 6.x.1950, *Munch* 343 (K; SRGH).

Lesotho, S. Africa (Natal, Orange Free State and the Transvaal) and Swaziland. In shallow sandy or peaty soils over rocks on mountain plateaux.

5. **Drosera natalensis** Diels in Engl., Pflanzenr. **IV**, 112: 93, fig. 31 G–J (1906).—Exell & Laundon in Bol. Soc. Brot., Sér. 2, **30**: 216–217, t. 2 (1956).—Oberm. in Fl. Southern Afr. **13**: 193 (1970). Type from S. Africa (Natal).

Perennial herb, acaulescent. Leaves in basal rosettes; lamina 0·4–2·0 × 0·2–1·1 cm., obovate-spathulate, glandular; stipules 0·2–0·5 cm. long, connate at the base, apex lacerated; petioles 0·4–1·5 cm. long, broadening into the lamina, sparsely to densely pilose. Peduncles 1–2, 10–45 cm. long, arising laterally from the rosette then curving to become erect, canaliculate, glandular. Inflorescence racemose, 3–13-flowered; pedicels 1–9 mm. long, glandular; bracts 1–2 mm. long, linear-elliptic, caducous. Sepals 5, 3–6 × 2–2·5 mm., connate at the base, ± elliptic, irregularly serrulate at the apex, glandular. Petals 5, 7–8 mm. long, white to pink. Stamens 5, filaments 3 mm. long. Ovary subglobose, glabrous; styles 3, 1 mm. long, 2-partite to the base with 2-fid apices. Seeds 0·4–0·6 × 0·1–0·2 mm., black, ellipsoid-fusiform, testa more or less foveolate.

Mozambique. Z: Pebane, fl. & fr. 24.x.1942, *Torre* 4653 (LISC). SS: Lagoa Pate, between Manhiça and Bilene, fl. & fr. 25.iii.1954, *Barbosa & Balsinhas* 5458 (LM). LM: Maputo, 15 m., fl. & fr. 22.x.1960, *Myre & Macêdo* 4063 (LM).

In SE. Africa from the eastern Cape, Natal and Transvaal to Mozambique, and in Madagascar. Lake shores.

6. **Drosera affinis** Welw. ex Oliv. in F.T.A. **2**: 402 (1871).—Diels in Engl., Pflanzenr. **IV**, 112: 88 (1906).—Laundon in F.T.E.A., Droseraceae: 4 (1959). TAB. **19** fig. A. Type from Angola.

  Drosera flexicaulis Welw. ex Oliv., tom. cit.: 403 (1871).—Diels, tom. cit.: 98 (1906).—Taton in F.C.B. **2**: 554 (1951). Type from Angola.

Perennial herb, caulescent. Stem 1–23 cm. long, bearing alternate leaves which are clustered towards the apex. Leaves mostly erect, but occasionally on the lower parts of the stem descending; lamina 0·3–3 × 0·2–0·5 cm., narrowly oblanceolate or narrowly oblanceolate-oblong, glandular around the margin of the upper surface, glabrous or sparsely pilose below; stipules 0·3–1·3 cm. long, apex lacerated; petioles 0·5–7 cm. long, very slender, glabrous or very sparsely pilose. Peduncles 1–4, 8–30 cm. long, arising laterally from the stem, then curving to become erect, canaliculate, glabrous. Inflorescence racemose, 3–13-flowered; pedicels 2–10 mm. long, glabrous or sparsely pilose; bracts 3–5 mm. long, linear to elliptic, caducous. Sepals 5, 5–7 × 1·5–2 mm., connate at the base, oblong-lanceolate, pubescent or pilose. Petals 5, 5–8 mm. long, white or purple. Stamens 5, filaments 5 mm. long. Ovary subglobose, glabrous; styles 3, 2·5 mm. long, 2-partite to the base. Seeds 0·7–0·9 × 0·2 mm., brownish-black, fusiform; testa reticulate with longitudinal and transverse ridges.

Zambia. B: 16 km. N. of Senanga, 1040 m., fl. 31.vii.1952, *Codd* 7301 (BM; K; PRE; SRGH). N: Mbala (Abercorn), Sunzu Hill, 1830 m., ix.1933, *Gamwell* 189 (BM). W: Mwinilunga Distr., Sinkabolo Swamp, 1200 m., fl. & fr. 20.xi.1962, *Richards* 17422 (K; SRGH). C: Mkushi R., 1680 m., 8.viii.1947, *Greenway* 7946 (EA; K). **Rhodesia.** C: Beatrice, fl. 15.xii.1922, *Eyles* 3811 (SRGH). E: Inyanga, Mare R., 1830 m., 27.x.1946, *Wild* 1532a (K; SRGH). **Malawi.** N: Mzimba Distr., Vipya Mts., 12.vi.1947, *Benson* 1290a pro parte (BM). **Mozambique.** N: Niandica-Maniamba, fl. & fr. 16.ix.1934, *Torre* 224 (COI).

Tanzania, Zaire (Katanga) and Angola. In marshes and swamps.

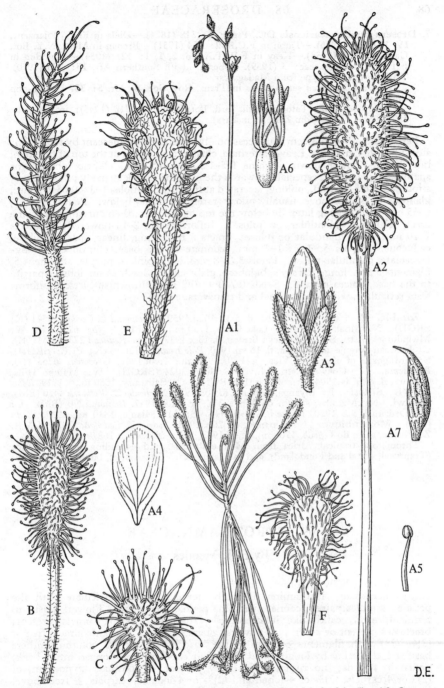

D.E.

Tab. 19. A.—DROSERA AFFINIS. A1, plant in flower (×⅔); A2, leaf (×5); A3, flower-
bud (×6); A4, petal (×6); A5, stamen (×6); A6, pistil (×6); A7, seed (×40), all
from *Milne-Redhead & Taylor* 10485. B.—DROSERA MADAGASCARIENSIS. Upper part
of leaf (×4) *Milne-Redhead & Taylor* 8490. C.—DROSERA BURKEANA. Upper part of
leaf (×4) *Milne-Redhead & Taylor* 10829. D.—DROSERA INDICA. Upper part of leaf
(×4) *Milne-Redhead & Taylor* 9807B. E.—DROSERA BEQUAERTII. Upper part of
leaf (×4) *Robinson* 2256. F.—DROSERA DIELSIANA. Upper part of leaf (×6) *Newman
& Whitmore* 38.

7. **Drosera madagascariensis** DC., Prodr. **1**: 318 (1824).—Diels in Engl., Pflanzenr. **IV**, 112: 98 (1906).—Taton in F.C.B. **2**: 554 (1951).—Brenan in Mem. N.Y. Bot. Gard. **8**: 436 (1954).—Keay in F.W.T.A. ed. 2, **1**, 1: 121 (1954).—Laundon in F.T.E.A., Droseraceae: 5 (1959).—Oberm. in Fl. Southern Afr. **13**: 200 (1970). TAB. **19** fig. B. Type from Madagascar.

Drosera ramentacea sensu Eyles in Trans. Roy. Soc. S. Afr. **5**: 357 (1916), non Burch. ex DC.

Drosera congolana Taton in Bull. Jard. Bot. Brux **17**: 310 (1945); in tom. cit.: 552 (1951). Type from Zaire (Kinshasa).

Perennial herb, caulescent, but occasionally apparently acaulescent by reduction; stem up to 25 cm. long. Leaves alternate, clustered and erect at the top of the stem, but deflexed lower down; lamina 0·5–1·5 × 0·2–0·5 cm., elliptic, obovate or spathulate, glandular, especially towards the margins above, sparsely pilose below; stipules up to 5 × 1 mm., oblong, lacerated at the apex; petioles 1–3 × 0·03–0·1 cm., glabrous or pilose above, usually pilose (rarely glabrous) below. Peduncles 1–2, 9–55 cm. long, arising laterally below the top of the stem, then curving to become erect, glabrous, glandular or pilose. Inflorescence 2–14-flowered; pedicels 1–10 mm. long, glandular or pilose; bracts 3 mm. long, linear, sparsely pilose, caducous. Sepals 5, 4–7 × 1–2 mm., connate at the base, elliptic or oblong-lanceolate, glandular-pilose. Petals 5, 5–8 mm. long, pink or purple. Stamens 5, filaments 5 mm. long. Ovary subglobose, glabrous; styles 3, 2 mm. long, 2-partite to the base, apices clavate. Seeds 0·7–0·9 × 0·2 mm., brownish-black, fusiform; testa reticulate, with longitudinal and transverse ridges.

Zambia. B: 5 km. S. of Kalabo, fl. & fr. 16.xi.1959, *Drummond & Cookson* 6541 (K; SRGH). **N**: Mbala (Abercorn), Lake Chila, 1760 m., fr. 9.i.1952, *Nash* 62 (BM). **W**: Mwinilunga Distr., SW. of Dobeka Bridge, fl. 13.x.1937, *Milne-Redhead* 2737 (BM; K). **C**: between Serenje and Mpika, fl. 16.vii.1930, *Pole Evans* 2912(19) (K; PRE; SRGH). **S**: 3rd gorge of Zambezi, Victoria Falls, 910 m., 1.viii.1941, *Greenway* 6263 (K). **Rhodesia. N**: Gokwe Distr., 14.ix.1949, *West* 3024 (SRGH). **W**: Matopo Hills, 1520 m., fl. & fr. iv.1904, *Eyles* 53 (BM; SRGH). **C**: Salisbury, 1580 m., x.1919, *Eyles* 1854 (K; SRGH). **E**: Inyanga, 1700 m., fl. 20.i.1931, *Norlindh & Weimark* 4450 (BM). **Malawi. N**: Mzimba, stream on Vipya Mt., fr. 12.vi.1947, *Benson* 1290 (BM). **C**: Mt. Dedza, fl. 5.ix.1950, *Wiehe* 633 (SRGH). **S**: Mt. Mlanje, 1830 m., 1919, *Shinn* (BM). **Mozambique. N**: Unango, fl. 26.v.1948, *Pedro & Pedrógão* 3936 (LMJ). **Z**: Gúruè, serra do Gúruè, 1700 m., fl. 4.i.1968, *Torre & Correia* 16879 (LISC).

Throughout tropical Africa from Mali eastwards, and extending into S. Africa (Transvaal, Natal and Pondoland), and in Madagascar. In marshes, bogs and swamps.

69. MYROTHAMNACEAE

By E. J. Mendes

Dioecious, wholly glabrous, balsamic-resinous shrubs; branches opposite. Leaves decussate; leaf-lamina digitately nerved, plicate, articulate with the petiole; stipules paired, connate with the petiole, persistent. Flowers sessile in terminal, erect, catkin-like inflorescences; perianth absent; bracts present; bracteoles present or absent. ♂ flowers: stamens 4 with filaments almost free or (3)4–5(?8) with filaments connate at the base into a central column, anthers basifixed, dehiscing by longitudinal slits, their connective prolonged into a beak; pollen in tetrads; no rudimentary ovary present. ♀ flowers: gynoecium of (2)3 or 4 carpels, (2)3- or 4-lobed and (2)3- or 4-locular; carpels ± free above, tapering into ± recurved styles; stigmas ± spathulate, oblong, papillose, lateral, facing inwards; ovules numerous, anatropous, axillary, 2-seriate; no staminodes present. Fruit a capsule, dehiscing by the adaxial sutures of the free apices of the carpels. Seeds numerous, small, pendulous; endosperm fleshy, copious; embryo minute.

A single genus, *Myrothamnus*, with 2 species, one only known from Africa south of the Sahara, the other (*M. moschatus* (Baill.) Baill.) from Madagascar.

MYROTHAMNUS Welw.

Myrothamnus Welw. in Journ. Linn. Soc., Bot. **3**: 155 (Feb. 1859); in Ann. Cons. Ultramar., Parte Não Off., Sér. 1, **1858**: 547 & 578, Nota 8 (Nov. vel Dec. 1859); in Trans. Linn. Soc., Bot. **27**: 22, t. 8 (1869).

Characters as for the family.

Myrothamnus flabellifolius Welw. in Journ. Linn. Soc., Bot. **3**: 155 (Feb. 1859); in Ann. Cons. Ultramar., Parte Não Off., Sér. 1, **1858**: 547 & 578, Nota 8 (Nov. vel Dec. 1859); in Trans. Linn. Soc., Bot. **27**: 23, t. 8 (1869) (" flabellifolia ").—Oliv. in F.T.A. **2**: 404 (1871) (" flabellifolia ").—Bak. in Trans. Linn. Soc., Bot. Sér. 2, **4**: 13 (1894) (" flabellifolia ").—Engl., Pflanzenw. Afr. **1**: 407, fig. 348* (1910); op. cit. **3**, 1: 289, fig. 187* (1915) (" (Sond.) Welw.").—Marloth, Fl. S. Afr. **2**: 32, fig. 18 & 19, t. 11 A (1925).—Niedenzu & Engl. in Engl. & Prantl, Pflanzenfam. ed. 2, **18a**: 264, fig. 152 C–D & 153* (1930).—Burtt Davy, F.P.F.T.: 430 (1932) (" (Sond.) Welw."). —Brenan, T.T.C.L.: 368 (1949); in Mem. N.Y. Bot. Gard. **8**: 436 (1954).— Dale & Greenway, Kenya Trees & Shrubs: 329 (1961).—F. White, F.F.N.R.: 67 (1962) (" (Sond.) Welw.").—Friedrich-Holzhammer in Prodr. Fl. SW. Afr. **51**: 1 (1966).—Mendes in C.F.A. **4**: 30 (1970).—Lisowski, Malaisse & Symoens in Bull. Jard. Bot. Nat. Belg. **40**: 255, fig. 1–2 (1970); in F.**C**.B., Myrothamnaceae: 2, t. 1 (1970). TAB. **20**. Type from Angola.

?*Cliffortia flabellifolia* Sond. in Harv. & Sond., F.C. **2**: 597 (1862). Type as above?

Myrothamnus flabellifolius subsp. *elongatus* H. Weim. in Bot. Not. **1936** (& in Medd. Lunds Bot. Mus. **26**): 454, fig. 1 sinister, 2 & 4 (distribution map) (1936) (" elongata "). Type: Rhodesia, Belingwe Distr., near Mnene, ♀ fl. ii.1931, *Norlindh & Weimarck* 5154 (LD).

Myrothamnus flabellifolius subsp. *robustus* H. Weim., tom. cit.: 459, fig. 3 & 4 (distribution map) (1936) (" robusta "). Type: Zambia, near Serenje, ♀ fl. vii.1930, *Pole Evans* 2880 (K).

A prostrate, ascending or erect shrub, much branched, usually 30–90 cm. tall (stated to grow up to 2m.); young branches tetragonous, the angles narrowly winged; branches soon becoming woody and subspinulose with the persistent stipules and petioles; older branches with bark fissuring longitudinally. Leaves decussately disposed along fast-growing long branches or ± congested on lateral short branches. Leaf-laminae (8)10–14(20) × (3)6–8(15) mm., ± rhombic, the apex (3)5–7(11)-crenate-dentate, the base entire and cuneate, digitately nerved (nerves sometimes 2-furcate near the apex), closely longitudinally pleated when dry, articulated with the petiole; petiole (1)2–3(5) mm. long, subvaginate; stipules subulate, connate with the petiole at the base and exceeding it by 1–3 mm. Inflorescences usually 2–3 cm. long, rarely up to 5 cm., terminal on the lateral short branches; each flower sessile in the axil of a single, elliptic to ovate, usually obtuse bract 1·2–1·8 × c. 1·0 mm.; bracts persistent after flower- and fruit-fall; ♀ inflorescences stouter than the ♂. ♂ flowers reduced to (3)5–6(?8) stamens; filaments filiform and connate at the base into a central column; anthers oblong, reddish, after dehiscence presenting an asymmetric X-shaped cross-section; connective prolonged into a short, ± curved beak. ♀ flowers zygomorphic, reduced to (2)3 green carpels connate only at the base and tapering into ± short, outwardly curved styles presenting purple, oblong, inwardly facing stigmas. Capsule coriaceous with carpels somewhat enlarged, deeply (2-)3-lobed, crowned by the persistent styles and stigmas. Seeds c. 0·5 mm. long., ovoid to tetragonal.

Botswana. SE: Lobatsi, fl. 29.x.1955, *McConnell* in SRGH 68917 (K; SRGH).
Zambia. N: c. 48 km. S. of Mpika, Nachkutu Cave, fl. 26.x.1967, *Simon & Williamson* 1219 (K; LISC; SRGH). C: near Serenje, fr. 15.vii.1930, *Pole Evans* 2880 (K; PRE; SRGH). E: Chipiri Hills near Chadiza, fl. 29.xi.1958, *Robson* 785 (BM; K; LISC; PRE; SRGH). **Rhodesia.** N: Mrewa, st. 28.i.1941, *Hopkins* in GHS 7897 (SRGH). W: Victoria Falls, fl. xii.1905, *Allen* 92 (K; SRGH). C: Gwelo, fl. 5.xi.1967, *Biegel* 1891 (K; LISC; PRE; SRGH). E: Inyanga Distr., N. of Mt. Dombo, fl. 25.xi.1965, *Simon* 516 (K; LISC; PRE; SRGH). S: Gwanda, near tributary of Mtshibizini R., 35 km. from Koodoovale Motel to Tuli Breeding Station, fr. 19.iii.1959, *Drummond* 5860 (K; LISC; PRE; SRGH). **Malawi.** N: Mzimba Distr., N. end of Vipya Plateau, fr. 1.v.1966, *Whellan* 2260 (K; SRGH). C: Dedza Distr., Chiwao Hill near Chongoni Forest School, fl. 14.iii.1963, *Salubeni* 6 (K; SRGH). S: Zomba Plateau, fr. 31.v.1946, *Brass* 16132 (K; PRE; SRGH). **Mozambique.** N: Amaramba, Mitucué Mts., 20 km.

* ♀ flower erroneously pictured with 4 carpels (figs. F & H).

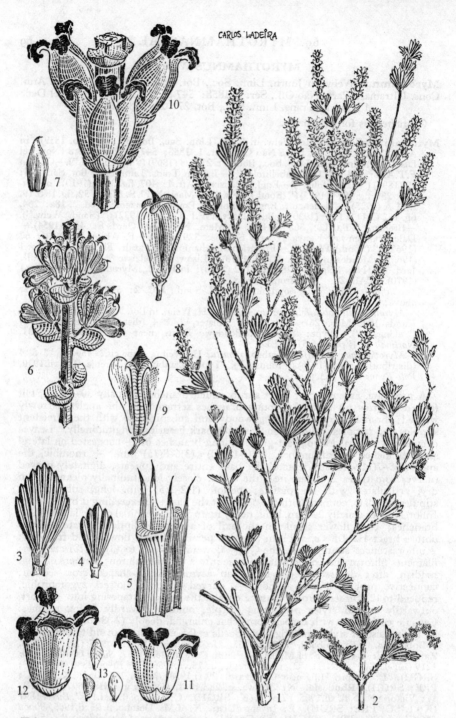

CARLOS LADEIRA

Tab. 20. MYROTHAMNUS FLABELLIFOLIUS. 1, branchlets with ♀ inflorescences (× ⅔);
2, branchlets with ♂ inflorescences (× 1); 3, leaf-lamina with its petiole and stipules,
showing bifurcate nerves (× 1); 4, a 5 crenate-dentate leaf (× 1); 5, a node with a
pair of persistent petioles and its stipules (× 3); 6, part of ♂ inflorescence, with
dehiscent stamens (× 8); 7, mature, unopened anther (× 8); 8, anther, after
dehiscence, lateral view showing filament-column (× 16); 9, anther, after dehiscence,
abaxial view showing filament-column (× 16); 10, part of ♀ fruiting inflorescence
(× 8); 11, ♀ flower, with bract, abaxial view (× 8); 12, dehiscing fruit, with bract,
lateral view (× 8); 13, seeds (× 16), 1, 4, 10, 12–13 from *Pereira, Sarmento & Marques*
1744, 2 from *Garcia* 454, 3, 5, 7 from *Angus* 814, 6, 8–9 from *Mendonça* 322, 11 from
Rushworth 337.

from Cuamba, fr. 15.ii.1964, *Torre & Paiva* 10602 (COI; EA; K; LISC; LMU; MO; SRGH). Z: Ile Mts., fl. 22.vi.1943, *Torre* 5563 (COI; LISC; LMU). T: Cabora-Bassa, Songo Plateau, fr. 21.i.1973, *Torre, Carvalho & Ladeira* 18818 (COI; FHO; LISC; LMU; MO). MS: Manica Distr., Báruè, between Vila Gouveia and Mungari, fl. 18.ix.1942, *Mendonça* 322 (COI; LISC; M). LM: Namaacha, fr. 9.i.1947, *Barbosa* 95A (LMA; SRGH).

Also known from Kenya, Tanzania, Zaire (Haut-Katanga), Angola, SW. Africa, Lesotho and S. Africa (Transvaal and Cape Prov.). Usually forming large stands on sunny rocky hills or along cracks and crevices in rocks; 500–1900 m. " Resurrection Plant."

Note: M. *flabellifolius* is phenotypically a rather variable species, reflecting in its habit and in the average size of leaves, internodes and inflorescences the local edaphic and climatic conditions of the year during which the gathering was obtained.

70. HAMAMELIDACEAE

By E. J. Mendes and M. P. Vidigal

Trees and shrubs, often with stellate indumentum. Leaves alternate or less often opposite, deciduous or persistent, simple, entire or not, pinnately or palmately nerved; stipules mostly paired, often deciduous. Inflorescences racemose, often spicate or capitate, sometimes very dense, terminal or axillary; bracts and bracteoles often present. Flowers bisexual or unisexual, actinomorphic or rarely zygomorphic, rarely without a perianth. Calyx-tube ± adnate to the base of the gynoecium; calyx-lobes 4–5 or more, imbricate or valvate. Petals 4–5 or more, sometimes absent in ♀ flowers, free, imbricate or valvate, rarely circinnate. Stamens 1-seriate, usually as many as and alternating with the petals, rarely fewer or indefinite, perigynous; filaments free; anthers 2-locular, opening lengthwise by slits or by valves; the connective often produced; staminodes sometimes present, alternating with the stamens. Disk absent or annular or of separate glands. Ovary inferior or nearly so, rarely sub-superior, (1)2(3)-carpellary and -locular; carpels often free at the apex; styles subulate, free, often recurved and persistent; stigmas terminal or lateral; ovules 1–many in each loculus, axillary, pendulous. Fruit a woody capsule, loculicidal or septicidal. Seeds mostly 1 per loculus, sometimes winged; endosperm thin; embryo straight.

A comparatively small family, mainly Asiatic, with c. 100 species corresponding to some 25 genera. 1 genus only in continental Africa and another in Madagascar, both endemic. Absent from S. America and the W. Indies.

TRICHOCLADUS Pers.

Trichocladus Pers., Syn. Pl. **2**: 597 (1807).

Evergreen trees or shrubs, mostly with conspicuous stellate indumentum. Leaves opposite or alternate, ovate-cordate to oblong, usually entire, sometimes peltate at the base, petiolate; stipules linear or subulate, inconspicuous. Flowers in small dense terminal or axillary capitula or spikes, bisexual or sometimes ♀ only. Calyx-tube adnate to the base of the gynoecium; calyx-lobes 4–5, valvate. Petals 4–5, linear-spathulate, greatly exceeding the calyx, with margins revolute, usually valvate, stated to be absent from ♀ flowers. Stamens as many as and alternating with the petals; filaments short and thick; anthers basifixed, usually mucronate or beaked, opening by lateral valves. Ovary inferior to almost superior; styles 2, free, often recurved, persistent; ovules 1 in each loculus, apical, pendulous. Fruit a woody loculicidally 2-valved capsule, with bony and elastic endocarp which in dehiscence separates from the epicarp. Seeds ellipsoid to oblong-ovoid, with a large apical hilar area; cotyledons large; endosperm scanty.

A small endemic African genus occurring from S. Africa to Ethiopia, probably with 5 species only.

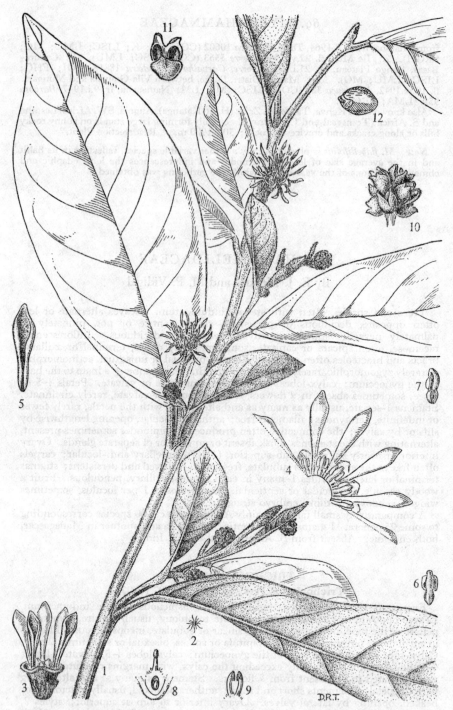

Tab. 21. TRICHOCLADUS ELLIPTICUS subsp. MALOSANUS. 1, fertile branch (×1); 2, stellate hair (×12); 3, flower (×4); 4, calyx and corolla opened out to show stamens (×4); 5, petal (×4); 6, stamen, front view (×6); 7, same, side view (×6); 8, longitudinal section of gynoecium (×9); 9, styles (×6); 10, infructescence (×1); 11, dehisced capsule (×2); 12, seed (×2), 1, 3–9 from *Rogers* 5, 2, 10–12 from *Thomas* 3617. From F.T.E.A.

Trichocladus ellipticus Eckl. & Zeyh., Enum. Pl. Afr. Austr. Extratrop.: 356 (1837).—
Sond. in Harv. & Sond., F.C. 2: 325 (1862).—Engl., Pflanzenw. Afr. 3, 1: 290 (1915)
pro parte.—Marloth, Fl. S. Afr. 2, 1: 25, t. 9 fig. B (1925).—Harms in Engl. & Prantl,
Pflanzenfam. ed. 2, 18a: 323 (1930) pro parte.—Hutch. in Kew Bull. 1933: 430
(1933).—Palmer & Pitman, Trees of S. Afr. 1: 671, top fig. on p. 670 (1972).
Type from S. Africa (Cape Prov.).

Subsp. **malosanus** (Bak.) Verdc. in Kew Bull. 24: 345 (1970); in F.T.E.A.,
Hamamelidaceae: 2, fig. 1 (1971). TAB. 21. Type: Malawi, Mt. Malosa, *Whyte*
(K, holotype).
    Trichocladus malosanus Bak. in Kew Bull. 1897: 266 (1897).—Hutch., loc. cit.—
Brenan, T.T.C.L.: 244 (1949).—Eggeling & Dale, Indig. Trees Uganda, ed. 2: 155,
t. 6 (1952).—F. White, F.F.N.R.: 67 (1962). Type as above.
    Trichocladus ellipticus var. *latifolius* Schweinf. ex Engl., Bot. Jahrb. 49: 456,
fig. 1/H-N* (1913) *nom. non rite publ.*
    Trichocladus ellipticus sensu Engl., loc. cit. pro parte et fig. 188/H-N.—Harms in
Engl. & Prantl, loc. cit. pro parte et fig. 169/H-N.—Germain in F.C.B. 2: 582
(1951).—Dale & Greenway, Kenya Trees & Shrubs: 233, t. 13 (1961).—Teixeira in
Bol. Soc. Brot., Sér. 2, 43: 162, t. 1-4 (1969); in C.F.A. 4: 29 (1970) non Eckl. &
Zeyh. sensu stricto.

A tree 6–10(14) m. high or a shrub 3–5 m. tall, sometimes with straggling
branches; young branchlets densely stellate-tomentose with greyish to coppery
hairs, older ones glabrescent. Leaves alternate; leaf-lamina 2·5–16(?28) × 1·5–
6(?12) cm., lanceolate or elliptic to oblanceolate, acute to acuminate at the apex,
cuneate to rounded or rarely subauriculate at the base, with entire margins, rather
coriaceous, glabrous above, densely stellate-tomentose beneath with greyish,
ochraceous or coppery hairs; petiole 0·4–1·8 cm. long, channelled above; stipules
c. 2 mm. long, subulate, soon caducous. Inflorescences up to 3·5 cm. in diameter,
capitate, axillary or terminal, on peduncles up to 1·2 cm. long. Bisexual flowers
(the only kind seen in F.Z. area): sessile and with 2 filiform c. 1·5 mm. long
bracteoles; calyx stellate-tomentose outside; calyx-tube c. 1 mm. deep and 2 mm.
in diameter at the mouth, campanulate, adnate to the base of the gynoecium;
calyx-lobes 5, c. 0·5 mm. long, triangular, apex obtuse, ± reflexed; petals 5,
0·8–1·8(2·2) × c. 0·15 cm., spathulate, the margins strongly revolute and thus
appearing linear, ± densely stellate-tomentose outside, fast growing and usually
soon caducous; stamens 5, the filaments up to 0·8 mm. long, obpyramidal, stout,
the anthers 1·0–1·5 mm. long and opening by 2 lateral valves, their connective
usually prolonged into a ± erect beak up to 0·5 mm. long or more, rarely shortly
apiculate; gynoecium stellate-tomentose, the styles up to 1·3 mm. long and
incurved, the stigmas lateral and facing each other. Capsule c. 7 × 5 mm. (slightly
compressed), woody, ± ellipsoid, sheathed by the persistent calyx-tube, loculicidally
2-valved, each valve bifurcate at the apex, usually 2-seeded. Seeds c. 5 × 3 mm.,
ivory-coloured, smooth, shining, ovoid to ellipsoid, showing apically 2 confluent
areas which are ± ovate and reddish-brown, each c. 2·5 × 1·6 mm., the adaxial one
with an elliptic hilum c. 1 mm. long near its apex.

Zambia. W: Ndola Distr., Kondola, fl. 14.vii.1952, *Holmes* H764 (FHO; K;
SRGH). C: Serenje Distr., Kundalila Falls, fl. 13.x.1963, *Robinson* 5698 (EA; K;
SRGH). **Rhodesia.** E: Melsetter Distr., Cashel, fl. 7.xi.1952, *Wild* 3888 (K; LISC;
SRGH). **Malawi.** N: Mzimba Distr., Mazamba, Luchilemu R., fl. 9.vii.1964, *Chapman*
2252 (FHO). C: Lilongwe Distr., Kasitu, Dzolanyama, fl. 4.xi.1962, *Chapman* 1724
(SRGH). S: Mt. Mlanje, Likulezi Valley, fl. 10.x.1957, *Chapman* 461 (BM; EA; FHO;
K; PRE). **Mozambique.** Z: Gúruè, near Namuli Peak, Malema R., fl. 8.xi.1967,
Torre & Correia 16013 (COI; LISC; LMU). MS: Beira Distr., Gorongosa Mts.,
fl. 18.x.1965, *Torre & Pereira* 12399 (COI; EA; K; LISC; LMU; MO; PRE).
Also known from Ethiopia, Uganda, Kenya, Tanzania, Zaire (Kivu and Haut-Katanga)
and Angola (N. Cuanza). Riverine forest or evergreen forest near waterfalls, 900–1800 m.

Note. Typical *T. ellipticus* is known to occur from the Cape to Natal; on average it has
smaller, more lanceolate and often more acuminate leaves.

* The figure given by Engler as representing this var. *latifolius* was afterwards
reproduced in Engl., Pflanzenw. Afr. 3, 1: fig. 188/H-N (1915) and in Engl. & Prantl,
Pflanzenfam. ed. 2, 18a: fig. 169/H-N (1930) without any varietal designation.

71. HALORAGACEAE

By E. J. Mendes

Herbs or undershrubs, often aquatic, monoecious, polygamous or dioecious. Leaves alternate, opposite or verticillate, pinnately or palmately nerved, sometimes very large, the submerged ones often much divided; stipules absent. Inflorescences axillary or terminal, the flowers usually small, arranged in cymes, fascicles, racemes, panicles or spikes, or solitary; bracts often present. Calyx-tube short, adnate to the ovary; calyx-lobes 2–4, mostly valvate and persistent, or absent. Petals 2–4, valvate or ± imbricate, or absent. Stamens 2, 3, 4 or 8, rarely 1 or absent; anthers basifixed, 2-celled, dehiscing lengthwise; filaments mostly filiform. Gynoecium 1–4-locular; ovules 1 in each loculus, pendulous from the apex; styles 1–4 or absent; stigmas papillose or plumose. Fruit a nutlet, indehiscent or breaking up into 2 or 4 single-seeded mericarps, or a drupe. Seeds with abundant fleshy or oily endosperm.

A small cosmopolitan family of 7 genera and some 120–150 species, rather rare in the tropics; usually found in damp or boggy places.

Leaves not radical, small, sessile or shortly petiolate, not palmately nerved:
    Leaves sessile or shortly petiolate, alternate or opposite, entire or shallowly dentate; flowers in axillary fascicles, the central one ☿, the others ♀; fruit an indehiscent nutlet - - - - - - - - - - - **1. Laurembergia**
    Leaves sessile, verticillate, pinnately nerved, the submerged ones pinnately divided into filiform segments; flowers arranged in terminal spikes or solitary and axillary; fruit separating into 4 (or fewer) 1-seeded nutlets - - **2. Myriophyllum**
Leaves borne on the rhizome, large, long-petioled, palmately nerved; flowers arranged in spikes disposed along the upper ¾ of a scape borne at the apex of the rhizome; fruit a drupelet - - - - - - - - - - - **3. Gunnera**

1. LAUREMBERGIA Berg.

Laurembergia Berg., Desc. Pl. Cap.: 359, t. 5 fig. 10 (Sept. 1767).—Schindl. in Engl., Pflanzenr. **IV**, 225: 61 (1905).—A. Raynal in Webbia **19**: 683 et seq. (1965).
    Serpicula L., Mant. Pl.: 16 (Oct. 1767).

Perennial low-growing herbs, sometimes weakly suffruticose plants, with a ± creeping rhizome and rooting at the lower nodes; stems usually branching at the base, often reddish. Leaves simple, opposite or in (3)4 rows, rarely verticillate or alternate, usually small; leaf-lamina entire to dentate, sessile or shortly petiolate. Flowers in axillary 3–11(15)-flowered fascicles, or rarely solitary; fascicles sometimes of 1–3 pedicellate central ☿ flowers with the others ♀ and sessile or subsessile, or sometimes having 1 long-pedicellate central ♂ flower with the others ♀ and sessile or subsessile, or sometimes having 1–3 long-pedicellate ♂ flowers in the upper leaf-axils and 1–7 sessile or subsessile ♀ flowers in the lower leaf-axils. Calyx-tube urceolate or ellipsoid, with longitudinal nerves which sometimes alternate with often strongly mamillate ribs; calyx-lobes 4, triangular, persistent. Petals 4, sometimes rudimentary or absent in ♀ flowers. Stamens 4 or 8, or absent. Gynoecium 4-locular, becoming 1-locular by resorption of septa; ovules 4; styles 4 or absent; stigmas papillose. Nutlets small, with or without 4 or 8 tuberculate longitudinal ribs. Seeds 1, pendulous; embryo cylindric.

A small genus occurring in the Maghreb, subtropical and tropical Africa, Madagascar and the Mascarene Is., India (SW. Deccan), Ceylon, Malaysia and eastern S. America. Represented in the F.Z. area by a single very variable subspecies belonging to subgen. *Serpiculastrum* A. Raynal.

Laurembergia tetrandra (Schott) Kanitz in Mart., Fl. Bras. **13**, 2: 378, t. 69 (1882).—Schindl. in Engl., Pflanzenr. **IV**, 225: 74 (1905).—A. Raynal in Webbia **19**: 693 (1965).—Boutique & Verdc. in F.T.E.A., Haloragaceae: 4 (1973). Type from Brazil.
    Haloragis tetrandra Schott apud Spreng. in Syst. Veg. ed. 16, **4**, 2, App.: 405 (1827). Type as above.

Suffrutescent usually decumbent or ± erect herb up to 30 cm. tall, or trailing and sometimes forming mats; stems ± branched, often reddish, densely

Tab. 22. LAUREMBERGIA TETRANDRA subsp BRACHYPODA var. BRACHYPODA. 1, habit (×2); 2, detail of flowering branch (×8); 3, ♀ flower (×20); 4, ♀ flower (×60); 5, fruit from ♀ flower (×30); 6, fruit from ♀ flower (×30), all from *Purseglove* 1745. From F.T.E.A.

pubescent-villous to entirely glabrous. Leaves opposite or alternate, rarely sub-
3-nate, mostly sessile, somewhat fleshy; leaf-lamina 5–15(20) × 0·5–5(7) mm.,
linear, linear-oblong, ovate, elliptic or obovate, rounded at the apex, ± cuneate at
base or narrowing into a short petiole, margins entire or 1–4-dentate towards the
apex. Fascicles of 7–11(15) flowers, of which the central one is ⚥ and has a filiform
pedicel 0·3–1·8 mm. long and the others are ♀ and sessile or almost so. Calyx-tube
slightly constricted at the apex, c. 0·3 mm. long, with 8 smooth or 3–4-mamillate
longitudinal ribs and ± distinct nerves alternating with the ribs; calyx-lobes
c. 0·2 mm. long, narrowly triangular, acute. Petals 4, 0·8–1·2 mm. long, oblong,
slightly cucullate, absent in ♀. Stamens 4, oppositisepalous; anthers linear;
filaments filiform. Styles short; stigmas capitate, papillose. Nutlets up to
0·9 × 0·6 mm., barrel-shaped, reddish-brown, with 8 smooth or (2)3–4-tuberculate
longitudinal ribs often confluent in pairs, tubercules whitish, glabrous to villous.

Subsp. **brachypoda** (Welw. ex Hiern) A. Raynal in Webbia **19**: 694, adnot. (1965).—
   Friedrich-Holzhammer & Schreib. in Prodr. Fl. SW. Afr., Haloragaceae: 1 (1967).—
   Boutique in F.C.B., Haloragaceae: 5, fig. 1 (1968).—Mendes in C.F.A. **4**: 31
   (1970).—Boutique & Verdc. in F.T.E.A., Haloragaceae: 6, fig. 2 (1973). Type from
   Angola.
      Laurembergia repens subsp. *brachypoda* (Welw. ex Hiern) Oberm. in Bothalia **11**:
   117 (1973). Type as above.

   Nutlets with 8 ribs ornamented with (2)3–4 usually not or only slightly confluent
tubercles.

Var. **brachypoda** A. Raynal, loc. cit. (1965); in Fl. Cameroun **5**: 136, t. 24 (1966).—
   Boutique, loc. cit., fig. 1A (1968).—Mendes, tom. cit.: 32 (1970).—Boutique &
   Verdc., loc. cit., fig. 2/1–6 (1973). TAB. **22**. Type as above.
      Serpicula repens sensu Oliv. in F.T.A. **2**: 405 (1871), non L. (1767).
      Tillaea aquatica sensu Bak. f. in Trans. Linn. Soc., Ser. 2, Bot. **4**: 13 (1894).
      Serpicula repens var. *brachypoda* Welw. ex Hiern, Cat. Afr. Pl. Welw. **1**: 332 (1896).
   Type as for *Laurembergia tetrandra* subsp. *brachypoda*.
      Laurembergia repens var. *brachypoda* (Welw. ex Hiern) Welw. ex Hiern, op. cit. **2**:
   482 (1901). Type as above.
      Laurembergia angolensis Schindl. in Engl., Pflanzenr. **IV**, 225: 72 (1905). Type
   from Angola.
      Laurembergia engleri Schindl., tom. cit.: 73, fig. 21 (1905).—Keay in F.W.T.A.
   ed. 2, **1**: 171 (1954). Types from Nigeria and Angola.
      Laurembergia villosa Schindl., loc. cit.—Keay, loc. cit. Type from Senegal.
      Laurembergia tetrandra subsp. *brachypoda* var. *numidica* sensu A. Raynal in Webbia
   19: 694, in adnot. (1965) pro parte.—Boutique in F.C.B., Haloragaceae: 6 pro max.
   parte, fig. 1B (1968).—Mendes, tom. cit.: 32 (1970) pro parte; non (Dur. ex Batt.)*
   A. Raynal (1965) sensu stricto.

   Plants glabrate to densely villous-pubescent. Leaf-lamina linear-oblong to
linear, rarely ovate or subovate, length/breadth ratio usually over 2·5, margins
entire or shallowly 1–2(3)-dentate near or towards the apex, not discolorous, sessile
or almost so, rarely narrowing into a petiole up to 2 mm. long (specially at the base
of stem). Fascicles 5–9(11)-flowered; ⚥ flowers with pedicels 0·5–0·7 mm. long,
pubescent. Petals c. 0·8 mm. long. Nutlets usually glabrous, sometimes densely
villous.

   Caprivi Strip. Singalamwe, fl. & fr. 31.xii.1958, *Killick & Leistner* 3211 (SRGH).
Botswana. N: Moremi Game Reserve, fl. & fr. 17.xi.1972, *Smith* 271 (SRGH).
Zambia. B: 24 km. ENE. of Mongu, fl. & fr. 10.xi.1959, *Drummond & Cookson* 6283
(K; LISC; PRE; SRGH). N: Kasama Distr., Chambesi R., fl. & fr. 7.iv.1961,
Richards 15020 (K; SRGH). W: Mwinilunga Distr., Matonchi R., fl. & fr. 23.x.1937,
Milne-Redhead 2918 (K; PRE). C: Serenje Distr., Lake Lusiwachi, fl. & fr. 15.x.1963,
Robinson 5737 (K; SRGH). E: Nyika Plateau near N. Rukuru waterfall, fl. & fr.
27.x.1958, *Robson* 403 (K). S: 5 km. E. of Choma, fr. 28.v.1955, *Robinson* 1269 (SRGH).
Rhodesia. W: Matobo Distr., Farm Quaringa, fl. & fr. xii.1958, *Miller* 5668 (SRGH).
C: Salisbury, Rua R. tributary, fl. & fr. vi.1916, *Eyles* 1337 (BM; K; PRE; SRGH).
E: Inyanga, Pungwe Heights S. of falls, fl. & fr. 19.xii.1955, *Chase* 5919 (BM; K; LISC;
PRE; SRGH). **Malawi.** N: S. Vipya, Luwawa Dam, fl. & fr. 13.i.1967, *Hilliard &*

---

* *Serpicula numidica* Dur. was originally published ex Batt. in Batt. & Trab., Fl. Algér.:
318 (1888) and not ex Batt. & Trab., Fl. Anal. Syn. Algér. Tunis.: 128 (1902) as A. Raynal
says (loc. cit.).

Burtt 4463 (K; SRGH). S: Ncheu Distr., Lower Kirk Range, Chipusiri, fl. & fr. 17.iii.1955, *E.M. & W.* 969 (BM; LISC; SRGH). **Mozambique.** Z: Gúruè Mts., near the Malema R. source, fl. & fr. 4.i.1968, *Torre & Correia* 16880 (LISC). MS: Manica Distr., Chimanimani Mts., between Skeleton Pass and the plateau, fl. 27.ix.1966, *Simon* 873 (K; LISC; PRE; SRGH). SS: Inhambane Distr., Vilanculos, fl. & fr. 27.viii.1968, *Gomes e Sousa & Balsinhas* 5109 (LMU; PRE). LM: Bela Vista, Santaca, Lake Lifuno, fl. & fr. 12.xii.1961, *Lemos & Balsinhas* 295 (BM; COI; LISC; LMA; PRE; SRGH).

Widespread in subtropical and tropical Africa from Senegal to Uganda and southwards to SW. Africa, Botswana, Transvaal and Natal. Grassy edges of swamps, sandy lake margins, on mud in grassland areas, moist pathsides etc.; 0–2200 m.

Note. The robust collections seen only from coastal and subcoastal southern Mozambique (SS and LM) are probably only a very distinct maritime ecoform (prostrate or decumbent plants up to 80 cm. in diameter, base of stems woody, stems branched, leaves from base of stems and branches up to 20 × 6 mm., elliptic-obovate with margins entire or 1-dentate, all parts of plant, including nutlets, densely pubescent-villous); they resemble phenotypically some forms of the S. African *Laurembergia repens* (L.) Berg., but have the flowers arranged in fascicles composed and disposed as in typical *L. tetrandra* subsp. *brachypoda.*

2. MYRIOPHYLLUM L.

Myriophyllum L., Sp. Pl. **2**: 992 (1753); Gen. Pl. ed. 5: 429 (1754).—Schindl. in Engl., Pflanzenr. **IV**, 225: 77 (1905).

Glabrous, mostly perennial, usually aquatic herbs, sometimes growing on mud. Stems free-floating or with rooted rhizomes, mostly branched. Submerged leaves in whorls of 3–6 and always pinnatisect into undivided segments; aerial leaves whorled, opposite or alternate, often dentate or entire or bract-like; stipules absent; leaf-bases often accompanied by 1–3 filiform to subulate, deciduous, stipule-like outgrowths. Flowers mostly sessile, 1 or 2 in the axil of a leaf or bract in terminal, emergent spikes or solitary in the lower leaf-axils, bisexual or unisexual, monoecious or sometimes polygamous (the upper flowers commonly ♂, the lower ♀, sometimes with ♀ ones in between), rarely dioecious; bracteoles 2, often inconspicuous. Calyx of 4 small lobes in ♂ flowers, minute in ♀. Petals in ♂ flowers 2 or 4, cucullate, caducous, in ♀ minute or absent. Stamens (2–7)8 or 1; staminodes absent in ♀ flowers. Gynoecium 2–4-locular, rudimentary or absent in ♂ flowers; styles 2 or 4, very short or absent; stigmas 2 or 4, subsessile or sessile, persistent. Fruit separating into 4 (or fewer by abortion) 1-seeded nutlets. Seeds pendulous; testa membranous; endosperm copious.

An almost cosmopolitan genus comprising some 40 species, represented in Africa south of the Sahara by two species only, probably both introduced.

Leaf-segments densely papillose, apex mucronulate; leaf-bases accompanied by deciduous stipule-like outgrowths 0·8–1·5 mm. long, linear, mostly recurved; flowers solitary in median and upper leaf-axils of stems usually floating at the surface - 1. *aquaticum*
Leaf-segments not papillose, apex not mucronulate; stipule-like outgrowths absent or inconspicuous; flowers in terminal emergent spikes 5–10 cm. long - 2. *spicatum*

1. **Myriophyllum aquaticum** (Vellozo) Verdc. in Kew Bull. **28**: 36 (1973).—Boutique & Verdc. in F.T.E.A., Haloragaceae: 7, fig. 3 (1973). TAB. **23.** Type from Brazil.
   Enydria aquatica Vellozo, Fl. Flum.: 57 (1825); Ic. Fl. Flum. **1**: t. 150 (1835). Type as above.
   Myriophyllum brasiliense Cambess. in St.-Hil., Fl. Bras. Merid. **2**: 252 (1829).— Schindl. in Engl., Pflanzenr. **IV**, 225: 88, fig. 25 & 28K (1905). Type from Brazil.

Aquatic herb, usually submerged, with the upper part floating at the surface or sometimes creeping on to sandy or muddy banks; rhizome rooted; stems up to 1·2 m. long, simple or sparsely branched at the base. Leaves all pinnatisect, in alternating whorls of (4)5–6, densely papillose (when dry simulating puberulence), usually much longer than internodes, their bases flanked by 2–3 linear, mostly recurved, deciduous stipule-like outgrowths; leaf-segments 8–16(30) on each side, linear-subulate, with a mucronulate apex; leaf-base dilated, almost concealing the stem at least at upper nodes. Flowers unisexual, solitary in the middle and upper leaf-axils; bracteoles subulate, with 1–2 lateral laciniae in ♂, laterally 1–2-dentate in ♀. Petals 4, c. 4 mm. long, absent in ♀. Stamens usually 8, rarely less; anthers

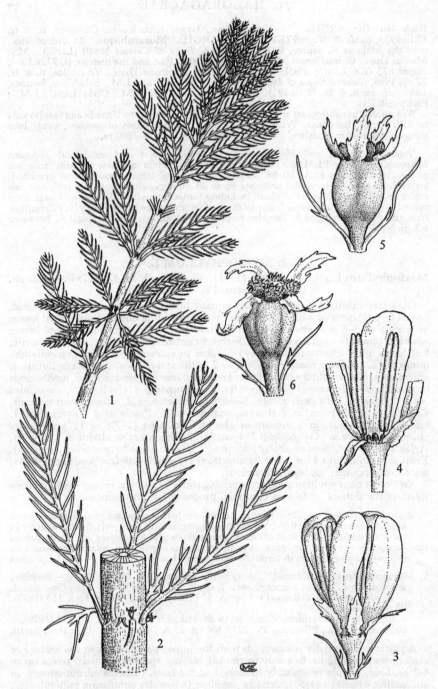

Tab. 23. MYRIOPHYLLUM AQUATICUM. 1, part of leaf stem with ♀ flowers (×1); 2, detail of same (×2); 3, ♂ flower (×12); 4, same with petal removed to show rudimentary styles (×12); 5, ♀ flower (×24); 6, same at a later stage (×24), 1, 5, 6 from *Kimani* in EA.14302, 2 from *Hanid* 1087, 3–4 from *Cuming* 164 (Chile). From F.T.E.A.

c. 3 mm. long; filaments up to 5 mm. long. Stigmas 4, shortly papillose. Fruit c. 1·8 × 1·2 mm., ovoid, longitudinally 4-sulcate, papillose-punctate.

Rhodesia. N: Lomagundi Distr., Endeavour Bridge, Hunyani R. S. of Darwendale, fl. 23.xi.1971, *Ellis* in GHS 224168 (SRGH). C: Salisbury Distr., L. McIlwaine, fl. 6.i.1970, *Marshall* in GHS 201016 (K; PRE; SRGH). S: Chiredzi Distr., Hippo Valley Estate, Section 18, st. 27.iii.1971, *Taylor* 166 (K; PRE; SRGH).

Native of S. America (Brazil, Peru, Chile, Argentina and Uruguay), naturalized or casual in parts of Europe, Africa (outside F.Z. area recorded from Kenya, Tanzania and S. Africa), Malaysia, Japan, Australia and N. America. Upland still, shallow waters near banks of rivers, lakes and ponds.

2. **Myriophyllum spicatum** L., Sp. Pl. 2: 992 (1753).—Eckl. & Zeyh., Enum. Pl. Afr. Austr. Extratrop.: 274 (1836).—Harv. in Harv. & Sond., F.C. 2: 572 (1862).— Schindl. in Engl., Pflanzenr. **IV**, 225: 90 (1905).—Marloth, Fl. S. Afr. **2**, 2: 231, fig. 149B (1925).—Burtt Davy, F.P.F.T.: 203 (1926).—Wild in Publ. Proj. Conj. C.C.T.A./C.S.A. **14**: 38, saltem pro parte, t. 14D (1964).—A. Raynal in Fl. Cameroun **5**: 132, t. 23 (1966).—Boutique & Verdc. in F.T.E.A., Haloragaceae: 7, nota (1973). Type from Europe.

Aquatic herb, usually submerged, occasionally creeping on to muddy banks; rhizome rooted; stems up to 2 m. long, usually branched. Submerged leaves pinnatisect, in alternating whorls of (3)4(5), glabrous, ± equalling internodes; leaf-segments 7–11 on each side, filiform, usually obtuse at apex; stipule-like outgrowths absent or inconspicuous. Flowers unisexual or bisexual in terminal emergent spikes up to 10 cm. long; spikes interrupted with whorls c. 1·5 cm. apart along a 4-angled rhachis; whorls of 4 obovate entire bracts (except the lowest which are pinnatisect or dentate); upper bracts shorter than the flowers; bracteoles broadly ovate to triangular, entire. Petals 4, c. 3 mm. long. Stamens (?2)8; anthers c. 2 mm. long; filaments c. 2 mm. long. Stigmas 4, spreading, papillose. Fruit subglobose, c. 3 mm. in diameter, 4-sulcate, usually with 8 longitudinal verrucose ridges or rarely smooth.

Botswana. N: Linyanti R., Hunter's Camp, st. 27.x.1972, *Gibbs Russell* 2196 (SRGH). **Zambia.** N: Mbala (Abercorn) Distr., L. Tanganyika, Mpulungu, fl. buds 20.x.1947, *Greenway & Brenan* 8245 (K; PRE). S: Mumbwa Distr., Kafue Hook Pontoon, st. 21.xi.1959, *Drummond & Cookson* 6743 (BM; K; LISC; PRE). **Rhodesia.** W: Victoria Falls, st. 28.xi.1949, *Wild* 3220 (BM; SRGH). **Malawi.** C: L. Nyasa, Nkhota Kota, st. 20.vi.1904, *Cunnington* 10 (BM). S: Mangochi (Fort Johnston) Distr., Monkey Bay, near jetty, st. 15.ii.1969, *Williams* 7 (SRGH).

Almost ubiquitous in the northern hemisphere; frequent in northern and southern Africa but rare in the tropics (Cameroon, Ethiopia, Tanzania, Malaysia, Philippines). Still, shallow waters near banks of rivers, lakes and ponds.

3. GUNNERA L.

Gunnera L., Mant. Pl.: 16 et 121 (1767); Syst. Nat. ed. 12: 587 et 597 (1767).— Schindl. in Engl., Pflanzenr. **IV**, 225: 104 (1905).

Perennial herbs, sometimes giant, with a creeping or erect rhizome. Leaves usually tufted near the apex of the rhizome, often very large; leaf-laminae reniform-cordate to ovate-truncate, simple to lobed, crenate to compound-dentate, rarely entire, palmatinerved; petiole equalling or longer than the lamina, sometimes provided with ochrea-like cataphylls. Inflorescences scapigerous, spicate, racemose or paniculate, borne on the rhizome. Flowers ebracteolate, bisexual or unisexual, monoecious or polygamous (♂ flowers towards the apex and ♀ ones towards the base, sometimes ☿ in between) or dioecious. Calyx-lobes 2(3), equal or unequal, or absent. Petals 2, small, longer than and alternating with the calyx-lobes, or absent. Stamens 2, oppositipetalous, or absent. Gynoecium 1-locular; styles 2, filiform, subulate or compressed, entirely papillose. ♂ flowers with calyx-tube reduced and styles vestigial or absent. ♀ flowers apetalous. Fruit an indehiscent nutlet or a ± globose to trigonous drupe. Seed 1; testa membranous.

A distinctly austral genus, comprising c. 50 species, represented in Central and S. America, eastern and southern Africa (1 species only), Madagascar, Malaysia, Tasmania, New Zealand (9 species) and the Hawaiian Is. (7 species).

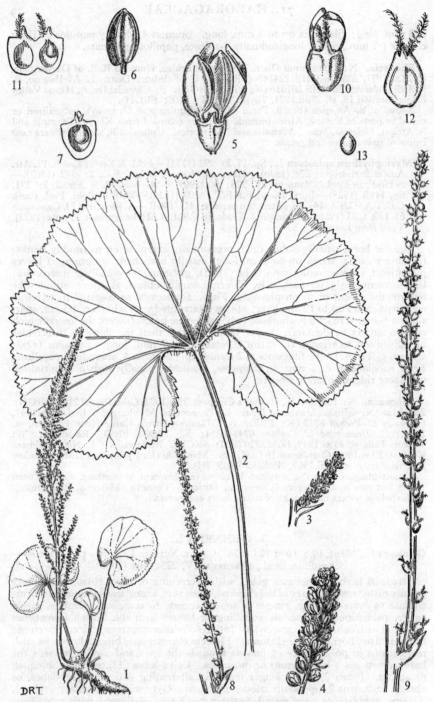

Tab. 24. GUNNERA PERPENSA. 1, habit (× ⅕); 2, leaf (× ½); 3, spike of ♂ flowers (× 1);
4, same (× 2); 5, ♂ flower (× 8); 6, stamen (× 8); 7, section of ovary from ♂ flower
(× 8); 8, spike of ♀ flowers (× 1); 9, same (× 2); 10, ♀ flower (× 8); 11, ovary from
same, opened out (× 8); 12, older ♀ flower (× 1); 13, ovule (× 8), all from *St. Clair-
Thompson* 799. From F.T.E.A.

Gunnera perpensa L., Mant. Pl.: 121 (1767); Syst. Nat. ed. 12: 597 (1767).—Sims in Curtis, Bot. Mag. **50**: t. 2376 (1823).—Harv. in Harv. & Sond., F.C. **2**: 571 (1862).— Oliv. in F.T.A. **2**: 406 (1871).—Schindl. in Engl., Pflanzenr. **IV**, 225: 116 (1905).— Marloth, Fl. S. Afr. **2**, 2: 231, fig. 149A, 150 & 151 (1925).—Burtt Davy, F.P.F.T.: 203 (1926).—Bader in Engl., Bot. Jahrb. **80**: 286, fig. 1 (1961).—Boutique in F.C.B., Haloragaceae: 2, t. 1 (1968).—Boutique & Verdc. in F.T.E.A., Haloragaceae: 3, fig. 1 (1973). TAB. **24**. Type from S. Africa (Cape Prov.).

Gunnera perpensa var. *angusta* Schindl., op. cit.: 117 (1905).—H. Perr. in Fl. Madag., Haloragaceae: 3, fig. 1/1–7 (1950). Type from Madagascar.

Rhizomatous herb, up to 1 m. tall; rhizome ± elongated, usually creeping, up to 3 cm. thick, yellow-fleshed. Leaves robust, tufted near the apex of rhizome; leaf-lamina 4–13(25) × 6–17(38) cm., reniform to subcircular, palmately nerved, margin irregularly dentate with glandular teeth, cordate at the base, appressed-hairy on both surfaces, particularly so beneath and along the nerves when young, glabrescent with age; petiole 15–75 cm. long, stout, appressed-hairy to glabrescent. Spikes (0·5)2–7(10) cm. long, slender, sparsely hairy, solitary or fasciculate, remotely to densely disposed usually only along the upper ¾ of scape, subtended by 4–10 mm. long triangular to linear and usually pubescent bracts; scapes usually solitary at the apex of rhizome, sparsely hairy or glabrescent, usually exceeding leaves. Flowers ♂, ♀ and ☿, ebracteolate; calyx-tube urceolate; calyx-lobes minute, triangular; petals c. 2 mm. long, linear-spathulate, glabrous or hairy at apex, caducous, or absent; stamens with anthers up to 1·5 mm. long, broadly ellipsoid, dorsiventrally compressed, and filaments c. 0·4 mm. long, or absent; styles up to 1 mm. long, filiform, plumose from the base, divergent, or rudimentary or absent. Drupelets c. 1·4 mm. in diameter, subglobose, slightly laterally compressed and thus ± 2-crested, glabrous, crowned by the calyx-lobes, sessile or on pedicels up to 1 mm. long.

Rhodesia. E: Melsetter Distr., 5 km. from Skyline on road to Umtali, immat. fr. 28.xi.1955, *Drummond* 5043 (K; LISC; SRGH). **Mozambique.** T: Ncheu–Neno road, Nankungwi stream, fr. 31.i.1959, *Robson* 1400 (K; LISC). MS: Manica Distr., Manica, Tsetsera Mts., fr. 9.ii.1955, *E.M. & W.* 326 (BM; LISC; SRGH).

Known also from Sudan, Ethiopia, Zaire, Rwanda, Uganda, Kenya, Tanzania and S. Africa (Transvaal and Cape Prov.). In cold or cool, continually moist localities, mainly along upland streambanks, above 1500 m.

Note. Recorded in cult. in Rhodesia as having a symbiotic *Nostoc* sp. inhabiting green zones in the roots.

72. RHIZOPHORACEAE
By A. R. Torre and A. E. Gonçalves

Monoecious trees, shrubs or, rarely, undershrubs; young branches glabrous or glabrescent with age. Leaves simple, petiolate, usually isomorphic, decussate (very rarely 3(4)-nate), with interpetiolar and caducous stipules, or rarely dimorphic, alternate, without stipules (*Anisophyllea*); nervation generally pinnate, rarely curvinerved and with 1–2 intramarginal nerves (*Anisophyllea*). Inflorescences simple or branched, axillary or, rarely, supra-axillary, of cymes (lax or condensed and clustered), racemes, panicles, fascicles or spikes, rarely a solitary flower. Flowers usually bisexual, rarely unisexual or polygamous, actinomorphic, with a pair of ± connate bracteoles, rarely ebracteolate. Calyx gamosepalous, usually ± adnate to the ovary, persistent in fruit; calyx-lobes (3)4–5(7) or 8–15 (*Bruguiera*), valvate. Petals as many as the calyx-lobes and alternating with them, free, often clawed, divided above to rarely entire, usually fleshy and conduplicate, caducous or rarely persistent. Disk fleshy, annular or ± cup-shaped, crenate, flat or lobed, rarely absent. Stamens usually twice as many as the petals, sometimes ∞ (indeterminate number), usually in 1 whorl, free, epipetalous or adnate to the calyx-tube, variously inserted in relation to the disk, when present; filaments usually subulate to ± filiform, sometimes very short; anthers 4-locular and dehiscing longitudinally or, rarely, multilocular and dehiscing by a large ventral valve (*Rhizophora*), dorsifixed, introrse. Ovary syncarpous, (1)2–4(5)-locular, inferior to superior; placentae axile, each with usually 2 (1 in *Anisophyllea*) pendulous, anatropous ovules; style usually single or 4–5(8) (*Anisophyllea*), ± persistent, with entire or ± lobed stigmas. Fruit a berry, a drupe (*Anisophyllea*) or a dry, septicidally dehiscent, 2–4-loculed capsule (*Cassipourea*). Seeds 1 or more, sometimes arillate (*Cassipourea*), often viviparous (mangrove species); embryo straight or curved and often with green cotyledons; endosperm present or rarely absent (*Anisophyllea*).

A family with c. 120 species, widespread mainly in tropical Africa, Asia, Oceania and America; of some economic importance, yielding edible fruits, tannin and timber.

Leaves alternate, usually dimorphic (the smaller, stipuliform leaves often caducous), with 2–4 thick lateral nerves ± parallel to the midrib, exstipulate; flowers bisexual, unisexual or polygamous, in simple or composite spikes; ovary inferior, (3)4(5)-locular; styles (3)4(8); fruit a drupe - - - - - **5. Anisophyllea**
Leaves decussate or rarely verticillate (3(4)-nate), isomorphic (without stipuliform leaves), with pinnate nervation, stipulate; flowers bisexual, solitary and pedicellate or in peduncled cymes or in fascicles; ovary inferior to superior, (1)2–4(5)-locular; style 1; fruit a berry or capsule:
  Plants viviparous, wholly glabrous, of muddy sea-shores and estuaries, usually in mangrove forests, often with aerial roots; flowers solitary and pedicellate or in peduncled cymes; ovary inferior to semi-inferior; fruit a berry:
    Flowers bracteolate, up to c. 1 cm. long, in peduncled cymes; calyx-lobes 4–5(6); hypocotyl very thick towards the apex and ± pointed:
      Leaves usually elliptic; calyx-lobes 4(5); petals entire; filaments very short; anthers multilocular, dehiscing by a large ventral valve; ovary 2-locular; hypocotyl not longitudinally ridged - - - - - **1. Rhizophora**
      Leaves usually obovate; calyx-lobes (4)5(6); petals not entire; filaments longer; anthers 4-locular, dehiscing longitudinally; ovary 3-locular; hypocotyl longitudinally ridged - - - - - - - **2. Ceriops**
    Flowers ebracteolate, at least 1·5 cm. long, solitary; calyx-lobes 8–15; hypocotyl ± isodiametric throughout or thicker submedially and blunt - **3. Bruguiera**
  Plants not viviparous, ± hairy, neither of muddy sea-shores or estuaries nor of mangrove forest, usually without aerial roots; flowers in fascicles or rarely solitary; ovary superior to subinferior; fruit a capsule - - - **4. Cassipourea**

1. RHIZOPHORA L.

Rhizophora L., Sp. Pl. 1: 443 (1753); Gen. Pl. ed. 5: 202 (1754).—Salvoza in Nat. Appl. Sci. Bull. 5: 179 (1936).

Wholly glabrous evergreen shrubs or trees of muddy sea-shores and estuaries, with stems supported by numerous aerial, adventitious, branched, capped prop-roots developed from the upper nodes; pneumatophores absent; branches stout, opposite. Leaves decussate, petiolate; leaf-lamina entire, leathery, usually black-dotted beneath, with nerves obscure below and visible or distinct above, midrib protruding at the tip into a caducous mucro; stipules large, lanceolate, ± red when fresh. Inflorescences cymose, simple or 2–3-dichotomously-branched, axillary, peduncled. Flowers bisexual, 2-bracteolate, shortly pedicellate. Calyx glabrous, leathery, surrounded at the base by a persistent cup-shaped pair of bracteoles, accrescent and reflexed in fruit; calyx-tube ± adnate to the ovary; calyx-lobes 4(5). Petals entire, not clawed, inserted at the mouth of the calyx-tube. Stamens 4 epipetalous plus 4 or 8 episepalous, inserted on the margin of the crenulate disk, sessile or subsessile; anthers narrowly oblong, triquetrous, pointed, with numerous pollen-sacs within a membranous epidermis, multilocular, opening by a large ventral valve. Ovary inferior to semi-inferior, 2-locular, each locule 2-ovulate; style obscure or up to 6 mm. long, with a simple or shortly 2-lobed stigma. Fruit a leathery, indehiscent, mostly 1-seeded (very rarely 2–3-seeded) berry, included in and ± adnate to the calyx-tube, with a granular or roughened surface; cotyledons connate into a fleshy body continuous with but set off from a clavate, elongate hypocotyl which perforates the apex of the fruit and falls out of it; endosperm absent.

A genus of c. 8 species, widespread mainly in tropical Africa, Asia, Oceania and America; represented along the E. coast of Africa and in Madagascar, Seychelles and Mauritius Is. by a single species.

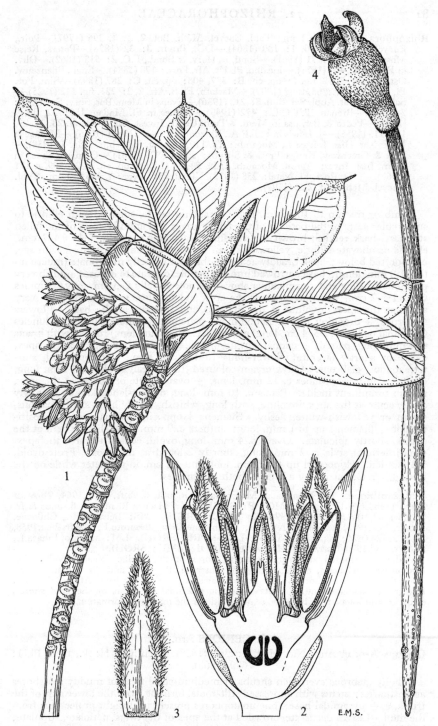

Tab. 25. RHIZOPHORA MUCRONATA. 1, flowering branch with lower leaves removed (× ⅔),
from *Schlieben* 2628; 2, longitudinal section of flower (× 2); 3, petal (× 4), both from
Elliott 263; 4, fruit and hypocotyl (× ⅔), from *Greenway* 4861. From F.T.E.A.

Rhizophora mucronata Lam., Tabl. Encycl. Méth. Bot. **2**, 2: t. 396 (1797).—Poir., Encycl. Méth. Bot. **6**, 1: 189 (1804).—DC., Prodr. **3**: 32 (1828).—Peters, Reise Mossamb. Bot. **1**: 71 (1861).—Sond. in Harv. & Sond., F.C. **2**: 513 (1862).—Oliv. in F.T.A. **2**: 407 (1871).—Ficalho, Pl. Ut. Afr. Port.: 178 (1884).—Engl., Pflanzenw. Ost-Afr. **A**: 7, 12, fig. 1; op. cit. **B**: 337, 408; op. cit. **C**: 287 (1895).—Sim, For. Fl. Port. E. Afr.: 65, t. 69 (1909).—Marloth, Fl. S. Afr. **2**, 2: 221, fig. 142 (1925).— Salvoza in Nat. Appl. Sci. Bull. **5**: 213 (1936).—Evans in Mem. Bot. Surv. S. Afr. **15**: 7 (1936).—Brenan, T.T.C.L.: 472 (1949).—Arènes in Fl. Madag. **150**: 32, fig. 9 (1954).—Pedro & Barbosa in Mem. e Trab. Centr. Invest. Cient. Algod. **23**, 2: 99, 104, 215 (1955).—J. Lewis in F.T.E.A., Rhizophoraceae: 2, fig. 1 (1956).—Macnae & Kalk, Nat. Hist. Inhaca I., Moçamb.: 13, 119, 150, fig. 5a, 11b, t. 3 fig. 3 (1958).— Dale & Greenway, Kenya Trees & Shrubs: 399, fig. 79 (1961).—Gomes e Sousa in Mem. Inst. Invest. Agron. Moçamb. **1**, 1: 54, 55 (1966); op. cit. **1**, 2: 599, t. 171 (1967).—J. H. Ross, Fl. Natal: 258 (1972). TAB. **25**. Type: t. 396 in Lam., Tabl. Encycl. Méth. Bot. **2**, 2 (1797).

Shrub or tree up to 10(18) m. high; branches terete, smooth, with elliptic to subcircular scars of the petioles of the fallen leaves and annular scars of the fallen stipules; bark reddish-brown to greyish. Leaf-lamina usually 7–15 × 4–7 cm., elliptic or obovate-elliptic, yellowish-green when young and dark green with age, cork-dotted below; apex broadly acute or obtuse, distinctly mucronate (mucro up to 8(10) mm. long); base narrowed into the petiole, cuneate to obtuse; midrib very prominent below and conspicuous above; petiole usually 2–3 cm. long; stipules 6–8 × 1·5–2 cm., ± striate towards the apex and rugulose towards the base, very acute at the apex, involute at the margins and truncate at the base. Flowers 4-merous, ± drooping, in 2–3-dichotomous, pauciflorous inflorescences; peduncles (principal peduncle 2·5–4 cm. long) and pedicels (c. 5 mm. long) robust, with bracts c. 4 mm. long, paired, ± connate, ovate-orbicular, obtuse to rounded at the apex, entire or shortly lacerate; flower-buds 10–14 × 5–7 mm., ellipsoid-ovoid, sub-tetragonous, glabrous. Calyx cream-coloured, thick; calyx-tube c. 4 mm. long, campanulate; calyx-lobes c. 12 mm. long, ± ovate, acute or obtuse at the apex, midrib prominent inside. Petals c. 10 mm. long, oblong-lanceolate or narrowly ovate, acute at the apex, involute, with long, whitish, woolly hairs at the margins, cream or yellowish-white, fleshy. Stamens 4 oppositisepalous and 4 oppositi-petalous; filaments up to 1 mm. long; anthers c. 9 mm. long, slightly bent at the angles, shortly apiculate. Ovary c. 4 mm. long, ovoid, conical towards the apex, semi-inferior; style c. 2 mm. long, shortly 2-lobed at the apex. Fruit ovoid, 1(2)-seeded. Hypocotyl up to 40 cm. long and 1·2 cm. in diameter while on the tree, ± terete, narrowing a little at both extremities.

Mozambique. N: Mogincual, Mogincual R. mouth, fl. & fr. 30.iii.1964, *Torre & Paiva* 11483 (J; K; LISC; MO; P; SRGH). Z: Macuze R. mouth, fl.-buds & fr. 23.viii.1962, *Wild & Pedro* 5894 (K; LISC; SRGH). MS: Beira Distr., Chiloane, fl. & fr. 25.x.1906, *Johnson* 20 (K). SS: Inhambane Distr., Bazaruto I., fl. & fr. 29.x.1958, *Mogg* 28728 (BM; K; LISC; LMA; LMU; PRE; SRGH). LM: Maputo, Inhaca I., fl. & fr. 6.vi.1970, *Correia & Marques* 1675 (COI; LMU; SRGH).
Widely distributed along the western shores of the Pacific Ocean, Ryukyu Is., Micronesia, Melanesia, northern coast of Australia, Polynesia and the Indian Ocean; along the E. African shores from near Massawa in the Red Sea to S. Africa (Durban). Evergreen woodlands and thickets along the intertidal mud-flats of sheltered shores, estuaries and inlets, mainly in the seaward side of the mangrove formation.

2. CERIOPS Arn.

Ceriops Arn. in Ann. Nat. Hist. **1**: 363 (1838).—Benth. & Hook., Gen. Pl. **1**: 679 (1865).

Wholly glabrous evergreen shrubs or medium-sized trees of muddy sea-shores and estuaries; stems with appressed stilt-roots, forming, with the lower part of the trunk, a ± pyramidal base; pneumatophores present as upright projections from the mud. Leaves decussate, crowded at the apex of the twigs, petiolate, stipulate; leaf-lamina entire, leathery, with the nerves obscure on both faces. Inflorescences (2)4–8(10)-flowered, cymose, condensed, solitary, axillary, towards the upper part of the branches. Flowers bisexual, subsessile to sessile, with cupular, partly connate pairs of bracts disposed beneath each flower and each cluster. Calyx glabrous, persistent; calyx-tube adnate to the ovary; calyx-lobes (4)5(6), spreading in fruit. Petals emarginate or truncate, fringed or with (2)3 long-stipitate clavate glands above, sometimes cohering from the middle towards the base by the uncinate-hairy margins, inserted at the mouth of the calyx-tube, each embracing 2 stamens. Disk shallowly (8)10(12)-lobed, cupuliform. Stamens twice as many as the petals, inserted in the sinuses of the disk; filaments unequal to ± equal in each pair; anthers ± ovoid. Ovary semi-inferior, 3-locular, each locule 2-ovulate; style simple, terete and conical below, minutely 2–3-lobed to entire at the apex. Fruit an ovoid, leathery, indehiscent, unilocular, usually 1-seeded berry, mostly superior; seed germinating on the plant, the radicle developing a clavate, ridged and sulcate hypocotyl tapering to the apex, which perforates the apex of the fruit and falls out of it.

A genus with 2 species, widespread mainly in tropical E. Africa, Asia and Oceania; represented along the E. coast of Africa, Madagascar, Comores and Seychelles Is. by a single species.

Ceriops tagal (Perr.) C. B. Robinson in Philipp. Journ. Sci. **3**, 5: 306 (1908).—Brenan, T.T.C.L.: 472 (1949).—Pedro & Barbosa in Mem. e Trab. Centr. Invest. Cient. Algod. **23**, 2: 99 (1955).—J. Lewis in F.T.E.A., Rhizophoraceae: 5, fig. 2 (1956).—Macnae & Kalk, Nat. Hist. Inhaca I., Mocamb.: 13, 119, 150, fig. 5b, 11b, t. 3 fig. 2 & 4 (1958).—Dale and Greenway, Kenya Trees & Shrubs: 399 (1961).—J. H. Ross, Fl. Natal: 258 (1972). TAB. 26. Type probably from the southern Philippines.

Rhizophora tagal Perr. in Mém. Soc. Linn. Paris **3**: 138 (1825). Type as above.

Ceriops candolleana Arn. in Ann. Nat. Hist. **1**: 364 (1838).—Oliv. in F.T.A. **2**: 409 (1871) " candolliana ".—Marloth, Fl. S. Afr. **2**, 2: 221 (1925). Type from Australia.

Ceriops mossambicensis Klotzsch in Peters, Reise Mossamb. Bot. **1**: 71 (1861). Type: Mozambique, Niassa, from Querimba to Moçambique, 11°–15° S., *Peters* (B, holotype †).

Shrub or small tree up to 6(15) m. high; branches terete or slightly compressed towards the extremities, smooth, with elliptic to ± deltate or subcircular scars of the petioles of the fallen leaves and annular scars of the fallen stipules; bark brown to grey. Leaf-lamina usually 4–9 × 2–5 cm., elliptic to obovate; apex rounded to retuse; base narrowed into the petiole; midrib rather prominent below; petiole usually 1–3 cm. long; stipules 1–2 cm. long, lanceolate, plicate-involute initially, very acute at the apex and truncate at the base. Inflorescences (3)4–8-flowered; principal peduncle up to 1·5 cm. long, very shortly or obsoletely 2–3-forked at the apex. Flowers 5-merous, subsessile; bracts and bracteoles c. 2 mm. long, ± deltate or ovate-circular, obtuse or subrounded at the apex; flower-buds c. 5 × 4 mm., ellipsoid-ovoid, subpentagonal, glabrous. Calyx reddish-brown; calyx-tube c. 2·5 mm. long; calyx-lobes c. 5 mm. long, triangular or ovate-triangular, acute at apex, fleshy, abaxial surface glossy, adaxial surface paler and mat, with a longitudinal thickened ridge. Petals 3·5 × 2 mm., oblong or obovate-oblong, truncate and bearing 3 distinct very short glandular appendages at the apex, shortly cohering a little below the middle by the uncinate-hairy margins, the base slightly narrower than the apex, white or cream. Disk composed of 5, ± separate, 2-lobed elements. Stamens 10; filaments 3–4 mm. long, ± filiform, narrowed and bent at the apex and slightly swollen at the base; anthers 0·5–0·7 mm. long, ovoid-sagittate, acute at the apex, ochraceous. Ovary c. 3 mm. long, ovoid, pale green; style c. 1·5 mm. long, thickened at the base and minutely 3-lobed at the apex. Hypocotyl extended up to 26 cm. from the fruit while on the tree, sharply longitudinally ridged, ± swollen subdistally.

Mozambique. N: Cabo Delgado Distr., Porto Amélia, Metuge, Bandar, fl. & fr. 19.xii.1963, *Torre & Paiva* 9618 (K; LISC; LUAI; WAG). Z: Macuze R. mouth, fl. & fr. 23.viii.1962, *Wild & Pedro* 5895 (K; LISC; SRGH). MS: Beira Distr., Chiloane, fl. & fr. 25.x.1906, *Johnson* 19 (K). SS: Inhambane Distr., Bazaruto I., fl. & fr. 29.x.1958, *Mogg* 28725 (BM; K; LISC; LMA; LMU; PRE; SRGH). LM: Inhaca I., fl. & fr. 6.vi.1970, *Correia & Marques* 1667 (COI; LISC; LMU; SRGH).

Widely distributed along the western shores of the Pacific Ocean, in Formosa, Micronesia, Melanesia, northern and western coasts of Australia, and round the Indian Ocean; along the E. African shores from Somali Republic to S. Africa (Durban). Evergreen woodlands and thickets along the intertidal mud-flats of sheltered shores, estuaries and inlets. The most inland of the Rhizophoraceous mangroves.

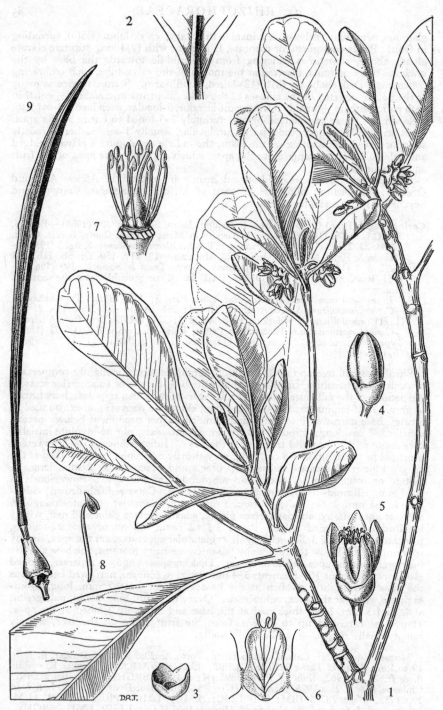

Tab. 26. CERIOPS TAGAL. 1, flowering branch with some leaves removed (×1); 2, underside of leaf (×2); 3, fused bracteoles (×3); 4, bud with bracteoles (×3); 5, flower (×3); 6, petal (×6); 7, pistil and stamens (×6); 8, stamen (×6); 9, fruit and hypocotyl (inverted) (×½), all from *Toms* 65. From F.T.E.A.

3. BRUGUIERA Lam.

Bruguiera Lam., Tabl. Encycl. Méth. Bot. **2**, 2: t. 397 (1797); Encycl. Méth. Bot. **4**, 2: 696 (1798).

Wholly glabrous and evergreen shrubs or trees of muddy sea-shores and estuaries, with geniculate pneumatophores arising from the mud, sometimes with aerial roots when young; roots and lower part of the trunk forming a pyramidal buttressed base. Leaves decussate, petiolate; leaf-lamina entire, leathery, usually black-dotted beneath, with the nerves obscure to visible below and visible or distinct above; stipules large, lanceolate. Inflorescences 1–few-flowered, cymose, axillary; peduncles usually arched outwards. Flowers bisexual, ebracteolate. Calyx glabrous, leathery, accrescent; calyx-tube adnate at the base to the ovary; calyx-lobes 8–15. Petals 2-fid and apically setigerous, involute at the base (each embracing a pair of stamens), not clawed, inserted at the mouth of the calyx-tube. Stamens twice as many as the petals, epipetalous; filaments unequal in length in each pair, filiform; anthers linear, dehiscing longitudinally. Ovary inferior, 2–4-locular, each locule 2-ovulate; style filiform, with 2–4 minute stigmatic lobes. Fruit a leathery, indehiscent, usually unilocular, mostly 1-seeded berry, included in and adnate at the base to the calyx-tube; seed germinating on the plant, the radicle developing a ± terete or obscurely ribbed, blunt hypocotyl, which perforates the fruit-apex and falls with the fruit; cotyledons minute and connate at the base.

A genus of c. 6 species, widespread mainly in tropical E. Africa, Asia and Oceania; represented along the E. coast of Africa and in Madagascar and Seychelles Is. by a single species.

Bruguiera gymnorrhiza (L.) Lam., Tabl. Encycl. Méth. Bot. **2**, 2: t. 397 (1797); Encycl. Méth. Bot. **4**, 2: 696 (1798) " gymnorhiza ".—Peters, Reise Mossamb., Bot. **1**: 72 (1861).—Sond. in Harv. & Sond., F.C. **2**: 514 (1862).—Engl., Pflanzenw. Ost-Afr. **A**: 8, 12; op. cit. **B**: 338, 404, 408; op. cit. **C**: 287 (1895).—Sim, For. Fl. Port. E. Afr.: 65, t. 70 (1909).—Marloth, Fl. S. Afr. **2**, 2: 221, 222, fig. 141 & 141b (1925).—Evans in Mem. Bot. Surv. S. Afr. **15**: 7 (1936).—Brenan, T.T.C.L.: 471 (1949).—Arènes in Fl. Madag. **150**: 34, fig. 10 (1954).—Pedro & Barbosa in Mem. e Trab. Centr. Invest. Cient. Algod. **23**, 2: 99, 104, 215 (1955).—J. Lewis in F.T.E.A., Rhizophoraceae: 6, fig. 3 (1956).—Dale & Greenway, Kenya Trees & Shrubs: 395 (1961).—Gomes e Sousa in Mem. Inst. Invest. Agron. Moçamb. **1**, 1: 54, 55 (1966); op. cit. **1**, 2: 600, t. 172 (1967).—J. H. Ross, Fl. Natal: 258 (1972). TAB. **27**. Type: Tab. 31, Rheede, Hort. Malab. **6** (1686) sub *Candel.*

Rhizophora gymnorhiza L., Sp. Pl. **1**: 443 (1753). Type as above.
Rhizophora conjugata L., loc. cit. Type: Ic. no. 279, Hermann Herb. (BM).
Bruguiera capensis Bl. in Mus. Bot. Lugd.-Bat. **1**: 137 (1850). Type from S. Africa.
Bruguiera cylindrica sensu Oliv. in F.T.A. **2**: 409 (1871) saltem pro parte.—Macnae & Kalk, Nat. Hist. Inhaca I., Moçamb.: 13, 119, 150, fig. 5c & 11b, t. 3 fig. 1 (1958) non Blume (1850).

Shrub or tree up to 7(15) m. high; branches terete, smooth with subcircular scars of the petioles of the fallen leaves and annular scars of the fallen stipules; bark fibrous, light brown, reddish-brown or grey. Leaf-lamina usually 6–12 × 2–6 cm., narrowly to broadly elliptic or sometimes obovate-elliptic; apex acute, subacute or shortly acuminate; margins somewhat revolute; base narrowed into the petiole, cuneate or subcuneate; midrib rather prominent below, lateral nerves slightly prominent above and just visible to obscure beneath; petiole usually 1–3 cm. long; stipules up to 4 × 1 cm., entire and involute, acute at the apex and truncate at the base. Flowers solitary, peduncles c. 1 cm. long, ± terete, arching; flower-buds c. 27 × 7 mm., narrowly ellipsoid, acute, strongly ribbed, glabrous. Calyx pinkish-green to reddish-brown; calyx-tube 1–2 cm. long, obconical; calyx-lobes (10)12(14), 1·5–2 cm. long, triquetrous, narrow, very acute at the apex. Petals c. 1·5 cm. long, ± oblong, conduplicate and costate inside, 2-lobed at the apex (the sinus with a median seta c. 4 mm. long and the lobes c. 5 mm. long, obtuse, apically 2–3-fimbriate), white or cream, soon turning brown, sparsely hirsute at the margins and densely hirsute at the base, coriaceous. Stamens (20)24(28), paired, oppositepetalous; filaments 6 or 8 mm. long; anthers c. 6 mm. long, mucronate at the apex. Ovary c. 4 mm. long, (2)3(4)-locular; style c. 20 mm. long, shortly (2)3(4)-lobed at the apex. Fruit a turbinate berry crowned by the calyx-lobes, unilocular, 1-seeded. Hypocotyl up to 15 cm. long and 1·2 cm. in

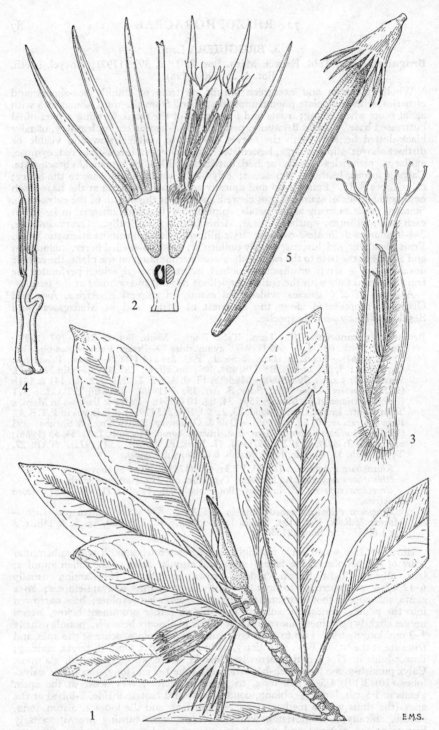

E.M.S.

Tab. 27. BRUGUIERA GYMNORRHIZA. 1, flowering branch with some leaves removed (× ⅔);
2, longitudinal section of flower with all but 2 petals and stamen-pairs removed (× 3);
3, petal with enclosed stamen-pair (× 6); 4, stamen-pair (× 6); 5, fruit and hypocotyl
(× ⅔), 1 & 5 from *Greenway* 5287, 2–4 from *Johnson* 22. From F.T.E.A.

diameter while on the tree, only narrowing a little at both extremities, shallowly and bluntly longitudinally ridged.

Mozambique. N: Cabo Delgado Distr., Porto Amélia, Metuge, Bandar, fl. & fr. 19.xii.1963, *Torre & Paiva* 9617 (COI; J; K; LISC; LMU; LUA; PRE; SRGH). Z: Macuze R. mouth, fr. 23.viii.1962, *Wild & Pedro* 5893 (K; LISC; SRGH). MS: Beira Distr.) near the port of Beira, fl. iii.1970, *Biegel* 3995 (SRGH). SS: Inhambane Distr., Bazaruto I., Gengareme Point, fr. 29.x.1958, *Mogg* 28726 (BM; K; PRE; SRGH). LM: Maputo, Inhaca I., fl. & fr. 6.vi.1970, *Correia & Marques* 1670 (COI; LISC; LMU; SRGH).

Widely distributed along the western shores of the Pacific Ocean, Ryukyu Is., Micronesia, Australia, Polynesia and in the Indian Ocean; along the E. African shores from Ethiopia to S. Africa (Natal and the Cape Prov.). Evergreen woodlands and thickets along the intertidal mud-flats of sheltered shores, estuaries and inlets, mainly towards the seaward side of mangrove formation.

4. CASSIPOUREA Aubl.

Cassipourea Aubl., Pl. Guian. **1**: 528 (1775).—Alston in Kew Bull. **1925**: 241 (1925).
Richaeia Thou., Gen. Nov. Madag.: 25 (1806).
Weihea Spreng., Syst. Veg. ed. 16, **2**: 559, 594 (1825) *nom. conserv.*
Dactylopetalum Benth. in Journ. Linn. Soc., Bot. **3**: 79 (1858).

Small shrubs to tall trees, without aerial roots. Leaves decussate or rarely verticillate, petiolate, stipulate; leaf-lamina ± elliptic to obovate or ovate, rarely circular, serrate (especially from near the middle towards the apex) to entire, leathery to membranous, glabrous to densely tomentose. Inflorescence 1–many-flowered, fasciculate, lax or congested, axillary. Flowers bisexual, pedicellate (pedicels articulate, with or without a pair of minute connate bracteoles, inserted usually at their base) to sessile. Calyx pilose or glabrous; calyx-tube ± campanulate, sometimes ± adnate to the ovary; calyx-lobes 4–7. Petals linear or spathulate, divided above (usually laciniate), clawed, glabrous to pilose, inflexed in bud, inserted on the margin of the disk or at the bottom of the calyx-tube (at the base and outside the disk). Disk deep, 8–45-lobed or low, fairly conspicuously lobed (or ? absent). Stamens 8–45, inserted on the margin of, or inside or outside, the disk; filaments filiform; anthers glabrous or rarely pubescent, dehiscing longitudinally. Ovary superior to subinferior, (1)2–4(5)-locular, each locule 2-ovulate, hairy to glabrous; style simple (very rarely divided), filiform to ± thick, minutely lobed to entire at the apex, often persistent. Fruit a fleshy or leathery, septicidally dehiscent capsule. Seeds 2–4, compressed or angular, arillate; testa leathery; endosperm fleshy; embryo straight; cotyledons flat.

A genus with c. 70 species, widespread in tropical Africa, Asia and America, including a few useful timber trees.

Calyx-tube elongate-campanulate; calyx-lobes shorter than calyx-tube, ± erect, glabrous
  or glabrescent outside; ovary 2(3)-locular; leaves glabrous (subgenus *Dactylopetalum*)
                                                     1. *gummiflua*
Calyx-tube not elongate-campanulate; calyx-lobes longer than calyx-tube, ± reflexed,
  ± hairy outside; ovary 3(4)-locular; leaves hairy at least on the midrib (subgenus
  Weihea):
  Leaves ± densely tomentose to pubescent on the inferior surface and tomentulose to
    glabrescent on the superior surface:
    Flowers with short pedicels (up to 4 mm. long), sericeous-puberulous; calyx-lobes
      4–6 mm. long; stamens c. 20; fruit white-hirsute at the apex - 2. *gossweileri*
    Flowers with usually long pedicels (up to 12 mm. long), hirsute; calyx-lobes
      (5)7–10(17) mm. long; stamens more than 25, up to 40(45); fruit ± densely
      puberulous - - - - - - - - - - 3. *mollis*
  Leaves ± sparsely, obscurely appressed-pubescent to glabrous on the inferior surface
    and usually glabrous on the superior surface:
    Flowers usually in dense axillary fascicles; leaves obovate - - 4. *obovata*
    Flowers solitary to few; leaves lanceolate, ± elliptic to obovate or ovate:
      Ovary densely sericeous; leaves elliptic to obovate, obtuse to acute and ± long-
        acuminate at the apex, entire or sparsely and scarcely serrate 5. *euryoides*
      Ovary glabrous to densely hairy; leaves lanceolate, ± elliptic to obovate or ovate,
        rounded, obtuse to acute and usually not or slightly acuminate, entire to
        serrate:

Leaves obovate, ovate, oblong or ± elliptic, entire or scarcely serrate towards
the apex, somewhat revolute; stipules ovate, subcordate to truncate at the
base; ovary glabrous or slightly hairy to hirsute at the apex
6. *mossambicensis*
Leaves elliptic, oblong or rarely obovate, serrate usually near the middle towards
the apex, rarely entire, not or very slightly revolute; stipules ovate,
lanceolate to oblong, truncate or subrounded at the base; ovary glabrous to
densely hirsute:
Style up to 1·5 mm. long; ovary densely hirsute; leaves discolorous; pedicels
2 mm. long - - - - - - - - 7. *fanshawei*
Style 3–5 mm. long; ovary glabrous to densely hirsute; leaves usually
concolorous; pedicels 2–5(9) mm. long - - - - 8. *malosana*

1. **Cassipourea gummiflua** Tul. in Ann. Sci. Nat. Bot., Sér. 4, **6**: 123 (1856).—Ind.
Kew. **2**: 1273 (1895) " gummifera ".—Alston in Kew Bull. **1925**: 272 (1925).—
J. Lewis in Kew Bull. **1955**: 147 (1955); in F.T.E.A., Rhizophoraceae: 15 (1956).—
Gomes e Sousa in Mem. Inst. Invest. Agron. Moçamb. **1**, 2: 601 (1967)
" gummiflora ". Type from Madagascar.

Tree (3)5–20(25) m. high, evergreen; branches terete, shortly appressed-
puberulous to glabrescent when young, glabrous with age; bark brown or grey,
smooth with scattered raised lenticels. Leaves decussate (sometimes verticillate)
on thick nodes; leaf-lamina (10·5)12–15 × (4)5–7·5 or 5–14(21) × 3–5(10) cm.,
elliptic-lanceolate, narrowly to broadly elliptic or oblong or rarely ovate or obovate,
glabrous, ± leathery; apex rounded to acuminate, obtuse or ± acute; margins
entire, shallowly sinuate to bluntly and shallowly remotely serrate usually above the
middle towards the apex; base rounded to cuneate, ± acute at its extremity;
petiole (6)8–15(17) mm. long, thick, pubescent when young and glabrous with age;
stipules 2–4 mm. long, ± triangular, thick, slightly puberulous to glabrous, soon
caducous. Inflorescences ± dense and congested fascicles. Flowers 4–5- or
5–6-merous; pedicels 1·5–4 mm. long, glabrous, articulate at or slightly above the
middle; bracteoles minute, puberulous to glabrous, or absent. Calyx glabrous;
calyx-tube 2–4 mm. long, elongate-campanulate; calyx-lobes 0·5–1·5 mm. long,
semi-circular to ± deltate, rounded or obtuse (rarely acute) and usually puberulous
at the apex, suberect. Petals linear-spathulate, laciniate at the apex, white,
glabrous, inserted outside at the base of the disk. Disk c. 1 mm. high, entire to ±
deeply lobed. Stamens 8–10(12) or (8)10–12 inserted along and outside below the
margin of the tubulous-campanulate ± lobed disk; filaments 4·5–7 mm. long (the
oppositipetalous longer than the alternipetalous); anthers 0·7–1·1 mm. long, ±
ellipsoid. Ovary 1–1·5 mm. long, obovoid to ovoid, 2(3)-locular, glabrous, sparsely
pubescent above the middle to regularly pubescent or densely white-hirsute mainly
above the middle, superior or semi-inferior; style 3–6 mm. long, glabrous, slightly
2(3)-lobed. Capsule up to 7 mm. long, ellipsoid, ± spherical to obovoid, glabrous
or puberulous mainly above the middle.

Var. **ugandensis** (Stapf) J. Lewis in Kew Bull. **1955**: 159 (1955); in F.T.E.A.,
Rhizophoraceae: 15 (1956).—Dale & Greenway, Kenya Trees & Shrubs: 396
(1961).—Mendes in C.F.A. **4**: 40 (1970). Type from Uganda.
Dactylopetalum ugandense Stapf in Journ. Linn. Soc., Bot. **37**: 515 (1906). Type
as above.
Cassipourea gummiflua sensu F. White, F.F.N.R.: 275 (1962).

Leaf-lamina (10·5)12–15 × (4)5–7·5 cm., usually narrowly to broadly elliptic;
apex rounded to acuminate or subacute; margins entire to shallowly sinuate or
bluntly and shallowly serrate usually above the middle; base rounded to cuneate
(acute at its extremity). Pedicels up to 2(3) mm. long. Calyx-lobes and petals
4–5(6). Stamens 8–10(12). Ovary glabrous, or very sparsely and irregularly
hirsute above. Capsule glabrous.

Zambia. W: Kitwe Distr., fl. 25.iii.1955, *Fanshawe* 2220 (EA; K; SRGH); Solwezi
Distr., near Solwezi Boma, fr. 12.ix.1952, *White* 3241 (BM; FHO; K; PRE).
Also in Cameroon, Angola (Malange), Zaire, Uganda and Kenya. Moist forest, gallery
forest, upland forest, rain forest and swamp forest, 1200–1500 m.

Var. **verticillata** (N. E. Br.) J. Lewis in Kew Bull. **1955**: 158 (1955); in F.T.E.A.,
Rhizophoraceae: 16 (1956).—Gomes e Sousa in Mem. Inst. Invest. Agron. Moçamb.
1, 2: 601, fig. 173 (1967).—J. H. Ross, Fl. Natal: 258 (1972). Type from S. Africa
(Natal).

Cassipourea verticillata N. E. Br. in Kew Bull. **1894**: 5 (1894).—Wood & Evans, Natal Pl. **3**, 4: 7, t. 276 (1899), republ. (1970).—Sim, For. Fl. Port. E. Afr.: 67 (1909).—Alston in Kew Bull. **1925**: 275 (1925).—Gomes e Sousa, tom. cit.: 602, 603 (1967). Type as above.

Cassipourea redslobii Engl., Bot. Jahrb. **54**: 368 (1917); Pflanzenw. Afr. **3**, 2: 673, fig. 299 (1921).—Alston, loc. cit.—Brenan, T.T.C.L.: 471 (1949). Type from Tanzania.

Cassipourea verticillata forma *decussata* Engl., tom. cit.: 369 (1917).—Brenan, op. cit.: 472 (1949). Type from Tanzania.

Leaf-lamina 5–14(21) × 3–5·5(10) cm., usually narrowly elliptic to more rarely broadly elliptic; apex obtuse to acuminate, rarely rounded; margins usually shallowly sinuate or bluntly and shallowly serrate above the middle, rarely entire; base cuneate to rarely rounded or truncate (acute at its extremity). Pedicels (1·5)2–3(4) mm. long. Calyx-lobes and petals (4)5–6. Stamens (8)10–12. Ovary pubescent especially above the middle. Capsule puberulous above (shortly and obscurely).

Rhodesia. E: Umtali Distr., Vumba Mts., fl.-buds 22.ii.1955, *Chase* 5484 (BM; COI; K; LISC; PRE; SRGH); Inyanga Distr., fr. 8.xi.1967, *Müller* 697 (K; PRE; SRGH). **Mozambique.** MS: Beira Distr., Cheringoma, 70 km. from Beira, fr. 9.vii.1941, *Torre* 3038 (COI; K; LISC; LMA; MO). LM: Maputo, Ponta do Ouro, fl. 18.xi.1944, *Mendonça* 2942 (BR; LISC; LMU; MO).
Also in Cameroun, Tanzania, Seychelles and Mascarene Is. and S. Africa (Natal). Upland forest, gallery forest, mixed dense forest, moist forest and rain forest, 1400–2000 m.

Note. Several sterile gatherings from Zambia (*Chapman* 2321 (FHO)), from Malawi (Vipya, *Chapman* 2226 (SRGH) and *Müller* 1586 (K; SRGH)) and from Mozambique (*Müller & Pope* 2043 (LISC) and *S. Hepker* s.n. (SRGH)) probably belong here.

2. **Cassipourea gossweileri** Exell in Journ. Bot., Lond. **66**, Suppl. Polypet.: 162 (1928).—Gossw. & Mendonça, Cart. Fitogeogr. Angola: 158 (1939).—Mendes in C.F.A. **4**: 39 (1970). Type from Angola.
Cassipourea sp. 1 sensu F. White, F.F.N.R.: 275 (1962).

Shrub or small tree up to 7·5 m. high, sometimes with scrambling branches; branches terete, branching ± at right-angles, smooth or scarcely rough, pubescent at first, later glabrescent; bark grey to brown, with lenticular streaks of paler grey. Leaf-lamina 1·5–4·5(7) × 1–2·5(4) cm., elliptic, leathery to papyraceous; apex rounded, obtuse or subacute to acute; margins entire, serrulate to serrate from the middle to the apex; base rounded to cuneate; venation conspicuous on both surfaces, the midrib prominent below; superior surface ± sparsely and minutely pubescent to glabrescent; inferior surface more densely pubescent; petiole 1–5 mm. long, sericeous-pubescent; stipules 3–5 mm. long, broadly ovate, subacute to obtuse at the apex, ± densely pubescent on abaxial surface. Flowers solitary, 5-merous; pedicels sericeous-pubescent, articulate above the middle; bracteoles 1–2·5 mm. long, conchiform, rounded at the apex and truncate at the base, ± pubescent outside. Calyx sericeous-pubescent outside and sparsely puberulous inside; calyx-tube 1–2 mm. long; calyx-lobes 4–6 mm. long, ovate-lanceolate or lanceolate. Petals c. 10 mm. long, spathulate, laciniate above, glabrous. Stamens c. 20; filaments up to 7 mm. long; anthers c. 1·5 mm. long, oblong-ellipsoid. Ovary c. 2 mm. long, ovoid, 3-locular, densely pubescent to pubescent at the apex only; style up to 5 mm. long, pubescent near the base, up to the middle or sometimes near its apex. Fruit up to 11 × 8 mm., ovoid to sub-spherical, densely sericeous-pubescent to sparsely pubescent (more densely near its apex).

Zambia. C: Lusaka Distr., Mt. Makulu, immat. fr. 26.xii.1959, *White* 6020 (FHO; K; SRGH). S: Kalomo Distr., Fulwa Forest, c. 6·4 km. SW. of Buwe Pool on Barotse Cattle Cordon Road, fl. 28.xi.1962, *Bainbridge* 609 (FHO; SRGH).
Also in Angola. Woodlands, savanna woodlands and tree-savannas, sometimes rupicolous or on sandy soils, up to c. 1200 m.

Note. Two specimens (*Martin* 504/33 (K) and *Mitchell* 17/13 (FHO; K; LISC)) present leaves larger than usual for this species, respectively up to 7·5 × 3·5 cm. and up to 8 × 4·5 cm.; we believe these leaves developed during particularly heavy rainy seasons.

3. **Cassipourea mollis** (R. E. Fr.) Alston in Kew Bull. **1925**: 257 (1925).—Brenan,
    T.T.C.L.: 472 (1949).—J. Lewis in F.T.E.A., Rhizophoraceae: 9 (1956).—
    F. White, F.F.N.R.: 276 (1962). Syntypes from Zambia: Mbala Distr., Mbala,
    R. E. Fries 1251 (LD, syntype) and Kalambo, between Mbala and Bismarckburg,
    R. E. Fries 1251a (B, syntype †; BM & K, photosyntypes).
        Weihea mollis R. E. Fr., Wiss. Ergebn. Schwed. Rhod.-Kongo-Exped. **1**: 165,
    t. 14 (1914). Syntypes as above.

Shrub or small tree up to 7 m. high; young branchlets puberulous to ± densely
tomentose, older branches glabrescent to glabrous; bark grey to brown, rough,
longitudinally ± deeply ridged, corky on the trunk and on the older branches.
Leaf-lamina 3–11 × 1·5–7 cm., ± broadly ovate or elliptic to oblong-elliptic,
subcircular to obovate, leathery to membranous; apex subacuminate, acute, subacute, obtuse, rounded or rarely retuse; margins entire or sharply serrate; base
cuneate, subcuneate, obtuse, truncate or rarely subcordate; superior surface
tomentulose to glabrescent, with conspicuous venation; inferior surface ± densely
tomentose to pubescent, mainly along the nerves which are prominent; petiole
usually 2–8 mm. long, sericeous-hirsute; stipules up to c. 9 mm. long, ±
triangular, brown or green, sericeous-tomentose, caducous. Inflorescences 1(few)-
flowered. Flowers (4)5(6)-merous; pedicels usually 3–8(rarely 14) mm. long,
hirsute, articulate above or rarely at the middle; bracteoles up to 5 mm. long,
conchiform, truncate at the apex, ± truncate at the base, densely appressed
pubescent to puberulous. Calyx greenish, ± sericeous-tomentose outside and ±
densely appressed pubescent to puberulous inside; calyx-tube 2–4 mm. long; calyx-
lobes (5)7–10(17) mm. long, ± broadly lanceolate. Petals 8–14 mm. long, spathulate,
laciniate above, white or cream, glabrous. Stamens (26)30–40; filaments 6–9 mm.
long; anthers up to 2 mm. long, oblong-ellipsoid. Ovary 1·5–3·5 mm. long, ovoid,
3(4)-locular, very densely and regularly hairy, superior; style up to c. 5–6 mm.
long, ± pubescent near the base or up to the middle or throughout, with 3(4)
stigmatic lobes. Fruit up to 14 mm. long, ovoid to ± spherical, ± densely
puberulous. Seeds dark brown to black; aril orange-coloured.

Zambia. N: Mbala Distr., fl. 19.ix.1949, *Bullock* 1011 (EA; K; LISC; SRGH).
C: Miengwe, fl. 3.xi.1955, *Fanshawe* 2582 (K; LISC). **E**: Fort Jameson Distr., Mfumu,
Asamfa, fr. 6.i.1959, *Robson* 1051 (BM; K; LISC; PRE; SRGH). **Malawi. N**:
Mzimba Distr., Mzimba, Katondo Road, fl. & immat. fr. 5.xi.1968, *Salubeni* 1194 (K;
LISC; SRGH). **C**: Kasungu Distr., Kasungu National Park, fl. 6.xii. 1970, *Hall-Martin*
1040 (SRGH). **Mozambique. T**: Cabora-Bassa, 5 km. from the dam, fr. 19.ii.1968,
Torre & Correia 17734 (COI; K; LISC; LMA).
    Also in Tanzania. Woodlands (mainly *Brachystegia*-woodland), savanna woodlands
and, sometimes, tree-savannas and thickets, on argillaceous-sandy, sandy and stony soils,
sometimes rupicolous, 600–1800 m.

4. **Cassipourea obovata** Alston in Kew Bull. **1925**: 261 (1925). Type: Mozambique,
    Cabo Delgado Distr., Macomia, near mouth of Msalu (Messalo) R., *Allen* 89 (2
    specimens) (K, syntypes).

Tree 6 m. high; branches pale grey, rough, pubescent when young, glabrous
with age. Leaf-lamina 3–5 × 1·8–3·5 cm., obovate, leathery; apex rounded,
truncate, obtuse or shortly acuminate and retuse at its extremity; margins entire
and somewhat revolute; base cuneate to ± attenuate into the petiole; midrib
straight, prominent below, and reticulation slightly prominent on both surfaces;
superior surface glabrous; inferior surface very sparsely appressed-pubescent
(except along the midrib); petiole 3–5 mm. long, glabrous or glabrescent; stipules
3 mm. long, broadly ovate, rounded at the apex and subcordate-truncate at
the base, externally glabrescent or glabrous (pubescent towards the apex).
Inflorescences ± dense and congested fascicles. Flowers 5–6-merous; pedicels
1·5–2 mm. long, ± pubescent, articulate near the apex; bracteoles c. 1·5 mm. long,
± conchiform, truncate or rounded (sometimes lacerate) at the apex, truncate at
the base, glabrous or glabrescent abaxially and puberulous at the margins. Calyx
appressed-pubescent outside and glabrous inside; calyx-tube c. 1 mm. long;
calyx-lobes 4–5 mm. long, lanceolate. Petals c. 5 mm. long, spathulate, laciniate
above, glabrous. Stamens c. 17; filaments 3–4 mm. long; anthers c. 1 mm. long,
oblong-ellipsoid. Ovary 1·5 mm. long, ovoid, 3–4-locular, sparsely hairy (more
densely towards the apex), semi-inferior; style 3–4 mm. long, hairy at and near the

base, persistent, with 3–4 stigmatic surfaces at the apex. Fruit unknown, stated to be edible.

Mozambique. N: Cabo Delgado Distr., iviacomia, near mouth of Msalu (Messalo) R., fl. xii.1911, *Allen* 89 (K).
Known only from the type collection, growing on sandy hills near the sea.

5. **Cassipourea euryoides** Alston in Kew Bull. **1925**: 254 (1925).—J. Lewis in Kew Bull. **1955**: 143 (1955); in F.T.E.A., Rhizophoraceae: 11 (1956).—Dale & Greenway, Kenya Trees & Shrubs: 396 (1961). Type from Kenya.
   Cassipourea honeyi Alston, loc. cit.—Gomes e Sousa in Mem. Inst. Invest. Agron. Moçamb. **1**, 2: 602 (1967). Type: Mozambique, Manica e Sofala, Beira Distr., Dondo, Siluvo Hills (? Chiluvo Mts.), *Honey* 757 (K, holotype).

Shrub or tree up to 15 m. high; young stem and branches very shortly and rather sparsely appressed-hirsute, glabrous with age; bark brown to grey, ± smooth, somewhat corky. Leaf-lamina mostly 4·5–7 × 2–3 cm., elliptic to obovate, papyraceous to leathery; apex obtuse to acute and ± acuminate; margins entire or remotely serrate (specially above the middle); base acute, ± long-cuneate; midrib ± curving towards the apex, lateral nerves and reticulation moderately conspicuous; superior surface shining and glabrous; inferior surface mat and very sparsely appressed pubescent to glabrous; petiole 1–6 mm. long, hirtellous to glabrous; stipules 3–5 mm. long, ovate to lanceolate, obtuse at the apex and truncate at the base, appressed pubescent to glabrescent outside, soon caducous. Inflorescences 1–3-flowered, ± lax. Flowers 4–5-merous; pedicels 1–2 mm. long, shortly hirsute, articulate near the apex; bracteoles 1–2 mm. long, conchiform, obovate, truncate to rounded at the apex, truncate at the base, slightly appressed pubescent on abaxial surface. Calyx densely to appressed pubescent outside and sparsely puberulous to appressed pubescent inside. Calyx-tube 1–1·5 mm. long; calyx-lobes 3–4 mm. long, narrowly triangular to ± ovate. Petals 3–6 mm. long, spathulate, laciniate above, white, glabrous. Stamens c. 20; filaments 2–4 mm. long; anthers 1 mm. long, ellipsoid-oblong. Ovary c. 1·5 mm. long, subspherical, 3–4-locular, densely white- to ± yellowish-sericeous throughout; style 2–2·5 mm. long, sparsely pubescent to glabrous, persistent. Fruit 5 mm. long, subspherical to ovoid, densely velvety with intermixed scattered longer hairs near the apex.

Rhodesia. E: Umtali Distr., Burma Valley, fl.-buds 30.xii.1959, *Chase* 7236 (BM; K; LISC; SRGH). S: Chibi Distr., st. 30.xii.1962, *Moll* 468 (K; SRGH).
Mozambique. N: Mozambique Distr., Nampula, Nassapo Mt., 23 km. from Nampula to Meconta, immat. fr. 13.i.1964, *Torre & Paiva* 9934 (COI; K; LISC; LMU; MO). Z: 10 km. from Gilé Mt., fl. 21.xii.1967, *Torre & Correia* 16691 (J; LISC; LMA; LUAI). MS: Vila Pery Distr., Chimoio, Chindaza Mts. near Tembe, fl. & fr. 14.iii.1948, *Garcia* 613 (BR; COI; K; LISC; LMA; LUA; SRGH).
Also in Kenya and Tanzania. Dry evergreen and deciduous forests, woodlands, savanna woodlands and tree-savannas, sometimes amongst rocks; up to c. 800 m.

6. **Cassipourea mossambicensis** (Von Brehm.) Alston in Kew Bull. **1925**: 261 (1925).—Mendes in C.F.A. **4**: 38 (1970).—J. H. Ross, Fl. Natal: 258 (1972). TAB. **28**. Type: Mozambique, Chidôro, Mayanda, Mossanzi, c. 60 m., *Johnson* 104 (K, holotype).
   Weihea mossambicensis Von Brehm. in Engl., Bot. Jahrb. **54**: 361 (1917). Type as above.

Small shrub 2–3(5) m. high; branches terete, grey or brownish, ± smooth, hairy when young, glabrous with age. Leaf-lamina 3–7·5(10) × 2–4·5 cm., ± elliptic or obovate, leathery; apex rounded, truncate, obtuse or shortly obtuse-acuminate; margins entire to sparsely serrate (2, 3 or more teeth) towards the apex and somewhat revolute; base ± cuneate to rounded; midrib straight, prominent below, the reticulation conspicuous on both surfaces; superior surface ± glossy, glabrous; inferior surface mat, sparsely appressed-pubescent to glabrate; petiole 3–6(8) mm. long, hairy to glabrescent; stipules 2·5–6 mm. long, ovate to ovate-triangular, obtuse to acute at the apex, subcordate-truncate at the base, ± pubescent to glabrous (with hairy margins towards the apex) outside, soon caducous. Inflorescences 1–3-flowered, ± lax. Flowers 5-merous; pedicels 1·5–4 mm. long, sparsely hairy, articulate near the apex; bracteoles c. 2 mm. long, conchiform, ± rounded at the apex, truncate at the base, slightly pubescent to

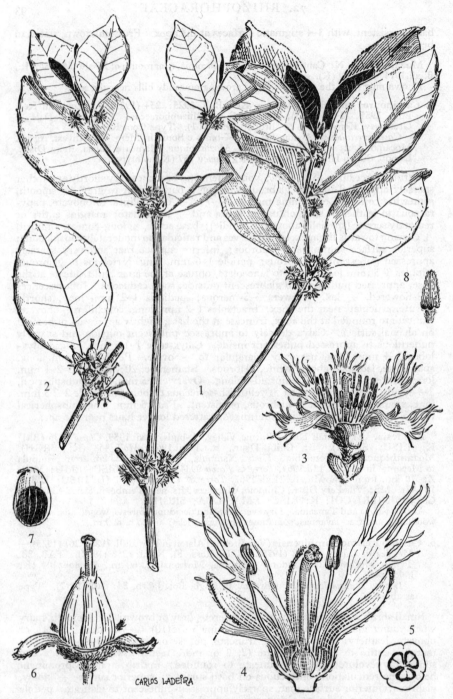

Tab. 28. CASSIPOUREA MOSSAMBICENSIS. 1, flowering branch (×1); 2, part of fruiting
branch (×1); 3, flower, side view, with pedicel and pair of bracts (×4); 4,
longitudinal section of flower (all but 2 stamens removed) (×8); 5, cross-section of
ovary (×16); 6, fruit with persistent calyx and style (×4); 7, seed with aril (×4);
8, stipule (×2), 1, 3–5, 8 from *Correia & Marques* 1844, 2, 6–7 from *Schlechter* 11656.

glabrous (except along the margins) abaxially and glabrous adaxially. Calyx green, sericeous outside and glabrous inside; calyx-tube 2 mm. long; calyx-lobes 4–5 mm. long, lanceolate or oblong-lanceolate. Petals 6–7 mm. long, spathulate, laciniate above, white to yellow, glabrous. Stamens 15–20; filaments 4–5 mm. long; anthers 1·5 mm. long, ellipsoid. Ovary up to 2 mm. long, ovoid to sub-spherical, 3–4-locular, glabrous or slightly and sparsely hairy to hirsute at the apex; style 5 mm. long, glabrous to pubescent near the base and puperulous towards the apex, persistent, sometimes lobulate, with 3–4 stigmatic surfaces 0·5 mm. long at the apex. Fruit c. 5 mm. long, ovoid to spherical, glabrous or slightly and sparsely hairy at the apex.

Mozambique. N: Cabo Delgado Distr., Mechanga, 5 km. from Mocímboa da Praia, fr. 6.vi.1961, *Gomes e Sousa* 4686 (COI; FHO; K; LMA; SRGH). SS: Inhambane Distr., between Mocodoene and Funhalouro, Xilaule forest, fl. 28.x.1947, *Barbosa* 609 (COI). LM: Maputo, between Bela Vista and Porto Henrique, fl. 19.xi.1944, *Mendonça* 2961 (COI; K; LISC; LUAI; MO; SRGH).

Also in S. Africa (Natal). Coastal, sublittoral, dry semi-deciduous forests, thickets, woodlands, savanna woodlands or tree-savannas, usually on sandy or sandy-argillaceous soils, up to 200 m.

Note. A specimen from Palma, Nangade (*Mendonça* 984 (K; LISC; LMA; MO; SRGH)) has 5(6)-merous flowers with pedicels 5–6 mm. long and narrowly obovate to oblong leaves which are c. ⅓ as wide as long; it seems to represent an extreme form of this species.

7. **Cassipourea fanshawei** Torre & Gonçalves in Garcia de Orta, Sér. Bot. **3**: 49, t. 1 (1976). Type: Zambia, Musondwa, *Fanshawe* 4897 (FHO, isotype; K, holotype).

Tree 4·5 m. high; branches terete, smooth, puberulous when young, glabrous with age; bark dark grey. Leaf-lamina 4–8 × 1·5–3 cm., elliptic to rarely sub-obovate, discolorous, leathery; apex obtuse and slightly acuminate; margins sparsely and inconspicuously serrate from below the middle towards the apex; base subcuneate, cuneate or attenuate into the petiole; superior surface green, shining and glabrous (except on the midrib and near the margins which are sparsely pubescent), reticulation dense and ± conspicuous; inferior surface green-yellowish, mat, sparsely appressed-pubescent (more densely so on midrib), midrib prominent, reticulation close and moderately prominent; petiole 3–5 mm. long, pubescent; stipules up to 4 mm. long, ovate, obtuse or rounded at the apex and truncate at the base, appressed-pubescent abaxially and puberulous adaxially. Inflorescences mostly 2-flowered. Flowers 4-merous; pedicels 2 mm. long, hirsute-pubescent, articulate near the apex; bracteoles 1 mm. long, conchiform, rounded at the apex, truncate at the base, puberulous abaxially and glabrous adaxially (pubescent at the base). Calyx appressed-pubescent outside and glabrous inside; calyx-tube c. 1·5 mm. long; calyx-lobes 3 mm. long, ± ovate, acute at the apex. Petals 8 mm. long, spathulate, laciniate above, glabrous, inserted on the margin of the disk outside the stamens. Disk hypogynous, shallowly cupuliform and flat at the bottom, weakly developed. Stamens 16–18, inserted on the margin of the disk; filaments 3 mm. long; anthers 1 mm. long, oblong-ellipsoid. Ovary 1–1·5 mm. long, subovoid, 3-locular, densely hirsute, slightly semi-inferior; style up to 1·5 mm. long, thick, white, glabrous, persistent, with 3 stigmatic surfaces. Fruit unknown.

Zambia. N: Musondwa, fl. 8.x.1958, *Fanshawe* 4897 (FHO; K; LISC). Known only from the type collection. Thickets.

8. **Cassipourea malosana** (Bak.) Alston in Kew Bull. **1925**: 258 (1925).—Burtt Davy & Hoyle, N.C.L.: 83 (1936).—Brenan, T.T.C.L.: 472 (1949).—J. Lewis in Kew Bull. **1955**: 144 (1955); in F.T.E.A., Rhizophraceae: 11 (1956).—Dale & Greenway, Kenya Trees & Shrubs: 398, fig. 78 (1961). Type: Malawi, Zomba Distr., Malosa Mt. near Zomba, *Whyte* s.n. (K, holotype).
Weihea malosana Bak. in Kew Bull. **1897**: 267 (1897). Type as above.
Weihea gerrardii Schinz in Bull. Herb. Boiss. **5**: 867 (1897).—Swynnerton in Journ. Linn. Soc., Bot. **40**: 67 (1911). Type from S. Africa (Natal).
Cassipourea gerrardii (Schinz) Alston in loc. cit.—Burtt Davy, F.P.F.T.: 250 (1926).—J. H. Ross, Fl. Natal: 258 (1972). Type as *Weihea gerrardii.*
Weihea elliottii Engl., Bot. Jahrb. **40**: 52, t. 1A (1907). Type from Kenya.
Cassipourea congoensis sensu Alston in Kew Bull. **1925**: 259 (1925) saltem pro parte

et sensu J. Lewis in F.T.E.A., Rhizophoraceae: 12, fig. 4(6) (1956) non DC., Prodr. **3**: 34 (1828) *nom. nud.*

  Cassipourea elliottii (Engl.) Alston, tom. cit.: 260, cum fig. p. 243 (1925).—Burtt Davy & Hoyle, loc. cit.—Evans in Mem. Bot. Surv. S. Afr. **22**: 283 (1948).— Brenan, loc. cit. Type as for *Weihea elliottii.*

Shrub or tree up to 25 m. high; young stem ± appressed-pubescent to glabrescent (first year twigs sparsely hirsute); branches hairy when young and glabrous with age; bark greyish-yellow, grey or brown. Leaf-lamina 3–10 × 1–5 cm., oblanceolate, lanceolate, narrowly to broadly elliptic, oblong, ovate or obovate; apex subacute, obtuse or short- to long-acuminate (rarely acute, rounded or retuse); margins serrate or rarely entire; base cuneate or obtuse, rarely rounded; superior surface ± shining, glabrous, with conspicuous venation; inferior surface mat, sparsely appressed-pubescent to glabrous (except on the midrib) and with ± prominent venation; petiole 2–7 mm. long, pubescent to glabrous; stipules truncate to subrounded at the base, abaxial surface densely appressed-pubescent to pubescent at the apex only, adaxial surface glabrous, soon caducuous. Inflorescences 1–4(6)-flowered, lax to ± congested. Flowers 4–5(6)-merous; pedicels 2–5(9) mm. long, pubescent, articulate near the apex; bracteoles 1·5–2·5 mm. long, conchiform, rounded or lacerate at the apex and truncate at the base. Calyx slightly pubescent to sericeous-pubescent outside and glabrous to slightly puberulous inside; calyx-tube 0·5–2 mm. long; calyx-lobes 3–6 mm. long, lanceolate, oblong or ovate-lanceolate. Petals 4–8 mm. long, spathulate, laciniate above, glabrous. Stamens 15–20; filaments 3–6 mm. long; anthers c. 1 mm. long, narrowly to broadly ellipsoid-oblong. Ovary 1–2 mm. long, ovoid to ± spherical, 3–4-locular, glabrous to densely hirsute; style 3–5 mm. long, glabrous to pubescent mainly towards the base, persistent. Fruit 6–10 mm. long, ± ellipsoid, ovoid to subspherical, sericeous-pubescent to glabrous.

Zambia. W: Kabompo Distr., on bank of Kabompo R. below " Boma ", fl. 20.xi.1952, *Holmes* 1005 (FHO; K). **Rhodesia.** E: Umtali Distr., Mt. Pene Forest, fl. 28.ix.1906, *Swynnerton* 1325 (BM; K; SRGH). **Malawi.** N: Nyika Distr., Nyika Plateau, fl. xi.1965, *Cottrell* in GHS 166164 (K; SRGH). C: Dedza Distr., halfway between Nchencherere and Chongoni, *Chapman* 1083 (FHO; K; SRGH). S: Zomba Distr., upper Mlunguzi above William's Falls, fr. 23.ix.1961, *Chapman* 1464 (FHO; K; SRGH). **Mozambique.** Z: Pebane, near the lighthouse, fl. & fr. 12.i.1968, *Torre & Correia* 17108 (BM; BR; LISC; WAG). MS: Vila Pery Distr., Manica, Rotanda, fl. 12.xi.1965, *Torre & Correia* 13061 (K; LISC; LMU; LUA; SRGH).

  Also in Ethiopia, Somali Republic, Uganda, Kenya, Tanzania, Swaziland and S. Africa (Transvaal and Natal). Moist forest, mixed forest, thickets, rain-forest and upland dry forest, 1000–2200 m., sometimes at lower altitudes.

5. ANISOPHYLLEA R. Br. ex Sabine

Anisophyllea R. Br. ex Sabine in Trans. Hort. Soc. **5**: 446 (1824).—Benth. & Hook., Gen. Pl. **1**: 683 (1865).

Undershrubs, shrubs or trees, usually glabrous, with young shoots often pubescent or pilose, without aerial roots. Leaves alternate, shortly petiolate, exstipulate, often with alternating reduced stipuliform leaves; leaf-lamina usually ± elliptic, entire, often inequilateral, with 2–6 lateral nerves curving up from near the base and soon becoming subparallel, leathery to papery, mostly ± glabrous with age. Inflorescences many-flowered, spike-like, racemose or paniculate, axillary or supra-axillary, solitary or serial, ± slender, ebracteate or with minute, brownish, pubescent, soon caducous bracts. Flowers small, unisexual, polygamous or more usually bisexual, sessile. Calyx pilose; calyx-tube campanulate, adnate to the ovary; calyx-lobes (3)4–5(7), ± deltate, acute at the apex, ± erect. Petals scarcely longer than the calyx-lobes, lobed or laciniate (rarely entire), involute, inserted at the mouth of the calyx-tube. Disk obscure, crenulate or lobed. Stamens twice as many as the petals, inserted at the mouth of the calyx-tube or sometimes the epipetalous ones adnate to the basal parts of the petals; filaments short, usually unequal in length, subulate or narrowly linear; anthers small, ellipsoid or ovoid, didymous. Ovary inferior, (3)4(8)-locular, each locule 1-ovulate; styles (3)4(8), short, subulate, ± thickened below, erect or recurved, ± free, arising from the epigynous disk. Fruit an ellipsoid to pyriform drupe, smooth or ridged, usually 1-locular and 1-seeded. Seed with a leathery testa; albumen thick and hard; embryo linear or slightly fusiform, longitudinally embedded in the seed; cotyledons minute or absent; endosperm absent.

A genus comprising c. 35 species, widespread in tropical Africa, Asia, Oceania and S. America.

Rhizomatous, usually many-stemmed undershrubs up to 0·8 m. high; leaf-lamina
    3–6 × 1–2·5 cm., ± elliptic to oblanceolate, ± apiculate at the apex, glabrous (or
    glabrate at the base) with age - - - - - - - 1. *quangensis*
Shrubs or trees up to 12(15) m. high; leaf-lamina usually 4–10 × 2–5·5 cm., ovate,
    narrowly to broadly elliptic, lanceolate or obovate, densely tomentose to glabrous
    with age:
    Young branchlets crimson, later yellowish-brown, ± densely tomentose; leaf-lamina
        densely tomentose to glabrescent (nerves yellow with long bristles below); petiole
        densely ferruginous-hairy; inflorescences moderately branched, stout, densely
        tomentose, patent-erect; calyx densely tomentose outside; hairs long, stout, meso-
        or pachydermatose, slightly twisted - - - - - - 2. *boehmii*
    Young branchlets brown or brownish-yellow or grey, tomentulose to glabrous; leaf-
        lamina glabrescent to glabrous; petiole glabrescent to glabrous; inflorescences
        slender, tomentulose to puberulous, ± drooping; calyx tomentulose to glabrate
        outside; hairs short, weak, leptodermatose, moderately twisted - 3. *pomifera*

1. **Anisophyllea quangensis** Engl. ex Henriques in Bol. Soc. Brot. **16**: 62, 76 (1899).—Mendes in C.F.A. **4**: 41 (1970). Type from Angola.

Rhizomatous, usually many-stemmed undershrub up to 0·8 m. high; branches tomentose when young and glabrescent with age (hairs short, slightly twisted, brownish-yellow); bark smooth or scarcely rough, brownish. Leaf-lamina 3–6 × 1–2·5 cm., ± elliptic to oblanceolate, rarely obovate, narrowed towards an acute or obtuse, ± apiculate, rarely rounded apex and cuneate to attenuate towards the petiole, or obtuse, symmetric or slightly asymmetric at the base, leathery, 5-nerved (the lower nerves close to the leaf margins); superior surface glossy, green or greenish-brown (tomentose to glabrescent when young; glabrous, or glabrate at the base, with age); inferior surface mat, yellowish-brown or yellowish-green (± tomentose when young, glabrous, or glabrate at the base, with age); petiole 1–4 mm. long, ± pubescent. Flowers 4(5)-merous, bisexual, in ± robust and upright, yellowish-brown, ± tomentose spikes (in the axils of the leaves borne on upper part of the branches); bracts 1·5–5 mm. long or more, subterete or ± foliaceous, tomentose. Calyx-lobes 2–3 mm. long, triangular to deltate, ± tomentose. Petals 4 mm. long, 3-lobed (central lobe slender and undivided, lateral lobes broader and 2–3-laciniate), white. Stamens with filaments ± 2 mm. long, subulate, slightly incurved; anthers 0·5 mm. long, broadly ellipsoid. Ovary 2 mm. long, (3)4-locular; styles (3)4, 2 mm. long, pyriform, narrowing towards the apex and reflexed. Drupe 3·5 × 1·5 cm., ellipsoid, narrowing towards both extremities, reddish, edible.

Zambia. W: Balovale Distr., Katuba Valley, fr. 25.x.1953, *Gilges* 18 (PRE; SRGH). Also in Cabinda (northern Angola) and Zaire (Haut-Kwango Distr.). Grassland on sandy soils, c. 1000 m.

2. **Anisophyllea boehmii** Engl., Pflanzenw. Ost-Afr. **C**: 287 (1895).—Engl. & Drude in Engl., Pflanzenw. Afr. **3**, 2: 675 (1921).—Brenan, T.T.C.L.: 470 (1949).—Duvign. & Dewit in Bull. Inst. Roy. Col. Belg. **21**: 925 (1950).—Duvign. in Lejeunia **16**: 108 (1952).—J. Lewis in F.T.E.A., Rhizophoraceae: 18, fig. 5 (1956).—F. White, F.F.N.R.: 275 (1962). TAB. 29. Type from Tanzania.

Evergreen shrub or tree up to 10 m. high with dense and close ramification; bark smooth to rough, brown to grey; young branchlets crimson, later yellowish-brown, ± densely tomentose (hairs long, mesodermatose, slightly twisted, reddish-brown to yellowish-brown or grey). Main leaf-lamina usually 5–10 × 2–5·5 cm. long, ovate, narrowly to broadly elliptic, lanceolate, obovate or, rarely, subcircular, narrowed to an acute or obtuse apex (sometimes rounded or, rarely, retuse or emarginate) and cuneate to obtuse, symmetric or slightly asymmetric at the base, ± leathery; superior surface ± glossy (purplish-pink or pale green to yellowish and ± densely tomentose when young, dark green to brownish-green or yellowish and pubescent to glabrescent, with age); inferior surface

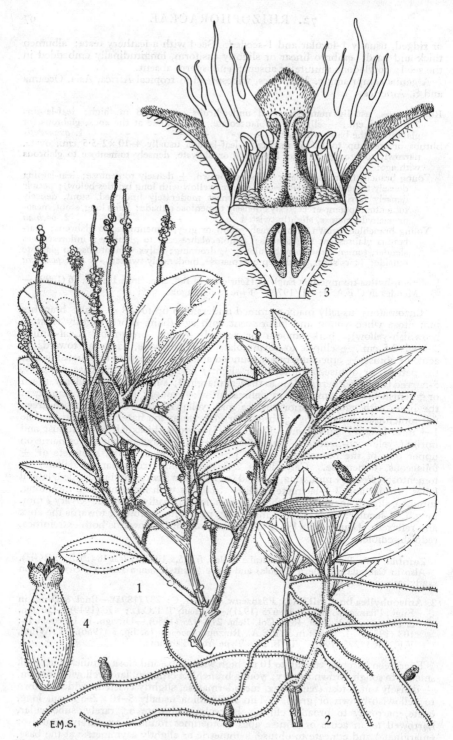

Tab. 29. ANISOPHYLLEA BOEHMII. 1, flowering branch (× ⅔); 2, fruiting branch (× ⅔); 3, longitudinal section of flower (× 20); 4, young fruit (× 4), 1 & 3 from *Watkins* 156, 2 & 4 from *Bullock* 3218.   From F.T.E.A.

mat (yellowish-green or brownish-yellow and densely tomentose when young, yellowish-green to brownish and densely to sparsely tomentose, with rather short, meso- or pachydermatose hairs, with age); midrib prominent, yellow, with long dense bristles below; petiole 2–5 mm. long, densely ferrugineous-hairy. Spikes 3–10(15) cm. long, stout, usually composite, often many-branched, yellowish-brown, densely tomentose (hairs strong or flexuous, meso-pachydermatose), ± remotely flowered, upright, usually in the axils of smaller (up to 3 cm. long), obovate and deciduous leaves (false bracts) at the base of the current year's growth. Flowers 4–5-merous, bisexual, densely tomentose; bracts ± 2 mm. long, subterete. Calyx-lobes 2 mm. long, triangular to deltate, pale green (tinged purplish-pink externally), densely tomentose outside and glabrous inside, their margins ciliolate. Petals 3 mm. long, 3-lobed (central lobe slender and undivided, lateral lobes broader and 2–3-laciniate), white, glabrous, papillose. Stamens with filaments 2 mm. long, subulate, incurved; anthers 0·3 mm. long, broadly ovoid. Ovary 1·5 mm. long, campanulate, 4–5-locular. Styles 4–5, 1·5 mm. long, pyriform with leptodermatose hairs (sometimes with the apex inflated and apiculate at the extremity) towards the base, divergent, bent outwards. Stigmas spherical with spherical papillae. Drupe 3·5 × 2 cm., ellipsoid, plum-coloured, edible.

Zambia. B: Mankoya Distr., Luampa-Kafue traverse, Namahanda, st. 15.x.1938, *Martin* 871 (FHO). N: Samfya, fl.-buds 26.vi.1956, *Fanshawe* 4153 (K). W: Ndola Distr., Ndola, fr. 19.x.1954, *Fanshawe* 1628 (K). C: Broken Hill Distr., Kabasa, fl. & immat. fr. viii.1934, *Trapnell* 1528 (K).
Also in Tanzania and Zaire. *Brachystegia* woodland on sandy or rocky soils, c. 1300 m.

3. **Anisophyllea pomifera** Engl. & Von Brehm. in Engl., Bot. Jahrb. **54**: 376 (1917).—
Engl. & Drude in Engl., Pflanzenw. Afr. **3**, 2: 677, fig. 301L (1921).—Burtt Davy & Hoyle, N.C.L.: 82 (1936).—Brenan, T.T.C.L.: 471 (1949).—Duvign. & Dewit in Bull. Inst. Roy. Col. Belg. **21**: 932 (1950).—J. Lewis in F.T.E.A., Rhizophoraceae: 18 (1956).—F. White, F.F.N.R.: 275 (1962). Type from Tanzania.

Evergreen or semi-evergreen shrub or tree up to 12(15) m. high; trunk up to c. 50 cm. in diameter; bark smooth or scarcely rough, light brown or grey, sometimes scaling off in small, thin scales; young branchlets brown, yellowish-brown or grey, tomentulose to glabrous (hairs minute, leptodermatose, moderately twisted, reddish-brown to yellowish-brown or grey). Main leaf-lamina usually 4–8 × 2–5 cm., ovate, ± elliptic, obovate or rarely subcircular, narrowed to an acute or obtuse apex (sometimes rounded or, rarely, retuse), cuneate to obtuse, symmetric or slightly asymmetric at the base, papery or often ± leathery, glabrescent to glabrous; superior surface ± glossy (orange-brown, yellowish-brown or light green when young; dark green or ± brownish with age); inferior surface mat (yellowish-green when young; green or yellowish with age); petiole 2–5(8) mm. long, glabrescent to glabrous. Spikes 3–13 cm. long, slender, tomentulose to puberulous (hairs leptodermatose, moderately twisted, dark brown with brownish-yellow contents), remotely flowered, ± drooping, in the axils of smaller (up to 3 cm. long) and usually deciduous leaves (false bracts) on the lower part of the branches. Flowers 4–5-merous, bisexual, glabrescent; bracts ± 1·5 mm. long, subterete. Calyx-lobes 2 mm. long, triangular to deltate, green, tomentulose to glabrate outside and glabrous inside, their margins ciliolate. Petals 3 mm. long, ± 3-lobed (central lobe slender and undivided, lateral lobes broader and 2–3-laciniate), white or yellowish, glabrous, papillose. Stamens with filaments 1·5 mm. long, subulate, incurved; anthers 0·5 mm. long, broadly ovoid. Ovary 2 mm. long, campanulate, 4–5(7)-locular; styles 4–5(7), 1·5 mm. long, pyriform, thickened and partially covered with sharp, short, hyaline leptodermatose and yellow meso-dermatose hairs towards the base, narrowed, reflexed, glabrous and papillose above. Drupe 4 × 2·5 cm., ellipsoid, plum-coloured, edible, flavoured like a nectarine.

Zambia. N: Mbala Distr., fr. 1.xi.1952, *Robertson* 195 (EA; K). W: Mwinilunga Distr., Kaundu, st. ix.1934, *Trapnell* 1580a (K). C: Chitambo Mission, *Stevenson* 66 (FHO). E: Fort Jameson, Forest Reserve, fl. 23.vi.1960, *Grout* 222a (FHO). S: Livingstone Distr., fr. 20.vi.1928, *Watt & Brandwyk* 2029 (PRE). **Malawi.** N: Mzimba Distr., Lundazi to Mzimba, 46 km., fl. 28.iv.1952, *White* 2505 (FHO; K; SRGH). C: Lilongwe Distr., Dzalanyama Forest, st. 29.iii.1962, *Adlard* (SRGH).
Also in Tanzania. Upland plateau thickets or *Brachystegia-Isoberlinia* woodlands on sandy or rocky soils, 600–1600 m.

73. COMBRETACEAE

By A. W. Exell

This account of the Combretaceae of the Flora Zambesiaca area was published in a slightly abbreviated form in Kirkia Vol. 7 pt. 2, pp. 159–252 (1970) through the kindness of Professor H. Wild.

I was indebted both in that and the present version to C. A. Stace for detailed descriptions and drawings of the scales and for general collaboration, to Thomas Müller of Salisbury Botanic Garden for descriptions of seedlings, to L. Grandvaux Barbosa and H. Wild for ecological notes and to F. White of the Forestry Herbarium, Oxford, where most of the work was done, for much help in every way.

Since the publication of the account in Kirkia, G. E. Wickens has published the Combretaceae of the Flora of Tropical East Africa (1973). Our two accounts have been largely correlated throughout and vary in a few comparatively minor cases where our opinions differ as regards treatment rather than fact. I am very grateful to him for pointing out a number of textual inaccuracies and omissions in my Kirkia account and for some valuable improvements and additional information which have been incorporated in the present version.

Lastly there has just appeared "Revisão de algumas Combretaceae de Moçambique" by Maria Telma Faria (in Mem. Inst. Invest. Agron. Moçamb. **4**: 1–175 (1973)) with copious citation of specimens and a very full list of vernacular names. The present account has profited considerably from this work.

Trees, shrubs, shrublets or climbers, rarely subherbaceous. Indumentum of unicellular compartmented (rarely non-compartmented) hairs, multicellular stalked glands and multicellular scales (in which the head consists of a multicellular plate only one cell thick). Leaves opposite, verticillate, spiral or alternate, exstipulate, simple and almost always entire (very rarely crenulate). Flowers bisexual or bisexual and male in the same inflorescence or very rarely dioecious, usually 4–5-merous, rarely slightly zygomorphic, in axillary or extra-axillary elongated or subcapitate spikes or racemes or in terminal or axillary panicles. Receptacle (calyx-tube or hypanthium) usually in two distinct parts, the lower receptacle surrounding and adnate to the inferior (except in *Strephonema* from West Africa) ovary and upper receptacle usually produced beyond it to form a short or long tube (perhaps a true calyx-tube in *Meiostemon*) terminating in the sepals (calyx-lobes), the latter sometimes scarcely developed. Sepals 4 or 5 (rarely 6 or 8) or almost absent, sometimes accrescent (*Calycopteris* in Asia). Petals 4 or 5 (rarely more) or absent, conspicuous or sometimes very small, usually inserted near the mouth of the upper receptacle (but near the disk in *Meiostemon*). Stamens usually twice as many as the sepals or petals but occasionally the same number (4 in *Meiostemon*), borne inside the upper receptacle with biseriate or more rarely uniseriate insertion (one whorl very rarely as staminodes), exserted or included; anthers dorsifixed, versatile (rarely adnate to the filaments, but not in our area). Disk intrastaminal, hairy or glabrous, sometimes inconspicuous or absent. Style usually free (attached to the upper receptacle for part of its length in *Quisqualis* and a very few species of *Combretum*). Ovary inferior (semi-inferior in *Strephonema*), 1-locular with usually 2 (up to 6) pendulous anatropous ovules of which only 1 usually develops. Fruit (a pseudocarp) very variable in size and shape, fleshy or dry, stipitate or sessile, usually indehiscent (occasionally tardily dehiscent) often variously winged or ridged, 1-seeded. Albumen absent. Cotyledons 2 (rarely 3 or 4) occasionally with their petioles connate almost to the apex.

Cosmopolitan in the tropics and subtropics.

In addition to the genera indigenous in our area, *Quisqualis indica* L., the Rangoon Creeper, with a very long upper receptacle and the style partially adnate to it, is sometimes grown in gardens.

The flowers in the Combretaceae are usually sessile or nearly so but appear to be pedicellate in several genera because the slender lower receptacle, containing the ovary, resembles a pedicel. The male flowers in *Terminalia* and *Pteleopsis* have long slender stalks, perhaps a sterile lower receptacle in origin. Sometimes there is a short true pedicel at the base of the lower receptacle.

The lobes at the apex of the upper receptacle (sometimes poorly developed but often quite long) probably represent the calyx and have usually been called calyx-lobes, which implies a gamosepalous calyx of which there is no convincing evidence except perhaps in *Meiostemon*. I thus now prefer to call them sepals.

The cotyledons show interesting and often striking differences in the way they unfold and in their behaviour during germination but as these characters are known in only a few of our species they can seldom conveniently be used for classification. If some of the cotyledon characters are adaptations to fire the differences may not be so fundamentally important in the classification as would at first sight appear. The union of the cotyledon petioles in a few species of *Combretum* is a very rare and unusual feature.

The compartmented hairs (requiring microscopic examination) known also as " combretaceous hairs " occur in all our species (unless entirely glabrous) and are known, apart from the Combretaceae, only in the Cistaceae and a few Myrtaceae so that they give a usually reliable method of identifying the family even from fragmentary material (see Stace in Journ. Linn. Soc., Bot. **59**: 233–234, text-figs. 1–12 (1965)).

Specimens with compound leaves or with stipules can confidently be excluded from the family.

Lower receptacle without adnate bracteoles; not mangroves:
  Flowers all bisexual; petals present (except in aberrant specimens); leaves usually opposite or verticillate, with scales or stalked glands (usually microscopic):
    Stamens 8–10; petals usually inserted near the apex of the upper receptacle*
      1. Combretum
    Stamens 4; petals inserted near the edge of the disk   -   -   **2. Meiostemon**
  Flowers usually andromonoecious (bisexual and male in the same inflorescence), if all bisexual then petals absent; leaves without scales or stalked glands; petals present or absent:
    Petals present; leaves ± opposite   -   -   -   -   -   - **3. Pteleopsis**
    Petals absent; leaves usually spiral or alternate (opposite or subopposite in *T. pteleopsoides*)   -   -   -   -   -   -   -   - **4. Terminalia**
Lower receptacle with 2 adnate bracteoles; mangroves   -   -   **5. Lumnitzera**

Note.—The scales may be completely hidden by a dense indumentum or conspicuous only at certain stages of development (usually in older leaves) so that microscopic examination may be necessary.

In the absence of flowers the key given below will enable most specimens to be generically identified with the exception that *Meiostemon* and *Combretum* can be separated only by flower characters.

Plant not a mangrove; fruit without traces of adnate bracteoles:
  Leaves opposite or subopposite or verticillate (very rarely alternate):
    Scales or stalked glands (usually microscopic) present   -   - **1. Combretum**
      2. Meiostemon
    Scales and stalked glands both absent:
      Fruit with a stipe at least 10 mm. long   -   -   -   -   - **3. Pteleopsis**
      Fruit with a stipe not exceeding 5 mm. long   -   -   -   - **4. Terminalia**
  Leaves spiral or alternate (sometimes in fascicles); fruits usually 2-winged, always with a sclerenchymatous layer in the pericarp   -   -   -   - **4. Terminalia**
Plant a mangrove; leaves spiral; fruits usually showing 2 adnate bracteoles or traces of them   -   -   -   -   -   -   -   -   -   - **5. Lumnitzera**

1. COMBRETUM Loefl.

Combretum Loefl., Iter Hisp.: 308 (1758) *nom. conserv.*

Trees, shrubs, shrublets or woody climbers, very rarely subherbaceous. Leaves opposite, verticillate or rarely alternate, usually petiolate (rarely subsessile), almost always entire; petiole sometimes persisting (especially in climbers) forming a ± hooked spine. Flowers always ⚥ (in our area), actinomorphic or slightly zygomorphic (not in our area), 4–5-merous, in elongated or subcapitate axillary or

* In *Combretum* Sect. *Hypocrateropsis* the upper receptacle is scarcely developed so that the petals are necessarily at the edge of the disk.

extra-axillary spikes or racemes or in terminal or terminal and axillary often leafy panicles. Receptacle usually clearly divided into a lower part (lower receptacle) surrounding and adnate to the ovary and an upper part (upper receptacle) varying from patelliform to infundibuliform and itself sometimes visibly differentiated into a lower part containing the disk (when present) and an often more expanded upper part. Sepals (calyx-lobes) 4–5 (rarely more), deltate to ± subulate or filiform, sometimes scarcely developed. Petals usually 4–5 (rarely absent, in aberrant specimens and up to 7 in occasional flowers), small and inconspicuous to showy and exceeding the sepals, of various colours. Stamens twice as many as the petals inserted in 1 or more usually 2 series inside the upper receptacle and usually exserted. Disk glabrous or hairy, with or without a free margin, sometimes inconspicuous or absent. Style free (in our species); stigma sometimes ± expanded. Ovary completely inferior. Fruit 4–5-winged, -ridged or -angled, sessile or stipitate, indehiscent or rarely tardily dehiscent; pericarp usually thin and papery, sometimes leathery, more rarely fleshy. Cotyledons various. Plants lepidote or with microscopic (and sometimes macroscopic) stalked glands.

Throughout the tropics (except in Australia and Pacific Is.) and extending to the subtropics.

Key to the subgenera

Scales present though sometimes inconspicuous or hidden by the indumentum; microscopic
    stalked glands absent (except in 3 American spp.); flowers usually 4-merous (petals
    usually not red) -   -   -   -   -   -   -   -   1. Subgen. *Combretum*
Scales absent; microscopic stalked glands present; flowers 5-merous or if 4-merous then
    petals usually red (in our area) -   -   -   -   -   - 2. Subgen. *Cacoucia*

For identification of the species it is of fundamental importance to distinguish between the two subgenera and the following notes may help.

1. Every species in our area has either scales (Subgen. *Combretum*) or microscopic stalked glandular hairs (Subgen. *Cacoucia*) but both these are sometimes inconspicuous (especially at various stages of development) or hidden by a dense indumentum or by glutinous secretions. If no scales are visible with a lens (the fruit is sometimes a good place to look for them) do not confidently assume that the specimen belongs to Subgen. *Cacoucia*. To be certain microscopic examination is necessary and is always required to see the microscopic stalked glandular hairs of Subgen. *Cacoucia*.

2. Most 4-merous species belong to Subgen. *Combretum* but some 4-merous red-flowered climbers and shrublets belong to Subgen. *Cacoucia* Sect. *Conniventia* and the 4-merous *C. andradae* belongs to Subgen. *Cacoucia* Sect. *Poivrea*.

3. All predominantly 5-merous specimens from our area (that is apart from occasional aberrant flowers) can be placed in Subgen. *Cacoucia* with some confidence.

4. Most (but not all) climbers belong to Subgen. *Cacoucia* especially those in which the persistent petioles (or their basal parts) form slightly hooked spines.

5. Species with macroscopic glandular hairs belong to Subgen. *Cacoucia*.

As it is naturally troublesome to have to make a microscopic examination the following key will separate the two subgenera, when flowers are present, in the great majority of cases.

Flowers and fruits 4-merous:
    Petals bright red -   -   -   -   -   -   -   -   - Subgen. *Cacoucia*
    Petals not bright red:
        Petals 4 mm. long or longer:
            Scales absent; flowers in handsome spikes   -   -   - Subgen. *Cacoucia*
            Scales very conspicuous -   -   -   -   -   - Subgen. *Combretum*
        Petals less than 3·5 mm. long; scales present but sometimes inconspicuous
                                                         Subgen. *Combretum*
Flowers and fruits 5-merous (except for occasional abnormal flowers or fruits)
                                                         Subgen. *Cacoucia*

1. Subgen. COMBRETUM

Scales present (sometimes inconspicuous or concealed by the indumentum). Flowers normally 4-merous.

The identification of the sections in Subgen. *Combretum* is not easy partly because there are a large number of small sections and no system (other than entirely artificial ones) has yet been found for grouping them in larger units. Scale-characters, though often the best, have been used as little as possible for primary keying because of the inconvenience to field workers. It is sometimes difficult to decide between 1-seriate and 2-seriate insertion of the stamens. I have classified as 2-seriate those species with the stamens clearly in 2 series usually at some distance from the margin of the disk. The species classified as 1-seriate are those apparently 1-seriate or very nearly so on superficial examination. It is also not always easy to see to what extent the margin of the disk is free in the small-flowered species. The expansion of the style at its apex may well be a good character but the stigma may wither soon after fertilization. The first key is based mainly on flowering material: the second is an attempt to separate the sections when only fruiting material is available.

Key to the sections of Subgen. Combretum (flowering material)

Upper receptacle little developed, almost flat; disk conspicuously visible, well developed; petals glabrous, linear-elliptic; stamens 1-seriate; scales conspicuous, large, usually at least (100)150μ in diam., divided by many radial and tangential walls

1. *Hypocrateropsis*

Upper receptacle cupuliform to infundibuliform; disk usually concealed within the upper receptacle and usually not visible in dried specimens without opening the flower; petals glabrous or hairy, of various shapes; stamens 1-seriate or 2-seriate; scales various (sometimes inconspicuous), if over 150μ in diam. either conspicuously wavy or scalloped at the margin or divided almost solely by radial walls:

Petals ciliate at the apex, 0·5–1·1(1·5) mm. long, obtriangular or obovate, often rather inconspicuous; stamens 1- or 2-seriate; scales various:

Stamens 1-seriate; fruit broad-winged - - - - - 8. *Ciliatipetala*
Stamens 2-seriate; fruit with very narrow (1 mm. wide) wings (C. *illairii*)

13. *Chionanthoida*

Petals glabrous (rarely with 2 or 3 hairs at the apex), usually (but not invariably) longer than 1 mm.; stamens 1- or 2-seriate:

Bark of branchlets peeling off in large ± cylindric or hemicylindric pieces leaving an exposed cinnamon-red surface (C. *psidioides* subsp. *glabrum*) - 8. *Ciliatipetala*

Bark of branchlets coming off in untidy irregular fibrous strips or threads:

Petals subreniform, c. 0·9 × 1·5 mm., broader than long; upper receptacle cupuliform, small, c. 1·2–1·5 × 1·2–1·5 mm.; stamens 2-seriate; scales of a simple 8-celled type - - - - - - - 2. *Combretastrum*

Petals obovate or subcircular (occasionally almost subreniform) to elliptic or spathulate, usually more than 1 mm. long and nearly always as long as broad or longer; stamens 1- or 2-seriate; scales various:

Stamens 1-seriate in insertion (or very nearly so) at or near the margin of the disk; scales conspicuous or not so:

Inflorescences of short (up to 3 cm. long) usually subcapitate or glomeruliform spikes; cotyledons (where known) borne above soil level:

Branchlets and rhachis not fuscous-pubescent:

Fruit with a fairly stout stipe up to 0·7 cm. long; petals obovate to elliptic; disk with no (or very short) free margin; style not expanded at the apex; scales divided by 8–16 primary radial walls alone; trees or shrubs - - - - - - - - 3. *Angustimarginata*

Fruit with a slender stipe 1–3 cm. long; petals ovate to subcircular; disk with a short free margin or the latter up to 1 mm. long; style often expanded at the apex (perhaps always but soon withering); spikes often subsessile, glutinous; scales of a simple 8-celled type, often with additional tangential and partial radial walls; scandent shrubs

4. *Macrostigmatea*

Branchlets and rhachis fuscous-pubescent; petals spathulate; stipe of fruit up to 14 mm. long - - - - - - - - 9. *Fusca*

Inflorescences of more elongated spikes (usually at least 5 cm. long); disk with a free margin; cotyledons (where known) usually with connate petioles and arising below soil level:

Petals subcircular (lamina sometimes almost subreniform); leaves and fruits often somewhat " metallic " in appearance; scales 80–180μ in diam., divided by many radial and tangential walls to give c. 16–40 marginal cells, conspicuous (except when hidden by the indumentum); style not expanded at the apex - - - 5. *Metallicum*

Petals obovate to spathulate; leaves and fruits not " metallic " in appearance, often glutinous (especially when young); style often slightly to considerably expanded at the apex; scales less than 80μ in diam. and with less than 16 marginal cells:

Fruits up to 2·5–3·5 × 2·5–3·5 cm., with stipe 0·5–0·7 cm. long; petals
obovate to spathulate; leaves often 3–4-verticillate; scales 50–65μ in
diam., usually divided by 8 radial walls alone   - 6. *Glabripetala*
Fruits (4)5(8) × (3·5)5(7) cm. with stipe up to 2·5 cm. long (elongating as
the fruit develops); petals obovate-spathulate to spathulate; leaves
usually opposite, sometimes 3-verticillate; scales 40–75μ in diam.,
usually with some tangential as well as radial walls

7. *Spathulipetala*
Stamens clearly 2-seriate in insertion; scales conspicuous, moderately to very
dense:
Principal lateral nerves 3–4(6) pairs; fruit usually with dark-reddish or
sometimes yellowish scales; upper receptacle somewhat constricted below
the apex - - - - - - - - - 10. *Breviramea*
Principal lateral nerves (4)5 or more pairs:
Petals up to 4 × 2 mm., emarginate; upper receptacle somewhat constricted
below the apex - - - - - - 11. *Aureonitentia*
Petals up to 2(3) mm. long, not emarginate; upper receptacle not constricted
below the apex; inflorescences of rather congested often subcapitate
spikes:
Scales on under surface of leaf very dense, often contiguous:
Upper receptacle cupuliform to very broadly infundibuliform; scales
over 100μ in diam., divided equally by radial and tangential walls,
silvery, not appearing impressed; fruit 4 winged
12. *Elaeagnoida*
Upper receptacle narrowly infundibuliform to tubular (*C. stocksii*, fruit
unknown) - - - - - - 13. *Chionanthoida*
Scales on under surface of leaf usually golden, moderately to fairly dense
but not or rarely contiguous, often appearing impressed, divided
almost solely by radial walls; upper receptacle broadly campanulate
to narrowly infundibuliform; fruit 4-angled or 4-winged
13. *Chionanthoida*

Key to sections of Subgen. Combretum (fruiting material)

Fruit broad-winged:
Stipe of fruit not more than 10 mm. long:
Scales conspicuous on mature leaves:
Scales often contiguous at least on the lower surface of the leaf:
Scales often contiguous on both surfaces of the leaf; stipe of fruit 2–3 mm. long
1. *Hypocrateropsis*
Scales contiguous only on the lower surface of the leaf:
Stipe of fruit 2–3(5) mm. long - - - - - - 8. *Ciliatipetala*
Stipe of fruit 5–10 mm. long:
Fruit brown, greyish-brown or reddish-grey to dark purple, often with a
" metallic " appearance - - - - - - 5. *Metallicum*
Fruit silvery- or golden- lepidote:
Rhachis tomentose - - - - - - 11. *Aureonitentia*
Rhachis lepidote but otherwise glabrous - - 12. *Elaeagnoida*
Scales sparse to rather dense on the lower surface of the leaf but not contiguous:
Principal lateral nerves 3–4(6) pairs - - - - 10. *Breviramea*
Principal lateral nerves (4)5 or more pairs:
Scales appearing somewhat impressed - - - 13. *Chionanthoida*
Scales not appearing impressed:
Reticulation not very prominent on the under surface of the leaf; scales
usually over 150μ in diam., circular, divided by many radial and
tangential walls - - - - - - 1. *Hypocrateropsis*
Reticulation usually very prominent on the under surface of the leaf; scales
rarely more than 120μ in diam. with up to only 11(16) usually concave
marginal cells and cells opaque - - - - 8. *Ciliatipetala*
Scales inconspicuous or rather inconspicuous on mature leaves (sometimes because
hidden by a dense indumentum):
Lower surface of mature leaf covered by a dense indumentum ± hiding the scales:
Principal lateral nerves 3–4(6) pairs - - - - - 10. *Breviramea*
Principal lateral nerves 5 or more pairs:
Fruit brown, greyish-brown or reddish-grey to dark purple, often with a
" metallic " appearance, not glutinous - - - - 5. *Metallicum*
Fruit usually pale, sometimes glutinous when young:
Fruit glutinous (especially when young), up to 3·5 cm. long
6. *Glabripetala*
Fruit not conspicuously glutinous, up to 2·5 cm. long:

Fruit up to 2·5 cm. long; reticulation of under surface of leaf usually prominent (occasionally ± concealed by the indumentum)
        8. *Ciliatipetala*
Fruit up to 2·3 cm. long; reticulation of under surface of leaf not prominent - - - - - - 3. *Angustimarginata*
Lower surface of mature leaf glabrous to pubescent or pilose with indumentum not entirely concealing the surface:
  Rhachis not fuscous-pubescent:
    Bark of branchlets peeling off in cylindric or hemi-cylindric flakes leaving an exposed cinnamon-red surface - - - - - 8. *Ciliatipetala*
    Bark of branchlets coming off in fibrous strips; fruit often glutinous (especially when young):
      Leaves usually 3–4-verticillate - - - - - 6. *Glabripetala*
      Leaves usually opposite - - - - - - 8. *Ciliatipetala*
  Rhachis fuscous-pubescent; scandent evergreen shrub or liane - 9. *Fusca*
Stipe of fruit (10)12–30 mm. long:
  Principal lateral nerves 3–4(6) pairs:
    Fruit densely and conspicuously reddish- or golden-lepidote - 10. *Breviramea*
    Fruit rather inconspicuously lepidote - - - - 4. *Macrostigmatea*
  Principal lateral nerves (4)5 or more pairs:
    Rhachis fuscous-pubescent - - - - - - - 9. *Fusca*
    Rhachis not fuscous-pubescent:
      Lateral nerves 6–10 pairs; stipe of fruit up to 5 mm. long, glutinous; petals subreniform - - - - - - - - 2. *Combretastrum*
      Lateral nerves 5–6(7) pairs; stipe of fruit more than 5 mm. long, pilose; petals spathulate - - - - - - - - - 9. *Fusca*
      Scales often contiguous, very conspicuous; stipe of fruit up to 12(15) mm. long; fruit up to 2–3·5 cm. long - - - - 12. *Elaeagnoida*
      Scales rather inconspicuous; stipe of fruit 10–30 mm. long:
        Fruit up to 3 cm. long - - - - - 4. *Macrostigmatea*
        Fruit ± 5 cm. long (up to 8 cm.):
          Scandent shrubs or lianes - - - - - 4. *Macrostigmatea*
          Small to medium-sized trees - - - - 7. *Spathulipetala*
Fruit angled or very narrowly winged - - - - - 13. *Chionanthoida*

1. Sect. HYPOCRATEROPSIS Engl. & Diels in Engl., Mon. Afr. Pflanz. **3**: 11 (1899).—Stace in Bot. Journ. Linn. Soc. **62**: 135, figs. 2–10 (1969).—Exell in Kirkia, **7**, 2: 167 (1970).—Wickens, F.T.E.A. Combret.: 11 (1973).

Flowers 4-merous. Upper receptacle almost flat, little developed. Petals linear-elliptic, glabrous. Stamens 8, 1-seriate, inserted at the margin of the disk. Fruit 4-winged. Cotyledons 2, borne above soil level (*C. celastroides*, *C. padoides*) or arising below soil level (*C. imberbe*, needing confirmation). Scales large, circular, usually at least 100μ in diam. (some rarely only 50μ) and mostly over 150μ, divided by many radial and tangential walls.

Disk glabrous; scales rarely contiguous - - - - - - 1. *celastroides*
Disk pilose at least on the margin:
  Leaves not very densely lepidote, scales rarely contiguous on mature leaves
        2. *padoides*
  Leaves very densely lepidote, scales mostly contiguous - - - 3. *imberbe*

1. **Combretum celastroides** Welw. ex Laws. in Oliv., F.T.A. **2**: 422 (1871).—Engl. & Diels in Engl., Mon. Afr. Pflanz. **3**: 12 (1899).—O. B. Mill., B.C.L.: 42 (1948).—Codd, Trees and Shrubs Kruger Nat. Park: 129 (1951).—Exell & Garcia in Contr. Conhec. Fl. Moçamb. **2**: 102 (1954) excl. specim. *Vincent* 44; C.F.A. **4**: 50 (1970).—Rattray in Kirkia, **2**: 89 (1961).—Stace in Mitt. Bot. Staatssamml. Münch. **4**: 12 (1961).—F. White, F.F.N.R.: 284 (1962).—Mitchell in Puku, **1**: 121 (1963).—Fanshawe & Savory in Kirkia, **4**: 190 (1964).—Boughey in Journ. S. Afr. Bot. **30**: 167 (1964).—Exell in Prodr. Fl. SW. Afr. **99**: 8 (1966); in Kirkia, **7**, 2: 167 (1970).—Wickens, F.T.E.A. Combret.: 11, fig. 3(1), 4(1) (1973). Type from Angola.

Small tree up to 9–12(20) m. tall or shrub, sometimes with sarmentose branches, or climber; crown rounded to flat-rounded; bark creamy-brown to grey, rough and very closely and shallowly fissured or smooth; branchlets tomentellous to pubescent, glabrescent. Leaves opposite or subopposite (rarely alternate or pseudo-verticillate); lamina 2·5–14 × 1–8 cm., subcoriaceous to papyraceous, narrowly to broadly elliptic or oblong-elliptic or obovate, usually shortly and rather bluntly acuminate at the apex and cuneate to rounded at the base, glabrous to fairly densely pubescent especially on the nerves below and often with tufts of hairs in the axils of the lateral nerves below, punctulate above, reddish- or golden- or silvery-lepidote below, scales usually not contiguous; lateral nerves 4–10 pairs, rather prominent below, slightly so above; petiole 2·5–20 mm. long. Inflorescences 5–15 cm. long, of terminal panicles of spikes with lateral spikes, often unbranched, in the axils of the upper leaves; rhachis tomentellous and sparsely lepidote; bracteoles 1–3 × 0·3–0·6 mm., caducous. Flowers yellow or cream or whitish, sweet-scented, sessile. Lower receptacle 2·5–6 × 0·3–1 mm. and, like the upper receptacle, rufous-lepidote otherwise glabrous. Sepals 2–2·5 × 1·5–2 mm., ovate-triangular, lepidote but otherwise glabrous. Petals 1·5–2·5 × 0·3 mm., linear-elliptic, glabrous. Stamen-filaments 1·5–2·5 mm. long; anthers 0·4–0·8 mm. long. Disk 2–5·5 mm. in diam., glabrous, with notches where the filaments are inserted, without distinct free margin. Style 1·5–2·8 mm. long, glabrous. Fruit 1·5–2·5 cm. in diam., subcircular to transversely broadly elliptic in outline, rufous- or golden-lepidote otherwise glabrous or nearly so, apical peg 0·5–2·5 mm. long, wings 6–10 mm. broad, stipe 1·5–3 mm. long articulated almost at the rhachis. Cotyledons 2 with petioles 5–6 mm. long, borne above soil level. Scales 150–300μ in diam. (as in the sectional description).

Zaire, Angola, Botswana, SW. Africa, Zambia, Rhodesia, Tanzania, Mozambique and northern Transvaal. For ecology see the subspecies.

Numerous intermediates between *C. celastroides* and *C. laxiflorum* have made it advisable to unite these two species. The three subspecies recognized form fairly homogeneous populations but there are some intermediates between subsp. *celastroides* and subsp. *laxiflorum*.

Flowers larger than in the other two subspecies, attaining the maximum dimensions in the
    specific description; leaves nearly glabrous; tufts of hairs in the axils of the lateral
    nerves absent or rare; scales usually reddish on the leaves; petiole up to 20 mm. long,
    rather stout; usually a small tree   -   -   -   -   -   subsp. *laxiflorum*
Flowers smaller, usually approaching the minimum dimensions in the specific description;
    tufts of hairs usually present in the axils of the nerves below; scales usually golden or
    silvery on the leaves; petiole up to 8 mm. long; usually a shrub or climber:
Leaves usually pubescent below (except in some intermediates); disk up to 4 mm.
    in diam. -   -   -   -   -   -   -   -   subsp. *celastroides*
Leaves almost glabrous below (except for tufts of hairs in the axils of the nerves and
    occasional pubescence on the midrib); disk 2–2·5 mm. in diam. - subsp. *orientale*

1a. **Subsp. laxiflorum** (Welw. ex Laws.) Exell in Bol. Soc. Brot. Sér. 2, **42**: 15 (1968); in Kirkia, **7**, 2: 168, t. 1 fig. 1a (1970).—Exell & Garcia in C.F.A. **4**: 51 (1970). TAB. **31** fig. G; TAB. **32** fig. 1a. Type from Angola (Cuanza Norte).
      Combretum laxiflorum Welw. ex Laws. in Oliv., F.T.A. **2**: 428 (1871).—Engl. & Diels in Engl., Mon. Afr. Pflanz. **3**: 11, t. 1 fig. D (1899).—Duvign. in Bull. Soc. Roy. Bot. Belg. **88**: 62, 70, fig. 1, 4A, 7(6) (1956).—F. White, F.F.N.R.; 284 (1962).—Boughey in Journ. S. Afr. Bot. **30**, 4: 167 (1964).—Liben in F.C.B., Combret.: 40 (1968).
      Combretum butayei De Wild. in Ann. Mus. Cong. Bot., Sér. 5, **1**: 196 (1904). Type from Zaire.

Usually a small tree. Leaves subcoriaceous or more rarely chartaceous, almost glabrous except for scales, nearly always devoid of tufts of hairs in the axils of the leaves below; lamina up to 14 × 8 cm. Disk up to 5·5 mm. in diam.

Zambia. B: Balovale Distr., 1070 m., fl. x.1953, *Gilges* 311 (K; PRE; SRGH). N: 45 km. E. of Mporokoso, 1430 m., fr. 13.iv.1961, *Phipps & Vesey FitzGerald* 3122 (BM; FHO; SRGH). W: 100 km. S. of Mwinilunga on the Kabompo road, fr. 3.vi.1963, *Loveridge* 760 (FHO; SRGH).
      Also in Zaire, Angola and Tanzania. *Brachystegia* woodland and evergreen thickets, often on termite mounds; c. 1000–1500 m. Occasional on copper bearing soils.
      There are intermediates with subsp. *celastroides* (appearing as almost glabrous forms of the latter) across Angola and Zambia where the areas of distribution meet or overlap. Two specimens of a climber from the Niassa Prov. of Mozambique (*Allen* 113 (K), 114 (K)) appear to be intermediate between subsp. *laxiflorum* and subsp. *orientale* having the larger and more coriaceous leaves of the former and the small flowers of the latter.

1b. Subsp. **celastroides.**—Exell in Bol. Soc. Brot., Sér. 2, **42**: 16 (1968); in Kirkia, **7**, 2: 168, t. 1 fig. 1b (1970).—Wickens in F.T.E.A., Combret.: 17, fig. 1(1a) 1973. TAB. **30** fig. A; TAB. **32** fig. 1b. Type as for the species.
Combretum patelliforme Engl. & Diels in Engl., Mon. Afr. Pflanzen. **3**: 12, t. 1 fig. C (1899) pro parte quoad specim. *Antunes* A 155 (B†).—Eyles in Trans. Roy. Soc. S. Afr. **5**, 4: 428 (1916).—Wild, Fl. Vict. Falls: 146 (1953). Syntypes from Angola and Mozambique: Delagoa Bay, *Schlechter* 11957 (B†; BM; COI; K).

Usually a shrub, sometimes climbing, occasionally a liane, rarely a small tree. Leaves papyraceous or more rarely chartaceous, pubescent especially on the nerves below; lamina usually c. 5–10 × 1–4·5 cm. Disk usually c. 2·5–3·5 mm. in diam. Scales usually golden or silvery (some reddish especially in intermediates with subsp. *laxiflorum*).

Caprivi Strip. C. 112 km. from Katima Mulilo on the road to Singalamwe, c. 1000 m., fl. 30.xii.1958, *Killick & Leistner* 3204 (M; PRE; SRGH). **Botswana.** N: Chobe Distr., Serondela, 910 m., fl. i.1952, *Miller* B 1274 (FHO; PRE). **Zambia.** B: Sesheke Distr., near Masese Forest Station, fr. 21.v.1952, *White* 1955 (BM; COI; EA; FHO; K). N: Lunzua Gorge, Lake Tanganyika, fr. 23.ii.1955, *Richards* 4616 (K). W: Chingola, fl. 3.ii.1958, *Fanshawe* 4243 (FHO; K; LISC). C: Lusaka Distr., near Chimbala Siding, young fl. 27.xii.1959, *White* 6036 (FHO). S: Livingstone Distr., Malanda Forest Reserve, fr. 10.iii.1960, *White* 7687 (FHO). **Rhodesia.** N: Urungwe Distr., Chirundu road, 520 m., fl. & fr. 26.ii.1953, *Wild* 4038 (BM; SRGH). W: Gwaai Forest Reserve, fr. xi.1937, *McGregor* A 2/38 (FHO; K; SRGH). S: Nuanetsi Distr., Gonakundzingwa, fr. 28.vi.1960, *Farrel* 223 (SRGH). **Mozambique.** T: Tete, banks of R. Zambeze, fr. 23.iii.1966, *Torre & Correia* 15321 (LISC).
Also in Angola and SW. Africa. Riverine forest, *Combretum-Commiphora* thickets and *Baikiaea* woodland; in hotter drier areas from c. 500–1200 m.
Myre & Paisana 2080 (LM) from Mozambique, Sul do Save, Alto Limpopo, not in flower and difficult to determine exactly, seems to be intermediate with subsp. *orientale* having the larger leaves of subsp. *celastroides* but very few hairs except for axillary tufts.
Generally at lower altitudes than subsp. *laxiflorum*, especially in regions where both subspecies occur.

1c. Subsp. **orientale** Exell in Bol. Soc. Brot. Sér. 2, **42**: 16 (1968); in Kirkia, **7**, 2: 169 t. 1 fig. 1c (1970).—Wickens, F.T.E.A. Combret.: 17 (1973).—Faria in Mem. Inst. Invest. Agron. Moçamb. **4**: 11 (1973). TAB. **32** fig. 1c. Type: Mozambique, Delagoa Bay, *Schlechter* 11957 (B†; BM, holotype; COI; K).
Combretum patelliforme Engl. & Diels in Engl., Mon. Afr. Pflanz. **3**: 12, t. 1 fig. C (1899) pro parte excl. specim. *Antunes* A 155 (B†).—Exell & Garcia in Contr. Conhec. Fl. Moçamb. **2**: 102 (1954). Syntypes from Angola and Mozambique (as above).
Combretum trothae Engl. & Diels, tom. cit.: 13, t. 2 fig. A (1899). Type from Tanzania.

Usually a shrub or climber. Leaves papyraceous or rarely chartaceous, nearly glabrous except for scales and for tufts of hairs in the axils of the nerves beneath; lamina up to c. 5 ×2 cm. Disk usually c. 2–2·5(3) mm. in diam. Scales usually golden or silvery (except in some intermediates).

Rhodesia. S: Nuanetsi Distr., Beacon 10 on Mozambique border, 300 m., fr. iv.1953, *Vincent* 206 (BM; SRGH). **Mozambique.** SS: Panda, fr. 25.ii.1955, *E.M. & W.* 594 (BM; LISC; SRGH). LM: between Magude and Panjane, fr. 27.i.1948, *Torre* 7223 (BM; K; LISC).
Also in Tanzania, Natal and possibly also in the Transvaal. Forest patches, deciduous woodland and shrub savanna; in hotter drier areas at low altitudes.

Specimens from the Niassa Prov. of Mozambique seem to be intermediates with subsp. *laxiflorum* (see p. 106).

2. **Combretum padoides** Engl. & Diels in Engl., Mon. Afr. Pflanz. **3**: 13, t. 2 fig. B (1899).—Exell & Garcia in Contr. Conhec. Fl. Moçamb. **2**: 103 (1954).—F. White, F.F.N.R.: 284 (1962).—Boughey in Journ. S. Afr. Bot. **30**, 4: 167 (1964).—Liben, F.C.B. Combret.: 40 (1968).—Exell in Bol. Soc. Brot. Sér. 2, **42**: 7 (1968); in Kirkia, **7**, 2: 170, t. 1 fig. 2 (1970).—Wickens, F.T.E.A. Combret. 17, fig. 4(2) (1973). —Faria in Mem. Inst. Invest. Agron. Moçamb. **4**: 13 (1973). TAB. **30** fig. B; TAB. **32** fig. 2. Syntypes from Tanzania and Mozambique: near Boroma, *Menyharth* (Z, lectotype).
Combretum tenuipes Engl. & Diels, tom. cit.: 13, t. 3 fig. B (1899).—Burtt Davy, F.P.F.T. **1**: 246 (1926). Type from the Transvaal.
Combretum homblei De Wild. in Fedde Repert. **13**: 196 (1914). Type from Zaire.

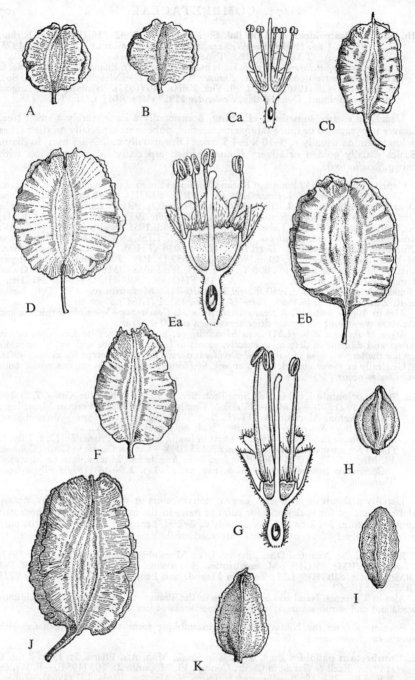

Tab. 30. A.—COMBRETUM CELASTROIDES subsp. CELASTROIDES, fruit (×1). B.—C. PADOIDES, fruit (×1). C.—C. UMBRICOLA, (a) flower, longitudinal section (×6); (b) fruit (×1). D.—C. SCHUMANNII, fruit (×1). E.—C. GILLETTIANUM, (a) flower, longitudinal section (×6); (b) fruit (×1). F.—C. FRAGRANS, fruit (×1). G.—C. PSIDIOIDES subsp. PSIDIOIDES, flower, longitudinal section (×6). H.—C. ILLAIRII, fruit (×1). I.—C. BUTYROSUM, fruit (×1). J.—C. XANTHOTHYRSUM, fruit (×1). K.—C. PISONIIFLORUM, fruit (×1). From F.T.E.A.

Combretum giorgii De Wild. & Exell apud Exell in Journ. Bot. **67**: 100 (1929). Type from Zaire.
Combretum minutiflorum Exell, op. cit. **68**: 245 (1930). Type from Tanzania.
Combretum celastroides sensu Exell & Garcia, tom. cit.: 102 (1954) pro parte quoad specim. Vincent 44 (BM).

Shrub or small tree usually up to c. 5 m. tall in our area (recorded up to 12 m. in Kenya) or liane; bark silvery-grey, smooth; branchlets puberulous to tomentellous when young, glabrescent. Leaves opposite or subopposite or occasionally alternate; lamina 3–10 × 3–4·5 cm. (up to 11·5 × 5·5 cm. in Tanzania), papyraceous, narrowly elliptic to elliptic (to broadly elliptic in Tanzania), rather abruptly acuminate at the apex and cuneate to rounded at the base, usually pubescent on the nerves below but sometimes almost glabrous (except for scales), tufts of hairs in the axils of the nerves beneath absent or rarely conspicuous, lepidote below; lateral nerves 5–8 pairs, rather prominent below, slightly so above; petiole 3–20 mm. long, slender, densely pubescent to almost glabrous, sometimes sparsely lepidote. Inflorescences up to c. 10 cm. long, of terminal panicles of spikes often with unbranched spikes in the axils of the upper leaves; rhachis tomentellous to pubescent, ± glabrescent; bracts not seen. Flowers sessile, yellow or white, lepidote but otherwise glabrous. Sepals 1 × 1·2 mm., broadly ovate, lepidote otherwise glabrous. Petals 1–2 mm. long, linear-elliptic, glabrous. Stamen-filaments 2 mm. long; anthers 0·4 mm. long. Disk 1·5–2 mm. in diam., pilose. Fruit 1·1–1·9 × 1–1·9(2) cm., subcircular in outline or somewhat broader than long, rufous-lepidote otherwise glabrous, apical peg absent or less than 0·5 mm. long, wings c. 7 mm. broad, stipe up to 2 mm. long. Cotyledons 2, 1–1·2 × 1·5–1·8 cm., subcircular to transversely elliptic or subreniform with petioles 2–4 mm. long, borne above soil level. Scales similar to those of C. celastroides but mature leaves with additional smaller scales as little as 50μ in diam. and with a simplified cellular pattern with down to c. 12 marginal cells.

Zambia. C: Feira Distr., 26 km. N. of Feira, fr. 30.v.1952, White 2907 (FHO; K). **Rhodesia.** E: Umtali Distr., Odzi R., Sun Valley Farm, 730 m., fl. 18.xii.1954, Chase 5359 (BM; SRGH). S: Ndanga, Chidumo Clinic, fl. i.1959, Farrell 21 (SRGH). **Malawi.** C: Mareli I., fl. 25.viii.1946, Gouveia & Pedro 1821 (LMJ). S: Fort Johnston, fr. 1954, Jackson 1363 (FHO). **Mozambique.** N: between Mueda and Negomano, fl. & fr. 4.iv.1960, Gomes e Sousa 4566 (COI; K; PRE). T: Tete, fr. 27.iii.1932, Vincent 44 (BM). MS: Báruè, km. 116 from Vila Gouveia, c. 750 m., fl. 20.xii.1965, Torre & Correia 13753 (LISC).
Also in Zaire, Kenya, Tanzania and the Transvaal. Deciduous tree/shrub savanna, dry rocky river banks and Combretum-Commiphora thickets. In hotter drier areas mainly at lower altitudes.

In flower this species can be distinguished from C. celastroides by the pilose disk (glabrous in the latter) but in the absence of flowers the distinction is not easy though the fruit is usually smaller and in most cases has no apical peg or a shorter one than in C. celastroides.
C. padoides is often described as a liane in Tanzania but in our area it is a scrambling shrub or small tree. Tanzanian specimens are more variable with several organs reaching larger dimensions than in our area.
Busse 1098 recorded by Exell and Garcia (loc. cit.) from " Planalto de Macondes " in the Cabo Delgado district of Mozambique probably came from Makondes, north of R. Rovuma in Tanzania, just outside our area.

3. **Combretum imberbe** Wawra in Sitz.-Ber. Math.-Nat. Akad. Wiss. Wien, **38**: 556 (1860).—Engl. & Diels in Engl., Mon. Afr. Pflanz. **3**: 14 (1899).—Monro in Proc. & Trans. Rhod. Sci. Ass. **8**, 2: 80 (1908).—Bak. f. in Journ. Linn. Soc. Bot. **40**: 67 (1911).—Eyles in Trans. Roy. Soc. S. Afr. **5**, 4: 428 (1916).—Steedman, Trees etc. S. Rhod.: 55 (1933).—Burtt Davy & Hoyle, N.C.L.: 38 (1936).—Brenan, T.T.C.L.: 138 (1949); in Mem. N.Y. Bot. Gard. **8**, 5: 437 (1954).—Wild, Guide Fl. Vict. Falls: 146 (1953); in Rhod. Agric. Journ. **50**, 5: 3, 4, 5 (1953); op. cit. **52**, 6: 13 (1955).—Exell & Garcia in Contr. Conhec. Fl. Moçamb. **2**: 103 (1954); C.F.A. **4**: 52 (1970).—Williamson, Useful Pl. Nyasal.: 38 (1955).—Story in Bot. Surv. S. Afr. **30**: 37, phot. 35 (1958).—Rattray in Kirkia, **2**: 85 (1961).—F. White, F.F.N.R.: 284 (1962).—Breitenb., Indig. Trees Southern Afr. **4**: 828 (1965).—Exell, Prodr. Fl. SW. Afr. **99**: 9 (1966); in Bol. Soc. Brot. Sér. 2, **42**: 7 (1968); in Kirkia, **7**, 2: 171, t. 1 fig. 3 (1970).—B. de Winter, M. de Winter & Killick, Sixty-six Transv. Trees: 122, cart. et phott. (1966).—Wickens, F.T.E.A. Combret.: 18, fig. 1(4), 4(4)

(1973).—Faria in Mem. Inst. Invest. Moçamb. **4**: 15 (1973). TAB. **31** fig. F; TAB. **32** fig. 3. Type from Angola (Benguela).
Argyrodendron petersii Klotzsch in Peters, Reise Mossamb., Bot. **1**: 101 (1861). Type: Mozambique, Sena, *Peters* (B†, holotype).
Combretum elaeagnoides sensu Laws. in Oliv., F.T.A. **2**: 426 (1871) pro parte quoad specim. Angol.
Combretum truncatum Welw. ex Laws. in Oliv., tom. cit.: 427 (1871).—Sim, For. Fl. Port. E. Afr.: 62, t. 62 fig. B (1909).—O. B. Mill., B.C.L.: 44 (1948). Syntypes from Angola and Mozambique: Lupata, *Kirk* (K).
Combretum primigenum Marloth ex Engl., Bot. Jahrb. **10**: 49 (1888).—Engl. & Diels in Engl., tom. cit.: 14, t. 3 fig. C (1899).—Eyles, loc. cit.—Steedman in Proc. Rhod. Sci. Ass. **26**: 10 (1927).—O. B. Mill., tom. cit.: (1948). Type from SW. Africa.
Combretum petersii (Klotzsch) Engl., Pflanzenw. Ost-Afr. C: 290 (1895).—Henkel in Proc. Rhod. Sci. Ass. **30**: 16 (1931). Type as for *Argyrodendron petersii*.
Combretum imberbe var. *dielsii* Engl. apud Engl. & Diels in Engl., tom. cit.: 14 (1899). Type from Tanzania.
Combretum imberbe var. *petersii* (Klotzsch) Engl. & Diels in Engl., tom. cit.: 14, t. 2 fig. C (1899).—Burtt Davy & Hoyle, N.C.L.: 38 (1936).—O. B. Mill., loc. cit.— Pardy in Rhod. Agric. Journ. **51**, 3: 171 cum phot. (1954).—Boughey in Journ. S. Afr. Bot. **30**, 4: 167 (1964). Type as for *Argyrodendron petersii*.
Combretum imberbe var. *truncatum* (Welw. ex Laws.) Burtt Davy, F.P.F.T. **1**: 246 (1926). Syntypes as for *C. truncatum*.

Tree up to 30 m. tall (usually 6–10 m.) or shrub; bark light-grey or whitish; branchlets often becoming spiny, glabrous. Leaves silvery, opposite or sub-opposite; lamina 2·5–8·5 × 1–3 cm., papyraceous to subcoriaceous, narrowly elliptic to elliptic-oblong, densely silvery-lepidote on both surfaces otherwise glabrous, scales often contiguous or overlapping, apex obtuse or rounded and often mucronate, base obtuse to narrowly cuneate; lateral nerves 4–7 pairs, slightly prominent on both surfaces, usually without domatia; petiole 4–10 mm. long, lepidote but otherwise glabrous or almost so. Inflorescences of spikes up to c. 10 cm. long, often forming a terminal panicle by suppression of the upper leaves, or unbranched lateral spikes up to 5 cm. long; rhachis lepidote, otherwise glabrous or sparsely puberulous; bracts absent or very caducous. Flowers sessile, yellowish, densely lepidote but otherwise glabrous or nearly so. Lower receptacle 2·5(3) × 0·25–0·3 mm., densely silvery- or rufous-lepidote, otherwise glabrous or nearly so; upper receptacle little developed, almost flat when the flowers are mature, densely silvery- or rufous-lepidote, otherwise glabrous or nearly so. Sepals 1·5 × 1·5 mm., ovate-triangular. Petals 1–1·2 × 0·4–0·6 mm., obovate to spathulate, glabrous. Stamen-filaments 1·5–2 mm. long; anthers 0·3–0·4 mm. long. Disk 2–2·5 mm. in diam., margin densely tomentose. Fruit 1·3–1·8 cm., subcircular to broadly ovate in outline, fairly densely to densely silvery-lepidote, apex pointed, apical peg very short (up to 0·5 mm.) to medium (1 mm.) or absent, wings up to 7 mm. broad, stipe 2–3 mm. long. Cotyledons 2, 15–18 × 15 mm., subcircular, with petioles 9–11 mm. long arising at or below soil level. Scales 120–300µ in diam., roughly circular, cells very numerous and small (marginal cells c. 40–100), divided as in the sectional description; cells beneath the scales with conspicuous round papillae.

Caprivi Strip. Linyanti, 915 m., fl. 27.xii.1958, *Killick & Leistner* 3144 (EA; FHO; PRE; SRGH). **Botswana.** N: Maun, fr. 2.vi.1930, *Van Son* in Herb. Transv. Mus. 28823 (BM; K; PRE). SE: near Mahalapye railway line, fl. 16.vii.1957, *de Beer* 524 (K; SRGH). **Zambia.** B: Sesheke Boma, fl. 27.xii.1952, *Angus* 1051 (FHO; K). N: Mpika, Luangwa Valley, *Michelmore* 632 (K). S: Katambora, fl. 2.xii.1935, *Gilges* 496 (PRE; SRGH). **Rhodesia.** N: Darwin Distr., 510 m., Mkumbura R., fl. 23.i.1960, *Phipps* 2386 (K; SRGH). W: Bulawayo Commonage, fl. 1932, *Pitt-Schenkel* 182 (FHO). C: Gwelo, Nahla Farm, fl. i, *Eyles* 7028 (K; SRGH). E: Melsetter Distr., Hot Springs Hotel, 610 m., fl. 9.xii.1951, *Chase* 4252 (BM; COI; LISC; SRGH). S: Victoria Distr., fl. 1909, *Monro* 646 (BM; SRGH). **Malawi.** S: Zomba Distr., fr. *Clements* 636 (FHO; K). **Mozambique.** N: Nampala Distr., Mutuáli, fr. 23.iii.1953, *Gomes e Sousa* 4083 (COI; K; LISC; PRE). Z: Mocuba, fr. 29.v.1949, *Andrada* 1535 (COI; LISC). T: between Km. 148 on the railway and Chineza Station, fr. 18.v.1948, *Mendonça* 4328 (BM; K; LISC). MS: Búzi, between R. Búzi and Grudja, fl. 11.xi.1941, *Torre* 3810 *Torre* 2746 (BM; K; LISC). LM: Goba, fl. 23.xii.1944, *Mendonça* 3472 (BM; K; LISC).
Also in Angola, SW. Africa, Tanzania and the Transvaal. Tree savanna (Botswana),

on sandy karroo soils with limestone outcrops, secondary *Baikiaea* woodland, *Acacia-Combretum-Terminalia* open woodland, mopane woodland, alluvial plains and edges of swamps, on basalt, grey loams and black cotton soils; at low altitudes.

The timber, yellow and easily worked, is used in turnery and also for fencing props, mine props and sleepers. Leaves much eaten by cattle.

C. imberbe is sometimes confused with *C. elaeagnoides* which is similarly densely conspicuously lepidote. The latter can be at once distinguished in flower by the much more developed cupuliform to infundibuliform upper receptacle, and in fruit by the much longer stipe (up to c. 12 mm. long).

2. Sect. COMBRETASTRUM Eichl. in Mart., Fl. Bras. **14**, 2: 115 (1867).—Keay in Kew Bull. **1950**: 252 (1950).—Exell in Journ. Linn. Soc. Bot. **55**: 108 (1953); in Kirkia, **7**, 2: 172 (1970).—Stace in Bot. Journ. Linn. Soc., Bot. **42**, 2: 143 (1969).—Wickens in F.T.E.A. Combret.: 19 (1973).

Sect. *Olivaceae* Engl. & Diels in Engl., Mon. Afr. Pflanzen **3**: 20 (1899).

Flowers 4-merous, small. Upper receptacle cupuliform to campanulate. Petals subreniform, glabrous. Stamens 8, biseriate. Disk very small, often inconspicuous. Fruit 4-winged (in our species) or 4-angled. Scales usually of a simple 8-celled type, 40–55(70)μ in diam., usually with the cuticle raised off the cell plate, sometimes rather inconspicuous.

4. **Combretum umbricola** Engl. [in Abh. Preuss. Akad. Wiss. **1894**: 17 (1894) *nom. nud.*] Pflanzenw. Ost-Afr. **C**: 288 (1895) (" umbricolum ").—Engl. & Diels in Engl., Mon. Afr. Pflanz. **3**: 23, t. 5 fig. D (1899).—Exell & Garcia in Contr. Conhec. Fl. Moçamb. **2**: 105 (1954).—Liben, F.C.B. Combret.: 56 (1968).—Exell in Bol. Soc. Brot., Sér. 2, **42**: 5 (1968); in Kirkia, **7**, 2: 173, t. 1 fig. 4 (1970).—Wickens, F.T.E.A., Combret.: 20, fig. 1(6), 3(6), 4(6) (1973).—Faria in Mem. Inst. Invest. Agron. Moçamb. **4**: 20 (1973). TAB. **30** fig. C; TAB. **32** fig. 4. Type from Tanzania (Usambara).

Small tree or shrub, sometimes scandent; branchlets densely pubescent when young, soon glabrescent, often silvery-lepidote when young. Leaves opposite; lamina 3·5–16 × 2–6·5 cm., chartaceous to subcoriaceous, elliptic to elliptic-oblong, minutely verrucose and usually shiny and glabrous above, minutely silvery-lepidote otherwise glabrous beneath, apex usually acuminate, base cuneate or obtuse to rounded; lateral nerves 6–10 pairs, usually slightly impressed above; petiole up to 9(13) mm. long, glabrescent. Inflorescence up to 20 cm. long, a terminal or axillary panicle of spikes; rhachis densely fuscous-pubescent when young, ± glabrescent; bracts 1–2 mm. long, linear, caducous. Flowers sessile, white, scented. Lower receptacle c. 1 mm. long, glabrous or pubescent; upper receptacle 1·2–1·5 × 1·2–1·5 mm., often petaloid, usually not conspicuously lepidote. Sepals very shallowly triangular or little developed. Petals 0·9 × 1·5 mm., subreniform, glabrous. Stamen-filaments 3–4 mm. long; anthers 0·5 mm. long; style 4 mm. long, glabrous. Fruit up to 2·8 × 2·3 cm., broadly elliptic in outline, minutely lepidote otherwise usually glabrous, wings 7 mm. broad, chartaceous, apical peg 1–2 mm. long, stipe up to 5 mm. long. Scales as for the section, frequent in the areoles of the lower surface of the leaf.

Mozambique. N: Malema, fr. 20.vi.1948, *Pedro & Pedrógão* 4341 (EA; LMJ). Z: Mopeia, between Naquexa and Cundine, fl. 22.ii.1905, *Le Testu* 700 (BM). Also in Zaire, Uganda and Tanzania. Riverine forest at lower altitudes.

3. Sect. ANGUSTIMARGINATA Engl. & Diels in Engl., Mon. Afr. Pflanzen **3**: 25 (1899) excl. spp. *C. prunifolium* et *C. kirkii*.—Stace in Journ. Linn. Soc., Bot. **62**: 155, figs. 119–125 (1969).—Exell in Kirkia, **7**, 2: 173 (1970).

Flowers 4-merous. Upper receptacle cupuliform to campanulate. Petals subcircular, obovate, narrowly obovate, spathulate or narrowly elliptic, glabrous. Stamens 8, 1-seriate or nearly so, inserted shortly above the margin of the disk. Disk glabrous with a pilose margin very shortly free (up to 0·5 mm.). Style not expanded. Fruit 4-winged. Cotyledons 2(3) arising above soil level. Scales inconspicuous, often obscured by the indumentum and/or glutinous secretions, more abundant on the upper surface of the leaf than the lower, c. 50–75μ in diam., not scalloped, delimited by 8–16 primary radial walls alone; cell-walls very thin.

Flowers pubescent to tomentose; leaves usually pubescent to tomentose beneath (but sometimes nearly glabrous), lateral nerves 6–8(12) pairs -   -   5. *erythrophyllum*
Flowers glabrous (except for scales) or very nearly so; leaves nearly glabrous (except for scales), lateral nerves 4–6 pairs - - - - - - - 6. *kraussii*

5. **Combretum erythrophyllum** (Burch.) Sond. in Linnaea, **23**: 43 (1850); in Harv. & Sond., F.C. **2**: 509 (1862).—Engl. & Diels in Engl., Mon. Afr. Pflanz. **3**: 26, t. 7 fig. D (1899).—Bak. f. in Journ. Linn. Soc. Bot. **40**: 67 (1911).—Eyles in Trans. Roy. Soc. S. Afr. **5**, 4: 427 (1916).—Boughey in Journ. S. Afr. Bot. **30**, 4: 167 (1964).— Breitenb., Indig. Trees Southern Afr. **4**: 836 (1965).—B. de Winter, M. de Winter & Killick, Sixty-six Transv. Trees: 120, cart. et phott. (1966).—Exell, Prodr. Fl. SW. Afr. **99**: 8 (1966); in Bol. Soc. Brot. Sér. 2, **42**: 7 (1968); in Kirkia, **7**, 2: 173, t. 1 fig. 5 (1970).—Wild in Kirkia, **7**, 1: 53 (1968).—Faria in Mem. Inst. Invest. Agron. Moçamb. **4**: 21 (1973).—Jacobsen in Kirkia, **9**, 1: 168 (1973). TAB. **32** fig. 5. Type from S. Africa.
    Terminalia? erythrophylla Burch., Trav. S. Afr. **1**: 400 (1822). Type as above.
    Combretum glomeruliflorum Sond. in Linnaea, **23**: 47 (1850); in Harv. & Sond., F.C. **2**: 509 (1862).—Engl. & Diels in Engl. tom. cit.: 26, t. 7 fig. B (1899).—Monro in Proc. & Trans. Rhod. Sci. Ass. **8**, 2: 80 (1908).—Eyles, tom. cit.: 428 (1916).— Burtt Davy, F.P.F.T. **1**: 247, fig. 35A–C excl. fig. 35B (1926).—Codd, Trees and Shrubs Kruger Nat. Park: 130, fig. 121a (1951). Type from S. Africa.
    Combretum riparium Sond., loc. cit.; in Harv. & Sond., tom. cit.: 511 (1862). Type from S. Africa.
    Combretum sonderi Gerr. ex Sond. in Harv. & Sond., loc. cit. Type from S. Africa.
    Combretum erythrophyllum var. *obscurum* Heurck & Müll. Arg. in Heurck, Obs. Bot. **2**: 237 (1871). Type from S. Africa.
    Combretum ligustrifolium Engl. & Diels ex Bak. f. in Journ. of Bot. **37**: 436 (1899) *nom. nud.*
    Combretum lydenburgianum Engl. & Diels in Engl., tom. cit.: 26, t. 7 fig. C (1899).— Eyles in Trans. Roy. Soc. S. Afr. **5**, 4: 428 (1916) ("lindenburgianum"). Type from the Transvaal.
    Combretum salicifolium sensu Monro in Proc. & Trans. Rhod. Sci. Ass. **8**, 20: 80 (1908).—Sim, For. Fl. Port. E. Afr.: 64 (1909).—Eyles, tom. cit.: 429 (1916).— Steedman in Proc. & Trans. Rhod. Sci. Ass. **24**: 17 (1925).
    Combretum glomeruliflorum var. *obscurum* (Heurck & Müll. Arg.) Burtt Davy, F.P.F.T. **1**: 37, 247 (1926). Type as for *C. erythrophyllum* var. *obscurum*.
    Combretum glomeruliflorum var. *riparium* (Sond.) Burtt Davy, loc. cit.—Steedman, Some Trees etc., S. Rhod.: 55 (1933). Type as for *C. riparium*.

    Small to large spreading tree; young branchlets tomentose with rather persistent indumentum, pinkish after shedding the bark. Leaves subopposite or occasionally 3-verticillate; lamina sometimes pale-yellow when young, usually c. 5 × 2 cm., elliptic, rather inconspicuously lepidote on both surfaces otherwise nearly glabrous above except for variable hairiness on the midrib and nerves, tomentose to nearly glabrous beneath, apex usually acute and mucronate, base cuneate; lateral nerves 6–8(12) pairs; petiole 1–4 mm. long, tomentose. Inflorescences usually of unbranched axillary spikes up to 2(3) cm. long, rarely forming short panicles by the suppression of leaves on short shoots; rhachis and peduncles usually tomentose more rarely glabrescent. Flowers cream or yellow, usually ± congested. Lower receptacle 1·5–2·5 × 1 mm., usually tomentose rarely pubescent; upper receptacle 2–2·5 × 3 mm., broadly campanulate, pubescent to densely pubescent, with some scales visible when the indumentum is not too dense. Sepals c. 1 × 1·3–1·5 mm., deltate to broadly deltate. Petals c. 1·5 × 1 mm., obovate, narrowly obovate, spathulate or narrowly elliptic, sometimes emarginate, unguiculate, glabrous. Stamen-filaments 5 mm. long; anthers 1 mm. long. Disk 1·5 × 1, cupuliform, glabrous, with a pilose margin free for c. 0·5 mm. Fruit 1·3 × 1 cm., subcircular to broadly elliptic in outline, lepidote and pubescent, apical peg absent, wings c. 5 mm. wide, slightly decurrent at the base, stipe up to 7 mm. long. Scales as for the section, almost always some with at least 12 primary walls.
    Figures by Engler and Diels (loc. cit.) show a markedly 2-seriate insertion of the stamens but in all the flowers I have dissected they appear to be very nearly 1-seriate and close to the margin of the disk. The petals are very variable in shape

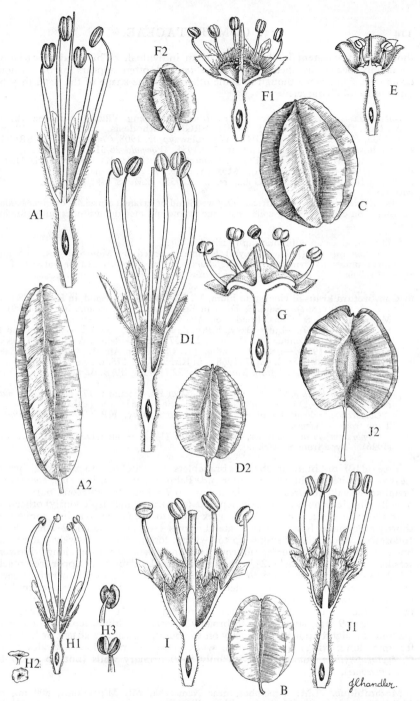

Tab. 31. A.—COMBRETUM PLATYPETALUM. A1, flower, longitudinal section (×4); A2, fruit (×1). B.—COMBRETUM MICROPHYLLUM, fruit (×1). C.—COMBRETUM LASIOCARPUM, fruit (×1). D.—COMBRETUM MOSSAMBICENSIS. D1, flower, longitudinal section (×2½); D2, fruit (×1). E.—MEIOSTEMON TETRANDRUS subsp. TETRANDRUS, flower, longitudinal section (×6). F.—COMBRETUM IMBERBE. F1, flower, longitudinal section (×6); F2, fruit (×1). G.—COMBRETUM CELASTROIDES subsp. LAXIFLORUM, flower, longitudinal section (×4). H.—COMBRETUM ELAEAGNOIDES. H1, flower, longitudinal section (×6); H2, 2 scales from lower receptacle (×12½); H3, 2 anthers (×24). I.—COMBRETUM SCHUMANNII, flower, longitudinal section (×6). J.—COMBRETUM ZEYHERI. J1, flower. longitudinal section (×4); J2, fruit (×½). From F.F.N.R,

though fairly constant in the flowers of an individual. This kind of variation commonly occurs in species of *Combretum* and often appears to be without taxonomic significance though in some other cases the shape of the petal may be quite a useful character.

Botswana. SE: Gaberones Distr., near Tlhokweng Village, Notwane R., fl. 11.xii.1961, *Yalala* 147 (FHO; LISC; SRGH). **Rhodesia.** N: Mazoe, 310 m., fl. xi.1906, *Eyles* 434 (BM; SRGH). W: Bulawayo, fr. 1908, *Chubb* 50 (BM; SRGH). C: Chilimanzi Tribal Trust Land, fr. 7.iii.1951, *Wormald* 26/51 (BM; SRGH). E: Nyahodi R., 1675 m., fl. 20.ix.1906, *Swynnerton* 701 (BM; K). S: Victoria Distr., fl. 1909, *Monro* 499 (BM; SRGH). **Mozambique.** Z: Quelimane, 120 m., fr. iii.1946, *Faulkner* 10 (FHO). LM: Goba, fr. 23.ii.1955, *E.M. & W.* 562 (BM; LISC; SRGH).

Also in S. Africa and SW. Africa. Dry woodland or savanna especially on river banks; occasional on copper-bearing soils; common at low and medium altitudes but occurring up to 1500 m.

Said to produce an antibiotic sap (Faria, loc. cit.).

Combretum caffrum (Eckl. & Zeyh.) Kuntze recorded by Monro, Eyles, Sim and Steedman under its synonym *C. salicifolium* may be a subspecies or variety of *C. erythrophyllum* but I have not seen this narrow-leaved plant from our area.

6. **Combretum kraussii** Hochst. in Flora, **27**, 2: 424 (1844).—Sond. in Harv. & Sond., F.C. **2**: 510 (1862).—Engl. & Diels in Engl., Mon. Afr. Pflanz. **3**: 28 (1899).— Monro in Proc. & Trans. Rhod. Sci. Ass. **8**, 2: 80 (1908)?—Sim, For. Fl. Port. E. Afr.: 63 (1909).—Burtt Davy, F.P.F.T. **1**: 246, fig. 35D–F (1926).—Exell & Garcia in Contr. Conhec. Fl. Moçamb. **2**: 137 (1954).—Palmer & Pitman, Trees of S. Afr.: 248 (1961).—Breitenb., Indig. Trees Southern Afr. **4**: 830 (1965).—Exell in Bol. Soc. Brot., Sér. 2, **42**: 7 (1968); in Kirkia, **7**, 2: 175, t. 1 fig. 6 (1970).—Faria in Mem. Inst. Invest. Agron. Moçamb. **4**: 22 (1973). TAB. **32** fig. 6. Type from S. Africa.

    Combretum lucidum E. Mey. ex Drège, Zwei Pfl.-Docum.; 174 (1843) *nom. nud.* non Bl. (1825).—Dummer in Gard. Chron., Ser. 3, **53**: 164 (1913).

    Combretum nelsonii Dummer, loc. cit.—Burtt Davy, F.P.F.T. **1**: 246 (1926). Type from S. Africa.

    Combretum woodii Dummer, tom. cit.: 181 (1913).—Burtt Davy, tom. cit.: 245 (1926). Type from S. Africa.

Tree 6(10) m. high or shrub; branchlets tomentellous to nearly glabrous. Leaves opposite, bright red in winter (*fide* Palmer & Pitman, loc. cit.), sometimes bractlike near the inflorescences; lamina up to 9 × 5 cm., chartaceous, narrowly to broadly elliptic or obovate-elliptic, lepidote (sometimes inconspicuously) otherwise nearly glabrous, often shiny above, apex acute to obtuse or rounded, sometimes shortly mucronate, base cuneate; lateral nerves 4–6 pairs; petiole 2–7 mm. long. Inflorescences of axillary spikes up to 1–3 cm. long; rhachis densely pubescent; bracts c. 2 mm. long, filiform, caducous; peduncle usually c. 1 mm. long. Flowers sessile. Lower receptacle 2–3 × 0·5 mm., viscid, glabrous; upper receptacle 1·5–2 × 3–4 mm., cupuliform to broadly campanulate, lepidote otherwise nearly glabrous. Sepals 1·5 × 2 mm., deltate. Petals 1–2 × 0·7–0·8 mm., spathulate, glabrous. Stamen-filaments 5–6 mm. long; anthers 1 mm. long. Disk with pilose margin free for c. 0·5 mm. Style 5 mm. long, rather stout. Fruit reddish, 1·3–2 × 1·3–1·9 cm., subcircular to broadly elliptic in outline, lepidote otherwise glabrous to slightly pubescent mainly on the body; apical peg absent or less than 0·5 mm. long, wings up to 8 mm. wide, stipe 4–7 mm. long. Scales as in *C. erythrophyllum* but usually with only 8–10 primary walls (but up to 10–16 in *C. nelsonii*).

Mozambique. LM: Libombos, near Namaacha, Mt. M'ponduine, 800 m., fr. 22.ii.1955, *E.M. & W.* 511 (BM; LISC; SRGH).

Also in S. Africa. Rocky hillsides, forest margins, rain-forest, grassland; at medium altitudes.

Said to cause dermatitis among forestry workers (Faria, loc. cit.).

The record by Monro (loc. cit.) is probably incorrect.

C. kraussii can usually be distinguished from *C. erythrophyllum* by the nearly glabrous flowers and fruits (apart from scales) and by the leaves being frequently more rounded at the apex and less narrowly cuneate at the base.

4. Sect. MACROSTIGMATEA Engl. & Diels in Engl., Mon. Afr.
Pflanz. **3**: 24 (1899).—Stace in Bot. Journ. Linn. Soc. **62**: 159 (1969).—
Exell in Kirkia, **7**, 2: 176 (1970).—Wickens, F.T.E.A., Combret.:
21 (1973).

Flowers 4-merous, glutinous, in subcapitate spikes. Upper receptacle cupuliform
to infundibuliform. Petals ovate, subcircular, obtriangular or obovate, glabrous.
Stamens 8, 1-seriate, inserted at the margin of the disk. Disk glabrous with only a
very short free margin or with a pilose margin free for c. 1 mm. Style sometimes
(perhaps always) with an expanded stigma (but the latter quickly withering).
Cotyledons 2, borne above soil level. Scales 45–85μ in diam., usually of a simple
8-celled type with the ± constant addition of a number of tangential and extra
radial walls giving up to 16 marginal cells.

This section is somewhat heterogeneous and could be divided into two subsections on
the character of the disk which is glabrous with only a very short free margin in *C.
schumannii* and has a pilose margin free for c. 1 mm. in *C. kirkii* and *C. gillettianum*.

Disk glabrous with only a very short free margin; anthers 0·6 × 0·5 mm.; fruit 2·5–3·5 cm.
   long; plant a small tree or shrub (occasionally said to be scandent)   7. *schumannii*
Disk with a pilose margin free for c. 1 mm.; anthers 1–1·2 × 0·8–1 mm.; fruit up to 5 cm.
   long; plant a liane:
   Fruit not decurrent at the base; stipe up to 2·5(3) cm. long - - - 8. *kirkii*
   Fruit decurrent at the base; stipe up to 1·2(2) cm. long - - - 9. *gillettianum*

7. **Combretum schumannii** Engl. [in Abh. Preuss. Akad. Wiss. 1894: 34 (1894) *nom.
nud.*] Pflanzenw. Ost-Afr. C: 289 (1895).—Engl. & Diels in Engl., Mon. Afr. Pflanz.
3: 24, t. 6 fig. C (1899).—O. B. Mill., B.C.L.: 43 (1948).—Brenan, T.T.C.L.: 138
(1949).—Dale & Greenway, Kenya Trees and Shrubs: 146 (1961).—Exell in Bol.
Soc. Brot., Sér. 2, **42**: 7 (1968); in Kirkia, **7**, 2: 177, t. 1 fig. 8 (1970).—Gillett &
McDonald, N.C.K.: 15 (1970).—Wickens, F.T.E.A. Combret.: 21, fig. 4(7) (1973).
TAB. **30** fig. D; TAB. **31** fig. I; TAB. **32** fig. 7. Type from Tanzania.
   Combretum macrostigmateum Engl. & Diels in Engl., loc. cit. Type from Tanzania.
   Combretum engleri Schinz apud De Wild. & Dur. in Bull. Herb. Boiss., Sér. 2, **1**:
878 (1901).—F. White, F.F.N.R.: 285, fig. 49 (1962) pro parte excl. specim. *Angus*
773 et *White* 3679.—Fanshawe & Savory in Kirkia, **4**: 190 (1964).—Exell, Prodr. Fl.
SW. Afr. **99**: 8 (1966); in Bol. Soc. Brot. Sér. 2, **42**: 7 (1968); in Kirkia, **7**, 2: 176,
t. 1 fig. 7 (1970).—Exell & Garcia, C.F.A. **4**: 53 (1970). Type from SW. Africa.
   Combretum myrtillifolium Engl. in Pflanzenw. Afr. **3**, 2: 695 (1921). Type as above.
Journ. S. Afr. Bot. **18**: 62 (1952). Non Engl. (1895). Type from SW. Africa.
   Combretum chlorocarpum Exell in Journ. of Bot. **66**, Suppl. Polypet.: 167 (1928).
Type from Angola.

Deciduous shrub or small tree up to 18 m. high (occasionally described as
scandent); bark grey, scaly; branchlets slender, usually pale, glabrous or slightly
pubescent and soon glabrescent. Leaves opposite; lamina 2–9(10) × 0·7–4·5 cm.,
papyraceous to chartaceous, narrowly elliptic to ovate or obovate, rather inconspicuously lepidote and verruculose above otherwise nearly glabrous except for
tufts of hairs in the axils of the lateral nerves beneath, somewhat shiny above, apex
rather blunt to rounded sometimes somewhat acuminate, base cuneate to rounded;
lateral nerves 3–5 pairs; petiole 2–8(10) mm. long, slender. Inflorescences of short
almost sessile glutinous lateral spikes, sometimes branched or fasciculate, often
appearing in the axils of fallen leaves with very young leaves of the current season;
rhachis sometimes up to 3–4 cm. long in fruit; bracts minute, caducous. Flowers
pale-yellow. Lower receptacle 1–1·5 mm. long, glutinous, glabrous; upper
receptacle 2–2·5 × 2·5–3 mm., broadly infundibuliform, glutinous, rather inconspicuously lepidote otherwise glabrous, crenulated to form 4 pouch-like swellings
alternating with the petals and in the position of the sepals which are otherwise
scarcely developed. Petals 1·6–2 × 1·8 mm., broadly ovate to subcircular, glabrous,
shortly unguiculate. Stamen-filaments 3–4·5 mm. long; anthers 0·6 mm. long.
Disk glabrous, margin scarcely produced. Style 5 mm. long, slightly expanded at
the apex. Fruit green, up to 3·5 cm. in diam., subcircular in outline, glutinous
when young, rather inconspicuously lepidote otherwise glabrous, apical peg absent
or very short, wings up to 1·2 cm. broad and rather thin, stipe up to 1·8 cm. long,
slender. Cotyledons 2, arising above soil level; petioles 4–5 mm. long. Scales
whitish- to reddish-translucent, 45–75μ in diam., rather sparse to frequent,

circular; cells delimited by 8(12) radial walls and also by tangential and extra radial walls; marginal cells (8)10–16.

Botswana. N: Chobe Distr., Serondela, fl. xi.1951, *Miller* B/1199 (PRE; SRGH). **Zambia.** B: Sesheke Distr., fl. 1911, *Macaulay* 223 (BM; K). N: Kalaba-Bulaya, fl. & fr. 20.x.1949, *Bullock* 1333 (K). S: Kalomo, Mulobezi, fr. 9.iv.1955, *E.M. & W.* 1442 (BM; LISC; SRGH). **Malawi.** S: Fort Johnston, 550 m., fr. 7.vii.1936, *Burtt* 5986 (BM; EA; K). **Mozambique.** Z: Maganja da Costa, Gobene, c. 20 m., fr. 12.ii.1966, *Torre & Gouveia* 14582 (LISC).

Also in Angola, SW. Africa, Zaire, Kenya and Tanzania. *Baikiaea* woodland and thickets on Kalahari Sand; locally common in dry deciduous *Baikiaea* and *Pteleopsis* forest; *Brachystegia* woodland and *Acacia-Combretum-Terminalia* tree savanna.

8. **Combretum kirkii** Laws. in Oliv., F.T.A. **2**: 429 (1871) pro parte excl. specim. Welw.—Engl. & Diels in Engl., Mon. Afr. Pflanz. **3**: 29 (1899) excl. var.? *C. tinctorum.*—Sim, For. Fl. Port. E. Afr.: 64 (1909).—Exell & Garcia in Contr. Conhec. Fl. Moçamb. **2**: 115 (1954).—Exell in Bol. Soc. Brot. Sér. 2, **42**: 7 (1968); in Kirkia, **7**, 2: 178, t. 1 fig. 9 (1970).—Faria in Mem. Inst. Invest. Agron. Moçamb. **4**: 23 (1973). TAB. **32** fig. 8. Type: Mozambique, Tete, *Kirk* (K, holotype).

  Combretum menyhartii Engl. & Diels in Engl., tom. cit.: 46, t. 14 fig. B (1899).— Schinz, Pl. Menyharth.: 432 (1906). Type: Mozambique, near Boroma, *Menyharth* (K; Z, holotype).

Liane climbing to 15 m.; stems up to 15 cm. in diam.; bark grey to silvery, rough and flaky. Leaves opposite; lamina 5–9 × 2·5–5 cm., papyraceous to chartaceous, obovate-elliptic or narrowly elliptic, rather inconspicuously lepidote otherwise nearly glabrous except for tufts of hairs in the axils of the lateral nerves beneath, apex blunt to acute and slightly acuminate, base cuneate; lateral nerves 5–7 pairs; petiole up to 12 mm. long, slender. Inflorescences of short axillary subcapitate spikes up to 2·5 cm. long; rhachis glabrous; bracts not seen; peduncle up to 1 cm. long. Flowers sessile, congested. Lower receptacle 3 mm. long, lepidote otherwise glabrous; upper receptacle 3·5–4 × 2·5–3 mm., infundibuliform, lepidote otherwise glabrous. Sepals 0·5 mm. long, acuminate. Petals 1·2–2 × 0·8– 1·5 mm., obovate to subcircular, unguiculate, glabrous. Stamen-filaments 3–5 mm. long; anthers 1–1·5 mm. long. Disk with a pilose margin free for 1 mm. Style 6 mm. long, slightly expanded at the apex. Fruit brownish- or yellowish-green or green, up to 5·5 × 4–5 cm., subcircular to elliptic in outline, lepidote otherwise glabrous, apical peg absent or very short, not decurrent at the base, wings 1·5–2 cm. broad and rather thin, stipe up to 2·5(3) cm. long, slender. Cotyledons 2, transversely elliptic, 3–3·5 × 5·5 cm., petioles 2–4 mm. long, arising above soil level. Scales c. 50–85µ in diam.

Zambia. S: Gwembe Valley, Sinazongwe-Mwemba road crossing, c. 500 m., fr. 23.vi.1961, *Bainbridge* 445 (FHO). **Rhodesia.** N: Kariba, fr. iv.1959, *Goldsmith* 23/59 (BM; EA; K; SRGH). **Malawi.** S: Monkey Bay, Nkapi turn-off, fl. & fr. 4.xii.1966, *Eccles* 53 (MAL). **Mozambique.** N: Ribáuè, fr. 26.viii.1967, *Macedo* 2658 (LMA). T: Tete, fr. ii.1860, *Kirk* (K).

Apparently mainly confined to the valley of the Zambezi and its tributaries. Dry low altitude thickets and riverine formations.

Very near to *C. schumannii* but a vigorous liane with much larger fruits which are sometimes confused with those of *C. zeyheri.*

9. **Combretum gillettianum** Liben in Bull. Jard. Bot. Brux. **35**, 2: 117 (1965); F.C.B., Combret.: 70, t. 6 (1969).—Exell in Bol. Soc. Brot. Sér. 2, **42**: 7 (1968); in Kirkia, **7**, 2: 178, t. 1 fig. 10 (1970).—Wickens, F.T.E.A. Combret.: 22, fig. 1(8), fig. 4(8) (1973). TAB. **30** fig. E; TAB. **32** fig. 9. Type: Zambia, Mpulungu, *Glover* in *Bredo* 6143 (BR, holotype).

  Combretum sp.—Hutch., Botanist in S. Afr.: 520 (1946).
  Combretum engleri sensu F. White, F.F.N.R.: 285 (1962) pro parte quoad specim. *Angus* 773 A et *White* 3679.

Shrub or small tree up to 4 m. high, or sometimes scandent; young branchlets purple, tomentellous (glabrous *fide* Liben). Leaves opposite; lamina up to 4(5·5) × 1·5(2) cm., papyraceous to chartaceous or subcoriaceous, narrowly elliptic, conspicuously lepidote when young but inconspicuously so when mature, otherwise nearly glabrous except for some indumentum on the nerves when young and tufts of hairs in the axils of the nerves beneath, apex rather blunt, base rounded; lateral nerves 3–5 pairs; petiole up to 3·5 mm. long, pubescent. Inflorescences of short subcapitate spikes in the axils of the young leaves or forming a short panicle by suppression of the upper leaves; rhachis tomentose; bracts 1·5 × 0·5 mm., obovate-elliptic, caducous. Lower receptacle 2–2·5 mm. long, tomentellous; upper receptacle 3·5–4(4·5) × 3–4 mm., cupuliform, appressed-pubescent often with 4 tomentellous nerves running up into the sepals. Sepals 1·5 × 1·2 mm., triangular. Petals 1·3 × 1 mm., obtriangular to obovate or subcircular, glabrous, scarcely exceeding the sepals. Stamen-filaments 3·5–4 mm. long; anthers 1·2 mm. long. Disk with a pilose margin free for c. 1 mm. Style 4·5 mm. long, somewhat swollen at the apex and apparently considerably expanded for a short period. Fruit up to 3(4) × 2·5(3) cm., tomentellous when young, not conspicuously lepidote, apical peg very short (up to c. 0·6 mm.), decurrent at the base into the up to 1·2(2) cm. long slender stipe. Scales like those of *C. schumannii* but tending to be smaller, 35(40)–60(70)μ in diam., and less subdivided, with 8–12 marginal cells.

Zambia. N: Mbala (Abercorn) Distr., 1·6 km. E. of Mpulungu, fl. 17.xi.1952, *Angus* 773 A (BM; FHO).
Also in Zaire and Tanzania. *Brachystegia allenii* woodland; deciduous thickets.

Although there is a superficial resemblance to *C. schumannii* there are several differences: (1) the much smaller petals scarcely exceeding the sepals; (2) the much larger anthers; (3) the disk with appreciable free margin; (4) the tomentellous fruit decurrent into the stipe; (5) the smaller less subdivided scales. The relationship with *C. kirkii* is probably closer but the latter has usually considerably larger fruits not decurrent into the stipe.

5. Sect. METALLICUM Exell & Stace in Bol. Soc. Brot., Sér. 2, **42**: 22 (1968).—Stace in Bot. Journ. Linn. Soc. **62**, 2: 145 (1969) —Exell in Kirkia, 7, 2: 179 (1970).—Wickens, F.T.E.A. Combret.: 23 (1973).

Sect. *Glabripetala* Engl. & Diels in Engl., Mon. Afr. Pflanzen **3**: 43 (1899) pro parte.

Flowers 4-merous. Upper receptacle campanulate at the base and cupuliform at the apex, clearly divided into two regions. Petals transversely elliptic to obovate or subcircular, usually glabrous (rarely with a few hairs at the apex). Stamens 8, 1-seriate, inserted at the margin of the disk. Disk with a free pilose margin. Fruit 4-winged, brown or greyish-brown or reddish-grey to dark purple, lepidote, glabrous (apart from the scales) to densely hairy, usually somewhat " metallic " in appearance. Cotyledons 2, arising below soil level on a stalk formed by the connate petioles (*C. collinum*) or above soil level with short free petioles (*C. dumetorum*). Scales 55–180μ in diam., usually extensively divided by many radial and tangential walls to give c. 10–40 marginal cells.

The great difference in scale-structure led us to separate this section from Sect. *Glabripetala*. Species of the latter section are, moreover, nearly always glutinous, especially when young.

10. **Combretum collinum** Fresen. in Mus. Senckenb. **2**: 153 (1837).—Engl., Pflanzenw. Ost-Afr. C: 290 (1895).—Engl. & Diels in Engl., Mon. Afr. Pflanz. **3**: 56 (1899).—Exell & Garcia in Contr. Conhec. Fl. Moçamb. **2**: 120 (1954); C.F.A. **4**: 57 (1970).—Okafor in Bol. Soc. Brot. Sér. 2, **41**: 137–150 (1967).—Exell in Bol. Soc. Brot., Sér. 2, **42**: 8, 10, t. 2 fig. 5–7 (1968); in Kirkia, **7**, 2: 179, t. 1 fig. 11 (1970).—Gillett & McDonald, N.C.K.: 14 (1970).—Wickens, F.T.E.A.: 24 (1973).—Jacobsen in Kirkia, **9**, 1: 168 (1973).—Faria in Mem. Invest. Agron. Moçamb. **4**: 25 (1973). TAB. **32** fig. 10; TAB. **34** fig. D. Type from Ethiopia.
As the very extensive synonymy has recently been published by Exell (loc. cit. 1970) and by Wickens (loc. cit. 1973) and the important literature concerning our area is listed in Exell (loc. cit.) the synonyms are repeated here, as far as they concern our area, in an abbreviated form.
C. binderanum Kotschy (1865); *C. mechowianum* O. Hoffm. (1880–1882); *C. coriaceum* Schinz (1888); *C. schinzii* Engl. & Diels (1899); *C. laeteviride* Engl. & Gilg (1903); *C. monticola* Engl. & Gilg (1903); *C. cognatum* Diels (1907); *C. bajonense* Sim (1909); *C. gazense* Swynnerton & Bak. f. (1911); *C. junodii* Dummer (1913); *C. album* De Wild. (1914); *C. angustilanceolatum* Engl. (1921); *C. griseiflorum* S. Moore (1921); *C. milleranum* Burtt Davy (1921); *C. tophamii* Exell ex Burtt Davy & Hoyle (1936) *nom. nud.*; *C. abercornense* Exell (1939); *C. burttii* Exell (1939);

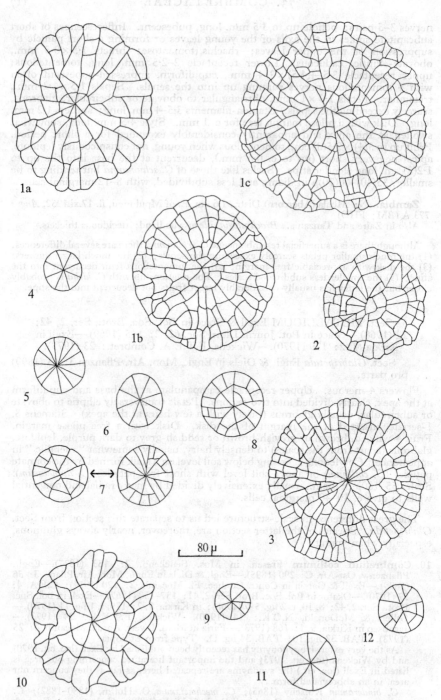

Tab. 32. Scales of COMBRETUM (dimensions indicated). The figures are numbered to correspond with the numbers of the taxa in the text. From C. A. Stace.

C. eylesii Exell (1939). Subspecies recently published by Okafor (loc. cit.) are discussed below.

Small semideciduous tree up to c. 17 m. high or shrub; crown rounded to flat; bark reddish-brown to grey or black or pale yellow to creamy-brown. Leaves opposite or alternate or verticillate; lamina up to 19(25) × 8(9) cm., very variable in shape, when dried brown or buff or golden-brown or silvery and usually somewhat " metallic " in appearance, densely tomentose to nearly glabrous (except for scales), conspicuously lepidote except when the scales are hidden by the indumentum; lateral nerves 6–14 pairs; reticulation usually (but not always) fairly prominent beneath; petiole usually up to 3(5) cm. long but very variable. Inflorescences of axillary spikes or panicles up to 10 cm. long, usually in the axils of bracts or reduced leaves on the current year's shoots. Flowers sessile, white or yellowish-green. Lower receptacle 2–4 mm. long, glabrous (except for scales) or hairy, lepidote; upper receptacle 3·5 × 2·5 mm., campanulate at the base and cupuliform at the apex, lepidote, glabrous (except for scales) to tomentose. Sepals 0·8 × 1·2 mm., broadly triangular. Petals 1·5–2·5 × 1–2·5 mm., transversely elliptic to obovate or subcircular, somewhat emarginate, unguiculate with the claw up to 1 mm. long, glabrous (rarely with a few hairs at the apex). Stamen-filaments 4–4·5 mm. long; anthers 0·9 mm. long. Disk 2 mm. in diam. with pilose margin free for c. 1 mm. Style 4 mm. long, glabrous or pubescent. Fruit brown or greyish-brown or reddish-grey to dark-purple, usually somewhat " metallic " in appearance, 2·5–5·5 × 2–5·5 cm., variable in shape, lepidote with scales often contiguous, glabrous (except for scales) to densely hairy, glossy or dull, wings variable in breadth, stipe up to 2 cm. long. Cotyledons 2, arising below soil level on a stalk formed by their connate petioles, free part of petiole 2 mm. long. Scales 80–180μ in diam., extensively divided by many radial and tangential walls to give 16–40 marginal cells; scales often contiguous on the lower epidermis of the leaf.

Caprivi Strip. Singalamwe, fr. 30.xii.1958, *Killick & Leistner* 3205 (PRE; SRGH). **Botswana.** N: between Nata and Maun, fl. ix.1946, *Miller* B 490 (FHO; PRE). SW: 80 km. NE. of Ghanzi, st. 27.vii.1955, *Story* 5052 (K; PRE). **Zambia.** B: Shangombo, 1035 m., fl. 8.viii.1952, *Codd* 7428 (BM; COI; EA; K; PRE; SRGH). N: Kalambo Falls, fr. 29.iii.1955, *E.M. & W.* 1327 (BM; LISC; SRGH). W: Solwezi, fl. 12.ix.1952, *Angus* 425 (BM; COI; FHO; K). C: Lusaka Distr., Mt. Makulu, fr. 29.iv.1956, *Angus* 1264 (BM; FHO; K). E: Fort Jameson, fl. 12.x.1950, *Gilges* 10 (SRGH). S: Livingstone Distr., Katambora, 885 m., fl. 15.vii.1955, *Gilges* 412 (LISC; PRE; SRGH). **Rhodesia.** N: Lomagundi Distr., Mangula, fr. 28.xi.1962, *Jacobsen* 1958 (PRE). W: Victoria Falls, 912 m., fl. & fr. 2.ix.1955, *Chase* 5741 (BM; COI; LISC; SRGH). C: Salisbury, 1465 m., fr. 30.vi.1931, *Brain* 5368 (COI; PRE; SRGH). E: Umtali, Parklands, fl. & fr. 7.x.1948, *Chase* 1505 (BM; LISC; SRGH). S: Fort Victoria Distr., Glenlivet, fl. & fr. 8.x.1949, *Wild* 2981 (BM; SRGH). **Malawi.** C: between Lilongwe and Salima, fr. 23.viii.1946, *Gouveia & Pedro in Pedro* 1795 (LMJ). S: Zomba, fr. 1959, *Banda* 1499 (BM). **Mozambique.** N: Mutuáli, Malema, fr. 5.iv.1962, *Lemos & Marrime in Lemos* 321 (BM; COI; K; LMJ). Z: Lugela, between Tacuane and R. Limboé, fr. 25.v.1949, *Andrada* 1501 (COI; LISC). T: Macanga, between Massanga and Casula, fr. 7.vii.1949, *Andrada* 1714 (COI; LISC). MS: Monte Umtereni, fl. 8.ix.1906, *Swynnerton* 587 (BM; K). SS: Massinga, between Funhalouro and Saúte, fr. 18.v.1941, *Torre* 2678 (BM; K; LISC). LM: between Magude and Uaniteze, fr. 4.v.1944, *Torre* 6569 (LISC).

Widespread in tropical, S. and SW. Africa. *Combretum-Pterocarpus-Pericopsis-Ostryoderris* tree savanna; *Acacia-Combretum-Terminalia* tree savanna; *Brachystegia-Julbernardia* woodland or savanna-woodland; *Baikiaea* woodland; tree savannas and woodlands of many types; deciduous and evergreen thickets; termite mounds.

A gum produced by this species is used medicinally by Africans (see Faria, loc. cit.).
This widely distributed and ecologically important aggregate species shows great variability in leaf-shape, indumentum and size of fruit. For some time the northern forms were known as *C. binderanum* and the southern forms as *C. mechowianum*, while *C. collinum* (the earliest name) remained isolated in NE. Africa. Okafor (loc. cit.) made a detailed study of the species from herbarium material and united *C. collinum*, *C. binderanum* and *C. mechowianum*. He recognized 11 subspecies but they are not always easy to separate and there are many intermediates in our area. Before accepting this system in our area it seems advisable to find out whether these subspecies have ecological significance and also to investigate germinations of seedlings. This latter has become of primary importance because *C. collinum* subsp. *dumetorum* (Exell) Okafor now appears to have a very different type of seedling from that of *C. collinum* and can scarcely be united with that species.

It is uncertain until further germination experiments have been made whether material from Barotseland which Okafor includes in his subsp. *dumetorum* really belongs here, so that I have not included *C. dumetorum* Exell in this Flora considering it to be only known with certainty from Angola.

Apart from *C. dumetorum*, Okafor recognized 6 subspecies from our area and the following key, extracted from his own key (loc. cit.), will enable ecologists and others to test the system in the field and to see whether the subspecies correspond with ecological entities and local distribution patterns. After recent experience with the seedlings of *C. dumetorum*, I am reluctant to accept the system until germinations have been carried out; but I am also reluctant to imply rejection of the work by putting the subspecies into synonymy.

Leaves distinctly hairy beneath, at least on the midrib:
  Fruit usually less than 3·6 cm. long, densely hairy, red scales if present usually not
    conspicuous; some leaves usually in whorls of 3 or 4   -   - subsp. *elgonense*
  Fruit usually more than 3·6 cm. long, usually without hairs but densely covered with
    conspicuous red scales; leaves usually opposite, subopposite or alternate with
    prominent reticulation beneath   -   -   -   -   - subsp. *gazense*
Leaves glabrous beneath or almost so:
  Fruit 3·8–5 cm. long, densely hairy, red scales conspicuous; inflorescence unbranched
    leaves variable as to insertion   -   -   -   -   -   - subsp. *suluense*
  Fruit usually more than 3·8 cm. long, glabrous, red scales conspicuous; inflorescence
    branched; leaves usually opposite, subopposite or alternate:
    Branchlets nearly always dark red; inflorescence-axis and lower receptacle densely
      covered with red scales; leaves drying golden-brown beneath subsp. *kwangense*
    Branchlets grey or brown; inflorescence-axis and lower receptacle with a sparse cover
      of red scales:
      Leaves usually acute to acuminate, silvery beneath with darker venation; fruit dark
        red-brown   -   -   -   -   -   -   -   - subsp. *taborense*
      Leaves usually rounded to subacute, yellowish beneath without conspicuously
        darker venation; fruit pale red-brown   -   -   - subsp. *ondongense*

The correct citations of these 6 subspecies with their basionyms and synonyms from our area are as follows: **subsp. elgonense** (Exell) Okafor (*C. elgonense* Exell; *C. abercornense* Exell); **subsp. gazense** (Swynnerton & Bak. f.) Okafor (*C. gazense* Swynnerton & Bak. f.; *C. bajonense* Sim; *C. eylesii* Exell; *C. mechowianum* subsp. *gazense* (Swynnerton & Bak. f.) Duvign.); **subsp. suluense** (Engl. & Diels) Okafor (*C. suluense* Engl. & Diels; *C. angustilanceolatum* Engl.; *C. griseiflorum* S. Moore; *C. milleranum* Burtt Davy); **subsp. kwangense** (Duvign.) Okafor (*C. mechowianum* subsp. *kwangense* Duvign.); **subsp. taborense** (Engl.) Okafor (*C. taborense* Engl.; *C. burttii* Exell; *C. mechowianum* subsp. *taborense* (Engl.) Duvign.); **subsp. ondongense** (Engl. & Diels) Okafor (*C. ondongense* Engl. & Diels).

For those who do not wish to enter into these complications the aggregate name *C. collinum* Fresen. is available and will serve most purposes.

6. Sect. GLABRIPETALA Engl. & Diels in Engl., Mon. Afr. Pflanz. **3**: 43 (1899) emend Exell & Stace apud Exell in Bol. Soc. Brot. Sér. 2, **42**: 22 (1968).—Stace in Bot. Journ. Linn. Soc. **62**: 158, figs. 129–135 (1969).—Exell in Kirkia, **7**, 2: 183, t. 1 fig. 12 (1970).—Wickens, F.T.E.A. Combret.: 28 (1973).

Flowers 4-merous. Leaves glutinous. Upper receptacle campanulate to infundibuliform at the base and cupuliform at the apex, clearly divided into two regions. Petals cuneate to spathulate or obovate, glabrous. Stamens 8, 1-seriate, inserted at the margin of the disk. Disk with a pilose free margin. Fruit 4-winged, glutinous. Cotyledons 2, arising below soil level and borne above ground on a long stalk formed by the connate petioles. Scales whitish or yellowish, usually rather inconspicuous, c. 50–65μ in diam., circular, not scalloped; cells delimited by 8(13) radial walls alone; cell-walls clear, cells transparent, more frequent on the upper surface of the leaf than on the lower.

11. **Combretum fragrans** F. Hoffm., Beitr. Kenntn. Fl. Centr.-Ost Afr.: 31 (1899).—Engl. & Diels in Engl., Mon. Afr. Pflanz. **3**: 51 (1899).—Exell in Kirkia, **7**, 2: 183, t. 1 fig. 12 (1970).—Wickens, F.T.E.A. Combret.: 29, fig. 2(11), 3(11), 4(11) (1973).—Faria in Mem. Inst. Invest. Agron. Moçamb. **4**: 32 (1973). TAB. **30** fig. F; TAB. **32** fig. 11; TAB. **34** fig. C. Syntypes from Tanzania.
  Combretum kilossanum Engl. & Diels in Notizbl. Bot. Gart. Berl. **2**: 193 (1898); in Engl., tom. cit.: 52 (1899).—Burtt Davy & Hoyle, N.C.L.: 39 (1936). Type from Tanzania.

Combretum albidiflorum Engl. & Diels in Engl., tom. cit.: 46, t. 14 fig. A (1899).
Type from Tanzania.
Combretum ghasalense Engl. & Diels in Engl., tom. cit.: 47, t. 15 fig. B (1899).—
Exell & Garcia in Contr. Conhec. Fl. Moçamb. **2**: 116 (1954).—Rattray in Kirkia,
2: 80 (1961).—Dale & Greenway, Kenya Trees and Shrubs: 145 (1961).—F. White,
F.F.N.R.: 285, fig. 50C (1962).—Boughey in Journ. S. Afr. Bot. **30**, 4: 167 (1964).—
Liben, F.C.B. Combret.: 75 (1968).—Exell in Bol. Soc. Brot., Sér. 2, **42**: 8, t. 2
fig. 10 (1968).—Wild in Kirkia, **7**, 1: 53 (1968).—Gillett & McDonald, N.C.K.: 14
(1970).—Jacobsen in Kirkia, **9**, 1: 168 (1973). Syntypes from the Sudan.
Combretum multispicatum Engl. & Diels in Engl., tom. cit.: 47, t. 15 fig. A (1899).
Type from the Sudan.
Combretum undulatum Engl. & Diels in Engl., tom. cit.: 48, t. 15 fig. C (1899).
Syntypes from the Sudan.
Combretum ternifolium Engl. & Diels in Engl., tom. cit.: 48, t. 14 fig. D (1899).—
Burtt Davy & Hoyle, loc. cit.—O. B. Mill., B.C.L.: 43 (1948).—Wild in Rhod.
Agric. Journ. **50**, 5: 6 (1953).—Brenan in Mem. N.Y. Bot. Gard. **8**, 5: 437 (1954).—
Williamson, Useful Pl. Nyasal.: 39 (1956).—Duvign. in Bull. Soc. Bot. Belg. **88**:
70 (1956). Type from Tanzania.
Combretum kamatutu De Wild. in Ann. Mus. Congo Belge, Bot. Sér. **4**: 215
(1903).—Diels in Engl., Bot. Jahrb. **39**: 497 (1907).—Engl., Pflanzenw. Afr. **3**, 2:
700 (1921) (" kamatula "). Type from Zaire.
Combretum tetraphyllum Diels in Engl., Bot. Jahrb. **39**: 499 (1907).—Monro in
Proc. & Trans. Rhod. Sci. Ass. **8**, 2: 80 (1908).—Eyles in Trans. Roy. Soc. S. Afr.
5, 4: 429 (1916).—Wild, Guide Fl. Vict. Falls: 146 (1953).—F. White, F.F.N.R.:
287 (1962). Type: Rhodesia or Zambia, Victoria Falls, *Engler* 2916 (B†, holotype).

Deciduous shrub or small tree up to c. 10(12) m. high; crown obovoid or
rounded; bark creamy-brown or grey-brown, vertically cracked, scaly, exuding
gum. Leaves opposite or very frequently 3–4-verticillate; lamina up to 20 × 9 cm.,
usually glutinous especially when young, ovate or ovate-elliptic, tomentose to nearly
glabrous (except for scales) and lepidote (often rather inconspicuously), apex acute
to rounded, base usually cuneate; lateral nerves 7–10 pairs, rather prominent
beneath; petiole up to 1·5 cm. long or sometimes very short (leaf subsessile),
leaving a prominent circular scar. Inflorescences of axillary spikes up to 7 cm.
long, sometimes branched, often appearing before the leaves in axils of fallen
leaves; rhachis usually densely pubescent; bracts 1·5 mm. long, deciduous.
Flowers sessile, greenish or dirty-white or yellow. Lower receptacle 1·5–3 mm.
long, usually tomentose; upper receptacle 2–3 × 2–3 mm., infundibuliform to
broadly campanulate at the base and cupuliform at the apex, pubescent to
tomentose. Sepals 1 × 1·5 mm., broadly triangular. Petals 2–3 × 1–1·5 mm.,
cuneate or spathulate to obovate, glabrous. Stamen-filaments 5–6 mm. long,
inserted at the margin of the disk; anthers 0·8–1 mm. long. Disk with a pilose free
margin. Fruit brown or reddish-yellow, 2·5–3·5 × 2·5–3 cm., glutinous, subcircular
to elliptic in outline, rather inconspicuously lepidote otherwise glabrous, apical peg
up to 3 mm. long, wings up to 12 mm. broad, stipe up to 5(7) mm. long.
Cotyledons 2, 2 × 3 cm., transversely elliptic, arising as in the sectional description.
Scales as for the section.

Botswana. N: Ngamiland, fr. i.1931, *Curson* 119 (PRE). **Zambia. B:** Balovale,
1065 m., fr. viii.1952, *Gilges* 176 (PRE; SRGH). **N:** Kasama Distr., Mungwe, fl.
8.x.1960, *Robinson* 3922 (SRGH). **W:** Mufulira, fl. 28 viii.1955, *Fanshawe* 2437 (K;
SRGH). **C:** Mt. Makulu, fl. 3.ix.1956, *Angus* 1403 (FHO). **E:** Fort Jameson, fl.
ix.1951, *Gilges* 104 (PRE; SRGH). **S:** Mazabuka Distr., near Municke R., fl. 2.viii.1952,
Angus 139 (BM; FHO; K). **Rhodesia. N:** Lomagundi, Silverside Mine, 1130 m.,
fl. 3.ix.1963, *Jacobsen* 2209 (PRE). **W:** Wankie, fl. 26.ix.1929, *Pardy* 4695 (FHO;
SRGH). **C:** Salisbury, fl. 5.x.1910, *Mundy* (K). **E:** Hondi Valley, fr. 27.xi.1948,
Chase 1279 (BM; COI; FHO; LISC; SRGH). **S:** Ndanga Distr., Chiredze R., fr.
15.x.1951, *Wormald* 89/51 (SRGH). **Malawi. N:** Rumphi, c. 1065 m., fr. ix.1953,
Chapman 177 (FHO). **C:** Dedza Distr., Mua-Livulezi Reserve, fr. 17.xi.1954, *Adlard* 199
(FHO). **S:** Balaka, fl. 7.x.1955, *Jackson* 1758 (FHO). **Mozambique. N:** 10 km. from
Mutuali on road to Lioma, fl. 8.xi.1953, *Gomes e Sousa* 4112 (K; LISC; PRE). **Z:**
Morrumbala, fl. 3.x.1949, *Andrada* 1939 (COI; LISC). **T:** between Tete and Zóbuè,
fl. 18.vi.1941, *Torre* 2907 (BM; K; LISC). **MS:** Chemba, near Maríngue, fr. 5.x.1944,
Mendonça 2357 (BM; LISC).
    Also from West Africa to the Sudan, Zaire, Kenya and Tanzania. Dry lowland tree
savanna with *Setaria* etc.; edges of dambos (seasonal swamps) with *Adansonia, Kirkia* and
Sterculia; plateau *Acacia-Combretum-Terminalia* open woodland; lake basin chipya (in

Zambia); mopane scrub; tree savannas; rocky places; heavy soils; from low to medium altitudes and occasional on copper-bearing soils.

This widespread and ecologically important species has usually been known as *C. ghasalense* in the northern part of its range and as *C. ternifolium* in the south. These were united under *C. ghasalense*. Recently Wickens examined authentic material of *C. fragrans*, an earlier name, which he found to be the same species, and Stace has confirmed that the scales are identical. The name must therefore unfortunately be changed again, it is to be hoped for the last time.

It is a curious fact that this common and widespread species, found throughout our area, has never been discovered in Angola, where one would certainly expect it to occur.

Although *C. fragrans* is very variable in indumentum and leaf-shape as well as in the arrangement of the leaves in 2's, 3's or 4's the species, perhaps because of its characteristic glutinous nature, is usually easily recognizable and there has not been the same tendency, as in most of the other very variable savanna species of *Combretum*, either to describe each form as a new species or to divide it into a number of intraspecific taxa.

7. Sect. SPATHULIPETALA Engl. & Diels in Engl., Mon. Afr. Pflanz. **3**: 58 (1899).—Stace in Bot. Journ. Linn. Soc. **62**: 158, figs. 126–128 (1969).—Exell in Kirkia, **7**, 2: 185 (1970).—Wickens, F.T.E.A. Combret.: 31 (1973).

Flowers 4-merous. Upper receptacle shortly infundibuliform. Petals obovate-spathulate to spathulate, glabrous. Stamens 8, 1-seriate, inserted at the margin of the disk. Disk with pilose margin free for 1·5–2 mm. Style with swollen apex (while functional) usually exserted before the stamens appear. Fruit 4-winged, large for the genus, with stipe up to 2·5 cm. long. Cotyledons 2, arising below soil level and borne above ground on a long stalk formed by the connate petioles, sometimes united to form a single peltate organ. Scales whitish or yellowish, c. 40–75µ in diam., circular but slightly convexly scalloped; cells delimited by 7–9 radial walls and also by a few tangential and extra radial walls; marginal cells 7–12; cell walls clear and cells transparent.

There is some evidence of relationship with Sect. *Macrostigmatea*.

12. **Combretum zeyheri** Sond. in Linnaea, **23**: 46 (1850); in Harv. & Sond., F.C. **2**: 511 (1862).—Engl. & Diels in Engl., Mon. Afr. Pflanz. **3**: 59 (1899).—Gibbs in Journ. Linn. Soc., Bot. **37**: 443 (1906).—Monro in Proc. & Trans. Rhod. Sci. Ass. **8**, 2: 80 (1908).—Sim, For. Fl. Port. E. Afr.: 63, t. 63 fig. A (1909).—Bak. f. in Journ. Linn. Soc., Bot. **40**: 68 (1911).—Eyles in Trans. Roy. Soc. S. Afr. **5**, 4: 429 (1916).— Steedman in Proc. & Trans. Rhod. Sci. Ass. **24**: 17 (1925); Some Trees, etc. S. Rhod.: 56 (1933).—Burtt Davy, F.P.F.T. **1**: 248 (1926).—Burtt Davy & Hoyle, N.C.L.: 39 (1936).—O. B. Mill., B.C.L.: 44 (1948).—Brenan, T.T.C.L.: 140 (1949); in Mem. N.Y. Bot. Gard. **8**, 5: 438 (1954).—Wild, Guide Fl. Vict. Falls: 146 (1953).—Exell & Garcia in Contr. Conhec. Fl. Moçamb. **2**: 120 (1954) excl. specim. *Barbosa* 2286; C.F.A. **4**: 66 (1970).—Pardy in Rhod. Agric. Journ. **51**, 1: 6 cum phot. (1954).—Duvign. in Bull. Soc. Roy. Bot. Belg. **88**: 70, 83, fig. 3A (1956); op. cit. **90**, 2: 194, 212 (1958).—Palgrave, Trees of Central Afr.: 135 cum phot. et tab. (1956).—Dale & Greenway, Kenya Trees and Shrubs: 147 (1961).—Palmer & Pitman, Trees S. Afr.: 249 cum fig. (1961).—Chapman, Veg. Mlanje Mount.: 35 (1962).—F. White, F.F.N.R.: 285, fig. 49M–N (1962).—Barclay-Smith in Kirkia, **4**: 32, 33 (1964).—Fanshawe & Savory in Kirkia, **4**: 190 (1964).—Boughey in Journ. S. Afr. Bot. **30**, 4: 167 (1964).—Breitenb., Indig. Trees Southern Afr. **4**: 833 cum fig. (1965).—B. de Winter & Killick, Sixty-six Transv. Trees: 124, cart. et phott. (1966).—Exell, Prodr. Fl. SW. Afr. **99**: 11 (1966); in Bol. Soc. Brot. Sér. 2, **42**: 8 (1968); in Kirkia, **7**, 2: 185, t. 1 fig. 13 (1970).—Liben, F.C.B. Combret.: 72 (1968).—Wild in Kirkia, **7**, 1: 53, t. 6 (1968); tom. cit. suppl.: 43, 45 (1970).— Gillett & McDonald, N.C.K.: 15 (1970).—Wickens, F.T.E.A. Combret.: 31, fig. 2(13), 3(13), 4(13) (1973).—Jacobsen in Kirkia, **9**, 1: 168 (1973).—Faria in Mem. Invest. Agron. Moçamb. **4**: 37 (1973). TAB. **31** fig. J; TAB. **32** fig. 12; TAB. **33**. Type from S. Africa.

　　Combretum tinctorum Welw. ex Laws. in Oliv., F.T.A. **2**: 430 (1871).—Engl. & Diels in Engl., tom. cit.: 30 (1899). Type from Angola.

　　Combretum teuszii O. Hoffm. in Linnaea, **43**: 132 (1882) (" theuschii ").—Engl. & Diels in Engl., tom. cit.: 60, t. 18 fig. G (1899). Type from Angola.

　　Combretum glandulosum F. Hoffm., Beitr. Kenntn. Fl. Centr.-Ost-Afr.: 33 (1889).—Engl. & Diels in Engl., tom. cit.: 59, t. 18 fig. C (1899). Non V. Sloot. (1919). Type from Tanzania.

Tab. 33. COMBRETUM ZEYHERI. 1, flowering branch (×1) *Fanshawe* 2437; 2, mature leaf, under side (×1) *Robinson* 289; 3, flower (×5) *Fanshawe* 2437; 4, flower, longitudinal section (×5) *Fanshawe* 2437; 5, fruit (×1) *Bullock* 1327.

Combretum oblongum F. Hoffm., tom. cit.: 34 (1889).—Engl. & Diels in Engl.,
tom. cit.: 60, t. 18 fig. D (1899).—R.E.Fr. in Wiss. Ergebn. Schwed. Rhod.-Kongo
Exped. **1**: 170 (1914).—Eyles in Trans. Roy. Soc. S. Afr. **5**, 4: 428 (1916).
Syntypes from Tanzania.
    Combretum bragae Engl., Pflanzenw. Ost-Afr. **C**: 289 (1895).—Engl. & Diels in
Engl., tom. cit.: 59 (1899).—Monro in Proc. & Trans. Rhod. Sci. Ass. **8**, 2: 80
(1908).—Eyles, tom. cit.: 427 (1916). Type: Mozambique, Beira, *Braga* 161
(B†, holotype).
    Combretum antunesii Engl. & Diels in Engl., Mon. Afr. Pflanz. **3**: 58, t. 18 fig. F
(1899). Type from Angola.
    Combretum sinuatipetalum De Wild. in Ann. Mus. Congo Belge, Bot. Sér. **4**: 215
(1903). Type from Zaire (Katanga).
    Combretum lopolense Engl. & Diels apud Diels in Engl., Bot. Jahrb. **39**: 502
(1907).—R.E.Fr., tom. cit.: 169 (1914).—Wild, Guide Fl. Vict. Falls: 146 (1953).
Type from Angola.
    Combretum rupicola Engl. apud Diels, tom. cit.: 503 (1907) non Ridl. (1890) *nom.
illegit.* Type from Tanzania.
    Combretum platycarpum Engl. & Diels apud Diels, loc. cit. Type from Tanzania.
    Combretum dilembense De Wild. in Fedde, Repert. **11**: 515 (1913). Type from
Zaire.
    Combretum calocarpum Gilg ex Dinter in Fedde, Repert. **16**: 169 (1919). Type
from SW. Africa.
    Combretum zeyheri var. *seineri* Engl., Pflanzenw. Afr. **3**, 2: 704 (1921). No type
cited.
    Combretum megalocarpum Exell ex Brenan, T.T.C.L.: 140 (1949). Type as for
C. rupicola Engl.
    Combretum zeyheri forma.—Wild in Kirkia, **7**, 1: 53 (1968).

Small to medium-sized deciduous tree up to 10(12) m. high or rarely a shrub;
crown rounded or flat-rounded; bark brown or grey-brown, smooth to scaly,
generally fissured; branchlets usually tomentose. Leaves opposite or 3-verticillate;
lamina up to 14(22) × 9(11) cm., chartaceous, broadly to narrowly elliptic or
obovate-elliptic or oblong-elliptic, usually tomentose (when young) or pubescent
to almost glabrous (except for scales), lepidote but scales rather inconspicuous,
apex usually rounded to obtuse, sometimes acute, base usually rounded, sometimes
slightly cordate; lateral nerves 5–12 pairs, prominent below; petiole up to 1 cm.
long. Inflorescences usually unbranched axillary spikes up to 8 cm. long including
the rather short tomentose peduncle; rhachis tomentose; bracts c. 2 mm. long,
caducous. Flowers sessile, yellowish. Lower receptacle 2–3 mm. long, tomentose;
upper receptacle c. 3 × 2·5–3 mm., shortly infundibuliform, pubescent and
lepidote. Sepals 1·5–2 × 1·5–2 mm., triangular. Petals yellow, 1·5–2·5 × 0·8–
1·2 mm., obovate-spathulate to spathulate, glabrous. Stamen-filaments 5–8 mm.
long, often reflexed at about 1 mm. below the anthers; anthers orange, 1·5 mm.
long. Disk 1·8 mm. in diam. with pilose margin free for 1·5–2 mm. Style c.
5 mm. long, glabrous, with swollen apex (while functional), usually exserted and
rather conspicuous before the stamens appear. Fruit usually c. 5·5 × 5 cm. (rarely
only 3–3·5 × 3–3·5 cm. and occasionally up to 10 × 8 cm.), subcircular in outline,
usually glabrescent, sometimes conspicuously lepidote on the body, apical peg very
short or absent, wings up to 3·5 cm. broad, stipe 1–3 cm. long, usually relatively
slender. Cotyledons 3·5–4 × 6 cm., subcircular to transversely elliptic, with free
petioles sometimes c. 2 mm. long (petioles united for the rest of their length) or the
two cotyledons united completely to form a single peltate organ c. 6 cm. in diam.,
arising as for the section. Scales as for the section.

Botswana. N: E. of Odiakwe, fr. 21.iii.1962, *de Beer & Yalala* 19 (SRGH). SE:
Kanye Hills, fr. 22.iv.1958, *de Beer* 687 (COI; LISC; SRGH). **Zambia.** B: Sefula,
fr. viii.1921, *Borle* 260 (FHO; PRE). N: Shiwa Nganda, 1675 m., fr. 22.vii.1938,
Greenway 5460 (EA; FHO; K). W: Mwinilunga Distr., 6·4 km. N. of Magowa Plains,
fl. & fr. 4.x.1952, *White* 3455 (BM; FHO; K). C: Lusaka, fl. 2.x.1960, *Dening* 47
(FHO). E: Fort Jameson Distr., 19 km. S. of Katete, fr. 24.iv.1952, *White* 2447 (BM;
FHO; K). S: Katambora, fl. & fr. 30.ix.1947, *Morze* 61 (FHO). **Rhodesia.** N:
Lomagundi Distr., Umboe Valley, fr. 19.ii.1964, *Wild* 6329 (FHO; SRGH). W:
Nyamandhlovu, fr. vii.1956, *Barrett* 106/56 (COI; LISC; SRGH). C: Hunyani R.,
Prince Edward Dam, fr. 16.v.1934, *Gilliland* 144 (BM; FHO; K; SRGH). E: Inyanga,
Lawley's Concession, fr. 19.ii.1954, *West* 3357 (SRGH). S: Fort Victoria Distr., fl. 1909,
Monro 509 (BM; SRGH). **Malawi.** N: 9·6 km. from Rumpi, fl. 20.x.1962, *Adlard* 509
(FHO; SRGH). C: between Lilongwe and Salima, fr. 23.viii.1946, *Gouveia & Pedro* in

Pedro 1793 (LMJ; PRE). S: Ncheu Distr., Sharpe Vale, fr. 19.vii.1958, *Jackson* 2236 (BM; FHO; LISC; SRGH). **Mozambique.** N: Maniamba, Macaloge, fr. 3.ix.1934, *Torre* 538 (COI). Z: Mocuba, Namagoa, fl. 12.x.1948, *Faulkner* 1654 (COI; K). T: Boroma, Msusa, fr. 27.vii.1950, *Chase* 2811 (BM; COI; SRGH). MS: Chimoio, Bandula, fr. 9.iii.1948, *Barbosa* 1146 (LISC). SS: Maringa, R. Save, fr. 27.vi.1950, *Chase* 2577 (BM; SRGH). LM: Maputo, Goba, 400 m., fr. 18.ii.1945, *Sousa* 33 (LISC: PRE).

Also in Angola, Zaire, Kenya, Tanzania, SW. Africa and S. Africa. Savanna woodland of *Brachystegia-Julbernardia*; *Cryptosepalum* dry forest and woodlands; lowland littoral forest with *Pteleopsis*, *Erythrophloeum* and *Brachystegia spiciformis*; *Pericopsis-Pterocarpus-Acacia* tree savanna; *Acacia-Combretum-Terminalia* tree savanna; copper limestone and lake basin chipya (Zambia); mopane termite mounds; scrub mopane; very tolerant of serpentiniferous and metalliferous soils. Very widespread from low to medium altitudes.

C. zeyheri is usually recognizable by its large yellowish-green fruits (sometimes drying orange) and in flower a characteristic feature is that the flowers are protogynous and the style (often drying black) often protrudes conspicuously from the flower-bud before it opens. There is considerable variability in leaf shape and indumentum as well as in the size of the fruit and it is mainly the various combinations of these features that has led to the considerable synonymy. The fruit size is usually about 5 × 5 cm. but fruits up to 10 × 8 cm. are met with while in one specimen (Botswana, Mochudi, *Shantz* 429 (PRE)) they are only 3–3·5 × 3–3·5 cm. and seem to be intermediate in size and shape between those of *C. zeyheri* and *C. schumannii*. There seems to be some correlation between increasing fruit size and increasing rainfall comparable to the leaf-shape cline in *C. molle*. In this connection it is interesting that very large-leaved specimens (up to 22 × 11 cm.) of *C. zeyheri* have been collected in Tanzania.

The only other African species of *Combretum* in our area which approach *C. zeyheri* in size of fruit are *C. kirkii* and large-fruited forms of *C. collinum*.

Both the unusual method of germination and the large fruits with rigid wings, better adapted for blowing along the ground than for transport by air, are thought to be adaptations for survival in areas subject to annual fires. This is discussed by Exell and Stace (in D. H. Valentine, Taxonomy, Phytogeography and Evolution: 316–321 (1972)).

8. Sect. **CILIATIPETALA** Engl. & Diels in Engl., Mon. Afr. Pflanz. **3**: 32 (1899) (" Ciliatopetalae ") excl. sp. *C. fulvotomentosum.*—Stace in Bot. Journ. Linn. Soc. **62**: 149, figs. 83–118 (1969).—Exell in Kirkia, **7**, 2: 187 (1970).—Wickens, F.T.E.A. Combret.: 32 (1973).

Flowers 4-merous. Upper receptacle cupuliform to broadly campanulate. Petals small (sometimes minute or even absent), obovate or obcuneate to sub-circular, sometimes emarginate, ciliate or pilose at the apex (glabrous only in *C. psidioides* subsp. *glabrum*). Stamens 8, 1-seriate in insertion or nearly so. Disk with a short free pilose margin. Fruit 4-winged. Cotyledons 2, with long petioles arising at or below soil level or (*C. albopunctatum*) with short petioles borne above soil level or (*C. viscosum*, not in our area) more or less connate. Scales somewhat variable, 40–120(130)μ in diam., ± circular, slightly or markedly scalloped at the margin, delimited by (7)8–c. 12 primary radial walls and often with additional tangential walls.

The species of this section are somewhat diverse, *C. albopunctatum* and *C. viscosum* (not in our area) as regards the germination and *C. psidioides* as regards the smaller scales and the characteristic exfoliation of the bark of the branchlets. Nevertheless *C. albopunctatum* and *C. viscosum* are both near to *C. apiculatum* while *C. psidioides* shows resemblance to *C. molle*. To complete the linkage, there are intermediates (and perhaps hybrids) between *C. apiculatum* and *C. molle* and the section seems to form a natural group, all the species having small (up to 1·5 × 1·5 mm. in *C. moggii*, from S. Africa) petals ciliate at the apex (except in *C. psidioides* subsp. *glabrum* which is almost glabrous throughout).

The small petals of Sect. *Ciliatipetala* seem to be vestigial having lost their role in insect-attraction which has been taken over by massing of the flowers, conspicuous stamens and scent.

Bark of the young branchlets coming off in untidy irregular fibrous strips or threads; scales (45)50–120μ in diam., usually with tangential walls additional to the radial ones: Scales mostly glistening; cotyledons borne well above soil level with petioles 2–4 mm. long; leaves up to 10 × 5·5 cm., typically obovate-elliptic, pubescent to sericeous-tomentose; plant a shrub of thickets, occasionally semi-scandent

13. *albopunctatum*

Scales usually not glistening; cotyledons arising at or below soil level with petioles up
    to 4–5 cm. long; leaves very variable; plant a small tree, shrub or climber:
Reticulation of under surface of leaf usually prominent (only subprominent in some
    intermediates); leaves up to 14(16) cm. long, typically sericeous-tomentose (but
    sometimes glabrous); fruit rarely exceeding 2 cm. in length or width; scales
    75(90)–120(130)μ in diam., cells opaque with walls appearing thick; plant a
    small tree or shrub - - - - - - - - 14. *molle*
Reticulation of under surface of leaf usually not prominent (subprominent in some
    intermediates); leaves up to 13 cm. long, glutinous when young; fruit up to
    3 × 2·5 cm., usually glutinous; scales (45)50–75(100)μ in diam., transparent and
    thin-walled:
Plant a small tree or shrub; leaves usually glabrous on the reticulation beneath
    (except in subsp. *leutweinii*) - - - - - 15. *apiculatum*
Plant a climber (or a scrambler if without support); leaves remaining hairy at
    maturity and pilose round the margins - - - 16. *acutifolium*
Bark of branchlets peeling off in large ± cylindric or hemicylindric pieces leaving an
    exposed cinnamon-red surface: reticulation of under surface of leaf prominent;
    scales glistening (but often obscured by the indumentum) c. 40–50μ in diam., cells
    very thin-walled and often obscured by glutinous secretions, delimited by 8 primary
    radial walls alone - - - - - - - - 17. *psidioides*

13. **Combretum albopunctatum** Suesseng. in Mitt. Bot. Staatssamml. München **1**: 336
    (1953).—F. White, F.F.N.R.: 435 (1962).—Boughey in Journ. S. Afr. Bot. **30**, 4:
    167 (1966).—Exell, Prodr. Fl. SW. Afr. **99**: 7 (1966); in Bol. Soc. Brot., Sér. 2, **42**:
    7 (1968); in Kirkia, **7**, 2: 188, t. 2 fig. 14 (1970). TAB. **35** fig. 13. Type from
    SW. Africa.
    Combretum sp.—O. B. Mill., B.C.L.: 44 (1948). *Miller* B/148.
    Combretum sp. 2.—F. White, F.F.N.R.: 287 (1962).

Thicket-forming shrub or small tree up to c. 3 m. high with tendency to climb;
bark grey or greyish-brown; young branchlets densely pubescent to tomentose.
Leaves opposite or subopposite; lamina up to 10 × 5·5 cm. (often only c. 4 × 2 cm.),
papyraceous to subcoriaceous, brownish-tomentose when young with the
indumentum almost concealing the scales, later pubescent to pilose beneath
especially on the midrib and nerves, lepidote mainly in the areolae, verruculose or
lepidote above, otherwise nearly glabrous, apex blunt to rounded and usually
mucronate, base subcordate; lateral nerves 5–7 pairs; petiole 3–5 mm. long,
pubescent. Inflorescences of axillary spikes up to 3 cm. long; rhachis nearly
glabrous; bracts 1·5 mm. long, filiform, caducous. Flowers sessile, yellow,
glutinous when young. Lower receptacle 1·7–2·5 mm. long, tomentose; upper
receptacle 2 × 2·5–3 mm., broadly campanulate, lepidote and pubescent. Sepals
0·5–1 × 1–1·5 mm., broadly deltate. Petals 1 × 0·8 mm., obovate, emarginate,
ciliate at the apex. Stamen-filaments 3·5–4 mm. long; anthers 1–2 mm. long.
Disk with very short pilose free margin. Style 3 mm. long, apex not expanded.
Fruit reddish-brown, usually satiny, 2–3 × 1·8–2·5 cm., subcircular to broadly
elliptic in outline, silvery lepidote with glistening scales, apical peg very short,
wings 8–10 mm. broad, stipe 2–7 mm. long. Cotyledons 2, 1–1·7 × 1·8–3 cm.,
transversely elliptic, arising above soil level; petioles 2–4 mm. long. Scales
whitish and usually glistening, c. 60–100μ in diam., deeply to shallowly bowl-
shaped and slightly convexly scalloped at each marginal cell, delimited by 8–c. 10
radial walls alone or also up to 6 tangential walls; cell-walls obscured by copious
glutinous secretions (except in very old leaves) and ± transparent apart from
secretions.

Caprivi Strip. Bagani Pontoon, fr. 19.i.1956, *de Winter* 4337 (K; M; PRE).
Botswana. N: Lower Ngwezumba R., fr. iii.1938, *Miller* B/200 (BM; FHO). **Zambia.**
B: Sesheke Distr., Masese, fr. 26.xii.1952, *Angus* 1043 (FHO; K). S: Gwembe Distr.,
Gwembe Valley, near Masuka R., fr. 30.iii.1952, *White* 2375 (BM; FHO; K).
Rhodesia. N: Sebungwe Distr., Kariangwe-Lubu R. junction area, fl. 16.xi.1956,
Lovemore 504 (K; SRGH). W: Wankie Tribal Trust Land, fr. v.1959, *Armitage* 68/59
(SRGH).
    Also from SW. Africa. *Baikiaea* woodland, *Colophospermum mopane* savanna woodland
(one record), dry deciduous tree savanna on grits and sandstones with *Commiphora* and
Combretum spp., thickets on alluvium; riverine thickets; sandveld; in hotter drier areas.

    An apparently distinct species with usually glistening white scales which has often been
confused with *C. apiculatum* and *C. molle* and occasionally with *C. erythrophyllum*. In
flower (but flowering material has very rarely been collected) it can usually be separated

from *C. apiculatum* subsp. *apiculatum* but it is not easy to distinguish from *C. apiculatum* subsp. *leutweinii* or some forms of *C. molle* without microscopic examination of the scales. At germination, however, it is recognizable by the cotyledons borne above soil level with short petioles. The young leaves on flowering specimens are very similar to those of *C. molle* at the same age. The resemblance to *C. erythrophyllum* is only superficial as the petals are elliptic and glabrous in the latter species. Some specimens have been less excusably named *C. obovatum*, a non-lepidote species belonging to Subgen. *Cacoucia*.

14. **Combretum molle** R. Br. [in Salt, Voy. Abyss. App.: lxiv (1814) *nom. nud.*] ex G. Don in Trans. Linn. Soc. Lond. **15**: 431 (1827) non Engl. & Diels (1899).— Brenan, T.T.C.L.: 137 (1949).—Wild, Guide Fl. Vict. Falls: 146 (1953); in Kirkia **7**, 1: 53 (1968); tom. cit. Suppl.: 41, 45, t. 2 fig. 15 (1970).—Exell & Garcia, Contr. Conhec. Fl. Moçamb. **2**: 106 (1954) excl. syn. *C. buchananii, C. puetense, C. brachypetalum* et *C. omahekae* et specim. *Mendonça* 396; C.F.A. **4**: 62 (1970).— Keay, F.W.T.A. ed. 2, **1**: 270 (1954).—Duvign. in Bull. Soc. Roy. Bot. Belg. **88**: 70 (1956).—Palgrave, Trees Central Afr.: 131 cum phot. et tab. (1956).—Chapman, Veg. Mlanje Mount.: 35 (1962).—F. White, F.F.N.R.: 287, fig. 50F (1962).— Boughey in Journ. S. Afr. Bot. **30**, 4: 167 (1964).—Barclay-Smith in Kirkia, **4**: 33 (1964).—Breitenb., Indig. Trees Southern Afr. **4**: 827 cum fig. (1965).—Stace in Bull. Brit. Mus. (Nat. Hist.) Bot. **4**, 1: 67 (1965).—Liben, F.C.B. Combret. 67 (1968) excl. syn. *C. griseiflorum.*—Exell in Bol. Soc. Brot. Sér. 2, **42**: 7 (1968); in Kirkia, **7**, 2: 189, t. 2 fig. 15 (1970).—Gillett & McDonald, N.C.K.: 14 (1970).—Wickens, F.T.E.A. Combret.: 33 (1973).—Jacobsen in Kirkia, **9**, 1: 168 (1973).—Faria in Mem. Inst. Invest. Agron. Moçamb. **4**: 45 (1973). TAB. **34** fig. F; TAB. **35** fig. 14; TAB. **36**. Type from Ethiopia.

Combretum gueinzii Sond. in Linnaea, **23**: 43 (1850); in Harv. & Sond., F.C. **2**: 509 (1862).—Engl. & Diels in Engl., Mon. Afr. Pflanz. **3**: 38, t. 12 fig. A (1899).— Eyles in Trans. Roy. Soc. S. Afr. **5**, 4: 428 (1916).—Burtt Davy & Hoyle, N.C.L.: 38 (1936).—Suesseng. & Merxm. in Proc. & Trans. Rhod. Sci. Ass. **43**: 27 (1951).— Pardy in Rhod. Agric. Journ. **49**, 2: 84 cum phot. (1952).—Brenan in Mem. N.Y. Bot. Gard. **8**, 5: 437 (1954). Type from Natal.

Combretum holosericeum Sond., tom. cit.: 44 (1850); in Harv. & Sond., tom. cit.: 510 (1862).—Laws. in Oliv., F.T.A. **2**: 430 (1871) pro parte.—Sim, For. Fl. Port. E. Afr.: 63, t. 62 (1909).—Burtt Davy, F.P.F.T. **1**: 247 (1926).—Steedman, Trees etc. S. Rhod.: 54 cum phot. (1933).—O. B. Mill., B.C.L.: 43 (1948). Type from S. Africa.

Combretum splendens Engl., Pflanzenw. Ost-Afr. C: 289 (1895).—R.E.Fr. in Wiss. Ergebn. Schwed. Rhod.-Kongo Exped. **1**: 167 (1914).—Steedman in Proc. & Trans. Rhod. Sci. Ass. **24**: 17, t. 17 (1925).—Burtt Davy & Hoyle, N.C.L.: 39 (1936). Syntypes from Tanzania and Malawi: *Buchanan* 859 (B†; K).

Combretum ulugurense Engl. & Diels in Engl., Mon. Afr. Pflanz. **3**: 35, t. 10 fig. A (1899).—Eyles in Trans. Roy. Soc. S. Afr. **5**, 4: 429 (1916). Type from Tanzania.

Combretum splendens var.? *nyikae* (Engl.) Engl. apud Engl. & Diels in Engl., tom. cit.: 37, t. 11 fig. E (1899).—R.E.Fr., tom. cit.: 165 (1914). Syntypes from Tanzania.

Combretum arbuscula Engl. & Diels in Warb., Kunene-Samb. Exped. Baum: 314 (1903).—Bak. f. in Trans. Linn. Soc., Bot. **40**: 67 (1911).—Eyles, tom. cit.: 427 (1916). Type from Angola.

Combretum atelanthum Diels in Engl., Bot. Jahrb. **39**: 494 (1907).—Eyles, loc. cit. Type: Rhodesia, near Salisbury, *Engler* 3098 (B†, holotype).

Combretum arengense Sim, For. Fl. Port. E. Afr.: 62, t. 63 fig. B (1909). Type: Mozambique, Maganja da Costa, *Sim* 5916 (probably = *Sim* 20902 (PRE)).

Combretum ellipticum Sim, tom. cit.: 63, t. 63 fig. D (1909). Type: Mozambique, without precise locality, *Sim* 6068 (PRE).

Combretum gueinzii var. *holosericeum* (Sond.) Exell ex Rendle in Journ. of Bot. **70**: 93 (1932). Type as for *C. holosericeum.*

Combretum gueinzii subsp. *eugueinzii* var. *holosericeum* (Sond.) Exell ex Burtt Davy & Hoyle, N.C.L.: 38 (1936). Type as above.

Combretum gueinzii subsp. *splendens* Exell ex Burtt Davy & Hoyle, loc. cit. *nom. nud.*
Combretum gueinzii subsp. *splendens* (Engl.) Exell ex Brenan, T.T.C.L.: 137 (1949). Type as for *C. splendens.*

The above synonymy accounts for most of the references to this species in our area. The remaining synonymy, mostly outside our area, follows in an abbreviated form.

C. *velutinum* DC. (1828); C. *trichanthum* Fresen. (1837); C. *vernicosum* Fenzl (1844) *nom. nud.*; C. *schimperanum* A. Rich. (1847); C. *punctatum* A. Rich. (1847) non Bl. nec Steud.; C. *quartinianum* A. Rich. (1847); C. *lepidotum* A. Rich. (1847) non Presl; C. *petitianum* A. Rich. (1847); C. *rochetianum* A. Rich. ex A. Juss. (1851);

C. gondense F. Hoffm. (1889); *C. tenuispicatum* (1894); *C. nyikae* Engl. (1894) *nom. nud.* (1895); *C. boehmii* Engl. (1894) *nom. nud.* (1895); *C. deserti* Engl. (1895); *C. nyikae* var. *boehmii* Engl. (1895); *C. microlepidotum* Engl. (1895); *C. lepidotum* var. *melanostictum* Welw. ex Hiern (1898); *C. hobol* Engl. & Diels (1899); *C. insculptum* Engl. & Diels (1899); *C. welwitschii* Engl. & Diels (1899); *C. welwitschii* var. *melanostictum* (Welw. ex Hiern) Engl. & Diels (1899); *C. galpinii* Engl. & Diels (1899); *C. obtusatum* Engl. & Diels (1899); *C. dekindtianum* Engl. (1902); *C. sokodense* Engl. (1907); *C. holtzii* Diels (1907); *C. ankolense* Bagshawe & Bak. f. (1908); *C. pretoriense* Dummer (1913) pro parte; *C. velutinum* var. *glabrum* Aubrév. (1950).

Semi-deciduous tree usually up to 8(17) m. high or shrub; crown rounded to flat-rounded; bark black-brown to grey-brown, rough, fissured; branchlets tomentose to almost glabrous with bark coming off in fibrous strips. Leaves opposite; lamina up to 14(16) × 9 cm., narrowly elliptic or narrowly ovate-elliptic or obovate or obovate-elliptic, typically densely grey-tomentose (often drying dark-velvety-brown) but almost glabrous in some forms, lepidote but scales often hidden by the indumentum, reticulation usually very prominent beneath (but less so in some apparently transitional forms), apex acute or obtuse and often mucronate or apiculate, base usually rounded to subcordate; lateral nerves 6–12 pairs; petiole usually 2–3 mm. long (up to 9 mm.) leaving after leaf-fall rather prominent circular projections on the stem. Inflorescences of axillary spikes up to 7(11) cm. long occasionally forming panicles by suppression of the upper leaves; rhachis tomentose; bracts 1–2 mm. long, filiform, caducous; peduncles 1–2 cm. long. Flowers sessile, yellow or greenish-yellow. Lower receptacle 1·5–2 mm. long, tomentose; upper receptacle 1·3–3 × 2–3 mm., campanulate, tomentose. Sepals 0·5–1 mm. long, broadly deltate. Petals 0·5–1 × 0·5–1 mm., sometimes minute or absent, irregularly obovate-deltate to reniform, ciliate at the apex. Stamen-filaments 5–6 mm. long, inserted at the edge of the disk; anthers 1 mm. long. Disk with a very short pilose free margin less than 0·5 mm. long. Style 5 mm. long, not expanded at the apex. Fruit 1·5–2(2·5) × 1·5–2(2·5) cm., subcircular to elliptic in outline, lepidote and tomentose (especially on the body) to nearly glabrous (except for scales), apical peg up to 1 mm. long (often much shorter), wings 5–7 mm. broad, stipe 2–3(5) mm. long. Cotyledons 2, 2·5 × 2·5–3·5 cm., subcircular to transversely elliptic, arising at or below soil level; petioles 3·5–4 cm. long. Scales ± contiguous on the areolae of the lower surface of the leaf, c. 75(90)–120(130)μ in diam., ± circular with marginal cells usually concave, usually with 8 radial walls and typically 8 (rarely 0) tangential walls and sometimes with many to several extra radial walls; marginal cells 8–11(16); cell-walls clear and usually very thick; cells usually opaque.

Botswana. N: Ngamiland, fl. xii.1930, *Curson* 811 (PRE). SE: Kanye Distr., 2·4 km. SE. of Pharing, fl. & fr. x.1947, *Miller* B/510 (FHO; K; LISC). **Zambia.** B: Balovale Distr., fr. viii.1952, *Gilges* 139 (SRGH). N: Lake Mweru, Nchelenje, fl. & fr. 4.x.1961, *Lawton* 771 (FHO). W: Mwinilunga Distr., 27 km. from Mwinilunga on the road to Kabompe, fr. 6.vi.1963, *Edwards* 648 (FHO; SRGH). C: Lusaka, 1310 m., fl. & fr. 7.x.1960, *Best* 239 (LISC; SRGH). E: Fort Jameson, fr. 30.viii.1929, *Burtt Davy* 1029/29 (BM; FHO). S: 48 km. E. of Mazabuka, fl. 23.xi.1931, *Stevenson* 182/31 (FHO). **Rhodesia.** N: Sebungwe Distr., Chicomba Vlei, fl. 9.xi.1951, *Lovemore* 161 (SRGH). W: Shangani Distr., Gwampa Forest Reserve, c. 915 m., fl. & fr. i.1957, *Goldsmith* 12/58 (SRGH). C: Salisbury, fl. & fr. 1917, *Eyles* 847 (BM; SRGH). E: S. of Penhalonga, fl. 1935, *Gilliland* 1229 (BM; FHO; PRE; SRGH). S: c. 8 km. NW. of Zimbabwe, fr. 7.v.1963, *Leach* 11670 (FHO; SRGH). **Malawi.** N: Vipya, between Rumpi Drift and Chikangawa, fr. 24.vi.1960, *Chapman* 784 (FHO; SRGH). C: Dedza, Chongoni Forest, fr. 16.viii.1960, *Chapman* 864 (BM; COI; FHO; SRGH). S: Mlanje Mt., 915–1065 m., fl. 21.ix.1957, *Chapman* 436 (FHO). **Mozambique.** N: c. 10 km. from Mutuáli, fr. 25.iii.1953, *Gomes e Sousa* 4084 (COI; LISC; PRE). Z: Mocuba, near Nhaluanda, fr. 27.v.1949, *Andrada* 1519 (COI; LISC). T: Vila Mouzinho, fr. 16.vii.1949, *Andrada* 1773 (COI; LISC). MS: Cheringoma, Inhaminga, fr. 2.vii.1941, *Torre* 3064 (BM; LISC). SS: Inharrime, Nhacoongo, fl. x.1935, *Gomes e Sousa* 1651 (BM; COI; K; LISC). LM: Maputo, fl. ix.1930, *Gomes e Sousa* 370 (COI).

Throughout the savanna and woodland regions of tropical and S. Africa and in Arabia. *Acacia-Combretum* savanna, *Brachystegia-Julbernardia* woodland and savanna woodland, *Combretum-Pterocarpus* savanna, *Acacia-Combretum-Terminalia* tree savanna, tree savanna with *Acacia nigrescens* (in Botswana), lowland forest margins with *Pteleopsis myrtifolia* and *Erythrophloeum suaveolens*, termite mounds, on heavy soils and on nickeliferous, cupriferous and graphitic soils; very widespread from low to medium altitudes.

Africans use both leaves and roots as an antidote to snake-bite (Faria, loc. cit.).

The 42 synonyms cited show the complexity of the problem provided by this very variable savanna species. This problem has been discussed by Exell and Garcia (loc. cit. 1954), by Exell (loc. cit. 1970) and most recently by Wickens (loc. cit. 1973). Nine of the principal forms were indicated by Exell (1970) and Wickens (1973) in the most recent survey recognized three principal forms in the F.T.E.A. area.

Very polymorphic aggregate species occur frequently not only in *Combretum* but in many other genera and families in the savanna regions of tropical Africa. They seem to be largely correlated in their areas of distribution with the occurrence of seasonal bush fires, the latter themselves correlated with an annual alternation of approximately six months' rainfall and six months' drought. Higher rainfall starts to produce forest and lower rainfall semi-desert conditions with insufficient vegetation to cause fierce fires. This problem was discussed recently by Exell and Stace (in Valentine, Taxonomy, Phytogeography and Evolution, 1972: 316–321). Whatever the effect that the fires may have on genetic instability, there is no doubt that the problem is a difficult one. These species-complexes have received in general the following kinds of treatment: (1) the chronologically earliest method was to describe each new combination of characters as a new species (hence the copious synonymy). This became impracticable and has now been largely discarded; (2) The various combinations have been given infra-specific epithets (as in *C. collinum* agg. by Okafor (loc. cit.) and in *Dichrostachys cinerea* by Brenan and Brummitt (F.Z. 3, 1: 37 et seq. (1970)) and in other cases; (3) the various combinations have been lettered A, B, C etc. as in *Clematopsis scabiosifolia* by Exell, Léonard and Milne-Redhead (in Bull. Roy. Soc. Bot. Belg. **83**: 407 et seq. (1951)) and in *Allophylus africanus* by Exell (F.Z. 2, 2: 506–508 (1966)). Wickens has followed this system to some extent (in F.T.E.A. Combret.: 35 (1973)) for *C. molle* agg. This method has certain practical advantages while our knowledge is still incomplete; (4) the complex may simply be treated as an aggregate species. *C. molle* is still largely at this stage. The obvious disadvantages lie in the fact that no assistance is given to ecologists if they find that they need a more precise classification.

The principal problem complexes in the African savanna regions in *Combretum* are: *C. collinum* agg., *C. molle* agg., *C. apiculatum* agg., *C. hereroense* agg. and the *C. paniculatum-microphyllum-platypetalum-oatesii* complex, the last-named though ecologically less important being one of the most difficult to deal with. To provide a real solution to the *C. molle* problem it would be necessary to study the species in the various ecological habitats throughout its range, to discover whether the various forms belong at all consistently to ecological niches and whether the same form belongs to the same niche in different regions, to conduct germination experiments with all the forms, to establish whether hybridization takes place (of which there is at present no definite evidence) and to conduct breeding experiments to find out to what extent the various forms breed true. In view of the immense range of the aggregate species this is probably more than the work of a lifetime and one is bound to question whether such an investigation will ever be carried out and even whether it is worth the enormous effort.

I have already mentioned the transequatorial cline (Exell, loc. cit. 1970) in which the ovate leaf-shape of the typical form in Ethiopia and the Sudan becomes a narrowly elliptic-ovate shape with much larger leaves in equatorial regions of high rainfall returning in the southernmost part of the range (Natal) to the same form found north of the equator. A less clumsy term for this might be a *holocline.**

In our area the variation in *C. molle* agg. is in general rather less complicated than in East Africa or Angola. Our main troubles are intermediates with *C. apiculatum*. It has in fact been suggested (L. E. Codd in litt. and Faria, loc. cit.) that these two species hybridize. C. A. Stace (to whom I am constantly grateful for examining *Combretum* scales) has always been able to identify apparent *molle-apiculatum* intermediates with one species or the other by the character of the scales but I know of no other method of identifying some of the more difficult specimens.

The best advice I can give to ecologists who need more precision than *C. molle* agg. is to make their own classification of any distinct-looking forms in the areas in which they are interested but not at present to give them more than a letter or number because an infra-specific classification burdened with Latin names may seem quite appropriate in one region but break down completely in another.

15. **Combretum apiculatum** Sond. in Linnaea, **23**: 45 (1850); in Harv. & Sond., F.C. **2**: 510 (1862).—Laws. in Oliv., F.T.A. **2**: 425 (1871).—Engl. & Diels in Engl., Mon. Afr. Pflanz. **3**: 42 t. 12 fig. C (1899).—Monro in Proc. & Trans. Rhod. Sci. Ass. **8**, 2: 80 (1908) (" ariculatum ").—Sim, For. Fl. Port. E. Afr.: 63 (1909).—Bak. f. in Journ. Linn. Soc., Bot. **40**: 68 (1911).—Eyles in Trans. Roy. Soc. S. Afr. **5**, 4: 427 (1916).—Burtt Davy, F.P.F.T. **1**: 245 (1926).—Steedman, Some Trees etc. S. Rhod.: 54 (1933).—Burtt Davy & Hoyle, N.C.L.: 38 (1936).—O. B. Mill., B.C.L.: 42 (1948).—Suesseng. & Merxm. in Proc. & Trans. Rhod. Sci. Ass. **43**: 27

---

* Wickens (loc. cit.) refers to it as an " amphiaequatorial palindrome ".

E

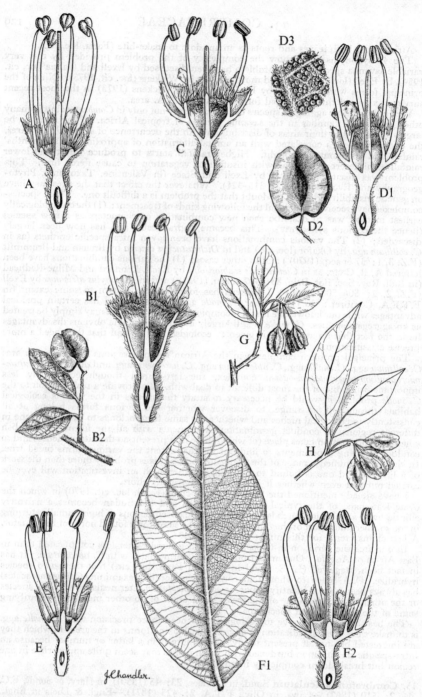

Tab. 34. A.—COMBRETUM GOSSWEILERI, flower, longitudinal section (×6). B.—
COMBRETUM HEREROENSE. B1, flower, longitudinal section (×6); B2, fruiting
branchlet (×½). C.—COMBRETUM FRAGRANS, flower, longitudinal section (×6).
D.—COMBRETUM COLLINUM. D1, flower, longitudinal section (×6); D2, fruit (×½);
D3, scales from lower leaf-surface (×15). E.—COMBRETUM APICULATUM subsp.
APICULATUM, flower, longitudinal section (×6). F.—COMBRETUM MOLLE. F1, leaf
(×½); F2, flower, longitudinal section (×6). G.—PTELEOPSIS ANISOPTERA, fruiting
branchlet (×½). H.—PTELEOPSIS MYRTIFOLIA, fruiting branchlet (×½). From
F.F.N.R.

(1951).—Wild, Guide Fl. Vict. Falls: 146 (1953); in Rhod. Agric. Journ. **52**, 6: 12, 16, 17 (1955); in Kirkia, **7**, 1: 49, 53 (1968); tom. cit. Suppl.: 40, 45 (1970).— Exell & Garcia in Contr. Conhec. Fl. Moçamb. **2**: 113 (1954); C.F.A. **4**: 61 (1970).— Duvign. in Bull. Soc. Roy. Bot. Belg. **90**, 2: 206 (1958).—Palmer & Pitman, Trees S. Afr.: 244 cum fig. (1961).—F. White, F.F.N.R.: 286, fig. 50E (1962).—Mitchell in Puku, **1**: 121 (1963).—Boughey in Journ. S. Afr. Bot. **30**, 4: 167 (1964).—Stace in Journ. Linn. Soc., Bot. **59**: fig. 26 (1965).—Breitenb., Indig. Trees Southern Afr. **4**: 823 (1965).—B. and M. de Winter & Killick, Sixty-six Transv. Trees: 118, cart. et phott. (1966).—Liben, F.C.B. Combret.: 65 (1968).—Exell in Bol. Soc. Brot., Sér. 2, **42**: 7 (1968); in Kirkia, **7**, 2: 195 (1970).—Gillet & McDonald, N.C.K.: 14 (1970).—Wickens, F.T.E.A., Combret.: 35, fig. 2(15), 5(15) (1973).—Jacobsen in Kirkia, **9**, 1: 168 (1973).—Faria in Mem. Invest. Agron. Moçamb. **4**: 55 (1973). Type from S. Africa.

Combretum apiculatum forma *sulphureum* Heurck & Müll. Arg. in Heurck, Obs. Bot. **2**: 229 (1871). Based on the same type as *C. apiculatum*.

Combretum apiculatum forma *viscosum* Heurck & Müll. Arg. in Heurck, tom. cit.: 231 (1871). Based on the same type as *C. apiculatum*. *Zeyher* 553 is cited both for this and for the preceding form.

Combretum apiculatum var. *parvifolium* Bak. f. in Journ. of Bot. **43**: 46 (1905).— Eyles in Trans. Roy. Soc. S. Afr. **5**, 4: 427 (1916). Type: Rhodesia, Bulawayo, *Eyles* 1094 (BM, holotype; SRGH).

Combretum glutinosum Wood in Trans. Phil. Soc. S. Afr. **18**, 2: 156 (1908) non Perr. ex DC. (1828). Type from S. Africa.

Combretum apiculatum var. *sulphureum* (Heurck & Müll. Arg.) Dummer in Gard. Chron. Ser. 3, **53**: 164 (1913). Type as for *C. apiculatum*.

Combretum apiculatum var. *viscosum* (Heurck & Müll. Arg.) Dummer, loc. cit.

Combretum apiculatum subsp. *boreale* Exell in Journ. of Bot. **67**: 46 (1929).—Burtt Davy & Hoyle, N.C.L.: 38 (1936).—Brenan, T.T.C.L.: 136 (1949). Type from Tanzania.

Small semi-deciduous tree 4–10 m. high, rarely a shrub (doubtfully recorded as scandent); crown open, flattened to rounded; bark greyish-black, deeply fissured, scaly; branchlets tomentose to glabrous (except for scales), lepidote; leaf-buds black or dark-brown. Leaves opposite; lamina 3–13 × 1·5–7·5 cm., glutinous when young, broadly to narrowly obovate-elliptic or oblong-elliptic or ovate to sub-circular, lepidote (sometimes inconspicuously so) and pubescent or pilose to glabrous (except for scales), apex usually apiculate or mucronate and usually twisted, base usually rounded to subcordate; lateral nerves 5–7 pairs, ± arcuate, reticulation rather inconspicuous to prominulous beneath; petiole up to 10 mm. long, leaving a prominent peg. Inflorescences of axillary spikes 3–7 cm. long; rhachis lepidote and otherwise glabrous or hairy; bracts 1·5 × 0·3 mm., caducous. Flowers sessile, yellow. Lower receptacle 2 mm. long, lepidote and otherwise hairy or glabrous; upper receptacle 3–4 × 2·5–3 mm., campanulate, lepidote and otherwise glabrous or pubescent. Sepals 0·5–1 mm. long, broadly deltate. Petals 1·2 × 1·2 mm., obtriangular, ciliate at the apex. Stamen-filaments 5 mm. long, anthers 1·2–1·5 mm. long. Disk with a pilose scarcely free margin. Style 5 mm. long, stout, not expanded at the apex. Fruit 2–3 × 1·5–2·5 cm., shiny (glutinous when young), subcircular to elliptic in outline, lepidote, otherwise glabrous or pubescent, apical peg up to 1 mm. long, wings up to 7 mm. broad, stipe 4–8 mm. long. Cotyledons 2, 2·5–2·8 × 3–4 cm., arising below soil level; petioles 4–5 cm. long. Scales yellowish-red, usually inconspicuous and often concealed by glutinous excretions, c. 45(50)–75(90)μ in diam., scalloped at each marginal cell; cells delimited by (7)8(10) radial walls and sometimes also up to 5 tangential walls and rarely with extra radial walls, marginal cells 7–10; cell-walls clear and thin, cells ± transparent.

Leaves glabrous (except for scales) when mature except for tufts of hairs in the axils of the
    lateral nerves beneath and/or with pilosity along the midrib and sometimes along the
    principal nerves beneath; scales 50–70μ in diam. - - - subsp. *apiculatum*
Leaves pubescent to pilose beneath even when mature; scales 60–100μ in diam.
                                                    subsp. *leutweinii*

15a. Subsp. **apiculatum**.—Exell in Mitt. Bot. Staatssamml. München **4**: 3 (1961); Prodr. Fl. SW. Afr. **99**: 7 (1966); in Bol. Soc. Brot. Sér. 2, **42**: 18 (1968); in Kirkia, **7**, 2: 194, t. 2 fig. 16a (1970).—Stace in Mitt. Bot. Staatssamml. München **4**: 13 (1961).—Exell & Garcia, C.F.A. **4**: 61 (1970).—Faria in Mem. Inst. Invest. Agron. Moçamb. **4**: 60 (1973). TAB. **34** fig. E; TAB. **35** fig. 15a.

Combretum buchananii Engl. & Diels in Engl., Mon. Afr. Pflanz. **3**: 40 (1899). Type: Malawi, *Buchanan* 1263 (B†, holotype; K).
Combretum apiculatum subsp. *boreale* Exell in Journ. of Bot. **67**: 46 (1929) pro parte.—Phillips in Journ. Ecol. **18**, 2: 211 (1930). Type from Tanzania.

Tree up to 10 m. high. Leaf-lamina up to 11–13 × 5–7·5, glabrous (except for scales) when mature except for tufts of hairs in the axils of the nerves beneath and occasionally pilosity along the midrib and some of the principal lateral nerves. Scales 50–75μ in diam.

Botswana. N: Ngamiland, near Tsodilo Hill, 1220 m., fr. 15.i.1963, *Banks* 39 (PRE). SE: Kanye, Polukwe Hill, 1220 m., fl. 13.xi.1948, *Hillary & Robertson* 504 (PRE). **Zambia.** B: Sesheke, fr. 28.xii.1952, *Angus* 1058 (FHO; K). N: Abercorn, 855 m., fr. 6.iii.1952, *Richards* 918 (K). C: Feira Distr., lower Luangwa Valley, fr. 30.v.1952, *White* 2910 (FHO; K). E: Chadiza Hill, fl. 1.xii.1958, *Robson* 794 (FHO). S: Gwembe Valley, Sinazeze, fr. 19.vi.1961, *Angus* 2927 (FHO; K; LISC). **Rhodesia.** N: Urungwe, Magunge, fr. 4.v.1955, *Schiff* 13 (COI; SRGH). W: Bulawayo Commonage, fr. 1932, *Pitt-Schenkel* 169 (FHO). C: Norfolk, Gwelo, fl. x.1933, *Steedman* 82 (FHO). E: Chipinga Distr., fl. 15.xi.1959, *Goodier* 642 (K; SRGH). S: Beitbridge Distr., Nulli Range, fr. 10.i.1961, *Leach* 10676 (BM; K; SRGH). **Malawi.** N: Fort Hill, 1675 m., fr. 6.vii.1936, *Burtt* 5960 (K). S: Monkey Bay, fr. 30.v.1937, *Lynes* 1402 (BM). **Mozambique.** N: Nampula, fr. 15.xii.1936, *Torre* 1099 (COI; LISC). Z: between Aguas Quentes and Morrumbala, fr. 28.vii.1949, *Barbosa & Carvalho* in *Barbosa* 3773 (LMJ). T: Cabora Bassa, fr. 25.vi.1949, *Barbosa & Carvalho* 3268 (LMA). MS: Chemba, fl. 10.xi.1946, *Pedro & Pedrógão* 118 (K; PRE). SS: Caniçado, fr. 18.vi.1947, *Pedrógão* 332 (K; PRE; SRGH). LM: Ressano Garcia, fr. 24.xii.1897, *Schlechter* 11911 (BM; BR; COI; K).

Also in Kenya, Tanzania, Angola, SW. Africa and S. Africa. Dry savanna and savanna woodland especially in rocky places; often with *Adansonia*, *Kirkia acuminata* and *Sterculia* spp.; secondary *Baikiaea* and Kalahari woodland; mopane termite mounds; *Acacia-Combretum-Terminalia* tree savanna; dambos and watershed grassland; tolerant of metalliferous soils; very common at low altitudes but occurring also at medium altitudes.

C. apiculatum subsp. *boreale* was described for plants (mainly from Tanzania) with longer relatively narrower leaves but it appears that this type of variation occurs in several species of *Combretum* in higher rainfall regions and little taxonomic importance should be attached to it. The type of subsp. *boreale* comes under typical *C. apiculatum* in the system now adopted. Various specimens named subsp. *boreale* are now considered to belong to subsp. *leutweinii*.

Intermediates or possible hybrids with *C. molle* are mentioned under that species.

15b. Subsp. **leutweinii** (Schinz) Exell in Mitt. Bot. Staatssamml. München **4**: 3 (1961); Prodr. Fl. SW. Afr. **99**: 8 (1966); in Bol. Soc. Brot. Sér. 2, **42**: 19 (1968); in Kirkia, **7**, 2: 195, t. 2 fig. 16b (1970).—Stace in Mitt. Bot. Staatssamml. München **4**: 13 (1961).—Faria in Mem. Inst. Invest. Agron. Moçamb. **4**: 68 (1973). TAB. **35** fig. 15b. Type from SW. Africa.
Combretum apiculatum var. *? pilosiuscula* Engl. & Diels in Engl., Mon. Afr. Pflanz. **3**: 43 (1899).—Schinz, Pl. Menyharth.: 431 (1906). Type: Mozambique, Boroma, *Menyharth* 892 (Z).
Combretum leutweinii Schinz apud De Wild. & Dur. in Bull. Herb. Boiss. **2**, 1: 878 (1901). Type as for *C. apiculatum* subsp. *leutweinii*.
Combretum kwebense N.E.Br. in Kew Bull. **1909**: 111 (1909).—O. B. Mill., B.C.L.: 43 (1948). Type: Botswana, Kwebe Hills, *Lugard* 48 (K, holotype).

Leaf-lamina up to 10 × 8 cm. but usually c. 4·5 × 3 cm. or smaller, densely to sparsely pubescent above and beneath, often rufous-tomentose on the nerves. Fruit often pubescent to pilose at least on the body. Scales 60–100μ in diam.

Botswana. N: Chobe Distr., Serondela, 915 m., fl. (fallen) xii.1951, *Miller* B/1265 (FHO). SE: Kgatla Distr., 32 km. NE. of Mochudi, *Miller* B/573 (FHO). **Zambia.** E: Petauke Distr., Beit Bridge, 425 m., fr. 18.iv.1952, *White* 2406 (FHO; K). **Rhodesia.** N: Sebungwe Distr., near Dett Salt Pan, fl. 27.xi.1956, *Lovemore* 509 (K; LISC; SRGH). E: Roadside Junction Tea Rooms, Melsetter road, fr. 1.i.1949, *Chase* 1389 (BM; SRGH). S: Ndanga, Chipinda Pools, fl. 3.xii.1959, *Goodier* 714 (K; LISC; SRGH). **Malawi.** S: Port Herald, fr. 1934, *Townsend* 40 (FHO). **Mozambique.** N: Amaramba, R. Lúrio, fl. 19.x.1948, *Andrada* 1421 (BM; COI; K; LISC; LMJ). T: between Tete and Chioco, fl. 27.ix.1942, *Mendonça* 474 (LISC). SS: Guijá, fl. 16.xi.1957, *Barbosa & Lemos* 8172 (K; LISC; LMJ).

Also in SW. Africa. Mainly in *Colophospermum mopane* savanna woodland but also common on cupriferous, nickeliferous and graphitic soils.

Subsp. *leutweinii* seems to occur mainly in the mopane belt. There are some intermediates but in general this subspecies can be satisfactorily distinguished from subsp. *apiculatum* by the indumentum on the lower surface of the leaves which remains until maturity, though then sometimes sparse, over the whole surface. Subsp. *leutweinii* may be confused with *C. albopunctatum* but the latter can usually be distinguished by its glistening white scales, especially on the fruit. The fruit of *C. albopunctatum* has shorter hairs than that of *C. apiculatum* subsp. *leutweinii* in which the indumentum on the body of the fruit is usually ± pilose.

16. **Combretum acutifolium** Exell in Journ. of Bot. **71**, Suppl. Polypet.: 232 (1933); in Bol. Soc. Brot., Sér. 2, **42**: 7 (1968); in Kirkia, **7**, 2: 195, t. 2 fig. 17 (1970).—F. White, F.F.N.R.: 286 (1962).—Liben, F.C.B. Combret.: 65 (1968).—Exell & Garcia, C.F.A. **4**: 60 (1970).—Wickens, F.T.E.A. Combret.: 36, fig. 2(16), 5(16) (1973). TAB. **35** fig. 16. Type from Angola.

Deciduous slender-stemmed climber or scrambling shrub; branchlets tomentose at first, soon glabrescent. Leaves opposite; lamina up to 7 × 3·5 cm., papyraceous, elliptic or broadly elliptic, rather sparsely pubescent to pilose on the principal nerves on both surfaces and sparsely pilose on the areolae below, densely ciliate on the margin, rather inconspicuously lepidote, apex acuminate and slightly mucronate, base rounded; lateral nerves 5–6 pairs; petiole up to 1·5 cm. long. Inflorescences of unbranched axillary spikes up to 3 cm. long; rhachis pubescent; bracts caducous. Flowers sessile, yellow. Lower receptacle 1–2 mm. long, glabrous or nearly so; upper receptacle c. 1·5 × 2 mm., broadly campanulate, sparsely and rather inconspicuously lepidote otherwise glabrous or nearly so. Sepals scarcely developed. Petals 1 × 1 mm., subcircular, ciliate at the apex, shortly unguiculate. Stamen-filaments 5 mm. long, inserted at or near the margin of the disk; anthers 0·5–0·6 mm. long. Disk c. 1 mm. in diam. with short free pilose margin. Style 3–4 mm. long, not expanded at the apex. Fruit greenish-yellow sometimes with red wings, up to 2·5 × 2·7 cm., subcircular in outline, glutinous, ± glabrous (except for scales), rather inconspicuously lepidote, apical peg very short (up to 0·5 mm. long) or absent, wings c. 10 mm. broad, stipe up to 7 mm. long. Scales 50–90(100)µ in diam., usually 8-celled with additional tangential and/or radial walls. Seedlings unknown.

Zambia. N: Abercorn Distr., Lake Tanganyika, Mpulungu, fl. 16.xi.1952, *Angus* 766 (BM; FHO; K). W: Mwinilunga, fl. 9.x.1952, *Angus* 584 (BM; FHO; K).
Also in Angola, Zaire (Katanga) and Tanzania. *Brachystegia-Julbernardia* woodland; Kalahari woodland; Itigi forest; thickets on shingle; at medium altitudes.
Very near to *C. apiculatum* but a climber or scandent shrub which up to now seems to be distinct. It is widespread in Angola extending beyond the Angolan frontiers into Katanga and western Zambia. The Abercorn record and the Tanzania localities are isolated from the rest of the range but the species may be found in other parts of Zambia.

17. **Combretum psidioides** Welw. in Ann. Cons. Ultram. **1856**, parte não official, Sér. 1: 249 (1856).—Engl. & Diels in Engl., Mon. Afr. Pflanz. **3**: 51 (1899).—Burtt Davy & Hoyle, N.C.L.: 39 (1936).—Exell & Garcia in Contr. Conhec. Fl. Moçamb. **2**: 112 (1954); C.F.A. **4**: 64 (1970).—Duvign. in Bull. Soc. Roy. Bot. Belg. **90**, 2: 194 et seq. (1958).—Stace in Mitt. Bot. Staatssamml. München **4**: 14 (1961).—F. White, F.F.N.R.: 286 (1962).—Boughey in Journ. S. Afr. Bot. **30**, 4: 167 (1964).—Fanshawe & Savory in Kirkia, **4**: 190 (1964).—Liben, F.C.B. Combret.: 66 (1968).—Exell in Bol. Soc. Brot., Sér. 2, **42**: 8 (1968); in Kirkia, **7**, 2: 196 (1970).—Wickens, F.T.E.A. Combret.: 37, fig. 5(17) (1973).—Faria in Mem. Inst. Invest. Agron. Moçamb. **4**: 70 (1973).
Combretum holosericeum sensu Laws. in Oliv., F.T.A. **2**: 430 (1871) pro parte quoad specim. Angol.

Semideciduous understorey tree up to 9 m. high or large bush; crown open, flattened to rounded; bark grey to black, deeply fissured or fairly smooth; branchlets usually tomentose when young with the bark peeling off in large strips or cylindrical or hemicylindrical pieces leaving a newly-exposed cinnamon-red surface. Leaves opposite; lamina 5–15(26) × 3–10(16) cm., elliptic or oblong-elliptic or oblong or ovate-elliptic or ovate-oblong, densely tomentose when young (except in subsp. *glabrum*), eventually pubescent to pilose or tomentose on the reticulation beneath and glabrous on the areolae (subsp. *psidioides*) or tomentose both on the reticulation and on the areolae (subsp. *dinteri*) or glabrous (subsp.

glabrum) apart from scales, rather inconspicuously lepidote, apex usually rounded or retuse, often mucronate or apiculate, base rounded or subcordate; lateral nerves 8–16 pairs; reticulation rather prominent beneath; petiole 3–10 mm. long. Inflorescences of rather dense usually tomentose axillary spikes up to 10 cm. long, usually appearing with the young leaves; rhachis usually tomentose; bracts 1–2 mm. long, subulate, caducous. Flowers sessile, yellow. Lower receptacle 1·2–1·3 mm. long, usually tomentose; upper receptacle 2·5 × 3·5 mm., broadly campanulate or cupuliform, usually tomentose. Sepals 1 × 1·5 mm., deltate. Petals 1·1 × 0·6 mm., obcuneate, ciliate at the apex (except in subsp. *glabrum*). Stamen-filaments 6 mm. long; anthers 1·1 mm. long. Disk 2 mm. in diam. with a very short free pilose margin. Style 6·5 mm. long, slightly expanded at the apex. Fruit usually crimson or pink, sometimes red-brown or tan, up to 3 × 3 cm., glutinous, subcircular or elliptic in outline, nearly glabrous except for rather inconspicuous scales, apical peg absent, wings up to 13 mm. broad, flexible and rather tenuous, stipe up to 10 mm. long. Cotyledons 2, 2–2·5 × 3–4 cm., transversely elliptic, arising below soil level; petioles 4–4·5 cm. long. Scales glistening, whitish or yellowish or reddish, often obscured by the indumentum, c. 40–50(70)μ in diam., circular, with 8 primary radial walls only, cell-walls very thin and obscured by glutinous secretions; cuticular membrane greatly raised from the cell-plate forming a ± spherical cavity.

This species was placed by Engler and Diels (op. cit.: 51) in Sect. *Glabripetala* but later transferred to Sect. *Ciliatipetala* by Exell and Garcia (loc. cit.).

Subsp. *kwinkiti* (De Wild.) Exell (*C. kwinkiti* De Wild.; *C. puetense* Engl. & Diels) in northern Angola and Zaire, with the lower surface of the leaf glabrous on the reticulation and tomentose to pubescent on the areolae, has not been recorded from our area but might well occur in Zambia.

Wickens (in Kew Bull. **26**, 1: 37–40 (1971)) has published a study of all five subspecies of *C. psidioides* with a map showing their distribution. The key given in Kirkia (loc. cit. 1970) has been completed from the key given by Wickens in the paper cited above.

Lower surface of leaf pubescent to tomentose; petals ciliate at the apex:
  Young leaves not glutinous; areolae on the lower surface of the leaf nearly glabrous (except for scales) to tomentose:
    Lower surface of leaf pubescent on the reticulation but sparsely pubescent to nearly glabrous (except for scales) on the areolae when mature -   subsp. *psidioides*
    Lower surface of leaf shortly tomentose on the reticulation and on the areolae
                           subsp. *dinteri*
  Young leaves glutinous and nearly glabrous (except for scales); areolae on the lower surface glabrous (except for scales) when mature; sides of the midrib and usually the principal lateral nerves ferruginous-pubescent; leaves attaining nearly maximum dimensions for the species -   -   -   subsp. *psilophyllum*
Lower surface of leaf glabrous; young leaves glutinous on the upper surface only; petals glabrous (without apical cilia) - - - - - - - subsp. *glabrum*

17a. Subsp. **psidioides.**—Exell in Mitt. Bot. Staatssamml. München **4**: 5 (1961); in Prodr. Fl. SW. Afr. **99**: 10 (1966); in Kirkia **7**, 2: 197, t. 2 fig. 18a (1970).—Stace in Mitt. Bot. Staatssamml. München **4**: 14 (1961); in Bull. Brit. Mus. (Nat. Hist.) Bot. **4**, 1: 27 (1965).—Exell & Garcia, C.F.A. **4**: 64 (1970).—Wickens in Kew Bull. **26**, 1: 38 (1971); in F.T.E.A. Combret.: 37, fig. 2(17a–d), fig. 3(17) (1973).—Faria in Mem. Inst. Invest. Agron. Moçamb. **4**: 71 (1973). TAB. **30** fig. G; TAB. **35** fig. 17a.

    Combretum grandifolium F. Hoffm., Beitr. Kenntn. Fl. Centr.-Ost-Afr.: 29 (1889).—Engl. & Diels in Engl., Mon. Afr. Pflanz. **3**: 39, t. 13 fig. B a–c (1899). Type from Tanzania.

    Combretum grandifolium var. *retusa* F. Hoffm., tom. cit.: 30 (1889).—Engl. & Diels in Engl., loc. cit. Type from Tanzania.

    Combretum grandifolium var. *eickii* Engl. & Diels in Engl., loc. cit. Type from Tanzania.

    Combretum brachypetalum R.E.Fr. in Wiss. Ergebn. Schwed. Rhod.-Kongo-Exped. **1**: 168, t. 1 fig. 4 et t. 13 fig. 4–6 (1914). Syntypes: Zambia, Bangweulu, *Fries* 773 (S), 953 (K; S), 953a (S), 965 (S).

    Combretum omahekae Gilg & Dinter ex Engl., Pflanzenw. Afr. **3**, 2: 698 (1921). Type from SW. Africa.

    Combretum psidioides subsp. *katangense* Duvign. in Bull. Soc. Roy. Bot. Belg. **88**: 70 (1956). No type cited but clearly equivalent to the type subspecies.

Leaf-lamina pubescent to densely pubescent or pilose on the reticulation beneath, sparsely pubescent or nearly glabrous (except for scales) on the areolae when mature. Petals ciliate at the apex.

Caprivi Strip. Singalamwe area, c. 1000 m., fr. 30.xii.1958, *Killick & Leistner* 3207 (M; PRE; SRGH). **Botswana.** SE: Mahalapye, near Lobatsi, st. 10.iii.1958, *de Beer* 674 (SRGH). **Zambia.** B: Mankoya Distr., near Luampa Mission, fr. 21.ii.1952, *White* 2011 A (BM; FHO). N: Fort Rosebery Distr., N. of Samfya Mission, fl. 7.x.1947, *Brenan & Greenway* in *Brenan* 8053 (FHO). W: Mwinilunga Distr., c. 6·5 km. N. of Kalene Hill Mission, fr. 16.vi.1963, *Edwards* 789 (COI; FHO; SRGH). C: Lusaka Distr., c. 185 km. E. of Lusaka on the Great East Road, fr. 16.iv.1952, *White* 2695 (BM; FHO; K). E: Tigone Dam, fl. 17.x.1958, *Robson & Angus* in *Robson* 142 (FHO). S: Mumbwa Distr., c. 9·5 km. from Nambwula Mission, fl. 17.ix.1947, *Brenan & Greenway* in *Brenan* 7868 (FHO). **Rhodesia.** N: Sebungwe Distr., fr. v.1950, *Davies* 568 (BM; SRGH). W: Victoria Falls, fr. 12.ii.1912, *Rogers* 13013 (PRE). E: Melsetter Distr., Martin Forest Reserve, fr. 24.ii.1952, *Mullin* 1/52 (BM; LISC; SRGH). **Malawi.** C: Dowa Distr., fr. 1934, *Topham* 982 (FHO; K). S: Zomba, fl. 1935, *Clements* 560 (FHO). **Mozambique.** N: Chamba, fl. 20.x.1948, *Andrada* 1428 (BM; COI; K; LISC). Z: between Molócuè and the Gilé, st. 15.x.1949, *Barbosa & Carvalho* 4447 (LMA). T: Macanga, between Furancungo and Régulo Bene, *Andrada* 1752 (COI; LISC). MS: Chimoio, Bandula, 700 m., fr. 27.ii.1948, *Garcia* 415 (BM; COI; K; LISC).

Also in Angola, Zaire, SW. Africa and Tanzania. Tree or shrub savanna especially on sandy soils; *Brachystegia* woodland: Kalahari woodland; rocky outcrops; hotter drier areas at medium altitudes.

17b. Subsp. **dinteri** (Schinz) Exell in Mitt. Bot. Staatssamml. München **4**: 3 (1961); Prodr. Fl. SW. Afr. **99**: 10 (1966); in Kirkia, **7**, 2: 198, t. 2 fig. 18b (1970).—Stace in Mitt. Bot. Staatssamml. München **4**: 14 (1961); in Bull. Brit. Mus. (Nat. Hist.) Bot. **4**, 1: 27 (1965).—Exell & Garcia, C.F.A. **4**: 66 (1970). TAB. **35** fig. 17b. Type from SW. Africa.
    Combretum dinteri Schinz apud De Wild. & Dur. in Bull. Herb. Boiss. **2**, 1: 877 (1901). Type as above.
    Combretum quirirense Engl. & Diels in Warb., Kunene-Samb.-Exped. Baum: 318 (1903). Type from Angola.

Leaf-lamina up to 10 × 5 cm., tomentellous beneath both on the reticulation and areolae. Petals ciliate at the apex.

Rhodesia. N: Gokwe Distr., fr. 9.iii.1962, *Bingham* 107 (FHO; LISC; SRGH). W: Bakumbusi, c. 80 km. N. of Bulawayo, fr. 4.vi.1947, *Keay* 21306 (BM; FHO; SRGH). E: Chipinga Distr., Giriwayo, 375 m., fr. 19.i.1957, *Phipps* 28b (SRGH).

Also in Angola and SW. Africa. *Baikiaea* woodland; *Colophospermum mopane* savanna woodland.

17c. Subsp. **psilophyllum** Wickens in Kew Bull. **26**, 1: 39 (1971); F.T.E.A. Combret.: 38, fig. 2(17e) (1973). Type from Tanzania.
    Combretum vanderystii De Wild., Ann. Mus. Cong., Bot. Sér. 5, **3**: 242 (1910). Syntypes from Zaire.
    Combretum anacardifolium Engl., *nom. nud.* in sched.
    Combretum psidioides subsp. *glabrum* Exell in Bol. Soc. Brot., Sér. 2, **42**: 18 (1968) pro parte quoad specim. *Richards* 11468.

Leaf-lamina 12–24 × 8–16 cm., ferrugineous-pubescent along the sides of the midrib and usually the principal lateral nerves as well but glabrous on the areolae (except for scales) when mature. Young leaves glutinous and appearing glabrous. Petals ciliate at the apex.

Zambia. N: Mbala (Abercorn) to Mpulungu road, c. 16 km. from Mpulungu, fl. 26.ix.1959, *Richards* 11468 (K).
Also in Zaire and Tanzania. *Brachystegia* woodland.

17d. Subsp. **glabrum** Exell in Bol. Soc. Brot., Sér. 2, **42**: 18 (1968) excl. specim. *Richards* 11468; in Kirkia, **7**, 2: 198, t. 2 fig. 18c (1970).—Wickens in Kew Bull. **26**, 1: 40 (1971). TAB. **35** fig. 17d. Type: Rhodesia, Wankie, *Pardy* in GHS 4564 (SRGH, holotype).

Rhodesia. N: Mazoe Valley, fr. xii.1932, *Seligman* (BM). W: Wankie, Fuller Siding, fl. 2.x.1929, *Pardy* in GHS 4565 pro parte (FHO; SRGH).

Known with certainty only from Rhodesia. Dry woodlands; Kalahari Sand.
It will be observed from the above that *C. psidioides* is one of the few savanna species of

Combretum that is amenable to normal taxonomic treatment; the species is easy to identify and up to the present there has been little difficulty in allocating the specimens to the various subspecies.

9. Sect. FUSCA Engl. & Diels in Engl., Mon. Afr. Pflanz. **3**: 76 (1899).—Exell in Bol. Soc. Brot., Sér. 2, **42**: 23 (1968); in Kirkia, **7**, 2: 198 (1970).—Stace in Journ. Linn. Soc., Bot. **62**: 160 (1969).—Macedo & Faria in Agron. Moçamb. **6**, 3: 211, fig. 1–2 (1972).—Wickens, F.T.E.A. Combret.: 38 (1973).

Sect. *Coriifolia* Engl. & Diels in Engl., tom. cit.: 75 (1899).

Flowers 4-merous, not glutinous. Upper receptacle elongate-campanulate, not obviously divided into two regions. Stamens 1-seriate. Disk without free margin. Fruit 4-winged. Cotyledons unknown. Branchlets and rhachis fuscous-pubescent. Scandent shrubs and lianes, rather inconspicuously lepidote. Scales (30)35–60μ in diam., circular in outline, convexly scalloped (or slightly so) at each marginal cell, divided by (7)8–16 primary radial walls alone.

18. **Combretum coriifolium** Engl. & Diels in Engl., Mon. Afr. Pflanz. **3**: 75, t. 22 fig. e (1899).—Brenan, T.T.C.L.: 134 (1949).—Liben, F.C.B. Combret.: 62 (1968).—Exell in Bol. Soc. Brot., Sér. 2, **42**: 23 (1968); in Kirkia, **7**, 2: 198 (1970).—Macedo & Faria in Agron. Moçamb. **6**, 3: 211, fig. 1–2 (1972).—Wickens, F.T.E.A. Combret.: 39, fig. 2 (1973).—Faria, Mem. Inst. Invest. Agron. Moçamb. **4**: 73 (1973). TAB. **35** fig. 18. Syntypes from Tanzania and Malawi: *Buchanan* 382 (B†; BM; K).

Combretum laurifolium Engl., Pflanzenw. Ost-Afr. **C**: 292 (1895) non Mart. (1839). Syntypes as above.

Combretum leiophyllum Diels in Engl., Bot. Jahrb. **39**: 506 (1907).—Brenan, T.T.C.L.: 141 (1949). Type from Tanzania.

Scandent evergreen shrub or liane climbing up to 30 m.; branchlets at first fuscous-pubescent, becoming glabrous. Leaves opposite; lamina up to 14 × 5·5 cm., coriaceous, oblong to oblong-elliptic, glabrous or nearly so except for tufts of hairs in the axils of the lateral nerves beneath and rather inconspicuously lepidote, apex slightly and rather bluntly acuminate, base obtuse to rounded; lateral nerves 5–6(8) pairs, rather broadly spaced; petiole up to 9 mm. long, rather stout. Inflorescences of short crowded axillary or terminal spikes sometimes branched by suppression of the upper leaves, c. 1 cm. long on peduncles c. 6 mm. long; rhachis fuscous-pubescent; bracts caducous. Flowers cream. Lower receptacle 1·5–2 mm. long, tomentose; upper receptacle 3 × 2 mm., elongate-campanulate. Sepals 1 × 1·2 mm., deltate. Petals 2–3 × 0·9 mm., spathulate, glabrous. Stamen-filaments up to 7·5 mm. long, 1-seriately inserted at or very slightly above the margin of the disk; anthers 0·9 mm. long. Disk 2 mm. in diam., with pilose not free margin. Fruit 4-winged, 2·8(3·5) × 2·5 cm., obovate to obovate-elliptic or subcircular, glutinous (especially when young), pilose to sparsely pilose, apex emarginate, apical peg absent or very short, wings up to 11 mm. broad, stipe up to 16 mm. long, slender, pilose. Cotyledons unknown. Scales 40–60μ in diam., some with up to 16 primary walls.

? **Rhodesia.** E: Gumiras, Sabi R., 425 m. alt., st. 21.iii.1960, *Farrell* 204 (SRGH). **Malawi.** Without precise locality, fl. 1891, *Buchanan* 382 (BM; K). **Mozambique.** N: Ribauè, Serra Mepaluè, fr. 7.ix.1968, *Macedo* 3560 (LMA); between Mueda and Chomba, fl. 25.ix.1948, *Pedro & Pedrógão* 5356 (LMJ). Z: Gúruè, c. 1800 m., fl. 28.ii.1966, *Torre & Correia* 14952 (LISC). MS: Vila Pery, between Matarara do Lucito and Gonda, fr. 6.vi.1949, *Pedro & Pedrógão* 6276 (LMA). Also in Zaire and Tanzania. Up to 1800 m. in submontane zone in high-rainfall areas, evergreen forest, gallery-forest along streams.

Farrell 204 is sterile and the identification uncertain. Description of the fruit and ecological information are from Macedo and Faria (loc. cit.).
Apparently near to *C. fuscum* Planch. from West Africa.

10. Sect. BREVIRAMEA Engl. & Diels in Engl., Mon. Afr. Pflanz. **3**: 61 (1899).—Stace in Journ. Linn. Soc., Bot. **62**: 148 (1969).—Exell in Kirkia, **7**, 2: 199 (1970).—Wickens, F.T.E.A. Combret.: 40 (1973).

Flowers 4-merous. Upper receptacle campanulate to infundibuliform. Petals spathulate to broadly obovate or obovate to subcircular, glabrous. Stamens

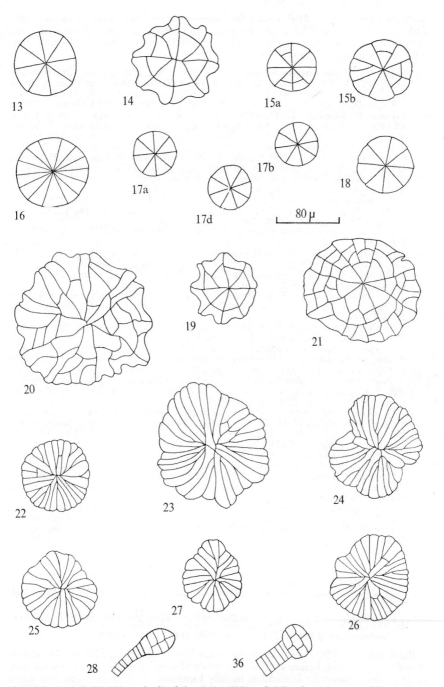

Tab. 35. Scales (13–27) and glandular hairs (28 and 36) of COMBRETUM (dimensions indicated). The figures are numbered to correspond with the numbers of the taxa in the text. From C. A. Stace.

2-seriate in insertion. Disk with a free margin. Fruit 4-winged. Leaves mostly with only 3–4(5) pairs of lateral nerves. Cotyledons 2, borne above soil level. Scales dense on the under surface of the leaf, (50)75–160μ in diam., cell-walls thin; cells clear and transparent.

19. **Combretum hereroense** Schinz in Verh. Bot. Verein. Brand. **30**: 245 (1888).— Engl. & Diels in Engl., Mon. Afr. Pflanz. **3**: 63 (1899).—Monro in Proc. & Trans. Rhod. Sci. Ass. **8**, 2: 80 (1908).—Eyles in Trans. Roy. Soc. S. Afr. **5**, 4: 429 (1916).—O. B. Mill. in Journ. S. Afr. Bot. **18**: 61 (1952).—Exell & Garcia in Contr. Conhec. Fl. Moçamb. **2**: 122 (1954); C.F.A. **4**: 56 (1970).—Rattray in Kirkia, **4**: 84 (1961).—Stace in Mitt. Bot. Staatssamml. München **4**: 12 (1961).—F. White, F.F.N.R.: 285, fig. 50B (1962).—Breitenb., Indig. Trees Southern Afr. **4**: 825 (1965).—Exell, Prodr. Fl. SW. Afr. **99**: 8 (1966): in Bol. Soc. Brot., Sér. 2, **42**: 8 (1968); in Kirkia, **7**, 2: 199, t. 2 fig. 20 (1970).—Wickens in Kew Bull. **25**, 3: 413 (1971); F.T.E.A. Combret.: 40, fig. 2(20), 3(20), 5(20) (1973).—Jacobsen in Kirkia, **9**, 1: 168 (1973).—Faria in Mem. Inst. Invest. Agron. Moçamb. **4**: 73 (1973). TAB. 34 fig. B; TAB. 35 fig. 19. Type from SW. Africa.

Combretum eilkeranum Schinz, tom. cit.: 246 (1888).—O. B. Mill., B.C.L.: 42 (1948).—Engl. & Diels in Engl., Mon. Afr. Pflanz. **3**: 61, t. 19 fig. A, a–c (1899) (" eilkeri "). Type from SW. Africa.

Combretum transvaalense Schinz in Bull. Herb. Boiss. **2**: 202 (1894).—Burtt Davy, F.P.F.T. **1**: 246 (1926).—Steedman, Some Trees, etc. S. Rhod.: 55 (1933).— O. B. Mill., B.C.L.: 43 (1948).—Brenan, T.T.C.L.: 139 (1949); in Mem. N.Y. Bot. Gard. **8**, 5: 438 (1954).—Codd, Trees and Shrubs Kruger Nat. Park: 133, t. 4b, fig. 23c–d (1951).—Palmer & Pitman, Trees S. Afr.: 249 cum fig. et phot. (1961). Type from the Transvaal.

Combretum usaramense Engl., Pflanzenw. Ost-Afr. **C**: 290 (1895).—Engl. & Diels in Engl., tom. cit.: 62, t. 19 fig. C (1899).—Brenan, T.T.C.L.: 139 (1949). Type from Tanzania.

Combretum prunifolium Engl. & Diels in Engl., tom. cit.: 29 (1899) pro parte. Type: Mozambique, Boroma, *Menyharth* 893 (Z).

Combretum sambesiacum Engl. & Diels in Engl., tom. cit.: 63, t. 19 fig. E (1899).— Schinz, Pl. Menyharth.: 432 (1906).—Gomes e Sousa, Pl. Menyharth: 319 (1936) (" zambesiacum "). Type: Mozambique, Boroma, *Menyharth* 892b pro parte (Z).

Combretum porphyrolepis Engl. & Diels in Engl., tom. cit.: 63 (1899) *nom. illegit.*— Bak. f. in Journ. Linn. Soc., Bot. **40**: 68 (1911). Syntypes from the Transvaal (including the type of *C. transvaalense*).

Combretum hereroense var. *villosissimum* Engl. & Diels in Engl., tom. cit.: 63, t. 19 fig. G (1899).—Wild, Guide Fl. Vict. Falls: 146 (1953).—Boughey in Journ. S. Afr. Bot. **30**, 4: 167 (1964).—Wickens in Kew Bull. **25**, 3: 414 (1971); F.T.E.A. Combret.: 41, fig. 2(20c) (1973). Type from the Transvaal.

Combretum rautanenii Engl. & Diels in Engl., tom. cit.: 64, t. 19 fig. D (1899).— O. B. Mill., B.C.L.: 43 (1948). Type from SW. Africa.

Combretum borumense Engl. & Diels in Engl., tom. cit.: 64, t. 19 fig. F (1899).— Schinz, Pl. Menyharth.: 431 (1906).—Bak. f., loc. cit. Type: Mozambique, Boroma, *Menyharth* 890 (Z, holotype).

Combretum rhodesicum Bak. f. in Journ. of Bot. **37**: 435 (1899); in Journ. Linn. Soc., Bot. **40**: 67 (1911).—Gibbs in Journ. Linn. Soc., Bot. **37**: 443 (1906).—Monro in Proc. & Trans. Rhod. Sci. Ass. **8**, 2: 80 (1908).—Eyles in Trans. Roy. Soc. S. Afr. **5**, 4: 428 (1916).—Henkel in Proc. & Trans. Rhod. Sci. Ass. **30**: 16 (1931).— Steedman, loc. cit.—Burtt Davy & Hoyle, N.C.L.: 39 (1936). Syntypes: Rhodesia, Bulawayo, *Rand* 582 (BM), 583 (BM).

Combretum transvaalense var. *bolusii* Dummer in Gard. Chron., Ser. 3, **51**: 201 (1913). Type from S. Africa.

Combretum transvaalense var. *ochrolepidotum* Dummer, loc. cit. Type from S. Africa.

Combretum transvaalense var. *villosissimum* (Engl. & Diels) Burtt Davy, F.P.F.T. **1**: 246 (1926). Type as for *C. hereroense* var. *villosissimum*.

Combretum sp. (*Curson* 114).—O. B. Mill., B.C.L.: 44 (1948).

Small tree up to 9(12) m. high or more rarely a shrub, often with arching stems, frequently flowering before the leaves; crown dense, irregular; bark creamy-grey to grey-black, fissured; branchlets usually tomentose to densely pubescent when young but often quickly glabrescent. Leaves opposite or subopposite, often borne on short shoots; lamina usually c. 2–7 × 1–4·5 cm., from very narrowly elliptic to very broadly obovate or subcircular (especially on sucker shoots when they may be as much as 5·5 cm. broad and broader than long), from densely tomentose to nearly glabrous (except for scales), densely golden-(rarely silvery-)lepidote beneath, scales usually contiguous but sometimes hidden by the indumentum, apex sometimes acute but often rounded or even retuse, base cuneate to rounded; lateral nerves usually only 3–4(5) pairs, prominent beneath; petiole up to 5 mm. long. Inflorescences of short rather congested spikes up to c. 3 cm. long, occasionally branched, axillary, often appearing on leafless branchlets in the axils of the scars of the fallen leaves; rhachis tomentose to glabrous (except for scales); bracts 3 mm. long, filiform, often somewhat persistent. Flowers sessile, yellow. Lower receptacle 2·3–3 mm. long; upper receptacle 2·5–3 × 2–3 mm., campanulate or shortly infundibuliform, lepidote and otherwise pubescent to nearly glabrous. Sepals 1–1·5 × 1–1·5 mm., deltate. Petals 1·5–2·5 × 1–2·5 mm., broadly to very broadly obovate or subcircular or very broadly transversely elliptic, emarginate, shortly unguiculate, glabrous. Stamen-filaments 3–4·5 mm. long; anthers 0·4–0·6 mm. long. Disk c. 2 mm. in diam. with pilose margin free for 0·7–0·8 mm. Style 3·5–4·5 mm. long, not expanded at the apex. Fruit usually c. 2 × 2 cm. but sometimes smaller and occasionally up to 3·5 × 3·5 cm., subcircular in outline, densely reddish- or golden-lepidote, apical peg absent or very short, wings c. 5(10) mm. wide, subcoriaceous, stipe up to 11 mm. long. Cotyledons 2, 1·5 × 2·5 cm., transversely elliptic, arising above soil level; petioles 10–11 mm. long. Scales irregularly undulate in outline, marginal cells 8–12 mostly retusely scalloped, delimited by usually 8 radial walls, 8 tangential walls and sometimes a few extra radial walls.

Caprivi Strip. E. of Cuanda R., 945 m., fl. 10.1945, *Curson* 1191 (PRE). **Botswana.** N: Ngamiland, c. 1220 m., plains near Tsodilo Hills, fr. 15.vii.1963, *Banks* 56 (PRE). SW: 24 km. W. of Ghanzi, fr. 24.iv.1963, *Ballance* 634 (FHO; SRGH). SE: 6·4 km. N. of Mahalapye, fr. 14.i.1960, *Leach & Noel* 60 (K; SRGH). **Zambia.** B: Sesheke Distr., Masese, fr. 25.i.1952, *White* 1960 (BM; FHO; K). C: Lusaka Distr., near Shimbala Siding, fr. 27.xii.1959, *White* 6037 (FHO). S: near Victoria Falls, fl. & fr. 13.ix.1957, *Angus* (FHO; K). **Rhodesia.** N: Darwin Distr., 3·2 km. from Mkumbura R., 520 m., fr. 23.i.1960, *Phipps* 2416 (K; SRGH). W: Bulawayo, fr. 18.ii.1912, *Rogers* 13085 (PRE). C: Gwelo, fl. & fr. x.1924, and v.1925, *Steedman* 47 (BM; FHO; K). E: Melsetter Distr., Junction Tea Rooms, fr. 2.i.1949, *Chase* 1467 (BM; SRGH). S: Lundi R., Chipinda Pools, fl. 12.ix.1961, *Goodier* 2 (BM; FHO; SRGH). **Malawi.** S: Zomba, Baloke Station, fr. 17.ii.1955, *Jackson* 1474 (BM; COI; FHO; K; LISC). **Mozambique.** N: Lago, Macaloge, fr. 3.ix.1934, *Torre* 539 (BM; COI; LISC). Z: Mocuba, fr. 27.v.1949, *Andrada* 1518 (COI; LISC). T: Marávia, between Chicoa and Chioco, fl. & fr. 25.ix.1942, *Mendonça* 416 (BM; LISC). MS: Manica do Lucite, fr. 19.v.1948, *Barbosa* 1493 (BM; LISC). SS: Massinga, Funhalouro, fr. 21.v.1941, *Torre* 2708 (BM; K; LISC). LM: Ressano Garcia, fr. 18.ii.1955, *E.M. & W.* 472 (BM; LISC; SRGH).

Also in Angola, SW. Africa, Uganda, Kenya, Tanzania and S. Africa. Mixed *Acacia* savanna; *Acacia-Combretum-Terminalia* tree savanna; *Brachystegia* woodland or savanna woodland; tree or shrub savanna with *Adansonia-Sterculia-Acacia nigrescens*; mopane woodland; thickets on termite mounds; dry deciduous shrub savanna with *Diplorhynchus* and *Pterocarpus brenanii*; edges of dambos; *secondary Baikiaea forest*; widespread and common from low to medium altitudes.

Extremely variable in leaf-shape and indumentum and sometimes in size of fruit. Wickens (in Kew Bull. **25**, 3: 413–416 (1971)) has recently divided this species into three subspecies of which only the type subspecies occurs in our area; so that if the proposed classification is accepted our material can all be called *C. hereroense* Schinz subsp. *hereroense*. In the same work Wickens further divides subsp. *hereroense* into var. *hereroense* and var. *villosissimum* Engl. & Diels. He separates these two varieties as follows:

Leaves glabrous or glabrescent; reticulation not prominent - - - var. *hereroense*
Leaves densely pubescent, hairs often concealing the scales; reticulation prominent
var. *villosissimum*

I have not found this division into varieties quite satisfactory but it should be checked by local workers throughout our area. In this aggregate species, as in many other savanna species where there is reticulate variation without much correlation of the characters, I feel that any infra-specific classification is a matter of approach. If one concentrates on the indumentum, as Wickens has largely done here, one arrives at taxa in which, for example, the leaf-shape is variable and *vice versa*.

11. Sect. AUREONITENTIA Exell & Stace in Bot. Journ. Linn. Soc. 62: 167 (1969).—Exell in Kirkia, 7, 2: 201 (1970).

Leaves with 5–10 pairs of lateral nerves. Flowers 4-merous. Upper receptacle infundibuliform. Stamens 2-seriate in insertion with comparatively long filaments.

Disk without free margin. Fruit 4-winged. Cotyledons unknown. Scales 100–190μ in diam., irregularly undulate in outline, divided rather irregularly by primary, secondary and partial radial walls; marginal cells 15–25, variously elongated; cells ± opaque with cell-walls thicker than in C. *hereroense* but still not resembling those of C. *molle*.

20. **Combretum aureonitens** Engl. & Gilg in Warb., Kunene-Samb.-Exped. Baum: 315 (1903).—Exell in Bol. Soc. Brot., Sér. 2, **42**: 8 (1968); in Kirkia, **7**, 2: 202, t. 2 fig. 21 (1970).—Exell & Garcia, C.F.A. **4**: 56 (1970). TAB. **35** fig. 20. Type from Angola.
    Combretum sp. 1.—F. White, F.F.N.R.: 287 (1962).

Small tree 5–6 m. high or sometimes scrambling; branchlets tomentose and rufous-lepidote. Leaves opposite or 3-verticillate; lamina up to 8 × 4·5–5 cm., elliptic or oblong-elliptic, densely to sparsely pubescent above and beneath, sometimes with tufts of hairs in the axils of the lateral nerves beneath, silvery- or golden-lepidote, densely so beneath with contiguous scales, apex and base rounded; lateral nerves 5–10 pairs; petiole 2–7 mm. long, tomentose. Inflorescences of axillary (rarely terminal) spikes 2–6 cm. long (including the peduncle). Flowers sessile, yellowish. Lower receptacle up to 3 mm. long, pubescent and lepidote, somewhat constricted towards the apex; upper receptacle 3·5–5·5 × 2·5–3·5 mm., infundibuliform, densely pubescent and lepidote. Sepals 1·5 × 1·5 mm., deltate. Petals up to 4 × 2 mm., obovate to spathulate, emarginate, glabrous. Stamen-filaments 8–10 mm. long; anthers 0·6 mm. long. Disk c. 2 mm. in diam., without free margin. Style 7–8·5 mm. long, pilose, not expanded at the apex. Fruit up to 3 × 2·8 cm., subcircular in outline, somewhat emarginate at the apex and cordate at the base, lepidote otherwise glabrous, apical peg absent, wings up to 10 mm. broad, stipe 5–10 mm. long. Cotyledons unknown. Scales dense, conspicuous, as for the section.

Zambia. N: Lake Mweru, Kafulwe Mission, fl. 2.xi.1952, *White* 3578 (BM; FHO; K).
Also in Angola. Semi-deciduous thicket; often on sand.

This species was placed near to C. *fulvotomentosum* (now in Sect. *Metallicum*) and C. *welwitschii* (Sect. *Ciliatipetala*) by Engler and Diels. It is closer to Sect. *Breviramea* but differs in the number of lateral nerves, spike not congested, no free margin to the disk, length of stamens and some scale characters.

12. Sect. **ELAEAGNOIDA** Engl. & Diels in Engl., Mon. Afr. Pflanz. **3**: 23 (1899).—Stace in Bot. Journ. Linn. Soc. **62**: 139, figs. 17–18 (1969).—Exell in Kirkia, **7**, 2: 202 (1970).

Flowers 4-merous, in dense often subcapitate axillary spikes. Upper receptacle cupuliform to broadly infundibuliform, densely lepidote. Petals narrowly obovate to spathulate, glabrous. Stamens 8, 2-seriate. Disk somewhat inconspicuous, glabrous, without free margin. Fruit 4-winged, silvery-lepidote. Cotyledons 2, borne above soil level. Scales 120–175μ in diam., sometimes contiguous; cells rather small and divided by numerous radial and tangential walls; marginal cells 16–30; central cells usually in an 8-celled rosette.

21. **Combretum elaeagnoides** Klotzsch in Peters, Reise Mossamb. Bot. **1**: 73 (1861).—
    Laws. in Oliv., F.T.A. **2**: 426 (1871) excl. specim. angol.—Engl., Pflanzenw. Ost-Afr. **C**: 290 (1895).—Engl. & Diels in Engl., Mon. Afr. Pflanz. **3**: 23, t. 8 fig. D (1899).—Sim, For. Fl. Port. E. Afr.: 62 (1909).—O. B. Mill., B.C.L.: 42 (1948).—Wild, Guide Fl. Vict. Falls: 146 (1953) (" elaeaginoides ").—Exell & Garcia in Contr. Conhec. Fl. Moçamb. **2**: 106 (1954).—Rattray in Kirkia, **2**: 89 (1961).—F. White, F.F.N.R.: 284, fig. 49, K (1962).—Boughey in Journ. S. Afr. Bot. **30**, 4: 167 (1964).—Fanshawe & Savory in Kirkia, **4**: 190 (1964).—Exell in Kirkia, **7**, 2: 202, t. 2 fig. 22 (1970).—Faria in Mem. Inst. Invest. Agron. Moçamb. **4**: 86 (1973). TAB. **31** fig. H; TAB. **35** fig. 21. Type: Mozambique, near Tete, *Peters* (B†; BM, fragm.; K, fragm.).
    Combretum prunifolium Engl. & Diels in Engl., tom. cit.: 28, t. 8 fig. B (1899) pro parte. Type: Mozambique, Boroma, *Menyharth* 893 (Z, holotype).
    Combretum stevensonii Exell in Journ. of Bot. **77**: 171 (1939).—O. B. Mill., B.C.L.: 44 (1948). Type: Zambia, Mazabuka, *Stevenson* 99 (BM, holotype; FHO).

Small deciduous tree up to 6·5 m. tall, or shrub; bark smooth, grey-brown to silver-grey; branchlets silvery-lepidote, otherwise glabrous. Leaves opposite or subopposite or rarely 3-verticillate; lamina 3·5–13 × 1–6 cm., chartaceous, narrowly elliptic or oblong-elliptic, densely silvery-lepidote on both surfaces with scales often contiguous especially beneath otherwise usually glabrous except for occasional tufts of hairs in the axils of the lateral nerves beneath, sometimes pubescent or sparsely pilose, apex acute to rounded, base usually somewhat cordate, sometimes distinctly so; lateral nerves 8–13 pairs; petiole 3–10 mm. long, lepidote otherwise glabrous or pubescent. Inflorescences of rather dense often subcapitate axillary spikes up to 2·5 cm. long, occasionally branched; rhachis densely lepidote otherwise glabrous or puberulous; bracts very small (c. 0·6 mm. long), caducous. Flowers sessile, greenish-yellow, densely lepidote otherwise glabrous or rarely puberulous. Lower receptacle 1·5–2·5 × 0·3–0·5 mm., lepidote otherwise glabrous or nearly so; upper receptacle up to 2 × 2 mm., cupuliform or broadly infundibuliform, lepidote otherwise glabrous or rarely puberulous. Sepals very shallowly triangular or merely small teeth. Petals 1·5–3 × 0·9–1·1 mm., narrowly obovate to spathulate, glabrous, apex rounded or obtuse, shortly unguiculate. Stamen-filaments 3·5–4·5 mm. long; anthers c. 0·5 mm. long. Disk 0·8–1 mm. in diam., glabrous, rather inconspicuous. Style 4·5–5 mm. long. Fruit 2–3·5 × 1–3·5 cm., subcircular or more rarely broadly elliptic in outline, silvery-lepidote (less densely on the wings), apical peg up to 1 mm. long or absent, wings up to 12 mm. broad, rigid, coriaceous, stipe slender and elongating to 12(15) mm. long. Cotyledons 2, 2–2·5 × 3–3·5 cm., transversely elliptic or subcircular to subreniform; petioles 1–3 mm. long. Scales as for the section.

Caprivi Strip. Between Katima Mulilo and Ngoma, c. 915 m., st. i.1959, *Killick & Leistner* 3309 (BM; K; PRE). **Botswana.** N: Tsotsoroga Pan, fr. 17.vi–8.vii.1930, *Van Son* in Herb. Transv. Mus. 28825 (BM; EA; PRE; SRGH). **Zambia.** B: Sesheke Distr., Katonga Forest Reserve near Masese, fr. 28.i.1952, *White* 1976 (BM; FHO; K). C: Kafue Flats, Chikumbe Hill, fr. 12.viii.1949, *Hoyle* 1117 (BR; FHO; K; LISC; NDO). E: Fort Jameson, fr. 11.v.1963, *Van Rensburg* 2127 (K; SRGH). S: Gwembe Distr., 3·2 km. S. of Sinazezi, fr. 2.iv.1952, *White* 2398 (BM; FHO; K). **Rhodesia.** N: Sebungwe Distr., Lubu R., fl. & fr. 2.xi.1951, *Lovemore* 121 (BM; SRGH). C: Gatooma Distr., fr. 11.iii.1954, *Cleghorn* 446 (K; SRGH). E: Inyanga Distr., Lawley's Concession, fr. 19.ii.1954, *West* 2398 (SRGH). **Mozambique.** T: between Tete and Boroma, fr. 5.v.1948, *Mendonça* 4091 (LISC). MS: Báruè, between Tambara and Mungári, fr. 3.ix.1943, *Torre* 5829 (BM; LISC).

Confined to our area in a belt stretching from northern Botswana to the Mozambique coast apparently following the valley of the Zambezi. In *Colophospermum mopane* shrub savanna; *Ostryoderris-Acacia sieberana-Combretum* spp. savanna; on Kalahari Sand in *Baikiaea* woodland; stony hillsides often forming thickets with *Commiphora* and *Combretum* spp. (Jesse thicket); occasional on copper-bearing soils; in hotter drier areas.

The fruit varies considerably in size. *C. stevensonii* was described from the large-fruited form but is scarcely specifically distinct. Specimens from the western end of the range have more oval fruits.

The type of *C. prunifolium*, *Menyharth* 893 (Z), consists of 3 specimens, the 2 upper ones (those figured by Engler and Diels) being *C. elaeagnoides* and the lower one *C. hereroense*. The material in the capsule is *C. elaeagnoides*.

13. Sect. CHIONANTHOIDA Engl. & Diels in Engl., Mon. Afr. Pflanz. **3**: 77 (1899).—Stace in Bot. Journ. Linn. Soc. **62**: 139, figs. 19–33, 164 (1969).—Exell in Kirkia, **7**, 2: 204 (1970).—Wickens, F.T.E.A. Combret.: 43 (1973).

Sect. *Meruensia* Engl. & Diels in Engl., tom. cit.: 20 (1899).

Flowers 4-merous. Upper receptacle campanulate to very narrowly infundibuliform or tubular, sometimes expanding slightly to a cupuliform apex. Petals narrowly obovate or obovate-spathulate or subcircular to (rarely) transversely elliptic, glabrous or ciliate (in *C. illairii*). Stamens 2-seriate, variously inserted. Disk usually inconspicuous without a free margin. Style not expanded at the apex. Fruit usually 4-angled, more rarely 4-winged. Cotyledons unknown. Scales (50)75–150μ in diam., often appearing impressed, ± circular in outline, divided by numerous radial walls with tangential walls absent or very few; marginal cells (15)25–35, radially elongated; usually conspicuous, dense to fairly sparse.

Several sections of *Combretum* contain species, otherwise clearly closely related, with very narrow-winged or angled fruits or with broad-winged fruits. The character seems to have little importance in delimiting sections.

Fruit 4-winged; leaves glabrous (except for scales) or nearly so:
  Leaves usually subcordate at the base; upper receptacle c. 1·5 mm. long, broadly
    campanulate - - - - - - - - - - 22. *gossweileri*
  Leaves obtuse to cuneate or somewhat rounded at the base, not subcordate; upper
    receptacle c. 4·5 mm. long, infundibuliform - - - 23. *xanthothyrsum*
Fruit 4-angled or 4-ridged or very narrowly 4-winged (unknown in *C. stocksii*); leaves
    glabrous (except for scales) to tomentose:
  Adult leaves conspicuously lepidote:
    Leaves glabrous (except for scales) or nearly so:
      Upper receptacle tubular to narrowly infundibuliform at the base expanding to a
        cupuliform apical part; scales on the lower surface of the leaf scarcely
        appearing depressed through eventually leaving depressions, rather dense
        (sometimes contiguous); fruit unknown - - - - 24. *stocksii*
      Upper receptacle narrowly infundibuliform throughout its length, expanding
        gradually towards the apex; scales on the lower surface of the leaf usually
        0·5–1 mm. apart, not contiguous, appearing somewhat impressed; fruit
        obovoid-ellipsoid to subglobose - - - - - - 25. *butyrosum*
    Leaves tomentose to rather densely pubescent and rather densely lepidote on the
      lower surface; fruit ovoid-prismatic to narrowly ovoid-prismatic
                                                              26. *pisoniiflorum*
  Adult leaves not conspicuously lepidote, coriaceous, glabrous (except for scales); petals
    ciliate at the apex - - - - - - - - - 27. *illairii*
Note. The fruit is unknown in *C. stocksii* which has been placed in the key on the assumption that it will be ridged or narrowly winged as in the majority of species in this section.

22. **Combretum gossweileri** Exell in Journ. of Bot. **66**, Suppl. Polypet.: 166 (1928);
    in Kirkia, **7**, 2: 204, t. 2 fig. 23 (1970).—F. White, F.F.N.R.: 285 (1962).—Lawton
    in Journ. Ecol. **52**: 472 (1964).—Liben, F.C.B. Combret.: 69, fig. 2D (1968).—
    Exell & Garcia, C.F.A. **4**: 68 (1968). TAB. **34** fig. A; TAB. **35** fig. 22. Type from
    Angola (Bié).
    Combretum subglomeruliflorum De Wild., Pl. Bequaert. **5**: 372 (1931). Type from
    Zaire.

Scrambling evergreen woody climber up to c. 5 m.; old stems fluted; bark grey-black to black-brown; branchlets albo-lepidote otherwise pubescent to nearly glabrous. Leaves opposite, often coloured and bract-like just below the inflorescences, somewhat glossy; lamina up to 10 × 4·5 cm., subcoriaceous, oblong or elliptic-oblong or obovate-oblong, usually glabrous (except for scales) or nearly so, rather densely lepidote but scales not contiguous and appearing somewhat impressed, apex blunt or rounded and sometimes shortly acuminate and mucronate, base usually subcordate or sometimes rounded; midrib somewhat impressed above, conspicuous beneath; lateral nerves 4–8 pairs; petiole up to 6 mm. long but often only 3 mm. Inflorescences of axillary (more rarely terminal) usually rather congested or even subcapitate spikes usually 1·5–2 cm. long (sometimes laxer and more elongate, up to 3·5 cm. long) on a tomentose to sparsely lepidote peduncle 1–3 cm. long or appearing paniculate by suppression of the upper leaves; rhachis tomentose to sparsely pubescent, lepidote; bracts 3 mm. long, filiform, caducous. Flowers dull-yellow or cream or greenish, scented. Lower receptacle 1–1·5 mm. long, tomentose to sparsely pubescent, lepidote; upper receptacle 1·5 × 2·5–3 mm., broadly campanulate, very sparsely pubescent, rather densely lepidote. Sepals 0·7 × 0·7 mm., deltate. Petals 1–2 × 1 mm., elliptic to broadly transversely elliptic, shortly unguiculate, glabrous. Stamen-filaments 5–7 mm. long; anthers 0·8–1 mm. long, apiculate at the apex. Style 4–5 mm. long. Fruit pale-green becoming pinkish, 2·3(3) × 2·4 cm., 4-winged, subcircular in outline, sparsely pubescent and lepidote, apical peg very short, wings up to 10 mm. broad, stipe 2–3 mm. long. Scales sometimes no more than 50–75μ in diam., cuticular membrane raised to form a secretion-filled cavity (otherwise as for the section).

Zambia. B: Balovale, 1065 m., fl. & fr. viii.1952, *Gilges* 152 (K; PRE; SRGH). N: Lake Bangweulu, Chiluwi I., fl. & fr. 13.v.1947, *Brenan & Greenway* in *Brenan* 8090 (FHO; K). W: Mwinilunga Distr., 96 km. S. of Mwinilunga on Kabompo road, fl. 1.vi.1963, *Loveridge* 717 (FHO; SRGH).

Also in Angola and Zaire. *Cryptosepalum* woodland; *Parinari* woodland; evergreen thicket; lake-basin chipya; occasional in *Marquesia* dry evergreen forest; termite mounds; at medium altitudes.

23. **Combretum xanthothyrsum** Engl. & Diels in Engl., Bot. Jahrb. **39**: 507 (1907).— Brenan, T.T.C.L.: 141 (1949).—Exell & Garcia in Contr. Conhec. Fl. Moçamb. **2**: 128 (1954).—Exell in Kirkia, **7**, 2: 205, tab. 2 fig. 24 (1970).—Wickens, F.T.E.A. Combret.: 46, fig. 2(25) (1973).—Faria in Mem. Inst. Invest. Agron. Moçamb. **4**: 85 (1973). TAB. **30** fig. J; TAB. **35** fig. 23. Type from Tanzania.
   Combretum stenanthoides Mildbr. in Notizbl. Bot. Gard. Berl. **14**: 105 (1938). Type from Tanzania.

Semiscandent evergreen shrub or liane (Tanzania). Leaves opposite; lamina up to 13(17·5) × 8 cm., elliptic to narrowly elliptic, rather densely lepidote, scales appearing somewhat impressed, otherwise glabrous, apex acute or acuminate, base usually rounded; lateral nerves 7–10 pairs; petiole 5–15 mm. long. Inflorescences axillary, sometimes paniculate by suppression of the leaves; spikes 3–5 cm. long; peduncle up to 2 cm. long, rufous-pubescent. Flowers whitish to pale-yellow, fragrant. Lower receptacle 1·5 mm. long, lepidote and pubescent; upper receptacle 4·5 × 2·5 mm., infundibuliform, lepidote and densely pubescent. Sepals 0·5 × 1·5 mm., broadly deltate. Petals 1·5–2·5 × 1·5 mm., broadly elliptic or subcircular, conspicuously emarginate, glabrous. Stamen-filaments 5–6 mm. long; anthers 0·5 mm. long. Disk inconspicuous. Style 6–7 mm. long. Fruit 4-winged, 3·5–5 × 2–3·5 cm., subcircular to oblong-elliptic in outline, lepidote otherwise glabrous or nearly so, apical peg up to 1 mm. long, wings up to 10 mm. broad and somewhat decurrent at the base, stipe up to 8 mm. long. Scales as for the section.

Mozambique. N: Nacala, near Fernão Veloso, fr. 15.x.1948, *Barbosa* 2419 (LISC); between Fernão Veloso and Quissangulo, fl. & immat. fr. 15.x.1948, *Barbosa* 2427 (LISC; LMJ).
   Also in Tanzania. Dry evergreen forest; dry thickets with *Guibortia schliebenii* and *Pseudoprosopis*; a littoral species in our area.

A vermifuge is produced from an infusion of the roots.

24. **Combretum stocksii** Sprague in Kew Bull. **1909**: 306 (1909).—Exell & Garcia in Contr. Conhec. Fl. Moçamb. **2**: 128 (1954).—Exell in Kirkia, **7**, 2: 205, t. 2 fig. 25 (1970).—Faria in Mem. Inst. Invest. Agron. Moçamb. **4**: 86 (1973). TAB. **35** fig. 24. Type: Mozambique, Macomia, *Stocks* 54 (K, holotype).

Scandent shrub; branchlets densely lepidote otherwise almost glabrous. Leaves opposite; lamina up to 5·5 × 2 cm., oblong-elliptic, rather densely lepidote above and below but scales usually not contiguous, not appearing impressed but leaving depressions, apex acute or shortly acuminate, base rounded to obtuse; lateral nerves 6–8 pairs; petiole up to 6 mm. long. Inflorescences of rather congested axillary spikes up to 3 cm. long; rhachis golden-lepidote; bracts c. 1·5 mm. long, filiform, caducous. Flowers creamy-white. Lower receptacle 1–1·4 × 0·8 mm., lepidote otherwise glabrous or nearly so; upper receptacle consisting of a very narrowly infundibuliform to tubular part at the base, 2–3 × 0·6–0·9 mm., expanding at the apex to a cupuliform part 1·5–2 × 1·5–2 mm., lepidote otherwise glabrous. Sepals 0·9 × 0·9 mm., deltate. Petals 2 × 0·8 mm., narrowly obovate, glabrous. Stamen-filaments 4·5–5 mm. long; anthers 0·5 mm. long. Disk inconspicuous. Style 8–9 mm. long. Fruit unknown. Scales as for the section.

Mozambique. N: Palma, near Nangade, fl. 20.x.1942, *Mendonça* 991 (BM; K; LISC). Known only from Mozambique. Dense evergreen forest.

25. **Combretum butyrosum** (Bertol. f.) Tul. in Ann. Sci. Nat., Sér. 4, **6**: 87 (1856).— Klotzsch in Peters, Reise Mossamb. Bot. **1**: 75 (1861).—Engl., Pflanzenw. Ost-Afr. C: 293 (1895).—Engl. & Diels in Engl., Mon. Afr. Pflanz. **3**: 81 (1899).—Monro in Proc. & Trans. Rhod. Sci. Ass. **8**, 2: 80 (1908).—Exell & Garcia in Contr. Conhec. Fl. Moçamb. **2**: 128, t. 10 (1954).—Exell in Kirkia, **7**, 2: 206, t. 2 fig. 26 (1970) — Wickens, F.T.E.A. Combret.: 44, fig. 2(22), 5(22) (1973).—Faria in Mem. Inst. Invest. Agron. Moçamb. **4**: 86 (1973). TAB. **30** fig. I; TAB. **35** fig. 25. Type: Mozambique, Inhambane, *Fornasini* (B†; BM, fragment; P).
   Sheadendron butyrosum Bertol. f., Ill. Piante Mozamb. Dissert.: 12 (" Scheadendron "), t. 4 fig. a–b (1850); in Mem. Accad. Sci. Ist. Bologna, **2**: 572, t. 40A, 40B (1850). Type as above.

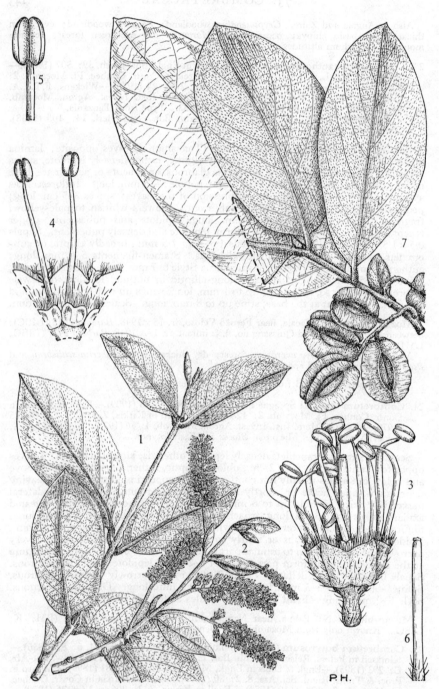

P.H.

Tab. 36. COMBRETUM MOLLE. 1, leafy branch (× ⅔); 2, flowering branch (× ⅔); 3, flower (×6); 4, part of flower opened out and 1 stamen removed to show petal (×6); 5, stamen (×12); 6, style (×6), 1–6 from *Eyles* 4500; 7, fruiting branch (× ⅔) *Rodin* 4392.

Woody climber up to 3–4 m. or prostrate shrub or shrublet; branchlets tomentose or tomentellous to densely pubescent; bark coming off in long ragged strips leaving a glabrous cinnamon-red exposed surface. Leaves opposite; lamina up to 8 × 4·5 cm., subcoriaceous when mature, oblong-elliptic to slightly obovate-oblong, glabrous (except for scales) except for pubescence along the midrib and a few hairs on the lateral nerves beneath, fairly densely lepidote beneath, scales appearing impressed and far from contiguous, rounded and often apiculate at the apex, subcordate at the base; lateral nerves 5–8 pairs; petiole up to 5 mm. long, pubescent. Inflorescences of rather dense axillary spikes 1·5–3 cm. long on densely pubescent peduncles up to 4 cm. long; rhachis tomentose to densely pubescent; bracts 2·5 mm. long, filiform, caducous. Flowers whitish or yellowish. Lower receptacle 1–1·5 mm. long, tomentose; upper receptacle 4–7 mm. long, infundibuliform, pubescent to nearly glabrous (except for scales) and lepidote. Sepals c. 1 × 1 mm., deltate. Petals 1·5–2 × 0·9, obovate-spathulate, glabrous. Stamen-filaments 4–5 mm. long, 2-seriately inserted near the mouth of the upper receptacle; anthers 0·8 mm. long. Disk inconspicuous. Style up to 9 mm. long. Fruit up to 3·7 × 2 cm., 4(5)-angled or very narrowly 4(5)-winged, ellipsoid or obovoid-ellipsoid to subglobose, usually tomentellous and lepidote (sometimes glabrous except for scales), subsessile. Scales as for the section.

Mozambique. SS: Massinga-Antiane, fl. 13.x.1945, *Pedro* 365 (LMJ; PRE); between Vilanculos and Mapinhame, fr. 17.xi.1941, *Torre* 3830 (BM; K; LISC).
Also from Kenya and Tanzania (*fide* Wickens). Tree or shrub savanna; evergreen forest; on sandy soils and Pleistocene coral deposits.

Evidently closely related to *C. pisoniiflorum* but differing in the more glabrous less densely lepidote leaves and the somewhat larger fruit tending to be obovoid rather than ovoid. The species is known in our area only from the S. of Mozambique so that if the species cited by Wickens (loc. cit.) is the same there is some discontinuity in the distribution. If the fruit figured by Wickens (fig. 5(22)) is compared with that figured by Exell and Garcia (loc. cit. t. 10) from near the type locality there is evidently some reason for doubt. It may be, of course, that *C. pisoniiflorum* may eventually have to be united with *C. butyrosum* in which case the aggregate species would cover discontinuities in the distribution and possibly discrepancies in the size and shape of the fruit.
"*Combretum butyracaceum*", evidently an error for *C. butyrosum* is mentioned by Monro (loc. cit.) as a species which may possibly occur in Rhodesia (which is unlikely).
C. butyrosum has oily fruits which are said to produce good butter but the species seems to be too rare to be of economic importance.

26. **Combretum pisoniiflorum** (Klotzsch) Engl., Pflanzenw. Ost-Afr. **C**: 293 (1895).— Exell & Garcia in Contr. Conhec. Fl. Moçamb. **2**: 127 (1954).—Exell in Bol. Soc. Brot., Sér. 2, **42**: 6 (1968); in Kirkia, **7**, 2: 207, t. 2 fig. 27 (1970).—Wickens, F.T.E.A. Combret.: 47, fig. 2(26), fig. 3(26), fig. 5(26) (1973).—Faria in Mem. Inst. Invest. Agron. Moçamb. **4**: 87 (1973). TAB. **30** fig. K; TAB. **35** fig. 26. Type: Mozambique, Sena, *Peters* (B†, holotype).
    Sheadendron molle Klotzsch in Peters, Reise Mossamb. Bot. **1**: 76 (1861). Type: Mozambique, Sena, *Peters* (B†, holotype).
    Sheadendron pisoniiflorum Klotzsch in Peters, tom. cit.: 77, t. 14 (1861) (*Combretum pisoniiflorum* in tab.). Type as for *C. pisoniiflorum*.
    Sheadendron pisoniiflorum var. *brachystachyum* Klotzsch, loc. cit. Type: Mozambique, Sena, *Kirk* (B†; K).
    Sheadendron pisoniiflorum var. *macrostachyum* Klotzsch, loc. cit. Type: Mozambique, Sena, *Peters* (B†; K).
    Combretum tetragonum Laws. in Oliv., F.T.A. **2**: 430 (1871).—Engl. & Diels in Engl., Mon. Afr. Pflanz. **3**: 80, t. 24 fig. F (1889).—Sim, For. Fl. Port. E. Afr.: 64 (1909).—Burtt Davy & Hoyle, N.C.L.: 39 (1936). Non *C. tetragonum* Presl (1836). Type: Mozambique, Sena, *Kirk* (K, holotype).
    Combretum molle (Klotzsch) Engl. & Diels in Engl., loc. cit. *nom. illegit.* non R.Br. ex G. Don (1827). Type as for *Sheadendron molle*.

Shrub up to 2–3(4) m. high, usually ± scandent, or liane; branchlets at first tomentose to densely pubescent; bark coming off in ragged strips leaving an exposed glabrous cinnamon-brown surface. Leaves opposite; lamina usually c. 4–4·8 × 2–3 cm. but up to 9·5 × 4 cm. (in Tanzania), membranous when young but becoming subcoriaceous, broadly elliptic or ovate-elliptic or narrowly oblong-elliptic, glabrescent above and shiny, tomentose (when young) to densely pubescent beneath even when mature, especially on the nerves and reticulation, rather densely lepidote beneath but most of the scales not quite contiguous, not appearing depressed but leaving depressions, apex shortly acuminate or rounded, base rounded to cordate; lateral nerves 4–7 pairs; petiole 2–4 mm. long, tomentose. Inflorescences of usually rather congested axillary spikes 3–5 cm. long including the tomentose peduncle; rhachis tomentose and lepidote; bracts c. 1·5 mm. long, filiform to ligulate, caducous. Flowers white or yellowish. Lower receptacle c. 1 mm. long; upper receptacle 3–5 mm. long, infundibuliform, pubescent and lepidote. Sepals 0·5–0·8 × 1 mm., broadly triangular. Petals 1·8–2 × 0·8–1·5 mm., narrowly obovate to spathulate, often emarginate. Stamen-filaments 4·5–6 mm. long, 2-seriately inserted in the upper half of the upper receptacle; anthers 0·4–0·5(0·7) mm. long. Disk inconspicuous. Style 9–10 mm. long. Fruit 2–2·5(3) × 1·2–1·5(1·7) cm., narrowly ovoid-prismatic to ovoid-prismatic, some-times acuminate at the apex, 4-angled or very narrowly 4-ridged or 4-winged, sometimes woody, tomentellous to glabrous (except for scales), lepidote but scales often hidden by the indumentum, dehiscing fairly readily along the angles. Scales with sometimes as few as 15–16 marginal cells, otherwise as for the section.

Malawi. S: Makoko, S. of Port Herald, fl. 1931, *Topham* 531 (BM; FHO). **Mozambique.** N: between Palma and Nangade, fr. 17.ix.1948, *Andrada* 1361 (BM; COI; LISC). T: Zumbo, M'Tamba, *Stocks* 59 (K). MS: Marromeu, Chupanga, fl. 19.x.1904, *Le Testu* 480 (BM). SS: 11·2 km. S. of Inhassoro turn-off, 75 m., fl. 7.x.1963, *Leach & Bayliss* 11864 (FHO; LISC; SRGH).

Also in Tanzania. *Brachystegia allenii* woodland; *Berlinia orientalis* woodland, often on sandy soil; tall thickets with *Landolphia-Dalbergia-Fernandoa-Combretum.*

Andrada 1361 and *Barbosa* 2150 (LISC) from the same region have fruits apparently dehiscing while still attached to the plant, which is rare in *Combretum.* This is clearly correlated with reduction of the wings as there would naturally be no advantage if a winged fruit, normal in the great majority of *Combretum* spp., were to dehisce on the plant. In some cases loss of wings seems to be correlated with distribution by water but here there is no evidence of this. I suggest as a hypothesis the following evolutionary sequence: loss of wings correlated with distribution by water (or possibly sometimes in dense forest), desiccation of the climate or perhaps spread to drier habitats, dehiscence of the fruit *in situ.*

Torre & Paiva 9865 (LISC) has lepidote but otherwise glabrous fruits but the leaves seem identical with those of *C. pisoniiflorum. Topham* 531 from Malawi has slightly larger more definitely emarginate petals and somewhat larger anthers than the Mozambique specimens.

27. **Combretum illairii** Engl. in Abh. Preuss. Akad. Wiss. **1894**: 15 (1894); Pflanzenw. Ost-Afr. **C**: 289 (1895).—Engl. & Diels in Engl., Mon. Afr. Pflanz. **3**: 80, t. 24 fig. D (err. fig. E on p. 80) (1899).—Brenan, T.T.C.L.: 135 (1949).—Exell in Bol. Soc. Brot., Sér. 2, **42**: 6 (1968); in Kirkia, **7**, 2: 208, t. 2 fig. 28 (1970).—Wickens, F.T.E.A. Combret.: 43, fig. 2(21), 5(21) (1973).—Faria in Mem. Inst. Invest. Agron. Moçamb. **4**: 89 (1973). TAB. **30** fig. H; TAB. **35** fig. 27. Type from Tanzania.
   Combretum hildebrandtii Engl., Pflanzenw. Ost-Afr. **C**: 289 (1895).—Engl. & Diels in Engl., tom. cit.: 79, t. 24 fig. E (err. D) (1899).—Brenan, loc. cit. Type from Tanzania.
   Combretum meruense Engl. [in Abh. Preuss. Akad. Wiss. **1894**: 34 (1894) *nom. nud.*] Pflanzenw. Ost-Afr. **C**: 291 (1895).—Engl. & Diels in Engl., tom. cit.: 20, t. 6 fig. B (1899). Syntypes from Tanzania.
   Combretum melchiorianum H. Winkl. in Fedde Repert. **18**: 123 (1922).—Brenan, T.T.C.L.: 136 (1949). Type from Tanzania.
   Combretum schliebenii Exell & Mildbr. in Notizbl. Bot. Gart. Berl. **14**: 106 (1938).— Brenan, loc. cit. Type from Tanzania.
   Combretum butyrosum sensu Dale & Greenway, Kenya Trees and Shrubs: 143 (1961).

Shrub 2–4 m. high or climber; branchlets pubescent and lepidote when young, glabrous and whitish when older. Leaves opposite, subopposite or 3-verticillate; lamina up to 14·5 × 6·5 cm., dark-green and shiny, coriaceous, oblong-elliptic, lepidote when young but appearing almost glabrous when mature with scales quite inconspicuous, apex acuminate, base rounded to subcordate; lateral nerves 5–12 pairs, very prominent beneath; petiole up to 5 mm. long, stout. Inflorescences of subcapitate axillary spikes c. 2(6) cm. long; peduncles 1–1·5 cm. long, lepidote and rufous-pubescent, stout and becoming very woody at the fruiting stage. Lower receptacle 2 mm. long, lepidote otherwise glabrous; upper receptacle 1·5–2·5 × 2·2 mm., campanulate, lepidote otherwise glabrous. Sepals scarcely developed. Petals 1·5 × 0·8 mm., obovate, ciliate at the apex. Stamen-filaments 4–5 mm. long; anthers 0·8 mm. long. Disk rather inconspicuous. Style 3·5–4 mm. long. Fruit 1·3–2·5 × 1–1·5 cm., 4-angled or very narrowly 4-winged, glabrous, with stiff wings up to 1 mm. broad, apical peg scarcely developed, stipe very short or fruit subsessile. Scales as for the section.

Mozambique. N: between Mucojo and Quiterajo, fr. 11.xi.1953, *Balsinhas* 76 (BM; LISC; LM; LMA); near Fernão Veloso, *Barbosa & Balsinhas* 5206 (BM; LISC; LM; LMA).
Also in Kenya and Tanzania. Coastal belt.

Subgen. CACOUCIA (Aubl.) Exell & Stace in Bol. Soc. Brot., Sér. 2, **40**: 10 (1966).—Exell in Kirkia, **7**, 2: 209 (1970).—Wickens, F.T.E.A. Combret.: 48 (1973).
Cacoucia Aubl., Pl. Guian. **1**: 450, t. 179 (1775).

Scales absent. Microscopic (and sometimes macroscopic) stalked glandular hairs always present. Flowers usually 5-merous (4-merous in Sect. *Conniventia* and *C. andradae*).

Key to the Sections of Subgen. Cacoucia

Flowers 4-merous:
Flowers red; petals ± connivent, glabrous - - - - 15. *Conniventia*
Flowers purplish; petals not connivent, pubescent (*C. andradae*) - 16. *Poivrea*
Flowers 5-merous:
Upper receptacle campanulate; petals hairy; fruit sessile or nearly so - 14. *Lasiopetala*
Upper receptacle elongate-campanulate to tubular, sometimes with an upper infundibuliform region and a lower subglobose or ± cylindric region containing the disk; petals hairy or glabrous; fruit with a stipe at least 2 mm. long - 16. *Poivrea*

14. Sect. LASIOPETALA Engl. & Diels in Engl., Mon. Afr. Pflanz. **3**: 65 (1899).—Exell in Kirkia, **7**, 2: 210 (1970).—Wickens, F.T.E.A. Combret.: 56 (1973).

Flowers 5-merous. Inflorescences of congested often subcapitate spikes. Upper receptacle campanulate. Petals hairy, exceeding the speals. Stamens 10, 2-seriate. Disk without free margin. Fruit 5(6)-winged or 5-angled, subsessile or very shortly stipitate. Cotyledons 3 (*C. obovatum*) arising at or below soil level and unfolding spirally. Woody climbers or scandent shrubs.

Fruit 5(6)-winged; leaves up to 6·5(8) cm. long - - - - - 28. *obovatum*
Fruit 5-angled; leaves 10–20 cm. long - - - - - - 29. *pentagonum*

28. **Combretum obovatum** F. Hoffm., Beitr. Kenntn. Fl. Centr.-Ost-Afr.: 28 (1889).—Engl., Pflanzenw. Ost-Afr. **C**: 291 (1895).—Engl. & Diels in Engl., Mon. Afr. Pflanz. **3**: 66, t. 20 fig. C (1899).—Brenan, T.T.C.L.: 134 (1949).—Exell & Garcia in Contr. Conhec. Fl. Moçamb. **2**: 125 (1954).—F. White, F.F.N.R.: 282 (1962).—Exell in Kirkia, **7**, 2: 210, t. 2 fig. 29 (1970).—Wickens, F.T.E.A. Combret.: 56, fig. 6(33), 7(33) (1973).—Faria in Mem. Inst. Invest. Agron. Moçamb. **4**: 90 (1973). TAB. **35** fig. 28; TAB. **37** fig. D. Type from Tanzania.
Combretum lasiocarpum sensu Exell & Garcia in Contr. Conhec. Fl. Moçamb. **2**: 135 pro parte quoad specim. *Torre* 3227 (BM; LISC).—sensu F. White, loc. cit.

Evergreen to semi-evergreen shrub, usually scandent, semi-scandent or sprawling, or ? small tree (one record); branchlets tomentose, tardily glabrescent. Leaves opposite or 3-verticillate, often white or whitish and bract-like below the inflorescences; lamina up to 6·5(10) × 3–4 cm., chartaceous, obovate or broadly obovate-oblong or obovate-elliptic, tomentose when young (usually ± glabrescent above) and remaining densely pubescent or pilose beneath when mature at least on the nerves and venation, apex usually rounded and occasionally rather bluntly acuminate, often mucronate, base usually rounded to subcordate; lateral nerves c. 6–8 pairs; petiole 5–12 mm. long, tomentose, sometimes indurating to form curved spines. Inflorescences of axillary congested or subcapitate spikes 2–4(6) cm. long or short panicles by reduction of the leaves to leaf-like bracts; rhachis tomentose; bracts 3–5 mm. long, filiform, caducous. Lower receptacle 2 mm.

long, tomentose; upper receptacle 2–3 × 2·5–3 mm., broadly campanulate, tomentose to densely pubescent. Sepals 1·5 × 2 mm., ovate-triangular. Petals 2 × 0·6–0·8 mm., narrowly oblong or obovate-oblong or spathulate, hairy. Stamen-filaments 6–7 mm. long; anthers 1 mm. long. Disk 2 mm. in diam., lobed, glabrous, margin scarcely free. Style 4·5 mm. long. Fruit 5(6)-winged, pale-brown, 3–3·5 × 2·5–3 cm., almost sessile, tomentose when young and remaining so at least on the body, apical peg absent or very short, wings 6–8 mm. broad, subcoriaceous, glabrescent towards the margins. Cotyledons 3, 1·5 × 2–3 cm., transversely elliptic to broadly obovate, with petioles 7–13 mm. long, arising at or below soil level.

Zambia. N: Isoka, fl. 21.xii.1962, *Fanshawe* 7200 (FHO; K). C: Lunsemfwa R., fr. 25.viii.1929, *Burtt Davy* 20918 (FHO). E: Nyamadzi R., fr. 25.iii.1955, *E.M. & W.* 1176 (BM; SRGH). S: Gwembe, fl. 23.xi.1955, *Bainbridge* 205/55 (FHO; SRGH). **Rhodesia.** N: Urungwe Distr., Tsuwiri R., fl. 22.xi.1957, *Goodier* 405 (COI; K; LISC; SRGH). **Mozambique.** T: between Chicoa and Chelima, fr. 30.vi.1949, *Barbosa & Carvalho* in *Barbosa* 3396 (LMJ).

Also in Tanzania. Thickets and semi-thickets; riverine thickets; *Colophospermum mopane* savanna woodland; low altitude valleys and escarpments.

It is interesting to note in this species the correlation between the production of 3 cotyledons and the occasional production of 6-winged fruits.

29. **Combretum pentagonum** Laws. in Oliv., F.T.A. **2**: 424 (1871).—Engl. & Diels in Engl., Mon. Afr. Pflanz. **3**: 102 (1899).—Sim, For. Fl. Port. E. Afr.: 61 (1909).—Brenan, T.T.C.L.: 134 (1949).—Exell & Garcia in Contr. Conhec. Fl. Moçamb. **2**: 124 (1954).—Exell in Bol. Soc. Brot., Sér. 2, **42**: 5 (1968); in Kirkia, **7**, 2: 211 (1970).—Wickens, F.T.E.A. Combret.: 57, fig. 7(34) (1973). TAB. **37** fig. E. Type: Mozambique, R. Rovuma, *Meller* (K).

Combretum wakefieldii Engl. in Pflanzenw. Ost-Afr. **C**: 291 (1895).—Engl. & Diels in Engl., tom. cit.: 65, t. 20 fig. B (1899). Type from Kenya (*fide* Wickens).

Combretum lasiopetalum Engl. & Diels in Engl., tom. cit.: 65, t. 20 fig. A (1899). Type from Tanzania.

Woody climber, branchlets fulvous-tomentose. Leaves opposite or subopposite; lamina 10–20 × 5·5–11·5 cm., papyraceous, elliptic to obovate-elliptic, almost glabrous above except for pubescence on the midrib and lateral nerves, densely fulvous-pubescent to fulvous-tomentose beneath especially when young, apex acuminate, base subcordate; lateral nerves 7–10(16) pairs; petiole up to 15 mm. long often forming a spine. Inflorescences of subcapitate spikes or racemes 3–6 cm. long (including the peduncle); rhachis fulvous-pubescent. Flowers sessile or sometimes very shortly pedicellate, cream to pink or maroon or crimson, sweet-scented. Lower receptacle 4·5–5·5 mm. long, constricted above the ovary, sericeous; upper receptacle 2–4 × 3–4 mm., campanulate, appressed-pubescent. Sepals little developed, broadly triangular. Petals 1·5–2·5 × 0·9–1·1 mm., obovate-oblong, hairy on the outside especially on the margins. Stamen-filaments 3·5–4(6) mm. long; anthers 1 mm. long, slightly apiculate. Disk 2 mm. in diam. with a pilose scarcely free margin. Style 4·5–5 mm. long, exserted before the stamens. Fruit 2–4·5 × 1·2–2·5 cm., 5-angled, ellipsoid to obovoid-ellipsoid with apex sometimes slightly acuminate, glabrous, subsessile or very shortly stipitate.

Zambia. N: Stevenson Road, fl. 1893–94, *Scott Elliot* 8402 (K). **Malawi.** C: Kaombe R., fl. 16.i.1952, *Steele* 114 (EA; K). **Mozambique.** N: Macondes, between Mueda and Nantula, 800 m., fl. & fr. 14.iv.1964, *Torre & Paiva* 9846 (LISC).

Also in Kenya, Tanzania and perhaps Zaire. *Brachystegia* woodlands and evergreen forest.

C. capitatum De Wild. & Exell from Zaire has a longer constriction separating the lower and upper receptacles but it may well prove to be the same species.

15. Sect. **CONNIVENTIA** Engl. & Diels in Engl., Mon. Afr. Pflanz. **3**: 69 (1899).—Exell in Kirkia, **7**, 2: 211 (1970).—Wickens, F.T.E.A. Combret.: 52 (1973).

Sect. *Parvulae* Engl. & Diels in Engl., tom. cit.: 67 (1899) pro parte.

Flowers 4-merous (more rarely 5-merous). Upper receptacle campanulate to tubular. Petals usually red, subcircular, exceeding the sepals, usually overlapping and connivent. Stamens 2-seriate. Disk rather inconspicuous, without free margin. Cotyledons 2, remaining below ground without unfolding. Climbers, shrubs, prostrate shrubs and shrublets.

Plant typically a climber or scrambler (sometimes a shrub in absence of support):
  Leaves usually 1½–2 times as long as broad, usually almost glabrous; branchlets rapidly glabrescent; inflorescences usually drying a fuscous or rufous colour but sometimes silvery; plant typically a strong evergreen climber of evergreen forest
    30. *paniculatum*
  Leaves usually subcircular (but sometimes more elongated), usually tomentose at least when young and retaining some indumentum; branchlets only tardily glabrescent; inflorescences usually drying a silvery colour but sometimes ± rufous; plant typically a deciduous climber or scrambler of drier vegetation types
    31. *microphyllum*
Plant a shrublet or small shrub very variable in leaf-shape, indumentum and shape and size of fruit; flowering on annual shoots; typically of savannas subject to seasonal burning - - - - - - - - - - 32. *platypetalum*

30. **Combretum paniculatum** Vent., Choix Pl.: sub t. 58 (1808).—Laws. in Oliv., F.T.A. 2: 425 (1871).—Engl. & Diels in Engl., Mon. Afr. Pflanz. 3: 70, t. 21 fig. C (1899) excl. var. *virgatum.*—Brenan, T.T.C.L.: 134 (1949).—Exell & Garcia in Contr. Conhec. Fl. Moçamb. 2: 127 (1954); C.F.A. 4: 69 (1970).—F. White, F.F.N.R.: 282, fig. 51 (1962).—Liben, F.C.B. Combret.: 34 (1968).—Exell in Bol. Soc. Brot., Sér. 2, **42**: 5 (1968); in Kirkia, **7**, 2: 212 (1970).—Wickens, F.T.E.A. Combret.: 53, fig. 6(29), 7(29) (1973) excl. subsp. *microphyllum.*—Faria in Mem. Inst. Invest. Agron. Moçamb. 4: 91 (1973). TAB. 37 fig. A–B. Type from Senegal.
  Combretum spinosum G. Don in Edinb. Phil. Journ. **11**: 345 (1824) *nom. illegit.* non Bonpl. (1809). Type from Sierra Leone.
  Combretum abbreviatum Engl., Pflanzenw. Ost-Afr. C: 292 (1895).—Engl. & Diels in Engl., tom. cit.: 72 (1899).—Bak. f. in Journ. Linn. Soc., Bot. **40**: 69 (1911).—Eyles in Trans. Roy. Soc. S. Afr. **5**, 4: 427 (1916). Syntypes from Tanzania.
  Combretum carvalhoi Engl., loc. cit.—Engl. & Diels in Engl., tom. cit.: 70, t. 21 fig. B (1899).—Brenan in Mem. N.Y. Bot. Gard. **8**, 5: 438 (1954). Type: Mozambique, Gorungosa, *Carvalho* (B†; COI).
  Combretum ramosissimum Engl. & Diels in Engl., tom. cit.: 72, t. 21 fig. D (1899). Syntypes from West Africa.
  Combretum buvumense Bak. f. in Journ. Linn. Soc., Bot. **37**: 152 (1905). Type from Uganda.
  Combretum unyorense Bagshawe & Bak. f. in Journ. of Bot. **46**: 5 (1908). Type from Uganda.
  Combretum seretii De Wild., Ann. Mus. Congo Belge, Bot. Sér. 5, **3**: 240 (1910). Type from Zaire.
  Combretum seretii var. *grandiflora* De Wild., loc. cit. Type from Zaire.
  Combretum bruneelii De Wild., tom. cit.: 234 (1910). Type from Zaire.
  Combretum thonneri De Wild., tom. cit.: 242 (1910) in obs.; Études Fl. Bangala & Ubangi: 241 (1911). Type from Zaire.
  Combretum thonneri var. *laurentii* De Wild., Ann. Mus. Congo Belge, Bot. Sér. 5, **3**: 241 (1910). Type from Zaire.
  Combretum lemairei De Wild. & Exell in Journ. of Bot. **67**: 143 (1929). Type from Zaire.
  Combretum paniculatum var. *vanderystii* De Wild., Pl. Bequaert. **5**: 356 (1931). Type from Zaire.

Usually a vigorous evergreen climber or a scrambling shrub in absence of support (occasionally a bush—perhaps intermediates or hybrids with *C. platypetalum*); bark grey to grey-black; branchlets usually rufous-tomentose at first but soon glabrescent. Leaves opposite; lamina up to 12(18) × 8(9·5) cm., usually c. 1½ times as long as broad, chartaceous, oblong-elliptic, usually glabrous or nearly so, apex rounded or acuminate, base obtuse to subcordate; lateral nerves 4–6(8) pairs; petiole up to 3 cm. long, the base often persistent and becoming spiny. Inflorescences terminal or axillary panicles; rhachis usually fuscous- or fulvous-tomentose; bracts c. 4 mm. long, very narrowly ovate-elliptic. Flowers red, 4(5)-merous, sessile or very shortly pedicellate. Lower receptacle up to 4(5) mm. long, usually somewhat constricted above and below the ovary, densely fulvous- or rufous-(more rarely silvery-)pubescent; upper receptacle 4–5 × 2–3 mm. (in our area), campanulate, densely fulvous- or rufous- (or more rarely silvery-) pubescent, sometimes nearly glabrous towards the apex. Sepals usually reduced to short teeth, sometimes triangular. Petals red, c. 2·5 × 2·5 mm., subcircular to ovate, usually somewhat overlapping at first and ± connivent. Stamen-filaments 7–8 mm. long, usually red; anthers 0·9 mm. long, red or purplish. Disk rather inconspicuous, without a free margin. Style 6–8 mm. long. Fruit 4(5)-winged, 2–2·5 × 1·5–2 cm. (in our area, sometimes larger elsewhere), subcircular to oblong-elliptic in outline, sparsely pubescent on the body, apical peg usually absent, wings up to 9 mm. broad, thin, stipe 5–10 mm. long.

Zambia. N: Abercorn Distr., Sunzu Hill., fl. 18.xi.1952, *Angus* 798 (BM; COI; FHO; K). **Rhodesia.** E: Melsetter Distr., Glencoe Forest Reserve, fl. 15.viii.1956, *McGregor* 47/56 (K; SRGH). **Malawi.** N: Livingstonia, Manchewe Valley, fl. 10.xi.1956, *Jackson* 2043 (FHO). C: Dedza, fl. 5.ix.1960, *Cartwright in Chapman* 911 (SRGH). S: Tuchila R., fl. 1.ix.1946, *Gouveia & Pedro* 1993 (LMJ; PRE). **Mozambique.** N: Macondes, between Mueda and Nangade, fl. 18.xi.1948, *Barbosa* 2216 (BM; K; LISC; LMJ). Z: Namarrói, near R. Mulumedi, fl. 18.viii.1949, *Andrada* 1890 (COI; LISC). MS: Macequece, 975 m., fl. 19.ix.1951, *Chase* 5288 (BM; COI; K; LISC; LMJ).
Widespread in the wetter regions of tropical Africa. Evergreen and semi-deciduous fringing forest; higher rainfall areas.

See under *C. microphyllum* and *C. platypetalum* for the relationship with those two species.
The pentamerous form is found in Ethiopia, Uganda, Kenya and Tanzania (Kilimanjaro) according to Wickens (loc. cit.) but occasional pentamerous flowers or 5-winged fruits can be found on otherwise tetramerous individuals.

31. **Combretum microphyllum** Klotzsch in Peters, Reise Mossamb. Bot. **1**: 74 (1861).—Laws. in Oliv., F.T.A. **2**: 427 (1871).—Engl., Pflanzenw. Ost-Afr. **C**: 292 (1895).—Engl. & Diels in Engl., Mon. Afr. Pflanz. **3**: 70, t. 21 fig. A (1899).—Schinz, Pl. Menyharth.: 432 (1906).—Monro in Proc. & Trans. Rhod. Sci. Ass. **8**, 2: 80 (1908).—Sim, For. Fl. Port. E. Afr.: 63, t. 62 fig. A (1909).—Bak. f. in Journ. Linn. Soc., Bot. **40**: 69 (1911).—Eyles in Trans. Roy. Soc. S. Afr. **5**, 4: 428 (1916).—Burtt Davy, F.P.F.T. **1**: 246 (1926).—Gomes e Sousa in Bol. Soc. Estud. Col. Moçamb. **4**, 26: 115 (1935).—Burtt Davy & Hoyle, N.C.L.: 39 (1936).—O. B. Mill., B.C.L.: 43 (1948).—Codd, Trees and Shrubs Kruger Nat. Park: 131, fig. 122a (1951).—Pardy in Rhod. Agric Journ. **49**, 2: 78, cum phot. (1952).—Brenan in Mem. N.Y. Bot. Gard. **8**, 5: 438 (1954).—Exell & Garcia in Contr. Conhec. Fl. Moçamb. **2**: 125 (1954).—F. White, F.F.N.R.: 282 (1962).—Boughey in Journ. S. Afr. Bot. **30**, 4: 167 (1964).—Exell in Bol. Soc. Brot., Sér. 2, **42**: 6, 7 (1968); in Kirkia, **7**, 2: 213 (1970).—Jacobsen in Kirkia, **9**, 1: 168 (1973).—Faria in Mem. Inst. Invest. Agron. Moçamb. **4**: 91 (1973). TAB. **31** fig. B; TAB. **37** fig. C. Type: Mozambique, rios de Sena and Tete, *Peters* (B†; BM, fragm.).
Combretum lomuense Sim, For. Fl. Port. E. Afr.: 62, t. 61 B (1909). Type: Mozambique, *Sim* 6393 (n.v.).
Combretum paniculatum subsp. *microphyllum* (Klotzsch) Wickens in Kew Bull. **26**, 1: 66 (1971); F.T.E.A. Combret.: 53 (1973). Type as for *C. microphyllum*.

A deciduous scandent or ± prostrate shrub; bark grey-brown or creamy-brown; branchlets tomentose when young and usually only tardily glabrescent, often pale. Leaves opposite; lamina usually up to c. 4·5 × 4 cm. and subcircular but sometimes up to 11 × 6 cm. and varying from ovate-oblong to obovate-oblong, chartaceous, usually greyish- or fulvous-tomentose when young and retaining some indumentum but sometimes glabrescent, apex usually rounded and often apiculate, base usually subcordate; lateral nerves 5–6 pairs; petiole up to 16 mm. long forming a spine. Inflorescences of large leafy or leafless terminal and axillary panicles; rhachis tomentose; bracts c. 2 mm. long, often transitional to small inflorescence-leaves which are in turn transitional to normal foliage. Flowers 4-merous, sessile or nearly so, red (due to red petals and stamens). Lower receptacle 3·5–4·5 mm. long, somewhat constricted above and below the ovary, densely greyish- or fulvous-pubescent; upper receptacle 2·5–3·5 × 1·5–3 mm., usually greyish- or silvery-sericeous. Sepals up to 0·9 × 1·2 mm. Petals red, 2–2·5 × 2 mm., subcircular, usually overlapping and connivent, glabrous. Stamen-filaments 10–13 mm. long; 0·9 mm. long. Style 8–10 mm. long. Disk rather inconspicuous, without a free margin. Fruit usually as in *C. paniculatum* (normally c. 2 × 2 cm.) but occasionally up to c. 5 × 3·3 cm.

Botswana. *Fide* O. B. Miller (B.C.L.: 43 (1948)). **Zambia.** N: Mpika, fl. 18.ix.1967, *Astle* 5101 (FHO; SRGH). C: Lusaka Distr., Luangwa R., near Beit Bridge, fl. 5.ix.1947, *Brenan & Greenway* in *Brenan* 7812 (FHO; K). E: Fort Jameson Distr., Luangwa Valley, 1065–1145 m., fl. x.1951, *Gilges* 101 (PRE; SRGH). S: Gwembe, fl. 19.ix.1955, *Bainbridge* 124/55 (FHO). **Rhodesia.** N: Sebungwe, fl. ix.1955, *Davies* 1420 (K; SRGH). W: Shangani Distr., 915 m., fl. xii.1956, *Goldsmith* 7/58 (SRGH). C: Marandellas Distr., Zwipadze, fl. 6.x.1951, *Corby* 742 (SRGH). E: Melsetter Distr., Sabi Valley, fl. 2.ix.1949, *Chase* 1738 (BM; COI; SRGH). S: Triangle Estate, Mtilikwe R., fl. & fr. 10.x.1951, *Greenhow* 73/51 (FHO). **Malawi.** C: Dedza Distr., Kabalika Village, fl. 25.viii.1954, *Adlard* 172 (BM; SRGH). S: Palm Beach, fl. & fr. 2.viii.1960, *Leach* 10382 (BM; FHO; SRGH). **Mozambique.** N: Lago Metangula, fr. 11.x.1942, *Mendonça* 748 (BM; LISC). Z: Mocuba, R. Licungo, fl. & fr. x.1942, *Torre* 4621 (LISC). T: Marávia, Chicoa, fl. & fr. 25.ix.1942, *Mendonça* 454 (BM; LISC). MS: Elephant Point, fl. 19.viii.1809, *Salt* (BM). SS: between Lumane and João Belo, fr. 23.x.1947, *Barbosa* 513 (LISC; LM). LM: Goba, fl. 13.ix.1961, *Lemos & Balsinhas* 189 (BM; COI; K; LISC; SRGH).

Also in Tanzania and the Transvaal. Woodland or savanna; thickets on alluvial soils in river valleys; hotter drier areas.

Wickens (locc. citt.) prefers to consider *C. microphyllum* as a subspecies of *C. paniculatum* and has done a great deal of careful and detailed work on this. For Flora Zambesiaca I have as far as possible maintained the status quo partly for reasons which I have already given (see Kirkia, loc. cit.) partly because I think Wickens's solution is only a part of the story, as of course he realizes himself. What seems to have happened (hypothesis of course) is that a forest climber, *C. paniculatum* or an ancestral form of it (perhaps originally 5-merous) gave rise, on desiccation of the climate, to a form (*C. microphyllum*) adapted to drier conditions, in general more hairy and more inclined to be shrubby, in East Africa and over most of the Fl. Zamb. area, extending southwards to the Transvaal. *C. paniculatum* (or its ancestor) also gave rise across the fire-swept savanna regions to a shrublet (*C. platypetalum*) showing the usual and well-known extreme variability of these savanna shrublets subject to annual burning (see Exell & Stace in Valentine, Taxonomy, Phytogeography and Evolution: 318 (1972)). We have intermediates between *C. paniculatum* and *C. microphyllum*, between *C. microphyllum* and *C. platypetalum* and (in Angola) between *C. paniculatum* and *C. platypetalum*, where their ranges touch or overlap. That is why I am disinclined to adopt the partial solution of reducing *C. microphyllum* to a subspecies of *C. paniculatum*. It is perhaps more difficult to explain why, in the case of the *paniculatum-platypetalum-microphyllum* complex, I have not decided to treat the complex as one aggregate species as I have advocated in the cases of *C. collinum*, *C. hereroense* and *C. molle*. The reason is that the latter three aggregate species are all predominantly savanna species usually showing, as far as we know at present, no great diversity in ecology; but in the first case we should have an aggregate extending from rain-forest to quite dry savanna conditions. Furthermore, although the flowers throughout the complex remain comparatively uniform and the leaf-shape and indumentum are only partially helpful, making identification in the herbarium difficult with, it must be admitted, a number of specimens difficult to place, it is nevertheless remarkable that in the field one is only rarely in any doubt. The position when working with the herbarium material is undoubtedly thrown out of perspective by the fact that the collectors know these species very well and will pass by hundreds of " normal " specimens but pick up the occasional intermediate or abnormal specimen just because it is puzzling.

32. **Combretum platypetalum** Welw. ex Laws. in Oliv., F.T.A. **2**: 433 (1871).— Engl. & Diels in Engl., Mon. Afr. Pflanz. **3**: 68 (1899).—Eyles in Trans. Roy. Soc. S. Afr. **5**, 4: 428 (1916).—Burtt Davy & Hoyle, N.C.L.: 39 (1936).—Stace in Mitt. Bot. Staatssamml. München **4**: 14 (1961).—F. White, F.F.N.R.: 280 (1962).— Liben in Bull. Jard. Bot. Brux. **35**, 2: 179, 180 (1965); F.C.B. Combret.: 39 (1968).—Exell in Bol. Soc. Brot., Sér. 2, **42**: 7 (1968); in Kirkia, **7**, 2: 214 (1970).— Wickens, F.T.E.A. Combret.: 54 (1973).—Faria in Mem. Inst. Invest. Agron. Moçamb. **4**: 97 (1973). TAB. **31** fig. A. Type from Angola.

Combretum katangensis De Wild. in Fedde Repert. **11**: 516 (1913). Type from Katanga.

Combretum zastrowii Dinter in Fedde Repert. **16**: 174 (1919).—O. B. Mill., B.C.L.: 44 (1948). Type from SW. Africa.

Shrublet usually c. 15–30 cm. high but occasionally a shrub up to 3 m. (probably intermediates with *C. paniculatum*) with a thick woody rhizome, often leafless when flowering on annually produced shoots; bark grey-brown to grey-black; branchlets tomentose to glabrous. Leaves opposite, subopposite or alternate or occasionally 3-verticillate; lamina from 1 × 1·4 cm. to 10 × 5 cm. extremely variable in size and shape (from subcircular to very narrowly elliptic) and in indumentum (glabrous to densely tomentose). Inflorescences and flowers (red) indistinguishable from those of *C. paniculatum*. Fruits as in *C. paniculatum* and *C. micorphyllum* but more elliptic and rarely subcircular, sometimes up to 5·5 × 3·5 cm. often with a distinct apical peg and stipe 2–6 mm. long.

Throughout our area and also in Zaire, Angola, SW. Africa and Tanzania.

The relationship between this species and *C. paniculatum* and *C. microphyllum* have been discussed under *C. microphyllum*. There remains the problem of extreme variability within the species itself for of all the savanna and grassland shrublets notorious for their difficulty this is one of the worst in its excessive variability. Forms appear with long narrow leaves or with narrow oblong fruits about twice the usual length and there is nothing to show whether they remain constant or not or whether they are thrown up from time to time by genetical disturbances caused by fire. Until a species like this one is properly studied in Africa by breeding experiments, effect of protection from fire, transplant experiments, edaphic variation in the individual and germination of seedlings we shall not understand the acute taxonomic problems with which these savanna species confront us. Owing to the time factor it will be much easier to study one of these shrublets than a savanna tree like *C. molle*. When the facts are known it will be time to consider a suitable nomenclatural system; but I am sure that it would be wrong to try to contort the facts to bring them within a rigid nomenclatural framework which may prove to be quite unsuitable. Meanwhile in the absence of the necessary facts the system here proposed is only a temporary makeshift.

Upper receptacle rather densely hairy to tomentose; leaves sparsely pubescent to tomentose:
    Leaves sparsely pubescent to rather densely pilose, up to 2½ times as long as broad
                                            subsp. *platypetalum*
    Leaves sericeous or tomentose up to 2–3 times as long as broad:
        Leaves sericeous-tomentose up to twice as long as broad - - subsp. *baumii*
        Leaves fulvous-tomentose, up to 3 times as long as broad, usually long-acuminate
                                          subsp. *virgatum*
Upper receptacle glabrous to sparsely pubescent; leaves glabrous to sparsely pubescent, up to 4 times as long as broad - - - - - - - subsp. *oatesii*

32a. Subsp. **platypetalum.**—Exell in Bol. Soc. Brot., Sér. 2, **42**: 24 (1968); in Kirkia, **7**, 2: 215 (1970).—Exell & Garcia, C.F.A. **4**: 71 (1968). Type as for the species.

Zambia. B: Sesheke, fl. 11.viii.1947, *Brenan & Keay in Brenan* 7670 (FHO; K). N: Kasama Distr., fl. 6.xi.1960, *Robinson* 4047 (K; SRGH). W: Solwezi, fl. 10.ix.1952, *Angus* 400 (BM; FHO; K). C: Serenje, fl. v.1930, *Lloyd* (BM). S: Mazabuka Distr., Siamambo Forest Reserve, fl. 7.vi.1952, *White* 2925 (BM; FHO).

Also in Angola, SW. Africa and Zaire. *Brachystegia* woodland; edges of dambos; copper and lake-basin chipya; Kalahari woodland; *Acacia-Combretum-Terminalia* tree savanna; termite mounds; open sandy ground; at medium altitudes.

This is the typical subspecies extremely common and widespread in Angola but in our area found only in Zambia.

32b. Subsp. **baumii** (Engl. & Gilg) Exell in Bol. Soc. Brot., Sér. 2, **42**: 24 (1968); in Kirkia, **7**, 2: 215 (1970).—Exell & Garcia, C.F.A. **4**: 72 (1968). Type from Angola.
    Combretum baumii Engl. & Gilg in Warb., Kunene-Samb.-Exped. Baum: 320 (1903). Type as above.
    Combretum arenarium Engl. & Gilg, tom. cit.: 318 (1903). Type from Angola.
    Combretum argyrochryseum Engl. & Gilg, tom. cit.: 320 (1903). Type from Angola.
    Combretum gnidioides Engl. & Gilg, tom. cit.: 319 (1903). Type from Angola.
    Combretum praecox De Wild. in Fedde Repert. **13**: 197 (1914).—R.E.Fr. in Wiss. Ergebn. Schwed. Rhod.-Kongo—Exped. **1**: 170 (1914). Type from Zaire.

Zambia. N: Mpongwe, fl. 2.ix.1963, *Fanshawe* 7964 (FHO; LISC). W: Mwinilunga, 48 km. on the road to Solwezi, fl. & fr. 18.ix.1952, *Angus* 482 (BM; FHO; K).
Also in Angola.

There are transitions between subsp. *baumii* and subsp. *virgatum*.

32c. Subsp. **virgatum** (Welw. ex Laws.) Exell in Bol. Soc. Brot., Sér. 2, **42**: 25 (1968); in Kirkia, **7**, 2: 216 (1970).—Exell & Garcia, C.F.A. **4**: 73 (1968). Type from Angola.

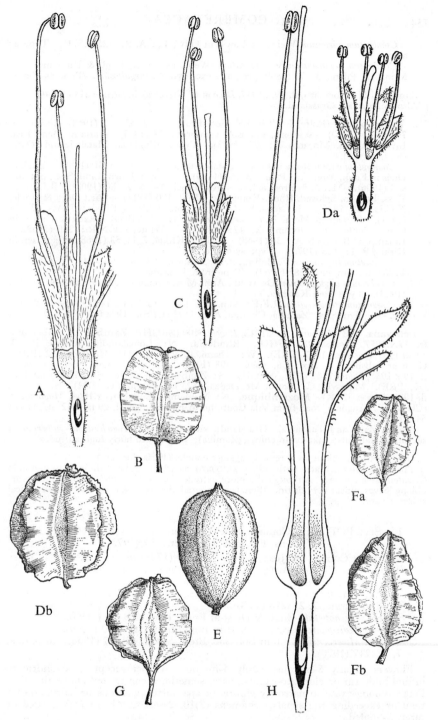

Tab. 37. A & B.—COMBRETUM PANICULATUM. A, flower, longitudinal section (×4);
B, fruit (×1). C.—C. MICROPHYLLUM, flower, longitudinal section (×4). D. (a & b)
—C. OBOVATUM, (a) flower, longitudinal section (×4); (b) fruit (×1). E.—C.
PENTAGONUM, fruit (×1). F. (a & b).—C. GOETZII, fruits (×1). G.—C. HOLSTII,
fruit (×1). H.—C. CONSTRICTUM, flower, longitudinal section (×4). From F.T.E.A.

Combretum virgatum Welw. ex Laws. in Oliv., F.T.A. **2**: 425 (1871). Type as above.
Combretum paniculatum var. *virgatum* (Welw. ex Laws.) Engl. & Diels in Engl., Mon. Afr. Pflanz. **3**: 71 (1899) pro parte excl. syn. *C. virgultosum*. Type as above.

Zambia. N: Abercorn, fl. 19.x.1947, *Brenan & Greenway* in *Brenan* 8149 (FHO; K). Also in Angola. Grassland.

33d. Subsp. **oatesii** (Rolfe) Exell in Bol. Soc. Brot., Sér. 2, **42**: 25 (1968); in Kirkia, **7**, 2: 216 (1970).—Wickens, F.T.E.A. Combret.: 54 (1973).—Faria in Mem. Inst. Invest. Agron. Moçamb. **4**: 97 (1973). Type: Rhodesia, Matabeleland, *Oates* (K, holotype).
Combretum oatesii Rolfe in Oates, Matabeleland, ed. 2: 399, t. 10 (1889).—Engl. & Diels in Engl., Mon. Afr. Pflanz. **3**: 68 (1899).—Gibbs in Journ. Linn. Soc., Bot. **37**: 443 (1906).—Monro in Proc. & Trans. Rhod. Sci. Ass. **8**, 2: 80 (1908).—R.E.Fr. in Wiss. Ergebn. Schwed. Rhod.-Kongo Exped. **1**: 170 (1914).—Burtt Davy & Hoyle, N.C.L.: 39 (1936).—Suesseng. & Merxm. in Proc. & Trans. Rhod. Sci. Ass. **43**: 28 (1951).—Brenan in Mem. N.Y. Bot. Gard. **8**, 5: 438 (1954).—Exell & Garcia in Contr. Conhec. Fl. Moçamb. **2**: 125 (1954).—F. White, F.F.N.R.: 280 (1962).—Liben, F.C.B. Combret.: 38 (1968).—Wild in Kirkia, **7**, 1: 53 (1968).—Jacobsen in Kirkia, **9**, 1: 168 (1973). Type as above.
Combretum turbinatum F. Hoffm., Beitr. Kenntn. Fl. Centr.-Ost-Afr.: 28 (1889).—F. White, tom. cit.: 282 (1962). Type from Tanzania.
Combretum angustifolium De Wild., Ann. Mus. Congo Belge, Bot. Sér. 4, **1**: 213 (1903). Type from Katanga.
Combretum stenophyllum R.E.Fr., tom. cit.: 170, t. 13 fig. 12–14 (1914).—F. White, loc. cit. Syntypes: Zambia, Chungu, *Fries* 1193 (UPS), 1193a (UPS).

Botswana. SE: Lobatsi, fr. x.1912, *Rogers* 6109 (SRGH). **Zambia.** N: Abercorn, fr. 17.x.1957, *Lawton* 282 (FHO). **Rhodesia.** N: Lomagundi, Mungwe R., fl. 20.xii.1961, *Jacobsen* 1577 (PRE). W: Khami, fl. & fr. xi.1903, *Marloth* 3392 (PRE). C: near Salisbury, fl. 2.ix.1956, *Angus* 1395 (FHO; K). E: Umtali Distr., Pounsley, fl. 19.x.1958, *Chase* 7362 (BM; SRGH). S: Zimbabwe, fl. 30.ix.1951, *Pole Evans* 4713 (K; PRE). **Malawi.** C: foot of Mt. Dedza, 1525 m., *Longfield* 62 (BM). S: Zomba, fl. 1901, *Purves* 32 (K). **Mozambique.** N: Maniamba, Cóbuè, fr. x.1964, *Magalhões* 6 (COI). T: Angónia, road from Vila Coutinho to Furancungo, fl. 29.ix.1942, *Mendonça* 542 (LISC).
Also in Zaire and Tanzania. Grassland; vleis (dambos); *Brachystegia-Julbernardia* savanna woodlands; lake-basin chipya (Zambia); medium to fairly high altitudes.

Used in Mozambique as a remedy against coughs (Faria, loc. cit.).
There are intermediates with subsp. *platypetalum* which make it difficult to maintain *C. oatesii* as a species. Although the more extreme forms of *C. turbinatum* with linear-oblong leaves appear very distinct there is every gradation between such forms and typical subsp. *oatesii*.

16. Sect. POIVREA (Comm. ex DC.) G. Don, Gen. Syst. **2**: 665 (1832).
—Wickens, F.T.E.A. Combret.: 57 (1973).

Poivrea Comm. [ex Juss., Gen. Pl.: 320 (1798) *in syn.* (" Pevraea "); ex Thou., Obs. Pl. Iles Austr. Afr.: 28 (1811) *nom. nud.*] ex DC., Prodr. **3**: 17 (1828). Type species: *Combretum coccineum* (Sonner.) Lam.
Sect. *Grandiflora* Engl. & Diels in Engl., Mon. Afr. Pflanz. **3**: 88 (1899).—Exell in Kirkia, **7**, 2: 216 (1970).
Sect. *Trichopetala* Engl. & Diels in Engl., tom. cit.: 92 (1899).
Sect. *Malegassica* Engl. & Diels in Engl., tom. cit.: 110 (1899).
Sect. *Mweroensia* Exell in Bol. Soc. Brot., Sér. 2, **42**: 27 (1968); in Kirkia, **7**, 2: 217 (1970).

Flowers usually 5-merous, rarely 4-merous. Upper receptacle cylindric to infundibuliform or elongate-campanulate, sometimes constricted above the disk. Petals oblong-ovate to narrowly elliptic or spathulate, glabrous or hairy, shorter than or exceeding the sepals. Stamens (8)10, 2-seriate. Fruit (4)5-winged or rarely 5-angled.

There is no doubt that Wickens (loc. cit.) is right in uniting sections *Grandiflora*, *Trichopetala* and *Malegassica* in Sect. *Poivrea*. This synthesis makes it advisable to add my recently described Sect. *Mweroensia* which becomes unnecessary with the union of all the other sections mentioned above.

Flowers 4-merous, hairy, not conspicuously glandular; petals much exceeding the sepals,
   pubescent outside - - - - - - - - - 33. *andradae*
Flowers 5-merous:
  Fruit 5-winged; upper receptacle not or only slightly constricted between the upper
    part and the lower part containing the disk:
    Petals glabrous, slightly shorter than the sepals; sepals caudate   34. *caudatisepalum*
    Petals hairy:
      Fruit with black stipitate glands 0·8 mm. long; petals not exceeding the sepals
                                    35. *mweroense*
      Fruit without macroscopic stipitate glands, hairy or glabrous:
        Upper receptacle hairy and/or conspicuously densely glandular:
          Upper receptacle hairy but not or only very sparsely conspicuously glandular:
            Fruit 2·5–3 × 2·5–3 cm.; rhachis of spike up to 5(7) cm. long
                                 36. *mossambicense*
            Fruit 3·5–4·5 × 4 cm.; rhachis of spike up to 9–10 cm. long
                                 37. *lasiocarpum*
         Upper receptacle conspicuously glandular - - - - 38. *goetzei*
        Upper receptacle glabrous or sparsely glandular - - - 39. *holstii*
  Fruit 5-angled; upper receptacle constricted between the upper and lower parts for
    5–6 mm.; margin of disk glabrous - - - - - - 40. *constrictum*

33. **Combretum andradae** Exell & Garcia in Contr. Conhec. Fl. Moçamb. **2**: 132, t. 11
(1954) (fig. C and D should be ×1 not ×2).—Exell in Bol. Soc. Brot., Sér. 2, **42**: 6
(1968); in Kirkia, **7**, 2: 218 (1970).—Wickens, F.T.E.A. Combret.: 60 (1973).—
Faria in Mem. Inst. Invest. Agron. Moçamb. **4**: 99 (1973). Type: Mozambique,
Amaramba, *Andrada* 1436 (BM, holotype; COI; LISC).
   Combretum lukafuense sensu Liben, F.C.B., Combret.: 27 (1968) quoad specim.
Mozamb.

Woody climber or shrub c. 2 m. high; branchlets at first tomentose but soon
glabrescent. Leaves subopposite; lamina up to 19 × 12·5 cm., membranous (at
least when young), broadly elliptic, densely minutely glandular above, tomentose
to densely pilose beneath, apex rounded or acuminate or apiculate, base subcordate
to cordate; lateral nerves 7–10 pairs; petiole up to 8 mm. long, tomentose.
Inflorescences of lateral spikes 4–7 cm. long, often in the axils of fallen leaves;
rhachis tomentose; bracts caducous (not seen). Flowers reddish-purple or
salmon-red, 4-merous. Lower receptacle c. 5 × 1·5 mm., densely pilose; upper
receptacle 12 × 3–6 mm., campanulate or infundibuliform, slightly constricted
above the part containing the disk, pubescent. Sepals 2–2·5 × 2·5 mm., broadly
triangular. Petals purple, 4–5 × 1–1·5 mm., oblong-spathulate or oblong-elliptic,
pubescent outside glabrous inside. Stamens 8, 2-seriate with filaments 18–22 mm.
long; anthers red, 1·5 × 1 mm. Disk 2–3 mm. in diam., glabrous, pilose margin
not free. Style c. 2·6 cm. long. Fruit 4-winged, c. 3·5(5) × 2·5(3) cm., elliptic or
elliptic-oblong in outline, apex and base truncate, velutinous, apical peg less than
1 mm. long, wings up to 10 mm. broad, stipe up to 7(10) mm. long.

Mozambique. N: Porto Amélia, road to Ancuabe near Metuge, fl. 6.ix.1948, *Andrada*
1325 (COI; LISC; LMJ); Nampula, fl. & fr. 18.x.1952, *Barbosa & Balsinhas* 5167
(BM; LISC; LM; LMJ).
  Also in Tanzania. Tree or shrub savanna; *Acacia-Lonchocarpus* woodlands; on clay
soils; at lower altitudes.

  The two specimens cited above are described as shrubs but the species is probably
always a putative climber.

34. **Combretum caudatisepalum** Exell & Garcia in Contr. Conhec. Fl. Moçamb. **2**:
130 cum fig. (1954).—Exell in Bol. Soc. Brot., Sér. 2, **42**: 6 (1968); in Kirkia, **7**, 2:
217 (1970).—Faria in Mem. Inst. Invest. Agron. Moçamb. **4**: 98 (1973). Type:
Mozambique, Memba, *Torre* 1352 (BM; COI, holotype; LISC).

Scandent shrub up to 3·5 m. high; branchlets purplish-brown, pubescent at first
later glabrescent. Leaves opposite; lamina up to 10 × 4 cm., chartaceous, obovate-
elliptic or oblong-elliptic, papillose on both surfaces otherwise glabrous except for
tufts of hairs in the axils of the lateral nerves beneath, shortly acuminate at the apex,
cordate at the base; lateral nerves 5–6 pairs; petiole up to 3 mm. long.
Inflorescences of elongated terminal and axillary spikes up to 10 cm. long; rhachis
rather sparsely pubescent and ± glabrescent; bracteoles 4 mm. long, linear-
elliptic, glabrous, caducous. Flowers reddish, subsessile. Lower receptacle 3 mm.

long, puberulous or glabrous; upper receptacle 8 × 3–4·5 mm., cylindric-infundibuliform, glabrescent. Sepals caudate-acuminate, equalling or slightly exceeding the sepals. Petals 3–4 × 1–1·5 mm., elliptic, apex acute, base truncate, glabrous. Stamen-filaments c. 12 mm. long; anthers 1·5 mm. long. Disk 2–2·5 mm. in diam., margin pilose, not free. Style 15 mm. long. Fruit 1·5 × 1·2 cm., elliptic in outline, glabrous, apical peg absent, wings 5 mm. broad, stipe short (up to 2 mm.).

Mozambique. N: Memba, fl. & fr. 17.v.1937, *Torre* 1532 (BM; COI; LISC). Not known elsewhere. Woodland; thickets.

35. **Combretum mweroense** Bak. in Kew Bull. **1895**: 290 (1895).—F. White, F.F.N.R.: 282 (1962).—Exell in Bol. Soc. Brot., Sér. 2, **42**: 6, 11 (1968); in Kirkia, **7**, 2: 217 (1970). Type: Zambia, Mweru Plateau, *Carson* 37 (BM, fragm.; K, holotype).

Scrambler or shrub up to c. 1·2 m. high, leafless when flowering; branchlets coarsely tomentose when young. Leaves opposite; lamina 4–5 × 2·5 cm., ovate, tomentose, acuminate at the apex, cordate at the base; lateral nerves 6–7 pairs; petiole 1 mm. long. Inflorescences of dense abbreviated subsessile lateral spikes up to 2·5 cm. long; rhachis coarsely tomentose with black stipitate glands. Flowers pink to reddish-yellow. Lower receptacle c. 3 mm. long, very densely sericeous-pilose; upper receptacle up to 12 mm. long, 2–2·5 mm. wide at the base and up to 4 mm. wide at the apex, tubular to·infundibuliform, sericeous-pilose, scarcely constricted but somewhat expanded at the apex. Sepals up to 4·5 mm. long, very narrowly triangular, with black stipitate glands (sometimes sparse) up to 0·8 mm. long. Petals 4–5 × 1·5 mm., narrowly elliptic, densely pilose, not exceeding the sepals. Stamen-filaments 14–20 mm. long; anthers orange-red, 0·9–1·1 mm. long. Disk rather fleshy, 2 mm. in diam., margin pilose, not free. Style 18–20 mm. long. Fruit up to 4·5 × 4 cm., subcircular to oblong in outline, densely pubescent and with black stipitate glands up to 0·8 mm. long, apical peg absent, wings c. 10 mm. broad, chartaceous, decurrent into the stipe, stipe up to 5 mm. long.

Zambia. N: Kalungwishi R., fl. 16.viii.1958, *Fanshawe* 4703 (FHO; K); fr. 16.viii.1958, *Fanshawe* 4704 (FHO; K). Not known elsewhere. Semi-deciduous thicket; in chipya woodland; semi-deciduous scrub.

36. **Combretum mossambicense** (Klotzsch) Engl., Pflanzenw. Ost-Afr. **C**: 292 (1895).—Engl. & Diels in Engl., Mon. Afr. Pflanz. **3**: 98, t. 26 fig. B (1899).—Bak. f. in Journ. Linn. Soc., Bot. **40**: 69 (1911).—Eyles in Trans. Roy. Soc. S. Afr. **5**, 4: 428 (1916).—Burtt Davy & Hoyle, N.C.L.: 39 (1936).—O. B. Mill., B.C.L.: 43 (1948).—Brenan, T.T.C.L.: 134 (1949); in Mem. N.Y. Bot. Gard. **8**, 5: 438 (1954).—Codd, Trees and Shrubs Kruger Nat. Park: 131, fig. 122b (1951).—Wild, Guide Fl. Vict. Falls: 146 (1953); in Rhod. Agric. Journ. **50**, 5: 4, 7 (1953).—Exell & Garcia in Contr. Conhec. Fl. Moçamb. **2**: 135 (1954); C.F.A. **4**: 78 (1970).—Stace in Mitt. Bot. Staatssamml. München **4**: 14 (1961).—F. White, F.F.N.R.: 282 (1962).—Fanshawe & Savory in Kirkia, **4**: 190 (1964).—Boughey in Journ. S. Afr. Bot. **30**, 4: 167 (1964).—Exell, Prodr. Fl. SW. Afr. **99**: 9 (1966); in Bol. Soc. Brot., Sér. 2, **42**: 6 (1968); in Kirkia, **7**, 2: 219, t. 2 fig. 37 (1970).—Liben, F.C.B. Combret.: 26 (1968).—Wickens, F.T.E.A. Combret.: 63, fig. 6(40), fig. 7(40) (1973).—Jacobsen in Kirkia, **9**, 1: 168 (1973).—Faria in Mem. Inst. Invest. Agron. Moçamb. **4**: 100 (1973). TAB. **31** fig. D; TAB. **35** fig. 36. Type: Mozambique, Sena, *Peters* (B†, holotype; BM, fragment).

Poivrea mossambicensis Klotzsch in Peters, Reise Mossamb. Bot. **1**: 78, t. 13 (1861). Type as above.

Poivrea glutinosa Klotzsch, tom. cit.: 79 (1861). Type: Mozambique, near Sena, *Peters* (B†, holotype).

Poivrea senensis Klotzsch, loc. cit. Type: rios de Sena, *Peters* (B†, holotype).

Combretum constrictum sensu Laws. in Oliv., F.T.A. **2**: 423 (1871) quoad syn. *Poivrea mossambicensis*.

Combretum ukambense Engl., Pflanzenw. Ost-Afr. **C**: 291 (1895). Syntypes from Kenya.

Combretum trichopetalum Engl., tom. cit.: 92 (1895).—Engl. & Diels in Engl., Mon. Afr. Pflanz. **3**: 97, t. 28 fig. C (1899).—Schinz, Pl. Menyharth.: 432 (1906).—Burtt Davy & Hoyle, N.C.L.: 39 (1936).—Gomes e Sousa in Bol. Soc. Estud. Col. Moçamb. **5**, 32: 319 (1936). Syntypes from Tanzania.

Combretum rigidifolium Welw. ex Hiern, Cat. Afr. Pl. Welw. **1**: 342 (1898). Type from Angola.

Combretum ischnothyrsum Engl. & Diels in Engl., tom. cit.: 96 (1899). Type: Mozambique, Tete, Boroma, *Menyharth* 888a (Z, holotype).
Combretum quangense Engl. & Diels in Engl., tom. cit.: 98 (1899). Syntypes from Zaire and Angola.
Combretum cataractarum Diels in Engl., Bot. Jahrb. **39**: 508 (1907).—Monro in Proc. & Trans. Rhod. Sci. Ass. **8**, 2: 80 (1908).—Eyles in Trans. Roy. Soc. S. Afr. **5**, 4: 427 (1916).—Burtt Davy, F.P.F.T. **1**: 248 (1926).—O. B. Mill., B.C.L.: 42 (1948). Syntypes: Victoria Falls, *Allen* (B†), *Engler* 2925 (B†).
Combretum floribundum N.E.Br. ex Diels, loc. cit. *in syn.* non Engl. & Diels (1899). Burtt Davy & Hoyle, N.C.L.: 38 (1936).
? *Combretum tomentosum* sensu Sim, For. Fl. Port. E. Afr.: 64 (1909).
Combretum detinens Dinter in Fedde Repert. **26**: 170 (1919). Type from SW. Africa.
Combretum migeodii Exell in Journ. of Bot. **67**: 48 (1929). Type from Tanzania.
Combretum armatum Phillips in Journ. Ecol. **18**: 217 (1930) *nom. nud.*

Small deciduous tree, shrub or woody climber, usually flowering before the leaves; bark grey to grey-brown; branchlets often pale, pubescent at first usually (but not always) soon glabrescent. Leaves opposite or subopposite; lamina up to 20 × 11 cm. but often only half that size, papyraceous to subcoriaceous, elliptic to elliptic-oblong or rarely subcircular, hairy when young, usually glabrescent but sometimes retaining a dense indumentum, apex usually acuminate, base rounded to cordate; lateral nerves 5–9 pairs, rather prominent beneath; petiole up to 5 mm. long, base forming a curved spine. Inflorescences of axillary spikes, sometimes subcapitate, up to 5(7) cm. long, often in the axils of fallen leaves; rhachis tomentose; bracts up to 4 × 2·5 mm., ovate, stalked, caducous. Flowers usually (4)5-merous, white or pinkish. Lower receptacle c. 5–6 mm. long, tomentose, often constricted above and below the ovary; upper receptacle up to 9(11) × 4 mm., hairy, upper part broadly infundibuliform, lower part surrounding the disk subglobose, slightly constricted between the two parts. Sepals 2 × 1·8 mm., deltate or triangular. Petals white or pinkish, 7–9 × 2–3·5 mm., elliptic, pilose, unguiculate. Stamen-filaments 16–17 mm. long; anthers orange-red, 1·7–1·8 mm. long. Disk c. 1·5 mm. in diam., pilose margin not free. Style 18 mm. long. Fruit (4)5-winged, up to 2–3 × 2–2·5 cm., elliptic to subcircular in outline, pubescent, apical peg very short, wings up to 10 mm. broad, stipe up to 10 mm. long.

Caprivi Strip. E. of Cuando R., fl. & fr. x.1945, *Curson* 1109 (PRE). **Botswana.** N: Matlapaning Drift, near Maun, 760 m., fl. 11.ix.1954, *Story* 4651 (PRE). **Zambia.** B: Sesheke, fl. ix.1921, *Borle* 287 (PRE). N: Kasama, fl. & fr., *Lawton* 764 (FHO). W: Ndola, fl. 14.viii.1952, *White* 3064 (FHO). C: near Beit Bridge, fl. 5.ix.1947, *Brenan & Greenway* in *Brenan* 7813 (FHO; K). E: 6·4 km. E. of Fort Jameson, 1220 m., fl. & fr. 1.x.1958, *King* 426 (SRGH). S: Livingstone, fl. 11.ix.1963, *Lawton* 1123 (FHO). **Rhodesia.** N: Lomagundi Distr., Mangula, fl. 30.ix.1962, *Jacobsen* 1790 (PRE). W: Victoria Falls, 885 m., fl. 17.ix.1954, *Greenway* 8804 (EA; FHO; K). C: Twenty Dales Estate, c. 19·2 km. E. of Salisbury, fl. ix.1946, *Wild, Gouveia & Pedro* in *Pedro* 2043 (LMJ). E: Melsetter Distr., Hot Springs, fl. & fr. 22.x.1948, *Chase* 939 (BM; COI; LISC; SRGH). S: Triangle Sugar Estate, Mtilikwe R., fl. 10.x.1951, *Kirkham* 52/51 (FHO; SRGH). **Malawi.** N: Rumpi, fl. viii.1953, *Chapman* 127 (FHO). C: Kaombe R., fl. & fr. 25.ix.1943, *Benson* 411 (PRE). S: Port Herald Distr., Lower Shire R., fl. & fr. 27.ix.1956, *Jackson* 2063 (BM; FHO; K). **Mozambique.** N: Nampula, fl. 15.x.1936, *Torre* 1337 (COI). Z: Namagoa, fl. 22.ix.1948, fr. 20.x.1948, *Faulkner* 285 (COI; K; SRGH). T: between Tete and Mandiè, fl. 5.ix.1946, *Gouveia & Pedro* 2018 (LMJ; PRE). MS: Chemba, Marіnguè, fl. 19.vii.1941, *Torre* 3108 (BM; LISC). SS: Olaria, c. 10 km. NW. of Caniçado, fl. 2.x.1963, *Leach & Bayliss* 11788 (PRE; SRGH). LM: Magude, fl. & fr. 30.xi.1944, *Mendonça* 3175 (LMA).

Widespread in Southern Tropical Africa and in SW. Africa and the Transvaal. Dry deciduous tree or shrub savanna; *Baikiaea* forest; thickets; termite mounds; hotter drier areas at low to medium altitudes.
Faria (loc. cit.) records flowers with up to 7 petals and 14 stamens.

37. Combretum lasiocarpum Engl. & Diels in Engl., Mon. Afr. Pflanz. **3**: 96 (1899). Schinz, Pl. Menyharth.: 432 (1906).—Exell & Garcia in Contr. Conhec. Fl. Moçamb. **2**: 135 (1954) excl. specim. *Torre* 3227.—Exell in Bol. Soc. Brot., Sér. 2, **42**: 6 (1968); in Kirkia, **7**, 2: 221 (1970). TAB. **31** fig. C. Type: Mozambique, Boroma, *Menyharth* 678 (Z, holotype).

Shrub 2 m. high or small tree; branchlets rather stout, tomentose at first eventually glabrescent. Leaves opposite; lamina up to 14 × 9·5 cm., broadly

elliptic to narrowly obovate-elliptic, chartaceous to subcoriaceous, glabrous or
nearly so, apex rounded or shortly acuminate, base subcordate; lateral nerves
5–10 pairs; petiole up to 1 cm. long, tomentose. Inflorescences of handsome
axillary spikes up to 9–10 cm. long, usually in the axils of fallen leaves; rhachis
tomentose; bracts (not seen) caducous. Flowers red or cream, 5-merous. Lower
receptacle up to 5 mm. long, tomentose; upper receptacle 9–12 × 2–3·5 mm. with
a broadly ellipsoid-cylindric basal part surrounding the disk then becoming slightly
constricted and with an infundibuliform upper part, tomentose. Sepals
1·5 × 2·5 mm., broadly deltate. Petals c. 4·5–5 × 1·5 mm., elliptic, pubescent
outside. Stamen-filaments 13 mm. long; anthers salmon-pink, 1·2 mm. long.
Fruit 5-winged, 3·5–4·5 × 2·5–4 cm., elliptic in outline, hairy, apical peg very short
or absent, wings 10 mm. broad, stipe 4–5 mm. long.

Mozambique. N: Nampula, fl. & fr. 18.x.1952, *Barbosa & Balsinhas* 5167 (BM;
LISC; LM; LMJ). T: near Boroma, *Menyharth* 678 (Z). MS: Mocuba, Namagoa,
fl. & fr. x.1946, *Faulkner* 14 pro parte (EA; K).
Known only from Mozambique. Dry deciduous tree or shrub savanna; at lower
altitudes.

Very close to *C. mossambicense* but with larger fruits and longer spikes. *Faulkner* 14
appears to be a mixture of *C. lasiocarpum* (EA; K) and *C. mossambicense* (BM; COI;
K; PRE), almost certainly different gatherings and no real evidence in favour of uniting
the two species except that the collector evidently thought they were the same.
This species has been at times confused with *C. obovatum* when in fruit and there is a
close resemblance at that stage though the flowers are quite different. In *C. lasiocarpum*
the fruits are distinctly stipitate but subsessile in *C. obovatum*.

38. **Combretum goetzei** Engl. & Diels in Engl., Mon. Afr. Pflanz. **3**: 97 (1899).—
Brenan, T.T.C.L.: 134 (1949).—Exell & Garcia in Contr. Conhec. Fl. Moçamb. **2**:
134 (1968).—Exell in Bol. Soc. Brot., Sér. 2, **42**: 6 (1968); in Kirkia, **7**, 2: 221
(1970).—Faria in Mem. Inst. Invest. Agron. Moçamb. **4**: 103 (1973). TAB. **37**
fig. F. Type from Tanzania.

Shrub up to 10 m. high, sometimes scandent; branchlets pale or reddish,
pubescent when young but soon glabrescent. Leaves opposite; lamina up to
8·5 × 4 cm., subcoriaceous, oblong-elliptic, almost glabrous when mature, apex
rounded or acuminate, base obtuse or rounded; lateral nerves 6–7 pairs, rather
prominent beneath; reticulation prominent above; petiole up to 5 mm. long
forming a curved spine when the plant is scandent. Inflorescences of lateral or
terminal spikes up to c. 10 cm. long; rhachis with dense macroscopic glands;
bracts 3 × 1 mm., narrowly elliptic, with stalked glands, caducous. Flowers white
(*fide* Chase) or pinkish, 5-merous. Lower receptacle 4–7 mm. long, glandular;
upper receptacle 4–6 × 3–4·5 mm., basal part (surrounding the disk) subglobose,
upper part campanulate-infundibuliform, slightly constricted between the two
parts, densely glandular but otherwise nearly glabrous. Sepals 2·5 × 2 mm.,
triangular. Petals (4)9 × 2–2·5 mm., unguiculate, pilose. Stamen-filaments
(6)18–19 mm. long; anthers 1·8 mm. long. Disk 2·5 mm. in diam., with pilose
margin not free. Style (6)20 mm. long. Fruit 5-winged, c. 3 × 2–2·5 cm., elliptic
or ovate-elliptic or broadly elliptic in outline, glabrous, somewhat glutinous, apical
peg c. 0·5 mm. long, stipe 2–3 mm. long, wings up to 9 mm. broad.

Mozambique. N: Ribauè, fl. & fr. 26.viii.1967, *Macedo* 2659 (LMA). T: Marávia,
R. Zambeze, fl. & fr. 25.ix.1942, *Mendonça* 410 (LISC).
Also in Tanzania. Woodland; deciduous shrub savanna.

According to Faria (loc. cit.) *Macedo* 2659 has much smaller flowers than usual.
Differs from *C. mossambicense* mainly in the densely glandular inflorescences but a few
intermediates occur (such as Mozambique, Tete, Boroma, *Chase* 2818 (BM)) and a
careful search often brings to light an occasional macroscopic gland in *C. mossambicense*.
C. goetzei is very little known; if the glabrous character of the fruit remains constant it may
provide a further distinction from *C. mossambicense*; but it is quite probable that *C. goetzei*
will eventually be considered to be a subspecies or variety of *C. mossambicense*.

39. **Combretum holstii** Engl., Pflanzenw. Ost-Afr. C: 291 (1895).—Engl. & Diels in
Engl., Mon. Afr. Pflanz. **3**: 95, t. 28 fig. A (1899).—Brenan, T.T.C.L.: 134 (1949).—
Exell & Garcia in Contrib. Conhec. Fl. Moçamb. **2**: 134 (1954); C.F.A. **4**: 79
(1968).—Liben, F.C.B. Combret. 24 (1968).—Exell in Bol. Soc. Brot., Sér. 2, **42**:

6 (1968); in Kirkia, **7**, 2: 222 (1970).—Faria in Mem. Inst. Invest. Agron. Moçamb. **4**: 103 (1973). TAB. **37** fig. G. Type from Tanzania.
Combretum affine De Wild. in Ann. Mus. Congo Belge **5**, 3: 233 (1910). Type from Zaire.
Combretum pynaertii De Wild., tom. cit.: 239 (1910). Type from Zaire.
Combretum landanaense De Wild. & Exell in Journ. of Bot. **67**: 179 (1929). Type from Zaire or Angola (*fide* Liben).

Woody climber (or small tree?); branchlets pale, glabrous. Leaves opposite or subopposite; lamina up to 20 × 8·5 cm., chartaceous, oblong-elliptic, minutely glandular but otherwise glabrous, apex acuminate, base rounded to cordate; lateral nerves 9–11 pairs, prominent below; petiole up to 10 mm. long, glabrous, basal part forming a curved spine. Inflorescences of axillary spikes (often in the axils of fallen leaves) up to 14 cm. long; rhachis glabrous; bracts c. 4 × 1–1·5 mm., caducous. Flowers crimson or purplish-red. Lower receptacle up to 5 mm. long, glandular but otherwise glabrous; upper receptacle 10–15 × 5–6 mm. with a lower subglobose part (surrounding the disk) and an upper infundibuliform part, slightly constricted between the two parts, sparingly glandular otherwise glabrous. Sepals 2–2·5 × 2–2·5 mm., deltate. Petals up to 6 × 2–2·8 mm., narrowly ovate-elliptic, shortly unguiculate, pubescent outside. Stamen-filaments 17–18 mm. long; anthers 2 mm. long. Disk 2–2·5 mm. in diam., margin pilose not free. Style 27–28 mm. long. Fruit 5-winged, up to 3·5 × 3 cm., broadly elliptic to subcircular in outline, glabrous, shiny, apical peg very short, wings up to 10 mm. broad, papyraceous, stipe up to 8 mm. long.

Mozambique. N: Porto Amélia, between Maate and Metuge, fl. 1.x.1948, *Andrada* 1392 (COI; LISC; LMJ). Z: between Nicuadala and Marral, fr. 27.vii.1942, *Torre* 4425 (BM; LISC). MS: near Chupanga, fl. viii.1858, *Kirk* (K). SS: Muchopes, 120 m., fr. 1905, *Sim* 20845 (PRE).
Also in Angola, Zaire and Tanzania. Woodland; riverine forest; mainly at low altitudes.

Balsinhas 116 (LMJ) from Mozambique, Porto Amélia, is described as a small tree 3 m. in height but it has the petioles persisting as curved spines, usually characteristic of a climber or potential climber.
This may prove to be no more than a glabrous variety of *C. mossambicense* and a few sparse glands also show a connection with *C. goetzei*. The style and the anthers are rather longer than any records for the other two species. The distribution is of a rather unusual type: northern Angola, Zaire, Tanzania and then throughout Mozambique as far south as Sul do Save but absent from Zambia and Malawi. This tends to give the taxon (of whatever rank it may be) some appearance of individuality.

40. **Combretum constrictum** (Benth.) Laws. in Oliv., F.T.A. **2**: 423 (1871) pro parte excl. syn. *Poivrea mossambicensis* Klotzsch et formae.—Engl. & Diels in Engl., Mon. Afr. Pflanz. **3**: 99, t. 26 fig. A (1899).—Sim, For. Fl. Port. E. Afr.: 61 (1909).—Gomes e Sousa, Bol. Soc. Estud. Col. Moçamb. **4**, 26: 102, 115 (1935).—Brenan, T.T.C.L.: 133 (1949).—Exell & Garcia, Contr. Conhec. Fl. Moçamb. **2**: 137 (1954).—Keay, F.W.T.A. ed. 2, **1**, 1: 273 (1954).—Dale & Greenway, Kenya Trees and Shrubs: 143 (1961).—Exell in Bol. Soc. Brot., Sér. 2, **42**: 6 (1968); in Kirkia, **7**, 2: 223 (1970).—Wickens, F.T.E.A. Combret.: 65, fig. 6(43), fig. 7(43) (1973).—Faria in Mem. Inst. Invest. Agron. Moçamb. **4**: 104 (1973). TAB. **37** fig. H. Type from Nigeria.
Poivrea constricta Benth. in Hook., Niger Fl.: 337 (1849). Type as above.
Combretum infundibuliforme Engl., Pflanzenw. Ost-Afr. **C**: 292 (1895). Syntypes from Zanzibar and Tanzania.
Combretum constrictum var. *tomentellum* Engl. & Diels in Engl., Mon. Afr. Pflanz. **3**: 100 (1899).—Brenan, T.T.C.L.: 134 (1949). Type from Tanzania.
? *Combretum bussei* Engl. & Diels in Engl., Bot. Jahrb. **39**: 509 (1907). Type from Tanzania. Wickens (loc. cit.) suggests this identification.
? *C. constrictum* var. *somalense* Pampan. in Bull. Soc. Bot. Ital. **1915**: 12 (1915). Syntypes from Somali Republic.

Shrub or climber; branchlets glabrescent. Leaves subopposite; lamina up to 6(12) × 3(6·5) cm., chartaceous to subcoriaceous, oblong to oblong-elliptic, glabrous or nearly so (or sometimes tomentose), apex rounded or acuminate, base rounded; lateral nerves 6–9 pairs; petiole 4–6 mm. long forming a slightly hooked spine at the base. Flowers red or pinkish, 5-merous. Inflorescences of subcapitate terminal or axillary spikes up to 6 cm. long; rhachis glabrous or tomentose;

bracts 6 × 2 mm., narrowly elliptic. Lower receptacle 5–6 mm. long, glabrous or tomentose; upper receptacle 10–18 mm. long and up to 7 mm. in diam. at the apex, constricted in the middle for 4–7 mm. to a diam. of 1·5 mm., broadly infundi-buliform at the apex, glabrous or tomentose. Sepals 4 × 3 mm., triangular. Petals 7–8 × 2·5 mm., narrowly elliptic, unguiculate, pilose. Stamen-filaments 17(25) mm. long, inserted in the apical part of the upper receptacle; anthers 0·8 mm. long. Disk 2 mm. in diam., margin glabrous not free. Style 25 mm. long. Fruit (not seen from our area) 5-angled, sessile, c. 2·5 × 1·2–1·5 cm., ellipsoid, glabrous, apical peg absent. Note: most organs described as glabrous or tomentose are glabrous in our area in the few specimens seen.

Mozambique. N: Mocímboa da Praia, R. Messalo, fl. 12.ix.1948, *Andrada* 1340 (BM; COI; K; LISC).

Also in Nigeria, ? Somali Republic, Kenya, Tanzania and Zanzibar. Edges of mangrove; seasonal swamps.

The fruits are probably distributed by water.

2. MEIOSTEMON Exell & Stace

Meiostemon Exell & Stace in Bol. Soc. Brot., Sér. 2, **40**: 18, t. 1 (1966).—Stace in Bot. Journ. Linn. Soc. **62**: 162, figs. 161–163 (1969).—Exell in Kirkia, 7, 2: 223 (1970).

Combretum sect. *Haplostemon* Exell in Journ. of Bot. **77**: 173 (1939).

Shrubs or more rarely small trees. Leaves opposite, petiolate, lepidote. Inflorescences of terminal and axillary panicles. Flowers small, 4-merous, sessile. Petals 4, inserted near the margin of the disk. Calyx-tube shallowly campanulate, slightly oblique. Stamens 4, 1-seriate, antipetalous, not exserted. Style scarcely exserted. Disk with pubescent scarcely free margin. Ovules 2. Fruit 4-winged. Cotyledons 2, unfolding spirally and arising above soil level (known only in *M. tetrandrus*). Scales c. 40–80μ in diam., circular in outline, slightly convexly scalloped at each marginal cell; cells delimited by 8–10 radial walls alone; cell-walls clear; cells transparent.

2 spp., one (with 2 subspecies) in Zambia and Rhodesia, the other (*M. humbertii* (Perrier) Exell & Stace) in Madagascar.

This genus, being lepidote, is evidently related to *Combretum* Subgen. *Combretum*. The suppression of one whorl of stamens, although its most obvious distinction, is not the only reason for treating it as a separate genus. It differs in the insertion of the petals at the margin of the disk. This occurs in *Combretum* Subgen. *Combretum* Sect. *Hypocrateropsis* as a matter of necessity because there is practically no development of the upper receptacle; but elsewhere in the genus as the upper receptacle evolves the place of insertion of the petals is carried up with it, the upper receptacle being, as its name implies, almost entirely of receptacular origin. In *Meiostemon* what we have regarded, perhaps by false analogy with *Combretum*, as the " upper receptacle " is probably really calycular tissue (in fact a true calyx-tube) with the petals left behind (so to speak) at the margin of the disk. As far as I can recollect (it is difficult to be absolutely certain in a large genus spread all over the tropics) there is no species of *Combretum* with this structure. Furthermore the distribution of *Meiostemon* in Zambia, Rhodesia and Madagascar (an unusual but not entirely unknown type of distribution) indicates relatively ancient origin.

Meiostemon tetrandrus (Exell) Exell & Stace in Bol. Soc. Brot., Sér. 2, **40**: 19, t. 1 (1966).—Exell in Bol. Soc. Brot., Sér. 2, **42**: 28 (1968); in Kirkia, 7, 2: 223 (1970).—Faria in Mem. Inst. Invest. Agron. Moçamb. **4**: 105 (1973). TAB. **38**. Type: Zambia, Mweru Wantipa, *Allen* 3 (BM, holotype; FHO).

Combretum tetrandrum Exell in Journ. of Bot. **77**: 172 (1939).—F. White, F.F.N.R.: 284 (1962).—Boughey in Journ. S. Afr. Bot. **30**, 4: 167 (1964). Type as above.

Deciduous scandent shrub up to 2–4 m. high or more rarely a small tree up to 5 m.; bark light-grey to blackish; branchlets reddish-brown, slender, at first minutely pubescent, glabrescent. Leaves opposite. Lamina up to 6 × 3 cm., papyraceous to chartaceous, elliptic to broadly elliptic, pubescent when very young but usually soon glabrescent except for pubescence on the midrib above, domatia and scales, lepidote (sometimes rather inconspicuously) on the lower surface, apex

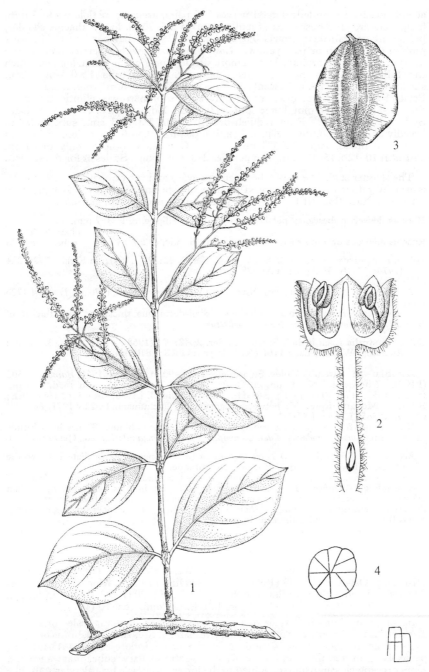

Tab. 38. MEIOSTEMON TETRANDRUS subsp. TETRANDRUS. 1, flowering branchlet (× ⅔)
Fanshawe 4896; 2, flower, longitudinal section (×10) *Fanshawe* 4896. M.
TETRANDRUS subsp. AUSTRALIS. 3, fruit (×1) *Bainbridge* 502; 4, scale (×270) from
a drawing by C. A. Stace.

acuminate, base rounded; lateral nerves 3–5 pairs, arcuate; petiole up to 5 mm. long, slender. Inflorescences of axillary spikes 2·5–9 cm. long; rhachis slender, pubescent or glabrous; bracts 0·8–2 mm. long, somewhat persistent. Flowers pinkish or greenish or pale-yellow. Lower receptacle with a part containing the ovary c. 1 mm. long produced into a more slender portion 0·5–1 mm. long on which the calyx-tube is borne somewhat obliquely. Calyx-tube 1 × 0·4–0·7 mm., shallowly campanulate, pubescent; lobes (sepals) 0·3 × 0·8 mm., rounded, reflexed. Petals 1 × 0·4–0·7 mm., ovate to broadly elliptic or narrowly elliptic, glabrous. Stamen-filaments 0·6 mm. long; anthers 0·7 mm. long. Disk c. 1·7 mm. in diam. Style 1·2 mm. long. Fruit reddish-brown, up to 1·3 × 1·3 cm., subcircular to broadly obovate in outline, glabrous, apical peg less than 0·5 mm. long, wings up to 5 mm. broad, stipe 0·5–1·5 mm. long. Cotyledons (known only in subsp. *australis*) 10–12 × 15–17 mm. with petioles 2–3 mm. long. Scales as for the section.

The slender stalk-like region between the ovary and the campanulate calyx-tube is usually rather more pronounced than in the specimen (*Angus* 708) figured by Exell and Stace (loc. cit.). The fruit in that plate is of subsp. *australis*.

Rhachis densely pubescent; petals broadly ovate to elliptic, c. 1 × 0·7 mm.
                                                                            subsp. *tetrandrus*
Rhachis glabrous or almost so; petals narrowly elliptic, c. 1 × 0·4 mm.   subsp. *australis*

Subsp. **tetrandrus.**—Exell in Bol. Soc. Brot., Sér. 2, **42**: 28 (1968); in Kirkia, **7**, 2: 224 (1970). TAB. **31** fig. E; TAB. **38** figs. 1–2.

Zambia. N: Mbala (Abercorn) Distr., Lufu Valley Escarpment, 1070 m., *Trapnell* 1726 (BM; K).
Known only from Zambia. Thickets; semi-deciduous thickets; Itigi thicket of *Bussea-Combretum* and *Commiphora-Combretum*.

Subsp. **australis** Exell in Bol. Soc. Brot., Sér. 2, **42**: 28 (1968). TAB. **9** fig. 3. Type: Rhodesia, Wankie, *Levy* 1161 (K, holotype; PRE; SRGH).

Zambia. S: Gwembe Valley, Sinazongwe area, 490 m., fr. 7.vii.1961, *Bainbridge* 502 (FHO). **Rhodesia.** N: Urungwe Distr., Zambezi Valley, Chirundu road, 520 m,. fr. 26.ii.1953, *Wild* 4039 (PRE; SRGH). W: Wankie, fl. i.1955, *Levy* 1161 (K; PRE; SRGH). **Mozambique.** T: between Changara and Cuchumano, fr. 22.v.1971, *Torre &* *Correia* 18565 (LMA; LMU)?
Known only from southern Zambia, Rhodesia and Mozambique. Dense low-altitude thicket with *Combretum elaeagnoides*, *Commiphora* spp., *Acacia nilotica*, etc. (Jesse thicket).

According to Faria (loc. cit.) *Torre & Correia* 18565 appears intermediate between the subspecies. Flowers are required from Mozambique.
Subsp. *australis* shows some transition towards the Madagascar species *M. humbertii* (Perrier) Exell & Stace. *Levy* 1161 (cited above) with inflorescences not more than 3·5 cm. long, petals barely 1 mm. long and leaves retaining some pubescence at time of flowering is clearly approaching *M. humbertii* which has inflorescences up to 3 cm. long, petals 0·5 mm. long and pubescent leaves.

3. PTELEOPSIS Engl.

Pteleopsis Engl. [in Abh. Preuss. Akad. Wiss. **1894**: 25 (1894) *nom. nud.*] Pflanzenw. Ost-Afr. **C**: 293 (1895).—Engl. & Diels in Engl., Mon. Afr. Pflanz. **4**: 2 (1900).—Exell & Stace in Bol. Soc. Brot., Sér. 2, **40**: 20 (1966).—Exell in Kirkia, **7**, 2: 225 (1970).—Wickens, F.T.E.A. Combret.: 70 (1973).

Small or medium-sized trees or occasionally shrubs, without scales or stalked glands. Leaves opposite or subopposite, petiolate, entire, almost glabrous or hairy. Flowers andromonoecious, ♀ and ♂ in the same inflorescence, (4)5-merous, pedicellate, in terminal and/or axillary or extra-axillary subcapitate racemes. Upper receptacle campanulate, joined to the lower receptacle by a slender stalk-like region; lower receptacle somewhat flattened. ♀ flowers: sepals deltate, little developed; petals (4)5, usually ± obovate; stamens (8)10, 2-seriate in our species; disk pilose with a short free margin; style not expanded at the apex. ♂ flowers usually towards the base of the inflorescence, as in the ♀ ones but with the ovary not developing and with a slender stalk replacing the lower receptacle (perhaps a true pedicel only towards the base); style present or vestigial. Fruit 2–5-winged, often decurrent into the comparatively long slender stipe. Cotyledons (where known) unfolding spirally and borne above soil level.

9–11 spp., all in tropical Africa.

The genus is intermediate in many respects between *Combretum* and *Terminalia* and is placed (Exell & Stace in Bol. Soc. Brot., Sér. 2, **40**: 20 (1966)) in the subtrib. *Pteleopsidinae* of the tribe *Combreteae*, between the subtribes *Combretinae* and *Terminaliinae*.

Lower receptacle glabrous or nearly so; fruit 2–4(5)-winged, glabrous:
  Fruit 2–3(4, very rarely 5)-winged, rarely with an apical peg and if present the latter very short; leaves usually shiny above and usually acuminate; petiole up to 10 mm. long - - - - - - - - - - - 1. *myrtifolia*
  Fruit 3–4(5)-winged, apical peg c. 1 mm. long; leaves usually (but not invariably) matt above, often mucronate or apiculate but not acuminate; petiole up to 6 mm. long 2. *anisoptera*
Lower receptacle rufous-sericeous; fruit (2)3-winged, pubescent - - 3. *barbosae*

The apical peg is, as in *Combretum*, a short projection at the apex of the fruit where the upper receptacle has broken off as the fruit develops.

There may be difficulty in separating *P. myrtifolia* and *P. anisoptera* in the above key (hybrids may occur). As they are ecologically important their separation is discussed in detail under *P. myrtifolia*.

1. **Pteleopsis myrtifolia** (Laws.) Engl. & Diels in Engl., Mon. Afr. Pflanz. **4**: 4, t. 1 fig. B (1900).—Bak. f. in Journ. Linn. Soc., Bot. **40**: 70 (1911).—Eyles in Trans. Roy. Soc. S. Afr. **5**, 4: 429 (1916).—Burtt Davy & Hoyle, N.C.L.: 39 (1936).— O. B. Mill., B.C.L.: 44 (1948).—Brenan, T.T.C.L.: 142 (1949).—Mogg in Macnae & Kalk, Nat. Hist. Inhaca I. Moçamb.: 150 (1958).—F. White, F.F.N.R.: 287, fig. 50H (1962).—Boughey in Journ. S. Afr. Bot. **30**, 4: 167 (1964).—Exell & Garcia, C.F.A. **4**: 82 (1970).—Exell in Kirkia, **7**, 2: 225 (1970).—Wickens, F.T.E.A. Combret.: 70, fig. 10 (1973).—Faria in Mem. Inst. Invest. Agron. Moçamb. **4**: 106 (1973). TAB. **34** fig. H; TAB. **39**. Syntypes: Mozambique, Lupata and Tete, *Kirk* (K).
  Combretum myrtifolium Laws. in Oliv., F.T.A. **2**: 431 (1871).—Sim, For. Fl. Port. E. Afr.: 64, t. 64 fig. C (1909). Syntypes as above.
  Pteleopsis variifolia Engl., Pflanzenw. Ost-Afr. **C**: 293 (1895). Syntypes from Tanzania.
  Pteleopsis stenocarpa Engl. & Diels in Engl., Mon. Afr. Pflanz. **4**: 5 (1900). Type from Angola.
  Pteleopsis obovata Hutch. in Kew Bull. **1917**: 232 (1917). Syntypes: Mozambique, R. Messalo, *Allen* 72 (K), 156 (K); Madanda Forest, *Dawe* 449 (K).

Deciduous tree up to 20 m. high (more often c. 10 m., said to reach 30 m. in Tanzania); wood red, very hard; bark grey or yellow-grey; crown rounded; young branchlets reddish-brown, slender, often pendulous, pubescent at first but soon glabrescent. Leaves opposite or subopposite; lamina dark-green, up to 9·5 × 3·5 cm. (sometimes as small as 1 × 0·6 cm.), usually shiny above, elliptic to very narrowly elliptic or obovate-elliptic, usually glabrous except for pubescence on the midrib, apex slightly acute or rather bluntly acuminate, base cuneate; lateral nerves 6–9 pairs; petiole up to 1 cm. long, rather slender, usually pubescent. Inflorescences of axillary subcapitate racemes up to 4·5 cm. long; rhachis slender, glabrous or sparsely pubescent; bracts 3 mm. long, filiform, caducous. Flowers ♀ and ♂ in the same inflorescence, white or yellow. ♀ flowers usually towards the apex of the inflorescence, pedicellate, 5-merous; lower receptacle c. 5 mm. long, flattened, glabrous; upper receptacle 2·5–3 × 2 mm., campanulate, glabrous, joined to the lower receptacle by a slender stalk-like portion 1–2 mm. long; sepals deltate; petals 1·5–2·5 × 1–2 mm., obovate to subcircular, shortly unguiculate, glabrous; stamens 2-seriate, antipetalous ones 4·5–5 mm. long, antisepalous ones 3·5–4 mm. long; anthers 0·7–0·8 mm. long; disk 2·5–3 mm. in diam. with a short free margin; style 5–6 mm. long. ♂ flowers usually towards the base of the inflorescence, ovary not developing but ♂ flowers otherwise similar to the ♀ ones (style present). Fruit 2–3(4, very rarely 5)-winged, 1–2·5 × 0·5–1·7 cm., very variable in size and shape, generally emarginate at the apex and usually without or with a very short apical peg, decurrent into the stipe and often unequal at the base; stipe up to 15 mm. long, very slender.

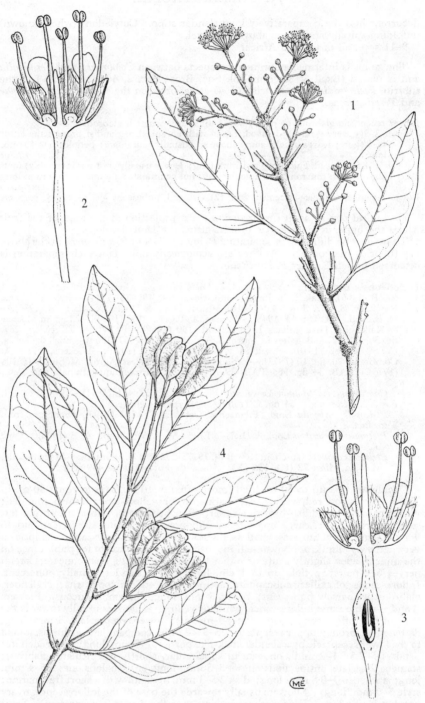

Tab. 39. PTELEOPSIS MYRTIFOLIA. 1, flowering branchlet (×1); 2, male flower, longitudinal section (×6); 3, bisexual flower, longitudinal section (×6); 4, fruiting branchlet (×1), all from *Fundi* 30. From F.T.E.A.

Botswana. N: Kazungula, fr. iv.1938, *Miller* B/194 (FHO). **Zambia.** S: Choma Distr., Sinazongwe, fl. 30.xii.1958, *Robson* 1024 (K; LISC; SRGH). **Rhodesia.** N: Sebungwe Distr., above Binga, 490 m., fr. ix, *Davies* 1043 (K; SRGH). W: Wankie, fl. & fr. 2.i.1952, *Lovemore* 233 (SRGH). E: Umtali Distr., Hot Springs, fr. 18.xii.1947, *Chase* 501 (BM; K; SRGH). S: Ndanga Distr., Umtilikwe R., fl. 26.i.1949, *Wild* 2753 (K; SRGH). **Malawi.** S: Zomba, 825–915 m., fl. xi.1915, *Purves* 242 (K). **Mozambique.** N: Nampula, Mureveia, fl. 20.xi.1948, *Andrada* 1465 (COI; K; LISC; LMJ). Z: between Milange and Mocuba, fl. 3.iii.1943, *Torre* 4880 (BM; K; LISC). T: Tete, fl. & fr. ii.1859, *Kirk* (K). MS: Chimoio, Garuso, fl. 16.ii.1948, *Mendonça* 3782 (BM; K; LISC). SS: Chibuto, 12 km. from Chibuto on the road to Alto Changane, fl. & fr. 12.ii.1959, *Barbosa & Lemos* 8383 (BM; K; LISC; SRGH). LM: Maputo, fr. 29.xi.1947, *Mendonça* 3571 (BM; K; LISC); Inhaca I., fl. 15.xii.1956, *Mogg* 26781 (BM; K).

Also in Angola, Kenya and Tanzania. Evergreen forest; riverine forest; *Brachystegia* woodland or savanna woodland; *Baikiaea* woodland; mopane woodland; *Acacia-Combretum-Terminalia* tree savanna; low to medium altitudes.

Produces good wood for furniture.

Small-leaved forms occur in the more southern drier parts of our area (e.g. *Mendonça* 3571 and *Torre* 6769 (BM; K; LISC)) but these forms are also very variable in leaf-shape and *P. obovata* Hutch. appears to be no more than one of them with mainly obovate leaves and often densely pilose.

Although *P. myrtifolia* and *P. anisoptera* differ in general by a number of characters it is difficult to find any single character which separates them infallibly and they probably hybridize. As the two species are important in our area I give some notes which should help to identify most specimens (the most difficult are young flowering specimens without fruits or mature leaves).

1. If some fruits are 2-winged: almost certainly *myrtifolia*.
2. If many of the fruits are 4-winged: probably *anisoptera*.
3. If the fruits are 1·5 cm. long or longer: probably *myrtifolia*.
4. If the fruits are 1 cm. long or shorter: probably *anisoptera*.
5. If the leaves are acuminate (even slightly so or some of them): almost certainly *myrtifolia*.
6. If the leaves are somewhat shiny on the upper surface: probably *myrifolia*.
7. If the leaves are matt on the upper surface: almost certainly *anisoptera*.
8. If the petioles are relatively long and slender: almost certainly *myrtifolia*.
9. If most of the fruits have distinct apical pegs: more probably *anisoptera* (but some fruits of *myrtifolia* have apical pegs).
10. If there are rather conspicuous conical glabrous buds 2·5–3 mm. long in the axils of the leaves (or fallen leaves): probably *anisoptera*.
11. If the specimen shows considerable variability in some of the above features and a puzzling mixture of characters: probably a hybrid.

The ecology of the two species is mainly similar but *P. myrtifolia* is recorded in our area in evergreen and riverine forests which is not the case for *P. anisoptera*.

2. **Pteleopsis anisoptera** (Welw. ex Laws.) Engl. & Diels in Engl., Mon. Afr. Pflanz. **4**: 4 (1900).—F. White, F.F.N.R.: 287, fig. 50G (1962).—Boughey in Journ. S. Afr. Bot. **30**, 4: 167 (1964).—Liben, F.C.B. Combret.: 5, fig. 1A (1968).—Exell & Garcia, C.F.A. **4**: 83 (1970).—Exell in Kirkia, **7**, 2: 227 (1970).—Wickens, F.T.E.A. Combret.: 72 (1973).—Jacobsen in Kirkia, **9**, 1: 168 (1973).—Faria in Mem. Inst. Invest. Agron. Moçamb. **4**: 116 (1973). TAB. 34 fig. G. Type from Angola.

    Combretum anisopterum Welw. ex Laws. in Oliv., F.T.A. **2**: 429 (1971). Type as above.

    Pteleopsis ritschardii De Wild., Contr. Fl. Katanga, Suppl. 2: 82 (1929). Syntypes from Katanga.

Deciduous tree up to 12(18) m. high or occasionally a shrub; crown conical, rounded or flat-rounded; bark grey to grey-brown with shallow longitudinal fissures; young branchlets at first pubescent, soon glabrescent. Leaves opposite or subopposite or sometimes alternate; lamina up to 7 × 3·5 cm., elliptic or narrowly elliptic or obovate-elliptic, at first densely sericeous-pilose (especially beneath), eventually pilose to pubescent or sometimes almost glabrous, matt or somewhat shiny above, apex not or scarcely acuminate usually apiculate or mucronate, base cuneate to rounded; lateral nerves 4–6 pairs, the lower pairs tending to form rather an acute angle with the midrib and to run parallel to the margin; petiole up to 6 mm. long; buds conical, nearly glabrous, tending to be conspicuous in the axils of the old leaves. Inflorescences of axillary subcapitate racemes up to 3 cm. long. Flowers white, ♀ and ♂ in the same inflorescence,

pedicellate. ♀ flowers: lower and upper receptacles as in *P. myrtifolia*; sepals
deltate; petals 1·5–2 × 0·8–1 mm., obovate, shortly unguiculate, glabrous; stamens
2-seriate, antipetalous ones 4·5 mm. long, antisepalous ones 3·5 mm. long; disk
pilose with a short free margin. ♂ flowers with pedicels 10 mm. long, similar to
the ♀ flowers but ovary not developing (style present but often shorter). Fruit
3–4(5)-winged, 1–1·8 × 0·6–1·2 cm., obovate to elliptic in outline, glabrous, apex
usually emarginate and with a distinct apical peg 0·5–3 mm. long, base decurrent
into the stipe and usually oblique, stipe up to 2 cm. long, slender.

   Zambia. B: Kabompo, 1035 m., fr. vii.1933, *Trapnell* 1220 (K). N: SW. of Lake
Tanganyika, fr. 21.vii.1930, *Hutchinson & Gillett* 3992 (K). W: Mwinilunga, 1370 m.,
fr. ix.1934, *Trapnell* 1599 (BM; K). C: Mt. Makulu, fr. 26.xii.1959, *White* 6013 (FHO;
K). S: Mazabuka Distr., between Pemba and Choma, fr. 22.i.1960, *White* 6330
(FHO; K). **Rhodesia.** N: Urungwe Distr., Sanyati-Cheroti area, fr. 6.i.1958, *Goodier*
527 (COI; K; SRGH). W: Wankie, 730 m., fr. 19.vi.1934, *Eyles* 8036 (BM; K;
SRGH). C: Gatooma, fr. 11.xi.1932, *Eyles* 7265 (K). **Mozambique.** T: Màgoé,
fr. 30.iv.1964, *Wild* 6543 (COI; FHO; LISC; SRGH).
   Also in Zaire, Angola and Tanzania. *Brachystegia* woodland; *Commiphora-Combretum*
thickets as an emergent; *Baikiaea* woodland; Kalahari woodland; *Acacia-Combretum-
Terminalia* tree savanna; graphitic slates.

   The relationship with *P. myrtifolia* is discussed under that species.

3. **Pteleopsis barbosae** Exell in Bol. Soc. Brot., Sér. 2, **42**: 28 (1968); in Kirkia, **7**,
   2: 228 (1970).—Faria in Mem. Inst. Invest. Agron. Moçamb. **4**: 116 (1973). Type:
   Mozambique, Mossuril, between Monapa and Lumbo, *Barbosa* 2466 (LISC,
   holotype).

   Tree up to 10 m. high; branchlets sparsely pilose when young, soon glabrescent.
Leaves opposite or subopposite; lamina up to 10 × 3·8 cm., papyraceous with ±
translucent dots, elliptic, glabrous except for a few reddish hairs on the midrib
beneath, apex blunt and slightly acuminate, base cuneate; lateral nerves 5–8 nerves
usually with 2 conspicuous subcircular glands (? extra-floral nectaries) with a
central pit situated on or near the 4th or 5th pair of nerves (from the base) c.
halfway between the midrib and the margin; petiole 5–7 mm. long, slender.
Inflorescences of terminal and axillary subcapitate racemes 2–4 cm. long; rhachis
pilose and sericeous-pubescent. Flowers white, ♀ and ♂ in the same inflorescence,
pedicellate, sericeous-pubescent. ♀ flowers: lower receptacle c. 3·5 mm. long,
sericeous-pubescent; upper receptacle 1·5 × 3 mm., broadly campanulate, sparsely
sericeous-pubescent, with a slender stalk-like portion at the base c. 0·5 mm. long;
disk pilose; style 3 mm. long. ♂ flowers similar to the ♀ ones but with the ovary
not developed and the style vestigial. Fruit (2)3-winged, 1–1·5 × 0·8–1·2 cm.,
obovate-elliptic in outline, appressed-pubescent, stipe up to 1·2 cm. long, slender,
densely appressed-pubescent.

   Mozambique. N: Nacala, 15 km. from Nacala Nova towards Nacala Velha, c. 10 m.
alt., fr. 3.xii.1963, *Torre & Paiva* 9390 (LISC).
   Unknown elsewhere. Tree or shrub savanna of *Acacia* spp. and in secondary bush;
at low altitudes.

   P. barbosae has rather thin translucent leaves which, when held up to the light and
looked at with a lens, show both a translucent reticulation and numerous pin-point
translucent dots. *P. diptera* Engl. & Diels and *P. hylodendron* Mildbr., both of which have
hairy flowers and fruits, differ from *P. barbosae* in that *P. diptera* has much thicker opaque
leaves and *P. hylodendron* does not show the gland-dots characteristic of *P. barbosae*.

4. TERMINALIA L.

Terminalia L., Syst. Nat. ed. 12, **2**: 674 (err. 638) (1767); Mant. Pl. Alt.: 21
     (1767) *nom. conserv.*

   Trees usually small in our area but sometimes very tall elsewhere, or rarely
shrubs, without scales or microscopic stalked glands. Leaves usually spirally
arranged, often crowded at the ends of the branches, sometimes on short shoots,
rarely opposite, petiolate or subsessile, usually entire but occasionally subcrenate,
often with 2 or more glands at or near the base of the lamina or on the petiole (but
not in our area). Flowers usually ♀ and ♂ in the same inflorescence (rarely all ♀),

usually in axillary spikes with ♂ flowers towards the apex and ♀ ones towards the base, rarely in terminal panicles; ♂ flowers stalked, stalks resembling pedicels but corresponding to the lower receptacle with abortion of the ovary; ♀ flowers sessile. Receptacle divided into a lower part (lower receptacle surrounding the inferior ovary) and an upper part, often scarcely developed, expanding into a shallow cup terminating in the sepals (or calyx-lobes). Petals absent. Stamens usually 10, exserted. Disk intrastaminal. Style free, not expanded at the apex. Ovary completely inferior. Fruit (pseudocarp) very variable in size and shape but usually 2-winged in our area, usually with an at least partially sclerenchymatous endocarp. Cotyledons (when known) spirally convolute.

Throughout the tropics and extending to the subtropics.

The flowers in *Terminalia* are remarkably uniform throughout the genus and rarely provide characters taxonomically useful.

This account is indebted to the work " A Revision of the African species of *Terminalia* " by M. E. Griffiths (in Journ. Linn. Soc., Bot. **55**: 818–907 (1959)) which has been largely drawn upon for descriptions, etc.

In addition to the species here described *T. catappa* L., *T. arjuna* (Roxb.) Wight & Arn. and *T. bellirica* (Gaertn.) Roxb. are sometimes grown in parks and gardens.

Key to the sections

Fruit ellipsoid-ovoid, 10–12(18) × 5 mm., not winged and only obscurely ridged (or very
    narrowly winged in one Madagascan species) - - - -   1. *Fatrea*
Fruit 2-winged or (4)5(6)-winged, at least 15 mm. long:
  Fruit 2-winged:
    Leaves in fascicles on short spur shoots; branches and branchlets often bearing spines
                                         2. *Abbreviatae*
    Leaves not in fascicles terminating short spur shoots though sometimes crowded
      towards the ends of lateral shoots:
      Bark on young branchlets purple-black, peeling off in cylindric or hemicylindric
        papery flakes leaving a reddish-brown or brown (later grey-brown) newly
        exposed surface - - - - - - - - - 3. *Psidioides*
      Bark on young branchlets not peeling off as above - - - 4. *Platycarpae*
  Fruit (4)5(6)-winged - - - - - - - - - 5. *Pteleopsoides*

1. Sect. FATREA (Juss.) Exell in Kirkia, **7**, 2: 229 (1970).

Fatrea Juss. in Ann. Mus. Nation. Hist. Nat. Paris **5**: 223 (1804).

Sect. *Myrobalanus* sensu Engl. & Diels in Engl., Mon. Afr. Pflanz. **4**: 9 (1900).

Leaves usually small, spirally arranged or alternate, often on spur shoots. Flowers all ♀. Fruit ellipsoid-ovoid, not or obscurely ridged. Small trees or shrubs.

1. **Terminalia boivinii** Tul. in Ann. Sci. Nat., Sér. 4, **6**: 95 (1856).—Capuron, Combrét. Arbust. ou Arboresc. Madagasc.: 83 (1967).—Exell in Kirkia, **7**, 2: 230 (1970).—Wickens, F.T.E.A. Combret.: 78 (1973).—Faria in Mem. Inst. Invest. Agron. Moçamb. **4**: 117 (1973). Type from Madagascar.

    Terminalia fatraea sensu Engl., Pflanzenw. Ost-Afr. **C**: 295 (1895).—Engl. & Diels in Engl., Mon. Afr. Pflanz. **4**: 9, t. 15 fig. D (1900).—Brenan, T.T.C.L.: 143 (1949).—Griffiths in Journ. Linn. Soc., Bot. **55**: 825, fig. 1 (1959).—Dale & Greenway, Kenya Trees and Shrubs: 152 (1961).

Shrub or small tree up to 12 m. high; wood very hard; bark smooth, greyish-brown; long shoots straight; lateral shoots terminating in a spur-shoot and with 4–6 lateral spurs 2–5 mm. long; further spur-shoots borne directly on the long shoots. Leaves alternate, petiolate; lamina pale-green and shiny above, somewhat brownish (when dried) beneath, 2·5–6 × 1–2·5 cm., chartaceous, narrowly obovate-elliptic to obovate, glabrous when mature; lateral nerves 6–7 pairs; petiole 1–5 mm. long. Inflorescences of lateral spikes 1·5–3·5 cm. long, sometimes subcapitate, often on spur-shoots; peduncle 1·5–1·8 cm. long, glabrous. Flowers yellowish, nearly sessile, glabrous or almost so. Sepals obtuse. Stamens 5 mm. long. Disk with 5–7 irregular lobes, pilose. Fruit pale-yellowish-green, 10–12 × 5 mm., ellipsoid-ovoid, acute at both ends, glabrous, obscurely ridged, very shortly stipitate.

Mozambique. N: R. Messalo, fl. *Allen* 96 (K). Z: Pebane, c. 20 m., fr. 8.iii.1966, *Torre & Correia* 15069 (LISC). SS: Bazaruto I., *Mogg* 28939 (K; SRGH).

Kenya, Tanzania, Zanzibar, northern Mozambique and Madagascar. Coastal tree or shrub savanna, coastal dunes and in mangrove (on Mafia I.); at low altitudes.

As Capuron (loc. cit.) has pointed out, this species, though very close to *T. fatraea* (Poir.) DC., differs in having smaller fruits without longitudinal ridges and appears to be distinct.

Although the fruits of *T. fatraea* and *T. boivinii* are typical " myrobalans " these species, small trees or shrubs with small leaves and spur-shoots are very different from the Asian species of Sect. *Myrobalanus* (Gaertn.) DC. which are large (often very large) trees with large leaves (often opposite or subopposite) not borne on spur-shoots and inflorescences often paniculate.

2. Sect. ABBREVIATAE Exell in Bol. Soc. Brot., Sér. 2, **42**: 30 (1968); in Kirkia, **7**, 2: 230 (1970).—Wickens, F.T.E.A. Combret.: 75 (1973).

Sect. *Platycarpae* Engl. & Diels in Engl., Mon. Afr. Pflanz. **4**: 17 (1900) pro parte.

Leaves in fascicles on short spur shoots; branchlets often bearing spines. Fruit 2-winged.

Lateral nerves not or scarcely prominent on the upper surface of the leaves:
Fruit 4–6·5 cm. long; spines occasionally present but only on the long shoots; long shoots usually straight or almost so - - - - - - 2. *prunioides*
Fruit 1·5–2·5 cm. long; spines present at the base of most spur shoots; long shoots usually zig-zag - - - - - - - - - - 3. *randii*
Lateral nerves prominent on the upper surface of the leaves; leaves pale-yellowish-green beneath; fruit (2)2·5–3·5 cm. long; long shoots usually zig-zag - 4. *stuhlmannii*

2. **Terminalia prunioides** Laws. in Oliv., F.T.A. **2**: 415 (1871).—Engl., Pflanzenw. Ost-Afr. **C**: 294 (1895).—Engl. & Diels in Engl., Mon. Afr. Pflanz. **4**: 22, t. 11 fig. A (1900).—Schinz, Pl. Menyharth.: 431 (1906).—Sim, For. Fl. Port. E. Afr.: 64 (1909).—Eyles in Trans. Roy. Soc. S. Afr. **5**, 4: 429 (1916).—Burtt Davy, F.P.F.T. **1**: 250 (1926).—O. B. Mill., B.C.L.: 44 (1948).—Wild, Guide Fl. Vict. Falls: 146 (1953).—Garcia in Contr. Conhec. Fl. Moçamb. **2**: 149 (1954).—Palgrave, Trees of Centr. Afr.: 144 cum tab. et phot. (1956).—Griffiths in Journ. Linn. Soc., Bot. **55**: 829, fig. 3 (1959).—Dale & Greenway, Kenya Trees and Shrubs: 153 (1961).—Palmer & Pitman, Trees S. Afr.: 251 (1961).—Letty, Wild Fl. Transv.: 232, t. 115 fig. 3 (1962).—F. White, F.F.N.R.: 291 (1962).—Boughey in Journ. S. Afr. Bot. **30**, 4: 167 (1964).—Roessler, Prodr. Fl. SW. Afr. **99**: 12 (1966).—Exell in Kirkia, **7**, 2: 231 (1970).—Exell & Garcia, C.F.A. **4**: 87 (1970).—Wickens, F.T.E.A. Combret.: 80, fig. 12(3) (1973).—Faria in Mem. Inst. Invest. Agron. Moçamb. **4**: 117 (1973). TAB. **40** fig. E. Syntypes: Mozambique, Tete, *Kirk* (K).

Terminalia porphyrocarpa Schinz in Verh. Bot. Ver. Brand. **30**: 242 (1888) *nom. illegit.* non F. Muell. ex Benth. (1864). Type from SW. Africa.

Terminalia rautanenii Schinz, tom. cit.: 243 (1888).—Engl. & Diels in Engl., Mon. Afr. Pflanz. **4**: 22, t. 10 fig. B (1900). Type from SW. Africa.

Terminalia holstii Engl. in Abh. Preuss. Akad. Wiss. **1894**: 34 (1894); Pflanzenw. Ost-Afr. **C**: 294 (1895).—Engl. & Diels in Engl., tom. cit.: 23, t. 11 fig. B (1900).—Brenan, T.T.C.L.: 145 (1949). Type from Tanzania.

Terminalia petersii Engl., Pflanzenw. Ost-Afr. **C**: 294 (1895). Type: Mozambique, Tete, *Peters* (B†).

Terminalia benguellensis Welw. ex Hiern, Cat. Afr. Pl. Welw. **1**: 339 (1898). Type from Angola.

Terminalia benguellensis var. *ovalis*, tom. cit.: 340 (1898). Type from Angola.

Small deciduous tree 7–15 m. high or shrub; crown narrow-ovoid to rounded; bark pale-grey to grey-black, deeply fissured; long shoots usually straight; lateral shoots ending in spur shoots, rarely with additional lateral spur shoots; spur shoots rarely borne directly on the long shoots; spines occasionally present but only on the long shoots. Leaves in fascicles on spurs, petiolate; lamina up to 7·5 × 3 cm., chartaceous, broadly obovate to elliptic-obovate, usually densely pubescent when young, ± glabrescent, apex rounded or emarginate or mucronate, base obtuse to cuneate; lateral nerves 3–5 pairs, ± impressed above; petiole 0·5–1·5(2·5) cm. long. Inflorescences of lateral spikes 5–8 cm. long; peduncle 1·7–2 cm. long, densely to sparsely pubescent. Flowers cream or white, glabrous or nearly so. Sepals triangular. Stamens 3–4 mm. long; anthers 0·5 mm. long. Disk pilose. Fruit purplish-brown or red, 4–6·5 × 2–3(5) cm., elliptic-oblong in outline, apex

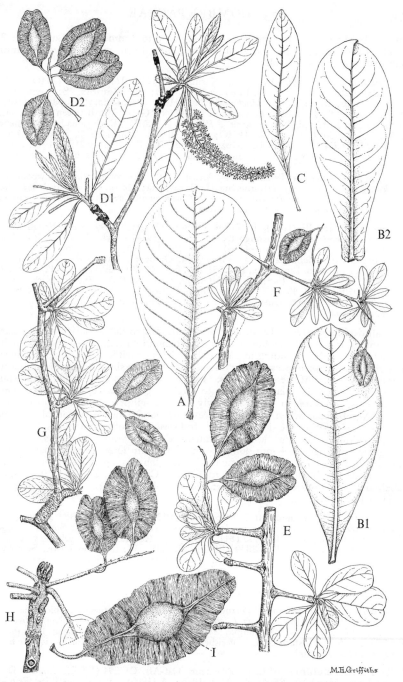

Tab. 40. A.—TERMINALIA ERICI-ROSENII, leaf ($\times \frac{1}{2}$). B.—TERMINALIA BRACHYSTEMMA. B1, leaf ($\times \frac{1}{2}$); B2, leaf ($\times \frac{1}{2}$). C.—TERMINALIA KAISERANA, leaf ($\times \frac{1}{2}$). D.— TERMINALIA SERICEA. D1, flowering branchlet ($\times \frac{1}{2}$); D2, fruits ($\times \frac{1}{2}$). E.— TERMINALIA PRUNIOIDES, fruiting branchlet ($\times \frac{1}{2}$). F.—TERMINALIA RANDII, fruiting branchlet ($\times \frac{1}{2}$). G.—TERMINALIA STUHLMANNII, fruiting branchlet ($\times \frac{1}{2}$). H.— TERMINALIA STENOSTACHYA, fruiting branchlet ($\times \frac{1}{2}$). I.—TERMINALIA MOLLIS, fruit ($\times \frac{1}{2}$). From F.F.N.R.

obtuse, deeply emarginate or mucronate: stipe up to 7 mm. long. Cotyledons 2, 10–11 × 18–20 mm., transversely elliptic to subreniform with petioles 3–4 mm. long, borne above soil level.

Botswana. N: Chobe Distr., near Kasinka, fr. 1938, *Miller* B/190 (FHO). SE: Bangwaketsi-Bakwena, fr. v.1930, *Van Son* 28818 (BM; K). **Zambia.** S: Livingstone Distr., Sala R., fr. 12.ii.1933, *Martin* 540/33 (FHO). **Rhodesia.** N: Mtoko Distr., Mkota Reserve, fr. 18.iv.1951, *Lovemore* 21 (FHO; SRGH). W: between Wankie and Victoria Falls, st. 13.viii.1946, *Gouveia & Pedro* 1645 (LMJ). E: Nyamadzi Settlement, fl. & fr. 21.x.1950, *Crook* M 206 (FHO; SRGH). S: Beitbridge, fl. & fr. 10.i.1961, *Leach* 10667 (COI; K; SRGH). **Mozambique.** T: Boroma, fr. 21.viii.1947, *Simão* 1494 (LM). SS: Alto Limpopo, Chicualacuala, fr. 17.iv.1955, *Myre* 2071 (LM). LM: Magude, fr. 28.i.1948, *Torre* 7235 (LISC).
   Also in Angola, Kenya, Tanzania, SW. Africa and the Transvaal. *Baikiaea* woodland; *Brachystegia* savanna woodland; mopane shrub savanna; *Acacia* tree or shrub savanna; *Acacia-Combretum-Terminalia* tree savanna; *Commiphora-Combretum* thickets; coastal savanna; most frequent in hotter drier areas.

3. **Terminalia randii** Bak. f. in Journ. of Bot. **37**: 435 (1899).—R.E.Fr. in Wiss. Ergebn. Schwed. Rhod.-Kongo Exped. **1**: 175 (1914).—Eyles in Trans. Roy. Soc. S. Afr. **5**, 4: 430 (1916).—O. B. Mill., B.C.L.: 45 (1948).—Pole Evans, Bot. Surv. S. Afr. Mem. **21**: 31 (1948).—Wild, Guide Fl. Vict. Falls: 146 (1953).— Griffiths in Journ. Linn. Soc., Bot. **55**: 839, fig. 7 (1959).—F. White, F.F.N.R.: 291 (1962).—Boughey in Journ. S. Afr. Bot. **30**, 4: 167 (1964).—Exell in Kirkia, **7**, 2: 232 (1970). TAB. **40** fig. F. Type: Rhodesia, Bulawayo, *Band* 325 (BM, holotype).
   Terminalia spinosa sensu Eyles, loc. cit.? This must be either *T. randii* or *T. stuhlmannii.*

Small deciduous tree 3–12 m. high or shrub; crown flat to flat-rounded; bark grey-brown or blackish, shallow-fissured in diamond pattern; long shoots usually zig-zag; lateral shoots ending in a spur shoot with or without lateral spur shoots; spur shoots also borne occasionally directly on the long shoots; spines frequently present at the base of the spur shoots. Leaves in fascicles at the ends of the spur shoots, petiolate; lamina pale-green to slightly glaucous, 1–2·5 × 0·7–1·2 cm., chartaceous, obovate to very narrowly obovate-elliptic, sparsely to densely sericeous-pubescent when young, glabrescent; lateral nerves inconspicuous on both surfaces; petiole 0·6–1·4 cm. long. Inflorescences of lateral spikes 3–3·5 cm. long; peduncle 1·5 cm. long, glabrous. Flowers white, glabrous. Sepals triangular. Stamens 3–4 mm. long; anthers 0·4 mm. long. Disk pilose. Fruit purplish, 1·5–2·5 × 1–1·2 cm., elliptic to oblong, glabrous, apex deeply emarginate, base cuneate to subtruncate, stipe 5–7 mm. long. Cotyledons 2, 7–10 × 10–15 mm., subcircular to obovate or subreniform or flabelliform, borne above soil level; petioles 1–2 mm. long.

Botswana. N: Kazungula, fr. iv.1936, *Miller* B/127 (BM; FHO). SE: Bamangwato Reserve, *Pole Evans* 3265 (13) (K). **Zambia.** S: Victoria Falls, near Songwe Gorge, 880 m., fr. 6.v.1963, *Bainbridge* 777 (FHO; SRGH). **Rhodesia.** N: Sebungwe Distr., Kana R., fr. 8.viii.1952, *Wild* 3849 (FHO; SRGH). W: Matobo Distr., Semokwe Reserve, fl. 1957, *McKay* in GHS 2360 (COI; SRGH). C: Hartley Distr., Poole Farm, fr. 25.iv.1945, *Hornby* 2464 (FHO; SRGH).
   Not known elsewhere. *Brachystegia* woodland; *Baikiaea* woodland; mopane savanna woodland or savanna; basalt outcrops; lower rainfall areas.

Allen 63 (K) said to be from " Portuguese East Africa " but otherwise unlocalized was identified by Griffiths (loc. cit.) as *T. randii* but from the distribution given above the occurrence of this species in Mozambique would be surprising since it has not been found in Malawi or in the eastern regions of either Zambia or Rhodesia. Nor has it ever been rediscovered in Mozambique.

4. **Terminalia stuhlmannii** Engl., Pflanzenw. Ost-Afr. C: 294 (1895).—Engl. & Diels in Engl., Mon. Afr. Pflanz. **4**: 22, t. 10 fig. A (1900).—R.E.Fr. in Wiss. Ergebn. Schwed. Rhod.-Kongo Exped. **1**: 175 (1914).—Eyles in Trans. Roy. Soc. S. Afr. **5**, 4: 430 (1916).—O. B. Mill., B.C.L.: 45 (1948).—Brenan, T.T.C.L.: 146 (1949).— Wild, Guide Fl. Vict. Falls: 146 (1953).—Garcia in Contr. Conhec. Fl. Moçamb. **2**: 149 (1954).—Griffiths in Journ. Linn. Soc., Bot. **55**: 840, fig. 8 (1959).—F. White, F.F.N.R.: 291 (1962).—Boughey in Journ. S. Afr. Bot. **30**, 4: 167 (1964).—Exell in Kirkia, **7**, 2: 232 (1970).—Wickens, F.T.E.A. Combret.: 82 (1973).—Faria in Mem. Inst. Invest. Agron. Moçamb. **4**: 120 (1973). TAB. **40** fig. G. Type from Tanzania.

Small deciduous tree up to 12 m. high or shrub, with flattened crown and horizontal branches; bark brownish-grey or whitish-grey, fissured, scaly or flaky; long shoots zig-zag, glabrous; spur shoots terminal or rarely lateral, on lateral shoots and occasionally on long shoots; spines sometimes present at the base of the spur shoots. Leaves clustered at the ends of spur shoots, subsessile or shortly petiolate; lamina glaucous-green above and yellowish-green beneath, 1·5–6 × 1–2(3) cm., coriaceous, obovate-elliptic to narrowly obovate-elliptic, apex obtuse to rounded, base cuneate, nearly glabrous; lateral nerves 4–5 pairs; petiole up to 4 mm. long, sometimes sericeous. Inflorescences up to 8 cm. long; rhachis and peduncle glabrous. Flowers whitish, glabrous. Sepals triangular. ♀ flowers with lower receptacle 5 mm. long. ♂ flowers with pseudopedicel 2·5–3 mm. long; filaments 4 mm. long; anthers 0·7 mm. long. Disk pilose. Fruit purple-pink or brown (2)2·3(4) × 1·6–1·8(3) cm., elliptic, glabrous. Cotyledons 2, 10–12 × 18–20 mm., half-moon-shaped to subreniform; petioles 2–4 mm. long, borne above soil level.

Caprivi Strip. Mpilila I., c. 915 m., fl. 13.i.1959, *Killick & Leistner* 3366 (BM; PRE; SRGH). **Botswana.** N: Kazungula, fr. 1936, *Miller* B/118 (FHO). **Zambia.** N: Munyamadzi R., fr. 1.x.1933, *Michelmore* 617 (K). S: near Livingstone, fl. & fr. 7.i.1953, *Angus* 1118 (COI; FHO; K). **Rhodesia.** N: Sebungwe Distr., Kariangwe, fl. 19.xii.1951, *Lovemore* 3540 (FHO; SRGH). W: Victoria Falls, fl. x.1930, *Pardy* in GHS 4757 (BM; FHO; SRGH). C: Gatooma, 1160 m., st. iv.1927, *Eyles* 4922 (FHO; K; SRGH). **Mozambique.** T: between Tete and Chicoa, fr. 25.vi.1949, *Andrada* 1640 (COI; LISC).
Also in Tanzania. *Acacia* savanna; *Brachystegia* woodland; mopane savanna or savanna woodland; margins of *Baikiaea* woodland; *Pterocarpus brenanii* tree savanna; hard-pan *Acacia* savanna (Tanzania); *Acacia-Combretum-Terminalia* open woodland; hotter drier areas.

Produces resin.

3. Sect. PSIDIOIDES Exell in Bol. Soc. Brot., Sér. 2, **42**: 30 (1968); in Kirkia, **7**, 2: 233 (1970).—Wickens, F.T.E.A. Combret.: 76 (1973).

Sect. *Platycarpae* Engl. & Diels in Engl., Mon. Afr. Pflanz. **4**: 17 (1900) pro parte.

Bark on young branchlets peeling off in cylindric or hemi-cylindric flakes leaving a reddish-brown or brown (later grey-brown) newly exposed surface. Fruits 2-winged.

The name of the section was chosen to indicate the close similarity to *Combretum psidioides* Welw. in the way the bark peels off.

Leaves tomentose or tomentellous, often sericeous (glabrescent when very old) or coarsely pilose:
Indumentum of leaves usually silvery-grey-sericeous:
Leaves up to 2–2·5 times as long as broad; lateral nerves subprominent
                                    5. *erici-rosenii*
Leaves at least 3 times as long as broad; lateral nerves rather inconspicuous
                                    6. *sericea**
Indumentum of leaves usually yellowish or brownish, tomentose or tomentellous or coarsely pilose; lateral nerves inconspicuous:
Leaves rather coarsely brownish-tomentose or -pilose - - - 7. *trichopoda**
Leaves yellowish- or brownish-tomentellous - - - - 8. *griffithsiana*
Leaves glabrous or pubescent or sparsely appressed-pilose:
Leaves mostly sessile or subsessile - - - - - - 9. *brachystemma*
Leaves petiolate:
Leaves glabrous or nearly so; petioles up to 3 cm. long, rather slender
                                  10. *kaiserana*
Leaves pubescent (occasionally) - - - - - - - 7. *trichopoda**

5. **Terminalia erici-rosenii** R.E.Fr. in Wiss. Ergebn. Schwed. Rhod.-Kongo-Exped. **1**: 173, t. 9 (1914).—Griffiths in Journ. Linn. Soc., Bot. **55**: 845, fig. 10 (1959).—F. White, F.F.N.R.: 288, fig. 52A (1962).—Exell in Kirkia, **7**, 2: 234 (1970). TAB. **40** fig. A. Type: Zambia, between Lake Bangweulu and Lake Tanganyika, *Fries* 1155 (UPS, holotype).

* Hybrids between these two species show intermediate characters.

Small semi-deciduous tree up to 12 m. high; crown rounded; bark grey-brown to grey-black, furrowed; branchlets stout, greyish-black, velutinous, with bark peeling off to reveal a reddish-brown striated newly exposed surface which later becomes greyish-brown. Leaves spirally arranged, petiolate; lamina up to 25 × 13 cm., coriaceous, broadly obovate to obovate-elliptic, silvery-(or cupreous-in dried specimens)sericeous-tomentose or -tomentellous on both surfaces, eventually glabrescent, somewhat glaucous, apex rounded to subtruncate or occasionally emarginate and usually mucronate, base cuneate; lateral nerves 9–18 pairs, subprominent; petiole 1–2 cm. long, stout. Inflorescences of lateral spikes 4–9 cm. long, usually in the axils of fallen leaves; rhachis tomentose; peduncle 2 cm. long, tomentose. Flowers creamy-white. Lower receptacle tomentose; upper receptacle sericeous-pilose. Sepals triangular-acuminate. Stamens 5 mm. long; anthers 0·5 mm. long. Disk pilose. Fruit 3·5–4 × 1·7–2·2 cm., elliptic-oblong, velutinous, apex emarginate, base subacute to truncate, stipe 5–7 mm. long.

Zambia. N: Chimbwi, Kawambwa, fl. 14.ix.1963, *Mutimushi* 386 (FHO); 45 km. E. of Mporokoso, 1435 m., fr. 13.iv.1961, *Vesey-FitzGerald* 3117 (BM; FHO; LISC; SRGH). W: Mufulira, fr. 22.vi.1957, *Fanshawe* 3333 (K; LISC).
Known only from Zambia. Shrub savanna at edges of dambos; laterite pans; chipya scrub; lake-basin chipya.

6. **Terminalia sericea** Burch. ex DC., Prodr. **3**: 13 (1828).—Sond. in Harv. & Sond., F.C. **2**: 508 (1862).—Laws. in Oliv., F.T.A. **2**: 416 (1871).—Engl. & Diels in Engl., Mon. Afr. Pflanz. **4**: 20, t. 13 fig. C (1900).—Monro in Proc. & Trans. Rhod. Sci. Ass. **8**, 2: 81 (1908).—Sim, For. Fl. Port. E. Afı.: 64, t. 64 A (1909).—Bak. f. in Journ. Linn. Soc., Bot. **40**: 70 (1911).—Eyles in Trans. Roy. Soc. S. Afr. **5**, 4: 30 (1916).—Steedman in Proc. & Trans. Rhod. Sci. Ass. **24**: 17 (1925): Some trees, etc. S. Rhod.: 56 (1933).—Burtt Davy, F.P.F.T. **1**: 249, fig. 37 (1926).—Henkel in Proc. & Trans. Rhod. Sci. Ass. **30**: 9, 14, 16, 19 (1931).—Burtt Davy & Hoyle, N.C.L.: 39 (1936).—O. B. Mill., B.C.L.: 45 (1948).—Gomes e Sousa, Dendrol. Mozamb. **1**: t. 239 (1951).—Suesseng. & Merxm. in Proc. & Trans. Rhod. Sci. Ass. **43**: 28 (1951).—Codd, Trees and Shrubs Kruger Nat. Park: 135, figs. 126e & f (1951).—Wild, Guide Fl. Vict. Falls: 146 (1953).—Garcia in Contr. Conhec. Fl. Moçamb. **2**: 147 (1954).—Palgrave, Trees of Centr. Afr.: 147 cum phot. et tab. (1956).—Williamson, Useful Pl. Nyasal.: 117 (1956).—Mogg in Macnae & Kalk, Nat. Hist. Inhaca I. Moçamb.: 15 (1958).—Griffiths in Journ. Linn. Soc., Bot. **55**: 853, fig. 14 (1959).—Rattray in Kirkia, **2**: 68 et seq. (1961).—Palmer & Pitman, Trees S. Afr.: 250 cum fig. (1961).—F. White, F.F.N.R.: 290, fig. 52E (1962).—Chapman, Veg. Mlanje Mountains: 37 (1962).—Fanshawe & Savory in Kirkia, **4**: 190 (1964).—Boughey in Journ. S. Afr. Bot. **30**, 4: 167 (1964).—Roessler, Prodr. Fl. SW. Afr. **99**: 12 (1966).—Liben, F.C.B. Combret.: 94 (1968).—Exell & Garcia, C.F.A. **4**: 89 (1970).—Exell in Bol. Soc. Brot., Sér. 2, **42**: 234 (1970).—Wickens, F.T.E.A. Combret.: 85, fig. 12(12) (1973).—Faria in Mem. Inst. Invest. Agron. Moçamb. **4**: 120 (1973). Non *T. sericea* Cambess. (1829). TAB. **40** fig. D. Type from S. Africa.
Terminalia angolensis O. Hoffm. in Linnaea, **43**: 131 (1881). Type from Angola.
Terminalia angolensis Welw. ex Ficalho in Bol. Soc. Geogr. Lisb. **2**: 708 (1882); Pl. Ut. Afr. Port.: 182 (1884) *nom. illegit.* non O. Hoffm. (1881). Type from Angola.
Terminalia fischeri Engl., Pflanzenw. Ost-Afr. **C**: 294 (1895). Type from Tanzania.
Terminalia nyassensis Engl., loc. cit.—Burtt Davy & Hoyle, N.C.L.: 39 (1936). Type: Malawi, Shire Highlands, *Buchanan* 189 (B†; BM; K).
Terminalia brosigiana Engl. & Diels in Notizbl. Bot. Gart. Berl. **2**: 191 (1898). Syntypes from Tanzania.
Terminalia sericea var. *angolensis* Hiern, Cat. Afr. Pl. Welw. **1**: 338 (1898).—Eyles in Trans. Roy. Soc. S. Afr. **5**, 4: 430 (1916). Syntypes from Angola.
Terminalia sericea var. *huillensis* Hiern, tom. cit.: 339 (1898). Syntypes from Angola.
Terminalia velutina sensu Eyles in Trans. Roy. Soc. S. Afr. **5**, 4: 430 (1916) non *T. velutina* Rolfe.
Terminalia bubu De Wild. & Ledoux in De Wild., Contr. Fl. Katanga, Suppl. **2**: 84 (1929). Syntypes from Zaire.

Small deciduous bushy tree 3–12(16) m. high or shrub; crown flat or rounded; bark grey-brown or pale-brown to grey-black, longitudinally fissured, fresh baık pale-rusty-red; branchlets with purplish-black bark peeling to reveal a light-brown newly exposed surface; young shoots sericeous-tomentose. Leaves spirally arranged, petiolate; lamina 5·5–12·5 × 1·5–4·5 cm., chartaceous, narrowly obovate-elliptic to narrowly elliptic, densely silvery-sericeous-tomentose (somewhat glabrescent when old), apex acute to rounded, base cuneate; lateral nerves usually inconspicuous beneath; petiole 2–10 mm. long. Inflorescences of lateral spikes 5–7·5 cm. long; peduncle 2·5–3 cm. long, densely sericeous. Flowers greenish-white. Lower receptacle sericeous-tomentose; upper receptacle sericeous but less densely so. Sepals triangular-acuminate. Stamens 4 mm. long; anthers 0·5 mm. long. Disk pilose. Fruit pinkish or purplish-brown, 3–4 × 1·7–2·5 cm.* broadly elliptic, apex obtuse to rounded and usually emarginate, base obtuse to subtruncate, finely tomentose, stipe 5–7 mm. long. Cotyledons 2, c. 2 cm. in diam., irregularly subcircular with petioles 1·5–2 cm. long, arising below soil level.

Caprivi Strip. Katima Mulilo, fr. 24.xii.1958, *Killick & Leistner* 3068 (K; PRE). **Botswana.** N: Kazungula, fl. x.1935, *Miller* B/60 (FHO). SE: Mochudi, fr. 1914, *Rogers* 6616 (K). **Zambia.** B: Sesheke Distr., fr. 24.xii.1952, *Angus* 1020 (FHO; K). N: Mkupa, fl. 7.x.1949, *Bullock* 1157 (K). C: Shimebala, fr. 30.iv.1961, *Angus* 2859 (SRGH). E: between Luangwa R. and Fort Jameson, st. 21.viii.1946, *Gouveia & Pedro* 1780 (LMJ). S: Namwala, fr. 19.iv.1963, *Van Rensburg* 2063 (K). **Rhodesia.** N: Mtoko Distr., Suskwe R., fl. 9.xi.1953, *Phelps* 74 (FHO; SRGH). W: Matopos, near World's View, fr. 13.iv.1955, *E.M. & W.* 1489 (BM; LISC; SRGH). C: 6·4 km. from Umvuma on Gwelo road, fr. 7.ii.1951, *Mullin* 3/51 (FHO; SRGH). E: Umtali Commonage, Palmerston, 1095 m., fr. 25.xi.1954, *Chase* 5342 (BM; COI; FHO; LISC; SRGH). S: Chibi Distr., near Lundi R., fl. 5.xii.1961, *Leach* 11313 (LISC; SRGH). **Malawi.** C: between Fort Manning and Lilongwe, fr. 22.viii.1946, *Gouveia & Pedro* 1786 (LMJ). S: Domasi, fr. 10.iii.1955, *E.M. & W.* 793 (BM; LISC; SRGH). **Mozambique.** N: Mutuali, 550 m., st. 6.vi.1947, *Pedro* 3443 (LMJ). Z: between Pebane and Mualama, fr. 5.x.1949, *Barbosa & Carvalho* 4301 (K; LM). T: between Tete and Cassamba, fr. 5.vii.1949, *Andrada* 1692 (COI; LISC). MS: Mazamba (Cheringoma), fr. 24.vii.1946, *Simão* 807 (LM). SS: Vilanculos, near Mapinhame, *Pedro* 368 (LMJ). LM: Marracuene, fr. 4.ix.1949, *Pedro* 4 (LMJ).

Also from Angola, Zaire, Tanzania, SW. Africa and S. Africa. *Combretum-Terminalia* savanna, often dominant or co-dominant; *Acacia-Combretum-Terminalia* tree savanna; *Brachystegia* savanna woodland; *Baikiaea* woodland; mopane savanna; lake basin chipya; deciduous thickets; margins of dambos; sandy soils; very widespread from low to medium altitudes.

The large number of references in the literature cited above are a measure of the great ecological importance of this species.

Specimens from the more northern parts of the area of distribution of *T. sericea* (including most of the specimens from Malawi) tend to have larger leaves, a more tawny (when dried) coarser indumentum and larger fruits. It is suggested that at least some of these may be *T. sericea* × *trichopoda* hybrids but the effect could sometimes be due to genes from other species or even to higher rainfall. The leaf and fruit characters are not always linked. Specimens which seem to be about halfway between *T. sericea* and *T. trichopoda* are cited as possible hybrids (below). A series of specimens collected by Meara at Plumtree in the western division of Rhodesia shows very clearly the transition between the two species.

6a. ? **Terminalia sericea × trichopoda** Exell in Kirkia, **7**, 2: 236 (1970).—Jacobsen in Kirkia, **9**, 1: 168 (1973).

Leaves with a darker indumentum than in *T. sericea* and with more prominent lateral nerves; fruit smaller than in *T. trichopoda*.

Zambia. N: between Kabwe and Kalangila, fr. *Bredo* 383 (FHO). E: 32 km. E. of Cachalolo, fr. 24.iii.1955, *E.M. & W.* 1168 (BM; LISC). S: Mazabuka Distr., fr. 17.iv.1929, *Stevenson* 4 (FHO). **Rhodesia.** N: Urungwe Distr., Katchera R., fr. 18.iv.1955, *Shiff* 7 (FHO; K; SRGH). W: near World's View, fr. 14.iv.1955, *E.M. & W.* 1499 (BM; LISC; SRGH). C: Salisbury Municipal Quarries, fl. 28.xi.1937, *Kelly Edwards* E 43/37 (FHO; SRGH). E: Umtali, Park Lands, fr. 12.i.1948, *Chase* 1180 (BM; LISC; SRGH). S: Ndanga Distr., Umtilikwe R., fr. 26.i.1949, *Wild* 2763 (SRGH). **Malawi.** S: Nyambi, fr. 26.iv.1955, *Jackson* 1644 (FHO).

Not known with certainty elsewhere though probably in Tanzania. There is no evidence that the ecology differs from that of *T. sericea*.

7. **Terminalia trichopoda** Diels in Engl., Bot. Jahrb. **39**: 514 (1907).—Eyles in Trans. Roy. Soc. S. Afr. **5**, 4: 430 (1916).—Palgrave, Trees of Centr. Afr.: 155 cum phot. et tab. (1956).—Griffiths in Journ. Linn. Soc., Bot. **55**: 858 (1959).—F. White, F.F.N.R.: 290 (1962).—Boughey in Journ. S. Afr. Bot. **30**, 4: 167 (1964).—Exell

* Somewhat larger dimensions given by Griffiths (loc. cit.) are probably from specimens I consider to be hybrids with *T. trichopoda*.

in Kirkia, **7**, 2: 236 (1970).—Wickens, F.T.E.A. Combret.: 85 (1973).—Faria in Mem. Inst. Invest. Agron. Moçamb. **4**: 127 (1973). Syntypes: Rhodesia, Matopos, *Engler* 2847a (B†) and Umtali, *Engler* 3142 (B†).

Small deciduous tree up to 9 m. high (up to 21 m. in Tanzania *fide* Wickens); bark dark-grey, deeply fissured; branchlets as in *T. sericea*. Leaves spirally arranged, petiolate; lamina up to 18 × 7 cm., obovate or narrowly obovate or narrowly obovate-elliptic, usually rather coarsely brown-tomentose or brown-pilose (rarely pubescent or ± sericeous in specimens tending towards *T. sericea*), apex usually acuminate, base cuneate; lateral nerves 10–14 pairs, fairly conspicuous beneath as is also the reticulation, petiole 15–20 mm. long. Inflorescences of axillary spikes 10–14 cm. long; peduncle 4–5 cm. long, brown-tomentose. Lower receptacle densely tomentose, not sericeous; upper receptacle less densely tomentose. Sepals triangular-acuminate. Stamens 4–5 mm. long; anthers 0·7 mm. long. Style 3·5 mm. long. Fruit purplish, up to 7·5 × 4·2 cm., broadly to narrowly oblong-elliptic, stipe 15–20 mm. long.

Botswana. N: Tati Distr., near Ramaquebane, fr. 15.viii.1963, *Yalala* 203 (SRGH). **Zambia.** N: Chilongowelo, 1465 m., fr. 8.ii.1952, *Richards* 751 (K). C: 16 km. S. of Lusaka, fr. 6.iv.1955, *E.M. & W.* 1410 (BM; LISC; SRGH). S: Kazungula, fr. 21.vii.1927, *Bourne* 21 (FHO); Bombwe, fl. 5.x.1932, *Martin* 300 (FHO). **Rhodesia.** N: Urungwe Distr., Cheoka Reserve, fr. 13.v.1952, *Whellan* 667 (FHO; SRGH). W: Matopos, near World's View, fr. 13.iv.1955, *E.M. & W.* 1493 (BM; LISC; SRGH). **Malawi.** C: Dedza Distr., Nganja Village, 610 m., 14.xi.1953, *Adlard* 6 (FHO). S: Fort Johnston, Kalombo, fr. 10.vi.1955, *Jackson* 1691 (FHO; K). **Mozambique.** T: between Changara and Mtoko, *Wild* 2669 (FHO; K). MS: Chimoio, Gondola, st. 7.x.1945, *Simão* 589 (LISC).

Also from Tanzania. Lake basin chipya (Zambia); scarp and plateau *Brachystegia* woodland; *Acacia-Combretum-Terminalia* tree savanna; dambo margins.

Differs from *T. sericea* in having on an average larger relatively broader leaves with more prominent lateral nerves, larger fruits and usually more tawny (at least when dried) indumentum. Both this species and *T. sericea* have been restricted as far as possible to what may be considered " typical " specimens and the intermediates have been placed in ? *T. sericea* × *trichopoda* (p. 173).

8. **Terminalia griffithsiana** Liben in Bull. Jard. Bot. Brux. **35**, 2: 181, t. 3 (1965); F.C.B. Combret.: 96, t. 9 (1968).—Exell in Kirkia, **7**, 2: 237 (1970). Type from Katanga.
    Terminalia sp. 6.—Griffiths in Journ. Linn. Soc., Bot. **55**: 858 (1959).
    Terminalia sp. 2.—F. White, F.F.N.R.: 292 (1962).

Shrub 4–8 m. high with a short thick twisted stem. Leaves spirally arranged, congested at the ends of the branchlets; lamina 9–22 × 6·5–14 cm., obovate, apex rounded or retuse, base cuneate and ± decurrent into the petiole, golden-yellow-tomentellous; lateral nerves 10–14 pairs, prominent. Inflorescences of axillary spikes 10–12 cm. long. Lower receptacle tomentellous; upper receptacle densely pubescent. Stamens 3 mm. long; anthers 0·5 mm. long. Disk densely pilose. Fruits 3–3·5 × 1·8 cm., elliptic, tomentellous, apex rounded or emarginate, base cuneate, stipe 2·5–3 mm. long.

Zambia. W: Mwinilunga Distr., 6·4 km. NW. of Kalene Hill Mission, fl. 20.ix.1952, *Holmes* 885 (FHO; K).
Also in Zaire. In *Brachystegia* woodland on Kalahari Sand and in chipya.

Differs from *T. erici-rosenii*, which it closely resembles, in having a rather fine brownish-yellow tomentum (not sericeous) but is nearly identical in other respects.

9. **Terminalia brachystemma** Welw. ex Hiern, Cat. Afr. Pl. Welw. **1**: 340 (1898).—Engl. & Diels in Engl., Mon. Afr. Pflanz. **4**: 20, t. 9 fig. C (1900).—Garcia in Contr. Conhec. Fl. Moçamb. **2**: 147 (1954).—Griffiths in Journ. Linn. Soc., Bot. **55**: 846, fig. 11 (1959).—F. White, F.F.N.R.: 290 (1962).—Boughey in Journ. S. Afr. Bot. **30**, 4: 167 (1964).—Liben, F.C.B. Combret.: 93 (1968).—Exell in Kirkia, **7**, 2: 237 (1970).—Exell & Garcia, C.F.A. **4**: 88 (1970).—Wickens, F.T.E.A. Combret.: 83, fig. 12(9) (1973).—Faria in Mem. Inst. Invest. Agron. Moçamb. **4**: 127 (1973). TAB. **40** fig. B1, B2. Type from Angola.
    Terminalia baumii Engl. & Gilg in Warb., Kunene-Samb.-Exped. Baum: 321 (1903). Type from Angola.

Small deciduous or semi-deciduous tree 5–8 m. high or shrub; crown flat-rounded; bark grey to dark-brown, longitudinally fissured; branchlets pubescent or glabrous with purplish-brown or purplish-black bark peeling off to reveal a light-brown newly exposed surface. Leaves spirally arranged, sessile or subsessile; lamina 9–15 × 5–7 cm., coriaceous, glaucous beneath, broadly obovate to obovate-elliptic, apex obtuse to rounded and sometimes shortly cuspidate, base narrowly cuneate and often when the leaf is quite sessile truncate at the point of insertion, glabrous or rather sparsely pubescent or sparsely appressed-pilose especially on the midrib; lateral nerves 10–20 pairs; petiole 0–2(4) mm. long. Inflorescences of axillary spikes 7·5–11 cm. long, often in the axils of the fallen leaves, glabrous or sparsely to densely sericeous. Flowers white. Lower receptacle glabrous or sparsely to densely sericeous; upper receptacle glabrous or sparsely pilose at the base. Sepals deltate- to triangular-acuminate. Stamens 3–3·5 mm. long; anthers 0·5 mm. long. Disk densely barbate. Fruit purplish or reddish-brown, 4–5·5 × 2·3–2·5 cm., elliptic to elliptic-oblong, apex obtuse to rounded and emarginate, base cuneate, stipe 5–7 mm. long.

Also in Angola, Zaire, Tanzania and the Transvaal.

Leaves glabrous; peduncle and receptacles glabrous; leaf-scars rounded or elliptic
subsp. *brachystemma*
Leaves pubescent (but old leaves sometimes glabrescent); peduncle tomentose; lower
receptacle sericeous-tomentose; leaf-scars deltate - - - subsp. *sessilifolia*

9a. Subsp. **brachystemma** Wickens, F.T.E.A. Combret.: 84 (1973).

Zambia. B: Balovale Distr., near Chitokaloki Mission, fl. 11.x.1952, *White* 3469 (BM; COI; FHO; K). W: Mufulira, fr. 20.vii.1957, *Fanshawe* 3363 (K; LISC). E: Fort Jameson, fr. 31.viii.1929, *Burtt Davy* 21044 (FHO). S: between Livingstone and Kalomo, fr. 15.viii.1946, *Gouveia & Pedro* 1664 (LMJ). **Rhodesia.** N: between Makuti and Kariba, fr. iv.1960, *Brewer* 6892 (CAH; FHO; SRGH). W: Matopos, near World's View, fr. 13.iv.1955, *E.M. & W.* 1490 (BM; LISC; SRGH). **Mozambique.** N: between Porto Amélia and Ancuabe, fl. immat. & fr. 24.viii.1948, *Barbosa* 1874 (LISC; LMJ). T: Macanga, between Angónia and Furancungo, fr. 15.v.1948, *Mendonça* 4246 (BM; LISC). MS: Cheringoma, fl. 30.xi.1944, Simão 301 (LISC).

Also in Zaire and the Transvaal. *Brachystegia allenii* woodland; *Burkea-Colophospermum-Dialium* savanna; *Baikiaea* woodland; mopane woodland; Kalahari Sand and other sandy soils; often on margins of dambos.

9b. Subsp. **sessilifolia** (R.E.Fr.) Wickens in Kew Bull. **25**: 184 (1971); F.T.E.A. Combret.: 84 (1973). Type: Zambia, Mbala Distr., Msisi, *Fries* 1328 (UPS). holotype).
Terminalia sessilifolia R.E.Fr. in Wiss. Ergebn. Schwed. Rhod.-Kongo-Exped. **1**: 174 (1914).—Griffiths in Journ. Linn. Soc., Bot. **55**: 851 (1959).—F. White in Griffiths, tom. cit.: 849 in nota (1959); F.F.N.R.: 290 (1962). Type as above.

Characters as in the key to the subspecies.

Zambia. N: Kawambwa, fl. 30.x.1952, *White* 3547 (FHO; K).
Also in Tanzania. Dry woodland; in swamp forest with *Bridelia micrantha, Rubus* and *Anthocleista kerstingii.*

The separation of the two subspecies is taken from Wickens (*locc. citt.*). In our area the position is considerably complicated by the occurrence of so many *brachystemma-sericea* hybrids described below.

9c. **Terminalia brachystemma × sericea** Exell in Kirkia, **7**, 2: 238 (1970).—Faria in Mem. Inst. Invest. Agron. Moçamb. **4**: 129 (1973).
Terminalia silozensis Gibbs in Journ. Linn. Soc., Bot. **37**: 444 (1906).—Monro in Proc. & Trans. Rhod. Sci. Ass. **8**, 2: 82 (1908).—Eyles in Trans. Roy. Soc. S. Afr. **5**, 4: 430 (1916).—Griffiths in Journ. Linn. Soc., Bot. **55**: 849 (1959).—Boughey in Journ. S. Afr. Bot. **30**, 4: 167 (1964). Type: Rhodesia, Matopos, *Gibbs* 277 (BM, holotype: K).

Intermediates (probably hybrids) between *T. brachystemma* and *T. sericea* with leaves usually smaller than in *T. brachystemma* and usually distinctly petiolate, and larger than in *T. sericea* and less densely hairy to almost glabrous. *Gibbs* 277, the type of *T. silozensis*, has, however, sessile or almost sessile leaves which makes a further complication.

Caprivi Strip. Andara Mission Station, fr. 22.ii.1956, *de Winter & Marais* 4794 (K; PRE). **Botswana.** N: Kazungula, fl. xi.1935, *Miller* B/86 (FHO). **Zambia.** W: Ndola, fr. 8.1.1955, *Fanshawe* 1782 (FHO; K; SRGH). C: Lusaka Distr., Rufunsa Rest House, fr. 3.v.1952, *White* 2913 (FHO). E: between Siwa Camp and Fort Jameson, st. 28.viii.1929, *Burtt Davy* 21006/29 (FHO). S: Mazabuka Distr., Siamambo Forest Reserve, fr. 9.vi.1952, *White* 2933 (FHO). **Rhodesia.** N: Urungwe Distr., fl. 3.x.1956, *Mullin* 67/56 (K; SRGH). W: Bulalima Mangwe, Plumtree, fl. ix.1954, *Meara* 89 (SRGH). C: Salisbury, Lower Hillside, fr. 20.iv.1934, *Gilliland* 164 (BM; FHO; SRGH). **Malawi.** Namakokwe R., fr. 14.ix.1929, *Burtt Davy* 21692/29 (FHO). **Mozambique.** N: between Palma and Cabo Delgado, fr. 16.ix.1948, *Barbosa* 2157 (LISC). Z: Mopeia, st. 29.vii.1942, *Torre* 4452 (LISC). T: between Fíngoè and Vila Vasco da Gama, fr. 13.viii.1941, *Torre* 3281 (LISC). MS: Chimoio, st. 7.ix.1942, *Mendonça* 160 (LISC).
Also in Zaire and Tanzania.

10. **Terminalia kaiserana** F. Hoffm., Beitr. Kenntn. Fl. Centr. Ost-Afr.: 26 (1889).—Engl., Pflanzenw. Ost-Afr. C: 294 (1895).—Engl. & Diels in Engl., Mon. Afr. Pflanz. 4: 19, t. 13 fig. D (1900).—Brenan, T.T.C.L.: 145 (1949).—Griffiths in Journ. Linn. Soc., Bot. 55: 849, fig. 12 (1959).—F. White, F.F.N.R.: 290 (1962).—Exell in Kirkia, 7, 2: 239 (1970).—Wickens, F.T.E.A. Combret.: 84, fig. 12(10) (1973). TAB. 40 fig. C. Type from Tanzania.
Terminalia holtzii Diels in Engl., Bot. Jahrb. 39: 513 (1907).—Griffiths in Journ. Linn. Soc., Bot. 55: 905 (1959). Type from Tanzania.

Small deciduous or semi-deciduous tree 7–10 m. high or shrub; crown flat-rounded or rounded; bark pale-brown to grey-black, fissured; branchlets sparsely sericeous-tomentose to glabrous with purplish-brown to purplish-black bark peeling off to reveal a light-brown newly exposed surface. Leaves spirally arranged, petiolate; lamina 7·5–12 × 2·5–3 cm., chartaceous, narrowly elliptic, sparsely sericeous when young but soon glabrescent, apex acute or shortly acuminate, base cuneate to narrowly cuneate; lateral nerves 10–16 pairs, slightly conspicuous above but rather inconspicuous beneath; petiole up to 3(3·5) mm. long, rather slender. Inflorescences of axillary spikes up to 10 cm. long; peduncle up to 3 cm. long, usually glabrous. Flowers cream or white. Lower receptacle glabrous or very sparsely pubescent; upper receptacle glabrous. Sepals deltate to triangular. Stamens 3 mm. long; anthers 0·4–0·5 mm. long. Disk pilose. Fruit purplish-brown, 4·5–5·5 × 2·5–3 cm., broadly elliptic, glabrous, apex obtuse to rounded and emarginate, base acute to subtruncate, stipe 5–7 cm. long.

Zambia. N: lower Lunzua Valley, fl. 20.x.1961, *Lawton* 786 (FHO). **Malawi.** N: 8 km. W. of Karonga, fr. 27.iv.1960, *Adlard* 340 (FHO).
Also in Burundi and Tanzania. *Brachystegia* woodland; *Acacia* shrub savanna; terrace alluvium; *Acacia-Combretum-Terminalia* tree savanna.

Up to the present this species, characterized by the longer more slender petioles remains satisfactorily distinct from *T. sericea* and *T. brachystemma* in our area but Wickens (F.T.E.A. Combret.: 86 (1973)) records *sericea-kaiserana* hybrids in Tanzania.

4. Sect. PLATYCARPAE Engl. & Diels emend. Exell in Bol. Soc. Brot., Sér. 2, 42: 31 (1968); in Kirkia, 7, 2: 239 (1970).—Wickens, F.T.E.A. Combret.: 77 (1973).

Sect. *Stenocarpae* Engl. & Diels in Engl., Mon. Afr. Pflanz. 4: 11 (1900).

Leaves spirally arranged, not borne in fascicles on short shoots; bark of branchlets not peeling off in the way characteristic of Sect. *Psidioides*; fruit 2-winged.

Branchlets developing a layer of cork in the second year which later becomes thick; flowers tomentose; fruit velutinous, 6·5–12 cm. long - - - 11. *mollis**
Branchlets not becoming very corky; flowers tomentose to glabrous; fruit puberulous to glabrous, up to 8·5 cm. long:
  Flowers and fruit glabrous; leaf-margins usually somewhat crenulate; fruit up to 3 cm. long - - - - - - - - - 13. *gazensis*
  Flowers (at least the lower receptacle) densely tomentose; leaf-margin occasionally somewhat crenulate in *T. sambesiaca* and *T. phanerophlebia*; fruit minutely

* Intermediates between this species and *T. stenostachya* may be hybrids.

puberulous to glabrous, usually over 3 cm. long (except sometimes in *T. phanerophlebia*):

Leaves elliptic to oblong-elliptic, generally rather rounded and somewhat asymmetric at the base and narrowed gradually towards the apex, usually somewhat glaucous beneath; older branchlets with very prominent leaf-scars   12. *stenostachya**
Leaves obovate or broadly obovate to obovate-elliptic, cuneate at the base and usually distinctly acuminate at the apex; leaf-scars usually not very prominent:
Tertiary nerves and veins scarcely impressed above; fruit nearly always more than 4 cm. long   -   -   -   -   -   -   -   -   14. *sambesiaca*
Tertiary nerves and veins impressed above; fruit 2·8–4 cm. long
15. *phanerophlebia*

11. **Terminalia mollis** Laws. in Oliv., F.T.A. **2**: 471 (1871) non Zoll. ex Teysm. & Binnendijk (1866) *nom. nud.* ex v. Sloot. (1924) nec Vidal (1885).—Keay, F.W.T.A. ed. 2, **2, 1, 1**: 279 (1954).—Griffiths in Journ. Linn. Soc., Bot. **55**: 871, fig. 19 (1959) excl. ref. Mendonça (1954).—F. White, F.F.N.R.: 291 (1962).—Stace in Bull. Brit. Mus. (Nat. Hist.) Bot. **4, 1**: 31 (1965).—Liben, F.C.B. Combret.: 98 (1968).— Exell & Garcia, C.F.A. **4**: 91 (1970).—Exell in Kirkia, **7, 2**: 240 (1970).—Wickens, F.T.E.A. Combret.: 88 (1973).—Jacobsen in Kirkia, **9, 1**: 168 (1973). TAB. **40** fig. I. Type from the Sudan.
Terminalia torulosa F. Hoffm., Beitr. Kenntn. Fl. Centr. Ost-Afr.: 27 (1889).— Engl. & Diels in Engl., Mon. Afr. Pflanz. **4**: 15, t. 5 fig. A (1900) excl. specim. Kirk et Buchanan. Type from Tanzania.
Terminalia dewevrei De Wild. & Dur. in Bull. Soc. Roy. Bot. Belg. **38**: 123 (1899). Type from Zaire.
Terminalia spekei Rolfe in Journ. Linn. Soc., Bot. **37**: 516 (1906). Syntypes from Uganda.
Terminalia mildbraedii Gilg ex Mildbr. in Wiss. Ergebn. Deutsch Zentr. Afr. Exped. 1907–1908, **2**: 581 (1913). Type from Tanzania.
Terminalia suberosa R.E.Fr. in Wiss. Ergebn. Schwed. Rhod.-Kongo Exped. **1**: 172, t. 4 (1914) non A. Chev. (1911) *nom. nud.* Syntypes: Zambia, Chirukuta, *Fries* 264 (UPS), 264a (UPS).
Terminalia kerstingii Engl., Bot. Jahrb. **39**: 511 (1907). Type from Togoland.
Terminalia reticulata Engl., loc. cit. *nom. illegit.* non Roth (1821). Type from Togoland.
Terminalia glandulosa De Wild., Contr. Fl. Katanga, Suppl. **2**: 86 (1929). Syntypes from Zaire.

Deciduous tree 5–13(20) m. high; crown rounded to flat-rounded; bark blackish-grey, deeply fissured; branchlets densely tomentose with grey-brown to dark-grey-brown bark becoming softly and thickly corky with age. Leaves spirally arranged, petiolate; lamina 16–37 × 7–19 cm., subcoriaceous, elliptic to obovate-oblong, densely brownish-tomentose (when dried), apex obtuse to rounded or sometimes acute, base obtuse to rounded or occasionally subcordate; lateral nerves 7–18 pairs, prominent beneath; petiole 3·5–5 cm. long, stout. Inflorescences of axillary spikes 8–17 cm. long; peduncle 1–2 cm. long. Flowers cream or greenish-white or pinkish, strong-smelling. Lower receptacle densely tomentose; upper receptacle sparsely tomentose, ± glabrous towards the apex. Fruit pale-yellowish-green, 6·5–12 × 2·5–5·5 cm., densely velutinous, stipe 5–7 cm. long. Cotyledons 2, 10–16 × 20–25 mm., transversely elliptic to subreniform with petioles 2–2·5 cm. long arising below soil level.

Zambia. N: Kasama Distr., 48 km. S. of Kasama, fl. 23.x.1949, *Hoyle* 1297 (FHO). W: Mwinilunga Distr., 6·4 km. NW. of Kalene Hill Mission, fl. 20.ix.1952, *Holmes* 892 (FHO). C: Mt. Makulu, fl. 17.x.1956, *Angus* 1421 (BM; COI; FHO; K). S: between Livingstone and Kabompo, fr. 15.viii.1946, *Gouveia & Pedro* in *Pedro* 1660 (LMJ).
Also in western tropical Africa, Angola, Zaire, Uganda, Kenya and Tanzania. *Brachystegia* woodland or savanna woodland; tall grass woodland on heavy soils; edges of dambos; low to medium altitudes.

As the distinction between *T. mollis* and *T. stenostachya* is mainly in the development of thick corky bark on the branchlets of the former, specimens which are not of an age to show this feature are necessarily difficult to classify. There also appear to be hybrids between the two species.

11a. ? **Terminalia mollis** × **stenostachya** Exell in Kirkia, **7, 2**: 241 (1970).—Faria in Mem. Inst. Invest. Agron. Moçamb. **4**: 130 (1973).

* Intermediates between this species and *T. mollis* may be hybrids.

Tree up to 18 m. high. Mostly with larger fruits than in *T. stenostachya* (i.e. up
to 8(10·5) × 4(5) cm.) and usually not showing the characteristic corky bark of
T. mollis.

Zambia. B: Samitondo Plain, fr. 30.xii.1938, *Martin* 925/38 (FHO). S: Mazabuka
Distr., Siamambo Forest Reserve, fr. 8.vi.1952, *White* 2927 (FHO). **Rhodesia.** N:
Mazoe Distr., Concession, fr. 8.v.1933, *Kelly Edwards* E 2/33 (FHO). W: Bulawayo, fr.
vii.1949, *Cuthbertson* 4 (FHO). C: Chilimanzi African Reserve, fr. 7.iii.1951, *Wormold*
25/51 (FHO).
   A specimen from Mozambique, Niassa, Mutuáli, fr. 1.iii.1953, *Gomes e Sousa* 4045 (K)
has been placed here by F. White (in sched.). The fruits are only up to 5·5 cm. long but
the specimen certainly tends towards *T. mollis.*
   Some of the Rhodesian references cited under *T. stenostachya* may refer in part to this
possible hybrid.

12. **Terminalia stenostachya** Engl. & Diels in Engl., Mon. Afr. Pflanz. **4**: 16, t. 7
   fig. B (1900).—Brenan, T.T.C.L.: 144 (1949).—Palgrave, Trees of Centr. Afr.:
   152 cum tab. et phot. (1956).—Griffiths in Journ. Linn. Soc., Bot. **55**: 888, fig. 24
   (1959).—F. White, F.F.N.R.: 291, fig. 52I (1962).—Chapman, Veg. Mlanje
   Mountains: 37 (1962).—Boughey in Journ. S. Afr. Bot. **30**, 4: 167 (1964).—Liben,
   F.C.B. Combret.: 101 (1968).—Exell in Kirkia, **7**, 2: 241 (1970).—Wickens,
   F.T.E.A. Combret.: 89 (1973).—Jacobsen in Kirkia, **9**, 1: 168 (1973).—Faria in
   Mem. Inst. Invest. Agron. Moçamb. **4**: 130 (1973). TAB. **40** fig. H. Syntypes:
   Malawi, *Buchanan* 352 (B†); Blantyre, *Buchanan* in Herb. Wood 6838 (B†).
      Terminalia torulosa Engl. & Diels in Engl., tom. cit.: 15 (1900) pro parte
   quoad specim. Kirk.—Eyles in Trans. Roy. Soc. S. Afr. **5**, 4: 430 (1916).—Steedman,
   Some Trees, etc. S. Rhod.: 56 (1933).—Burtt Davy & Hoyle, N.C.L.: 39 (1936).—
   Suesseng. & Merxm. in Proc. & Trans. Rhod. Sci. Ass. **43**: 28 (1951).—Wild, Guide
   Fl. Vict. Falls: 146 (1953).
      Terminalia rhodesica R. E. Fr. in Wiss. Ergebn. Schwed. Rhod.-Kongo-Exped. **1**:
   172, t. 13 (1914).—Burtt Davy & Hoyle, loc. cit. Syntypes: Zambia, Victoria Falls,
   Fries 78 (UPS), 78a (UPS).
      Terminalia mollis sensu Garcia in Contr. Conhec. Fl. Moçamb. **2**: 145 (1954)
   quoad specim. cit.—Griffiths, tom. cit.: 871 (1959) pro parte quoad ref. Mendonça.—
   Palgrave, tom. cit.: 140 cum tab. et phot. (1956).
      Terminalia sp. Brenan in Mem. N.Y. Bot. Gard. **8**, 5: 437 (1954).

   Small deciduous or semi-deciduous tree 5–12(20) m. high or shrub 4 m. high;
crown rounded to flat-rounded; bark dark-brown to grey-black, fissured, rough,
lattice-like; branchlets densely tomentose when young, glabrescent, with greyish-
brown fibrous bark and prominent leaf-scars. Leaves spirally arranged, petiolate;
lamina 11–18(26) × 4·5–5·5(11·5) cm., coriaceous, elliptic to oblong-elliptic,
tomentose when young, ± glabrescent and somewhat glaucous, sometimes bullate
above, apex acute to rounded or subcordate: lateral nerves 8–12 pairs, prominent
beneath; petiole 1·5–3 cm. long. Inflorescences of axillary spikes 10–16 cm. long;
peduncle 2·5–4 cm. long, tomentose. Flowers cream, strong smelling. Lower
receptacle tomentose; upper receptacle sparsely pubescent. Sepals deltate to
triangular-acuminate. Stamens 3 mm. long; anthers 0·5–0·6 mm. long. Disk
barbate. Fruit dark-red or crimson or yellowish-red, (3)4·5–6·2 × 2·4–3 cm.,
oblong to narrowly elliptic, minutely puberulous, apex obtuse to rounded, base
obtuse to subtruncate, stipe 5–6 mm. long.

   Zambia. 40 km. N. of Isoka, fr. 11.viii.1965, *Lawton* 1235 (FHO). C: Lusaka Distr.,
Great East Road, 192 km. from Lusaka, *White* 2696 A (FHO; K). E: Fort Jameson,
fr. 25.iv.1952, *White* 2466 (FHO). S: Chilanga, fl. 6.i.1966, *Lawton* 1350 (FHO).
Rhodesia. N: Trelawney Research Station, 1340 m., fl. x.1940, *Brain* 10993 (FHO;
LMJ; SRGH). W: Wankie Distr., Fuller Forest Reserve, fr. iii.1960, *Armitage* 63/60
(SRGH). C: Featherstone, 1370 m., fr. 26.iii.1947, *Cairns* 11/47 (FHO; SRGH).
E: Umtali Distr., Darlington, fl. 12.xi.1954, *Chase* 5327 (BM; FHO; LISC; SRGH).
S: Sabi-Lundi Junction, fr. 9.vi.1950, *Wild* 3471 (FHO; K; LISC; SRGH). **Malawi.**
N: Ifumbo, 610–1220 m., fl. 1939, *Lewis* 40 (FHO). C: Fort Manning, fl. 16.xii.1955,
Adlard 137 (FHO). S: Mtwice, fr. 10.iii.1955, *E.M. & W.* 774 (BM; LISC; SRGH).
Mozambique. N: Mutuáli, st. 29.v.1947, *Pedro* 3301 (LMJ). Z: Mocuba Distr.,
Namagoa, fl. 1944, *Faulkner* 337 (BM; COI; EA; K; LMJ; PRE; SRGH). T:
between Tete and Massamba, fr. 5.vii.1949, *Andrada* 1696 (COI; LISC). MS: between
Dombe and Sanguene, fl. 28.x.1952, *Pedro* 4505 (LMJ).
   Also in Zaire and Tanzania. *Brachystegia* woodland; low altitude tree savanna; dry
tree savanna with *Acacia nigrescens*; Lake basin chipya; *Acacia-Combretum-Terminalia*
savanna; scrub mopane; on heavy soils such as rich red soils and olei soils.

13. **Terminalia gazensis** Bak. f. in Journ. Linn. Soc., Bot. **40**: 69 (1911).—Garcia in Contr. Conhec. Fl. Moçamb. **2**: 145 (1954).—Griffiths in Journ. Linn. Soc., Bot. **55**: 892, fig. 25 (1959).—Boughey in Journ. S. Afr. Bot. **30**, 4: 167 (1964).—Exell in Kirkia, **7**, 2: 242 (1970).—Faria in Mem. Inst. Invest. Agron. Moçamb. **4**: 135 (1973). Type: Mozambique, Mossurize Distr., *Swynnerton* 152a (BM, holotype).

Small tree 7–17 m. high; bark dark-brown, fissured; branchlets tomentellous when young, glabrous with grey-brown fibrous bark. Leaves spirally arranged, petiolate; lamina 6–14 × 3·5–7 cm., papyraceous to chartaceous, obovate to elliptic, pubescent to shortly pilose or almost glabrous, apex usually acuminate, margin usually crenulate, base cuneate and sometimes narrowed; lateral nerves 5–8 pairs, prominulous above, somewhat prominent beneath; petiole up to 4 cm. long, usually rather slender. Inflorescences of axillary spikes 3·5–10 cm. long; rhachis appressed-pubescent; peduncle up to 3·5 cm. long, appressed-pubescent. Flowers white, foetid, glabrous. Sepals deltate to triangular. Stamens 4 mm. long; anthers 0·6 mm. long. Disk barbate. Fruit red or greenish-yellow, 2·5–3 × 1·5–2 cm., oblong-elliptic, glabrous, apex subtruncate and emarginate, base rounded to subtruncate, stipe 4–5 mm. long.

Rhodesia. N: Kariba, fl. & fr. i.1959, *Goldsmith* 1/59 (K; SRGH). C: Selukwe Distr., Ferny Creek, 1220 m., fl. 8.xii.1953, *Wild* 4295 (FHO; SRGH). E: Chipinga Distr., Chirinda Forest margin, 1525 m., fl. i.1962, *Goldsmith* 14/62 (FHO; LISC; SRGH). S: Belingwe Distr., south tip of Great Dyke, fl. immat. 8.xii.1960, *Wild* 6399 (FHO; SRGH). **Malawi.** S: Blantyre, Mpatamanga Gorge, fl. immat. 8.xii.1960, *Chapman* 1072 (FHO; K). **Mozambique.** N: between Porto Amélia and Mecúfi, fl. 27.x.1942, *Mendonça* 1093 (LISC). MS: Chimoio Distr., Monte Garuso, fl. i.1948, *Mendonça* 3584 (BM; K; LISC).
Not known elsewhere. *Brachystegia* woodland at medium altitudes or in tree savanna at lower altitudes.

E.M. & W. s.n. (LISC) growing in a garden at Marracuene, Mozambique, has glabrous fruits c. 6 × 2·5 cm. Apart from the much larger fruits (possibly the effect of cultivation) it closely resembles *T. gazensis* but might be *T. gazensis* × *sambesiaca*.
Typically this species has nearly glabrous leaves usually with crenulate margins (a very unusual feature in the Combretaceae). Griffiths (loc. cit.) has included specimens such as *Wild* 4344 (FHO; SRGH) = GHS 44793 from the Belingwe Distr. with thicker very hairy leaves and entire margins which may be a distinct serpentine tolerant species or subspecies.

13a. ? **Terminalia gazensis** × **sambesiaca.**

Flowers glabrous or almost so. Fruit up to 5 × 2 cm., almost glabrous. Leaf-margin slightly crenate. Petiole up to 3·5 cm. long.

Mozambique. N: between Meconta and Corrane, fr. 18.i.1964, *Torre & Paiva* 10048 (LISC).

Several other specimens such as *Torre & Paiva* 9288 (LISC) and 9485 (LISC), *Andrada* 1252 (LISC) and possibly *Simão* 1269 (LISC; LM) with petioles up to 7·5 cm. long probably come here.

14. **Terminalia sambesiaca** Engl. & Diels in Engl., Mon. Afr. Pflanz. **4**: 13, t. 4 A (1900).—Schinz, Pl. Menyharth.: 431 (1906).—Garcia in Contr. Conhec. Fl. Moçamb. **2**: 145 (1954).—Chapman & White, Evergreen Forests of Malawi: 99 (1970).—Exell in Kirkia, **7**, 2: 243 (1970).—Wickens, F.T.E.A. Combret.: 91 (1973).—Faria in Mem. Inst. Invest. Agron. Moçamb. **4**: 135 (1973). Type: Mozambique, near Boroma, *Menyharth* 613 (Z, holotype).
    Terminalia brownii sensu Laws. in Oliv., F.T.A. **2**: 415 (1871) pro parte quoad specim. Kirk.—Sim, For. Fl. Port. E. Afr.: 64 (1909).—Wild, Guide Fl. Vict. Falls: 146 (1953).
    Terminalia aemula Diels in Engl., Bot. Jahrb. **39**: 511 (1907).—Brenan, T.T.C.L.: 143 (1949).—Garcia, loc. cit. Type from Tanzania.
    ? *Terminalia obovata* Sim, For. Fl. Port. E. Afr.: 65, t. 64 fig. B (1909) *nom. illegit.* non Cambess. (1829) nec Steud. (1841). Type: Mozambique, Maganja da Costa, *Sim* 5672 (n.v.).
    Terminalia hildebrandtii sensu Burtt Davy & Hoyle, N.C.L.: 39 (1936).
    Terminalia kilimandsharica sensu Griffiths in Journ. Linn. Soc., Bot. **55**: 897, fig. 27 (1959) pro parte.—Boughey in Journ. S. Afr. Bot. **30**, 4: 167 (1964).

Large tree up to 25(39) m. high; bark greyish, smooth to slightly rough and fissured; young branchlets tomentose, glabrescent, with fibrous bark. Leaves spirally arranged, petiolate; lamina up to 18 × 13 cm., papyraceous to chartaceous, elliptic to broadly elliptic or obovate-elliptic, pubescent to pilose especially on the nerves and reticulation beneath; apex rounded and acuminate, margin sometimes crenulate, base cuneate to obtuse or rounded; lateral nerves 8–11 pairs, rather prominent beneath; petiole up to 4 cm. long, tomentose. Inflorescences of axillary or occasionally terminal spikes up to 15 cm. long; peduncle up to 6 cm. long, tomentellous. Flowers white. Lower receptacle tomentose; upper receptacle pilose at the base, almost glabrous towards the apex. Sepals triangular. Stamens 5 mm. long; anthers 0·5 mm. long. Disk barbate. Fruit up to 7(9) × 3(4·5) cm., elliptic, pubescent, apex subtruncate and sometimes emarginate, base narrowed into the stipe, stipe up to 15 mm. long. Cotyledons 2, 1–1·8 × 2–3 cm., transversely elliptic or half-moon-shaped, borne above soil level, with petioles 3–5 mm. long.

Zambia. N: Lunzua Valley, *Richards* (K). **Rhodesia.** N: Urungwe Distr., fl. 6.xii.1952, *Lovemore* 326 (FHO; SRGH). C: Charter Distr., 21 km. N. of Lalapansi on Great Dyke, fl. immat. 17.i.1962, *Wild* 5620 (FHO; SRGH). E: Umtali Distr., Laurenceville, fr. 21.xii.1952, *Chase* 4746 (BM; COI; FHO; LISC; SRGH). **Malawi.** C: Dedza Distr., Mua Livulezi, fr. 23.xii.1953, *Adlard* 20 (FHO). S: Mpatamanga Gorge, fr. 28.ii.1961, *Richards* 14487 (K; SRGH). **Mozambique.** N: between Mogincual and Quixaxe, fr. 27.vii.1948, *Pedro & Pedrógão* 4707 (LMJ). T: between Tete and Chicoa, fr. 25.vi.1949, *Barbosa & Carvalho* 3275 (LISC; LMJ). MS: Báruè, between Vila Gouveia and Changara, 750 m., fl. 20.xii.1965, *Torre & Correia* 13749 (LISC).
Also in Kenya and Tanzania. Evergreen forest; lowland fringing forest (in Malawi); savanna woodland with *Adansonia* and *Sterculia*; rocky outcrops; sometimes on serpentine soils.

I am indebted to F. White for pointing out that this species was misidentified by Griffiths (loc. cit.) as *T. kilimandsharica* Engl. & Diels. The leaves of *T. sambesiaca* tend to be obovate-elliptic-acuminate and rather larger and the fruits are usually smaller. Both species might be reasonably considered as subspecies of *T. brownii* Fresen.
Pedro & Pedrógão 4707 (cited above) has nearly glabrous fruits only 4–5·5 cm. long but appears to belong to this species as the fruits are unlike those of *T. gazensis* and much larger. *Andrada* 1252 (COI; LMJ) from Quixaxe is the same. *Pedro & Pedrógão* 3914 (LMJ) from Mozambique, Metangula, has fruits about the same size but tomentellous and *White* 2820 (FHO; K) from the northern province of Malawi is somewhat similar. All these specimens may be intermediates between *T. sambesiaca* and *T. brownii* (see note by White apud Griffiths in Journ. Linn. Soc., Bot. **55**: 895 (1959)).

15. **Terminalia phanerophlebia** Engl. & Diels in Engl., Mon. Afr. Pflanz. **4**: 19, t. 12 fig. C (1900).—Burtt Davy, F.P.F.T. **1**: 249 (1926).—Codd, Trees and Shrubs Kruger Nat. Park: 134, figs. 126a and b (1951).—Garcia, Contr. Conhec. Fl. Moçamb. **2**: 146 (1954).—Griffiths in Journ. Linn. Soc., Bot. **55**: 900, fig. 28 (1959).—Exell in Kirkia, **7**, 2: 244 (1970).—Faria in Mem. Inst. Invest. Agron. Moçamb. **4**: 140 (1973). Syntypes from the Transvaal and Mozambique: Delagoa Bay, *Junod* 141 (Z). The designation of *Galpin* 884 (K) as lectotype by Griffiths (loc. cit.) was incorrect as at least two syntypes cited by the original authors, *Galpin* 884 (Z) and *Junod* 141 (Z) exist.

Small tree up to 7 m. high or shrub; branchlets appressed-pubescent, glabrescent. Leaves spirally arranged, petiolate; lamina 3–10·5 × 1·5–4·5 cm., chartaceous, obovate to obovate-elliptic, sparsely pubescent or pilose to almost glabrous, apex acuminate, margin slightly crenulate or serrate-crenulate, base cuneate or narrowed; midrib appressed-pubescent beneath; lateral nerves 5–7 pairs, somewhat impressed above, prominent beneath; reticulation impressed above; petiole 1–2·5 cm. long. Inflorescences of axillary spikes 7–10 cm. long; peduncle 1·5–2 cm. long, appressed-pubescent. Flowers white. Lower receptacle tomentose; upper receptacle pubescent at the base, glabrous towards the apex. Sepals deltate. Stamens 3 mm. long; anthers 0·5 mm. long. Disk barbate. Fruit greenish-yellow, 2·8–4 × 1·8–2·7 cm., broadly elliptic-oblong, minutely pubescent, apex acute to rounded or truncate, base subtruncate, stipe 2–3 mm. long.

Mozambique. SS: Massingir, fr. 22.v.1972, *Myre et al.* 5784 (LMA). LM: between Goba and Catuane, fr. 29.iii.1953, *Pedro* 4056 (LMJ).
Also in the Transvaal, Swaziland and Natal. Gallery forest; *Androstachys* woodland; low altitude shrub savanna.

5. Sect. PTELEOPSOIDES Exell in Bol. Soc. Brot., Sér. 2, **42**: 31 (1968); in Kirkia, **7**, 2: 244 (1970).

Bark of young branchlets fibrous. Leaves opposite or subopposite. Flowers in capitulate spikes on very short spur shoots. Fruit (4)5(6)-winged.

16. **Terminalia pteleopsoides** Exell in Bol. Soc. Brot., Sér. 2, **42**: 31 (1968); in Kirkia, **7**, 2: 245 (1970). Type: Zambia, Nsama, *Fanshawe* F 4929 (FHO, holotype).

Small semi-deciduous tree up to 9 m. high but often stunted; crown flat-rounded; bark grey, shallow-furrowed; branchlets often rather stout, tomentose, with bark peeling off in longitudinal fibrous strips. Leaves opposite or subopposite, petiolate; lamina up to 8 × 3 cm., obovate-elliptic to narrowly elliptic, tomentose or velutinous when young, glabrescent when old, apex slightly acuminate, base cuneate to rounded; lateral nerves 4–7 pairs; petiole up to 3 mm. long. Inflorescences of subcapitate spikes c. 2 cm. long borne on short spur shoots with the young leaves. Flowers white or pale-cream or yellow, sweet-scented; rhachis densely pilose; bracts 3 mm. long, linear, densely pilose. ♀ flowers: lower receptacle 3 mm. long, tomentose; upper receptacle 1–1·5 × 2–2·5 mm., cupuliform, pilose, slightly zygomorphic; sepals deltate, acute, 2 dorsal ones slightly larger; stamens 4–6 mm. long; anthers 0·6–0·9 mm. long, shortly apiculate; disk pilose; style 7 mm. long, pilose. ♂ flowers similar but with the ovary not developing and with a pseudopedicel 8 mm. long. Fruit (4)5(6)-winged, up to 3·5 × 2 cm., elliptic in outline, tomentellous, apex blunt with a short apical peg, base decurrent into the stipe (or fruit subsessile), stipe up to 5 mm. long but often scarcely distinguishable from the decurrent base of the fruit.

Zambia. N: Mporokoso Distr., fl. 11.xi.1958, *Fanshawe* F 4793 (FHO; K).

Not known elsewhere. *Combretum* woodland; *Acacia-Combretum-Terminalia* woodland; plateau woodland; *Brachystegia* scrub on laterite; edges of pans and dambos.

The (4)5(6)-winged fruit is unlike that of any other known African species of *Terminalia* and is reminiscent of *Pteleopsis* in which genus the specimens had been placed in herbaria. It appears, however, to be a true *Terminalia* as shown by the absence of petals, the pericarp of the fruit with a highly sclerenchymatous layer, the inflorescences produced on short shoots and the usually stout branchlets. Understanding of the true systematic position of this remarkable species must await a worldwide reclassification of the sections of *Terminalia* but *T. pteleopsoides* seems so far removed from the rest of the genus in Africa that it would not be surprising if it eventually merited subgeneric status.

5. LUMNITZERA Willd.

Lumnitzera Willd. in Ges. Naturf. Fr. Berl. Neue Schr. **4**: 186 (1803).

Mangroves. Small evergreen trees or shrubs. Leaves spirally arranged, sessile or subsessile, fleshy-coriaceous. Flowers ♀, 5-merous, actinomorphic, red, white, pink or yellow, in short terminal spikes or racemes. Receptacle not externally differentiated into an upper and lower part but produced beyond the inferior ovary to form a tube bearing 2 adnate persistent bracteoles and terminating in a 5-lobed persistent calyx (or 5 sepals). Petals 5, caducous. Stamens (5)10, 2-seriate. Disk inconspicuous. Style filiform, persistent, not adnate to the wall of the receptacle, not expanded at the apex. Ovules 2–5. Fruit indehiscent, compressed-ellipsoid and obtusely angled, ± woody, crowned by the persistent calyx. Cotyledons unknown.

Two species: one in E. Africa, S. Africa (Natal), Madagascar, tropical Asia and Polynesia; the other in tropical Asia, N. Australia and Polynesia.

Lumnitzera racemosa Willd. in Ges. Naturf. Fr. Berl. Neue Schr. **4**: 187 (1803).—Laws. in Oliv., F.T.A. **2**: 418 (1871).—Engl. & Diels in Engl., Mon. Afr. Pflanz. **4**: 34 (1900).—Brenan, T.T.C.L.: 142 (1949).—Mogg in Macnae & Kalk, Nat. Hist. Inhaca I. Moçamb.: 150 (1958).—Dale & Greenway, Kenya Trees and Shrubs: 147, fig. 29 (1961).—Stace in Bull. Brit. Mus. (Nat. Hist.) Bot. **4**, 1: 40, 46, 50, 67 (1965); in New Phytol. **65**: 310, fig. 1a, d (1966).—Exell in Kirkia, **7**, 2: 245 (1970).—Wickens, F.T.E.A. Combret.: 93, fig. 13 (1973). TAB. **41**. Type from India.

Small tree up to 9 m. high or shrub; bark rough, reddish-brown; young branchlets reddish or grey, sometimes appressed-pubescent at first, soon

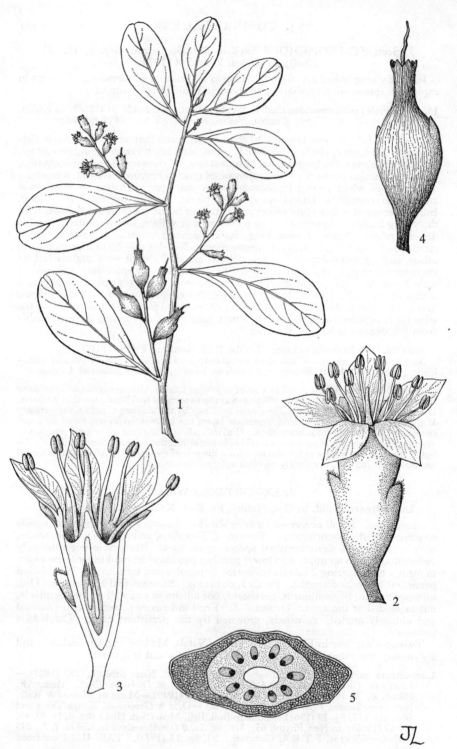

Tab. 41. LUMNITZERA RACEMOSA var. RACEMOSA. 1, fertile branchlet (× ⅔) *Greenway* 4957;
2, flower (×6); 3, longitudinal section of flower (×6), 2–3 from *Vaughan* 461; 4,
young fruit (×4); 5, transverse section of fruit (×6), 4–5 from *Greenway* 4957.

glabrescent. Leaf-lamina 2–9 × 1–3 cm., narrowly obovate or narrowly obovate-elliptic or narrowly elliptic, cuneate to the subsessile base or sometimes narrowed to appear subpetiolate. Inflorescences of short axillary spikes 2–7 cm. long; rhachis glabrous. Flowers sessile or nearly so. Receptacle 6–8 mm. long, tubular or narrowly urceolate, laterally compressed, glabrous or pubescent, usually contracted just above the middle at the insertion of the 2 broadly ovate 1·5 mm. long sometimes ciliolate opposite or subopposite adnate bracteoles. Sepals 0·8–1 mm. long, broadly ovate-acuminate, gland-tipped or eglandular. Petals white or cream (? sometimes pink) or yellow (var. *lutea*), 4 × 1 mm., narrowly elliptic or narrowly obovate-elliptic, glabrous. Stamens 10, equalling or slightly exceeding the petals. Style 6–7 mm. long, glabrous. Fruit 10–12 × 3–5 mm., appressed-pubescent or glabrous; pericarp with a well-developed inner layer of sclerenchyma with a spoke-like appearance in cross-section.

Var. **racemosa.**—Exell in Fl. Males. Ser. 1, **4**: 589 (1954); in Kirkia, **7**, 2: 246 (1970).—Wickens, tom. cit.: 95 (1973).—Faria in Mem. Inst. Invest. Agron. Moçamb. **4**: 141 (1973). TAB. **41**. Type as above.

Petals white or cream (? sometimes pinkish).

Mozambique. N: R. Mepanga, 5 km. N. of Mocímboa da Praia, fl. immat. 10.xi.1960, *Gomes e Sousa* 4624 (COI; K). Z or MS: R. Zambeze, fl. 28.v.1858, *Kirk* (K). SS: Daimane, near Inhambane, fl. iii.1938, *Gomes e Sousa* 2097 (COI; K; LISC). LM: Marracuene, fl. & fr. 1.vi.1959, *Barbosa & Lemos* in *Barbosa* 8542 (COI; K; LISC; LMJ).

Kenya to Natal, Madagascar, tropical Asia, N. Australia and Polynesia. Mangrove swamps, usually on the landward side.

L. racemosa var. *lutea* (Gaud.) Exell, with yellow flowers, is confined to Timor.

74. MYRTACEAE

By F. White*

Trees, shrubs or rhizomatous, geoxylic suffrutices. Leaves usually opposite or subopposite, rarely alternate, ternate, or in fours, simple, entire, often coriaceous, pellucid-punctate. Stipules absent. Inflorescence axillary or terminal, paniculate, thyrsoid, umbellate, cymose, racemose, or flowers solitary or fasciculate. Flowers mostly bisexual, sometimes unisexual by abortion, actinomorphic, partly or completely epigynous. Sepals (0)4–5, often persistent, rarely accrescent, sometimes fused to form an operculum, often with punctate glands. Petals (0)4–5, imbricate, free or coherent to form a calyptra or fused to form an operculum, often with punctate glands. Stamens usually numerous, free or basally connate; filaments often coiled or folded in the bud; anthers basifixed or dorsifixed, connective sometimes with an apical gland. Ovary inferior or half-inferior, with (1)2–5(10) locules; septa sometimes incomplete; placentation axile or parietal and then often with deeply intruded placentae; ovules 1 to numerous per locule or placenta; style 1, punctiform, capitate, funnel-shaped or shallowly 2–3-lobed at the apex. Fruit a berry, drupe or loculicidal capsule which only dehisces near the apex, rarely nut-like and indehiscent. Seeds 1 to many per locule; endosperm absent or scanty; embryo straight or incurved.

A large, almost exclusively tropical and subtropical family, well represented in tropical America and Asia and in Australia, but poorly represented in tropical Africa. More than 100 generic and more than 3000 specific names have been published.

Much remains to be done on the taxonomy of this family. In many cases the distinctions between currently accepted genera and species are weak. Some of the problems are man-made but the group is intrinsically difficult.

Some species, especially *Syzygium guineense*, and, to a lesser extent, *Eugenia*

* In collaboration with F. K. Kupicha.

capensis, show a remarkable capacity to transgress important chorological and ecological boundaries and so occur in several quite distinct vegetation types. Their habit in different habitats is often strikingly different, though differences in other characters may be slight. At first sight it appears that variation is continuous or even chaotic. A careful analysis of the abundant material now at hand, combined with much more field-evidence than is normally available, shows, however, that most specimens occupy distinct noda in the apparent continuum. Several of these noda can be usefully recognized as subspecies which are of ecogeographical significance. The differences between any two closely related subspecies are often slight, though the differences between more distantly related noda may be quite appreciable. Collectively each of these species shows a much wider range of variation in morphology, habit and ecology than is usual in more normal species. Because of the overall complexity of the situation it is virtually impossible to describe the pattern verbally. Special methods are needed. In the present instance silhouette diagrams have been prepared to show the range of variation in leaf-shape of the subspecies and the differences between them.

There is strong circumstantial evidence both for apomixis and hybridization in Africa, at least for *Syzygium*. Certain American and Asiatic species of *Eugenia* and *Syzygium* are known to be apomictic (van der Pijl in Rec. Trav. Bot. Nederl. **31**: 113–187 (1934), summarized by Gustafsson, *"Apomixis in Higher Plants"* (1946). In the Asiatic species *Syzygium jambos*, apomixis is associated with polyembryony and up to 13 embryos can be found in a single seed. By the time the embryo-sac is mature most of the nucellar tissue, with the exception of the nucellar cap, has disappeared. The cells in the nucellar cap gradually separate from each other and each has the potential of forming an embryo. The adventitious embryos can only develop after pollination and fertilization since they are dependent on nutriment from the endosperm. The fertilized egg-cell does not divide or forms only a few-celled embryo which does not develop further. In *S. malaccensis*, also from Asia, the embryo arises from the integumental epithelium. More than 50% of the seeds in this species examined by van der Pijl contained 2 or more embryos. Both *S. jambos* and *S. malaccensis* include apomictic and sexual strains. Apomictic *S. malaccensis* is self-sterile, whereas apomictic *S. jambos* includes both self-sterile and self-fertile forms. The sexual species examined by van der Pijl, including *S. cumini*, are self-sterile. There is no experimental evidence for apomixis in Africa, but polyembryony has been observed in *Syzygium guineense* subsp. *guineense* (*Hoyle* 1313a, *White* 3273) and subsp. *huillense* (*White* 2139); these taxa are presumably apomictic, at least in part. Merrill & Perry (in Journ. Arn. Arb. **19**: 208 (1938)) state that 4 Chinese species of *Syzygium*, in addition to *S. jambos*, are ordinarily polyembryonic.

Syzygium cordatum and *S. guineense* are connected by a complex chain of intermediates, presumably of hybrid origin (see p. 194).

Economically the family is important as a source of timber, fuel, essential oils and edible fruits. Many species are widely grown in Africa for these purposes and for ornament. Several have become naturalized in the F.Z. area, and others, which are naturalized elsewhere in Africa, are likely to become so. For these reasons, more exotic species are included in this account, and are dealt with more fully, than is usual for Flora Zambesiaca.

Heteropyxis is sometimes included in Myrtaceae, but more often is placed in its own monogeneric family. The latter course is followed here.

Fruit a berry:
Filaments straight in bud; cultivated -   -   -   -   -   -   **1. Feijoa**
Filaments bent inwards in bud:
    Ovary 4–5-locular; fruit many-seeded; cultivated and naturalized   **2. Psidium**
    Ovary 1–2(3)-locular; fruit few-seeded:
        Placentation (at least in upper part) appearing parietal; cultivated   **3. Myrtus**
        Placentation axile:
            Flowers axillary, solitary or in fascicles or in very short racemes, rarely in 2–3-flowered cymules; flower-buds subglobose, receptacle not gradually narrowed into a pedicel-like base above the articulation, not prolonged beyond summit of ovary as an " upper receptacle "; bracteoles persistent; calyx-lobes distinct, ciliolate; petals completely free, subpersistent, ciliolate; filaments and style scarcely longer than the petals; cotyledons partly or completely fused; indigenous and cultivated -   -   -   **4. Eugenia**

Flowers in terminal or, rarely, lateral, 5- or more, usually many-flowered and
corymbose cymes; flower-buds pyriform; receptacle gradually narrowed
into a pedicel-like base above the articulation, distinctly prolonged beyond
summit of ovary as an " upper receptacle "; bracteoles minute, fugacious;
calyx usually forming an indistinctly lobed rim-like extension to the upper
receptacle; petals usually cohering, and falling together as a calyptra at
anthesis; filaments and style c. twice as long as the petals; cotyledons free;
indigenous, cultivated and naturalized -   -   -   - **5. Syzygium**

Fruit a capsule:
Flowers solitary in the axils of leaves or bracts, often in spikes:
Filaments shorter than the petals -   -   -   -   - **6. Leptospermum**
Filaments much longer than the petals:
Filaments free   -   -   -   -   -   -   -   - **7. Callistemon**
Filaments united into bundles opposite the petals   -   -   - **8. Melaleuca**
Flowers in cymes or capitula:
Perianth of unmodified, free sepals and petals:
Filaments united into bundles opposite the petals   -   -   - **9. Tristania**
Filaments free; flowers fused in capitula -   -   -   - **10. Syncarpia**
Perianth modified in whole or in part to form an operculum which falls off entire
when the stamens expand (see p. 206) -   -   -   - **11. Eucalyptus**

1. FEIJOA Berg.

Feijoa Berg. in Linnaea **29**: 258 (1958).

A small genus with 2 species in Brazil. *F. sellowiana* Berg. is grown in Rhodesian
gardens for ornament or fruit. It is a shrub or small tree with small elliptic leaves,
whitish-tomentose beneath, and large solitary flowers with conspicuous red
filaments.

2. PSIDIUM L.

Psidium L., Sp. Pl. **1**: 470 (1753); Gen. Pl. ed. 5: 211 (1754).

Trees or shrubs. Inflorescence axillary, 1–3-flowered. Calyx unlobed and
almost or completely concealing the corolla before anthesis, subsequently splitting
irregularly into 4–5-lobes, persistent. Petals 4–5. Stamens numerous, free.
Ovary imperfectly 4–5-locular, with 4–5 intrusive, parietal placentae; ovules
numerous. Fruit a berry with numerous angular seeds.

More than 100 species, mostly in tropical America; a few in Oceania. Not
indigenous to tropical Africa, but the following two species are widely planted and
are often naturalized.

Branchlets quadrangular; leaves hairy beneath, elliptic or oblong-elliptic, apex rounded
or acute, base rounded   -   -   -   -   -   -   -   - 1. *guajava*
Branchlets terete; leaves glabrous beneath, obovate, apex cuspidate, base cuneate
2. *cattleianum*

1. Psidium guajava L., Sp. Pl. **1**: 470 (1753).—Williamson, Useful Pl. Nyasal.: 101
(1956).—F. White, F.F.N.R.: 303 (1962).—Boutique in F.C.B., Myrtaceae: 31
(1968).—Amshoff in C.F.A. **4**: 94 (1970).—J. H. Ross, Fl. Natal: 261 (1972).
Type said to come from India.

Small tree up to 10 m. tall. Bark smooth, pale brown, peeling over large areas;
branchlets quadrangular. Leaf-lamina up to 13 × 7 cm., elliptic or oblong-elliptic,
apex rounded or acute, base rounded; lower surface densely puberulous; lateral
nerves in c. 16 pairs, parallel, prominent beneath. Flowers white, solitary, axillary.
Calyx completely enclosing the young flower-bud; lobes c. 0·9 × 0·5 cm., in open
flower reflexed, lingulate, whitish-tomentellous inside. Petals c. 1·3 × 0·8 cm.,
oblong-elliptic. Stamens c. 1 cm. long. Style c. 1 cm. long, slightly capitate at
apex. Berry up to 10 cm. long, globose, ovoid or pyriform, usually yellow, with
white, yellow or crimson flesh.

Zambia. B: Balovale Distr., near Chavuma, planted, fl. fr. 13.x.1952, *White* 3479
(FHO). C: Lusaka Distr., near University of Zambia campus, naturalized, fl. 7.xi.1972,
Strid 2477 (FHO). **Rhodesia.** N: Lomagundi Distr., Umboe, naturalized, fl. 4.xi.1969,
Jacobsen 4035 (SRGH). W: Bulawayo, naturalized, fr. iii.1962, *Millen* 8188 (SRGH).
E: Melsetter Distr., Cashel to Umvumvumvu R., naturalized, fl. 10.xi.1952, *Wild* 3879
(SRGH). S: Zimbabwe, naturalized, st. 22.v.1951, *Thompson* 61/51 (SRGH). **Malawi.**

N: Nkhata Bay, near Limpasa R., naturalized, fl. 21.x.1966, *Pawek* 226 (SRGH). S: Blantyre Distr., near Soche Hill College, 1130 m., naturalized, fr. 7.ii.1970, *Brummitt* 8440 (FHO; K). **Mozambique.** Z: between Mocuba and Quelimane, cultivated, fl. 27.v.1949, *Barbosa & Carvalho* 2900 (SRGH). T: Angonia Distr., near Lipidzi Mission, cultivated, fr. 8.iii.1964, *Correia* 187 (LISC). MS: Chimoio Distr., Amatongas Mission, cultivated, fr. 25.i.1948, *Mendonça* 3719 (LISC). SS: Bazaruto I., naturalized, fl. 24.x.1958, *Mogg* 28623 (LISC). LM: Inhaca I., cultivated, fl. 3.x.1957, *Mogg* 27642 (LISC).

The " Guava " is a native of tropical America. Widely planted for its edible fruit, which is eaten both raw and cooked and is rich in vitamin C. Within the F.Z. area it is naturalized in several places.

2. **Psidium cattleianum** Sabine in Trans. Roy. Hort. Soc. **4**: 317, t. 11 (1821).—
    Schroeder in Journ. Arn. Arb. **27**: 314 (1946).—Brenan & Greenway, T.T.C.L.: 378 (1949).—F. White, F.F.N.R.: 303 (1962).—Amshoff in Fl. Gab. **11**: 5 (1966).—
    J. H. Ross, Fl. Natal: 261 (1972). Type from a plant cultivated in China.
    Psidium littorale Raddi, Opusc. Sci. **4**, 5: 254, t. 7 fig. 2 (?1822).—Merrill & Perry in Journ. Arn. Arb. **19**: 199 (1938). Type from Brazil.

Evergreen shrub or small tree up to 8 m. tall; branchlets terete. Leaf-lamina up to 7·5 × 4 cm., apex cuspidate, base cuneate, lower surface glabrous; lateral nerves in c. 9 pairs, scarcely prominent beneath. Flowers white, solitary or in pairs, axillary. Calyx not quite enclosing the young flower-bud; lobes c. 0·2 × 0·3 cm., hemi-orbicular, spreading, glabrous inside. Petals c. 0·5 × 0·5 cm., suborbicular. Satmens c. 0·5 cm. long. Style c. 0·5 cm. long, distinctly capitate at apex. Berry up to 4 cm. in diameter, obovoid to globose, dark purple red.

Zambia. N: Mbala (Abercorn) Distr., Sunzu Farm, cultivated, st. 18.xi.1952, *White* 3716 (FHO). **Rhodesia.** C: National Botanic Gardens, Salisbury, cultivated, fl. & fr. 11.ii.1971, *Biegel* 344 (SRGH). **Malawi.** S: Zomba Plateau, naturalized, fr. 7.ix.1961, *Chapman* 1452 (SRGH). **Mozambique.** LM: Jardim Vasco da Gama, Lourenço Marques, cultivated, fr. 29.i.1971, *Balsinhas* 1786 (SRGH).

The " Strawberry Guava ", a native of S. America, is relatively uncommonly grown for its edible fruit. It is naturalized locally.

Evidence that *P. cattleianum* was published earlier than *P. littorale* is discussed by Schroeder (loc. cit.). The taxonomy of this species and its relatives is somewhat confused. Diploid and tetraploid races (2n =44 & 88) of *P. cattleianum* are known. Some flowers are cleistogamous and others on the same plant open normally.

3. MYRTUS L.

Myrtus L., Sp. Pl. **1**: 471 (1753); Gen. Pl., ed. 5: 212 (1754).

A medium-sized genus with c. 100 species, mostly in the tropics and subtropics, though none is indigenous to tropical Africa. *M. communis* L., the " Myrtle ", is grown as a garden shrub or hedge plant in Zambia and Rhodesia. It is an aromatic evergreen shrub with glossy, dark green foliage; the small, rigid, lanceolate leaves end in a horny point; the flowers are small, white, solitary and axillary.

4. EUGENIA L.

Eugenia L., Sp. Pl. **1**: 470 (1753); Gen. Pl. ed. 5: 211 (1754).

Trees, shrubs, or rhizomatous, geoxylic suffrutices. Flowers bisexual, or ♂ by abortion, axillary, solitary or in fascicles, or in very short racemes, rarely in 2–3-flowered cymules. Flower-buds subglobose. Bracteoles conspicuous and persistent at base of receptacle. Receptacle sharply differentiated from the usually elongate and slender pedicel, in bisexual flowers completely fused to the inferior ovary, in ♂ flowers usually deeply concave; upper receptacle absent or poorly developed. Calyx of 4–5 well-developed, free, ciliolate sepals. Petals 4–5, ± ovate, ciliolate, free, persistent. Stamens numerous, in ♂ flowers inserted on the rim of the receptacle; filaments free, shorter than or scarcely longer than the petals. Ovary 2-locular; placentation axile; ovules numerous. Style as long as or longer than the filaments, punctiform, funnel-shaped, capitate or shortly 2-lobed at the apex, in ♂ flowers rudimentary and less than 0·1 cm. long. Fruit a berry with 1–3 seeds. Cotyledons partly or completely fused.

A large genus with probably several hundred species, mostly in tropical and subtropical America; relatively few in Africa and Asia, where it is largely replaced by *Syzygium*.

In the F.Z. area the flowers are of two kinds. All have well-developed stamens, the anthers of which appear to be functional, but some flowers have a very short style, less than 0·1 cm. long. Sometimes it is lacking altogether. According to Wood (Natal Plants **3**: 7, 1900) the short style represents a developmental stage and afterwards lengthens. Our own observations do not confirm this. Those individuals showing well-developed styles in the open flower show equally well-developed styles in the bud before it has opened, and this feature is always associated with well-developed ovules. In short-styled flowers ovules are not present or are very rudimentary. Styles of intermediate length were never seen. *Eugenia* in our area clearly shows sexual dimorphism. Bisexual flowers are sometimes, though by no means always, accompanied by flowers which appear to be functionally ♂. According to Amshoff (Act. Bot. Neerl. **7**: 58, 1958) the W. African species, *E. calophylloides* DC., is always short-styled. If this is indeed so perhaps *E. calophylloides* is apomictic. This same phenomenon should be looked for in our area. Long-styled flowers in some taxa, e.g. *E. capensis* subsp. *chirindensis*, are very poorly represented in herbaria. The same is true for *E. capensis* subsp. *capensis* in S. Africa.

Sepals not or scarcely longer than broad, erect or spreading; indigenous:
    Leaves if more than 3 times as long as broad then subacuminate at apex   1. *capensis*
    Leaves 3–15 times as long as broad, apex obtuse - - - - 2. *malangensis*
Sepals much longer than broad, strongly reflexed; exotic - - - 3. *uniflora*

1. **Eugenia capensis** (Eckl. & Zeyh.) Sond. in Harv. & Sond., F.C. **2**: 522 (1862).—Sim, For. Fl. Port. E. Afr.: 68 (1909).—Palmer & Pitman, Trees of S. Afr. **3**: 1669 cum tab. & photogr. (1972).—J. H. Ross, Fl. Natal: 260 (1972). TAB. **42**. Type from S. Africa.
    Memecylon capense Eckl. & Zeyh., Enum. Pl. Afr. Extratrop.: 274 (1836). Type as above.

Evergreen shrub or small tree 1–5(12) m. tall, or a rhizomatous suffrutex up to 25 cm. tall. Leaf-lamina 1 × 1–11 × 5·5 cm., chartaceous to coriaceous, very variable in shape and size, suborbicular to lanceolate or lanceolate-elliptic, apex broadly rounded to acuminate, base subcordate to narrowly acute, glabrous. Flowers axillary, solitary or in 2–9-flowered fascicles, rarely in 2–3-flowered cymules or (subsp. *gracilipes*) in racemes up to 1·5 cm. long; pedicels 0·3–0·9(2·8) cm. long; bracteoles c. 0·1 cm. long, ovate-deltate, persistent; receptacle and ovary c. 0·25 × 0·2 cm., obconical. Sepals 4, c. 0·25 × 0·25 cm., hemi-orbicular, margin ciliolate. Petals 4, 0·4–0·55 × 0·3–0·4 cm., lingulate or lingulate-deltate, margin ciliolate. Stamens 0·3–0·5 cm. long. Style 0·4–0·6 cm. long; style-head 0·1–0·3 cm. in diameter, sometimes punctiform (not in the F.Z. area) more often capitate or capitate-2-fid, orbicular, square, elliptic or rectangular in outline. Fruit c. 1·5 × 1·5 cm., purple-black, subglobose.

From Kenya and Zaire, southwards to S. Africa (Cape Province). In forest and thicket; 0–2150 m.

E. capensis is a polytypic species with 8 subspecies, 5 of which occur in the F.Z. area and the remainder in S. Africa. Apart from some trifling differences in size, the flowers and fruits seem to be uniform. The subspecies differ chiefly in habit and habitat and in the shape, size and texture of their leaves. Although the differences between related subspecies are slight, and collectively their variation forms a continuum, they are almost entirely ecogeographically distinct.

Leaves broadly rounded to subacute at apex:
    Leaves coriaceous, up to 4·6 × 3 cm., base usually subcordate or broadly rounded
        subsp. *capensis*
    Leaves chartaceous, 5·5 × 2·3–8·5 × 4·5 cm., base subacute   subsp. *aschersoniana*
Leaves subacuminate to attenuate-acuminate at apex:
    Leaves broadest in the upper ½; apex subacuminate; style-head slightly capitate, scarcely wider than the style - - - - - - subsp. *albanensis*
    Leaves broadest at or near the middle; apex bluntly subacuminate; style-apex 2-fid, more than twice as wide as style - - - - subsp. *nyassensis*
    Leaves broadest in lower ½; apex attenuate-acuminate; style-head subcapitate, scarcely twice as broad as style - - - - - - - subsp. *gracilipes*

Subsp. **capensis**.
  Eugenia capensis var. *major* Sond. in Harv. & Sond., F.C. **2**: 523 (1862). TAB. **42**,
  fig. A. Type from S. Africa.

Shrub or small tree, usually 2–4 m. tall, sometimes less. Leaf-lamina up to
4·6 × 3 cm., coriaceous, suborbicular to elliptic or oblanceolate-elliptic, apex usually
broadly rounded, sometimes subacute, base usually subcordate or broadly rounded,
sometimes cuneate. Flowers solitary or in 1–5-flowered fascicles. Style-head
capitate or 2-fid.

Mozambique. SS: Praia de Zavora, fl. 27.ii.1955, *E. M. & W.* 686 (BM; LISC;
SRGH). LM: Inhaca I., west coast, fr. 15.vii.1959, *Mogg* 31449 (BM; K; SRGH).
Also in S. Africa. In sand-dune thicket and (more rarely) in forest or wooded grassland,
0–200 m.

Mendonça 2951 from Ponta do Ouro, Mozambique is intermediate between subsp.
capensis and the S. African subsp. *gueinzii* (Sond.) F. White. In leaf-shape it more closely
resembles the former, but in its rhizomatous suffruticose habit it agrees with the latter.

Subsp. **aschersoniana** (F. Hoffm.) F. White in Kirkia, **10**: 403 (1977). Type from Tanzania.
  Eugenia aschersoniana F. Hoffm., Beitr. Fl. Centr.-Ost-Africa: 35 (1889).—Brenan
  & Greenway, T.T.C.L.: 376 (1949). TAB. **42** fig. C. Type as above.

Shrub or small tree 1·5–6 m. tall. Leaf-lamina (5·5 × 2·3)6·5 × 3–8·5 × 4·5 cm.,
chartaceous, ± elliptic, tapering from near the middle to the subacute apex and
base, larger and more tapered than in subsp. *capensis* and less pointed at the apex
than in subsp. *chirindensis*. Flowers in 3–9-flowered fascicles. Style-head
capitate.

Mozambique. N: Metónia Distr., between Mandimba and Vila Cabral, Lucumezi R.,
fl. 8.x.1942, *Mendonça* 677 (LISC). Z: Maganja da Costa Distr., Gobene forest, fl.
10.i.1968, *Torre & Correia* 17044 (LISC).
Also in Kenya, Tanzania, Zanzibar and Pemba I. In coastal forest and sand-dune
thicket, 0–50 m.

Intermediates between subspp. **capensis** and **aschersoniana**. TAB. **42** fig. B.
  Eugenia mossambicensis Engl. in Notizbl. Bot. Gart. Berl. **2**: 289 (1899). Type:
  Mozambique, Beira, fl. v.1895, *Schlechter* s.n. (B†, holotype; BM, drawing).

Shrub 1–2 m. tall. Leaf-lamina subcoriaceous, variable in shape but ±
intermediate in shape and size between subsp. *capensis* and subsp. *aschersoniana*.

Mozambique. MS: Mavita, fl. 25.x.1944, *Mendonça* 2575 (LISC); Beira Distr.,
7 km. N. of Macuti Beach, fl. 8.ix.1962, *Noel* 2465 (K; LISC; SRGH).
Occurring in a narrow belt between the areas of subsp. *capensis* and subsp. *aschersoniana*.

Subsp. **albanensis** (Sond.) F. White in Kirkia, **10**: 402 (1977). TAB. **42** fig. F. Type from
  S. Africa.
  Eugenia albanensis Sond. in Harv. & Sond. F.C. **2**: 522 (1862).—J. H. Ross,
  Fl. Natal: 260 (1972). Type as above.

Shrub 1–2·5 m. tall, or small tree up to 8 m. tall, elsewhere a rhizomatous
geoxylic suffrutex. Leaf-lamina 3 × 0·9–8 × 3 cm., subcoriaceous, oblanceolate or
obovate, apex shortly acuminate, base cuneate. Flowers solitary or in long,
pedunculate, 2–3-flowered cymules, borne only towards the base of the shoot,
usually in the axils of reduced leaves or bracts. Peduncles (or pedicels where the
flowers are solitary) 0·5–1·5 cm. long. Bracteoles c. 0·25 cm. long.

Mozambique. SS: Inhambane, Panda, fr. immat. 14.x.1968, *Balsinhas* 1385 (LISC).
LM: Costa do Sol, fr. 6.xi.1963, *Balsinhas* 664 (K; LISC).
Also in S. Africa. In coastal forest and sand-dune thicket.

Over the greater part of its geographical range subsp. *albanensis* is a rhizomatous
suffrutex. At its northern limits in the extreme S. of Mozambique and in adjacent parts
of Tongaland it also occurs as a shrub or small tree.

Subsp. **nyassensis** (Engl.) F. White in Kirkia, **10**: 403 (1977). TAB. **42** fig. D. Type:
  Malawi. without precise locality, fl., *Buchanan* 146 (B†, holotype; BM, drawing; K,
  isotype).
  Eugenia nyassensis Engl. in Notizbl. Bot. Gart. Berl. **2**: 290 (1899). Type as
  above.

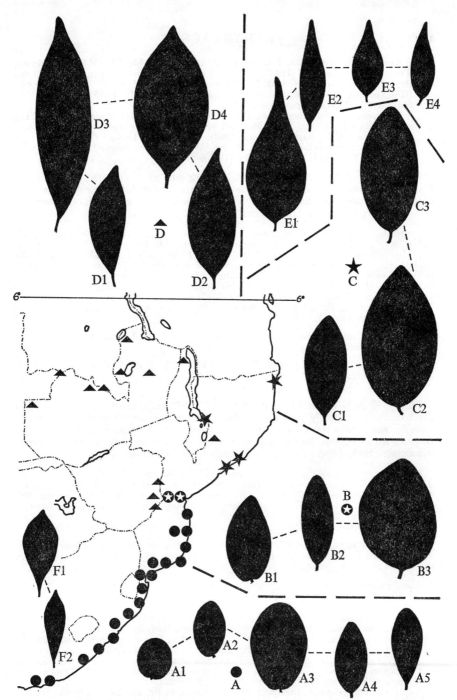

Tab. 42. Leaf-shape in EUGENIA CAPENSIS (all × ½). A.—Subsp. CAPENSIS. A1, *Mogg* 31449; A2, *Barbosa & Balsinhas* 5076; A3, *Gomes e Sousa* 2335; A4, *Mendonça* 1929; A5, *Mogg* 30897. B.—Intermediates between subsp. CAPENSIS and subsp. ASCHERSONIANA. B1, *Mendonça* 2575; B2, *Schlechter* s.n.; B3, *Noel* 2465. C.— Subsp. ASCHERSONIANA. C1, *Torre & Correia* 15162; C2, *Mendonça* 677; C3, *Torre & Correia* 17115. D.—Subsp. NYASSENSIS. D1, *Fanshawe* 1471; D2, *Buchanan* 146; D3, *Holmes* 880; D4, *Fanshawe* 1598. E.—Subsp. GRACILIPES. E1, *Wild et al.* 6602; E2, *Chapman* 968; E3, *Armitage* 68/55; E4, *Torre & Correia* 16429. F.—Subsp. ALBANENSIS. F1, *Balsinhas* 664; F2, *Balsinhas* 1385.

The geographical range of subsp. *gracilipes* interdigitates with the southern part of that of subsp. *nyassensis* and is not shown on the map; that of subsp. *albanensis* partly overlaps with subsp. *capensis* and is not shown.

Eugenia chirindensis Bak. f. in Journ. Linn. Soc., Bot. **40**: 70 (1911).—Steedman, Trees etc. S. Rhod.: 57 (1933). Types: Rhodesia, Chirinda Forest, 1125–1220 m., fl. xii.1905, *Swynnerton* 128 (BM, holotype; K, isotype), fl. x.1906, *Swynnerton* 443 (BM, paratype; K, isoparatype); Chipete Forest Patch, 1160 m., fl. x.1906, *Swynnerton* 1343 (BM, paratype).
? Eugenia bukobensis Engl., tom. cit.: 289 (1899).—Brenan & Greenway, T.T.C.L.: 377 (1949).—F. White, F.F.N.R.: 302 (1962), (including variants A, B & C).—Boutique in F.C.B., Myrtaceae: 25 (1968). Types from Tanzania.
Eugenia aschersoniana sensu R. E. Fr., Wiss. Ergebn. Schwed. Rhod.-Kongo-Exped. **1**: 175 (1914) non F. Hoffm. (1889).

Shrub or small tree 2·5–5(12) m. tall. Bark pale grey, smooth. Leaf-lamina 4 × 2–11 × 5·5 cm., chartaceous, broadest at or near the middle, apex nearly always bluntly subacuminate, base usually cuneate, sometimes rounded. Flowers in (1)3–9-flowered fascicles. Style-head 2-fid.

Zambia. B: Balovale Distr., Zambezi R., Chavuma, fl. 13.x.1952, *White* 3502 (BM; FHO; K). N: Lake Bangweulu, Samfya, fl. 19.viii.1952, *Angus* 240 (BM; FHO; K). W: Mwinilunga Distr., Zambezi R., 7 km. NW. of Kalene Hill, fl. 20.ix.1952, *Holmes* 880 (FHO; K). E: Nyika Plateau, 1·5 km. SW. of Rest House, 2150 m., fl. 15.xi.1958, *Robson* 616 (BM; K; LISC; SRGH). **Rhodesia.** E: Chinyamacheri Peak, 1750 m., fr. 24.viii.1951, *Chase* 3901 (BM; K; LISC; SRGH). **Mozambique.** Z: Serra do Gúruè, 1700 m., fl. immat. 6.xi.1967, *Torre & Correia* 15927 (LISC).
Also in Tanzania, Pemba I., Zanzibar and Zaire. In submontane, montane, fringing and dry evergreen forest, 1050–2150 m.

The leaves of subsp. *nyassensis* are more pointed than those of subsp. *aschersoniana* but much less attenuate than those of subsp. *gracilipes*.

Subsp. **gracilipes** F. White in Kirkia, **10**: 403 (1977). TAB. **42** fig. E. Type: Mozambique, Ribáuè, serra Mepáluè, 1600 m., fl. 9.xii.1967, *Torre & Correia* 16429 (LISC, holotype).

Shrub 1–2 m. tall. Leaf-lamina 3·6 × 1·5–8 × 3·5 cm., chartaceous, broadest near the base, and gradually tapering to the attenuate-acuminate apex, base usually cuneate, sometimes rounded. Flowers solitary or in 2–5-flowered fascicles or in racemes up to 1·5 cm. long; pedicels usually much more slender than in subsp. *nyassensis*. Style-head subcapitate, scarcely twice as wide as style.

Rhodesia. E: Melsetter Distr., Glencoe Forest Reserve, slopes of Mt. Pene, fl. 24.xi.1955, 1370 m., *Drummond* 4987 (K; LISC; SRGH). **Malawi.** N: Nyika Plateau, Luselo Forest, fr., *Salubeni* 354 (K). S: Soche Mt., fl. 6.x.1960, *J. D. Chapman* 968 (FHO; SRGH). **Mozambique.** N: Serra de Ribáuè, 1450 m., fl. 28.i.1964, *Torre & Paiva* 10294 (LISC).
Only known from the F.Z. area. In evergreen forest, 395–1600 m.

Differs from subsp. *nyassensis* in its markedly attenuate-acuminate leaves and less expanded style-head. It tends to occur at lower altitudes but the extent to which the two subspecies are ecogeographically distinct is uncertain.

2. **Eugenia malangensis** (O. Hoffm.) Niedenzu in Engl. & Prantl, Pflanzenfam. **3**, 7: 81 (1893).—Boutique in F.C.B., Myrtaceae: 26, t. 2 (1968).—Amshoff in C.F.A. **4**: 96 (1970). TAB. **43**. Type from Angola.
Myrtopsis malangensis O. Hoffm. in Linnaea **43**: 134 (1881). Type as above.
Eugenia angolensis Engl. in Notizbl. Bot. Gart. Berl. **2**: 288 (1899).—Bak. f. in Journ. Linn. Soc., Bot. **40**: 71 (1909).—R. E. Fr., Wiss. Ergebn. Schwed. Rhod.-Kongo-Exped. **1**: 175 (1914).—Brenan & Greenway, T.T.C.L.: 376 (1949).—F. White, F.F.N.R.: 302 (1962). Type from Angola.
Eugenia marquesii Engl., tom. cit.: 290.—Boutique in op. cit.: 24 (1968).—Amshoff in C.F.A. **4**: 97 (1970). Type from Angola.
Eugenia poggei Engl., tom. cit.: 289 (1899).—R. E. Fr., loc. cit.—F. White, loc. cit. Type from Angola.
Eugenia laurentii Engl., loc. cit. Type from Zaire.
Eugenia stolzii Engl. & von Brehm. in Engl., Bot. Jahrb. **54**: 330 (1917).—Brenan & Greenway, tom. cit.: 377 (1949). Type from Tanzania.

Extensively rhizomatous, geoxylic suffrutex; stems 15–50 cm. tall, caespitose, usually unbranched and burnt back to ground level each year, occasionally escaping fire and then sparsely branched. Leaves opposite or in 3s or 4s, rarely alternate. Leaf-lamina 4 × 1·2–7 × 0·6 or 11 × 3 cm., broadest near the middle, narrowly

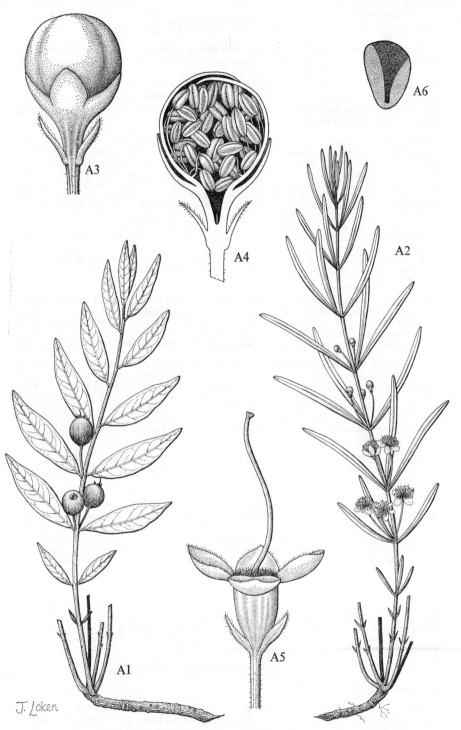

Tab. 43. EUGENIA MALANGENSIS. 1, broad-leaved variant in fruit (× ½) *Chapman* 263;
2, narrow-leaved variant in flower (× ½) *Lawton* 1536; 3, flower-bud (× 5) *Hoyle*
1211; 4, half flower-bud in median section (× 5) *Norman* R4; 5, immature fruit
(× 5) *Lawton* 1536; 6, cotyledon, showing region of fusion (lightly stippled) (× 3)
Brenan 8027.

elliptic or narrowly rhombic to linear, apex obtuse, base cuneate. Flowers usually solitary, in leaf-axils or, when borne near base of shoot, in the axils of bracts, rarely in very short, 2–9-flowered racemes; pedicels 0·6–2 cm. long; bracteoles 0·1–0·2 cm. long, deltate, persistent. Sepals 4, 0·15 × 0·25–0·3 × 0·3 cm., deltate or hemi-orbicular-deltate, ciliolate. Petals 4, 0·6–0·7 × 0·4 cm., lingulate, ciliolate. Style 0·5 cm. long; style-head 0·1 cm. in diameter, capitate. Fruit c. 1·5 × 1·5 cm., purple-black, subglobose.

Zambia. B: Balovale Distr., Chavuma, fl. 5.vii.1954, *Gilges* 397 (K; SRGH). N: Mbala (Abercorn) Distr., Kambole escarpment, 1800 m., fl. fr. 12.ix.1960, *Richards* 13230 (K). W: Ndola Distr., Mwekera Rest House, near Nkana, fl. 27.ix.1947, *Brenan & Greenway* 7968 (FHO; K). C: Nyika Plateau, 9 km. SW. of Rest House, 2150 m., fr. immat. 25.x.1958, *Robson* 342 (BM; K; LISC; SRGH). S: Siamambo Forest Reserve, near Choma, fl. 4.vi.1952, *White* 2920 (FHO). **Rhodesia.** C: Rusapi Distr., St. Faith's Mission Farm, Makoni, 1525 m., fl. 24.iv.1957, *Norman* R4 (K; LISC). E: Chipinga, fl. 27.x.1950, *Crook* M236 (FHO; K; LISC; SRGH). **Malawi.** N: Vipya Plateau, spurs above Luwawa Dam towards Kawendama, fr. 27.xii.1955, *J. D. Chapman* 263 (BM; FHO; K). **Mozambique.** MS: Rotanda, fl. & fr. 26.x.1944, *Mendonça* 2627 (LISC).

Also in Angola, Zaire and Tanzania. In grassland with other suffrutices at the edge of seasonally waterlogged grassland (dambos) and in secondary grassland replacing woodland and forest, 1000–2150 m.

In nearly all of the specimens of *E. malangensis* examined the flowers and fruits are borne on the current year's shoots which had sprouted, usually following fire, from the underground rhizome. When plants escape burning, branching occurs, and it seems that they can survive as low shrubs, for a short time. Since individuals more than 0·5 m. tall have not been reported it would appear that this species has only a limited capacity for upward growth.

E. marquesii Engl. is maintained as a species by Boutique and by Amshoff. It differs chiefly in having more than one flower in the leaf-axils, usually borne in short racemose inflorescences. It occurs sporadically throughout most of the range of *E. malangensis*. Most of the multiflorous individuals appear to have escaped burning the previous season and are producing flowers in the leaf-axils for a second time. If this is so, it would appear that multiflory is no more than a developmental condition and hence unworthy of taxonomic recognition. *E. marquesii* is provisionally regarded here as a synonym of *E. malangensis*.

3. **Eugenia uniflora** L., Sp. Pl. **1**: 470 (1753).—Brenan & Greenway, T.T.C.L.: 377 (1949).—Amshoff in Fl. Gab. **11**: 23 (1966); in C.F.A. **4**: 95 (1970). Type from Ceylon.

Shrub or small tree up to 7 m. tall. Leaf-lamina up to 6 × 3 cm., ovate or ovate-elliptic, apex bluntly subacuminate, base rounded; lower surface glabrous; lateral nerves in 7–9 pairs, indistinct; petiole 0·2 cm. long. Flowers 1·5–2·5 cm. long, white, solitary or few together, axillary; pedicels very slender. Sepals 4, 0·45 × 0·3 cm., ovate-deltate, strongly reflexed. Petals c. 0·6 × 0·3 cm., obovate-elliptic, strongly reflexed. Stamens 0·5 cm. long. Style 0·6 cm. long, punctiform. Berry 1·5–2·5 cm. in diameter, depressed-globose, crimson.

Zambia. W: Kitwe, fr. 10.vii.1967, *Fanshawe* s.n. (SRGH). **Rhodesia.** C: Salisbury, fr. 30.v.1932, *Arnold* in GHS 5850 (SRGH). E: Penhalonga, fl. 17.ix.1966, *Müller* 424 (SRGH). **Malawi.** S: Zomba, fr. 1.xii.1972, *Salubeni* 1880 (SRGH). **Mozambique.** LM: Lourenço Marques, fr. 11.iii.1971, *Balsinhas* 1791 (SRGH).

The "Pitanga Cherry" is a native of Brazil, but is now widely planted in the tropics, both as a hedge plant and for its edible fruit.

SPECIES INSUFFICIENTLY KNOWN

4. **Eugenia sp. 1.**

Tree 15 m. tall. Leaf-lamina up to 15 × 6 cm., elliptic or oblanceolate-elliptic, apex acuminate, base cuneate. Inflorescence sometimes 2-flowered; peduncle c. 1 cm. long, supra-axillary, inserted ± halfway along the internode. Fruit c. 2·5 cm. in diameter, subglobose.

Mozambique. MS: Dombe, Upper R. Lucite, Haroni-Makurupini Forest, 400 m., fr. 4.xii.1964, *Wild, Goldsmith & Müller* 6646 (FHO; K; LISC; SRGH).

5. Eugenia sp. 2.
   Eugenia sp. 1; F. White, F.F.N.R.: 302 (1962).

Suffrutex c. 20 cm. tall. Leaf-lamina (immature) up to 4 × 2 cm., elliptic or lanceolate-elliptic, apex acute to subacuminate, base cuneate, densely and softly pubescent on both surfaces. Flowers in short 2–5-flowered, leafless racemes borne towards the base of the shoot, or solitary in axils of bracts; pedicels up to 4·5 cm. long. Flower-buds c. 1 cm. in diameter.

Zambia. S: Mumbwa, fl. buds, 1911, *Macaulay* s.n. (K).

5. SYZYGIUM Gaertn.
Syzygium Gaertn., Fruct. **1**: 166, t. 33 (1788).

Trees, shrubs or rhizomatous, geoxylic suffrutices. Inflorescence usually a terminal, many-flowered, often compound and corymbose cyme, more rarely (*S. cumini*) a lateral cyme from the axils of fallen leaves. Bracteoles very small and inconspicuous, fugacious. Flowers bisexual, pyriform in bud. Receptacle gradually narrowed towards base to form a stout " pseudopedicel " above the articulation and expanded distally to form a rim-like " upper receptacle " beyond the apex of the ovary, which bears the stamens on its margin. Calyx forming an indistinct, irregularly lobed extension to the upper receptacle (African species) or distinctly lobed (*S. jambos*), eciliolate. Petals 4, suborbicular or obovate, caducous, cohering and falling together as a calyptra (African species) or falling separately (*S. jambos*), eciliolate. Stamens numerous, free, more than twice as long as the petals, very conspicuous, white or cream or bright red (not in F.Z. area). Ovary 2-locular; placentation axile; ovules numerous. Style slightly longer than the filaments, punctiform. Fruit a berry, usually 1-seeded. Cotyledons large, completely free.

In the past, *Syzygium* has frequently been united with *Eugenia* as was done by Lawson in F.T.A. Most authors dealing with African species, however, have kept the two genera apart, though often on unconvincing characters. A recent study by Schmid (in Amer. Journ. Bot. **59**: 423–436, 1972), based on 31 American and Old World (though not including African) species, has convincingly shown that the 2 genera are almost totally different in the structure of their flowers and seeds. In addition to the 9 major characters used to key out the genera in this work (p. 184) Schmid mentions several others, mostly anatomical. As Schmid points out, few of these characters taken singly are absolutely diagnostic, but any one species of *Eugenia* or *Syzygium* would only be anomalous for a small minority of characters. For African species the 9 characters used in the key are absolutely constant. Certain Asiatic species which are cultivated and, in some cases, naturalized in the F.Z. area are however anomalous in a few features.

Petals falling separately, not forming a calyptra; filaments c. 4 cm. long; planted and
   naturalized - - - - - - - - - - - 6. *jambos*
Petals cohering to form a calyptra; filaments up to 1·5 cm. long:
   Inflorescence mostly lateral from axils of fallen leaves on older branchlets; planted and
     naturalized - - - - - - - - - 5. *cumini*
   Inflorescence mostly terminal, sometimes in axils of uppermost leaves, never from axils
     of fallen leaves; indigenous:
     Leaves subsessile, deeply cordate, amplexicaul; petiole up to 0·2 cm. long; stems
       quadrangular and winged; calyx + upper receptacle 0·35–0·5 cm. long; filaments
       (0·8)1·1–1·5 cm. long; fruit urceolate - - - - - 1. *cordatum*
     Leaves at most subcordate, not amplexicaul; petiole usually more than 0·2 cm. long;
       stems mostly not quadrangular, unwinged; calyx + upper receptacle 0·15–0·3 cm.
       long; filaments 0·35–1·1 cm. long; fruit ellipsoid or subglobose:
       Pneumatophores present; leaves broadest near the base - - 3. *owariense*
       Pneumatophores absent; leaves broadest at or near the middle:
         Leaves subcordate or broadly rounded at the base - - - 2. *masukuense**
         Leaves cuneate at the base - - - - - - - 4. *guineense*

1. Syzygium cordatum Hochst. ex Krauss in Flora **27**: 425 (1844).—Sond. in Harv. &
   Sond., F.C. **2**: 521 (1862).—R. E. Fr., Wiss. Ergebn. Schwed. Rhod.-Kongo-Exped.

* See also *S. cordatum × guineense*, p. 194.

G

1: 176 (1914).—Burtt Davy, F.P.F.T. **1**: 241 (1926).—Brenan & Greenway, T.T.C.L.: 379 (1949).—Codd, Trees & Shrubs Kruger Nat. Park: 136, t. 128 fig. c (1951).—Eggeling & Dale, Indig. Trees Uganda: 273 (1952).—Wild, Guide Fl. Vict. Falls: 145 (1953); S. Rhod. Bot. Dict.: 129 (1953).—Brenan in Mem. N.Y. Bot. Gard. **8**: 439 (1954).—Williamson, Useful Pl. Nyasal.: 114 (1955).—Palgrave, Trees of Central Afr. 283 cum tab. & photogr. (1957).—Gomes e Sousa, Dendrol. Moçamb. **1**: 232, cum tab. (1960); Dendrol. Moçamb. Estudo Geral, **2**: 612, t. 183 (1966).— Dale & Greenway, Kenya Trees & Shrubs: 335 (1961).—F. White, F.F.N.R.: 303 (1962).—B. & M. de Winter & Killick, Sixty-six Transvaal Trees: 130 cum photogr. (1966).—Boutique in F.C.B., Myrtaceae: 4, t. 1 (1968).—Amshoff in C.F.A. **4**: 100 (1970).—Palmer & Pitman, Trees of S. Afr. **3**: 1675 cum tab. & photogr. (1972).— J. H. Ross, Fl. Natal: 260 (1972). TAB. **44**; TAB. **45*** fig. A. Type from S. Africa.

Syzygium cordifolium Klotzch in Peters, Reise Mossamb. Bot.: 63, t. 11 (1861) non Walp. (1843). Type: Mozambique, *Klotzsch* s.n. (B†, holotype).

Eugenia cordata (Hochst. ex Krauss) Laws. in F.T.A. **2**: 438 (1871) ("*cordatum*") pro parte excl. specim. *Meller* s.n.—Sim., For. Fl. Port. E. Afr.: 67 (1909).—Bak. f. in Journ. Linn. Soc., Bot. **40**: 71 (1911).—Steedman, Trees etc. S. Rhod.: 56 (1933). Type as for *S. cordatum*.

Small or medium-sized evergreen tree up to 20 m. tall; foliage very dense, slightly glaucous. Bark dark brown, rough, flaking. Young stems quadrangular and slightly winged. Leaf-lamina 4–8 × 2·2–13·5 × 7 cm., lanceolate-elliptic to oblong-elliptic or suborbicular, apex broadly rounded to subacute or, very rarely very shortly subacuminate, or emarginate, base deeply cordate and amplexicaul; petiole up to 0·2 cm. long. Receptacle (including pseudopedicel) + calyx 0·6–0·9 cm. long; calyx + upper receptacle 0·35–0·5 cm. long; filaments (0·8)1·1–1·5 cm. long. Fruit c. 1·8 × 0·9 cm., purple-black, urceolate, persistent calyx + upper receptacle 0·3–0·4 cm. long × 0·5 cm. wide.

Botswana. N: Kwando R., near Caprivi Strip, fl. 13.xi.1974, *Smith* 1183 (SRGH). **Zambia.** B: Mongu Distr., edge of Barotse Plain, Namushakende, fl. 21.vii.1961, *Angus* 3014 (FHO; K; SRGH). N: Lake Bangweulu, Samfya Mission, fl. 19.viii.1952, *White* 3089 (BM; FHO; K). W: Ndola, fr., *Fanshawe* 1705 (FHO; K; LISC; SRGH). C: Kabwe Distr., Mpunde, 53 km. NW. of Kabwe, 1170 m., fl. 1.x.1972, *Kornaś* 2220 (FHO). E: Petauke Distr., Kacholola, fl. 21.x.1967, *Mutimushi* 2189 (FHO). S: Zambezi R., Katombora, fl. 22.viii.1947, *Brenan & Greenway* 7731 (BM; FHO; K). **Rhodesia.** N: Mazoe Distr., Umvukwe Range, fl. 4.ix.1950, *Leach & Bayliss* 10477 (FHO; SRGH). W: Victoria Falls, "rainforest", fl. xi.1959, *Armitage* 193/59 (SRGH). C: Lower Hillside, Salisbury, fl. 20.v.1934, *Gilliland* 181 (FHO; K). E: Umtali Distr., Manyika Bridge, Odzani R., fl. 10.ix.1965, *Chase* 8306 (FHO; SRGH). S: Kyle National Park, fr. 9.ix.1970, *Basera* 178 (SRGH). **Malawi.** N: Misuku Hills, fl., *Lewis* 27 (FHO). C: Kota-kota Distr., Ntchisi Mt., 1330 m., fl. 5.viii.1946, *Brass* 17129 (FHO; K). S: Cholo Mt., 1200 m., fl. 26.ix.1946, *Brass* 17825 (BM; K). **Mozambique.** N: Ribáuè Distr., Serra de Mepálùè, 1550 m., fr. 9.xii.1967, *Torre & Correia* 16424 (LISC). Z: Gúruè, 700–1400 m., fl. 30.ix.1941, *Torre* 3535 (LISC). MS: Serra de Vumba, fr. 2.i.1948, *Barbosa* 791 (LISC). SS: Lower Limpopo, Inhamissa, fl. 16.xii.1957, *Macedo* 16 (LISC). LM: between Marracuene and Manhica, fl. 9.xii.1940, *Torre* 2253 (LISC).

From Uganda and Zaire southwards to Angola, the F.Z. area, and S. Africa. Usually in swamp forest, but also in moist gullies in montane forest (in Malawi and Mozambique) and in wooded, seasonally waterlogged grassland in Rhodesia and on the sandy coastal plain of southern Mozambique, 10–1550 m.

The fruit is edible and the wood is used for making mortars and, after water-seasoning, as construction material, especially for boat planking.

Intermediates, presumably of hybrid origin (see below), between *Syzygium cordatum* and *S. guineense* have frequently been collected, and the two species are connected by a complete range of intermediates. Since, in such situations, it is impossible to know for certain what were the original limits of variation of the parental species, decisions as to their delimitation are inevitably somewhat arbitrary.

1 × 4. Syzygium cordatum × S. guineense.—F. White, F.F.N.R.: 303 (1962) pro parte excl. specim. *White* 3029 et *Morze* 44. TAB. **45** fig. B.

Syzygium intermedium Engl. [in Sitz.-Ber. Königl. Preuss Akad. Wiss. Berl. **52**: 888 (1906) *nom. nud.*; ex R. E. Fr., Wiss. Ergebn. Schwed. Rhod.-Kongo-Exped. **1**: 177 (1914) *nom. illegit.* (*nom. provis.*)] ex Engl. & von Brehm. in Engl., Bot. Jahrb. **54**: 339 (1917).—Brenan & Greenway, T.T.C.L.: 380 (1949).—Wild, Guide Fl.

* TAB. **45** in back-cover pocket.

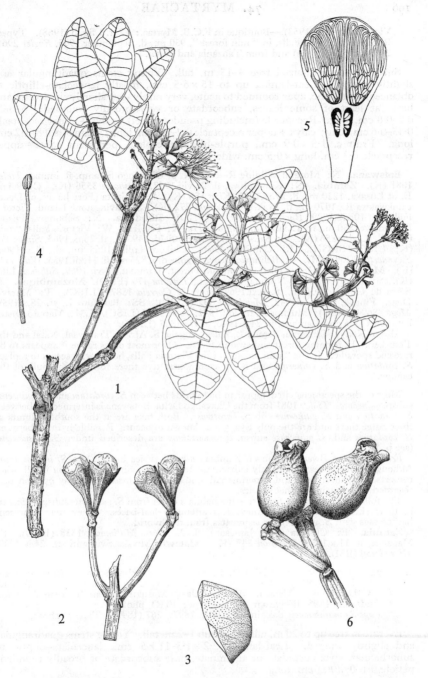

Tab. 44. SYZYGIUM CORDATUM. 1, flowering branchlet (× ½) *Kesler* 106; 2, flower-buds
(× 2); 3, calyptra (× 3); 4, stamen (× 5); 5, flower bud, longitudinal section (× 5)
2–5 from *Devillé* 49; 6, fruits (× 2) *Devillé* 485. By permission of the Editors of
F.C.B.

Vict. Falls: 145 (1953).—Boutique in F.C.B. Myrtac.: 12, t. 1 fig. h (1968). Types: Rhodesia, Victoria Falls, in " rain forest ", 930 m., fl. September 1905, *Engler* 2961, 2965a (B†, syntypes) and from Tanzania and Natal.

Small or medium-sized tree 4–15 m. tall. Young stems quadrangular and slightly winged. Leaf-lamina up to 15 × 6·5 cm., mostly lanceolate-elliptic to oblanceolate-elliptic, apex rounded to acute, very rarely very shortly subacuminate, base (at least of some leaves, subcordate or very broadly rounded; petiole 0·2–0·6 cm. long. Receptacle (including pseudopedicel) + calyx + upper receptacle 0·45–0·6 cm. long; calyx + upper receptacle 0·2–0·3 cm. long; filaments 0·8–1·2 cm. long. Fruit c. 1·8 × 0·9 cm., purple-black, urceolate, persistent calyx + upper receptacle 0·2 cm. long × 0·5 cm. wide.

Botswana. N: Moremi Wildlife Reserve, N. of Okavango Swamp, fl. immat. *Tinley* 1081 (K). **Zambia.** N: Kawambwa, fl. 26.viii.1957, *Fanshawe* 3556 (K). C: 9 km. E. of Lusaka, 1220 m., fl. 27.ix.1955, *King* 156 (K). E: Chipata (Fort Jameson) Distr., Lunkwakwa R., 1070 m., fl. 18.ix.1960, *Wright* 272 (K). S: Livingstone Island, Victoria Falls, st. 10.viii.1929, *Burtt Davy* 20506 (FHO). **Rhodesia.** N: Sebungwe Distr., Karyangwe, 850 m., fr. 15.xi.1958, *Phipps* 1477a (K; SRGH). W: Victoria Falls, " rain forest ", fl., *Allen* 56 (K). C: Salisbury Distr., Hatfield, 1495 m., fl. 26.ix.1965, *Simon* 474 (K; LISC; SRGH). E: Chipinga Distr., Chikore Hills, 1065 m., fl. 11.xi.1906, *Swynnerton* 1309 (BM). S: Belingwe Distr., Ngobe dip, 825 m., fl. 11.xii.1953, *Wild* 4345 (K). **Malawi.** C: Dedza Distr., Chongoni Forest, fl. immat. 14.vii.1969, *Salubeni* 1367 (K). S: Blantyre Distr., Ntunda, fl. immat., *Topham* 479 (FHO). **Mozambique.** Z: Serra do Gúruè, 1200 m., fl. 4.xi.1967, *Torre & Correia* 15870 (LISC). T: Marávia Distr., Fíngoè, fl. 24.ix.1942, *Mendonça* 392 (LISC). SS: Bazaruto I., st. 28.x.1958, *Mogg* 28694 (LISC); Muchopes, fr. 26.i.1941, *Torre* 2549 (LISC). LM: Maputo Distr., between Goba and Catuane, fl. 24.x.1940, *Torre* 1863 (LISC).

Also in Kenya, Tanzania, Pemba I., Zaire, Angola, S. Africa (Transvaal, Natal and the Transkei). In fringing forest and swamp forest. Over most of its range it appears to be rare and sporadic. In the " rain forest " of the Victoria Falls, however, it seems to replace *S. cordatum* and *S. guineense* and is the commonest tree there forming the bulk of the canopy.

Most of the specimens cited appear to be hybrids between *S. cordatum* and *S. guineense* subsp. *guineense*. *Tinley* 1081 from the Okavango Delta, Botswana is intermediate between *S. cordatum* and *S. guineense* subsp. *barotsense*. Both taxa are at the southern limits of their range there and are the only Syzygiums known to occur. Possible hybrids between *S. cordatum* and *S. guineense* subsp. *afromontanum* are described under *S. masukuense* (see p. 197).

Fries 727 from Bwana Mkubwa, Zambia, is cited (Fries loc. cit.) as *S. intermedium*. Although some leaves are slightly subacuminate it agrees with *S. cordatum* in all other respects, and, until field and experimental studies can be made, is in my opinion best regarded as belonging to that species.

The following specimens from riverine habitats differ from *S. guineense* subsp. *guineense* in their shorter petioles and much less attenuate leaf-bases. They may represent backcrosses with *S. guineense* or segregates from the hybrid.

Zambia. E: Chipata (Fort Jameson), fl. 2.x.1966, *Mutimushi* 1538 (FHO). S: Mapanza, fl. 11.x.1953, *Robinson* 349 (K). **Malawi.** Mwaka Syunguti, fl. 24.ix.1930, *Migeod* 950 (BM).

2. **Syzygium masukuense** (Bak.) R. E. Fr., Wiss. Ergebn. Schwed. Rhod.-Kongo-Exped. **1**: 177 (1914).—Brenan in Mem. N.Y. Bot. Gard. **8**: 439 (1954). TAB. **45**, fig. C–D. Type: Malawi, Misuku hills (" Masuku plateau "), 1980–2130 m., fl.-buds July 1896, *Whyte* s.n. (K, holotype; FHO, phot.).
Eugenia masukuensis Bak. in Kew Bull. **1897**: 267 (1897). Type as above.

Evergreen tree up to 20 m. tall or a shrub 1–2 m. tall. Young stems quadrangular and slightly winged. Leaf-lamina 2·2 × 1·3–11 × 5 cm., lanceolate-elliptic to suborbicular, apex cuspidate or acuminate, base subcordate or broadly rounded; petiole 0·1–0·6(0·8) cm. long.

Only known from the F.Z. area where it is virtually confined to montane habitats. In the protologue Baker states that *S. masukuense* is intermediate between *S. guineense* (" owariense ") and *S. cordatum*. It is possible that *S. masukuense* originated by hybridization between *S. cordatum* and *S. guineense* subsp. *afromontanum*. This hypothesis is supported by the fact that *S. masukuense* subsp. *masukense* only occurs where the ranges of these two taxa overlap, and most collectors state that it is rare or occurs as isolated individuals. Present evidence, however, is inconclusive since *S. masukuense* is by no means exactly intermediate between its putative parents and is not known to be connected to *S. cordatum* by other intermediates. This is in contrast to *S. cordatum* and *S. guineense* subsp. *guineense* which are more-or-less completely connected by a chain of intermediates (see p. 194). Other considerations weigh against the supposed hybrid origin of *S. masukuense*. *S. masukuense* subsp. *pachyphyllum* only differs from the nominate subsp. in having smaller, sometimes proportionately broader, less acuminate leaves and the dividing line between them is somewhat arbitrary. It occurs at higher altitudes than *S. cordatum* in the F.Z. area. Its fruit shows none of the features of that of *S. cordatum*. Unfortunately the fruit of subsp. *masukuense* is still unknown. The possibility that *S. masukuense* subsp. *pachyphyllum* may have originated by hybridization between *S. masukuense* and small leaved variants (Group B) of *S. guineense* subsp. *afromontanum* cannot be ruled out.

Subsp. **masukuense**. TAB. **45** fig. C.

Small tree 6–15 m. tall. Crown sometimes very wide-spreading. Bark dark, roughly scaly. Leaf-lamina 6·5 × 3·0–11 × 5 cm., lanceolate-elliptic or elliptic, apex distinctly acuminate, base broadly rounded or subcordate; petiole 0·2–0·6(0·8) cm. long. Receptacle (including pseudopedicel) + calyx 0·45–0·6 cm. long; calyx + upper receptacle 0·2–0·25 cm. long; filaments 0·6–0·8 cm. long. Fruit unknown.

Rhodesia. C: Salisbury Distr., Gomokurire-Chinamora Reserve, fl. 22.viii.1954, *Meyer* 2 (K). E: Umtali Distr., Mt. Nhuri, Mtasa S. Reserve, fl. 31.vii.1955, *Chase* 5986 (K; LISC). **Malawi.** N: Mugesse Forest Reserve, Misuku Hills, st. 3.ix.1952, *Chapman* 8 (FHO); Vipya Plateau, Chikangawa, 1675 m., fl. 1.x.1964, *Chapman* 2266 (FHO). C: Kota-kota Distr., Ntchisi Mt., 1600 m., fl. 26.vii.1946, *Brass* 16966 (K). S: Zomba Plateau, Mlunguzi R., fl. 25.ix.1961, *Chapman* 1466 (FHO; K; SRGH).
Confined to the F.Z. area. In montane forest on well-drained slopes, montane fringing forest and secondary montane grassland. Mostly between 1350 and 1750 m.

Subsp. **pachyphyllum** F. White in Kirkia, **10**: 404 (1977). TAB. **45** fig. D. Type: Rhodesia, Melsetter Distr., Chimanimani Mts., Stonehenge, 1710 m., among quartzite crags, fl. 28.xii.1959, *Goodier & Phipps* 309 (K, holotype; SRGH, isotype).

Tree up to 20 m. tall, or, on exposed crests, a shrub 1–2 m. tall. Young stems quadrangular and slightly winged. Leaf-lamina up to 6·5 × 4·2 cm., usually much smaller, oblong-elliptic to suborbicular, apex cuspidate or shortly acuminate, base subcordate; petiole 0·1–0·2 cm. long. Receptacle (including pseudopedicel) + calyx 0·3–0·4 cm. long; calyx + upper receptacle 0·15–0·2 cm. long; filaments 0·3–0·7 cm. long. Fruit c. 1·4 × 1·5 cm., purple-black, subglobose; persistent calyx 0·1 × 0·25 cm.

Rhodesia. E: Inyanga Distr., Pungwe-Mtendere divide, common at upper limits of forest in Pungwe Gorge, fl. 28.xii.1965, *West* 7078 (FHO; SRGH). **Malawi.** N: N. Vipya Plateau, near summit of Chimaliro, fl. 16.ii.1965, *Chapman* 2317 (FHO). **Mozambique.** MS: Báruè Distr., Serra de Choa, 1600 m., fr. 7.ix.1943, *Torre* 5858 (LISC).
Confined to the F.Z. area, where it is plentiful on the mountains of Rhodesia and adjacent parts of Mozambique. It occurs as a medium-sized tree in kloofs towards the upper limit of montane forest, and as a shrub or stunted tree on exposed rocky ridges, 1675–2135 m.
The solitary record from Chimaliro rock in N. Malawi suggests that this taxon may be of polytopic origin.
This subspecies is superficially similar in appearance to *S. sclerophyllum* Brenan which is confined to certain mountains in Tanzania and Kenya. The leaves of the latter, however, differ in being proportionately broader and more coriaceous and in having markedly recurved margins and a less well-developed apiculus.

3. **Syzygium owariense** (Beauv.) Benth. in Hook., Niger Fl.: 359 (1849).—Keay in F.W.T.A. ed. 2, **1**: 240, t. 95 fig. C (1954).—F. White, F.F.N.R.: 304 (1962).—Amshoff in Fl. Gab. **11**: 13 (1966); in C.F.A. **4**: 109 (1970).—Boutique in F.C.B. Myrtaceae: 9, t. 1 fig. E (1968). Type from Nigeria.
Eugenia owariensis Beauv., Fl. Oware Benin 2: 20, t. 70 (1810). Type as above.

Medium-sized or tall evergreen tree up to 30 m. tall, but usually smaller. Bole straight. Bark grey, smooth, or rough and flaking. Roots with abundant knee-shaped pneumatophores. Young stems ±terete. Leaf-lamina 11 × 3–13 × 6 cm., lanceolate, gradually tapered from near the base to a conspicuous, elongate, subcaudate acumen, base cuneate to subtruncate. Receptacle (including pseudo-pedicel) + calyx 0·4–0·65 cm. long; calyx + upper receptacle 0·2–0·25 cm. long; filaments 0·4–1·1 cm. long. Fruit c. 1·4 × 1 cm., purple-black, ellipsoid; persistent calyx 0·15 × 0·4 cm.

Zambia. N: Mpika Distr., Muchinga Escarpment, 48 km. S. of Shiwa Ngandu, fr. 29.xi.1952, *White* 3797 (BM; FHO; K). C: Serenje, fl. 27.ix.1961, *Fanshawe* 6722 (FHO; K; SRGH). **Rhodesia.** Without locality, st., *Gilliland* 1341 (FHO). **Malawi.** N: S. Vipya Plateau, Mzuzu, fr. i.1956, *Chapman* 265 (BM; FHO; K). C: Kota-kota Distr., Ntchisi Forest, st. 7.v.1961, *Chapman* 1303 (FHO). S: Mt. Mlanje, Lukulezi stream, fl. 11.ix.1957, *Chapman* 430 (BM; FHO; K). **Mozambique.** N: Mandimba Distr., Serra da Massangulo, fl. 12.x.1942, *Mendonça* 791 (LISC). Z: Gúruè Distr., Mt. Namuli, fl. 13.viii.1949, *Andrada* 1870 (LISC).

From Sierra Leone to Uganda and southwards to the F.Z. area, where it only occurs in fringing forest but always in swampy places. Confined to higher-rainfall areas, especially above 1350 m. It is particularly characteristic of swamp forest surrounding springs at the sources of small perennial streams. It often occurs in almost pure stands. Associates include: *Ocotea usambarensis, Podocarpus latifolius, Dacryodes edulis* and *Xylopia aethiopica*.

4. **Syzygium guineense** (Willd.) DC., Prodr. **3**: 259 (1828).—R. E. Fr., Wiss. Ergebn. Schwed. Rhod.-Kongo-Exped. **1**: 176 (1914).—Burtt Davy, F.P.F.T. **1**: 241 (1926).—O. B. Mill., B.C.L.: 45 (1948); in Journ. S. Afr. Bot. **18**: 65 (1952).—Brenan & Greenway, T.T.C.L.: 380 (1949).—Codd, Trees & Shrubs Kruger Nat. Park: 136, t. 128 figs. a & b (1951).—Pardy in Rhod. Agric. Journ.: 76 cum photogr. (1952).—Eggeling & Dale, Indig. Trees Uganda: 273, tt. 12 & 60 (1952).—Wild, Guide Fl. Vict. Falls: 145 (1953); S. Rhod. Bot. Dict.: 130 (1953).—Brenan in Mem. N.Y. Bot. Gard. **8**: 439 (1954).—Williamson, Useful Pl. Nyasal.: 428 (1955).—Palgrave, Trees of Centr. Afr.: 287 cum. tab. & photogr. (1957).—Keay in F.W.T.A. ed. 2, **1**: 240, t. 96 (1954).—Gomes e Sousa, Dendrol. Moçamb. **1**: 235 cum tab. (1960); Dendrol. Moçamb. Estudo Geral **2**: 613, t. 184 (1966).—Dale & Greenway, Kenya Trees & Shrubs: 335, t. 20 (1961).—F. White, F.F.N.R.: 303 (1962).—Amshoff in Fl. Gab. **11**: 13 (1966); in C.F.A. **4**: 104 (1970).—Friedrich-Holzhammer in Merxm., Prodr. Fl. SW. Afr. **97**: 1 (1966).—Boutique in F.C.B. Myrtac.: 13 (1968).—Palmer & Pitman, Trees of S. Afr. **3**: 1681 cum tab. et photogr. (1972).—J. H. Ross, Fl. Natal: 260 (1972). TAB. **45** fig. E–H. Type from W. Africa.

   Calyptranthes guineensis Willd., Sp. Pl. **2**: 974 (1800). Type as above.
   Eugenia guineensis (Willd.) Baill. ex Laness., Pl. Ut. Col. Fr.: 822 (1886).—Sim. For. Fl. Port. E. Afr.: 67 (1909). Type as above.
   Eugenia owariensis sensu Bak. f. in Journ. Linn. Soc., Bot. **40**: 71 (1911).—sensu Steedman, Trees etc. S. Rhod.: 56 (1933) non Beauv. (1810).

Small or medium-sized (elsewhere a large) evergreen tree or a suffrutex. Leaf-lamina 4 × 2–14 × 7 cm., very variable in shape, broadest at or near the middle, apex obtuse to acuminate, base cuneate; petiole 0·2–2·2 cm. long. Receptacle (including pseudopedicel) + calyx 0·35–0·65 cm. long; calyx + upper receptacle 0·15–0·25 cm. long; filaments 0·35–0·9 cm. long. Fruit subglobose or ellipsoid, 1·3–3·5 × 1·2–2·5 cm.

Widespread in Africa from Sénégal to Somalia and southwards to SW. Africa, Botswana, S. Africa (the Transvaal and Eastern Cape Province).

Syzygium guineense, as interpreted here, is one of the most widespread African tree species. It certainly occurs in a greater range of vegetation types and shows a larger variety of growth forms than any other African plant. Outside our area it is widely distributed in the rain forests of the Guineo-Congolian Region and the montane forests of the Afromontane Region, as well as in riverine forest and woodland in the Sudanian Region. In habit it ranges from a lofty forest tree 30 m. or more tall with large plank buttresses to a rhizomatous suffrutex less than 30 cm. tall, but in other respects its structure varies little. The flowers are essentially uniform and the fruits show only slight difference in shape. Variation in leaf-shape and size at first sight appears to be continuous, but in the field much of this variation is seen to be closely correlated with ecology and habit. Even Engler as long ago as 1921 (Pflanzenw. Afr. **3**, 2: 738) accepted *S. guineense* as a widespread transgressive species, the limits of which roughly correspond to those adopted here.

Primarily on the basis of leaf-shape 11 subspecies can be recognized for Africa as a whole. Some of these are connected by intermediates, and others, were it not for ecogeographic weighting, might appear to be arbitrarily delimited, but the overwhelming majority of specimens occupy noda which can be well characterized in ecogeographical terms. The subspecies are thus biologically significant: not merely typological abstractions; they have objective reality.

The species as a whole is possibly the most notorious ecological and chorological transgressor in Africa, but each subspecies is confined to a single major habitat or range of closely related habitats within a single major chorological unit, or is only slightly transgressive to about the same degree as its " normal " associates. The only exception to this is subsp. *guineense* which occurs in fringing forest and in woodland without showing any taxonomically significant differences. There are good reasons for believing that at least some of the complexity of *S. guineense* may be associated with apomixis. Apomixis, associated with nucellar polyembryony, has been reported for the Asiatic species, *S. jambos*. Polyembryony has also been found to occur in *S. guineense* subsp. *guineense* and subsp. *huillense*. The nature of this has not been studied developmentally, but, since polyembryony in Angiosperms is usually of the nucellar type, and is thus associated with at least facultative apomixis, it is quite likely that polyembryonic *S. guineense* is also apomictic. *S. guineense* and its relatives have much to offer the experimental taxonomist.

Trees; leaf-lamina often acuminate; petiole (0·6)0·8–2·2 cm. long:
  Leaf apex with a narrow, delicate acumen:
    Acumen long; bark on bole rough, brownish; fruit purple-black, subglobose or
      slightly ellipsoid, c. 1·8 × 1·5 cm. - - - - - subsp. *afromontanum*
    Acumen short; bark on bole smooth, pale grey or greyish-white; fruit cerise, ovoid
      or ellipsoid, c. 3·5 × 2·5 cm. - - - - - - subsp. *barotsense*
  Leaf apex non-acuminate, subacuminate or with a broad-based, coarse acumen
                                      subsp. *guineense*
Suffrutex; leaf-lamina obtuse to subacuminate; petiole 0·2–0·5(0·6) cm. long
                                        subsp. *huillense*

Subsp. **afromontanum** F. White, F.F.N.R.: 455, 303 (1962).—Boutique in F.C.B., Myrtac.: 14, t. 1 fig. B (1968).—Amshoff in C.F.A **4**: 107 (1970).—Chapman & White, Evergreen Forests of Malawi: 44, photogr. 21 (1970). TAB. **45** fig. F. Type: Zambia, Ndola, fl. 14.viii.1952, *White* 3058 (FHO, isotype; K, holotype).

Medium-sized, evergreen tree, in F.Z. area rarely more than 20 m. tall, rarely a small tree or a shrub. Bole often short. Crown wide-spreading when tree is growing in open. Bark pale brown, rough. Young stems more-or-less terete or quadrangular and winged. Receptacle (including pseudopedicel) + calyx 0·4– 0·55 cm. long; calyx + upper receptacle 0·2–0·25 cm. long; filaments 0·35–0·7 cm. long. Fruit c. 1·8 × 1·5 cm., purple-black, subglobose or very slightly ellipsoid, persistent calyx c. 0·15 × 0·3 cm.
Widespread on the mountains and uplands of Africa.

Over most of its range this subspecies is very uniform. At its upper altitudinal limits on certain mountains in the F.Z. area it is replaced by a variant (Group B, see below) which has quadrangular branchlets and much smaller more sclerophyllous leaves and is also often of greatly reduced stature. Group B is connected by intermediates on the one hand to Group A (subsp. *afromontanum* sensu stricto) and on the other hand to *S. masukuense* subsp. *pachyphyllum*.

Group A (subsp. *afromontanum* sensu stricto).

Small or medium-sized tree; young stems subterete. Leaf-lamina 5·5 × 2– 13 × 4 cm., more-or-less elliptic but markedly attenuate towards the distinctly and acutely acuminate apex and the narrowly cuneate base.

Zambia. B: Kalabo Distr., Njobositu Forest, fr. 12.xi., *Rea* 142 (K). N: Lake Bangweulu, Chilubi I., fl. 31.viii.1933, *Michelmore* 559 (K). W: Ndola 1280 m., fl. 12.ix.1938, *Greenway & Miller* 5662 (FHO). C: Serenje Distr. Kundalila, fl. 1.ix.1966, *Fanshawe* 9790 (K; LISC). **Rhodesia.** C: Salisbury, on granite kopje, fl. 1.vi.1927, *Eyles* 4952 (FHO; SRGH). E: Inyanga Distr., Rodel Farm, 1920 m., fl. 26.ii.1966, *Chase* 8381 (FHO; LISC; SRGH). **Malawi.** N: Vipya Plateau, Kawendama, 2135 m., fr. 6.xi.1964, *Chapman* 2286 (FHO). C: Kota-kota Distr., Ntchisi Mountain, 1500 m., fl. vii.1949, *Brass* 17027 (BM; FHO; K). S: Mlanje Mountain, Lichenya path, 1675 m., fr. 31.xii.1956, *Chapman* 335 (FHO). **Mozambique.** Z: Gúruè, 1300 m., fr. 9.xi.1967, *Torre & Correia* 16031 (LISC). MS: Báruè, Serra de Choa, 1600 m., fr. 10.xii.1965, *Torre & Correia* 13533 (LISC).

From the Sudan Republic to the F.Z. area and Angola and Zaire (Katanga). In primary and secondary montane forest and, on certain mountains, in forest fringing perennial streams below the lower limit of montane forest as on Mt. Mlanje, where it occurs with *Syzygium cordatum, Parkia filicoidea, Newtonia buchananii* and *Adina microcephala.* Away from the main mountainous massifs it occurs in a wide range of habitats mostly above 1280 m. including " kloof " forest on rocky hills, termite mounds in *Brachystegia taxifolia* woodland, and forest on levées fringing the Chambeshi River. It is one of the commonest and most characteristic species of dry evergreen forest (*mateshi*) of the Zambezi-Zaire watershed. The fruit is edible. 915–2150 m.

Note. In riverine habitats mostly below 1280 m. Plants which are intermediate between subsp. *afromontanum* and subsp. *barotsense* are found (see below), but they are not necessarily of hybrid origin.

Group B.

Tree up to 20 m. tall or shrub as low as 1·3 m. tall. Young stems quadrangular and winged. Leaves up to 5·5 × 2·2 cm.

Rhodesia. E: Chimanimani Mts., in bed of Bundi R., below Turret Towers, 1675 m., fr. immat. 5.iv.1969, *Kelly* 66 (FHO; K). **Mozambique.** MS: Mt. Gorongosa, Mt. Nhandore, 1740 m., fr. 21.x.1965, *Torre & Pereira* 12487 (LISC).
Confined to the F.Z. area. In montane forest and scrub forest. On Gorongosa with *Widdringtonia nodiflora, Philippia* and *Podocarpus latifolius.*

Intermediates between subspp. **afromontanum** and **barotsense.**

Leaves less acuminate than those of subsp. *afromontanum* and more acuminate than those of subsp. *barotsense.* Fruit c. 1·8 × 1·2 cm., purple-black, ellipsoid.

Zambia. B: 16 km. N. of Kalabo, fl. 14.xi.1959, *Drummond & Cookson* 6471 (K; LISC; SRGH). N: Mpika Distr., Lufila R., Luangwa Valley, fl. 15.x.1957, *Savory* 222 (K; SRGH); Mbala (Abercorn) Distr., Sansia Falls, Kalambo R., 1350 m., fr. 28.iii.1957, *Richards* 8909 (K). W: Kitwe, riverbanks, fl. 17.xii.1958, *Fanshawe* 5044 (FHO). E: Lundazi, fl. 20.x.1967, *Mutimushi* 2219 (FHO). **Rhodesia.** N: Lomagundi Distr., Sinoia, Hunyani R., fl. x.1926, *Rand* 283 (BM). C: Wedza Mt., fl. 18.ix.1964, *West* 4966 (FHO). E: Chipinga Distr., Sabi R., between Birchenough Bridge and Chisumbanje, fl. 19.x.1963, *Plowes* 2351 (FHO; SRGH). **Mozambique.** N: between Mocímboa da Praia and Quiterajo, fl. immat. 12.ix.1948, *Barbosa* 2114 (LISC). Z: Maganja da Costa, fr. 12.xi.1966, *Torre & Correia* 14519 (LISC). MS: Serra de Choa, 1300–1400 m., fl. 10.vii.1941, *Torre* 3000 (LISC).
Only known from the F.Z. area and Angola. These intermediates have a scattered distribution and mainly occur between the main areas occupied by subspp. *barotsense* and *afromontanum* but some specimens which are indistinguishable have been collected in Mozambique at considerable distances from the area of subsp. *barotsense.*

Apart from *Drummond & Cookson* 6471, which was collected in woodland at the edge of a grassy plain on Kalahari sand, all the records are from the banks of streams. In all cases the leaves are more distinctly acuminate than those of subsp. *barotsense* but less so than those of subsp. *afromontanum.* These intermediates appear to combine the smooth greyish-white bark of *barotsense* with the purple fruit of *afromontanum.* In shape the fruit more closely resembles *barotsense.*

Intermediates between subsp. **afromontanum** Group B and **S. masukuense** subsp. **pachyphyllum.**

Rhodesia. E: Umtali Distr., Tsetsera Mts., Wengezi R., 1830 m., fl. 26.i.1965, *Chase* 8251 (FHO; SRGH). **Mozambique.** MS: Serra Zuira, planalto Tsetsera, 1800 m., fr. 6.xi.1965, *Torre & Pereira* 12726 (LISC).

Subsp. **barotsense** F. White, F.F.N.R.: 455, 304 (1962).—Amshoff in C.F.A. **4**: 108 (1970). TAB. **45** fig. G. Type: Zambia, Livingstone Distr., Katombora, fl. 16.viii.1947, *Brenan & Greenway* 7758 (FHO, isotype; K, holotype).

Small evergreen tree up to 12 m. tall, usually with several stems from near base. Bole sometimes with aerial roots like those of *Ficus* in lower 2 m. Bark on bole and branches smooth, pale grey or greyish-white; bark on branchlets pale brown or grey, almost smooth. Young stems ±terete, not winged or angled. Leaf-lamina 5 × 2–10 × 3·5 cm., more-or-less elliptic, apex very shortly acuminate, base cuneate. Receptacle (including pseudopedicel) + calyx 0·5–0·65 cm. long; calyx + upper receptacle 0·2–0·25 cm. long; filaments 0·6–0·9 cm. long. Fruit c. 3·5 × 2·5 cm., cerise, ovoid or ellipsoid; persistent calyx 0·3 × 0·5 cm.

Caprivi Strip. Zambezi R., Katima Mulilo, 915 m., fr. immat. 8.i.1959, *Killick &
Leistner* 3316 (K; SRGH). **Botswana.** N: Zambezi R., Kazungula, fl. iv.1935, *Miller*
B/72 (FHO). **Zambia.** B: near Mankoya Boma, fr. 23.ii.1952, *White* 2139 (FHO).
S: Palm Island, Victoria Falls, fl. 20.xi.1949, *Wild* 3116 (FHO). **Rhodesia.** W: Wankie
Distr., Zambezi R., Kazungula, fr. iv.1955, *Davies* 1091 (LISC; SRGH).
Also in Angola and Zaire. Confined to the upper Zambezi basin and adjacent regions
where it is the most abundant and most characteristic tree of fringing forest, often
occurring almost pure, sometimes in strips only one tree wide. At times of flood it may be
submerged to a depth of several metres.

Although in leaf shape subsp. *barotsense* closely approaches subsp. *afromontanum* and
the two are connected by intermediates, all observant collectors from the time of Kirk
(1860) have commented on its distinctive appearance in the field, which is based largely on
the smooth whitish bark and bright green foliage.

Subsp. **guineense.**—Boutique in F.C.B., Myrtaceae: 13, t. 2 fig. A (1968).—Amshoff in
C.F.A. **4**: 105 (1970). TAB. **45** fig. E. Type from Ghana.
    Syzygium guineense var. *macrocarpum* Engl., Pflanzenw. Afr. **3**, 2: 738 (1921).—
Brenan, loc. cit.—Keay, tom. cit.: 241, t. 95 fig. F (1954).—Amshoff in Fl. Gab. **11**:
14 (1966) (" *macrocarpon* "). Type unknown.
    Syzygium guineense var. *guineense* Keay in F.W.T.A. ed. 2, **1**: 240, t. 95 fig. D,
t. 96 (1954).—Brenan in Mem. N.Y. Bot. Gard. **8**: 439 (1954). Type as above.
    Syzygium owariense sensu Williamson, Useful Pl. Nyasal.: 115 (1955).
    Syzygium guineense subsp. *macrocarpum* (Engl.) F. White, F.F.N.R.: 455, 304
(1962).—Boutique, tom. cit.: 16, t. 2 fig. E (1968).—Amshoff, tom. cit.: 106 (1970).
Type as above.

Small evergreen tree (3)5–10 m. tall. Bark on bole dark brown, rough, scaly;
bark on branchlets dark grey or ferrugineous, minutely flaking. Young stems
terete, not or only slightly angled. Leaf-lamina 7 × 2·6–14 × 7 cm., elliptic, ovate,
lanceolate or lanceolate- or ovate-elliptic, apex obtuse to shortly acuminate, base
mostly cuneate; petiole (0·6)1–2·2 cm. long. Receptacle (including pseudopedicel) + calyx 0·35–0·5 cm. long; calyx + upper receptacle 0·15–0·2 cm. long;
filaments 0·55–0·7 cm. long. Fruit 1·3 × 1·3–2·5 × 2 cm., purple-black, subglobose
or ellipsoid; persistent calyx 0·1 × 0·3–0·2 × 0·5 cm.
From Sénégal to Ethiopia and southwards to the F.Z. area, Angola and S. Africa.
In woodland and in fringing forest, 20–1750 m.
In leaf- and fruit-shape and in ecology, this is the most variable of the 4 subspecies in our area. On the basis of leaf-shape 3 major variants can be recognized—
one (Group B) occurs in woodland; another (Group C) is confined to fringing
forest, whereas the third (Group A) occurs both in woodland and fringing forest.
Fruit shape is partly correlated with ecology but partly cuts across groupings based
on leaf-shapes. There is a marked tendency for riverine plants both in Groups A
and C to have ellipsoid fruits, and woodland plants, both in Groups A and B, to
have subglobose fruits, but intermediates and exceptions occur.

Group A (subsp. *guineense* sensu stricto). TAB. **45** fig. E 1–3.

Leaves 9 × 4–14 × 7·5 cm., apex suddenly and shortly but distinctly acuminate.
Fruit 1·5 × 1·5–2·5 × 2 cm., ellipsoid (chiefly in riverine plants) or subglobose
(chiefly in woodland plants).

Zambia. N: Mbala (Abercorn) Distr., Chilongowelo Escarpment, 1500 m., fl.
30.viii.1960, *Richards* 13171 (K; SRGH). W: Mwinilunga Distr., Zambezi rapids,
fl. 18.v.1969, *Mutimushi* 3184 (K). C: Handsworth Park, Lusaka, 1280 m., fl. 15.ix.1960,
Best 227 (LISC; SRGH). E: Chipata (Fort Jameson), fl. 31.viii.1929, *Burtt Davy* 21035
(FHO). S: 19 km. N. of Choma, *White* 6756 (FHO; K). **Rhodesia.** N: Goromonzi,
Chindamora Reserve, fl. immat. 9.ix.1960, *Rutherford-Smith* 50 (LISC); Shamva Hills,
The Vale, fl. fr. 26.i.1973, *White* 10001 (FHO). E: Makurupini Forest, Haroni, fr.
4.xii.1964, *Wild* 6638 (FHO). **Malawi.** N: Nkhata Bay Distr., 8 km. E. of Mzuzu,
1280 m., fr. immat. 6.xii.1969, *Pawek* 3041 (K). C: Bunda forest, Lilongwe, 1065 m.,
fr. 10.xii.1962, *Chapman* 1763 (FHO). S: Zomba, fl.-buds, *Clements* 791 (FHO).
Mozambique. Z: between Quelimane and Marral, fl. immat. 17.ix.1941, *Torre* 3448
(LISC). T: Zóbuè, 900 m., fr. 21.x.1941, *Torre* 3685 (LISC). MS: Serra de Choa,
1400 m., fl. 9.xii.1965, *Torre & Correia* 13425 (LISC).
From Sénégal to Ethiopia and southwards to S. Africa (Natal), 100–1675 m. In
woodland, wooded grassland and fringing forest.

Group B. TAB. **45** fig. E 4–8.

Leaves 7 × 3·5–14 × 8·5 cm, apex rounded or subacute or very shortly cuspidate. Fruit 1·3 × 1·3–2·2 × 2·2 cm., subglobose.

Zambia. B: near Luena R., 15 km. ESE. of Mankoya, fr. 21.xi.1959, *Drummond &
Cookson* 6717 (K; SRGH). N: near L. Bangweulu, 25 km. S. of Samfya Mission, fl. 18.viii.1952, *White* 3079 (FHO; K). W: Mwinilunga to Solwezi km. 131, fr. 16.ix.1952, *White* 3273 (FHO). C: 15 km. ENE. of New Serenje, fl. 11.x.1949, *Hoyle* 1276 (FHO). E: Tigone Dam, Lundazi to Chama 2 km., fl. 17.x.1958, *Robson* 139 (BM; K; LISC). S: Choma, fl. 2.x.1955, *Bainbridge* 144 (FHO; K; SRGH). **Rhodesia.** N: Mazoe, fl. ix.1906, *Eyles* 403 (BM). C: Salisbury to Lusaka 15 km., fl. 19.x.1960, *Rutherford-Smith* 301 (K). E: Inyanga Distr., near Nyatsanga, 915–975 m., fr. 16.i.1967, *Biegel* 1771 (FHO). S: Chibi Distr., 15 km. N. of Lundi R., fl. 5.xii.1961, *Leach* 11312 (FHO). **Malawi.** N: Chitipa Distr., Misuku Hills, fl. 22.xii.1966, *Pawek* 665 (SRGH). C: Dedza Distr., Chongoni Forest, fl. 9.ix.1960, *Chapman* 915 (FHO; K; SRGH). S: Palombe Plain, fr. x.1929, *Clements* 145 (FHO). **Mozambique.** N: Imala to Mocuburi 9 km., 400 m., fr. immat. 16.i.1964, *Torre & Paiva* 10000 (LISC). Z: Pebane, 10–20 m., fr. immat. 12.i.1968, *Torre & Correia* 17107 (LISC). T: Moatize Distr., Zóbuè to Metengobalama 16 km., 900 m., fr. 11.i.1966, *Correia* 379 (LISC). MS: Rotanda to Mavita 4 km., 800 m., fl. immat. 30.x.1965, *Correia* 282 (LISC). SS: Massinga Distr., Funhalouro, fl. 21.vi.1941, *Torre* 2703a (LISC).

Also in Tanzania, Zaire, Angola, S. Africa (Transvaal). In woodland and wooded grassland, especially in *Brachystegia* woodland on rocky hills and at edges of seasonally waterlogged grassy depressions (dambos) in *Brachystegia* woodland. Sometimes occurring on low termite mounds, e.g. on the flood plain of the Chambeshi R., 20–1750 m.

Group C. TAB **45** fig. E 9–12.
   Eugenia fourcadei Dümmer in Gard. Chron., Ser. 3, **52**: 152 (1912). Type from S. Africa (Natal).
   Syzygium fourcadei (Dümmer) Burtt Davy, F.P.F.T.: 50, 241 (1926).—J. H. Ross, Fl. Natal: 260 (1972). Type as above.

Leaves 6·3 × 2·6–13 × 4 cm., nearly always narrower than those of Group B and more gradually tapered towards the acute or subacute apex. Fruit c. 1·8 × 1·6 or c. 2·4 × 1·7 cm., subglobose or ellipsoid.

Zambia. B: Balovale Distr., Kasisi R., 1065 m., fl. 4.ix.1953, *Gilges* 222 (K). N: Mporokoso Distr., Chisi swamp forest, 900 m., fl. 24.ix.1956, *Richards* 6274 (K). W: Chingola, fl. 18.x.1955, *Fanshawe* 2544 (K). C: Kabwe Rural Distr., Chisamba, 1200 m., fl. 6.x.1972, *Strid* 2303 (FHO). E: Chipata (Fort Jameson), fl. 2.x.1966, *Mutimushi* 1538 (K). S: Choma-Namwala 48 km., fl. 2.viii.1952, *White* 3029 (BM; FHO; K). **Rhodesia.** E: Inyanga Distr., Pungwe R., fl. 8.ix.1963, *Plowes* 2348 (FHO; K; SRGH). S: Hippo Pools, fr. 3.ii.1973, *White* 10107 (FHO). **Malawi.** N: Nyungwe, 490 m., fr. immat. 28.x.1930, *Migeod* 988 (BM). C: above Benga on Visanza road, 760 m., *Chapman* 1743 (FHO). **Mozambique.** N: Vila Cabral, Meponda, fl. 9.ix.1958, *Monteiro* 43 (LISC).

Also in Angola and S. Africa (Natal). In fringing forest, 400–1300 m.

Subsp. **huillense** (Hiern) F. White, F.F.N.R.: 304, 455 (1962).—Boutique in F.C.B., Myrtaceae: 17, t. 2 fig. F (1968) pro parte. TAB. **45** fig. H. Type from Angola.
   Eugenia guineensis var. *huillensis* Hiern, Cat. Afr. Pl. Welw. **1**: 359 (1898) pro parte quoad specim. *Welwitsch* 4401. Type as above.
   Syzygium huillense (Hiern) Engl., Bot. Jahrb. **54**: 339 (1917).—Milne-Redhead in Kew Bull. **1947**: 24 (1947).—Amshoff in Acta Bot. Neerl. **9**: 406 (1959); in C.F.A. **4**: 101 (1970) pro parte excl. inter alia specim. *Welwitsch* 4399 & 4400. Type as above.
   Syzygium mumbwaense Greenway in Kew Bull. **1928**: 196 (1928) pro parte excl. specim. *Bourne* 91. Type: Zambia, Mumbwa, Chibuluna vlei, fl., *Macaulay* 995 (K, holotype).

Rhizomatous geoxylic suffrutex up to 0·6 m. tall, but often no more than 0·3 m., usually flowering on unbranched current year's shoots arising directly from the rootstock. Flowering shoots usually markedly curved, terete. Bark on branchlets dark brown. Leaf-lamina 4 × 2–11 × 6 cm., very variable in shape, ovate or obovate to lanceolate-elliptic or oblanceolate-elliptic, apex obtuse to subacute, base usually cuneate, rarely rounded; petiole 0·2–0·5(0·6) cm. long. Receptacle (including pseudopedicel) + calyx 0·5–0·6 cm. long; calyx + upper receptacle 0·2–0·25 cm. long; filaments 0·4–0·7 cm. long. Fruit c. 3 × 2 cm., purple-black, ellipsoid or obovoid; persistent calyx 0·15 × 0·6 cm.

Zambia. B: Balovale, fl. 12.viii.1952, *Gilges* 170 (K; SRGH). N: Kawambwa Distr., between Luwingu-Kawambwa road junction and Luongo R. pontoon, fl. 15.x.1947, *Brenan & Greenway* 8116 (FHO; K). W: Mwinilunga Distr., N. of Kalene Hill, fl. 24.ix.1952, *Angus* 534 (BM; FHO). C: Mumbwa to Kabwe (Broken Hill) 96 km., fl. immat. 19.ix.1947, *Brenan & Greenway* 7891 (FHO; K). S: Namwala Distr., Naminwe dambo, fl. 21.x.1963, *Lawton* 1129 (FHO). **Rhodesia.** C: Delport road E. of Salisbury, fl. 11.xi.1960, *Rutherford-Smith* 349 (FHO; K; LISC; SRGH). S: 1·6 km. NW. of Buhera, fl. 20.iv.1969, *Biegel* 2926 (SRGH). **Malawi.** N: Vipya Plateau, Luwawa Dam, fl. 28.xi.1961, *Chapman* 1486 (FHO; SRGH).

Also in Angola and Zaire. In edaphic, seasonally waterlogged grassland with other suffrutices, usually on sandy soils, especially at the edges of dambos and on Kalahari sand. Also in secondary, fire-maintained grassland. The suffrutex, *Parinari capensis*, appears to be its most constant associate, 900–1400 m. The fruit is edible.

One of the most remarkable features of the African flora, especially in the Zambezian Region, is the occurrence of pairs of species which differ markedly in habit, but scarcely at all in other structural features. Most commonly one member of the pair is a tree and the other is a dwarf rhizomatous geoxylic suffrutex. The taxonomic relationships of such species is sometimes controversial and the suffrutex is often regarded by some authors as a variety or subspecies of the larger plant whereas other authors would give it specific rank. Reasons are given elsewhere in this volume (p. 37) for giving specific rank to *Parinari capensis*, the most widespread and characteristic suffrutex in the Zambezian Region. It would appear, however, that the claims of the suffrutex *Syzygium* to full specific status are less strong and that it would be better treated as a subspecies of *S. guineense*. The differences between the suffrutex and some variants of subspecies *guineensis* are very slight and some populations occur which include trees referable to subsp. *guineensis*, colonies of the suffrutex, and apparent intermediates between them.

The characteristic habitat of subsp. *huillense* is seasonally waterlogged grassland, from which trees are normally excluded, or, if present, are found only on termite mounds. In such places subsp. *huillense* can form extensive populations from which subsp. *guineense* is quite absent. Several collectors have described this situation in their field-notes. It is illustrated by Barbosa in Carta Fitogeográfica de Angola (1970: phot. 17.2).

Other collectors have commented on the extreme rarity of subsp. *guineense* in populations of subsp. *huillense*, e.g. *Brenan & Greenway* 7891 (" subsp. *huillense* is common here and uniform; a single tree of subsp. *guineense* ('S. owariense') in same locality "), or have recorded evidence that even when the 2 subspecies occur together they appear to remain distinct. A photograph in FHO taken by Hoyle (x. 54–5) at the edge of a dambo in Serenje District, Zambia, shows colonies of subsp. *huillense* and clumps of subsp. *guineense*, occurring both as mature trees and 2 m. high coppice, in close proximity. Hoyle remarked that subsp. *huillense* flowers near the ground, whereas subsp. *guineense* does not flower until it reaches a considerable size. It would appear that over much of its range subsp. *huillense* is genetically dwarf and subsp. *guineense* normally flowers only as a small tree. If this were always so, the claims of the suffrutex to specific rank would be almost as great as those of *Parinari capensis*.

Plants intermediate in habit between subsp. *huillense* and subsp. *guineensis* have been recorded from several localities. Some which bear flowers or fruits are genuine intermediates and are cited below. Others, however, appear to be no more than frost or fire-suppressed individuals of subsp. *guineense*. *Strid* 2135 resembles subsp. *huillense* in leaf-shape and petiole length, but is said to be a large tree.

Some genuine intermediates are cited below. It would appear that they are relatively rare. Both subsp. *huillense* and the variants of subsp. *guineense* that it comes into contact with at the edges of dambos sometimes have polyembryonic seeds, in this case, an almost certain indication of apomixis. This might explain the rarity of intermediate forms.

Intermediates between subsp. **huillense** and subsp. **guineense**.

Zambia. B: Balovale Distr., Manyinga, st. 10.xi.1932, *Gilges* 275 (K). N: Mbala (Abercorn), 1500 m., fl. 1.x.1956, *Richards* 6329 (K). C: near Lusaka, 1300 m., fl. 14.ix.1972, *Strid* 2135 (FHO). S: Namwala Distr., Choma to Namwala 112 km., fl. immat. 23.vi.1952, shrub 1 m. tall, *White* 2965 (FHO; K); Choma Distr., Siamambo Forest Reserve, fr. 6.ii.1960, shrub 1·6 m. tall, *White* 6855 (FHO; K; SRGH). **Rhodesia.** C: Salisbury to Lusaka 16 km., fl. 19.x.1960, shrub 1·5 m. tall, *Rutherford-Smith* 301 (K; LISC; SRGH). S: 1·6 km. NW. of Buhera, 1200 m., fl. 20.iv.1969, *Biegel* 2926 (SRGH). **Malawi.** C: Lilongwe, fl. 6.xi.1951, shrub 1·3–1·6 m. tall, *Jackson* 629 (FHO).

5. **Syzygium cumini** (L.) Skeels in U.S. Dept. Agr. Bur. Pl. Ind. Bull.: 25 (1912).—Brenan & Greenway, T.T.C.L.: 379 (1949).—Amshoff in Fl. Gab. **11**: 17 (1966). Type from Ceylon.

Myrtus cumini L., Sp. Pl. **1**: 471 (1753). Type as above.
Eugenia jambolana Lam., Encycl. Méth. Bot. **3**: 198 (1789). Type from tropical Asia.

Evergreen tree 6–15 m. tall. Bark rough, dark. Leaf-lamina up to 13 × 5 cm., elliptic or oblanceolate-elliptic, apex subacuminate or shortly acuminate, base cuneate. Inflorescence a lateral, lax panicle, mostly borne in the axils of fallen leaves on older stems, very rarely also axillary or terminal. Receptacle (including pseudopedicel) + calyx c. 0·5 cm. long; calyx + upper receptacle 0·2 cm. long; filaments 0·5–0·6 cm. long. Fruit up to 2·4 × 2 cm., ellipsoid or subglobose.

Zambia. W: Kitwe, cultivated, fl. 1.x.1971, *Fanshawe* I/12 (SRGH). **Rhodesia.** N: Goromonzi Distr., Ewanrigg National Park, cultivated, fr. 20.vii.1970, *Biegel & Müller* 3243 (K; SRGH). E: Chipinga Distr., Chikore Mission, naturalized, fl. xi.1962, *Goldsmith* 232/62 (BM; K; LISC; SRGH). **Mozambique.** Z: Namacurra Distr., Nicuadala to Campo 85 km., 40 m., naturalized, fr. 2.ii.1966, *Torre & Correia* 14392 (LISC). T: Mutarara, Zambezi Valley, cultivated, fl. 22.ix.1971, *Haffern* 6 (SRGH). MS: Rotanda, cultivated, fr. 26.x.1944, *Mendonça* 2630 (LISC). SS: Vilanculos, cultivated, fl. 19.x.1965, *Mogg* 28601 (BM; K; LISC; SRGH). LM: Jardim Vasco da Gama, cultivated, fl. 29.x.1972, *Balsinhas* 2446 (K; SRGH).

The " Jambolan ", which is probably a native of tropical Asia and is widely planted for its edible fruit, is sometimes naturalized.

6. **Syzygium jambos** (L.) Alston, Hand. Fl. Ceyl. **6**, Suppl.: 115 (1931).—Amshoff in Fl. Gab. **11**: 17 (1966). Type from Asia.

    Eugenia jambos L., Sp. Pl. **1**: 470 (1753). Type as above.
    Jambosa jambos (L.) Millsp., Publ. Field Mus. Nat. Hist. Chicago, Bot. Ser. **2**: 80 (1900).—Brenan & Greenway, T.T.C.L.: 378 (1949). Type as above.

Evergreen tree up to 15 m. tall. Leaf-lamina up to 20 × 5·5 cm., narrowly lanceolate, apex attenuate-subacuminate, base cuneate; lateral nerves widely spaced, up to 1 cm. apart. Inflorescence a ± 5–6-flowered terminal corymb; pedicels c. 1 cm. long. Receptacle (including pseudopedicel) + calyx c. 1·8 cm. long, upper receptacle 0·2–0·4 cm. long. Calyx-lobes 4, 0·5 × 0·8 cm. long, hemi-orbicular. Petals 4, c. 1·8 × 1·2 cm., obovate, acute, caducous, not calyptrate. Filaments c. 4 cm. long. Style c. 4·5 cm. long. Fruit up to 4·5 cm. in diameter, pyriform, white or yellowish, tinged with rose.

Rhodesia. C: Salisbury Distr., Borrowdale, cultivated, fl. 19.viii.1968, *Biegel* 2870 (SRGH). E: Chipinga Distr., Chirinda forest, naturalized, fl. xi.1962, *Goldsmith* 223/62 (FHO; LISC; SRGH). **Malawi.** Zomba, cultivated, fl. fr. 20.vi.1961, *Chapman* 1376 (FHO; SRGH). **Mozambique.** Z: Mocuba, cultivated, fl. 27.viii.1948, *Barbosa & Carvalho* 3812 (SRGH). LM: Viveiros Municipais, cultivated, fl. 4.ix.1973, *Balsinhas* 2566 (LISC).

The " Rose Apple ", a native of tropical Asia, is widely planted in Africa for its edible fruit, and is sometimes naturalized.

6. LEPTOSPERMUM J. R. & G. Forst.

Leptospermum J. R. & G. Forst., Char. Gen. Pl.: 71, t. 36 (1776).

A small genus with c. 50 species in Malesia, Australia, New Zealand and the Caroline Islands. The following species are grown in the F.Z. area for ornament.

Leptospermum laevigatum F. Muell. Shrub 2–6 m. tall. Branchlets pendulous. Leaves up to 3 × 0·6 cm., oblanceolate-oblong, very shortly mucronate, rigid, glabrous, glaucous. Flowers white. Capsule opening by c. 9 slits. This species is extensively and sometimes abundantly naturalized in South Africa.

Leptospermum petersonii F. M. Bailey. Shrub c. 2·5 m. tall. Leaves up to 5 × 0·5 cm., linear-lanceolate, apex muticous. Flowers white. Cultivated in Rhodesia.

Leptospermum scoparium J. R. & G. Forst. Shrub or small tree. Leaves up to 1 cm. long, sharp pointed, narrowly oblanceolate-elliptic. Flowers white or carmine. Capsule opening by c. 5 slits. It is the commonest indigenous shrub in New Zealand, and also occurs in Australia.

7. CALLISTEMON R. Br.

Callistemon R. Br. in App. Flind. Voy. **2**: 547 (1814).

A small genus of c. 25 species indigenous to Australia and New Caledonia. Several species are planted for their showy crimson flowers. These are the familiar and appropriately named " Bottle-Brush " plants. Their flowers occur in dense or interrupted spikes and have numerous far-exserted filaments giving a bottle-brush appearance. The axis of the inflorescence grows on beyond the flowers and continues to produce leaves. The capsules are woody and do not fall, so that the older stems consist of segments covered with old fruits alternating with bare segments.

Several species are in cultivation in the F.Z. area. Their taxonomy is difficult and their nomenclature is uncertain.

8. MELALEUCA L.

Melaleuca L., Mant. Pl. Alt. **1**: 14 (1767).

A medium-sized genus with c. 100 species in Australia and the Pacific. The following are planted in the F.Z. area for ornament. Some are similar in appearance to *Callistemon* but their flowers are whitish.

Melaleuca armillaris Sm. Small tree c. 6 m. tall. Leaves up to 3 × 0·1 cm., linear, ericoid, falcate, bright green, apex mucronate. Flowers cream, in dense, short, leafless lateral spikes, or crowded at the base of the previous year's leafy shoots. Cultivated in Rhodesia.

Melaleuca genistifolia Sm. Shrub or small tree. Leaves c. 1·5 × 0·2 cm., linear-lanceolate, rigid, somewhat glaucous. Flowers cream, in lax, interrupted spikes. Capsules small, c. 0·3 cm. in diam., with 5 short teeth. Cultivated in Rhodesia.

Melaleuca hypericifolia Sm. Tall shrub. Leaves up to 3 × 0·7 cm., lanceolate-elliptic, rigid, somewhat glaucous, apex subacute to mucronate, margin recurved. Flowers dull red, in dense spikes at the base of leafy branches. Cultivated in Zambia and Rhodesia.

Melaleuca lateritia Otto. Shrub. Leaves c. 2 × 0·15 cm., linear, ericoid, apex muticous. Flowers scarlet, in spikes at the base of leafy shoots. Cultivated in Rhodesia.

Melaleuca leucadendron (L.) L.* The " Cajuput Tree " or " Punk Tree ". Tree. Bark, thick, spongy, laminated, whitish. Leaves c. 5·5 × 1·5 cm., elliptic, longitudinally nerved, silvery-sericeous when young, later glabrous. Flowers whitish in spikes which at first appear terminal but are later surmounted by a leafy branch. Capsule truncate. Cultivated in Zambia, Rhodesia, Malawi and Mozambique.

Melalauca styphelioides Sm. Small tree. Leaves up to 1·5 × 0·4 cm., rigid, lanceolate, spinous-acuminate. Flowers cream, in dense spikes, at base of leafy shoots. Capsule with 5 subulate teeth. Cultivated in Zambia.

9. TRISTANIA R. Br.

Tristania R. Br. in Ait., Hort. Kew., ed. 2, **4**: 417 (1812).

A small genus with c. 50 species in South-East Asia, Australia and the Pacific.

Tristania conferta R. Br. " Brisbane Box ", a tall evergreen tree. Leaves c. 15 × 5 cm., lanceolate or elliptic-lanceolate, subacuminate. Flowers solitary or in 2–3-flowered cymes c. 2·5 cm. in diameter. Bundles of stamens longer than the petals. Capsule campanulate, with c. 6 longitudinal veins; opening by 3 valves. Has been planted by Forest Departments, but probably has no advantages over certain eucalypts.

10. SYNCARPIA Ten.

Syncarpia Ten. in Ind. Sem. Hort. Neap.: 12 (1839); in Ann. Sci. Nat. Bot., Sér. 2, **13**: 381 (1840).

A small Australian genus of c. 5 species.

Syncarpia glomulifera (Sm.) Niedenzu. " Turpentine Tree ". Tall evergreen tree. Leaves up to 12 × 5·5 cm., elliptic, apex and base acute or subacute,

* According to Blake (in Contrib. Queensl. Herb. **1**, 1968) 6 other species have been confused with *Melaleuca leucadendron*. Further collecting is necessary to establish the identity of the plant grown in the F.Z. area.

whitish-tomentose beneath. Flowers white, in globose heads; receptacle-tubes united at the base. Fruiting heads c. 1·5 cm. in diameter. Because of the great durability of its wood, this species is under trial as a plantation tree.

11. EUCALYPTUS L'Hérit.

Eucalyptus L'Hérit., Sert. Angl.: 18 (1788).

Evergreen trees and shrubs. Juvenile leaves mostly horizontal, opposite, cordate and sessile. Adult leaves sclerophyllous, mostly vertical, 2-facial, alternate, petiolate, usually elongate and narrowly lanceolate. Inflorescence fundamentally a dichasial cyme showing various degrees of expansion, aggregation or compaction. The unit inflorescence or " umbel " is in fact a condensed and usually partly reduced dichasium. Receptacle-tube obconical, campanulate or oblong, adnate to the ovary, truncate and entire after the operculum has fallen or with 4 minute teeth. Perianth variously modified (see below) to form an operculum which falls off entire when the filaments expand. Stamens numerous in several series, free or very rarely very shortly fused at the base into 4 clusters. Ovary inferior, 3–6-locular; placentation axile; ovules numerous in each locule; style subulate, rarely clavate, with a small truncate, capitate or rarely peltate stigma. Fruit usually hard and woody, consisting of the more-or-less enlarged truncate receptacle-tube enclosing the capsule, which is always adnate to the receptacle-tube although often readily separable from it, opening by as many valves as there are locules.

A large genus with c. 500 species in Australia and Tasmania and 2 or 3 species in the Indo-Malesian region. Many species are planted in the warmer parts of the world for timber, firewood, essential oils, shelter or ornament. Well over 50 species have been introduced into the F.Z. area and some of these are widely planted and contribute a great deal to the appearance of the landscape. In a work such as this it is only possible to include the most widely planted species together with some others, which, although not widely planted at present, appear to hold considerable promise for forestry purposes. In using the key it should be borne in mind that certain species which are grown in the F.Z. area, but have not been included here, may " key out ". To make quite sure that the specimen does belong to one of the included species it must also be carefully checked against the descriptions and illustrations.

The nature of the operculum has been much disputed. Recent work (summarized by Pryor & Johnson, A classification of the Eucalypts, 4–6, 1971) has clarified many points. The operculum is often a double structure; in some species the outer operculum is retained until shortly before anthesis when it falls as an outer cap just before the inner one is shed; in other species the outer operculum is shed early in the development of the bud, either as a small entire cap or as four small segments with their edges adpressed though not actually fused. In some other species the two opercula are separable only with care or by anatomical investigation and fall together at anthesis. In still other species only one operculum is found which is accompanied by four outer separate teeth. When two opercula are present, the outer and inner are interpreted as homologous respectively with the sepaline and petaline whorls of other Myrtaceous flowers. In one group of species there is no trace of a second operculum. Recent work has shown that the surviving operculum is wholly calycine, and that the corolline whorl has been lost.

Valves of capsule deeply enclosed, not visible at the mouth:
  Bark rough or subfibrous, persistent on trunk:
    Filaments red; flowers long-pedicellate; fruit > 2 cm. long - - 1. *ficifolia*
    Filaments white; flowers subsessile, in axillary umbels; fruit <1 cm. long
          6. *botryoides*
  Bark smooth and flaking; filaments white or cream; flowers distinctly pedicellate in terminal and axillary pseudo-racemes of umbels:
    Operculum slightly apiculate, not rostrate - - - - - 2. *citriodora*
    Operculum conspicuously rostrate - - - - - - 3. *maculata*
Valves of capsule clearly visible at the mouth:
  Inflorescence terminal, paniculate:
    Inflorescence ± as long as the leaves; operculum hemispherical, half the length of receptacle-tube - - - - - - - - 4. *cloeziana*
    Inflorescence much shorter than the leaves; operculum conical, ± equalling the receptacle-tube - - - - - - - - 10. *paniculata*

Inflorescence axillary:
Pedicels distinct, slender, longer than and sharply differentiated from the receptacle-tube - - - - - - - - - - - 9. *camaldulensis*
Pedicels scarcely differentiated from the receptacle-tube:
Operculum shorter than broad - - - - - - - 5. *grandis*
Operculum much longer than broad:
Fruit cyathiform, c. 1·5 cm. long - - - - - - 7. *robusta*
Fruit subglobose, c. 0·7 cm. long - - - - - - 8. *tereticornis*

1. **Eucalyptus ficifolia** F. Muell., Fragm. **2**: 85 (1860).—F. White, F.F.N.R.: 296 (1962).—Blakely, Key Eucalypts, ed. 3: 88 (1965).—Barrett & Mullin in Rhod. Bull. For. Res. **1**: 46 (1968). TAB. **46** fig. A. Type from Australia.

Small tree up to 9 m. tall. Bark rough and persistent throughout. Leaves up to 11 × 4 cm., broadly lanceolate or ovate-lanceolate. Pedicels 2–3 cm. long. Umbels c. 5-flowered; buds 1·5–2 cm. long, clavate to pyriform; operculum hemispherical, 2–3 times shorter than the receptacle-tube. Filaments scarlet or flame-coloured. Fruit 2–3·5 × 2–3 cm., urceolate, very thick and hard, very large; valves deeply enclosed.

Zambia. W: Kitwe, fr. 26.vi.1970, *Fanshawe* 10824 (SRGH). **Rhodesia.** W: Matopos Park, st. x.1911, *Curator, Matopos Nat. Park* in GHS 2004 (SRGH).
The " Red Flowering Gum ", which as a wild plant has a very restricted range on the S. coast of Western Australia, is widely planted in southern Africa for ornament.

2. **Eucalyptus citriodora** Hook. in Mitch. Journ. Trop. Austral.: 235 (1848).—F. White, F.F.N.R.: 294 (1962).—Blakely, Key Eucalypts, ed. 3: 96 (1965).—Barrett & Mullin in Rhod. Bull. For. Res. **1**: 36 (1968). TAB. **46** fig. B. Type from Australia.

Tall tree. Bark smooth and deciduous throughout. This species is very closely related to *E. maculata* Hook. and is sometimes treated as a variety of it. It differs chiefly in the higher percentage of citronellal, and hence the more fragrant lemon-scented leaves, and in its non-rostrate operculum and somewhat smaller fruits. Its leaves are often narrower.

Zambia. B: Mongu, fl. & fr. 18.ii.1952, *White* 2095 (FHO). W: Kitwe, fl. & fr. 12.vii.1970, *Fanshawe* 10856 (SRGH). C: Lusaka, fl. viii.1954, *Cooling* 89 (FHO). S: Siamambo Forest Reserve, fl. immat. 8.vii.1954, *Cooling* 74 (FHO). **Rhodesia.** W: Matopos Park, st. x.1910, *Curator, Matopos Park* s.n. (SRGH). **Malawi.** C: Chongoni Forest, fl. & fr. 14.iv.1971, *Salubeni* 1547 (SRGH). **Mozambique.** T: Zumbo, fr. 28.v.1952, *White* 2891 (FHO; SRGH). SS: Macia, fr. 17.vii.1947, *Pedro & Pedrógão* 1481 (SRGH).
The " Lemon Scented Gum " is a native of central Queensland. The wood is similar to that of *E. maculata* and there are reasonable prospects that these two species will produce a better quality saw timber than most other eucalypts.

3. **Eucalyptus maculata** Hook., Ic. Pl. **7**: 619 (1844).—F. White, F.F.N.R.: 294 (1962).—Blakely, Key Eucalypts, ed. 3: 97 (1965).—Barrett & Mullin in Rhod. Bull. For. Res. **1**: 56 (1968). TAB. **46** fig. C. Type from Australia.

Tall tree. Bark smooth, purple when old, yellowish when newly exposed, flaking in large patches to give the tree a mottled appearance. Leaves up to 30 × 6 cm., lanceolate, with numerous, almost transverse lateral nerves. Inflorescence of terminal and axillary pseudo-racemes of 3–5-flowered umbels. Pedicels 0·5–1 cm. long; buds up to 1·2 × 0·8 cm., obovoid-apiculate; operculum hemispherical, distinctly rostrate, c. ½ as long as receptacle-tube. Fruit up to 1·8 × 1·4 cm., broadly urceolate; valves deeply enclosed.

Zambia. W: Kitwe, fr. 16.vii.1970, *Fanshawe* 10871 (SRGH). C: Lusaka, fl.-buds viii.1954, *Cooling* 88 (FHO).
The " Spotted Gum " is widely distributed in eastern Australia from southern Queensland to Victoria. In S. Africa *E. maculata* is regarded as one of the most useful and valuable of the heavier timbers and is used for many purposes.

4. **Eucalyptus cloeziana** F. Muell., Fragm. **11**: 44 (1878).—F. White, F.F.N.R.: 295 (1952).—Blakely, Key Eucalypts, ed. 3: 178 (1965).—Barrett & Mullin in Rhod. Bull. For. Res. **1**: 39 (1968). Type from Australia.

Medium-sized tree. Bark dark brown, flaky-fibrous, persistent throughout, of the stringybark type. Leaves up to 12 × 3 cm., lanceolate, acuminate. Inflorescence

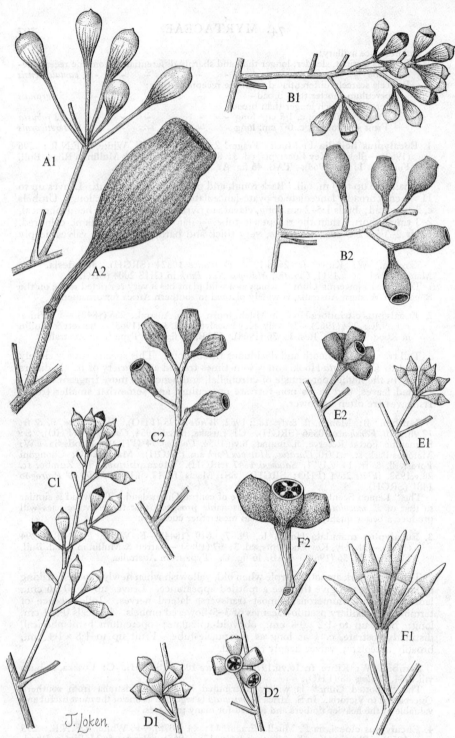

Tab. 46. A.—EUCALYPTUS FICIFOLIA. A1, flower-buds (×1) *Mitchell* s.n.; A2, fruit (×1) *S. Afr. For. Dept.* 7029. B.—E. CITRIODORA. B1, flower-buds (×1) *Cooling* 89; B2, fruits (×1) *White* 2095. C.—E. MACULATA. C1, flower-buds (×1 *Cooling* 88; C2, fruits (×1) *Dillon* 9 MS. D.—E. GRANDIS. D1, flower-buds (×1) *FHO* 93058; fruits (×1) *FHO* 93058. E.—E. BOTRYOIDES. E1, flower-buds (×1) *Wigg* FH 1743; E2, fruits (×1), *Wigg* FH 1743. F.—E. ROBUSTA. F1, flowers (×1), *Savory* 184; F2, fruits (×1) *Savory* 184.

a broad, terminal, many-flowered panicle, sometimes longer than the leaves; branches robust. Umbels 4–6-flowered; peduncles stout, 0·5–1 cm. long; buds 0·5 × 0·5 cm., clavate to globular-ovoid, pedicellate; operculum thick, hemispherical, ½ the length of the receptacle-tube. Fruit 0·9 × 1 cm., hemispherical; valves shortly exserted.

Zambia. S: Siamambo Forest Reserve, st. 22.iv.1958, *Savory* 170 (FHO). **Rhodesia.** C: Salisbury, Forestry Research Centre, st. vii.1967, *Müller* 1743 (SRGH). " Queensland " or " Gympie Messmate " is widespread in Queensland. It has few equals as a telephone or transmission pole because of its strength, excellent form and amenability to preservative treatment, and is under trial in Rhodesia and Zambia.

5. **Eucalyptus grandis** Hill ex Maiden in Journ. Proc. Roy. Soc. N.S.W. **52**: 501 (1918).—F. White, F.F.N.R.: 296 (1952).—Blakely, Key Eucalypts, ed. 3: 98 (1965).—Barrett & Mullin in Bull. Rhod. Bull. For. Res. **1**: 49 (1968). TAB. **46** fig. D. Type from Australia.

Tall tree. Bark smooth, deciduous, white, glaucous. Leaves up to 15 × 2 cm., narrowly lanceolate, attenuate-acuminate. Umbels axillary, 3–10-flowered; peduncles c. 1–1·2 × 0·25 cm., compressed; pedicels c. 0·3 cm. long; buds c. 1 × 0·5 cm., pyriform; operculum conical, shortly apiculate to rostrate, shorter than receptacle-tube. Fruit 0·7–0·8 × 0·6–0·8 cm., glaucous, cyathiform, rather thin, slightly contracted at the orifice; valves 4–6, rather thin, inserted just below the orifice.

Zambia. B: Sesheke, fr. *Bourne* s.n. (FHO). C: Serenje, fr. 1.xi.1958, *Savory* 46 (FHO). S: Siamambo Forest Reserve, fl. 8.vii.1954, *Cooling* 80 (FHO). **Rhodesia.** E: Inyanga, fr. xi.1957, *Miller* 4702 (SRGH). **Malawi.** C: Lilongwe Distr., Bunda Forest, fl. & fr. 29.vi.1962, *Chapman* 1658 (SRGH). " Toolur " occurs naturally in Queensland and New S. Wales, along a coastal belt 160 km. wide from 17°–32° S. At the northern end of its range it ascends to 750 m. The closely related *E. saligna* Sm., " Sidney Blue Gum " or " Saligna ", extends from 28°–35° S. and mostly occurs at higher altitudes (up to 1220 m.). There is a considerable area where the ranges of the 2 species overlap, and here the populations are variable and intermediate, though the extremes are somewhat different. It appears that most of the stands of *E. grandis/saligna* in southern Africa are of hybrid origin, though closer to the former. Professional foresters tend to use the name " Saligna " for trees of both these species and their presumed hybrids. In this work no attempt is made to distinguish between them. " Saligna " has been planted in the F.Z. area since long before the beginning of this century and is now one of the most widely planted and familiar exotic trees.

6. **Eucalyptus botryoides** Sm. in Trans. Linn. Soc., Bot. **3**: 286 (1797).—F. White, F.F.N.R.: 295 (1952).—Blakely, Key Eucalypts, ed. 3: 100 (1965).—Barrett & Mullin, Rhod. Bull. For. Res. **1**: 32 (1968). TAB. **46** fig. E. Type from Australia.

Medium-sized or large tree. Bark subfibrous and persistent on trunk and main branches. Leaves up to 12 × 4 cm., lanceolate, filiform-acuminate, green on both surfaces. Umbels axillary and also borne laterally on the internodes, 6–10-flowered; peduncles 1–1·5 × 0·25 cm., compressed; buds c. 1·2 × 0·4 cm., subsessile, ± fusiform, conspicuously glandular; operculum conoidal, slightly apiculate, ½ the length of the receptacle-tube. Fruit up to 0·9 × 0·8 cm., barrel-shaped, sessile; valves distinctly included.

Rhodesia. C: Salisbury, fr. vii.1967, *Müller* 1745 (SRGH). " Bangalay " or " Southern Mahogany " occurs naturally in a narrow coastal strip from central New S. Wales to eastern Victoria. It was one of the earliest eucalypts to be introduced into Rhodesia and has been widely planted, mainly in farm plantations.

7. **Eucalyptus robusta** Sm., Specimen Bot. New Holl.: 39 (1783).—F. White, F.F.N.R.: 294 (1962).—Blakely, Key Eucalypts, ed. 3: 100 (1965).—Barrett & Mullin in Rhod. Bull. For. Res. **1**: 77 (1968). TAB. **46** fig. F. Type from Australia.

Tall tree. Bark rough, subfibrous, persistent throughout. Leaves up to 17 × 6 cm., dark green above, pale green beneath, broadly lanceolate, attenuate-acuminate. Umbels mostly axillary, 5–10-flowered; peduncles 2–3 × 0·6 cm., compressed and strap-shaped; pedicels very stout, gradually merging into the receptacle; flower buds c. 2·8 × 0·7 cm. (including the pedicel), fusiform; operculum conoidal-rostrate, at least as long as the receptacle. Fruit c. 1·5 × 1 cm., cyathiform; valves 3–4, inserted just below the orifice.

Zambia. B: Mongu, fr. 18.ii.1952, *White* 2097 (FHO). S: Siamambo Forest Reserve, fl. & fr. 22.iv.1958, *Savory* 184 (FHO). **Rhodesia.** W: Matopos Park, st. x.1910, *Curator Matopos Park* 2017 (SRGH). C: Chilimanzi, Mtao Forest, fr. iv.1962, *Barrett* s.n. (SRGH). **Malawi.** S: Mlanje Distr., Likabula Stream, fl. iv.1960, *Lewis* 177 (SRGH). **Mozambique.** SS: Banhamine, fl. 19.iii.1948, *Torre* 7541 (LISC). LM: Inhaca I., fr. 8.vii.1956, *Mogg* 31259 (SRGH).

The " Swamp Magogany " is found along the E. coast of Australia in a narrow belt from southern Queensland to southern New S. Wales. It has been widely planted but grows relatively slowly and is not regarded as an important exotic. In 1935 it was planted in Itawa Swamp near Ndola in Zambia as an anti-malarial measure. Near Melsetter in Rhodesia trees 36 m. tall can be seen. *E. robusta* regenerates more freely than most eucalypts in Rhodesia, and dense thickets of saplings may be found near old stands.

8. **Eucalyptus tereticornis** Sm., Specimen Bot. New Holl.: 41 (1793).—F. White, F.F.N.R.: 300 (1952).—Barrett & Mullin in Rhod. Bull. For. Res. **1**: 80 (1968). TAB. **47** fig. A. Type from Australia.

Eucalyptus umbellata sensu Blakely, Key Eucalypts, ed. 3: 135 (1965) non (Gaertn.) Domin.

Tall tree. Bark smooth and irregularly blotched throughout or sometimes with a little rough, flaky, persistent bark at base. Leaves up to 20 × 2·5 cm., narrowly lanceolate, apex attenuate-acuminate. Umbels axillary, 5–12-flowered; peduncles 0·5–1·2 cm. long, almost terete; pedicels 0·3–0·4 cm. long; flower-buds 1·2–1·6 cm. long, ± fusiform; operculum conical, usually 2–3-times larger than the small cupular receptacle. Fruit c. 0·7 × 0·7 cm., subglobose; valves 3–4, strongly exserted.

Zambia. W: Kitwe, fl. & fr. 12.vii.1970, *Fanshawe* 10863 (SRGH). S: Siamambo Forest Reserve, fl. 22.iv.1958, *Savory* 145 (FHO). **Rhodesia.** C: Salisbury, Forestry Research Centre, fl. vii. 1967, *Müller* 1740 (SRGH).

The " Forest Red Gum " is indigenous to New Guinea and along the eastern coast of Australia. Because of its tolerance of harsh conditions and the quality of its timber, it is regarded as a potentially important species.

9. **Eucalyptus camaldulensis** Dehn., Cat. Pl. Hort. Camald., ed. 2: 20 (1832).— F. White, F.F.N.R.: 293 (1962).—Blakely, Key Eucalypts, ed. 3: 140 (1965).— Barrett & Mullin in Rhod. Bull. For Res. **1**: 34 (1968). TAB. **47** fig. B. Type from Australia.

Tall tree. Bark on upper part of bole smooth, dull white or ash-coloured, deciduous in long strips, at base of bole rough, blackish, exuding gum, flaking in plates c. 3–5 cm. in diam. Leaves up to 25 × 1·5 cm., narrowly lanceolate, falcate, apex narrowly acute. Umbels axillary, 5–10-flowered; peduncles 1–1·5 cm. long, terete; pedicels 0·5–0·7 cm. long, slender, longer than and sharply differentiated from the receptacle tube; flower-buds c. 0·8 × 0·5 cm., subglobose and rostrate; operculum markedly rostrate, 1·5–2 times as long as the receptacle. Fruit c. 0·7 × 0·7 cm., subglobose; valves usually 4, strongly exserted.

Zambia. B: Balovale Distr., near Chavuma, fr. 13.x.1952, *White* 3488 (FHO). W: Kitwe, fl. & fr. 22.vii.1970, *Fanshawe* 10892 (SRGH). **Malawi.** C: Dedza Arboretum, st. 3.viii.1968, *Salubeni* 1129 (SRGH).

" Murray " or " River Red Gum ", is widely distributed in Australia. It has long been cultivated in the F.Z. area and is now one of the most widely planted exotic trees., Although planted in practically every farm plantation in Rhodesia, it is not a major species in the principal forests devoted to eucalypts. It is considered more suitable for marginal areas, shelter belts, farm woodlots and firewood plots. In Zambia it is superior to other eucalypts on sites subject to considerable fluctuations in the water-table.

10. **Eucalyptus paniculata** Sm. in Trans. Linn. Soc., Bot. **3**: 287 (1797).—F. White, F.F.N.R.: 298 (1952).—Blakely, Key Eucalypts, ed. 3: 269 (1965).—Barrett & Mullin in Rhod. Bull. For. Res. **1**: 66 (1968). TAB. **47** fig. C. Type from Australia.

Tall tree. Bark dark grey, hard, deeply furrowed, persistent throughout. Leaves lanceolate, apex narrowly acute. Inflorescence terminal, paniculate, much shorter than the leaves. Umbels 3–9-flowered; peduncles 1–2 cm. long, subterete; pedicels 0·3–0·5 cm. long; buds 0·7 × 0·5 cm., broadly fusiform; operculum conical, subapiculate, ± equalling the receptacle-tube. Fruit 0·7 × 0·5 cm., hemispherical-obconoidal; valves small, inserted near mouth.

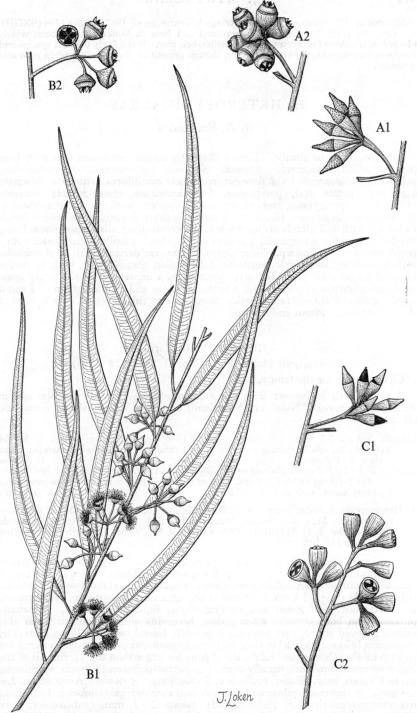

J. Loken

Tab. 47. A.—EUCALYPTUS TERETICORNIS. A1, flower-buds (×1), *Savory* 145; A2, fruits
(×1) *FHO* 18537. B.—E. CAMALDULENSIS. B1, flowering branchlet (×½) *Cooling*
86; B2, fruits (×1) *White* 3488. C.—E. PANICULATA. C1, flower-buds (×1) *C. T.*
White 6088; C2, fruits (×1) *Dillon* 8 MS.

Rhodesia. C: Salisbury, Forestry Research Centre, fr. vii.1967, *Müller* 1744 (SRGH). " Grey Ironbark " is found in Queensland and New S. Wales. It has been widely planted in Rhodesia for amenity or as a shelter-belt tree. It is a fairly tolerant species and produces an exceptionally strong timber, though growth in plantations is often slow and irregular.

75. HETEROPYXIDACEAE

By A. Fernandes

Aromatic trees or shrubs. Leaves alternate, simple, attenuate into a ± long petiole, pellucid-punctate. Flowers dioecious by abortion, actinomorphic, yellowish, sweet-scented. ♂ flowers: receptacle cyathiform, with (4)5 imbricate segments; petals 4(5), perigynous, free, imbricate, gland-dotted; stamens (4)5–8(9–10), perigynous, free, exserted, (4)5 opposite to the petals, the others to the receptacle segments, filaments straight; anthers dorsifixed, dehiscing longitudinally, with 1–3 glands on the back of the connective; disk perigynous, lining the lower ½ of the resceptacle, glabrous or ± pilose; pistillode turbinate, with a sessile or short style and a papillose stigma. ♀ flowers: perianth as in ♂; staminodes (4)5–8, minute; ovary free, superior, 2(3)-locular, gland-dotted, with ∞ ovules inserted on an axile placenta; style inserted in a depression of the ovary apex, cylindric, glabrous or pilose, with a large, papillose and capitate stigma. Fruit a small loculicidal 2(3)-valved capsule. Seeds ± straight, the outer ones winged at the extremities, without endosperm.

HETEROPYXIS Harv.

Heteropyxis Harv., Thes. Cap. 2: 18, t. 128 (1863).

Characters as for the family.

A genus with 3 species: 2 in our area (one extending into Angola) and in S. Africa (Transvaal, Natal and Swaziland) and 1 in Transvaal and Swaziland (*H. canescens* Oliv.).

Capsule 3–4 × 2–2·5 mm., ellipsoid; stamens (5–6)7–8(9–10); leaf-lamina (3–5)6–14 × (1)2–4 cm., usually with several (up to 10) lateral nerves prominent beneath; outer seeds c. 3 × 0·5 mm. - - - - - - - - - 1. *dehniae*
Capsule 2–3 × 1·75–2 mm., globose or broadly ellipsoid; stamens (4)5–7(8); leaf-lamina (2·5–4)4·5–7(8–9) × (0·6)1–2·5 cm., with or without 1 pair of lateral nerves prominent beneath; outer seeds c. 1·75 × 0·75 mm. - - - - - 2. *natalensis*

1. **Heteropyxis dehniae** Suesseng. in Proc. Trans. Rhod. Sci. Ass. **43**: 27 (1951).— Fernandes in Mitt. Bot. Staatss. München **10**: 223 (1971).—Martins in Garcia de Orta, Sér. Bot. **3**, 1: 51 (1976). TAB. **48**; TAB. **49**. Type: Rhodesia, Marandellas, *Dehn* 188b (M, holotype).

Tree up to 20 m. tall, with pale bark and pendent foliage, the leaves, young shoots and inflorescences dotted with oil-cavities. Leaf-lamina (3–5)6–14 × (1)2–4 cm., usually broadly lanceolate, sometimes ovate, attenuate at the base into a canaliculate petiole to 1·5–2·5 cm. long and at the apex into an acute tip, membranous to chartaceous, dull green and shining above, light green below, ± patent-pubescent on both surfaces when young, becoming almost glabrous when old; midrib grooved above, very prominent beneath, lateral nerves 2–10 on each side, prominent below, reticulation close and very conspicuous, usually with a gland dot in each mesh. Cymes paniculate, the ♂ panicles larger than the ♀; rhachis of the inflorescence and pedicels usually densely ± patent-pubescent; bracteoles minute, up to 1·5 mm. long, pilose; pedicels c. 2 mm. long. ♂ flower: receptacle c. 2 × 2·5 mm., ± appressed-pilose and gland-dotted outside; calyx-lobes c. 1 × 1 mm., irregularly semi-circular, gland-dotted; petals 2 × 2 mm., subcircular, very slightly unguiculate, reticulately nerved and gland-dotted; stamens (5–6)7–8(9–10), inserted near the margin of the disk; filaments 2·5–3 mm. long; anthers c. 1 mm. long, the connective with 1–3 glands on the back; disk c. 1 mm. broad, usually densely pubescent; pistillode turbinate, c. 1 × 1 mm., ± wrinkled, gland-dotted

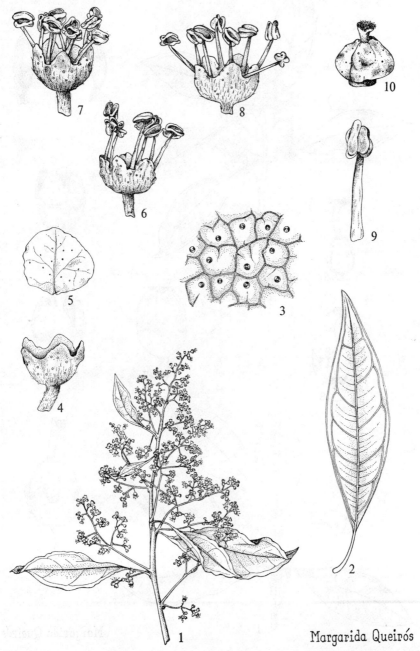

Margarida Queirós

Tab. 48. HETEROPYXIS DEHNIAE (♂ plant). 1, inflorescence (× ½); 2, inferior surface of
leaf (× ½); 3, detail of same showing reticulate venation and glands (× 12); 4, calyx
(× 6); 5, petal (× 12); 6–8, flowers with varying numbers of stamens (× 6); 9, stamen
(× 12); 10, pistillode (× 12), 2–3 from *Garcia* 879, the rest from *Chase* 1954.

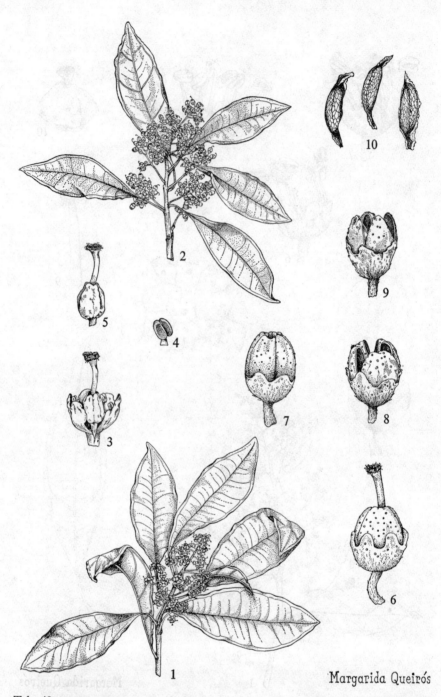

Margarida Queirós

Tab. 49. HETEROPYXIS DEHNIAE (♀ plant). 1, flowering branch (×½); 2, fruiting branch (×½); 3, flower showing pistil and staminodes (×6); 4, staminode (×12); 5, pistil (×6); 6, not fully matured fruit (×6); 7, unopened capsule (×6); 8, opened capsule with 3 valves stemming from a 3-carpellate ovary (very rare case) (×6); 9, old capsule with 4 valves showing loculicidal and septicidal dehiscence (×6); 10, seeds in frontal and lateral views (×12), 1–5 from *Chase* 5463, 6–10 from *Torre* 4266.

and ± pilose; style very short, with a capitate stigma. ♀ flower: perianth as in male; staminodes similar to the stamens but very small (c. 1 mm. long); ovary 3·2 × 1·5 mm., ovoid, gland-dotted and sometimes pilose; style up to 2·5 mm. long, glabrous or patent-pilose; stigma capitate, papillose, reddish. Capsule 3–4 × 2–2·5 mm., ellipsoid, loculicidal, 2(3)-valved, gland-dotted, sometimes pubescent. Seeds slightly curved, the outer ones longer and narrower, 2–3 × 0·5 mm., winged at both ends.

Rhodesia. N: Mazoe, Iron Mask Hill, 1320–1500 m., ♂ fl. i.1915, *Eyles* 607 (BM; K; PRE; SRGH). W: Farm Besna Kobila, ♂ fl. i.1953, *Miller* 1489 (K; SRGH); ♀ fl. xii.1953, *Miller* 2029 (K; LISC; SRGH); fr. iv.1954, *Miller* 2316 (K; LISC; SRGH). C: Gwelo, Salemore Farm, ♂ fl. 10.ii.1950, *Harvie* 7/50 (SRGH); Domboshawa, fr. 16.ii.1947, *Wild* 1666 (BM; EA; SRGH). E: Umtali Commonage, ♂ fl. iii.1946 & fr. 22.vi.1946, *Chase* 195 (BM; LISC; SRGH); Umtali, S-eastern slope of Quagga Hill, ♀ fl. 30.i.1955, *Chase* 5463 (BM; COI; K; LISC; LMA; SRGH). S: Chibi, Sibanajene Hills, fr. iii.1955, *Davies* 1788 (K; LISC; LMA; SRGH); Belingwe near Mnene, ♂ fl. 27.ii.1931, *Norlindh & Weimarck* 5195 (COI; LD) et ♀ 5192 (COI; LD). **Malawi.** C: Dedza, road to Sosola Rest House, Mua-Livulezi Forest, st. 22.iii.1962, *Adlard* 456 (SRGH). S: Shire Highlands, ♂ fl. 8.iv.1906, *Adamson* 317 (BM; K); Shire Highlands, ♀ fl. & fr. s.d., *Buchanan* 88 (63) (K). **Mozambique.** N: Ribáuè Mt. (Melapuè), ♂ fl. 23.iii.1964, *Torre & Paiva* 11343 (BR; LISC; LMU; P; PRE). Z: Lioma, south of the Mossolua Mts., ♀ fl. 22.iii.1954, *Gomes e Sousa* 4248 (COI; K; LISC; LMA; PRE); Gúruè, 24 km. on the road to Lioma, ♂ fl. 25.ii.1966, *Torre & Correia* 14862 (B; BM; LISC; LUA; LUAI). MS: Inyamatshira Mt. Range, ♂ fl. 29.i.1950, *Chase* 1954 (BM; COI; K; LISC; SRGH); Mossurize, near Espungabera Catholic Mission, fr. 8.vi.1942, *Torre* 4266 (LISC).

Also in northern Transvaal and Angola. Slopes and summits of hills between granite or quartzite rocks, stream-banks, galleries and *Brachystegia* forest, savanna, bush clumps, etc.

2. **Heteropyxis natalensis** Harv., Thes. Cap. **2**: 18, t. 128 (1863).—Hook. f. in Benth. & Hook. f., Gen. Pl. **1**: 785 (1867).—Burtt Davy, F.P.F.T. **2**: 468 (1932).—Stern & Brizicky in Bull. Torrey Bot. Club **85**: 116, fig. 1 a–d (1958).—Melchior in Engl., Syll. Pflanzenfam. **2**: 351 (1964).—Fernandes in Mitt. Bot. Staatss. München **10**: 228 (1971). Syntypes from Natal.

Tree up to 10 m. high or shrub, with whitish bark. Leaf-lamina (2·5)4·5–7(8–9) × (0·6) 1–2·2 cm., narrowly to broadly lanceolate, attenuate at the base into a canaliculate petiole up to 1 cm. long and at the apex into an acute tip, membranous to chartaceous, dull green above, light green below, sparsely appressed pilose on both surfaces, becoming glabrous when old; midrib grooved above, very prominent beneath, lateral nerves inconspicuous or sometimes with 1 prominent pair, reticulation close and very conspicuous, usually with a gland-dot in each mesh. Cymes paniculate, the ♂ panicles longer and broader than the ♀; inflorescence axis and pedicels ± sparsely appressed-pilose; bracteoles minute, up to 1 mm. long, carinate, appressed-pilose and gland-dotted outside; pedicels c. 2 mm. long. ♂ flower: receptacle 1–1·5 × 1·5–1·75 mm., sparsely pilose and gland-dotted outside; calyx-lobes c. 0·75 mm., triangular, obtuse; petals 1·5–2 × 1–2·5 mm., ovate, slightly unguiculate, reticulately nerved, gland-dotted, inserted at the mouth of the receptacle-tube; stamens (4)5–7(8), inserted near the margin of the disk; filaments c. 2 mm. long, sometimes slightly unequal; anthers c. 1 mm. broad, ± pilose; pistillode c. 0·75 × 1 mm., turbinate, ± wrinkled, gland-dotted outside; style up to 0·5 mm. long, with a capitate stigma. ♀ flower: perianth as in ♂; staminodes c. 1 mm. long; ovary subglobose, c. 1·25 in diam., usually bilobate, gland-dotted outside; style c. 2 mm. long, glabrous; stigma capitate, papillose. Capsule subglobose or ellipsoid, 2–3 × 1·75–2 mm., gland-dotted. Seeds ± straight, the outer ones c. 1·75 × 0·75 mm., ± winged at both ends.

Rhodesia. E: Umtali, St. David's School, ♀ fl. 12.i.1963, *Chase* 7934 (K; LISC; SRGH). **Mozambique.** MS: Chiluvo, ♂ fl. iii–iv, *Junod* 1090 (K). LM: Namaacha near Canada Dry factory, fl. 27.iii.1957, *Barbosa & Lemos* 7548 (COI; LISC); Namaacha, fr. 4.vii.1957, *Barbosa & Lemos* 7582 (K; LISC; LMA); Goba, ♂ fl. 18.iii.1945, *Sousa* 93 (BM; COI; FHO; LISC; LMA; PRE; SRGH).

Also in Transvaal, Swaziland and Natal. Slopes and tops of hills, riverine forests, woodlands, thickets, etc.

76. BARRINGTONIACEAE

By A. Fernandes

Evergreen trees or shrubs. Leaves alternate, exstipulate, usually large, simple, not gland-dotted. Flowers showy, actinomorphic or zygomorphic, bisexual, solitary or in fascicles on the old wood or in terminal racemes. Receptacle ± crateriform. Calyx (2–3)4–5-lobed, the lobes sometimes calyptriform, covering the petals and stamens in bud. Petals 4 or 0 (*Napoleonaea*), free or somewhat adnate to the staminal ring. Stamens ∞, in several verticils, all fertile or only some fertile and accompanied by numerous antherless staminodes, the outer series of staminodes united and forming a radiate or cup-shaped pseudocorolla; filaments usually united into a ring of varying depth; anthers basifixed, opening by lateral slits; staminal disk sometimes lobed. Ovary inferior or semi-inferior, 2(3)–5-locular; style very long and filiform or short and expanded into a broad peltate disk; ovules 1–∞ per loculus, anatropous. Fruit woody, fibrous or fleshy, indehiscent or operculate at the top, crowned by the persistent calyx-segments. Seeds without endosperm; embryo entire or divided.

A family of the tropical and subtropical regions of the Old World.

Calyx closed in bud and splitting later into 2–4(5) lobes; petals 4, free; stamens very numerous, connate at the base in 3–8 whorls, the inner one antherless; ovary 2–4-locular; style filiform, very long; inflorescence racemose   -   **1. Barringtonia**

Calyx 5-lobed, the lobes valvate; petals 0; antherless staminodes and fertile stamens ± connate at the base in 4 series, the outer one of staminodes forming a large radiate or campanulate pseudocorolla; ovary 5-locular; style short and expanded into a broad peltate disk; flowers solitary or in fascicles on the old wood -   **2. Napoleonaea**

1. BARRINGTONIA J. R. & G. Forst.

Barringtonia J. R. & G. Forst., Char. Gen. Pl.: 75, t. 38 (1776) *nom. conserv.* *Huttum* Adans., Fam. Pl.: 88 (1763).

Evergreen trees usually tall, more rarely shrubs. Flowers in long terminal or lateral racemes or spikes. Calyx ± 4-lobate or 2-4 irregularly laciniate. Petals 4, free from each other but somewhat adnate to the staminal ring. Stamens numerous, multiseriate, with the filiform filaments of each whorl connate at the base into a short ring; anthers small. Disk intrastaminal. Style filiform, very long, usually contorted; stigma small, simple. Ovary 2–4-locular; ovules 2–8 in each locule, horizontal, pendulous, 2-seriate. Drupe fibrous, usually 1-seeded; embryo entire.

A genus of c. 40 species of the tropical and subtropical regions of E. Africa, Madagascar, S. Asia, Malaysia, Australia and the Pacific Is.

Barringtonia racemosa (L.) Roxb., Hort. Beng.: 52 (1814).—Spreng. in L., Syst. Veg. ed. 16, **3**: 127 (1826).—DC., Prodr. **3**: 288 (1828).—Wight, Ic. Pl. Ind. Or. **1**: t. 152 (1839).—Hook. in Curtis, Bot. Mag. **67**: t. 3831 (1841).—Lawson in F.T.A. **2**: 438 (1871).—Medley Wood, Hand. Fl. Natal: 49 (1907).—Engl., Pflanzenw. Afr. **3**, 2: 658, fig. 290 (1921).—Marloth, Fl. S. Afr. **2**, 2: 218, fig. 140 (1925).—Knuth in Engl., Pflanzenr. **IV**, 219: 17 (1939).—Garcia in Estud. Ens. Doc. Junta Invest. Ultramar **12**: 157 (1954).—Perrier in Fl. Madag., Lecythid.: 4, fig. 2 (1954).—Cufod. in Bull. Jard. Bot. Brux. **29**, Suppl.: 613 (1959).—Dale & Greenway, Kenya Trees & Shrubs: 243, fig. 48 (1961).—Gomes e Sousa in Mem. Inst. Invest. Agron. Moçamb. **2**, 1: 597, fig. 170 (1967).—Payens in Blumea **15**: 192 (1967). TAB **50**. Type from India.

Eugenia racemosa L., Sp. Pl. **1**: 471 (1753).—Lam., Encycl. Méth. Bot. **3**: 197 (1789). Type as above.

Butonica caffra Miers in Trans. Linn. Soc., Bot. **1**: 78 (1875). Type from S. Africa.

Barringtonia caffra (Miers) Knuth, tom. cit.: 19 (1939). Type as above.

A much-branched riparian shrub or small to medium-sized tree, 2–20(27) m. high. Leaves crowded at the ends of the branches, obovate-oblong or obovate-oblanceolate, 14–38(42) × 4–14(16) cm., acute or acuminate but sometimes rounded

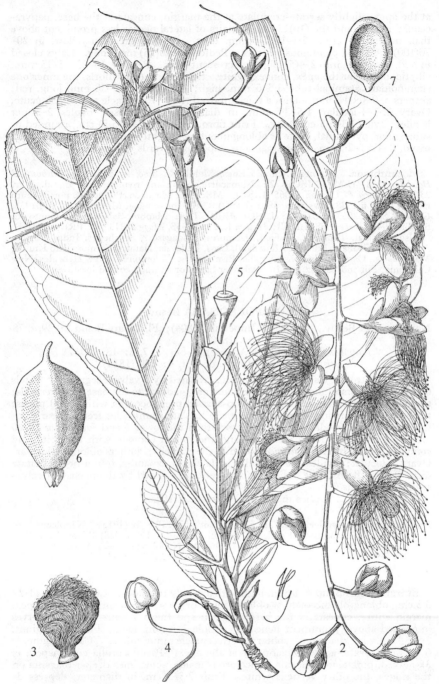

Tab. 50. BARRINGTONIA RACEMOSA. 1, branch with leaves ($\times\frac{1}{2}$); 2, inflorescence ($\times\frac{1}{2}$); 3, flower with calyx and petals removed to show stamens ($\times$ 1); 4, stamen ($\times$ 10); 5, ovary and style ($\times$ 1); 6, fruit ($\times$ 1); 7, seed ($\times$ 1), all from *Torre & Correia* 14065.

at the apex, slightly serrate-crenulate at the margin, cuneate at the base, papyraceous; midrib and the (10)13–18(20) pairs of lateral nerves less prominent above than beneath; petiole 2·5–10(15) mm. long. Flowers greenish-yellow, in 20–70(100) mm. long pendulous racemes. Pedicels 3–16(25) mm. long. Calyx closed in bud, rupturing into 2–4(5) elliptic accrescent lobes. Petals 4, 15–25 × 5–15 mm., elliptic, obtuse at the apex, convex, white. Stamens in (5)6 whorls, the inner one staminoidal; staminal-tube 3·5–6 mm. high; filaments 22·5–38 mm. long, red; anthers c. 0·8 × 0·8 mm. Disk a thick grooved ring. Receptacle tetragonal, acute. Ovary (2)3–4-locular, 1·5–3·5 mm. in diameter, ± globose; ovules 2–3 per locule; style (2)3–5·5 cm. long. Fruit fibrous, 5–7(9) × 2–4(5·5) cm., rounded or subangular, pyramidal, ovate or oblong in outline, slightly winged, 1-seeded. Seed ovoid, 2–4 × 1–2·5 cm., subtetragonous, tapering towards the apex.

Mozambique. N: near António Enes, Meluli R., fl. & fr. 5.xi.1952, *Barbosa & Balsinhas* 5237 (BM; LISC). Z: Namacurra, Muijaiana rivulet, c. 30 m., fl. & fr. 23.i.1966, *Torre & Correia* 14065 (LISC). MS: Vila Machado, Pungué R. port, Craro rivulet, fl. 26.ii.1948, *Mendonça* 3803 (LISC). SS: Inhambane, Mutamba R., fl. i.1938, *Gomes e Sousa* 2063 (COI; K; LISC; PRE). LM: Maputo, Salamanga, Maputo R., fl. 23.iv.1948, *Torre* 7688 (LISC); Inhaca I., 24.ix.1958, *Mogg* 28313 (K; PRE; SRGH).
Also in Natal, Tanzania, Mafia I., Comores, Madagascar, Seychelles, India, Ceylon, Andamans, Nicobars, Burma, S. Thailand, Laos, Malaysia, Hainan, Formosa, Riu Kiu Is., Marianas Is., Carolinas Is., N. Australia, Bismarck Is., Solomon Is., New Hebrides, New Caledonia, Fiji and Samoa Is. In tidal river banks or swampy coastal localities.

2. NAPOLEONAEA Beauv.

Napoleonaea Beauv., Napoléone Impériale (1804); Fl. Oware Benin 2: 29, t. 78 (1810).
" *Napoleona* "—Heine in Adansonia, Sér. 2, 7, 2: 137 (1967).
Belvisia Desv. in Journ. de Bot. 4: 130 (1814).

Shrubs or trees. Calyx-tube turbinate, crowned by 5 valvate segments. Petals 0. Antherless staminodes and fertile stamens ± adnate to the base in 4 verticils; the outer one forming a radiate or cup-shaped pseudocorolla with 20–40 pleats; the second comprising numerous narrowly ligulate staminodes free or nearly so; the third of staminodes connate for more than the lower ½ and forming a 20–60 lobed cup with the lobes inflexed; and the fourth connate with the third and composed of intermixed staminodes and fertile stamens; anthers oblong, 1-locular. Ovary 5-locular; ovules 2-6 per cell; style short, expanded into a broad peltate disk. Drupe fleshy, resembling a pomegranate, crowned by the persistent calyx-segments. Seeds few.
A genus of c. 15 species mainly of W. tropical Africa.

Napoleonaea gossweileri Bak. f. in Cat. Talbot's Nig. Pl.: 32 (1913) " Napoleona ".— Greves in Journ. Bot., Lond. **66**, Suppl. Polypet.: 176 (1928) " Napoleona ".— Knuth in Engl., Pflanzenr. **IV**, 219: 72 (1939) " Napoleona ".—F. White, F.F.N.R.: 274 (1962) " Napoleona ".—A. & R. Fernandes in C.F.A. **4**: 113 (1970). TAB. **51**. Type from Angola.

Evergreen shrub, up to 1·8 m. high with a woody rootstock. Leaves 3–10 × 1·2–3·5 cm., oblong-oblanceolate or oblong, blunt or sometimes acuminate at the apex, margin entire, cuneate at the base; midrib and the 6–7 pairs of lateral nerves impressed above, prominent beneath; petiole 3–5 mm. long. Flowers fragrant, 3·5–5 cm. in diameter, solitary. Receptacle muriculate. Calyx-segments 6 × 4 mm., ovate, acute, biglandular at the apex. Pseudocorolla cup-shaped, c. 35-nerved, pink along the ribs, concolorous inside, sometimes deeper magenta on the ridges, the other verticils white. Fruit 2·5–3 cm. in diameter, depressed-globose, verrucose.

Zambia. B: Sesheke Distr., c. 3·2 km. NW. of Masese Valley, fl. 11.viii.1947, *Brenan* 7675 (BM; EA; K); Mongu, 14·5 km. S. of Hamushakende, fr. 22.vii.1961, *Angus* 3029 (LISC; SRGH).
Also in Angola. Mixed woodlands on sandy slopes.

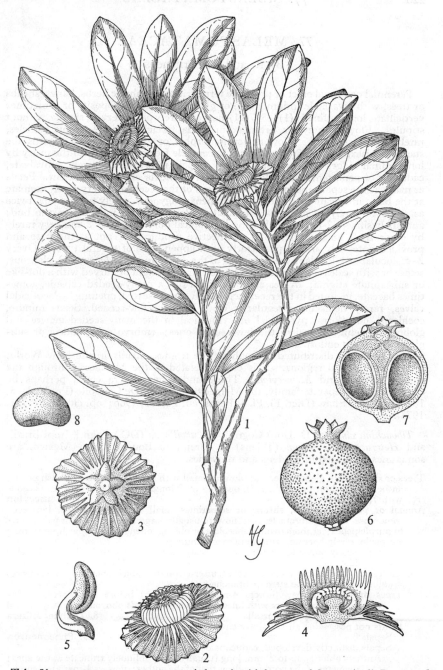

Tab. 51. NAPOLEONAEA GOSSWEILERI. 1, branch with leaves and flowers (× ½) *Drummond & Cookson* 6559; 2, flower (× ¾); 3, flower seen from below (× ¾); 4, flower in vertical section (× ¾), 2–4 from *Rea* 123; 5, stamen (× 4); 6, fruit (× ¾); 7, fruit in longitudinal section (× ¾); 8, seed (× ¾), 5–8 from *Fanshawe* 8058.

77. MELASTOMATACEAE

By R. and A. Fernandes

Perennial or annual erect or prostrate terrestrial or epiphytic herbs, lianes, shrubs or trees, with a variable indumentum or glabrous. Leaves opposite or sometimes verticillate, longitudinally (1)3–5–7(11)-nerved, rarely penninerved, simple, without stipules. Flowers in cymes, arranged in heads, fascicles or in ± ample panicles, rarely flowers solitary. Flowers actinomorphic or zygomorphic as regards the stamens, bisexual, (3)4–5(6)-merous. Receptacle free or adherent to the ovary by longitudinal ridges or ± adnate to it. Sepals imbricate or rarely subvalvate, caducous or persistent, rarely 0; intersepalar segments sometimes present. Petals as many as the sepals, inserted on the margin of the receptacle, very rarely connate at the base, imbricate or contorted. Stamens epigynous, rarely perigynous, twice as many or rarely as many as the petals; filaments free, inflected in the bud; anthers basifixed, 2-locular, dehiscing by 1 apical pore (rarely by 2 and very rarely by 4) or by 2 longitudinal slits; connective produced or not at the base and provided anteriorly and/or posteriorly with appendages of several types. Ovary 2–∞-locular or 1-locular by suppression of the dissepiments (*Memecylon*), glabrous, setose or with scales at the apex; style 1, terminal, straight or curved with a dot-like or subcapitate stigma; ovules anatropous. Fruit a many-seeded capsule, sometimes baccate, included in the receptacle, rarely semiexserted, opening by loculicidal valves, rarely dehiscing irregularly, or a berry, 1–few-seeded. Seeds minute, cochleate, cuneiform, pyramidal or cylindrical in the many-seeded genera and globose or hemispherical in the few-seeded ones; embryo subcylindric or subglobose, without endosperm.

A large family distributed throughout the tropics, mainly in the New World, rare in subtropical regions. It is clearly related to the Myrtaceae through the genera *Mouriri* and *Memecylon*. These and other genera could perhaps be considered as a separate family, the Memecylaceae; cf. De Candolle (Prodr. **3**: 5, 1828) and Airy Shaw (Dict. Fl. Pl. & Ferns, ed. 7: 712, 1966; op. cit. ed. 8: 731, 1973).

Tibouchina viminea (D. Don) Cogn. [= *T. urvilleana* (DC.) Cogn.], from Brazil, and *Heterocentrum roseum* (Triana) A. Braun & Bouché, from Mexico, are sometimes cultivated in gardens and nurseries.

Trees or shrubs with glabrous leaves; flowers small with usually whitish or bluish petals; anthers dolabriform (axe-shaped), opening by 2 longitudinal slits; fruit drupaceous with 1(2) large seeds - - - - - - - - **1. Memecylon**
Annual or perennial herbs, shrubs or sometimes small trees with setose, lanate or tomentose, rarely glabrous leaves; flowers usually large and showy with pale mauve to purple petals; anthers ovoid, lanceolate or oblong, opening by 1 apical pore; fruit a capsule, rarely baccate, with numerous minute seeds:
Stemless; usually only a leaf arising from a scaly tuber as well as the ± long-peduncled scorpioid cyme - - - - - - - **10. Haplophyllophorus**
Caulescent; leaves more or less numerous, inserted on stem and branches; inflorescences axillary or ending the stem or branches:
Trees or large shrubs; flowers 4-merous, produced before the leaves; pedo-connectives* of anthers with linear appendages; seeds obovoid or subpyramidal with hyaline papillae dorsally - - - - **8. Dichaetanthera**
Plants not possessing all the above characters:
Sepals 0 - - - - - - - - - - **9. Phaeoneuron**
Sepals distinctly developed, caducous or persistent:
Anthers oblong, up to 2 mm. long (usually less), obliquely truncate at the apex; diameter of the pore as wide as that of the anther; flowers 4-merous; annual - - **2. Antherotoma**
Anthers longer, attenuate to the apex; diameter of the pore narrower than that of the anther; flowers 4–5(6)-merous; perennial, rarely annual:
Intersepalar segments absent, rarely present and then 1-setose; flowers 5-merous with persistent sepals:

* With Jacques-Félix (in Bull. Inst. Fr. Afr. Noire **15**: 237, 1953), we call the prolongation of the connective below the anther-loculi the *pedoconnective*.

Flowers not surrounded by persistent large bracts, in panicles, sometimes
contracted; small trees or large shrubs
7. Dissotis Subgen. **Dissotidendron**
Flowers solitary or in capitate inflorescences, surrounded by large persistent
bracts; perennial herbs or small shrubs:
Stamens equal, with slightly attenuate not rostrate anthers; bristles of
receptacle in 1–several, complete or incomplete rings; fruit a baccate
capsule, irregularly bursting   -   -   -   - **3. Tristemma**
Stamens usually very unequal, with rather attenuate-rostrate anthers;
bristles of receptacle, if present, not in rings; fruit a valved capsule
6. Melastomastrum
Intersepalar segments present, if absent then receptacle glabrous; sepals early
or tardily caducous, rarely persistent:
Stamens equal:
Anthers erostrate, opening by an extrorse pore; pedoconnective short,
with two divergent, thick, obtuse, anterior appendages; intersepalar
segments subulate   -   -   -   -   - **4. Pseudosbeckia**
Anthers rostrate, opening by a terminal introrse pore; pedoconnective
inconspicuous with thinner anterior appendages   - **5. Osbeckia**
Stamens usually very unequal, the longer ones with long pedoconnectives
provided anteriorly with a ± deeply furcate appendage, the shorter
ones with short pedoconnectives anteriorly bitubercled (if stamens
subequal*, then pedoconnectives conspicuous at least on the longer
ones); sepals not or tardily caducous   -   -   - **7. Dissotis**

1. MEMECYLON L.

Memecylon L., Sp. Pl. **1**: 349 (1753); Gen. Pl., ed. 5: 166 (1754).

Trees or shrubs usually everywhere glabrous, with the young branches narrowly
4-winged, 4-gonous or terete. Leaves opposite, entire, longitudinally 1–3(5)-
nerved, shortly petiolate. Flowers 4-merous, small or minute, in fascicles or in
cymes forming panicle-like or corymbose inflorescences. Bracts minute.
Receptacle broadly campanulate, hemispherical or cupular, with the limb ±
dilated, truncate, 4-toothed or -lobed. Petals broadly ovate, obovate, circular-
subdeltate, obtuse or apiculate, whitish or bluish. Stamens 8, equal, sometimes 4
with abortive anthers; filaments filiform; anthers axe-shaped, with 2 anterior
thecae dehiscing by 2 longitudinal slits and usually with a gland on the concave
back. Ovary 1-locular, wholly adherent to the receptacle, sunken or convex at the
apex; style filiform; ovules 2–12(20), on a central placenta. Fruit an ovoid,
globose or ellipsoid 1(2)-seeded berry, crowned by the persistent calyx. Seeds
large; cotyledons crumpled or not so.
C. 320 species in the Old World tropics and subtropics, with c. 70 species in Africa.

Leaf-lamina longitudinally 1-nerved:
Leaf-lamina roundish, truncate or emarginate at the apex and cuneate towards the base:
Cymes 3(2–1)-flowered; flower-buds obtuse, c. 4 mm. wide -   -   1. *torrei*
Cymes many-flowered; buds acute, up to 3 mm. wide:
Flower-buds c. 5 ×3 mm., with membranous, contorted petals; receptacle hemi-
spherical; leaf-lamina 2·3–8·5(11·5) ×1–5(6·5) cm.   -   - 2. *flavovirens*
Flower-buds c. 3 ×2·25 mm., with coriaceous, not contorted petals; receptacle
obconical; leaf-lamina 1·5–4·5 ×0·5–2·7 cm.   -   -   - 3. *insulare*
Leaf-lamina ± acuminate at the apex:
Leaf-lamina cuneate or acute at the base; flowers not in fascicles:
Cymes many-flowered, in panicles up to 4 cm. long; leaf-lamina 4·5–12 ×2–6 cm.;
petiole 3–7 mm. long   -   -   -   -   -   - 4. *myrianthum*
Cymes 1–3-flowered, up to 2 cm. long, not paniculate; leaf-lamina
(2·5)3–6 ×(1·2)1·5–3·2 cm.; petiole c. 4 mm. long   -   - 5. *natalense*
Leaf-lamina slightly cordate or roundish at base, 7–18 ×3–6·5 cm.; petiole 1·5–2 mm.
long; flowers in dense many-flowered axillary fascicles -   6. *erythranthum*
Leaf-lamina longitudinally 3–5-nerved:
Inflorescences distinctly pedunculate, up to 5 cm. long; reticulation of the leaf-lamina
inconspicuous:
Cymes many-flowered, up to 2·5 cm. long, in dense clusters; receptacle 2 ×3 mm.,
with lobes c. 1 ×2 mm., obtuse, not truncate; tree or shrub -   7. *sapinii*
Cymes (1)2-flowered, in axillary panicles up to 5 cm. long; receptacle 2 ×1·5 mm.,
with the lobes 0·5–1 mm., truncate; a shrub or sometimes a liane
8. *zambesiense*
* Osbeckioid forms.

Inflorescences sessile or subsessile; reticulation of the leaf-lamina conspicuous especially
  beneath:
Reticulation of the inferior leaf surface very close:
  Leaf-lamina 3·8–10·5 × 2–5·2 cm., ± acuminate; fruits pedicellate
                                                                 9. *sansibaricum* var. *buchananii*
  Leaf-lamina (1·5)2–3·5 × 1–2·5 cm., not acuminate, obtuse or roundish at the apex;
    fruits sessile   -   -   -   -   -   -   -   10. *sessilicarpum*
Reticulation of the inferior leaf surface not so close; leaf-lamina elliptical to widely
  elliptical (3·5–9 × 2–4·5 cm.), ovate (3–9 × 1·5–5 cm.) or almost circular
  (5–7 × 4·5–6·5 cm.)   -   -   -   -   -   -   -   11. *sousae*

1. **Memecylon torrei** A. & R. Fernandes in Bol. Soc. Brot., Sér. 2, **46**: 63, t. 1 (1972).
TAB. 52. Type: Mozambique, Mozambique Distr., Moma, *Torre & Correia* 17229
(COI; K; LISC, holotype; SRGH).

Shrub c. 3 m. high. Young branches narrowly 4-winged, yellowish, the old ones
inconspicuously 4-gonous or terete, covered with greyish and irregularly fissured
bark; nodes thickened; internodes 0·5–1·5 cm. long. Leaf-lamina 4–7 × 2–4 cm.,
elliptic or obovate, obtuse or rounded at the apex, cuneate at the base, coriaceous,
discolorous, green on the upper face, yellowish-green beneath, dull on both faces,
longitudinally 1-nerved (sometimes with a pair of thin nerves close to the leaf
margin), the midrib impressed on the upper face, prominent below, the transverse
nerves inconspicuous on both faces. Cymes up to 1·5 cm. long, 1–3-flowered,
axillary or at the nodes of leafless branches; inflorescence-axis up to 4 mm. long,
the peduncle and the pedicel each c. 2 mm. long. Flower-buds obtusely apiculate,
c. 5 × 4 mm. Receptacle campanulate, c. 3 mm. high and 4 mm. in diam. at the
apex. Receptacle-lobes broadly triangular, 1 × 2·5 mm. Petals (in bud)
2·5 × 3·25 mm., whitish, irregularly rhombic, truncate at the base. Anthers (in
bud) 1·75 mm. long, with a gland on the back. Ovules 6. Fruit unknown.

Mozambique. N: Moma, road to Naburi 15 km. from R. Ligonha, 17.i.1968, *Torre &*
Correia 17229 (COI; K; LISC; SRGH).
Known only from Mozambique. On termitaria surrounded by mangrove formation.

2. **Memecylon flavovirens** Bak. in Kew Bull. **1897**: 268 (1897).—Gilg in Engl., Mon.
Afr. Pflanzen. **2**: 44 (1898).—R. E. Fr., Wiss. Ergebn. Schwed. Rhod.-Kongo.
Exped. **1**: 182 (1914).—Engl., Pflanzenw. Afr. **3**, 2: 769 (1921).—Brenan, T.T.C.L.:
311 (1949); in Mem. N.Y. Bot. Gard. **8**: 441 (1954).—A. & R. Fernandes in Mem.
Soc. Brot. **11**: 61 (1956); in C.F.A. **4**: 119 (1970).—F. White, F.F.N.R.: 310
(1962). Type: N. Malawi, *Whyte* (K, holotype).
    Memecylon cyanocarpum Gilg in Engl., Bot. Jahrb. **30**: 366 (1901).—Brenan,
loc. cit. (1949). Type from Tanzania.
    Memecylon gossweileri Gilg ex Engl., loc. cit. Type from Angola (Malanje).

A much-branched tree up to 7·5 m. or a shrub up to 5 m. Stem with
grey-brown or dark-grey, deeply fissured bark, breaking off in flakes; young
shoots 4-gonous, grey-green or whitish-grey; nodes thickened. Leaf-lamina
2·3–8·5(11·5) × 1–5(6·5) cm., obovate-cuneate, rarely obovate-oblong or elliptic,
roundish, truncate or emarginate (rarely subacute) at the apex, coriaceous,
subconcolorous (yellowish-green, turning deep-yellow or brown-yellow on drying,
but bluish-black or glaucous in the young leaves), dull on both faces or somewhat
shining above, with 1 longitudinal nerve prominent below and the transverse
nerves very oblique, slightly raised beneath, sometimes also above, frequently
inconspicuous on both faces; petiole ± absent or up to 4 mm. long, stout. Cymes
up to 12 mm. long, many-flowered, mainly on the upper leafless part of the
branches, solitary or in fascicles, sessile or with peduncle up to 8 mm. long;
pedicels 2–6 mm. long, deep red-purple. Flower-buds c. 5 × 3 mm., conical and
very pointed at the top. Receptacle 1–2 mm. high, hemispherical, contracted above
the base, with the limb 2·5–3 mm. in diam., saucer-shaped, sublobulate, green.
Petals 4 × 3·5 mm., subdeltate, acuminate, very acute, white, sweet-scented.
Filaments 4–5 mm. long, white or pale violet; anthers pale to bright blue before
dehiscing, with a red gland on the back. Style 6–8 mm. long, pale violet. Fruit
ovoid-ellipsoid, 12–14 × 6–8 mm., or subglobose and up to 15 mm. in diam., at first
green or pale yellow, turning blue-black at maturity.

Zambia. B: Zambezi (Balovale), Chavuma, fl. & fr. 14.x.1952, *White* 3509 (BM;
COI; FHO). N: Mbala, Kalambo Falls, fl. & fr. 14.xi.1952, *White* 3645 (COI; FHO;

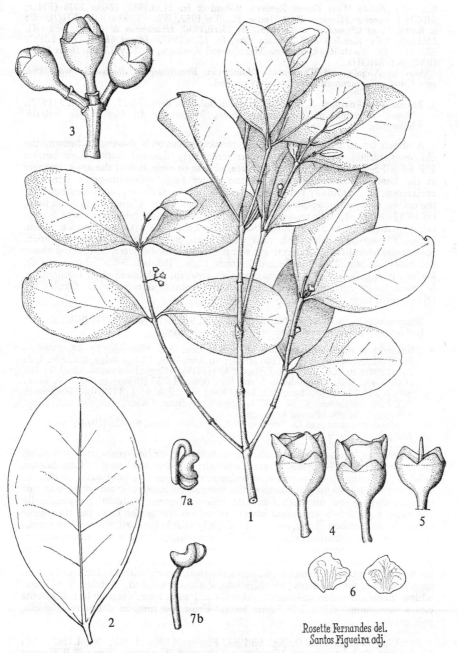

Rosette Fernandes del.
Santos Figueira adj.

Tab. 52. MEMECYLON TORREI. 1, flowering branch (×½); 2, underside of leaf (×1); 3, inflorescence (×3); 4, flowers (×3); 5, receptacle after shedding petals and stamens (×3); 6, petals (×3); 7, stamens, (a) in bud, (b) in flower (×6), all from *Torre & Correia* 17229.

K). W: Ndola West Forest Reserve, fl.-bud & fr. 21.xi.1959, *Angus* 2076 (FHO; SRGH); Serenge, Kanona Rest House, fl. 30.xi.1952, *White* 3800 (COI; FHO). C: c. 8 km. E. of Chiwefwe, c. 1350 m., fr. 15.vii.1930, *Hutchinson & Gillett* 3688 (K). **Malawi.** N: Kandoli Hill, Nkhata Bay, fr. 13.ix.1963, *Adlard* 551 (COI; LISC; SRGH). C: North Road, between Mzimba and Kasungu, fr. 23.viii.1946, *Brass* 17381 (BM; K; SRGH).

Also in Angola and Tanzania. Mainly in *Brachystegia–Julbernardia* woodland, sometimes on escarpments, in dry stony ground.

3. **Memecylon insulare** A. & R. Fernandes in Bol. Soc. Brot., Sér. 2, **46**: 65, t. 3 (1972). Type: Mozambique, Inhambane Distr., Magaruque I., *Guy* (COI; SRGH, holotype).

A shrub up to 2 m. high. Young branches narrowly 4-winged, brown, the old ones terete, with greyish brown irregularly fissured bark. Leaf-lamina 1·5–4·5 × 0·5–2·7 cm., elliptic or obovate, obtuse or roundish at the apex, cuneate at the base, discolorous (green on the upper face, yellowish-green beneath), coriaceous, longitudinally 1-nerved, the nerve impressed above, prominent beneath, the transverse ones inconspicuous; petiole c. 2 mm. long. Cymes up to 2 cm. long, up to 12-flowered; inflorescence-axis up to 8 mm. long; peduncle up to 6 mm. long; pedicels up to 3 mm. long, bracteolate at the base with bracteoles c. 2 mm. long. Flower-buds biconical, c. 3 × 2·25 mm., acutely apiculate at the apex. Receptacle obconical, c. 2 mm. high and c. 3·5 mm. in diameter, contracted above the base; lobes c. 2 mm. wide and 0·5 mm. high. Petals c. 3·5 × 2·5 mm., thick, keeled, triangular, with a scarious whitish margin. Stamens with filaments c. 5·5 mm. long; anthers c. 2 mm. long, with a minute dorsal gland. Ovules 11; style 8 mm. long. Fruit unknown.

Mozambique. SS: Vilanculos, Magaruque I., 9.xi.1958, *Mogg* 28999 (SRGH). Known only from Mozambique. On recent sandstones.

4. **Memecylon myrianthum** Gilg in Engl., Mon. Afr. Pflanzen. **2**: 44 (1898).— De Wild., Miss. Laurent. **1**: 166 (1905); in Ann. Mus. Congo Belge, Bot. Sér. 5, **2**: 333 (1908); in Bull. Jard. Bot. Brux. **5**: 379 (1919).—Engl., Pflanzenw. Afr. **3**, 2: 768 (1921).—A. & R. Fernandes in Mem. Soc. Brot. **11**: 59 (1956); in Bol. Soc. Brot., Sér. 2, **34**: 84 (1960); tom. cit.: 198 (1960); in C.F.A. **4**: 118 (1970).—Wickens in F.T.E.A., Melastom.: 87 (1975). Syntypes from Gabon, Angola and Zaire; lectotype: Angola (Pungo Andongo).

Memecylon claessensii De Wild. in Bull. Jard. Bot. Brux. **4**: 426 (1914); op. cit. **5**: 378 (1919). Type from Zaire.

A tree up to 25 m. tall or a shrub. Stem with grey or brownish-grey ± fissured bark; branchlets 4-gonous, green, smooth; nodes thickened. Leaf-lamina 4·5–12 × 2–6 cm., elliptic to ovate, acuminate, the acumen ± long, acute or obtuse, acute at the base, slightly revolute at the margin, coriaceous or subcoriaceous, not shining, light-green on both faces or yellowish-green beneath, longitudinally 1-nerved, the midrib impressed above, raised on the under face, the transverse nerves inconspicuous above, sometimes slightly visible beneath, the tertiary nerves and reticulation inconspicuous; petiole 3–7 mm. long. Cymes in axillary, solitary or 2-nate, ± branched many-flowered panicles up to 4 cm. long; peduncle 0·5–1·3 cm. long; pedicels 2–4 mm. long. Flower-buds c. 1·5 mm. high, acutely conical. Receptacle 0·5 mm. high, hemispherical, green, with the limb c. 0·5 mm. high and 1·5–2 mm. in diam., campanulate, shallowly 4-lobed. Petals c. 2 × 2 mm., white, subsemicircular, apiculate. Anthers c. 1 mm. long, deep violet; filaments c. 1·5 mm. long. Style 2·5–3 mm. long. Fruit 6–8 mm. in diameter, globose, pale green.

Zambia. N: Kasama, fr. juv. vii.1964, *Fanshawe* 8808 (FHO; K; LISC). W: Mwinilunga, Sinkabolo dambo, Matonchi, c. 1350 m., fl. 22.xii.1969, *Simon & Williamson* 1951 (COI; SRGH). Also in Gabon, Zaire, Angola, Uganda and Tanzania. In riparian forests.

5. **Memecylon natalense** Markgr. in Notizbl. Bot. Gart. Berl. **11**: 1078 (1934).— A. & R. Fernandes in Bol. Soc. Brot., Sér. 2, **46**: 64, t. 2 (1972) descr. ampl. Type from S. Africa (Natal).

Memecylon australe Gilg & Schlecht. ex Engl., Pflanzenw. Afr. **3**, 2: 768 (1921) non F. Muell. ex Triana (1872). Type as above.

A shrub or tree up to 7 m. high. Young branches slender, narrowly 4-winged, brown, the old ones inconspicuously 4-gonous, covered with greyish longitudinally fissured bark; nodes thickened; internodes 0·5–3 cm. long. Leaf-lamina (2·5)3–6 × (1·2)1·5–3·2 cm., ovate, acuminate at the apex, the acumen up to 8 mm. long, obtuse or sometimes rather acute, cuneate at the base, subcoriaceous, yellowish-green on the upper face, lighter beneath, longitudinally 1-nerved, the midrib impressed above, prominent on the under face; transverse nerves and reticulation conspicuous on both faces; petiole up to 4 mm. long. Cymes up to 2 cm. long, axillary, usually 3-flowered; inflorescence-axis 5–7 mm. long; peduncle 1–7 mm. long; pedicels 1–4 mm. long. Flower-buds c. 5 mm. long, biconical, subacute at the apex. Receptacle c. 4 mm. high and 4·6 mm. in diameter, obconic; lobes c. 2·75 mm. wide and 1 mm. high. Petals 5 × 4 mm., subrhombic, acute at the apex, whitish, subcoriaceous. Stamens with filaments c. 6 mm. long and anthers c. 2·5 mm. long, with a dorsal, elliptic, yellowish gland, all fertile or sometimes 4 fertile and 4 sterile. Ovules 2–8; style 10 mm. long. Fruit subglobose, c. 9 mm. in diameter, crowned with the calyx-lobes.

Malawi. S: Mlanje, Ruo Gorge, c. 1000 m., fr. 1.ix.1970, *Müller* 1474 (COI; SRGH). **Mozambique.** N: Ribáuè, Mepáluè Mt., 1600 m., fl. 9.xii.1967, *Torre & Correia* 16431 (COI; K; LISC; LMU; PRE; SRGH). Z: Gúruè Mt., slope of the mountain, near R. Malema, 1700 m., fr. 6.xi.1967, *Torre & Correia* 15956 (COI; K; LISC; SRGH). Also in Natal. Montane mist forests.

6. **Memecylon erythranthum** Gilg in Engl., Mon. Afr. Pflanzen. **2**: 45, t. 10C (1898).—Engl., Pflanzenw. Afr. **3**, 2: 769 (1921).—Brenan, T.T.C.L.: 311 (1949).—A. & R. Fernandes in Mem. Soc. Brot. **11**: 60 (1956); in Bol. Soc. Brot., Sér. 2, **34**: 201 (1960) excl. specim. *Greenway* 4814 et *Zimmermann* 6704 et 6705.—Wickens in F.T.E.A., Melastom.: 84 (1975). Type from Tanzania.

A tree up to 8 m. high or a shrub. Branchlets terete, brownish; nodes rather thickened. Leaf-lamina 7–18 × 3–6·5 cm., ovate or oblong-ovate, roundish or more frequently slightly cordate at the base, somewhat contracted at the apex into a ± long and obtuse acumen, slightly revolute at the margin, coriaceous, discolorous or subconcolorous, slightly shining above, dull beneath, longitudinally 1-nerved (sometimes with a very slender nerve running parallel to and near the margin), transverse nerves numerous, very slender, faintly visible on both faces, straight, ± at right angles to the midrib, tertiary nerves and reticulation inconspicuous; petiole 1·5–2 mm. long. Flowers in fascicles of 20–40 at the nodes of leafless old branches (fide Gilg); pedicels 2–3 mm. long. Receptacle saucer-shaped, shallowly 4-lobed. Petals pink to scarlet. Fruit c. 10 mm. in diameter, subglobose, black at maturity (red when immature).

Rhodesia. E: Stirling, southern Melsetter, st. x.1912, *Swynnerton* 190 (BM). Also in Tanzania. Lower storey of rain forests.

7. **Memecylon sapinii** De Wild., Comp. Kasai: 378 (1910).—A. & R. Fernandes in Bol. Soc. Brot., Sér. 2, **30**: 185, t. 24 & 25 (1956); in Mem. Soc. Brot. **11**: 54 (1956); in C.F.A. **4**: 121 (1970).—F. White, F.F.N.R.: 310 (1962). Type from Zaire.
Memecylon angolense Exell in Journ. of Bot. **67**, Suppl. Polypet.: 180 (1929).—Gossw. & Mendonça, Cart. Fitogeogr. Angol.: 112, 159 (1939). Type from Angola.

A much-branched tree up to 9 m. tall with a narrow elongated crown or a shrub. Stem with corky, greyish, deeply fissured bark; branchlets 4-gonous, brownish, smooth. Leaf-lamina 4·5–14 × 2·2–5·5(7·5) cm., elliptic to narrowly elliptic, rarely ovate or ovate-lanceolate, ± acuminate at the apex, the acumen blunt or acute, cuneate or acute at the base, coriaceous or subcoriaceous, discolorous, dark green and slightly shining above, lighter green and dull beneath (brown, reddish brown or green above, green or yellowish-green beneath on drying), longitudinally 3-nerved, the nerves impressed on the upper face, raised beneath, the transverse and tertiary nerves either visible or inconspicuous, reticulation inconspicuous; lateral longitudinal nerves slightly arched between the insertions with the transverse ones towards the distal ⅓; petiole 2–7 mm. long. Cymes up to 2·5 cm. long, many-flowered, in dense axillary clusters or, frequently, on the leafless nodes; peduncles 3–13 mm. long, compressed, blackish; pedicels 6–13 mm. long, blackish. Receptacle c. 2 × 3 mm., hemispherical; limb 3–3·5 mm. in diameter, sub-campanulate, 4-lobed, the lobes blunt-triangular, black on drying like the

H

receptacle. Petals c. 2 × 3 mm., obovate, obtuse, deep violet. Stamens with violet filaments c. 5–6 mm. long, and yellowish anthers c. 2 mm. long. Style c. 7 mm. long. Fruit 8–10 × 7–8·5 mm., subglobose, green when young, black at maturity.

Zambia. B: Zambezi (Balovale), near Chavuma, fr. 13.x.1952, *White* 3481 (BM; COI; FHO; K). W: Mwinilunga, fl. 10.vi.1963, *Edwards* 716 (COI; K; LISC; SRGH).
Also in Angola and Zaire. Fringing forests and woodlands on grey compact Kalahari sand.

8. **Memecylon zambesiense** A. & R. Fernandes in Bol. Soc. Brot., Sér. 2, **43**: 304, t. 16 (1969). Type: Zambia, Mwinilunga Distr., *Angus* 528 (BM; COI, holotype; FHO; K).
    Memecylon sp. 1—F. White, F.F.N.R.: 312 (1962).

A shrub up to 4 m. high, sometimes a liane. Young branches indistinctly 4-gonous, becoming subcylindric, with dark-brown bark. Leaf-lamina 6–14 × 3·2–7 cm., narrowly ovate to broadly elliptic, acuminate at the apex, the acumen itself short (c. 0·5 cm. long), wide and ± obtuse or sometimes ± acute, narrowed or cuneate at the base, coriaceous, green or brown-green (when dried), longitudinally 3-nerved (sometimes 5-nerved but if so, then the external pair thinner, close to the margin and arched between the transverse venules), the nerves impressed on the upper face, raised below; lateral longitudinal nerves arched between the insertions with the transverse ones towards the distal ⅓; petiole up to 1 cm. long. Cymes (1)2-flowered, arranged in axillary panicles up to 5 cm. long; inflorescence-axis up to 3 cm. long, 4-gonous; peduncle 2–5 mm. long; pedicels very short. Petals 2 × 1·25 mm., obovate, clawed. Receptacle c. 2 × 1·5 mm., broadly campanulate; lobes 0·5–1 mm. long, truncate. Ovules 2. Fruit (immature) globose, black.

Zambia. W: Mwinilunga Distr., near Kalene Hill Mission, fr. juv. 20.ix.1952, *White* 3304 (COI; FHO; K).
Known only from Zambia. In gallery forests of the Zambesi.

9. **Memecylon sansibaricum** Taub. in Engl., Pflanzenw. Ost-Afr. **C**: 296 (1895).— Gilg in Engl., Mon. Afr. Pflanzen. **2**: 40 (1898).—Engl., Pflanzenw. Afr. **3**, 2: 765 (1921).—A. & R. Fernandes in Bol. Soc. Brot., Sér. 2, **34**: 195 (1960); op. cit. **43**: 302 (1969).—Wickens in F.T.E.A., Melastom.: 81 (1975). Type from Zanzibar.

Var. **buchananii** (Gilg) A. & R. Fernandes in Bol. Soc. Brot., Sér. 2, **46**: 66 (1972)— Wickens, op. cit. 82. TAB. **53**. Type: Malawi, *Buchanan* 141 (B†, holotype; BM; K. lectotype.).
    Memecylon buchananii Gilg in Engl., loc. cit.; in loc. cit.—A. & R. Fernandes, tom. cit.: 76, t. 15, 195 (1960).—F. White, F.F.N.R.: 310 (1962). Type as above.
    Memecylon sansibaricum var. *sansibaricum* sensu A. & R. Fernandes in Bol. Soc Brot., Sér. 2, **43**: 302 (1969) pro parte.

A tree up to 11 m. tall or a much branched shrub. Stem with the bark purplish-grey to purplish-brown, falling off in thin flakes; branchlets 4-gonous, smooth, grey or whitish-grey; nodes thickened. Leaf-lamina 3·8–10·5 × 2–5·2 cm., elliptic, broad-elliptic or ovate, ± acuminate or subcaudate at the apex, the acumen ± oblong, obtuse or acute, usually acute or cuneate or rarely roundish at the base, subcoriaceous to stiffly coriaceous, dark, mid- or yellowish-green and glossy above, pale green and mat beneath (on drying brownish on both faces or darker above than beneath, or dark- to yellowish-green above and reddish-brown beneath), longitudinally 3-nerved, the nerves impressed above, raised beneath; lateral longitudinal nerves arched between the insertions with the transverse ones towards the upper half; transverse nerves oblique, slightly visible on both sides; reticulation very close, conspicuous mainly beneath, not raised; petiole 3–6 mm. long. Cymes up to c. 1 cm. long, very shortly pedunculate, umbel-like, many-flowered, in axillary fascicles or on leafless nodes; bracts numerous at the peduncle-tip, keeled, reddish-brown; pedicels 6–9 mm. long. Receptacle c. 1–1·25 mm., subhemi-spherical, brown (on drying), with the limb c. 3 mm. in diameter, saucer-shaped, distinctly 4-lobate, the lobes ovate, roundish. Petals misty-purple or blue. Anthers c. 0·75 mm. long. Style 6–8 mm. long. Fruit c. 10 mm. in diameter, globose, blue-black at maturity (pale green when immature).

Rosette Fernandes del.
et Santos Figueira adj.

Tab. 53. MEMECYLON SANSIBARICUM var. BUCHANANII. 1, flowering branch (×1); 2, inflorescence (×2); 3, petal (×4); 4, stamen (×4), all from *Buchanan* sn.

228   77. MELASTOMATACEAE

Zambia. N: Kundabwika Falls, fl. 7.x.1958, *Fanshawe* 4882 (FHO; K). W: Mwinilunga, 6·4 km. N. of Kalene Hill Mission, fl. 20.i.1952, *Angus* 494 (BM; COI; FHO; K). **Malawi.** S: Mlanje, Lichenya Forest Reserve, near Mabuka Court, c. 600 m., st. 7.ix.1970, *Müller* 1561 (COI; SRGH). **Mozambique.** Z: Pebane, near the lighthouse, 20 m., st. 8.iii.1966, *Torre & Correia* 15090 (LISC). T: Zóbuè Mt., st. 3.x.1942, *Mendonça* 630 (BR; COI; LISC; M).

Also in Tanzania. Sand-dunes and laterite outcrops, thickets on shores of sea and lakes, fringing forests, " mushitus " and rain forests.

10. **Memecylon sessilicarpum** A. & R. Fernandes in Garcia de Orta, Sér. Bot. **2**: 55, t. 1 (1974). Type: Moçambique Distr., António Enes, *Torre & Correia* 17355 (COI; K; LISC, holotype; SRGH).

A small tree or a shrub up to 7 m. high. Young branches narrowly 4-winged, reddish-brown, ± cylindric when older, covered with greyish ± irregularly fissured bark; nodes thickened; internodes 0·5–2·5 cm. long. Leaf-lamina (1·5)2–3·5 × 1–2·5 cm., ovate or elliptic, obtuse or roundish at the apex, roundish or slightly cuneate at the base, dark green above, yellowish-green or bright green beneath, longitudinally 3-nerved, the longitudinal nerves ± raised on both faces, the laterals arched between the insertions with the transverse ones above the middle of leaf; reticulation close and very conspicuous mainly on the under side; petiole c. 1 mm. long. Flowers unknown. Fruits sessile, clustered, sometimes at the nodes of the leafless part of branches, crowned by the reddish, 1·5 mm. long receptacle lobes.

Mozambique. N: António Enes, shore, bud 17.x.1965, *Mogg* 32306 (LISC). Known only from Mozambique. Coastal forests on sandstone.

11. **Memecylon sousae** A. & R. Fernandes in Bol. Soc. Brot., Sér. 2, **46**: 67, t. 4 (1972). Type: Beira, Cheringoma, Chinizina, *Gomes e Sousa* 4380 (COI, holotype; K). *Memecylon sp. 2 et 3*—A. & R. Fernandes in An. Junta Invest. Ultramar **10**: 11 (1955).

Shrub c. 3 m. high. Young branches acutely 4-gonous, yellowish-brown or grey, the old ones subcylindric, covered with greyish striate bark; nodes much thickened; internodes 0·8–5·5 cm. long. Leaf-lamina elliptic to broadly elliptic (3·5–9 × 2–4·5 cm.) or ovate (3–9 × 1·5–5 cm.) to almost circular (5–7 × 4·5–6·5 cm.), chartaceous or coriaceous, rounded or usually shortly acuminate at the apex, the acumen wide and obtuse, rounded to ± cuneate at the base, brownish-green on the superior surface, green beneath, longitudinally 3(5)-nerved; midrib usually impressed above, prominent beneath; lateral nerves thinner than the midrib, arched between the insertion with the transverse ones along the distal ⅓; marginal nerves (when present) thin, arched between the transverse nerves from the base to the apex or only along the distal ½; transverse nerves slender, oblique; reticulation conspicuous, somewhat open; petiole 2–5 mm. long. Flowers in leaf-axils or at the nodes of leafless branches, arranged in dense sessile fascicles; bracts small, scarious; pedicels c. 4 mm. long, articulate below the flower. Young fruit globose, crowned with the c. 0·75 mm. long receptacle-lobes, becoming obovoid (9 × 7 mm.) and black at maturity.

Mozambique. MS: Beira, Cheringoma, near R. Chinizina, st. 22.x.1949, *Pedro & Pedrógão* 8861 (COI). LM: Inhaca I., fr. i.1956, *Davidson* 156 (SRGH). Known only from Mozambique. In open forests or dense coastal bush on sandy soils.

Memecylon sp. A.

Malawi. N: Uzumara forest, c. 2000 m., 21.i.1964, *Chapman* 2231 (FHO). This specimen has a reticulation of the inferior leaf surface similar to that of *M. sousae*. However, the plant is a large tree up to 25 m. high and the form of leaf is different.

Memecylon sp. B.

Mozambique. SS: Madanda forest, c. 130 m., fr. juv. 5.xii.1906, *Swynnerton* 1074 (BM). This specimen resembles *M. sousae* in the reticulation of the inferior leaf surface, but the leaves are smaller and glossy above.

Memecylon sp. C.

Mozambique. N: Macomia, from Quiterajo to Mocímboa da Praia, between Ingoane and R. M'salo, st. 12.ix.1948, *Pedro & Pedrógão* 5170 (COI; LMA).

This specimen was originally cited (A. & R. Fernandes in An. Junta Invest. Ultramar **10**, 3 : 1, 1955) as insufficient material for identification. Later on (A. & R. Fernandes in Bol. Soc. Brot., Sér. 2, **43**: 301, t. 15, 1969), it was referred to *M. microphyllum* Gilg ex Engl., species known from E. Usambara. A comparison with the type of this species has shown that the leaves of the Mozambique specimen are much larger and with different reticulation of the inferior leaf surface. As the specimen is sterile, the material is insufficient for us to decide about its identification. It may belong to *M. mouririifolium* Brenan.

Memecylon sp. D.

Mozambique. N: António Enes, at the shore, st. 20.x.1965, *Mogg* 32462 (LISC). As the material has neither flowers nor fruits, identification is not possible.

Memecylon sp. E.

Malawi. S: Mlanje, Ruo Gorge, ± 1000 m., st. 1.ix.1970, *Müller* 1472 (COI; SRGH). Probably a new species.

2. ANTHEROTOMA (Naud.) Hook. f.

Antherotoma (Naud.).) Hook. f. in Benth. & Hook., Gen. Pl. **1**: 745 (1867).

Delicate annual herbs with slender indistinctly 4-gonous stem. Leaves opposite. Flowers 4-merous. Receptacle broadly campanulate. Sepals long-persistent; intersepalar segments present. Petals obovate. Stamens 8, equal, straight, yellow; pedoconnective somewhat long with anterior bilobed appendage at the base; anthers obovate-oblong, dehiscing at the truncate apex by a large pore. Ovary 4-locular, the lower half adnate to the receptacle; style filiform; stigma punctiform. Capsule 4-valved. Seeds cochleate.

A genus with two or three species in tropical Africa and Madagascar.

Antherotoma naudinii Hook. f. in Benth. & Hook. f., Gen. Pl. **1**: 745 (1867); in F.T.A. **2**: 444 (1871).—Triana in Trans. Linn. Soc. Lond. **28**: 57, t. 4 fig. 43 [1871] (1872).—Gilg in Engl., Mon. Afr. Pflanzen. **2**: 9, t. 1 fig. F (1898).—Bak. in Journ. Linn. Soc., Bot. **40**: 72 (1901).—De Wild. in Ann. Mus. Congo Belge, Bot. Sér. 5, **2**: 327 (1908).—Th. & H. Dur., Syll. Fl. Cong.: 208 (1909).—Engl., Pflanzenw. Afr. **3**, 2: 746, fig. 317D (1921).—Perrier in Fl. Madag., Melastom.: 7, t. 1 fig. 6, 6', 6", 7, 8 (1951).—Brenan in Mem. N.Y. Bot. Gard. **8**: 439 (1954).—Jacques-Félix in Ic. Pl. Afr. **3**: t. 51 (1955).—A. & R. Fernandes in An. Junta Invest. Ultramar **10**, 3: 47 (1955); in Mem. Soc. Brot. **11**: 8 & 66 (1956); in Kirkia **1**: 69 (1961); in C.F.A. **4**: 124 (1970).—Binns, H.C.L.M.: 67 (1968).—Wickens in F.T.E.A., Melastom.: 9 (1975). TAB. **54**. Type from Madagascar.

    Osbeckia antherotoma Naud. in Ann. Sci. Nat. Bot., Sér. 3, **13**: 50, t. 6 fig. 10 (1850); op. cit. **14**: 56 (1850).—Cogn. in A. & C. DC., Mon. Phan. **7**: 330 (1891).—Hiern Cat. Afr. Pl. Welw. **1**, 2: 363 (1898). Type as above.

    Antherotoma antherotoma (Naud.) Krasser in Engl. & Prantl, Pflanzenfam. **3**, 7: 154 (1893) *nom. illegit.*

    Dissotis kundelungensis De Wild. in Ann. Mus. Congo Belge, Bot. Sér. 4, **2**: 117 (1913). Type from Zaire.

Stem up to 38 cm. high (usually less), unbranched or with long ascending branches, laxly leafy, sparsely appressed-setulose. Leaf-lamina 0·5–4·5(5–7) × 0·1–1·8(2) cm., increasing in size upwards, ovate to oblong, acute at the apex, contracted or acute at the base, entire or crenulate, the teeth ending in a bristle, membranous, concolorous, yellowish-green or pale green and sparsely subappressed-pilose on both faces, more densely so on the nerves beneath, 3–5-nerved; petiole 3–6(8) mm. long, setose. Flowers in terminal capitate cymes, surrounded at the base by 2–4(5) leaves; pedicels c. 2·5 mm. long, setulose; bracts subequalling the pedicels, setulose. Receptacle c. 2 mm. high, pale green, with sparse, brownish red or purplish bristles inserted towards the summit at the apex of little appendages. Sepals c. 1·5 mm. long, triangular-subulate, stellate-setose at the apex like the intersepalar segments. Petals 5 × 3·5 mm., obovate, pale mauve to pink. Anthers (0·5)0·75–1·75 mm. long, with non-undulate cells; pedoconnective 0·5–2 mm. long; filaments c. 2 mm. long. Ovary with a ring of bristles surrounding the base of the style. Fructiferous receptacle c. 3 × 2·75 mm., roundish at the base, a little contracted above the capsule. Seeds very minutely tuberculate.

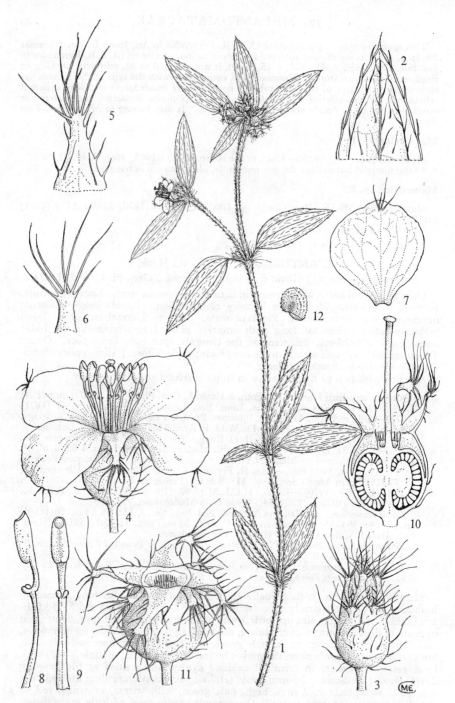

Tab. 54. ANTHEROTOMA NAUDINII. 1, flowering branch (×1); 2, part of leaf, inferior surface (×6); 3, flower bud (×8); 4, flower (×6); 5, calyx-lobe (×10); 6, intersepalar appendage (×10); 7, petal (×6); 8, 9, stamen in lateral and frontal view respectively (×12); 10, longitudinal section of flower (×8); 11, fruiting calyx, with included capsule (×8); 12, seed (×20), all from *Milne-Redhead & Taylor* 9493. From F.T.E.A.

Zambia. N: Kasama Distr., Chisimba Falls, fl. & fr. 20.v.1962, *Robinson* 5196 (K; SRGH). W: Mufulira, 1200 m., fl. & fr. 5.v.1934, *Eyles* 8135 (BM; K; SRGH). C: 13 km. SE. of Lusaka, fl. & fr. 20.iv.1957, *Noak* 208 (K; SRGH). E: Lundazi Distr., Kangampande Mt., Nyika Plateau, 2100 m., fl. & fr. 3.v.1952, *White* 2570 (COI; FHO; K). S: between Landlers Corner and Mumbwa, fl. 19.iii.1963, *Van Rensburg* 1702 (K; SRGH). **Rhodesia.** N: Umvukwe Mt. near Dawson, fl. & fr. 29.iv.1948, *Rodin* 4473 (SRGH). W: Matobo, Farm Besna Kobila, 1440 m., fl. & fr. v.1956, *Miller* 3567 (SRGH). C: Marandellas, fl. & fr. 4.iv.1950, *Wild* 3275 (K; SRGH). E: Umtali, Vumba, Norsland Farm, fl. & fr. 8.iii.1949, *Chase* 1356 (BM; COI; SRGH). S: Victoria, Zimbabwe, fl. & fr. 4.v.1962, *Drummond* 7934 (SRGH). **Malawi.** N: Nkhata Bay, Kandoli Hill, fl. 24.iv.1960, *Eccles* 24 (SRGH). C: Ntchisi Mt., 1400 m., fl. & fr. 24.vi.1946, *Brass* 16914 (K; SRGH). S: Namwera Escarpment, Jalasi, 1120 m., fl. & fr. 15.iii.1955, *E. M. & W.* 907 (BM; LISC; SRGH). **Mozambique.** N: Vila Cabral, Massangulo Mt., 1450 m., fl. & fr. 25.ii.1964, *Torre & Paiva* 10795 (COI; LISC; LMU; PRE). T: Macanga (Furancungo), 1140–1265 m., fl. & fr. 17.iii.1966, *Pereira, Sarmento & Marques* 1812 (K). MS: Bárùe, Choa Mt., 1400 m., fl. & fr. 28.iii.1966, *Torre & Correia* 15482 (BR; COI; EA; LISC; M).

Widespread in tropical Africa and Madagascar. In open forests, woodlands, shady places in damp thickets, grasslands, dambos, humid granite outcrops, cultivated land, etc.

3. TRISTEMMA Juss.

Tristemma Juss., Gen. Pl.: 329 (1789).

Annual or perennial herbs or shrubs, with sharply 4-angular or -winged stems and branches. Leaves petiolate, opposite. Flowers 5-merous, sessile or subsessile, in terminal heads surrounded by leaves and leafy bracts; inner bracts coriaceous, chaffy or membranous. Receptacle ovoid or ovoid-oblong, glabrous or densely setose on the upper half or with 1–6 rings of bristles. Sepals persistent, ± reflexed, ciliate at the margin; intersepalar segments 0. Petals mauve, pink or blue-violet, rarely white. Stamens 10, subequal; anthers lanceolate, all similar, yellow; pedoconnective short with an anterior 2-lobed or 2-fid appendage at the base. Ovary 5-locular, adnate to the receptacle all around to above the middle. Capsule 5-valved, with fleshy placentas, usually splitting irregularly. Seeds numerous, cochleate.

A genus with c. 15 species in tropical Africa, Madagascar and Mascarene Islands.

Branches and petioles with patent bristles 2–5 mm. long; leaves up to 12·5 ×6 cm., acuminate, very acute; anthers 5–6·5 mm. long - - - - 1. *acuminatum*
Branches and petioles with appressed or subappressed shorter bristles; leaves (4)7–15(22) ×(2)3–8(13) cm., not so acuminate; anthers 4–5 mm. long
2. *mauritianum*

1. **Tristemma acuminatum** A. & R. Fernandes in Bol. Soc. Brot., Sér. 2, **30**: 184, t. 21–23 (1956); in Mem. Soc. Brot. **11**: 43 & 95 (1956). Type from Tanzania (Usambara).

A perennial herb up to 2 m. high. Branches 4-angled, sulcate on the faces, hairy, the hairs 2–5 mm. long, patent, reddish yellow, denser and longer at the nodes. Leaf-lamina up to 12·5 ×6 cm., elliptic or ovate-elliptic, acuminate, the acumen very acute, acute or roundish at the base, entire, membranous, dark green and somewhat densely subappressed-strigose above, yellow-greenish and sparsely strigose beneath, longitudinally 5–7-nerved, the nerves prominent and covered with long and subpatent hairs, the transverse ones arched; petiole 1–3 cm. long, furrowed above, ± densely and patently hairy. Flowers 4-12, in compact shortly pedunculate heads; outer bracts subfoliaceous, strigose on nerves and apex, the inner ones membranous, obovate, nearly glabrous. Receptacle c. 8 ×4 mm., with 1–2 usually interrupted rings of bristles at the top. Sepals c. 8 ×4 mm., triangular, very acute. Petals c. 15 ×12 mm., obovate, blunt, lilac to blue-violet. Anthers 5–6·5 mm. long; pedoconnective c. 0·75 mm. long with the anterior basal appendage deeply lobed. Fructiferous receptacle c. 10 ×7·5 mm.

Mozambique. N: 5 km. from Chombe on the road to Negomano, c. 800 m., fl. & fr. 13.iv.1964, *Torre & Paiva* 11892 (COI; K; LISC).
Also in Tanzania. Marshy places in clayey red soil of forest floors.

2. **Tristemma mauritianum** J. F. Gmel., Syst. Pl. **2**: 693 (1791).—Wickens in F.T.E.A., Melastom.: 17 (1975). TAB. **55**. Type from Mauritius.

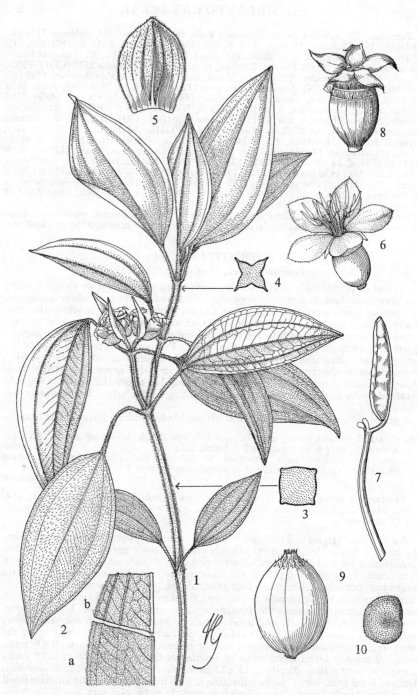

Tab. 55. TRISTEMMA MAURITIANUM. 1, flowering branch (×½); 2, part of leaf, (a) superior, (b) inferior surface (×1); 3, cross-section of stem, lower part (×3); 4, same, upper part (×3); 5, bract (×3), 1–5 from *Angus* 2623; 6, flower (×1); 7, stamen, lateral view (×8); 8, fruiting calyx (×6), 6–8 from *Richards* 15488 & 4026; 9, fruit (×3), from *Zenker* 2725; 10, seed (×20), from *Richards* 15488.

Tristemma virusanum Juss., Gen. Pl.: 329 (1789) *nom. inval.*—Humbert in Fl. Madag., Melastom.: 4, tab. 1, 1–5 (1951). Type as above.
Tristemma incompletum R. Br. in Tuckey, Narrat. Exped. River Zaire, App.: 435 (1818).—Gilg in Engl., Mon. Afr. Pflanzen. 2: 25 (1898).—Engl., Pflanzenw. Afr. 3, 2: 754 (1921).—Robyns & Tournay, Fl. Parc Nat. Alb. 1: 675 (1948).—A. & R. Fernandes in An. Junta Invest. Ultramar 10, 3: 50 (1955); in Mem. Soc. Brot. 11: 42 & 94 (1956); in Bol. Soc. Brot., Sér. 2, 34: 191 (1960); in Kirkia 1: 77 (1961); in C.F.A. 4: 127 (1970). Type from Zaire.
Melastoma albiflorum G. Don, Gen. Hist. 2: 764 (1832). Type from Sierra Leone.
Tristemma schumacheri Guill. & Perr. in Guill., Perr. & A. Rich., Fl. Senegamb. Tent. 1: 311 (1833).—Benth. in Hook., Niger Fl.: 354 (1849).—Naud. in Ann. Sci. Nat. Bot., Sér. 3, 13: 298, t. 6 fig. 6 (1850).—Hook. f. in F.T.A. 2: 446 (1871).— Triana in Trans. Linn. Soc. Lond. 28: 56 [1871] (1872).—Cogn. in A. & C. DC., Mon. Phan. 7: 361 et 1180 (1891).—Krasser in Engl. & Prantl, Pflanzenfam. 3, 7: 154, fig. 70Y (1893).—Taub. in Engl., Pflanzenw. Ost-Afr. C: 295 (1895). Syntypes from Senegal and Gambia.
Tristemma albiflorum (G. Don) Benth., op. cit.: 353 (1849).—Cogn., op. cit.: 362 (1891).—Hiern, Cat. Afr. Pl. Welw. 1, 2: 365 (1898). Type as for *Melastoma albiflorum.*
Tristemma schumacheri var. *grandifolium* Cogn., op. cit.: 361 (1891). Type from Angola.
Tristemma incompletum var. *grandifolium* (Cogn.) Hiern, op. cit.: 364 (1898). Type as above.
Tristemma grandifolium (Cogn.) Gilg, op. cit.: 26, t. 1 fig. N (1898).—Engl., Pflanzenw. Afr. 3, 2: 755 (1921).—A. & R. Fernandes, tom. cit.: 43 (1956); tom. cit.: 192 (1960); tom. cit.: 128 (1970). Type as above.
Tristemma grandifolium var. *congolanum* De Wild. in Ann. Mus. Congo Belge, Bot. Sér. 5, 2: 329 (1908).—Th. & H. Dur., Syll. Fl. Cong.: 206 (1909). Type from Zaire.

An erect or climbing succulent perennial herb or a lax shrub up to 1·2(2·5) m. high, dark green, sometimes suffused reddish. Branches sharply 4-angled or slightly 4-winged, with furrowed faces, sparsely strigose, with short stiff appressed purplish bristles, denser towards the apex. Leaf-lamina (4)7–15(22) × (2)3–8(13) cm., elliptic or ovate-elliptic, acute at the apex, cuneate or rarely roundish at the base, entire, membranous, dark green above, lighter beneath, ± strigose on both faces, the hairs short, appressed on the upper face, sparser and subappressed on the lower, longitudinally 5–7-nerved, these nerves rather slender, impressed above, prominent beneath, the transverse ones ± obliquely ascending, conspicuous mainly beneath; petiole 1·2–4(5) cm. long, furrowed above, strigose. Flowers numerous, in dense headed inflorescences; outer bracts completely foliaceous or with a foliaceous apex, acute, dorsally appressed-strigose, the inner ones subcircular to obovate, apiculate to obtuse, coriaceous to chaffy or scarious, glabrous. Receptacle 6–9 × 4–5 mm., ovoid, with 1(2) rings of bristles on the upper part, the lower ring, when present, usually incomplete. Sepals 3·5–5 × 2–3 mm., triangular, acute. Petals 6–10 mm. long, pale mauve or pink. Anthers 4–5 mm. long; pedoconnective 0·5–0·75 mm. with the basal appendage deeply 2-lobed, the lobes obtuse. Fructiferous receptacle 15–16 × 10–12 mm., purple or purplish-brown.

Zambia. N: Kawambwa Distr., 19 km. S. of Kawambwa on the Mansa road, fl. 30.x.1952, *Angus* 674 (COI; FHO; K). W: Mwinilunga, Samahina, 12 km. E. of Kalene Hill, fl. & fr. 16.xii.1963, *Robinson* 6083 (K; SRGH). **Rhodesia.** E: Melsetter, fl. & fr. 10.xi.1957, *Ball* 716 (LISJC; SRGH); Melsetter, near junction of Lusitu R. and Haroni R., c. 350 m., fl. & fr. 25.xi.1955, *Drummond* 5016 (K; LISC; SRGH). **Mozambique.** Z: Metolola, fl. & fr. 9.ix.1949, *Barbosa & Carvalho* 4004 (COI; LMA); Massingire, Metolola Mts., fl. & fr. 23.v.1943, *Torre* 5372 (COI; EA; LISC). MS: Buzi R. on road to Gogoi, c. 600 m., fl. & fr. i.1962, *Goldsmith* 16/62 (K; LISC; SRGH); Gorongosa, Vila Paiva de Andrade, fl. & fr. 5.x.1944, *Mendonça* 2374 (LISC).
Widespread in tropical Africa, Madagascar and Mauritius. Fringing forests on banks of rivers.

4. PSEUDOSBECKIA A. & R. Fernandes

Pseudosbeckia A. & R. Fernandes in Bol. Soc. Brot., Sér. 2, 30: 167 (1956).

Shrubs with the young branches 4-gonous. Leaves opposite. Flowers 5-merous. Bracts caducous. Sepals tardily deciduous; intersepalar segments subulate.

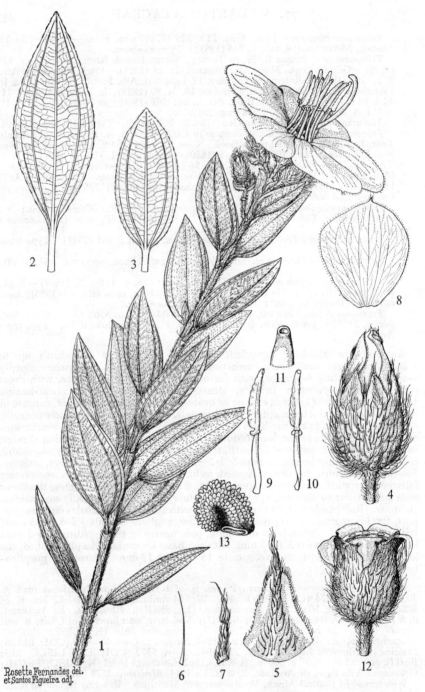

Tab. 56. PSEUDOSBECKIA SWYNNERTONII. 1, flowering branch (×1); 2 & 3, leaves showing variation in shape (×1); 4, flower-bud (×2); 5, sepal (×4); 6, bristle of the receptacle (×4); 7, intersepalar appendage (×4); 8, petal (×1); 9, stamen in lateral view (×2); 10, stamen in posterior view (×2); 11, apex of anther, posterior view (×2); 12, fructiferous receptacle (×2); 13, seed (×15), 1, 4–13 from *Wild* 2861, 2–3 from *Pedro & Pedrógão* 7204.

Rosette Fernandes del.
et Santos Figueira adj.

Petals obovate. Stamens 10, equal; filaments straight; anthers oblong-lanceolate, straight, yellow, with rather undulate loculi and one extrorse pore; pedoconnective very short with two anterior divergent obtuse appendages and two posterior minute prominences at the base. Ovary 5-locular, adnate to the receptacle by 10 longitudinal septa; style filiform; stigma punctiform. Capsule 5-valved. Seeds cochleate, covered with rather elongate tubercles arranged in curved parallel lines. Monotypic; known only in our area.

Pseudosbeckia swynnertonii (Bak. f.) A. & R. Fernandes in Bol. Soc. Brot., Sér. 2, **30**: 168, t. 1 (1956); in Mem. Soc. Brot. **11**: 10 & 68 (1956). Type: Rhodesia, Chimanimani Mts., *Swynnerton* 2085 (BM, holotype; K). TAB. **56**.

    Osbeckia swynnertonii Bak. f. in Journ. Linn. Soc., Bot. **40**: 71 (1911). Type as above.

    Dissotis swynnertonii (Bak. f.) A. & R. Fernandes in Bol. Soc. Brot., Sér. 2, **28**: 71 (1954); in Garcia de Orta **2**: 184 (1954). Type as above.

A shrub up to 1·2 m. tall. Young branches 4-gonous, ± densely bristly, the bristles whitish or whitish-yellow, often densely leafy upwards, the oldest ones subterete, glabrescent, greyish; nodes somewhat thickened, long-setose. Leaves opposite. Leaf-lamina 1·5–5 × 0·7–2·5 cm., elliptic, oblong-ovate or ovate, acute at the apex, acute or rarely roundish at the base, subentire or crenulate at the margin, somewhat rigid, very discolorous (dark green above, yellowish-green beneath), appressed-setose on the upper face, the bristles short, whitish, each prolonged on the leaf surface by a raised line, subpatently setose on the under face, more dense on nerves, longitudinally (3)5-nerved, the nerves impressed above, prominent beneath, the transverse ones and the reticulation inconspicuous on the upper face, ± visible on the under one; petiole 4–8 mm. long, setose. Flowers in terminal few-flowered cymes or solitary; pedicels 3–5 mm. long, setose; bracts linear-oblong to ovate, acute, bristly outside, caducous. Receptacle 8–11 × 7–9 mm., campanulate, densely setose, the bristles long, ascending, simple, whitish. Sepals 6–6·5 mm. long, triangular-ovate, obtuse, setose on the back, long-setose at apex, ciliate at the margin, tardily deciduous; intersepalar segments linear, setose, ending in a long white rather strong bristle. Petals 2·5–3 × 2·4–2·7 cm., purple, ciliate at the margin. Anthers c. 8 mm. long; basal anterior appendages of the pedoconnective 0·5–1 mm. long. Fructiferous receptacle up to 11 × 8 mm., widely campanulate. Capsule slightly shorter than the receptacle, spherical, strigose on the upper part. Seeds c. 1 mm. in diameter.

Rhodesia. E: Melsetter Distr., Chimanimani Mts., below Ben Nevis, 1290 m., fl. & fr. 6.vii.1957, *Phipps* 702 (COI; K; SRGH). **Mozambique**. MS: Martin's Falls, 1350 m., fl. & fr. 28.viii.1964, *Whellan* 2139 (COI; SRGH).

Known only from the Chimanimani Mts. Rocky slopes, *Brachystegia* woodlands and along rivers.

5. OSBECKIA L.

Osbeckia L., Sp. Pl. **1**: 345 (1753); Gen. Pl. ed. 5: 162 (1754).

Perennial erect or ascending herbs or shrubs, setose, with the branches usually 4-gonous. Leaves opposite, 3–7-nerved, entire or serrulate. Flowers 4–5-merous, solitary, capitate or paniculate. Bracts caducous. Receptacle ovoid, urceolate or subglobose, usually covered by simple bristles or stellate-setose or penicillate appendages. Sepals caducous or persistent. Petals obovate, ciliate at the margin. Stamens 10, equal; anthers linear-subulate, rostrate, opening by a terminal introrse pore; connective not or slightly produced, with the appendages not anteriorly bitubercled. Ovary 4–5-locular, adhering to the receptacle by 8–10 longitudinal septa, with a crown of bristles at the apex. Capsule 4–5-valved, included in the receptacle. Seeds numerous, minute, cochleate.

About 60 species in Asia, Australia, Mauritius and Madagascar, and 5 in Africa.

Osbeckia congolensis Cogn. ex Büttn. in Verh. Bot. Ver. Prov. Brand. **31**: 95 (1889).— Cogn. in A. & C. DC., Mon. Phan. **7**: 314 (1891).—Krasser in Engl. & Prantl, Pflanzenfam. **3**, 7: 156 (1893).—Gilg in Engl., Mon. Afr. Pflanzen. **2**: 6 (1898).— De Wild. & Dur. in Ann. Mus. Congo Belge, Bot. Sér. 1, **1**: 23, t. 12 fig. 1–9 (1898).— Engl., Pflanzenw. Afr. **3**, 2: 744 (1921).—Exell in Journ. Bot., Lond. **66**, Suppl. Polypet.: 176 (1928).—Keay in F.W.T.A. ed. 2, **1**: 249 (1954).—Jacques-Félix in

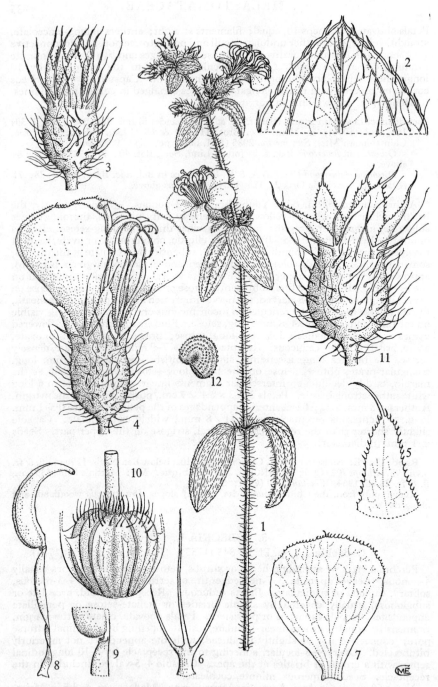

Tab. 57. OSBECKIA CONGOLENSIS. 1, flowering branch (×1); 2, part of leaf, underside
(×6); 3, flower bud (×4); 4, flower with 1 petal removed (×4); 5, calyx-lobe
(×8); 6, intersepalar appendage (×10); 7, petal (×4); 8, stamen, side view (×6);
9, detail of same showing connective appendages (×10); 10, ovary (×8); 11, fruiting
calyx (×6), 1–11 from *Lind* 2329; 12, seed (×20), from *Haswell* 81.   From F.T.E.A.

Ic. Pl. Afr. **3**: t. 52 (1955).—R. & A. Fernandes in C.F.A. **4**: 129 (1970).—Wickens in F.T.E.A., Melastom.: 11, t. 14 (1975). TAB. **57**. Type from Zaire.

Osbeckia congolensis var. *robustior* Cogn. ex Büttn., loc. cit.—Cogn., op. cit.: 314 et 1177 (1891).—Dur. & De Wild. in Bull. Soc. Roy. Bot. Belg. **37**: 114 (1898).— De Wild., Miss. Laurent. **1**: 163 (1905). Type from Zaire.

Perennial erect or ascending many-stemmed herb up to 20–80 cm. high. Stems 4-gonous, with appressed or ± patent long bristles; nodes somewhat thickened, long-bristly. Leaves decussate, usually reflexed; leaf-lamina 1·5–6 × 0·6–1·5 cm., oblong or lanceolate, acute or subacute at the apex, ± conspicuously crenulate at the margin, base cuneate, membranous, dull green above, lighter beneath, ± appressed long-hairy on both faces, longitudinally 3–5-nerved, nerves inconspicuous above, prominent below; petiole 2–3(5) mm. long, densely bristly. Flowers 5-merous, in terminal and lateral leafy false panicles up to 12 cm. long. Bracts caducous, scarious. Receptacle 4–6 mm. long, campanulate, ± densely appressed setose. Sepals 3 mm. long, oblong-triangular, persistent, ciliate at the margin, alternating with linear simple or sparsely branched long-bristly appendages. Petals c. 7 × 5 mm., pinkish mauve, shortly ciliate at the upper margin. Stamens straight, up to 10 mm. long; anthers c. 4 mm. long; pedoconnective very short with 2 anterior protuberances; filaments 4–6 mm. long. Ovary ovoid, 3 × 2·5 mm., the upper ½ free, with a crown of bristles at the top surrounding the style-base; style sigmoid. Capsule c. 5 mm. long, ovoid. Seeds c. 0·6 mm. long, papillose.

Zambia. W: Mwinilunga, fl. & fr. 16.v.1969, *Mutimushi* 3330 (K).
Widespread in tropical Africa. Watershed plains.

6. MELASTOMASTRUM Naud.

Melastomastrum Naud. in Ann. Sci. Nat. Bot., Sér. 3, **13**: 296 (1850).

Perennial herbs or shrubs with 4-gonous branches. Leaves petiolate, opposite, entire. Flowers 5-merous, subsessile, in terminal many-flowered heads surrounded by persistent bracts at the apices of the branches or in 1–2(3)-flowered terminal cymes or flowers axillary and solitary. Receptacle glabrous or setose at the base or on the upper half (the bristles not in rings) or covered everywhere by stellate or palmate bristles. Sepals persistent; intersepalar segments 0. Petals obovate, mauve, purple or violet. Stamens 10, very unequal (subequal in osbeckioid forms); pedoconnectives of the anthers of the longer stamens with a deeply bilobed anterior appendage (the lobes club-shaped) and a small posterior spur at the base; appendage of the pedoconnectives of the shorter stamens similar but smaller. Ovary 5-locular, adnate to the receptacle by 10 longitudinal septa. Capsule 5-valved. Seeds cochleate.
Species 6, in tropical Africa.

Flowers in terminal usually many-flowered heads; leaf-lamina 5–13(16) × 1·5–6·5(7·5) cm., longitudinally 5–7-nerved; petiole 0·5–3 cm. long - - - 1. *capitatum*
Flowers in 1–2(3)-flowered terminal cymes or axillary and solitary; leaf-lamina usually smaller and relatively narrower (2·5–11 × 1·2–4·5 cm.), longitudinally 3–5-nerved; petiole usually shorter than above, not more than 1·2 cm. long - 2. *segregatum*

1. **Melastomastrum capitatum** (Vahl) A. & R. Fernandes in Garcia de Orta **2**: 278 (1954); in Mem. Soc. Brot. **11**: 14 & 69 (1956); in Kirkia **1**: 69 (1961); in C.F.A. **4**: 130 (1970).—Jacques-Félix in Bull. Mus. Natn. Hist. Nat. Paris, Sér. 3, Bot. no. 270: 57, fig. 3 (1974).—Wickens in F.T.E.A., Melastom.: 20, t. 7 fig. 12 (1975). Type a plant cultivated at Kew?

Melastoma capitatum Vahl, Eclog. Amer.: 45 (1797).—DC., Prodr. **3**: 199 (1828). Type as above.

Melastoma capitatum G. Don, Gen. Syst. **2**: 764 (1832) *nom. illegit.* Type from Sierra Leone.

Tristemma erectum Guill. & Perr. in Guill., Perr. & A. Rich., Fl. Senegamb. Tent. **1**: 312 (1833). Type from Senegal.

Heterotis capitata Benth. in Hook., Niger Fl.: 352 (1849). Based on *Melastoma capitatum* G. Don non Vahl.

Melastomastrum erectum (Guill. & Perr.) Naud. in Ann. Sci. Nat. Bot., Sér. 3, **13**: 296, t. 5 fig. 4 (1850). Type as for *Tristemma erectum*.

Dissotis capitata (Benth.) Hook. f. in F.T.A. **2**: 449 (1871).—Cogn. in A. & C. DC., Mon. Phan. **7**: 365 (1891).—Taub. in Engl., Pflanzenw. Ost-Afr. C: 295 (1895).—

Gilg in Engl., Mon. Afr. Pflanzen. 2: 13 (1898).—Th. & H. Dur., Syll. Fl. Cong.: 208 (1909).—Thonn., Fl. Pl. Afr.: t. 115 (1915).—Engl., Pflanzenw. Afr. 3, 2: 747 (1921).—Robyns & Tournay, Fl. Parc Nat. Alb. 1: 673 (1948). Based on *Melastoma capitatum* G. Don.
Tristemma capitatum (Vahl) Triana in Trans. Linn. Soc. Lond. 28: 56, t. 4 fig. 41d [1871] (1872). Type as for *Melastomastrum capitatum.*
Dissotis erecta (Guill. & Perr.) Dandy in F. W. Andr., Fl. Pl. Anglo-Egypt. Sudan 1: 192, fig. 110 (1950).—Keay in F.W.T.A. ed. 2, 1: 259 (1954). Type as for *Tristemma erectum.*

An erect or suberect perennial herb or a shrub 40–60(90) cm. high. Branches obtusely 4-gonous and deeply furrowed upwards, cylindric towards the base, sparsely strigillose, the bristles short, appressed, longer at the nodes; internodes 4–17 cm. long. Leaf-lamina 5–13(16) × 1·5–6·5(7·5) cm., ovate to narrowly elliptic, acute at the apex, acute or roundish at the base, membranous to somewhat rigid, dark green above, paler beneath, ± appressed-strigose on both faces, the bristles on the upper face prolonged by white raised lines, longitudinally 5–7-nerved, the longitudinal nerves impressed above, prominent beneath, the transverse ones inconspicuous above, ± conspicuous beneath, very oblique; petiole 0·5–3 cm. long, slender, somewhat compressed, strigose. Flowers ± numerous, in dense heads surrounded by the upper leaves and by foliaceous bracts; inner bracts closely sheathing the receptacle, ovate to ovate-lanceolate, scarious, ciliate at margin, glabrous or strigose on the back along the median line. Receptacle c. 10 × 5 mm., cylindric or ovoid-cylindric, surrounded by long whitish bristles at the very base. Sepals 6–7 × 2·5–3 mm., triangular-subulate, ciliate; intersepalar segments 0. Petals 20 × 15 mm., widely obovate, light mauve to violet. Long stamens: anthers 7–8 mm. long, lanceolate-subulate, purple; pedoconnective 5–6 mm. long with the 2 anterior appendages 2 mm. long. Short stamens: anthers 6–7 mm. long, lanceolate, yellow; pedoconnective c. 0·75 mm. long. Fructiferous receptacle c. 13–15 × 6–7 mm., urceolate, cream-coloured to greyish. Seeds c. 0·75 mm. long, indistinctly tuberculate.

Zambia. N: Kawambwa Distr., Kalungwishi R., 1050 m., fl. & fr. 26.iv.1957, *Richards* 9444 (COI; K). W: Mwinilunga, Ikelenge, fl. & fr. ii.1960, *Pinhey* 9 (COI; SRGH); Kitwe, Misaka, fr. 28.iii.1958, *Fanshawe* 4373 (K).
Also in W. tropical Africa, Zaire, Uganda and Tanzania. On fringes of riparian woodland, semi-evergreen rain forests, swamps and fire clearings.

2. **Melastomastrum segregatum** (Benth.) A. & R. Fernandes in Mem. Soc. Brot. 11: 12 & 68, t. 1 (1956); in Kirkia 1: 69 (1961); in C.F.A. 4: 130 (1970).—F. White, F.F.N.R.: 310 (1962).—Jacques-Félix in Bull. Mus. Natn. Hist. Nat. Paris, Sér. 3, Bot. no. 270: 74, fig. 10 (1974).—Wickens in F.T.E.A., Melastom.: 22, t. 7 fig. 1–11 (1975). TAB. 58. Syntypes from Nigeria.
Heterotis segregata Benth. in Hook., Niger Fl.: 350 (1849). Syntypes as above.
Dissotis segregata (Benth.) Hook. f. in F.T.A. 2: 448 (1871).—Cogn. in A. & C. DC., Mon. Phan. 7: 363 (1891).—Taub. in Engl., Pflanzenw. C: 295 (1895).—Gilg in Engl., Mon. Afr. Pflanzen. 2: 12 (1898).—Th. & H. Dur., Syll. Fl. Cong.: 212 (1909).—R. E. Fr., Wiss. Ergebn. Schwed. Rhod.-Kongo-Exped. 1: 177 (1914).—Brenan, T.T.C.L.: 310 (1949). Syntypes as above.
Tristemma segregatum (Benth.) Triana in Trans. Linn. Soc. Lond. 28: 56, t. 4 fig. 41c [1871] (1872). Syntypes as above.
Dissotis minor Gilg in Engl., tom. cit.: 12, t. 2 fig. C (1898).—Engl., Pflanzenw. Afr. 3, 2: 747 (1921). Type from Tanzania.
Tristemma schlechteri Gilg in Schlecht., West Afr. Kautsch Exped.: 302 (1900) *nom. nud.*
Melastomastrum schlechteri A. & R. Fernandes in Bol. Soc. Brot., Sér. 2, 29: 48, t. 2 (1955). Type from Cameroon.

An erect much-branched shrub up to 2 m. high. Branches cylindric at the base, 4-gonous and furrowed towards the apex, shortly appressed-setose; nodes slightly thickened, long-setose; internodes generally rather short (1·1–3·5 cm. long), occasionally longer (up to 10 cm.). Leaf-lamina 2·5–11 × 1·2–4·5 cm., ovate, ovate-lanceolate to lanceolate, ± acute at the apex, contracted and roundish at the base, rigid or rarely membranous, ± discolorous, sparsely appressed-strigose on both faces, the bristles longer and denser on the nerves beneath, longitudinally 3–5-nerved, the longitudinal nerves impressed above, rather prominent beneath, the transverse ones usually inconspicuous on both faces; petiole 0·4–1·2 cm. long,

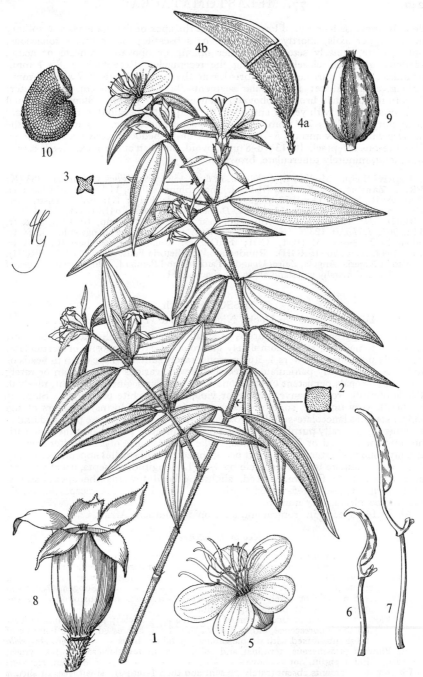

Tab. 58. MELASTOMASTRUM SEGREGATUM. 1, flowering branch (×½); 2, cross-section of stem at median part (×3); 3, same, upper part (×3); 4, leaf, (a) inferior, (b) superior surface (×1); 5, flower (×1); 6 & 7, small and large stamens respectively (×3), 1–7 from *Davies* sn; 8, fruiting calyx (×3); 9, fruit (×2); 10, seed (×25), 8–10 from *Reynolds* 129135.

densely appressed-setose. Flowers 1–2(3) at the apex of the branches and solitary in the upper axils, shortly pedunculate or subsessile; outer bracts foliaceous, oblanceolate, densely setose, the inner oblong or obovate, truncate or acute, scarious, glabrous, closely sheathing the receptacle. Receptacle c. 8 × 4 mm., cylindric, green, surrounded by bristles at the base. Sepals 6–7 × 2·5–3 mm., oblong, acute, ciliate; intersepalar segments 0. Corolla 2–3 cm. in diameter; 20 × 16 mm., petals broadly obovate, mauve-pink to purple. Stamens unequal (subequal and all with yellow anthers in the frequent osbeckioid forms), the longer with purple anthers 7–9 mm. long and pedoconnective c. 5–6 mm. long, the shorter with yellow anthers 5–7(8) mm. long and a pedoconnective c. 1 mm. long. Fructiferous receptacle 10–13 × 6–8 mm., ovoid, purplish at the apex. Seeds c. 0·75 mm. long, minutely tuberculate, brownish.

Caprivi Strip. Mpililia I., c. 900 m., fl. & fr. 15.i.1959, *Killick & Leistner* 3383 (K; PRE). **Zambia.** B: Zambezi (Balovale), near Shingi village, 11 km. N. of Chavuma, Zambezi R., fl. & fr. 14.x.1952, *Angus* 637 (BM; COI; FHO; K); Kalabo Distr., near Rest house Kalabo, fl. & fr. 13.xi.1959, *Drummond & Cookson* 6418 (COI; K; LISC; SRGH). N: Luapula Distr., Mbereshi, near Luapula R. swamp, 1050 m., fl. & fr. 11.i.1960, *Richards* 12326 (K); Mansa Distr., Samfya, Lake Bangweulu, fl. & fr. 20.viii.1952, *Angus* 251 (BM; COI; FHO). S: Katambora, Zambezi R., fl. & fr. 14.iv.1949, *West* 2909 (SRGH). **Rhodesia.** W: Victoria Falls, *Allen* 111 (K; SRGH).
Also in Nigeria, Angola, Zaire, Uganda, Tanzania and Pemba I. On riparian fringes of vegetation and marshy grassland along the rivers and lakes.

7. DISSOTIS Benth.

Dissotis Benth. in Hook., Niger Fl.: 346 (1849) *nom. conserv.*

Annual or perennial herbs, shrubs or small trees, usually ± pilose, pubescent, lanate or tomentose, rarely completely glabrous. Leaves opposite or verticillate, generally petiolate. Flowers 4–5(6)-merous, rarely solitary, more often in heads or in cymes, frequently paniculate. Receptacle with various indumentum or rarely glabrous. Sepals persistent or caducous; intersepalar segments present, rarely 0. Petals obovate, rose, mauve, purple or violet, rarely white or orange. Stamens 8–10(12), very unequal, sometimes subequal (osbeckioid forms); anthers of the longer stamens lanceolate-subulate, rostrate, 1-porose, with introrse pore and undulate thecae, generally purple, with a long pedoconnective provided at the base with an anterior 2-lobed or 2-fid appendage like a spur, those of the shorter stamens usually yellow, shorter, with a short pedoconnective and smaller appendage. Ovary 4–5-locular, adnate to the receptacle by 8–10(12) longitudinal septa, usually bristly at the top; style filiform, curved, slightly dilated towards the apex; stigma punctiform. Capsule 4–5-valved, coriaceous, included in the receptacle. Seeds minute, cochleate, very numerous.
About 140 species in tropical and subtropical southern Africa.

Key to the subgenera

Sepals persistent or tardily caducous; intersepalar segments present or absent:
  Intersepalar segments present or rarely absent; annual or perennial herbs or suffrutices:
    Flowers 4-merous in heads surrounded by the uppermost leaves; bracts small, persistent - - - - - - - - - Subgen. *Osbeckiella*
    Flowers 5-merous or rarely 4-merous, neither in heads nor surrounded by the uppermost leaves; bracts larger, caducous:
      Flowers 5-merous in racemiform or ± condensed panicles; indumentum, at least on the inflorescence, either only of minute ± appressed stellate hairs or of these intermixed with simple capitate-glandular hairs - Subgen. *Argyrella*
      Flowers (4)5-merous, terminal and solitary or in few-flowered ± lax cymes; indumentum not as above - - - - - - Subgen. *Heterotis*
  Intersepalar segments absent (rarely present and then 1-setose); small trees or shrubs
    Subgen. *Dissotidendron*
Sepals ± soon caducous; intersepalar segments usually present:
  Fructiferous receptacle distinctly costate; pedoconnective appendage of the longer stamens thick, bilobed; receptacle-bristles minutely spinulose
    Subgen. *Dupineta*
  Fructiferous receptacle usually not costate; pedoconnective-appendages of the longer stamens 2-lobed or deeply bifid; receptacle-bristles or -hairs smooth
    Subgen. *Dissotis*

Subgen. OSBECKIELLA A. & R. Fernandes in Bol. Soc. Brot., Sér. 2,
43: 285 (1969).

Bristles of the stems and branches ± short, appressed or subappressed, usually
conspicuously bulbous at the base, whitish:
Anthers of the longer stamens suddenly contracted into a short beak and with
pedoconnective-appendage divided into 2–3 subulate lobes; annual, frequently
with unbranched stems - - - - - - - - 1. *tenuis*
Anthers of the longer stamens tapering into a beak and with the pedoconnective-
appendage divided into 2 obtuse lobes; perennial or rarely annual, with stems
frequently branched:
Stems herbaceous and usually not much branched towards the top; internodes below
the floral leaves very long (up to 30 cm.); leaf-lamina 0·5–7·5 × 0·3–1·8 cm.,
linear-lanceolate to ovate; heads dense and usually many-flowered; capsule
broadly ovoid - - - - - - - - 2. *debilis*
Stems woody and branched towards the top; internodes below the floral leaves short
(0·3–4 cm.); leaf-lamina 1–5 × 0·2–0·4(0·7) cm., narrowly lanceolate; heads
2–4-flowered or flower solitary; capsule subglobose - - 3. *angustifolia*
Bristles of the stems and branches long and ± dense, usually patent, not conspicuously
bulbous at base, often fulvous - - - - - - - 4. *phaeotricha*

Subgen. HETEROTIS (Benth.) A. & R. Fernandes in Bol. Soc. Brot.,
Sér. 2, 43: 286 (1969).

Receptacle glabrous; sepals longer than the receptacle; inflorescence-axis and branches
with glandular hairs - - - - - - - - 5. *glandulosa*
Receptacle with simple bristles and ± long appendages long stellate-setose at apex; sepals
as long as or shorter than the receptacle; inflorescence-axis and branches without
glandular hairs - - - - - - - 6. *rotundifolia* var. *prostrata*

Subgen. ARGYRELLA (Naud.) A. & R. Fernandes in Bol. Soc. Brot.,
Sér. 2, 43: 287 (1969).

Only one species in F.Z. area - - - - - - - - 7. *canescens*

Subgen. DUPINETA (Raf.) A. & R. Fernandes in Bol. Soc. Brot.,
Sér. 2, 43: 288 (1969).

Receptacle with simple bristles only; all leaves distinctly petiolate (petiole 1–6 cm.);
lamina ovate to lanceolate, round or subcordate at the base, longitudinally 3–5(7)-
nerved - - - - - - - - - - - 8. *hensii*
Receptacle with simple bristles on the lower part and appendages ending in a stellate-setose
disk on the median and upper parts; upper leaves subsessile; lamina ovate-cordate,
longitudinally 7–11-nerved - - - - - - - - 9. *brazzae*

Subgen. DISSOTIDENDRON A. & R. Fernandes in Bol. Soc. Brot.,
Sér. 2, 43: 289 (1969).

Plant everywhere (except old parts) densely lanate (hairs long, thin and crispate)
10. *lanata*
Plant not lanate:
Receptacle 4–5(6) mm. wide, cylindric, attenuate towards the base, conspicuously
10-ribbed, contracted below the mouth in fruit - - - - 11. *caloneura*
Receptacle slightly wider, campanulate or cylindric-campanulate, rounded at the base,
not ribbed and not contracted at the apex in fruit:
Flowers in very compact short inflorescences; sepals 3–5 × 5–6 mm., semicircular or
broadly ovate; receptacle-bristles simple or 2–few-fid, patent or ascending,
sometimes glandular-capitate, usually inserted on short protuberances; leaves
not simultaneous with the flowers - - - - - - 12. *melleri*
Flowers in loose or not very compact panicles up to 24 × 20 cm.; sepals 6–8 × 4–6 mm.,
oblong-semicircular to oblong; receptacle-bristles, if present, simple, appressed,
long, the uppermost sometimes inserted on short prominences; leaves and
flowers simultaneous - - - - - - - 13. *johnstoniana*

Subgen. DISSOTIS.—A. & R. Fernandes in Bol. Soc. Brot., Sér. 2, 43:
289 (1969).

Flowers solitary or rarely 2–3 at the end of branches, surrounded by the uppermost leaves
only or by these and large caducous scarious bracts; receptacle and leaves sericeous-
silvery, the hairs rather slender and long:

Flowers surrounded by the uppermost leaves only; receptacle 6–8 × 5–6 mm.
                                                        14. *cryptantha*
Flowers surrounded by the uppermost leaves and by large caducous scarious bracts;
    receptacle 10–12 × 9–12 mm.:
    Leaf-lamina up to 7·5 × 2 cm., ± spreading, not folded along the median nerve
                                                        15. *speciosa*
    Leaf-lamina up to 6 × 1·6 cm., arched and reflexed, folded along the median nerve
                                                        16. *simonis-jamesii*
Flowers in cymes usually grouped in spiciform, racemiform or paniculate, rarely capitate
    inflorescences; indumentum not as above:
    Receptacle glabrous or with simple appressed sparse or not very dense bristles:
        Petiole up to 8 mm. long; leaf-lamina elliptic to subcircular, crenulate; flowers
            5–6(7)-merous; stamens of the two verticils subequal but with pedoconnectives
            well developed - - - - - - - - - 17. *gilgiana*
        Petiole 0; leaf-lamina linear-lanceolate to ovate-lanceolate, entire or indistinctly
            crenulate; flowers 4- or 5-merous; stamens of the two verticils very unequal:
            Flowers 4-merous - - - - - - - - - 18. *gracilis*
            Flowers 5-merous:
                Glabrous everywhere; leaf-lamina 4–12·5 × 1–4·5 cm.; receptacle 8–9 × 4–5 mm.;
                    intersepalar segments absent - - - - - - 19. *welwitschii*
                Appressed-setose; leaf-lamina relatively narrower, up to 11·5 × 1·5 cm.;
                    receptacle somewhat larger, up to 10 × 6 mm.; intersepalar segments
                    present, linear - - - - - - - - 20. *anchietae*
    Receptacle ± densely scaly, setose or hairy, the bristles or hairs inserted on the
        receptacle surface or on ± long appendages:
        Receptacle densely covered with appressed very minute echinulate-branched scales
                                                        21. *romiana*
    Receptacle with a different indumentum:
        Receptacle densely covered with bristles 1–2·5 mm. long, flat at the base, simple or
            divided, appressed, inserted directly on the receptacle:
            Flowers in very congested panicles; indumentum of stems, branches and leaves
                ± dense; inferior surface of leaf-lamina usually without orange capitate
                hairs; receptacle long-urceolate with long stout divided bristles
                                                        22. *trothae*
            Flowers in long lax panicles; indumentum of stems, branches and leaves looser
                than above; inferior surface of leaf-lamina with orange capitate hairs;
                receptacle shortly urceolate with usually shorter slender simple bristles
                                                        23. *pulchra*
        Receptacle covered with appendages:
            Receptacle ovoid, densely covered with patent or reflexed nipple-shaped
               appendages, 1–few-setose at the top; leaf-lamina glabrous or sparsely
              appressed-hairy - - - - - - - - 24. *pachytricha*
            Receptacle-appendages not as above; leaf-lamina usually ± densely hairy or
            setose:
                Receptacle-appendages short, ± flat, pectinate or subpectinate-setose at the
                apex:
                    Leaf-lamina 5–7-nerved, 3–14·5 × 1–5·5 cm., with the superior surface
                      nearly smooth and ± densely subappressed-setose - 25. *princeps*
                    Leaf-lamina 3-nerved, up to 17·5 × 3·8 cm., with the superior surface
                      covered with conic or pyramidal short appendages each ending in a
                      bristle - - - - - - - - - 26. *peregrina*
            Receptacle-appendages ± elongated, cylindric or slightly compressed, dilated
              into a stellate-setose head or disk at the apex:
                Leaf-lamina not membranous; stem and nerves of the inferior leaf surface
                  scaly-setose, the scales fimbriate or pectinate:
                    Leaves opposite with lamina 1·8–10·5 × 0·6–3·5 cm., oblong-lanceolate to
                      ovate-oblong, round or subcordate at the base; stem appressed-scaly
                                                        27. *falcipila*
                  Leaves 2–3-nate with lamina 2·3–11·5 × 1·8–5 cm., ovate-lanceolate or
                      ovate, cordate at the base; stem ± patently scaly-setose
                                                        28. *denticulata*
            Leaf-lamina membranous; stems and leaves hirsute with simple bristles:
                Flowers 5-merous in terminal very dense subcapitate inflorescences
                  surrounded by the two uppermost leaves; stamens subequal; leaf-
                  lamina 2–14 × 1·5–5 cm., broadly ovate to lanceolate; petiole very
                  short or absent - - - - - - - 29. *densiflora*
                Flowers 4–5-merous in ± congested inflorescences not surrounded by the
                  uppermost leaves; stamens very unequal or subequal; leaf-lamina
                  1·5–9 × 0·8–2·5 cm., narrowly lanceolate to ovate-lanceolate; petiole
                2–10 mm. long - - - - - - 30. *senegambiensis*

1. **Dissotis tenuis** A. & R. Fernandes in Bol. Soc. Brot., Sér. 2, **43**: 291, t. 1 (1969).
TAB. **59**. Type: Zambia, Mwinilunga, ± 6·4 km. N. of Kalene Hill Mission,
Edwards 802 (COI, holotype; LISC; SRGH).

Annual herb up to 20 cm. high. Stem slender, simple or branched at the base,
the branches rather elongate and erect; stems and branches terete, dark reddish,
± appressed-setose, the bristles whitish and bulbous at the base, leafy at the base
and below the inflorescence, leafless elsewhere. Leaf-lamina 2–15 × 1·5–9 mm.,
ovate, rather acute at the apex, entire or slightly crenulate at the margin, attenuated
into a petiole 1·5 mm. long or contracted at the base, membranous, discolorous
(reddish on the upper face, light-green beneath), 3-nerved, ± densely subappressed-
setose, the setae white and long. Flowers 4-merous, small, shortly (c. 3 mm. long)
pedicellate, 1–10 in terminal cymes usually arranged in heads leafy at the base.
Bracts minute, persistent, long-setose. Receptacle c. 3 × 2·5 mm., subhemispherical, reddish, sparsely setose, the bristles simple, whitish. Intersepalar
segments c. 2 mm. long, linear, stellate-setose at the top, the bristles up to 3 mm.
long. Sepals c. 3 × 2 mm., persistent or very tardily deciduous, triangular-oblong,
ciliate at the margin, stellate-setose at the top, the bristles up to 2·5 mm. long.
Petals up to 7 × 6 mm., broadly obovate, penicillate-setose at the apex, shortly
unguiculate, intensely orange (fide collect.) or purple. Stamens 8, very unequal.
Longer stamens: anthers 2 mm. long, suddenly contracted into a c. 0·5 mm. long
beak; pedoconnective 2 mm. long, very arched and produced at the base into a
2·5 × 1 mm., 2–3-fid appendage; filaments 2·5 mm. long. Shorter stamens:
anthers 2 mm. long, oblong, obtuse at the top, with a very short pedoconnective
produced into a c. 0·25 mm. long appendage. Ovary sparsely setose at the top;
style 5 mm. long, strongly curved towards the apex; stigma papillose. Seed c.
0·5 mm. in diameter, cochleate, straw-yellow, with minute tubercles disposed in
curved, parallel lines.

Zambia. W: Mwinilunga Distr., Zambesi Rapids, c. 6 km. N. of Kalene Hill, fl.
20.ii.1975, *Hooper & Townsend* 255 (K).
Known also in NW. Tanzania. In bogs overlying rocks.

2. **Dissotis debilis** (Sond.) Triana in Trans. Linn. Soc. Lond. **28**: 58, t. 4 fig. 44a, b
[1871] (1872).—Cogn. in A. & C. DC., Mon. Phan. **7**: 367 (1891).—Gilg in Engl.,
Mon. Afr. Pflanzen. **2**: 14, t. 2 fig. D (1898).—Th. & H. Dur., Syll. Fl. Cong.: 209
(1909).—Engl., Pflanzenw. Afr. **3**, 2: 748 (1921).—Burtt Davy, F.P.F.T. **1**: 243
(1926).—Jacques-Félix in Bull. Mus. Hist. Nat. Paris, Sér. 2, **7**: 372 (1935).—
Brenan in Mem. N.Y. Bot. Gard. **7**: 439 (1954).—Garcia in Est. Ens. Doc. Junta
Invest. Ultramar **12**: 156 (1955).—A. & R. Fernandes in Mem. Soc. Brot. **11**: 15 &
70 (1956); in Bol. Soc. Brot., Sér. 2, **34**: 182 (1960); in Kirkia **1**: 70 (1961); in
C.F.A. **4**: 137 (1970).—Binns, H.C.L.M.: 68 (1968).—Wickens in F.T.E.A.,
Melastom.: 35, t. 8 fig. 2 (1975). Type from S. Afr. (Transvaal).
  Osbeckia debilis Sond. in Linnaea **23**: 47 (1850) non Naud. (1850). Type as above.
  Osbeckia phaeotricha var. *debilis* (Sond.) Sond. in Harv. & Sond., F.C. **2**: 519
(1862). Type as above.
  Dissotis penicillata Gilg in Engl., loc. cit.—Engl., loc. cit. Type from Angola
(Huíla).
  Dissotis paludosa Gilg in Engl., loc. cit. Type from Zaire.
  Dissotis bangweolensis R. E. Fr. in Wiss. Ergebn. Schwed. Rhod.-Kongo-Exped.
1911–12, **1**: 178 (1914). Type: Zambia, Lake Bangweulu, Kamindas, Kapata, *Fries*
936 (UPS, holotype).

A perennial or rarely annual usually many-stemmed plant 10–95 cm. tall. Stems
erect, ascending or decumbent, simple or ± branched, 4-gonous and herbaceous
towards the apex, cylindric and somewhat woody at the base, green or pink to dark
red, sparse to ± densely hairy, the hairs white, appressed to subpatent and usually
bulbous at the base; basal internodes short, those below the floral leaves very long
(up to 30 cm.); nodes with dense, ± stiff bristles. Leaves opposite, the floral ones
closely approximate (pseudo-verticillate) and shortly petiolate; leaf-lamina
0·5–7·5 × 0·3–1·8 cm., linear-lanceolate to ovate (the lowermost the widest), acute or
subobtuse at apex, contracted or attenuate at the base, crenulate-setulose (each
tooth ending in a small bristle), membranous, concolorous or nearly so (of a
somewhat lighter green above), appressed-setose on both faces, the setae sparse to
± dense, mainly on the nerves, longitudinally 3–5-nerved, the nerves impressed
above, raised beneath, the transverse nerves and reticulation inconspicuous;

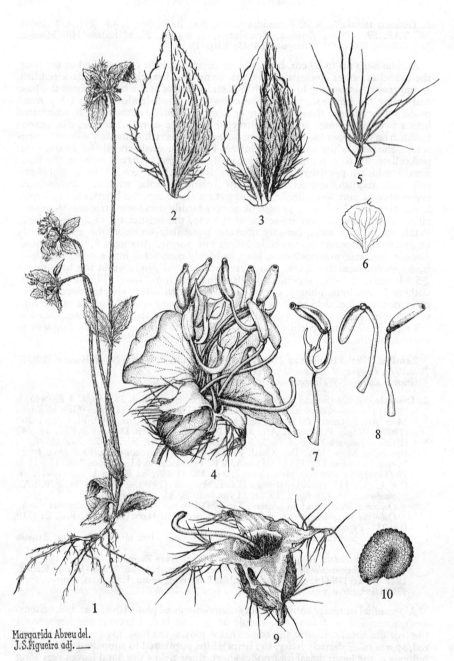

Margarida Abreu del.
J.S. Figueira adj.

Tab. 59. DISSOTIS TENUIS. 1, habit (×1); 2, leaf, superior surface (×3); 3, leaf, underside (×3); 4, flower with 1 petal removed (×6); 5, intersepalar segment (×12½); 6, petal (×3); 7, long stamen (×6); 8, two short stamens (×6); 9, receptacle after petals and stamens are shed (×6); 10, seed (×35), all from *Edwards* 802.

petiole 2–5 mm. long. Flowers in contracted cymes arranged in terminal (rarely axillary) dense, many- or sometimes few-flowered heads 1–3·5 cm. in diameter, surrounded by the uppermost leaves; pedicels up to 5·5 mm. long; bracts 2·5 mm. long, lanceolate or elliptic, acute, scarious or with a foliaceous tip, pale green or pink with ciliate margin, persistent. Receptacle 3–4 × 3 mm., ovoid, green to bright red, sparsely setose, the bristles white, simple, or sometimes towards the summit of receptacle inserted in pairs or in clusters at the apex of short appendages. Sepals 4–5 × 2–2·5 mm., oblong-triangular, obtuse or acute with a subterminal stellate- or penicillate-setose appendage; intersepalar segments c. 2 mm. long, penicillate-setose, persistent. Petals 7·5–10 mm. long, mauve to deep pink or rosy-purple. Anthers of the longer stamens 4–4·5 mm. long, pedoconnective 4–6 mm. long with the basal appendage 2–3 mm. long, spathulate and ± deeply bilobed; anthers of the shorter stamens similar to those of longer ones but shorter and with a c. 1 mm. long minutely appendiculate pedoconnective. Fructiferous receptacle 3–5 × 3–3·5 mm., broadly campanulate, ovoid or subspherical. Capsule slightly shorter than the receptacle, sparsely setose on the margins of the valves and with a cup-shaped, marginally setose crown around the insertion of the style.

A very polymorphic species. We consider it probable that var. *pusilla* and var. *prostrata* are only ecophenes induced mainly by fires.

Perennial with a woody rootstock:
  Stems erect or ascending:
    Leaves ovate to broadly elliptic (at least the lower ones), 0·8–4·5 ×0·3–1·7 cm., contracted at base; stems simple or few-branched:
      Stems usually more than 10 cm. high, simple or few-branched; lower leaves ovate, the upper ones oblong or lanceolate, 1–4·5 ×0·5–1·8 cm.; heads usually dense (10–30 flowers on robust stems) - - - - - var. *debilis*
      Stems c. 10 cm. high, simple and weak; leaves elliptic to broadly elliptic, 0·8–1·5 ×0·5–1 cm.; heads few-flowered - - - - var. *pusilla*
    Leaves linear-lanceolate, 3·5–7·5 ×0·5–1·8 cm., attenuate at the base; stems usually many-branched:
      Stems and branches ± sparsely setose; leaves up to 14 mm. wide; heads compact; pedicels up to 3 mm. long - - - - - - var. *lanceolata*
      Stems and branches densely sericeous; leaves up to 18 mm. wide; heads lax; pedicels up to 5·5 mm. long - - - - - - var. *pedicellata*
  Stems decumbent, rooting at the nodes; inflorescence usually few-flowered
                                var. *prostrata*
Annual; stems simple or few-branched - - - - - var. *postpluvialis*

Var. debilis.

Zambia. B: near Shangombo, 1020 m., fl. & fr. 9.viii.1952, *Codd* 7465 (BM; COI; K; PRE; SRGH). N: Kawambwa, Timnatushi Falls, 1260 m., fl. & fr. 19.iv.1957, *Richards* 9328 (COI; K). W: Mwinilunga Distr., 26 km. W. of R. Kabompo, fl. & fr. 11.iv.1930, *Milne-Redhead* 1114 (K). C: Lusaka, near Mt. Makalu Research Station, fl. & fr. 20.vii.1956, *Angus* 1375 (K; LISC; SRGH). **Rhodesia.** N: Shamva, 900 m., fl. & fr. 21.xii.1921, *Eyles* 3795 (SRGH). W: Shangani Distr., Gwampa Forest Reserve, c. 900 m., fl. & fr. i.1956, *Goldsmith* 75/56 (K; LISC; SRGH). C: Charter Distr., on Enkeldorn-Gutu road, fl. & fr. 4.xii.1960, *Leach & Chase* 10539 (COI; SRGH). E: Melsetter, fl. 1.x.1919, *Walters* 2737 (K; SRGH). S: Belingwe Distr., near Shabani, fl. & fr. 24.iv.1954, *Plowes* 1716 (COI; SRGH). **Malawi.** N: Mzimba, fl. & fr. 12.vi.1947, *Benson* 1308 (BM). C: Dedza, 1700 m., fl. 2.ii.1959, *Robson* 1424 (K; LISC). S: Phalombe plain, near Kundwelo village, 630 m., fl. & fr. 29.vii.1956, *Newman & Whitmore* 298 (BM; SRGH). **Mozambique.** N: Túnguè, between Quionga and Palma, fl. & fr. 21.x.1942, *Mendonça* 1030 (COI; LISC; LMA). T: Angónia, between Furancungo and Vila Coutinho, fl. & fr. 29.ix.1942, *Mendonça* 537 (K; LISC; LMA; SRGH; WAG). MS: Manica, Mavita, Murôruè, fl. & fr. 22.iv.1948, *Barbosa* 1520 (COI; EA; LISC; PRE). SS: Homoíne, 7 km. from Jacubécua, on Inharrime road, fl. & fr. 9.xii.1944, *Mendonça* 3343 (COI; K; LISC; LMA; LMU). LM: between Manhiça and Bilene, Pati lagoon, fl. & fr. 25.iii.1954, *Barbosa & Balsinhas* 5457 (COI; LMA).

Widespread in tropical and in S. Africa. In swamps, moist places, rather dry sandy soils and as a weed of cultivated land.

Var. **pusilla** (R. E. Fr.) A. & R. Fernandes in An. Junta Invest. Ultramar **10**: 19 (1955); in Mem. Soc. Brot. **11**: 19 (1956); in Bol. Soc. Brot., Sér. 2, **34**: 183 (1960); in C.F.A. **4**: 138 (1970). Type: Zambia, Luesa R., *Fries* 593 (UPS, holotype).
  Dissotis pusilla R. E. Fr., Wiss. Ergebn. Schwed. Rhod.-Kongo-Exped. **1**: 178 (1914).—Engl., Pflanzenw. Afr. **3**, 2: 748 (1921). Type as above.

246 77. MELASTOMATACEAE

Zambia. B: road to Kaoma, fl. & fr. 1.x.1957, *West* 3486 (K). N: Mansa, near Samfya, L. Bangweulu, fl. & fr. 28.viii.1952, *Angus* 314 (K). W: Mwinilunga Distr., source of Matonchi dambo, fl. & fr. 26.x.1937, *Milne-Redhead* 2963 (K; LISC). C: c. 10 km. N. of Kabwe, fl. & fr. 23.ix.1947, *Brenan* 7934 (K). S: Machili, fl. & fr. 21.ix.1969, *Mutimushi* 3783 (K).
Also in Angola. Ecology as for the type variety.

Var. **prostrata** A. & R. Fernandes in Bol. Soc. Brot., Sér. 2, **28**: 181, t. 1 (1954); in Mem. Soc. Brot. **11**: 72 (1956); in C.F.A. **4**: 139 (1970). Type from Angola.

Zambia. B: Senanga Distr., W. of R. Zambezi, fl. 28.vi.1964, *Verboom* 11223 (K). N: Kapalala, Luapula R., fl. 3.ix.1953, *Fanshawe* 253 (K). C: c. 5 km. E. of Lusaka, 1200 m., fl. & fr. 15.vi.1958, *Best* 141 (K).
Also in Angola. Ecology as for the type variety.

Var. **lanceolata** (Cogn.) A. & R. Fernandes in Bol. Soc. Brot., Sér. 2, **29**: 49, t. 3 (1955); in Mem. Soc. Brot. **11**: 19 (1956); in Bol. Soc. Brot., Sér. 2, **34**: 183 (1960); in Kirkia **1**: 70 (1961); in C.F.A. **4**: 139 (1970).—Wickens in F.T.E.A., Melastom.: 36 (1975). Type from Angola.
  Dissotis lanceolata Cogn. in A. & C. DC., Mon. Phan. **7**: 366 (1891).—Hiern, Cat. Afr. Pl. Welw. **1**, 2: 366 (1898).—De Wild. in Ann. Mus. Congo Belge, Bot. Sér. 5, **3**: 224 (1910). Type as above.
  Dissotis debilis var. *debilis* sensu A. & R. Fernandes in An. Junta Invest. Ultramar **10**, 3: 67 (1955) pro parte.

Caprivi Strip. Mashi R., fl. & fr. 31.x.1962, *Fanshawe* 7103 (K; SRGH). **Zambia,** B: Kaoma, Boma, fl. & fr. 23.ii.1952, *White* 2123 (COI; FHO). N: Mporokoso Distr., Nsama, 1200 m., fl. & fr. 3.iv.1957, *Richards* 8994 (COI; SRGH). C: Serenje Distr., Lake Lusiwasi, 1200 m., fl. & fr. 6.iv.1961, *Richards* 14961 (K; SRGH). W: Ndola. between Kitwe and Forest Department Research Station, 1250 m., fl. & fr. 28.iv.1961, *Linley* 141 (COI; SRGH). E: Fort Jameson-Nsefu-Luangwa road, fl. & fr. 11.v.1963, *van Rensburg* 2119 (K; SRGH). S: Mumbwa Distr., Chunga, Kafue National Park, fl. & fr. 27.iii.1963, *Mitchell* 18/44 (COI; LISC; SRGH). **Rhodesia.** E: Umtali, Odzani River Valley, 1914, *Teague* 207 (K). **Malawi.** C: Nkhota Kota Distr., 1500 m., fl. & fr. 7.viii.1946, *Shortridge* 17393 (K). S: Blantyre, 1100 m., fl. & fr. 18.vi.1946, *Brass* 16353 (K; SRGH). **Mozambique.** N: Marrupa, 36 km. from Maúa on road to Marrupa, 600 m., fl. 19.ii.1964, *Torre & Paiva* 10666 (COI; K; LISC; LMU); Nampula near CICA Experimental Station, fl. & fr. 11.xi.1952, *Barbosa & Balsinhas* 5258 (COI; LISC; LMA). Z: Maganja da Costa, km. 42 on the road to the beach, c. 15 m., fl. & fr. 15.ii.1966, *Torre & Correia* 14660 (BR; EA; LISC; M; P).
Also in Angola, Zaire, Tanzania and S. Africa. Ecology as for the type variety.

Var. **pedicellata** A. & R. Fernandes in Kirkia **1**: 70 (1961). Type: Zambia, Northern Prov., Mpulungu, *Richards* 907 (K, holotype).

Zambia. N: Mpulungu, near Mission, 840 m., fl. & fr. 6.iii.1952, *Richards* 907 (K). Known only from Zambia. In open marshy localities among grass.

Var. **postpluvialis** (Gilg) A. & R. Fernandes in Garcia de Orta **2**: 171 (1954); in An. Junta Invest. Ultramar **10**, 3: 19 (1955); in Bol. Soc. Brot., Sér. 2, **34**: 183 (1960).—Wickens in F.T.E.A., Melastom.: 36 (1975). Type from the Sudan.
  Osbeckia postpluvialis Gilg in Engl., Mon. Afr. Pflanzen. **2**: 6 (1898).—Engl., Pflanzenw. Afr. **3**, 2: 744 (1921).—Hutch. in Kew Bull. **1921**: 371 (1921). Type as above.

Mozambique. N: Nampula, fl. & fr. 20.v.1935, *Torre* 798 (COI; LISC).
Also in the Sudan and Uganda. Grasslands and open woodlands.

Observations in the field are needed to decide if var. *postpluvialis* is annual or only represents the first year growth of the other varieties.

The specimens *Fanshawe* 3523 (K), *Richards* 491 (K), 9328 (COI; K), 9333 (COI; K) and 9333B (K) from the Northern Province of Zambia, which we have identified as *D. debilis* var. *debilis*, are remarkable because the cauline and floral leaves are unusually large (up to 4·5 × 1·7 cm.), 5-nerved, and sometimes tinged with red, the hairs of the appendages of the receptacle and of the sepals are long and the appendages of the pedoconnectives are very large and ± deeply 2-lobed distally. They may represent another variety of this variable species.

The following specimens are probably hybrids between *D. debilis* and *D. phaeotricha*: **Zambia.** N: Mbala Distr., Saisi R., c. 1500 m., fl. & fr. 22.i.1970, *Sanane* 1052 (K); Upper end of Lake Chila, 1500 m., fl. 29.i.1955, *Richards* 4281 (K); Lunzua Agricultural Station, 1800 m., fl. 25.iii.1955, *Richards* 5165 (K); Chisinga Ranch near Luwingu, c. 1560 m., fl. & fr. 27.iv.1961, *Astle* 533 (K). W: Solwezi, c. 1350 m., *Robinson* 3462 (K).

3. **Dissotis angustifolia** A. & R. Fernandes in Bol. Soc. Brot., Sér. 2, **28**: 181, t. 2 & 3 (1954); in An. Junta Invest. Ultramar **10**, 3: 16 (1955). Type: Mozambique, Cabo Delgado, near Ingoane, *Barbosa* 2073 (COI, holotype; LISC; LMA).

A shrublet up to 45 cm. high, with a thickened woody rootstock. Stem erect, woody, rather branched towards the apex, covered with irregularly fissured bark; branches erect or ascending, slender, 4-gonous, reddish, with short sparse whitish appressed bristles bulbous at the base, glabrescent towards the base; nodes with a ring of long bristles; internodes 0·3-4 cm. long. Leaves 1–5 × 0·2–0·4(0·7) cm., narrowly lanceolate, acute at the apex, crenulate at the margin, narrowed into a 2 mm. long petiole at the base, rigid, patent, green on both faces and with bristles as on branches mainly on the nerves and along the margin, 3-nerved, the lateral nerves inconspicuous on superior surface, they and the midrib prominent beneath. Flowers 4-merous, solitary or 2–4 in heads at the end of the branches. Bracts minute, green or yellow, acute, ciliate along the margin, persistent. Pedicels 2–3 mm. long, appressed-setose. Receptacle 3–4 × 2–2·5 mm., cylindric-campanulate, appressed-setose, provided towards the summit with reddish, very shortly pedicellate appendages, setose at apex. Sepals 3 × 1·5 mm., persistent or tardily caducous, obtuse, reddish, ciliate along the margin, long stellate-setose at the apex, alternating with 1 mm. long intersepalar segments stellate-setose at the apex. Petals 1 × 1 cm., rose-violet. Stamens 8, unequal. Longer stamens: anthers 4 mm. long with a pedoconnective 5–6 mm. long, arched and furnished at the base with a 2 mm. long, furcate appendage; filaments 5 mm. long. Shorter stamens: anthers 3·5 mm. long, with a very short pedoconnective provided at the base with a minute, bituberculate appendage; filament 6 mm. long. Style c. 12 mm. long. Fructiferous receptacle 4 × 3 mm., constricted towards the apex. Capsule 4-valved, 4-angular above, setose at the top. Seeds c. 0·4 mm. long, cochleate, yellowish, with minute papillae arranged in curved lines.

Mozambique. N: António Enes, fr. 18.x.1965, *Gomes e Sousa* 4871 (COI; LISC). Known only from Mozambique. Moist plains near the sea coast.

4. **Dissotis phaeotricha** (Hochst.) Hook. f. in F.T.A. **2**: 451 (1871).—Triana in Trans. Linn. Soc. Lond. **28**: 58 [1871] (1872).—Cogn. in A. & C. DC., Mon. Phan. **7**: 367 (1891).—Krasser in Engl. & Prantl, Pflanzenfam. **3**, 7: 156 (1893).—Taub. in Engl., Pflanzenw. Ost-Afr. C: 295 (1895).—Gilg in Engl., Mon. Afr. Pflanzen. **2**: 14 (1898).—Engl., Pflanzenw. Afr. **3**, 2: 748 (1921).—Garcia in Est. Ens. Doc. Junta Invest. Ultramar **12**: 156 (1954).—A. & R. Fernandes in An. Junta Invest. Ultramar **10**, 3: 21 (1955); in Mem. Soc. Brot. **11**: 20 & 72 (1956); in Bol. Soc. Brot., Sér. 2, **34**: 183 (1960); in Kirkia **1**: 71 (1961); in C.F.A. **4**: 142 (1970).—Binns, H.C.L.M.: 68 (1968).—Wickens in F.T.E.A., Melastom.: 36, t. 8 fig. 3 (1975). Type from Natal.
  Osbeckia phaeotricha Hochst. in Flora **27**: 424 (1844); in Walp., Repert. **5**: 708 (1846).—Sond. in Harv. & Sond., F.C. **2**: 519 (1873) pro parte excl. var. Type as above.
  Argyrella phaeotricha (Hochst.) Naud. in Ann. Sci. Nat. Bot., Sér. 3, **13**: 300 (1849). Type as above.
  Dissotis villosa Hook. f. in F.T.A. **2**: 450 (1871). Type from Guinea.
  Osbeckia zambesiensis Cogn. in Bol. Soc. Brot. **7**: 226 (1889); in A. & C. DC., Mon. Phan. **7**: 331 (1891).—Taub. in Engl., loc. cit.—Engl., op. cit. 744 (1921). Type from Mozambique, R. Zambezi, *Carvalho* 10 (BR; COI, holotype).
  Dissotis phaeotricha var. *zambesiensis* (Cogn.) A. & R. Fernandes, tom. cit.: 25 (1955); tom. cit.: 21 (1956). Type as above.

Erect or sometimes ascending or prostrate, perennial herb up to 60 cm., with the stems often arising from a woody rootstock. Stems 1 to several, simple or ± branched towards the apex, obtusely 4-angled, reddish, more or less densely covered with long and whitish ± patent bristles. Leaves 1–6 × 0·5–2 cm., subcoriaceous, the lower ovate or ovate-oblong, the upper ovate-oblong, oblong or lanceolate, acute or subacute, roundish at the base, with cartilaginous ± conspicuously serrate margin, dark green above, paler beneath, 3–5-nerved, ± densely hairy on both faces; petiole up to 4 mm. long, densely setose. Flowers 4-merous, in dense heads surrounded by the uppermost leaves; bracts numerous, variable in size, scarious, bristly mainly at the margin; pedicels up to 5 mm. long. Receptacle up to 6 mm. long, reddish, covered with simple bristles and appendages stellate-setose at the apex; intersepalar segments stalked, the stalks c. 4·5 mm. long, penicillate-setose at the dilate apex, the setae c. 4 mm. long. Sepals triangular, as long as or shorter than the receptacle, with an appendage similar to the intersepalar segments at the apex. Petals usually 17 × 15 mm., obovate, mauve-pink. Stamens usually unequal or sometimes subequal, the longer ones with filaments 6 mm. long, anthers 2·5–5 mm. long with a strongly curved pedoconnective 5·5 mm. long provided with a furcate appendage 2·5 mm. long, the shorter ones with filaments 5·25 mm. long, anthers 3·5 mm. long with a pedoconnective c. 0·5 mm. long, 2-tubercled at the base. Fructiferous receptacle up to 7 × 4 cm., urceolate. Capsule subglobose, with an apical corona of bristles. Seeds numerous, c. 0·5 mm. long, cochleate, reddish.

Indumentum of the stems not very dense, of fulvous, ± rigid, ± patent bristles; flowers up to 2 cm. in diameter; anthers of the longer stamens 2–3·5 mm. long; receptacle 4–6 mm. high - - - - - - - - - - - - var. *phaeotricha*
Indumentum of the stems very dense, of long, whitish, weak, ± coiled subappressed hairs; flowers up to 3 cm. in diameter; anthers of the longer stamens c. 5 mm. long; receptacle c. 6 mm. high - - - - - - - - var. *villosissima*

Var. **phaeotricha**.

Zambia. N: Kawambwa, fl. & fr. 22.viii.1957, *Fanshawe* 3523 (M; SRGH). W: Solwezi, 1350 m., fl. & fr. 10.iv.1960, *Robinson* 3507 (K; SRGH). **Rhodesia.** E: Melsetter, fl. & fr. 1.x.1919, *Wallis* 2737 (SRGH). **Malawi.** N: Nkata Bay, Limpasa dambo, fl. & fr. 31.vii.1960, *Eccles* 30 (BM; COI; K; LD; SRGH). S: Zomba Distr., Matawale stream, fl. & fr. 30.vii.1937, *Lawrence* 419 (K). **Mozambique.** N: Mogincual, between Quixaxe and Liupo, fl. & fr. 8.x.1948, *Barbosa* 2489 (COI; LISC; LMA); Meconta, between Corrane and Muatua, 180 m., fl. 31.iii.1964, *Torre & Paiva* 11505 (COI; EA; K; LISC; LD; PRE). Z: Lugela-Mocuba, Namagoa Estate, 60 m., fl. & fr. v.1943, *Faulkner* 295 (COI; K; PRE; SRGH; UPS). MS: between Beira and Manga, 8 km. from Beira, fl. & fr. 6.xi.1946, *Pedro & Pedrógão* 21 (COI; LMA). SS: Vilanculos, 8 km. from Inhassoro on the road Inhassoro-National Road no. 1, fl. & fr. 20.xi.1968, *Fernandes, Fernandes & Pereira* 210 (BR; COI; LISC; LMU; PRE). LM: Delagoa Bay, 1893, *Junod* 334 (P).
Widespread in tropical and S. Africa. Swamps and moist places.

Var. **villosissima** A. & R. Fernandes in Kirkia **1**: 71 (1961).—Wickens in F.T.E.A., Melastom.: 37 (1975). Type: Zambia, Mporokoso Distr., Mweru-Wa-Ntipa, near Bulaya Chishyela, *Richards* 9065 (COI; K, holotype).

Zambia. N: Mweru-Wa-Ntipa, Chishyela dambo, 915 m., fl. & fr. 11.viii.1962, *Tyrer* 380 (BM; SRGH).
Also in Zaire and Tanzania. Swamps and moist places.

Hybrids between *D. debilis* (2) and *D. phaeotricha* (4) may occur (see page 246).

5. **Dissotis glandulosa** A. & R. Fernandes in Bol. Soc. Brot., Sér. 2, **43**: 293, t. 3 (1969). TAB. **60**. Type: Zambia, Mwinilunga, source of Zambesi R., *Robinson* 5990 (SRGH, holotype).

Perennial herb up to 20 cm. high probably woody at the base. Stems erect, branched; stems and branches reddish, glabrous and narrowly winged towards the base, widely winged towards the apex, the wings sparsely setose; internodes 1–1·5 cm. long. Leaves opposite, sessile, broadly elliptic (2–3 × 1–1·5 cm.) or lanceolate (up to 4·5 × 1·3 cm.), rounded or somewhat obtuse at the apex, obscurely denticulate at margin, the teeth setose, ± cuneate at the base, membranous, discolorous (yellow-green above and pale yellow-green beneath), 3- or 5-nerved, the nerves thin, more raised on the under side than above. Flowers 5-merous, in few flowered (4-flowered in the specimen seen) cymes at the top of stems and branches; peduncles, bracts, pedicels, sepals and margin of petals with capitate-glandular reddish bristles c. 0·7 mm. long; bracts caducous, membranous, reddish, conspicuously nerved; pedicels c. 2 mm. long. Receptacle 6 × 4 mm., campanulate, reddish, conspicuously longitudinally nerved, glabrous; sepals c. 6 × 2 mm., narrowly triangular, distinctly longitudinally nerved, with capitate-glandular hairs dorsally and with eglandular bristles along the margin. Petals 15 × 6 mm., bright rose, obovate, with capitate-glandular hairs along the margin. Stamens 10, unequal; filaments c. 6 mm. long, yellow; anthers of the longer stamens 5·5 mm. long, reddish yellow, with a strongly arched pedoconnective c. 8 mm. long provided with an arched appendage c. 3 mm. long, dilated above the middle and slightly bilobed or shortly and obtusely apiculate at apex; anthers of the shorter stamens

c. 4 mm. long, yellow, with the pedoconnective c. 1 mm. long, arched, appendiculate at the base, the appendage entire, dilated at apex. Fruit unknown.

Zambia. W: Mwinilunga, source of Zambesi R., fl. 13.xii.1963, *Robinson* 5990 (SRGH).
Known only from Zambia. In moist sandy woodland.

6. **Dissotis rotundifolia** (Sm.) Triana in Trans. Linn. Soc. Lond. **28**: 58 [1871] (1872).—Cogn. in A. & C. DC., Mon. Phan. **7**: 369 (1891).—Taub. in Engl., Pflanzenw. Ost-Afr. **C**: 295 (1895).—Gilg in Engl., Mon. Afr. Pflanzen. **2**: 15 (1898).—Keay in F.W.T.A. ed. 2, **1**: 257, fig. 101 (1954).—A. & R. Fernandes in Mem. Soc. Brot. **11**: 23 & 74 (1956); in Bol. Soc. Brot., Sér. 2, **34**: 184 (1960); in C.F.A. **4**: 143 (1970).—Binns, H.C.L.M.: 62 (1968).—Wickens in F.T.E.A., Melastom.: 39, t. 8 fig. 6 (1975). Type from Sierra Leone.

Var. **prostrata** (Thonn.) Jacques-Félix in Adansonia **11**: 548 (1971). Type from Ghana.
Melastoma prostratum Thonn. in Schumacher, Beskr. Guin. Pl.: 220 (1827); in Kongel. Dansk. Vid. Selsk. Naturvid. Math. Afh. **3**: 240 (1828). Type as above.
Heterotis prostrata (Thonn.) Benth. in Hook., Niger Fl.: 349 (1849). Type as above.
Osbeckia zanzibarensis Naud. in Ann. Sci. Nat. Bot., Sér. 3, **13**: t. 7 fig. 5 (1849); op. cit. **14**: 55 (1850). Type from Zanzibar.
Lepidanthemum triplinervium Klotzsch in Peters, Reise Mossamb. Bot. **1**: 64 (1861). Type: Mozambique, Boror, *Peters* (B†, holotype).
Dissotis prostrata (Thonn.) Hook. f. in F.T.A. **2**: 452 (1871).—Triana in Trans. Linn. Soc. Lond. **28**: 58 [1871] (1872).—Oliv. in Trans. Linn. Soc. Lond. **29**: 73, t. 39 (1873).—Cogn. in A. & C. DC., loc. cit.—Krasser in Engl. & Prantl, Pflanzenfam. **3**, **7**: 156, fig. 70 R, S (1893).—Taub. in Engl., loc. cit.—Garcia in Est. Ens. Docum. Junta Invest. Ultramar **12**: 155 (1954). Type as for *Dissotis rotundifolia* var. *prostrata*.

A perennial procumbent or rarely scandent, branched herb up to 60–90 cm. long. Stem rooting at the nodes, fleshy, angular, hollow, the oldest ± woody and subterete; branches ascending, appressed-setulose or patently-pilose; nodes with stiff erect bristles. Leaves opposite; leaf-lamina 0·9–5·7 × 0·6–2·8 cm., narrowly to broadly elliptic, ovate or subcircular, acute or obtuse at apex, acute at base, entire, dull green above and pale green or whitish beneath, slightly fleshy, appressed-setose on both faces, the bristles short and sparse, longitudinally 3(5)-nerved, the longitudinal nerves slender, impressed above, raised beneath, the lateral ones and the reticulation inconspicuous on both faces; petiole 0·2–3 cm. long, slender, compressed, sulcate above, shortly setose. Cymes few-flowered or flowers solitary; bracts ovate, membranous, ciliate on back and margin, shorter than the receptacle, deciduous; pedicels 4–5 mm. long. Receptacle c. 6 mm. long, ovoid or cylindric, densely to ± sparsely setose and with ± numerous ± long patent, linear appendages usually setose all around and with several long radiate pink or purple bristles at the top or sometimes with only a few or a single terminal bristle. Sepals 4–9 × 1–1·5 mm., linear-oblong, ciliate at the margins and with a subapical appendage similar to those of the receptacle, persistent; intersepalar segments like the uppermost receptacular appendages. Petals 1–2 cm. long, obovate, pale mauve to purple. Anthers of the longer stamens 6–8 mm. long, pedoconnective 5·5–7 mm. with the anterior appendage bilobed (lobes obtuse) up to c. 2 mm. long and a minute posterior spur; anthers of the shorter stamens 4–6 mm. long, pedoconnective c. 1 mm. long, very shortly appendiculate at the base. Fructiferous receptacle 6–11 × 6–7·5 mm., ovoid to ovoid-ellipsoid. Capsule subequalling the receptacle, setose on the upper part. Seeds cochleate, ridged along the back and with a fringe of long papillae around the hilum.

Zambia. N: Kawambwa Distr., Kafulwe, L. Mweru, 960 m., fl. 24.iv.1957, *Richards* 9428 (K). W: Chingola, fr. 15.v.1957, *Fanshawe* 3284 (K; SRGH). **Rhodesia.** E: Melsetter, Haroni R., 390 m., fl. 24.iv.1962, *Wild* 5733 (COI; K; SRGH). **Malawi.** N: Nkhata Bay, Vizara Estate near Chinteche road, fl. & fr. 18.ii.1960, *Adlard* 381 (COI; SRGH). S: Zomba Distr., Mlomba, fl. & fr. 2.ii.1955, *Jackson* 1440 (K; LISC). **Mozambique.** N: 4 km. from Montepuez on the Montepuez-Namuno road, c. 430 m., fl. & fr. 25.xii.1963, *Torre & Paiva* 9664 (LISC; LMU; PRE; WAG); 14 km. from Palma on the Palma-Nangade road, fl. 10.iv.1964, *Torre & Paiva* 12132 (BR; COI; K; LISC; LMU; PRE). Z: Boror, near Nhamacurra, Maganja da Costa road, fl. & fr. 26.iii.1943, *Torre* 4976 (LISC); Pebane, km. 75 on the road to Mualama, c. 20 m., fl.

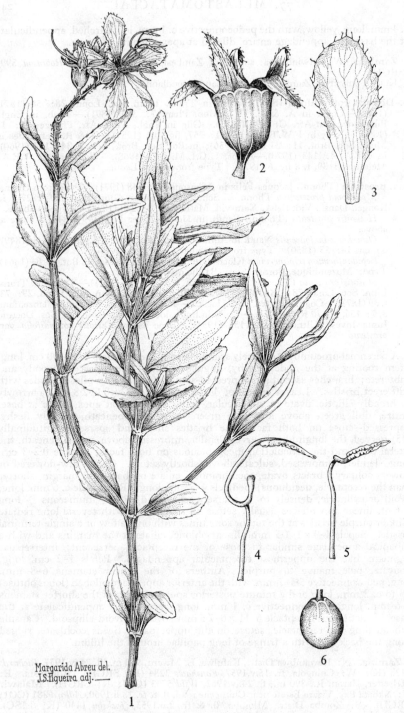

Tab. 60. DISSOTIS GLANDULOSA. 1, upper part of plant (×1); 2, receptacle after shedding of petals and stamens (×3); 3, petal (×3); 4, long stamen (×3); 5, short stamen (×3); 6, ovary (×3), all from *Robinson* 5990.

10.iii.1966, *Torre & Correia* 15155 (LISC; LMA; LUA; SRGH). MS: Inhaminga, Inhansato, 50 m., fl. & fr. 13.iii.1956, *Gomes e Sousa* 4292 (COI; K; LMA).
Widespread in tropical Africa. In moist or damp places in forests and along streams, sometimes on rocks or creeping and climbing among boulders and in open *Brachystegia* woodland.

7. **Dissotis canescens** (E. Mey. ex Grah.) Hook. f. in F.T.A. **2**: 453 (1871).—Jacques-Félix in Bull. Inst. Fr. Afr. Noire **15**: 979 (1953).—Keay in F.W.T.A. ed. 2, **1**: 258 (1954).—Garcia in Est. Ens. Docum. Junta Invest. Ultramar **12**: 155 (1954).— A. & R. Fernandes in An. Junta Invest. Ultramar **10**, 3: 30 (1955); in Mem. Soc. Brot. **11**: 26 & 80 (1956); in Bol. Soc. Brot., Sér. 2, **34**: 186 (1960); in Kirkia **1**: 73 (1961); in C.F.A. **4**: 145 (1970).—F. White, F.F.N.R.: 306 (1962).—Binns, H.C.L.M.: 57 (1968).—Wickens in F.T.E.A., Melastom.: 40, t. 8 fig. 7 (1975). Type grown at the Royal Botanic Garden, Edinburgh from material of unknown origin.
   Osbeckia canescens E. Mey. ex Grah. in Edinb. Nat. Phil. Journ. **28**: 399 (1840); in Curtis, Bot. Mag. **66**: t. 3790 (1840). Type as above.
   Osbeckia umlaasiana Hochst. in Flora, Jena **27**: 424 (1844).—Sond. in Harv. & Sond., F.C. **2**: 518 (1862). Type from S. Africa (Natal).
   Osbeckia incana E. Mey. ex Walp., Repert. **5**: 708 (1846) *nom. illegit.* Based on *Osbeckia umlaasiana.*
   Argyrella incana Naud. in Ann. Sci. Nat. Bot., Sér. 3, **13**: 300, t. 6 fig. 7 (1849). Based on *Osbeckia umlaasiana.*
   Argyrella canescens (E. Mey. ex Grah.) Harv., Gen. Afr. Pl.: 113 (1868). Type as for *Dissotis canescens.*
   Dissotis incana (Naud.) Triana in Trans. Linn. Soc. Lond. **28**: 58, t. 4 fig. 44d [1871] (1872).—Cogn. in A. & C. DC., Mon. Phan. **7**: 370 (1891).—Krasser in Engl. & Prantl, Pflanzenfam. **3**, 7: 156, fig. 70 T–V (1893).—Taub. in Engl., Pflanzenw. Ost-Afr. **C**: 295 (1895).—Gilg in Engl., Mon. Afr. Pflanzen.: 17 (1898).—Th. & H. Dur., Syll. Fl. Cong.: 211 (1909).—R. E. Fr., Wiss. Ergebn. Schwed. Rhod.-Kongo-Exped. **1**: 180 (1914).—Engl., Pflanzenw. Afr. **3**, 2: 749 (1921).—Burtt Davy, F.P.F.T.: 243 (1926).—Hutch. & Dalz. in F.W.T.A. **1**: 212 (1927). Based on *Osbeckia umlaasiana.*
   Dissotis angolensis Cogn. in A. & C. DC., op. cit.: 371 (1891).—Gilg in Engl., loc. cit.; in Warb., Kunene-Samb. Exped. Baum: 323 (1903).—Hiern, Cat. Afr. Pl. Welw. **1**: 367 (1898).—Engl., op. cit.: 750 (1921).—Jacques-Félix in Bull. Mus. Hist. Nat. Paris, Sér. 2, **8**: 108 (1936); loc. cit.—R. & A. Fernandes, tom. cit.: 147 (1970). Type from Angola.
   Tristemma verdickii De Wild. in Ann. Mus. Congo Belge, Bot. Sér. 4, **1**: 219 (1903). Type from Zaire.
   Dissotis canescens var. *transvaalensis* A. & R. Fernandes in Bol. Soc. Brot., Sér. 2, **28**: 187, t. 7 (1954); in Mem. Soc. Brot. **11**: 32 (1956). Type from S. Africa (Transvaal).

A perennial herb or shrub 0·25–1·8 m. high. Stems erect, ± branched or simple, 4-gonous, purple or brownish, tomentose with branched hairs, glabrescent. Leaves opposite, subsessile or with a petiole up to 5 mm. long; leaf-lamina 1·4–7·5 × 0·3–2·6(3) cm., linear-oblong, oblong, ovate-oblong, sometimes ovate, roundish, obtuse or acute at apex, round or subcordate at the base, entire or rarely serrulate at margin, somewhat membranous or rigid, dark green, brownish-green to greyish-green on the superior surface, yellowish-, whitish- or greyish-green below, with sparse minute stellate hairs above and ± tomentose with branched hairs beneath, longitudinally 3–5(7)-nerved, the nerves impressed above, prominent on the undersurface, the transverse nerves inconspicuous, except beneath in sparsely tomentose leaves. Flowers in dense axillary cymes, forming compact short terminal inflorescences or ± long and narrow, rarely wide panicles; bracts caducous, ovate, obtuse or acute, concave, purplish, stellate-tomentose outside, the stellate hairs sometimes intermixed with simple capitate-glandular ones; pedicels 2–3 mm. long. Receptacle 4·5–7 × 3–4·5 mm., ovoid or subspherical, purplish, covered by minute stellate hairs (or sometimes branched hairs towards the summit of receptacle), intermixed with simple capitate-glandular ones, occasionally the latter absent or rarely the only ones present, glabrescent with age; intersepalar segments linear, ending in a glandular-tipped bristle, or, in a few cases, absent. Sepals 4–7 × c. 2 mm., triangular-oblong, acute, with the outer face stellate-tomentose, the inner one glabrous and purple, persistent. Corolla ± 3·5 cm. in diameter, pink to bright purple. Anthers of the longer stamens 6–7 mm. long; pedoconnective 6–8 mm. long, with the appendage c. 2·5 mm. long, linear-spathulate; anthers of the shorter stamens 4–6 mm. long; pedoconnective c. 1 mm. long, shortly appendiculate. Fructiferous receptacle c. 7 × 5–6 mm., crowned by the persistent sepals. Seeds c. 0·75 mm. long, ridged.

Zambia. N: Mbala (Abercorn) Distr., Kambole area, 16 km. from Kambole-Mbala road, 1500 m., fl. & fr. 29.viii. 1956, *Richards* 5987 (COI; K; SRGH). W: Mwinilunga Distr., Lisombo R., fl. & fr. 8.vi.1963, *Loveridge* 858 (COI; SRGH). E: Lunkwakwa, Chitipa, fl. & fr. 28.ix.1966, *Mutimushi* 1494 (K; SRGH). C: Mkushi Hotel, fl. & fr. 5.viii.1959, *West* 4003 (LISC; SRGH). **Rhodesia.** N: Mazoe Distr., 1290 m., fl. & fr. i.1907, *Eyles* 509 (BM; SRGH). C: Marandellas, Digglefold, fl. & fr. 29.ii.1946, *Corby* 15 (COI; SRGH). E: Melsetter, Chimanimani, upper Bundi plain, fl. & fr. 1.ii.1957, *Phipps* 341 (COI; K; LISC; SRGH). S: Buhera, Sabi R., fl. & fr. iii.1954, *Davies* 708 (SRGH). **Malawi.** N: Nkhata Bay, Chombe Estate, fl. & fr. 6.ix.1955, *Jackson* 1747 (K; LISC). C: Dzalanyama Forest Reserve near Chiungiza, 1550 m., fl. & fr. 9.ii.1959, *Robson* 1526 (K; LISC). S: Mlanje Distr., Mchese Mt., c. 750 m., fl. & fr. 6.vi.1958, *Chapman* 583 (FHO; K; SRGH). **Mozambique.** N: Maniamba, fl. & fr. 27.viii.1934, *Torre* 526 (COI; LISC). Z: between Moemba and Mobede, fl. & fr. 23.v.1949, *Andrada* 1483 (COI; LISC). T: Marávia, between Fíngoè and Vila Vasco da Gama, 800–1000 m., fl. & fr. 13.viii.1941, *Torre* 3276 (COI; K; LISC; LMU; SRGH). MS: Báruè, Choa Mt., 21 km. from Vila Gouveia, 1400 m., fl. 28.iii.1966, *Torre & Correia* 15434 (BR; EA; LISC; P; PRE). LM: Peto forest, near Ponta do Ouro, fl. & fr. 27.xii.1948, *Gomes e Sousa* 3913 (COI; K; LISC; SRGH).

Widespread in tropical and S. Africa. In marshy situations in clearings, woodlands and forests mainly along streams.

8. **Dissotis hensii** Cogn. in A. & C. DC., Mon. Phan. 7: 372 (1891).—De Wild. & Dur. in Ann. Mus. Congo Belge, Bot. Sér. 1, 1: 18, t. 10 (1898).—Gilg in Engl., Mon. Afr. Pflanzen. 2: 19 (1898).—Engl., Pflanzenw. Afr. 3, 2: 751 (1921).—A. & R. Fernandes in Mem. Soc. Brot. 11: 34 (1956); in C.F.A. 4: 149 (1969). Type from Zaire.

A perennial herb up to 1 m. high with slender 4-gonous scabrid stem and branches; bristles sparse, short, appressed. Leaf-lamina 5·5–12·5 × 2–5·4 cm., ovate to oblong-ovate or lanceolate, acute at apex, round or subcordate at base, serrulate-setose, membranous, light green above, pale green beneath, sparsely setose on both faces, the bristles short and subappressed, longitudinally 3–5(7)-nerved, the nerves slender, impressed above, prominent below, the transverse nerves only slightly visible below; petiole 1–4(6) cm. long, slender, strigillose. Flowers sessile in terminal lax panicles; bracts shorter than the receptacle, ovate, acute, thin, pale green, dorsally setose, ciliate, caducous. Receptacle c. 4 × 2·5 mm., cylindric, sparsely setulose, the bristles short, appressed, scabrid. Sepals 3 × 1·5 mm., lanceolate, acute, setose, caducous; intersepalar segments short, setose at the top, caducous. Petals 7–8 mm. long, pink. Anthers of the longer stamens c. 4 mm. long; pedoconnective c. 3 mm. long with a c. 1 mm. long appendage; anthers of the shorter stamens c. 4 mm. long; pedoconnective 0·75 mm. long with a very short appendage. Fructiferous receptacle 6–8 × 4–5 mm., ovoid, rather longer than the capsule, glabrescent, the longitudinal and transverse nerves distinctly raised. Capsule subspherical, very sparsely setose at the apex, the bristles weak and short. Seeds c. 1 mm. long, indistinctly tuberculate.

Zambia. W: Mwinilunga Distr., Lisombo R., fr. 10.vi.1963, *Loveridge* 884 (COI; SRGH); Mwinilunga Distr., Ikelenge, fl. ii.1960, *Pinhey* 8A (COI; SRGH). Also in Angola and Zaire. In swampy ground in " mushitu ".

9. **Dissotis brazzae** Cogn. in A. & C. DC., Mon. Phan. 7: 372 (1891).—De Wild. & Dur. in Ann. Mus. Congo Belge, Bot. Sér. 1, 1: 29, t. 15 (1898).—Th. & H. Dur., Syll. Fl. Cong.: 208 (1909).—Jacques-Félix in Bull. Mus. Hist. Nat. Paris, Sér. 2, 7: 370 (1935).—Keay in F.W.T.A. ed. 2, 1: 258 (1954).—A. & R. Fernandes in Mem. Soc. Brot. 11: 34 & 85, t. 6 (1956); in C.F.A. 4: 148 (1970).—Wickens in F.T.E.A., Melastom.: 41, t. 9 fig. 1 (1975). Type from Gabon (Franceville).
Dissotis multiflora sensu Gilg in Engl., Mon. Afr. Pflanzen. 2: 18, t. 2 fig. F (1898) pro parte quoad syn. *D. brazzae* Cogn.—Engl., Pflanzenw. Afr. 3, 2: 751 (1921) pro eadem parte.—Hutch. & Dalz. in F.W.T.A. ed. 1, 1: 212 (1927) pro eadem parte.
Osbeckia multiflora sensu Hiern, Cat. Afr. Pl. Welw. 1, 2: 364 (1898) non Sm. (1813).

A many-stemmed perennial herb up to 1·5 m. high. Stems stout, scarcely branched, sharply 4-gonous or winged towards the apex, green, hirsute, glabrescent. Leaf-lamina 3·5–11 × 1·5–5 cm., ovate-cordate, acute, entire or minutely crenulate, membranous to somewhat rigid, dark green above, pale green beneath, sparsely setose on both faces with short, subappressed bristles, longitudinally 7–11-nerved, the nerves impressed above, prominent below, transverse nerves inconspicuous above, slightly visible beneath; petiole up to 3 cm. long but the upper leaves subsessile. Flowers sessile in dense cymes arranged in pyramidal sometimes very large or ± contracted panicles; bracts 3–5 mm. long, ovate-lanceolate, acute, green or pellucid, setose on the outer side, ciliate, caducous. Receptacle c. 6 mm. long, ovoid-oblong, green, shortly and patently setose (the bristles ± dense and spinulose) and with sparse elongate-cylindric patent caducous appendages, dilated at the apex into a multiradiate setose disk. Sepals 2·5–3 mm. long, triangular-subulate, green, ciliate at the margins and apex, early caducous; intersepalar segments like the receptacular appendages and also caducous. Petals c. 12·5 mm. long, pink-mauve. Longer stamens with anthers 7–8 mm. long, pedoconnectives 3·5–4 mm. long with appendages dilated towards the notched apex; shorter stamens with anthers 6–7 mm. long, pedoconnectives c. 1 mm. long with short bilobed appendages. Fructiferous receptacle 7–8 mm. long, oblong-urceolate, glabrescent, longitudinally 20-nerved, the nerves somewhat raised. Capsule ovoid, densely setose at the apex with the bristles long and rigid. Seeds c. 0·75 mm. long, indistinctly tuberculate.

Zambia. N: Kawambwa Distr., road to Nchelengi, 1200 m., fl. & fr. 20.iv.1957, *Richards* 9379 (COI; K). W: Mwinilunga Distr., Lisombo R., fr. 13.vi.1963, *Loveridge* 944 (K; LISC; SRGH).
From the Guinea Republic to Angola, Zaire, Uganda, Kenya and Tanzania. Open grassy woodlands.

10. **Dissotis lanata** A. & R. Fernandes in Bol. Soc. Brot., Sér. 2, **43**: 294, t.4 (1969).
Type: Malawi, Northern Prov., Mafinga Mt., *Robson* 533 (K, holotype; LISC).

Shrub up to 1·5 m. high. Stems erect, 6-gonous, at first covered with a very dense woolly tomentum, later becoming glabrous and corky, the bark longitudinally and transversely fissured. Branches thick, very nodose, with short or sometimes elongate internodes, the oldest ones corky like the stems, the younger ones, like the petioles, inflorescence-axis, pedicels, back of bracts, receptacle and sepals, densely floccose-tomentose (hairs long, thin, crisped). Leaves petiolate, disposed at the apex of the branches of the current year; lamina narrowly elliptic (4–7 × 1·4–2 cm.) or elliptic (up to 7·5 × 3·5 cm.) and subacute to subobtuse at the apex, or ovate (up to 11·5 × 5 cm.) and obtuse at the apex, round or cordate at the base, green and with conic projections each ending in a bristle on the upper face, very densely floccose-tomentose on the under side, longitudinally 5-nerved, the nerves impressed above, prominent below but concealed by the dense tomentum. Flowers 5-merous, arranged in dense umbel-like cymes at the apices of the branches; bracts c. 5 mm. long, triangular, reddish on the inner face, caducous; pedicels 4-gonous, 5–8 mm. long. Receptacle 11 × 7 mm., campanulate, slightly constricted below the sepals, longitudinally 10-ribbed (ribs concealed by the tomentum), floccose-tomentose, the hairs of the tomentum intermixed with short strong bristles dilated at the base. Sepals 6–7 × 4 mm., triangular, reddish within, ciliolate at margin, persistent, alternating with intersepalar seta-like segments dilated at the base and helicoidally curled at the apex. Stamens 10, very unequal. Longer stamens: anthers 11 mm. long, red; pedoconnective 1 cm. long, arched, appendiculate at the base (appendage c. 2 mm. long, 2-fid, with oblong and obtuse segments) and shortly spurred behind; filaments 16 mm. long. Shorter stamens: anthers 9 mm. long, yellow, pedo-connective c. 1 mm. long, appendiculate at base (appendage bilobed, the lobes c. 2 mm. long, oblong, obtuse and curved upwards); filaments 12·5 mm. long. Ovary 9 mm. long, 5-angular and setose above. Style 2·2 cm. long, red; stigma punctiform. Fructiferous receptacle subhemispherical, becoming glabrous, crowned with the persistent and reflexed sepals. Capsule somewhat exserted. Seeds c. 0·75 cm. in diameter, cochleate, with minute papillae arranged in curved lines.

Malawi. N: Mafinga Mt., fl. 26.viii.1958, *Lawton* 460 (K).
Known only from Northern Prov. of Malawi. In skeletal soils of the slopes of mountains in the upper limit of *Brachystegia* woodland.

11. **Dissotis caloneura** Gilg ex Engl., Pflanzenw. Afr. **3**, 2: 749 (1921).—A. & R. Fernandes in Bol. Soc. Brot., Sér. 2, **30**: 171, t. 3, 4 (1956); in Mem. Soc. Brot. **11**:

77 (1956); in Bol. Soc. Brot., Sér. 2, **34**: 59 & 185 (1960).—F. White, F.F.N.R.: 308 (1962).—Wickens in F.T.E.A., Melastom.: 42, t. 9 fig. 2 (1975). Type from Zaire.

A shrub or a small tree up to 3·5(4) m. tall. Stems with the bark soft, pale grey, fissured and coming off in strips when old; branches bluntly 4-gonous, sulcate along the faces, strigose (bristles with a bulbous base) towards the apex and subterete, sometimes very swollen and glabrescent below, leafy only for a rather short length below the inflorescence; internodes short. Leaves opposite, erect; leaf-lamina 3·5–11 × 1·5–6·2(9) cm., broadly ovate to ovate-lanceolate, cordate or roundish at the base, shortly acuminate at the apex, entire or crenulate, dull green to light green and minutely bullate above, pale green and reticulate-foveolate beneath, somewhat thin, ± bristly on both faces (bristles sparser and shorter beneath), rarely subglabrous, longitudinally 5–7(9)-nerved, with the longitudinal nerves, the transverse ones and the reticulation impressed on the upper surface, prominent on the lower one; petiole 0·7–2·5 cm. long, strigose. Flowers in terminal, leafy, pyramidal, contracted to lax panicles up to 14 × 7 cm.; pedicels 2·5–4 mm. long in flower, up to 8 mm. long in fruit; bracts ovate, c. 5 mm. long, very soon caducous. Receptacle 8–10(13) × 4–5(6) mm., cylindric, attenuate towards the base, sometimes slightly contracted at the apex, 10-ribbed, green suffused with purple. Sepals 3–5 × 4–5 mm., semicircular, red, ciliolate, persistent. Petals c. 2 × 2 cm. (rarely 3·5 × 2·8 cm.), purplish-mauve, ciliolate. Longer stamens: anthers 11–13(17) mm. long; pedoconnective 8–11(13) mm. long, with the appendage deeply bilobed (the lobes spathulate or cochleate) and a small posterior protuberance at the base. Shorter stamens: anthers 8–10(12) mm. long; pedoconnective 0·5–1 mm. long, also appendiculate at base. Fructiferous receptacle up to 15 × 9 mm., ellipsoid, 10-ribbed, crowned by the reflexed or patent sepals. Capsule equalling or slightly longer than the receptacle, shortly bristly on the upper ⅓.

Receptacle and sepals glabrous; pedicels setose at base - - - var. *caloneura*
Receptacle with short, weak, patent or ascending bristles; sepals with some bristles mainly
    along the median line; pedicels setose throughout - - - var. *pilosa*

Var. **caloneura**.
    Dissotis venulosa Hutch., Botanist in S. Afr.: 512 (1946). Type: Zambia near Mbala, Lake Chila, *Hutchinson & Gillett* 3887 (K, holotype).

Zambia. N: Misamfu, 6·4 km. N. of Kasama, fl. 4.iv.1961, *Angus* 2643 (FHO; K; SRGH); Kasama, fr. 11.xi.1952, *Angus* 743 (BM; COI; FHO; K); Abercorn Distr., Kambole Escarpment, 1500 m., fl. 4.vi.1957, *Richards* 9987 (COI; K); 10 km. E. of Kasama, fl. 6.iv.1962, *Robinson* 5077 (K; SRGH).
    Also in Zaire, Rwanda, Burundi and Tanzania. In exposed situations among rocks.

Var. **pilosa** A. & R. Fernandes in Kirkia **1**: 73, t. 7 (1961). Type: Zambia, Luanshya, *Fanshawe* 3118 (K, holotype).

Zambia. W: Luanshya, fl. 29.iii.1957, *Fanshawe* 3123 (K). C: Mkushi, fl. 27.iii.1961, *Angus* 2526 (FHO; K; SRGH).
    Known only from Zambia. Exposed situations on quartzite or sandstone rocks.

The specimen from Zambia, E: Lundazi, Kangampande Mt., Nyika Plateau, c. 2100 m., 6.v.1952, *White* 2736 (COI; FHO; K) is sterile, so it is impossible to know whether it belongs to var. *caloneura* or var. *pilosa*.

12. **Dissotis melleri** Hook. f. in F.T.A. **2**: 453 (1871).—Triana in Trans. Linn. Soc. Lond. **28**: 58 [1871] (1872).—Cogn. in A. & C. DC., Mon. Phan. **7**: 371 (1891).—Taub. in Engl., Pflanzenw. Ost-Afr. C: 295 (1895).—Gilg in Engl., Mon. Afr. Pflanzen. **2**: 18 (1898).—Engl., Pflanzenw. Afr. **3**, 2: 750 (1921).—A. & R. Fernandes in An. Junta Invest. Ultramar **10**, 3: 32 (1955); in Mem. Soc. Brot. **11**: 25 (1956); in Bol. Soc. Brot., Sér. 2, **30**: 172, t. 6, 7 (1956); op. cit. **34**: 185 (1960).—Wickens in F.T.E.A., Melastom.: 44 (1975). Type: Malawi, Chiradzulu, Manganja range, ix.1861, *Meller* (K, holotype).
    Dissotis whytei Bak. f. in Kew Bull. **1897**: 267 (1897). Type: Malawi, Zomba Mt., 1200–1818 m., *Whyte* (K, holotype).

A small tree up to 6 m. tall, with 1 or sometimes 2–3 main trunks, and a light open crown, or a large shrub, generally without leaves when in flower. Branches bluntly 4-gonous, ± deeply sulcate, scabrid (covered with small appressed bristles bulbous at base), thickened at nodes, the oldest glabrescent and with grey and irregularly fissured bark; internodes below the inflorescence usually very short. Leaves opposite; lamina 4·5–16 × 1·5–10 cm., ovate-lanceolate, acute at apex, roundish or somewhat acute at base, entire, chartaceous, discolorous (light to dark green above, yellow-green beneath), longitudinally 5–7-nerved, the longitudinal and transverse nerves and the reticulation impressed above making the upper face slightly bullate, reticulation prominent below; upper face sparsely to densely appressed-setose, the lower also setose, the bristles short, bulbous at base and very dense mainly on the reticulation (those on the nerves up to 3·5 mm. long); petiole up to 2·5 cm. long, strigose. Flowers 5-merous, in terminal usually short and very compact panicles, leafy at the base; bracts 3–6 × 2·5–3·5 mm., ovate or lanceolate, acute, densely setose outside, soon caducous; pedicels 2–3 mm. long. Receptacle 7–12 × 5–8 mm., cylindric-campanulate, round at the base, green to purplish, with ± sparse patent or ascending bristles, these simple or 2–3-branched from the base, usually inserted on short protuberances, sometimes red-capitate-glandular. Sepals 3–5 × 5–6 mm., semicircular or broadly ovate, oblique and apiculate at the top, glabrous, ciliate at margin, persistent; intersepalar segments absent. Petals up to 2·5 × 2 cm., obovate, ciliate, purple. Longer stamens: anthers 9–12 mm. long; pedoconnective 16–18 mm. long with the appendage deeply bilobed (lobes obtuse) and with a very short posterior spur at the base; filaments 14–15 mm. long. Shorter stamens: anthers 7–9 mm. long; pedoconnective 1·5–2·5 mm. long, appendiculate at base (appendage shortly bilobed); filaments 12 mm. long. Ovary ellipsoid, setose at the apex. Seeds shortly papillose-tuberculate, the papillae arranged in curved lines.

Receptacle with bristles not capitate-glandular; reticulation of the leaf underside obscure
- - - - - - - - - - var. *melleri*

Receptacle with at least some bristles capitate-glandular, the glands red; reticulation of the leaf underside prominent - - - - - - - - var. *greenwayi*

Var. **melleri**.

Zambia. E: Nyika Plateau, near Govt. Rest House, fl. 24.ix.1950, *Benson* N.R. 172 (BM). **Malawi.** N: below western boundary of Mugesse, fl. 1.ix.1952, *Chapman* 3 (FHO). C: Dedza, fl. 3.vi.1961, *Chapman* 1345 (COI; SRGH). S: Zomba Plateau, *Clements* 666 (FHO). **Mozambique.** N: Moçambique Distr., Ribáuè Mt., Mepalué, 1350 m., fr. 25.i.1964, *Torre & Paiva* 10253 (COI; K; LISC). T: Zóbuè Mt., fl. 3.x.1942, *Mendonça* 592 (COI; K; LISC; SRGH; WAG).
Known also from Tanzania.

Var. **greenwayi** (A. & R. Fernandes) A. & R. Fernandes in Bol. Soc. Brot., Sér. 2, **46**: 68 (1972).—Wickens in F.T.E.A., Melastom.: 46, fig. 13 (1975). TAB **61**. Type from Tanzania.
Dissotis greenwayi A. & R. Fernandes in Bol. Soc. Brot., Sér. 2, **30**: 172, t. 8 (1956); op. cit. **34**: 185 (1960); in Mem. Soc. Brot. **11**: 25 & 97 (1956). Type as above.

Zambia. E: Nyika Plateau, fl. 19.ix.1960, *Coxe* 33 (SRGH). **Malawi.** N: Walindi Forest, Misuku Hills, 2000 m., fl. 12.xi.1958, *Robson* 582 (K; LISC). C: Dedza, Chongoni Forest, fl. 24.ix.1960, *Chapman* 925 (FHO; LISC; SRGH).
Also in Tanzania. Exposed situations on cliffs, mountain slopes in high grassland and rock crevices.

13. **Dissotis johnstoniana** Bak. f. in Trans. Linn. Soc. Lond., Ser. 2, **4**: 14, t. 2 fig. 13–17 (1894).—Taub. in Engl., Pflanzenw. Ost-Afr. C: 295 (1895).—Gilg in Engl., Mon. Afr. Pflanzen. 2: 16 (1898).—Engl., Pflanzenw. Afr. **3**, 2: 749 (1921).—Brenan in Mem. N.Y. Bot. Gard. **8**: 440 (1954).—A. & R. Fernandes in Mem. Soc. Brot. **11**: 24 & 76 (1956); in Kirkia **1**: 72 (1961).—Chapman, Veg. Mlanje Mt. Nyasal.: 47, 54 (1962).—Binns, H.C.L.M.: 68 (1968). Type from Malawi, Mlanje Mt., *Whyte* 74 (K, holotype).

A much-branched straggling shrub 1·5–3 m. high. Main stems up to 10 cm. in diameter at the base; bark grey, falling off in thin papery flakes; branches bluntly 4-gonous, sulcate and striate along the faces, light brown and glabrescent proximally, purplish and setose distally, leafy only along a rather short length below the inflorescence; nodes somewhat thickened, long-setose. Leaves opposite; lamina 5·5–14 × 2·7–7 cm., ovate to ovate-lanceolate, contracted into an acute acumen or attenuate towards the apex, cordate, subcordate, rounded or rarely acute at the base,

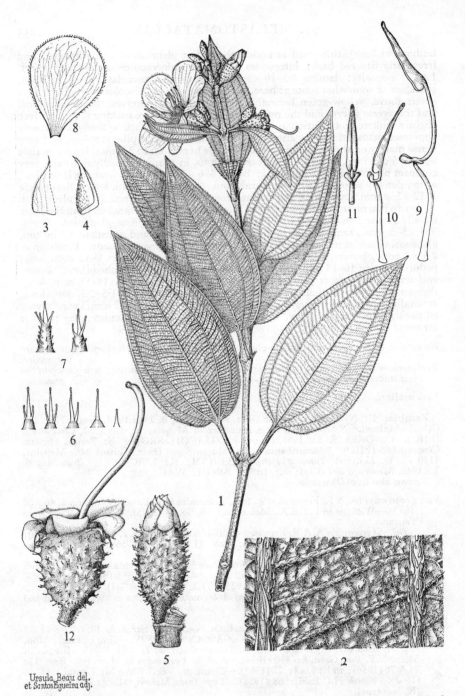

Ursula Beau del.
et Santos Figueira adj.

Tab. 61. DISSOTIS MELLERI var. GREENWAYI. 1, flowering branch (×½); 2, part of leaf underside (×6); 3, ½ of a bract seen from inside (×2); 4, bract. lateral view (×2); 5, flower-bud (×2); 6, scales and bristles of the lower and median parts of receptacle (×6); 7, appendages of upper part of receptacle (×6); 8, petal (×1); 9, long stamen in lateral view (×2); 10, short stamen in lateral view (×2); 11, upper part of short stamen in anterior view (×2); 12, receptacle after petals and stamens are shed (×2), all from *Greenway* 8410.

minutely denticulate-ciliate at margin, dark green to purplish and rough and ±
setose above, light- to yellowish-green and reticulate and more sparsely setose
beneath, longitudinally 5–7-nerved, the longitudinal and transverse nerves and the
reticulation impressed above, raised beneath; petiole up to 2·5(3) cm. long, strigose.
Flowers 5-merous, in terminal pyramidal panicles, up to 24 × 20 cm., leafy at the
base; bracts 6–7 × 6–7 mm., ovate, acute, purple, very soon caducous; pedicels
c. 2 mm. long in flower and up to c. 7 mm. long in fruit. Receptacle 6–9 × 5–8 mm.,
campanulate, roundish at the base, often purple. Sepals 6–8 × 4–6 mm., ciliate at
margin, persistent, reflexed in fruit; intersepalar segments absent. Petals mid- to
deep-blackish-purple. Longer stamens: anthers 8–10 mm. long, not much
attenuate at the apex; pedoconnective c. 10 mm. long, with the appendage
1–1·5 mm. long and deeply 2-lobed (lobes obtuse). Shorter stamens: anthers
5–8 mm. long; pedoconnective c. 0·75 mm. long with a similar appendage at base.
Fructiferous receptacle 10–14 × 9–10 mm. Capsule subequalling the receptacle,
setose towards the apex. Seeds shortly papillose-tuberculate.

Receptacle and sepals glabrous; sepals 6–7 × 4–6 mm., subsemicircular or shortly oblong,
    round or oblique at the apex; inflorescence usually lax, up to 24 × 20 cm.; reticulation
    of leaf-lamina loose -   -   -   -   -   -   -   var. *johnstoniana*
Receptacle and sepals ± densely strigose with simple long white bristles, all inserted on
    the epidermis or the uppermost at the top of short to slightly elongate prominences;
    sepals c. 8 × 4·5 mm., oblong, acute; inflorescence usually shorter and denser than
    above; reticulation of leaf-lamina generally somewhat closer -   -   var. *strigosa*

Var. **johnstoniana**.

Malawi. S: Mlanje Mt., Nayawani Forest, fl. & fr. 23.viii.1956, *Newman & Whitmore*
535 (BM; SRGH). **Mozambique. Z**: Gúruè Mt., between Nuirre and Loloè, 2100 m.,
fl. 24.ix.1944, *Mendonça* 2277 (COI; LISC).
Known only from Malawi and Mozambique. Slopes of high mountains either in rocky
places, or in grassy edges of forests, or in mixed evergreen undergrowths.

Var. **strigosa** Brenan in Mem. N.Y. Bot. Gard. **8**: 440 (1954).—A. & R. Fernandes in
Mem. Soc. Brot. **11**: 25 (1956).—Chapman, Veg. Mlanje Mt. Nyasal.: 54, 58
(1962).—Binns, H.C.L.M.: 68 (1968). Type: Malawi, Mlanje Mt., Chambe
Plateau, 2100 m., fl. & fr. 9.vii.1946, *Brass* 16757 (K, holotype; SRGH).

Malawi. S: Mlanje Mt., Thuchila plateau, fl. 21.vii.1956, *Newman & Whitmore* 100
(BM; COI; LISC; SRGH).
Only on Mlanje Mt. Ecology as for var. *johnstoniana*.

Intermediates between the type variety and var. *strigosa* occur, e.g. *Brass* 16861 (K;
SRGH).

14. **Dissotis cryptantha** Bak. f. in Kew Bull. **1894**: 345 (1894).—Gilg in Engl., Mon.
Afr. Pflanzen. **2**: 17 (1898).—Engl., Pflanzenw. Afr. **3**, 2: 750 (1921).—A. & R.
Fernandes in Mem. Soc. Brot. **11**: 83 (1956); in Bol. Soc. Brot., Sér. 2, **34**: 186
(1960); op. cit. **43**: 296 (1969).—Wickens in F.T.E.A., Melastom.: 56 (1975).
Type: Malawi, *Buchanan* 625 (BM; K, holotype).
    Dissotis spectabilis Gilg in Engl., Bot. Jahrb. **30**: 366 (1901).—Engl., loc. cit.—
Brenan & Greenway, T.T.C.L.: 310 (1949).—A. & R. Fernandes in An. Junta
Invest. Ultramar **10**, 3: 35 (1955); in Mem. Soc. Brot. **11**: 32 (1956). Type from
Tanzania.

A much-branched shrub up to c. 2 m. tall. Young branches 4-gonous, ± densely
yellowish-brown-hairy, the hairs short to ± long, patent. Leaves opposite,
subsessile or shortly petiolate, the petiole up to 3 mm. long, setose. Leaf-lamina
up to 4·7 × 1·4 cm., oblong to elliptic- or ovate-oblong, acute or obtuse and
apiculate or sometimes folded at apex, roundish at the base, entire, submembranous
to papyraceous, concolorous or of a lighter green beneath, sericeous on both faces
but more densely so beneath, the indumentum of the oldest leaves sparser and
shorter, longitudinally 3–5-nerved, the nerves impressed above, prominent below,
the transverse nerves and reticulation neither raised nor visible. Flowers 5-merous,
surrounded by the uppermost leaves, solitary or in 3-flowered cymes, forming ±
dense terminal inflorescences; pedicels very short or absent; bracts absent.
Receptacle 6–8 × 5–6 mm., subhemispheric or subcampanulate, densely whitish-
yellow sericeous above, the bristles inserted at the roundish top of short flat
appendages. Sepals 7–13 × 2·5–3·5 mm., somewhat broadened towards the blunt

I

or notched apex, outside sericeous, inside glabrous and red, late deciduous; intersepalar segments 2·5–4 mm. long, rectangular to spathulate, sericeous, stiff-bristly at apex. Petals c. 3 × 2·5 cm., violet. Longer stamens: anthers 9–11 mm. long; pedoconnective 15–18 mm. long with the appendage c. 1·5 mm. long and filaments 8 mm. long; shorter stamens: anthers 8 mm. long; pedoconnective c. 2 mm. long with the appendage 1 mm. long and filaments 8 mm. long. Fructiferous receptacle c. 7 × 7 mm., subspherical, contracted at the mouth; capsule shorter than the receptacle, spherical, setose on the upper part and with a crown of short setae at the apex.

Zambia. N: Mbala (Abercorn) Distr., Lumi R. near Customs Post, 1680 m., fl. 12.vi.1957, *Richards* 10089c (COI; K). **Malawi.** N: Lusangadzi Forest, fl. 15.v.1958, *Chapman* 585 (FHO). S: Mlanje, Likabula Forest Station, 810 m., fl. & fr. 18.v.1958, *Chapman* 572 (FHO). **Mozambique.** N: Unango, fl. 26.v.1948, *Pedro & Pedrógão* 3933 (COI; LMA); Metónia, Massangulo Mt., fl. & fr. 12.x.1942, *Mendonça* 789 (COI; K; LISC; SRGH). Z: Shire Highlands, Zambesiland, fl. 6.iv.1906, *Adamson* 69 p.p. (K). Also in Tanzania. Swampy plains and along rivers.

15. **Dissotis speciosa** Taub. in Engl., Pflanzenw. Ost-Afr. **C**: 295 (1895).—Gilg in Engl., Mon. Afr. Pflanzen. **2**: 18, t. 3C (1898).—Engl., Pflanzenw. Afr. **3**, 2: 750 (1921).—Wickens in F.T.E.A., Melastom.: 57 (1975). Type from Uganda.
    Dissotis macrocarpa Gilg in Engl., Mon. Afr. Pflanzen. **2**: 18 (1898).—Th. & H·Dur., Syll. Fl. Cong.: 211 (1909).—Engl., Pflanzenw. Afr. **3**, 2: 750 (1921).—Brenan & Greenway, T.T.C.L.: 309 (1949).—A. & R. Fernandes in Mem. Soc. Brot. **11**: 32 & 83 (1956); in Bol. Soc. Brot., Sér. 2, **34**: 187 (1960); in Kirkia **1**: 74 (1961). Type from Uganda.
    Dissotis helenae Busc. & Muschl. in Engl., Bot. Jahrb. **49**: 479 (1913).—Gilg in Engl., Bot. Jahrb. **53**: 371 (1915). Type: Zambia, Northern Prov., between Lakes Bangweulu and Tanganyika, *von Aosta* 1346 (B†, holotype).

Similar to *D. cryptantha* Bak. f. Differs by: leaves up to 7·5 × 2 cm., usually not so densely sericeous; flowers larger, surrounded by the upper leaves and by ovate bracts; floriferous receptacle 10–12 × 9–12 mm.; sepals up to 15 mm. long, more constricted at the base; intersepalar segments up to 6 mm. long; petals 3–3·5 × 2·5–2·7 cm.; anthers of the longer stamens c. 12–13 mm., those of the shorter ones c. 9–11 mm. long; sericeous hairs of the receptacle longer (up to 10 mm. long), all inserted on flat appendages, successively longer from the base to the apex of the receptacle; fructiferous receptacle up to 12–13 mm. long.

Zambia. N: Mbala, Saisi R., 1500 m., fl. & fr. 15.iv.1959, *Richards* 11218 (K; SRGH). E: Nyika Plateau, 1800–2100 m., vi.1846, *Whyte* (K). **Malawi.** N: Kaningina, Lusangadzi dambo, c. 1500 m., fl. & fr. v.1954, *Chapman* 295 (BM; SRGH). Also in Sudan, Zaire, Uganda, Kenya and Tanzania. In marshy places mainly along rivers.

The following specimens: **Zambia.** N: Lake Chila, 1500 m., fl. 15.iii.1955, *Richards* 4936 (K); Mporokoso Distr., near Mporokoso, 1050 m., fl. 27.iv.1957, *Richards* 9458 (COI; K; SRGH) are somewhat intermediate between *D. speciosa* and *D. pachytricha*. The receptacle-protuberances are like those of *D. speciosa* but not so flat and with the bristles not so long and silvery; the sepals are more like those of *D. pachytricha* but are hairy on the back; the leaves are also more like those of *D. pachytricha*, but are provided with sparse short hairs; etc. Thus we consider these specimens as probable hybrids between the two species.

16. **Dissotis simonis-jamesii** Busc. & Muschl. in Engl., Bot. Jahrb. **49**: 481 (1913).—Gilg, loc. cit.—F. White, F.F.N.R.: 306 (1962).—A. & R. Fernandes in Bol. Soc. Brot., Sér. 2, **46**: 68 (1972). Type: Zambia, Northern Prov., Lake Bangweulu, 1500 m., *von Aosta* 927 (B†, holotype).
    Dissotis degasparisiana Busc. & Muschl., tom. cit.: 480 (1913).—Gilg, loc. cit.—A. & R. Fernandes in Mem. Soc. Brot. **11**: 84 (1956); in Kirkia **1**: 74 (1961). Type: Zambia, L. Bangweulu, *von Aosta* 1135 (B†, holotype).

A shrub or shrublet 0·5–2 m. tall. Branches 4-gonous, very densely leafy towards the apex and ± sericeous with long very thin white hairs, leafless and glabrescent proximally; bark brownish or grey, ± fissured. Leaves opposite, subsessile to shortly petiolate. Leaf-lamina 2–6 × 0·7–1·6 cm., ovate to ovate-lanceolate or elliptic, acute at the apex, contracted at the base, entire, folded along the midrib or rarely flat, arched and reflexed, rigid, concolorous, sometimes reddish, ± sericeous (hairs white, shining, very thin, long, straight, denser towards the lamina-base), longitudinally 5–7-nerved, the longitudinal nerves slightly raised beneath, the transverse nerves and reticulation inconspicuous on both faces; petiole ± absent to 4 mm. long, sericeous. Flowers 5-merous, solitary at the apices of branches, surrounded by the uppermost rather broad leaves and by large, sub-circular, apiculate, ± scarious, purple or brown, caducous bracts sericeous outside; pedicels c. 3 mm. long. Receptacle c. 12 × 10 mm., subhemispherical, densely sericeous, the long thin white hairs inserted on appressed ± flat appendages, successively longer from the base to the receptacle summit (here c. 3 mm. long). Sepals ± 17 × 5 mm., linear-oblong or oblong, contracted at the base, somewhat acute, dorsally sericeous, late caducous; intersepalar segments c. 5 mm. long, spathulate, sericeous on both faces and tipped with long bristles. Petals 4–5 × 3–3·7 cm., bright magenta to purple. Longer stamens: anthers 13–16 mm. long; pedoconnective c. 19 mm. long with the basal appendage 3–5 mm. long. Shorter stamens: anthers 11–13 mm. long, pedoconnective c. 4 mm. long, with basal appendage c. 3 mm. long. Fructiferous receptacle c. 15 mm. in diameter, subspherical. Capsule c. 15 mm. in diameter, globose.

Zambia. N: 72 km. S. of Mbala (Abercorn) on Kasama road, fl. 30.iii.1955, *E. M. & W.* 1336 (BM; SRGH).

Known only in the Northern Province of Zambia. In swamps.

17. **Dissotis gilgiana** De Wild. in Ann. Mus. Congo Belge, Bot. Sér. 4, **1**: 217 (1903).—A. & R. Fernandes in Bol. Soc. Brot., Sér. 2, **28**: 186, t. 6 (1954); in Kirkia **1**: 74 (1961) non Hutch. & Dalz. (1927). Type from Zaire.
    Dissotis wildemaniana Gilg ex Engl. in Engl., Pflanzenw. Afr. **3**, 2: 749 (1921). Type from Zaire.

A perennial herb up to 45 cm. tall (usually shorter) with numerous stems forming a ± dense clump arising from a woody rootstock. Stems upright, simple or few-branched, distinctly 4-gonous, green, sparsely setose with short bristles bulbous at the base; internodes up to 10 cm. long. Leaves opposite, scattered; lamina 1·5–3·5 × 0·8–3 cm., elliptic or ovate to subcircular, attenuate at the base into an indistinct petiole or contracted into a petiole up to 8 mm. long, acute or obtuse at the apex, crenulate at the margin, each tooth ending in a falcate bristle, a little darker or reddish-green above, light green beneath, somewhat rigid, sparsely appressed-setose (bristles short, whitish) on both faces (more densely so above), longitudinally 3–5(7)-nerved, the longitudinal nerves slightly impressed above, prominent beneath, the transverse nerves and the reticulation usually incon-spicuous. Flowers 5–6(7)-merous, in 3-flowered cymes, rarely solitary, arranged in subcorymbiform inflorescences or in leafy panicles up to 10 cm. long; pedicels 0·5–2 mm. long; bracts lanceolate, acute, scarious, whitish-pink to red, ciliolate, tardily caducous. Receptacle c. 6–8 mm. long, subhemispherical or subcam-panulate, green to deep red, appressed-setose (bristles sparse to somewhat dense, short, whitish). Sepals 5–7 × 2·5–3·5 mm., oblong-lanceolate, obtuse or acute, red, dorsally appressed-setose, ciliolate at margin, caducous; intersepalar segments 2·5 mm. long, shortly triangular, bristle-tipped. Petals c. 2 cm. long, pale mauve to bright pink. Longer stamens: anthers 6–7 mm. long, bright yellow; pedo-connective 2–2·5 mm. long with the basal appendage deeply bilobed, the lobes curved upwards. Shorter stamens: very similar to the longer ones in colour and shape but a little smaller and with a pedoconnective c. 1 mm. long. Fructiferous receptacle 6–8 × 7–8 mm., subhemispherical. Capsule slightly shorter than the receptacle, spherical, setose at the apex.

Zambia. W: Mwinilunga, Lunga R., fl. 23.xi.1937, *Milne-Redhead* 3366 (K); Mwinilunga, 16 km. W. of Kakomo, fl. & fr. 29.ix.1952, *Angus* 571 (BM; COI; FHO; K).

Also in Katanga. In dambos, vegetation fringing riparian " mushitus " and sometimes in *Brachystegia-Julbernardia* woodlands.

18. **Dissotis gracilis** Cogn. in A. & C. DC., Mon. Phan. **7**: 366 (1891).—Gilg in Engl., Mon. Afr. Pflanzen. **2**: 14 (1898).—Engl., Pflanzenw. Afr. **3**, 2: 748 (1921).—Th. & H. Dur., Syll. Fl. Cong.: 210 (1909).—A. & R. Fernandes in Mem. Soc. Brot. **11**: 22 (1956); in Kirkia **1**: 71 (1961); in C.F.A. **4**: 141 (1970). Type from Angola.

A perennial herb up to c. 80 cm. tall. Stems several, erect, straight, simple or few-branched, slender, 4-gonous, green, turning dark red with age, glabrous or sparsely hairy (the hairs short, appressed); internodes rather long. Leaves opposite, sessile; lamina 1·5–4·7(5·7) × 0·2–1 cm., linear-oblong or linear, the lower ones ovate or elliptic, all acute or the lowermost and median obtuse or suboptuse, attenuate at base, entire or indistinctly crenulate, somewhat rigid, concolorous (yellow-green) or slightly discolorous (sometimes reddish above with age), glabrous or sparsely appressed-setose (bristles weak, short) mainly along the nerves beneath, longitudinally 3-nerved, the longitudinal nerves slender, impressed above, prominent beneath, the transverse nerves and reticulation inconspicuous. Flowers 4(5)-merous, solitary or 3–several in terminal or axillary (in uppermost axils) cymes, sometimes forming lax terminal narrow panicles up to 22 cm. long, usually shorter; primary and secondary peduncles longer or subequalling the axillary leaf, slender, dark purple; bracts 4–9 mm. long, ovate-lanceolate, acute, red, glabrous, ciliate at the margin, caducous; pedicels c. 2 mm. long. Receptacle 4·5–6 × 3–4 mm., campanulate, sparsely setose (the bristles short, slender, white, appressed). Sepals 5·5–7 × 2–3 mm., triangular-lanceolate or oblong, acute or obtuse, rose-purplish, dorsally appressed-pilose, ciliate at the margin, deciduous; intersepalar segments 3–4 mm. long, linear-subulate, ending in a bristle. Petals 1–1·8 cm. long, narrowly obovate, mauve to purple-violet. Long stamens with anthers 4–5 mm. long, pedoconnective 6–7 mm. long, the anterior appendage 2–3 mm. long, ± deeply 2-lobed or 2-fid; short stamens with anthers 3–4 mm. long, pedoconnective c. 1 mm. long, minutely 2-tuberculate at the base. Fructiferous receptacle slightly contracted below the mouth, glabrescent. Capsule ovoid, equalling the receptacle, with a crown of bristle-tipped scales at the apex.

Zambia. B: Mongu Distr., Kande plain, fr. 20.vii.1961, *Angus* 3003 (K). N: Mporokoso Distr., near Mwita R., 1350 m., fl. & fr. 1.vi.1957, *Richards* 8961 (COI; K). W: Mwinilunga, Sinkabolo dambo, fl. ii.1938, *Milne-Redhead* 3770, 3770A (K).
Also in Angola and Zaire. In swamps along rivers.

19. **Dissotis welwitschii** Cogn. in A. & C. DC., Mon. Phan. **7**: 371 (1891).—Gilg in Engl., Mon. Afr. Pflanzen. **2**: 18 (1898).—Hiern, Cat. Afr. Pl. Welw. **1**, 2: 367 (1898).—Engl., Pflanzenw. Afr. **3**, 2: 750 (1921).—A. & R. Fernandes in C.F.A. **4**: 150 cum photogr. (1970). Type from Angola.

A perennial herb, 0·3–1·0 m. high, glabrous throughout. Stems numerous, erect, upright, unbranched, 4-gonous or slightly 4-winged, densely leafy, pale green to reddish. Leaves opposite, sessile, upright; lamina 4–12·5 × 1–4·5 cm., ovate-lanceolate to oblong-lanceolate, acute or suboptuse at the apex, round or subcordate at the semi-amplexicaul base, entire, papyraceous, somewhat discolorous, shining and light green above, yellowish-green below, often turning reddish with age, longitudinally 3–5-nerved, the longitudinal nerves subimpressed above, prominent beneath, the transverse nerves and reticulation inconspicuous. Flowers 5-merous, in terminal pyramidal lax to somewhat dense panicles up to 20 cm. long; pedicels 2–10 mm. long; bracts 14–18 × 12–18 mm., broadly ovate, concave, emarginate, purple, scarious, caducous. Receptacle 8–9 × 4–5 mm., cylindric, dark red or purple. Sepals 6–7 × 2–3 mm., ovate-oblong, truncate, emarginate or bilobed at the apex, very thin, purplish, caducous; intersepalar segmencs absent. Petals c. 2 × 1·6 cm., obovate, bright pink to purplish-violet. Longer stamens: anthers 9–10 mm.; pedoconnective 9–11 mm. long. Shorter stamens: anthers c. 6 mm. long; pedoconnective c. 1 mm. long. Fructiferous receptacle 8–9 × 5·5–7 mm., urceolate, contracted above the capsule, usually very dark purple. Capsule subspherical. Seeds minutely tuberculate.

Zambia. N: Luwingu, fl. & fr. 15.v.1958, *Angus* 1947 (K; LISC).
Also in Angola. Marshy places in clearings and evergreen " mushitu ".

20. **Dissotis anchietae** A. & R. Fernandes in Bol. Soc. Brot., Sér. 2, **28**: 189, t. 10 (1954); in C.F.A. **4**: 150, t. 8 (1970). Type from Angola.

Similar to *D. welwitschii* (19) but stems, leaves, receptacle and sepals with sparse, simple, ± long and appressed bristles; leaf-lamina up to 11·5 × 1·5 cm., linear-lanceolate to lanceolate; inflorescences rather smaller, with fewer flowers; floriferous receptacle slightly larger, up to 10 × 6 mm.; sepals also larger, c. 10 × 4 mm.; intersepalar segments present, 4 mm. long, linear, setose at the apex; anthers, pedoconnectives and respective appendages longer; fructiferous receptacle somewhat larger, up to 12 × 8·5 mm.; capsule also larger.

Zambia. N: Luwingu, fr. 29.v.1964, *Fanshawe* 8718 (FHO; K).
Also in Angola. Marshy places.

21. **Dissotis romiana** De Wild., Comp. Kasai: 375 (1919).—A. & R. Fernandes in Bol.
Soc. Brot., Sér. 2, **30**: 175, t. 10 (1956); in Kirkia **1**: 75 (1961).—F. White,
F.F.N.R.: 308 (1962). Type from Zaire.

A shrub or small tree up to c. 3 m. high. Young branches striate, scabrous with
minute brownish echinulate scales, the oldest greyish, glabrescent. Leaves
opposite, petiolate; lamina up to 18 × 6 cm., ovate-lanceolate to ovate, very
attenuate and acute at the apex, round or subcordate at the base, crenulate at the
margin, somewhat discolorous, light or dark green and with short slightly raised
white lines ending in a minute pectinate bristle on the upper face, pale- or yellowish-
green and with pectinate or flabellate bristles beneath, longitudinally 5–7-nerved,
the longitudinal nerves impressed above, very prominent (especially the midrib)
and densely scaly (scales minutely echinulate-setulose) beneath, the transverse
nerves and the reticulation inconspicuous above, slightly visible beneath; petiole
up to 4·5 cm. long, rough-scaly. Flowers 5-merous, in panicles up to 20 × 12 cm.,
terminal, pyramidal, somewhat lax; pedicels 1–2 mm. long in flower. Bracts
12–17 × 8–9 mm., ovate to oblong-ovate, acuminate, purplish, shortly setose,
caducous. Receptacle 10–12 × 5 mm., cylindric, densely scaly, the scales very
minute, appressed, echinulate. Sepals 9–10 mm. long and c. 3 mm. wide at base,
lanceolate, attenuate, acute, pink or purple, scaly like the receptacle, caducous;
intersepalar segments linear, dilated at the apex into a stellate-setose disk, caducous.
Petals c. 2 cm. long, purple. Longer stamens: anthers 11–14 mm. long;
pedoconnective c. 22 mm. long with the bifid appendage c. 1 mm. long. Shorter
stamens: anthers 10–12 mm. long; pedoconnective c. 2·5 mm. long with the
appendage less than 1 mm. long. Fructiferous receptacle 12–13 × 8 mm., urceolate,
somewhat contracted above the capsule. Capsule c. 6 mm. in diameter, subspherical, appressed-setose, the bristles short, simple. Seeds c. 0·5 mm. long,
papillose.

Zambia. N: Mbala (Abercorn) Distr., Kambole area, Calongora R., 1050 m., fl. & fr.
25.viii.1956, *Richards* 5972 (COI; K; SRGH). W: Mwinilunga Distr., Lunga R.,
9·6 km. W. of Kakomo, fl. & fr. 30.ix.1952, *Angus* 575 (COI; FHO; K).
Also in Zaire. In swampy forests along rivers.

22. **Dissotis trothae** Gilg in Engl., Mon. Afr. Pflanzen. **2**: 19, t. 2 fig. B (1898).—Engl.,
Pflanzenw. Afr. **3**, 2: 751 (1921).—A. & R. Fernandes in Mem. Soc. Brot. **11**: 35 &
86 (1956); in Bol. Soc. Brot., Sér. 2, **34**: 167 & 188 (1960); in Kirkia **1**: 75 (1961).—
F. White, F.F.N.R.: 310 (1962).—Wickens in F.T.E.A., Melastom.: 55 (1975).
Type from Burundi.
Dissotis mildbraedii Gilg in Mildbr., Wiss. Ergebn. Deutsch. Zentr.-Afr. Exped.
1907–1908, **2**: 583 (1913). Type from Burundi.
Dissotis grandiceps Kraenzlin in Viert. Nat. Ges. Zurich **76**: 149 (1931). Type
from Zaire.

A tall perennial herb or shrub 0·9–3·6 m. high. Floriferous branches terete,
brownish and scaly (scales minute, dilated and spinulose at the base, fringed at
margins and shortly setose at top) proximally, sulcate, pale green or yellow and also
scaly (the scales prolonged into ± patent bristles) apically. Leaves 3-nate; lamina
3–15·5 × 1·6–7 cm., ovate-lanceolate or ovate, cordate or subcordate at base,
attenuate, acute or subobtuse at the apex, serrulate at the margin (teeth prolonged
from the base by branched bristles), very discolorous (light to dull green above,
pale- to yellowish-green beneath), covered on the upper face with short subconic
prominences, each provided with raised whitish lines converging from base to
summit into a simple or sometimes basally branched bristle, foveolate-reticulate
and soft-pubescent (hairs rather long, thin, white, subappressed) beneath,
longitudinally 5–7-nerved, the longitudinal nerves impressed above, prominent and
scaly beneath (scales ovate or ovate-lanceolate, setose at apex), the transverse
nerves and reticulation raised on the under face; petiole 0·5–3 cm. long,
subcylindric, scaly. Flowers 5-merous, in terminal, very compact panicles; bracts
12–13·5 × 7–9 mm., ovate, scarious, purple inside, densely and appressedly white-
scaly outside, caducous; pedicels up to 2 mm. long. Receptacle 9–11 × 4·5–6·5 mm.,
oblong-cylindric, densely setose, the bristles 1–2·5 mm. long, flat at the base, white,

usually divided. Sepals 8–9 × 3–4 mm., oblong-lanceolate, glabrous and purple inside, densely setose outside, with long bristles at the apex; intersepalar segments 2–3 × 1–1·5 mm., united with the sepals by a circumscissile, caducous receptacle rim. Petals c. 3 × 2·5 cm., obovate, mauve to bright purple, ciliate at margin. Longer stamens: anthers 10–14 mm. long; pedoconnective 18–20 mm. long with the bifid basal appendage curved c. 1 mm. long. Shorter stamens: anthers 10–13 mm. long; pedoconnective 2·5–4 mm. long, 2-appendiculate at the base. Fructiferous receptacle 10–11 × 6–7·5 mm., very constricted and prolonged above the capsule by the 5-undulate-dentate rim. Capsule subspherical, 6–7 mm. in diameter, or widely ovoid, c. 8 × 6 mm., appressed-setose on the upper ⅓. Seeds minutely tuberculate.

Zambia. N: Mbala (Abercorn), Lusakafu (Lungu), 1650 m., fl. & fr. 23.vi.1950, *Bullock* 2950 (EA; K; SRGH).
Also in Zaire, Rwanda, Burundi, Uganda and Tanzania. In marshy places of dense or open bushy woodlands and forests mainly on the banks of lakes and rivers.

23. **Dissotis pulchra** A. & R. Fernandes in Bol. Soc. Brot., Sér. 2, **29**: 54, t. 8 (1955); in Mem. Soc. Brot. **11**: 87 (1956). Type: Melsetter, Chimanimani Mts., *Plowes* 1216 (COI, holotype; SRGH).

Similar to *D. trothae* (22), but differing in the looser and shorter indumentum of stems, branches and leaves; in the presence of an under layer of irregular short orange, usually capitate hairs in the indumentum of the inferior leaf surface; in the longer and looser panicles; in the rather shorter usually not divided bristles of the receptacle; and in the fructiferous receptacle not prolonged and only slightly constricted above the capsule.

Rhodesia. E: Melsetter, " The Corner ", fl. 21.ix.1965, *West* 6813 (COI; SRGH).
Mozambique. MS: Valley of Wizards, 1650 m., fl. 29.viii.1964, *Whellan* 2152 (COI; SRGH).
Known only in the Chimanimani region. Along streams, sometimes in rock crevices.

24. **Dissotis pachytricha** Gilg ex R. E. Fr., Wiss. Ergebn. Schwed. Rhod.-Kongo-Exped. **1**: 180, t. 13 fig. 7–11 (1914).—Engl., Pflanzenw. Afr. **3**, 2: 750 (1921).—Brenan & Greenway, T.T.C.L.: 310 (1949).—A. & R. Fernandes in Mem. Soc. Brot. **11**: 33 & 85 (1956); in Kirkia **1**: 74 (1961).—F. White, F.F.N.R.: 306 (1962).—Wickens in F.T.E.A., Melastom.: 53 (1975). Syntypes from Zambia, between Malolo and Kitwe [Katwe], N. from Luwingu, *R. E. Fries* 1203 (UPS), Zaire and Tanzania.

An erect rather branched shrub up to 1·8 m. high. Stems cylindric and glabrescent towards the base; branchlets 4–6-gonous, slender, short or not very long, borne at ± 45° to the main stems, sparsely and shortly appressed-setose. Leaves 2–3-nate; lamina 2·4–8 × 0·5–2·1 cm., lanceolate to elliptic or oblong, acute at the apex, acute or somewhat contracted and roundish at base, entire or minutely crenulate at the margin, papyraceous (rigid on drying), usually discolorous (darker green and sometimes shining on the upper face, bright- to yellow-green on the under one), glabrous or sparsely appressed-setose (the bristles slender) above, sparsely appressed-setose beneath mainly on nerves, longitudinally 3-nerved, the longitudinal nerves impressed above, raised beneath, the transverse nerves and reticulation inconspicuous on both faces; petiole 4–6 mm. long, somewhat slender, yellowish-green. Cymes with few to several 5-merous flowers at the end of the stems and branches, rarely flowers solitary; inner bracts shorter than the receptacle, widely ovate to subcircular, scarious, brownish and ciliate, outer ones ovate-lanceolate, acute, with a scarious base and leafy apex, all caducous; pedicels c. 2 mm. long. Receptacle 9–10 × 7–8 mm., ovoid, densely covered with nipple-shaped, sometimes slightly compressed, reddish to brown, patent (eventually reflexed) appendages which are successively longer from the base to the receptacle-apex and bear 1 to few white, short subulate bristles at the top. Sepals up to ± 10 × 6 mm., oblong-ovate, contracted at the base, unequally bilobed (the lobes obtuse) or emarginate at the apex, deep pink- to reddish-brown with a submedian sparsely ciliate keel and glabrous widely-scarious margins, often persistent in fruit, eventually falling; intersepalar segments ± 5 mm. long, erect, similar to the uppermost receptacle-appendages. Petals 2–3 × 1·5–2 cm., mauve to bright purple. Longer stamens: anthers 10–13 mm. long; pedoconnective 15–20 mm. long with the basal appendage c. 3 mm. long. Shorter stamens: anthers 8–10 mm. long, pedoconnective c. 4 mm. long with the appendage c. 2·5 mm. long. Fructiferous receptacle 12–15 × 8–9 mm., ovoid to subspherical, contracted at the mouth. Capsule ¾ as long as receptacle, sparsely setose on the margins of valves and with a crown of short bristles swollen at base around the insertion of style. Seeds c. 1 mm. long, minutely papillose-tuberculate.

Zambia. N: Mbala (Abercorn), 1670 m., fl. & fr. 29.iii.1955, *E. M. & W.* 1231 (BM; LISC; SRGH).
Also in Zaire and Tanzania. In damp ground near streams and lakes and in grasslands, open bushes or woodlands.

25. **Dissotis princeps** (Kunth) Triana in Trans. Linn. Soc. Lond. **28**: 57 [1871] (1872).—Cogn. in A. & C. DC., Mon. Phan. **7**: 375 (1891).—Taub. in Engl., Pflanzenw. Ost-Afr. **C**: 295 (1895) pro parte (excl. syn. *D. violacea* Gilg).—Gilg in Engl., Mon. Afr. Pflanzen. **2**: 22 (1898).—Engl., Pflanzenw. Afr. **3**, 2: 753 (1921).— Burtt Davy, F.P.F.T.: 243, fig. 34 (1926).—Brenan, T.T.C.L.: 310 (1949); in Mem. N.Y. Bot. Gard. **8**: 441 (1954).—Garcia in Est. Ens. Docum. Junta Invest. Ultramar **12**: 157 (1954).—A. & R. Fernandes in An. Junta Invest. Ultramar **10**, 3: 38 (1955); in Bol. Soc. Brot., Sér. 2, **30**: 177 (1956); op. cit. **34**: 190 (1960); in Mem. Soc. Brot. **11**: 37 & 90 (1956); in Kirkia **1**: 76 (1961); in C.F.A. **4**: 154 (1970).—Chapman, Veg. Mlanje Mt. Nyasal.: 47 (1962).—F. White, F.F.N.R.: 308 (1962).—Binns, H.C.L.M.: 68 (1968).—Wickens in F.T.E.A., Melastom.: 51 (1975). Type probably from Mozambique, *?* Galvão da Silva (P).
Rhexia princeps Kunth in Humb. & Bonpl., Mon. Melastom. **2**: 122, t. 46 (1823). Type as above.
Osbeckia? princeps (Kunth) DC., Prodr. **3**: 140 (1828).—Naud. in Ann. Sci. Nat. Bot., Sér. 3, **14**: 54 (1850). Type as above.
Osbeckia eximia Sond. in Linnaea **23**: 48 (1850); in Harv. & Sond., F.C. **2**: 518 (1862).—D. Dietr., Fl. Univ., N.F.: t. 88 (1849–55). Syntypes from S. Africa (Natal).
Dissotis eximia (Sond.) Hook. f. in F.T.A. **2**: 454 (1871) pro parte. Syntypes as above.
Dissotis verticillata De Wild. in Bull. Jard. Bot. Brux. **5**: 79 (1915). Type from Zaire.

A perennial herb or a much-branched shrub up to 3(4) m. tall. Old branches subcylindric and striate, the younger ones angular, ± scabrous or setose with small branched scale-like bristles. Leaves 2–3(4)-nate, frequently reflexed, petiolate; lamina 3–14·5 × 1–5·5 cm., lanceolate to ovate-lanceolate, roundish to subcordate (sometimes cordate) at the base, acute at the apex, crenulate at the margin, ± dark or light green above, pale green, whitish or yellowish below, subappressed-setose on both faces but more densely so beneath (the bristles ± long, simple or divided from the base), longitudinally 5–7-nerved, the longitudinal nerves impressed on the upper face, prominent on the lower one, the transverse ones numerous, 1 mm. distant, usually inconspicuous above, ± raised beneath like the reticulation; petiole 0·5–4 cm. long, scaly-setose. Flowers 5-merous, subsessile in dense cymes, arranged in short terminal very compact inflorescences or in racemiform or pyramidal panicles up to 23 × 16 cm.; bracts shorter than the receptacle, broad-ovate, obtuse or acute, purple and appressed-setulose outside, caducous. Receptacle 6–10 × 5–6 mm., covered by numerous ± long flat appendages pilose all around, tipped by several ± long ascending white or yellowish-white bristles. Sepals 7–10 × 2·5–3·5 mm., oblong, oblique-truncate or acute and setose or with a setose appendage at the apex, ciliate at the margin, purplish, dorsally hairy, caducous; intersepalar segments rectangular or linear, ± long-bristly at the top. Petals c. 2·5 × 2 cm., lilac to dark purple or violet, rarely white. Longer stamens: anthers 9–12 mm. long; pedoconnective 15–20 mm. long, with the basal appendage c. 1·5 mm. long. Shorter stamens: anthers 8–10 mm. long; pedoconnective c. 3 mm. long with the appendage shorter than above. Fructiferous receptacle 8–12 × 6–8·5 mm., scarcely to somewhat contracted above the capsule. Capsule shorter than the receptacle, broadly ovoid to spherical, appressed-setose at the apex and with a crown of long bristles surrounding the insertion of the style.

Receptacle-appendages elongate with dense, long and strong bristles; inflorescence
usually dense - - - - - - - - - - - - var. *princeps*
Receptacle-appendages short with not very dense, short and weak bristles; inflorescence
usually loose - - - - - - - - - - - var. *candolleana*

Var. **princeps**.

Zambia. N: 72 km. S. of Mbala (Abercorn) on Kasama road, fl. & fr. 30.iii.1955, *E. M. & W.* 1352 (BM; LISC; SRGH). C: Mkushi, fl. & fr. 5.viii.1959, *West* 4008 (SRGH). E: Chitipa (Fort Jameson), fl. & fr. 23.v.1951, *Gilges* 106 (SRGH). **Rhodesia.** W: Bulawayo Distr., fl. 1930, *Cheesman* 13 (BM). C: Makoni Distr., Forest Hill, 1440 m., fl. & fr. vi.1917, *Eyles* 793 (BM; K; SRGH). E: Umtali, Imbeza Forest Estate, fl. & fr. 15.v.1949, *Chase* 1643 (BM; COI; LISC; SRGH). S: E. of Zimbabwe R., fr. 3.vii.1930, *Hutchinson & Gillett* 3381 (BM; K; SRGH). **Malawi.** N: Nyika Rest House, near the Shire R., fl. & fr. 20.vi.1960, *Chapman* 773 (COI; FHO; SRGH). C: Dedza Distr., Chongoni forest, fr. 20.xi.1960, *Chapman* 1052 (COI; FHO; K; SRGH). S: Zomba Plateau, fl. & fr. 5.vi.1946, *Brass* 16254 (BM; K; SRGH). **Mozambique.** N: Malema, Murralelo, c. 650 m., fl. & fr. 19.iii.1964, *Torre & Paiva* 11271 (COI; K; LISC; LMU; SRGH). Z: Quelimane Distr., Lugela, Tacuane, fl. & fr. viii, *Faulkner* 36 (BR; COI; EA; K; PRE; SRGH). T: Angónia, Domué Mt., fl. & fr. 9.iii.1964, *Torre & Paiva* 11073 (LISC; LMA). MS: Gorongosa, Unora Mt., 800 m., fl. & fr. 6.v.1964, *Torre & Paiva* 12255 (BR; LISC; LUA; M; WAG).

Tropical Africa and S. Africa. Upland rain-forest, gallery forest and wooded grassland.

Var. **candolleana** (Cogn.) A. & R. Fernandes in Bol. Soc. Brot., Sér. 2, **29**: 56, t. 9, 10 (1955); in Mem. Soc. Brot. **11**: 40 & 92 (1956); in Kirkia **1**: 76 (1961); in C.F.A. **4**: 155 (1970).—Wickens in F.T.E.A., Melastom.: 52 (1975). Type from Angola.
  Dissotis candolleana Cogn. in A. & C. DC., Mon. Phan. **7**: 375 (1891).—Gilg in Engl., Mon. Afr. Pflanzen. **2**: 19 (1898).—Hiern, Cat. Afr. Pl. Welw. **1**, 2: 367 (1898).—Engl., Pflanzenw. Afr. **3**, 2: 751 (1921).—Jacques-Félix in Bull. Mus. Hist. Nat. Paris, Sér. 2, **7**: 372 (1935).—Brenan & Greenway, T.T.C.L.: 309 (1949).—Binns, H.C.L.M.: 67 (1968). Type as above.
  Dissotis muenzneri Gilg ex Engl., loc. cit.—Brenan, loc. cit. Type from Tanzania.

Zambia. W: Mwinilunga, Lisombo R., fl. & fr. 8.vi.1963, *Edwards* 684 (COI; K; LISC; SRGH). C: between Lusaka and Rufunza, Maswero Hills, fl. & fr. 20.viii.1946, *Gouveia & Pedro* in *Pedro* 1726 (COI; LMA). E: Nyika Plateau, Rest House on path to N. Rukuru Waterfall, 1250 m., fl. & fr. 27.x.1958, *Robson* 396 (K; LISC). **Rhodesia.** E: Melsetter, near Busi R., 1050 m., fl. & fr. 15.v.1953, *Chase* 4977 (BM; COI; SRGH). **Malawi.** N: Nkhata Bay, Limpasa dambo, fl. & fr. 31.vii.1960, *Eccles* 31 (BM; COI; K; SRGH). C: Ntchisi Distr., Ntchisi Mt., 1400 m., fl. & fr. 24.vii.1946, *Brass* (K; SRGH). S: Mlanje Mt., 1891, *Whyte* 103 (BM). **Mozambique.** Z: Namuli Peaks, near Vila Junqueiro, 1020 m., fl. & fr. 25.vii.1962, *Leach & Schelpe* 11464 (K; LISC; SRGH). T: Macanga, between Casula and Furancungo, fl. & fr. 9.vii.1949, *Barbosa & Carvalho* 3524 (COI; LMA). MS: Maronga Forest, fl. 13.viii.1945, *Simão* 454 (LMA).

Tropical Africa, Swaziland and Natal. In swampy grasslands, along streams and sometimes in woodlands and savannas.

26. **Dissotis peregrina** A. & R. Fernandes in Bol. Soc. Brot., Sér. 2, **30**: 180, t. 15 (1956).—F. White, F.F.N.R.: 308 cum tab. (1962). Type: Zambia, Kawambwa Distr., Kawambwa-Mansa road, 30.x.1952, *Angus* 676 (COI, holotype; FHO).
  Dissotis kassneri Gilg ex De Wild. in Ann. Mus. Congo Belge, Bot. Sér. 4, **2**: 117 (1918) *nom. nud.*

A perennial herb or shrub up to c. 2 m. high. Branches terete, covered with very appressed scales up to 1·5 mm. long. Leaves opposite or 3-nate; lamina up to 17·5 × 3·8 cm., oblong-lanceolate, round or subcordate at the base, subacute at the apex, crenate-serrate at the margin (each tooth ending in a fringed bristle), coriaceous, concolorous or slightly discolorous (light or dull green above, paler beneath), longitudinally 3(5)-nerved, the longitudinal nerves impressed above, raised beneath as well as the numerous transverse ones and the reticulation; upper face of lamina covered with short conic or pyramidal prominences ornamented by raised white lines converging into a short apical bristle; under face foveolate, ± hairy and with the longitudinal nerves covered by appressed scales similar to the cauline ones but longer; petiole 1–2 cm. long, robust, scaly. Flowers 5-merous, in terminal somewhat lax panicles; bracts 8–9 × 6–7 mm., ovate, obtuse, scaly outside, caducous; pedicels c. 1 mm. long. Receptacle c. 8 × 5 mm., cylindric-campanulate, covered with very short appendages pectinate-setose at the apex. Sepals 8–11 × 3–4 mm., oblong, emarginate or truncate at the apex, purplish, dorsally scaly-setose at the apex; intersepalar segments 2·5 × 1·5 mm., oblong, slightly dilated and setose near the apex. Fructiferous receptacle c. 10 × 8 mm., ovoid-cylindric. Capsule c. 8 × 6 mm., setose on the upper half, and with a crown of white fringed-pectinate scales around the insertion of the style. Seeds c. 0·75 mm. long, minutely papillose-tuberculate.

Zambia. N: Kawambwa, fl. & fr. 23.viii.1957, *Fanshawe* 3582 (K).
Also in Zaire. On fringe of riverine " mushitu ", mainly in swampy places.

27. **Dissotis falcipila** Gilg in Engl., Mon. Afr. Pflanzen. **2**: 23, t. 3 fig. A (1898).—
Th. & H. Dur., Syll. Fl. Cong.: 210 (1909).—Engl., Pflanzenw. Afr. **3**, 2: 753
(1921).—A. & R. Fernandes in Bol. Soc. Brot., Sér. 2, **30**: 178 (1956); in Mem. Soc.
Brot. **11**: 41, t. 5 (1956); in C.F.A. **4**: 157 (1970).—F. White, F.F.N.R.: 308 (1962).
Type from Zaire.
  Dissotis verdickii De Wild. in Ann. Mus. Congo Belge, Bot. Sér. 4, **1**: 218 (1903).—
Th. & H. Dur., tom. cit.: 213 (1909). Type from Zaire.
  Dissotis crenulata sensu R. E. Fr., Wiss. Ergebn. Schwed. Rhod.-Kongo-Exped.
1: 181 (1914) non Cogn. (1898).
  Dissotis angusii A. & R. Fernandes in Bol. Soc. Brot., Sér. 2, **30**: 176, t. 14 (1956).—
F. White, loc. cit. Type: Zambia, Mansa Distr., Samfya, near Lake Bangweulu,
Angus 292 (COI, holotype; FHO; K).

A shrub or perennial herb up to 1·8 m. high. Branches subcylindric, densely
appressed-scaly, the scales c. 2 mm. long, dilated at the base, acute at apex,
fimbriate at margin, brownish on the oldest branches and white on the youngest
and on the inflorescence-axis. Leaves opposite; lamina 1·8–10·5 × 0·6–3·5 cm.,
oblong-lanceolate or ovate-oblong, round or subcordate at the base, acute at the
apex, denticulate-serrate at the margin, each tooth ending in a bristle, the upper
face dull green, with short subconic prominences each ending in a white appressed
bristle, the under face whitish or pale green, ± densely lanate (hairs long, crispate,
very slender), somewhat thick, longitudinally 3–5-nerved, the longitudinal nerves
(covered beneath with scales similar to those of the branches but longer or ending
in a ± long seta), the very close, parallel, straight, transverse ones and the
reticulation impressed above and much raised beneath, usually (longitudinal nerves
excepted) completely hidden by the indumentum; petiole 0·2–2 cm. long, scaly-
setose. Flowers 5-merous, subsessile, in short terminal compact panicles
4 × 3·5–4 cm., the pedicels up to 2 mm. long in fruit; bracts 10–13(15) × 6–
8(17) mm., subcircular or broadly ovate, round or acute at the apex, reddish or
brownish, dorsally appressed-setose, caducous. Receptacle 7–9 × 5–6 mm.,
campanulate, densely and shortly setulose and with patent ± long terete capitate
appendages, these setose throughout and stellate-setose at the apex (the bristles
white, simple, those of the apex stouter and longer); intersepalar segments similar
to the upper appendages of receptacle but longer. Sepals c. 6 × 3·5 mm., oblong,
oblique-truncate at the top, dorsally appressed-setose, with an appendage stellate-
setose at the apex, caducous. Petals c. 16 mm. long, obovate, purple. Longer
stamens: anthers 8–12 mm. long; pedoconnective 13–15 mm. long with the
appendage short. Shorter stamens: anthers 8–9 mm. long; pedoconnective
3–4 mm. long with a small appendage. Fructiferous receptacle 10–11 × 6–8 mm.,
ovoid-campanulate. Capsule subspherical or ovoid, a little shorter than the
receptacle, densely setose on the upper half and with a crown of long white bristles
at the apex.

Zambia. B: Zambezi Distr., Chavuma, fl. & fr. 4.viii.1959, *Gilges* 196 (K; SRGH).
N: Bangweulu, fl. & fr. 19.ix.1911, *R. E. Fries* 733 (K; UPS); Samfya, fl. & fr. 2.x.1953,
Fanshawe F.339 (LISC). W: Mwinilunga Distr., Lisombo R., fl. & fr. 8.vi.1963,
Loveridge 859 (COI; K; SRGH).
Also in Angola and Zaire. In swamps along the rivers.

28. **Dissotis denticulata** A. & R. Fernandes in Bol. Soc. Brot., Sér. 2, **29**: 57, t. 11
(1955); op. cit. **30**: 177 (1956); in Mem. Soc. Brot. **11**: 41 & 93 (1956); in Kirkia
1: 76 (1961).—F. White, F.F.N.R.: 308 cum tab. (1962).—Wickens in F.T.E.A.,
Melastom.: 52 (1975). TAB. 62. Type: Zambia, Shiwa Ngandu, *Greenway* 5765
(EA; K; PRE, holotype).

A shrub up to 2·5 m. high or a perennial herb. Branches 2–3-nate, bluntly
6-gonous or terete, ± densely and patently setose (the bristles whitish to rusty,
usually long, pectinate-branched from a bulbous base) towards the apex, the oldest
scabrous. Leaves 2–3-nate; lamina 2·3–11·5 × 1·8–5 cm., ovate-lanceolate or
ovate, cordate at base (those of the flowering branches the smallest and most deeply
cordate), subacute at the apex, conspicuously denticulate at the margin, discolorous,
dark or brownish-green above and whitish or pale-green below (the old leaves often
turning bright red), rather thick, rough on the upper face with subconic

prominences each ending in a simple white rather long bristle, densely covered on the under side by long thin silvery hairs, longitudinally 5–7-nerved, the longitudinal nerves impressed above, raised and appressed-scaly-setose beneath (the scales with a thick laterally dentate-setose base ending in a long bristle); petiole 0·2–2·5 mm. long, scaly-setose. Flowers 5-merous, subsessile in short compact terminal panicles; bracts up to 9 × 8 mm., subcircular, obtuse, caducous. Receptacle 8–10 × 6–7 mm., cylindric, densely setose (bristles simple or divided from the base) and with many ± patent linear appendages up to 2·5 mm. long, ending in a small stellate-setose disk (the bristles of the terminal disks ascending, numerous, up to 2·5 mm. long, rather stout, whitish); intersepalar segments like the uppermost receptacle-appendages but longer. Sepals c. 9 × 5 mm., oblong, obliquely truncate, dorsally minutely appressed-setose and with a stellate-setose appendage at the apex, caducous. Petals c. 2·8 × 2 cm., purple or rose. Longer stamens: anthers 10–13 mm. long; pedoconnective c. 15 mm. long with the basal appendage c. 1 mm. long; filaments 7–8 mm. long. Shorter stamens: anthers 8–12 mm. long; pedoconnective c. 3 mm. long with the basal appendage 0·5 mm. long; filaments c. 10 mm. long. Fructiferous receptacle up to 11 × 10 mm., broadly cylindric or subspherical. Capsule ovoid, ± equalling the receptacle, appressed-setose on the upper half.

Zambia. N: Mansa (Fort Rosebery) Distr., c. 26 km. S. of Samfya, Lake Bangweulu, fl. & fr. 18.viii.1952, *Angus* 238 (BM; COI; FHO; K). C: Chiwefwe, fl. & fr. 1.v.1957, *Fanshawe* 3224 (K).
Also in Zaire and Tanzania. In swampy grasslands, forests and river banks.

29. **Dissotis densiflora** (Gilg) A. & R. Fernandes in Garcia de Orta **2**: 180 (1954); in Mem. Soc. Brot. **11**: 90 (1956); in Bol. Soc. Brot., Sér. 2, **34**: 189 (1960); in C.F.A. **4**: 158 (1970).—Wickens in F.T.E.A., Melastom.: 50 (1975). Lectotype: Malawi, Shire Highlands, *Buchanan* 484 (K).
  Osbeckia densiflora Gilg in Engl., Mon. Afr. Pflanzen. **2**: 8 (1898), excl. specim. *Buchanan* 114.—Engl., Pflanzenw. Afr. **3**, 2: 746 (1921). Type as above.

An annual or perennial herb up to 2 m. tall. Stem erect or ascending, simple or with a few branches, 4(6)-gonous, flat or sulcate on the faces, pale green, ± hirsute with long, whitish or yellowish hairs; internodes usually rather elongate, sometimes up to 18 cm. long; nodes long-setose. Leaves opposite or, rarely, 3-nate, shortly petiolate to subsessile; lamina 2–14 × 1–5·5 cm., the lowermost the smallest, broadly ovate and obtuse, the median ones the largest, ovate or ovate-lanceolate and acute, the uppermost the narrowest, lanceolate, very acute, all subcordate or roundish at the base, crenulate at the margin, membranous, light green and sparsely appressed-setose on both faces, longitudinally 3–5(7)-nerved, the longitudinal nerves slender, impressed above, raised below, the transverse ones inconspicuous on the upper face, somewhat visible on the under one, very slender and oblique, the reticulation inconspicuous; petiole up to 8 mm. long, slender, long-hirsute. Flowers 5-merous, in terminal very dense capitate inflorescences surrounded by the uppermost leaves, or also in axillary not so dense ones; pedicels ± 0·5 mm. long; bracts 2·5–3 mm. long, oblong, pale, ciliate at the margin. Receptacle 7–10 × 4–6 mm., cylindric, covered with linear, patent or reflexed appendages up to 2·5 mm. long sparsely setose all around and stellate-setose (the bristles long ± numerous, strong, whitish or rosy to dark purple) at the slightly dilated apex and sometimes also with bristles inserted directly on the receptacle. Sepals 5–8 × 1·75–3 mm., oblong, glabrous or dorsally ± appressed-setose, ciliate at the margin and long-setose at the roundish apex, caducous; intersepalar segments like the uppermost receptacular appendages. Petals 13–17(24) × 11–15 mm., pale mauve to purple. Stamens subequal with anthers 5–7(10) mm. long, pedo-connective c. 1 mm. long with a c. 0·5 mm. long, bituberculate appendage at the base. Fructiferous receptacle up to 11 × 6 mm., urceolate-campanulate, slightly contracted above the capsule. Capsule c. 5 mm. in diameter, spherical, setose on the upper part and with an apical crown of bristles.

Zambia. N: Kasama Distr., Chief Mwamba's Village, 1320 m., fl. & fr. 25.ii.1962, *Richards* 16165 (K; SRGH). W: Mwinilunga Distr., S. of Matonchi Farm, fl. & fr. 19.ii.1938, *Milne-Redhead* 4632 (K). **Malawi.** N: Uzumara Forest Reserve, 1800 m., fl. & fr. iii.1953, *Chapman* 129 (BM; SRGH). C: Ntchisi Distr., Ntchisi, 1440 m., fl. & fr. 20.ii.1944, *Benson* 284 (K). S: Namasi, fl. & fr. 1897, *Cameron* (K). **Mozambique.**

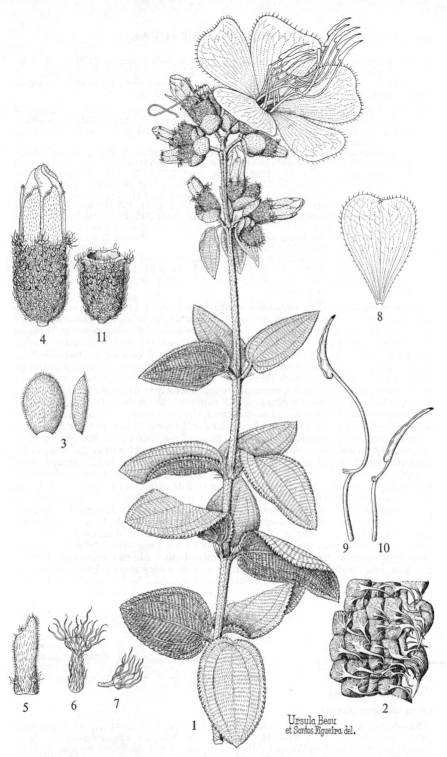

Ursula Beau
et Santos Figueira del.

Tab. 62. DISSOTIS DENTICULATA. 1, flowering branch (×1); 2, marginal part of leaf seen from above (×8); 3, bract, dorsal and lateral view (×2); 4, flower before anthesis (×2); 5, sepal (×2); 6, intersepalar segment (×6); 7, receptacle appendage (×6); 8, petal (×1); 9, long stamen (×2); 10, short stamen (×2); 11, fructiferous receptacle (×2), all from *Greenway* 5765.

N: 50 km. from Vila Cabral on the Vila Cabral-Maniamba road, c. 1350 m., fl. & fr. 29.ii.1964, *Torre & Paiva* 10902 (BR; COI; K; LISC; LMU; LUA; SRGH).

Also in Angola, Zaire and Tanzania. In shady situations along rivers and woodlands and in grasslands.

30. **Dissotis senegambiensis** (Guill. & Perr.) Triana in Trans. Linn. Soc. Lond. **28**: 58 [1871] (1872).—A. & R. Fernandes in Garcia de Orta **2**, 3: 284 (1954); in Bol. Soc. Brot., Sér. 2, **46**: 69 (1972).—Wickens in F.T.E.A., Melastom.: 48 (1975). Type from Senegal.

 Osbeckia senegambiensis Guill. & Perr. in Guill., Perr. & A. Rich., Fl. Senegamb. Tent. **1**: 310 (1833).—Benth. in Hook., Niger Fl.: 346 (1849).—Naud. in Ann. Sci. Nat. Bot., Sér. 3, **14**: 60 (1850).—Hook. f. in F.T.A. **2**: 443 (1871).—Cogn. in A. & C. DC., Mon. Phan. **7**: 334 (1891).—Krasser in Engl. & Prantl, Pflanzenfam. **3**, 7: 156 (1893).—Gilg in Engl., Mon. Afr. Pflanzen. **2**: 8 (1898).—Engl., Pflanzenw. Afr. **3**, 2: 746 (1921).—Hutch. & Dalz. in F.W.T.A. **1**: 208 (1927).—P. Sousa in An. Junta Invest. Colon. **1**: 74 (1946).—F. W. Andr., Fl. Pl. Anglo-Egypt. Sudan **1**: 194 (1950).—Keay in F.W.T.A. ed. 2, **1**: 249 (1954).—Type as above.

 Osbeckia abyssinica Gilg, loc. cit.—Engl., loc. cit.—Brenan in Mem. N.Y. Bot. Gard. **8**: 439 (1954). Type from Ethiopia.

 Dissotis irvingiana Hook. var. *irvingiana* forma *abyssinica* (Gilg) A. & R. Fernandes, tom. cit.: 179 (1954); in Mem. Soc. Brot. **11**: 36 & 87 (1956); in Bol. Soc. Brot., Sér. 2, **34**: 188 (1960); in Kirkia **1**: 75 (1961).—Chapman, Veg. Mlanje Mt. Nyasal.: 33 (1962). Type as above.

An annual, biennial or perennial, much-branched, erect herb up to 2 m. tall. Branches 4-gonous, pale green to dull red, sparsely to densely hirsute with pale spreading bristles up to 7 mm. long, longer at the nodes. Leaves opposite, those of the main stem patent or folded along the midrib and reflexed; lamina 1·5–9 × 0·8–2·5 cm., narrowly lanceolate to elliptic-oblong or ovate-lanceolate, rarely ovate, acute at the apex, roundish and contracted at the base, subentire or crenate, subconcolorous (both faces light green) or discolorous (dark- or yellow-green above, lighter beneath), somewhat rigid, sparsely to densely appressed-setose above, subhirsute below, longitudinally 3–5-nerved, the nerves slender, impressed on the upper face, slightly raised on the lower, the transverse nerves and reticulation inconspicuous; petiole 2–10 mm. long, slender, hirsute. Flowers 4- or 5-merous, subsessile, in terminal and axillary cymes, those of the main stem often approximate and usually forming congested inflorescences; bracts shorter than the receptacle, brownish, scarious, ciliate, late deciduous. Receptacle 5–7 × 4 mm., cylindric, long whitish- or purplish-hirsute, covered with cylindric, patent or reflexed usually long appendages somewhat dilated and long stellate-setose at the apex or rarely the appendages short and with 1–few bristles at the top. Sepals 4–6 × 2–3 mm., oblong, stellate-setose at the apex, ciliate at the margin, caducous; intersepalar segments like the uppermost receptacle-appendages but longer. Petals 12–17 × 8–13 mm. long, obovate, pale to light purple. Longer stamens of dissotidoides forms: anthers 7–8 mm. long, pedoconnective 5·5–8·5 mm. long with 2-lobate, 1–2 mm. long appendage at base; filaments c. 8 mm. long. Shorter stamens: anthers 5–8 mm. long, pedoconnective c. 1 mm. long with a very short 2-lobed appendage at the base; filaments 6 mm. long. Stamens of osbeckioides forms subequal, those opposite to the sepals with the pedoconnective slightly longer. Fructiferous receptacle 8–9 × 5–7 mm., ovoid to urceolate. Capsule shorter than the receptacle, setose at the apex on the valve-margins and with a ring of strong bristles around the style-insertion.

Flowers 5-merous - - - - - - - - var. *senegambiensis*
 Stamens subequal - - - - - - - forma *senegambiensis*
 Stamens very unequal - - - - - - - forma *irvingiana*
Flowers 4-merous - - - - - - - - - var. *alpestris*

Var. **senegambiensis**.

forma **senegambiensis**.

Rhodesia. E: La Rochelle, Imbeza Valley, 1200 m., fl. & fr. 11.iv.1955, *Chase* 5544 (BM; K; LISC; LMA; SRGH). **Malawi.** N: Mugesse Forest Reserve, fl. & fr. 2.ix.1952, *Chapman* 15 (BM). S: Mlanje Mt., near Likabula Forest Station, fl. & fr. 10.v.1958, *Chapman* 608 (K; SRGH). **Mozambique.** Z: Gúruè, Gúruè Mt., 1300 m., fl. 21.ii.1966, *Torre & Correia* 14720 (COI; LISC). MS: Manica, Vumba Mt., fl. & fr. 25.iii.1948, *Garcia* 721 (LISC); Gorongosa, Unora Mt., c. 800 m., fl. 6.v.1964, *Torre & Paiva* 12249 (K; LISC; LMU; P; WAG).

forma **irvingiana** (Hook.) A. & R. Fernandes in Bol. Soc. Brot., Sér. 2, **46**: 69 (1972).
Type from Nigeria.
  Dissotis irvingiana Hook. in Curtis, Bot. Mag. **85**: t. 5149 (1859).—Hook. f.,
tom. cit.: 453 (1871).—Triana, loc. cit.—Cogn., op. cit.: 375 (1891).—Krasser, loc.
cit.—Gilg, op. cit.: 20 (1898).—Engl., op. cit.: 752 (1921).—Jacques-Félix in Bull.
Mus. Hist. Nat. Paris, Sér. 2, **7**: 370 (1935).—Keay, tom. cit.: 259 (1954).—A. & R.
Fernandes in Garcia de Orta **2**: 179 (1954); in Mem. Soc. Brot. **11**: 35 & 87 (1956);
in Bol. Soc. Brot., Sér. 2, **34**: 188 (1960); in Kirkia **1**: 75 (1961). Type as above.
  Dissotis irvingiana var. *irvingiana* forma *irvingiana*—A. & R. Fernandes in An. Junta
Invest. Ultramar **10**, 3: 36 (1955). Type as above.

Malawi. Without locality or date, fl. & fr., *Whyte* 78 (K).

Var. **alpestris** (Taub.) A. & R. Fernandes in Bol. Soc. Brot., Sér. 2, **46**: 69 (1972).
Syntypes from Tanzania.
  Dissotis alpestris Taub. in Engl., Pflanzenw. Ost-Afr. **C**: 295 (1895).—Gilg in
Engl., Mon. Afr. Pflanzen. **2**: 20 (1898).—Engl., Pflanzenw. Afr. **3**, 2: 752 (1921).
Syntypes as above.
  Dissotis cincinnata Gilg, loc. cit.—Engl., loc. cit.—Brenan & Greenway, T.T.C.L.:
309 (1948) Type from Tanzania.
  Dissotis irvingiana var. *alpestris* (Taub.) A. & R. Fernandes in Garcia de Orta **2**:
179 (1954); in Mem. Soc. Brot. **11**: 37 & 89 (1956). Syntypes as for *Dissotis alpestris*.
  Dissotis irvingiana var. *alpestris* forma *alpestris*—A. & R. Fernandes in An. Junta
Invest. Ultramar **10**, 3: 36 (1955); in Mem. Soc. Brot. **11**: 89 (1956); in Bol. Soc.
Brot., Sér. 2, **34**: 189 (1960). Syntypes as for *Dissotis alpestris*.
  Dissotis irvingiana var. *alpestris* forma *osbeckioides* A. & R. Fernandes in An. Junta
Invest. Ultramar **10**: 38 (1955). Type: Mozambique, Manica, Vumba Mt., *Pedro &
Pedrógão* 6989 pro parte (COI, holotype; LMA).

Malawi. N: Misuku Plateau, 1950–2100 m., fl. & fr. viii.1896, *Whyte* (K).
Mozambique. MS: Vumba Mt., *Pedro & Pedrógão* 6989 pro parte (COI, LMA).
Distribution of species: widely distributed throughout tropical Africa. Moist places,
forest edges, grassland, banks of rivers.

8. DICHAETANTHERA Endl.

Dichaetanthera Endl., Gen. Pl.: 1215 (1840).—Jacques-Félix in Bull. Soc.
Bot. Fr. **102**: 37 (1955).

Small trees or shrubs, strigillose with rigid bristles. Branches ± distinctly
4-gonous. Leaves opposite, subentire. Flowers 4-merous, pedicellate, 2-bracteate,
arranged in terminal panicles, produced before the leaves; bracts caducous. Sepals
short, persistent; intersepalar segments absent. Petals obovate or oblong.
Stamens 8, unequal; filaments filiform; anthers linear-subulate, sigmoid,
1-porose, with undulate thecae; pedoconnective rather long with two linear
anterior appendages at the base. Ovary 4-locular, adnate to the receptacle by
8 septa; style filiform, somewhat thickened distally; stigma punctiform. Capsule
4-valved. Seeds very numerous, obovoid or subpyramidal with hyaline papillae on
the back; hilum basal.

Bristles of receptacle up to 2 mm. long, dense, distinctly spinulose; leaf-lamina densely
  tomentose beneath and setose above, the bristles up to 3 mm. long, subulate
                                                                    1. *rhodesiensis*
Bristles of receptacle shorter, smooth or slightly scabrous; leaf-lamina not tomentose
  beneath; bristles of the upper face shorter:
Sepals up to c. 2 mm. long in flower, somewhat accrescent; receptacle 10–13 × 9–11 mm.
  in fruit, with appressed bulbous bristles, erect and rather hardened in fruit; capsule
  subequalling the fructiferous receptacle, truncate at the top; leaf-lamina sparsely
  setose on both faces, the bristles minute -   -   -   -   - 2. *corymbosa*
Sepals 4–6 mm. long in flower, reflexed and not accrescent in fruit; receptacle
  8–9 × 7–8 mm. in fruit, with arched-ascending bristles not bulbous at the base;
  capsule ± exserted, acute; leaf-lamina densely setose beneath and ± sparsely
  setose above, the bristles up to c. 1 mm. long -   -   -   - 3. *erici-rosenii*

1. **Dichaetanthera rhodesiensis** A. & R. Fernandes in Bol. Soc. Brot., Sér. 2, **30**: 182,
t. 18, 19 & 20 (1956); in Kirkia **1**: 77 (1961).—F. White, F.F.N.R.: 306 (1962).
Type: Zambia, Barotseland, Zambezi, Shingi, *Angus* 628 (BM; COI, holotype;
FHO).

A small much-branched tree up to 3 m. tall, with stout trunk and rounded crown.

Young branches 4-gonous, densely setose (bristles up to 3·5 mm. long, yellow-rusty, spinulose, dilated at base), the oldest indistinctly 4-gonous, glabrescent, with the bark thick, pale grey, longitudinally fissured and transversely cracked; nodes much thickened on the old branches. Leaves crowded at the top of the branchlets; lamina up to 15·5 × 11·7 cm., ovate or ovate-elliptic, acute at apex and base, entire, rather thick, discolorous, bright yellow-green above, paler beneath, setose on the upper face (bristles up to 3 mm. long, subulate, appressed) and densely tomentose on the under one, the 5–7 longitudinal nerves impressed above, raised beneath, the transverse ones slightly impressed above, prominent beneath, reticulation inconspicuous; petiole up to 1 cm. long, appressed-setose. Panicles up to 14 × 13 cm., ± dense, with the axis, branches and pedicels densely appressed-setose; bracts lanceolate or ovate-lanceolate, setose outside. Receptacle ± 7 × 6 mm., cylindric-campanulate, densely setose, the bristles up to 2 mm. long, subulate, dilated at the base, spinulose, arched-ascending. Sepals ± 5 × 5 mm., roundish, dorsally setose, marginally ciliate. Petals c. 16 × 13 mm., bright pink or mauve, marginally ciliate. Long stamens: anthers c. 11 mm. long; pedoconnective c. 9 mm. long with the two basal appendages c. 4 mm. long. Short stamens: anthers c. 10 mm. long; pedoconnective c. 3 mm. long with a c. 3·5 mm. long appendage. Fructiferous receptacle 7–10 × 7–9 mm., campanulate, with the persistent sepals neither much thickened nor accrescent. Capsule exserted, setose on the upper half.

Zambia. B: Zambezi (Balovale) Distr., c. 9·6 km. S. of Shingi village, 1·6 km. N. of the Zambezi R., fr. 13.x.1952, *Angus* 628 (BM; COI; FHO; K). N: Kapata Falls, fl. & fr. 6.viii.1952, *Gilges* 158 (SRGH). W: Mwinilunga Distr., Kalene Hill, fr. 25.ix.1952, *Angus* 553 (BM; COI; FHO; K).
Known only in Zambia. On lateritic soils and rocky top of hills.

2. **Dichaetanthera corymbosa** (Cogn.) Jacques-Félix in Bull. Soc. Bot. Fr. **102**: 38 (1955).—A. & R. Fernandes in Mem. Soc. Brot. **11**: 45 (1956); in Kirkia **1**: 77 (1961); in C.F.A. **4**: 160 (1970).—F. White, F.F.N.R.: 305 (1962).—Wickens in F.T.E.A., Melastom.: 14 (1975). Type from Central African Republic.
    Barbeyastrum corymbosum Cogn. in A. & C. DC., Mon. Phan. **7**: 376 (1891).— Gilg in Engl., Mon. Afr. Pflanzen. **2**: 23, t. 1 fig. G (1898).—Krasser in Engl. & Prantl, Pflanzenfam. **3**, 7: 56 (1898).—Engl., Pflanzenw. Afr. **3**, 2: 753, fig. 317E (1921). Type as above.
    Sakersia corymbosa (Cogn.) Jacques-Félix in Bull. Inst. Fr. Afr. Noire **15**: 1001 (1953). Type as above.
    Sakersia laurentii Cogn. ex De Wild. & Dur. in Ann. Mus. Congo Belge, Bot. Sér. 2, **1**: 23 (1899).—De Wild. & Dur., op. cit. **1**: 135, t. 68 (1900).—Engl., op. cit.: 759 (1921).—Jacques-Félix in Bull. Mus. Hist. Nat. Paris, Sér. 2, **8**: 113 (1936). Syntypes from Zaire.

A tree up to 18 m. tall or a shrub; bole 25–40 cm. in diameter; crown narrow, sparse; bark light brown or grey-brown, thin, loosely and shallowly longitudinally furrowed; young branches obscurely 4-gonous, thickened at nodes, densely appressed-setose (bristles bulbous at base, longer and stiffer at the nodes), the oldest subterete, glabrescent. Leaf-lamina 6–12 × 2·5–5·2 cm., ovate, rounded or subcordate at the base, acute at the apex, entire or minutely crenulate, coriaceous, very discolorous, dark green above, pale or light green beneath, sparsely setose on both sides, the bristles very short and with a bulbous base, longer beneath on the nerves, the 5–7 longitudinal nerves impressed above, raised below, the transverse ones inconspicuous on the upper face, visible, slender and slightly prominent on the under one; petiole 0·9–2·5 cm. long, densely appressed-setose. Panicles up to 15 × 15 cm., 40–50-flowered, not very dense, with the axis, branches and pedicels (up to 5–6 mm. long) appressed-setose; bracts lanceolate, c. 5 × 2 mm., acute. Receptacle c. 7 × 5 mm., campanulate, setose, the bristles appressed, short, subulate, bulbous at base. Sepals c. 2 × 3 mm., widely triangular or semicircular, confluent at the base, obtuse, red, ciliolate. Petals 10–14 × 8–10 mm., obovate, magenta or deep pink with darker nerves. Stamens subequal with yellow or pale lemon anthers 8–10 mm. long, crimson pedoconnective ± 4 mm. long and crimson appendages 2–2·5 mm. long. Fructiferous receptacle 10–13 × 9–11 mm., campanulate; sepals persistent, hardened and somewhat accrescent. Capsule c. 10 × 9 mm., subequalling the receptacle, truncate at the top, appressed-setose on c. the upper half. Seeds 1–1·25 mm. long.

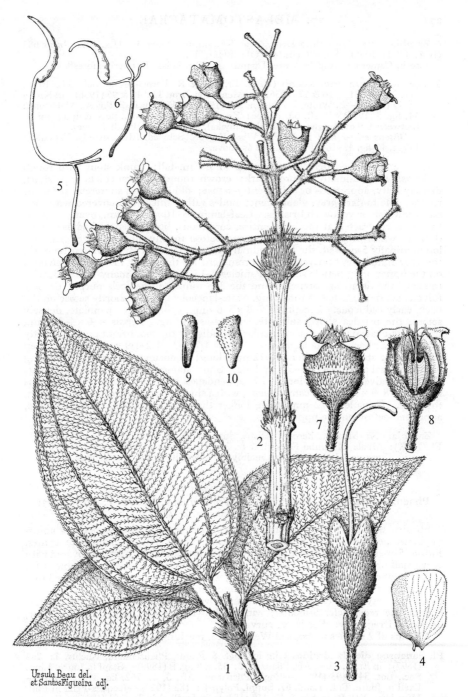

Ursula Beau del.
et SantosFigueira adj.

Tab. 63. DICHAETANTHERA ERICI-ROSENII. 1, upper part of branch showing underside of leaves (×1); 2, fructiferous branch (×1); 3, receptacle with pedicel and bracteoles after petals and stamens are shed (×2); 4, petal (×1); 5, long stamen (×2); 6, short stamen (×2); 7, fructiferous receptacle (×2); 8, same, showing opened capsule (×2); 9, anterior view of seed (×c. 20); 10, seed in lateral view (×c. 20), 1–2, 6–10 from *Angus* 797; 3–5 from *Robertson* 170.

Zambia. N: Kawambwa Distr., near Kawambwa Boma, fr. 31.x.1952, *Angus* 683 (BM; COI; FHO; K), 686 (BM; COI; FHO; K).
Also in Cameroon, Angola, Zaire, Uganda and Tanzania. In riverine forests.

3. **Dichaetanthera erici-rosenii** (R. E. Fr.) A. & R. Fernandes in Bol. Soc. Brot., Sér. 2, **30**: 181, t. 16 & 17 (1956); in Mem. Soc. Brot. **11**: 46 & 95 (1956); in Kirkia **1**: 77 (1961).—F. White, F.F.N.R.: 306 (1962).—Wickens in F.T.E.A., Melastom.: 14, fig. 5 (1975). TAB. **63**. Type: Zambia, Northern Prov., near Kalungwini R. (between Lakes Bangweulu and Tanganyika), *Fries* 1154 (UPS, holotype).
  Dissotis erici-rosenii R. E. Fr., Wiss. Ergebn. Schwed. Rhod.-Kongo-Exped. **1**: 179, t. 13 fig. 15–18 (1914). Type as above.

A much-branched deciduous tree up to 7·5 m. tall; trunk stout with rough yellowish longitudinally fissured bark; crown rounded; branchlets very short, densely leafy, appressed- or subpatently-setose; old branches subterete, striate or rugose, pale to dark grey, glabrescent; nodes rather thickened, covered with long stiff brownish or yellowish bristles. Leaf-lamina 5–10 × 2·7–7 cm., ovate, acute at the apex, contracted at the base, entire, coriaceous, light green and setose above (bristles white or pale yellow), paler and setose mainly on the nerves beneath, longitudinally 5-nerved, the longitudinal nerves impressed above, raised beneath, the very numerous transverse ones prominent on the under face, inconspicuous on the upper one; petiole setose. Panicles 7–13 × 9–13 cm., many-flowered, with the axis, the divaricate branches and the 3–5 mm. long pedicels purple, densely fulvous-setose; bracts 4–5 mm. long, ovate-lanceolate, acute, shortly setose on the back, early caducous. Receptacle 7–8 × 5–6 mm., cylindric-campanulate, densely setose, the bristles short, subulate, arched-ascending. Sepals 4–6 × 4–5 mm., triangular-ovate or oblong-semicircular, roundish at the top, bright purple, shortly setose on the back, ciliate at margin. Petals 1·7–2·2 × 1·2–1·9 cm., obovate, deep pink. Long stamens: anthers 11–12 mm. long; pedoconnective 12–17 mm. long with the basal appendages 4–5 mm. long. Short stamens: anthers 8–9 mm. long; pedoconnective 3–4 mm. long with appendages c. 3 mm. long. Fructiferous receptacle 8–9 × 7–8 mm., campanulate, with a sinuous ridge all around at ½–⅔ from the base. Capsule c. 12 mm. high, longer than the receptacle, ellipsoid, acute, appressed-setose on the upper half.

Zambia. N: Mbala (Abercorn) Distr., Sunzu Hill, fr. 18.xi.1952, *Angus* 797 (COI; FHO; K); Mbala, Kalambo Falls, fl. & fr. 21.x.1947, *Brenan* 8180 (FHO; K).
Also in Tanzania. In rocky places mainly by waterfalls and in woodlands.

9. PHAEONEURON Gilg

Phaeoneuron Gilg in Engl. & Prantl, Pflanzenfam., Nachtr. **1**: 267 (1897); in Engl., Mon. Afr. Pflanzen. **2**: 34 (1898).

Shrubs, undershrubs or perennial herbs. Leaves opposite, sometimes anisophyllous, petiolate. Inflorescences of terminal and axillary paniculate cymes. Flower 5-merous. Receptacle obovoid, shallowly undulate at margin, covered with very small brown hairs. Sepals 0. Petals broadly ovate, ± oblique. Stamens 10, equal or subequal; anthers linear, straight, with the pedoconnective produced into a posterior thick ± square appendage and provided with 2 anterior subglobose tubercles. Style longer than the stamens; ovary adnate by septa to the inferior half of the receptacle, 5-locular, with multiovulate placentas. Fruit capsular, dehiscing irregularly. Seeds ∞, curved.
A genus of 2 species in tropical West Africa, reaching E. Sudan and W. Zambia.

Phaeoneuron dicellandroides Gilg in Engl. & Prantl, Pflanzenfam., Nachtr. **1**: 267 (1897); in Engl., Mon. Afr. Pflanzen. **2**: 35, t. 8 fig. B (1898).—Stapf in Journ. Linn. Soc., Bot. **34**: 493 (1900).—Engl., Pflanzenw. Afr. **3**, 2: 762, fig. 318H (1921).—Exell in Journ. Bot., Lond. **67**, Suppl. Polypet.: 182 (1929).—Jacques-Félix in Bull. Mus. Hist. Nat. Paris, Sér. 2, **8**: 114 (1936); in Ic. Pl. Afr. **3**: t. 61 (1955).—Keay in F.W.T.A. ed. 2, **1**: 247 (1954).—R. & A. Fernandes in C.F.A. **4**: 161 (1970). TAB. **64**. Syntypes from Cameroon and Central Africa.
  Medinilla africana Cogn. ex De Wild. & Th. Dur. in Ann. Mus. Congo Belge, Bot. Sér. 2, **1**: 24 (1898). Type from Zaire.
  Phaeoneuron moloneyi Stapf in Curtis, Bot. Mag. **126**: t. 7729 (1900); in Journ. Linn. Soc., Bot. **34**: 494 (1900). Type from Nigeria.
  Dicellandra gracilis A. Chev., Expl. Bot. Afr. Occid. Fr.: 276 (1920) *nom. nud.*

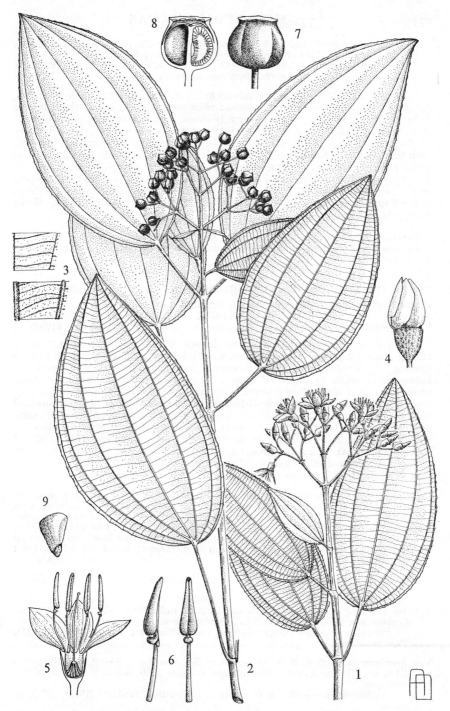

Tab. 64. PHAEONEURON DICELLANDROIDES. 1, flowering branch ($\times\frac{1}{2}$); 2, fruiting branch ($\times\frac{1}{2}$), 1–2 from *Richards* 3347 and *Le Testu* 544; 3, marginal part of leaf, superior (top) and inferior (lower) surface ($\times1$) *Richards* 3347; 4, flower bud ($\times2$) *Le Testu* 9077; 5, longitudinal section of flower ($\times2$); 6, stamens in lateral and anterior view ($\times3$), 5–6 from *Le Testu* 9077 and *Talbot* sn; 7, fruit ($\times2$); 8, same in longitudinal section to show placentation ($\times2$); 9, seed ($\times10$), 7–9 from *Le Testu* 544.

Phaeoneuron gracile Hutch. & Dalz. in F.W.T.A. **1**: 206 (1927) pro parte; in Kew
Bull. **1928**: 220 (1928) pro parte. Syntypes from Ivory Coast.
   Dicellandra gracilis (Hutch. & Dalz.) A. Chev. in Bull. Mus. Hist. Nat. Paris,
Sér. 2, **4**: 679 (1932) pro parte. Syntypes as above.

Perennial herb, undershrub or shrub up to 2·5 m. high, with the young branches,
petioles, peduncles and pedicels covered with minute rusty hairs. Leaf-lamina
5–20 × 2–10 cm., oblong, ovate to broadly ovate, ± acute at apex, denticulate at
margin, acute, subacute or obtuse to cordate at base, membranous, minutely hairy
on the nerves mainly on the underface, 5–7-nerved, the nerves prominent beneath,
with the numerous transverse nerves parallel, at right angles to the longitudinal
ones; petiole 1·5–8 cm. long, canaliculate above. Panicle many-flowered; pedicels
2–3 cm. long. Receptacle c. 5 × 5 mm. Sepals 0. Petals 10–12 × 7–8 mm., broadly
ovate, ± oblique at top, violet or rose. Stamens with filaments and anthers
c. 5·5–6 mm. long. Style 12–15 mm. long. Fruit globose, creamy-brown. Seeds
yellowish, c. 1 mm. long.

Zambia. W: Mwinilunga, fr. 19.v.1969, *Mutimushi* 3370 (K; SRGH).
   Extends from West tropical Africa to E. Sudan, Zaire and W. Zambia. Swampy places
in interior of "mushitus".

10. HAPLOPHYLLOPHORUS (Brenan) A. & R. Fernandes

Haplophyllophorus (Brenan) A. & R. Fernandes in Bol. Soc. Brot., Sér. 2, **46**:
70 (1972).
   Amphiblemma Sect. *Haplophyllophora* Brenan in Kew Bull. **1953**: 85 (1953).

Acaulescent herbs with the leaves (usually one, sometimes 2) and inflorescences
arising from a bulbil-bearing rhizome, the bulbils densely covered with stellate
bristles. Leaves longitudinally 5–9-nerved, petiolate. Flowers 5(4)-merous, in
subscorpioid single or branched cymes. Receptacle turbinate, hispid. Sepals
triangular, persistent. Petals ovate, obovate or oblong. Stamens 10(8), very
unequal; anthers of long stamens dehiscing by a terminal broad oblique pore and
with a long pedoconnective anticously furnished with an upcurved clavate, entire
or bilobed appendage; anthers of short stamens abortive. Ovary 5(4)-locular,
adherent to the receptacle nearly to the top; style columnar with a dilated
subcapitate stigma. Capsule turbinate, crowned by 5 scales. Seeds very small,
straight, straw-coloured with a lateral appendage at the top.
   A small genus of 2 species from Angola, Zaire and Zambia.

Long stamens with filaments 2–3 mm. long and yellow anthers 1·5–1·7 mm. long, furnished
   with bilobed 1–1·5 mm. long appendages; leaves with setulose long hairs on both
   faces; petiole 0·5–1·8 cm. long; inflorescence 1–3-flowered   1. *acaulis* var. *brevipes*
Long stamens with filaments 1·5–2 mm. long and violet anthers 1–1·3 mm. long, furnished
   with entire 0·5–0·7 mm. long appendages; leaves with setulose long hairs intermixed
   with minute ones (less than 0·1 mm. long) on both faces; petiole 4·5–9·5 cm. long;
   inflorescence up to 10-flowered -   -   -   -   -   -   -   2. *seretii*

1. **Haplophyllophorus acaulis** (Cogn.) A. & R. Fernandes in Bol. Soc. Brot., Sér. 2,
46: 70 (1972). Type from Angola.
   Amphiblemma acaule Cogn. in Bol. Soc. Brot. **11**: 89 (1893).—Gilg in Engl., Mon.
Afr. Pflanzen. **2**: 29 (1898).—Brenan in Kew Bull. **1953**: 87 (1953).—R. & A.
Fernandes in C.F.A. **4**: 164 (1970). Type as above.
   Cincinnobotrys acaulis (Cogn.) Gilg in Engl., Pflanzenw. Afr. **3**, 2: 757 (1921).
Type as above.

Var. **brevipes** (Brenan) A. & R. Fernandes in Bol. Soc. Brot., Sér. 2, **46**: 70 (1972).
TAB. **65**. Type: Zambia, Mwinilunga, Luakera Falls, *Milne-Redhead* 4348 (K,
holotype).
   Amphiblemma acaule var. *brevipes* Brenan in Kew Bull. **1953**: 87 (1953). Type as
above.

Perennial acaulous herb with a short rhizome. Leaf-lamina 1·5–7 × 1–6·5 cm.,
broadly ovate-cordate, obtusely to subacutely acuminate at the apex, serrulate at the
margin, the teeth ending in a bristle, rather rugose, green above, pinkish beneath,
with scattered long setulose hairs on both sides, longitudinally 5–7-nerved;
transverse nerves conspicuous, forming with the longitudinal ones a very open

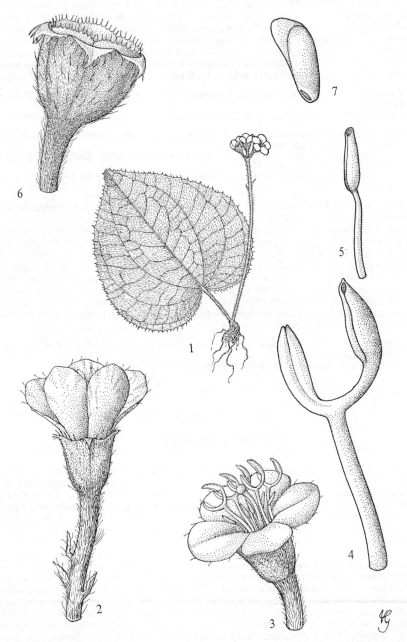

Tab. 65. HAPLOPHYLLOPHORUS ACAULIS var. BREVIPES.   1, habit (× ⅔); 2, flower in lateral
view (×4); 3, flower from above showing stamens and style (×4); 4 & 5, long and
short stamens respectively (×18); 6, fruiting calyx (×6); 7, seed (×32), all from
Milne-Redhead 4348.

reticulation; petiole 0·5–1·8 cm. long. Peduncle 3–7·5 cm. long, pinkish, hairy mainly upwards, with some minute scales, 1–3-flowered. Receptacle c. 3·5 × 3 mm., turbinate, densely covered with appressed hairs. Sepals c. 1 mm. long, hairy. Petals c. 5 × 3 mm., pale pink. Long stamens: filaments 2–3 mm. long, white; anthers 1·5–1·7 mm. long, yellow; pedoconnective crimson with bilobed appendage 1–1·5 mm. long. Small stamens: filaments 1 mm. long; anthers c. 1 mm. long, ± abortive. Fructiferous receptacle c. 4 × 3 mm. Seeds 0·75 mm. long, obovate.

Zambia. W: N. of Mwinilunga, Luakera Falls, fl. & fr. 25.i.1938, *Milne-Redhead* 4348 (K).
Known only from Zambia. On damp mossy rocks in shade of evergreen vegetation.

2. **Haplophyllophorus seretii** (De Wild.) A. & R. Fernandes in Bol. Soc. Brot., Sér. 2, **46**: 71 (1972). Type from Zaire.
    Cincinnobotrys seretii De Wild. in Ann. Mus. Congo Belge, Bot. Sér. 5, **2**: 330, t. 89 (1908). Type as above.
    Amphiblemma seretii (De Wild.) Brenan in Kew Bull. **1953**: 88 (1953). Type as above.

Perennial acaulous herb. Leaf-lamina 4·5–8 × 4·5–7 cm., cordate-ovate, shortly acuminate at the apex, serrulate-dentate at margins, green above, probably pinkish below, with bristles intermixed with very minute hairs c. 0·1 mm. long on both sides, with 7–9 longitudinal nerves, these forming a very open reticulation with the conspicuous transverse ones; petiole 2–11 cm. long. Peduncle 4–15 cm. long, hairy, with scattered small scales. Inflorescence 1–5(–10)-flowered. Receptacle c. 3 × 2·5 mm., hairy. Sepals c. 2 mm. long. Petals c. 5 mm. long, dull pink. Long stamens: filaments 2 mm. long; anthers 2 mm. long, violet; pedoconnective c. 1·5 mm. long, curved upwards, with a spur c. 1·75 mm. long, attenuate to the extremity. Short stamens: anthers abortive, 1·5 mm. long, yellow. Fructiferous receptacle 4 × 3 mm. Seeds minute, obovate.

Zambia. N: Mporokoso Distr., Lumangwe Falls, Kalungwishi R., 900 m., fl. & fr. 9.i.1960, *Richards* 12313 (K).
Also in Zaire. On steep rocky banks on peaty soil.

78. LYTHRACEAE
By A. Fernandes

Annual or perennial (sometimes marshy or aquatic) herbs, suffrutices, shrublets, shrubs or trees. Leaves simple, entire, 1-nerved or ± distinctly penninerved, decussate, sometimes verticillate, rarely alternate; stipules 0 or 2–10 or more, small, subulate, axillary. Flowers bisexual, actinomorphic or very rarely zygomorphic, (3)4–5(6–16)-merous, homomorphic or often di-trimorphic. Inflorescence various from single axillary flowers to fasciculate, cymose or paniculate; pedicels usually bracteolate. Calyx persistent, tubular, urceolate or campanulate, the lobes alternating often with ± developed appendages in the sinuses. Petals inserted in the calyx-tube, as many as and alternating with the calyx-lobes, sometimes few or 0, often clawed, membranous, corrugated and imbricate in the bud, equal or sometimes unequal. Stamens numerous (32–200) in the primitive woody genera, often as many as the calyx-lobes, sometimes fewer (2 or 1) in the more advanced herbaceous ones, equal or sometimes very unequal (in heterostylous plants), inserted on the calyx-tube; filaments free; anthers 2-locular, bent inwards in the bud, dorsifixed and versatile, rarely basifixed (*Pleurophora* and *Cranea*, not from F.Z. area), dehiscing by longitudinal slits. Disk absent or very small, cupular or unilateral. Ovary usually free, sessile or stipitate, 2–6-locular or 1-locular (*Cryptotheca*); placentation axile, rarely basal, sometimes the central axis not reaching to the top of the ovary; style absent, short or elongate and flexuous; stigma capitate or punctiform, rarely 2-lobed; ovules numerous to 2, small, anatropous, ascending. Fruit mostly a capsule included in the calyx-tube or ± exserted, indehiscent or opening loculicidally or septicidally by valves or by a transverse lid or irregularly, with the placentas forming a central column. Seeds

2–numerous, various in form, small, sometimes winged; embryo straight, with endosperm; cotyledons flat or rarely convolute.

A large family of c. 500 species in the tropics and subtropics, rare in the temperate regions. *Lagerstroemia indica* L. (*Simão* 56/48, LISC) is sometimes cultivated as a small ornamental tree.

Trees or shrubs:
  Dissepiments of the ovary complete and then the placenta continuous with the style; flowers 4-merous; stamens (4)8(12); fruit indehiscent; seeds not winged
  1. **Lawsonia**
  Dissepiments of the ovary interrupted above the placenta, the latter not continuous with the style; flowers 5–6-merous; fruit dehiscent; seeds winged:
    Small trees; flowers 5–6-merous, homomorphic, arranged in panicles; calyx turbinate, membranous, with folded very short appendages; stamens 5–6; ovary 2-locular; capsule included; seeds thinly winged; leaves minutely pellucid-punctate, with a large gland on the underside of the rounded apex
    2. **Galpinia**
    Shrubs, villous-tomentose; flowers 6-merous, sometimes dimorphic-heterostylous, axillary, solitary or rarely geminate; calyx campanulate, coriaceous, with short not folded appendages; stamens 12 or 18; ovary 1-locular; capsule slightly exserted; seeds with a thick wing; leaves not pellucid-punctate, subacute or obtuse at the apex and without a gland below - - - - - - 3. **Pemphis**
Herbs or suffrutices:
  Dissepiments of the ovary complete and then placenta continuous with the style; flowers in axillary ± pedunculate cymes or in contracted umbellate or capitate cymes enveloped by two large bracteoles; capsule dehiscing by a small apical operculum, the lower part persisting, subseptifragal or disrupting irregularly
  4. **Nesaea**
  Dissepiments of the ovary incomplete and then placenta not continuous with the style; flowers axillary and solitary or in ± pedunculate cymes or in glomerules at the expanded base of leaves; capsule dehiscing by valves or disrupting irregularly:
    Capsule disrupting irregularly with the walls not transversely striate (under the microscope); hygrophilous plants but usually not aquatic:
      Ovules and seeds numerous, small; flowers 3–5-merous; anthers yellow; capsule with relatively thick walls - - - - - - - 5. **Ammannia**
      Ovules and seeds 2–5, larger than above; flowers (3)4(5)-merous; anthers violet; capsule with very thin walls - - - - - 6. **Hionanthera**
    Capsule dehiscing by valves transversely striate (under the microscope); usually aquatic plants - - - - - - - - 7. **Rotala**

1. LAWSONIA L.

Lawsonia L., Sp. Pl. **1**: 349 (1753); Gen. Pl. ed. 5: 166 (1754).

Shrub or small tree. Leaves decussate, elliptic to oblanceolate, entire. Flowers 4-merous, bisexual, in lax leafy terminal panicles. Calyx turbinate with the tube short; lobes spreading and broadly ovate; appendages absent. Petals broadly ovate, cordate at base, very much corrugated in bud. Stamens exserted, 4 opposite to the sepals or often 8 in pairs also opposite to the sepals or rarely 9–12 ternate; filaments filiform-subulate; anthers broadly oblong. Ovary sessile, subglobose, 2–4-locular; style thick, continuous with placentas, persistent, a little longer than the stamens; stigma capitate. Fruit globose, indehiscent or disrupting irregularly. Seeds many, thick, 4-sided; cotyledons flat.

A monospecific genus.

Lawsonia inermis L., Sp. Pl. **1**: 349 (1753).—Koehne in Engl., Bot. Jahrb. **4**: 36 (1883); in Engl. & Prantl, Pflanzenfam. **3**, 7: 15, fig. 6 (1891); in Engl., Pflanzenr. **IV**, 216: 270, fig. 59 (1903).—Gilg in Engl., Pflanzenw. Ost-Afr. **C**: 286 (1895).—Th. & H. Dur., Syll. Fl. Cong.: 218 (1909).—De Wild. in Ann. Mus. Congo Belge, Bot. Sér. 5, **3**: 452 (1912).—Engl., Pflanzenw. Afr. **3**, 2: 655, fig. 288 (1921).—Fernandes & Diniz in Garcia de Orta **4**, 3: 388 (1956).—Boutique in F.C.B., Lythraceae: 25 (1967). TAB. **66**. Syntypes from Egypt and India.
  Lawsonia spinosa L., Sp. Pl. **1**: 349 (1753) non Lour. (1790). Type from India.
  Lawsonia alba Lam., Encycl. Méth. Bot. **3**: 106 (1789); Tab. Encycl. Méth. Bot. **2**: 435, t. 296 (1792).—DC., Prodr. **3**: 91 (1828).—Hiern in F.T.A. **2**: 483 (1871).—Perrier in Humbert, Fl. Madag., Lythraceae: 23, fig. III, 7–11 (1954). *Nom. superfl.*, based on *Lawsonia inermis*.
  Rotantha combretoides Bak. in Journ. Linn. Soc., Bot. **25**: 317 (1890). Syntypes from Madagascar.

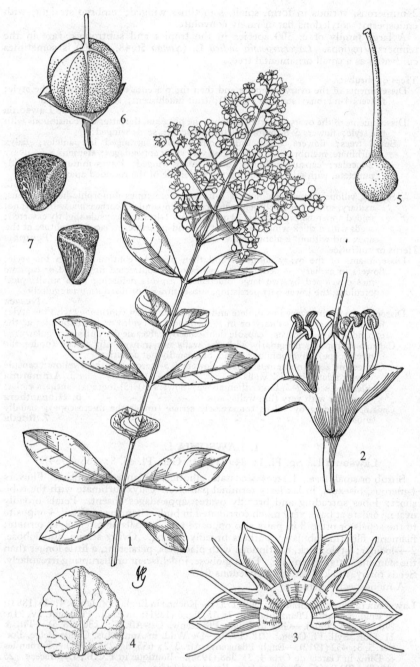

Tab. 66. LAWSONIA INERMIS. 1, flowering branch (× ⅔); 2, flower (×8); 3, calyx opened
to show filaments (×6); 4, petal (×8); 5, pistil (×8); 6, capsule (×4), 1–6 from
Rodrigues de Carvalho 1884–1885; 7, seeds (×6), from *Mendonça* 2351.

Evergreen shrub or small tree 2–7 m. high, glabrous, with the young branches 4-angular and the old ones terete, rigid and often spinescent. Leaves 1·2–6·7 × 0·5–2·7 cm., elliptic, obovate to oblanceolate, acute at the apex, sometimes mucronulate, narrowed into a short petiole, pinnately nerved, submembranous. Panicles 5–25 cm. long. Flowers whitish, scented; pedicels 2–4 mm. long; bracts 0·4–0·7 mm. long, linear, caducous. Calyx-tube 1–1·7 mm. high, with the lobes 2–3 mm. long, ovate-triangular. Petals 1·5–2·5 mm. long, slightly clawed, yellowish. Stamens with the filaments c. 5 mm. long. Style c. 3 mm. long. Fruits 4–6 × 4·5–8 mm. Seeds 2–2·6 mm. long.

Mozambique. N: Mossuril and Cabaceira, fl. & fr. 1884–85, *Carvalho* (COI). T: Tete, cultivated in a garden, 130 m., fl. 31.xii.1965, *Torre & Correia* 13992 (LISC). MS: Sena, Chemba, cultivated in a garden, fl. & fr. 4.x.1944, *Mendonça* 2351 (LISC). LM: Maputo, Jardim Municipal Vasco da Gama, fl. 20.xi.1945, *Gomes e Sousa* (COI).
Native of western tropical Asia, N. Africa and probably of the eastern coast of Africa. In alluvial soils along rivers and in villages.

Widespread by culture in the tropical regions of the world. Used mainly by the Arabs as a producer of dyes.

2. GALPINIA N. E. Br.
Galpinia N. E. Br. in Kew Bull. **1894**: 345 (1894).

Small tree. Leaves decussate, shortly petiolate. Flowers 5–6-merous in terminal panicles with 4-gonous branches. Calyx-tube campanulate, the lobes triangular, valvate, alternating with plicate appendages in the sinuses. Petals caducous, corrugated in bud. Stamens 6, opposite to the petals, circinate in bud, inserted below the middle of the calyx-tube, exserted; anthers dorsifixed, curved. Ovary subsessile, depressed cone-shaped, incompletely 2-locular; style subulate, a little shorter than the stamens, with a dot-like stigma; ovules many, erect or radiating on a basal placenta. Capsule ± exserted, thinly membranous, tipped by the style, dehiscing irregularly; seeds flat, winged all around.
A monospecific genus.

Galpinia transvaalica N. E. Br. in Kew Bull. **1894**: 346 (1894).—Oliv. in Hook., Ic. Pl. **24**: t. 2375 (1895).—Koehne in Engl., Bot. Jahrb. **29**: 164 (1900); in Engl. & Prantl, Pflanzenfam. Nachtr. **2**: 48 (1900); in Engl., Pflanzenr. IV, 216: 184, fig. 29 (1903).—Engl., Pflanzenw. Afr. **3**, 2: 649, fig. 284 (1921).—Burtt Davy, F.P.F.T.: 196, fig. 24 (1926).—Hutch., Botanist in S. Afr.: 339, fig. pag. 377 (1946).— Fernandes & Diniz in Garcia de Orta **4**, 3: 389 (1956).—Gomes e Sousa, Dendrol. Moçamb. **2**: 557, t. 168 (1967). TAB. 67. Type from the Transvaal.

Small tree 3–7 m. high, glabrous, the branches 4-gonous. Leaves (2)5–9 × (1)2–4·5 cm., oblong-obovate or elliptic-obovate or elliptic-oblong, rounded at the often plicate apex, margin narrowly recurved, base ± cuneate, coriaceous, densely and very minutely pellucid-dotted (at least under the microscope), pinnately nerved, the midrib with a gland on the under side just below the apex; petiole 3–4 mm. long. Panicles terminal, 6–12 cm. long. Calyx-tube c. 2·5 × 4 mm., the lobes triangular, erect, as long as the tube. Petals 4 mm. long, ovate-lanceolate, acute, shortly clawed, caducous, creamy whitish. Stamens exserted; filaments c. 8 mm. long; anthers subcircular, c. 1 mm. in diam. Ovary c. 2 mm. in diameter; style 6 mm. long. Capsule up to 5 mm. in diameter, globose. Seeds c. 2·5 mm. in diameter, subcircular, with a very thin wing.

Rhodesia. S: Nuanetsi, Malangwe R., SW. Mateke Hills, c. 615 m., fr. 6.v.1958, *Drummond* 5668 (COI; K; LISC; PRE; SRGH). **Mozambique.** LM: Maputo, Polana slope, fl. 22.i.1960, *Lemos & Balsinhas* 23 (COI; LISC; PRE; SRGH); Inhaca I., Ponta Ponduine, 0–200 m., fr. 8.vii.1958, *Mogg* 28053 (K; LISC; LMU; PRE).
Also in Transvaal and Swaziland. In dune scrub, littoral forests, evergreen thickets and savannas on argillaceous-sandy or rocky soils.

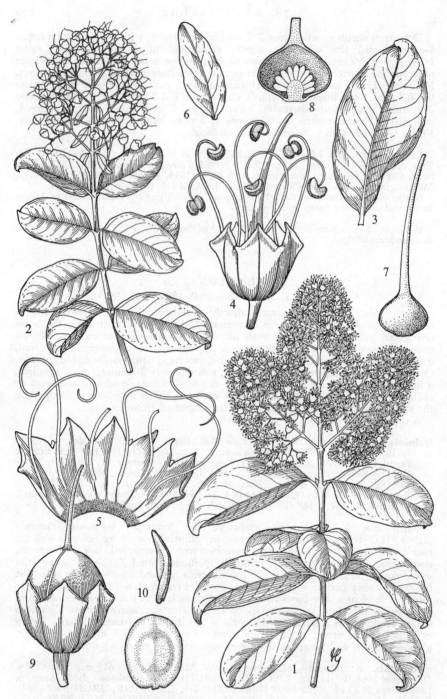

Tab. 67. GALPINIA TRANSVAALICA. 1, flowering branch (× ⅔); 2, fruiting branch (× ⅔); 3, leaf, inferior surface (×1); 4, flower (×8); 5, calyx opened to show insertion of stamens (×10); 6, petal (×8); 7, pistil (×8); 8, longitudinal section of ovary (×6); 9, fruit within calyx (×6); 10, seeds (×6), 1, 3–7 from *Myre & Balsinhas* 263, 2, 8–10 from *Balsinhas* 1226.

3. PEMPHIS J. R. & G. Forst.

Pemphis J. R. & G. Forst., Char. Gen. Pl.: 67, t. 34 (1776).

Shrublets, shrubs or small trees, silky-pilose, with the young branches sub-angular, the older ones terete. Leaves decussate, oblong-lanceolate, entire. Flowers axillary, solitary, rarely geminate, pedicellate, the pedicels 2-bracteolate at the base, the bracteoles very soon caducous. Flowers 6-merous, regular. Calyx turbinate-campanulate, coriaceous, with the lobes widely 3-angular and the appendages ± short, horn-shaped. Petals 6, obovate, corrugated. Stamens 12–18, 2- or 1-seriately inserted a little below the middle of the calyx-tube, subequal or alternately unequal; filaments filiform; anthers broadly elliptic. Ovary globose, shortly stipitate, 3-locular; style short or elongate; stigma 2-lobed; ovules numerous, erect. Capsule subglobose, ± enclosed in the calyx-tube, becoming almost 1-locular, dehiscing by a transverse slit at the middle (circumscissile). Seeds numerous, erect, imbricate, obcuneate, winged, the wing thickened.

Genus with 2 species, one littoral and widespread, the other confined to the low mountains of SW. Madagascar.

Pemphis acidula J. R. & G. Forst., Char. Gen. Pl.: 68, t. 34 (1776).—DC., Prodr. 3: 89 (1828).—Hiern in F.T.A. 2: 482 (1871).—Koehne in Engl., Bot. Jahrb. 3: 133 (1882); in Engl., Pflanzenr. IV, 216: 185, fig. 30B (1903).—Gilg in Engl., Pflanzenw. Ost-Afr. C: 285 (1895).—Engl., Pflanzenw. Afr. 3, 2: 651, fig. 285 (1921).—Perrier in Fl. Madag., Lythraceae: 20 (1954).—Fernandes & Diniz in Garcia de Orta 4, 3: 389 (1956).—Gomes e Sousa, Dendrol. Moçamb. 2: 555, t. 166 (1967).—Lewis in Proc. Roy. Soc., B, 178 (1050): 79–94 (1971). TAB. 68. Syntypes from the Pacific Islands.

Lythrum pemphis L. f., Suppl. Pl.: 249 (1781).—Lam., Tabl. Encycl. Méth. Bot. 2: 524, t. 408 fig. 2 (1794) *nom. illegit.* Based on *Pemphis acidula.*

Melanium fruticosum Spreng., Syst. Veg. 2: 455 (1825) *nom. illegit.* Based on *Pemphis acidula.*

Shrub up to 6 m. or small tree up to 11 m. high; nodes rather thickened, with conspicuous scars after leaf fall. Leaves 1–3·2 × 0·3–1·3 cm., ovate-oblong to linear-lanceolate, subacute or obtuse at the apex, narrowed to the base, 1-nerved, the nerve prominent beneath. Pedicels 5–13 mm. long; bracteoles c. 4 mm. Flowers heterostylous: short-styled flower with stamens sub-2-seriately inserted, alternately unequal, the longer ones slightly exceeding the calyx lobes, and the style not longer than the ovary; long-styled flower with the stamens not exceeding the sinuses and the style twice as long as the ovary. Calyx-tube 4–7 mm. long, 12-grooved, the lobes 2 × 2 mm., erect. Petals 6 × 3 mm., white or violet (*fide* Gomes e Sousa), obovate, shortly clawed. Stamens 12. Ovary c. 2 mm. in diameter. Capsule c. 5 mm. in diameter. Seeds up to 3 × 2 mm., reddish.

Mozambique. N: Palma, Cabo Delgado lighthouse, c. 10 m., fl. & fr. 17.iv.1964, *Torre & Paiva* 12123 (LISC); Tambuze I., fl. & fr. 2.v.1959, *Gomes e Sousa* 4451 (COI; K; LMA). SS: Vilanculos, Santa Carolina I., fr. 20.xi.1968, *Fernandes, Fernandes & Pereira* 179 (COI; LMU); Bazaruto I., Ponta Gengorime, fl. & fr. 29.x.1958, *Mogg* 28731 (K; LISC).

Widespread along the coast of E. Africa, Asia and Australia. Dunes, coral rocks, mangrove fringes and mangrove swamps.

4. NESAEA Commers. ex Juss.

Nesaea Commers. ex Juss., Gen. Pl.: 332 (1789).

Annual or perennial herbs or shrublets, rarely shrubs, glabrous, pilose or tomentose. Stems erect, usually 4-gonous. Leaves decussate or subdecussate, rarely 3-nate or alternate, entire, sessile or shortly petiolate, obscurely penninerved. Inflorescences lax or dense axillary cymes, sometimes capituliform and with 2–4 large bracteoles at the base, or flowers solitary. Flowers bisexual, sometimes heterostylous, 4–8-merous, bracteolate, the bracteoles very variable in size. Calyx turbinate-campanulate, campanulate, tubular, urceolate or semiglobose, herbaceous, with the tube 8–16-ribbed; appendages 0 or very short or longer than the calyx-lobes. Petals 0–8, deciduous, inserted in the throat of the calyx, sessile or clawed. Stamens 4–23, in 1 or 2 verticils, sometimes geminate or 3-nate, inserted at $\frac{1}{6}-\frac{1}{2}$ of the tube length, equal or alternately longer; filaments filiform; anthers dorsifixed,

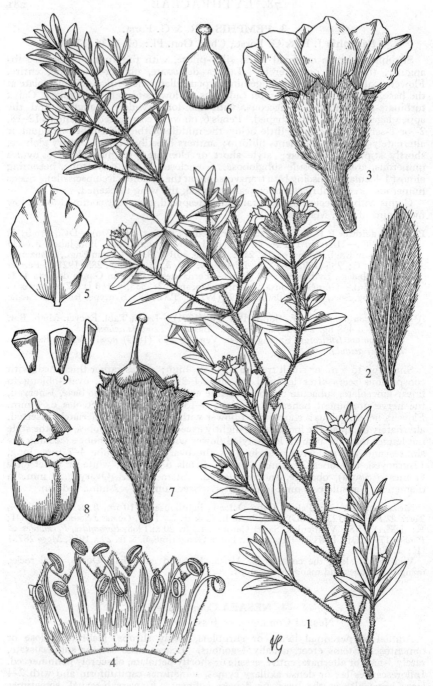

Tab. 68. PEMPHIS ACIDULA. 1, flowering branch (× ⅔) from *Pedro & Pedrógão* 5199; 2, leaf (×2); 3, lateral view of flower (×6); 4, calyx opened to show insertion of stamens (×6); 5, petal (×8); 6, pistil of a short-styled flower (×8); 7, capsule within calyx (×4); 8, dehisced capsule (×4); 9, seeds (×4), 2–9 from *Torre & Paiva* 12123.

didymous or oblong. Ovary sessile, 2–5-locular; ovules numerous; style 0 or short or very long; stigma capitate or punctiform. Capsule globose or ellipsoid, opening by an apical operculum, the lower part subsepticidal or dehiscing irregularly. Seeds ∞, small.

A genus with c. 50 species, mainly in Africa and Madagascar. Some species occur in tropical Asia, Australia and N. and Central America.

Operculum thin, tardily recognizable and soon becoming detached from the placentas; calyx subglobose, campanulate or rarely suburceolate:
  Style 0 or up to at most 1·5 mm. long; stamens usually as many as the calyx-lobes rarely up to 8 in some flowers (*N. crassicaulis*) or always 8 (*N. moggii*); petals present or absent:
    Dichasia many-flowered, producing very dense sessile or shortly pedunculate glomerules; flowers (4)5(6)-merous; stigma sessile or subsessile; appendages erect, twice as long as the calyx-lobes - - - - - 1. *sarcophylla*
    Dichasia 1–10(15)-flowered, sessile or ± pedunculate, ± loose; flowers 4(5)-merous; appendages shorter:
      Outer bracteoles shorter than the pedicels or attaining at most the middle of the calyx:
        Style attaining at most the ovary length:
          Style and stigma up to 0·3 mm. long; leaves and calyx usually minutely hairy; dichasia (1–3)5–15-flowered; calyx 1–1·5 mm. long; stamens 4; stem not-thickened - - - - - - - 2. *ondongana*
          Style and stigma 0·5–1 mm. long; leaves and calyx always glabrous; dichasia 1–7(many)-flowered; calyx 2–2·5 mm. long; stamens 4–6(7–8); stem usually thickened - - - - - - - 3. *crassicaulis*
        Style 1·5 mm. long, exceeding the ovary length; leaves 4–7 × 2–2·5 mm., elliptic or slighter obovate; calyx c. 2·5 mm. long, turbinate-campanulate; petals purple; stamens 8 - - - - - - - 4. *moggii*
      Outer bracteoles longer than the pedicels, attaining or exceeding the top of the calyx:
        Outer bracteoles not exceeding the tips of the flowers; upper leaves cordate or subcordate at the base; dichasia 3–few-flowered; plant usually minutely hairy - - - - - - - - - 5. *aspera*
        Outer bracteoles conspicuously exceeding the tips of the flowers; upper leaves roundish or attenuated into a short petiole at the base; dichasia 3–11-flowered; plant glabrous - - - - - - 6. *robinsoniana*
  Style much longer than 1·5 mm., rarely shorter and if so stamens exserted or leaves sagittate-cordate at the base or dichasia surrounded by very large bracteoles; stamens 4 or 8 or 10–16(23); petals present or sometimes absent:
    Calyx-appendages 0 or minute or as long as the calyx-lobes or if longer then leaves sagittate-cordate:
      Outer bracteoles oblong or linear, up to 2 mm. long:
        Flowers 5–8-merous, solitary or in 2–3(5)-flowered dichasia; stamens twice as many as the calyx-lobes; calyx-appendages 0 or minute:
          Leaves obtuse at the apex; calyx ± puberulous; flowers 6-merous, solitary or in 2–3-flowered dichasia, with dimorphic heterostyly - 7. *rigidula*
          Leaves acute at the apex; calyx glabrous; flowers homomorphic:
            Shrublet 10–40 cm. high, arising from a woody rootstock and branched from the base; dichasia (1–2)3-flowered; flowers 6–7(8)-merous
                                                                   8. *heptamera*
            Shrub usually 1 m. high, sometimes up to 2·25 m., branched upwards; dichasia 1–6-flowered; flowers (4)5(6)-merous - - 9. *fruticosa*
        Flowers 4-merous in 1–many-flowered dichasia; stamens 4 or 8; calyx-appendages 0 or sometimes exceeding the lobes:
          Stamens 8; plants glabrous; petals 2–3 mm. long:
            Shrublet virgately much branched; leaves 4–17 × 1–3(5) mm.; dichasia 1–3(5)-flowered; calyx-lobes equalling c. ⅓ of the tube length; ovary ellipsoid - - - - - - - - 10. *schinzii*
            Herb with ascending branches; leaves up to 30 × 10 mm.; dichasia 2–7(many)-flowered; calyx-lobes equalling ⅓ of the tube length
                                                                   11. *ramosissima*
          Stamens 4 (4–8 in *N. dinteri* subsp. *elata*); plants glabrous or hairy; petals 0·5–1·5 mm. long, sometimes 0:
            Annual glabrous herbs; petals 4, pink or pinkish; stamens (when 4) opposite to the calyx-lobes:
              Erect herb up to 65 cm. high, usually branched at the middle, rarely from the base; leaves 5–45 × 1·5–7 mm., hastate-cordate at the base; appendages as long as the calyx lobes; petals pink, c. 1 × 0·75 mm.; style c. 2·25 mm. long - - - - 12. *dinteri* subsp. *elata*

Herb up to 15 cm. high much branched from the base to the apex, the
    lower branches procumbent; leaves 5–22 × 1–2·5 mm., ± auriculate
    at the base; appendages very short; petals pinkish, c. 0·5 mm. in
    diameter; style 0·75–1 mm. long -   -   -   -   13. *gazensis*
Perennial hairy undershrub with woody rootstock; leaves 10–16 × 2·5–
    5 mm., deeply sagittate-cordate at the base; appendages twice as long
    as the calyx-lobes; petals 0–4, yellowish, scarcely 1 mm. long; stamens
    opposite to the petals   -   -   -   -   -   14. *passerinoides*
Outer bracteoles larger, foliaceous, up to 3 × 1·5 mm., the inner ones many,
    purplish; leaves 10–20 × 2·5–6 mm., lanceolate, cordate at the base; dichasia
    7–10-flowered; petals c. 2 × 1·25 mm., obovate; stamens 8; style 3·5 mm.
    long; whole plant flushed purple   -   -   -   -   15. *purpurascens*
Calyx-appendages longer than the lobes, usually ciliate; dichasia surrounded by large
    bracteoles:
Dichasia in axillary leafy racemes, long or short, or in very dense terminal heads;
    involucral bracteoles large, oblong or spathulate:
Involucral bracteoles oblong, acuminate, c. 5·5 × 4 mm.; dichasia axillary,
    3–7-flowered, at the end of peduncles 1–6 mm. long; appendages c.
    1·75 mm. long, glabrous or ciliolate at the apex; style c. 4 mm. long;
    flowers homomorphic -   -   -   -   -   -   - 16. *tolypobotrys*
Involucral bracteoles expanded upwards, broadly spathulate and shortly
    acuminate; dichasia sessile, aggregated at the ends of stems and branches in
    large very dense heads; appendages c. 2 mm. long, ciliolate; calyx 3·5–
    4·5 mm. long; flowers dimorphic-heterostylous   -   -   17. *linearis*
Dichasia in axillary heads; involucral bracteoles 2–4, very large, rounded or
    kidney-shaped, abruptly acuminate:
Appendages 0·75 mm. long, scarcely exceeding the lobes; stamens and style
    included; leaves 10–30 × 1·5–2·5 mm., narrowly linear-lanceolate
                                                                18. *rautanenii*
Appendages much longer (see however *N. pedroi*) than the lobes; stamens and
    style exserted:
Leaves neither cordate nor truncate at the base:
Leaves 10–55 × 4–25 mm., oblong or elliptic to elliptic-lanceolate; dichasia
    many-flowered; peduncles 7–55 mm. long; flowers 5–6(7)-merous;
    stamens usually as many as the calyx-lobes; style 6–8 mm. long
                                                                19. *radicans*
Leaves 8–30 × 1·5–6(10) mm., linear or narrowly oblong; peduncle
    1–15 mm. long:
Flowers homomorphic or dimorphic-heterostylous; involucral bracteoles
    7–9 × 6–8 mm.:
Flowers homomorphic; root not tuberous; leaves linear or lanceolate-
    oblong:
Annual erect herb 5–35 cm. high; appendages 1–1·5 mm. long; style
    2·5–3·5 mm. long; leaves linear or lanceolate-oblong
                                                                20. *erecta*
Suffruticose herb up to 70 cm. high, much branched; appendages
    1·75 mm. long; style c. 6·5 mm. long; leaves linear
                                                                21. *ramosa*
Flowers dimorphic-heterostylous; perennial herb with tuberous root;
    leaves spathulate   -   -   -   -   -   - 22. *spathulata*
Flowers trimorphic-heterostylous; involucral bracteoles 2·4 × 5 mm.;
    appendages 0·75 mm. long   -   -   -   -   - 23. *pedroi*
Leaves (at least the upper ones) cordate or truncate at the base:
Leaves 5–20(25) × 5–10 mm., cordate at the base; dichasia ± 5-flowered;
    peduncle 5–27 mm. long; involucral bracteoles 4–7 × 4–7 mm.;
    appendages c. 1 mm. long; style 1·5–2 mm. long   -   24. *cordata*
Leaves up to 9 × 3 mm., truncate, roundish or slightly cordate at the base;
    dichasia 1–3-flowered; peduncle 1·5–4 mm. long; involucral bracteoles
    c. 2·5 × 1·25 mm.; appendages longer than the lobes; style c. 1·35 mm.
    long   -   -   -   -   -   -   -   -   - 25. *pygmaea*
Operculum thick, early recognizable, persisting for some time attached to the placentas;
    calyx narrowly tubular-urceolate; stamens 8, included -   -   - 26. *drummondii*

1. **Nesaea sarcophylla** (Welw. ex Hiern) Koehne in Engl., Bot. Jahrb. **3**: 328 (1882);
    in Engl., Pflanzenr. **IV**, 216: 228, fig. 43E (1903).—Hiern, Cat. Afr. Pl. Welw. **1**:
    375 (1898).—Engl., Pflanzenw. Afr. **3**, 2: 651 (1921).—Fernandes & Diniz in Garcia
    de Orta **6**: 104 (1958).—Fernandes in C.F.A. **4**: 182 (1970). Type from Angola.
    Ammannia sarcophylla Welw. ex Hiern in F.T.A. **2**: 479 (1871). Type as above.

Annual (or perennial?) decumbent herb up to 40 cm. long. Stems 4-gonous, scabrous at the angles, thick (c. 3 mm.), rooting at the lower nodes. Leaves up to 40 × 15 mm., decussate, ovate-lanceolate, obtuse at the apex, auriculate-cordate at the base, glaucous, 1-nerved, the nerve impressed above and ± prominent and minutely hairy beneath. Dichasia many-flowered, producing very dense sessile or shortly pedunculate glomerules; bracteoles lanceolate, ciliate, ± equalling the calyx. Flowers (4)5(6)-merous. Calyx 1·75–2 mm. long, semiglobose; lobes equalling ⅓ of the tube length; appendages erect, twice as long as the calyx-lobes, ciliolate. Petals 0. Stamens 1, 2 or 3, opposite the lobes. Ovary globose, 2–3-locular; stigma sessile. Capsule globose, not or scarcely exserted. Seeds c. 0·4 mm. long, brownish.

Mozambique. SS: Limpopo, 48 km. on the road Vila Teixeira Pinto to Mabuiapanse, fl. & fr. 20.viii.1969, *Correia & Marques* 1099 (COI; LMU).
Also in Angola and S. Africa. In moist mud or sand along rivers and in marshy places.

2. **Nesaea ondongana** Koehne in Mém. Herb. Boiss. **10**: 78 (1900); in Engl., Bot. Jahrb. **29**: 165 (1900); in Engl., Pflanzenr. **IV**, 216: 225 (1903).—Engl., Pflanzenw. Afr. **3**, 2: 651 (1921).—Fernandes & Diniz in Garcia de Orta **6**: 104 (1958).— Pohnert & Roessler in Prodr. Fl. SW. Afr. **95**: 7 (1966).—Fernandes in C.F.A. **4**: 182 (1970); in Garcia de Orta, Sér. Bot. **2**: 78, t. 1 (1975). Type from SW. Africa.

Annual, prostrate or ascending, sometimes erect, glabrous or minutely hairy herb, up to 25 cm. high (probably more in subsp. *orientalis*). Stems branched at the base, 4-gonous, the branches slender, prostrate or ascending. Leaves 5–25 × 1–6 mm., decussate, narrowly elliptic or narrowly lanceolate, obtuse or subacute at the apex, margin scabrous, all cuneate at the base or the upper ones rounded or cordate and the lower ones cuneate, 1-nerved. Dichasia axillary, 1–15-flowered, sessile or nearly so; pedicels 1–4 mm. long, slender, 4-gonous, glabrous or minutely hairy, bracteolate at the base, the bracteoles c. 1 mm. long, linear, scarious. Flower 4-merous. Calyx 1–1·5 mm. long, campanulate; lobes triangular, equalling c. ¼ of the calyx-tube length; appendages shorter or scarcely longer than the lobes, without cilia or ciliolate at the apex. Petals 0. Stamens 4, opposite to the lobes, inserted a little below the middle of the calyx-tube and attaining the apex of the lobes. Ovary c. 0·6 mm. in diameter, globose; stigma sessile or style and stigma up to 0·3 mm. long. Capsule globose, exceeding the calyx lobes. Seeds many, c. 0·45 mm. long, brownish.

Erect, decumbent or prostrate herbs, branched at the base rarely about the middle
- - - subsp. *ondongana*
Style and stigma shorter than 0·3 mm.; dichasia usually 1–10-flowered:
  Erect or decumbent; calyx c. 1·5 mm. long; appendages as long as the calyx-lobes;
  capsule c. 1·5 mm. in diam. - - - - - - var. *ondongana*
  Prostrate; calyx c. 1 mm. long; appendages shorter than the calyx-lobes; capsule
  c. 1 mm. in diam. - - - - - - - var. *evansiana*
Style and stigma 0·30 mm. long; dichasia usually 5–15-flowered - var. *beirana*
Stronger plants much and densely branched upwards - - - subsp. *orientalis*

Subsp. **ondongana**.
Var. **ondongana**. TAB. 69.
  Ammannia baccifera subsp. *intermedia* Koehne in Engl., Bot. Jahrb. **41**: 79 (1908). Type: Mozambique, Lourenço Marques, Matola, *Quintas* 101 (COI, holotype).
  Ammannia intermedia (Koehne) Fernandes & Diniz in Garcia de Orta **4**, 3: 406 (1956). Type as above.

Caprivi Strip. c. 11 km. N. of Katima Mulilo on road to Ngoma, fl. 22.xii.1958, *Killick & Leistner* 3025 (PRE; SRGH). **Botswana.** N: Mumpswe Pan, 40 km. NNW. of mouth of Nata R., fl. & fr. 21.iv.1957, *Drummond & Seagrief* 5169 (K; PRE; SRGH). SE: Molepolole, Khungwane, 101 m., fl. & fr. 14.vi.1955, *Story* 4886 (PRE). **Zambia.** S: Mazabuka, edge of Kafue Flats, fl. & fr. 7.iv.1955, *E. M. & W.* 1430 (BM; K; LISC). **Rhodesia.** N: Kariba, Sengwe, Lake Kariba, 508 m., fl. & fr. 19.xii.1964, *Mitchell* 913 (COI; K; SRGH). W: Wankie, Makwa Pan, fl. & fr. 17.iv.1972, *Gibbs Russell* 1596 (COI). S: Nuanetsi R., Mabalanta, Gona-re-zhou Game Reserve, fl. & fr. 3.vi.1971, *Ngoni* 162A (COI). **Mozambique.** SS: R. Limpopo, Dumela, fl. & fr. 30.iv.1961, *Drummond & Rutherford-Smith* 7619 (COI; SRGH). LM: Costa do Sol, fl. & fr. 11.xi.1963, *Balsinhas* 673 (K; LISC; LMA); idem, *Grosvenor* 695 (COI).
Also in Angola and SW. Africa. In swamps, seasonal dambos, moist sandy soils, moist grassy plains, marshy ground along rivers and river-beds.

Var. **evansiana** (Fernandes & Diniz) Fernandes in Bol. Soc. Brot., Sér. 2, **48**: 115 (1974).

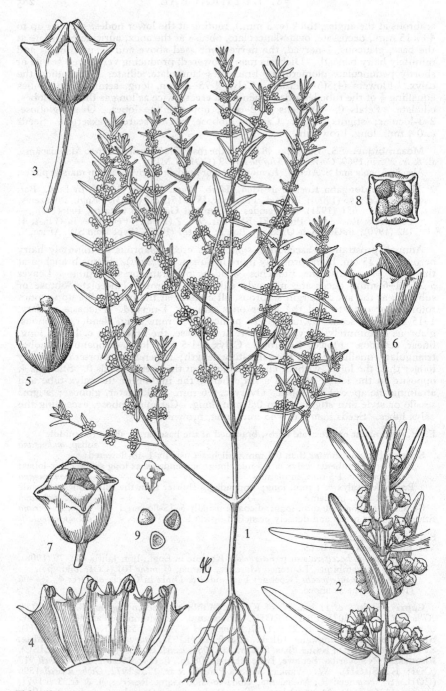

Tab. 69. NESAEA ONDONGANA subsp. ONDONGANA var. ONDONGANA. 1, habit (×1); 2, upper part of inflorescence (×4); 3, flower-bud (×24); 4, calyx opened (×24); 5, pistil (×24); 6, capsule within calyx (×18); 7, capsule dehisced within calyx (×18); 8, capsule from above (×12); 9, seeds and operculum (×12), all from *Quintas* 101.

Type: Botswana, Kachikau, Chobe R., *Pole Evans* 4187 (IFAN; K; PRE; SRGH, holotype).
Ammannia evansiana Fernandes & Diniz in Bol. Soc. Brot., Sér. 2, **31**: 155, t. 6 (1957). Type as above.

Caprivi Strip. Grootfontein-Nord, Andara, 1939, *Volk* 2104 (M). **Botswana. N:** 58 km. N. of Kachikau, on road to Kasane, Chobe R., 11.vii.1937, *Pole Evans* 4187 (IFAN; K; PRE; SRGH). **Zambia. B:** 13 km. from Nangweshi, 1088 m., fl. & fr. 23.vii.1952, *Codd* 7176 (BM; EA; K; PRE; SRGH). **S:** Livingstone Distr., Katambora to Kasangulu, 960 m., fl. & fr. 25.viii.1947, *Greenway & Brenan* 7976 (EA; PRE; SRGH). **Mozambique. SS:** Caniçado, 74 km. from Maginge on the road to Mabuiapanse, fl. & fr. 29.viii.1969, *Correia & Marques* 1337 (COI; LMU).
Also in Angola and SW. Africa. Ecology as for the type variety.

A specimen from Zambia, B: 30 km. W. of Kalabo on Likongo road, fl. & fr. 14.xi.1959, *Drummond & Cookson* 6874 (SRGH), has a style c. 0·25 mm. long.

Var. **beirana** Fernandes in Bol. Soc. Brot., Sér. 2, **48**: 115, t. 1 (1974). TAB. **70**. Type: Mozambique, Beira, N. of Macuti Beach, *Noel* 2488 (K; LISC; SRGH, holotype).
Ammannia senegalensis sensu Fernandes & Diniz in Garcia de Orta **4**, 3: 405 (1956) pro parte quoad specim. *Chase* 2489 (SRGH).

Zambia. S: Namwala, Umbuzu on the Ngabo to Kwichila road, fl. & fr. 18.iv.1963, *van Rensburg* 2041 (K; SRGH). **Rhodesia. W:** Victoria Falls, fl. & fr. 13.vii.1952, *Codd* 7065 (BM; K; PRE; SRGH). **S:** Sabi-Lundi Junction, Chitsa's Kraal, 256 m., fl. & fr. 4.vi.1950, *Wild* 3336 (K; PRE; SRGH); Ndanga Distr., Chitsa's Kraal, Sabi R., fl. & fr. 5.vi.1950, *Chase* 2270 (BM; SRGH). **Malawi. N:** Songwe to Karonga, fl. vii.1896, *Whyte* 59 (K). **Mozambique. MS:** Marínguà, Sabi R., 192 m., fl. & fr. 25.vi.1950, *Chase* 2489 (BM; K; SRGH).
Known only from the F.Z. area. On sandbanks at edges of swamps and beside rivers.

Subsp. **orientalis** Fernandes in Bol. Soc. Brot., Sér. 2, **48**: 115, t. 2 (1974). Type: Rhodesia, Chipinga, Chibuwe Pan, fl. & fr. vi.1972, *Gibbs Russell* 2083 (COI, holotype).

Rhodesia. E: Chipinga, Chibuwe Pan, fl. & fr. vi.1972, *Gibbs Russell* 2083 (COI). Known only from Rhodesia. Habitat unknown; probably edges of pans and rivers.

3. **Nesaea crassicaulis** (Guill. & Perr.) Koehne in Engl., Bot. Jahrb. **3**: 324 (1882); in Engl., Pflanzenr. **IV**, 216: 225, fig. 43A (1903).—R. E. Fr. in Wiss. Ergebn. Schwed. Rhod.-Kongo-Exped. **1**: 165 (1914).—Engl., Pflanzenw. Afr. **3**, 2: 651 (1921).— Fernandes & Diniz in Garcia de Orta **6**: 104 (1958).—Pohnert & Roessler in Prodr. Fl. SW. Afr. **95**: 6 (1966).—Boutique in F.C.B., Lythraceae: 18 (1967).—Fernandes in C.F.A. **4**: 182 (1970). Type from Senegambia.
Ammannia crassicaulis Guill. & Perr. in Guill., Perr. & A. Rich., Fl. Senegamb. Tent. **1**: 303 (1833).—Hiern in F.T.A. **2**: 479 (1871).—Wild, Guide Fl. Vict. Falls: 144 (1953). Type as above.

Annual or perennial glabrous prostrate creeping semi-succulent herb, 7–50 cm. (up to 2 m. when aquatic) long. Stems sub-4-gonous, usually thick, rooting at the lower nodes. Leaves 5–60 × 3–10 mm., decussate, sessile, obovate-oblong to oblanceolate or lanceolate-elliptic, obtuse or roundish at the apex, acute or cuneate-attenuated at the base, 1-nerved, nerve impressed above, prominent beneath. Dichasia loose, 1–7(many)-flowered, sessile or pedunculate, the peduncle 0·5–2(10) mm. long; bracteoles 0·5–1 mm. long, linear-lanceolate; pedicels 2–6 mm. long. Flowers 4-merous. Calyx 2–2·5 mm. long, almost globose; lobes 0·5–0·6 mm. long, triangular; appendages spreading, equalling the lobes. Petals 4, 1–1·5 mm. long, broadly elliptic to subcircular, mauve, sometimes 0. Stamens 4–6(7–8), included. Ovary c. 1 mm. in diam., globose, 2-locular; style 0·5–1 mm. long. Capsule c. 2 mm. in diameter, subglobose, equalling the calyx-lobes. Seeds c. 0·5 mm. long, brownish, concave-convex.

Caprivi Strip. Singalamwe, 1056 m., fl. & fr. 31.xii.1958, *Killick & Leistner* 3215 (COI; PRE). **Botswana. N:** Soda Pan, Upper Maragha Valley near Mushu, 36 km. WSW. of Maun, fl. & fr. 20.iii.1965, *Wild & Drummond* 7185 (COI; K; PRE; SRGH). **Zambia. B:** between Mongu and Lealui, fl. & fr. 12.xii.1959, *Drummond & Cookson* 6362 (COI; K; LISC; PRE; SRGH). **N:** Mporokoso, Chishyela Dambo, Mweru-wa-Ntipa, 992 m., fl. 5.viii.1962, *Tyrer* 288 (BM; SRGH). **W:** Ndola, Kafubu, fl. & fr. 6.vi.1964, *Mutimushi* 725 (K; SRGH). **S:** Kafue Flats W. of Mazabuka, fl. & fr. v.1962,

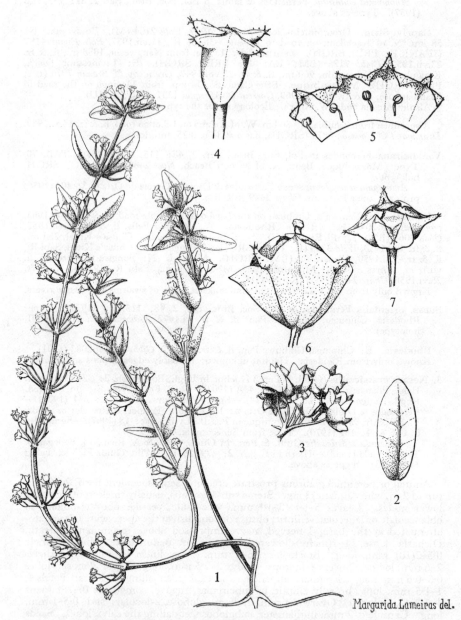

Tab. 70. NESAEA ONDONGANA var. BEIRANA. 1, habit (×1); 2, leaf (×3); 3, dichasium
(×3); 4, flower-bud (×12½); 5, calyx opened (×6); 6, capsule within calyx
(×12½); 7, capsule dehisced within calyx (×6), all from *Noel* 2488.

Margarida Lameiras del.

Brockington 5 (SRGH). **Rhodesia.** W: Wankie, Victoria Falls, fl. & fr. 22.x.1959, *Wild* 4850 (K; LISC; SRGH). **Mozambique.** SS: Bilene, S. Martinho, Parque Flores, fl. & fr. 17.viii.1965, *Balsinhas* 941 (COI; LISC; LMU). LM: Inhaca I., 37 km. from Lourenço Marques, Ponta Rasa swamp, 0–200 m., fl. & fr. 8.vii.1958, *Mogg* 28028 (K; LMA; LMU; PRE; SRGH); Maputo, Lagoon Satine, fl. & fr. 10.vii.1971, *Correia & Marques* 2135 (LMU).

Widespread in tropical Africa. At edges of swamps, pools, along edges and beds of rivers, in moist grasses near waterfalls, sometimes submerged, etc.

4. **Nesaea moggii** Fernandes in Bol. Soc. Brot., Sér. 2, **48**: 116, t. 3 (1974). TAB. **71.**
   Type: Mozambique, António Enes, Nantangula Beach, *Mogg* 32410 (LISC, holotype).

Decumbent or sometimes erect glabrous herb up to 8 cm. high. Creeping stems subterete, rooting at the lower nodes, the erect ones narrowly 4-winged. Leaves 4–7 × 2–2·5 mm., decussate, elliptic or slightly obovate, rounded at the apex, with margin minutely scabrous, tapering to the base into a short petiole, yellowish green, penninerved. Dichasia 1(3)-flowered, sessile or with peduncle up to 2 mm. long; pedicels 3–4 mm. long, slender, bracteolate at the base, the bracteoles c. 1 mm. long, linear. Flowers 4-merous. Calyx 2·5 mm. long, turbinate-campanulate, 8-ribbed; lobes 1·5 mm. broad and 0·5 mm. high, triangular; appendages spreading, as long as the calyx-lobes. Petals c. 3 × 3 mm., purple, roundish, veined. Stamens 8, inserted slightly above the bottom of the calyx-tube, the episepalous with filaments c. 2·5 mm. long, exceeding the lobes by c. 1 mm., the epipetalous with filaments c. 1·5 mm. long, included. Ovary c. 1·25 × 1·1 mm., 2-locular, ovoid; style c. 1·5 mm. long, attaining the apex of the calyx-lobes. Capsule globose, included or slightly exserted. Seeds c. 0·5 mm., brownish, concave-convex.

Mozambique. N: António Enes, near Nantangula Beach, c. 2 m., fl. & fr. 20.x.1965, *Mogg* 32410 (LISC).
Known only from Mozambique. Freshwater swamps near the coast.

5. **Nesaea aspera** (Guill. & Perr.) Koehne in Engl., Bot. Jahrb. **3**: 327 (1882); in Engl., Pflanzenr. **IV**, 216: 226 (1903).—Hiern, Cat. Afr. Pl. Welw. **1**: 375 (1898).—Engl., Pflanzenr. Afr. **3**, 2: 651 (1921).—Fernandes & Diniz in Garcia de Orta **6**: 105 (1958).—Pohnert & Roessler in Prodr. Fl. SW. Afr. **95**: 5 (1966).—Fernandes in C.F.A. **4**: 183 (1970). Type from Senegambia.
   Ammannia aspera Guill. & Perr. in Guill., Perr. & A. Rich., Fl. Senegamb. Tent **1**: 304 (1833). Type as above.

Annual erect or ascending herb, 4–25 cm. high. Stem 4-gonous, glabrous or minutely scabrous or hispidulous, branched at the base, the lower branches procumbent. Leaves 7–27 × 3–10 mm., decussate, lanceolate, oblong-lanceolate or elliptic, all subacute at the apex, the upper ones cordate or subcordate, the lower ones cuneate-attenuate at the base, subglaucous, sometimes minutely hispidulous beneath on the nerve and at the margin. Dichasia 3–few-flowered, sessile or ± long-pedunculate; pedicel of the middle flower up to 4 mm. long; bracteoles almost as long as the calyx, oblong-lanceolate, whitish, with green or greenish nerve. Flowers 4(5–6)-merous. Calyx c. 2 mm. long, sometimes hirtellous; lobes ⅓–¼ of the tube length; appendages erect, c. twice as long as the lobes, ciliolate at the top. Petals 0 or very rarely present. Stamens 4(5–6), episepalous, with the filaments c. 1 mm. long, inserted below the middle of calyx-tube and attaining the lobe-apices. Ovary 2-locular; style and stigma as long as ⅓–½ of the ovary length (c. 0·5 mm.). Capsule c. 1·5 mm. in diameter, scarcely exceeding the calyx-lobes. Seeds c. 0·25 mm., brownish.

Rhodesia. W: Wankie National Park, 2nd Ngamo Pan, ± 100 km. SE. of Main Camp, fl. & fr. 17.iv.1972, *Grosvenor* 709 (COI). C: Gwelo Distr., Mlezu School Farm, 29 km. SSE. of Que Que, 1344 m., fl. & fr. 2.ii.1966, *Biegel* 873 (COI; K; SRGH). S: near Lundi R. bridge, fl. & fr. 2.v.1962, *Drummond* 7852 (COI; K; SRGH).
Mozambique. SS: Dumela, R. Limpopo, fl. & fr. 30.iv.1961, *Drummond & Rutherford-Smith* 7620 (COI; K; SRGH) and 7621 (COI; K; SRGH).
In tropical Africa. In moist sand or mud at the edge of rivers, dams and pools.

Petals were seen in the specimen from **Rhodesia**, C: Salisbury, Prince Edward Dam, 1536 m., fl. & fr. 14.vii.1931, *Brain* 5499 (SRGH).

K

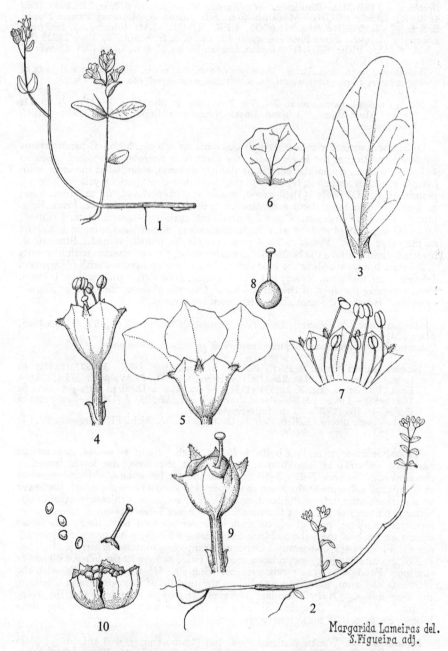

Margarida Lameiras del.
S.Figueira adj.

Tab. 71. NESAEA MOGGII. 1, 2, habit (×1); 3, leaf (×6); 4, flower (without petals), pedicel and bracteoles (×6); 5, flower with petals (×6); 6, petal (×6); 7, calyx opened (×6); 8, pistil (×6); 9, capsule within calyx; 10, dehisced capsule, operculum with style and seeds (×6), all from *Mogg 32410*.

6. **Nesaea robinsoniana** Fernandes in Bol. Soc. Brot., Sér. 2, **48**: 116, t. 4 (1974).
TAB. 72. Type: Zambia, 95 km. E. of Kasama, fl. 6.v.1962, *Robinson* 5137 (COI;
EA; K; SRGH, holotype).

Erect or decumbent glabrous annual herb, up to 20 cm. high. Stem subterete
towards the base, 4-gonous or 4-winged above, branched sometimes from the base.
Leaves 10–25 × 2–6 mm., decussate, narrowly lanceolate, acute at the apex, margin
slightly scabrous, rounded or attenuate into a short petiole at the base, 1-nerved.
Dichasia axillary, 3–11-flowered, pedunculate, the peduncles 1–10 mm. long; outer
bracteoles c. 6 × 1·25 mm., foliaceous with the basal part slightly scarious, recurved,
sometimes folded at the midrib; inner bracteoles c. 2·25 × 0·5 mm., linear, greenish
upwards, becoming shorter and completely scarious inwards. Flowers 4–5-merous,
pedicellate, the pedicels c. 1·5 mm. long. Calyx 1·5–2 mm., turbinate-campanulate;
lobes c. 1 mm. broad and 0·5 mm. high, deltate, inflected inwards; appendages
spreading, c. 1 mm. long, curved upwards, glabrous or ciliolate at the apex.
Petals 0. Stamens as many as and opposite the calyx-lobes, inserted slightly
below the middle of calyx-tube, attaining the apex of the calyx-lobes. Ovary
0·75 × 0·5 mm., ovoid, 2-locular; style and stigma c. 0·5 mm. long. Capsule
globose, included. Seeds c. 0·25 mm. in diameter, yellowish, concave-convex.

Zambia. N: 95 km. E. of Kasama, fl. & fr. 6.v.1962, *Robinson* 5137 (COI; EA; K;
SRGH).
Known only from Zambia. Muddy places.

7. **Nesaea rigidula** (Sond.) Koehne in Engl., Bot. Jahrb. **3**: 333 (1882); op. cit. **29**: 166
    (1900); in Mém. Herb. Boiss. no. 20: 24 (1900); in Engl., Pflanzenr. **IV**, 216: 235
    (1903); in Warb., Kunene-Samb.-Exped. Baum: 312 (1903).—Engl., Pflanzenw.
    Afr. **3**, 2: 652 (1921).—Wild, Guide Fl. Vict. Falls: 145 (1953).—Fernandes & Diniz
    in Garcia de Orta **6**: 108 (1958).—Pohnert & Roessler in Prodr. Fl. SW. Afr. **95**: 7
    (1966).—Fernandes in C.F.A. **4**: 184 (1970). Type from the Transvaal.
    Lythrum rigidulum Sond. in Linnaea **23**: 42 (1850); in Harv. & Sond., F.C. **2**: 516
    (1862). Type as above.
    Nesaea mucronata Koehne in Verh. Bot. Ver. Prov. Brand. **30**: 249 (1888); in
    Engl., tom. cit.: 236 (1903).—Engl., tom. cit.: 653 (1921).—Fernandes & Diniz,
    loc. cit. Type from SW. Africa.

Shrublet erect or decumbent, 8–35 cm. high, glabrous or shortly hairy. Stems
branched, 4-gonous or narrowly 4-winged upwards. Leaves 3–25 × 1·5–7 mm.,
decussate, oblong to lanceolate, obtuse or subacute at the apex, cordate at the base,
glaucous, 1-nerved. Flowers 6-merous, dimorphic-heterostylous, solitary or in
(1)2–3-flowered dichasia; pedicels c. 2–10 mm. long, puberulous; bracteoles
1–1·5 mm. long, oblong. Calyx 4–6 mm. long, turbinate-campanulate or
cupuliform; lobes equalling c. ⅓ of the calyx-tube length; appendages very short
or 0. Petals as long as the calyx, white, rose, salmon, orange or red. Stamens 12.
Long-styled flowers: episepalous stamens exceeding the calyx-lobes, the epipetalous
ones c. as long as the lobes; style greatly exceeding the stamens. Short-styled
flowers: episepalous and epipetalous stamens exserted; style equalling or scarcely
exceeding the lobes. Ovary obovoid or oblong, 2-lobular. Capsule included.
Seeds 0·3 mm. in diameter, brownish, concave-convex.

Caprivi Strip. Mpilila I., 960 m., fl. & fr. 12.i.1959, *Killick & Leistner* 3323 (PRE;
SRGH). **Botswana.** N: Kangwa (Xanwe), 27 km. NE. of Aha Hills, 12.iii.1965,
Wild & Drummond 6935 (SRGH). SW: Ghanzi Commonage, 992 m., fl. & fr. 10.i.1970,
Brown 7568 (COI; K). **Zambia.** B: Senanga, fl. & fr. 18.x.1961, *Fanshawe* 6743 (K;
LISC). S: Namwala, Maschila, fl. & fr. 16.x.1963, *van Rensburg* 2541 (K; SRGH).
Rhodesia. W: Wankie, near Makalolo I. Pan, Main Camp of Wankie National Park,
fl. & fr. 11.xii.1968, *Rushworth* 1348 (COI; PRE).
Also in Angola, SW. Africa and S. Africa. Edges of marshes and rivers, flooded
grassland areas, alluvial plains, cracks in limestone, etc.

8. **Nesaea heptamera** Hiern in F.T.A. **2**: 472 (1871).—Koehne in Engl., Bot. Jahrb. **3**:
    335 (1882); in Engl., Pflanzenr. **IV**, 216: 235 (1903).—Boutique in F.C.B.,
    Lythraceae: 20, t. 2 (1967). Type: Malawi, Zomba and E. end of Lake Shirwa,
    Meller (K, holotype).

Shrublet erect, 10–40 cm. high, from a woody rootstock, glabrous. Stems
4-gonous or narrowly winged, branched from the base, ± reddish. Leaves
10–45 × 2–10 mm., decussate, linear-lanceolate, sessile, acute at the apex, rounded,

truncate or subcordate at the base, glaucous, subcoriaceous, 1-nerved. Dichasia (1–2)3-flowered; peduncle slender up to 15 mm. long, bracteolate at the top, the bracteoles 1·5–2 mm. long, linear-lanceolate; pedicels of the lateral flowers 1–2 mm. long, bracteolate at the middle, the bracteoles 0·5–1·3 mm. long, linear-lanceolate. Flowers 6–7(8)-merous. Calyx-tube 3–4·7 mm. long, cupuliform; lobes 0·8–1 mm. long, deltate, acute; appendages very short. Petals 2·5–5 mm. long, obovate, orange, pink with red midline, pale red or red. Stamens 12–14(16), the episepalous 6–7 mm. long, the others 4–5 mm. long, all exserted; filaments crimson; anthers yellow. Ovary 1–1·5 mm. in diam., globose; style 8–9 mm. long, yellow, exceeding the stamens. Capsule 3–3·5 mm. in diameter, globose. Seeds c. 0·5 mm. long, ovate, concave-convex.

Zambia. N: 120 km. E. of Kasama, Chambeshi R., fl. & fr. 11.x.1960, *Robinson* 3943 (EA; K; SRGH). C: NW. Mkushi, fl. & fr. 24.ix.1957, *Fanshawe* 3713 (SRGH). E: Petauke, Chimutengo, fl. & fr. 5.x.1966, *Mutimushi* 1465 (K; SRGH). S: Mumbwa Distr., S. of Chenobi, 1120 m., fl. & fr. 16.ix.1947, *Greenway & Brenan* 8093 (EA; K; PRE). **Rhodesia.** N: Lomagundi, Alaska Smelter Copper Mine General Office, 1216 m., fl. & fr. 26.xi.1965, *Jacobsen* 2772 (PRE). W: Sebungwe, Chebina spring, fl. & fr. x.1948, *Whellan* 389 (SRGH). C: Salisbury, between Avondale West and Mabelreign, 1480 m., fl. & fr. 23.x.1935, *Drummond* 4921 (K; LISC; PRE; SRGH). E: Umtali, East Commonage, 1152 m., fl. & fr. 8.x.1959, *Chase* 7177 (BM; K; LISC; PRE; SRGH). **Malawi.** C: Dedza, N. A. Kachere, fl. & fr. 6.xi.1958, *Jackson* 2259 BM; K; PRE; SRGH). S: Zomba Hill, 640 m., xi.1900, *Manning* 37 (K). **Mozambique.** MS: Manica, Chimoio, 19 km. on the road Vila Pery to Manica, c. 700 m., fl. & fr. 24.xi.1965, *Torre & Correia* 13242 (COI; K; LISC; MO; P); Báruè, 6 km. from the crossing Vila Gouveia-Tete on the road to Macossa, c. 400 m., fl. & fr. 11.xii.1965, *Torre & Correia* 13561 (BR; LISC; LMU; M).

Also in Zaire and Tanzania. Margins of dambos and vleis, grasslands, seasonally burnt wet grasslands, edge of termite mounds, degraded *Brachystegia* woodlands, etc.

9. **Nesaea fruticosa** Fernandes & Diniz in Bol. Soc. Brot., Sér. 2, **31**: 159, t. 10 & 11 (1957). Type from Tanzania.

Shrub usually c. 1 m. high, sometimes up to 2·25 m., glabrous. Stem greyish, much branched upwards, the branches 4-gonous or narrowly winged. Cauline leaves 30–35 × 9–12 mm., decussate, elliptic-lanceolate, acute at the apex, margin entire, subcordate at the base, glossy green, 1-nerved, the nerve impressed above, prominent beneath; leaves of branches similar to the cauline ones but smaller (usually 4–6 × 1·5–3 mm.). Dichasia many, axillary, 1–6-flowered; peduncles c. 2 mm. long; pedicel of the middle flower up to 5 mm. long, those of the laterals c. 2 mm. long; bracteoles c. 0·5 mm. long, lanceolate. Flowers homomorphic, (4)5(6)-merous. Calyx c. 4 mm. long, tubular-campanulate, 8, 10 or 12-ribbed; lobes short, triangular, reddish, scarious at the apex; appendages very short. Petals c. 3 mm. long, pink, oblong, corrugate, with the median nerve thick beneath. Stamens twice as many as the calyx-lobes, inserted a little above the base of the calyx-tube (the epipetalous ones inserted slightly above the episepalous ones), the episepalous ones 7 mm. long, c. ½ exserted, the epipetalous ones c. 5·3 mm. long and c. ⅓ exserted. Ovary 2 × 1·6 mm., ellipsoid, 3-locular; style 7 mm. long, exceeding the stamens. Capsule c. 3 mm. in diameter, globose, included in the calyx-tube. Seeds numerous, c. 0·5 mm. long.

Zambia. N: Isoka Distr., Pangala area, fl. 12.viii.1965, *Lawton* 1240 (K; SRGH). E: Lundazi, fl. & fr. 18.x.1967, *Mutimushi* 2248 (K); Nsadzu R. at Katete-Chadiza crossing, 1000 m., 8.x.1958, *Robson* 24 (BM; K; LISC; PRE).

Also in Tanzania. Along rivers.

10. **Nesaea schinzii** Koehne in Verh. Bot. Ver. Prov. Brand. **30**: 250 (1888); in Engl., Bot. Jahrb. **22**: 151 (1895); in Journ. Bot., Lond. **40**: 69 (1902); in Engl., Pflanzenr. IV, 216: 239 (1903).—Engl., Pflanzenw. Afr. **3**, 2: 654 (1921).—Wild, Guide Fl. Vict. Falls: 145 (1953).—Fernandes & Diniz in Garcia de Orta **6**: 109 (1958).— Pohnert & Roessler in Prodr. Fl. SW. Afr. **95**: 7 (1966).—Boutique in F.C.B., Lythraceae: 20 (1967).—Fernandes in C.F.A. **4**: 186 (1970). Type from SW. Africa.

Erect or sometimes decumbent glabrous shrublet, up to 65 cm. high, usually with a strong root. Stems 4-gonous or narrowly winged, much branched from the base (the branches also branched), herbaceous to ± woody. Leaves 4–17 × 1–3(5) mm.,

Margarida Lameiras del.

Tab. 72. NESAEA ROBINSONIANA. 1, habit (×1); 2, dichasium with peduncle and bracteoles (×6); 3, 4-merous flower (×6); 4, 5, 5-merous flowers (×6); 6, calyx of a 4-merous flower opened (×6); 7, idem of a 5-merous flower (×6); 8, pistil (×6); 9, dehisced capsule within calyx and seeds (×6); 10, dehisced capsule and operculum (×6), all from *Robinson* 5137.

sometimes those of the lower part of stems up to 25 × 7 mm., linear, lanceolate or
oblong, tapering and subacute at the apex, margins revolute, broadened and
subhastate-cordate at the base, sessile or very shortly petiolate, 1-nerved. Dichasia
1–3(5)-flowered; peduncles 1–13 mm. long; bracteoles 0·5–1 mm. long, lanceolate.
Flowers 4-merous, trimorphic-heterostylous; pedicels 0·3–3 mm. long. Calyx
1·5–3 mm. long, campanulate; lobes c. 0·5 mm. high, triangular; appendages
shorter than the calyx-lobes, sometimes 0. Petals 2–2·5 mm. long, elliptic-obovate,
red-purple. Stamens 8, inserted slightly above the bottom of the calyx tube (the
epipetalous ones a little higher). Long-styled flowers: episepalous stamens 2·5 mm.
exserted, the epipetalous 3·5 mm. exserted; style 3·5 mm. exserted. Short-styled
flowers: episepalous stamens c. 3·5 mm. exserted, the epipetalous c. 2·5 mm.
exserted; style scarcely exceeding the lobes. Middle-styled flower: episepalous
stamens c. 3·5 mm. exserted, the epipetalous included (as long as the lobes);
style c. 0·5 mm. exserted. Ovary c. 1 mm. long, ellipsoid, shortly stipitate.
Capsule 1·5–2 × 0·8–1 mm., ellipsoid, exserted. Seeds c. 0·4 mm. long, brownish,
concave-convex.

Zambia. B: Sesheke Distr., Masese valley, fl. 11.viii.1947, *Brenan & Keay* 7692 (K).
C: Lusaka, University campus, fl. 4.ix.1972, *Strid* 2055 (K). **Rhodesia.** W: Bulawayo,
Matopos road, 1434 m., fl. 8.iii.1964, *Best* 395 (K; LISC; SRGH). E: Sabi Valley,
Honde Dip, fl. 26.ix.1947, *Whellan* 248 (SRGH). S: Gwanda, 8 km. N. of Hwali Store,
fl. & fr. 20.iii.1959, *Drummond* 5886 (COI; EA; K; LISC; PRE; SRGH).
Mozambique. SS: Gaza, R. Limpopo, Dumela, fl. 30.iv.1961, *Drummond & Rutherford-
Smith* 7626 (COI; K; SRGH).
Also in Angola, Zaire, Rwanda, Uganda, Kenya, Tanzania, S. Africa (Transvaal) and
SW. Africa. Damp ground near the outflow of dams, river banks, river beds, dambos,
savannas, sandveld, etc.

11. **Nesaea ramosissima** Fernandes & Diniz in Bol. Soc. Brot., Sér. 2, **29**: 98, t. 13
    (1955); in Garcia de Orta **4**, 3: 392, t. 1 (1956). Type: Mozambique, Amaramba,
    Mandimba, *Pedro & Pedrógão* 4139 (COI, holotype; EA; LMA).

Annual (or biennal?) erect herb, c. 40 cm. high. Stem thick (c. 5 mm. in
diameter), sub-4-gonous and unbranched below, 4-gonous and much branched
above; floriferous branches ascending, simple or few-branched, narrowly 4-winged,
the wings serrulate. Stem-leaves c. 30 × 10 mm., decussate, lanceolate, subacute at
the apex, margin revolute, auriculate at the base, the auricles roundish, inconspicuously penninerved, the midrib impressed above, prominent beneath; leaves
of the branches similar to those of stems, but smaller. Dichasia 2–7(many)-
flowered, up to 12 mm. long; pedicels 1–8 mm. long, 4-gonous; bracteoles c.
1 mm. long, scarious, linear. Flowers 4-merous (heterostylous?). Calyx c. 2 mm.
long, broadly campanulate, 8-nerved, the nerves prominent; lobes erect, equalling
c. ⅓ of the tube-length; appendages very short, curved. Petals 4, 2·5 × 2·5 mm.,
subcircular, reddish-purple, clawed at the base, slightly emarginate at the apex.
Stamens 8, inserted slightly below the middle of the calyx-tube, exserted at
anthesis, the episepalous exceeding the lobes by c. 2 mm., the epipetalous by
c. 1 mm. Ovary c. 1 mm. in diameter, globose; style c. 5 mm. long, exceeding
by c. 1·5 mm. the episepalous stamens (long-styled flower?). Capsule globose,
equalling the calyx-lobes. Seeds c. 0·15 mm. long.

Mozambique. N: Amaramba, Mandimba R., fl. & fr. 3.v.1948, *Pedro & Pedrógão*
4139 (COI; EA; LMA). Known only from Mozambique. In swamps, on river banks.

12. **Nesaea dinteri** Koehne subsp. **elata** Fernandes in Bol. Soc. Brot., Sér. 2, **48**: 122,
    t. 9 (1974). TAB. **73**. Type: Zambia, Mumbwa, Chunga, Kafue National Park,
    Mitchell 18/50 (COI; LISC, holotype; PRE; SRGH).

Annual erect glabrous herb up to 65 cm. high. Stem ± thick, subterete and
usually leafless below, branched at the middle, the branches ascending, 4-winged,
the wings serrulate. Leaves 5–45 × 1·5–7 mm., decussate, linear to lanceolate,
obtuse at the apex, margin revolute, hastate-cordate at the base, minutely scabrous
especially on the upper face, 1-nerved. Dichasia (1)3–many-flowered; primary
peduncle 2–5 mm. long, slightly scabrous, bracteolate at the apex, the bracteoles
c. 0·75 mm. long, linear, the secondary and tertiary peduncles shorter; pedicels c.
1·5 mm. long, minutely scabrous. Flowers 4-merous. Calyx 1·5–2 mm. long,

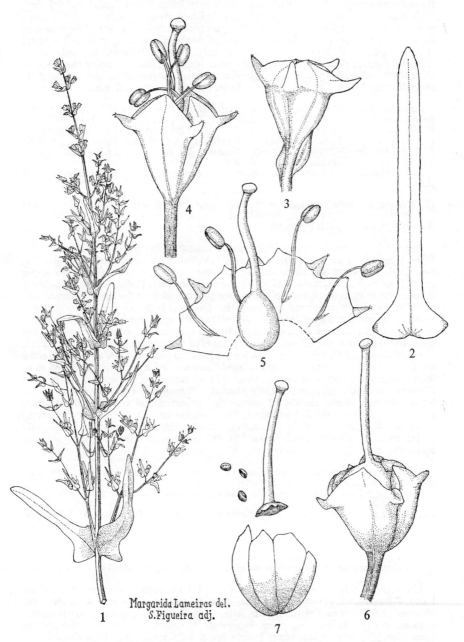

Tab. 73. NESAEA DINTERI subsp. ELATA. 1, upper part of stem (×1); 2, leaf (×2); 3, flower-bud (×12½); 4, flower without petals (×12½); 5, calyx opened (×12½); 6, capsule within calyx (×12½); 7, dehisced capsule, operculum with style and seeds (12½), all from *Mitchell* 18/50.

reddish, minutely scabrous, turbinate, 8-ridged; lobes 1 mm. broad and 0·5 mm. high, triangular; appendages spreading, c. 0·5 mm. long. Petals c. 1 × 0·75 mm., obovate, clawed, reddish-purple or dark blue (*fide* Astle). Stamens 4 (4–8 or 8 in some plants), opposite to the calyx-lobes, with the filaments purple and anthers yellow, inserted a little below the middle of calyx-tube, c. 1·5 mm. exserted. Ovary c. 1 × 0·75 mm., ellipsoid; style c. 2·25 mm. long, purple, exceeding the anthers. Capsule 1·5–2 mm. in diameter, globose, reddish, included in calyx or slightly exceeding the lobes. Seeds c. 0·4 mm. long, brownish, rounded, concave-convex.

Zambia. N: Mpika Distr., Luangwa valley, Mfuwe, fl. & fr. 15.v.1969, *Astle* 5996 (K). S: between Mumbwa and Namgoma, fl. & fr. 21.iii.1963, *van Rensburg* 1773 (K; SRGH). **Rhodesia.** W: Victoria Falls, 960 m., iii.1918, *Eyles* 1294 (BM; SRGH); Nyamandhlovu Pasture Station, fr. 10.vi.1955, *Plowes* 1859 (SRGH). C: Hartley, Poole Farm, fl. & fr. 3.v.1964, *Hornby* 3442 (PRE; SRGH). **Malawi.** C: Mangochi (Ft. Johnston) Distr., Mwera, fl. & fr. 1.vi.1955, *Jackson* 1668 (K).
Also in Tanzania. In wet heavy clayey soils, drying mud, damp roadsides, flooded plains, etc.

13. **Nesaea gazensis** Fernandes in Bol. Soc. Brot., Sér. 2, **49**: 9, t. 1 (1975). Type: Mozambique, Gaza Distr., Dumela, R. Limpopo, *Drummond & Rutherford-Smith* 7623 (COI; SRGH, holotype).

Annual herb, up to 15 cm. high, much branched from the base to the apex. Stem 4-gonous in the lower part, 4-winged upwards; branches numerous, slender, 4-winged, the wings slightly scabrous, the lower branches decumbent, the upper ones ± ascending. Leaves 5–22 × 1–2·5 mm., decussate, linear, ± curved, obtuse at the apex, margin involute, slightly to broadly auriculate at the base, 1-nerved, the nerve impressed above, prominent beneath. Dichasia 1–8-flowered, loose; primary peduncle slender, c. 4 mm. long, 2-bracteolate at the apex, the bracteoles c. 1·5 mm. long, linear, scarious; secondary peduncles c. 2 mm. long, 2-bracteolate also at the apex, the bracteoles c. 0·75 mm. long; pedicel of the middle flower c. 1 mm. long, those of the lateral flowers c. 2·5 mm. long. Flowers 4-merous. Calyx c. 1·25 mm. long, campanulate, longitudinally 8-ribbed; lobes c. 0·5 mm. broad and 0·35 mm. long, triangular; appendages very short (c. 0·15 mm. long). Petals c. 0·5 mm. in diameter, pinkish, rounded, slightly clawed. Stamens 4, opposite to the calyx-lobes, inserted at the middle of calyx-tube, exserted and reaching the stigma; filaments c. 1 mm. long and anthers c. 0·25 mm. broad. Ovary c. 0·75 × 0·5 mm., obovate, 2-locular; style 0·75–1 mm. long, crowned by a capitate stigma. Capsule c. 1·5 mm. in diameter, subglobose, slightly exserted. Seeds c. 0·25 mm. in diameter, brownish, concave-convex.

Mozambique. SS: Gaza Distr., Dumela, R. Limpopo, fl. & fr. 30.iv.1961, *Drummond & Rutherford-Smith* 7623 (COI; SRGH).
Known only from the F.Z. area. On mud in riverbeds.

14. **Nesaea passerinoides** (Welw. ex Hiern) Koehne in Engl., Bot. Jahrb. **3**: 338 (1882); in Verh. Bot. Ver. Prov. Brand. **30**: 250 (1888); in Journ. Bot., Lond. **40**: 68 (1902); in Engl., Pflanzenr. IV, 216: 237 (1903).—Hiern, Cat. Afr. Pl. Welw. **1**: 377 (1898).—Engl., Pflanzenw. Afr. **3**, 2: 653 (1921).—Fernandes & Diniz in Garcia de Orta **6**: 110 (1958).—Fernandes in C.F.A. **4**: 186 (1970). Type from Angola.
Ammannia passerinoides Welw. ex Hiern in F.T.A. **2**: 480 (1871). Type as above.

Undershrub with woody rootstock, up to 1·2 m. high. Stem erect, reddish, hairy, 4-gonous, simple or ± branched. Leaves 10–16 × 2·5–5 mm., subappressed, alternate, decussate or 3–4-nate, lanceolate, acute at the apex, margin very revolute, sagittate-cordate at the base, hairy, 1-nerved. Dichasia 1–7-flowered, the flowers homomorphic, subsessile; outer bracteoles c. 1·25 mm. long, boat-shaped, the inner ones shorter, hairy. Calyx 2·5–3 mm. long, light green striped wine-red, 4-gonous, hairy; lobes very short, connivent; appendages twice as long as the lobes. Petals 0–4, c. 1 mm. long, yellowish. Stamens 4, rarely 8 (fide Hiern), opposite to the petals, inserted about the middle of the calyx-tube, with filaments c. 1·5 mm. long, reaching the stigma and exceeding the appendages. Ovary ellipsoid, 2-locular; style exserted, c. 1·5 mm. long, deep wine-red. Capsule ellipsoid, dark red, exceeding the calyx. Seeds c. 0·3 mm. long, brownish, concave-convex.

Botswana. SE: Pharing, near Kanye, 1280 m., fl. & fr. iii.1947, *Miller* B/496 (PRE).
Rhodesia. N: Gokwe R., Fly Gate, Chehandirisi R., fl. & fr. 23.iv.1964, *Bingham* 1294 (K; LISC; SRGH). W: Matopos, 1376 m., 9.iii.1932, *Brain* 8621 (SRGH). C: Salisbury, between Avondale West and Mabelreign, 1480 m., fl. & fr. 23.x.1955, *Drummond* 4915 (K; PRE; SRGH). S: Victoria, Chenyati Stream, Kyle National Park, fl. & fr. 27.iv.1971, *Basera* 364 (K).
Also in Angola. Grasslands, margins of vleis and termite mounds, on black clayey or sandy soils.

15. **Nesaea purpurascens** Fernandes in Bol. Soc. Brot., Sér. 2, **48**: 120, t. 7 (1974). Type: Zambia, Chandola, Fort Jameson, 26.ix.1966, *Mutimushi* 1482 (K; SRGH, holotype).

Decumbent, purple-flushed herb, the erect part up to 30 cm. high. Stem rooting at the creeping portion, branched, the branches narrowly winged, minutely hairy. Leaves 10–20 × 2·5–6 mm., decussate, lanceolate, flushed purple, obtuse at the apex, cordate at the base, minutely hairy, penninerved, the midrib impressed above, prominent beneath. Dichasia 7–10-flowered, pedunculate, the peduncle 4–10 mm. long, papillose; outer bracteoles c. 3 × 1·5 mm., foliaceous, reddish, boat-shaped, with the midrib scabrous on the underside, the inner ones numerous, c. 2·5 × 0·5 mm., purplish, scarious, also boat-shaped. Flowers 4-merous, with c. 0·5 mm. long pedicels. Calyx 2 mm. long, turbinate-campanulate, flushed purple, conspicuously ribbed, papillose; lobes 1 mm. broad and 0·5 mm. high, deltate; appendages c. 0·5 mm. long, spreading, horn-shaped, minutely papillose. Petals c. 2 × 1·25 mm., obovate, purple. Stamens 8, inserted slightly below the middle of the calyx-tube, exserted. Ovary c. 1 mm. in diameter, globose; style c. 3·5 mm. long, greatly exceeding the stamens. Capsule c. 1·5 mm. in diameter, globose, included. Seeds c. 0·3 mm. long.

Zambia. E: Chandola, Chipata (Fort Jameson), fl. & fr. 26.ix.1966, *Mutimushi* 1482 (K; SRGH).
Known only from Zambia. On dried muddy places by dams.

16. **Nesaea tolypobotrys** Koehne in Engl., Bot. Jahrb. **22**: 151 (1895); in Engl., Pflanzenr. **IV**, 216: 232 (1903).—Engl., Pflanzenw. Afr. **3**, 2: 651 (1921). Syntypes from S. Africa.

Perennial?, entirely glabrous, erect herb, up to 60 cm. high. Stem subterete below, 4-gonous upwards, dark brown, branched approximately from the middle. Leaves 18–85 × 8–16 mm., decussate, sessile, narrowly to broadly lanceolate, obtuse at the apex, margin entire, slightly scabrous, obtuse or cuneate at the base, 1-nerved. Dichasia axillary, 3–7-flowered, 5–10 mm. long; peduncles 1–6 mm. long; outer bracteoles c. 5·5 × 4 mm., oblong, ± long-acuminate, whitish, conspicuously nerved, with a green keel; inner bracteoles up to 2·5 mm. long, linear, scarious. Flowers 4-merous, pedicellate, the pedicels c. 1 mm. long. Calyx c. 3 mm. long, campanulate becoming subglobose, 8-nerved; lobes c. 0·5 mm. long, triangular; appendages c. 1·75 mm. long, erect, curved inwards, completely glabrous or ciliolate at apex. Petals pink. Episepalous stamens c. 5 mm. long, the epipetalous c. 3 mm., all exserted. Ovary globose, 2-locular; style c. 4 mm. long, exceeding the stamens. Capsule c. 1·5 mm. in diameter, globose, included. Seeds c. 0·5 mm. long, yellowish, concave-convex.

Mozambique. LM: Rikatla, fl. & fr. vii.1917, *Junod* 321 (PRE).
Also in S. Africa (Natal, Kaffraria, Ungulubi). On sandy flats.

17. **Nesaea linearis** Hiern in F.T.A. **2**: 475 (1871).—Koehne in Engl., Bot. Jahrb. **3**: 333 (1882); in Engl., Pflanzenr. **IV**, 216: 233, fig. 45C (1903).—Engl., Pflanzenw. Afr. **3**, 2: 651, fig. 286C (1921).—Garcia in Est. Ens. Doc. Junta Invest. Ultramar **12**: 159 (1954).—Fernandes & Diniz in Garcia de Orta **4**, 3: 392 (1956). Syntypes from Mozambique, *Forbes* (K); mouth of Zambezi, xiii.1862, *Kirk* (K).

Annual, erect, glabrous herb, up to 1 m. high. Stem usually simple and subterete below, 4-gonous and branched upwards. Leaves 20–80 × 3–10 mm., decussate, linear, acute at the apex, obtuse, acute or cuneate at the base, rigid, 1-nerved. Dichasia capituliform, large, subglobose, at the ends of stems and branches, very dense, many-flowered and sessile, surrounded at base by several imbricate bracteoles, the bracteoles reniform, acuminate, venose. Flowers 4–6-merous, dimorphic-heterostylous, in subsessile clusters (the flowers also subsessile

in the clusters), surrounded by oblong bracteoles expanded upwards and shortly
acuminate at the apex. Calyx-tube 3–4 mm. long, narrowly campanulate,
membranous, 8–12-nerved; lobes c. 0·5 mm. high, deltate, acute; appendages
erect or arching, much longer than the lobes (c. 2 mm.), ciliate at the top. Petals
c. 4·5 mm. long, purple, narrowly obovate, caducous. Stamens as many as the
calyx-lobes, inserted near the base of the calyx-tube (the epipetalous on a higher
level). Short-styled flowers: filaments of long stamens 7 mm. long, exceeding the
stigma by 2·5 mm.; filaments of short stamens 5·5 mm., exceeding the stigma by
1 mm.; style c. 4·2 mm. long. Long-styled flowers: filaments of long stamens
6 mm. long, exceeding the appendages by 2 mm.; filaments of short stamens 4 mm.
long, scarcely attaining the apex of the appendages; style 7·5 mm. long. Ovary
1·7 mm. long, ellipsoid, 2-locular. Capsule and seeds not seen.

Mozambique. N: Moma, Chalana, fl. & fr. 13.vii.1948, *Pedro & Pedrógão* 4474 (EA).
Z: between Quelimane and Nicuadala, fl. & fr. 25.vii.1942, *Torre* 4417 (COI; LISC;
MO; PRE). MS: 29 km. NW. of Nova Sofala, fl. 21.vi.1961, *Leach & Wild* 11125
(COI; K; LISC; PRE); Cheringoma, R. Muanza, fl. & fr. 22.vii.1967, *Moura* 195
(COI; LISC; LMU).
Known only from Mozambique. Wet clayey soils.

18. **Nesaea rautanenii** Koehne in Bull. Herb. Boiss. **6**: 750 (1898); in Engl., Bot. Jahrb.
   29: 165 (1900); in Engl., Pflanzenr. **IV**, 216: 231 (1903).—Engl., Pflanzenw. Afr. **3**,
   2: 651 (1921).—Pohnert & Roessler in Prodr. Fl. SW. Afr. **95**: 3 (1966). Type from
   SW. Africa.

Erect or procumbent annual herb, up to 30 cm. high. Stem rooting at nodes,
simple or few branched, 4-gonous and sparsely hairy like the branches. Leaves
10–30 × 1·5–2·5 mm., decussate, narrowly linear-lanceolate, acute or obtuse at the
apex, margin entire, obtuse and ciliate at the base, sometimes sparsely pilose,
1-nerved, the nerve impressed above, prominent beneath. Dichasia 1–7-flowered,
pedunculate, the peduncles 1–9 mm. long; involucral bracteoles c. 5 × 4 mm.,
subcircular, suddenly acuminate, the acumen 1–2 mm. long, conspicuously nerved,
flushed red, with the margin ciliolate and with some hairs on the underside of
midrib, the other bracteoles c. 3 mm. long, scarious and villous on the underside
of the midrib. Flowers 4-merous. Calyx 3–3·5 mm. long, rose-coloured; lobes
c. 1·5 mm. broad and 0·5 mm. high, deltate; appendages c. 0·75 mm., erect,
ciliolate at the apex. Petals purple. Stamens 4, episepalous, inserted below the
middle of the calyx-tube, included. Ovary c. 1 mm. in diameter, subglobose; style
c. 1 mm. long, included. Capsule included. Seeds c. 0·4 mm. long, yellowish,
subcircular.

Zambia. B: Sesheke, fl. & fr. 18.vi.1963, *Fanshawe* 7859 (SRGH).
Also in SW. Africa. In sandy flood plains.

19. **Nesaea radicans** Guill. & Perr. in Guill., Perr. & A. Rich., Fl. Senegamb. Tent. **1**:
   306, t. 70 (1833).—Hiern in F.T.A. **2**: 474 (1871); Cat. Afr. Pl. Welw. **1**: 376
   (1898).—Koehne in Engl., Bot. Jahrb. **3**: 330 (1882); in Engl., Pflanzenr. **IV**, 216:
   231 (1903).—Gilg in Engl., Pflanzenw. Ost-Afr. **C**: 285 (1895).—Grandidier, Atlas
   Hist. Pl. Madag.: t. 380 (1896).—R. E. Fr., Wiss. Ergebn. Schwed. Rhod.-Kongo-
   Exped. **1**: 165 (1914).—Engl., Pflanzenw. Afr. **3**, 2: 651 (1921).—Wild, Guide Fl.
   Vict. Falls: 145 (1953).—Garcia in Est. Ens. Doc. Junta Invest. Ultramar **12**: 158
   (1954).—Perrier in Fl. Madag., Lythracées: 7 (1954).—Fernandes & Diniz in Garcia
   de Orta **4**, 3: 394 (1956); op. cit. **6**: 107 (1958).—Boutique in F.C.B., Lythraceae:
   17 (1967). Type from Senegambia.

Perennial herb, rarely ± woody below, procumbent or erect, 20–120 cm. long,
glabrous or canescent with crispate hairs, sometimes glabrescent. Stem rooting at
lower nodes especially when aquatic, subterete below, 4-gonous above, green to
brownish, ± branched. Leaves 10–55 × 4–25 mm., decussate, sessile or with petiole
1–3 mm. long, oblong, elliptic, elliptic-lanceolate or elliptic-ovate, rarely ovate,
rounded or subacuminate at apex, rounded or narrowed at base, the margin sub-
cartilaginous, sometimes ciliate; lamina dark green, 1-nerved, sometimes distinctly
penninerved. Dichasia usually many-flowered, borne at the end of 4-gonous
peduncles 7–55 mm. long, surrounded by 2–4 foliaceous involucral bracteoles, these
3–7 × 3–7 mm., cordate, venose, shortly acuminate, glabrous or ciliate at margin.
Flowers 5–6(7)-merous, pedicellate, the pedicels 0·5–3 mm. long; inner bracteoles

3–4 mm. long, linear. Calyx 3–4 mm. long, subturbinate-campanulate, 10–12 (14)-ribbed; lobes c. 0·5 mm. long, deltate; appendages 1–5 mm. long, erect, ciliate at the apex. Petals 2·5–4 mm., obovate, mauve. Stamens usually twice as many as the calyx-lobes, inserted near the bottom of calyx-tube, the episepalous rather longer than the others, all exserted. Ovary 1–1·5 mm. in diameter, subglobose, 2–3-locular; style 6–8 mm. long, exceeding the stamens. Capsule 1·5–2 mm. in diameter, subglobose. Seeds 0·4 mm. long, yellowish, concave-convex.

Plant glabrous or nearly so - - - - - - - - var. *radicans*
Plant canescent with crispate hairs to glabrescent - - - - var. *floribunda*

Var. **radicans**.

Zambia. B: Ilonga, fl. 18.vii.1962, *Fanshawe* 7007 (K; SRGH). N: Mporokoso, Bulaya, Chishyela dambo, Mweru-wa-Ntipa, 992 m., fl. 11.viii.1962, *Tyrer* 386 (BM; SRGH). C: Mount Makulu Research Station near Chilanga, 1312 m., fl. 14.x.1960, *Coxe* 109 (COI; SRGH). W: Solwezi, fl. & fr. 26.vii.1964, *Fanshawe* 8846 (K; SRGH). E: Lundazi, fl. & fr. 17.x.1967, *Mutimushi* 2232 (K). S: Mazabuka, Veterinary Research Station, fl. 18.vii.1963, *van Rensburg* 2359 (K; SRGH). **Rhodesia.** W: Victoria Falls, S. bank in front of Main Falls, fl. 30.viii.1947, *Greenway & Brenan* 8025 (EA; K; PRE). C: Gwelo, Mlezu Government School, 1344 m., fl. & fr. 2.ii.1966, *Biegel* 874 (SRGH). E: Odzi, Tsungwesi R., 1056 m., fl. & fr. 31.v.1936, *Eyles* 8596 (K). **Mozambique.** N: Macomia, Ingoane, fl. 12.xi.1948, *Barbosa* 2062 (K; LISC; LMA; MO); Mogincual, between Quixaxe and Liupo, fl. & fr. 18.x.1948, *Barbosa* 2490 (LISC; LMA; M). Z: Quelimane, Nicoadala, Licuári, fl. & fr. 8.viii.1970, *Balsinhas* 1760 (LMA). SS: between Vilanculos and Macovane, st., 24.xi.1942, *d'Orey* 26 (LISC). LM: Lourenço Marques, 1917–18, *Junod* 321 (LISC).
Widespread in tropical Africa and Madagascar. In swamps, river banks and river beds, moist grasslands, in the spray of the waterfalls, etc.

Var. **floribunda** (Sond.) Fernandes in Bol. Soc. Brot., Sér. 2, **48:** 117 (1974). Type from S. Africa (Cape Prov.).
Nesaea floribunda Sond. in Harv. & Sond., F.C. **2:** 517 (1862).—Hiern in F.T.A. **2:** 474 (1871); Cat. Afr. Pl. Welw. **1:** 376 (1898).—Koehne in Engl., Bot. Jahrb. **3:** 331 (1882); in Warb., Kunene-Samb.-Exped. Baum: 312 (1903); in Engl., Pflanzenr. **IV**, 216: 231 (1903).—Thonner, Blütenpflanz. Afr.: t. 111 (1908).—Wild, Guide Fl. Vict. Falls: 145 (1953).—Fernandes & Diniz in Garcia de Orta **4**, 3: 394 (1956).—Pohnert & Roessler in Prodr. Fl. SW. Afr. **95:** 6 (1966).—Boutique in F.C.B., Lythraceae: 16 (1967).—Fernandes in C.F.A. **4:** 189 (1970). Type as above.

Caprivi Strip. West region, fr. 10.iv.1946, *Kruger* (PRE). **Zambia.** B: Sioma, along Zambezi R., fl. 21.vii.1921, *Borle* 226 (PRE; SRGH). N: (Abercorn) Mbala Distr., Sandpits, 1500 m., fl. & fr. 20.xi.1943, *Richards* 18420 (K; LISC). W: Chingola, fl. & fr. 10.v.1968, *Mutimushi* 2605 (K). C: Serenje, fl. 12.viii.1965, *Fanshawe* 9246 (K; SRGH). E: Petauke, Kacholola, fl. & fr. 21.x.1967, *Mutimushi* 2188 (K). S: Livingstone, fl. 10.ix.1962, *Fanshawe* 7017 (SRGH). **Rhodesia.** W: Victoria Falls, fl. & fr. 8.vii.1930, *Hutchinson & Gillett* 3452 (BM; K; SRGH). C: Salisbury, E. Commonage, 1468 m., fl. & fr. ii.1917, *Eyles* 666 (BM; K; SRGH). E: Umtali Commonage, fl. 24.x.1949, *Chase* 1800 (BM; SRGH). S: Beitbridge, Shashi R., 9·7 km. NW. of junction with Limpopo, fl. 6.v.1959, *Drummond* 6094 (SRGH). **Malawi.** N: Fort Hill, 1120–1280 m., fl. & fr. vii.1896, *Johnston* (K). S: Zomba, Mpanda Stream, fl. & fr. 11.vi.1963, *Salubeni* 53 (COI; SRGH). **Mozambique.** Z: Mocuba, Namagoa, fl. & fr. v.1943, *Faulkner* 89 (COI; K; PRE; SRGH). MS: road to Vila de Manica, 160 m., fl. 24.ix.1950, *Chase* 2902 (BM; COI; K; LISC; SRGH). SS: Lindela (Inhambane)-Maxixe junction, ± 32 m., fl. 4.x.1963, *Leach & Bayliss* 11820 (COI; K; LISC; SRGH). LM: Namaacha, on the crossroad to Vila Pouca, fl. & fr. 16.vi.1971, *Pereira* 1914 (LMU).
Widespread in tropical and southern Africa and Zanzibar. Ecology as for the type variety.

20. **Nesaea erecta** Guill. & Perr. in Guill., Perr. & A. Rich., Fl. Senegamb. Tent. **1:** 305, t. 69 (1833).—A. Rich., Fl. Abyss. **1:** 280 (1847).—Hiern in F.T.A. **2:** 474 (1871); Cat. Afr. Pl. Welw. **1:** 376 (1898).—Koehne in Engl., Bot. Jahrb. **3:** 331 (1882); in Engl., Pflanzenr. **IV**, 216: 231 (1903).—Engl., Pflanzenw. Afr. **3**, 2: 651 (1921).—Wild, Guide Fl. Vict. Falls: 144 (1953).—Fernandes & Diniz in Garcia de Orta **4**, 3: 396 (1956); op. cit. **6:** 107 (1958).—Boutique in F.C.B., Lythraceae: 17 (1967).—Fernandes in C.F.A. **4:** 188 (1970). Type from Senegambia.
Nesaea humilis Klotzsch in Peters, Reise Mossamb. Bot.: 68 (1862). Type: Mozambique, Rios de Sena, *Peters* (B, holotype†; K).
Nesaea racemosa Klotzsch, loc. cit. in adnot. Type from Madagascar.

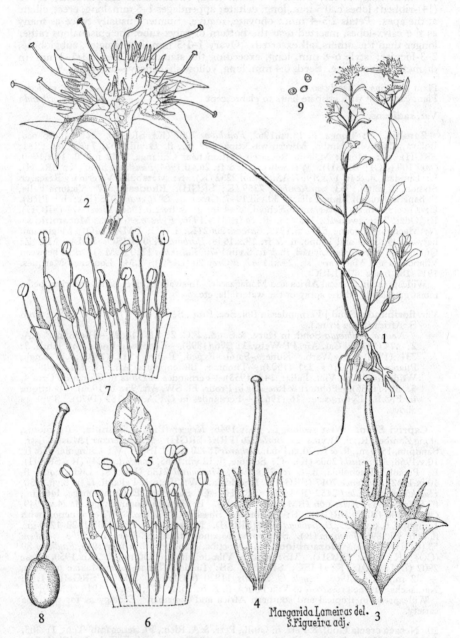

Tab. 74. NESAEA SPATHULATA. 1, habit (×½); 2, dichasium (×3); 3, 5-merous flower
with bracteoles (×6); 4, 6-merous flower (×6); 5, petal (×6); 6, calyx of a
5-merous flower opened (×6); 7, calyx of a 6-merous flower opened (×6); 8, pistil
of a dolichostylous flower (×6); 9, seeds (×6), all from *Lea* (PRE).

Annual erect or ascending herb, 5–35 cm. high. Stem simple or ± branched from the base, 4-gonous or narrowly winged to the top. Leaves 8–30 × 1·5–6(10) mm., decussate, sessile, linear to narrowly oblong, acute or obtuse at the apex, subacute or subobtuse at the base, 1-nerved. Dichasia capituliform, solitary and axillary, 5–7m(any)-flowered, surrounded at the base by 2 broadly cordate-acuminate, 7·5–8 × 6–7·5 mm., greenish, nervose involucral bracteoles; peduncles 3–15 mm. long, glabrous or with patent hairs. Flowers 4–6-merous; inner bracteoles 3–4 × 1 mm., linear-lanceolate. Calyx 2·5–3·5 mm., campanulate; lobes c. 0·5 mm. high, triangular; appendages 1–1·5 mm. long, linear, erect, ± bent inwards, ciliate or glabrous. Petals 2·5–3 × 2–2·5 mm., obovate, clawed, rose-coloured or violet. Stamens as many or twice as many as the calyx-lobes, exserted. Ovary 1–2 mm. in diameter, subglobular, 2–4-locular; style exserted 2·5–3·5 mm., exceeding the stamens. Capsule c. 2·5 mm. in diameter, included. Seeds c. 0·3 mm. long, brownish.

Zambia. E: Lupande, Longamusi R., fl. & fr. 17.viii.1963, *Bingham* 795 (SRGH). S: Namwala, Kafue National Park, fl. & fr. 14.v.1963, *Mitchell* 18/92 (COI; EA; PRE; SRGH). **Rhodesia.** W: Shangani, Connemara Mine near Hunter's Road, *Goldsmith* 123/55 (LISC; PRE; SRGH). C: Hartley, Poole Farm, fl. & fr. 29.iii.1955, *Hornby* 3383 (K; SRGH). S: Nuanetsi, Malangwe R., SW. Mateke Hills, 640 m., fl. & fr. 5.v.1958, *Drummond* 5564 (COI; K; PRE). **Mozambique.** N: António Enes, fr. 20.x.1965, *Mogg* 22430 (LISC). Z: Mocuba, Namagoa, 64 m., fl. & fr. iv–v.1943, *Faulkner* 290 (K; PRE; SRGH). MS: Chemba, between Tambara and Mitondo, at the Muíre and Zambeze junction, c. 170 m., fl. & fr. 15.v.1971, *Torre & Correia* 18439 (COI; LMA; LMU). SS: c. 8 km. W. of Inhacoro, 32 m., fl. & fr. 7.x.1963, *Leach & Bayliss* 11869 (COI; SRGH). LM: Namaacha, near dam of Canada-Dry, fl. & fr. 17.v.1971, *Marques* 2275 (LMU).
Widespread in tropical Africa. Moist places especially along streams.

21. **Nesaea ramosa** Fernandes in Bol. Soc. Brot., Sér. 2, **48**: 118, t. 5 (1974). Type: Mozambique, Inhambane, 9 km. on the road Vilanculos to Mambone, *Torre & Pereira* 12330 (COI; K; LISC, holotype; SRGH).

Shrubby herb up to 70 cm. high (or more?). Stem subterete, much branched c. 10 cm. above the base; branches 4-gonous below, narrowly winged upwards, the wings scabrous. Leaves 15–25 × 1·5–3 mm., decussate, linear, acute at the apex, sometimes curved and folded along the midrib, roundish or attenuate at the base, 1-nerved, the nerve impressed above, slightly prominent beneath, margin scabrous. Dichasia numerous, axillary, 3–10-flowered, pedunculate, surrounded at base by 2 involucral bracteoles; peduncles 1–15 mm. long, pubescent; involucral bracteoles c. 7 × 8 mm., subsemicircular, suddenly acuminate, the acumen 2·2 mm. long, conspicuously nerved, the midrib prominent beneath, the other bracteoles 4–4·5 × 1–1·5 mm., scarious, all apiculate and with the margin minutely ciliate. Flowers (4)5(6)-merous. Calyx-tube c. 3 mm. long, turbinate-campanulate, 8-, 10- or 12-ribbed; lobes 0·75 mm. broad and 0·5 mm. long, triangular, apiculate; appendages c. 1·75 mm. long, erect, densely ciliolate. Petals 3–4 × 2–3 mm., purple, broadly elliptic, clawed, with a conspicuous midrib, very soon caducous. Stamens usually as many as the calyx-lobes, inserted c. 0·5 mm. above the bottom of calyx-tube, the episepalous c. 5 mm. long and 2·5 mm. exserted (1·5 mm. above the lobes and c. 1 mm. above appendages), the epipetalous c. 3·5 mm. long, slightly exceeding the lobes (0·5 mm.). Ovary c. 1·5 × 1 mm., ellipsoid, 2-locular; style c. 6·5 mm. long, c. 4·5 mm. exserted (above the lobes). Capsule c. 2·75 × 2 mm., ellipsoid, brownish. Seeds c. 0·3 mm. long, brownish, elliptic.

Mozambique. SS: c. 5 km. S. of Vilanculos turn-off, c. 160 m., fl. & fr. 6.x.1963, *Leach & Bayliss* 11844 (COI; K; LISC; SRGH).
Known only from Mozambique. Vlei grasslands, low lands and open forests of *Brachystegia spiciformis*, *Strychnos* and *Phoenix* in black or grey soils.

22. **Nesaea spathulata** Fernandes in Bol. Soc. Brot., Sér. 2, **48**: 119, t. 6 (1974). TAB. **74**. Type: Mozambique, Manica e Sofala Distr., Mucarangue near Buzi, *Lea* 78 (K; PRE, holotype).

Perennial herb 15–65 cm. high, reddish below, greyish upwards, glabrous. Stems of 2 kinds arising from a tuberous root: short stems up to 15 cm. high (or more?), branched, tetragonous or narrowly 4-winged, floriferous upwards; long stems stronger, much branched upwards and floriferous. Leaves of short stems

15–35 × 3–8 mm., spathulate, obtuse or subacute at the apex, margin entire, longly attenuate at the base, 1-nerved, the midrib impressed above prominent beneath; leaves of long branches up to 35 × 2 mm., linear. Dichasia many-flowered, axillary and at the ends of stems and branches, the axillary ones pedunculate, the peduncle 3–10 mm. long; involucral bracteoles c. 9 × 6 mm., broadly ovate, suddenly acuminate, the acumen 2–3 mm. long, conspicuously nerved, the other bracteoles c. 5 × 2·5 mm., spathulate, shortly acuminate at apex and long attenuate at base. Flowers 5–6-merous, dimorphic-heterostyled. Calyx 3–4 mm. long, turbinate-campanulate, 10–12-ribbed; lobes c. 1 mm. broad and 0·5 mm. high, triangular; appendages c. 1·5 mm. long, erect, ciliate. Petals c. 3 × 1·75 mm., pinkish-purple, lanceolate, clawed, conspicuously nerved. Stamens 10 or 12, inserted c. 1 mm. above the bottom of the tube; anthers c. 0·75 × 0·5 mm. Ovary c. 1·5 × 1 mm., ovoid, slightly stipitate, 2-locular. Long-styled flowers: episepalous stamens c. 6 mm. long, exceeding the lobes by c. 3 mm., the epipetalous c. 4 mm. long, attaining the appendages-top; style c. 7 mm. long. Short-styled flowers: episepalous stamens c. 7 mm. long, the epipetalous c. 5 mm. long; style c. 2·5 mm. long, attaining the appendages-top. Immature capsule c. 2·5 mm. long, ovoid, included. Immature seeds c. 0·25 mm.

Mozambique. MS: Manica e Sofala Distr., Mucarangue near Buzi, 32 m., fl. 19.x.1935, *Lea* 78 (K; PRE).
Known only from Mozambique. In dambos on black soils.

23. **Nesaea pedroi** Fernandes & Diniz in Bol. Soc. Brot., Sér. 2, **29**: 96, t. 11 (1955); in Garcia de Orta **4**, 3: 396, t. 3 (1956). Type: Mozambique, Niassa, Moma, *Pedro & Pedrógão* 4523 (COI, holotype; LMA).

Annual erect glabrous herb, up to 30 cm. high. Stem simple or ± branched, 4-gonous or narrowly 4-winged. Leaves 8–30 × 1·5–2 mm., decussate, sessile, ± spreading or ascending, linear, acute at the apex, margin minutely serrulate, attenuate at the base. Dichasia capituliform, few-flowered, surrounded by two involucral bracteoles 4 × 5 mm., venulose, broadened upwards and ending in a short acumen; peduncles up to 3 mm. long. Flowers 4–5-merous, trimorphic-heterostyled; inner bracteoles membranous, ± narrowly lanceolate, as long as the calyx, with a ciliate margin and a conspicuous midrib. Calyx c. 3·5 mm. long, turbinate-campanulate, 8- or 10-ribbed; lobes curved inwards, broadly deltate, apiculate, ¼ as long as the tube or shorter; appendages c. 0·75 mm. long, inflected, ciliolate. Petals c. 3·5 × 2·5 mm., purple, caducous, obovate, clawed. Stamens 8 or 10, inserted slightly above the bottom of calyx-tube, the epipetalous a little higher. Long-styled flowers: episepalous stamens long-exserted, the others exceeding twice the length of the appendages; style c. 7 mm. long, exceeding the episepalous stamens. Middle-styled flowers: episepalous stamens very much exserted, the others exceeding twice the length of the appendages; stigma c. 1 mm. below the anthers of the episepalous stamens. Short-styled flowers: episepalous stamens long-exserted, the others exceeding the appendages by 1·5 mm.; stigma a little below the anthers of the epipetalous stamens. Ovary elliptic-ovoid, 2-locular, shortly stipitate. Capsule and seeds not seen.

Mozambique. N: Palma, fl. 14.viii.1916, *Pires de Lima* 28 (PO). Known only from Mozambique. In marshy places.

24. **Nesaea cordata** Hiern in F.T.A. **2**: 475 (1871); Cat. Afr. Pl. Welw. **1**: 376 (1898).— Oliv. in Trans. Linn. Soc. Lond. **29**: 74, t. 40B (1873).—Koehne in Engl., Bot. Jahrb. **3**: 332 (1882); in Engl., Pflanzenr. **IV**, 216: 232 (1903).—Engl. Pflanzenw. Afr. **3**, 2: 651 (1921).—Fernandes & Diniz in Garcia de Orta **4**, 3: 392 (1956); op. cit. **6**: 106 (1958).—Pohnert & Roessler in Prodr. Fl. SW. Afr. **95**: 6 (1966).—Boutique in F.C.B., Lythraceae: 15 (1967).—Fernandes in C.F.A. **4**: 187 (1970). Syntypes from Upper Guinea (Niger), Uganda and Angola.
Nesaea cordata forma *villosa* Koehne in Engl., op. cit. **29**: 166 (1900). Type: Malawi, *Buchanan* 423 (B, holotype†).

Annual erect herb, 4–26 cm. high. Stem slender, simple to much branched, 4-gonous or narrowly winged, glabrous, scabrous or hairy. Leaves 5–20(25) × 5–10 mm., decussate, sessile, lanceolate to broadly ovate-lanceolate, acute or subacuminate at the apex, margin whitish, cordate at the base (sometimes only the upper leaves), glabrous to ± hairy, 1-nerved. Dichasia c. 5-flowered; peduncles

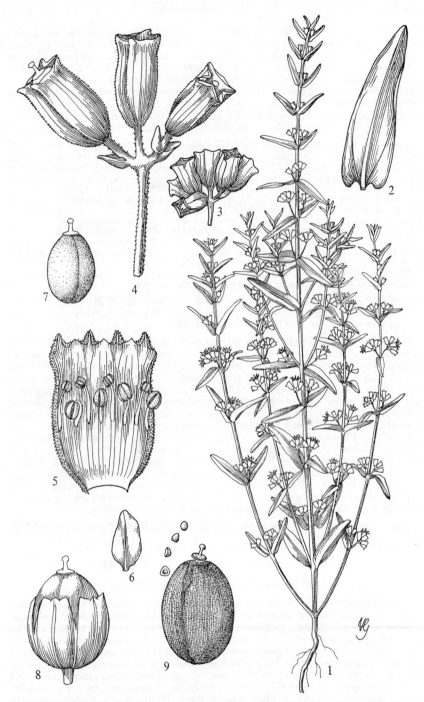

Tab. 75. NESAEA DRUMMONDII. 1, habit ($\times\frac{2}{3}$); 2, leaf ($\times 3$); 3, 6-flowered dichasium
($\times 3$); 4, 3-flowered dichasium ($\times 10$); 5, calyx opened ($\times 16$); 6, petal ($\times 20$);
7, pistil ($\times 16$); 8, capsule within calyx (8); 9, dehiscing capsule and seeds ($\times 8$),
all from *Wild & Drummond* 7019.

5–27 mm. long, usually rather shorter than the leaves; involucral bracteoles 4–7 × 4–7 mm., broadly cordate, acuminate, cymbiform, ± hairy outside; inner bracteoles 1·5–3·3 mm. long, elliptic to linear, ciliate at the margin. Flowers 4(6)-merous. Calyx 1·75–2 mm. long, campanulate, 8- or 12-ribbed; lobes c. 0·5 mm. high, triangular; appendages c. 1 mm. long, erect, ciliate at the apex. Petals 1·5–2 mm. long, obovate, rose-coloured, mauve or pink, almost as long as the calyx-tube. Stamens 4–6, rarely 8–12, exserted. Ovary c. 1 mm. in diameter, subglobular, 2(3)-locular; style 1·5–2 mm. long. Capsule 1·5–1·7 mm. in diameter, subglobose, included. Seeds c. 0·25 mm. long, brownish, concave-convex.

Zambia. B: Masese, fl. 14.iii.1961, *Fanshawe* 6429 (K; SRGH). N: 45 km. NE. of Isoka, fl. & fr. 23.v.1962, *Robinson* 5236 (K). S: Choma, Mapanza, 1020 m., fl. & fr. 13.iv.1958, *Robinson* 2839 (K; PRE; SRGH). **Rhodesia.** N: Gokwe, Lengwe Research Station, fl. & fr. 16.iv.1969, *Jacobsen* 614 (COI). W: Matobo, Farm Besna Kobila, 1536 m., fl. & fr. iv.1955, *Miller* 2757 (K; LISC; PRE; SRGH). C: Marandellas, fl. & fr. 9.iv.1948, *Corby* 70 (BM; K; SRGH). S: Gwanda, Doddieburn Ranch, ± 800 m., fl. & fr. 11.v.1972, *Pope* 760 (COI; K). **Malawi.** S: Mlanje Distr., Likabula Plain, 780 m., fl. & fr. 15.vi.1962, *Richards* 16702 (K). **Mozambique.** N: Malema, Mutuáli, Estação Experimental do CICA, fl. & fr. 24.iv.1961, *Balsinhas & Marrime* 437 (BM; K; LISC; LMA; PRE). Z: Chinde, viii.1907, anonymous 78 (COI). MS: near Garuso, fl. & fr. iv.1935, *Gilliland* 1839 (K). SS: R. Limpopo, Dumela, fl. & fr. 30.iv.1961, *Drummond & Rutherford-Smith* 7616 (K; SRGH).

Widespread in tropical Africa. Moist places among grasses and sedges, mud on riverbeds, shallow wet soil over rocks, near running water, etc.

25. **Nesaea pygmaea** Fernandes & Diniz in Bol. Soc. Brot., Sér. 2, **29**: 97, t. 12 (1955); in Garcia de Orta **4**, 3: 394, t. 2 (1956). Type: Mozambique, between Nampula and Corrane, *Torre* 1365 (COI, holotype; LISC).

Annual slender erect glabrous herb, up to 6 cm. high. Stem simple or little branched, 4-gonous or narrowly 4-winged. Leaves up to 9 × 3 mm., decussate, ovate-lanceolate, subacute or acute at the apex, truncate or roundish at the base, 1-nerved, the lower ones and those of the middle shorter than the internodes, the upper ones as long as or longer. Dichasia capituliform, 1–3-flowered, at the apex of stem and in the axils of upper leaves; peduncles 1·5–4 mm. long; involucral bracteoles 2, c. 2·5 × 1·25 mm., ovate-lanceolate, greenish, conspicuously veined, curved at the apex; inner bracteoles linear, membranous, exceeding the middle of the calyx. Flowers 4-merous. Calyx 1·5 mm. long, turbinate-campanulate, 8-ribbed; lobes deltate, curved inwards, ¼ as long as the calyx-tube; appendages c. 0·75 mm. long, erect, much longer than the lobes, not ciliate. Petals 1 × 0·75 mm., obovate, clawed, emarginate at apex, wine-coloured. Stamens 4, opposite to the calyx-lobes, inserted slightly below the middle of calyx-tube, exceeding the appendages. Ovary subglobose, 2-locular; style c. twice the ovary length, scarcely exceeding the stamens. Capsule scarcely exceeding the calyx-lobes.

Mozambique. N: Malema, Inago Mt., fl. & fr. 20.iii.1964, *Torre & Paiva* 11279 (BM; COI; K; LISC; SRGH).

Known only from Mozambique. Over rocks in dry places.

26. **Nesaea drummondii** Fernandes in Bol. Soc. Brot., Sér. 2, **48**: 124, t. 10 (1974). TAB. **75**. Type: Botswana, Chadum Valley, SW. Africa border, 16 km. W. of Knau Knau, *Wild & Drummond* 7019 (COI; K; SRGH, holotype).

Annual erect herb 5–35 cm. high. Stem branched from the base or a little above, 4-winged like the ascending branches, the wings conspicuously scabrous. Leaves 7–45 × 1·5–10 mm., decussate, broadly lanceolate, obtuse or subacute at the apex, margin scabrous and ± involute, cordate at the base and contracted a little above, greyish green or dark green, 1-nerved, the nerve impressed above, prominent and scabrous beneath. Dichasia (1)3–many-flowered, pedunculate, the peduncle 1–8 mm. long; bracteoles up to 1·25 mm. long, scarious, with a green midrib and scabrous margin. Flowers 4(5)-merous. Calyx 3–3·5 mm. long, narrowly tubular, a little constricted above, reddish, conspicuously longitudinally ribbed, minutely scabrous especially along ribs; lobes c. 0·75 mm. broad and 0·25 mm. high, deltate; appendages spreading, c. 0·5 mm. long, a little longer than the lobes. Petals c. 1·5 × 0·75 mm., ovate-lanceolate, with a conspicuous midrib, caducous, violet. Stamens 8(10), included, inserted in 2 rows at c. ⅓ of the calyx-tube length; anthers c. 0·2 mm. long, globose. Ovary 1 × 0·6 mm., ellipsoid, 2-locular; style (with stigma) c. 0·25–0·5 mm. long, included. Capsule included, many-seeded subsepticidal; operculum thick, persisting attached to the voluminous placentas. Seeds c. 0·25 mm. in diam., brownish.

Botswana. N: Chadum Valley, SW. Africa border, 16 km. W. of Knau Knau, fl. & fr. 14.iii.1965, *Wild & Drummond* 7019 (COI; K; SRGH). **Rhodesia.** N: Kariba, Chirundu, Zambesi R., 24.iv.1961, *Rutherford-Smith* 688 (SRGH). W: Gwaai R. near Dahlia on road to Gwaai River Hotel, fl. & fr. 16.iv.1972, *Grosvenor* 681 (COI). **Mozambique.** SS: Dumela, R. Limpopo, 30.iv.1961, *Drummond & Rutherford-Smith* 7628 (COI; K; SRGH); Caniçado, 31 km. on the road Vila Pinto Teixeira to Balule, fl. & fr. 26.viii.1969, *Correia & Marques* 1236 (LMU).
Also in SW. Africa. In wet grassy places and on mud and sand in riverbeds.

5. AMMANNIA L.

Ammannia L., Sp. Pl. **1**: 119 (1753); Gen. Pl. ed. 5: 55 (1754).

Annual (sometimes biennial?) glabrous (rarely with the calyx minutely hairy) herbs. Stems erect or decumbent, 4-gonous or 4-winged, simple or ± branched. Leaves decussate, sessile, entire, 1-nerved. Dichasia (1)3–many-flowered, subsessile or pedunculate; bracteoles small, membranous, whitish. Flowers 4(5–8)-merous, never heterostylous. Calyx campanulate or urn-shaped, 8-ribbed; appendages absent or very short. Petals 4–8, very caducous or absent, small, obovate, spathulate or rounded. Stamens as many as or twice the number of calyx-lobes, rarely fewer, with the filaments inserted at ¼–½ of the calyx-tube length; anthers didymous. Ovary incompletely 2–4(5)-locular, sometimes unilocular; style 0 or ± long, not continuous with placentas; stigma capitate; ovules numerous, axile or on the septa. Capsule globose or ellipsoid, included in the calyx or ± exserted, thinly membranous, dehiscing transversally and irregularly; seeds numerous, very small, angular.
A subtropical and tropical genus of c. 20 species, found in wet habitats.

Style longer than 0·3 mm.; leaves usually all auriculate-cordate at the base; petals present:
  Style 0·5–1·8 mm. long; dichasia very loose, with peduncles 4–18 mm. long
                                        1. *auriculata*
  Style 0·3–1 mm. long; dichasia more dense than above, with peduncles 1·5–4 mm. long:
    Plant unbranched or with a few very short branches, up to 120 cm. high; dichasia relatively loose with peduncles c. 4 mm. long; stamens 4–8; capsule c. 3 mm. in diameter, scarcely exceeding the calyx-lobes - - - - 2. *elata*
    Plant much branched, up to 60 cm. high; dichasia usually dense with peduncles 1·5–3 mm. long; stamens 4; capsule 1·5–1·8 mm. in diameter, much exceeding the calyx-lobes - - - - - - - - - - 3. *prieuriana*
Style absent or up to 0·3 mm. long; leaves very rarely all auriculate-cordate at the base; petals frequently absent:
  Fructiferous calyx with appendages short but distinct; petals 4; calyx tube 1·5–2 mm. long - - - - - - - - - - - 4. *wormskioldii*
  Fructiferous calyx without appendages; petals 0; calyx tube 1–1·2 mm. long
                                        5. *baccifera*

1. **Ammannia auriculata** Willd., Hort. Berol. **1**: 7, t. 7 (1803).—DC., Prodr. **3**: 80 (1828).—Koehne in Bot. Jahrb. **1**: 244 (1880); op. cit. **4**: 389 (1883); in Engl. & Prantl, Pflanzenfam. **III**, 7: 7 (1898); in Engl., Pflanzenr. **IV**, 216: 45, fig. 5B (1903).—Gilg in Engl., Pflanzenw. Ost-Afr. **C**: 285 (1895).—Engl., Pflanzenw. Afr. **3**, 2: 645 (1921).—Burtt Davy, F.P.F.T. **1**: 198 (1926).—Fernandes & Diniz in Garcia de Orta **4**, 3: 404 (1956).—Pohnert & Roessler in Prodr. Fl. SW. Afr. **95**: 2 (1966).—Boutique in F.C.B., Lythraceae: 22 (1967).—Fernandes in C.F.A. **4**: 175 (1970). Type from Africa.

Annual erect simple or ± branched herb up to 65 cm. high. Stem up to 5 mm. thick and subterete below, tapering upwards, the upper part of stem and the branches 4-gonous, narrowly winged to the apex, the wings sometimes scabrous. Leaves 15–80 × 3–14 mm., linear or sublanceolate, attenuate to the subacute apex, margin entire, sometimes minutely scabrous, 1-nerved, the nerve impressed above, prominent below, cordate-auriculate at the base (sometimes the lower ones cuneate). Dichasia (2)3–15-flowered, loose, with the peduncle 4–10(18) mm. long; bracteoles lanceolate. Flowers 4-merous, with pedicel 0·5–3 mm. long. Calyxtube campanulate, 1–1·5 mm. long; lobes 4, deltoid, attaining ⅓–½ of the calyx-tube length; appendages 4, ± distinct, up to 0·5 mm. long. Petals obovate, 0·6–0·7 mm. long, pink. Stamens 4–8, exserted. Ovary 1–1·5 mm. in diameter, globose; style 0·5–1·8 mm. long, filiform. Capsule 2–3·5 mm. in diameter, globose, equalling or slightly exceeding the lobes. Seeds c. 0·4 mm. long, brownish, concave-convex.

Botswana. N: Isodilo Hills, fl. & fr. 8.ii.1964, *Guy* 176/64 (SRGH). Zambia. N: Kasama, 95 km. E. of Kasama, fl. & fr. 6.v.1962, *Robinson* 5139 (K; SRGH). W: Ndola, fl. & fr. 3.iv.1954, *Fanshawe* 1056 (EA; K; LISC; SRGH). C: Luangwa Valley, fl. & fr. 26.xi.1965, *Astle* 4129 (K; SRGH). S: Mapanza, Choma, fl. & fr. 20.iv.1958, *Robinson* 2846 (PRE; SRGH). Rhodesia. N: Gokwe, Mbumbuze R., fl. & fr. 17.iii.1964, *Bingham* 1323 (K; LISC; SRGH). W: Shangani Reserve, fl. & fr. 1.iv.1951, *West* 3159 (SRGH). C: Umswewe, fl. & fr. 20.iv.1921, *Borle* 179 (K; PRE). S: Beitbridge, Sentinel Ranch, fl. & fr. v.1967, *Symes* 67/M/31 (COI). Malawi. N: Kondowe to Karonga, 1920–2880 m., fl. & fr. vii.1896, *Whyte* 32 (K). Mozambique. N: Mandimba, fl. & fr. 31.v.1948, *Pedro & Robinson* 4139 (EA). Z: Quelimane, Nicoadala, Licuare, bank of R. Limare, fl. & fr. 12.v.1971, *Balsinhas* 1859 (LMA). T: Tete, right bank of R. Zambezi, fl. & fr. 16.x.1965, *Neves Rosa* 30 (LMA). MS: Beira, Chemba, confluence of R. Muira with R. Zambezi, fl. & fr. 15.v.1971, *Torre & Correia* 18439 (LISC), 18441 (LISC). SS: Caniçado, near Maguge, R. Limpopo, 23.viii.1969, *Correia & Marques* 1171 (COI; LISC; LMU). LM: Namaacha, bank of R. Movene, 5.vii.1967, *Marques* 204 (COI; LMU; PRE).

Widespread in tropical and subtropical Africa, Asia, Australia and America. In humid sands and alluvial plains near water.

2. **Ammannia elata** Fernandes in Bol. Soc. Brot., Sér. 2, 48: 125, t. 11 (1974). TAB. 76.
Type: Mozambique, Zambezia, Mathilde, Dumbazi R., *Le Testu* 749 (BM, holotype).

Erect annual herb up to 1·2 m. high. Stem unbranched or with a few very short branches, thick (c. 6 mm. in diameter), exfoliating longitudinally, 4-gonous below, 4-winged upwards. Leaves up to 60 × 8 mm., decussate, narrowly lanceolate, attenuated to the subacute apex, almost hastate-cordate at the base, obscurely penninerved, the midrib impressed above prominent beneath. Dichasia not very dense, c. 20-flowered; primary peduncle 4-gonous, c. 4 mm. long, the secondary and the tertiary c. 2 mm.; pedicels 1–1·5 mm. long; bracteoles minute, c. 1·25 mm. long, linear, acute. Calyx-tube 1·75 mm. long, campanulate, with 8 longitudinal ribs; calyx-lobes 4, c. 1 × 1 mm., triangular; appendages 4, shorter than the calyx-lobes. Petals 4, c. 1 × 0·75 mm., pinkish, caducous, subcircular. Stamens 4–8; filaments c. 1·5 mm. long; anthers c. 0·5 mm. long. Ovary c. 1 mm. in diameter, globose, ovules numerous; style 0·5 mm. long. Capsule c. 3 mm. in diameter, globose, usually slightly exceeding the calyx-lobes. Seeds c. 0·5 mm. long, brownish, angular.

Mozambique. Z: Mathilde, R. Zambezi, fl. & fr. 2.vi.1908, *Le Testu* 749 (BM). Known only from Mozambique. Marshy places of river banks.

3. **Ammannia prieuriana** Guill. & Perr. in Guill., Perr. & A. Rich., Fl. Senegamb. Tent. 1: 303 (1833).—Koehne in Engl., Bot. Jahrb. 1: 248 (1880); in Engl., Pflanzenr. IV, 216: 48 (1903).—Engl., Pflanzenw. Afr. 3, 2: 646 (1921).—Fernandes & Diniz in Garcia de Orta 4, 3: 405 (1956).—Boutique in F.C.B., Lythraceae: 23 (1967).—Fernandes in C.F.A. 4: 175 (1970). Type from Senegambia.
Ammannia multiflora sensu Fernandes & Diniz, loc. cit. non Roxb. (1820).
Ammannia senegalensis sensu Fernandes & Diniz, tom. cit. 406 pro parte quoad specim. *Torre* 521 et *Pedro & Pedrógão* 4424.

Annual usually branched herb 10–60 cm. high. Stems and branches 4-gonous, widely winged at the summits. Leaves 15–70 × 1·5–11 mm., linear or oblong attenuated to the apex, almost hastate-cordate at the base, sometimes broader above the middle. Dichasia dense, 3–many-flowered; peduncle 1·5–3 mm. long. Flowers pedicellate, the pedicels 1–4 mm. long. Calyx-tube c. 1 mm. long, campanulate; lobes 4, deltate; appendages 4, short, horn-shaped. Petals 4, 0·5–0·6 mm. long, obovate, pinkish, caducous. Stamens 4, inserted ± at the middle of the calyx-tube. Ovary 0·7–1 mm. in diameter, globose; style 0·5–1 mm. long. Capsule 1·5–1·8 mm. in diameter, globose, exserted. Seeds c. 0·4 mm. long.

Zambia. B: Masese, fl. & fr. 3.v.1961, *Fanshawe* 6528 (K; SRGH). N: SW. of Mweru-wa-Ntipa, Mofwe Dambo, c. 912 m., fr. 30.vii.1962, *Tyrer* 150 (BM; SRGH). W: Mwekara, fl. & fr. 28.iv.1962, *Fanshawe* 6802 (SRGH). C: 9·5 km. E. of Chisamba,

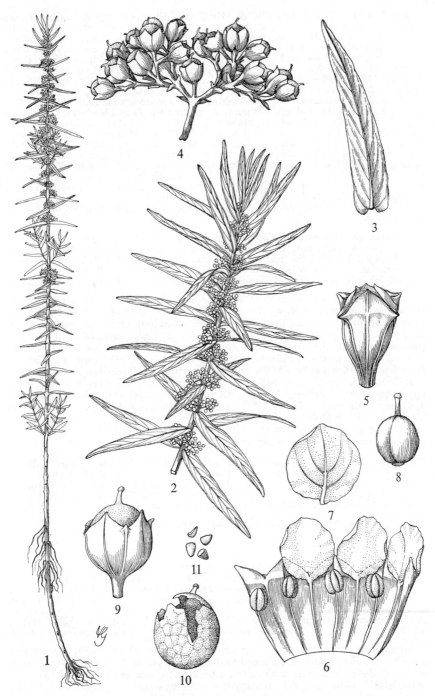

Tab. 76. AMMANNIA ELATA. 1, habit ($\times \frac{1}{5}$); 2, upper part of plant ($\times \frac{1}{2}$); 3, leaf, superior surface ($\times 1$); 4, dichasium ($\times 4\frac{1}{2}$); 5, flower-bud ($\times 15$); 6, flower opened to show petals and insertion of stamens ($\times 15$); 7, petal ($\times 20$); 8, pistil ($\times 15$); 9, capsule within calyx ($\times 9$); 10, dehisced capsule ($\times 9$); 11, seeds ($\times 9$), all from *Le Testu* 749.

fl. & fr. 19.v.1957, *Best* 130 (K; SRGH). E: Chipata Distr., Chilongozi, Luangwa Game Reserve, Luangwa R., 600 m., fl. & fr. 11.x.1960, *Richards* 13325 (K; SRGH). S: Choma, fl. & fr. 16.iv.1963, *van Rensburg* 1969 (K; SRGH). **Rhodesia.** N: Urungwe, near Mogororo R., fl. & fr. viii.1957, *Goodier* 310 (K; SRGH). W: Bulawayo, 1440 m., fl. & fr. vii.1956, *Miller* 3618 (K; LISC; SRGH). C: Gwelo, Mlezu School Farm, 29 km. SSE. of Que Que, 1344 m., fl. & fr. 16.vi.1966, *Biegel* 1243 (COI; SRGH). E: Inyanga, Honde Valley, 832 m., fl. & fr. 17.iv.1958, *Phipps* 1084 (COI; SRGH). S: Nuanetsi, Malangwe R., SW. Mateke Hills, 640 m., fl. & fr. 5.v.1958, *Drummond* 5582 (COI; SRGH). **Malawi.** N: between Mpata and Tanganyika Plateau, vii.1896, *Whyte* (K). C: Benga, Nkhota-Kota, 470 m., fl. & fr. 2.ix.1946, *Brass* 17488 (BM; K; PRE; SRGH). S: Ciwalo, Kalembe, Mangochi, fl. & fr. 2.vi.1955, *Jackson* 1671 (K). **Mozambique.** N: Meconta, Corrane, fl. & fr. 11.vii.1948, *Pedro & Pedrógão* 4424 (COI; EA). Z: Lugela-Mocuba, Namagoa Estate, fl. & fr. vii, *Faulkner* 9 (COI; EA; K; PRE; SRGH). T: Kabagahangwa Kraal, Mazoe R., 320 m., fl. & fr. 22.ix.1948, *Wild* 2581 (K; SRGH). MS: Beira, Chemba, Mitondo, c. 170 m., fl. & fr. 1971, *Torre & Correia* 18445 (LISC). SS: between Macontene and Chibuto, near R. Zenge, fl. & fr. 30.vi.1950, *Myre* 889 (COI; LMA). LM: Namaacha, km. 28·7 on national road no. 3, R. Changalane, fl. & fr. 1.viii.1967, *Marques* 2139 (COI; LMU).

Widespread in tropical Africa. Sands, banks of water courses, edges of swamps, marshes, etc.

4. **Ammannia wormskioldii** Fisch. & Mey. in Ind. Sem. Horti Petrop. **7**: 42 (1841).—
Koehne in Engl., Bot. Jahrb. **1**: 256 (1880); op. cit. **4**: 391 (1883); in Engl., Pflanzenr. **IV**, 216: 53, fig. 5L (1903).—Engl., Pflanzenw. Afr. **3**, 2: 646 (1921).—
Fernandes & Diniz in Garcia de Orta **6**: 100 (1958).—Pohnert & Roessler in Prodr. Fl. SW. Afr. **95**: 3 (1966).—Fernandes in C.F.A. **4**: 177 (1970). Type from Zaire.

Annual (sometimes perennial?) erect or ascending herb up to 50 cm. high, rooting at the lower nodes. Stem unbranched or little-branched, 4-gonous. Leaves narrowly lanceolate to elliptic-obovate, 10–32(60) × 2·5–8(12) mm., acute at the apex, the upper ones subcordate at the base, the lower ones attenuate-cuneate, pinnately nerved, the midrib impressed above, prominent below. Dichasia 3–many-flowered, sessile or subsessile, dense; bracteoles 0·4–0·5 mm. long, linear; fructiferous pedicels as long as the calyx. Calyx-tube 1·5–2 mm. long, campanulate, semiglobose in fruit. Calyx-lobes 4, equalling ⅓ to ½ the length of the tube; appendages 0 or very short. Petals 4, white, ± as long as the calyx-lobes. Stamens 4, as long as the calyx. Ovary 0·7–1·2 mm. in diameter, globose; stigma sessile to subsessile (the style sometimes up to 0·5 mm. long). Capsule 1·5–2·5 mm. in diameter, globose, covered by the calyx to ⅔ of its length. Seeds c. 0·5 mm. long, brownish, pyriform, concave-convex.

Zambia. N: Mbala (Abercorn) Distr., Lumi River Marsh, 1680 m., fl. & fr. 31.v.1957, *Richards* 9937 (K). **Rhodesia.** N: Miami, fl. & fr. iv.1926, *Rand* 33 (BM). W: Nyamandhlovu, Gwaai R., fl. & fr. 19.iv.1973, *Gibbs Russell* 1703 (COI). S: Nuanetsi, Malangwe R., SW. Mateke Hills, 672 m., fl. & fr. 7.v.1958, *Drummond* 5679 (COI; K; LISC; PRE; SRGH). **Malawi.** N: Kondowe to Karonga, 640–1920 m., vii.1896, *Whyte* (K).

Also in Angola, SW. Africa and Zaire. Edges of marshes and swamps.

5. **Ammannia baccifera** L., Sp. Pl. **1**: 120 (1753); op. cit., ed. 2, **1**: 175 (1762).—
Hiern in F.T.A. **2**: 478 (1871); Cat. Afr. Pl. Welw. **1**: 374 (1898).—Koehne in Engl., Bot. Jahrb. **1**: 258 (1880); op. cit. **4**: 391 (1883); in Engl., Pflanzenr. **IV**, 216: 53, fig. 5M (1903).—Engl., Pflanzenw. Afr. **3**, 2: 647 (1921).—Keay in F.W.T.A. ed. 2: 165 (1954).—Fernandes & Diniz in Garcia de Orta **6**: 101 (1958).—Pohnert & Roessler in Prodr. Fl. SW. Afr. **95**: 2 (1966).—Fernandes in C.F.A. **4**: 177 (1970). Type from China.

Ammannia attenuata Hochst. ex A. Rich., Tent. Fl. Abyss. **1**: 278 (1847).—
Koehne, loc. cit.—Fernandes & Diniz, op. cit. **4**, 3: 406 (1956). Type from Ethiopia.

Annual erect or ascending ± branched herb up to 80 cm. high. Stems and branches reddish, 4-angular, winged to the apex. Leaves 7–70 × 1·5–10(16) mm., elliptic to linear, attenuate at the apex, attenuate-cuneate or obtuse to subcordate at the base. Dichasia (1–2)3- to many-flowered, lax or dense; peduncles 1(2) mm. long; bracteoles linear-lanceolate; pedicels 1–2·5(4) mm. long. Calyx-tube 1–1·2 mm. long, turbinate-campanulate; lobes 4; appendages 0 or inconspicuous. Petals 0. Stamens 4, as long as the calyx-lobes or shorter. Ovary 0·7–1·2 mm. in diameter, globose; style 0·1–0·3 mm. long. Capsule 1–2·5 mm. in diameter, globose, ½ to ¾ included in the calyx. Seeds c. 0·4 mm. in diameter, brownish.

78. LYTHRACEAE 309

Leaves attenuate-cuneate at the base; dichasia lax; capsule 1–2 mm. in diameter; stems
and branches slender - - - - - - - - - subsp. *baccifera*
Leaves obtuse to subcordate at the base; dichasia dense; capsule (1·5)2–2·5 mm. in
diameter; stems and branches more robust - - - subsp. *aegyptiaca*

Subsp. **baccifera**.

Zambia. W: Kitwe, Mindolo, fl. & fr. 28.iv.1963, *Mutimushi* 290 (K; SRGH). C:
Lusaka, Chikupa Estate, fl. & fr. 12.iv.1963, *van Rensburg* 1900 (K; SRGH). S: Machili,
fl. & fr. 25.iii.1961, *Fanshawe* 6462 (LISC; SRGH). **Rhodesia.** N: Lomagundi, fl. &
fr. 6.i.1963, *Jacobsen* 2048 (PRE). S: Nuanetsi R., gorge upstream from Buffalo Bend,
fl. & fr. 29.iv.1961, *Drummond & Rutherford-Smith* 7579 (K). **Malawi.** N: Karonga,
L. Nyasa, fl. & fr. xi.1887, *Scott Elliot* (K). **Mozambique.** Z: without precise locality,
Carvalho (COI). MS: islands of R. Zambezi, fl. & fr. 1946, *Simão* 660 (LMA).
SS: Gaza, R. Limpopo, Dumela, fl. & fr. 30.iv.1961, *Drummond & Rutherford-Smith* 7627
(COI; K; SRGH). LM: Estação Agrícola de Umbelúzi, fl. & fr. 13.vi.1946, *Gomes e
Sousa* 3327 (LISC).
Tropical Africa, Asia and Australia (fide Koehne, loc. cit.). In moist sands and marshes.

Subsp. **aegyptiaca** (Willd.) Koehne in Engl., Bot. Jahrb. **1**: 258 (1880); op. cit. **4**: 391
(1883); in Engl., Pflanzenr. IV, 216: 55 (1903).—Boutique in F.C.B., Lythraceae:
24 (1962). Type from Egypt.
Ammannia aegyptiaca Willd., Hort. Berol. **1**: 6, t. 6 (1803); Enum. Hort. Berol. **1**:
167 (1809). Type as above.

Zambia. N: Mansa (Fort Rosebery), fl. & fr. 4.vii.1957, *Fanshawe* 3409 (K).
Rhodesia. W: Matobo, Farm Besna Kobila, 1504 m., fl. & fr. ii.1953, *Miller* 1562
(SRGH). C: Salisbury, i.1918, *Eyles* 6884 (K; SRGH).
Also in Europe (introduced), Egypt, Zaire, Angola, Sudan, Ethiopia and Asia. On
sandy river banks.

6. HIONANTHERA Fernandes & Diniz

Annual or perennial aquatic or terrestrial glabrous herbs. Leaves decussate,
sessile, linear, widened at the base, canaliculate above. Flowers (3)4(5)-merous, in
axillary sessile many-flowered dense dichasia surrounded by the widened bases of
leaves; outer bracteoles 2, ± as long as the glomerules, the inner ones numerous,
subulate, membranous, whitish. Calyx campanulate, scarious, 8-nerved below,
4-nerved above the insertion of the stamens, with short appendages in the sinuses
and a nectariferous ring at the bottom. Petals persistent, corrugated. Stamens 4
(sometimes 3 or 5), opposite to the lobes; anthers dorsifixed, they and the pollen
violet. Ovary sessile or shortly stipitate, with the dissepiments interrupted above
the column of the placenta (the placenta then not continuous with the style),
incompletely 2-locular, 2–5-ovulate; style ± long. Capsule thinly membranous,
dehiscing irregularly. Seeds few, concave-convex, relatively large, dark violet.
Tropical African genus with 4 species.

Aquatic plants, caespitose; stems simple or little-branched:
 Stems thick, c. 4 mm. in diam.; leaves up to 6 cm. long, longer than the internodes,
  erect or ascending, attenuate to the apex; outer bracteoles foliaceous; flowers
  (3)4(5)-merous; petals violet; ovary 2–4-ovulate; style 2 mm. long
                                                                    1. *mossambicensis*
 Stem slender; leaves up to 2·5 cm. long, the lower ones spreading and shorter than the
  internodes, the upper ones erect or ascending, all narrow and acuminate; outer
  bracteoles not foliaceous, usually violet; flowers 4-merous; petals pale lilac; ovary
  2-ovulate; style ± 1·5 mm. long - - - - - - - 2. *graminea*
Terrestrial plants, not caespitose; stems branched from the base or only further up:
 Plant procumbent; leaves up to 4·5 cm. long, the lower ones spreading, the upper ones
  ascending or erect, acute; flowers 4-merous; calyx campanulate, almost truncate;
  appendages short, triangular; petals white; ovary usually 2-ovulate; style c. 2 mm.
  long - - - - - - - - - - - 3. *torrei*
 Plant erect; upper leaves c. 4 × 0·2 cm., spreading-ascending, acute or subacute at apex;
  flowers 4(5)-merous; calyx campanulate with the lobes deltate, well developed and
  with inconspicuous appendages; petals lilac; ovary (4)5-ovulate; style c. 1–
  1·5 mm. long - - - - - - - - - - 4. *garciae*

1. **Hionanthera mossambicensis** Fernandes & Diniz in Bol. Soc. Brot., Sér. 2, **29**: 91,
 t. 6 (1955); in Garcia de Orta **4**, 3: 398, t. 4 (1956). TAB. 77. Type: Mozambique,
 Nampula, *Torre* 719 (COI, holotype; LISC).

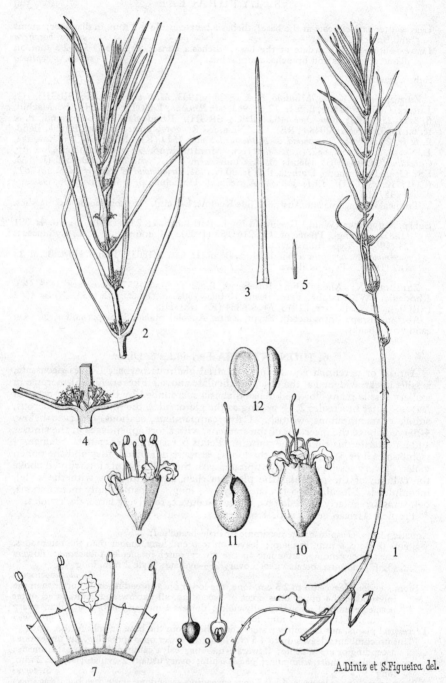

A.Diniz et S.Figueira del.

Tab. 77. HIONANTHERA MOSSAMBICENSIS. 1, habit (×½); 2, upper part of stem (×1); 3, leaf (×1); 4, part of stem showing floral glomerules; 5, outer bracteole (×4); 6, lateral view of flower (×10); 7, calyx opened to show appendages of calyx, one petal, stamens and nectariferous ring (×10); 8, pistil (×10); 9, longitudinal section of the ovary (×10); 10, fructiferous calyx (×10); 11, capsule dehiscing (×10); 12, seeds (×10), all from *Torre* 719.

Aquatic caespitose herb up to 45 cm. long (probably longer). Stem simple or few branched, 4 mm. in diameter, 4-gonous, rooting and leafless below, leafy above; internodes up to 4·5 cm. long, very short towards the apex. Leaves up to 6 × 0·3 cm., attenuated to an acute apex, expanded at the base, ascending or erect, distinctly 1-nerved. Flowers (3)4(5)-merous in many-flowered very dense glomerules; outer bracteoles up to 3·5 × 0·75 mm., foliaceous, linear, the inner ones minute (c. 1 mm.), linear, thinly membranous, whitish; pedicels up to 1 mm. long. Calyx 1·5 × 1 mm., campanulate; lobes short or very short, c. 1 mm. wide, slightly apiculate; appendages short, scarcely exceeding the lobes. Petals 1 mm. long, violet, corrugated, oblong, sometimes slightly emarginate at the apex. Stamens usually 4, 2–2·25 mm. long, inserted a little below the middle of the tube and exserted for c. 1·5 mm. Ovary c. 0·5 mm. long, obovoid, 2–4-ovulate; style 2 mm. long, almost as long as the stamens. Capsule c. 2 mm. long, ellipsoid, scarcely exserted. Seeds often 2, c. 1·75 × 1 mm., dark violet.

Mozambique. N: Nampula, 7.ii.1936, *Torre* 719 (COI; LISC).
Known only from Mozambique. In swamps.

2. **Hionanthera graminea** Fernandes & Diniz in Bol. Soc. Brot., Sér. 2, **29**: 92, t. 7 (1955); in Garcia de Orta **4**, 3: 398, t. 5 (1956). Type: Mozambique, Nampula, *Torre* 715 (COI, holotype; LISC).

Caespitose herb up to 40 cm. long. Stems yellowish, slender, flexible, simple, leafless and rooting below, leafy towards the apex; internodes up to 6 cm. long, very short towards the apex. Leaves up to 2·5 cm. long, linear, acuminate at the apex, widened and reddish at the base, folded or canaliculate above, keeled beneath, the lower ones patent, the upper ones ascending. Flowers 4-merous in dense many-flowered glomerules; outer bracteoles 3 × 1 mm., linear, acute at the apex, widened at the base, membranous, canaliculate above, keeled beneath, usually violet, the inner ones subulate, whitish or rose. Calyx up to 1·5 mm. long, campanulate, lilac or rose; lobes short, broadly deltate, apiculate; appendages short, as long as the lobes. Petals c. 1 mm. long, oblong, pale lilac, corrugated. Stamens 4, exserted c. 1·5 mm. Ovary 0·5 mm. long, obovoid, 2-ovulate; style ± 1·5 mm. long. Capsule c. 0·5 mm. long, exserted. Seeds 2, c. 1·5 × 0·75 mm., brownish, concave-convex.

Mozambique. N: Nampula, fl. & fr. 7.ii.1935, *Torre* 715 (COI; LISC).
Known only from Mozambique. In swamps.

3. **Hionanthera torrei** Fernandes & Diniz in Bol. Soc. Brot., Sér. 2, **29**: 93, t. 8 (1955); in Garcia de Orta **4**, 3: 398, t. 6 (1956). Type: Mozambique, Nampula, *Torre* 1212 (COI, holotype; LISC).

Annual procumbent herb, 25 cm. long or more. Stems c. 2·5 mm. in diameter, ascending, rooting towards the base; internodes up to 4 cm. long, shorter towards the apex. Leaves up to 4·5 × 0·3 cm., acute at the apex, widened at the base, distinctly 1-nerved. Flowers 4-merous in very dense many-flowered glomerules; outer bracteoles up to 3 × 1·5 mm., lanceolate, acute at the apex, thinly membranous, usually rose-pink(?); inner bracteoles similar to the outer ones but smaller. Calyx 1·5 mm. long, campanulate, almost truncate; lobes very short, apiculate; appendages short. Petals 4, c. 1·25 mm. long, oblong, white. Stamens 4, exceeding the calyx-lobes by c. 1 mm. Ovary c. 0·5 mm. long, obovoid, 2-ovulate; style c. 2 mm. long, equalling or slightly exceeding the stamens. Capsule ellipsoid, scarcely exceeding the calyx-lobes. Seeds 2, c. 1 × 0·5 mm., dark-violet, concave-convex.

Mozambique. N: Nampula, 21.ii.1937, *Torre* 1212 (COI; LISC).
Known only from Mozambique. In the soil upon rocks.

More material is needed to decide whether *H. mossambicensis* and *H. torrei* are indeed distinct, or whether they are merely aquatic and terrestrial forms, respectively, of a single species.

4. **Hionanthera garciae** Fernandes & Diniz in Bol. Soc. Brot., Sér. 2, **29**: 94, t. 9 (1955); in Garcia de Orta **4**, 3: 398, t. 7 (1956). Type: Mozambique, Manica e Sofala Distr., Serra de Bandula, near Chimoio, *Garcia* 790 (COI; K; LISC, holotype; MO; P; PRE; SRGH).

Annual or perennial erect herb up to 25 cm. high. Stem not or somewhat thickened, unbranched and rooting at the nodes towards the base, simple or often branched further up, the branches ascending, decussate; internodes up to 3·5 cm. long, very short towards the apex. Basal leaves 2, c. 17 × 4 mm., opposite, at the lowest node between the roots, oblong, rounded at the apex, membranous, 1-nerved; leaves of second node c. 6 × 1·5 mm., linear, thinly membranous; upper cauline leaves and those of the branches up to 4 × 0·2 cm., linear, acute or subacute at the apex, widened at the base, suberect or erect, 1-nerved. Flowers 4(5)-merous, in many-flowered glomerules; outer bracteoles c. 2 × 0·5 mm., linear, acute at the apex, thinly membranous, whitish or pale rose; inner bracteoles similar to the outer ones but progressively smaller and thinner. Calyx 1·25 mm. long, broadly campanulate; lobes deltate, equalling c. ⅓ of the tube; appendages inconspicuous. Petals 0·75 mm. long, lilac, corrugated. Stamens 4, with the filaments 1–2 mm. long, exceeding the lobes; anthers 0·25 mm. long. Ovary obovoid, 2–5-ovulate; style 1–1·5 mm. long, equalling the anthers. Capsule scarcely exserted, with the walls thinly membranous. Seeds 2–4, c. 1 × 0·75 mm., dark-brown, ovate, curved.

Rhodesia. C: Gwelo, Mlezu School Farm, 1344 m., 26.ii.1966, *Biegel* 936 (COI; SRGH). E: Umtali Distr., Tscuzo Purchase Area, Ruwara, 1920 m., ii.1961, *Davies* 44 (SRGH). S: Victoria Distr., *Monro* 911 (BM; SRGH). **Mozambique.** MS: Serra de Bandula, near Chimoio, 28.iii.1948, *Garcia* 790 (COI; LISC; MO; P; PRE; SRGH).
Known only from Rhodesia and Mozambique. In mud of drying pools and moist humus-rich soil upon rocks.

7. ROTALA L.
Rotala L., Mant. Pl. Alt.: 175 (1771).

Annual or rarely perennial glabrous herbs, aquatic or in marshy places. Stems terete or 4-gonous, striate, often rooting at the nodes. Leaves decussate or verticillate, rarely alternate, sessile or subsessile. Flowers small, actinomorphic, homostylous or heterostylous, 3–6-merous, axillary, solitary, or in terminal racemes or spikes, very rarely in axillary umbels; bracteoles 2, rarely absent. Calyx-tube semiglobose, campanulate or urceolate-tubular, scarious, very rarely herbaceous, the lobes triangular, alternating with ± developed appendages or these often absent; nectariferous ring present at base. Petals persistent, sometimes caducous or absent, equal or unequal. Stamens 1–6, opposite to the sepals. Ovary sessile or substipitate, incompletely 2–4-locular (dissepiments interrupted above the placentas); ovules ± numerous, small; style absent or ± developed, simple. Capsule septicidally 2–4-valved, cartilaginous, the walls horizontally striate (under the microscope). Seeds small, concave-convex.
A genus of c. 40 species of warm regions.

Leaves decussate:
  Flowers 4-merous in axillary (1)3–12-flowered umbels; pedicels 0·4–0·5(1·3) mm. long; calyx-appendages absent; petals 0; stamens 1–3, included; capsule 3-valved; leaves 5–25 × 1–5 mm., narrowly lanceolate or narrowly oblong, obtuse at the apex, narrowed to the base into a petiole - - - - - - - 1. *serpiculoides*
  Flowers solitary and axillary throughout the plant or in racemes or spikes sometimes capitate at the apex of stems and branches:
    Flowers solitary and axillary throughout the plant:
      Appendages of the calyx very conspicuous:
        Flowers 3-merous; calyx campanulate; capsule subglobose, 3-valved; bracteoles almost as long as the calyx:
          Robust erect aerial plant, with the main stem almost woody and densely branched; leaves of the main stem 4–12 × 2–5 mm., those of branches c. 3 × 2 mm. - - - - - - - 2. *juniperina*
          Creeping plant, with the main stem prostrate up to 4 mm. thick, producing erect shoots; leaves of the prostrate stem 9–20 × 4–8 mm., those of the erect shoots 4–10 × 1–4 mm. - - - - - 3. *decumbens*
        Flowers 4-merous; calyx tubular, tetragonal; capsule ellipsoid, 4-valved; bracteoles shorter than the calyx; petals 4, roundish - 4. *fluitans*
      Appendages of the calyx very short or absent:
        Flowers pedicellate, the pedicels up to 2 mm. long; petals narrowly triangular or 0; stamens 2–4, included; capsule 4-valved; leaves 4–15 × 2–5 mm. 5. *submersa*

Flowers sessile or subsessile:
  Stamens usually fewer than the calyx-lobes (2 or 3, very rarely 1 or 4); petals
    0 or ± developed:
    Capsule 3-valved:
      Bracteoles ± as long as the calyx; stamens inserted ± at the middle of
        the calyx-tube:
        Flowers (4)5(6)-merous; leaves linear or subulate, obtuse or mucronate
          at the apex, 3–5 × 0·3–0·4 mm. -   -   -   -   18. *pusilla*
        Flowers (3)4-merous; lower leaves c. 4 × 1 mm., the median narrowly
          elliptic, 4–7 × 1·25–2 mm., the upper ones ovate or broadly elliptic,
          3–4 × 1·75–2 mm. -   -   -   -   -   -   6. *minuta*
      Bracteoles c. 0·5 mm. long, distinctly shorter than the calyx; stamens
        inserted at the base of the calyx-tube -   -   -   7. *gossweileri*
    Capsule 2-valved:
      Style up to 0·5 mm. long, straight:
        Aerial and submerged leaves very narrow, linear to narrowly lanceolate;
          calyx 4-merous; petals 0; stamens 2; capsule exceeding the
          calyx-lobes -   -   -   -   -   -   -   8. *capensis*
        Aerial leaves not as above:
          Capsule globose or subglobose:
            Aerial leaves broadly ovate, cordate at the base, 3·5–5 × 2–3 mm.,
              the submerged ones up to 2 cm. long, narrowly linear, usually
              recurved at apex; calyx c. 2 mm. long; petals 0 or 2–4;
              stamens 1–2(3–4); seeds c. 0·5 mm. -   - 9. *heterophylla*
            Aerial leaves lanceolate or ovate-lanceolate, up to 9 × 2·5 mm., the
              lower ones linear, c. 9 mm. long; calyx c. 1 mm. long;
              petals 0; stamens 2–3; seeds c. 0·25 × 0·15 mm.
                                         12. *milne-redheadii*
          Capsule obovoid-ellipsoid or ellipsoid:
            Caespitose or procumbent herb 8–20 cm. long, simple or ±
              branched, with the leaves densely imbricate at the top of stems
              and branches; upper leaves 2–3(5) × 1·5–2(3) mm.; stamens
              2–3(4); style c. 0·5 mm. long   10. *heteropetala* var. *engleri*
            Robust herb 15–35 cm. high, much branched to the apex, the
              branches very floriferous; leaves 3–10 × 1·5–8 mm.; stamens
              2(3); style 0·1–0·2 mm. long -   -   - 11. *congolensis*
          Style longer than 0·5 mm., recurved; leaves ovate-oblong, the upper ones
            up to 10 × 2·5 mm., slightly cordate at base; stamens 2(3); seeds
            c. 1 × 0·3 mm. -   -   -   -   -   -   -   13. *wildii*
  Stamens always 4, opposite to the calyx-lobes; petals usually developed:
    Stamens included; petals distinctly cordate, 1–1·25 × 1–1·25 mm.; sub-
      merged leaves linear-lanceolate (up to 3 × 1 mm.), the aerial ones
      subcircular (3 × 2·5 mm.), cordate at base; stems 2–4 mm. in diam.
                                         14. *cordipetala*
    Stamens exserted; petals slightly cordate, sometimes clawed, c.
      1·25 × 1 mm.; submerged leaves lanceolate (c. 5 × 1·5 mm.) or linear
      (c. 10 × 0·5 mm.), the aerial ones broadly ovate (3 × 2 mm.), slightly
      cordate; stems c. 2 mm. in diameter -   -   -   15. *nashii*
  Flowers in racemes or spikes sometimes capitate at the apex of the stems and branches
                                         17. *hutchinsoniana*
Leaves verticillate or sometimes leaves decussate and verticillate on the same plant,
  rarely plants with some branches with decussate and other branches with verticillate
  leaves:
  Leaves not always all verticillate:
    Appendages of the calyx conspicuous -   -   -   -   -   -   3. *decumbens*
    Appendages of the calyx absent:
      Stamens 2–3; petals 0; bracteoles as long as the calyx; capsule 3-valved
                                         18. *pusilla*
      Stamens 4; petals 4; bracteoles shorter than the calyx; capsule 2-valved:
        Flowers dimorphic-heterostylous, in racemes or spikes sometimes capitate at the
          apex of stems and branches -   -   -   -   -   17. *hutchinsoniana*
        Flowers neither in racemes nor in spikes at apex of stems and branches:
          Submerged leaves lanceolate (c. 5 × 1·5 mm.) or linear (c. 10 × 0·5 mm.), the
            aerial ones broadly ovate (3 × 2 mm.); calyx 2·5 mm. long; ovary ellipsoid, c. 1 × 0·5 mm.; flowers homomorphic
                                         15. *nashii*
          Submerged leaves linear, up to 11 × 0·5–1 mm., the aerial ones 3·5–6 × 2–
            3·5 mm.; bracteoles c. 1 mm. long; calyx c. 3 mm. long; ovary sub-
            globose, c. 1 × 0·75 mm.; flowers with dimorphic heterostyly
                                         16. *longistyla*

Leaves always all verticillate:
  Stamens (1)2(3); petals 0; bracteoles as long as the calyx; capsule 3-valved; leaves
    narrowly linear, 3–5 × 0·3–0·4 mm.   -   -   -   -   -   18. *pusilla*
  Stamens 4; petals 4; bracteoles shorter than the calyx; capsule 2-valved; aerial
    leaves not narrowly linear:
      Aerial leaves 3·5–6 × 1·5–2·5 mm., oblong or ovate, rounded at the base; submerged
        leaves when present linear or narrowly lanceolate up to 11 × 0·5–1 mm.;
        bracteoles 0·75–1 mm. long; usually a short-stemmed plant growing under
        water spray -   -   -   -   -   -   -   -   - 16. *longistyla*,
      Aerial leaves 2·5–10 × 1–3 mm., lanceolate to oblong or ovate, subcordate at the
        base; submerged leaves very narrow to hair-like, 5–25 mm. long; bracteoles
        0·2–0·3(0·75) mm. long; usually a long-stemmed water plant
                                                          19. *myriophylloides*

1. **Rotala serpiculoides** Welw. ex Hiern in F.T.A. **2**: 469 (1871); Cat. Afr. Pl. Welw. **1**:
   371 (1898).—Koehne in Engl. Bot. Jahrb. **1**: 158 (1880); in Engl., Pflanzenr. **IV**,
   216: 33, fig. 1M (1903).—Engl., Pflanzenw. Afr. **3**, 2: 643 (1921).—Fernandes &
   Diniz in Garcia de Orta **6**: 92 (1958).—Boutique in F.C.B., Lythraceae: 4 (1967).—
   A. Raynal in Adansonia, N.S. **7**: 544 (1967).—Fernandes in C.F.A. **4**: 168 (1970).
   Type from Angola.

Annual weak erect or sometimes decumbent herb up to 30 cm. long. Stem
simple, sometimes ± branched. Leaves 5–25 × 1–5 mm., decussate, sublinear or
narrowly oblong, obtuse at the apex, narrowed at the base into a conspicuous
petiole. Flowers (3)4-merous in axillary sessile (1)3–12-flowered umbels; pedicels
0·4–0·5(1·3) mm. long; bracteoles linear-subulate, as long as or longer than the
pedicels. Calyx 0·75–1·3 mm. long, campanulate, without appendages; lobes
0·4–0·7 mm. long, triangular. Petals 0. Stamens 1–3, included. Ovary 0·5–1 mm.
long, ellipsoid; style very short. Capsule 1·5 mm. long, ellipsoid, 3-valved. Seeds
c. 0·4 mm. long, brownish, concave on one side, convex on the other.

Zambia. C: Mkushi Distr., Fiwila Mission, 1320 m., fl. & fr. 1932, *Hewitt* 6 (BM).
S: Siamambo, Choma, fl. & fr. 20.iii.1958, *Robinson* 2817 (COI; K; PRE; SRGH).
Rhodesia. C: Salisbury, Gatooma road, fl. & fr. 12.i.1946, *Wild* 1062 (SRGH).
Malawi. S: Zomba, W. of Lake Chilwa, 600 m., fl. & fr. 21.vi.1962, *Robinson* 5398 (K).
Also in Angola, Zaire, Sudan, Rwanda, Kenya and Tanzania. In damp grasslands,
dambos and swamps.

2. **Rotala juniperina** Fernandes in Bol. Soc. Brot., Sér. 2, **48**: 126, t. 12 (1974).
   TAB. 78. Type: Malawi, Mlanje Distr., Likabula Forest, *Robinson* 5353 (K;
   SRGH, holotype).

Erect annual herb, up to 20 cm. high. Stem thick, almost woody at the base,
profusely branched from the base to the apex, the branches 4-gonous or 4-winged,
densely leafy; internodes up to 4 mm. long, shorter towards the apex. Leaves of
the main stem 4–12 × 2–5 mm., opposite, ovate-lanceolate, subacute at the apex,
rounded at the base, those of the branches much smaller, c. 3 × 2 mm., ovate-
oblong, canaliculate on the upper face and ending in a recurved point, contracted at
the base into a short petiole, penninerved. Flowers 3-merous, axillary, solitary,
sessile, distributed throughout the plant; bracteoles 2, as long as the calyx (c.
1·5 mm.), linear, scarious. Calyx 1·5 × 1 mm., campanulate, scarious; tube
c. 1 mm. long; appendages 3, c. 0·75 mm. long, curved; lobes c. 1 mm. broad
and c. 0·5 mm. long. Petals 3, c. 0·5 × 0·25 mm., persistent, elliptic, slightly
apiculate. Stamens 3, included, inserted slightly below the middle of the calyx-
tube. Ovary c. 0·75 mm. in diameter, stipitate, globose, 3-locular; style c.
0·25 mm. long. Capsule c. 1·5 mm. long, subglobose, 3-valved. Seeds c. 0·5 mm.
long, whitish.

Malawi. S: Mlanje Distr., 16 km. NW. of Likabula Forest Depot, 700 m., fl. & fr.
15.vi.1962, *Robinson* 5353 (K; SRGH).
Only known from Malawi. Muddy places.

3. **Rotala decumbens** Fernandes in Bol. Soc. Brot., Sér. 2, **48**: 127, t. 13 (1974). Type:
   Zambia, Kabulamwanda dam, 128 km. N. of Choma, *Robinson* 723 (K, holotype).

Creeping, (?) perennial herb 30 cm. long or longer. Main stem 4-gonous, up to
4 mm. thick, with internodes 8–25 mm. long, rooting at the nodes and sending up
erect shoots c. 13 cm. high. Leaves of the creeping stem 9–20 × 4–8 mm., opposite,
ovate-lanceolate, subacute at the apex, margin entire, rounded or slightly cordate at

Tab. 78. ROTALA JUNIPERINA. 1, habit ($\times \frac{2}{3}$); 2, upper part of branch ($\times 6$); 3, leaf ($\times 3$); 4, flower with bracteoles ($\times 16$); 5, calyx opened ($\times 16$); 6, petal ($\times 27$); 7, pistil ($\times 20$); 8, capsule within calyx ($\times 20$); 9, capsule dehisced ($\times 20$); 10, seeds ($\times 20$), all from *Robinson* 5353.

the base, penninerved, midrib impressed above, prominent beneath; leaves of the erect shoots decussate or 3-nate, similar to those of the creeping stem but smaller (from 4 ×1 mm. to 10 ×4 mm.), densely crowded towards the end of the stem. Flowers 3-merous, solitary, axillary, scattered along the erect shoots; bracteoles as long as the calyx-tube (c. 1·25 mm.), linear, scarious. Calyx c. 1·5 mm. long, campanulate; lobes c. 0·6 mm. broad and 0·35 mm. high; appendages c. 0·75 mm. long, ± curved. Petals 3, c. 0·5 ×0·3 mm., elliptic. Stamens 3, opposite the calyx-lobes, inserted at the middle of the calyx-tube, reaching the apex of the calyx-lobes; anthers c. 0·25 mm. long. Ovary stipitate, subglobose, c. 1 ×0·75 mm., 3-locular; style (and stigma) c. 0·25 mm. long. Capsule subglobose, c. 1·5 ×1·25 mm., 3-valved, scarcely exserted. Seeds many, c. 0·5 mm. long, whitish, concave on one side, convex on the other.

Zambia. S: Kabulamwanda Dam, 128 km. N. of Choma, 1088 m., fl. & fr. 24.iv.1954, *Robinson* 723 (K).
Known only from Zambia. Muddy bottoms of shallow irrigation channels.
This species very much resembles *R. juniperina* in the floral characters, and we suggest that *R. juniperina* may be a land form and *R. decumbens* the acquatic form of the same species.

4. **Rotala fluitans** Pohnert in Mitt. Bot. Staatss. München **1**: 448 (1954).—Pohnert & Roessler in Prodr. Fl. SW. Afr. **95**: 9 (1966).—Fernandes in C.F.A. **4**: 169 (1970). Type from SW. Africa.
    Rotala tetragonocalyx Fernandes & Diniz in Bol. Soc. Brot., Sér. 2, **29**: 88, t. 2 (1955); in Garcia de Orta **6**: 93 (1958). Type from Angola.

Creeping annual herb, 40 cm. long or more. Stems leafy or leafless on the lower part, emitting slightly 4-winged aerial branches up to 12 cm. long. Leaves decussate, the cauline up to 20 × 11 mm., sessile, widely elliptic, rounded at the apex, cordate at the base, thinly membranous, with the midrib somewhat conspicuous, those of the branches 2·5–6 × 1·5–5·5 mm., also sessile, broadly elliptic or subcircular, rounded at the apex, cordate at the base, inconspicuously penninerved, smaller and crowded to the top. Flowers 4-merous, subsessile, solitary, axillary; bracteoles 2, c. 1 mm. long, subulate, scarious. Calyx-tube 2–2·5 mm. long at anthesis, tubular, 4-gonous, somewhat constricted at the mouth, when fructiferous up to 3·5 mm. long; lobes 4, very short, almost truncate; appendages c. 1 mm. long, subulate, erect, usually incurved. Petals 1 ×0·75 mm., subcircular, slightly emarginate at the apex, exserted 0·5 mm. beyond the calyx-tube mouth. Stamens 4, inserted a little below the middle of the calyx-tube, included; anthers c. 0·25 mm. Ovary c. 1·5 mm. long, ellipsoid; style very short, slightly longer than the capitate stigma. Capsule 4-valved, 3·5 mm. long, ellipsoid, as long as the calyx-tube. Seeds numerous, c. 0·5 mm. long.

Caprivi Strip. Ngamiland, E. of the Cuando R., 1023 m., fr. x.1945, *Curson* 1013 (PRE). **Botswana.** N: Chobe R., 3·2 km. N. of Kasane, fl. & fr. 13.vii.1937, *Erens* 416 (K; PRE; SRGH). **Zambia.** B: 128 km. N. of Senanga, fl. & fr. 2.viii.1952, *Codd* 7345 (K; PRE). S: Zambezi R., above Victoria Falls, fl. & fr. 17.viii.1937, *Mogg* 14592 (PRE). **Rhodesia.** N: Sebungwe Distr., Zambezi R., 330 m., fr. 8.ix.1959, *Mitchell* 1 (COI; SRGH).
Also in Angola, Zaire and SW. Africa. In water and on riverside mud or sandbanks.

5. **Rotala submersa** Pohnert in Mitt. Bot. Staatss. München **1**: 449 (1954).—Pohnert & Roessler in Prodr. Fl. SW. Afr. **95**: 9 (1966).—Fernandes in Bol. Soc. Brot., Sér. 2, **48**: 128 (1974). Type from SW. Africa.
    Rotala pedicellata Fernandes & Diniz in Bol. Soc. Brot., Sér. 2, **31**: 152, t. 2 (1957). Type: Zambia, Mapanza, 1115 m., *Robinson* 622 (K; SRGH, holotype).

Annual herb. Stem prostrate, sometimes ascending towards the apex, usually thick, rooting at the nodes and emitting erect 4-winged branches. Leaves 4–15 × 2–5 mm., decussate, sessile, oblong-elliptic, rounded at the apex, rounded or slightly cordate at the base, inconspicuously penninerved, thinly membranous, the midrib conspicuous beneath. Flowers 4-merous, axillary, solitary, with pedicels up to 2 mm. long, bracteolate at the apex, the bracteoles subulate, scarious, c. ½ as long as the tube. Calyx c. 2·5 mm. long, tubular (when fructiferous almost campanulate), 4-gonous, without appendages; lobes plicate at the sinuses, c. ¼ as long as the tube. Petals c. 1 ×0·25 mm., linear to narrowly lanceolate or 0. Stamens 2–3(4), inserted at ⅓ of the tube, included. Ovary 0·75 ×0·5 mm.,

4-locular, longitudinally 4-lobed; style very short. Capsule 2 × 1 mm., 4-valved, ellipsoid, included. Seeds numerous, 0·4–0·75 × 0·25–0·5 mm.

Petals 0; seeds c. 0·4 × 0·25 mm. - - - - - - - var. *submersa*
Petals present, linear to narrowly lanceolate, c. 1 × 0·25 mm.; seeds 0·75 × 0·5 mm.
var. *angustipetala*
Var. **submersa**.

Botswana. N: 196·5 km. N. of Maun, 19·5 km. N. of Jovenega Pan, fl. & fr. 7.vii.1937, *Erens* 345 (PRE). **Zambia.** B: Masese, fl. & fr. 3.v.1961, *Fanshawe* 6535 (SRGH). S: Mazabuka, Mapanza, Choma, c. 1155 m., fl. & fr. 13.iv.1958, *Robinson* 2837 (COI; EA; K; PRE; SRGH). **Rhodesia.** N: Kariba, Sengwa, fl. & fr. 18.xi.1964, *Jarman* 80 (COI; K). W: Wankie, fl. & fr. iv.1932, *Levy* 24 (K; PRE). C: Gatooma, Salope Farm, fl. & fr. 12.iv.1971, *Conway* 34/71 (COI; K). S: Nuanetsi R., upstream from Buffalo Bend, fl. & fr. 28.iv.1961, *Drummond & Rutherford-Smith* 7580 (COI; SRGH). Also in SW. Africa. By pools in riverbeds, savannas or creeping in mud.

Var. **angustipetala** Fernandes in Bol. Soc. Brot., Sér. 2, **48**: 128, t. 14 (1974). Type: Zambia, Mbala (Abercorn), *Michelmore* 434a (K, holotype).

Zambia. N: Mbala (Abercorn), *Michelmore* 434a (K). Known only from Zambia. Ecology as for the type variety.

6. **Rotala minuta** Fernandes & Diniz in Bol. Soc. Brot., Sér. 2, **31**: 151, t. 1 (1957). Type: Zambia, Mufulira, *Eyles* 8343 (K; SRGH, holotype).

Dwarf annual erect or ascending herb up to 3 cm. high. Stem branched at the base, the branches 4-gonous, rooting at the lower nodes; internodes up to 5·5 mm. long, shorter towards the apex. Lower leaves c. 4 × 1 mm., linear, the median 4–7 × 1·25–2 mm., narrow-elliptic, subobtuse at the apex, rather narrowed to the base, the upper ones 3–4 × 1·75–2 mm., ovate, rounded or subcordate at the base. Flowers (3)4-merous, axillary, solitary, subsessile; bracteoles 2, thinly membranous, linear-lanceolate, almost as long as the calyx-tube. Calyx c. 2 mm. long, campanulate; lobes c. 0·75 mm. long, triangular; appendages very short, sometimes absent. Petals 0. Stamens (1)2, inserted below the middle of the tube, included. Ovary 3-locular, globose, very shortly stipitate; style half as long as the ovary. Capsule ellipsoid, 3-valved, not exceeding the lobes. Seeds c. 0·5 mm. in diameter, subcircular, concave on one side, convex on the other.

Zambia. W: Mwinilunga, Kalenda Plain, Matonchi, 16.iv.1960, *Robinson* 3578 (K; SRGH). Known only from Zambia. In damp places and shallow water in lateritic dambos.

7. **Rotala gossweileri** Koehne in Engl., Bot. Jahrb. **42**, Beibl. 97: 48 (1909).—Exell in Journ. Bot., Lond. **67**, Suppl. Polypet.: 61 (1929).—Fernandes & Diniz in Garcia de Orta **6**: 95 (1958).—A. Raynal in Adansonia, N.S. **7**: 541 (1967).—Fernandes in C.F.A. **4**: 172 (1970). Type from Angola (Malanje).

Annual caespitose erect herb 10–15 cm. high, simple or few-branched towards the apex. Stems simple, 4-gonous. Leaves 2–4 × 1·5–3 mm., decussate, broadly ovate, obtuse at the apex, subcordate at the base. Flowers 4(5)-merous, solitary, axillary on the upper part of the stems, sessile; bracteoles c. 0·5 mm. long, linear, scarious. Calyx 1·5–2 mm. long, campanulate; lobes ± as long as the tube, triangular; appendages 0. Petals 0. Stamens 2, inserted at the base of the calyx-tube, with the filaments as long as the tube. Ovary c. 0·75 mm. in diameter, globose; style 0·3–0·5 mm., a little more than ⅓ the length of the ovary. Capsule 1·5 mm. long, subglobose, scarcely longer than the calyx-tube, 3-valved. Seeds c. 0·5 mm. long, obovate, brownish.

Zambia. N: Mbala Distr., Mbala Township, Mbulu R., fl. & fr. 18.v.1968, *Sanane* 132 (K; LISC). C: Serenje, Bolelo R., 1370 m., fl. & fr. 24.vi.1963, *Symoens* 10429 (K). **Rhodesia.** N: Sebungwe, Chicomba Vlei, fl. & fr. 12.viii.1957, *Whellan* 525 (BM; SRGH). Also in Senegal, Central African Republic and Angola. Pools and swamps, in water or on mud.

8. **Rotala capensis** (Harv.) Fernandes & Diniz in Bull. Jard. Bot. Brux. **27**: 105 (1957); in Garcia de Orta **6**: 94 (1958).—Boutique in F.C.B., Lythraceae: 8 (1967).— A. Raynal in Adansonia, N.S. **7**: 540 (1967).—Fernandes in C.F.A. **4**: 170 (1970). Type from S. Africa.

Suffrenia capensis Harv., Thes. Cap. **2**: 56, t. 189 (1863). Type as above.
Rotala filiformis (Bellardi) Hiern in F.T.A. **2**: 468 (1871) pro parte quoad syn.
Suffrenia capensis Harv.
Rotala filiformis forma *hiernii* Koehne in Engl., Bot. Jahrb. **1**: 168 (1880) pro parte
quoad specim. huill.; in Engl., Pflanzenr. **IV**, 216: 37 (1903) pro ead. parte.— Hiern,
Cat. Afr. Pl. Welw. **1**: 372 (1898) pro parte quoad specim. *Welwitsch* 2338.

Annual caespitose herb up to 5 cm. long when growing on the ground, longer
when submerged. Stem unbranched, reddish. Leaves 5–15 × 0·5–1 mm.,
decussate, all narrowly oblong or the upper ones linear-lanceolate, attenuate to the
apex, 1-nerved. Flowers 4-merous, axillary, solitary, sessile or subsessile;
bracteoles 0·3–0·5 mm. long, subulate. Calyx 0·8–1 mm. long, campanulate; lobes
0·2–0·3 mm. long, triangular. Petals 0. Stamens 2, included, inserted at the base
of the calyx-tube. Ovary 0·4–0·5 mm. long, ellipsoid; style 0·1–0·2 mm. long.
Capsule c. 1·25 mm. long, ellipsoid, 2-valved, exserted. Seeds c. 0·5 mm. long,
light brown, concave-convex.

Zambia. N: Kasama, 105 km. E. of Kasama, fl. & fr. 6.v.1962, *Robinson* 5160 (K;
SRGH). W: Kasempa, 7 km. E. of Chizera, fl. & fr. 27.iii.1961, *Drummond & Rutherford-
Smith* 7435A (COI; SRGH). **Rhodesia.** W: Matopos, iii.1902, *Eyles* 1045 (SRGH).
C: Salisbury, Cleveland Dam, fl. & fr. 19.v.1953, *Wild* 4114 (LISC; PRE; SRGH).
S: Bikita, Turgwe-Dafana confluence, 1050 m., fl. & fr. 5.v.1969, *Biegel* 3029 (COI; K).
Also in Cameroon, Angola and S. Africa. Submerged in pools, streams and swamps or
on stones, margins of streams, etc.

9. **Rotala heterophylla** Fernandes & Diniz in Bull. Jard. Bot. Brux. **27**, 1: 106, t. 3
(1957).—Boutique in F.C.B., Lythraceae: 10 (1967).—A. Raynal in Adansonia,
N.S. **7**, 4: 540 (1967).—Fernandes in C.F.A. **4**: 171, t. 18 (1970). Type from
Angola.
Ammannia heterophylla Welw. in sched.
Rotala filiformis sensu Hiern in F.T.A. **2**: 468 (1871) pro parte quoad specim.
Welwitsch 2341 et 2342.
Rotala filiformis forma *typica* Koehne in Engl., Bot. Jahrb. **1**: 168 (1880); in
Engl., Pflanzenr. **IV**, 216: 37 (1903) pro parte quoad specim. angol. et botsw.
Rotala filiformis forma *hiernii* sensu Hiern, Cat. Afr. Pl. Welw. **1**: 372 (1898) pro
parte quoad specim. *Welwitsch* 2341 et 2342.

Annual partially submerged herb, 2·5–35 cm. long. Stems slender, simple or few
branched, rooting at the nodes on the lower part, 4-gonous; internodes 2–10 mm.
long, sometimes shorter than the leaves. Leaves decussate, sessile or subsessile,
heteromorphic, the lower ones (5)10–20 × 0·5–0·8(3) mm., linear, truncate,
emarginate or slightly bifid at the apex, rounded at the base, those of the middle
narrowly ovate, emarginate at the apex and the upper ones 3·5–5 × 2–3 mm., ovate,
obtuse at the apex, cordate at the base, obscurely penninerved. Flowers (3)4-
merous, axillary, solitary, sessile; bracteoles 0·2–0·7 mm. long, linear-subulate,
scarious. Calyx c. 2 mm. long, campanulate; calyx-lobes equal or unequal,
0·5–0·7 mm. high, triangular; appendages 0. Petals 0 or 2–4, up to 0·75 × 0·25 mm.,
subulate, elliptic or oblong. Stamens 1–2(3–4), included, inserted at the middle of
the calyx-tube. Ovary 0·8 × 0·6 mm., ellipsoid or globose; style 0·3–0·4 mm. long.
Capsule 1–2 mm. long, reddish, globose or subglobose, 2-valved. Seeds c. 0·3 mm.
long, brownish, concave-convex.

Zambia. B: Masese, fl. & fr. 3.v.1961, *Fanshawe* 6534 (SRGH). N: 25 km. of
Kasama, near Lukulu R., fl. & fr. 7.v.1961, *Robinson* 4637 (EA; K; SRGH). W:
Kasempa, 7 km. E. of Chizera, fl. & fr. 27.iii.1961, *Drummond & Rutherford-Smith* 7435
(COI; SRGH). S: Mazabuka, Choma, fl. & fr. 19.iii.1958, *Robinson* 2798 (COI; K;
PRE; SRGH). **Rhodesia.** N: Urungwe, Chipani Pan, fl. & fr. 13.v.1952, *Whellan* 656
(SRGH). C: Salisbury Distr., Parktown, fl. & fr. 18.iv.1944, *Greatrex* 18412 (BM;
SRGH). **Malawi.** N: Nkhata Bay Distr., Vipya, 59 km. S. of Mzuzu, 1792 m., fl. & fr.
16.iv.1967, *Pawek* 982 (COI).
Also in Mali, Cameroon, Angola, Central African Republic, Zaire and Tanzania. In
pools, swamps, damp muddy soils, drying muds on lake shores, shallow pools on rocky
outcrops, etc.

10. **Rotala heteropetala** Koehne in Engl., Bot. Jahrb. **22**: 149 (1895); in Engl.,
Pflanzenr. **IV**, 216: 38 (1903).—Engl., Pflanzenw. Afr. **3**, 2: 645 (1921). Type from
Ethiopia.
Ammannia elatinoides A. Rich., Tent. Fl. Abyss. **1**: 279 (1847) non DC. (1825).

Var. **engleri** Koehne in Engl., Bot. Jahrb. **39**: 663 (1907); op. cit. **41**: 76 (1907).—
R. E. Fr., Wiss. Ergebn. Schwed. Rhod.-Kongo-Exped. **1**: 164 (1914).—Fernandes
in Garcia de Orta, Sér. Bot. **2**: 77 (1975). Type: Rhodesia, Victoria Falls, *Engler*
2983 (B, holotype†).

Annual or perennial (fide collector), 8–20 cm. high, caespitose, procumbent or
erect herb. Stems fleshy, whitish or crimson, ± branched from the base. Leaves
decussate, sessile, the upper ones c. 2–3(4–5) × 1·5–2(3) mm., usually densely
imbricate, broadly ovate, obtuse or subacute at the apex, cordate at the base,
obscurely penninerved. Flowers 4-merous, axillary, solitary, sessile; bracteoles 2,
c. 0·5 mm. long, subulate, scarious. Calyx c. 1·5 mm. long, campanulate; tube
c. 1 mm. long; lobes c. 0·75 mm. broad and 0·5 mm. high, deltate, apiculate;
appendages 0. Petals 1, subulate, or 1, oblong, or 2 narrowly oblong, or 4 of which
2 oblong and 2 subulate, sometimes 0. Stamens 2–3(4), inserted slightly above the
middle of the calyx-tube, included. Ovary c. 0·75 mm. long, subglobose, 2-locular;
style c. 0·5 mm. long, with a capitate stigma. Capsule obovate-ellipsoid to ellipsoid,
2-valved, shortly stipitate. Seeds c. 0·5 mm. long, brownish.

Zambia. B: Mongu, Kande Lake c. 13 km. NE. of Mongu, fl. & fr. 11.xi.1959,
Drummond & Cookson 6349 (COI; K; PRE; SRGH). N: Bangweulu swamps,
Ncheta I., fl. & fr. 3.ix.1969, *Verboom* 2638 (K). S: Namwala, Kafue National Park,
Bird Pool, Nhala, Ngama, fl. & fr. 14.v.1963, *Mitchell* 18/91 (COI; SRGH). **Rhodesia.**
W: Victoria Falls, fl. & fr. 38–1947, *Greenway & Brenan* 8031 (EA; K; PRE). C:
Salisbury, Parktown, fl. & fr. 11.iv.1944, *Greatrex* (SRGH).
Also in Angola. Swamps, spray zone of waterfalls, muds or sands on the shores of lakes
and streams, in drying pans, etc.

11. **Rotala congolensis** Fernandes & Diniz in Bol. Soc. Brot., Sér. 2, **29**: 89, t. 3
(1955).—Boutique in F.C.B., Lythraceae: 8 (1967). Type from Zaire.

Annual erect or ascending herb, 15–35 cm. long. Stems thick, up to 3 mm. in
diam. at the base, simple or branched, densely leafy, reddish. Leaves
3–10 × 1·5–8 mm., decussate, sessile, ovate to widely ovate, rounded at the apex,
cordate at the base, distinctly penninerved. Flowers 4-merous, axillary, solitary,
subsessile; bracteoles 0·3–0·5 mm. long, subulate. Calyx c. 1·5 mm. long,
campanulate; lobes deltate, c. ½ as long as the tube. Petals often 0, rarely 3, of
which 2 are lanceolate and shorter than the calyx-lobes and 1 subulate, even smaller
(0·2–0·3 mm. long). Stamens 2(3), included, inserted slightly below the middle of
the calyx-tube. Ovary 0·5–0·7 mm. long, 2-locular, globose, sessile; style
0·1–0·2 mm. long. Capsule 1·5 mm. long, ellipsoid, 2-valved. Seeds c. 0·35 mm.
long, brownish, ellipsoid.

Rhodesia. W: Matobo Distr., Farm Besna Kobila, 1500 m., st. iii.1953, *Miller* 1651
(SRGH). C: Rusape, Mona, fl. & fr. 3.i.1954, *Dehn* 1003 (BM; SRGH).
Also in Zaire and Tanzania. Edges of streams, running water, damp grasslands.

12. **Rotala milne-redheadii** Fernandes & Diniz in Bol. Soc. Brot., Sér. 2, **33**: 21 (1959).
Type: Zambia, Mwinilunga Distr., Kalenda Plain, *Milne-Redhead* 4414 (COI; K,
holotype).

Annual erect herb up to 11 cm. high. Roots slender, whitish, fasciculate. Stem
simple or branched at the base or towards the apex, thick, terete, green; internodes
up to 8 mm. long, shorter than the leaves, very short towards the apex. Leaves up
to 9 × 2·5 mm., decussate, sessile, the lower ones linear, those of the middle and the
uppermost ones lanceolate or ovate-lanceolate, obtuse and sometimes emarginate
at the apex, roundish at the base, inconspicuously penninerved. Flowers 4-merous,
axillary, solitary, reddish, subsessile, 2-bracteolate at the base, the bracteoles
scarious, subulate, c. half as long as the calyx. Calyx c. 1 mm. long, campanulate;
appendages 0; lobes c. 0·5 mm. broad and 0·45 mm. long, triangular, apiculate. Petals 0.
Stamens 2–3, inserted ± at the middle of the calyx-tube, not exceeding the apices
of the calyx-lobes. Ovary c. 0·5 mm. in diameter, globose, 2-lobed, 2-locular;
style c. 0·2 mm. long, with the stigma capitate. Capsule (immature) c. 0·6 mm. in
diameter, globose. Seeds c. 0·25 × 0·15 mm.

Zambia. W: Mwinilunga Distr., Kalenda Plain, 30.i.1937, *Milne-Redhead* 4414
(COI; K).
Known only from Zambia. In shallow waters and peaty soils of dambos.

More material is needed to decide whether this species is distinct from *R. dinteri* Koehne (in Mém. Herb. Boiss. no. 20: 24, 1900); the latter occurs in SW. Africa.

13. **Rotala wildii** Fernandes in Bol. Soc. Brot., Sér. 2, **40**: 128, t. 15 (1974). TAB. **79**.
Type: Rhodesia, Mtoko, Makate Ruins, *Wild* 5662 (COI; K; SRGH, holotype).

Annual partly submerged and partly aerial herb up to 30 cm. long. Stems decumbent and rooting at the nodes in the lower part, ascending towards the apex, ± branched, the branches ascending; internodes up to 2 cm. long, becoming shorter towards the apex where the leaves are crowded. Leaves decussate, green, the lower ones very variable in size, the upper ones up to 10 × 2·5 mm., ovate-oblong, obtuse at the apex, slightly cordate at the base, obscurely penninerved, the midrib very distinct. Flowers 4-merous, axillary, solitary; bracteoles c. 0·75 mm. long, linear, scarious. Calyx campanulate, c. 1·5 mm. long, scarious; tube c. 1 mm. long; lobes c. 0·5 × 0·75 mm., apiculate; appendages 0. Petals 3, one lanceolate, one ovate and one linear, or all absent. Stamens 2(3), as long as the lobes, inserted a little below the middle of the calyx-tube. Ovary ellipsoid, c. 0·75 × 0·5 mm.; style as long as the ovary (0·75 mm.), curved; stigma capitate, very papillose. Capsule broadly ellipsoid, c. 2 × 1·5 mm., stipitate, 2-valved. Seeds brownish, c. 1 mm. long and 0·3 mm. broad.

Rhodesia. N: Mtoko, Makate Ruins, 1320 m., fl. & fr. 15.ii.1962, *Wild* 5662 (COI; K; SRGH).
Known only from Rhodesia. Granite kopjes.

14. **Rotala cordipetala** R. E. Fr. in Fedde, Repert. **12**: 541 (1913); Wiss. Ergebn. Schwed. Rhod.-Kongo-Exped. **1**: 164, t. 10, fig. 2, 3 and text fig. 13d, e (1914).— Engl., Pflanzenw. Afr. **3**, 2: 645 (1921). Type: Zambia, Lake Bangweulu, Kasomo, *R. E. Fries* 656 (UPS, holotype).

Perennial rhizomatose herb, with erect stems up to 50 cm. high. Stems thick, 2–4 mm. in diameter at the base, simple, reddish towards the apex; lower internodes c. 1·5 cm. long, the upper ones very short, the leaves densely imbricate and crowded at the summit. Leaves 1–3 × 1–2·5 mm., decussate, ± reddish, sessile, the submerged ones up to 3 × 1 mm., oblong, ovate or rounded-ovate, the upper ones 3 × 2·5 mm., subcircular, obtuse or rounded at the apex, cordate at the base, obscurely penninerved. Flowers 4-merous, c. 2 mm. long, solitary in the axils of the upper leaves, sessile; bracteoles c. 0·5 mm. long, subulate, scarious. Calyx c. 2 mm. long, campanulate; calyx-tube c. 1·5 mm. long; appendages 0. Petals 4, 1–1·25 × 1–1·25 mm., persistent, ovate or subcircular, rounded or emarginate at the apex, cordate at the base. Stamens 4, inserted a little below the middle of the calyx-tube, included. Ovary c. 1 mm. in diameter, globose; style as long as the ovary. Capsule not seen.

Zambia. N: Bangweulu Lake, Kasomo, fl. 19.ix.1911, *R. E. Fries* 656 (UPS).
Known only from Zambia. In water on sandy ground.

15. **Rotala nashii** Fernandes in Bol. Soc. Brot., Sér. 2, **48**: 129, t. 16 (1974). Type: Zambia, Mbala (Abercorn), Lake Chila, *Nash* 65 (BM, holotype).

Annual partly submerged herb c. 20 cm. high. Stems c. 2 mm. in diameter; lower internodes up to 12 mm. long, the upper ones very short. Leaves decussate on some stems, 3-nate on others, the submerged ones lanceolate (c. 5 × 1·5 mm.), or linear (c. 10 × 0·5 mm.), the aerial ones 3 × 2 mm., broadly ovate, obtuse or rounded at the apex, contracted into a petiole c. 0·5 mm. long, obscurely penninerved. Flowers 4-merous, solitary, axillary, sessile; bracteoles c. 0·5 mm. long, subulate, scarious. Calyx c. 2·5 mm. long, campanulate; tube c. 1·5 mm. long; lobes triangular, c. 1 × 1 mm., apiculate; appendages 0. Petals c. 1·25 × 1 mm., pink-mauve, broadly ovate, slightly rounded at the apex, shortly clawed and cordate at the base. Nectariferous ring of 4 free scales opposite to the petals. Stamens 4, opposite the lobes, inserted a little below the middle of the calyx-tube; filaments c. 3 mm. long, exserted. Ovary c. 1 × 0·5 mm., ellipsoid, 2-locular, shortly stipitate; style c. 0·75 mm. long; stigma capitate. Capsule 2 × 1·5 mm., ellipsoid, 2-valved. Seeds c. 0·6 × 0·3 mm. long, elliptic in outline.

Zambia. N: Mbala (Abercorn) Distr., Lake Chila, 1600 m., 15.iii.1955, *Richards* 4998 (K).

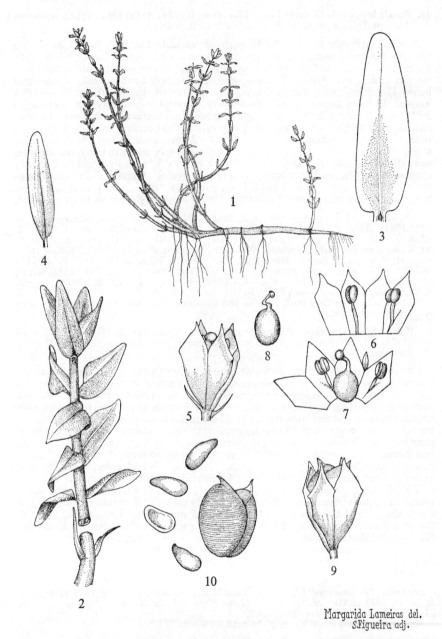

Margarida Lameiras del.
S.Figueira adj.

Tab. 79. ROTALA WILDII. 1, habit (×½); 2, lower and upper parts of a branch (×3); 3, upper leaf (×6); 4, leaf of the median region (×6); 5, flower with bracteoles (×12½); 6, calyx opened (×12½); 7, flower opened showing one petal and pistil (×12½); 8, pistil (×12½); 9, capsule within calyx (×12½); 10, dehiscing capsule and seeds (×12½), all from *Wild* 5662.

16. **Rotala longistyla** Gibbs in Journ. Linn. Soc., Bot. **37**: 445 (1906). Type: Rhodesia, Victoria Falls, fl. & fr. ix.1905, *Gibbs* 170 (BM, holotype).

A small straggling perennial(?) herb of variable length. Stems ascending towards the apex, rooting at the prostrate lower part. Aerial leaves 3·5–6 × 1·5–3·5 mm., decussate or 3(4–5)-nate, sessile, oblong, obtuse or subacute at the apex, margin entire, rounded at the base, green, obscurely penninerved; submerged leaves 11 × 0·5–1 mm., usually shorter than the internodes, linear, the upper ones longer. Flowers 4-merous, axillary, solitary, dimorphic-heterostylous; bracteoles 0·75–1 mm. long. Calyx 3 × 1·5 mm., campanulate, scarious, with a nectariferous reddish ring of free scales opposite the petals at the bottom; lobes 4, 1 × 1 mm., triangular, acute; appendages 0. Petals c. 2 × 2 mm., pink, broadly ovate, rounded at the apex, ± persistent and ± clawed. Stamens 4, opposite to the lobes, inserted near the middle of the calyx-tube. Ovary c. 1 × 0·75 mm., subglobose, shortly stipitate, 2-locular, each loculus with numerous ovules. Capsule 2-valved. Long-styled flowers: stamens c. 1 mm. long, included; style 3 mm. long, exserted. Short-styled flowers: stamens exserted, 3 mm. long; style 1 mm. long, included.

Botswana. N: near Thamalakane R. bridge, 8 km. NE. of Maun, Matlapeneng, fl. 19.iii.1965, *Wild & Drummond* 7157 (COI; SRGH). **Zambia.** B: Kalabo, fl. 17.xi.1959, *Drummond & Cookson* 6563 (COI; K; PRE; SRGH, long-styled form). S: Kafue National Park, Namwala, Nkala R., Ngoma, fl. & fr. 29.v.1963, *Mitchell* 18/96 (SRGH, short-styled form). **Rhodesia.** W: Victoria Falls, fl. 18.xi.1949, *Wild* 3188 (SRGH, short-styled form); idem, fl. 13.vii.1952, *Codd* 7063 (K; PRE, long-styled form); idem, 960 m., fl. & fr. v.1904, *Eyles* 124 (BM; K; SRGH). C: Salisbury, Gwebi, fl. 11.x.1960, *Rutherford-Smith* 229 (SRGH).
Known only from Botswana, Zambia and Rhodesia. In water of rivers and in grass sprayed by waterfalls.

17. **Rotala hutchinsoniana** Fernandes in Bol. Soc. Brot., Sér. 2, **48**: 130, t. 17 & 18 (1974). Type: Zambia, 1·6 km. from Abercorn, *Hutchinson & Gillett* 3872 (BM, holotype).

Creeping annual or more probably perennial herb with underground runners. Stems up to 8 cm. (or longer), 4-gonous, reddish, rooting at the nodes; internodes up to 10 mm. long, becoming progressively shorter towards the apex. Leaves decussate or 3-nate, sessile or attenuate into a very short petiole, the lower ones variable in form and size from triangular (3 × 1 mm.) to broadly ovate (3 × 2·5 mm.), the upper ones ovate (c. 6 × 3 mm.), obtuse at the apex, slightly cordate at the base, penninerved. Flowers 4-merous, axillary, sessile, in spikes at the apices of stems and branches, dimorphic-heterostylous; bracteoles minute, c. 0·25 mm. long, scarious, subulate. Calyx campanulate; lobes c. 1 × 1·25 mm., deltate; tube 1·5 mm. long; nectariferous ring with 4 free scales opposite the petals. Petals 4, c. 2 × 1·5 mm., persistent, deep pink, ovate, subacute or obtuse at the apex, roundish or attenuate towards the base, slightly clawed. Stamens 4, inserted just above the base of the tube, with anthers c. 0·5 mm. long. Long-styled flowers: stamens included; style c. 3·5–4 mm. long, exceeding the lobes by c. 2 mm. Short-styled flowers: stamens with filaments c. 4 mm. long, long-exserted; style included, c. 1 mm. long. Ovary c. 1·25 × 1 mm., broadly ellipsoid. Capsule 2-valved, c. 2·25 × 1·5 mm., included, ellipsoid. Seeds c. 0·5 mm. long, brownish, elliptic in outline.

Zambia. N: near small lake 1·6 km. from Mbala (Abercorn), fl. 19.vi.1930, *Hutchinson & Gillett* 3872 (BM). **Malawi.** N: Nyika Plateau, Lake Kaulime, 2150 m., fl. & fr. 24.x.1958, *Robson & Angus* 325 (LISC).
Known only from the F.Z. area. In mud on shores of lakes.

18. **Rotala pusilla** Tul. in Ann. Sci. Nat. Bot., Sér. 4, **6**: 128 (1856).—Perrier in Fl. Madag., Lythracées: 11 (1954).—Fernandes & Diniz in Garcia de Orta **6**: 92 (1958).—Fernandes in C.F.A. **4**: 168 (1970). Type from Madagascar.
    Rotala mexicana subsp. *pusilla* (Tul.) Koehne in Engl., Pflanzenr. **IV**, 216: 30 (1903); in Engl., Bot. Jahrb. **41**: 75 (1907). Type as above.

Annual creeping or erect tufted moss-like herb, 3–10 cm. high. Stems filiform, pink. Leaves 3–5 × 0·3–0·4 mm., often 3–4-nate or some opposite and the others 3–4-nate, all linear or subulate, obtuse or mucronate at the apex, sessile, green. Flowers (4)5(6)-merous, c. 1 mm. long, reddish, sessile, solitary in the axils;

bracteoles as long as the calyx or a little longer, linear, scarious. Calyx-tube c. 0·5 mm. long, campanulate; lobes c. 0·5 mm. long, triangular, apiculate. Petals 0. Stamens (1)2(3), inserted at the middle of the calyx-tube; nectariferous scales free, narrowly linear, up to $\frac{1}{3}$ or $\frac{1}{2}$ of the tube length; stigma subsessile. Capsule c. 1 mm. long, exserted, globose, 3-valved. Seeds c. 0·5 mm. long, ellipsoid.

Zambia. S: between Mazabuka and Kaleya, fl. & fr. 15.iv.1963, *van Rensburg* 1914 (K; SRGH). Rhodesia. N: Lomagundi, 1320 m., fl. & fr. vii.1953, *Ward* 1400 (SRGH). C: Hunyani Poort, fl. & fr. 2.v.1952, *Wild* 3812 (K; LISC; SRGH). Malawi. S: Mangochi (Fort Johnston), Ciwalo V., fl. & fr. 2.vi.1953, *Banda* 105 (BM; K). Mozambique. N: Mutuáli, near Malema road, fl. & fr. 29.v.1947, *Pedro* 3326 (LMA).
  Also in W. tropical Africa, Central African Republic, Angola and Madagascar. In water and wet mud, sometimes as a weed in rice fields.

19. **Rotala myriophylloides** Welw. ex Hiern in F.T.A. 2: 469 (1871); Cat. Afr. Pl. Welw. 1: 371 (1898).—Koehne in Engl., Bot. Jahrb. 1: 154 (1880); in Engl., Pflanzenr. IV, 216: 31, fig. 2C (1903); in Warb., Kunene-Samb.-Exped. Baum: 313 (1903).—Engl., Pflanzenw. Afr. 3, 2: 643, fig. 281C (1921).—Fernandes & Diniz in Garcia de Orta 6: 91 (1958).—Boutique in F.C.B., Lythraceae: 12 (1967).— Fernandes in Bol. Soc. Brot., Sér. 2, 48: 131 (1974). Type from Angola.
  Rotala longicaulis Fernandes & Diniz in Bol. Soc. Brot., Sér. 2, 29: 87, t. 1 (1955); in Garcia de Orta 4: 404, t. 8 (1956). Type: Mozambique, Niassa, Vila Cabral, *Torre* 96 (COI, holotype; LISJC).

Perennial caespitose usually partially submerged herb, 10–60 cm. long or more. Stems various, arising from a creeping rhizome, rooting at the nodes, crimson. Leaves (3)4(6)-nate, the aerial ones 2·5–10 × 1–3 mm., lanceolate to oblong or ovate, obtuse at the apex, subcordate at the base, the submerged 5–25 mm. long, very narrow to capillary. Flowers (3)4(5)-merous on the upper parts of the stems and branches, axillary, sessile or with pedicels up to 1 mm. long, dimorphic-hetero-stylous; bracteoles 0·2–0·3(0·75) mm. long, linear-subulate. Calyx up to 2·5 mm. long, campanulate; tube 1–1·7 mm. long; lobes c. 0·5–1 mm. high, triangular; appendages 0. Petals elliptic or ovate, shortly clawed, pink-mauve; nectariferous ring slightly lobed. Long-styled flowers: stamens inserted ± at the middle of the tube, scarcely attaining the lobes-apex; style 3·5 mm. long, very much exserted. Short-styled flowers: stamens long-exserted, exceeding the calyx-lobes by 1·5 mm.; style included, 0·7–1 mm. long. Ovary 0·8–1 mm. long, ellipsoid, 2-locular. Capsule c. 1·5 mm. long, ellipsoid, 2-valved. Seeds many.

Botswana. N: Thamalakane R., upstream from Crocodile Camp, 23° 30′ E., 19° 55′ S., fl. & fr. 22.i.1972, *Gibbs-Russell & Biegel* 1365 (COI; K). Zambia. N: L. Bangweulu, fl. 27.v.1964, *Fanshawe* 8694 (K; SRGH); Mbala (Abercorn) Distr., Lake Chila, fl. & fr. 4.viii.1949, *Greenway* 8374 (EA; K; PRE). S: c. 5 km. E. of Choma, 1319 m., fl. 28.v.1955, *Robinson* 1270 (K; SRGH). Rhodesia. N: Goromonzi, 40 km. of Salisbury, along the Umtali road, 1650 m., fl. 25.iv.1967, *Rushworth* 662 (COI; PRE; SRGH). C: Cleveland Dam, fl. & fr. 5.v.1934, *Gilliland* 30 (BM; K; SRGH). Mozambique. N: Vila Cabral, *Torre* 96 (COI; LISJC).
  Also in Angola and Zaire. In slowly moving waters, at the edges of streams or in swamps.

SPECIES INSUFFICIENTLY KNOWN

Rotala cataractae Koehne in Engl., Bot. Jahrb. 39: 663 (1907).—Engl., Pflanzenw. Afr. 3, 2: 645 (1921). Type: Rhodesia, W: Victoria Falls, grassy margin of the rain forest, sprayed rocks, 930 m., 12–13.ix.1905, *Engler* 2990 (B, holotype†).

We have not seen the type of this species which was destroyed in Berlin, but we think that it is probably a gathering of plants of *R. longistyla* Gibbs (16) with decussate leaves and short-styled flowers.

79. OLINIACEAE

By B. Verdcourt

Trees or shrubs. Leaves simple, opposite or ternate, at first sight apparently without stipules but vestiges are present. Flowers ☿, regular, in terminal or axillary cymes. Calyx-tube joined to the ovary and produced above it as a cylindrical tube the margin of which is sinuate or minutely 4–5-toothed and ultimately deciduous. Petals 4–5, imbricate, oblong-linear to spathulate, inserted on the margin of the calyx-tube, pubescent at the base, alternating with 4–5 small valvate more or less incurved pubescent coloured scales which close the throat in the young flower. Stamens 4–5, inserted on the calyx-tube, alternating with the petals, inserted below the scales and at first hidden by them; filaments very short; connective thickened; anthers 2-celled, dehiscing inwardly; alternating with the stamens are 4–5 2-lobed hairy swellings at the base of the petals and usually interpreted as staminodes. Ovary inferior, 3–5-locular with (1)2–3 pendulous ovules in each locule; placentation axile; style simple with a clavate stigma. Fruit a 3–5-celled false drupe with 1 seed per cell. Seeds with spiral or convolute embryos; endosperm absent.

A single genus *Olinia*. The interpretation of the floral structure varies and there is a good deal to be said for considering the " calyx-tube " to be a prolongation of the receptacle, the " petals " as calyx-lobes and the " scales " as the true petals. This is in fact the interpretation of Fernandes and several previous workers. Since, however, the application of specific terms to these organs has little real meaning I have retained the interpretation which a user of the Flora would probably make on a first inspection. The floral anatomy has been discussed by Rao & Dahlgren (Bot. Not. **122**: 160–171 (1969)) and they agree to a large extent with the interpretation I have followed. The affinities of the family are usually stated to lie with the Lythraceae and there is indeed some resemblance in structure, but recent workers suggest that the true affinities are with the Thymelaeaceae and Penaeaceae. Discussions on this matter have been published by Fernandes (Comptes Rendus IVe Réunion A.E.T.F.A.T. **1960**: 283–288 (1962)), Weberling (Engl., Bot. Jahrb. **82**: 119–128 (1963)) and Mujica & Cutler (Kew Bull. **29**: 93–123 (1974)). It is interesting to note that Takhtadzhyan (Syst. Phyl. Magnoliophytorum: 304 (1966)) combines the above ideas since although putting the Oliniaceae some little way after the Lythraceae he places the Penaeaceae after Oliniaceae and not with the Thymelaeaceae.

OLINIA Thunb.

Olinia Thunb. in Roemer, Arch. Bot. **2**, 1: 4 (1800) *nom. conserv.*
Plectronia L., Syst. Nat. ed. 12, **2**: 138, 183 (1767); Mant. Pl. Alt.: 6, 52 (1767).
Tephea Del. in Rochet d'Héricourt, Second Voyage . . . Mer Rouge: 340 (1846).

Characters as for the family.

A small genus usually estimated to contain c. 10 species but there are probably only 5 and some of these are scarcely worth more than subspecific recognition.

Petals 1–2·5 mm. wide, 2·3–5 mm. long, rarely under 2 mm. long - 1. *rochetiana*
Petals much narrower, mostly under 0·5(1) mm. wide, usually only 1–1·5(2) mm. long
2. *vanguerioides*

1. **Olinia rochetiana** Juss. in Comptes Rendus Acad. Sci. Paris **22**: 812 (1846).—Liben in Bull. Jard. Bot. Nat. Belg. **43**: 235 (1973); in Fl. Afr. Centr., Oliniaceae: 1, t. 1 (1973).—Verdcourt in F.T.E.A., Oliniaceae: 2, t. 1 (1975). TAB **80**A. Type from Ethiopia.
   Tephea aequipetala Del. in Rochet d'Hericourt, Second Voyage . . . Mer Rouge: 340 (1846). Type from Ethiopia.
   Olinia cymosa sensu Hiern in F.T.A. **2**: 485 (1871) non Thunb. (1800).
   Olinia usambarensis Gilg in Engl., Bot. Jahrb. **19**: 278 (1894); in Engl. & Prantl, Pflanzenfam. **3**, 6a: t. 74 fig. A–G (1894).—Engl., Pflanzenw. Afr. **3**, 2: 624, t. 277 fig. A–G (1921).—Robyns, Fl. Parc Nat. Alb. **1**: 650, t. 67 (1948).—Brenan, T.T.C.L.: 395 (1949).—Eggeling & Dale, Indig. Trees Uganda ed. 2: 290, t. 14 (1952).—Brenan in Mem. N.Y. Bot. Gard. **8**: 441 (1954).—Dale & Greenway, Kenya Trees & Shrubs: 350, t. 22 (1961).—F. White, F.F.N.R.: 269 (1962). Type from Tanzania.
   Olinia volkensii Engl., Pflanzenw. Ost-Afr. **C**: 285 (1895).—Brenan, loc. cit. Syntypes from Tanzania.
   Olinia macrophylla Gilg in Wiss. Ergebn. Zweit. Deutsch. Zentr.-Afr.-Exped. 1910–1911, **2**: 575 (1913). Type from Zaire.
   Olinia ruandensis Gilg, loc. cit. Type from Rwanda.
   Olinia discolor Mildbr. in Notizbl. Bot. Gart. Berl. **11**: 669 (1932).—Brenan, loc. cit. Type from Tanzania.

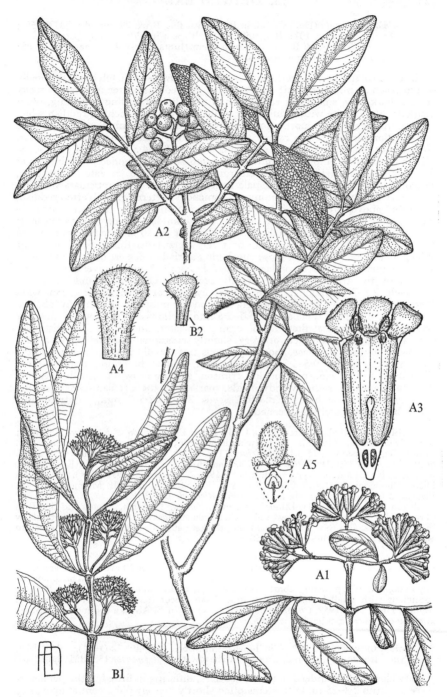

Tab. 80. A.—OLINIA ROCHETIANA. A1, flowering shoot (× ⅔) *Robson* 619; A2, fruiting shoot (× ⅔) *Chapman* 1217; A3, longitudinal section of flower (×4); A4, petal (×8); A5, scale and stamen from inside (×10), A3–A5 from *Robson* 619. B.—OLINIA VANGUERIOIDES. B1, flowering shoot (× ⅔); B2, petal (×8), both from *Ball* 20.

Olinia aequipetala (Del.) Cufod. in Bull. Jard. Bot. Brux. **29**, Suppl.: 603 (1959); in Öst. Bot. Zeitschr. **107**: 106, 109 (1960). Type as for *Tephea aequipetala*.
Olinia huillensis A. & R. Fernandes in Mem. Junta Invest. Ultramar, Sér. 2, **38**: 15, t. 1 (1963). Type from Angola.

Shrub, small tree or less often a large tree, mostly (0·9)4–16 m. tall but occasionally said to reach 27 m., evergreen; often a small gnarled bushy tree in exposed rocky places; bark grey, smooth or rough, sometimes slightly peeling or flaking, often fissured; blaze white with reddish or purplish border. Branchlets reddish when very young, later pale, mostly squarish, which together with the characteristic leaf venation renders sterile twigs easily identifiable. Leaves opposite or ternate, often bright red when young; leaf-lamina 0·6–12 × 0·5–4·5 cm., broadly rounded-elliptic to narrowly elliptic or rounded-diamond-shaped, rounded to acuminate at the apex, the actual tip blunt or emarginate, rounded to cuneate at the base, glabrous or minutely puberulous, pale beneath with the close reticulate venation characteristically darker and forming a conspicuous pattern; midrib often reddish, impressed above, prominent beneath; petiole 2–9 mm. long, reddish. Inflorescences 1·5–7·5 cm. in diameter, globose or pyramidal, usually many-flowered; branches often puberulous; pedicels very short or up to c. 1 mm. long; bracts 1·5–6 × 1–3·5 mm., very deciduous or rarely persistent, thin veined, ovate or oblong, cucullate, pubescent, those enveloping a triad of flowers or single terminal flower, larger than the secondary bracts, 1–3·5 × 1–2 mm., enveloping lateral flowers of the triad. Flowers sweetly scented. Calyx-tube and combined receptacle 2·5–7 mm. long, more or less cylindrical when mature, somewhat narrowed to the base, mostly crimson, glabrous or puberulous, minutely lobed or undulate at the apex. Petals 2·3–5 × 1–2·5 mm., yellowish-cream, white, pink, or crimson, often white at first and becoming red later, linear-oblong to distinctly obovate-spathulate, rounded or very slightly emarginate at the apex, always narrowest towards the base, mostly glabrous but usually pubescent towards the base inside. Scales 0·5–1·5 × 0·5–1·25 mm., yellow becoming crimson-pink. Fruit 0·5–1 × 0·5–1 cm., pink or dull crimson, globose or ovoid, usually speckled with pale lenticels, rather woody inside, marked by the circular scar remaining after the calyx-tube has fallen off. Seeds 2·5–3 × 2·5–3 mm., subtrigonous-ovoid, brown, minutely shagreened.

Zambia. N: Mbala (Abercorn), L. Chila, fl. 5.ii.1958, *Robson & Fanshawe* 499 (BM; (K; LISC; PRE; SRGH). W: Mwinilunga Distr., R. Zambesi above rapids 8 km. from Kalene Mission, fl. 15.ii.1962, *Richards* 17233 (K). E: Nyika Plateau, 1·6 km. SW. of rest house, fl. 15.ii.1958, *Robson & Fanshawe* 619 (BM; K; LISC; PRE; SRGH). **Malawi.** N: 24 km. on Nyika road, fr. 13.ii.1956, *Chapman* 289 (FHO; K). C: Dedza Distr., Chongoni, fl. 31.x.1960, *Chapman* 996 (K; SRGH). S: Mt. Mlanje, Luchenya Plateau to Chambe Basin path, fr. 13.vi.1962, *Richards* 16672 (K).

Also in E. Zaire, Rwanda, Ethiopia, Uganda, Kenya, Tanzania, Angola and S. Africa (Transvaal). Evergreen rain forest and thicket often at edges and upper limits, particularly in rocky gullies, also *Brachystegia* woodland, thicket patches on termite mounds, mushitu and riverine forest patches; 1200–2150 m.

This is a variable species particularly in the size of the leaves and petals. Certain specimens from W. Zambia and Angola have smaller petals and have been separated by Fernandes as a distinct species but varietal rank is the most they deserve. The Transvaal populations would probably come here also. In rocky exposed places over 2000 m. specimens with small leaves not exceeding 3 cm. occur (e.g. Malawi, N: E. side of Mafingi Hills, 22.xi.1952, *Chapman* 56 (FHO; K)). In E. Africa a common gall causes swelling of the calyx-tube, but it remains smooth.

2. **Olinia vanguerioides** Bak. f. in Journ. Linn. Soc., Bot. **40**: 72 (1911). TAB **80**B.
Type: Rhodesia, Gazaland, Umswirizwi headwaters, *Swynnerton* 158 (BM; holotype)

Very closely allied indeed to the last species, differing in little but the petal size. Tree or shrub 2·4–25 m. tall; stems often shortly densely pubescent at apex, less often quite glabrous. Leaves 2·7–11·5 × 1·5–4·5 cm., very much as in previous species with characteristic dark pellucid venation against the glaucous pale under-surface, elliptic to elliptic-lanceolate, very often pubescent when young particularly on the midrib both above and below; petioles often red. Flowers in mostly dense many-flowered inflorescences, the peduncles branched and pedicels mostly densely shortly pubescent; bracts very small. Calyx-tube and combined receptacle 3–6 mm.

long, more or less cylindrical, narrowed to the base, mostly pale, glabrous. Petals 1–1·5(2) mm. long, up to 0·5(1) mm. wide, yellow or white, mostly narrow, with a few hairs inside at base but scales densely pubescent so that the whole centre of the flower appears choked with hairs. Fruit 6 mm. in diameter, ?white, globose.

Rhodesia. N: Mazoe, Umvukwes, Ruorka Ranch, fl. 16.xii.1952, *Wild* 3928 (K; SRGH). W: Matobo, fl. xii.1953, *Miller* 1994 (K; SRGH). C: Marandellas, Delta, fl. 5.iv.1950, *Wild* 3310 (K; SRGH). E: Melsetter, Bridal Veil Falls, fl. xii.1952, *Ball* 20 (K; SRGH). S: Bikita, S. cliff face of Mt. Horne, fl. 9.v.1969, *Biegel* 3082 (K; SRGH). **Mozambique.** MS: Chimanimani Mts. above R. Bundi, *Whellan* 2203* (SRGH).

Not known elsewhere. Margins of evergreen forest and on bare rocky ground beyond, particularly on granite, also in streamside gallery forest; 990–1560 m.

S. African (Transvaal), E. Zambian and Angolan populations of *O. rochetiana* are to a certain extent intermediate with *O. vanguerioides* but the latter is, I feel, distinct enough to recognize at specific rank. Very many sheets show a very characteristic gall affecting the calyx-tube and receptacle which becomes grooved and bears numerous small rather elongate wart-like tubercles. This fact has had some influence on my maintaining the taxon at specific rank. This gall has never been seen on *O. rochetiana* and the smooth galls found on that species have not been seen on *O. vanguerioides*. *Wild* 3606 (Rhodesia, Chimanimani Mts., Mt. Peza, 1800 m., 15.x.1950, among crags) has small leaves not exceeding 3 × 1·5 cm. and the flowers are pathologically deformed; it may be a variant of this species.

80. SONNERATIACEAE

By A. Fernandes

Trees or shrubs, with normal and sometimes aerial roots (pneumatophores). Leaves opposite, petiolate, coriaceous, entire, exstipulate. Flowers bisexual, actinomorphic, solitary or in 3s, axillary or terminal. Calyx-tube campanulate, thickly coriaceous, 4–8-lobed, the lobes valvate. Petals 4–8 or 0. Stamens 12 to numerous, inserted on the calyx-tube usually in several series; filaments free; anthers reniform, medifixed, opening longitudinally. Ovary free or adnate at the base to the calyx-tube, 4–many-locular; septa thin; style long; stigma subcapitate; ovules numerous, embedded in thick axile placentas, ascending. Fruit a many-seeded berry or a capsule. Seeds without endosperm; embryo with short leafy cotyledons.

A small family of 2 genera, included by Bentham & Hooker in the Lythraceae, distributed along tropical coasts from East Africa to Australia. Only the genus *Sonneratia* is represented in our area. *Duabanga grandiflora* (Roxb. ex DC.) Walp. is grown in the Forest Nurseries of Rhodesia.

SONNERATIA L. f.

Sonneratia L. f., Suppl. Plant.: 38 (1781) *nom. conserv.*

Glabrous trees or shrubs with coriaceous leaves. Calyx-lobes 4–8, acute, valvate. Petals as many as the calyx-lobes, broad to very narrow. Stamens ∞, inserted in several rows on a perigynous ring at the mouth of the calyx-tube. Ovary adnate to the calyx-tube at the base, depressed-globose, multilocular; ovules numerous in each loculus; style straight. Fruit a multilocular berry, free from the calyx-tube and eventually stipitate, with the loculi many-seeded. Seeds curved, angular, with a thick coriaceous testa; cotyledons shorter than the radicle.

Sonneratia alba Sm. in Rees, Cycl. 33, no. 2 (1816).—DC., Prodr. **3**: 231 (1828).—Engl., Pflanzenw. Afr. **3**, 2: 655 (1921).—Pires de Lima in Brotéria, Sér. Bot., **19**: 141 (1921).—Brenan, T.T.C.L.: 592 (1949).—Garcia in Est. Ens. Doc. Junta Invest. Ultram. **12**: 159 (1954).—Dale & Greenway, Kenya Trees & Shrubs: 539, fig. 98 (1961).—Gomes e Sousa, Dendrol. Moçamb. **2**: 556, t. 167 (1967).—Sangai in F.T.E.A. Sonneratiaceae: 1 (1968). TAB. **81**. Type: Indonesia, Amboina,

---

* *Whellan* 2203 appears to be the same variant as *Wild* 3606; in view of this additional specimen it may prove correct to treat this as a separate variety but neither is adequate and both seem to have deformed flowers due, perhaps, to some galling agent.

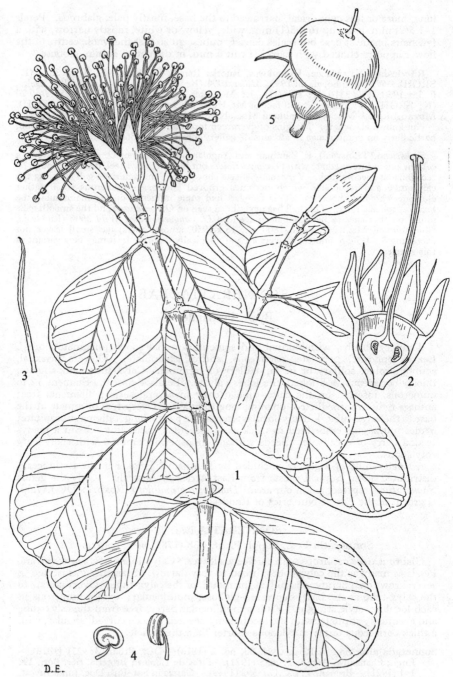

Tab. 81. SONNERATIA ALBA. 1, flowering branch (×⅔) *Drummond & Hemsley* 3234; 2, longitudinal section of flower with stamens removed (×1), from a drawing by *Kirk*; 3, petal (×2); 4, anthers, two aspects (×3) both from *Drummond & Hemsley* 3234; 5, fruit (×⅔) *Mogg* 28730. From F.T.E.A.

Rhumphius, illustration of *Mangium caseolare album* in Herb. Amb. **3**: 111, t. 73 (1743).
Sonneratia mossambicensis Klotzsch in Peters, Reise Mossamb., Bot. **1**: 66, t. 12 1861). Type: Mozambique, 14–18° L.S., *Peters* (B†, holotype; K, isotype).
Sonneratia acida sensu Hiern in F.T.A. **2**: 483 (1871) non L. f.
Sonneratia acida var. *mossambicensis* (Klotzsch) Matiei in Boll. Ort. Palermo, **7**: 108 (1908). Type as for *Sonneratia mossambicensis*.
Sonneratia caseolaris Engl., Pflanzenw. Ost-Afr. **C**: 286 (1895).—Williams, Useful Orn. Pl. Zanzib. Pemba: 449 cum icon. (1949).

An evergreen mangrove shrub or tree, 3–15 m. tall, with spreading branches and many thick finger-like pneumatophores. Old branches covered with a dark-grey or greyish-brown rough and fissured bark, the young ones terete, swollen at the nodes. Leaf-lamina 3–12·5 × 1·7–9 cm., obovate-circular to broadly obovate-elliptic or sometimes elliptic, rounded or emarginate at the apex, cuneate at the base, somewhat waved at the margin, coriaceous, yellow-green; midrib impressed above, prominent beneath, with 11–14 pairs of lateral nerves; petiole stout, 3–15 mm. long. Flowers odoriferous, rather large, 1–3 together, terminal. Bud ellipsoid, c. 3 × 2 cm. Calyx-tube c. 1·5 cm. long, thick, campanulate, angular, the angles alternating with the 6–8 lobes; lobes reddish inside, green outside, 12–20 × 6–9 mm., lanceolate, erect in the flower, ± reflexed in fruit. Petals 13–20 × 0·5–1·25 mm., ligulate, white or dorsally tinged with pink, caducous. Stamens ∞, the filaments up to 3·5 cm. long, inflexed in bud; anthers c. 1·5 mm. long. Ovary 14–18-locular; style 4·5–6·5 cm. long. Fruit c. 3·8 cm. in diam., surrounded at the base by the persistent calyx.

Mozambique. N: Tunguè, Palma, fl. 15.ix.1948, *Andrada* 1351 (COI; LISC; LMA); Mossuril, Cabaceira, fr. 14.xii.1948, *Pedro & Pedrógão* 4762 (EA). Z: R. Luabo, mouth of Pankeis Mucelo, fl. 6.vi.1858, *Kirk* (K). MS: Beira, fl. 12.ix.1965, *Balsinhas* 947 (COI). SS: Inhambane, Mangue, N. of Maxixe, fr. 14.i.1954, *Schelpe* 4432 (BM). LM: R. Umbeluzi, fl. 8.ii.1910, *Galpin* 7888 (PRE).
East coast of Africa (from south of Mozambique) and neighbouring islands, SE. Asia to N. Australia, S. Ryu Kyu Is., Micronesia, Solomon Is., New Hebrides and New Caledonia. Frequent in salty muds of mangrove swamps.

81. ONAGRACEAE

By P. H. Raven

Herbs, shrubs, or rarely trees with alternate or opposite, rarely whorled, simple, entire, lobed or pinnatifid leaves; stipules minute or absent. Flowers actinomorphic or zygomorphic, ☿ or rarely unisexual, 4- or 5-(rarely 2-, 3-, or 6-)merous, borne in the axils of usually reduced foliage leaves or in a more or less distinct inflorescence, usually a corymb or raceme. Floral tube present or absent. Sepals free or very rarely connate. Petals free (very rarely fused to the sepals), or sometimes absent. Stamens usually twice as many as the sepals and in 2 sets, 1 opposite the sepals and 1 opposite the petals, or as many as the sepals, sometimes reduced in number to 2; anthers versatile but sometimes attached near the base, dehiscing longitudinally. Ovary inferior. Style single; stigma variously lobed, discoid, elongate, or capitate. Fruit a loculicidally or more rarely irregularly dehiscent capsule, sometimes indehiscent, or a berry, 1–many-seeded; placentation axile. Seeds small, anatropous, lacking endosperm.

Seeds with a tuft of long hairs at the chalazal end; petals purplish or white, always present
                                             3. Epilobium
Seeds lacking hairs; petals yellow, white or rarely purplish, or sometimes absent:
Floral tube present; sepals caducous - - - - - - **1. Oenothera**
Floral tube absent; sepals persistent in fruit - - - - - **2. Ludwigia**

In addition to the above genera, *Gaura lindheimeri* Engelm. ex A. Gray, with white petals and few-seeded, indehiscent fruits, has been collected as a cultivated plant in Zambia; and species of *Fuchsia*, with berries, are occasionally cultivated in the F.Z. area.

330   81. ONAGRACEAE

1. OENOTHERA L.

Oenothera L., Sp. Pl. 1: 346 (1753); Gen. Pl. ed. 5: 163 (1754).

Annual, biennial, or perennial herbs. Leaves alternate, entire to pinnatifid. Stipules absent. Flowers borne in the axils of often reduced leaves, often clustered in an inflorescence near the ends of the stems. Bracteoles absent. Floral tube present. Sepals 4, caducous. Petals 4, caducous, yellow, white, or rose-purple, with contorted aestivation. Stamens 8; anthers attached near the middle. Pollen shed singly. Stigma deeply 4-lobed. Ovary 4-locular; ovules in 1–many rows in each locule, the seeds free. Capsule loculicidal. Seeds lacking hairs.

C. 110 species, all native to N. and S. America, but widely naturalized in the Old World.

Petals yellow; capsule cylindric:
Sepals 5–8 mm. long; petals 3–7 mm. long:
  Seeds coarsely and conspicuously pitted; leaves sinuate-dentate or sinuate-pinnatifid, or rarely subentire - - - - - - - - 1. *laciniata*
  Seeds indistinctly and shallowly pitted; leaves sinuate-dentate - 3. *indecora*
Sepals 15–22 mm. long; petals 20–32 mm. long - - - - 2. *stricta*
Petals white or rose-purple; capsule clavate:
  Petals white; flowers opening near sunset; plants hirsute - - 4. *tetraptera*
  Petals rose-purple; flowers opening near sunrise; plants mostly strigulose 5. *rosea*

1. **Oenothera laciniata** Hill in Ait., Hort. Kew: 172, pl. 6 (1768).—Munz in Amer. Journ. Bot. 22: 654 (1935). Type from seeds obtained near Charleston, S. Carolina.

Erect, often rank, annual, usually with many branches from the base, these often decumbent and up to 40 cm. long. Plants strigulose and densely villous above, especially in the inflorescence. Leaves 1–8 cm. long, sinuate-dentate or sinuate-pinnatifid, more rarely subentire, lanceolate; lower leaves petiolate, the upper ones sessile. Inflorescence erect, the villous buds individually deflexed from it before opening; flowers borne in the axils of much reduced leaves. Flowers opening near sunset. Floral tube 8–10 mm. long. Sepals 6–8 mm. long, with slender free tips c. 1 mm. long. Petals 5–7 mm. long, bright yellow, fading reddish. Anthers 2–3 mm. long, the filaments 3–6 mm. long. Style surrounded by the anthers at anthesis; stigma with linear lobes c. 2 mm. long. Capsule cylindric, 1·5–3 cm. long, not winged, subsessile. Seeds c. 1 mm. long, light brown, obovoid, coarsely and conspicuously pitted. Chromosome number $n = 7$ (ring of 14 at meiotic metaphase I). Autogamous.

Rhodesia. E: Chipinga, Melsetter Distr., fl. ix.1935, *Eyles* 8462 (K; SRGH). Native in N. and S. America, from the U.S.A. to Ecuador.

2. **Oenothera stricta** Ledeb. ex Link, Enum. Hort. Berol. 1: 377 (1821).—Munz in Amer. Journ. Bot. 22: 661 (1935). TAB. 82, figs. B1–B9. Type from Chile.
    Oenothera nocturna Harv. in Harv. & Sond., F.C. 2: 506 (1862).—Burtt Davy, F.P.F.T. 1: 202 (1926). Type from S. Africa.

Perennial herb, often flowering the first year, with several stout stems 30–100 cm. tall from a taproot. Plants subglabrous below, villous above. Leaves sparsely serrulate, very narrowly oblanceolate, those in the basal rosette up to 20 cm. long, gradually narrowed to an indistinct petiole; cauline leaves 2–8 cm. long, sessile. Inflorescence erect, the flowers in the axils of the much reduced upper leaves. Flowers opening near sunset. Floral tube 1·5–2·8 cm. long, slender. Sepals 15–22 cm. long, with free tips c. 2 mm. long. Petals 20–32 mm. long, bright yellow, fading reddish. Anthers 7–8·5 mm. long, the filaments 12–15 mm. long. Style surrounded by the anthers at anthesis; stigma with linear lobes 4–5 mm. long. Capsule cylindric, 2–3 cm. long, 3–4 mm. thick, enlarged upward, not winged, sessile or on a stout pedicel up to 3 mm. long. Seeds c. 1·5 mm. long, brown, obovoid, in 1 row in each locule of the capsule. Chromosome number $n = 7$ (ring of 14 at meiotic metaphase I). Autogamous.

Rhodesia. N: Mtoko, fl. & fr. iv.1956, *Davies* 1926 (K; SRGH). C: Salisbury, fl. 1.iii.1919, *Eyles* 1533 (BM; SRGH). E: Chimanimani Hotel, Melsetter Distr., fl. 22.ix.1960, *Phipps* 2844 (K; LISC; SRGH). **Mozambique.** LM: Inhaca I., fl. & fr. 15.ix.1959, *Mogg* 31640 (K).
Native to southern Chile and Argentina; cultivated for ornament and widely naturalized elsewhere.

Tab. 82. A.—OENOTHERA ROSEA. A1, flowering branch (×½) *Drummond* 4887; A2, flower (×1½) *Brooke* 233. B.—OENOTHERA STRICTA. B1, habit (×½) *Phipps* 2844; B2, flower bud (×1) *Chase* 2960; B3, flower partly dissected (×1); B4, style and stigma (×2); B5, anther, 2 aspects (×2), B3–B5 from *Munz* 169; B6 & B7, fruit before and at beginning of dehiscence (×1) *Chase* 2960; B8, seed in ventral view (×12); B9, same in lateral view (×8), B8–B9 from *Goldsmith* 126/68.

3. **Oenothera indecora** Cambess. in A. St. Hilaire, Fl. Bras. Mer. **2**: 268 (1830).—Munz in Amer. Journ. Bot. **22**: 658 (1935). Type from Brazil.

Erect annual or biennial with several branches from the base, these up to 60 cm. tall, everywhere covered with fine, appressed pubescence. Leaves 1–6 cm. long, sinuate-dentate, lanceolate or narrowly ovate, sessile. Inflorescence erect, the buds also erect; flowers borne in the axils of much reduced leaves. Flowers opening near sunset. Floral tube 8–15 mm. long. Sepals 5–6 mm. long, with slender free tips c. 1 mm. long. Petals 3–6 mm. long, bright yellow, fading reddish. Anthers c. 2 mm. long, the filaments 4–6 mm. long. Style surrounded by the anthers at anthesis; stigma with linear lobes 2–3 mm. long. Capsule 15–22 mm. long, cylindric, not winged, subsessile. Seeds c. 1 mm. long, light brown, obovoid, indistinctly and shallowly pitted. Chromosome number $n = 7$ (ring of 14 at meiotic metaphase I). Autogamous.

Rhodesia. C: Selukwe, Ferny Creek, fl. 8.xii.1953, *Wild* 4281 (K; LISC; SRGH). E: Melsetter, fr. 30.x.1957, *Whellan* 1443 (SRGH).
Native of S. America. Weedy places and about cultivated fields.

4. **Oenothera tetraptera** Cav., Ic. Pl. **3**: 40, t. 279 (1796).—Burtt Davy, F.P.F.T.: 202 (1926).—R. & A. Fernandes in C.F.A. **4**: 202 (1970). Type from Mexico.

Perennial herb, often flowering the first year, with numerous stems commonly 15–40 cm. tall from a somewhat woody caudex. Plants covered with long spreading hairs throughout, and also with shorter appressed hairs. Leaves 3–10 cm. long, oblanceolate or elliptical, the basal ones usually sinuate-pinnatifid, the cauline ones irregularly sinuate-pinnatifid in outline to entire, much reduced in size; petioles mostly less than 1 cm. long. Inflorescence erect, the flowers in the axils of much reduced leaves. Flowers opening near sunset. Floral tube 8–10 mm. long. Sepals 2–3 cm. long, usually coherent and deflexed to one side in anthesis. Petals 2·5–3·5 cm. long or as short as 1·4 cm. late in the season, white, fading purplish, broadly obovate. Anthers 5–6 mm. long, the filaments 1–1·5 cm. long. Style held above the anthers at anthesis; stigma with linear lobes 4–8 mm. long. Capsule clavate, 10–15 × 6–8 mm., with prominent wings 2–3 mm. wide, the base gradually narrowed into a hollow, ribbed stipe 5–25 mm. long. Seeds c. 1·3 mm. long, light brown, obovoid, in 2 rows in each locule. Chromosome number $n = 7$. Self compatible.

Rhodesia. C: Salisbury, fl. & fr. 10.ix.1955, *Drummond* 4872 (K; LISC; SRGH). E: Hillside Golf Course, Umtali, fl. 4.xi.1958, *Chase* 7015 (SRGH).
Native from Texas to northern S. America. In F.Z. area, introduced and weedy locally around Salisbury, Rusape and Umtali. Common in S. Africa.

5. **Oenothera rosea** L'Hérit. ex Ait., Hort. Kew. ed. 1, **2**: 3 (1789).—Burtt Davy, F.P.F.T. **1**: 202 (1926).—Wild, Common Rhod. Weeds: 36, fig. 36 (1955). TAB. **82** figs. A1–A2. Type grown from seeds collected in Peru.

Perennial herb, often blooming the first year, with numerous stems commonly 20–50 cm. tall from a somewhat woody caudex. Plants strigulose throughout, more densely so in the inflorescence. Leaves 2–5 cm. long, oblanceolate to narrowly obovate, entire to somewhat pinnatifid at the base of the blade, acute, the base narrowly cuneate; petioles 0·4–3 cm. long, distinct. Inflorescence nodding before anthesis, the flowers up to c. 1 cm. long, borne in the axils of much reduced leaves. Flowers opening near sunrise. Floral tube 3·5–6·5 mm. long, slender. Sepals 5–8 mm. long, commonly coherent and deflexed to one side in anthesis. Petals 4·5–8 mm. long, bright purplish-rose. Anthers 2·5–4 mm. long, on slender filaments 4–6 mm. long. Style surrounded by the anthers at anthesis; stigma c. 2 mm. long, with linear lobes. Capsule clavate, 8–10 × 3–4 mm., strigulose, narrowly winged, the base passing gradually into a hollow, ribbed stipe 5–20 mm. long. Seeds c. 0·6 mm. long, brown, oblong-ovoid, in 2 rows in each locule. Chromosome number $n = 7$ (ring of 14 at meiotic metaphase I). Autogamous.

Zambia. W: Buchi, Kitwe, fl. & fr. 15.xi.1969, *Mutimushi* 3827 (K; SRGH). Rhodesia. N: near Mazoe Dam, fl. 2.ix.1960, *Rutherford-Smith* 39 (K; SRGH). W: Hillside Dam, Bulawayo Distr., fl. & fr. ix.1957, *Miller* 4588 (SRGH). C: Mt. Hampden, Salisbury Distr., fl. 6.x.1955, *Drummond* 4887 (LISC; SRGH). E: Umtali, fl. & fr. 26.ix.1955, *Chase* 5807 (BM; K; SRGH).
Weed of American origin, now widespread in all warm parts of the world.

2. LUDWIGIA L.

Ludwigia L., Sp. Pl. **1**: 118 (1753); Gen. Pl. ed. 5: 55 (1754).
Isnardia L., Sp. Pl. **1**: 120 (1753); Gen. Pl. ed. 5: 56 (1754).
Jussiaea L., Sp. Pl. **1**: 388 (1753); Gen. Pl. ed. 5: 183 (1754).
Prieurea DC., Prodr. **3**: 58 (1828).

Slender herbs, erect or creeping and rooting at the nodes, to large shrubs. Underwater parts often swollen and spongy or bearing inflated white spongy pneumatophores. Leaves alternate or opposite, mostly entire. Stipules absent or reduced, deltate. Flowers borne singly, clustered or arranged in an inflorescence. Bracteoles lacking or conspicuous, usually 2, at or near the base of the ovary. Floral tube absent. Sepals 3–7, persistent after anthesis. Petals as many as the sepals or absent, caducous, yellow, with contorted aestivation. Stamens as many or twice as many as the sepals; anthers versatile. Pollen shed in tetrads or singly. Disk (summit of the ovary) flat to conical, often with depressed nectaries surrounding the bases of the episepalous stamens. Stigma hemispherical or capitate, the upper $\frac{1}{2}$–$\frac{2}{3}$ receptive, often lobed, the number of lobes corresponding to the number of locules. Ovary with a number of locules equal to the number of sepals; ovules pluriseriate or uniseriate in each locule, in one species (*L. hyssopifolia*) uniseriate below, pluriseriate above; if uniseriate, the seeds sometimes embedded in powdery or woody endocarp from which they detach easily or with difficulty. Dehiscence of the capsules irregular, by a terminal pore, or by flaps separating from the valve-like top. Seeds subcircular or elongate, the raphe usually easily visible and in some species equal or nearly equal in length to the seed.

The reasons for uniting *Ludwigia*, *Jussiaea* and *Isnardia* are reviewed in Reinwardtia **6**: 327–427 (1963). The resulting genus comprises some 75 species, mostly of tropical distribution and better represented in the New World than in the Old.

Stamens twice as many as the sepals:
  Seeds free, not embedded in endocarp, pluriseriate:
    Raphe equal in length to the diameter of the seed - - - 4. *octovalvis*
    Raphe not more than $\frac{1}{4}$ the diameter of the seed:
      Plants subglabrous; sepals 2–6 mm. long - - - - - 1. *erecta*
      Plants densely and coarsely pubescent; sepals (4)6–14 mm. long 2. *stenorraphe*
  Seeds embedded in endocarp, uniseriate:
    Seeds firmly embedded in woody coherent endocarp, pendulous, appearing as bumps in the capsule wall c. 1·5 mm. apart; narrow wings not running down to the ovary from the sinuses between the sepals - - - - 9. *stolonifera*
    Seeds loosely embedded in horse-shoe-shaped pieces of endocarp, horizontal, appearing as bumps in the capsule wall c. 0·5 mm. apart; a narrow wing extending down from the sinuses between each pair of sepals to the upper portion of the ovary - - - - - - - - 7. *leptocarpa*
Stamens as many as the sepals:
  Leaves opposite; petals absent - - - - - - 10. *palustris*
  Leaves alternate; petals present:
    Sepals 3 (rarely 4 or 5); stems creeping and rooting at the nodes; capsules normally tapering to the apex - - - - - - - 6. *senegalensis*
    Sepals 4, rarely 5; stems erect; capsules truncate at the apex:
      Seeds free, not embedded in endocarp, pluriseriate:
        Sepals 6–13 mm. long; petals 10–15 × 10–16 mm.; capsule 2–4·3 cm. long
            3. *jussiaeoides*
        Sepals 1·3–3·5 mm. long; petals 1–3 × 0·7–2 mm.; capsule 0·3–1·9 mm. long
            5. *perennis*
      Seeds more or less firmly embedded in endocarp at maturity, uniseriate
            8. *abyssinica*

1. **Ludwigia erecta** (L.) Hara in Journ. Jap. Bot. **28**: 292 (1953).—A. & R. Fernandes in Garcia de Orta **5**: 113 (1957); op. cit. **7**: 487 (1959); in C.F.A. **4**: 192 (1970). Type from America, perhaps Colombia, cultivated in Europe.
    Jussiaea erecta L., Sp. Pl. **1**: 388 (1753).—Brenan in F.T.E.A., Onagraceae: 12 (1953); in F.W.T.A. ed. 2, **1**: 169 (1954); in Mem. N.Y. Bot. Gard. **8**: 442 (1954).—Perrier in Fl. Madag., Oenotheracées: 21 (1950).—Binns, H.C.L.M.: 70 (1968).
    Isnardia discolor Klotzsch in Peters, Reise Mossamb. Bot.: 70 (1861). Type: Mozambique, the Zambezi between Sena and the Lupata Mts., *Peters* (B†).

Subglabrous erect herb from 3 cm. to more than 3 m., sometimes more or less woody at base, freely branched, the stems sharply angled from the decurrent leaf-bases. Leaves 2–13 × 0·2–4·5 cm., lanceolate to elliptic, rarely ovate, narrowly cuneate at the base, the apex acuminate to acute, rarely obtuse; main veins 16–27 on each side of midrib; petiole 2–15 mm. long. Flowers solitary in upper axils. Bracteoles c. 0·5 mm. long. Sepals 4, 2–6 × 1–1·5 mm., lanceolate-acuminate. Petals 3·5–5 × 2–2·5 mm., obovate. Stamens 8, subequal; filaments c. 1·5 mm. long; anthers c. 0·6 mm. long, shedding pollen directly on the stigma at anthesis. Pollen shed in tetrads. Disk not elevated, with a sunken, white-hairy nectary around the base of each epipetalous stamen. Style 0·5–1 mm. long; stigma globose, 1–1·1 mm. thick, its upper ⅔ receptive. Capsule 10–19 × 2–2·5 mm., glabrous, rarely puberulent, pale brown with 4 prominent dark brown ribs, sharply 4-angled with 4 nearly flat walls, irregularly and readily loculicidal, subsessile or on a pedicel up to 2 mm. long. Seeds 0·3–0·4(0·5) × 0·2–0·3 mm., pluriseriate in each locule of the capsule, free, pale brown, minutely cellular-pitted, elongate-obovoid; raphe c. ⅕ the diameter of the body. Chromosome number n = 8.

Botswana. N: Lake Ngami, fl. 3.i.1963, *Smithers* (SRGH). Zambia. N: Kangele R., 6·4 km. from Nsama on track between Nsama and L. Chisi, fr. 20.vii.1962, *Tyrer* 48 (BM). S: Sinazongwe, Mazabuka Distr., fr. 10.xii.1960, *Crozier & Mwanza* 8195 (SRGH). Rhodesia. N: Mtoko Distr., Mkoto Reserve, Mazoe R., fl. 11.viii.1950, *Whellan* 461 (K; SRGH). W: Wankie, fr. 21.vi.1934, *Eyles* 8061 (BM; K; SRGH). S: Sabi-Lundi junction, fr. 6.vi.1950, *Wild* 3419 (K; SRGH). Malawi. N: Karonga, fr. 31.vii.1952, *Williamson* 51 (BM). C: Benga, Nkhota Kota, fr. 2.ix.1946, *Brass* 17486 (K; SRGH). S: Tangadzi R. c. 27·7 km. W. of Chiromo, S. Shire Valley, fr. 16.vii.1958, *Seagrief* 3067 (SRGH). Mozambique. N: Mutuali near road to Malema, fr. 5.vi.1947, *Pedro* (PRE). Z: Margins of the Zambezi, 1884–1885, *Carvalho* (COI). T: Tete, Missão Católica de Boroma, margins of the Zambezi, fr. 22.ix.1942, *Mendonça* 347 (LISC). MS: Maringa, R. Save, fr. 25.vi.1950, *Chase* 2502 (BM; K; SRGH).

Native in the tropics of the New World, probably introduced but found throughout tropical Africa from Mauritius and the Sudan to Angola and Mozambique, also in Madagascar, the Seychelles and the Mascarene Is., from 0–1100 m.

2. **Ludwigia stenorraphe** (Brenan) Hara in Journ. Jap. Bot. **28**: 294 (1953). Type from Nigeria.

Jussiaea stenorraphe Brenan in Kew Bull. **8**: 164 (1953); in F.T.E.A., Onagraceae: 10 (1953); in F.W.T.A. ed. 2, **1**: 169 (1954).—Chapman, Veg. Mlanje Mts., Malawi: 38 (1962).—Binns, H.C.L.M.: 170 (1968). Type as above.

Jussiaea stenorraphe Brenan var. *stenorraphe* loc. cit.; loc. cit.; loc. cit. Type as above.

Ludwigia stenorraphe (Brenan) Hara var. *stenorraphe*.—A. & R. Fernandes in Garcia de Orta **5**: 471 (1957); **7**: 487 (1959). Type as above.

Robust suffruticose herb or shrub 1–3 m. tall, clothed everywhere with more or less dense erect or appressed pubescence. Leaves 2–13 × 0·2–3·8 cm., narrowly linear to oblanceolate, gradually narrowed to the base, the apex acute or subacute; main veins on each side of midrib 10–20; petiole 0–4 mm. long, rarely longer. Sepals 4, (4)6–13 × 1·5–5 mm., lance-deltoid, puberulent or hirsute, often turning reddish after anthesis. Petals 6–16 × 4–16 mm., ovate or suborbicular. Stamens 8, the epipetalous ones shorter; filaments 2–5 mm. long; anthers 0·75–2 mm. long, extrorse and shedding pollen outward, not on the stigma. Pollen shed in tetrads. Disk raised up to 2 mm., each epipetalous stamen surrounded at its base by a sunken, densely white-hairy nectary. Style 2–6 mm. long; stigma globose, 1·5–2 mm. in diameter, often slightly elevated above the anthers at anthesis. Capsule 10–40 × 1·5–4 mm., thin walled, puberulent or hirsute, brown with 8 dark brown ribs, readily and irregularly loculicidal, terete; pedicel 1–10(20) mm. long. Seeds 0·75–0·8 × 0·4 mm., pluriseriate in each locule of the capsule, free, pale brown, oblong-ellipsoid; raphe c. ⅛ the diameter of the body.

Endemic to tropical Africa, from Senegal and the southern Sudan to northern Angola, southern Malawi and Zambia. Represented in the F.Z. area by 3 subspecies.

Subsp. **stenorraphe.**—R. & A. Fernandes in C.F.A. **4**: 192 (1970). Type as for species.

Plant more or less densely pubescent, with long erect hairs. Sepals 6–9(10) mm. long, up to 4 mm. wide. Petals 9·5–13 × 6–12 mm., obovate. Disk elevated up to 1·5 mm. Style 2–4(6) mm. long. Capsule c. 2–3 cm. long; pedicel 3–10 mm. long.

Zambia. W: near Lwamukunyi R., Mwinilunga Distr., fl. 22.viii.1930, *Milne-Redhead* 966 (K). **Malawi.** S: Mlanje, fr. 6.vi.1958, *Chapman* H/633 (BM; K; PRE; SRGH).

Senegal, Mali, and the southern Sudan to northern Angola, southern Malawi and Zambia. Low swampy places, up to 1200 m.

Subsp. **macrosepala** (Brenan) Raven in Reinwardtia **6**: 353 (1963). Type from Uganda.
Jussiaea stenorraphe var. *macrosepala* Brenan in Kew Bull. **8**: 167 (1953); in F.T.E.A., Onagraceae: 11 (1953).—Binns, H.C.L.M.: 70 (1968). Type as above.

Plant densely covered with long spreading hairs. Sepals 10–14 mm. long, up to 5 mm. wide. Petals 11–14 × 10–15 mm., suborbicular. Disk elevated up to 1·5 mm. Style 3·5–5 mm. long. Capsule 2–4 cm. long; pedicel 0·5–2 cm. long.

Zambia. W: Ndola, fl. 25.iv.1954, *Fanshawe* 1139 (K). **Malawi.** N: Nyika Plateau, Mwanemba, fl. 2.iii.1903, *McClunie* 127 (K).

Uganda S. to Zambia and Malawi and in coastal Tanzania. Swamps and wet places, 600–2500 m.

Most of the collections from the F.Z. area—such as *Mutimushi* 748 (K), from Ndola, Zambia—are intermediate in flower size between subsp. *macrosepala* and subsp. *stenorraphe*. Zambia and Malawi lie between the principal areas of these 2 geographically distinct races.

Subsp. **speciosa** (Brenan) Raven in Reinwardtia, **6**: 352 (1963). Type: Mozambique, Namagoa Estate, Lugela Distr., Zambezia, *Faulkner* 801 (K, holotype; SRGH, isotype).
Jussiaea stenorraphe var. *speciosa* Brenan in Kew Bull. **8**: 167 (1953); in F.T.E.A., Onagraceae: 11 (1953). Type as above.
Ludwigia stenorraphe var. *speciosa* (Brenan) A. & R. Fernandes in Garcia de Orta **5**: 113 (1957). Type as above.

Plant densely covered with long spreading hairs. Sepals 9–14 mm. long, up to 4·5 mm. wide. Petals c. 16 × 15–16 mm., suborbicular. Disk elevate 1·25–2 mm. Style 5–6 mm. long. Capsule c. 2·5 cm. long; pedicel 9–12 mm. long.

Mozambique. Z: between Régulo Ságura and Namarroi, fl. 17.ix.1949, *Barbosa & Carvalho* 4111 (COI).

Outside the F.Z. area, this very distinctive entity—which very probably deserves specific status—is known only from Rufiji, Tanzania; it thus occupies 2 disjunct areas at low elevations on the E. African coast. It grows in swamps.

3. **Ludwigia jussiaeoides** Desr. in Lam., Encycl. Méth. Bot. **3**: 614 (1792).—Perrier in Fl. Madag., Oenotheracées: 8 (1950).—Garcia in Contr. Conh. Fl. Moçamb. **2**: 160 (1956).—A. & R. Fernandes in Garcia de Orta **5**: 113 (1957). Type from Mauritius.
Jussiaea jussiaeoides (Desr.) Brenan in Kew Bull. **8**: 163 (1953); in F.T.E.A., Onagraceae: 12 (1953). Type as above.
Ludwigia prostrata sensu Perrier, Fl. Madag., Oenotheracées: 9 (1950).

Tall herb, sometimes woody near the base, up to 3 m. tall, puberulent or strigulose especially on the young parts. Leaves 2·5–13 × 0·2–2·5 cm., lanceolate or narrowly lanceolate, minutely ciliate-pubescent, narrowly cuneate at the base, the apex acute; main veins on each side of the midrib 11–17; petiole 2–20 mm. long. Sepals 4, 6–13 × 1·4–3·5 mm., lanceolate-deltate. Petals 10–15 × 10–16 mm., broadly obovate. Stamens 4; filaments 2·5–4 mm. long; anthers 2–3 mm. long, extrorse and not shedding pollen directly on the stigma. Pollen shed in tetrads. Disk 1·5–2·5 mm. high, conical, with a depressed white-hairy nectary surrounding the base of each petal. Style 3·5–5 mm. long; stigma 1–2 mm. thick, globose, 4-lobulate. Capsule 20–34 × 2–3 mm., thin-walled, puberulent, terete, pale brown with 8 darker brown ribs, readily and irregularly loculicidal; pedicel 2–8 mm. long. Seeds 0·5–0·6 × 0·3–0·4 mm., pluriseriate in each locule of the capsule, free, pale brown, obovoid; raphe c. ⅛ the diameter of the body. Chromosome number $n = 24$.

Mozambique. N: 4·8 km. W. of Lumbo, Niassa Distr., fl. & fr. 21.v.1961, *Leach & Rutherford-Smith* 10942 (K; SRGH).

From the Cherangani Hills, Kenya to the Niassa Distr. of Mozambique, also Madagascar, the Comores, Seychelles and Mauritius. Damp ground in swamps and about standing water, 0–1400 m.

4. **Ludwigia octovalvis** (Jacq.) Raven in Kew Bull. **15**: 476 (1962).—Binns, H.C.L.M.: 70 (1968). TAB. **83** fig. B. Type from the West Indies.

Oenothera octovalvis Jacq., Enum. Syst. Pl. Ins. Carib.: 19 (1760). Type as above.

Jussiaea pubescens L., Sp. Pl. ed. 2, **1**: 555 (1762). Type from Jamaica (?).

Jussiaea angustifolia Lam., Encycl. Méth. Bot. **3**: 331 (1789).—Burtt Davy, F.P.F.T.: 201 (1926).—Garcia in Contr. Conh. Fl. Moçamb. **2**: 159 (1958). Type probably from the Moluccas.

Jussiaeae linearis Hochst. in Flora **27**: 425 (1844). Type from S. Africa.

Jussiaea angustifolia var. *linearis* (Hochst.) Harv. in Harv. & Sond., F.C. **2**: 504 (1862). Type as above.

Jussiaea suffruticosa var. *linearifolia* Hassler in Fedde, Repert. **12**: 277 (1913).— Brenan in F.T.E.A., Onagraceae: 15 (1953). Type from Paraguay.

Ludwigia pubescens (L.) Hara, Journ. Jap. Bot. **28**: 293 (1953).—A. & R. Fernandes in Garcia de Orta **5**: 474 (1957). Type as for *Jussiaea pubescens.*

Ludwigia pubescens var. *linearifolia* (Hassler) A. & R. Fernandes in tom. cit.: 115 (1957). Type as above.

Robust well-branched herb, sometimes woody at the base or even shrubby, up to 4 m. tall, subglabrous, puberulent, or densely villous. Leaves 0·7–14·5 × 0·1–4 cm., linear to subovate, narrowly or broadly cuneate at base, the apex attenuate; main veins 11–20 on each side of the midrib; petioles up to 10 mm. long. Bracteoles reduced or up to 1 mm. long. Sepals 4, 3–15 × 1–7·5 mm., ovate or lanceolate. Petals 3–17 × 2–17 mm., broadly obovate or cuneate, emarginate. Stamens 8, the epipetalous ones shorter; filaments 1–4 mm. long; anthers 0·5–4 mm. long, extrorse but soon crumbling and shedding pollen directly on the stigma. Pollen shed in tetrads. Disk slightly raised, with a white-hairy sunken nectary surrounding the base of each epipetalous stamen. Style 1·5–3·5 mm. long; stigma 1·2–3 mm. across, subglobose, shallowly 4-lobed. Capsule 17–45 × 2–8 mm., thin-walled, terete, pale brown with 8 darker ribs, readily and irregularly loculicidal; pedicel up to 10 mm. long. Seeds 0·6–0·75 × 0·5–0·7 mm., pluriseriate in each locule of the capsule, free, brown, rounded, including the raphe which is equal in size to the body of the seed and evenly transversely ridged. Chromosome number $n = 16$ (in F.Z. area).

Throughout the tropics of the world, and represented in F.Z. area by 3 subspecies.

Subsp. **octovalvis.** Type as for species.

Subglabrous or with sparse or dense appressed pubescence. Leaves 3–14·5 × 0·4–4 cm., lanceolate or narrowly lanceolate to narrowly ovate. Sepals (6)8–13 mm. long. Petals 5–16 × 4–17 mm. Anthers 1·3–2 mm. long. Chromosome number $n = 16$.

Zambia. N: Bulaya, Chishyela Dambo, NE. of Mweru-wa-Ntipa, fl. & fr. 11.viii.1962, *Tyrer* 370 (BM; SRGH). E: Petauke Dam, fl. & fr. 6.xii.1958, *Robson* 853 (K). S: Magoye, Kafue Basin, fl. & fr. 27.iii.1963, *van Rensburg* 1839 (K; SRGH). **Rhodesia.** N: College Farm near Mazoe, fl. 1.v.1962, *Noel* 2420 (SRGH). W: Khami, Bulawayo Distr., fl. & fr. 25.iv.1946, *Wild* 1067 (K; SRGH). C: Poole Farm, Hartley, fl. 16.iii.1948, *Hornby* 2913 (SRGH). E: Umtali, fr. 12.iv.1946, *Chase* 14737 (SRGH). S: Malangwe R., SW. Mateke Hills, Nuanetsi Distr., fl. 6.v.1958, *Drummond* 5616 (COI; PRE; SRGH). **Malawi.** S: Kapininjote, Ft. Johnston Distr., fl. 27.v.1954, *Jackson* 1326 (K). **Mozambique.** N: Eráti, 6·4 km. from Namapa on road to Odinepa, fl. 12.xii.1963, *Torre & Paiva* 9525 (LISC). Z: Namagoa, Mocuba Distr., fl. iii–v.1943, *Faulkner* 299 (SRGH). MS: Chimoio, Garuso, fl. 8.i.1948, *Barbosa* 816 (LISC). LM: Lourenço Marques, fl. & fr. 30.xi.1897, *Schlechter* 11536 (K).

Tanzania, Mozambique, Malawi, Rhodesia and eastern S. Africa. Moist places, often near cultivated fields, from 0–1500 m.

Widespread throughout the tropics of the world.

African populations referred to this subspecies might have originated as hybrids between subsp. *sessiliflora* and subsp. *brevisepala*; their morphological and geographical relationships lend some credence to this hypothesis. Whatever their origin, they constitute a relatively stable series of populations occupying a wide area at the present time.

Subsp. **sessiliflora** (Micheli) Raven in Kew Bull. **15**: 476 (1962). Type from Brazil.

Jussiaea suffruticosa L., Sp. Pl. **1**: 388 (1753).—Perrier in Fl. Madag., Oenotheracées: 20 (1950).—Brenan in F.T.E.A., Onagraceae: 14 (1953).—Garcia in

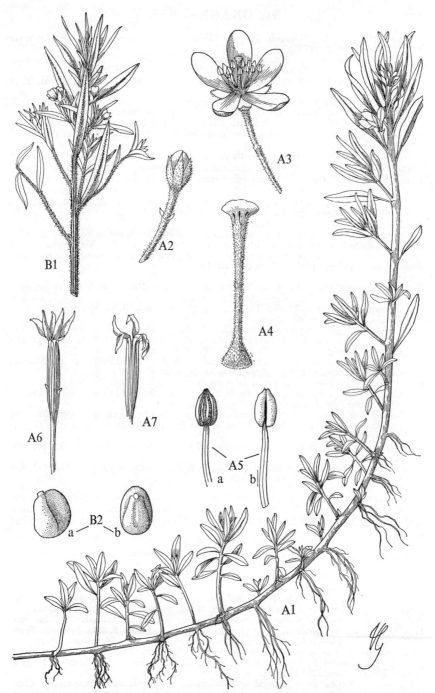

Tab. 83. A.—LUDWIGIA STOLONIFERA. A1, habit (×½); A2, flower bud (×1½), A1–A2
from *Robson & Fanshawe* 679; A3, flower (×1½); A4, style and stigma (×6);
A5, stamen, (a) ventral (b) dorsal view (×6), A3–A5 from *Pope* 154; A6, fruit
(×1); A7, fruit splitting (×1), A6–A7 from *Robson & Fanshawe* 679. B.—LUDWIGIA
OCTOVALVIS. B1, flowering branch (×½) *van Rensburg* 858; B2, seed, (a) lateral,
(b) ventral view (×16) *Eyles* 120.

338 81. ONAGRACEAE

Contr. Conh. Fl. Moçamb. **2**: 160 (1956). Type from India. Not *Ludwigia suffruticosa* Walt. (1788).
Jussiaea villosa Lam., Encycl. Méth. Bot. **3**: 331 (1789).—Peters, Reise Mossamb. Bot.: 70 (1861).—Binns, H.C.L.M.: 70 (1968). Type from India.
Jussiaea octonervia f. *sessiliflora* Micheli in Mart., Fl. Bras. **13**, 2: 171 (1875). Type as above.
Jussiaea suffruticosa var. *sessiliflora* (Micheli) Hassler in Bull. Soc. Bot. Genève **2**, 5: 271 (1913).—Brenan, op. cit.: 16 (1953). Type as above.
Ludwigia pubescens var. *sessiliflora* (Micheli) Hara in Journ. Jap. Bot. **28**: 292 (1953).—A. & R. Fernandes in Garcia de Orta **5**: 114 (1957). Type as above.

More or less densely covered with spreading pubescence, at least in the upper parts. Leaves 2–10 × 0·8–4 cm., lanceolate to subovate. Sepals 6–15 mm. long. Petals 6–17 × 5–17 mm. Anthers 1·2–4 mm. long. Chromosome number $n = 16$.

Rhodesia. E: Gungunyana Forest Reserve, Melsetter Distr., fl. x.1961, *Goldsmith* 87/61 (SRGH). **Mozambique.** MS: Chimoio between R. Búzi and Vila Pery, fl. & fr. 2.vi.1941, *Torre* 2824 (LISC; SRGH).
Common in Natal, Rhodesia, Mozambique, possibly introduced in Zanzibar (Pemba) and in the Cape Prov. of S. Africa, Madagascar (common), Seychelles, Mauritius. Widespread in the rest of the tropics of the world. Damp and swampy places, often by rivers, streams, lakes and in marshes, from 0–c. 1000 m.

Subsp. **brevisepala** (Brenan) Raven in Kew Bull. **15**: 476 (1962).—R. & A. Fernandes in C.F.A. **4**: 144 (1970). Type from Cameroun.
Jussiaea linearis Willd., Sp. Pl. **2**: 575 (1800).—Peters, Reise Mossamb. Bot.: 70 (1861). Type from W. Africa.
Jussiaea suffruticosa var. *linearis* (Willd.) Oliv. ex Kuntze, Rev. Gen. Pl. **1**: 251 (1891).—Brenan in F.T.E.A., Onagraceae: 15 (1953); in F.W.T.A. ed. 2, **1**: 169 (1954).—Binns, H.C.L.M.: 70 (1968). Type as above.
Jussiaea didymosperma Perrier in Notul. Syst. **13**: 148 (1947); in Fl. Madag., Oenotheracées: 23 (1950). Type from Madagascar.
Jussiaea suffruticosa L. var. *brevisepala* Brenan in Kew Bull. **8**: 168 (1953); op. cit.: 14 (1953); loc. cit.; in Mem. N.Y. Bot. Gard. **8**: 441 (1954).—Binns, loc. cit Type as for *Ludwigia octovalvis* subsp. *brevisepala*.
Ludwigia pubescens var. *brevisepala* (Brenan) Hara in Journ. Jap. Bot. **28**: 294 (1953).—A. & R. Fernandes in Garcia de Orta **5**: 474 (1957); op. cit. **7**: 489 (1959). Type as above.
Jussiaea suffruticosa var. *pseudo-linearis* Brenan, op. cit.: 169 (1953); loc. cit. Type from Togo.
Ludwigia pubescens var. *linearis* (Willd.) A. & R. Fernandes, op. cit.: 115, 471, 474 (1957). Type as for *Jussiaea linearis*.

More or less densely covered with spreading pubescence, especially on the younger parts. Leaves 2–9 × 0·1–1·5 cm., linear or lanceolate. Sepals 3–6 mm. long. Petals 3–8 × 2–4·5 mm. Anthers 0·5–1·2 mm. long. Chromosome number $n = 16$.

Zambia. N: Marsh below Nimkole Village, fl. 23.ii.1955, *Richards* 4633 (K). W: Mindolo, Kitwe, fl. & fr. 24.iv.1963, *Mutimushi* 289 (SRGH). C: Golden Valley, Chisamba, fl. 27.ii.1933, *Michelmore* 684 (K). S: Mazabuka, fl. 29.iii.1963, *van Rensburg* 1858 (SRGH). **Rhodesia.** N: 8 km. NE. of Mtoko, fl. & fr. 21.viii.1956, *Davies* 2093 (LISC; SRGH). W: Deka R., Wankie, fl. & fr. 21.vi.1934, *Eyles* 8063 (BM; K; SRGH). S: Gorge upstream from Buffalo Bend, Nuanetsi R., Nuanetsi Distr., fr. 28.iv.1961, *Drummond & Rutherford-Smith* 7562 (K; SRGH). **Malawi.** N: Karonga, L. Malawi, fl. ix.1887, *Scott* (K). C: Benga, Nkhota Kota Distr., fl. & fr. 2.ix.1946, *Brass* 17485 (K). S: Lower Mwanza R., Chikwawa Distr., fr. 6.x.1946, *Brass* 18014 (K). **Mozambique.** N: between Montepuez and Balama, fl. 29.viii.1948, *Barbosa* 1926 (LISC). Z: Mocuba, fl. & fr. 5.vi.1949, *Gerstner* 7099 (K; PRE). T: Marávia, Fíngoè, fl. 11.viii.1941, *Torre* 3246 (LISC).
Endemic to Africa, from Senegal and the Sudan to Angola and Mozambique, Cape Verde Is.; S. Tomé, northwest coast of Madagascar, where probably introduced. Moist. ground around lakes and along rivers and streams, from 0–2700 m.

Subsp. *brevisepala* and subsp. *octovalvis* intergrade where they come into contact, some of the intermediates resembling subsp. *sessiliflora* in their long sepals and erect pubescence, but without the broad leaves of the latter, which remains very distinct. One such intergrade was cited as subsp. *sessiliflora* in error: Raven in Reinwardtia **6**: 363 (1963), Rhodesia, W: Cataract I., Victoria Falls, fl. & fr. viii.1909, *Rogers* 3283 (K).

81. ONAGRACEAE

81. ONAGRACEAE 339

5. **Ludwigia perennis** L., Sp. Pl. **1**: 119 (1753); excl. verba falsa " foliis oppositis ".—
A. & R. Fernandes in Garcia de Orta **5**: 114 (1957). Type from Ceylon.
  Lugwigia nesaeoides Perrier in Notul. Syst. **13**: 141 (1947); in Fl. Madag.,
Oenotheracées: 10 (1950). Type from Madagascar.
  Ludwigia humbertii Robyns & Lawalrée in Bull. Jard. Bot. Brux. **18**: 291 (1947).—
Robyns & Tournay, Fl. Parc Nat. Alb. **1**: 681 (1948). Type from Zaire.
  Jussiaea perennis (L.) Brenan in Kew Bull. **8**: 163 (1953); in F.T.E.A., Onagraceae:
13 (1953); in F.W.T.A. ed. 2, **1**: 169 (1954). Type as above.

Annual herb up to 1 m. tall, subglabrous or minutely puberulent on younger
parts. Leaves 1–11 × 0·3–2·7 cm., narrowly elliptical to lanceolate, narrowly
cuneate at the base, the apex subacute; main veins on each side of the midrib 6–12;
petiole 2–15 mm. long, winged. Sepals 4, rarely 5, (1·3)2–3·5 × (0·5)0·7–1·8 mm.,
deltate, glabrous or minutely puberulent. Petals 1–3 × 0·7–2 mm., elliptical.
Stamens usually as many as the sepals, rarely more; filaments 0·3–0·7 mm. long;
anthers 0·5–0·7 × 0·5–0·7 mm., shedding pollen directly on the stigma at anthesis.
Pollen shed in tetrads. Disk slightly elevated, glabrous. Style 0·7–1·5 mm. long;
stigma 0·4–0·5 mm. thick, globose. Capsule 3–16(19) mm. long, thin-walled,
glabrous or puberulent, terete, pale brown, readily and irregularly loculicidal;
capsule sessile or on a pedicel up to 6 mm. long, often more or less nodding. Seeds
0·3–0·5 × 0·2–0·25 mm., pluriseriate in each locule of the capsule, free, brown with
fine brown lines, ellipsoid-rounded; raphe very narrow and inconspicuous.
Chromosome number $n = 8$.

Zambia. S: Mazabuka, edge of Kafue Flats, fr. 7.iv.1955, *E.M. & W.* 1424 (BM;
LISC; SRGH). **Rhodesia.** N: c. 8 km. N. of Gokwe, fl. & fr. 25.iii.1963, *Bingham* 584
(SRGH). S: Malangwe R., SW. Mateke Hills, Nuanetsi Distr., fl. & fr. 6.v.1958,
Drummond 5590 (K; SRGH). **Malawi.** S: W. of L. Chilwa, Zomba Distr., fl. & fr.
21.vi.1962, *Robson* 5404 (K; SRGH). **Mozambique.** Z: Namagoa, Mocuba Distr.,
fl. 1943, *Faulkner* 298 (K; PRE; SRGH). MS: Lion's Creek, fl. & fr. 7.iv.1898,
Schlechter 12189 (COI; K). SS: Limpopo, between Caniçado and Saúte, fr. 15.v.1948,
Torre 7817 (LISC).
  Throughout the tropics of the Old World; in Africa from Senegal, the region of Lake
Chad, and the Sudan to eastern Zaire and northern Natal and northwestern Madagascar.
Wet places, as on flood plains and about lakes, 0–1200 m.

6. **Ludwigia senegalensis** (DC.) Troch. in Mém. Inst. Fr. Afr. Noire **2**: 378 (1940).—
R. & A. Fernandes in C.F.A. **4**: 195 (1970). Type from Senegal.
  Prieurea senegalensis DC., Prodr. **3**: 58 (1828). Type as above.
  Ludwigia pulvinaris Gilg in Warb., Kunene-Samb.-Exped. Baum: 324 (1903); in
F.W.T.A. **1**: 146 (1927).—A. & R. Fernandes, Garcia de Orta **7**: 490 (1959). Type
from Angola.
  Jussiaea pulvinaris (Gilg) Brenan in Kew Bull. **8**: 163 (1953); in F.W.T.A. ed. 2,
1: 169 (1954). Type as above.
  Jussiaea senegalensis (DC.) Brenan, tom. cit.: 164 (1953); loc. cit. Type as above.
  Ludwigia pulvinaris subsp. *lobayensis* Raven in Reinwardtia **6**: 373 (1963). Type
from Zaire.

Low herb, creeping and rooting at the nodes, the stems mostly 5–35 cm. long,
glabrous or minutely puberulent. Leaves 0·5–3 × 0·3–1 cm., narrowly lanceolate to
broadly ovate, narrowly cuneate at the base, the apex acute or obtuse; veins
obscure; petioles 1–2 mm. long. Sepals 3, rarely 4 or even 5, 1–3 mm. long,
deltate-acute. Petals 2–2·5 × 0·5–1·5 mm., linear to obdeltate, acute. Stamens as
numerous as the sepals; filaments 0·8–1·5 mm. long; anthers surrounding the
stigma and shedding pollen directly on it at anthesis. Pollen shed in tetrads. Disk
conspicuously elevated, glabrous. Style c. 2 mm. long; stigma globose, c. 0·8 mm.
thick. Capsule 3·5–9 × 1·3–2 mm., plump, thin-walled, pale brown, readily and
irregularly loculicidal, subsessile. Seeds showing through the capsule wall, in
approximately 2 rows in each locule, 0·5–0·7 mm. long, free, light to dark brown,
ovoid; raphe narrow and inconspicuous. Chromosome number $n = 8$.

Zambia. B: Zambesi, Sanduala pontoon, Mongu Distr., fr. 12.ix.1959, *Drummond &
Cookson* 6371 (K; LISC; PRE; SRGH). N: Ft. Rosebery, fl. & fr. 5.v.1964, *Fanshawe*
8527 (K). S: Victoria Falls, Livingstone Distr., 0·9 km. W. up river, 27.viii.1947,
Greenway & Brenan 7991 (K). **Rhodesia.** W: Cataract I., Victoria Falls, fl. & fr.
viii.1909, *Rogers* 5286 (K).
  Endemic to Africa, from Senegal and the southern Sudan to southern Angola and the
Zambesi. Wet places, sometimes completely submerged, 0–1100 m.

Having had the opportunity of growing several strains of this species and studying their characteristics in the experimental garden, I am now convinced that the 3 taxa I recognized in 1963 (in Reinwardtia **8**: 371–373) are best regarded as belonging to 1 somewhat polymorphic species, in which the differences are maintained by consistent autogamy and sometimes even cleistogamy combined with efficient vegetative reproduction.

7. **Ludwigia leptocarpa** (Nutt.) Hara in Journ. Jap. Bot. **28**: 292 (1953).—A. & R. Fernandes in Garcia de Orta **5**: 116, 474 (1957); op. cit. **7**: 490 (1959); in C.F.A. **4**: 196 (1970). Type from the U.S.A.

Jussiaea leptocarpa Nutt. in Gen. N. Amer. Pl. **1**: 279 (1818).—Brenan in F.T.E.A., Onagraceae: 16 (1953); in F.W.T.A. ed. 2, **1**: 169 (1954). Type as above.

Jussiaea pilosa Kunth, Nov. Gen. et Sp. **6**: 101, t. 532 (1823).—R. E. Fr., Wiss. Ergebn. Schwed. Rhod.-Kongo-Exped. **1**: 182 (1914).—Perrier in Fl. Madag., Oenotheracées: 17 (1950).—Mogg in Macnae & Kalk, Nat. Hist. Inhaca I.: 150 (1958).—Binns, H.C.L.M.: 70 (1968). Type from Colombia.

Jussiaea seminuda Perrier, op. cit.: 18 (1950). Type from Madagascar.

Robust hairy plants to 3 m. tall, often somewhat woody below, reclining at base but erect and well-branched, with erect floating pneumatophores arising from roots under water. Leaves 3·5–18 × 1–4 cm., long-hairy, broadly lanceolate, narrowly cuneate at base, the apex acuminate; main veins on each side of midrib 11–20; petiole 0·2–3·5 cm. long. Bracteoles at base of ovary absent or rarely present, narrowly deltate. Sepals 5, rarely 4, 6 or 7, 5·5–11 × 1·5–3 mm., deltate-acuminate, long-hairy, with a narrow wing running down from the sinus between adjacent sepals to the apical portion of the ovary. Petals 5–11 × 4–8 mm., obovate. Stamens twice as many as the sepals; filaments 2–4 mm. long, the epipetalous ones shorter; anthers 1·2–1·6 mm. long, extrorse and thus not shedding pollen directly on the stigma. Pollen shed in tetrads. Disk slightly elevated, the base of each epipetalous stamen surrounded by a depressed nectary densely covered with matted white hairs. Style 3–4·5 mm. long, glabrous; stigma 2–2·5 mm. across, c. 1 mm. high, globose, the upper ⅔ receptive. Capsule 15–50 × 2·5–4 mm., relatively thin-walled, long-hairy, terete, dull light brown, with prominent ribs over the locules and less prominent ones over the septa, marked on the outside with bumps c. 0·5 mm. apart, corresponding to the position of the seeds, slowly and irregularly loculicidal; pedicels 2–20 mm. long. Seeds 1–1·2 mm. long, uniseriate in each locule of the capsule, horizontal, shiny pale brown, finely pitted, obovoid; raphe much narrower than the body of the seed; each seed loosely embedded in an easily detached horseshoe-shaped segment of firm pale brown endocarp c. 1–1·5 mm. thick and c. 1 mm. high. Chromosome number $n = 24$ (in African populations).

Caprivi Strip. Katima Mulilo area, fr. 24.xii.1958, *Killick & Leistner* 3077 (M; PRE). **Botswana.** N: Mokankanyane, Ngamiland, fl. & fr. xii.1930, *Curson* 754 (PRE). **Zambia.** B: Namushakende, Mongu Distr., fl. & fr. 21.viii.1961, *Angus* 3015 (K; SRGH). N: Luaba lagoon, L. Bangweulu, fl. 3.ii.1959, *Watmough* 219 (SRGH). C: Lukanga, Kabwe Distr., fl. 19.viii.1958, *Seagrief* 3180 (SRGH). E: Chilongozi, Ft. Jameson Distr., fl. 11.x.1960, *Richards* 13327 (K; SRGH). S: Lochinvar, Mazabuka, fl. 25.iv.1962, *Mitchell* 13199 (SRGH). **Rhodesia.** N: Kariba Lake shore, Binga Distr., fl. 3.xii.1960, *Boughey* 7528 (SRGH). S: Zimbabwe, fr. 11.viii.1929, *Rendle* 267 (BM). **Malawi.** N: Vua, fl. 17.vii.1952, *Williamson* 39 (BM). C: 3·2 km. S. of L. Nyasa Hotel, Salima Distr., fl. & fr. 9.viii.1951, *Chase* 3877 (K; SRGH). S: Lower Shire Valley, Elephant Marsh, fl. ix.1956, *Robertson* 2 (K). **Mozambique.** MS: on the mainland near Expedition I., fl. & fr. xii.1859, *Kirk* (K). SS: between Vila de João Belo and Chongoene, fr. 24.vii.1947, *Barbosa* 317 (COI). LM: Marracuene near Bobole, fl. 2.x.1957, *Barbosa & Lemos* 7920 (COI; K).

In Africa from Senegal and the vicinity of L. Chad to L. Tana, Ethiopia, and S. to the interior of Angola and Zululand, and Madagascar. In the New World from the south-eastern U.S.A. and the W. Indies to Peru and Argentina. Often abundant in marshes, along streams, rivers and lakes, up to 1300 m.

8. **Ludwigia abyssinica** A. Rich., Tent. Fl. Abyss. **1**: 274 (1848).—A. & R. Fernandes in Garcia de Orta **5**: 116 (1957); op. cit. **7**: 494 (1959); in C.F.A. **4**: 198 (1970). Type from Ethiopia.

Ludwigia prostrata sensu Oliv. in F.T.A. **2**: 491 (1871).—Robyns & Tournay, Fl. Parc Nat. Alb. **1**: 681 (1948).

Ludwigia jussiaeoides sensu Harv. in Harv. & Sond., F.C. **2**: 505 (1894) pro parte.—Burtt Davy, F.P.F.T. **1**: 201 (1926).

Ludwigia parviflora sensu R.E.Fr., Wiss. Ergebn. Schwed. Rhod.-Kongo-Exped. **1**: 182 (1914).

Jussiaea abyssinica (A. Rich.) Dandy & Brenan in F. W. Andr., Fl. Pl. Anglo-Egypt. Sudan **1**: 145 (1950).—Brenan in F.T.E.A., Onagraceae: 18 (1953); in F.W.T.A. ed. 2, **1**: 170 (1954); in Mem. N.Y. Bot. Gard. **8**: 441 (1954).—Williamson, Useful Pl. Nyasal.: 71 (1955).—Binns, H.C.L.M.: 70 (1968). Type as for *Lugwigia abyssinica*.

Stout succulent herb, sometimes woody at the base, up to 3 m. tall, well branched, glabrous except for minute hairs on the midribs and margins of young leaves; stems usually somewhat reddish. Leaves 2–13 × 0·5–3·5 cm., lanceolate or broadly elliptical, narrowly cuneate at the base, the apex subacute; main veins on each side of midrib 13–22; petiole 2–20 mm. long. Flowers clustered on short axillary shoots also bearing reduced leaves. Sepals 4, 1·7–3 × 0·4–1 mm., lanceolate-ovate, mucronate, usually with reddish margins. Petals 1·5–3·5 × 1·2–2·6 mm., nearly circular in outline. Stamens 4; filaments 0·8–1·2 mm. long; anthers c. 0·5 × c. 0·8 mm., weakly attached to the filament, shedding pollen directly on the stigma at anthesis. Pollen shed in tetrads. Disk elevated c. 0·5 mm., with a depressed nectary fringed with short hairs surrounding the base of each petal. Style 0·5–0·8 mm. long; stigma c. 1 mm. thick, c. 0·5 mm. high, depressed-globose. Capsule 10–20 × 1·2 mm., relatively thin-walled, glabrous, terete, light brown, at first thin-walled and torulose, but as the endocarp swells and hardens, becoming smooth; pedicels 0·5–3 mm. long. Seeds 0·6–0·75 × 0·4–0·5 mm., uniseriate in each locule of the capsule, diagonal, brown, obovoid; raphe inconspicuous, each seed loosely but completely embedded in an easily detached piece of soft powdery endocarp 0·6–1 mm. long, 0·5–0·7 mm. wide. Chromosome number *n* = 24.

Zambia. B: near Senanga, fr. 29.vii.1952, *Codd* 7226 (BM; COI; K; PRE; SRGH). N: Mbala (Abercorn), fl. 2.v.1952, *Richards* 1736 (K). W: Kitwe, fl. 16.iv.1961, *Linley* 135 (SRGH). C: Walamba, fr. 22.v.1954, *Fanshawe* 1231 (K; SRGH). S: Katombora, Livingstone Distr., fr. 7.vii.1956, *Gilges* 647 (SRGH). **Rhodesia.** W: Victoria Falls, fl. & fr. vi.1959, *Armitage* 76/59 (SRGH). E: below Mtarazi Falls, Inyanga Distr., fl. & fr. 19.iv.1958, *Phipps* 1153 (K; SRGH). **Malawi.** N: Nkhata Bay Distr., fr. 2.x.1960, *Eccles* 44 (SRGH). S: Blantyre, *Buchanan* 80 (K). **Mozambique.** N: Massangulo, fl. iii.1933, *Gomes e Sousa* 1286 (COI). Z: Murrumbala, fl. 30.xii.1858, *Kirk* (K). T: Angónia, Posto Zootécnico, valley of R. Maué, fl. 13.v.1948, *Mendonça* 4212 (LISC) MS: Chiniziua, Beira Distr., fl. 30.x.1957, *Gomes e Sousa* 4430 (COI; K; PRE).

Endemic to Africa, from Senegal and the southern Sudan to Angola and Zululand and Madagascar. Common in swampy situations up to 1900 m.

The leaves were said by Williamson (loc. cit.) to be edible (Mzimba Distr. of Malawi).

9. **Ludwigia stolonifera** (Guill. & Perr.) Raven in Reinwardtia **6**: 390 (1963). TAB. **83** fig. A. Type from Senegal.

Jussiaea diffusa Forsk., Fl. Aegypt-Arab.: 210 (1775).—Burtt Davy, F.P.F.T. **1**: 201 (1926).—Garcia in Contr. Conh. Fl. Moçamb. **2**: 160 (1956). Type from Egypt.

Jussiaea stolonifera Guill. & Perr., Fl. Senegamb. Tent.: 292 (1833). Type as above.

Jussiaea alternifolia E. Mey. ex Peters, Reise Mossamb. Bot.: 69 (1861). Type from S. Africa.

Jussiaea repens sensu Robyns & Tournay, Fl. Parc Nat. Alb. **1**: 680 (1948).—Perrier in Fl. Madag., Oenotheracées: 16 (1950).

Jussiaea diffusa subsp. *albiflora* Perrier, tom. cit.: 15 (1950). Type from Madagascar.

Jussiaea repens var. *diffusa* (Forsk.) Brenan in Kew Bull. **8**: 171 (1953); in F.T.E.A., Onagraceae: 19 (1953); in F.W.T.A. ed. 2, **1**: 170 (1954). Type as for *Jussiaea diffusa*.

Ludwigia adscendens var. *diffusa* (Forsk.) Hara in Jour. Jap. Bot. **28**: 291 (1953).—A. & R. Fernandes in Garcia de Orta **5**: 116 (1957); op. cit. **7**: 491 (1959); in C.F.A. **4**: 197 (1970). Type as above.

Herb with prostrate or ascending stems, rooting at the nodes, with conspicuous white erect fusiform mucronate pneumatophores arising in clusters at the nodes of the floating stems and from the roots; plants more or less densely villous to glabrous. Leaves 2–9 × 0·5–1·7(2·3) cm. on flowering stems, broader on floating non-flowering branches, dark green, shining, narrowly lanceolate to narrowly elliptical, narrowly cuneate at the base, the apex acute; main veins 6–12 on each side of the midrib; petioles 0·2–2 cm. long. Flowers borne singly in upper leaf axils. Bracteoles c. 1 mm. long, deltate. Sepals 5, 5–14 × 1·5–2·8 mm., deltate acuminate. Petals 7–18 × 4–10 mm., lemon-yellow with a darker spot at the base, obovate, rounded at the apex. Stamens 10, the epipetalous ones slightly shorter; filaments 2·5–4 mm. long; anthers 1·2–8 mm. long, extrorse and not shedding pollen on the stigma at anthesis. Pollen grains shed individually. Disk slightly elevated, with a depressed white-hairy nectary surrounding the base of each epipetalous stamen. Style 3–8 mm. long, densely long-hairy to just below stigma; stigma 1·5–2 mm. across, 1–1·2 mm. high, golden-yellow, elevated above the anthers at anthesis, the upper receptive. Capsule 10–30 × 3–4 mm., glabrous or villous, light brown, with 10 conspicuous darker brown ribs, terete, the seeds evident between the ribs as bumps c. 1·5 mm. apart; capsules thick-walled, very tardily and irregularly dehiscent; pedicel 0·5–2 cm. long. Seeds 1·1–1·3 mm. long, uniseriate in each locule of the capsule, pale brown, more or less vertical, firmly embedded in coherent cubes of woody endocarp 1·2–1·5 × 1–1·2 mm., the endocarp firmly fused to the capsule wall. Chromosome number $n = 16$.

Botswana. N: Junction of the Thamalakane and Botletle Rs., fl. ix.1896, *Lugard* 3 (K). SE: 20 km. S. of Gaberones, fl. 19.i.1960, *Leach & Noel* 223 (K; SRGH). **Zambia.** B: Omboya R., 32·2 km. N. of Boma, Kalabo, fl. & fr. 11.ix.?, *Rea* 144 (K). N: Cascalawa, L. Tanganyika, Mbala (Abercorn) Distr., 16.vii.1960, *Richards* 12874 (K). W: Ndola, fl. 23.x.1954, *Fanshawe* 1636 (K). C: 15 km. E. of Chisamba, fl. 19.v.1957, *Best* 129 (SRGH). E: Chilongozi, Luangwa R., Ft. Jameson Distr., 11.x.1960, *Richards* 13328 (K; SRGH). S: Victoria Falls, fl. & fr. 21.xi.1949, *Wild* 3139 (K; SRGH). **Rhodesia.** N: Kariba Gorge, fl. & fr. ix.1960, *Goldsmith* 104/60 (K). W: Headwaters Maelele/Lubu Rs., Wankie Distr., fl. 5.x.1958, *Lovemore* 553 (K; SRGH). C: Poole Farm, Hartley, fl. & fr. 4.iv.1954, *Hornby* 3344 (SRGH). E: Chiribira Falls, fl. 8.vi.1950, *Wild* 3418 (K; SRGH). S: Umzingwane R. crossing 48·3 km. NE. of Tuli Police Camp, Beitbridge Distr., fl. 11.v.1959, *Drummond* 6119 (K; SRGH). **Malawi.** N: Mwanemba, Nyika Plateau, fl. ii–iii.1903, *McClounie* 129 (K). C: Nkhota Kota Distr., fl. & fr. 17.x.1947, *Benson* 385 (PRE). S: James Lagoon 9·6 km. N. of Chiromo, Port Herald Distr., fl. 22.iii.1960, *Phipps* 2614 (K; PRE; SRGH). **Mozambique.** T: Missão Católica de Boroma, fl. & fr. 22.ix.1942, *Mendonça* 357 (LISC). MS: Macuti Beach, Beira Distr., fl. 10.ix.1962, *Noel* 2482 (LISC; SRGH). SS: between Vila de João Belo and Macia, fr. 21.ix.1948, *Myre & Carvalho* 260 (COI). LM: Marracuene, Bobole, fl. 15.xii.1961, *Lemos & Balsinhas* 314 (BM; COI; K; LISC).
Nearly throughout Africa N. of 30° Lat., except for the desert, and in the Near East from Israel and Lebanon E. to Iraq. Wet places, especially along rivers and lakes where often growing floating on water, sometimes forming huge masses, up to 1900 m.

10. **Ludwigia palustris** (L.) Ell., Sketch Fl. S. Carolina & Georgia **1**: 211 (1817).—
Burtt Davy, F.P.F.T. **1**: 201 (1926).—A. & R. Fernandes in Garcia de Orta **7**: 490 (1959); in C.F.A. **4**: 195 (1970). Type presumably from Europe.
Isnardia palustris L., Sp. Pl. **1**: 120 (1753). Type as above.

Entirely glabrous herb, up to 0·5 m. long or perhaps sometimes longer, creeping and rooting at the nodes, with opposite leaves, the stems at most ascending-decumbent, well-branched and forming mats. Leaves 0·7–4·5 × 0·4–2·3 cm., broadly elliptical or subovate, broadly cuneate and abruptly narrowed to a broadly winged petiole, the apex subacute; main veins 4–8 on each side of midrib. Flowers axillary and usually paired. Bracteoles lacking or minute, up to 1 mm. long. Sepals 4, 1·4–2 × 0·8–1·8 mm., deltate-acute. Petals 0. Stamens 4, green; filaments 0·5–0·6 mm. long; anthers 0·4–0·6 mm. across, 0·2–0·4 mm. high, shedding pollen directly on the stigma at anthesis. Pollen grains shed singly. Disk elevated c. 0·3 mm., glabrous, bright green. Style 0·5–0·7 mm. long, pale green; stigma 0·25–0·4 mm. thick, globose. Capsule (2)2·5–5 × 2–3 mm., dull light brown, elongate-globose, obscurely 4-angled, smooth and somewhat corky-walled, but fairly readily and irregularly loculicidal, with a broad green band 0·4–0·5 mm. wide on each of the angles of the capsule terminating at or well below the summit. Seeds 0·6–0·9 × c. 0·3 mm., pluriseriate in each locule of the capsule, free, light-brown, elongate-ovoid; raphe very narrow. Chromosome number $n = 8$.

Botswana. N: Toten, fl. & fr. 18.iii.1965, *Wild & Drummond* 7150 (SRGH). **Zambia.** B: Shangombo, edge of Mashi R., fl. 8.viii.1952, *Codd* 7441 (BM; K; PRE; SRGH). **Rhodesia.** Salisbury, fl. 20.ix.1952, *Wild* 3861 (K; LISC; MO; SRGH).
Widespread from temperate N. America to northern S. America, in temperate Europe

and N. Africa, including Socotra, and in temperate southern Africa from southern Angola, Botswana, Zambia and the vicinity of Salisbury S. to the Cape; also introduced into Hawaii, New Zealand and Australia. Along the margins of lakes, along streams and in wet places up to 1300 m.

Rare and widely scattered, and possibly introduced into the F.Z. area.

3. EPILOBIUM L.

Epilobium L., Sp. Pl. **1**: 347 (1753); Gen. Pl. ed. **5**: 164 (1754).—Raven in Bothalia **9**: 309–333 (1967).

Perennial herbs, often flowering the first year, rarely annual. Leaves opposite, usually alternate above, toothed or entire. Stipules absent. Flowers in a terminal inflorescence (in F.Z. area) or borne in the axils of the leaves. Bracteoles lacking. Floral tube present (in F.Z. area) or nil. Sepals 4, not persistent. Petals 4, rose-purple or white (yellow in 1 N. American species), emarginate. Stamens 8; anthers versatile. Pollen shed in tetrads, singly in a few non-African species. Stigma clavate, globose, or deeply 4-lobed. Ovary 4-locular; ovules pluriseriate in each locule, the seeds free. Dehiscence of capsule loculicidal. Seeds with a prominent tuft of hairs, the coma.

Approximately 200 species, well represented at relatively high latitudes and elevations, and occurring on all continents except Antarctica. There are 10 native species in Africa.

Stigma deeply 4-cleft:
  Stems clothed with spreading pubescence; leaves subsessile, distinctly clasping at the
    base - - - - - - - - - - - 1. *hirsutum*
  Stems clothed with appressed pubescence; leaves petiolate, usually rounded to obtuse
    at the base - - - - - - - - - 2. *capense*
Stigma entire - - - - - - - - - - - 3. *salignum*

1. **Epilobium hirsutum** L., Sp. Pl. **1**: 347 (1753).—Burtt Davy, F.P.F.T. **1**: 201 (1926).—R. & A. Fernandes in C.F.A. **4**: 200 (1970). Type from Europe.
  Epilobium hirsutum var. *villosissimum* Koch, Syn. Fl. Germ. Helv. ed. **1**: 240 (1835) " villosissima ".—Brenan in F.T.E.A., Onagraceae: 2 (1953).—A. & R. Fernandes in Garcia de Orta **7**: 493 (1959). Type presumably from N. Central Europe.

Robust herb, 0·2–2·5 m. tall, the subligneous stems sometimes persistent; rhizome stout, producing thick white underground runners with very scattered cataphylls; plants more or less white-pubescent, densely covered, especially in the inflorescence, with long spreading trichomes. Leaves 2–12 × 0·5–4 cm., mostly opposite, alternate above, oblong-lanceolate, densely hairy, acute, sessile, clasping at the base, coarsely toothed; rosette leaves longer and less hairy than later leaves; young leaves to 20 cm. long, more glabrous; petiole up to 5 cm. Inflorescence with an admixture of glandular trichomes, erect in bud; flowers erect in bud. Floral tube 2·5–3 mm. across, c. 1–1·5 mm. deep. Sepals 6–10 × 2–2·5 mm., apiculate. Petals 6–16 × 6–15 mm., deeply notched, bright purplish-rose. Anthers 1·5–2 mm. long; the filaments of the longer 3·5–6 mm. long, of the shorter 2–3 mm. long. Styles 6–10 mm. long; stigma deeply 4-lobed, held above the anthers at anthesis, lobes 1·5–2·5 mm. long. Capsules 3–8 mm. long, densely villous, on pedicels 2–12 mm. long; seeds 0·9–1·15 mm. long, dark brown or even coppery, oblong-obovoid, acute at the base, coarsely papillose, the coma 5–7 mm., dull white. Chromosome number $n = 18$.

Rhodesia. N: Mazoe, 1370 m., fl. & fr. ii.1907, *Eyles* 524 (BM; BOL; K; PRE; SAM; SRGH). C: Mann's Farm, Salisbury Distr., fl. 8.i.1950, *Greatrex* (K; LISC; PRE; SRGH). E: near Cheshire, Inyanga, 1300 m., fl. & fr. 3.ii.1931, *Norlindh & Weimarck* 4795 (BM; BR; PRE).
N. Africa, including the Canaries and Cape Verde Is., and S. through E. Africa to S. Africa (Cape Province). From 0–2600 m.

2. **Epilobium capense** Buch. ex Hochst. in Flora **27**: 425 (1844).—Binns, H.C.L.M.: 70 (1968). TAB. **84**. Type from S. Africa.
  Epilobium flavescens E. Mey. ex Harv. in Harv. & Sond., F.C. **2**: 506 (1862).—Burtt Davy, F.P.F.T.: 201 (1926). Type from S. Africa.
  Epilobium bojeri Hausskn. in Öst. Bot. Zeitschr. **29**: 90 (1879).—Perrier in Fl. Madag., Oenotheracées: 3 (1950). Type from Madagascar.
  Epilobium sp. 2—Brenan in F.T.E.A., Onagraceae: 4 (1953).

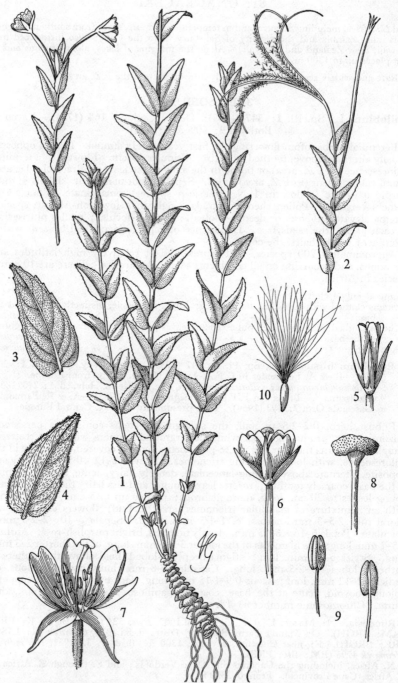

Tab. 84, EPILOBIUM CAPENSE. 1, habit (×½); 2, branch with mature fruit (×½); 3, leaf, superior surface (×1); 4, leaf, inferior surface (×1); 5, bud (×2); 6, flower (×1½); 7, flower with petal removed to show style and stamens (×2); 8, part of style with stigma (×6); 9, stamen, 2 aspects (×4); 10, seed (×4), 1, 3, 4 from *Robson 245*, 2 & 10 from *Stohr N44*, 5–9 from *Chase 5883*.

Perennial herb, 0·1–1·2 m. tall; underground stems vertical or nearly so, densely invested with thick white fleshy rounded scales c. 4 × 2–10 mm. after first year of growth, the new shoots scaly and arising from this region; plants strigulose with some glandular trichomes in the more densely pubescent inflorescence; stems with weakly marked elevated lines decurrent from the margins of the petioles. Leaves 2–5 × 0·4–2·5 cm., opposite near the base, alternate above, the margins and veins and sometimes the entire surface finely strigulose, narrowly ovate to narrowly lanceolate, mucronate to long-acuminate at the apex, rounded to obtuse or more rarely subcordate at the base, weakly or more often coarsely serrate with prominent forward-directed teeth, especially in the upper ½; petioles 1–2·5 mm. long, short but distinct. Inflorescence densely strigose with an admixture of glandular trichomes, erect in bud; flowers erect or somewhat drooping in bud. Floral tube 2–2·5 mm. across, 1·1–1·5 mm. deep, usually long-ciliate at the mouth with trichomes mostly 0·6–0·8 mm. long. Sepals 4·2–10 × 1·2–2·5 mm., narrowly oblong, acuminate or apiculate. Petals obovate, 6–16 × 3–10·5 mm., bright rose-purple, paler purplish, creamy or white, especially in smaller flowered forms, the notch c. ⅛ the length of the petal. Anthers white, 1·3–2·5 × c. 0·8–1 mm.; filaments pale rose to white, those of the longer stamens 2·5–8 mm. long, of the shorter ones 1·5–5 mm.; pollen yellow. Styles 5–15 mm. long, pale rose to white, glabrous; stigma 4-lobed usually deeply so, white, the lobes 0·7–2 mm. long, papillose and receptive within, held far above the anthers in larger-flowered plants but reached by them in smaller-flowered ones. Capsules 3–9 cm. long, densely strigulose with an admixture of glandular trichomes, erect, on a pedicel 1–6 cm. long; seeds 1·3–1·6 mm. long, brown, the coma c. 5–7 mm., dingy white. Chromosome number $n = 18$.

Zambia. E: Nyika Plateau, 2100 m., fl. 27.xi.1955, *Lees* 102 (K). **Rhodesia.** E: Umtali Distr., Engwa, 1980 m., fl. 2.ii.1955, *E.M. & W.* 114 (BM; LISC; SRGH). **Malawi.** N: Nyika Plateau, 1350 m., fl. 10.x.1947, *Benson* 1382 (SRGH). C: Dedza Mt. slopes, fl. 22.viii.1956, *Banda* 285 (BM; SRGH). S: Ncheu Distr., Lower Kirk Range, Chipusiri, 1460 m., fl. 17.iii.1955, *E. M. & W.* 965 (BM; LISC; SRGH).

South-western Tanzania S. along the W. side of L. Malawi to the eastern Cape Province, whence W. to the vicinity of Cape Town; mountains of central Madagascar, 900–2100 m.

3. **Epilobium salignum** Hausskn. in Öst. Bot. Zeitschr. **29**: 90 (1879).—Perrier in Fl. Madag., Oenotheracées: 4 (1950).—Brenan in F.T.E.A., Onagraceae: 5 (1953); in F.W.T.A. ed. 2, **1**: 166 (1954); in Mem. N.Y. Bot. Gard. **8**: 441 (1954).— A. & R. Fernandes in Garcia de Orta **5**: 111 (1957); op. cit. **7**: 493 (1959).—Binns, H.C.L.M.: 70 (1968).—R. & A. Fernandes in C.F.A. **4**: 200 (1970). Type from Madagascar.
Epilobium neriophyllum Hausskn. in Abh. Naturw. Verein Bremen **7**: 19 (1880).— Burtt Davy, F.P.F.T.: 202 (1926).—Robyns & Tournay, Fl. Parc Nat. Alb. **1**: 682 (1948). Type from S. Africa.
Epilobium perrieri H. Lév., Rév. Geogr. Bot. **27**: 3 (1917).—Perrier, loc. cit. Type from Madagascar.

Perennial herb 0·2–1·6 m. tall, the stems sometimes persistent and subligneous; strongly rhizomatous, the rhizomes lacking scales, long-spreading and giving rise to new leafy shoots; plants evenly strigulose, sometimes sparsely so with faintly marked lines running down from the margins of the pedicels. Leaves 2·8 × 0·3–2 cm., mostly opposite or subopposite, alternate in the inflorescence, the margins and veins and sometimes the entire surface finely strigulose, very narrowly to narrowly elliptic, rarely almost lanceolate, acute, rarely acuminate, narrowly cuneate to attenuate at the base, very rarely rounded, weakly serrulate or rarely serrate towards the base; petiole 1–8 mm. long, distinct. Inflorescence to 30 cm. long, densely strigulose, erect in bud, the bracts usually not much reduced; flowers nodding in bud and when they first open, later erect. Floral tube 1·5–2·3 × 0·8–2 mm., its mouth glabrous or long-ciliate. Sepals 3·5–8·5 × 1·2–2·2 mm., apiculate. Petals 5–15 × 2–7 mm., narrowly obovate, at first white or cream, then rose-pink following pollination, the notch 0·2–2 mm. deep. Anthers 0·8–2 mm. long, brownish, fading rose; filaments white, those of the longer stamens 2·5–8 mm. long, those of the shorter ones 1·5–5 mm. Styles white, 4·2–10 mm. long; stigma 1·8–4 × 1·2–1·8 mm., white, usually clavate, rarely subcapitate, the longer anthers usually just reaching the base of the stigma. Capsules 3–7 mm. long, densely strigulose, erect, on a pedicel 0·8–4·5 cm. long;

seeds 1–1·35 × 0·35–0·6 mm., light brown or tan, oblong-obovoid, obtuse at the base, very minutely pitted, the coma c. 5–9 mm. long, copious, white. Chromosome number $n = 18$.

Zambia. N: Saisi R. Marsh, Mbala (Abercorn) Distr., 1500 m., fl. 27.ii.1957, *Richards* 8381 (K). W: Solwezi Distr., 1350 m., fl. & fr. iv.1960, *Robinson* 3461 (BR; K; M; SRGH). E: Lundazi Distr., Nyika Plateau, 2100 m., fl. 3.i.1959, *Richards* 10416 (K). S: Mochipapa, 8 km. E. of Choma, 1300 m., fl. & fr. 29.v.1955, *Robinson* 1279 (BR; K; SRGH). **Rhodesia.** N: Miami, fl. & fr. v.1926, *Rand* 141 (BM). W: Matopos, fr. vi.1919, *Rogers* 5257 (SRGH). C: Prince Edward Dam, Salisbury, fl. & fr. 28.ii.1937, *Eyles* 8943 (SRGH). E: Inyanga, 1950 m., fl. 29.i.1931, *Norlindh & Weimarck* 4710 (BM). **Malawi.** N: Kondowe to Karonga, 600–1800 m., fl. & fr. vii.1896, *Whyte* (K). S: Zomba Plateau, 1500 m., fl. & fr. 7.vi.1946, *Brass* 16314 (BM; BR; K; SRGH; US). **Mozambique.** N: Vila Cabral, fr. 23.v.1939, *Torre* 97 (COI; LISC).

From eastern Zaire and the vicinity of Mt. Elgon in Uganda S. in the mountains of E. Africa to the eastern Cape Prov., Cameroon, Angola, throughout Madagascar. Moist places, 500–3000 m.

82. TRAPACEAE
By J. P. M. Brenan

Aquatic floating herbs. Leaves alternate, floating, in rosettes, only present at the upper nodes; stipules small, scarious, cleft to base and thus apparently more than two per leaf; petiole spongy and more or less inflated about the middle; lamina rhombic to deltate. Adventitious roots (?) submerged, paired but not opposite, one from either side of petiole or leaf-scar, chlorophyllose and thus leaf-like, pinnatisect into many filiform segments; in those from upper nodes segments shorter or absent. Flowers solitary, from upper axils, pedunculate, hermaphrodite, regular. Sepals, petals and stamens 4, the latter perigynous. Petals white. Ovary half-inferior, bilocular; ovules pendulous, one per loculus. Fruit a one-seeded, top-shaped drupe; pericarp soon disappearing; endocarp very hard, variously 2–4-horned, the horns derived from the persistent sepals.

Clearly related to and probably derived from the Onagraceae, but so distinct in its morphology that it is better regarded as a separate family. There is only one genus. Miki (in Proc. Japan. Acad. **35**: 289–294 (1959)) considers *Trapa* (with the fossil genus *Hemitrapa*) to constitute a distinct family Hydrocaryaceae, derived from Lythraceae not Onagraceae. The evidence does not seem satisfactory and Trapaceae is the conserved name of the family.

There has been much controversy over the nature of the adventitious roots mentioned above. Thus some authors have considered them leaves, others roots. The most recent work is by Couillault in Bull. Soc. Bot. France **119**: 177–198 (1973), who summarises previous work and supports the view that they are roots. I am grateful to my colleague Miss M. Gregory, on the staff of the Jodrell Laboratory, Royal Botanic Gardens, Kew, for help on this problem.

1. TRAPA L.
Trapa L., Gen. Pl. ed. 5: 56 (1754).

Characters of the family.

Trapa natans L., Sp. Pl. **1**: 120 (1753).—Hegi, Ill. Fl. Mitteleuropa **5**, 2: 884 (1925).—Brenan in F.T.E.A. Trapac.: 1 (1953). TAB. **85**.* Type presumably from Europe.

Annual. Leaves with lamina often broader than long, 1·5–5 cm. × 1·5–8 cm., toothed in upper part, glabrous above, beneath varying from almost glabrous to densely hairy all over. Sepals about 4–7 cm. long. Petals obovate-oblong to obovate, about 8–16 mm. long. Fruit very variably 2–4-horned.

* The illustration depicts var. *africana* which differs from var. *bispinosa* by having slimmer horns on the fruit.

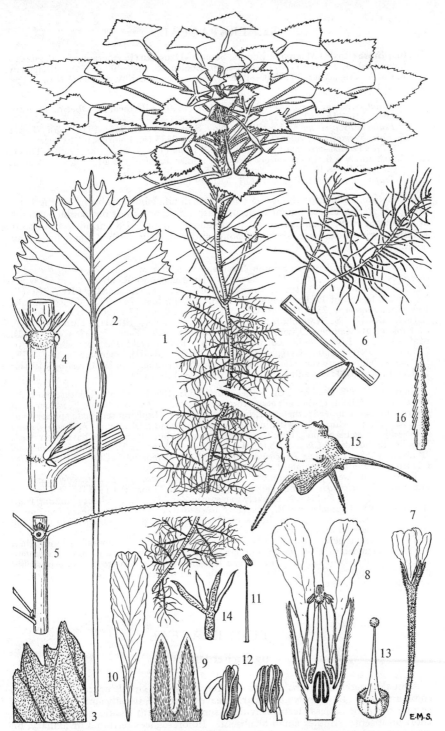

Tab. 85. TRAPA NATANS. 1, part of plant showing floating leaves and submerged adventitious roots ($\times\frac{1}{3}$); 2, floating leaf ($\times\frac{2}{3}$); 3, marginal part of leaf, underside, natural size; 4, nodes showing stipules ($\times 3$); 5, nodes showing first stage of adventitious roots ($\times 1\frac{1}{2}$); 6, nodes showing fully developed adventitious roots ($\times 1\frac{1}{2}$); 7, flower, natural size; 8, flower cut longitudinally ($\times 3$); 9, two sepals ($\times 3$); 10, petal ($\times 3$); 11, stamen ($\times 3$); 12, anther, 2 views ($\times 12$); 13, ovary and disk ($\times 3$); 14, ovary and calyx enlarging after anthesis, natural size; 15, endocarp of fruit, natural size; 16, apex of one of the horns of the fruit ($\times 3$). From F.T.E.A.

Var. **bispinosa** (Roxb.) Makino in Iinuma, Somoku-Dzusetzu (Iconography ... Plants ... Japan), ed. 3, **1**: 137 (1907).—Dubois in Zooléo **26**: 399–400 (1954).—Brenan in F.T.E.A. Trapaceae: 3 (1953).—R. & A. Fernandes in C.F.A. **4**: 203 (1970). Type from India.

Trapa bispinosa Roxb., Fl. Ind. **1**: 449 (1820).—Bremek. & Oberm. in Ann. Transv. Mus. **16**: 427 (1935).
Trapa austroafricana V. Vassiljev in Nov. Syst. Pl. Vasc. **1965**: 192 (1965). Type: Zambia, Western Province, Chingola, *Fanshawe* 2535 (K, holotype; SRGH).
" *Trapa austro-africana* var. *vassii* Auct." Binns, H.C.L.M.: 100 (1968). This is almost surely a fictitious variety based on a misreading of the author's abbreviation " V. Vassil.", probably from *Jackson* 2115 from Malawi (K).
Trapa natans sensu Binns, H.C.L.M.: 99 (1968).

Pyrene (i.e. endocarp) about 3–5 cm. across in all, 2-horned, horns arising from the upper angles, erecto-patent to arcuate-ascending or almost horizontal, straight or somewhat curved, conical, or attenuate above, about 1–1·8 cm. long and about 4–7 mm. wide near base, sharp at point and reflexedly barbed for a little way below it.

Botswana. N: Okavango Swamp, Txatxanika Lagoon, st. 1.iii.1972, *Gibbs Russell & Biegel* 1491 (K; PRE; SRGH). **Zambia.** B: Luanginga R. about 6 km. NW. of Sandaula Pontoon, st. 17.xi.1959, *Drummond & Cookson* 6585 (K; LISU; PRE; SRGH). N: Bwali, lagoon of L. Bangweulu, st. 11.xi.1969, *Verboom* 2641 (K). W: Chingola, 18.x.1955, *Fanshawe* 2535 (K; SRGH). E: Lundazi, st. 17.x.1967, *Mutimushi* 2285 (LISC; SRGH). S: Katombora, st. 5–11.xi.1949, *West* 3048 (SRGH). **Rhodesia.** W: S. bank of R. Zambesi, Livingstone District, 17.i.1929, *Young in Moss* 482 (BM). **Malawi.** N: Livingstonia, *Laws* 5 (K). S: Port Herald District, Chiromo, fr. 22.iii.1960, *Phipps* 2602 (COI; K; LMA; PRE; SRGH). **Mozambique.** MS: R. Lucite, Mufo, between Matarara do Lucite and Dombe, 14.x.1953, *Pedro* 4310 (K; LMA; PRE). LM: Incanine, 14.i.1898, *Schlechter* 12030 (BM; K).

Gams, in Veröff. Geobot. Inst. Rübel, Zürich **33**: 108–115 (1958), has discussed the two divergent schools of thought about the taxonomy of *Trapa*: one taking *T. natans* L. in a very wide sense indeed; the other, mainly Russian, multiplying segregate species to at least 25, a trend which Gams deplores as unnecessary and misleading. *T. austroafricana* V. Vassiljev was based on a type from the Flora area, but other names have been used: Vassiljev has identified *Chase* 5088 (COI; LISC) as *T. insperata* V. Vassiljev; *Drummond & Cookson* 6585 (LISC) as *T. acicularis* V. Vassiljev; *Phipps* 2602 (COI; LISC) & *Kirk* (K) from Mozambique, Luabu R. as *T. congolensis* V. Vassiljev.
These species were poorly defined. Fruits of *T. insperata* and *T. austroafricana* were unknown. *T. acicularis* is what I defined (in F.T.E.A. Trapac.: 1 (1953)) as *T. natans* var. *africana* Brenan and is very unlikely to occur in Zambia. An excellent gathering from Botswana, N: Thamalakane R., *Biegel & Russell* 3714 (SRGH) appears to show a complete transition between the leaves of *T. congolensis* and *T. austroafricana* (and the latter was separated on its leaves).
To sum up, the views of Gams seem wholly justified. There may possibly be more than one taxon included in *T. natans* in southern Africa, but the evidence is quite inadequate and at present unconvincing. The alleged leaf-differences appear to be more probably phenotypic, due to differences in the aquatic environment well-known to affect the vegetative parts of aquatic plants.

83. TURNERACEAE
By R. Fernandes

Annual or perennial herbs, shrubs or rarely trees with simple or stellate hairs and sometimes setae, frequently glandular. Leaves alternate, usually simple, subentire, serrate to more or less deeply divided, sometimes with sessile glands; petiole short or absent; stipules small or absent. Inflorescences terminal or axillary, of racemes, panicles or cymes, or flowers solitary and axillary. Flowers ☿, actinomorphic, sometimes heterostylous. Calyx usually with a short to long tube and 5 lobes, imbricate in aestivation. Petals 5, usually brightly coloured, contorted in aestivation, adnate to the calyx-tube forming a short to fairly long hypanthium, sometimes with a ligule at the base. Stamens with the filaments more or less adnate to the calyx-tube; anthers 2-celled, dorsifixed, introrse, dehiscing longitudinally. Ovary superior, 1-locular and usually several-ovulate, rarely 1-ovulate; ovules anatropous. Styles 3, with fimbriate or laciniate stigmas. Fruit a 3-valved capsule. Seeds 1 to several with a unilateral aril; testa hard, reticulate-alveolate or longitudinally striate-sulcate.

A tropical and subtropical family of c. 100 species mainly from America, with 7 genera in Africa.

Flowers solitary, axillary:
　　Annual herb; flowers small (up to 4·5 mm. long), nearly concealed in the axils of the leaves; leaves aggregated at the ends of branches; capsules up to 2·5 mm. long, smooth, pendulous; seeds curved　-　-　-　-　-　3. **Hyalocalyx**
　　Shrub; flowers larger, not as above; leaves scattered along the branches; capsule larger (5–6 mm. long), tubercled in F.Z. species, erect; seeds straight or slightly curved　-　-　-　-　-　-　-　-　-　-　-　**1. Piriqueta**
Flowers in racemes or panicles:
　　Shrub; capsule 1-seeded; seed straight, longitudinally striate-sulcate, submuricate
　　　　　　　　　　　　　　　　　　　　　　　　　　　　2. Stapfiella
　　Annual or perennial herbs; capsule many-seeded; seeds straight or curved, regularly reticulate-alveolate:
　　　　Capsule linear (not less than 3·5 times as long as wide), usually constricted between the seeds; seeds straight; petals with a basal ligule　-　**4. Tricliceras**
　　　　Capsule ellipsoid, not constricted between the seeds; seeds curved; petals eligulate
　　　　　　　　　　　　　　　　　　　　　　　　　　　　5. Streptopetalum

1. PIRIQUETA Aubl.

Piriqueta Aubl., Hist. Pl. Guiane Fr. **1**: 298, t. 117 (1775).

Annual or perennial herbs, shrubs or trees with simple or stellate hairs. Leaves with some impressed submarginal glands towards the base, sometimes glands also at the apex of the marginal teeth; stipules caducous or absent. Flowers usually solitary and axillary, with the pedicel articulate at the apex of a 2–4-bracteolate free peduncle. Sepals connate below the middle or free or nearly so. Petals shortly clawed, inserted at the throat of the calyx-tube; corona of 10 free or connate, fimbriate or laciniate scales at the base of the petals. Stamens with the filaments usually glabrous, inserted nearly at the base of the calyx; anthers rarely apiculate or mucronate. Ovary many-ovulate; styles with usually penicillate stigmas. Capsule globose to elliptic, smooth or tubercled, 3-valved. Seeds pyriform, straight or curved, with hard, reticulate-alveolate testa; aril crenate, lacerate or entire; cotyledons shorter than the radicle.

A mainly American genus with 1 species in the F.Z. area.

Piriqueta capensis (Harv.) Urb. in Jahrb. Königl. Bot. Gart. Berl. **2**: 78 (1883).—Gilg in Engl. & Prantl, Pflanzenfam. **3**, 6a: 62 (1893); op.cit. ed. 2, **21**: 464 (1925).—Burtt Davy, F.P.F.T. **1**: 119 (1926).—Garcia in Est. Ens. Docum. Junta Invest. Ultramar **12**: 161 (1954).—A. & R. Fernandes in Bol. Soc. Brot., Sér. 2, **35**: 160, t. 7 (1961); in Mem. Junta Invest. Ultramar, Sér. Bot. 2, **34**: 22 (1962).—J. H. Ross, Fl. Natal: 251 (1972). TAB. **86**. Type from S. Africa (Transvaal).
　　Turnera capensis Harv. in Harv. & Sond., F.C. **2**: 599 (1862); Thes. Cap. **2**: 25, t. 140 (1863). Type as above.

An undershrub 30-100 cm. high, or sometimes a perennial herb. Stems patently pilose, the hairs frequently short and somewhat rigid; old stems without leaves, often purplish, young ones usually densely foliate. Leaves 1–4·5 × 0·5–1·5 cm., elliptical to oblong or oblong-ovate to rarely obovate, obtuse or the narrowest ones acute at the apex, remotely serrate, frequently cuneate at the base, somewhat thick, subconcolorous, hispidulous or with subappressed hairs on both surfaces, with 1–3(4) suborbicular, submarginal impressed glands on each side near the base; lateral nerves obsolete above, slender and fairly prominent beneath, the midrib prominent on both surfaces; petiole 2–4 mm. or 0. Peduncle 4–20(25) mm. long, hispidulous; pedicel (1·25)5–10 mm. long, also hispidulous. Flowers usually heterostylous. Sepals 7–13 mm. long, lanceolate, acute or attenuate-cuspidate, hispidulous or with subappressed hairs, connate for 0·5–0·8 mm. at the base. Petals 6–12 × 2–6 mm., oblong-ovate, rounded at the top, cuneate at the base, golden-yellow. Longistylous flowers: filaments 3–4(4·5) mm. long; anthers 1–1·5 mm. long, oblong; styles (3·5)4–5·5(6) mm. long. Brevistylous

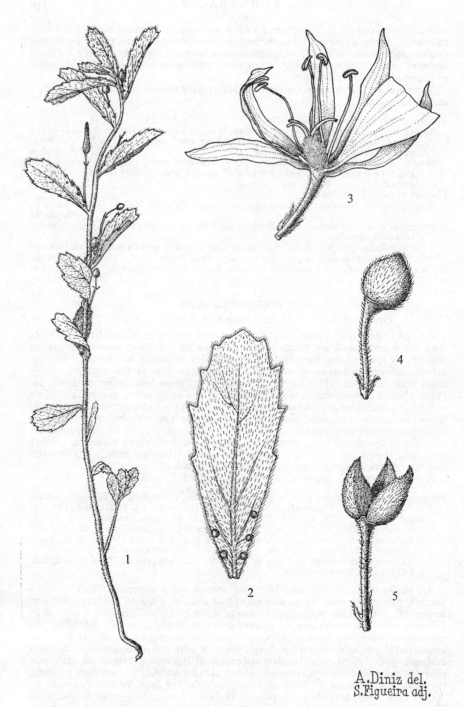

A.Diniz del.
S.Figueira adj.

Tab. 86. PIRIQUETA CAPENSIS. 1, habit (× ⅔); 2, leaf (×2); 3, flower, partly dissected
(×4); 4, fruit (×4); 5, open capsule (×4), all from *Mendonça* sn.

flowers: filaments (6)6·5–8 mm. long; anthers 0·75–1·5 mm. long; styles 2 mm. long. Capsule 4–7 mm. long, ovoid, hairy, brownish-green. Seeds 3–3·5 mm. long, obovoid-oblong, slightly curved, roundish at the top, reticulate-alveolate; aril whitish, equalling c. ⅔ of the seed, subentire or crenulate.

Rhodesia. S: Beitbridge, c. 8 km. N. of Shashi-Limpopo confluence, fl. (longistylous) & fr. 22.iii.1959, *Drummond* 5927 (EA; LISC; PRE; SRGH). **Mozambique.** SS: Gaza, Caniçado, between Massingir and Transvaal frontier, fl. (homostylous) 13.xi.1970, *Correia* 1940 (LMU). LM: Maputo, Goba near R. Maiuana, fl. (longistylous and brevistylous) & fr. 2.xi.1960, *Balsinhas* 157 (COI; LISC; LMA; PRE).

Also in S. Africa (Transvaal and Natal). In open forests, bushland and stony savannas.
The majority of plants from Mozambique differ from the Rhodesian ones as follows: the leaf-indumentum hairs are longer and softer and subappressed, not patent; the leaves are sessile, not usually attenuate into a short petiole; the teeth of the leaf margin are larger, more acute and more closely spaced, and the glands are nearer the leaf margin. Among the 10 Mozambican gatherings, 6 comprise only longistylous plants, 2 are mixtures of brevistylous and longistylous plants, 1 includes only brevistylous plants and the Caniçado specimen is homostylous.

2. STAPFIELLA Gilg

Stapfiella Gilg in Mildbr., Wiss. Ergebn. Deutsch. Zentr.-Afr. Exped. 1907–1908, 2: 571 ut Flacourtiaceae (1913).—J. Lewis in Kew Bull. 8: 282 (1953); in F.T.E.A., Turneraceae: 4 (1954).

Shrubs. Leaves petiolate, usually elliptic and serrate at the margin, frequently glandular-punctate; stipules 0. Flowers in terminal and axillary panicles or racemes. Calyx connate below the middle, with 5 tubercles within. Petals eligulate, white, whitish or nearly translucent, inserted at the mouth of the calyx-tube. Filaments filiform, inserted at the mouth of the calyx-tube. Ovary 1-locular, 1-ovulate, the ovule basal; stigmas fringed. Capsule ellipsoid or clavate, shortly rostrate, dehiscing septicidally into 3 valves from the apex downwards. Seed obovoid, straight, longitudinally striate-reticulate.
An endemic genus of tropical Africa with 6 species.

Stapfiella zambesiensis R. Fernandes in Bol. Soc. Brot., Sér. 2, 49: 13, t. 1 (1975). TAB. 87. Type: Zambia, 8 km. E. of Kasama, *Robinson* 4720 (COI; K, holotype; SRGH).

A very much branched undershrub or shrub up to 2 m. high. Old branches black or rarely somewhat rusty-black, glabrescent, the young ones short, slender, greenish or yellowish-green and rather pubescent with yellowish, straight or curved, antrorse hairs, densely leafy. Leaves 1·5–10 × 0·5–2·5 cm., elliptic, ± attenuate to the acute or subacute apex and to the base, slightly crenate-serrate at the margin, the teeth mucronulate (but not hooked), chartaceous, not shining, pubescent on both surfaces, more densely so along the midrib and nerves, punctate beneath with black glands; petiole slender, up to 2 cm. long. Raceme up to 7-flowered, rarely 1-flowered, pubescent, without tubercle-based hairs; pedicels 4–5 mm. long, very slender, hispidulous or both hispidulous and pubescent. Buds subspherical, pubescent and sometimes also hispidulous (hairs short, very slender). Calyx-lobes 2·25–2·5 mm. long, oblong, with hyaline margin, black-punctate on the back. Petals 3–5·5 mm. long, obtuse, white. Ovary densely pubescent; styles ± 0·6–0·75 mm. long. Capsule 2·5–3 mm. long, ovoid-ellipsoid, pale green, puberulous, without black glands, sometimes pendulous; valves c. 1·5 mm. broad, subelliptic; fructiferous peduncle c. 5 mm. long, sometimes deflexed. Seed 2·25–2·75 mm. long, yellowish-brown; aril less than half the length of the seed.

Leaves up to 2 × 0·7 cm., not or slightly attenuate to both extremities; petiole up to 0·25 cm. long; young branches very densely leafy with very short internodes
forma *zambesiensis*
Leaves 1·5–10 × 0·5–2·5 cm., rather attenuate to both extremities and usually with the apex very acute; petiole up to 2 cm. long; young branches not very densely leafy, with relatively long internodes - - - - - - forma *grandifolia*

Forma zambesiensis

Zambia. N: 8 km. E. of Kasama, fl. & fr. 4.xii.1961,*Robinson*4720 (COI; K; SRGH). Known only from Zambia. In thickets and margin of mushitus near streams.

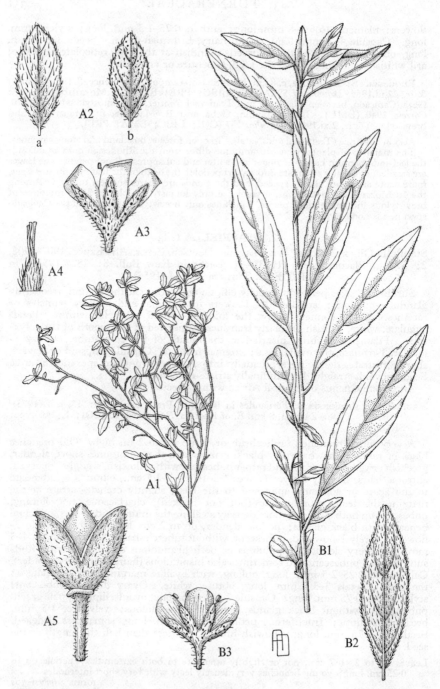

Tab. 87. A.—STAPFIELLA ZAMBESIENSIS forma ZAMBESIENSIS. A1, habit (×½); A2, leaf,
(a) superior, (b) inferior surface (×2); A3, flower (×8); A4, part of ovary with
styles (×15); A5, dehisced capsule (×8), all from *Mutimushi* 431. B.—STAPFIELLA
ZAMBESIENSIS forma GRANDIFOLIA. B1, branch (×½); B2, leaf (×2); B3, flower
(×8), all from *Fanshawe* 3581.

Forma **grandifolia** R. Fernandes in Bol. Soc. Brot., Sér. 2, **49**: 14, t. 2 (1975). Type: Zambia, Chilongowelo, *Richards* 49 (K, holotype).

Zambia. N: Kawambwa, 23.viii.1957, *Fanshawe* 3581 (K). Known only from Zambia. In mushitus near streams and margins of swamp forest.

Perhaps an ecological form, conditioned by wet soil and shade.

3. HYALOCALYX Rolfe

Hyalocalyx Rolfe in Journ. Linn. Soc., Bot. **21**: 257 (1884).

Annual herbs. Leaves exstipulate, petiolate. Flowers solitary in the axils of the upper very closely arranged leaves, heterostylous. Sepals hyaline-membranous, connate in the lower half. Petals and stamens adnate to the base of the calyx-tube; filaments linear, flattened below; anthers cordate-ovate. Ovary ovoid, with 3 parietal placentas; styles 3, filiform, straight, with laciniate-flabellate stigmas. Capsule 3-valved, opening loculicidally from the apex. Seeds oblong-obovoid, curved, with rigid, reticulate-alveolate testa; aril entire; albumen abundant, oily.

A monotypic genus from Madagascar, SE. and E. Africa.

Hyalocalyx setifer Rolfe in Journ. Linn. Soc., Bot. **21**: 258, t. 7 (1884) " *setiferus* ".—Vatke in Abh. Naturwiss. Ver. Bremen **9**: 116 (1887).—Gilg. in Engl. & Prantl, Pflanzenfam. **3**, 6a: 62 (1893); op. cit. ed. 2, **21**: 464 (1925).—Perrier in Arch. Bot. Caen **4**, 1: 7 (1930).—Mildbr. in Notizbl. Bot. Gart. Berl. **13**: 282 (1936) " setiferus ". —Perrier in Fl. Madag., Turneraceae: 4, fig. 1 (1950) " *setiferus* ".—Drummond & Meikle in Kew Bull. **5**: 335 (1951).—J. Lewis in F.T.E.A., Turneraceae: 7, fig. 3 (1954).—A. & R. Fernandes in Mem. Junta Invest. Ultramar, Sér. 2, **34**: 21 (1962). TAB. **88**. Type from Madagascar (Nossi-bé Isle, a longistylous plant).
Turnera setifera (Rolfe) Baill. in Bull. Soc. Linn. Paris **1**: 582 (1886). Type as above.
Hyalocalyx dalleizettei Capitaine in Bull. Herb. Boiss., Sér. 2, **8**: 252, t. 5 (1908). Type from Madagascar (near Tananarive, a brevistylous plant).

An annual herb covered all over with long, appressed or subappressed yellowish hairs denser on young parts especially on petioles and nerves of the leaves. Stem up to 27 cm. tall, erect, usually much-branched from the base, rarely simple; lower branches frequently as long as or longer than the main stem and decumbent or patent, the others patent or ascending, few-branched or unbranched, all terete, brownish to dark purplish. Leaves 1·5–4 × 0·5–1·2 cm., elliptic to elliptic-oblanceolate, attenuate-acute to subacute at the apex, cuneate at the base, entire in the proximal $\frac{1}{3}$–$\frac{1}{2}$, serrate distally, very scattered along the stem and branches but very closely arranged and forming subcapitate aggregates at their extremities; lateral nerves ascending, somewhat prominent above like the midrib; petiole 1–5 mm. long. Peduncle 1–2 mm. long and straight in flower, up to 5(6) mm. long and suddenly deflexed at the top in fruit, densely long-setose (the setae ascending) mainly towards the apex; bracteoles 2, below the flower, very small, opposite. Flowers narrowly cylindrical, concealed by the subtending leaves; calyx 2–2·5 mm. long; lobes oblong, rounded at the apex and with 1–3 subapical setae. Petals 4–6 mm. long, narrowly oblanceolate-spathulate, obtuse, pale yellow to orange, glabrous, falling off with the calyx and the stamens after pollination. Stamens with somewhat unequal filaments and short, cordate, pale yellow anthers. Ovary narrowly ovoid, subglabrous; styles of unequal length (shorter in the brevistylous flowers, longer in the longistylous ones). Capsule 2–2·5 × 1·5 mm., ellipsoid-cylindrical to subglobose, smooth, papillose at the apex, reflexed, with 3–10 seeds. Seeds 1–1·5 × 0·7 mm., slightly curved, at first pale, then dark brown, later nearly black; aril cordate, thin, subhyaline, c. half as long as the seed.

Mozambique. N: 15 km. from Mocímboa da Praia to R. Messalo, fl. & fr. 13.v.1959, *Gomes e Sousa* 4467 (BR; COI; K; LMA; PRE); Serra de Ribáuè (Mepáluè), c. 800 m., fr. 23.iii.1964, *Torre & Paiva* 11338 (LISC). Z: Lugela, Mnobede road, fr. 6.iv.1948, *Faulkner* 234 (BR; COI; K).
Also in Tanzania and Madagascar. In open woodlands (*Brachystegia* open forest), along roadsides and on cultivated land, mainly in dry places on sandy soils.

M

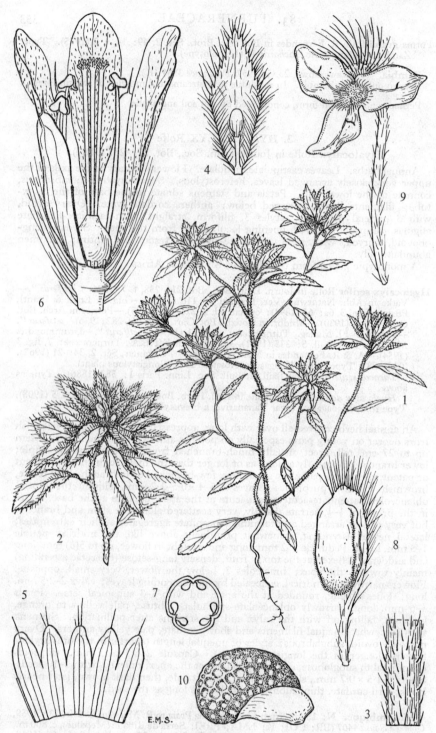

Tab. 88. HYALOCALYX SETIFER. 1, habit (× ⅔); 2, branch-apex (× 1½); 3, segment of stem (× 8); 4, bract and bud (× 4); 5, opened calyx (× 8); 6, partly dissected flower (× 16); 7, diagrammatic transverse section of ovary; 8, capsule (× 8); 9, dehisced capsule (× 8); 10, seed (× 32), all from *Faulkner 234*. From F.T.E.A.

4. TRICLICERAS Thonn. ex DC.

Tricliceras Thonn. ex DC., Pl. Rar. Jard. Genève: 56 (1826).
Wormskioldia Thonn. in Schumach. & Thonn., Beskr. Guin. Pl.: 165 (1827).

Annual or perennial, caulescent or acaulescent herbs, puberulous to hirsute and often also setose. Leaves sessile or shortly petiolate, not or ± divided, usually exstipulate. Inflorescences scapose or axillary, of 1–many-flowered racemes; bracteoles single or paired. Flowers erect, shortly pedicellate, homostylous or heterostylous. Calyx-tube cylindrical, hairy within near the base and usually with 5 tubercles above the insertion of the stamens. Petals pale yellow to scarlet or brick-red, rarely white, inserted below the throat of the calyx-tube, spathulate to obovate, cuneate and clawed, with a small ligule opposite to the insertion. Stamens with subhypogynous filaments, usually 3 longer and 2 shorter; anthers oblong, dorsifixed. Ovary 1-locular, many-ovulate, with parietal placentas; styles 3 with entire or lobed stigmas. Capsule usually linear or narrowly ellipsoid, siliquiform, not or ± constricted between the seeds, rostrate, smooth, with irregular loculicidal dehiscence. Seeds numerous, straight, reticulate-foveolate, with 2-pored pits.

A tropical and S. African genus with 14 species.

Perennials; petals usually not less than 25 mm. long:
 Stems, if present, without setae; leaves not lobed:
  Pedicels deflexed in fruit; capsule pendulous or spreading, up to 3 cm. long, fusiform-trigonous; seeds c. 4 × 1·75 mm.; acaulescent or with short pubescent stem - - - - - - - - - - 1. *brevicaule*
  Pedicels not deflexed; capsule erect, up to 6·5 cm. long, cylindrical; seeds c. 6 × 2 mm.; caulescent with glabrous stem - - - 2. *mossambicense*
 Stems usually with setae; leaves usually more or less deeply lobed or divided at least towards the base:
  Ovary densely setose; capsule more or less setose - - - 5. *schinzii*
  Ovary and capsule glabrous or nearly so:
   Capsule pendulous or spreading; peduncle up to 38 cm. long, much exceeding the upper leaves; raceme with 3–14 flowers; stem with dark purple non-glandular setae - - - - - - - - 3. *longepedunculatum*
   Capsule erect; peduncle 4·5–16 cm. long, usually not exceeding the upper leaves; raceme with 1–6 flowers; stem with yellow glandular setae - 4. *hirsutum*
Annuals; petals up to 24 mm. long:
 Ovary and capsule more or less hairy; pedicels erect in fruit; setae of the stem cylindrical, swollen in the basal ⅓–½ - - - - - - 6. *glanduliferum*
 Ovary and capsule glabrous; pedicels deflexed in fruit; setae of the stem, if present, with or without a short, bulbous, not cylindrical base:
  Leaves irregularly 2-pinnatifid, with narrow acute segments - 7. *tanacetifolium*
  Leaves undivided or with a few lobes:
   Setae of the stem more than 0·5 mm. long (up to 1·5 mm.), not bulbous at the base; leaves with 2 distinct, reflexed auricles at the base - 11. *auriculatum*
   Setae of the stem not more than 0·5 mm. long, nearly reduced to the bulbous base, sometimes very sparse; leaves not as above:
    Stem patently hairy; leaves with a gland at the usually rounded apex
      10. *elatum*
    Stems appressed hairy; leaves acute, without apical gland:
     Racemes 1–4-flowered; leaves up to 4·5 cm. broad, usually irregularly lobed (lobes obtuse) in the proximal ⅓–½ - - - - 8. *lobatum*
     Racemes (4)5–9-flowered; leaves not more than 1·6 cm. broad, unlobed (sometimes with some acute teeth at the base) - - 9. *lanceolatum*

1. Tricliceras brevicaule (Urb.) R. Fernandes in Bol. Soc. Brot., Sér. 2, **49**: 16 (1975). Type from Zanzibar.
 Wormskioldia brevicaulis Urb. in Jahrb. Königl. Bot. Gart. Berl.**2**: 51 (1883).—Gilg in Engl. & Prantl, Pflanzenfam. **3**, 6a: 61 (1893); op. cit. ed. 2, **21**: 463 (1925).—Engl., Pflanzenw. Afr. **3**, 2: 593 (1921).—J. Lewis in Kew Bull. **8**: 285 (1953); in F.T.E.A., Turneraceae: 13, t. 4, fig. 1–6 (1954). Type as above.

A perennial herb with a long, thick, straight or contorted woody rootstock. Stems 1 to few, usually short to almost absent. Leaves crowded at the top of the stem, erect to spreading, somewhat discolorous (paler beneath), more or less densely scabrous on both surfaces, rarely subglabrous when young, finally glabrescent, sessile; midrib and lateral nerves slightly raised on the inferior surface with longer hairs than the areas between the veins. Flowers 2–7(9) per raceme, heterostylous; peduncles 2–14·5 cm. long, with white patent hairs and patent yellowish setae, the latter 0·5–1·5 mm. long and somewhat dilated towards the base; bracts 1–4 mm. long, subacuminate; pedicels 6–16(18) mm. long, patently hairy and setose. Calyx 12–18 mm. long, greenish, pubescent and shortly setose; lobes 5–8 mm. long, linear. Petals 25–40 × 10–18 mm., broadly obovate-cuneate, usually bright orange (somewhat paler on the abaxial surface), sometimes yellow. Stamens unequal, the 3 longer ones 14–16 mm., the other two 12–14 mm. long; anthers c. 1·75 mm. long. Styles of the longistylous flowers 14–16 mm. long, exceeding by c. 4 mm. the anthers of the longest stamens; styles of the brevistylous ones 8·5–10 mm. long, not reaching the anthers of the shorter stamens. Ovary 4–5 × 1–1·5 mm., ellipsoid, acute, puberulous or subglabrous. Fruiting pedicels recurved, accrescent. Capsule 1·5–3 × 0·3–0·5 cm., fusiform-trigonous, tapering to both extremities, sometimes slightly falcate, with smooth faces and raised ribbed angles, glabrous or puberulous, at a right angle or oblique to the peduncle; beak 2–4 mm. long. Seeds 3·5–4 × 1·75 mm., obovoid.

Usually caulescent; leaves not less and usually more than 4 times as long as broad, elliptic, acute, erect; inflorescence usually shorter than the subtending leaf
var. *brevicaule*
Acaulescent; adult leaves up to 4 times as long as broad, obovate to broadly elliptic, usually obtuse or rounded at the apex, for the most part prostrate, rosulate, the inner ones erect-patent; inflorescence frequently longer than the subtending leaf
var. *rosulatum*

Var. **brevicaule.**

Usually caulescent, the stem 3–10(35?) cm. long, pubescent but not setiferous. Leaves up to 22 × 4 cm., elliptic to narrowly elliptic, numerous, erect, crowded at the top of the stem (those below the terminal tuft rather smaller), acute or acuminate, bluntly serrate or 2-serrate at the margin. Peduncles 3–10 cm. long, (2)3(6)-flowered; pedicels 10–18 mm. long.

Zambia. Without precise locality, 11.xii.1954, *Richards* 3580 (BR). N: Mbala (Abercorn), ii.1935, *Gamwell* 228 (BM).
Also in Kenya and Tanzania (incl. Zanzibar). On sandy soil in open bushland and grassland, 1000–1200 m.

Var. **rosulatum** (Urb.) R. Fernandes in Bol. Soc. Brot., Sér. 2, **49**: 16 (1975). Type from Tanzania.
　　Wormskioldia rosulata Urb. in Notizbl. Bot. Gart. Berl. **4**: 173 (1905). Type as above.
　　Wormskioldia brevicaulis var. *rosulata* (Urb.) J. Lewis in Kew Bull. **8**: 285 (1953); in F.T.E.A., Turneraceae: 14, t. 4, fig. 7–8 (1954).—A. & R. Fernandes in Mem. Junta Invest. Ultramar, Sér. 2, **34**: 17 (1962).—Binns, H.C.L.M.: 101 (1958). Type as above.

Acaulescent. Leaves 4–13, up to 16·5 × 6 cm., rosulate, applied to the soil or erect-patent, usually obovate, sometimes oblong-obovate to elliptical or sub-orbicular, rounded at the apex or sometimes shortly acuminate and subacute, cuneate to contracted at the base, with 2-crenate margin. Peduncles 2–14·5 cm. long, with (3)5(11) flowers; pedicels 7–10 mm. long in flower, up to 15 mm. in fruit.

Zambia. N: Mbala (Abercorn) Distr., Kawe R. Gorge, 1200 m., fl. 14.ii.1959, *Richards* 10905 (SRGH). E: Lundazi, c. 6·5 km. W. of Chikomeni, Lukusuzi National Park, 800 m., fl. 3.xii.1970, *Sayer* 494 (SRGH). **Malawi.** S: Ncheu, Mphepo Zinai, 23.xi.1967, *Salubeni* 899 (LISC). **Mozambique.** N: Marrupa, 13.vi.1948, *Pedro & Pedrógão* 4296 (EA; LMA). Z: Gúruè, between Mutuáli and Lioma, c. 26 km. from Mutuáli, c. 650 m., 10.ii.1964, *Torre & Paiva* 10500 (LISC). MS: road between Dondo and Inhaminga, c. 10 km. N. of Muanza R., fl. 4.xii.1971, *Pope & Müller* 514 (SRGH).
Also in Kenya and Tanzania. In open forest, savannas, roadsides, in crevices of granitic rocks, on argillaceous and sandy soils.

2. **Tricliceras mossambicense** (A. & R. Fernandes) R. Fernandes in Bol. Soc. Brot., Sér. 2, **49**: 22 (1975). TAB. **89**. Type: Mozambique, Inhambane, between Quissico and Chiducoane, *Mendonça* 3308 (LISC, holotype).
　　Wormskioldia mossambicensis A. & R. Fernandes, op. cit. **35**: 158, t. 5 (1961); in Mem. Junta Invest. Ultramar, Sér. 2, **34**: 18 (1962).—J. H. Ross, Fl. Natal: 251 (1972). Type as above.

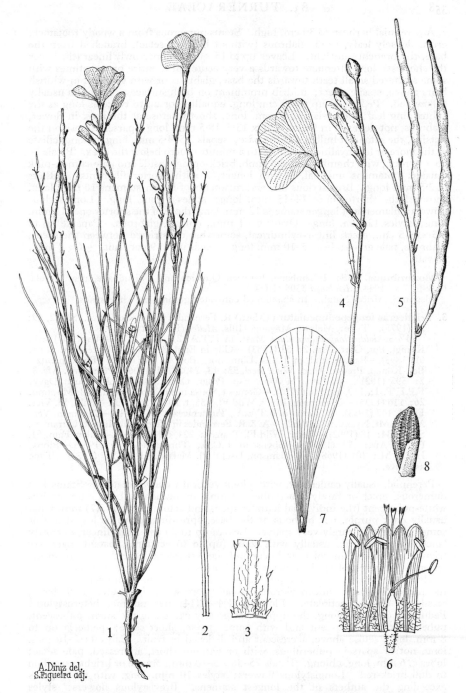

Tab. 89. TRICLICERAS MOSSAMBICENSE. 1, habit ($\times\frac{1}{2}$); 2, leaf ($\times\frac{1}{2}$); 3, base of leaf ($\times3$); 4, inflorescence ($\times1$); 5, fruits ($\times1$); 6, calyx opened showing stamens, ovary and styles ($\times2$); 7, petal with ligule ($\times2$); 8, seed with aril ($\times3$), all from *Mendonça* 3308.

A perennial herb up to 30 cm. high. Stems numerous from a woody rootstock, erect, densely leafy, rigid, glabrous (without hairs or setae), branched from the base, the branches upright. Leaves up to 15 × 0·2 cm., narrowly linear (the lower the shortest), long-attenuate towards a very acute apex, entire or sometimes with some scattered small teeth towards the base, glabrous or with sparse thin whitish hairs above, sessile, erect; midrib prominent on both surfaces. Racemes usually 4-flowered. Peduncles up to 19 cm. long, equalling or up to twice as long as the subtending leaf; pedicels c. 2·5 mm. long, shortly setose at the apex in flower, glabrous, not deflexed in fruit. Calyx 13·5–18·5 mm. long, sparsely setose on the outside, the setae not bulbous at the base; sepals 3·5–4·5 mm. long, oblong, ciliate at the more or less hyaline margin, with a minute mucro below the apex. Petals ± 35 × 10 mm., with rhombic-obovate limb, brick-orange inside and pale salmon-pink outside. Stamens unequal, 3 with longer, 2 with shorter filaments; anthers c. 2·5 mm. long. Brevistylous flowers: filaments of longer stamens 16 mm. long, those of the shorter ones 14–15 mm. long; styles 8 mm. long. Longistylous flowers: filaments of longer stamens 12 mm. long, those of the shorter ones 11 mm. long; styles 12 mm. long. Ovary 6 × 1 mm., oblong, glabrous. Capsule up to 6·5 × 0·35 cm., erect, linear-cylindrical, somewhat constricted between the seeds, glabrous, pale green; beak 5–10 mm. long. Seeds up to 9 per fruit, ± 6 × 2 mm., clavate.

Mozambique. SS: Inhambane, between Quissico and Chicudoane, near Zavala, fl. & fr. 7.xii.1944, *Mendonça* 3308 (LISC).
Also in S. Africa (Natal). In abandoned cultivated ground, on white sands.

3. **Tricliceras longepedunculatum** (Mast.) R. Fernandes in Bol. Soc. Brot., Sér. 2, **49**: 21 (1975). Type: Malawi, Maganja Hills, *Meller* (K, holotype).
      Wormskioldia longepedunculata Mast. in F.T.A. **2**: 502 (1871).—Urb. in Jahrb. Königl. Bot. Gart. Berl. **2**: 53 (1883).—Gilg in Engl. & Prantl, Pflanzenfam. **3**, 6a: 61 (1893).—Britten et al. in Trans. Linn. Soc., Bot. **4**: 14 (1894).—Engl. in Sitz.-Ber. Königl. Preuss. Akad. Wiss. Berl. **52**: 13, 24, 28, 31 (1906); Pflanzenw. Afr. **3**, 2: 593 (1921).—P. Lima in Asoc. Esp. Progr. Cienc.: 63 (1923).—Burtt Davy, F.P.F.T. **1**: 119, fig. 11a (1926).—Gomes e Sousa in Bol. Soc. Estud. Col. Moçamb. **26**: 116 (1935).—Martineau, Rhod. Wild Fl.: 55, t. 20 fig. 2 (1953).—Wild, Fl. Vict. Falls: 143 (1953).—J. Lewis in F.T.E.A., Turneraceae: 14 (1954).—Chapman, Veg. Mlanje Mt. Nyasal.: 33 (1962).—A. & R. Fernandes in Mem. Junta Invest. Ultramar., Sér. 2, **34**: 18 (1962).—Letty, Wild Fl. Transv.: 224, t. 111 (1962).—Riley, Fam. Fl. Pl. S. Afr.: 59 (1963).—Robyns in F.C.B., Turneraceae: 9 (1967).—Binns, H.C.L.M.: 101 (1968).—Williamson, Useful Pl. Malawi, ed. 2: 127 (1972). Type as above.

Perennial, usually caulescent, with a long vertical woody rootstock. Stems 1 to numerous, erect or rarely procumbent, simple or branched, both more or less white-pubescent (the individual hairs crisped) and setose, setae 1–5(7) mm. long, usually dark purple, not bulbous at the base, spreading, more or less dense or sometimes sparse, rarely very sparse. Leaves up to 20 cm. long, linear, narrowly lanceolate or elliptic, usually with some (up to 6) scattered, patent, narrowly triangular teeth or lanceolate to linear and acute segments near the base or in the lower half, the rest of the lamina undivided but serrate or serrulate at the margin, rarely all the lamina undivided, pubescent or puberulous to glabrescent, setose on the nerves beneath, subconcolorous, usually somewhat rigid, erect and dense, sessile to shortly petiolate. Flowers (1)4–12(14) per raceme, heterostylous. Peduncles 6–38 cm. long, the uppermost greatly exceeding the stem, pubescent, puberulous or glabrescent and with some sparse, short setae; pedicels up to 8 mm. long, setose above, accrescent and deflexed in fruit. Calyx 11·5–16 mm. long, not or sparsely puberulous, with or without short, appressed, pale setae; lobes c. 6 mm. long, oblong. Petals 25–38 × 5–10 mm., orange or bright vermilion to dull brick-red. Longistylous flowers: styles 10 mm. long, with the stigmas exceeding the anthers of the longest stamens. Brevistylous flowers: styles 3·5 mm. long. Ovary 4·5–5·5 × 1 mm., pubescent. Capsule 4–8·5 × 0·1–0·15 (0·2) cm., pendulous, with or without short, sparse setae. Seeds 2·5–3·3 × 1·2–1·7 mm., obovoid.

Usually not more than 60 cm. high; lateral segments of the leaves entire, not longer than 2(2·7) cm. and up to 2½–4½ times as long as the width of the rhachis; seeds 2·5–3(3·5) mm. long - - - - - - - - var. *longepedunculatum*

83. TURNERACEAE 359

Taller (the stem up to 1 m. high) and stronger; lateral segments of the leaves sometimes dentate, up to 4 cm. long, up to 6 times as long as the width of the rhachis, narrower and more acute than above; seeds up to 2 mm. long, a little narrower than above

var. *eratense*

Var. **longepedunculatum.**
Wormskioldia longepedunculata var. *integrifolia* Urb. in Engl., Bot. Jahrb. **15**: 160 (1893). Type: Malawi, near Blantyre, *Last* (B, holotype†; K).
Wormskioldia longepedunculata var. *bussei* Urb. in Notizbl. Bot. Gart. Berl. **4**: 174 (1905). Type from Tanzania.

Caprivi Strip. fl. & fr. 26.xii.1958, *Killick & Leistner* 3101 (PRE). **Botswana.** SE: near Mahalapye, fl. 21.xii.1957, *de Beer* 531 (BR; PRE; SRGH). **Zambia.** B: Kataba, fl. & fr. 13.x.1960, *Fanshawe* 5837 (BR; K; SRGH). N: Isoka Distr., c. 32 km. SE. of Tunduma, fl. & fr. 1.i.1959, *Richards* 10381 (EA; K; SRGH). C: Kapiri Mposhi-Kanona road, c. 140 km. from Kanona, fl. 5.vii.1960, *Richards* 12827 (K). E: Lundazi, Tigone Dam, c. 3·2 km. on Choma road, fl. 19.xi.1958, *Robson* 661 (K; LISC; SRGH). S: Namwala, Mala and Kabulamwando, fl. & fr. 17.xii.1962, *van Rensburg* 1110 (K; PRE; SRGH). **Rhodesia.** N: Lomagundi, Mangula, fl. juv. 3.xii.1961, *Jacobsen* 1543 (PRE). W: Matobo, Farm Besna Kobila, fl. xi.1954, *Miller* 2531 (LISC; PRE; SRGH). C: Marandellas, Dozmeri Farm, Macheke, ii.1964, *Strang* 2285 (SRGH). E: Inyanga, fl. 29.x.1930, *Fries, Norlindh & Weimarck* 2388 (BM; BR; LD). S: Ndanga Distr., c. 50 km. E. of Bikita, fl. 21.x.1930, *Fries, Norlindh & Weimarck* 2164 (LD). **Malawi.** N: Iponjoro to Matipa, fl. 19.xi.1952, *Chapman* 38 (BM). C: base of Mt. Dedza, fl. 16.x.1937, *Longfield* 48 (BM). S: Fort Johnston, fl. juv. 22.xii.1954, *Jackson* 1433 (K; LISC; SRGH). **Mozambique.** N: between Ribáuè and Malema, fl. 24.vii.1962, *Leach & Schelpe* 11443 (K; LISC; SRGH). Z: Namacurra on road to Vila Maganja da Costa, fl. & fr. 25.i.1966, *Torre & Correia* 14109 (LISC). T: Moatize, between Zóbuè and Metengo-Balame, fl. & fr. 10.i.1966, *Correia* 352 (LISC). MS: Manica, Dombe, near Escola de Matindiro, fl. & fr. 18.xi.1965, *Pereira & Marques* 717 (COI; LISC; LMU; SRGH).
Also in Tanzania and S. Africa (Transvaal). In *Brachystegia* woodland, in tree savannas, pastures, roadsides, cultivated land, abandoned ground, etc., on red and black sandy-argillaceous soils, stony and dry places, etc., 40–1700 m.

Specimens from Mozambique, Pebane (*Torre* 4682; *Torre & Correia* 15135, 17183) have thick, woody stems sometimes branching from the base, a dense indumentum of short, crisped or straight hairs interspersed with very short, sparse setae, and more slender peduncles than the type. Perhaps they should be considered at subspecific level.

Var. **eratense** R. Fernandes in Bol. Soc. Brot., Sér. 2, **49**: 21, t. 6 (1975). Type: Mozambique, Eráti, Estação Experimental de Namapa, *Lemos & Macuácua* 12 (BM; COI, holotype; K; LISC; PRE).

Mozambique. N: Eráti, at the margins of Namapa R., fl. & fr. 5.iv.1964, *Torre & Paiva* 11645 (LISC).
Known only from Mozambique.

4. **Tricliceras hirsutum** (A. & R. Fernandes) R. Fernandes in Bol. Soc. Brot., Sér. 2, **49**: 18, t. 3–5 (1976). Type: Mozambique, Manica e Sofala, Cheringoma, *Pedro & Pedrogão* 40 (LMA, holotype).
Wormskioldia schinzii Urb. var. *hirsuta* A. & R. Fernandes, op. cit. **35**: 160, t. 6 (1961); in Mem. Junta Invest. Ultramar, Sér. 2, **34**: 15, 20 (1962). Type as above.

A perennial, somewhat viscid herb, 17–40 cm. high. Stems 1 to several from a woody tortuous rootstock, erect, simple, whitish, subrigid, striate, setose (the setae up to 4 mm. long, spreading, ending in a minute gland, the upper ½–⅔ yellowish, translucent and very thin, the lower part whitish or pinkish, somewhat rigid, provided with long white hairs, not bulbous at the base) and also hairy (the hairs very slender, long, white, more or less dense towards the stem-apex). Leaves up to 15 × 3 cm., usually erect (lower leaves smaller, the lowermost sometimes only 10 × 4 mm.), oblong to oblong-elliptic or lanceolate to linear-lanceolate in outline, lobed, pinnatifid or coarsely serrate-dentate in the lower ½–⅔ (the lobes or segments triangular-oblong to lanceolate, spreading or slightly recurved, acute or obtuse, mucronate, usually entire) or sometimes only with some teeth near the base (the teeth broadly triangular), the upper part of the lamina more or less attenuate, acute or subacute, entire, subentire or shallowly serrate-dentate, contracted at the base, membranous, slightly discolorous (paler beneath), hispidu-lous on both surfaces, sessile; midrib and lateral nerves (ascending and parallel to the margin) raised on both surfaces but more so and sparsely setose beneath.

Peduncles 4·5–16 cm. long, erect, slender, usually not exceeding the upper leaves 1–4(6)-flowered, setose and hairy; pedicels 1·5–3 mm. long in flower, 4–7 mm. long in fruit, erect, densely setose. Flowers heterostylous. Calyx 13–17 mm. long, sparsely and weakly setose and hairy; lobes c. 4 mm. long, oblong. Petals 3– 3·3 cm. long with limb 1·2–1·4 cm. broad, salmon-pink to orange-red or bright red. Longistylous flowers: filaments of the longer stamens ± 12·5 mm. long, those of the shorter ones ± 10 mm. long; styles c. 11·5 mm. long, with the stigmas above the upper anthers. Brevistylous flowers: filaments of the longer stamens 14– 15 mm. long, those of the shorter ones 12–13 mm. long; anthers 2–2·5 mm. long; styles c. 5 mm. long. Ovary glabrous. Capsule up to 7 × 0·2 cm., not or slightly constricted between the seeds, glabrous, erect; beak ± 5 mm. long. Seeds c. 3 mm. long; aril subequalling ¾ of the seed.

Rhodesia. N: Kariba Distr., Tshete Gorge, L. Kariba, 10.i.1963, *Whellan* 2014 (SRGH). E: Chipinga, Sabi Valley Experimental Station, xii.1959, *Soane* 187 (SRGH). S: Ndanga, Gudus, Sangwe Reserve, c. 430 m., fl. & fr. i.1960, *Farrell* 121 (BR; LISC; PRE; SRGH). **Malawi.** S: Chickwawa, Lengwe National Park, fl. 14.xi.1970, *Hall-Martin* 1126 (SRGH). **Mozambique.** MS: 24 km. from Vila de Manica on road to Vila Pery, c. 900 m., fl. 25.xi.1965, *Torre & Correia* 13274 (LISC). SS: between Massangena and Corone, c. 4 km. from Massangena, fl. 8.xi.1969, *Myre & Duarte* 5053 (LMA).

Known only from Rhodesia, Malawi and Mozambique. In mopane woodland, mixed savanna woodland and grassland.

5. **Tricliceras schinzii** (Urb.) R. Fernandes in Bol. Soc. Brot., Sér. 2, **49**: 23, t. 7 (1975). Type from Angola.
   Wormskioldia schinzii Urb. in Engl., Bot. Jahrb. **15**: 159 (1893).—A. & R. Fernandes in Bol. Soc. Brot., Sér. 2, **35**: 160 (1961); in Mem. Junta Invest. Ultramar, Sér. 2, **28**: 15 (1961); in C.F.A. **4**: 211 (1970). Type as above.

A perennial herb, 11·5–47 cm. high (incl. the racemes). Stems 1-several from a long, woody, noduled rootstock, 4–25 cm. long, dark brownish when dry, erect, simple or sometimes branched from the base, more or less setose like the leaves, peduncles, pedicels, calyx and capsule, the setae spreading, not bulbous at the base, minutely capitate-glandular. Leaves more or less dense, usually attenuate to the acute apex subcontracted or attenuate into a subpetiolar base 1–5 mm. long, setose especially on the inferior surface; midrib and nerves raised on both surfaces. Peduncles up to 26 cm. long, usually exceeding the upper leaves, 2–7-flowered; pedicels 5–10 mm. long, deflexed in fruit. Calyx (9)11–15 mm. long. Petals obovate, cuneate, salmon to orange-red. Anthers 2–2·5 mm. long. Ovary densely setose. Capsule setose, pendulous.

Setae slender, yellow, translucent, shining and rather thin in the upper half, those of the stem up to 2 mm. long, usually shorter than the diameter of the stem; leaves erect, without white hairs on the midrib and nerves of superior surface or, if hairs present, then sparse; plant very viscid on stem and leaves   -   subsp. *schinzii* var. *juttae*
Setae more robust, not shining, translucent only at the mucro-like apex, those of the stem up to 4 mm. long, usually equalling or longer than the diameter of the stem; leaves usually spreading with very dense, long, white hairs on the midrib and nerves of the superior surface; plant not viscid   -   -   -   -   -   subsp. *laceratum*

Subsp. **schinzii** var. **juttae** (Dinter & Urb.) R. Fernandes in Bol. Soc. Brot., Sér. 2, **49**: 23, t. 8 (1975). Syntypes from SW. Africa (Omaheke).
   Wormskioldia juttae Dinter & Urb. in Fedde, Repert. **13**: 153 (1914).—Tikovsky & Schreiber in Prodr. Fl. SW. Afr. **88**: 4 (1968). Syntypes as above.
   Wormskioldia rehmii Suesseng. in Mitt. Bot. Staatss. München **1**: 55 (1950). Type from SW. Africa (Grootfontein).
   Wormskioldia glandulifera sensu Dinter in Fedde, Repert. **25**: 50 (1928) non Klotzsch (1862).
   Wormskioldia schinzii sensu auct. pro parte, non Urb. (1893).

Indumentum of slender, spreading setae, opaque in the lower ½–⅓, translucent, shining and very thin in the upper ½–⅔, rather dense towards the stem-apex, without intermixed hairs or rarely with some white hairs on the midrib of the superior leaf surface. Leaves up to 14 × 2·5 cm., usually erect, lanceolate, sometimes elliptic, more or less attenuate towards the acute apex, somewhat attenuate to the base, ± lobed from the apex to the base (the lobes up to 1 cm. long, but usually shorter, oblong or lanceolate, obtuse or acute, again shallowly dentate to serrate-dentate, ascending or the lower spreading), sometimes the leaves only distinctly lobed near the base and the greater part of the margin shallowly lobed or crenate-dentate or all the margin serrate-dentate or crenate, rigid, concolorous, light green (even on drying), rather setose, with the setae similar to those of the stem but more slender. Petals 17–25 × 7–8 mm. Styles of the longistylous flowers 10 mm. long, those of the brevistylous ones 3 mm. long. Capsule up to 5 × 0·3–0·4 cm., not constricted between the seeds, densely setose. Seeds 3–5 mm. long.

Botswana. N: Chobe, Nungwe Valley, 960 m., fl. 2.i.1966, *Henry* 60 (SRGH). SE: c. 93 km. from Lobatsi to Ghanzi, ix.1967, *Lambrecht* 347 (PRE; SRGH). **Rhodesia.** W: Wankie National Park, Makwa road, Main Camp, fl. 26.x.1968, *Rushworth* 1237 (SRGH).
Also in SW. Africa. In woodland and scrub on deep greyish-white Kalahari sand.

The fruits of typical Angolan *T. schinzii* are known only in the young state, and so their final shape and size cannot be appreciated. If more Angolan material with fruits in good condition should prove that they are different from those of var. *juttae*, then this would be considered better as a subspecies.

Subsp. **laceratum** (Oberm.) R. Fernandes in Bol. Soc. Brot., Sér. 2, **49**: 24, t. 9–11 (1975). Type from S. Africa (Transvaal).
   Wormskioldia schinzii var. *schinzii* sensu A. & R. Fernandes in Mem. Junta Invest. Ultramar, Sér. 2, **34**: 15 (1962).—sensu M. Diniz in Rev. Cienc. Biol. **3**: 52 (1970) non Urb. (1893).
   Wormskioldia lacerata Oberm. in Bothalia **11**, 3: 288 (1974). Type as for *Tricliceras schinzii* subsp. *laceratum*.

Setae rather thick and long (those of the stem and midrib of the leaves up to 4 mm. long), subequalling or exceeding the diameter of the stem, spreading, whitish or brownish on drying, not very rigid, attenuate from the not bulbous base to the translucent mucro-like apex, this ending in a very small capitate gland, sometimes intermixed with long, white hairs. Leaves frequently spreading, up to 19·5 × 6 cm. (the lower ones rather smaller, sometimes only 0·6 × 0·4 cm.), lanceolate to elliptic in outline, pinnatilobed from the base nearly to the apex (the lobes or segments up to 2·5 cm. long, usually triangular-oblong to narrowly lanceolate, spreading or somewhat recurved, very acute, dentate, more or less scattered to nearly contiguous, sometimes the lobes obtuse or rounded), with the rhachis 0·3–1·5 cm. broad, usually dentate-serrate between the segments, light green or yellowish-green on drying, membranous or slightly rigid, setose (densely so mainly on the midrib beneath towards the petiolar base) and densely hairy above on the midrib and nerves (hairs long, dense, soft, thin, white, subappressed and retrorse). Calyx densely setose, sometimes also hairy. Petals 22–30 mm. long and with limb up to 13 mm. broad. Anthers 2·5 mm. long. Immature capsule 4 cm. long, patently setiferous.

Mozambique. SS: Caniçado, 16 km. from Lagoa Nova to Mapuoanguene, fl. 20.xi. 1970, *Correia* 2064 (LMU). LM: between Manhiça and Incomati Valley, fl. & fr. juv. 22.x.1947, *Barbosa* 459 (COI; LISC; LMA; LMU; PRE; SRGH).
Also in S. Africa (Transvaal). In open forests, bushy savannas and roadsides, on argillaceous-sandy soils.

Besides the characters cited in the key, subsp. *laceratum* differs from subsp. *schinzii* by the less rigid, more deeply divided leaves with less prominent nerves, and by the usually larger flowers with more brightly coloured petals. It may prove to be a distinct species.
   A collection from Mozambique (SS: Gaza, Bilene, fl. 5.xi.1969, *Correia & Marques* 1385 (COI; LISC; LMU; SRGH)) differs from all other members of the genus in having stipulate leaves; the stipules are foliaceous and dentate or linear.

6. **Tricliceras glanduliferum** (Klotzsch) R. Fernandes in Bol. Soc. Brot., Sér. 2, **49**: 18 (1975). Syntypes: Mozambique, Rios de Sena, Tete and Boror, *Peters* (EA, isosyntypes).
   Wormskioldia glandulifera Klotzsch in Peters, Reise Mossamb. Bot. **1**: 146, t. 26 (1861).—Mast. in F.T.A. **2**: 503 (1871).—Urb. in Jahrb. Königl. Bot. Gart. Berl. **2**: 49 (1883).—Gilg in Engl. & Prantl, Pflanzenfam. **3**, 6a: 61 (1893).—Engl., Pflanzenw. Afr. **3**, 2: 593 (1921).—Gilg in Engl. & Prantl, op. cit. ed. 2, **21**: 463 (1925).—Burtt Davy, F.P.F.T. **1**: 119 (1926).—Bremek. in Ann. Transv. Mus. **15**, 2: 248 (1933).—Martineau, Rhod. Wild Fl.: 55, t. 20 fig. 3 (1953).—A. & R.

Fernandes in Mem. Junta Invest. Ultramar, Sér. 2, **34**: 16 (1962).—Letty, Wild
Fl. Transv.: 224 (1962).—Tikovsky & Schreiber in Prodr. Fl. SW. Afr. **88**: 4
(1968). Syntypes as above.

Annual, erect herb up to 40 cm. high. Stem branched to simple with a double
indumentum of soft, whitish, shining hairs and of more or less dense, yellowish,
spreading, up to 1 mm. long setae with swollen, blackish base. Leaves 2·5–
11·5 × 1–3 cm., ovate-lanceolate to lanceolate, attenuate towards the acute apex,
cuneate at the base, more or less incised-serrate to subpinnatifid, the teeth or
segments again serrate, the main ones with a terminal seta (rarely the lower leaves
very small, 0·6–1·1 × 0·3–0·55 cm., oblong-ovate, obtuse, entire), membranous,
darkening on drying, densely hairy on both faces, more so on the nerves, the
midrib also setose towards the base; petiole up to 1 cm. long, hairy like the
stem. Flowers 2-5(7) per raceme, heterostylous or homostylous; peduncles
2–10 cm. long; pedicels 3–6 mm. long, not deflexed in fruit; peduncles and
pedicels hairy like the stem. Calyx 10–15·5 mm., patently hairy and with weak
setae; lobes 2·5–5 mm. long, oblong, subobtuse. Petals up to 23 × 12·5 mm.,
obovate-cuneate, yellow to orange. Filaments of the longest stamens (7)9–13 mm.
long, those of the shorter ones 5·5–10 mm. long; anthers c. 2·3 mm. long. Ovary
3·5–5 mm. long, hairy; styles 10 mm. long and subequalling the longest stamens
in the longistylous flowers or 3 mm. long and shorter than the shortest stamens
in the brevistylous flowers. Capsule 3–6 cm. long, erect on straight pedicels,
shortly and patently hairy; beak up to 6 mm. long. Seeds 3–4 × 1 mm.

Botswana. N: Chobe, Kasane, 960 m., fl. & fr. vii.1966, *Mutakela* 93 (SRGH).
Zambia. E: near Changwe, between Petauke (Old Boma) and Mwape, fl. & fr. 16.xii.1958,
Robson 957 (BM; LISC; PRE; SRGH). **Rhodesia.** N: Urungwe, Zambesi Escarp-
ment, on Salisbury-Lusaka main road, fl. & fr. 31.i.1958, *Drummond* 5393 (BR; COI;
EA; LISC; PRE; SRGH). W: Wankie, Matetsi, iii.1918, *Eyles* 1288 (BM; SRGH).
E: Umtali, banks of Odzi R., Sun Valley Farm c. 6·5 km. from Junction Tea Room, c.
770 m., fl. & fr. 18.xii.1954, *Chase* 5366 (BM; BR; PRE; SRGH). S: c. 29 km. NNW.
of Beitbridge, fl. & fr. 14.i.1963, *Leach* 11584 (LISC; PRE; SRGH). **Malawi.** S:
Nsanje (Port Herald), between Thangadzi and Liladje Rs., 960 m., fl. & fr. 25.iii.1960,
Phipps 2693 (PRE; SRGH). **Mozambique.** Z: Zambezia, vii.1859, *Kirk* (K). T:
Tete, on road to Changara, fl. & fr. 13.ii.1968, *Torre & Correia* 17534 (LISC). MS: Vila
de Sena, fl. & fr. 1884–1885, *Carvalho* (COI). SS: Gaza, Massangena, fl. & fr. 20.ii.1968,
Magalhães 157 (COI).
Also in SW. Africa and S. Africa (Transvaal). In open forest, tree- and bush-savannas,
grassland, roadsides, or as a weed in old African gardens, on sandy, rocky soils, etc.

7. **Tricliceras tanacetifolium** (Klotzsch) R. Fernandes in Bol. Soc. Brot., Sér. 2, **49**: 25
(1975). Type: Mozambique, Zambezia, Boror, *Peters* (B, holotype†; K).
    Wormskioldia tanacetifolia Klotzsch in Peters, Reise Mossamb. Bot. **1**: 147
(1861).—Urb. in Jahrb. Königl. Bot. Gart. Berl. **2**: 51 (1883).—Engl., Pflanzenw.
Afr. **3**, 2: 593 (1921).—Burtt Davy, F.P.F.T. **1**: 119 (1926).—Garcia in Est. Ens.
Doc. Junta Invest. Ultramar **12**: 161 (1954).—A. & R. Fernandes in Mem. Junta
Invest. Ultramar, Sér. 2, **34**: 16 (1962).—Letty, Wild Fl. Transv.: 224 (1962).
Type as above.

An annual herb. Stem erect, up to 1 m. high, branched sometimes from the
base, the branches ascending, often the lower ones as long as or longer than the
stem, rather leafy upwards, with a double indumentum of very thin, arcuate,
antrorse or subpatent hairs and of numerous, yellowish, capitate-glandular, up
to 1·5 mm. long setae with bulbous, blackish-violet base, finally scabrous from the
persistent bases of the setae. Leaves 1·5–13 × 1–12 cm., ovate, triangular or
rhombic in outline, pinnatifid, with the longer segments again partite and the
shorter ones irregularly and ± deeply dentate, all usually rather attenuate and
acute like the terminal part of the lamina, with the rhachis also irregularly dentate
between the segments, membranous, light green but paler beneath, not darkening
on drying, subglabrous or pubescent on the midrib above, sparsely setose beneath,
more densely so towards the base mainly on the nerves, with the main teeth
glandular-setose at the apex; petiole 0·1–2·3 cm. long. Flowers 2–7 per raceme,
heterostylous; peduncles 3–24 cm. long, slender, with indumentum as on the
stem but hairs more sparse and patent and the setae not so dense and with smaller
bulbous bases; pedicels 2–7·5 mm. long, densely setose at the apex, deflexed in
fruit. Calyx 11–16 mm. long, glabrous or sparsely and weakly setose towards the

top; lobes c. 6 mm. long, linear-lanceolate, obtuse. Petals c. 24 mm. long, pale
yellow to orange. Styles 10–11·5 mm. long in the longistylous flowers, with the
stigmas above the anthers of the longer stamens and 2–3 mm. long in the brevistylous ones, with the stigmas below the anthers of the shorter stamens. Ovary
c. 3·5 mm. long, glabrous. Capsule (0·7)1–4(5) cm. long, glabrous. Seeds 2–2·5 ×
1–1·2 mm., obovoid.

Rhodesia. N: Sinoia, Zwimba Reserve, 1440 m., fl. & fr. iii.1961, *Davies* 2904
(SRGH). E: Umtali, banks of Odzi R., Sun Valley Farm, Wengesi R., c. 6 km. W. of
Junction Tea Room, c. 770 m., fl. & fr. 18.xii.1954, *Chase* 5355 (BM; BR; COI; LISC;
PRE; SRGH). S: Chibi, c. 72·5 km. S. of Fort Victoria on Beitbridge road, fl. & fr.
15.iii.1967, *Rushworth* 331 (LISC; SRGH). **Mozambique.** Z: Namacurra, 26 km.
from Nicuadala on road to Campo, fl. (brevistylous) & fr. 1.ii.1966, *Torre & Correia* 14313
(LISC). MS: Gorongosa National Park, fl. & fr. 10.vi.1966, *Macedo* 2149 (LMA). SS:
Govuro, between the road No. 1 and Mambone, on the new road, fl. & fr. 15.xii.1969,
Myre, Duarte & Rosa 5192 (LMA).
Also in S. Africa (Transvaal). In *Brachystegia* open forest, secondary savannas,
stony places, roadsides, cultivated ground, on small granite outcrops with shallow soil
patches (" rock mats "), in wet vlei at foot of granite kopjes, fissures of rocks, sandy soil,
etc., from 50 to 1440 m.

8. **Tricliceras lobatum** (Urb.) R. Fernandes in Bol. Soc. Brot., Sér. 2, **49**: 21 (1975).
Syntypes from Sudan, Kenya and Angola.
Wormskioldia lobata Urb. in Jahrb. Königl. Bot. Gart. Berl. **2**: 52 (1883).—Gilg in
Engl. & Prantl, Pflanzenfam. **3**, 6a: 61 (1893); op. cit. ed. 2, **21**: 463 (1925).—Engl.
in Sitz.-Ber. Königl. Preuss. Akad. Wiss. Berl. **52**: 13 (1906); Pflanzenw. Afr. **3**,
2: 593 (1921).—Thonn., Fl. Pl. Afr.: t. 105 (1915).—Broun & Massey, Fl. Sudan:
56 (1929).—F. W. Andr., Fl. Pl. Anglo-Egypt. Sudan **1**: 29, fig. 22 (1950).—
Martineau, Rhod. Wild Fl.: 55 (1953).—Garcia in Est. Ens. Doc. Junta Invest.
Ultramar **12**: 162 (1954).—J. Lewis in F.T.E.A., Turneraceae: 10 (1954).—A. & R.
Fernandes in Mem. Junta Invest. Ultramar, Sér. 2, **28**: 14 (1961); op. cit. **34**: 15
(1962); in C.F.A. **4**: 210 (1970).—W. Robyns in F.C.B., Turneraceae: 7–8 (1967).—
Binns, H.C.L.M.: 101 (1968).—Tikovsky & Schreiber in Prodr. Fl. SW. Afr. **88**:
4 (1968). Syntypes as above.

An annual herb. Stem 5–100 cm. high, erect, simple or branched, more or less
crisply puberulous and also sparsely setose, the setae short (up to 0·5 mm. long),
yellow or brown with swollen base. Leaves heteromorphic, acute at the apex,
shallowly crenate-serrate or subentire at the margin, membranous, slightly
discolorous, subpatently hairy, rarely subglabrous above, the 2–3 lower ones up to
5 × 1·3 cm., elliptic, unlobed and petiolate (petiole up to 4 mm. long), the others
larger (up to 15 × 3·5(4·5) cm.), lanceolate or rarely ovate-lanceolate, irregularly
lobed in the lower ½–⅔ (lobes 1–4, blunt, frequently with a sessile gland at the top)
and with a pair of very small glandular auricle-like lobes at the base, subsessile to
sessile, rarely all the leaves unlobed. Flowers 1–4 per raceme; peduncles slender,
straight, 2–9 cm. long in flower, up to 20 cm. long in fruit, with indumentum as
on the stem but more sparse and also with patent, soft white hairs up to 2 mm.
long, denser towards the base of the peduncles; bracteoles c. 1 mm. long,
pubescent; pedicels 3(4) mm. long. Calyx hairy and sometimes also setose on
the outside; tube 9–10·5 mm. long, hairy inside on the lower 4 mm.; lobes 3–5 mm.
long, oblong, usually obtuse, with hyaline, ciliolate margin. Petals 15–18(20) ×
3·5–5 mm., obovate-cuneate, pale yellow to orange, rarely white; ligules 1·5 mm.
long. Stamens with 9–13 mm. long filaments, 2 shorter and 3 longer; anthers
c. 2·5 mm. long. Ovary c. 4·5 × 1 mm., narrowly ellipsoid, glabrous; styles 8–
10 mm. long. Capsule 4–7 cm. long and 1–1·5 mm. in diameter, moniliform, at
right angles to the peduncle or deflexed, glabrous; beak 2·5 mm. long. Seeds 2–
2·5 × 1 mm., obovate or obovate-oblong; aril up to ⅔ the length of the seed.

Caprivi Strip. c. 915 m., fl. 20.xii.1958, *Killick & Leistner* 3005 (PRE; SRGH).
Zambia. B: Mongu Lealui, fl. & fr. 10.i.1960, *Gilges* 954 (SRGH). N: Mbala (Aber-
corn), fl. 6.i.1955, *Bock* 157 (COI; PRE). E: near Changwe, between Petauke and
Mwape, fl. 16.xii.1958, *Robson* 958 (BM; LISC; PRE; SRGH). S: Mazabuka,
c. 6·5 km. from Chirundu bridge on Lusaka road, fl. & fr. 6.ii.1958, *Drummond* 5502
(BR; LISC; PRE; SRGH). **Rhodesia.** N: Urungwe, Zambesi Escarpment, main
Salisbury-Lusaka road, fl. 31.i.1958, *Drummond* 5392 (BR; LISC; PRE; SRGH). W:
Bulawayo, xii.1897, *Rand* 13 (BM). C: Hartley, Poole Farm, fl. & fr. 3.iii.1948, *Hornby*
2910 (PRE; SRGH). E: Umtali, Muhuni (Muhenye?) Purchase area, fl. & fr. i.1960,

Davies 2680 (SRGH). S: Chibi Kopje, near Madzivire, fl. & fr. 30.xii.1962, *Moll* 478 (LISC; SRGH). **Malawi.** N: c. 13 km. W. of Karonga, 25.ii.1953, *Williamson* 179 (BM). C: Bunda Hill, Lilongwe, 1250 m., fl. & fr. 7.ii.1959, *Robson* 1494 (BM; LISC; PRE; SRGH). S: Shire R., near Liwonde Ferry, 475 m., fr. 13.iii.1955, *E. M. & W.* 836 (BM; LISC; SRGH). **Mozambique.** N: Cabo Delgado, Macondes, between Nantulo and Mueda, c. 310 m., fl. & fr. 30.xii.1963, *Torre & Paiva* 9778 (LISC). MS: 24 km. from Manica on road to Vila Pery, fl. 25.xi.1965, *Torre & Correia* 13277 (LISC).

Widespread through tropical Africa (Sudan, Uganda, Kenya, Tanzania, Zaire, Angola, SW. Africa, etc.). Common in open situations in mopane woodland, fringing forests, secondary savannas, on rocky outcrops, sand soils, etc., or as a weed in old African gardens, up to 1500 m.

9. **Tricliceras lanceolatum** (A. & R. Fernandes) R. Fernandes in Bol. Soc. Brot., Sér. 2, **49**: 20 (1975). Type: Mozambique, Mutuáli, Malema road, *Gomes e Sousa* 4222 (COI, holotype; LMA).

Wormskioldia lanceolata A. & R. Fernandes, op. cit. **35**: 156, t. 3 (1961); in Mem. Junta Invest. Ultramar, Sér. 2, **34**: 15 (1962). Type as above.

Annual herb. Stem up to 60 cm. high, erect, simple or branched from the base, puberulous (hairs very short and thin, arcuate, antrorse, subappressed) and sparsely setose (setae c. 0·5 mm. long, yellowish towards the apex, brownish at the swollen base). Leaves all similar, up to 15 × 1·6 cm., oblong-lanceolate to linear-lanceolate, unlobed, tapering towards the base and the very acute apex, shallowly undulate and with minute teeth at the margin, sometimes with some triangular teeth near the base and a pair of sessile glands just at the base, thinly membranous, glabrous everywhere or sparsely puberulous towards the base; petiole 1–2 mm. long to absent. Flowers (4)5–9 per raceme; peduncles up to 25 cm. long, sparsely hairy at the base, glabrous upwards; pedicels c. 2 mm. long in flower, up to 4 mm. long in fruit. Flower-buds acute. Calyx sparsely setose, without hairs; tube 10 mm. long; lobes 5 mm. long, oblong-linear, acute, with a brown seta below the apex. Petals c. 22 mm. long, orange, attenuate-cuneate towards the base, adnate to the calyx-tube for 3·5 mm., the limb c. 9 × 5 mm., rhombic-obovate. Brevistylous flowers: filaments of the longer stamens ± 12 mm. long, those of the shorter ones ± 9 mm. long; anthers 3 mm. long. Ovary 3 mm. long; styles 3 mm. long. Capsules up to 5·5 × 0·15 cm., glabrous. Seeds 2·25–3 × 1–1·25 mm.

Mozambique. N: Mutuáli, on Malema road, 5 km. from Cotton Station, fl. & fr. 25.ii.1954, *Gomes e Sousa* 4222 (COI; LMA). MS: Corone, Inhaminga, fl. & fr. 20.iv.1956, *Gomes e Sousa* 4310 (COI; LISC).

Known only from Mozambique. In open *Brachystegia* forest, on sandy or clayey-sandy soils, near the coast.

10. **Tricliceras elatum** (A. & R. Fernandes) R. Fernandes in Bol. Soc. Brot., Sér. 2, **49**: 16 (1975). Type: Mozambique, between Corrane and Nametil, *Torre* 1273 (COI, holotype; LISC).

Wormskioldia elata A. & R. Fernandes in Bol. Soc. Brot., Sér. 2, **35**: 155, t. 1 & 2 (1961); in Mem. Junta Invest. Ultramar, Sér. 2, **34**: 15 (1962). Type as above.

An annual herb. Stem up to 1 m. high, erect, simple or branched, more or less hirsute and with very sparse yellow or brownish setae swollen towards the base. Leaves all similar, lanceolate and up to 12 × 4·5 cm., or elliptic and up to 8·5 × 3·7 cm., subacute, obtuse or rounded at the apex, shallowly crenulate-serrate, provided below the middle with 2 pairs of subopposite blunt lobes, the basal lobes rather smaller, usually all with a blackish sessile gland at the apex as that of the apex of the lamina, membranous, slightly discolorous, hispidulous, all sessile or the lower ones very shortly petiolate. Flowers 6–11 per raceme; peduncles up to 30 cm. long, hispid and very sparsely setose; bracteoles 1–2 mm. long; pedicels 2·5–3·5 mm. long, hairy. Calyx 11·5–13 mm. long, patently hairy on the outside; tube c. 9·5 mm. long, hairy inside below the insertion of the petals; lobes c. 3·5 mm. long, oblong, subacute. Petals 19–20 × 3 mm., bright scarlet, slightly paler abaxially; ligules acute. Longistylous flowers: filaments of the longer stamens 7–8 mm. long, those of the shorter ones 5·5–6 mm. long; anthers 2–2·5 mm. long; styles c. 8 mm. long. Brevistylous flowers: filaments of the longer stamens 8–11 mm. long, those of the shorter ones 6·5–9 mm. long; styles 3·5 mm. long. Ovary 3 × 0·4 mm., oblong, linear, glabrous. Capsule up to 5·5 cm. long; beak up to 6 mm. long. Seeds 2–2·5 mm. long.

Mozambique. N: Imala, between Muecate and Alide, near Mt. Melita, fl. & fr. 15.i.1964, *Torre & Paiva* 9975 (LISC).
Known only from Mozambique. In savanna, xerophytic scrub and rupideserta, on sandy soils.

11. **Tricliceras auriculatum** (A. & R. Fernandes). R. Fernandes in Bol. Soc. Brot., Sér. 2, **49**: 15 (1975). Type: Mozambique, Namina, *Pedro & Pedrógão* 3238 (EA; LMA, holotype).
   Wormskioldia auriculata A. & R. Fernandes in Bol. Soc. Brot., Sér. 2, **35**: 157, t. 4 (1961); in Mem. Junta Invest. Ultramar, Sér. 2, **34**: 16 (1962). Type as above.

An annual herb. Stem up to 90 cm., prostrate to erect, simple or with few, spreading branches, purplish, puberulous (hairs thin, curved, antrorse) and sparsely setose (setae up to 1·5 mm. long, not bulbous at the base, spreading, purplish towards the base, yellowish towards the apex). Leaves all similar, up to 7·5 × 0·9 cm., narrowly lanceolate, apex acute, sometimes subcaudate, regularly serrulate at the margin, subcontracted above the base and provided there with a pair of spreading, linear or linear-spathulate, obtuse auricles up to 7 × 2 mm., somewhat fleshy or membranous, glabrous or hairy on superior surface towards the base, sparsely setose beneath on the midrib, obsoletely penninerved; petiole up to 2 mm. long, setose. Flowers 4–6 per raceme; peduncles 8–11 cm. long, glabrous or sparsely hairy; bracteoles c. 1·5 mm. long; pedicels up to 4·5 mm. long in fruit. Calyx sparsely and shortly setose and without hairs on the outside; tube c. 6 mm. long, whitish-pubescent inside in the lower ⅓; lobes 2·5–4 mm. long, oblong-lanceolate, obtuse. Petals 22 × 8·5 mm., yellow. Filaments of the longer stamens c. 8 mm. long, those of the shorter ones c. 6 mm. long; anthers 1·3–1·5 mm. long. Ovary 3–4 mm. long, oblong; styles 7·5 mm. long, with the stigmas projecting beyond the anthers of the longer stamens. Capsule up to 4 × 0·15 cm., glabrous, deflexed; beak c. 2·5 mm. long. Seeds 9–15 per capsule, c. 2·5 mm. long, oblong-obovoid.

Mozambique. N: c. 35 km. E. of Ribáuè, fl. & fr. 17.v.1961, *Leach & Rutherford-Smith* 10912 (LISC; SRGH).
Known only from Mozambique. On granitic rocks.

5. STREPTOPETALUM Hochst.

Streptopetalum Hochst. in Flora **24**: 665 (1841).

Annual or perennial herbs, pubescent and setose. Leaves petiolate or sessile, sometimes heteromorphic, without stipules or basal glands. Inflorescences axillary, 1–many-flowered, usually a secund raceme with free peduncle. Flowers hetero- or homostylous, on 2-bracteolate, not deflexed, sometimes accrescent pedicels. Sepals adnate into a cylindrical tube, 15-nerved, internally hairy for 3–6 mm. from the base and with 5 linear-elliptic tubercles above the insertion of the stamens. Petals yellow to orange, inserted at the throat of the calyx-tube, obovate to spathulate, without ligules. Stamens perigynous, inserted at the base of the calyx-tube; filaments filiform or winged in the lower ⅓; anthers oblong. Ovary 1-locular, with parietal placentas; styles straight with multipartite or lobulate stigmas. Capsules erect, loculicidal. Seeds numerous, arcuate, reticulate-alveolate, with 2-pored pits; aril unilateral, shorter than or equalling the seed.

An African genus mainly of tropical E. and Central Africa, 1 species extending into S. Africa. 5 species known to date.

Perennial; leaves sessile, rigid; petals 18–20 × 10 mm.   -   -   1. *luteoglandulosum*
Annual; leaves shortly petiolate, membranous; petals smaller:
  Homophyllous: all leaves unlobed, with serrulate or subentire margins; capsule nearly as long as wide; setae of the stem up to 5 mm. long, somewhat rigid.
                                  2. *wittei*
  Heterophyllous: lower leaves more or less lobed, the others serrate; capsule distinctly longer than wide; setae of the stem up to 1 mm. long, somewhat soft
                                  3. *serratum*

1. **Streptopetalum luteoglandulosum** R. Fernandes in Bol. Soc. Brot., Sér. 2, **49**: 25, t. 12 & 13 (1975). TAB. **90**. Type: Zambia, Luapula, *Richards* 12403 (BR; EA; K, holotype; SRGH).

A perennial herb up to 45 cm. high. Stems numerous, erect, simple or more of less branched, the branches also erect, densely leafy, with a double indumentum of

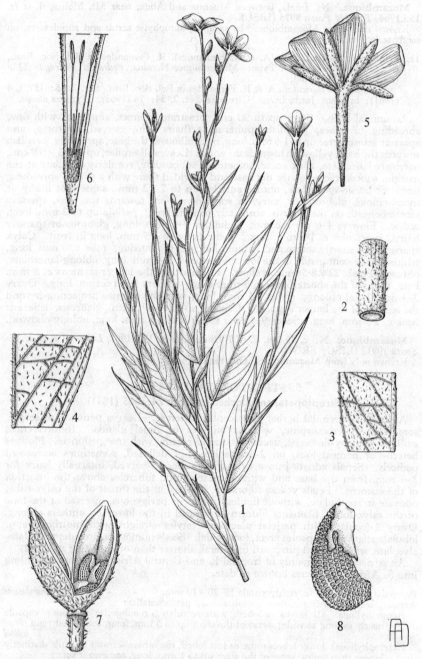

Tab. 90. STREPTOPETALUM LUTEOGLANDULOSUM. 1, habit (× ½); 2, segment of stem
(× 3); 3, leaf, superior surface (× 2); 4, leaf, inferior surface (× 2); 5, flower (× 1½);
6, part of flower in longitudinal section (× 1½); 7, open capsule (× 4); 8, seed (× 8),
all from *Richards* 12403.

very short, subpatent whitish hairs and short yellow setae up to 0·5 mm. long with a
swollen translucent base and a very thin, glandular, caducous apex. Leaves up
to 8 × 2·8 cm. (the uppermost and the lower ones rather smaller), sessile, all
similar, elliptic or ovate-elliptic to narrowly elliptic or linear-lanceolate, attenuate
or somewhat contracted towards a very acute, frequently subcaudate acumen,
entire or minutely denticulate-glandular at the margin, acute or subcontracted
at the base, light green, concolorous, rigid in the dry state, with a similar indumentum as on the stem but looser; midrib, lateral nerves and reticulum somewhat
raised on both faces but a little more so beneath. Racemes 4–25 cm. long,
terminal, with the lower internodes rather elongate. Flowers heterostylous;
pedicels up to 7 mm. long in fruit, erect, subclavate. Calyx pubescent and also
sparsely glandular-setose; tube 10–13 mm. long, with 5 dense tufts of white hairs
inside 4 mm. distant from the base; lobes 10–11 × 1·5–2 mm., oblong, obtuse, with
white, membranous margin. Petals c. 18 × 10 mm., obovate, bright yellow or
orange. Stamens with c. 10·5 mm. long filaments in the brevistylous flowers and
c. 8·5 mm. long filaments in the longistylous ones; anthers 2–2·5 mm. long.
Styles 3–3·5 mm. long in the brevistylous flowers, with distinctly branched
stigmas reaching the middle of the filaments, and 8·5–10 mm. long in the longistylous flowers with shortly branched stigmas exceeding the anthers. Capsule
5–7 × 3–4·5 mm., ellipsoid or ovoid, slightly muricate from the persistent bases
of the setae. Seeds 3–3·5 mm. long, arcuate-clavate, finally blackish; aril ½ as
long as to almost equalling the seed.

Zambia. N: Kawambwa, 15.ii.1957, *Fanshawe* 4039 (BR; K).
Known only from Zambia. In grassland on sandy soils.

2. **Streptopetalum wittei** Staner in De Wild. & Staner, Contr. Fl. Katanga, Suppl. **4**:
68 (1932).—Robyns in F.C.B., Turneraceae: 11 (1967). Type from Zaire (Katanga).

Annual (or also perennial?) herb, 20–70 cm. high. Stem erect, simple or
branched, with a double indumentum of very short, thin, white, somewhat
crisped or curved subappressed antrorse hairs and of yellowish or golden-yellow,
up to 5 mm. long, patent, more or less dense setae, some of these, towards the
upper part of the stem as well as on the raceme-axes and pedicels, with a very swollen
vesicular, yellow base. Leaves all similar, 5–15 × 0·3–4 cm., oblong-lanceolate or
elliptic to narrowly lanceolate or narrowly elliptic, attenuate towards the acute
apex (the uppermost with the apex nearly caudate), serrulate, the teeth somewhat
scattered and mucronulate, cuneate at the base, membranous, pubescent and
setiferous, more densely so on the nerves; petiole up to 2 mm. long. Flowers
heterostylous. Racemes 10–25 cm. long, 6–15-flowered; pedicels thick, up to
4 mm. long in fruit. Calyx-tube 7·5–13 mm. long, with vesiculose setae on the
outside and hairy within in the lower 2–3 mm.; lobes (2·5)4–6 mm. long, oblong-
lanceolate, acute. Petals 8–14 × 8–10 mm., obovate, unguiculate, pale yellow or
lemon-yellow. Stamens with filaments 10–12 mm. long in the brevistylous
flowers and 5–6 mm. long in the longistylous ones; anthers c. 1·5 mm. long. Ovary
ovoid, vesiculose-setose; styles 3 mm. long in the brevistylous flowers or 10 mm.
long in the longistylous ones. Capsule elliptic-ovoid (5–6 × 3–4 mm.), or sub-
spheric (4 × 4 mm.), setose, the setae bulbous. Seeds 2·5–3 mm. long, rather
arcuate, finally blackish; aril longer than ½ of the seed, sinuate at the margin.

Zambia. N: Luapula Distr., Mweru, c. 1200 m., fl. & fr. iii.1932, *Walter* 25 (K).
Also in Zaire (Katanga). In woodlands, grasslands and roadsides, in rocky semi-
shady situations, between 700 and 1200 m.

A very poisonous plant, especially dangerous to children.

3. **Streptopetalum serratum** Hochst. in Flora **24**, 2: 666 (1841).—Urb. in Jahrb.
Königl. Bot. Gart. Berl. **2**: 56 (1883).—Gilg in Engl. & Prantl, Pflanzenfam. **3**, 6a:
61 (1893); op. cit. ed. 2, **21**: 463 (1925).—Engl. in Sitz.-Ber. Königl. Preuss. Akad.
Wiss. Berl. **52**: 13 (1906).—Burtt Davy, F.P.F.T. **1**: 119 (1926).—J. Lewis in
F.T.E.A., Turneraceae: 17, t. 5 fig. 1–5 (1954).—Cufod. in Bull. Jard. Bot. Brux. 29,
3 (Suppl.): 598 (1959).—Tikovsky & Schreiber in Prodr. Fl. SW. Afr. **88**: 2
(1968).—Clavarino in Webbia **23**: 360, fig. 2 & 3 (1969). Type from Ethiopia.
Wormskioldia serrata (Hochst.) Walp., Repert. **5**: 782 (1846).—Mast. in F.T.A. **2**:
502 (1871). Type as above.
Wormskioldia abyssinica A. Rich., Tent. Fl. Abyss. **1**: 299 (1847) *nom. illegit.*

Streptopetalum serratum var. *latifolium* Pirotta in Ann. Ist. Bot. Roma **8**, 2: 259 (1904). Syntypes from Ethiopia.

Streptopetalum serratum var. *angustifolium* Pirotta, tom. cit.: 260 (1904). Syntypes from Ethiopia.

An annual herb. Stem up to 50 cm. high, erect, usually simple, with a double indumentum of short, thin, patent hairs and of soft, up to 1 mm. long setae which are slightly swollen at the base. Leaves heteromorphic, the lower ones 1·5–6 × 0·5–1 cm., rhombic-cuneate, pinnatifid to pinnatilobed with obtuse lobes, the median and upper ones larger, up to 14 × 4 cm., lanceolate, unlobed, attenuate to the acute apex, the greater part of the margin distinctly 2-serrate or in uppermost leaves minutely serrulate, cuneate and entire at the base, thinly, subappressedly and shortly hairy on both faces, more densely and subpatently so on the nerves. Peduncles up to 2 cm. long, pubescent and also setose; pedicels c. 1 mm. long, accrescent, hairy like the peduncles. Racemes (1)2–12-flowered, up to 9 cm. long in fruit. Flowers homostylous. Calyx pubescent and setose on the outside; tube 8–13 mm. long, densely hairy within on the lower 3–4 mm.; lobes 2–4 mm. long, obtuse or sometimes obsolete. Petals 11–16 × 2–4 mm., obovate-cuneate, yellow to orange. Stamens with filaments 9–12(13) mm. long and anthers c. 1·8 mm. long. Ovary oblong, setose, 12–15-ovulate; styles 8–10 mm. long. Capsule (5)7–12 × 3–4 mm., ellipsoid, setose, the setae with swollen and sometimes blackish base. Seeds 2·75–3 × 0·8–0·9 mm., clavate, finally nearly black; aril as long as or slightly shorter than the seed, oblong, crenate at the margin.

Botswana. N: near Tsessebe, fl. & fr. 8.iii.1965, *Wild & Drummond* 6816 (K; LISC; PRE; SRGH). SE: road from Francistown to Gaberones, fl. & fr. ix.1967, *Lambrecht* 304 (BR; K; SRGH). **Zambia.** N: Mporokoso, Kundabwika Falls on Kalungwishi road, c. 1184 m., fl. & fr. 14.iv.1961, *Phipps & Vesey-FitzGerald* 3171 (SRGH). S: Livingstone, i.1910, *Rogers* 7523 (K). **Rhodesia.** N: Urungwe, Mensa Pan, c. 17·5 km. SSE. of Chirundu Bridge, 460 m., fl. & fr. 30.i.1958, *Drummond* 5357 (BR; K; PRE; SRGH). W: Wankie National Park, Main Camp, c. 1050 m., fl. & fr. 26.i.1969, *Rushworth* 1456 (COI; K; SRGH). E: Umtali, c. 69 km. S. of Umtali on Chipinga road, fl. & fr. 6.i.1969, *Biegel* 2741 (K; LISC; PRE; SRGH). S: Buhera, c. 32 km. from Birchenough Bridge, fl. & fr. 6.ii.1964, *Masterson* 337 (SRGH).

From Ethiopia to S. Africa (Transvaal) and SW. Africa, except Mozambique. In mopane woodland, *Commiphora* veld, thorn-scrub, roadsides, etc., on sandy soil among rocks.

84. PASSIFLORACEAE

By R. & A. Fernandes

Trees, shrubs or herbs, erect or more usually tendrillous climbers, sometimes with a napiform root. Leaves alternate, rarely opposite, entire or variously lobed or 3–7-foliolate, sessile or petiolate, often with glands at the margin and on the abaxial surface of the lamina and at the apex of the petiole. Stipules 0 or 2, caducous or persistent. False stipules present in some species of *Basananthe*. Tendrils, when present, axillary (sterile peduncles) or at the apex of the floriferous peduncle. Flowers solitary, in racemes or in paniculate cymes, pedicellate, the pedicel often articulate and 3-bracteolate. Flowers ☿ or unisexual or polygamous, actinomorphic. Hypanthium ± developed, broad or narrow. Calyx-tube short or ± elongate, coriaceous or herbaceous; lobes 3–∞ (3–6 in F.Z. species), valvate or imbricate, persisting with the tube. Petals 0 or as many as the calyx-lobes, inserted at the base or mouth of the calyx-tube or in an intermediate position, the margin entire or fimbriate. Corona annular, single or double, rarely 0, tubular and often fimbriate at the margin with the laciniae erect or radiant, or reduced to a rim of hairs. Stamens 5, rarely 6–10 or ∞; filaments free or partially connate into a tube, inserted at the base of the flower or on an androgynophore; anthers oblong or linear, basifixed or dorsifixed, introrse, 2-locular, dehiscing by longitudinal slits, sometimes apiculate. Ovary superior, 3–5(6)-carpellar, 1-locular, often stipitate (borne on an androgynophore) or rarely subsessile; ovules many or few, on 3–5(6) parietal placentas, pendulous, anatropous, with the funicle elongate; styles 1 or

3–5, with capitate or clavate, sometimes fimbriate stigmas. ♂ flowers often with a vestigial ovary. ♀ flowers usually with staminodes. Disk-glands sometimes present. Fruit baccaceous or a 3–5-valved capsule, usually many-seeded. Seeds 1–∞, on long funicles, ovate, compressed, rarely oblong or tumid, mostly arillate, with a furrowed and ridged seed-coat; endosperm fleshy, more or less abundant. Embryo large, with leafy cotyledons and cylindrical radicle.

A family of c. 550 species and 18 genera of the tropical and subtropical regions, mainly in S. America.

Trees or shrubs, without tendrils; leaves ± coriaceous, sometimes glandular-dentate; flowers 5-merous, ☿, with a well-developed corona:
- Stamens 10–16; petiole with 2 glands at the apex - - - - **1. Viridivia**
- Stamens 5; petiole without glands at the apex - - - - **2. Paropsia**

Shrubs, climbers or perennial herbs, usually with tendrils; leaves often membranous, sometimes coriaceous; flowers ☿ or unisexual; corona single or double or sometimes wanting:
- Flowers ☿, with a well-developed corona (or sometimes 2); petiole with or without glands at the apex:
  - Sepals 3(4); petals (2)4; stamens 6(8); style single, very short, with a (3)–4-lobed stigma; seeds smooth; petiole with glands at the apex; tendrillous hetero-phyllous shrub - - - - - - - - - **3. Schlechterina**
  - Sepals 5–6; petals 5 or 0; stamens 5–6; styles 3(4), each one with a capitate stigma; seeds alveolate-reticulate; petiole without glands at the apex; herbs or climbers:
    - Flowers small, whitish or greenish, in axillary ± long-pedunculate few-flowered inflorescences, the peduncle often ending in a ± well-developed tendril **4. Basananthe**
    - Flowers large and showy (passion-flowers), axillary and solitary; strong climbers, all introduced in the F.Z. area - - - - - - **6. Passiflora**
- Flowers unisexual, ☿ or polygamous; corona absent or poorly developed (comprising a laciniate rim or membrane or a row of filaments or hairs); leaves usually with glands at the apex of the petiole or near the base of the lamina - **5. Adenia**

1. VIRIDIVIA J. H. Hemsl. & Verdc.

Viridivia J. H. Hemsl. & Verdc. in Hook., Ic. Pl. **36**: t. 3555 (1956).

Small tree without tendrils. Flowers ☿, usually developing before leaves, in dense racemes at the end of short branches; bracts small, carinate, acute, soon caducous. Sepals 4, imbricate, sericeous outside, 3–7-nerved. Petals 4, smaller than the sepals, 1-nerved. Corona shortly tubular, irregularly fimbriate and with clavate whitish glands. Stamens 10–16, with filaments free and hairy; anthers oblong. Ovary globose, stipitate, 1-locular; styles 4–6 with fleshy ± kidney-shaped stigmas; ovules ± 50, anatropous, inserted in 2 rows on 4–5 placentas. Capsule subglobose, stipitate. Seeds ovoid, compressed, included in a cupulate aril.

A monospecific genus.

Viridivia suberosa J. H. Hemsl. & Verdc. in Hook., Ic. Pl. **36**: t. 3555 (1956).—F. White, F.F.N.R.: 268 (1962).—de Wilde in F.T.E.A., Passifloraceae: 9 (1975). TAB. **91**. Type: Zambia, Kasama Distr., c. 42 km. on road from Chambezi R. to Kasama, c. 5 km. N. of Lukulu Rest Camp, *Hoyle* 1310 (FHO; K, holotype).

Small tree or shrub up to 8 m. high producing flowers before leaves. Old branches covered with fleshy longitudinally fissured cork, the young ones hairy, the hairs short, stiff, golden-yellow or reddish-brown. Leaves alternate, with the lamina 7·5–17(20) × 4–7·5(9·5) cm., ovate or elliptic, rarely elliptic-oblong or lanceolate, acute or rounded at the apex, denticulate at the margin, broadly cuneate, rounded or cordate at the base, hairy on both faces but more so below, the hairs short, stiff and spreading; midrib and lateral nerves (7–11) ± prominent on both sides, the veins conspicuously reticulate beneath; petiole up to 5(10) mm. long, covered by short hairs and provided with 2 glands at the apex. Inflorescence up to 2 cm. long, ± 7-flowered, occurring at the end of the branches; pedicels (4)7–18 mm. long, hairy; bracts 5·5 × 3·5 mm., ovate, soon caducous. Flowers ☿. Sepals up to 1·9 × 1·3 cm., unequal, elliptic or broadly ovate, obtuse at the apex, rounded at the base, densely sericeous outside. Petals up to 18 × 6 mm. with claw c. 4 mm. long, narrowly elliptic-spathulate, greenish-white or yellowish-white, glabrous. Corona up to 3 mm. high, densely hairy. Stamens 10–16, with

Tab. 91. VIRIDIVIA SUBEROSA. 1, part of leafy branchlet (×1) *Richards* 5327; 2, 3, flowering branchlets (×1); 4, longitudinal section of flower (×3); 5, fruit, with wall partly removed (×1); 6, seed, with part of aril removed (×2), 2 from *Bullock* 1347, 3–6 from *Hoyle* 1310. Reproduced by permission of the Bentham-Moxon Trustees.

filaments 1·2 cm. long and anthers 2·5–3·5 mm. long, elongate-oblong, with the connective minutely apiculate. Ovary 4–5 mm. in diameter, globose, densely hairy, borne on a gynophore c. 5 mm. long; styles up to 3 mm. long; stigmas c. 2·5 mm. broad, fleshy. Capsule 3·3 × 3–4 cm., rounded at the apex, contracted at the base into a stipe c. 1 cm. long, minutely pubescent. Seeds up to 9 mm. long, included in a cup-shaped aril with serrulate margin.

Zambia. N: Kawambwa, fl. 25.v.1957, *Fanshawe* 3639 (BR; K; SRGH); Abercorn Distr., Inono R., fl. & fr. 1.ix.1956, *Richards* 6063 (BR; K). E: Fort Jameson, Machinje Hills, st. 15.v.1965, *Mitchell* 2973 (K; SRGH).
Known also from Tanzania. Woodlands, riverine forests and rocky slopes.

2. PAROPSIA Noronha ex Thou.

Paropsia Noronha ex Thou., Hist. Vég. Is. Austr. Afr.: 59, t. 19 (1805).

Trees or shrubs, usually with a short rust-coloured indumentum. Leaves alternate, coriaceous, ± hairy, petiolate; stipules very small, caducous. Inflorescences in terminal panicles or flowers solitary or in pairs, rarely 3–5, contemporary with or appearing before leaves. Flowers ☿. Calyx-tube very short. Sepals (4)5, imbricate, downy, marcescent. Petals (4)5, also imbricate, narrower and thinner in texture than the sepals. Corona consisting of numerous laciniae disposed in 1 row, free or connate below, regularly distributed or arranged in 5 oppositipetalous bundles. Stamens 5, alternate with the petals, with the lower part of the filaments connate and adherent to the gynophore or to the ovary; anthers linear or subovate-oblong, cordate at the base, dorsifixed, 2-locular, dehiscing by 2 longitudinal slits. Ovary sessile or shortly stipitate, ovoid, 1-locular, with 3–5 multiovulate parietal placentas; styles (2)3(5), free, sometimes hairy; stigmas subreniform or capitate. Capsule subglobose, 3(5)-valved, the valves membranous, downy. Seeds ovoid, compressed, enveloped by a fleshy cup-shaped aril, with a crustaceous testa. Embryo straight, surrounded by fleshy endosperm; cotyledons leafy.
A tropical genus with 10 species in Africa and Madagascar and 1 in Sumatra and Malaysia.

Ovary at anthesis sparsely hairy or glabrous; styles glabrous at the base; flowers usually
appearing before leaves - - - - - - - - - 1. *braunii*
Ovary at anthesis densely hairy or tomentose; styles hairy at the base; flowers
contemporary with the leaves:
  Leaves oblong, gradually attenuate to the apex, glabrous or nearly so on the upper face,
    sparsely and shortly hairy beneath especially on midrib and nerves; pedicels
    6–10 mm. long; sepals at anthesis 15–20 mm. long - 2. *grewioides* var. *orientalis*
  Leaves oblong-elliptic or ovate-oblong, rarely broadly elliptic or ovate, shortly and
    broadly acuminate at the apex, initially tawny-tomentose on both faces, glabrescent
    above with maturity; pedicels up to 5 mm. long; sepals 10–12 mm. long
                                          3. *brazzeana*

1. **Paropsia braunii** Gilg in Engl., Bot. Jahrb. **40**: 472 (1908); in Engl. & Prantl, Pflanzenfam. ed. 2, **21**: 415 (1925).—Engl., Pflanzenw. Afr. **3**, 2: 572 (1921).—Brenan & Greenway, T.T.C.L. **2**: 448 (1949).—A. & R. Fernandes in Garcia de Orta **6**, 2: 248 (1958).—Sleumer in Bull. Jard. Bot. Nat. Belg. **40**, 1: 57, fig. 2 (1970).—de Wilde in F.T.E.A., Passifloraceae: 8 (1975). Type from Tanzania.
Paropsia schliebeniana Sleumer in Notizbl. Bot. Gart. Berl. **12**: 475 (1935).—Brenan & Greenway, loc. cit.—A. & R. Fernandes, loc. cit. Type from Tanzania.
Paropsia sp.—A. & R. Fernandes, loc. cit.

Small tree or shrub up to 10 m. high, much-branched, the young branches rusty-tomentose or -velvety, the old ones covered with grey bark, slightly scaly, usually without leaves when in flower. Leaf-lamina (5)7–12 × (2·5)4·5–6(7) cm., broadly oblong or elliptic, ± obtuse, attenuate at the apex, margin remotely denticulate, broadly cuneate or roundish and slightly unequal at the base, papery, with ± dense, yellowish, soft and spreading hairs along the midrib and nerves on the upper surface and ± densely rusty-tomentose beneath; midrib and the 5–7 pairs of lateral nerves flat on the upper side, prominent on lower side which is conspicuously reticulate-veined; petiole 5–6 mm. long. Flowers appearing before the leaves, cream-coloured, in pedicellate fascicles of 3–5, borne in the axils of cucullate caducous bracts, arranged along the branches and forming a terminal tawny-tomentose panicle up to 32 cm. long; pedicels 6–8(9) mm. long, yellowish-tomentose. Calyx campanulate; sepals 6–8 × 3·5 mm., oblong, yellowish, silky outside at least along the median nerve, glabrous inside. Petals 7–8(10) × 3 mm., oblong, yellowish, glabrous. Corona c. 2 mm. high, with the laciniae connate in the lower ⅔ but free apically. Stamens with filaments 6–7 mm. long and anthers 2–3 mm. long. Ovary sparsely hairy or glabrous at anthesis, with a gynophore 1–1·5 mm. high; styles 1–1·5 mm. long, glabrous. Capsule c. 1·5 × 1 cm., ellipsoid, sparsely hairy.

Mozambique. N: Macondes, from Macomia towards Chai, fl. 30.ix.1948, *Andrada* 1382 (COI; LD; LISC; LMU; MO); between Meconta and Corrane, 220 m., fr. 18.i.1966, *Torre & Paiva* 10052 (COI; LD; LISC; LMU; MO). Z: Mopeia, Zambezi R., fl. buds 28.vii.1942, *Torre* 4435 (EA; FHO; K; LISC; LMA; SRGH; WAG). MS: Buzi, Mucheve Forest Reserve, fr. 4.xi.1967, *Carvalho* 957 (LISC). Also in Tanzania. In scrub and forests of various types.

2. **Paropsia grewioides** Welw. ex Mast. in F.T.A. **2**: 505 (1871).—Warb. in Engl. & Prantl, Pflanzenfam. **3**, 6a: 27 (1893).—Gilg in Engl., Bot. Jahrb. **40**: 472 (1908); in Engl. & Prantl, Pflanzenfam. ed. 2, **21**: 415 (1925).—A. & R. Fernandes in Garcia de Orta **6**, 4: 653 (1958); in C.F.A. **4**: 213 (1970).—Sleumer in Bull. Jard. Bot. Nat. Belg. **40**, 1: 63, fig. 4 (1970). Type from Angola.

Var. **orientalis** Sleumer, tom. cit.: 65 (1970).—de Wilde in F.T.E.A., Passifloraceae: 9 (1975). TAB. **92**. Type: Mozambique, Niassa, Cabo Delgado, between Mueda and Mocímboa do Rovuma, *Pedro & Pedrógão* 5312 (COI, holotype; EA; LMA; SRGH).

Paropsia grewioides auct. non Welw. ex Mast.—Verdc. in Kew Bull. **11**: 449 (1957).—A. & R. Fernandes in Garcia de Orta **6**, 2: 248, t. 1 (1958).

Small tree or shrub 4·5–12 m. high with the branches shortly and densely yellowish-tomentose. Leaf-lamina (4)7–11(13) × 1·5–3·5 cm., oblong, with the apex subacute or acuminate, margin regularly serrate-crenate, the teeth 1–1·5 mm. deep, curvate and hairy at the apex; midrib and the 10–14 pairs of lateral nerves flat or nearly so on the upper surface, prominent on lower surface which is conspicuously reticulate-veined; petiole 6–8 mm. long, slender, yellowish-hairy. Flowers axillary, creamy-greenish, solitary, pedicellate, the pedicels 6–10(15) mm. long, slender, tawny-hairy. Sepals 15–20(25) × 4–6(10) mm., oblong-lanceolate, densely yellowish-tomentose outside and sparsely so inside. Petals resembling the sepals but thinner and shorter, greenish-yellow or cream-coloured. Corona c. 4 mm. high. Stamens with filaments 8–10 mm. long and linear anthers c. 3 mm. long. Ovary yellow-tomentose; gynophore 1·5 mm. high, glabrous; styles 4–5 mm. long, slender, hairy. Fruit and seeds not seen.

Mozambique. N: between Mueda and Mocímboa do Rovuma, 20 km. from Mueda, fl. 23.ix.1948, *Pedro & Pedrógão* 5312 (COI; EA; LMA; SRGH). Also in Tanzania. In scrub and forests.

Var. *grewioides* is distributed from Cameroon to northern Angola; it has smaller flowers and larger fruit than var. *orientalis*.

3. **Paropsia brazzeana** Baill. in Bull. Soc. Linn. Paris **1**: 611 (1886).—Warb. in Engl. & Prantl, Pflanzenfam. **3**, 6a: 27 (1893).—Gilg in Engl., Bot. Jahrb. **40**: 472 (1908); in Engl. & Prantl, Pflanzenfam. ed. 2, **21**: 415 (1925).—Th. & H. Dur., Syll. Fl. Cong.: 222 (1909).—Engl., Pflanzenw. Afr. **3**, 2: 572, fig. 253D–F (1921).—A. & R. Fernandes in Garcia de Orta **6**, 4: 653 (1958); in C.F.A. **4**: 214 (1970).—F. White, F.F.N.R.: 268 (1962).—Sleumer in Bull. Jard. Bot. Nat. Belg. **40**, 1: 70, fig. 6 (1970). Type from Congo-Brazzaville.

Paropsia reticulata Engl., Bot. Jahrb. **14**: 391 (1892).—Warb., loc. cit., fig. 9D–F. —De Wild. & Th. Dur. in Ann. Mus. Congo Belge, Bot. Sér. 2, **1**, 2: 24 (1899); in Bull. Herb. Boiss. Sér. 2, **1**: 22 (1900).—Th. & H. Dur., loc. cit.—Gilg, tom. cit.: 414, fig. 183D–F (1925). Type from Angola.

Paropsia reticulata var. *ovatifolia* Engl., loc. cit.—Th. Dur. & Schinz in Mém. Cour. Acad. Roy. Sci. Belg. in 8°, **53**, 4: 140 (1896). Type from Zaire.

Paropsia reticulata var. *proschii* Briq. in Ann. Conserv. Jard. Bot. Genève **6**: 1 (1902). Type: Zambia, Barotseland, *Prosch* 35 (G, holotype).

Paropsia argutidens Sleumer in Fedde, Repert. **45**: 13 (1938). Type: Zambia, Solwezi Distr., between Lunsala and Mwafwe, *Milne-Redhead* 726 (BR; K, holotype).

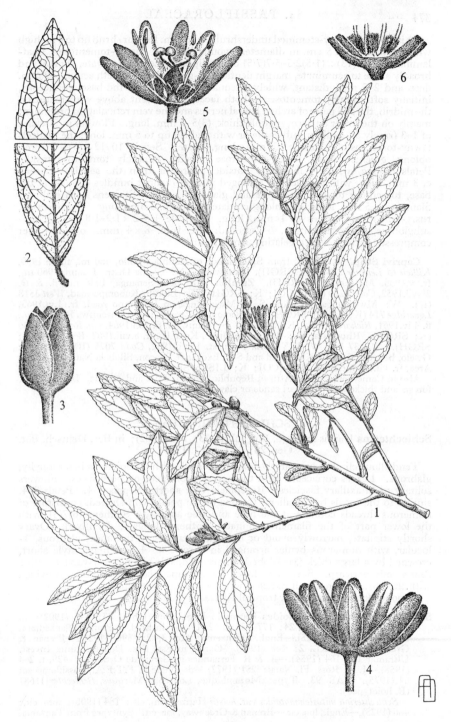

Tab. 92. PAROPSIA GREWIOIDES var. ORIENTALIS. 1, branchlets (×½); 2, leaf showing superior (top) and inferior surface (×1½); 3, flower bud (×2); 4, flower in lateral view (×1½); 5, flower with facing sepal and petals removed (×1½); 6, base of flower partly dissected to show filament attachment (×2), all from *Pedro & Pedrógão* 5312.

Rhizomatous many-stemmed undershrub up to 1 m. high or shrub up to 4 m. high with a trunk up to 5 cm. in diameter; young branches tawny-tomentose. Leaf-lamina (3)6–10(13) × (1·5)2·5–5·7(7·5) cm., elliptic-lanceolate to ovate, shortly and broadly acute to acuminate, margin denticulate-glandular, the teeth acute, 1–2 mm. deep and 2–5 mm. distant, widely attenuate to rounded at the base, coriaceous, initially softly tawny-tomentose on both faces, glabrescent above with maturity, the midrib, the 5–8 pairs of arched lateral nerves and the vein reticulation prominent mainly on the lower surface; petiole thick, (3)5–7 mm. long. Flowers in fascicles of 1–3 (rarely 5), axillary, subsessile or with pedicel up to 5 mm. long, the pedicels tawny-tomentose; bracts c. 2 mm. long, ovate. Sepals 10–12 × (3)4–5 mm., oblong, greenish-white, tawny-tomentose outside, minutely tomentose inside. Petals oblong, white, puberulous outside, narrower than the sepals. Corona c. 3 mm. high, with the laciniae arranged in 5 ± distinct bundles connate at the base, tomentose towards the apex and glabrous below. Stamens with glabrous filaments c. 5 mm. long and oblong-cordate anthers c. 2 mm. long. Ovary sessile, rusty-tomentose; styles c. 1 mm. long. Capsule 1·5–2 × 1·2–1·8 cm., ovoid-subglobose, rusty-tomentose, 6–7-seeded. Seeds c. 8 × 4 mm., ovoid, rather compressed, brown with a gelatinous orange aril.

Caprivi Strip. c. 53 km. from Singalamwe to Katima Mulilo, 960 m., fl. 3.i.1959, *Killick & Leistner* 3281 (K; SRGH). **Botswana.** N: Chobe Distr., Lesuma, 960 m., fr. v.1966, *Mutakela* 55 (SRGH). **Zambia.** B: near Senanga, 1088 m., fl. & fr. 29.vii.1952, *Codd* 7220 (BM; COI; K; SRGH). N: Lukulu-Kabompo road, *West* 3518 (K). W: Mwinilunga, 96·5 km. S. of Mwinilunga on Kabompo road, fr. 1.vi.1963, *Loveridge* 714 (K; LISC; SRGH). C: Kanona to Lusaka, near Lunsemfwa R., 1280 m., fl. 4.iv.1961, *Richards* 14929 (K; SRGH). S: Namwala, fl. 16.i.1964, *van Rensburg* 2786 (K; SRGH). **Rhodesia.** N: Sebungwe Distr., Kana R., fr. 8.viii.1962, *Wild* 3850 (K; SRGH). W: 37 km. W. of Victoria Falls, st. 15.vii.1952, *Codd* 7075 (BM; K). C: Gwelo, between the new Kana road and SW. Exchange Beacon, Silobela Native Purchase Area, fr. vii.1956, *Cowan* 14/56 (COI; K; LISC; SRGH).

Also in Cameroon, Central African Republic, Congo, Zaire and Angola. In woodland, forests and thickets on Kalahari sands or clayey soils.

3. SCHLECHTERINA Harms

Schlechterina Harms in Engl., Bot. Jahrb. **33**: 148 (1902); in Ber. Deutsch. Bot. Ges. **24**: 177, t. 12 (1906).

Tendrillous climber or sparsely-branched shrub with conspicuous heterophylly, glabrous. Leaves coriaceous with the nerves prominent on both faces. Flowers solitary or in axillary fascicles. Receptacle very short. Sepals 3–4. Petals 2–4, similar to the sepals but smaller. Corona inside petals, cup-shaped, consisting of numerous threads connate at the base and free apically. Stamens (6)7(8), with the lower part of the filaments connate; anthers elongate, dorsifixed. Ovary shortly stipitate, narrowly ovoid or oblong, attenuate to the apex, glabrous, 1-locular, with numerous ovules arranged in 2 rows on 4 placentas; style short, crowned by a large thick (3)–4-lobed stigma. Capsule oblong, 3–4-valved, with a thin woody pericarp. Seeds enveloped by a campanulate aril; embryo provided with thin endosperm, small radicle and thick ovate cotyledons.

A monospecific genus of E. tropical Africa.

Schlechterina mitostemmatoides Harms in Engl., Bot. Jahrb. **33**: 148 (1902); in Ber. Deutsch. Bot. Ges. **24**: 177–184, t. 12 (1906); in Engl. & Prantl, Pflanzenfam. ed. 2, **21**: 485 (1925).—Engl., Pflanzenw. Afr. **3**, 2: 596 (1921).—Brenan & Greenway, T.T.C.L. **2**: 448 (1949).—Garcia in Est. Ens. Docum. Junta Invest. Ultramar **12**: 164 (1954).—A. & R. Fernandes in Garcia de Orta **6**, 2: 429, t. 2–4 (1958).—J. H. Ross, Fl. Natal: 251 (1972).—de Wilde in F.T.E.A., Passifloraceae: 64 (1975). TAB. **93**. Type: Mozambique, Lourenço Marques, *Schlechter* 11681 (B, holotype†).

Schlechterina mitostemmatoides var. *holstii* Harms, tom. cit.: 184 (1906); tom. cit.: 486 (1925).—Engl., loc. cit.—Brenan & Greenway, loc. cit. Syntypes from Tanzania.

Climber or sparsely-branched shrub up to 3 m. high, glabrous, showing conspicuous heterophylly. Leaves coriaceous, with the nerves and veins prominent on both faces, those of saplings and juvenile shoots 10–21 × 0·4–1·1 cm., linear, pinnatipartite with the segments obliquely triangular; leaves on sterile adult

Tab. 93. SCHLECHTERINA MITOSTEMMATOIDES. 1, habit (× ⅔) *Faulkner* 1773 & *R. M. Graham* B467; 2, flower (×4); 3, stamens (×6); 4, pistil (×8), all from *Padwa* 791; 5, fruit (×⅔); 6, seed (×3), both from *Hornby* 656. From F.T.E.A.

branches 4–10(12) × 1–4·5(6) cm., entire, elliptic or oblong, acute or sometimes acuminate at the apex, cuneate at the base, or irregularly pinnatipartite or divided to the midrib; leaves of floriferous branches 4–7 × 1·5–3·5 cm., elliptic, acute or obtuse at the apex, entire or with a few fairly deep divisions, strongly undulate at the margin; petiole 7–12 mm. long, with a pair of glands at the top. Tendrils axillary, slender to stout, 3–10(15) cm. long. Flowers c. 1·5 cm. long and c. 2 cm. in diameter, white, odoriferous, solitary or in axillary fascicles; pedicels c. 5 mm. long. Hypanthium very short, c. 5 mm. wide. Sepals 7–9 × 3–5 mm., ovate or broadly oblong, obtuse or acute, with very distinct reticulation. Petals similar to the sepals, but smaller, narrower and of a thinner texture. Corona with the tube 2·5–3 mm. high and the threads c. 4 mm. long. Stamens with filaments c. 10 mm. long, shortly connate at the base; anthers c. 4 mm. long. Ovary c. 2·5 × 2 mm., ovoid, on a swollen gynophore c. 3 mm. long; style c. 1·5 mm. long, thick, crowned by a (3)–4-lobed stigma c. 2·5 mm. in diameter. Capsule 4·5–5 × 2·5 cm., many-seeded. Seeds c. 6·5 × 4 × 2·5 mm., pyriform.

Mozambique. N: between Mocímboa da Praia and Mueda, 15.ix.1948, *Pedro & Pedrógão* 5210 (COI; LMA). Z: Maganja da Costa, Gobene forest, c. 20 m., fr. 12.ii.1966, *Torre & Correia* 14575 (LD; LISC; MO). MS: Gorongosa, 27.ix.1943, *Torre* 3972 (COI; LD; LISC; LMA; MO). SS: Chibuto, fl. 12.ii.1942, *Torre* 3970 (BM; BR; COI; LD; LISC; LMA; MO; SRGH). LM: Maputo, near Bela Vista, fr. 20.xi.1940, *Torre* 2105 (LISC).
Also in Tanzania and S. Africa (Natal). In forests, shrubby savannas, thickets and littoral scrub on sandy or black soils.

4. BASANANTHE Peyr.*

Basananthe Peyr. in Bot. Zeit. **17**: 101 (1859).
Tryphostemma Harv., Thes. Cap. **1**: 32, t. 51 (1859).

Annual or perennial erect, prostrate or ascending herbs, rarely shrubs, with or without tendrils. Leaves alternate, entire, serrate or dentate or ± deeply lobed, sessile or petiolate. Stipules small, linear; some species provided with false stipules, sometimes foliaceous, developed from supra-axillary buds. Tendrils axillary or absent. Inflorescence a 1–3-flowered sessile or pedunculate cyme, sometimes with a simple tendril or a mucro-like prolongation in place of the terminal flower; bracteoles small, linear; pedicels slender. Flowers ☿, small, campanulate, whitish or greenish, stipitate, the stipe indistinctly articulate to the pedicel. Hypanthium narrow or shallowly cup-shaped. Sepals 5(6), oblong to lanceolate, imbricate, free. Petals 0 or 1–2 or (4)5(6), oblong to lanceolate, free, narrower and shorter than the sepals. Outer corona inserted at the bottom of the hypanthium, consisting of a barrel-shaped tube crowned with a ring of free threads. Inner corona shorter than the outer one and represented by a cup-shaped membrane with entire or irregularly lobulate margin. Stamens 5(6–9), with the free filaments inserted on the inside of the inner corona; anthers basifixed, subsagittate, 2-locular. Ovary ellipsoid, ovoid or obovoid, sessile or shortly stipitate, 1-locular; ovules several or few, inserted on 3(4) placentas; styles 3–4, free or connate at the base, each with a capitate stigma. Capsule ellipsoid, 3(4)-valved. Seeds 1 or few, ellipsoid to reniform, ± compressed, arillate, with a coriaceous, usually rugose, blackish testa; embryo straight with horny endosperm and foliaceous cotyledons.
A genus of c. 25 species in tropical and southern Africa.

* Dr. W. J. J. de Wilde has recently produced a revision of this genus (in Blumea **21**: 327–356, 1973) as well as a monograph of *Adenia* (in Meded. Landb. Wag. **71**, 18: 1–281, 1971). Dr. de Wilde has examined not only material from the F.Z. area but also from other regions (when the species are not confined to the F.Z. area) and his descriptions are excellent and very ample. For these reasons we have followed his descriptions and our keys have been adapted from those in his works.
On account of several circumstances we were unable to see any material from PRE and only saw part of the SRGH collection. In general we cite specimens which we have observed; however, when certain regions were not represented in material available to us but were reported by Dr. de Wilde to have yielded relevant gatherings, we have referred to the specimens cited by this author, marking them with an asterisk.

False stipules present:
  Leaves not lobed (margins entire, serrate or dentate):
    Leaf-margin only with a few minute teeth towards the base   -   1. *pseudostipulata*
    Leaf-margin conspicuously serrate-dentate   -   -   -   -   -   2. *apetala*
  Leaves 3–9-lobed:
    Sepals 2·5–4·5 mm. long, with keeled nerves; leaves 3–7(9)-lobed   -   3. *pedata*
    Sepals 5–7·5 mm. long, without keeled nerves; leaves 3-lobed   -   4. *triloba*
False stipules absent:
  Petals absent or 1–2:
    Petiole 0–5 mm. long; leaves unlobed; tendrils usually absent:
      Leaves more than 2·5 cm. long (up to 15 cm.); sepals 4·5–8·5 mm. long:
        Leaves lanceolate or linear-lanceolate, up to 15 × 1·2 cm., with the margin
          serrulate-denticulate or subentire; tube of the outer corona c. 3 mm. high;
          inner corona 1·5–2·5 mm. high   -   -   -   -   5. *longifolia*
        Leaves broader, sometimes obovate or subcircular, up to 8 × 5 cm.; tube of the
          outer corona 1–2 mm. high; inner corona 1–1·5 mm. high
                                                                      6. *sandersonii*
      Leaves 0·7–2·5 cm. long, suborbicular on the creeping shoots, elliptic-acute on the
        erect ones; sepals 3–5 mm. long   -   -   -   -   7. *parvifolia*
    Petiole 5–45 mm. long or more; leaves deeply 3-lobed; tendrils present
                                                                      8. *phaulantha*
  Petals 4 or 5:
    Styles free or connate for up to ¼ of their length; sepals 3–14 mm. long; bracteoles
      0·5–4(5) mm. long:
      Plant with tendrils, annual or perennial; stem up to 3 m.; leaves 3–5-lobed,
        rarely unlobed   -   -   -   -   -   -   -   9. *hanningtoniana*
      Plant without tendrils, perennial, arising from a woody rootstock, with erect or
        ± decumbent shoots; leaves unlobed:
        Shoots erect; leaves pubescent on both faces; disk well developed
                                                                      10. *holmesii*
        Shoots ± decumbent, short; superior surface of leaves glabrous; disk absent
          or inconspicuous:
          Inferior leaf surface, petiole, stem and flowers pubescent; stems prostrate or
            ascending, up to 50 cm. long; leaf-lamina 3–7 × 1·5–4 cm.
                                                                      11. *reticulata*
          Inferior leaf surface papillate (sometimes the papillae rather long and
            resembling hairs); petiole, stem and flowers glabrous or ± long-papillate;
            stems prostrate, up to 10 cm. long; leaf-lamina 3–19 × 1·5–13 cm.
                                                                      12. *baumii*
    Styles connate for more than ¼ of their length; sepals 7–15 mm. long; bracteoles
      3–7 mm. long   -   -   -   -   -   -   -   - 13. *heterophylla*

1. **Basananthe pseudostipulata** de Wilde in Blumea **21**, 2: 337, fig. 1c & 5 (1973).
Type from Zaire (Katanga).

Glabrous perennial herb with a woody rootstock. Stems up to 70 cm. long, erect or
decumbent. Leaf-lamina 1·5–6·5 × (0·1)0·2–1·5(3) cm., elliptic to lanceolate, acute
and mucronate at the apex, margin subentire (with only a few teeth in the lower ½),
acute at the base. Stipules 1·5–2 mm. long. False stipules 0·3–2 cm. long, oblong
to lanceolate, acute. Tendrils absent or 1–3 cm. long. Cymes 1–2-flowered;
peduncle 0·4–2 cm. long; bracteoles 1–1·5 mm. long. Flowers glabrous with stipe
3–6 mm. long. Hypanthium 1·5–2 mm. broad. Sepals 4–6 mm. long, subobtuse
at the apex. Petals absent or present, 3–4·5 mm. long. Outer corona with tube
1–1·5 mm. high and the threads 1–1·25 mm. Inner corona 0·75–1 mm. high, cup-
or funnel-shaped. Stamens 5, with the filaments 2–2·5(3) mm. long and the anthers
0·5–0·75 mm. long. Ovary c. 0·75 mm.; styles 2·5–3·5 mm. long, free. Capsule
1·1(1·5) × 0·6 cm., ellipsoid, light green. Seed c. 5 × 3 mm., solitary.

Rhodesia. E: Melsetter, Tarka Forest Reserve E. of Bunga, 1000 m., fl. x.1971,
Goldsmith 28/71 (BR; LMA; M; SRGH). **Mozambique.** MS: Manica, Mavita,
Mocuta Mt., 1588 m., fl. 13.xii.1965, *Pereira & Marques* 1072 (COI; LISC; LMU).
Also in Zaire (Katanga). In steppes, savannas and forests.

2. **Basananthe apetala** (Bak. f.) de Wilde in Blumea **21**, 2: 338, fig. 1e & 5 (1973); in
F.T.E.A., Passifloraceae: 56 (1975). Syntypes: Malawi, *Buchanan* 1063 (K); *Scott
Elliot* 8661 (K); Zomba Mts., *Whyte* (BM; K).
    Tryphostemma apetalum Bak. f. in Trans. Linn. Soc. Lond. Ser. 2, **4**: 14, t. 3
fig. 7–11 (1894).—Hutch. & Pearce in Kew Bull. **1921**: 263 (1921).—Engl., Pflanzenw.
Afr. **3**, 2: 599 (1921).—Harms in Engl. & Prantl, Pflanzenfam. ed. 2, **21**: 488

(1925).—Norlindh in Bot. Notis. **1934**: 107 (1934).—A. & R. Fernandes in Garcia de Orta **6**, 2: 251 (1958). Syntypes as above.

    Tryphostemma apetalum var. *serratum* Bak. f. in Journ. Bot., Lond. **37**: 437 (1899).—Engl. in Sitz.-Ber. Königl. Preuss. Akad. Wiss. Berl. **52**: 27, 28, 30 (1906); loc. cit. fig. 265 (1921).—Harms, tom. cit.: 487, fig. 222 (1925).—A. & R. Fernandes, loc. cit. Syntypes: Rhodesia, Mazoe Distr., *Eyles* 412, 461 (BM); *Rand* 1347 (BM); Umtali, *Teague* 477 (K); Victoria Distr., *Munro* 1548 (BM).

Erect, perennial, glabrous herb up to 70 cm. high, with a woody rootstock. Leaf-lamina 1·5–10 × 0·5–2(4·5) cm., elliptic to lanceolate, obtuse or acute at the apex, margin ± conspicuously serrate, the teeth up to 1 mm. deep, mucronate, acute at the base; petiole 1–4(6–12) mm. long. Stipules 2–3 mm. long. False stipules asymmetrical, (0·5)1–4 cm., acute at both extremities, ± serrate at the margin. Axillary tendrils up to 8 cm. long. Cymes 1–2-flowered; peduncles 1–5 cm. long; bracteoles 1·5–3 mm. long, filiform; inflorescence tendrils up to 6 cm. long, slender. Flowers pale yellow with stipe 2–5 mm. long. Hypanthium c. 2·5 mm. broad. Sepals 5, the inner ones ± petaloid, 4–5·5 mm. long, obtuse at the apex. Petals absent. Outer corona with the tube c. 2 mm. high and the threads 1–1·75 mm. Inner corona 0·75–1 mm. high, cup-shaped. Stamens 5, with the filaments 1·5–2·5 mm. long and the anthers 0·75–1 mm. long. Ovary 0·75–1·25 mm. long; styles 3(4), 0·75–1·25 mm. long, free. Capsule 1·5–2 cm. long, 1–3-seeded. Seeds c. 7 mm. long.

    Rhodesia. N: Mazoe, N. of Jumbo, c. 1280 m., fl. & fr. 13.i.1958, *Phipps* 845 (SRGH). W: Maschambas, fl. ix.1931, *Myres* 149 (K). C: Salisbury Distr., Hatcliffe, c. 22·5 km. NNE. of Salisbury on road to Domboshawa, 1555 m., fl. & fr. 26.ix.1955, *Drummond* 4882 (BR; K; LISC; LMA; SRGH). E: Umtali Distr., East Commonage, 1152 m., fl. & fr. 4.xii.1953, *Chase* 7920 (SRGH). S: Victoria, fl., *Monro* 1548 (BM). **Malawi.** C: Dedza, Chongoni Forest, fl. 14.vii.1969, *Salubeni* 1368 (SRGH). S: Zomba and Shire Highlands, *Whyte* (BM; K). **Mozambique.** MS: Manica, Rotanda, near Messambize R., c. 1000 m., fl. & fr. 17.xi.1965, *Torre & Correia* 12991 (COI; LD; LISC; MO).

    Also in Tanzania. In open forests, savannas, scrub, grasslands, burnt places, etc.

3. **Basananthe pedata** (Bak. f.) de Wilde in Blumea **21**, 2: 333, fig. 1a & 5 (1973). TAB. **94**. Type: Rhodesia, Shashi R., *Rand* 67 (BM, holotype).

    Tryphostemma pedatum Bak. f. in Journ. Bot., Lond. **37**: 436 (1899).—Engl. Pflanzenw. Afr. **3**, 2: 600 (1921).—Hutch. & Pearce in Kew Bull. **1921**: 263 (1921).— Harms in Engl. & Prantl, Pflanzenfam. ed. 2, **21**: 488 (1925). Type as above.

    Tryphostemma schlechteri Schinz in Viert. Nat. Ges. Zürich **55**: 243 (1911).— Engl., loc. cit.—Hutch. & Pearce, tom. cit.: 260 (1921).—Schreiber in Prodr. Fl. SW. Afr. **89**: 2 (1968). Type from S. Africa (Transvaal).

    Tryphostemma arenophilum Pott in Ann. Transv. Mus. **5**: 234 (1915). Type from S. Africa (Transvaal).

    Tryphostemma harmsianum Dinter in Fedde, Repert. **24**: 304 (1928). Type from SW. Africa.

Annual or biennial, aromatic, glabrous, erect or decumbent herb up to 50 cm. high. Stem simple or branched from the base, the branches ascending. Leaf-lamina deeply pedately 3–7(9)-lobed, the lobes (0·5)1–8 cm. long, elliptic to linear, obtuse to acuminate at the apex, margin dentate-mucronate, the teeth up to 1·5 mm., light green; leaf-base subcordate, decurrent into a 2·5–15 mm. long petiole. Persistent cotyledons 2·2–2·7 × 1–1·5 cm., broadly elliptic, with petiole 5–9 mm. long. Stipules 2–6 mm. long. False stipules 5–20 mm. long, foliaceous, asymmetrical, acute-acuminate and mucronate at the apex. Tendrils absent. Cymes 1–2-flowered, with peduncle up to 2 cm. long; bracteoles 2–4 mm. long. Flowers glabrous, with stipe 1·5–3·5 mm. long. Hypanthium 1–2 mm. broad. Sepals 2·5–4·5 mm. long, with 2(3) keeled or winged green nerves. Petals 2–4·5 mm. long. Outer corona with the tube (0·5)0·75–1·25 mm. high and the threads (0·5)0·75–1 mm. long. Inner corona 0·33–0·75 mm. high, cup-shaped. Stamens 5, with the filaments 1–1·5 mm. long and the anthers 0·5–1 mm. long. Ovary 0·5–1·5 mm. long; styles free, 0·60–2·25 mm. long. Capsule 7–10(14) mm. long, ellipsoid to narrowly pyriform, shortly pedicellate, 1-seeded. Seed c. 7 × 4 mm., subreniform, alveolate, blackish.

    Botswana. N: 78 km. N. of Aha Hills, SW. African border, fl. 13.iii.1965, *Wild & Drummond* 7001 (COI; K; SRGH). SW: Okwa Valley, 5 km. from Lobatsi to Ghanzi, fr. 11.v.1969, *Brown* 6060 (SRGH). SE: Mahalapye, *van Rensburg* B4071 (PRE*).

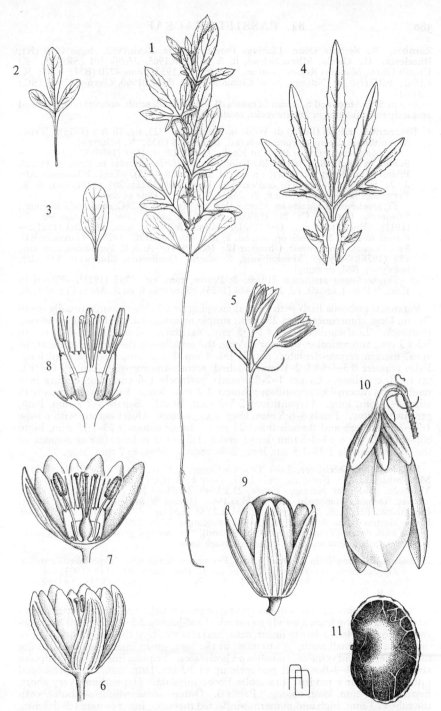

Tab. 94. BASANANTHE PEDATA. 1, habit (×½); 2, lower leaf (×½); 3, persistent cotyledon
(×½), 1–3 from *Wild* 4657; 4, upper leaf and false stipules (×½); 5, inflorescence
(×5), 4–5 from *Angus* 1078; 6, flower in lateral view (×8); 7, flower in longitudinal
section (×8); 8, base of flower in longitudinal section showing disc, stamen attachment, inner and outer coronas and pistil (×10), 6–8 from *Chase* 4770; 9, flower with
developing fruit in lateral view (×5); 10, same with dehiscing capsule (×4); 11,
seed (×4), 9–11 from *Angus* 1078.

Zambia. B: Sesheke Distr., Lichinga Forest, fl. & fr. 20.xii.1952, *Angus* 1078 (K).
Rhodesia. C: Gwelo, Mlezu School, fl. & fr. 22.iii.1965, *Molife* 101 (SRGH). E:
Umtali Distr., Maranka Reserve, 800 m., fl. & fr. 10.ii.1953, *Chase* 4770 (BM; COI; K;
LISC; SRGH). S: Ndanga, N. of Chipinda Pools, fl. 17.i.1960, *Goodier* 826 (LISC;
SRGH).
Also in SW. Africa and S. Africa (Transvaal). Woodland, scrub, savanna and grassland
on sandy soils, fissures in granite rocks, roadsides, etc.

4. **Basananthe triloba** (Bolus) de Wilde in Blumea **21**: 335, fig. 1b & 5 (1973). Type:
Mozambique, Delagoa, Puzeen's Kraal, *Bolus* 7606 (BM; K, holotype).
    Tryphostemma trilobum Bolus in Hook., Ic. Pl. **19**: t. 1838 (1889) ("triloba").—
Schinz in Engl., Bot. Jahrb. **15**, Beibl. 33: 2 (1892).—Harms in Engl. & Prantl,
Pflanzenfam. **3**, 6a: 81 (1893); op. cit. ed. 2, **21**: 488 (1925).—Engl., Pflanzenw. Afr.
3, 2: 599 (1921).—Hutch. & Pearce in Kew Bull. **1921**: 261 (1921).—A. & R.
Fernandes in Garcia de Orta **6**, 2: 252, t. 7 (1958). Type as above.
    Tryphostemma schinzianum Harms [loc. cit. fig. 25A & 26C nom. nud.] in Engl.,
Pflanzenw. Ost-Afr. **C**: 281 (1895).—Bak. f. in Journ. Linn. Soc., Bot. **40**: 73
(1911).—Engl., tom. cit.: 600 (1921).—Hutch. & Pearce, tom. cit.: 263 (1921).—
Harms in Engl. & Prantl, op. cit. ed. 2, **21**: 488, fig. 217C (1925).—Garcia in Est
Ens. Docum. Junta Invest. Ultramar **12**: 165 (1954).—A. & R. Fernandes, tom. cit.:
253 (1958). Type: Mozambique, Zambezia, Quelimane, *Stuhlmann* 835 (B†,
holotype; BM, drawing).
    Tryphostemma sagittatum Hutch. & Pearce, tom. cit.: 262 (1921).—Harms in
Engl. & Prantl, op. cit. ed. 2, **21**: 488 (1925). Syntypes from S. Africa (Transvaal).

Perennial glabrous herb with several creeping or climbing shoots, usually up to
70 cm. long, sometimes up to 250 cm., simple or branched, arising from a woody
rootstock. Leaf-lamina 1·5–8 × 1–7·5 cm., sagittate, 3-lobed, the lobes up to
5·5 × 3 cm., suborbicular, ovate or oblong, the middle one the largest, acute at the
apex, margin serrate-dentate; petiole 0·1–4 cm. long. Stipules 3–8 mm. long.
False stipules 0·5–3·5 × 0·2–1 cm., 2-lobed, serrate-mucronate. Axillary tendrils
up to 10 cm. long. Cymes 1–2-flowered; peduncle 1–6 cm. long, ending in a
tendril or a mucro-like appendage; bracts 2–3 mm. long. Flowers glabrous with
stipe 4–7 mm. long. Hypanthium 2·5–3 mm. broad. Sepals 5–7·5 mm. long,
greenish-yellow. Petals 4–6·5 mm. long, concolorous. Outer corona with a tube
1·75–2 mm. high and threads 0·5–1·25 mm. Inner corona 1·25–1·75 mm. high,
cup-shaped. Ovary 1–1·5 mm. long; styles (1)2·5–4 mm. long, free or connate at
the base. Capsule 1·25–1·5 cm. long, 2–3-seeded. Seeds c. 7 mm. long.

Rhodesia. E: Melsetter, above Tembioi Gorge, fl. 13.ix.1956, *Taylor* 1748 (SRGH).
Mozambique. N: Nampula, 14.iv.1937, *Torre* 1371 (COI; LISC). Z: Quelimane,
Moebede, road near Namagoa Estate, 23.x.1948, *Faulkner* 308 (BR; COI; K). MS:
Manica, Dombe, Machongo Mt., fl. 26.xi.1965, *Pereira & Marques* 944 (LMU). SS:
Inhambane, Panda, fl. 14.vii.1958, *Balsinhas* 1384 (LMU). LM: Manhiça, near José
Maria Martins Farm, fl. 22.x.1969, *Correia & Marques* 1348 (LMU).
Also in S. Africa (Transvaal, Natal, Swaziland). In swamps, grassland, scrub, riparian
bush and open forests on sandy, stony and black soils.

5. **Basananthe longifolia** (Harms) R. & A. Fernandes, comb. nov. Type from Tanzania.
    Tryphostemma longifolium Harms in Engl., Bot. Jahrb. **33**: 149 (1902).—Engl.,
Pflanzenw. Afr. **3**, 2: 598 (1921).—Hutch. & Pearce in Kew Bull. **1921**: 264 (1921).—
A. & R. Fernandes in Garcia de Orta **6**, 2: 252, t. 5 (1958). Type as above.

Perennial, glabrous herb with ascending or erect usually unbranched stems up to
60 cm. long arising from a woody rootstock. Leaf-lamina 2·5–15 × 0·5–1·5 cm., narrowly elliptic or lanceolate to linear, acute and mucronate at the apex, margin entire
or with a few small teeth, ± attenuate at the base, green, membranous, with the
midrib, nerves and veins conspicuous on both faces. Stipules linear. False stipules
absent. Cymes 1–3-flowered; peduncle up to 3·5 cm. long, sometimes prolonged
into a cirrhus or a tendril; bracteoles linear-subulate. Hypanthium very short.
Sepals 5, c. 7 mm. long, oblong. Petals 0. Outer corona cylindric-tubular, with
the tube c. 3 mm. high and numerous inflected threads. Inner corona 1·5–2·5 mm.
high, broadly infundibuliform. Stamens 5, with the filaments inserted on the inner
side of the inner corona. Ovary ovoid, shortly stipitate; styles 3, filiform, shortly
connate at the base, provided with capitellate stigmas; ovules usually 3. Capsule
not seen.

Mozambique. N: Marrupa, fl. 13.vi.1948, *Pedro & Pedrógão* 4299 (COI; LMA).
Also in Tanzania. In sandy soils.

6. **Basananthe sandersonii** (Harv.) de Wilde in Blumea **21**: 339, fig. 2a–b (1973); in F.T.E.A., Passifloraceae: 57 (1975) pro parte. Type from S. Africa (Natal).

  Tryphostemma sandersonii Harv., Thes. Cap. **1**: 33, t. 51 (1859); in Harv. & Sond., F.C. **2**: 499 (1862).—Harms in Engl. & Prantl, Pflanzenfam. **3**, 6a: 81 (1893); op. cit. ed. 2, **21**: 488 (1925).—Engl., Pflanzenw. Afr. **3**, 2: 598 (1921).—Hutch. & Pearce in Kew Bull. **1921**: 265 (1921).—J. H. Ross, Fl. Natal: 251 (1972). Type as above.

  Tryphostemma natalense Mast. in Trans. Linn. Soc. Lond. **27**: 639 (1871) *nom. illegit.*

  Tryphostemma friesii Norlindh in Bot. Notis. **1934**: 107, fig. 9 & 10 (1934).— A. & R. Fernandes in Garcia de Orta **6**, 2: 252 (1958). Type: Rhodesia, Inyanga, *F. N. & W.* 3112 (BM; BR; PRE; SRGH; UPS, holotype).

Perennial glabrous herb, with woody rootstock producing 1–several erect or prostrate simple or ± branched shoots. Leaf-lamina unlobed, 2·5–8 × 2·5–5 cm., suborbicular, broadly ovate, obovate or elliptic, rounded or obtuse at apex, remotely and acutely denticulate especially towards the base, rounded or cuneate at the base, green on the upper surface, glaucous beneath; midrib and nerves prominent beneath, conspicuously and closely net veined on both surfaces; petiole 0–5 mm. long. Stipules 1·5–5 mm. long. False stipules absent. Cymes 1–3-flowered; peduncle up to 4·5 cm. long, often with a tendril in place of the terminal flower; bracteoles 1–4 mm. long. Flowers pale yellow with stipe 3–17 mm. long. Hypanthium 2–4 mm. broad. Sepals 5–7, 4·5–8·5 mm. long, obtuse at the apex. Petals 0, rarely 2. Outer corona blue, with the tube 1–2 mm. high and the threads 0·75–1·5 mm. long. Inner corona 1–1·5 mm. high, cup-shaped, lobulate. Stamens 5, inserted at the inner corona, with filaments 1·5–4 mm. long and anthers 0·75 -1·25 mm. long. Ovary 1–1·5 mm. long, obovoid; styles 3–4, free, 3–4 mm. long. Capsule 1·5–2 × 0·7 cm., pendulous, obovoid, 1–4-seeded. Seeds 6–10 mm. long.

Rhodesia. E: Umtali Distr., between Henkel's Gap and SE. boundary of Stapleford Forest Reserve, 1760 m., fl. 25.ix.1952, *Chase* 4660 (BM; SRGH). **Mozambique.** MS: Barué, Vila Gouveia, Choa Mt., near Chapéu, fl. & fr. 17.ix.1942, *Mendonça* 305 (COI; LD; LISC; LMU; MO).
Also in S. Africa (Transvaal, Natal and Cape Prov.) and Swaziland. In grassland, savanna, open scrub and forest edges on sandy and rocky soils.

7. **Basananthe parvifolia** (Bak. f.) de Wilde in Blumea **21**: 340, fig. 2d & 5 (1973). Type: Rhodesia, Melsetter, *Swynnerton* 1415 (BM, holotype).

  Tryphostemma parvifolium Bak. f. in Journ. Linn. Soc., Bot. **40**: 73 (1911).— Engl., Pflanzenw. Afr. **3**, 2: 598 (1921).—Hutch. & Pearce in Kew Bull. **1921**: 265 (1921).—Harms in Engl. & Prantl, Pflanzenfam. ed. 2, **21**: 488 (1925). Type as above.

  Tryphostemma humile Dandy in Kew Bull. **1927**: 251 (1927). Type: Rhodesia, Melsetter, *Walters* 2727 (K, holotype; SRGH).

Perennial glabrous herb, erect, prostrate or with both prostrate and erect shoots, the prostrate shoots up to 50 cm., the erect ones up to c. 15 cm., arising from a woody rootstock. Leaves simple, alternate, overlapping on the main branches, thinly fleshy, glaucous, those of the prostrate shoots suborbicular, 0·7–2·5 cm. in diameter, cordate at the base, those of the erect ones ovate to oblong-lanceolate, 0·75–2 cm. long, acute at the apex, margin entire or with a few minute teeth at the base; petiole 0·5–2 mm. long. Stipules paired, 1–2 mm. long, filiform. False stipules absent. Tendrils absent. Cymes 1–2-flowered, the flowers small, whitish with stipe 6–10 mm. long; peduncle up to 1 cm. long; bracteoles c. 1·5 mm. long. Hypanthium c. 1·5 mm. broad. Sepals 5(6), (3)4–5 mm. long, subobtuse at the apex. Petals 0. Outer corona blue, with the tube 1–1·5 mm. high and threads 1(1·5) mm. Inner corona c. 1 mm. high, cup-shaped. Stamens 5, with filaments 2–2·25 mm. long and anthers 0·5–0·75 mm. long. Ovary c. 1 mm. long; styles 3–4, free, 1·5–2·5 mm. long. Capsule c. 1 cm. long, with thin pericarp, 1-seeded. Seed c. 7 × 6 mm., roundish.

Rhodesia. E: Chipinga, Fortuna Farm, 1120 m., st. 4.v.1959, *Chase* 7127 (BM; SRGH).
Known only from E. Rhodesia. In grassland and open forest.

8. **Basananthe phaulantha** (Dandy) de Wilde in Blumea **21**: 341, fig. 2e & 5 (1973); in F.T.E.A., Passifloraceae: 58 (1975). Type from Tanzania.

    Tryphostemma phaulanthum Dandy in Kew Bull. **1927**: 251 (1927). Type as above.

Annual, glabrous, erect, simple or short-branched herb up to 60 cm. high. Leaf-lamina up to 6 × 5 cm., deeply 3-lobed, subcordate at the base, the lobes 1–5 × 0·5–2 cm., elliptic to oblong, obtuse or acute and shortly mucronate at the apex, margin serrate-mucronate; petiole (0·5)1–4·5 cm. long, slender. Stipules c. 2 mm. long. False stipules absent. Tendrils 1–3 cm. long, slender. Cymes 1–2-flowered, the flowers greenish with stipe 1–2·5 mm. long, peduncle up to 0·75 mm. long ending in a slender tendril up to 1 cm. long; bracteoles c. 1 mm. long. Hypanthium c. 0·75 mm. broad. Sepals 1·5–2 mm. long, subacute to obtuse at the apex. Petals 0. Outer corona with the tube c. 1 mm. high and threads c. 0·75 mm. long. Inner corona 0·25–0·35 mm. high, cup-shaped. Stamens 5, with the filaments 0·5–0·65 mm. long and anthers c. 0·5 mm. long. Ovary c. 0·75 mm. long; styles c. 0·25 mm. long, free. Capsule c. 7 × 5 mm., ovoid, 1-seeded. Seed 4–5 mm. long.

Zambia. N: Mupamadzi R., *Astle* 4419 (K). E: Chipata Distr., Chizombo, 4 km. SE. of Mfuwe, fl. 30.xii.1968, *Astle* 5404 (K). **Malawi.** C: N. of Chitala on Kasache road, 700 m., fl. & fr. 12.ii.1959, *Robson* 1570 (BM; BR; K; LISC; SRGH).
Also in Tanzania. Along watercourses on clay soils.

9. **Basananthe hanningtoniana** (Mast.) de Wilde in Blumea **21**: 342, fig. 3a–c & 6 (1973); in F.T.E.A., Passifloraceae: 59 (1975). Syntypes from Kenya and Tanzania.

    Tryphostemma hanningtonianum Mast. in Hook., Ic. Pl. **15**: t. 1484 (1885).—Engl., Bot. Jahrb. **14**: 390 (1891); Pflanzenw. Afr. **3**, 2: 599 (1921).—Harms in Engl. & Prantl, Pflanzenfam. **3**, 6a: 81 (1893); op. cit. ed. 2, **21**: 488 (1925); in Engl., Pflanzenw. Ost-Afr. **C**: 280 (1895).—Hutch. & Pearce in Kew Bull. **1921**: 261 (1921). Syntypes as above.

    Tryphostemma niloticum Engl., tom. cit.: 389 in obs. (1891); op. cit. **15**: 577 (1893); tom. cit.: 598 (1921).—Harms, loc. cit. (1893); tom. cit.: 281 (1895).—Hutch. & Pearce, loc. cit. Type from Sudan.

    Tryphostemma volkensii Harms in Engl., Bot. Jahrb. **19**, Beibl. 47: 40 (1894); in Engl., Pflanzenw. Ost-Afr. **C**: 281 (1895); in Engl. & Prantl, loc. cit.—Hutch. & Pearce, loc. cit. Type from Tanzania.

    Tryphostemma hanningtonianum var. *latilobum* Harms in Engl., Pflanzenw. Ost-Afr. **C**: 280 (1895) (" latiloba "). Type from Tanzania.

    Tryphostemma latilobum (Harms) Engl., op. cit.: 599 (1921). Type as above.

    Tryphostemma snowdenii Hutch. & Pearce, loc. cit. fig. 1.—Harms, loc. cit. ed. 2 (1925). Type from Kenya.

    Tryphostemma stolzii Engl. & Harms in Engl., Pflanzenw. Afr. **3**, 2: 599 (1921); in Notizbl. Bot. Gart. Berl. **8**: 291 (1923).—Harms, loc. cit. ed. 2 (1925).—Garcia in Est. Ens. Docum. Junta Invest. Ultramar **12**: 165 (1954).—A. & R. Fernandes in Garcia de Orta **6**, 2: 252 (1958). Syntypes from Tanzania.

    Tryphostemma foetidum Lebrun & Taton in Bull. Jard. Bot. Brux. **18**: 283 (1947). Type from Zaire.

Annual or perennial, glabrous or pubescent erect or climbing herb 0·1–3 m. tall. Leaf-lamina 1–13 × 1–13 cm., 3–5-lobed, rarely simple, glabrous or hairy (sometimes only beneath), truncate or cordate at the base; lobes up to 8·5 × 3 cm., ovate, elliptic, lanceolate or obovate, obtuse or acute at the apex, margin entire or ± remotely dentate; petiole 0·25–8 cm. long, sometimes with a few glandular teeth at the apex. Persistent cotyledons up to 2 × 0·8 cm., elliptic, with slender petiole c. 6 mm. long. Stipules 3–15 mm. long. False stipules absent. Tendrils 1–15 cm. long, slender. Cymes 1–2-flowered, the flowers greenish-cream with stipe 2–10 mm. long; peduncle 1–10 cm. long; bracteoles 0·5–4 mm. long, sometimes caducous or absent. Hypanthium 1·5–4 mm. broad. Sepals 3–8 mm. long, obtuse or acute at the apex, glabrous or hairy. Petals 2–6 mm. long. Outer corona bluish, with the tube 1·5–2·25 mm. high and threads 1·5–2·25 mm. long. Inner corona 0·75–1·5 mm. high, cup-shaped. Stamens 5(9), with filaments 2–3 mm. long and sagittate anthers 1–1·5(2·5) mm. long. Ovary 0·65–2 mm. long; styles 3(4), free, 1–3 mm. long. Capsule 1·5–1·7 × 0·7 cm., ellipsoid, 1–5-seeded. Seeds 6–7 × 4 mm., subreniform, blackish.

Zambia. B: Barotse, Senanga Distr., *H.J.A.R.* (K*). E: Kapatamoyo, near Ft. Jameson, 1150 m., fl. 5.i.1959, *Robson* 1035 (BM; K; LISC; SRGH). S: Kalomo, Mulobezi, fl. & fr. 9.iv.1955, *E. M. & W.* 1440 (BM; LISC; SRGH). **Rhodesia.** N:

Mtoko Distr., Nyagoko R., 800 m., fl. & fr. 14.ii.1962, *Wild* 5647 (BR; COI; K; LISC; M; SRGH). W: Wankie Distr., Wankie Game Reserve, fl. 22.ii.1956, *Wild* 4783 (SRGH). **Malawi.** C: Dedza Distr., *Banda* 616 (SRGH*). **Mozambique.** Z: Gurué, R. Licungo, c. 1000 m., fl. & fr. 7.iv.1943, *Torre* 5101 (BR; LD; LISC; LMA; MO). T: Cabora Bassa, Songa Mt., c. 900 m., fl. & fr. 31.xii.1965, *Torre & Correia* 13958 (COI; LISC; LMU).
Also in Sudan, Ethiopia, Zaire, Uganda, Kenya and Tanzania. Woodland, savanna and thickets on soils of various types.

10. **Basananthe holmesii** R. & A. Fernandes in Bol. Soc. Brot. Sér. 2, **50**: 165, t. 1 & 2 (1976). Type: Zambia, Mwinilunga, *Holmes* 1320 (K, holotype).

Perennial, erect villous herb, arising from a woody rootstock. Stem up to 25 cm. high, branched. Tendrils absent. Leaves 2–6 × 0·3–2·4 cm., narrowly or broadly lanceolate, acute and ± long-acuminate, the acumen black, margin entire, the base attenuate into a petiole up to 4 mm. long, densely cinereous-hairy on both faces, the hairs ± appressed, pinnately nerved, the nerves ± concealed by the indumentum. Stipules up to 12 mm. long, linear, hairy. False stipules absent. Cymes 1–2-flowered; peduncle 1–2 cm. long, hirsute; bracteoles c. 5 mm. long, linear, hairy. Flowers pubescent, with hirsute stipe up to 7 mm. long. Hypanthium 4–5 mm. wide. Sepals 5, up to 14 × 2·5 mm., hairy outside, glabrous inside, obtuse at the apex, 3-nerved. Petals 4, up to 10 × 2 mm., thin, shorter than the sepals. Outer corona with a tube c. 2·5 mm. high and threads c. 3·75 mm. Disk c. 0·5 mm. high. Inner corona c. 1 mm. high, cup-shaped. Stamens 5, with the filaments c. 4·5 mm. long and sagittate anthers c. 4·5 × 0·6 mm. Ovary ellipsoid; gynophore c. 1 mm. long; styles 3, c. 6 mm. long, connate for c. 1·5 mm. at the base. Capsule c. 2·2 × 1 cm., ellipsoid. Seeds unknown.

Zambia. W: Mwinilunga, fl. 7.xi.1955, *Holmes* 1320 (K).
Known only from Zambia. In calcareous soils or on red sandy loam derived from limestone.

11. **Basananthe reticulata** (Bak. f.) de Wilde in Blumea **21**: 350, fig. 4f & 7 (1973). Type from Angola.
  Tryphostemma reticulatum Bak. f. in Hutch. & Pearce in Kew Bull. **1921**: 264 (1921).—A. & R. Fernandes in Garcia de Orta **6**, 4: 664, t. 9 (1958); in C.F.A. **4**: 227 (1970). Type as above.

Perennial prostrate or ascending herb with stems up to 50 cm. long, arising from a woody rootstock. Leaf-lamina 3–7 × 1·5–4 cm., simple, ovate or elliptic or oblong, rounded, obtuse or subacute at the apex, entire-margined, rounded to acute at the base, glabrous on the upper surface, shortly pubescent beneath; petiole 0·25–1 cm. long. Stipules 2–6 mm. long. False stipules absent. Tendrils absent. Cymes 1–6(8)-flowered, the flowers pubescent with stipe 5–14 mm. long; peduncles 0–0·75 cm. long; bracteoles 1·5–4 mm. long. Hypanthium 2–2·5 mm. broad. Sepals 5–7 mm. long, obtuse at the apex. Petals 5–7 mm. long. Outer corona with tube 1–2 mm. high and threads 2–3·5 mm. long. Inner corona 1–1·5 mm. high, cup-shaped. Stamens 5, with filaments 2·5–3 mm. long and anthers 1–1·5 mm. Ovary 0·65–1 mm. long; styles free, 3–4 mm. long. Capsule c. 1·5 cm. long, pubescent. Seeds unknown.

Zambia. N: Mporokoso to Mkupa, fl. 25.x.1949, *Bullock* 1363 (K).
Also in Angola. In woodland and scrub on sandy soils.

12. **Basananthe baumii** (Harms) de Wilde in Blumea **21**: 350 (1973). Type from Angola.
  Tryphostemma baumii Harms in Warb., Kunene-Samb.-Exped. Baum: 310 (1903). Type as above.

Perennial producing 1 or a few prostrate shoots up to 10 cm. long from a woody rootstock. Leaves 3–12(19) × 1·5–6(13) cm., simple, oblong to broadly ovate or obovate or subcircular, obtuse to acute at the apex, entire-margined, acute at the base, stiffly coriaceous, glabrous on the upper surface, densely white-papillate or shortly hairy along the veins beneath; midrib impressed above, prominent (and sometimes hairy especially at base) on the lower surface; lateral nerves and vein-reticulation also prominent. Stipules 3–8 mm. long, linear, sometimes ± spathulate or dissected. False stipules absent. Tendrils absent. Cymes 1–5-flowered, sessile

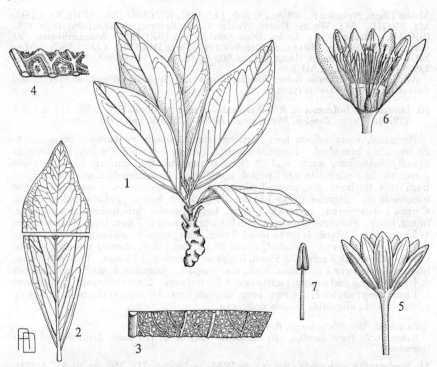

Tab. 95. BASANANTHE BAUMII var. CAERULESCENS. 1, habit (× ½); 2, leaf showing superior
(top) and inferior surface (× ½); 3, section of leaf undersurface showing venation
(×2); 4, marginal section of leaf (×6), 1–4 from *Drummond & Cookson* 6457;
5, flower in lateral view (×3); 6, flower with facing sepals and petals removed (×4)
5–6 from *Sambesi Expedition* 391; 7, stamen (×4) from *Drummond & Cookson* 6491.

or with peduncle up to 1·5 cm. long, the flowers greenish-white, glabrous or
sparsely papillate or hairy, with stipe 7–13(15) mm. long; bracteoles 2–5 mm. long.
Hypanthium 1–2 ×2–3 mm., cup-shaped. Sepals 5·5–9 mm. long, ± obtuse.
Petals 5–8·5 mm. long. Outer corona with tube 1–1·75 mm. high and threads
(1·5)2–3·5 mm. long. Inner corona 1–2 mm. high, cup-shaped. Stamens 5–6,
with filaments 2·5–4 mm. long and anthers 1–1·5 mm. long. Ovary c. 1 mm. long;
styles 3–4, (2)3–6 mm. long, free. Capsule 1–1·5 cm. long, purplish-green,
oblong-globose or ellipsoid, 2–3-seeded. Seeds 6–7 mm. long, subreniform.

Leaves (in sicco) greenish above, greyish-white or greyish-green beneath; nerves on the
underside of leaves yellowish-brown to purplish-brown; peduncle 5–15 mm. long
var. *baumii*
Leaves (in sicco) glaucous or purplish above, greenish-white to purplish-grey beneath;
nerves on the underside of leaves reddish to purplish; peduncle up to 3 mm. long
var. *caerulescens*

Var. **baumii**—de Wilde, loc. cit.: fig. 4g–h & 7.
    Tryphostemma baumii Harms, loc. cit.; in Engl. & Prantl, Pflanzenfam. ed. 2, **21**:
488 (1925).—Hutch. & Pearce in Kew Bull. **1921**: 264 (1921) pro parte.—Engl.,
Pflanzenw. Afr. **3**, 2: 598 (1921).—A. & R. Fernandes in Garcia de Orta **6**, 4: 664,
t. 11 (1958); in C.F.A. **4**: 228 (1970). Type as above.
    Tryphostemma mendesii A. & R. Fernandes in Bol. Soc. Brot., Sér. 2, **32**: 86, t. 3
(1958); in Garcia de Orta **6**, 4: 554, t. 12 (1958); in C.F.A. **4**: 228 (1970). Type from
Angola.

    Zambia. N: Mpika, between Mushinga Escarpment and Great N. Road, fl. & fr.
24.ix.1966, *Lawton* 1464 (K; SRGH).
    Also in Angola. Dry secondary forest, woodland and scrub on sandy soils.

84. PASSIFLORACEAE

Var. **caerulescens** (A. & R. Fernandes) de Wilde in Blumea **21**: 351 (1973). TAB. **95**.
Type from Angola.
Tryphostemma caerulescens A. & R. Fernandes in Bol. Soc. Brot., Sér. 2, **32**: 85,
t. 2 (1958); in Garcia de Orta **6**, 4: 664, t. 13 (1958); in C.F.A. **4**: 228 (1970). Type
as above.

Zambia. B: Kalabo Distr., 65 km. N. of Kalabo, fl. 14.xi.1959, *Drummond & Cookson*
6457 (COI; K; SRGH). W: Mwinilunga Distr., Cha Mwana (Chibara's) Plain, fl.
14.x.1937, *Milne-Redhead* 2764 (COI; K).
Also in Angola. In dry woodland and open sandy ground.

The following specimen has unusually large leaves, up to 19 × 13 cm.: Zambia, W:
Mwinilunga, *Milne-Redhead* 3357 (COI; K).

13. **Basananthe heterophylla** Schinz in Verh. Bot. Ver. Prov. Brand. **30**: 252 (1888).—
de Wilde in Blumea **21**: 353, fig. 4k & 6 (1973). Type from SW. Africa.
Tryphostemma heterophyllum (Schinz) Engl., Bot. Jahrb. **14**: 338 (1891) in obs.;
op. cit. **15**: 577 (1893); Pflanzenw. Afr. **3**, 2: 599 (1921).—Hutch. & Pearce in
Kew Bull. **1921**: 261 (1921).—Harms in Engl. & Prantl, Pflanzenfam. ed. 2, **21**: 488
(1925).—Schreiber in Prodr. Fl. SW. Afr. **89**: 2 (1968). Type as above.

Annual or biennial glabrous herb. Stem up to 50 cm. high, erect, simple or
sometimes branched at the base, terete, fleshy and smooth. Leaf-lamina up to
0·8(1·4) cm., oblong-lanceolate to linear, obtuse or acute at the apex, margin
8 × 8 cm., usually deeply 3–5-lobed, sometimes unlobed, the lobes 2–7 × 0·2–
serrate-setaceous-dentate; leaf-base subacute; petiole 1–5 cm. long, winged,
with glandular teeth. Persistent cotyledons 2–2·25 × 1–1·5 cm., with a petiole
1–1·25 cm. long. Stipules 3–6 mm. long. False stipules absent. Tendrils
absent. Cymes 1–2-flowered, the flowers cream-coloured, with stipe 2–4 mm.
long; peduncle 2·5–15 mm. long; bracteoles 3–7 mm. long. Hypanthium
2·5–4 mm. wide. Sepals 7–15 mm. long, acute, the outer ones with 2 prominent
submarginal nerves. Petals 5–7 mm. long, ± acute. Outer corona with tube
1–1·5 mm. high and threads 2·5–3·5 mm. Inner corona 0·5–0·75 mm. high,
cup-shaped. Stamens 5, the filaments 4–5 mm. long, the anthers 2–2·25 mm. long.
Ovary 1·5–2 mm. long; styles 2·5–6 mm. long, connate at the base for 1–3·5 mm.
Capsule 1·5–1·75 cm. long, 1–4-seeded. Seeds 7–8 mm. long.

Botswana. SW: Ghanzi, *Brown* 7927 (K; SRGH). **Rhodesia.** S: Gwanda,
Bandani Hill near Tuli Experimental Station, fl. 16.xii.1964, *Norris Rogers* 486 (SRGH).
Also in SW. Africa. On granite kopjes.

5. ADENIA Forsk.*

Adenia Forsk., Fl. Aegypt.-Arab.: 77 (1775).—de Wilde in Meded. Landb.
Wag. **71**, 18: 50 (1971).
Modecca Lam., Encycl. Méth. Bot. **4**: 208 (1797).
Kolbia Beauv., Fl. Oware Benin **2**: 91, t. 120 (1807).
Blepharanthes J. E. Sm., Gram. Bot.: 188 (1821) *nom. illegit.*
Paschanthus Burch., Trav. Int. S. Afr. **1**: 543 (1822).
Microblepharis (W. & A.) Roem., Syn. Mon., 2 Peponiferae: 133, 200 (1846).
Erythrocarpus Roem., op. cit.: 133, 204 (1846).
Clemanthus Klotzsch in Peters, Reise Mossamb. Bot.: 143 (1861).
Ophiocaulon Hook. f. in Benth. & Hook. f., Gen. Pl. **1**: 813 (1867).
Machadoa Welw. ex Benth. & Hook. f., op. cit.: 814 (1867).
Keramanthus Hook. f. in Curtis, Bot. Mag.: t. 6271 (1876).
Jaeggia Schinz in Verh. Bot. Ver. Prov. Brand. **30**: 253 (1888).
Echinothamnus Engl., Bot. Jahrb. **14**: 383 (1891).

Herbaceous to ligneous perennial climbers with tendrils, sometimes erect
herbs or shrubs or small trees usually without tendrils, arising from a rootstock or
tuber or provided with a thick main stem. Leaves sessile or ± long-petiolate,
simple and entire, variously lobed to palmately partite or rarely palmately
compound, glabrous or pubescent. Glands (0)1–2 at the base of the lamina, at or
near the top of the petiole, sometimes with others on the lower surface or at the

* See note on p. 376.

N

margin of the lamina. Stipules minute, narrowly triangular or reniform. Tendrils axillary (sterile tendrils). Inflorescences axillary, cymose, the median or the first three flower(s) often replaced by tendril(s) (inflorescence tendrils). Bracts and bracteoles minute, triangular to subulate. Flowers dioecious or rarely monoecious, ⚥ or polygamous, campanulate or urceolate to tubular or infundibuliform, usually greenish or yellowish, with a stipe articulate at the base. Hypanthium saucer- or cup-shaped or tubular. Sepals (4)5(6), free or ± long-connate, imbricate, persistent. Petals (4)5(6), free or ± adnate to the calyx-tube included in the calyx, usually fimbriate or laciniate. Corona annular or consisting of 5 cup-shaped parts or of a laciniate rim or membrane or a row of filaments or hairs, or 0. Disk-glands 5, ± strap-shaped, truncate or capitate, inserted at or near the base of the hypanthium, alternating with the petals, or sometimes absent. ♂ flowers: stamens (4)5(6), hypogynous or perigynous, with the filaments free or partially connate into a tube and the anthers basifixed, oblong to linear, acute or obtuse, often apiculate or mucronate, 2-locular, with 2 longitudinal slits; vestigial ovary minute. ♀ flowers: usually smaller than the ♂ ones; staminodes ± subulate; ovary superior, subsessile or shortly stipitate, globular to oblong, 1-locular, with 3(5) parietal placentas, usually with numerous anatropous 2-tegumented ovules; styles 3(5), free or partially connate; stigmas usually reniform to subglobular, laciniate or plumose or densely woolly-papillate. Capsule stipitate, 3(5)-valved, the pericarp coriaceous to somewhat fleshy, greenish, yellowish or bright red. Seeds ± compressed, with crustaceous pitted testa, enclosed in a membranous to pulpy aril; endosperm horny; embryo large, straight, with foliaceous cotyledons.

A tropical genus distributed through Africa, Madagascar, S. India, Ceylon and Malaysia, with c. 93 species.

Sepals and petals free, inserted on the rim of the hypanthium:
    Corona membranous or consisting of hairs, rarely absent; glands of the disk present, rarely absent; stigmas not sessile; glands of the lamina-base 1 or 2, never on a special spathulate appendage; usually trees or shrubs:
        Stem rarely thorny; leaves simple or 5-partite or (2)3–5-foliolate; corona consisting of laciniae or of woolly or not woolly hairs:
            Glands at the lamina-base 1; leaves simple or (2)3-foliolate, the leaflets sessile or nearly so - - - - - - - - - 1. *fruticosa* subsp. *simplicifolia*
            Glands at the lamina-base 2(4), separate (sometimes approximate); leaves 5-partite:
                Hypanthium 2–3 mm. wide; corona of non-woolly hairs; disk-glands absent; climber, sometimes shrub-like - - - - - 2. *glauca*
                Hypanthium 3–5 mm. wide; corona of woolly hairs; disk-glands present; tree
                                                  3. *karibaensis*
        Stem thorny; leaves simple with obscure reticulation; corona consisting of a few stiff hairs - - - - - - - - - - - 4. *spinosa*
    Corona absent or a fleshy rim; disk-glands absent; stigmas sessile or subsessile; glands at the lamina-base 1, on a spathulate to semicircular median appendage; subligneous climbers:
        Leaf surfaces with a very distinct reticulation:
            Leaves broadly ovate to orbicular, unlobed; nerves arching towards the apex of the lamina - - - - - - - - - 5. *stolzii*
            Leaves broadly ovate, orbicular or ± triangular, often bluntly 3(5)-lobed; nerves ± straight - - - - - - 6. *gummifera* var. *gummifera*
        Leaf surfaces not very distinctly reticulate; plant usually covered in wax; leaves rhomboid to 3-lobed, cuneate to subtruncate at the base
                                6. *gummifera* var. *cerifera*
Sepals partially connate into a calyx-tube; petals free or ± adnate to the calyx-tube:
    Filaments inserted 4–5 mm. above the base of the hypanthium; corona and disk-glands absent - - - - - - - - - - - 23. *repanda*
    Filaments inserted at the base of the hypanthium or on an androgynophore; corona and disk-glands usually present:
        Leaf-lamina digitately divided to the base, the lobes or leaflets usually ± distinctly stalked:
            Hypanthium of the ♂ flowers 8–13 mm. wide; tendrils present - 9. *stenodactyla*
            Hypanthium narrow, up to c. 5 mm. wide:
                  Anthers straight, free (also in bud) - - - - 17. *mossambicensis*
                Anthers ± curved, connate at the apex - - - - - 16. *digitata*
        Leaf-lamina simple or ± deeply lobed, not compound:
            Plants with tendrils:
                Corona absent; petals inserted above the middle of the calyx-tube:
                    Leaves glabrous; ♂ flowers 20–25 mm. long (incl. stipe); petals entire or

84. PASSIFLORACEAE 387

irregularly dentate at the apex; anthers c. 2·5 ×1 mm., not apiculate;
filaments c. 5·5 mm. long - - - - - 12. *zambesiensis*
Leaves pubescent especially along nerves; ♂ flowers 25–35 mm. long (incl.
stipe); petals fimbriate; anthers (3)4–7 × 0·5–1 mm., with apiculum up to
0·1 mm. long; filaments 10–21 mm. long - - - 13. *schliebenii*
Corona present, consisting of hairs; petals inserted at the same level as the
corona or up to the middle of the calyx-tube:
Glands at the base of the leaf-lamina 1 or 2, on two free or ± adnate auricles
or on a peltate or subspathulate appendage:
Plant pubescent; hypanthium 5–12 mm. wide; leaf-margin dentate
14. *stricta*
Plant glabrous; hypanthium 2–6 mm. wide, tapering to the base:
Calyx-tube of ♂ flowers 5–12 mm. wide; petals 30–45 mm. long,
fimbriate; ♂ flowers (with stipe) 40–75 mm. long
10. *dolichosiphon*
Calyx-tube of ♂ flowers 3–5 mm. wide; petals 5–9 mm. long; ♂ flowers
(with stipe) 13–30 mm. long:
Leaves cordate, truncate or hastate at the base - - 11. *hastata*
Leaves acute to rounded at the base 15. *lanceolata* subsp. *scheffleri*
Glands at the base of the leaf-lamina 2, on two free auricles lateral to the apex
of the petiole; hypanthium 5–15 mm. wide:
Anthers obtuse, not apiculate; petals of the ♂ flower broadly spathulate,
shortly fimbriate, those of the ♀ flower lanceolate, fimbriate only near
the apex; leaves entire to shallowly 3–5(7)-lobed 7. *panduriformis*
Anthers ± acute, apiculate, the apiculum (1)1·5–2·5 mm. long; petals
spathulate; leaves unlobed, rarely shallowly lobed
8. *rumicifolia* var. *rumicifolia*
Plants without tendrils, 0·1–1·5 m. high:
Leaves glabrous, unlobed, with entire margin; calyx-lobes subentire to serrulate:
Leaves lanceolate to linear, 0·2–5(6) cm. broad:
Flowers (1)3–5, erect-patent:
Flowers (incl. the stipe) (35)40–52 × 2–5 mm.; petals inserted at c. 4 mm.
below the throat of the calyx-tube; leaves linear, (8)12–20 × 0·2–0·4
cm.; anthers 4 ×1 mm. with a 0·5 mm. long not papillate apiculum
18. *erecta*
Flowers 10–26 × 3–7·5 mm.; petals inserted at the same level as the
corona; apiculum of anthers usually papillate - 19. *goetzei*
Flowers 1(2–3), pendent - - - - - - 20. *tuberifera*
Leaves ovate to suborbicular, 7–14 × 5–12 cm.; anthers with a papillate
apiculum - - - - - - - - 21. *ovata*
Leaves pubescent, unlobed, with dentate margin, or lobed or partite; calyx-lobes
(at least three of them) long woolly-fimbriate at the margin
22. *volkensii*

1. **Adenia fruticosa** Burtt Davy, F.P.F.T. **1**: 36, 221 (1926).—Bremek., Vegetations-
bilder **23**, 3: 6, t. 18 (1932); in Ann. Transv. Mus. **15**, 2: 248 (1933).—Liebenberg
in Bothalia **3**, 4: 528, 532, 538, t. 1 & 2 (1939).—Dyer et al., Wild Fl. Transv.: 225
(1962).—de Wilde in Meded. Landb. Wag. **71**, 18: 69 (1971).—J. H. Ross, Fl. Natal:
251 (1972). Type from S. Africa.

Subsp. **simplicifolia** de Wilde, tom. cit.: 71, fig. 6 (1971). Type: Rhodesia, Melsetter
Distr., ♂ fl., *Chase* 1321 (BM; K; LISC; SRGH, holotype).

Climber with trunk-like softly woody base up to 2 m. tall, producing at apex
numerous liane-like stems up to 5 m. long. Leaf-lamina 1–6 × 0·75–5 cm., simple
or (2)3-foliolate, ovate or subcircular in outline, subcordate at base subcoriaceous,
green or grey, or glaucous-green, finely punctate or not, with (1)3–5 main
nerves from the base; petiole (0·3)0·5–2·5 cm. long; leaflets 1–5 × 1–3(4) cm.,
suborbicular to elliptic, ovate or rhomboid, ± obtuse at the apex, entire, acute to
rounded at the base; petiolules very short or 0. Gland at the apex of the petiole
single, 1–3 mm. in diameter, on a fleshy, ± upward-curved lobe 1·5–3 mm. in
diameter. Stipules c. 0·5 mm. long, narrowly triangular. Sterile tendrils 3–12 cm.
long, simple, sometimes breaking off and leaving a thorn. Cymes 2–5-flowered
in ♂, 1–3-flowered in ♀, solitary in the axils of normal or much reduced leaves on
short shoots up to 1 cm. long; peduncle up to 1 cm. long; fertile tendrils 0; bracts
and bracteoles 1–2·5 mm. long, triangular to oblong, ± serrulate; pedicels 1–6 mm.
long in ♂, 1–4 mm. long in ♀ flower. ♂ flower: 13–15 × 4–4·5 mm. (incl. the
3–4 mm. long stipe), campanulate, 5(6)-merous; hypanthium 2 × 2·5–4 mm.,
broadly cup-shaped; calyx-tube 0; sepals 8–10 mm. long, lanceolate, obtuse,

subentire; petals 7–8 × 1·5–2 mm., oblong to lanceolate, acute, 3-nerved, serrulate; corona comprising a set of fine dense hairs 0·5–1 mm. long; stamens with filaments 3–4 mm. long connate at the base and anthers 5–5·5 × 0·75 mm., subobtuse at the apex; disk-glands 0·5–0·75 mm. long. ♀ flower: 7–8 × 2·5–3·5 mm. (incl. the 0·5–0·75 mm. long stipe), campanulate; hypanthium 1–1·5 mm. long; calyx-tube 0; sepals 5–6 mm. long; petals 3–4 × 0·75 mm.; corona-hairs c. 0·5 mm. long; staminodes 2–2·5 mm., connate at the base for c. 1 mm.; disk-glands 0·25–0·5 mm. long; pistil 5–6·5 mm. long; gynophore 1·5–2 mm. long; ovary 1·5–3 × 1·5–2·5 mm.; styles connate at the base for 0·5–0·75 mm., with arms 0·5–1 mm. long; stigmas c. 1·5 mm. in diameter. Capsule 1–2 × 0·8–1·8 cm., ellipsoid, with smooth coriaceous pericarp, 3–6-seeded. Seeds c. 6–6·5 × 6–6·5 × 2·5 mm., broadly orbicular, pitted.

Rhodesia. E: Melsetter Distr., Muwushu, Sabi Valley, 800 m., fr. x.1958, *Davies* 2526 (K; SRGH). S: Nuanetsi, between Chitove Falls and border, c. 300 m., fl. 14.ix.1967, *Müller* 667 (K; SRGH).
Also in S. Africa (Transvaal). In dry savannas and thorn bushveld in rocky places and near hot springs.

2. **Adenia glauca** Schinz in Engl., Bot. Jahrb. **15**, Beibl. 33, 1: 1 (1892).—Engl., Pflanzenw. Afr. **3**, 2: 605 (1921).—Burtt Davy, F.P.F.T. **1**: 222 (1926).—Steyn, Toxicol. Pl. S. Afr.: 314, fig. 43b, 46, 47b (1934).—Liebenberg in Bothalia **3**, 4: 523, 532, 539, fig. 6–8, t. 4–5 (1939).—Dyer et al., Wild Fl. Transv.: 225, t. 112 (1962).—Watt & Breyer-Brandwijk, Med. & Pois. Pl. S. & E. Afr. ed. 2: 828 (1962).—de Wilde in Meded. Landb. Wag. **71**, 18: 73, fig. 1f & 7 (1971). Syntypes from S. Africa (Transvaal).
Modecca glauca Schinz ex Engl., op. cit. **14**: 393 (1891) *nom. nud.*

Climber, sometimes shrub-like, up to 3·5 m. high, with the basal part c. 1 × 0·4 m. above ground, thickened and ± fleshy, emitting numerous stems up to 3 m. long. Leaf-lamina 2–12 × 2·5–12 cm., 5-partite to the base, suborbicular in outline, membranous to coriaceous, greyish, glaucous or purplish-grey, sometimes punctate, with 5 main nerves from the base; petiole (0·5)1–5 cm. long; leaf-segments 1–7 × 0·5–3·5(4·5) cm., ovate to oblong, obtuse, sometimes retuse at the apex, margin entire, acute or subacute at the base, with 4–8 pairs of lateral nerves and rather distinct reticulation. Glands of the lamina-base 2, 1–2 mm. in diameter, contiguous, on two ± connate, ± fleshy, brownish auricles; lamina-glands 1–3(5), c. 0·25 mm. in diameter, apical or subapical at the extremities of the nerves. Stipules 1–1·5 mm. long, triangular, acute. Sterile tendrils up to 10 cm. long. Cymes 2–5-flowered in ♂, 1–3-flowered in ♀; peduncles 1(2) cm. long. Bracts and bracteoles 0·5–2 mm. long, triangular. Pedicels 2–20 mm. long. Tendrils (0)1, 2–6 cm. long. ♂ flower: (13)15–30 × 2–4 mm. (incl. the 3–7 mm. long stipe), infundibuliform; hypanthium (1)2–4 × 2–3 mm.; calyx-tube 0; sepals (8)10–18(20) mm. long, lanceolate, obtuse, entire; petals 5(6), 6–11 × 1·5–2·5 mm., lanceolate, acute to obtuse at the apex, narrowed to the base, 1–3-nerved, serrulate; corona a fringe of fairly thick hairs 0·5–1 mm. long; stamen-filaments (3)4·5–7 mm. long, ± long-connate at the base, inserted at the base of the hypanthium; anthers 3–5·5 × 1 mm., apiculate; disk-glands 0; vestigial ovary 0·5–1·25 mm. long (incl. the gynophore). ♀ flower: (9)10–15 × 2–3 mm. (incl. the 1·5–4 mm. long stipe), tubular-infundibuliform; hypanthium 1·5–2 mm. long; calyx-tube 0; sepals 7–9 mm. long, lanceolate, obtuse, subentire; petals 4–5 × 0·5–1 mm., lanceolate, acute, 1-nerved, irregularly serrulate at the apex; corona of sparse thick hairs; staminodes 2–3·5 mm. long, connate at the base; disk-glands 0; pistil 6–8 mm. long; gynophore 1–2 mm. long; ovary 2·5–4·5 × 2·5–4 mm., subglobose; styles shortly connate at the base, with the arms 1–1·5 mm. long; stigmas c. 1·5 mm. in diameter, subglobose, papillate. Capsule 1·8–2·5 × 1·5–2 cm., ellipsoid, with the pericarp thickly coriaceous, 3–5-seeded. Seeds 5·5–7·5 × 5·5–7·5 × 2·5–3 mm., orbicular, pitted.

Botswana. SE: c. 3 km. E. of Pharing, ♂ fl. ix.1947, *Miller* B/508 (K).
Also in S. Africa. In rocky places in dry bushveld.

3. **Adenia karibaensis** de Wilde in Meded. Landb. Wag. **71**, 18: 75 (1971). TAB. **96**.
Type: Rhodesia, Rukowakuona Escarpment, 640–960 m., ♂ fl., *Wild & Barbosa* 5907 (BR; COI; K; LISC; M; SRGH, holotype).

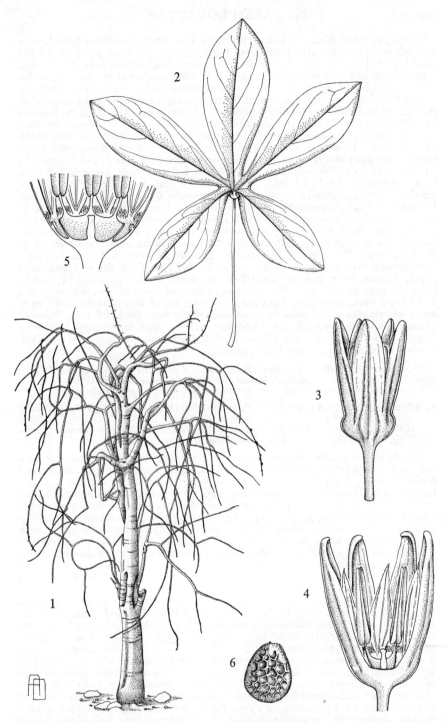

Tab. 96. ADENIA KARIBAENSIS. 1, habit (×$\frac{1}{4.5}$) from *Wild & Barbosa* 5907 (based on photograph); 2, leaf (×$\frac{1}{2}$) from *Wild* 5765; 3, lateral view of flower (×3); 4, partly dissected ♂ flower (×4); 5, longitudinal section of flower-base showing vestigial ovary and stamen attachments (×5); 6, seed (×5), 3–6 from *Bingham* 1595.

Tree up to 6·5 m. high. Trunk up to 2·5 × 0·5 m., erect, sometimes forked, soft-wooded, with smooth grey lenticellate bark, towards the apex producing half-climbing or drooping branches up to 3 m. long. Leaf-lamina 3–16 × 3–15 cm., 5-partite almost to the base, suborbicular in outline, cordate at the base, ± coria-ceous, greyish- or purplish-green; leaf-segments 2–11 × 0·8–6 cm., ovate or elliptic to oblong-lanceolate, obtuse at the apex, margin entire, acute to rounded at the base, with distinct reticulation; petiole 1–8(9) cm. long. Glands at the lamina-base 2(4), 1–3 mm. in diameter, separate or contiguous, on 2 ± connate auricles. Stipules 1·5–2 mm. long, narrowly triangular to linear. Sterile tendrils up to 12 cm. long. Cymes in the axils of normal or ± reduced leaves, the latter crowded on short shoots up to 3 cm. long, up to 20-flowered in ♂, 1–3-flowered in ♀; peduncle up to 2·5 cm. long; tendrils 0–1, 1–6 cm. long; pedicels 2–8 mm. long. ♂ flower: 14–17 × 4 mm. (incl. the 2·5–6 mm. long stipe); hypanthium 1·5–2 × 3–5 mm., broadly cup-shaped; calyx-tube 0; sepals 8–10 mm. long, lanceolate, obtuse, subentire; petals 6–7 × 1·5–2 mm., lanceolate, acute, 3–5-nerved, serrulate; corona comprising a set of fine woolly hairs 0·5–1 mm. long; stamen-filaments 2–3 mm. long, connate at the base for 1–1·5 mm., inserted at the base of the hypanthium; anthers 4·5–5 × 1 mm., obtuse, with apiculum up to 0·1 mm. long; disk-glands 0·5–1 mm. long; vestigial ovary c. 1 mm. long on a gynophore 0·5–1 mm. long. ♀ flower: c. 9 × 3–4 mm. (incl. the c. 3 mm. long stipe), campanulate; hypanthium 1·5–2 mm. long, broadly cup-shaped; calyx-tube 1(2) mm. long; sepals 2–3 mm. long, elliptic-oblong, obtuse, entire; petals 4–5 × 1–1·5 mm., subentire, 3-nerved; corona a set of woolly hairs c. 0·5 mm. long; staminodes c. 2·5 mm. long, connate at the base for 1–1·5 mm., inserted at the base of the hypanthium; disk-glands c. 0·5–1 mm. long; pistil unknown. Capsule 1·5–2 × 1·2–1·5 cm., ellipsoid, with coriaceous pericarp, 4–10-seeded. Seeds 7 × 7 × 2·5 mm., suborbicular, pitted.

Rhodesia. N: Kariba Distr., Bumi Escarpment, ♂ fl. 12.xi.1966, *Cannell* 13 (SRGH); Sipolilo, N. of Chenanga Camp, fl. in bud 16.x.1965, *Bingham* 1595 (SRGH).
Known only in Rhodesia. In rocky savanna, schist outcrops and sandstone.

4. **Adenia spinosa** Burtt Davy, F.P.F.T. **1**: 36 & 222 (1926).—Bremek. in Ann. Transv. Mus. **15**, 2: 249 (1933).—Liebenberg in Bothalia **3**, 4: 523, 532, 533, t. 3 fig. 9 (1939).—Dyer et al., Wild Fl. Transv.: 215 (1962).—de Wilde in Meded. Landb. Wag. **71**, 18: 76, fig. 7 (1971). Type from S. Africa (Transvaal).

Woody shrub or small tree with massive, irregular, light green, fleshy or hard bole up to 0·50 × 2·5 m., producing slender, greyish-green, very thorny unbranched fertile stems c. 1 m. long, sometimes climbing by means of the thorns, with internodes 0·5–4 cm. long. Leaf-lamina 1–3·5 × 0·7–3 cm., ovate to elliptic, obtuse or sometimes retuse at the apex, margin entire or somewhat remotely shallowly lobulate and undulate, rounded or cordate at the base, ± coriaceous, greyish-glaucous green, with the midrib, lateral nerves and vein reticulation distinct on both surfaces but more so below; petiole 0·2–0·7 cm. long. Glands of the lamina-base 1 or 2, 1–2·5 mm. in diameter, subreniform to orbicular, sessile at the apex of the petiole; lamina-gland 1, apical or subapical, c. 0·25 mm. in diameter, at the end of the midrib. Stipules c. 0·5 mm. long, triangular, acute. Sterile tendrils 3·5–6 cm. long, axillary, soon becoming sharp patent thorns up to 4 cm. long. Cymes axillary, 2–6-flowered in ♂, 1–3-flowered in ♀; peduncle c. 0·5 cm. long; pedicels 0–2 mm. long in ♂, 0–1 mm. long in ♀. ♂ flower: (10)12–24 × 2·5–4 mm. (incl. the 1·5–3 mm. long stipe), tubular-campanulate; hypanthium 1–3 mm. long, cup-shaped; sepals (8)9–18 mm. long, free, lanceolate, obtuse, entire; petals (6)8–10 × 1–2 mm., lanceolate, acute, serrulate, 1–3-nerved; corona consisting of a few stiff hairs near the insertion of the petals; stamens with the filaments 2·5–5·5 mm. long, connate at the base, inserted at or up to 1·5 mm. above the base of the hypanthium; anthers 4–5 × 0·75–1 mm., obtuse, with apiculum up to 0·2 mm. long; disk-glands 0 or reduced to callose pads up to 0·25 mm. in diameter; vestigial ovary 0·5–1 mm. long on a gynophore 0·5–1 mm. long. ♀ flower: smaller than the ♂; hypanthium 1–1·5 mm. long; sepals 5–8 mm. long, lanceolate, obtuse; petals 2·5–3·5 × 0·5–0·75 mm., lanceolate, 1-nerved; corona 0 or consisting of a few stiff hairs; staminodes c. 1·5 mm. long, connate at the base, inserted at the base of the hypanthium; disk-glands 0; pistil 4–6(8) mm. long; gynophore 1–2 mm. long; ovary 2–4(5) × 2–3·5 mm., subglobular; styles connate for c. 1 mm. at the base, with the free arms c. 0·5 mm. long; stigmas c. 1 mm. in diameter, papillate. Capsule 1·4–2·2 × 1·2–1·8 cm., subglobular, with thinly coriaceous pericarp, 3–6-seeded. Seeds c. 6·5 × 7 × 2–2·5 mm., suborbicular, pitted.

Rhodesia. S: Beitbridge, Sentinel Ranch, ♂ fl. 2.ii.1973, *Ngoni* 173 (K; SRGH). Also in S. Africa (Transvaal). Dry rocky places.

5. **Adenia stolzii** Harms in Fedde, Repert. **11**: 35 (1913).—de Wilde in Acta Bot. Neerl. **17**: 131, fig. 2g (1968); in Meded. Landb. Wag. **71**, 18: 271, fig. 43 (1971); in F.T.E.A., Passifloraceae: 54 (1975). Type from Tanzania.

Subligneous climber up to 20 m. Leaf-lamina 4–12 × 3·5–10 cm., orbicular to ovate, obtuse to acute at the apex, margin entire or rarely with lobes up to 7·5 mm, deep, cordate at the base, chartaceous, brownish-green above, greyish-glaucous and sometimes punctate beneath, with 3(5) main nerves from the base and with 1(2) pairs of nerves from the midrib ascending towards the apex, distinctly reticulate; petiole 1·5–11 cm. long. Gland at the lamina-base single, 1–2 mm. in diameter, on a spathulate median appendage 1·5–2·5 mm. long; lamina-glands (0)2–4, 0·5–1 mm. in diameter, near the axils of the upper lateral nerves; marginal glands c. 0·25 mm. in diameter, up to 6 on each side. Stipules 0·5–1 mm. long, reniform, lacerate. Sterile tendrils 10–25 cm. long, simple. Cymes in the axils of normal leaves or sometimes in the axils of ± reduced leaves borne on short shoots up to 10 cm. long; peduncles 0·25–5 cm. long, up to 15-flowered in ♂ and 1–3-flowered in ♀; tendrils 0 or 1, 1–2 cm. long; bracts and bracteoles 0·5–1·5 mm. long, narrowly triangular, acute, serrulate; pedicels 0·5–5(12) mm. long in ♂, 0·5–4 mm. in ♀. ♂ flower: 12–15 × 2–4(5) mm. (incl. the 2–3 mm. long stipe), campanulate; hypanthium 1–1·5 × 2–4(5) mm., shallowly cup-shaped; calyx-tube 0; sepals 8·5–13 × 2–3 mm., lanceolate, obtuse, subentire at the margin, remotely punctate; petals 8–13 × 2–3·5 mm., lanceolate, obtuse, 3–5-nerved, serrulate in the upper ½, sparsely punctate; corona 0 or a low ring c. 0·1 mm. wide; stamen-filaments 2–3 mm. long, connate at the base for 1–2 mm., inserted at the base of the hypanthium; anthers 4·5–5 × 1(1·5) mm., obtuse; disk-glands 0; vestigal ovary, incl. the gynophore, 1–1·5 mm. long. ♀ flower: c. 10 × 4 mm. (incl. the c. 0·5 mm. long stipe), campanulate; hypanthium c. 1 mm. long, saucer-shaped; calyx-tube 0; sepals 8–9 × 3 mm., lanceolate, subobtuse, subentire; petals c. 3·5 × 0·33–0·5 mm., linear, subacute, 1-nerved, entire; corona 0; staminodes c. 0·5 mm. long, free; disk-glands 0; pistil 7–7·5 mm. long; gynophore c. 0·5 mm. long; ovary c. 5·5 × 3 mm., ellipsoid; style 0–0·5 mm. long, with stigmas c. 1·5 mm. in diameter, reniform, with papillate margins. Capsule 4(4·5) × 2–2·5 cm., ovate-oblong or subfusiform, with smooth coriaceous pericarp, c. 30-seeded. Seeds c. 4–5 × 2·5 × 2 mm., ± ellipsoid, pitted.

Zambia. E: Lundazi, Nyika, at entrance to Chowo Forest, c. 2300 m., ♂ fl. 5.vii.1971, *Pawek* 4986 (K). **Malawi.** N: Willindi Forest Reserve, fr., *Chapman* 230 (FHO*). Also in Kenya and Tanzania. In montane forest and scrub.

6. **Adenia gummifera** (Harv.) Harms in Engl. & Prantl, Pflanzenfam. **3**, 6a, Nachtr. 1: 255 (1897); op. cit. ed. 2, **21**: 490 (1925).—Burtt Davy in Ann. Transv. Mus. **3**: 121 (1912); F.P.F.T. **1**: 222 (1926).—Engl., Pflanzenw. Afr. **3**, 2: 602 (1921).— R. E. Fr. in Notizbl. Bot. Gart. Berl. **8**: 567 (1923).—Summerhayes in Trans. Linn. Soc. Lond., Ser. 2 Zool. **19**: 279 (1931).—Henkel, Woody Pl. Natal. Zulul.: 110 (1934).—Norlindh in Bot. Notis. **1934**: 107 (1934).—Liebenberg in Bothalia **3**, 4: 523, 532, 535, fig. 10–11 (1939).—Brenan & Greenway, T.T.C.L. **2**: 447 (1949).— Garcia in Est. Ens. Doc. Junta Invest. Ultramar **12**: 162 (1954).—A. & R. Fernandes in Garcia de Orta **6**, 2: 256, t. 10–11 (1958).—Watt & Breyer-Brandwijk, Med. & Pois. Pl. S. & E. Afr. ed. 2: 828 (1962).—F. White, F.F.N.R.: 267 (1962).—Dyer et al., Wild Flowers of the Transv.: 225 (1962).—de Wilde in Acta Bot. Neerl. **17**: 131, fig. 2h (1968); in Meded. Landb. Wag. **71**, 18: 261 (1971); in F.T.E.A., Passifloraceae: 51 (1975).—J. H. Ross, Fl. Natal: 251 (1972). Type from S. Africa.
    Modecca gummifera Harv. in Harv. & Sond., F.C. **2**: 500 (1862). Type as above.
    Ophiocaulon gummifer Mast. in F.T.A. **2**: 518 (1871).—Harms in Engl., Pflanzenw Ost-Afr. **C**: 281 (1895) "gummiferum".—R. E. Fr. in Wiss. Ergebn. Schwed. Rhod.-Kongo-Exped.: 157 (1914) "gummiferum". Syntypes: Mozambique, Zambezia, Shamo, *Kirk* (K); Tete, Kongone, between Tete and sea coast, *Kirk* (K).
    Ophiocaulon cissampeloides sensu Bak., Fl. Maurit. Seychell.: 106 (1877); in Journ. Linn. Soc., Bot. **40**: 74 (1910) non (Hook) Mart. (1871).
    Adenia rhodesica Suess. in Trans. Rhod. Sci. Ass. **43**: 13 (1951). Type: Rhodesia, Marandellas, *Dehn* 696/52 (BR; M, holotype; SRGH).

Subligneous climber up to 30 m. long and up to 10 cm. thick at the base. Leaf-lamina (1)1·5–11 × (1)1·5–11 cm., entire or ± deeply 3(5)-lobed (lobes obtuse), orbicular to ovate or rhomboid in outline, obtuse or retuse, rarely subacute at apex, margin entire, cordate to truncate or cuneate at the base, membranous, dark green above, greyish-green or glaucous beneath, sometimes punctate, 3-nerved from the base and with a pair of nerves from the midrib ending in marginal glands; reticulate venation fine, distinct or indistinct; petiole (1)1·5–11 cm. long. Glands at lamina-base single, 0·5–1·5 mm. in diameter, situated on a median circular to spathulate appendage 1–3 mm. long; lamina-glands 0–4, 0·5–1(2) mm. in diameter, contiguous to the axils of the nerves; marginal glands c. 0·5 mm. in diameter, 3–7 on each side. Stipules 0·5(1) mm. long, broadly rounded to triangular, lacerate. Sterile tendrils 5–20 cm. long, simple or 3-fid. Cymes with peduncle (0·5) 1–12(16) cm., up to 35-flowered in ♂, 2–6-flowered in ♀; bracts and bracteoles 0·5–1 mm. long, narrowly triangular, acute, ± serrulate; pedicels 2–10(15) mm. long in ♂, 3–5(10) mm. in ♀; tendril 0 or 1, 1–4 cm. long. ♂ flower: (11)12–18(20) × 2–4 mm. (incl. the 2–8 mm. long stipe), ± campanulate; hypanthium 1–2(2·5) × 2–4 mm.; calyx-tube 0; sepals (7)8 × 2–3 mm., lanceolate, subobtuse, laciniate, punctate; petals (6)8–11 × 1·5–2 mm., lanceolate, obtuse, 3-nerved, laciniate-serrulate in the upper ⅔, remotely punctate; corona 0; stamen-filaments (1)2–3·5 mm. long, up to 2 mm. connate at the base, inserted at base of hypanthium; anthers (3)4–6 × 1–1·25 mm., obtuse, with apiculum up to 0·1 mm. long; disk-glands 0; vestigial ovary 0·5–1(2) mm.; gynophore 0·25–0·5(1·5) mm. long. ♀ flower: 5·5–8 × 2–2·5(3) mm. (incl. the 0·33–0·5 mm. long stipe), ± campanulate; hypanthium c. 0·5 mm. long, saucer-shaped; calyx-tube 0; sepals (4)4·5–6·5 × 1·5–2 mm., oblong, obtuse to subacute, entire, punctate; petals 2–4·5 × 0·33–0·5 mm., lanceolate-linear, suboblong, 1–3-nerved, entire or finely serrulate at the apex, punctate or not; staminodes c. 0·5 mm. long, free; corona 0; disk-glands 0; pistil 3·5–6 mm. long; gynophore 0·33–0·5 mm. long; ovary 3–4·5 × 2–3·5 mm., ovoid; style 0–0·5 mm. long; stigmas subreniform, laciniate-papillate, each 1–1·5 mm. in diameter. Capsule 2·5–4(4·5) × 1·75–3 cm., sub-globose to ovoid or ellipsoid, sometimes ± 3(6)-angular, with woody coriaceous pericarp, 30–50-seeded. Seeds 3·5–5·5 × 3–4 × 2 mm., ± ovoid, pitted.

Leaves green, 2·5–11 × 2·5–11 cm. with the reticulation of the lower surface very distinct, cordate to truncate at the base; stems green or greyish-green, ± pruinose

var. *gummifera*

Leaves glaucous, 1–4·5 × 1–4·5 cm. with the reticulation of the lower surface indistinct, cuneate to subtruncate at the base; stems pruinose, with a whitish waxy cover

var. *cerifera*

Var. **gummifera.**—de Wilde in Meded. Landb. Wag. **71**, 18: 263, fig. 41 (1971).

Zambia. B: *Gilges* 270 (SRGH). **N:** Kasama, 1280 m., ♂ fl., *Astle* 1099 (SRGH). **W:** Kabompo R., ♂ fl. 6.x.1952, *Angus* 613 (BM; BR; K); fr. 7.x.1952, *Angus* 617 (BM; BR; K). **E:** Chadiza, 850 m., ♀ fl. & fr. 28.xi.1958, *Robson* 772 (BM; K; SRGH). **S:** Mazabuka Distr., Kafue Gorge, ♀ fl. & fr. 24.xi.1963, *Robinson* 5858 (K; M; SRGH). **Rhodesia. N:** Goromonzi, Chinamora Reserve, fl. 15.xi.1970, *Ngoni* 79 (K; SRGH). **W:** Victoria Falls, S. bank of Zambesi R., fl. xii.1913, *Rogers* 13053 (K). **C:** Marandellas, Inverneil Estates, Heydon, 1600 m., fr. 7.i.1973, *Simon* 2343 (SRGH). **E:** Inyanga Distr., Pungwe Valley, 1400 m., ♂ fl. 18.xii.1930, *F. N. & W.* 3922 (BM; BR; LD; P; SRGH). **S:** Victoria Distr., Kyle National Park Headquarters area, st. 7.v.1970, *Basera* 85 (SRGH). **Malawi. N:** Chitipa Distr., Misuku Hills, Mughesse Forest, 1606 m., ♂ fl. 28.xii.1972, *Pawek* 6207 (SRGH). **C:** Dedza, Bembeke Mission, fr. 21.ii.1961, *Chapman* 1155 (COI; SRGH). **S:** Mlanje Mt., Lukulezi Valley, 1340–1600 m., ♂ fl. 15.x.1957, *Chapman* 469 (K). **Mozambique. N:** Cabo Delgado, Porto Amélia, st. *Myre & Macedo* 3517 (LMA; SRGH). **Z:** Morrumbala, near M'bobo, ♀ fl. & fr., *Torre* 5326 (COI; LD; LISC; LMA). **T:** Chicoa, 900 m., fr., *Torre & Correia* 13957 (LISC; LMA; P; SRGH). **MS:** Manica, Vumba Mt., ♂ fl. 2.i.1948, *Barbosa* 795 (EA; K; LISC; WAG). **SS:** João Belo, Chipenhe, Chachuene Forest, ♂ fl. 13.xi.1957, *Barbosa & Lemos* 8097 (COI; LISC; LMA). **LM:** Goba, near Maiuna R., fr. 3.xi.1960, *Balsinhas* 172 (COI; K; LISC); Inhaca I., between the Lighthouse and Ponta Torres, ♂ fl. 12.vi.1970, *Correia & Marques* 1781 (COI; LMU).

Also in the Somali Republic, Uganda, Zaire, Kenya, Tanzania (incl. Zanzibar), Seychelles, S. Africa (incl. Swaziland). In forest, scrub, savanna, on stony slopes, termitaria, in littoral bush, etc.

Var. **cerifera** de Wilde in Meded. Landb. Wag. **71**, 18: 264, fig. 41 (1971). Type:

Zambia, Mbala (Abercorn) Kalambo Falls, 850 m., *Robson* 490 (BM, holotype; BR; K; LISC).
Adenia sp. 1—F. White, F.F.N.R.: 268 (1962).

Zambia. N: Mpulungu, Lake Tanganyika, ♂ fl. 17.xi.1952, *White* 3685 (BR; K). Known only from Zambia. On rocky slopes of *Brachystegia* woodland.

7. **Adenia panduriformis** Engl., Bot. Jahrb. **14**: 376 (1891); Pflanzenw. Afr. **3**, 2: 604 (1921).—Harms in Engl., Bot. Jahrb. **15**: 573 (1893); in Engl. & Prantl, Pflanzenfam. **3**, 6a: 84 (1893); op. cit. ed. 2, **21**: 491 (1925).—de Wilde in Meded. Landb. Wag. **71**, 18: 152, fig. 22 (1971); in F.T.E.A., Passifloraceae: 33 (1975). TAB. **97** fig. A.
Type: Mozambique, between Tete and Kauvabatta (Cabora Bassa), ♂ fl., *Kirk* (B†, holotype; K, lectotype).

Woody climber up to 10 m., with strong, ± terete, fibrous, not knobbly stems. Fertile branches ± pruinose, the internodes 5–15 cm. long. Leaf-laminas 4–13 × 3–12 cm., entire or 3–5(7)-lobed, the lobes obtuse, orbicular to ovate, acute at the apex with acumen 3–5 mm. long, entire at the margin, ± deeply cordate or rarely truncate at the base membranous, brown or blackish when dry, with 5(10) main nerves from the base or penninerved with 3–5 pairs of nerves and distinct reticulation; petiole 0·5–6 cm. long. Glands of the lamina-base 2, 1–2 mm. in diameter, on 2 auricles 1·5–3 mm. in diameter lateral to the apex of the petiole. Stipules 0·5–1·5 mm. long, oblong, acute. Sterile tendrils up to 20 cm. long. Cymes with peduncles up to 2 cm. long, up to 10-flowered in ♂, 1–5-flowered in ♀; pedicel 5–20(35) mm. long in ♂, 5–15 mm. long in ♀; tendril 0 or 1, 2–15 cm. long. ♂ flower: 15–22 × (8)10(14) mm. (incl. the 0·5–1·5 mm. long stipe), broadly tubular-campanulate, greenish-yellow; hypanthium 3–5 × (8)10(14) mm.; calyx-tube 4–8 mm. long; calyx-lobes 4–10 mm. long, narrowly triangular, subacute; petals 7–8 × 3–4 mm., spathulate, rounded to subacute at the apex, 3–5-nerved, dentate-fimbriate at the margin, inserted at the level of the corona; corona comprising membranous appendages 0·5(1) mm. long; stamen-filaments 2·5–3 mm. long, shortly connate at the base, inserted at base of hypanthium; anthers 7·5–10 × 1–1·5 mm., obtuse, not apiculate; disk-glands c. 1 mm. long; vestigial ovary c. 1 mm. long incl. gynophore. ♀ flower: similar to the ♂, (10)15–20 × 7(10) mm. (incl. the 0–1 mm. long stipe); hypanthium 2–3 mm. long; calyx-tube 2–5 mm. long; calyx-lobes 7–14 mm. long, narrowly triangular, acute; petals 5·5–7 × 0·5–1 mm., linear-lanceolate, acute, crenulate-fimbriate towards the apex, 1–3-nerved; staminodes 2–3 mm. long, shortly connate at the base; corona and disk-glands as in the ♂; gynophore 0·5–1 mm. long; ovary 4–5 × 3·5–4·5 mm., subglobose; styles connate at the base for 1·5–2 mm., with the arms 1·5–2 mm. long; stigmas c. 4 mm. in diameter, subglobular, densely woolly-papillate. Capsule 2·25–3·5(4) × 2–3·5 cm., subglobose, 3-ribbed, greenish turning yellow, blackish when dry, with thick, smooth, coriaceous pericarp 1·5–2·5 mm. thick, 20–45-seeded. Seeds c. 6 × 4·5(5) × 2 mm., ovoid-ellipsoid, pitted.

Zambia. C: Katondwe, ♀ fl. & fr. 5.xii.1964, *Fanshawe* 9040 (K). E: Jumbe, Luangwa Valley, ♂ fl. 24.xi.1966, *Mutimushi* 1620 (BR; K; SRGH). S: Bombwe, ♂ fl., *Martin* 387 (K). **Rhodesia.** N: Mtoko, c. 8 km. N. of Ruenya R., ♂ fl. xii.1965, *Wild* 7480 (K; LISC; SRGH). E: Melsetter, Dokodoko, Hot Springs, 640 m., ♂ fl. 16.xi.1952, *Chase* 4702 (BM; COI; LISC; SRGH). S: Beitbridge, Chikwarakwara, 480 m., ♂ fl. xi.1960, *Davies* 2855 (SRGH). **Mozambique.** N: António Enes, Boila, 60 km. on the road to Nametil, c. 80 m., fr. 25.i.1968, *Torre & Correia* 17382 (LISC; MO). T: Estima-Candôdo, fr. 25.i.1972, *Macedo* 4675 (LISC; LMU; SRGH). MS: Chemba, ♂ fl., *Surcouf* (P).
Also in Tanzania. Dry open forests and thickets.

8. **Adenia rumicifolia** Engl. & Harms in Engl., Pflanzenw. Afr. **3**, 2: 603 (1921).—Harms in Notizbl. Bot. Gart. Berl. **8**: 296 (1923).—de Wilde in Acta Bot. Neerl. **17**: 292 (1968); in Meded. Landb. Wag. **71**, 18: 154 (1971); in F.T.E.A., Passifloraceae: 34, fig. 7 (1975).—A. & R. Fernandes in C.F.A. **4**: 221 (1970). Type from Tanzania.
Adenia megalantha Harms in Engl., Pflanzenw. Ost-Afr. A: 92 (1895) *nom. nud.*
Adenia lobata sensu Engl., Pflanzenw. Ost-Afr. B: 216 (1895).—R. E. Fr., Wiss. Ergebn. Schwed. Rhod.-Kongo-Exped.: 157 (1914).—A. & R. Fernandes in Garcia de Orta **6**, 2: 257 (1958); op. cit. **6**, 4: 659 (1958) pro parte.—F. White, F.F.N.R.: 268 (1962) non Jacq. (1809).
Adenia lobata var. *grandiflora* R. E. Fr., loc. cit.—Engl., tom. cit.: 604 (1921).—Harms in Engl. & Prantl, Pflanzenfam. ed. 2, **21**: 491 (1925). Type: Zambia, L. Bangweulu, ♂ fl., *Fries* 1048 (UPS, holotype).

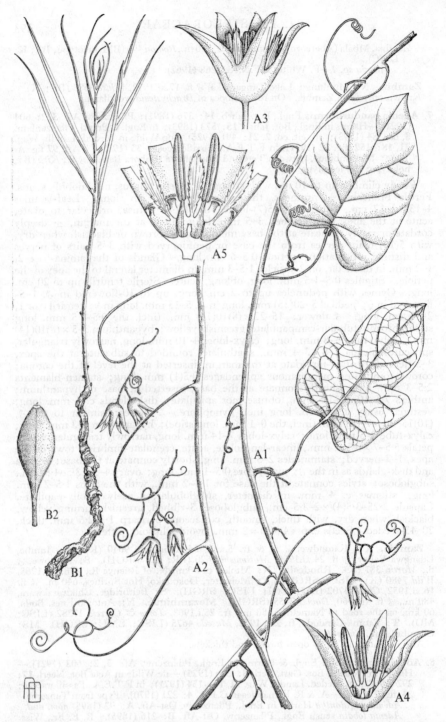

Tab. 97. A.—ADENIA PANDURIFORMIS. A1, leafy branch (× ½) *Macedo* 4675; A2, flowering branch (× ½); A3, lateral view of ♂ flower (× 2); A4, ♂ flower with facing sepal and petal removed (× 2); A5, longitudinal section of ♂ flower showing vestigial ovary, stamen attachments disk-glands and 3 petals (× 2), A2–A5 from *Mutimushi* 1620. B.—ADENIA ERECTA. B1, habit (× ½); B2, young fruit (× 1); both from *Milne-Redhead* 2926.

Var. **rumicifolia.**—de Wilde in Meded. Landb. Wag. **71**, 18: 156, fig. 24 (1971).

Woody climber up to 20 m., with terete or 3–5-angular, sometimes knobbly stems. Fertile branches 2–6 mm. in diameter, the internodes 4–10 cm. long. Leaf-lamina 3·5–15(20) × 2·5–10(14)cm., ovate or suborbicular to oblong, acute to rounded at the apex with acumen up to 2 cm. long, entire or rarely shallowly lobed or with sinuate margin, acute to shallowly cordate or hastate at the base, membranous or coriaceous, dark green to blackish above, paler beneath, with 5(7) main nerves from the base and with 1–6 pairs of ascending nerves from the midrib; petiole 1·5–10 cm. long. Glands of the lamina-base 2, 1–3 mm. in diameter, on 2 auricles 1·5–6 mm. in diameter lateral to the apex of the petiole; lamina-glands 0–6 on either half of the blade. Stipules 0·5–1 mm. long, triangular. Sterile tendrils simple or 3-fid, up to 25 cm. long. Cymes up to 10-flowered in ♂, 2–6-flowered in ♀; peduncle 0–8 cm. long; bracts and bracteoles 1–2 mm. long, triangular; pedicels 3–40 mm. long in ♂, 3–15 mm. in ♀; fertile tendrils 0–1. ♂ flower: 15–35 × 10–18 mm. (incl. the 0·5–2 mm. long stipe), broadly tubular-campanulate, pale yellow; hypanthium 2·5–5 × 10–18 mm., saucer-shaped; calyx-tube (5)7–15 mm. long; calyx-lobes 7–14 mm. long, triangular, crenulate-laciniate at the margin; petals up to 10 × 5 mm., broadly spathulate, lacerate-fimbriate at the margin, inserted at the same level as the corona; corona of woolly hairs; stamen-filaments 2–6 mm. long, connate for 0·5–1·5 mm. at the base, inserted at the base of the hypanthium; anthers (6)7–12 × 1–2 mm., acute, with apiculum (1)1·5–2·5 mm. long; disk-glands 2–3 mm. long; vestigial ovary with the gynophore 0·5–1 mm. long. ♀ flower: similar to the ♂; pistil 8–17 mm. long; gynophore (0·5)1–2(2·5) mm. long; ovary (3·5)4–8 × 2·5–4·5 mm., ovoid to ellipsoid, 3-ribbed; styles free or connate at the base for up to 2 mm., with the arms 1·5–5 mm. long; stigmas 2·5–5 mm. in diameter, reniform, yellowish-green, papillate. Capsule 3–6 × 1·5–3 cm., shortly pyriform, the attenuate part corresponding to ⅓ or less of the length of the fruit, blackish when dry, 40–150-seeded. Seeds 3·5–5(6) × 3–4 × 2 mm., ellipsoid to suborbicular, pitted.

Zambia. N: Kawambka, L. Mweru near Kafulwe Mission, ♀ fl. & fr. 2.xi.1952, *White* 3576 (BR; K). W: Mwinilunga Distr., Matonchi Farm, ♂ fl. 4.x.1938, *Milne-Redhead* 2543 (BM; BR; K). **Rhodesia.** E: Inyanga, Nyamingura R., E. Inyangani, 1025 m., fl. 26.x.1959, *Chase* 7194 (BM; K; LISC; SRGH). **Malawi.** N: Rumphi Distr., Lura Escarpment, Chinguliro, c. 6·5 km. up from lake, 1184 m., fr. 22.iv.1972, *Pawek* 5156 (K; SRGH). **Mozambique.** N: Malema, Murripa Mt., 40 km. on the Entre Rios-Ribáuè road, ♀ fl. & fr. 15.xii.1967, *Torre & Correia* 16523 (COI; LD; LISC; LMA; MO). Z: Milange, 800 m., ♂ fl. 10.x.1942, *Torre* 4573 (BR; K; LISC; LMA; MO). MS: Manica, pathside to Josi R. Falls, 1312 m., fr. 16.vi.1962, *Chase* 7764 (K; SRGH).

Also in Guinea-Bissau, Guinea, Sierra Leone, Cameroon, Zaire, Sudan, Ethiopia, Uganda, Kenya, Tanzania and Angola. Forest edges and gallery forests, woodlands, thickets and savannas.

9. **Adenia stenodactyla** Harms in Notizbl. Bot. Gart. Berl. **8**: 297 (1923).—de Wilde in Meded. Landb. Wag. **71**, 18: 163, fig. 26 (1971); in F.T.E.A., Passifloraceae: 37 (1975). Type from Tanzania.

    Adenia angustisecta Engl. & Harms. in Engl., Pflanzenw. Afr. **3**, 2: 605 (late 1921) *nom. illegit.* non Burtt Davy (Sept. 1921).

    Adenia stenodactyla var. *kondensis* Harms, tom. cit.: 298 (1923); op. cit. **13**: 426 (1936). Type from Tanzania.

Herbaceous climber up to 2·5 m., arising from a tuberous rootstock. Branches often greyish-glaucous, striate, with internodes 2–10 cm. long. Leaf-lamina (2)4–10(25) × 4–20(25) cm., deeply 5-parted or 5-foliolate, suborbicular in outline, membranous, green above, greyish-green sometimes with purplish spots beneath, with 5 main nerves from the base, the nerves prominent, sometimes reddish; leaflets 2–16(25) × 0·2–2(9) cm., oblong to linear, subobtuse to acute and with acumen up to 1·5 cm. long, margin entire or deeply 3–8(16)-lobed, narrowing gradually at the base, pinnately nerved with the fine reticulation somewhat distinct, petiolules up to 1·5 cm. long; petiole 0·5–5 cm. long. Glands of the lamina-base 2(4), 1–1·5 mm. in diameter, on 2 separate or contiguous wart-like appendages; lamina-glands (0)2–4(16), c. 1 mm. in diameter, submarginal or scattered. Stipules c. 1 mm. long, linear. Sterile tendrils up to 12 cm. long.

Cymes with peduncles 1–8 cm. long, up to 10-flowered in ♂, 1–3-flowered in ♀, the flowers cream-coloured; bracts and bracteoles 1–3 mm. long, lanceolate, acute; pedicels 3–12 mm. long in ♂, 2–8 mm. in ♀; tendril (0)1, 2–6 cm. long. ♂ flower: 25–40 × 8–13 mm. (incl. the 4–8 mm. long stipe), broadly urceolate; hypanthium 4–7 × 8–13 mm.; calyx-tube 12–20 mm. long; calyx-lobes 4–8 mm. long, ovate-triangular, obtuse, lacerate-denticulate; petals 6–10 × 2·5–4 mm., oblong-spathulate, obtuse at the apex, shortly unguiculate, 5-nerved, lacerate-dentate in the upper ½, inserted at the same level as the corona or up to 2 mm. above this; corona consisting of slender sometimes branched hairs 0·5–1·5 mm. long; stamen-filaments 4–7·5 mm. long, connate at the base, inserted at the base of the hypanthium; anthers 6–8(10) × 1(1·5) mm., obtuse; disk-glands 3–3·5 mm. long; vestigial ovary c. 1 mm. long; gynophore 0–1 mm. long. ♀ flower: 15–18 × 5–8 mm. (incl. the 2–3(4) mm. long stipe), tubular-campanulate; hypanthium 2–2·5 × 5–8 mm., broadly cup-shaped; calyx-tube 6–8 mm. long; calyx-lobes 5–7 mm. long, ovate-triangular, subacute, entire; petals c. 6 × 2·5 mm., oblong-spathulate, obtuse, denticulate-lacerate in the upper ½, inserted at the same level as the corona, 5-nerved; corona similar to that of the ♂ flower; staminodes c. 4 mm. long, connate at the base; disk-glands 0·75–1 mm. long; pistil 9–10 mm. long; gynophore c. 2 mm. long; ovary c. 4 × (2·5)3 mm., ellipsoid, 3-ribbed; styles 2–2·5 mm. long, connate at the base; stigmas sessile, woolly-papillate. ♀ flowers similar to the ♂ but with a well-developed pistil. Capsule (2·5)3–6 × 2–3·5 cm., ellipsoid, with smooth coriaceous pericarp, 40–60-seeded. Seeds 5–6·5 × 5–6 mm., obovate, pitted.

Zambia. N: Mbale (Abercorn) Distr., Kawimbe to Mbale, 1740 m., ♂ fl. 1.xii.1959, *Richards* 11851 (K; SRGH). **Mozambique.** Z: Morrumbala, 22 km. from the crossing between the roads Campo-Nicuadala and Morrumbala, c. 100 m., fr. 29.xii.1967, *Torre & Correia* 16796 (LISC).
Also in Tanzania. In grassland, savannas, scrub, edges of woodlands, rocky hillsides on various types of soils.

10. **Adenia dolichosiphon** Harms in Notizbl. Bot. Gart. Berl. **13**: 425 (1936).—de Wilde in Meded. Landb. Wag. **71**, 18: 165, fig. 26 (1971); in F.T.E.A., Passifloraceae: 38 (1975). Type from Tanzania.

Herbaceous climber up to 5 m., with a turnip-shaped tuber. Leaf-lamina 2–10 × 3–8·5 cm., simple, unlobed, ovate to triangular or lanceolate, apex obtuse to acute with mucro 0·5–1 mm. long, margin entire, cordate to truncate or hastate at the base, membranous, greenish above, paler below, not punctate, with 3–5(7) main nerves from the base, 1–2(4) pairs of nerves from the midrib and rather indistinct reticulation; petiole 0·5–5·5 cm. long. Glands of the lamina-base 2, 1·5–3 mm. in diameter, contiguous, yellowish, borne on the subspathulate or ± 2-lobed 2–5 mm. wide peltate lamina-base; glands of the lamina 0. Stipules 1–2 mm. long, narrowly triangular. Sterile tendrils up to 15 cm. long. Cymes sessile or with peduncle up to 5 cm. long, 1–15-flowered in ♂, 1–3-flowered in ♀; bracts and bracteoles 1–2·5 mm. long, narrowly triangular to oblong, acute; pedicels 10–30 mm. long in ♂ flower. ♂ flower: (35)40–75 × 5–9 mm. (incl. the 4–9 mm. long stipe), tubiform; hypanthium 1·5–3 × 3–6 mm., tapering, cup-shaped; calyx-tube 30–55 mm. long; calyx-lobes 5–7·5 mm. long, ovate-oblong, subobtuse, crenulate or fimbriate at the margin; petals 30–45 × 0·5(0·75) mm., linear, acute, 1–3-nerved, fimbriate, inserted at the same level as the corona or up to 2 mm. above it; corona-hairs 1–1·5 mm. long, slender, ramified; stamen-filaments 7–10 mm. long, connate at the base, inserted at the base of the hypanthium or on an androgynophore up to 2 mm. long; anthers 7–8·5 × 1 mm., obtuse, with apiculum c. 0·75 mm. long; disk-glands 1·5–2·5 mm. long; vestigial ovary c. 0·5 mm. long. ♀ flower unknown. Capsule 4–5 × 4 cm., subglobose, with thick coriaceous pericarp, 15–20-seeded. Seeds c. 7 × 5 mm., ovate, pitted.

Mozambique. N: Cabo Delgado, Macondes, 12 km. from Nantulo to Mueda, c. 350 m., ♂ fl. 30.xii.1963, *Torre & Paiva* 8786 (B; COI; LD; LISC; LMU; MO). Z: Gurué, 26 km. from Mutuáli to Lioma, 650 m., fr. 10.ii.1964, *Torre & Paiva* 10498 (COI; LD; LISC; MO). MS: Vila Machado, ♂ fl. 26.ii.1948, *Mendonça* 3782) BR; EA; J; K; LISC; LMA; M; P; SRGH; WAG).
Also in Tanzania. In *Brachystegia* woodland and shrubby vegetation on clayey-sandy soils.

11. **Adenia hastata** (Harv.) Schinz in Engl., Bot. Jahrb. **15**, Beibl. 33, 1: 3 (1892).—
Burtt Davy in Ann. Transv. Mus. **3**: 121 (1912).—Engl., Pflanzenw. Afr. **3**, 2: 603
(1921).—Harms in Notizbl. Bot. Gart. Berl. **8**: 295 (1923); in Engl. & Prantl,
Pflanzenfam. ed. 2, **21**: 491 (1925).—Burtt Davy, F.P.F.T. **1**: 221 (1926).—
Liebenberg in Bothalia **3**: 519, 532, 536, fig. 4–5 (1939).—A. & R. Fernandes in
Garcia de Orta **6**, 2: 256, t. 9 (1958).—Watt & Breyer-Brandwijk, Med. & Pois. Pl.
S. & E. Afr. ed. 2: 828 (1962).—de Wilde in Meded. Landb. Wag. **71**, 18: 168
(1971).—J. H. Ross, Fl. Natal: 251 (1972). Type from S. Africa.
 Modecca hastata Harv., Thes. Cap. **2**: 43, t. 167 (1863). Type as above.
 Adenia schlechteri Harms in Engl., Bot. Jahrb. **33**, 1: 150 (1902). Type from
S. Africa (Transvaal).

Herbaceous climber up to 4 m., arising from a tuberous rootstock. Branches
slender, greyish-green, striate, with internodes up to 5 cm. long. Leaf-lamina
1·5–10(14) × 1·5–10(13) cm., broadly ovate to ± hastate, apex obtuse to acute
and with mucro up to 2 mm. long, margin entire, cordate to truncate at the base,
membranous (herbaceous) to coriaceous, greyish-brown to blackish, often punctate,
with 3–7 main nerves from the base, 1–2 pairs of nerves from the midrib and
indistinct reticulation; petiole 0·5–5(10) cm. long. Glands at the lamina-base
1–2, 1–3(5) mm. in diameter, contiguous, borne on the ± semiorbicular or 2-lobed
up to 4 mm. wide peltate lamina-base; lamina-glands 0–2(4), up to 5 mm. in
diameter, at or near the apex. Stipules 1·5–2 mm. long, narrowly triangular,
acute. Sterile tendrils up to 12 cm. long. Cymes with peduncles 0·5–4(11) cm.
long, up to 12-flowered in ♂, 1–3-flowered in ♀; bracts and bracteoles 1–3 mm.
long, narrowly triangular; pedicels 2–10 mm. long in the ♂, 1–2 mm. long in the
♀. ♂ flower: (13)15–30 × (3)4–5 mm. (incl. the 2–6 mm. long stipe), tubular-
infundibuliform; hypanthium 2–3 × 3–6 mm., cup-shaped; calyx-tube (5)7–
15 mm. long; calyx-lobes 4–7 mm. long, ovate to elliptic, densely fimbriate at the
margin; petals 5–7(9) × 0·33–1 mm., linear-lanceolate, 1–3-nerved, entire to
densely fimbriate, inserted (1)2–8 mm. above the corona; corona hairy, the hairs
sometimes branched; stamen-filaments 3–7 mm. long, connate for 2·5 mm. at the
base, inserted at the base of the hypanthium or on an androgynophore; anthers
4–7 × 1 mm., obtuse, with apiculum up to 0·33 mm. long; disk-glands 1–1·5 mm.
long; vestigial ovary c. 1 mm. long, with gynophore up to 1 mm. long. ♀ flower:
(8)10–18 × 5–7(8) mm. (incl. the 0·5–2·5 mm. long stipe), campanulate-
infundibuliform; hypanthium 1–2 mm. long; calyx-tube 3–6 mm. long; calyx-
lobes 4–7 mm. long, ovate, crenulate-laciniate at the margin; petals 3–5(6) × 0·1–
0·33(0·5) mm., linear, acute, 1-nerved, entire to ± fimbriate, inserted 1·5–3 mm.
above the corona; corona consisting of slender hairs 1–1·5 mm. long; disk-glands
0·33–0·75 mm. long; pistil 7–9 mm. long; gynophore 1–1·5 mm. long; ovary
(2)2·5–4·5 × (1·5)2–4 mm., ellipsoid to subglobose; styles connate at the base, with
the arms 1·5–2·5 mm. long and stigmas palmately branched, papillate. Capsule
2–3·5(6) cm. in diameter, subglobose, with coriaceous pericarp, 5–25-seeded.
Seeds 7(8) × 5·5–6·5 × 3 mm., ovoid to ellipsoid, pitted.

Var. **hastata**.—de Wilde in Meded. Landb. Wag. **71**, 18: 169, fig. 26 (1971).

Leaves without glands at the apex. Glands of the lamina-base single or 2,
contiguous on a single, semiorbicular, not lobed appendage. Petals entire or
remotely irregularly serrate-fimbriate only at the base. Anthers at anthesis
reaching the throat of the calyx-tube or nearly so.

Mozambique. SS: Gaza, Mabalane, 192 m., ♂ fl. 1.x.1963, *Leach & Bayliss* 11776
(BR; COI; K; LISC; SRGH). LM: Magude, near Motaze, ♀ fl. & fr. 6.xi.1944,
Mendonça 2783 (BR; LD; LISC; LMA; MO).
Also in S. Africa. Woodlands and shrubby savannas on alluvial clays or rocky or sandy
soils.

12. **Adenia zambesiensis** R. & A. Fernandes in Bol. Soc. Brot. Sér. 2, **50**: 166, t. 3 & 4
(1976). Type: Mozambique, 21 km. from Mocuba to Maganja da Costa, *Torre &
Correia* 16975 (LISC, holotype).

Herbaceous, glabrous climber c. 1 m. long (or more ?), arising from a woody
rootstock. Stem slender, yellowish-green, sulcate, with internodes 2·5–7 cm. long.
Leaf-lamina 2–5 × 2–5 cm., (3)5-lobed, broadly ovate to suborbicular in outline,
cordate at the base, membranous, green on the superior surface, light green beneath,

3(5)-palmatinerved and with inconspicuous reticulation; lobes unlobed or shallowly 3-lobed, obtuse or rounded at the apex, entire-margined, ± deeply constricted at the base, the median up to 4 × 2 cm., the intermediate ones up to 2·5 × 1·5 cm., the basal ones up to 1 × 0·9 cm.; petiole 1–4 cm. long, canaliculate above. Glands of the lamina-base 2, c. 0·75 mm. in diameter, sessile; lamina-glands minute, scattered, sometimes submarginal. Sterile tendrils axillary, up to 5 cm. long, slender. Stipules c. 1 mm. long, lanceolate. Inflorescences usually 1-flowered and with a tendril; peduncle 2–3 cm. long, slender. Bracteoles c. 1 mm. long. ♂ flower 20–25 × 3–4 mm. (incl. 8–9 mm. long stipe); hypanthium 3–4 mm. wide; calyx-tube (with hypanthium) 8–9 × 3–4 mm.; calyx-lobes 4–5 × 3 mm., thin, ± obtuse at the apex, with reticulate nerves; petals 2–2·5 × 1·25–1·5 mm., ovate or broadly elliptic, very thin, apiculate or irregularly dentate at the apex, 1-nerved, inserted c. 6 mm. above the hypanthium-base; corona 0; stamen-filaments c. 5·5 mm. long, connate at the base into a tube c. 1–1·5 mm. long and inserted at the base of the hypanthium; anthers c. 2·5 × 1 mm., erect, obtuse and without apiculum at the apex; disk-glands c. 1 mm. long, spathulate, inserted c. 1·5 mm. above the base of the hypanthium; vestigial ovary c. 0·5 mm. long, on a gynophore c. 0·5 mm. long. ♀ flower and fruit unknown.

Mozambique. Z: 21 km. from Mocuba to Maganja da Costa, c. 150 m., ♂ fl. 9.i.1968, *Torre & Correia* 16975 (LISC).
Known only from Mozambique. Open forest with grassy understorey on sandy soils.

13. **Adenia schliebenii** Harms in Notizbl. Bot. Gart. Berl. **13**: 426 (1936).—A. & R. Fernandes in Garcia de Orta **6**, 2: 257, t. 13 (1958).—de Wilde in Meded. Landb. Wag. **71**, 18: 175, fig. 26 (1971); in F.T.E.A., Passifloraceae: 40 (1975). Type from Tanzania.

Herbaceous climber up to 5 m., arising from a tuberous rootstock. Fertile branches 2–5 mm. in diameter, greenish, glabrous, with internodes 5–15 cm. long. Leaf-lamina 5–12 × 4–13 cm., entire to deeply 3–5-lobed, suborbicular, acute-acuminate and mucronate at the apex, cordate to subtruncate at the base, membranous, brownish-green on the upper surface, paler beneath, usually densely reddish-punctate, pubescent especially on the nerves, with 5 main basal nerves, up to 4 pairs of nerves from the midrib and indistinct reticulation; lobes 1–8 cm. long, elliptic to oblong, usually constricted at the base, remotely denticulate; petiole 2–6·5 cm. long. Glands of the lamina-base 2, on 2 auricles 1–1·5 mm. in diameter, lateral to the apex of the petiole. Stipules c. 1 mm. long, triangular. Sterile tendrils up to 15 cm. long. Cymes 3–15-flowered in ♂; peduncle 0·5–6 cm. long; bracts and bracteoles 3–6 mm. long, dentate-laciniate; pedicels 4–10 mm. long, glabrous. ♂ flower: 25–35 × (6)8–10 mm. (incl. the 3–6 mm. long stipe), tubular-infundibuliform; hypanthium and calyx-tube (15)18–24 mm. long; calyx-lobes 4–5·5 mm. long, ovate-triangular, subacute, serrulate at the margin; petals 3–5 × 0·75–1(1·5) mm., lanceolate, subacute, 3-nerved, fimbriate, inserted 11–15 mm. above the base of the hypanthium; corona 0; stamen-filaments 10–21 mm. long, connate for 1–2 mm. at the base, inserted at the base of the hypanthium; anthers (3)4–7 × 0·5–1 mm., obtuse, with apiculum up to 0·1 mm. long; disk-glands c. 1·5 mm. long; vestigial ovary 0·5–1 mm. long; gynophore 0·25–1 mm. long. ♀ flower and capsule unknown.

Mozambique. N: Cabo Delgado, Msalu R., ♂ fl. iii.1912, *Allen* 150 (K).
Known only from Tanzania and northern Mozambique. In forests and scrub.

14. **Adenia stricta** (Mast.) Engl., Pflanzenw. Afr. **3**, 2: 605 (1921).—A. & R. Fernandes in Garcia de Orta **6**, 2: 258, t. 14–15 (1958).—de Wilde in Meded. Landb. Wag. **71**, 18: 176, fig. 28 (1971). Type: Mozambique, vicinity of Tete, ii.1860, *Kirk* (K, lectotype).
    Modecca stricta Mast. in F.T.A. **2**: 515 (1871). Syntypes from Mozambique, Tete, *Kirk* (K); Malawi, Murchison Falls, *Meller* (K).

Trailing herb 0·5–2·5 m. long with a tuberous rootstock, ± pubescent in all parts except the flowers. Fertile branches 3–7 mm. in diameter, greyish-green, with internodes 2–8 cm. long. Leaf-lamina 3–15 × (2)3–15 cm., entire to 3–7-lobed, ovate-oblong to suborbicular in outline, acute at the apex with mucro up to 1·5 mm. long, cordate to subtruncate at the base, brownish-green on the upper surface, paler beneath, not punctate, ± pubescent mainly along the nerves, with 3–5(7) main nerves from the base, (1)2–5 pairs of nerves from the midrib and indistinct reticulation; lobes (1)2–10 × 1–5(6) mm., ovate to spathulate-oblong, sinuate-dentate; petiole 2–7(15) cm. long. Glands of the lamina-base 2, on 2 auricles connate over the apex of the petiole; submarginal glands 0–6 and marginal ones minute, borne on the teeth. Stipules 1–2 mm. long, triangular, acute. Sterile tendrils up to 10 cm. long. Cymes 5–10-flowered in ♂, 1–2-flowered in ♀; peduncle 0·5–5 cm. long; bracts and bracteoles 2–6 mm. long, lanceolate and sometimes dentate; pedicels 5–20 mm. long in ♂, 2–10 mm. long in ♀, pubescent; tendrils (0)1, 2–6 cm. long. ♂ flower: 25–37 × (7)9–13 mm. (incl. the 3–6 mm. long stipe), tubiform to ± urceolate; hypanthium 3–5 × 5–12 mm., cup-shaped; calyx-tube 12–20 mm. long; calyx-lobes c. 8 mm. long, triangular to oblong, obtuse to subacute, fimbriate; petals 7–17 × 1·5 mm., lanceolate to linear, subacute, 3-nerved, fimbriate, inserted on the calyx-tube 4–5 mm. above the corona; corona consisting of fine hairs 1·5–2 mm. long; stamen-filaments 3·5–7 mm. long, free, inserted at the base of the hypanthium or on an androgynophore up to 1·5 mm. long; anthers 10–12 × 1–1·5 mm., obtuse, with apiculum up to 0·1 mm. long, reddish-dotted; disk-glands 1·5–2 mm. long; vestigial ovary 0·5–1 mm., on a gynophore 0·5–1·5 mm. long. ♀ flower: 15–18 × 10(15) mm. (incl. the 1·5–2 mm. long stipe), campanulate; hypanthium c. 2 mm. long, cup-shaped; calyx-tube 4–6 mm. long; calyx-lobes 9–10 mm. long, oblong, obtuse, with the margin fimbriate; petals c. 7 × 1–1·5 mm., lanceolate, acute, 3-nerved, fimbriate, inserted 1–2 mm. above the corona; corona of fine hairs 1–1·5 mm. long; staminodes c. 3 mm. long, free; disk-glands 1–1·5 mm. long; pistil 7–10 mm. long with a gynophore 0·5–1 mm. long; ovary 3·5–5 × 3·5–4·5 mm., subglobose; styles 2–2·5 mm. long, free; stigmas c. 3 mm. in diameter, subreniform, woolly-papillate. Capsule c. 4·5 × 3·5 cm., ovoid-ellipsoid, with coriaceous pericarp, c. 40-seeded. Seeds c. 7·5 × 5·5–6 × 3 mm., obovoid, pitted.

Rhodesia. N: Urungwe, Mensa Pan, 17·5 km. SE. of Chirundu bridge, 460 m., ♂ fl. 29.i.1958, *Drummond* 5349 (K; SRGH). **Malawi.** S: Murchison Falls, *Meller* (K). **Mozambique.** N: Malema, 10 km. from Estação Algodoeira do Mutuáli, ♂ fl. 25.ii.1954, *Gomes e Sousa* 4226 (LMA). T: 6 km. on Changara road, 130 m., ♂ fl. 26.xii.1965, *Torre & Correia* 13812 (LISC); Chicoa, Serra de Sanga, ♀ fl. 30.xii.1965, *Torre & Correia* 13891 (COI; LD; LISC; LMU).

Known only from Rhodesia, Malawi and Mozambique. Open woodlands on sandy or stony soils.

15. **Adenia lanceolata** Engl., Bot. Jahrb. **14**: 378 (1891); Pflanzenw. Afr. **3**, 2: 603 (1921).—Harms in Engl., Bot. Jahrb. **15**: 572 (1893); in Engl. & Prantl, Pflanzenfam. **3**, 6a: 84 (1893).—F. W. Andr., Fl. Pl. Anglo-Egypt. Sudan **1**: 163 (1950).— de Wilde in Meded. Landb. Wag. **71**, 18: 185 (1971); in F.T.E.A., Passifloraceae: 43 (1975). Syntypes from the Sudan.

Climber or suberect branched herb up to 6 m., arising from a tuberous rootstock up to 30 cm. in diameter. Leaf-lamina 1–15 × 0·5–5 cm., ovate, obovate or lanceolate, apex acute-acuminate to broadly obtuse or sometimes ± truncate, usually with mucro 0·5–1 mm. long, margin entire, acute to rounded at the base, membranous to coriaceous, greenish or brownish above, grey to glaucous beneath, punctate or not, with 3–5 main nerves from the base, 2–5 pairs of nerves from the midrib and usually distinct reticulation; petiole 0·2–3 cm. long. Lamina-base glands 1 or 2, 0·5–2 mm. in diameter, contiguous or separate, on a wart-like or subspathulate median appendage or on 2 ± connate or separate auricles; lamina-glands 0 or 1–3 pairs, usually approximate to the axils of the nerves. Stipules 1–1·5 mm. long, linear. Sterile tendrils 2–10 cm. long. Cymes 2–20(50)-flowered in ♂, 1–3(5)-flowered in ♀; peduncle up to 2(3) cm. long; bracts and bracteoles 1–2 mm. long, narrowly triangular to lanceolate, entire or serrulate; pedicels 2–20 mm. long in ♂, 2–10 mm. long in ♀; tendril 0–1, 3–7(10) cm. long. ♂ flower: (13)15–27 × 3–5 mm. (incl. the (2)4–10 mm. long stipe), tubular-infundibuliform; hypanthium 1·5–3·5 × 2–5 mm., cup-shaped; calyx-tube (4)5–10 mm. long; calyx-lobes 5–8(10) mm. long, oblong to lanceolate, ± obtuse at the apex, serrulate-fimbriate at the margin; petals 5·5–9 × (1)1·5–2 mm., lanceolate, acute, 3-nerved, serrulate, inserted at the same level as the corona or up to 2 mm. above; corona consisting of fine hairs 0·33–1 mm. long; stamen-filaments 2–4·5 mm. long, connate at the base, inserted at the base of the hypanthium or on an androgynophore up to 3 mm. long; anthers 3·5–5 × (0·5)0·75–1 mm. with apiculum up to 0·2 mm. long; disk-glands 0·75–1·5 mm. long; vestigial ovary 0·5–0·75 mm. on a gynophore up to 0·75 mm. long. ♀ flower: 10–22 × (3)3·5–6 mm. (incl. the 1–7 mm. long stipe), tubular-campanulate; hypanthium 1–2·5(3) × 3–5 mm., cup-shaped; calyx-tube 2·5–7·5 mm. long; calyx-lobes 3–6(8) mm. long, oblong, obtuse to acute, serrulate-laciniate; petals 3–7·5 × 0·5–1·5 mm., lanceolate, acute, 1–3-nerved, entire or shallowly denticulate in the upper ½, inserted at the level of the corona; corona comprising hairs 0·5–1 mm. long; staminodes 2·5–4 mm. long, connate for up to 1·5 mm. at the base, inserted at the base of the hypanthium; disk-glands 0·5–1 mm. long; pistil 6–11 mm. long; gynophore 1·5–3 mm. long; ovary 3·5–6 × 2·5–4 mm., ovoid-ellipsoid; styles connate at the base, with the arms 0·5–1 mm. long, ending in subglobular papillate stigmas. Capsule 2–4 × 1·5–2·5 cm., oblong-obovoid, with ± thickly coriaceous pericarp, 10–30-seeded. Seeds 5–7 × 4·5–6 × 2·5–3 mm., ovate to orbicular, pitted.

Subsp. **scheffleri** (Engl. & Harms) de Wilde in Blumea **17**: 180 (1969); in Meded. Landb. Wag. **71**, 18: 187, fig. 30 (1971); in F.T.E.A., Passifloraceae: 45 (1975). Type from Kenya.
    Adenia scheffleri Engl. & Harms in Engl., Pflanzenw. Afr. **3**, 2: 603 (1921). Type as above.

Leaf-lamina 1–6(8·5) × 0·5–2·5(5) cm. Lamina-base glands 1–2, on a single median wart-like or subspathulate appendage, or on 2 ± connate auricles 1·5–3 mm. in diameter. Cymes up to 50-flowered in ♂, 1–3(5)-flowered in ♀. ♂ flower: 17–26 mm. long (incl. the 3–7·5 mm. long stipe). ♀ flower 10–22 mm. long (incl. the 1–6 mm. long stipe). Capsule (2)2·5–4 cm. long.

    Zambia. N: Isoka, fr. 23.xii.1962, *Fanshawe* 7213 (K). E: Lundazi, Tigone Dam, c. 3 km. on Choma road, 1200 m., ♂ fl. 19.xi.1958, *Robson* 663 (BM; BR; K; LISC; SRGH); ♀ fl. & fr. 19.xi.1958, *Robson* 677 (BM; BR; K; LISC; SRGH). **Malawi.** N: Rumphi Distr., Rumphi R. junction, 1088 m., ♂ fl. 6.xi.1966, *Pawek* 281 (SRGH). C: 77 km. N. of Lilongwe, S. of Mtiti R., 1140 m., ♂ fl., *Gillett* 17515 (EA*).
    Also in Kenya and Tanzania. In savannas, woodlands, termitaria and rocky places.

16. **Adenia digitata** (Harv.) Engl., Bot. Jahrb. **14**: 375 (1891).—Burtt Davy in Ann. Transv. Mus. **3**: 121 (1912); F.P.F.T. **1**: 221 (1926).—Engl., Pflanzenw. Afr. **3**, 2: 605 (1921).—Harms in Engl. & Prantl, Pflanzenfam. ed. 2, **21**: 491 (1925).—Steyn, Toxicol. Pl. S. Afr.: 310, fig. 43a, 44, 45, 47a (1934).—Henkel, Woody Pl. Nat. Zulul.: 118 (1934).—Liebenberg in Bothalia **3**, 4: 527, 530–532, 541, fig. 1–3, 14–17, t. 6–36 (1939).—A. & R. Fernandes in Garcia de Orta **6**, 2: 259 (1958).—Dyer et al., Wild Fl. Transv.: 225 (1962).—Watt & Breyer-Brandwijk, Med. & Pois. Pl. S. & E. Afr. ed. 2: 826, fig. 218 (1962).—de Wilde in Meded. Landb. Wag. **71**, 18: 188, fig. 30, 31a–h (1971); in F.T.E.A., Passifloraceae: 45 (1975). Type from S. Africa (Zululand).
    Modecca digitata Harv., Thes. Cap. **1**: 8, t. 12 (1859); in Harv. & Sond., F.C. **2**: 500 (1862). Type as above.
    Clemanthus senensis Klotzsch in Peters, Reise Mossamb. Bot.: 143 (1862). Type: Mozambique, Manica e Sofala, Rios de Sena, *Peters* (B, holotype†).
    Modecca senensis (Klotzsch) Mast. in F.T.A. **2**: 513 (1871).—Schinz in Bull. Trav. Soc. Bot. Genève **1891**: 67 (1891).—Hook. f. in Curtis, Bot. Mag. Ser. 3, **57**: t. 7763 (1901). Type as above.
    Adenia senensis (Klotzsch) Engl., loc. cit. (1891).—Harms in Engl., Bot. Jahrb. **15**: 573 (1893); in Engl. & Prantl, Pflanzenfam. **3**, 6a: 84 (1893); op. cit. ed. 2, **21**: 491 (1925).—Bak. f. in Journ. Linn. Soc., Bot. **40**: 73 (1910).—Engl., loc. cit. (1921).—Burtt Davy, loc. cit. (1925).—Norlindh in Bot. Notis. **1934**: 106 (1934).—Garcia in Est. Ens. Doc. Junta Invest. Ultramar **12**: 163 (1954).—A. & R. Fernandes, tom. cit.: 258 (1958). Type as above.
    Adenia stenophylla Harms in Engl., op. cit. **26**: 238 (1899).—Engl., loc. cit. (1921).—
    Adenia stenophylla Harms in Engl., op. cit. **26**: 238 (1899).—Engl., loc. cit.— Burtt Davy, op. cit.: 222 (1926). Type from S. Africa (Transvaal).
    Adenia multiflora Pott in Ann. Transv. Mus. **5**: 235 (1917).—Harms in Notizbl. Bot. Gart. Berl. **8**: 298 (1923).—Burtt Davy, op. cit.: 221 (1926). Type from S. Africa (Transvaal).
    Adenia angustisecta Burtt Davy in Kew Bull. **1921**: 280 (1921); op. cit.: 222 (1926). Type from S. Africa (Transvaal).
    Adenia buchananii Harms in Engl., loc. cit. Type: Malawi, southern province, fl., *Buchanan* (B, holotype†; K).

    Subherbaceous climber 0·2–3 m., arising from a tuber c. 10 × 6 cm. Stems usually annual, the fertile ones 1·5–5 mm. in diameter, greenish or greyish, and with inter-nodes 1·5–12 cm. long. Leaf-lamina 4–18 × 3–17 cm., deeply (3)5-partite or -foliolate, suborbicular or 5-gonal in outline, cordate at the base, membranous to coriaceous, green to yellowish-green above, yellowish, greyish or glaucous beneath, with 5 main nerves from the base; leaflets very variable in form and size, 1·5–15 × 0·75–4(7) cm., entire to ± deeply (2)3–5(10)-lobed, ovate or obovate to linear, rounded or acute sometimes mucronate at the apex, attenuated into a petiolule up to 2 cm. long, with 2–12 pairs of nerves and indistinct reticulation; petiole 1–9 cm. long. Glands of the lamina-base 2, 0·5–1·5 mm. in diameter, on 2 separate auricles 1–2 mm. in diameter situated at the transition between petiole and lamina; other glands: 2–4, 0·5–1·5 mm. in diameter, on the basal part of the lamina between the insertion of the leaflets, and 0–8 on each leaflet, irregularly scattered or submarginal. Stipules 1–3 mm. long, triangular, acute.   Sterile tendrils up to 15 cm. long. Cymes (1)5–20(60)-flowered in ♂, 1–10-flowered in ♀, the flowers cream, yellow or orange, purplish-green at the base; peduncle stout, up to 7 cm. long; bracts and bracteoles 1·5–3 mm. long, lanceolate, acute; pedicels 1–15 mm. long in ♂ flowers, 1–6 mm. long in ♀; tendril 1, 2–10 cm. long. ♂ flower: (14)20–38 × 3–10 mm. (incl. the 3–12(15) mm. long stipe), tubular-infundibuliform; hypanthium (1)2–3·5 × 2–4(5) mm., cup-shaped; calyx-tube (5)8–12 mm. long; calyx-lobes (4)7–11 mm. long, ovate or oblong to lanceolate, obtuse, dentate-fimbriate at the margin; petals 6–12 × (1)1·5–2·5 mm., lanceolate, 1(3)-nerved, dentate or fimbriate at the margin, inserted at the level of the corona or up to 5 mm. above it; corona consisting of short hairs, or rarely absent; stamen-filaments 3·5(9–12) mm. long, connate at the base for (1·5)2·5–7 mm., inserted on an androgynophore 1–3·5 mm. long; anthers 3–6 × 1(1·5) mm., ± curved inward, clinging together by the 0·2 mm. long papillate apicula; disk-glands 0·5–1·5 mm. long; vestigial ovary 0·5–1 mm. long, on a gynophore up to 1·5 mm. long. ♀ flowers 20–25 mm. long. ♀ flower: 15–26 × 5–6 mm. (incl. the 2–7 mm. long stipe), tubular-infundibuliform; hypanthium 2–4 × 2–4 mm.; calyx-tube 4–8 mm. long; calyx-lobes 5–7 mm. long, ovate to oblong, obtuse, entire; petals 2–7 × 0·35–1 mm., lanceolate to linear, acute, 1(3)-nerved, entire or serrulate to the apex, inserted at the same level as the corona or up to 4 mm. above it; corona as in ♂ flower; disk-glands c. 1 mm. long; pistil 8–12 mm. long; gynophore 2–4 mm. long; ovary (4)5–6 × (2)3–4 mm., ovoid to oblong; styles connate for 1–1·5 mm. at the base with arms 1–1·5 mm. long and stigmas 2–3 mm. in diameter, subreniform, woolly-papillate. Capsule (2·5)3–5·5(7·5) × (1·5)2–3·5(4) cm., ovoid to ellipsoid, greenish-yellow, with smooth coriaceous pericarp up to 4 mm. thick, (10)20–60-seeded. Seeds 5–8 × 4·5–6·5 × 3 mm., ovoid to ellipsoid, pitted.

Botswana. SW: Kgalagadi, c. 16 km. NW. of Lephepe village, st. 8.iii.1969, *Chiwita* 532 (SRGH). SE: Mahalapye, fr., *Yalala* 345 & 353 (SRGH*). **Zambia.** B: Mankoya near Luena R., 14·5 km. ESE. of Mankoya, ♂ fl. 21.xi.1959, *Drummond & Cookson* 6704 (COI; K; LISC; SRGH). C: Lusaka, ♂ fl., *Fanshawe* 4107 (K). S: Namwala, Ngoma, Kafue National Park, Nakalombwe, fr. 7.xii.1961, *Mitchell* 11/77 (SRGH). **Rhodesia.** N: Gokwe, ♂ fl. 28.x.1963, *Bingham* 866A (LISC; SRGH). W: Wankie, near Wankie National Park, ♂ fl. 26.xi.1968, *Rushworth* 1296 (SRGH). C: Marandellas, 12 km. E. of Marandellas on Umtali road, ♂ fl. 21.xi.1970, *Biegel* 3420 (K; SRGH); fr. 21.xi.1970, *Biegel* 3421 (SRGH). E: Umtali Golf Course, Commonage, 1152 m., ♂ fl. 26.x.1951, *Chase* 4148 (BM; BR; COI; LISC; SRGH). S: Chiredzi, Chuanja Hills, Gona-Re-Zhou Game Reserve, ♂ fl. 14.xi.1971, *Sherry* 496 (SRGH). **Malawi.** S: Zomba, Lambulira, fr. 10.i.1958, *Jackson* 2129 (K; SRGH). **Mozambique.** N: Ribáuè, ♂ fl., ♀ fl. & fr., *Torre & Paiva* 10180 (LD; LISC; LMA; MO). Z: Lugela-Mocuba, Namagoa Estate, ♂ & ♀ fl. & fr. 11.xii.1943, *Faulkner* 266 (BR; COI; K; SRGH). T: Tete to Lupata, *Kirk* (K). MS: Gorongosa, Chitengo, ♂ fl. 10.xi.1965, *Balsinhas* 1051 (COI; LMA). LM: Maputo, Manhoça, between Zitundo and Ponta do Ouro, fr. 5.xi.1968, *A. & R. Fernandes & Pereira* 70 (COI; LMU); Inhaca I., fr. 5.iv.1944, *Mogg* (K; SRGH).
    Also in Tanzania, S. Africa (Swaziland) and Angola. In savannas, woodlands, bush-lands, rocky places, grasslands, termitaria, on Kalahari sand and red clayey soils.

17. **Adenia mossambicensis** de Wilde in Meded. Landb. Wag. **71**, 18: 194, fig. 32 (1971). Type: Mozambique, Eráti, 2 km. from Alua on Mejuco road, 450 m., ♂ fl., *Torre & Paiva* 9544 (LISC, holotype).

Herbaceous climber up to 2 m., arising from a tuberous rootstock. Fertile branches greyish-green, with internodes 1–10 cm. long. Leaf-lamina 3–6 × 2–4

(5) cm., 3–5-foliolate, ovate-oblong in outline, cordate at the base, membranous, green on upper surface, glaucous green and not punctate below; median leaflet 2·5–5 × 1–1·5(2·5) cm., unlobed or pinnatilobed with up to 8 lobes, ovate to oblong, obtuse or acute at the apex, acute at the base, petiolulate, the petiolule 0·5(1) cm.; basal leaflets much smaller (1 -2 cm. long), unlobed, subsessile; petiole 0·5–2·5 cm. long. Lamina-base glands 2, 0·5–1 mm. in diameter, wart-like; lamina-glands c. 0·5 mm. in diameter, up to 6 on the median leaflet, submarginal. Stipules c. 1 mm. long, lanceolate, withering. Sterile tendrils up to 10 cm. long. Cymes with peduncle 1·5–4 cm. long, up to 12-flowered in ♂; bracts and bracteoles 1–1·5 mm. long, lanceolate, acute; pedicels 5–15 mm. long; tendril 1, 1–5 cm. long. ♂ flower: (20)25–30 × 3–4(5) mm. (incl. the 9–11 mm. long stipe), tubular, cream-coloured; hypanthium 3–5 × 3–4 mm., cup-shaped; calyx-tube 4–7 mm. long; calyx-lobes 5–8 mm. long, oblong, obtuse, entire or nearly so; petals 5–7 × 1 mm., lanceolate, ± obtuse, 3-nerved, irregularly dentate near the apex, inserted at the same level as the corona or up to 1 mm. above; corona-hairs fine, 0·5–1 mm. long, branched; stamen-filaments 4·5–6 mm. long, connate at the base, inserted at the base of the hypanthium or on an androgynophore up to 2 mm. long; anthers c. 4–5 × 0·75 mm., obtuse, with an apiculum up to 0·2 mm. long; disk-glands (0·5)1·5–2 mm. long; vestigial ovary c. 0·5 mm. long; gynophore c. 0·5 mm. long. ♀ flower and capsule unknown.

Mozambique. N: Eráti, 2 km. from Alua on Mejuco road, 450 m., ♂ fl. 13.xii.1963, *Torre & Paiva* 9544 (LISC).
Known only from Mozambique. Granite rocks.

18. **Adenia erecta** de Wilde in Meded. Landb. Wag. **71**, 18: 199, fig. 32 (1971). TAB. **97** fig. B. Type: Zambia, Mwinilunga Distr., Mujileshi R., ♂ fl., *Richards* 16959 (K, holotype).

Erect unbranched herb up to 40 cm. high, the annual stems arising from an elongate rootstock. Leaf-lamina 8–20 × 0·2–0·4 cm., linear, attenuate at the apex, margin entire, acute at the base, ± coriaceous, ± glaucous-green, not punctate beneath, penninerved with indistinct reticulation; petiole 0–0·1 mm. Glands of the lamina-base 2, 1–1·5 mm. in diameter, on 2 inconspicuous auricles; lamina-glands 10–20, 0·5–1 mm. in diameter, in a single row on either side of the midrib. Stipules 1·5–2 mm. long, triangular-lanceolate, acute. Tendrils absent. Cymes sessile or with a peduncle up to 5 mm. long, 1–3-flowered in ♂; bracts and bracteoles 1·5–2 mm. long, narrowly triangular to linear, fimbriate at the margin; pedicels 1–3 mm. long. ♂ flower: (35)40–52 × 3–5 mm. (incl. the 10–15 mm. long stipe), narrowly tubular-infundibuliform; hypanthium and calyx-tube 20–23 mm. long; calyx-lobes 10–15 mm. long, lanceolate, obtuse, dentate; petals c. 4 × 0·66 mm., lanceolate, obtuse, 1-nerved, laciniate-fimbriate near the apex, inserted c. 4 mm. below the throat of the calyx-tube; corona consisting of hairs 0·5–1 mm. long, scattered from 5–10 mm. above the base of the hypanthium; stamen-filaments 7–9 mm. long, connate at the base, inserted on an androgynophore 1·5–2 mm. long; anthers 4 × 1 mm., obtuse, with apiculum c. 0·5 mm. long; disk-glands 2–3 mm. long; vestigial ovary (incl. gynophore) 1–1·5 mm. long. ♀ flower not known. Immature capsule 2 × 1·1 cm., ellipsoid, with a peduncle c. 1 cm. long.

Zambia. W: Mwinilunga Distr., S. of Matonchi Farm, c. 800 m., ♂ fl. 7.xi.1937, *Milne-Redhead* 3144 (K); immat. fr. 24.x.1937, *Milne-Redhead* 2926 (K).
Known only from Zambia. Grassland at edges of rivers and in *Brachystegia* woodlands.

19. **Adenia goetzei** Harms in Engl., Bot. Jahrb. **30**: 360, t. 14 (1902); in Engl. & Prantl, Pflanzenfam. ed. 2, **21**: 491 (1925).—Engl., Pflanzenw. Afr. **3**, 2: 603 (1921).— de Wilde in Meded. Landb. Wag. **71**, 18: 200, fig. 32 (1971); in F.T.E.A., Passifloraceae: 48 (1975). Type from Tanzania.

Erect, simple or few-branched herb up to 35 cm. high, arising from a subspherical tuber 5 -10(20) cm. in diameter. Stems annual, with internodes 0·5–3(7) cm. long, reddish near apex. Leaf-lamina 7–22 × (0·75)1–6·5 cm., oblong-lanceolate to lanceolate-linear, acute at the apex with mucro c. 0·5 mm. long, margin entire, acute to long-attenuate and inconspicuously auricled at the base, membranous to subcoriaceous, reddish-green or glaucous-green, finely purplish-brown spotted beneath, penninerved, with 4–6 pairs of ascending main nerves and fairly distinct reticulation; petiole (0)1–2(4) mm. long. Glands of the lamina-base 2, 1–2 mm. in diameter, on 2 ± inwards curved auricles c. 2 mm. in diameter; lamina-glands 0–20, scattered. Stipules 1–2·5 mm. long, narrowly triangular, entire or 2–5-dentate or -partite. Tendrils 0. Cymes sessile or with a peduncle up to 15 mm. long, 1–5-flowered, the flowers yellowish; bracts and bracteoles 1–2·5 mm. long, narrowly triangular, sometimes serrulate; pedicel 3–5 mm. long in the ♂ flower and 2–4 mm. long in the ♀. Flowers polygamous. ♂ flower: 10–26 × 3–6(8) mm. (incl. the 3–7 mm. long stipe), tubular-infundibuliform; hypanthium 1·5–3 mm. long; calyx-tube 5–8(9) mm. long; calyx-lobes 3–7(8) mm. long, oblong, obtuse, fimbriate or denticulate or entire; petals 6–10 × 0·75–2 mm., lanceolate-linear, acute, 1–3-nerved, fimbriate in the upper ⅔ or only near the apex, inserted at the same level or up to 3 mm. above the corona, this consisting of a dense row of hairs 0·75–2 mm. long; stamen-filaments 4–7 mm. long, connate at the base, inserted at the base of the hypanthium or on an androgynophore up to 3 mm. long; anthers 3–5 × 0·75–1 mm., obtuse, with a ± blunt or acute, ± papillate apiculum 0·2–0·5 mm. long; disk-glands 0·5–0·75 mm. long; vestigial ovary 1–1·5 mm. long (incl. gynophore). ♀ flower: 12–26 × 3–8 mm. (incl. 2·5–8 mm. long stipe), tubular-infundibuliform, slightly larger than the ♂ flower, yellowish-green; hypanthium (0·5) 1–2·5 mm. long; calyx-tube 4–8(10) mm. long; calyx-lobes 4–8(9) mm. long, oblong, obtuse or acute; petals 5–12 × 0·5–1·75 mm., lanceolate-linear, acute, fimbriate, inserted at the same level as the corona; corona as in ♂; stamen-filaments 3–7 mm. long, free or shortly connate at the base, inserted at the base of the hypanthium or on an androgynophore up to 1 mm. long; anthers 2·5–4 (4·5) × 0·5–1 mm., sometimes partly abortive towards the apex, with a blunt, often papillate apiculum 0·5–1 mm. long; pistil 7–10(12) mm. long; gynophore 2(4) mm. long; ovary 3·5–5(6) × 2·5–4(4·5) mm., ellipsoid-oblong; styles ± free or shortly connate at the base, with arms 0·75–1·5 mm. long ending in subglobular papillate stigmas. ♀ flower resembling the ♂ and the ♀ flowers but 16–22 mm. long (incl. the 2·5–5 mm. long stipe), and with petals 5–7 × 0·5–1 mm. and ± free staminodes c. 2 mm. long. Capsule 3–4(4·5) × 2–2·5 cm., pendent at maturity, ellipsoid to obovoid, 3–6-ribbed, with coriaceous pericarp, 30–40-seeded. Seeds 4–4·5 × 3·5 × 2–2·5 mm., ovoid to subglobular, pitted.

Zambia. N: L. Mweru region NE. of Chiengi, Mweru-wa-Ntipa, fr. 13.x.1949, *Bullock* 1266A (K); ♂ fl. 15.x.1940, *Bullock* 1290 (K). C: Lusaka, fl. 17.xi.1965, *Fanshawe* 9423 (SRGH); ♀ fl. 8.xii.1968, *Fanshawe* 10487 (K). S: Mumbwa, *Macaulay* 958 (K). **Rhodesia.** N: Urungwe Reserve, Mgunje, ♂ & ♀ fl. 23.ii.1953, *Wild* 4238 (K; LISC; SRGH); Lomagundi Distr., 1280 m., ♀ fl. & fr. 14.xii.1964, *Wild &* *Drummond* 6673 (K; LISC; M; SRGH).
Also in Zaire and Tanzania. In savannas, *Brachystegia* woodlands, shrubby vegetation and stony places (granite kopjes).

20. **Adenia tuberifera** R. E. Fr. in Wiss. Ergebn. Schwed. Rhod.-Kongo-Exped. **1**: 158, t. 12 fig. 3–8 (1914).—Engl., Pflanzenw. Afr. **3**, 2: 603 (1921).—de Wilde in Meded. Landb. Wag. **71**, 18: 207, fig. 32 (1971). Type: Zambia, Kalambo Falls, 900–1500 m., *Fries* 1353 (K; UPS, holotype; Z).

Erect herb up to 60 cm., arising from a turnip-shaped tuber up to 12 cm. in diameter. Stems annual, simple or branched, with internodes 0·5–1·5 cm. long. Leaf-lamina (2)4–7(9) × 0·5–1·5 cm., lanceolate-linear, acute or subobtuse at the apex with mucro up to 1·5 mm. long, entire at the margin, acute to rounded at the base, membranous, greyish-glaucous, punctate beneath, with 5–13 pairs of nerves and indistinct reticulation; petiole 0·1–0·3 cm. long. Glands of the lamina-base 2, 0·5–1 mm. in diameter on the margin at the transition to the petiole; glands of the lamina 0. Stipules c. 1 mm. long, linear. Tendrils 0. Cymes sessile or with a peduncle 0·3–3 cm. long, 1(2–3)-flowered in ♂ and ♀, the flowers pendent; bracts and bracteoles c. 1 mm. long, narrowly triangular, acute; pedicels 10–20 mm. long in ♂, 5–10 mm. long in ♀. ♂ flower: 4–5-merous, 16–23 × 2–4·5 mm. (incl. the 5–7 mm. long stipe), tubular-infundibuliform; hypanthium 1–1·5 mm. long; calyx-tube 7–10 mm. long; calyx-lobes (3)4–6 mm. long, oblong to lanceolate, denticulate at the margin; petals 7–10 × 0·75–1 mm., linear-lanceolate, acute, 1-nerved, fimbriate in the upper ½, inserted at the same level as the corona or 1 mm. above it; corona consisting of hairs 0·5–1 mm. long; stamen-filaments 3–3·5 mm. long, connate at the base, inserted at the base of the hypanthium; anthers

2·5–4 × 0·75–1 mm., obtuse, with apiculum up to 0·2 mm. long; disk-glands
c. 0·5 mm. long, truncate; vestigial ovary c. 0·5 mm. long and gynophore c. 0·5 mm.
long. ♀ flower: 11–14 × 2–4 mm. (incl. the 2·5–4 mm. long stipe), ± tubiform;
hypanthium c. 1 mm. long; calyx-tube 4–5 mm. long; calyx-lobes 3–4 mm. long,
oblong, obtuse to subacute, entire; petals 4–5·5 × 0·3–0·75 mm., lanceolate-linear,
acute, 1-nerved, entire or shortly laciniate towards the apex, inserted at the same
level as the corona, this consisting of fine hairs 0·5–0·75 mm. long; staminodes
1·5–2 mm. long, free or shortly connate at the base; disk-glands 0·33–0·5 mm. in
diameter; pistil c. 7 mm. long; gynophore 1–2 mm. long; ovary 3–5 × 1·5–2·5 mm.,
ellipsoid-oblong, faintly 6-ribbed; styles c. 1 mm. long, free or connate; stigmas
subglobular, woolly-papillate. Capsule 4–4·5 × 2·5 cm., fusiform, ± 3-angular,
with thickly coriaceous pericarp, c. 30-seeded. Seeds c. 5 × 3·5–4 × 2·5 mm., ±
ovate, pitted.

Zambia. N: Mbala (Abercorn) Distr., Kalambo Falls, 1050 m., ♂ fl. 15.xi.1960, *Richards*
13565 (K); Abercorn Distr., Mpulungu road, 1536 m., ♀ fl. 22.x.1954, *Richards* 2103 (K);
Abercorn Distr., Kambole Escarpment, 1500 m., fr. 12.xi.1964, *Richards* 19259 (K).
Known only from Zambia. Open woodland and stony places in dry forests.

21. **Adenia ovata** de Wilde in Meded. Landb. Wag. **71**, 18: 204, fig. 32, 33 (1971).
Type: Zambia, Mufulira, *Fanshawe* 2516 (BR; K, holotype).

Herb 5–10 cm. high, with napiform warty tuber. Stems annual or partly
perennial, 1–3 cm. long, with internodes 0·2–1 cm. long; leaves ± appressed to
the ground. Leaf-lamina 7–14 × 5–12 cm., broadly ovate to subcircular, broadly
rounded and sometimes minutely apiculate at the apex, margin entire, broadly
rounded and shortly cuneate at the base, subcoriaceous, brownish- or reddish-
green or greyish-green above, paler, greyish-green and densely purplish-red
mottled beneath, with (3)5(7) main nerves from the base and with 3–4(7) pairs
of nerves from the midrib and distinct reticulation; petiole 0·1–0·4 cm. long.
Glands of the lamina-base 0 or 2, 0·5–1 mm. in diameter, on the margin near the
insertion of the petiole; glands of the lamina 0. Stipules 2–4 mm. long, narrowly
triangular, acute, entire or remotely dentate or dissected. Tendrils 0. Cymes
1–10-flowered in ♂, ⚥ and ♀, the flowers cream; peduncle up to 0·4 cm. long;
bracts and bracteoles 1·5–3·5 mm. long, narrowly triangular to linear, acute,
denticulate; pedicels 2–5 mm. long in the ⚥ and ♀ flower. ♂ flower: buds up to
10 mm. long; anthers as in the ⚥ flower, 3–3·5 mm. long incl. the blunt, papillate
apiculum c. 1 mm. long; vestigial ovary 1–1·5 mm. long. ⚥ flower: (14)18–
22 × 4–7(8) mm. (incl. the (4)5–7 mm. long stipe), tubular-infundibuliform;
hypanthium 1·5–2 mm. long; calyx-tube 6–8 mm. long; calyx-lobes 3–5(7) mm.
long, ovate-oblong, obtuse, fimbriate; petals 7–9 × 1–1·5 mm., lanceolate, subacute,
(1)3-nerved, fimbriate-papillate in the upper ½, inserted at the same level as the
corona, this consisting of hairs 1–2 mm. long; stamen-filaments 4–5 mm. long, ±
long-connate at the base, inserted at the base of the hypanthium; anthers (1·5)2–
4 × 0·75–1·25 mm. incl. the obtuse papillate apiculum 0·5–1·25 mm. long; disk-
glands 0·33–1 mm. long; pistil 7–10 mm. long; gynophore 1·5–3 mm. long;
ovary 4–5 × 2–3 mm., oblong-ellipsoid, usually finely warty; styles connate for
c. 1 mm., the arms 1–1·5 mm., ending in subglobular woolly-papillate stigmas.
♀ flower: (12)17–19 × 4–6(8) mm. (incl. the 4–5(6) mm. long stipe), tubular-
infundibuliform; hypanthium c. 1·5 mm. long; calyx-tube 5–6 mm. long; calyx-
lobes 3·5–5 mm. long, oblong, obtuse, subentire; petals (4)5–6 × 1–1·5(2) mm.,
obovate to oblanceolate, obtuse, (1)3-nerved, finely fimbriate in the upper ½,
inserted at the same level as the corona, this consisting of hairs 1–1·5 mm. long;
staminodes c. 4 mm. long, with the filaments shortly connate at the base and
vestigial anthers c. 1 mm. long; disk-glands c. 0·5 mm. long; pistil 7–8 mm. long;
gynophore 1–1·5 mm. long; ovary c. 4 × 2 mm., ellipsoid-oblong; styles connate
at the base, with the arms 1–1·5 mm. long, ending in subglobular woolly-papillate
stigmas. Capsule 1·5–2·5 × 1 cm., obovoid-oblong, subacute at the apex, with
smooth coriaceous pericarp. Seeds unknown.

Zambia. N: Kasama Distr., Chosi Flats, 1200 m., ⚥ fl. 14.xii.1974, *Richards* 19374
(K). W: Kitwe, immat. fr. 3.xii.1968, *Mutimushi* 2857 (K).
Known only from Zambia. In *Brachystegia* woodland on lateritic and sandy soils.

22. **Adenia volkensii** Harms in Engl., Pflanzenw. Ost-Afr. **C**: 281 (1895).—Engl., Pflanzenw. Afr. **3**, 2: 606 (1921).—Harms in Engl. & Prantl, Pflanzenfam. ed. 2. **21**: 492 (1925).—Verdc. & Trump, Common Pois. Pl. E. Afr.: 37, fig. 3 (1969).— de Wilde in Meded. Landb. Wag. **71**, 18: 182, fig. 28–29 (1971); in F.T.E.A., Passifloraceae: 43 (1975).—R. & A. Fernandes in Bol. Soc. Brot. Sér. 2, **50**: 168, t. 5 (1976). Type from Tanzania.

Erect perennial herb with annual shoots 20–50 cm. high, arising from a tuberous rootstock c. 5·5 cm. in diameter, without tendrils, ± pubescent, rarely glabrous Leaves 3–16 × 3–14 cm., entire or 3–7-lobed or -partite, apex acute with mucro up to 1 mm. long, margin remotely and very shallowly dentate or the teeth approximate and deeper, all ending in a black glandular mucro up to 1 mm. long, base cordate to truncate, membranous to coriaceous, brownish-green above, greyish-green and sometimes punctate beneath, pubescent especially along the nerves and veins, with 3–5(7) main nerves from the base and 2–5 pairs of nerves from the midrib, reticulation distinct or not, the nerves and veins sometimes reddish on the lower surface; lobes 2–12 × (0·25)0·5–6 cm., oblong to lanceolate, acute at the apex, contracted at the base; petiole 1·5–10 cm. long. Glands of the lamina-base 2, 0·5–1·5 mm. in diameter, on 2 suborbicular auricles at the top of the petiole; lamina-glands 2–6, submarginal. Stipules 0·5–1·5 mm. long, triangular, acute, gland-dotted. Cymes with peduncle up to 0·5 cm. long, 1–6-flowered in ♂, 1–2(3)-flowered in ♀ and ♀; bracts and bracteoles 2·5–10 mm. long, lanceolate, acute, usually dentate and gland-dotted. Flowers pubescent or glabrous. ♂ flower: 20–35(45) × 10–18(20) mm. (incl. the 2–8 mm. long stipe), broadly urceolate; pedicels 5–25 mm. long, ± pubescent; hypanthium broadly cup-shaped; calyx-tube 12–20 × 7–12 mm. at throat; calyx-lobes (3)4–7(9) mm. long, ovate, triangular, obtuse at the apex, 2 lobes with the margin ciliate and the others densely woolly-fimbriate, the hairs branched, 2–4 mm. long; petals 10–14 × 0·5–2 mm., lanceolate-linear, acute, 3(5)-nerved, densely woolly-fimbriate at the margin, the hairs branched, 3–5 mm. long, inserted at the same level as or up to 5 mm. above the corona, this consisting of branched hairs 1·5–3 mm. long; stamen-filaments 3–5·5 mm. long, free, inserted at the base of the hypanthium; anthers 8–12 × 1–1·5 mm., obtuse, with apiculum c. 0·1 mm. long; disk-glands 1·5–3 mm. long; vestigial ovary (incl. gynophore) 0·5–1·5 mm. long. ♀ flower: c. 26 × 10 mm. (incl. the c. 2 mm. long stipe), broadly tubiform, pubescent; pedicels up to 15 mm. long, pubescent; hypanthium c. 8 mm. wide; calyx-tube 18 × 10 mm., densely reddish-dotted; calyx-lobes c. 8 × 5 mm., ovate-triangular, obtuse, longitudinally reddish-dotted, 2 lobes with the margins ciliate and the others with the margins long-woolly-laciniate as in ♂ flower; petals 12–13 mm. long, linear-lanceolate, 3-nerved, margins fimbriate and inserted a little above the level of the corona, this as in ♂; stamen-filaments c. 9 mm. long, free, inserted at the base of the hypanthium; anthers c. 5·5 × 0·75 mm., with apiculum up to 0·5 mm. long, densely reddish-dotted; disk-glands 3–3·5 mm. long; gynophore c. 4 mm. long; ovary 4–4·5 × 3·5 mm., ellipsoid or subglobose; styles 3·5 mm. long, connate; stigma capitate, woolly-haired. ♀ flowers 16–22 × 8–11 mm. (incl. the 0·25–1·5 mm. long stipe), broadly tubiform; pedicels 5–10 mm. long, pubescent; hypanthium 1·5–2 mm. long, cup-shaped; calyx-tube 7–10 mm. long; calyx-lobes 5–8 × 0·5–2·5 mm., oblong, ± obtuse, fimbriate as in ♂ and ♀ flowers; petals 8–12 × 0·3–0·5 mm., linear, 1-nerved, sparsely fimbriate, inserted 1–2·5 mm. above the corona, this as in other flowers; staminodes 2·5–5 mm. long, up to 1·5 mm. connate; disk-glands 1–2 mm. long; pistil 10–14(16) mm. long; gynophore 1–2·5 mm. long; ovary 4–5·5(8) × 3–4(6) mm., ellipsoid; styles connate for 2–4 mm., the arms c. 1 mm. long; stigmas c. 4 mm. in diameter, much-branched, subreniform, ± woolly-papillate. Fruit 3·5–5·5 × 3–4·5 cm., subglobose to ellipsoid, with coriaceous pericarp, 15–30-seeded. Seeds 8–9 × 7–7·5 × 3–4 mm., ovate, pitted.

Some flowers of the ♀ plants have reduced ♂ and ♀ organs. In one of the flowers examined we found 4 anthers: one 5·5 mm. long, two 4 mm. long and one 3 mm. long with a long apiculum. The fifth stamen was reduced to the filament. The ovary was not as large as in the ♀ flower and the stigma was capitate. Perhaps this plant may also produce ♀ flowers.

Malawi. C: Kasungu Distr., Chamama, 1000 m., ♀ fl. 16.i.1959, *Robson* 1227 (K).

Also in Uganda, Kenya and Tanzania. On termitaria in *Brachystegia* woodlands, scrub and rocky places.

23. **Adenia repanda** (Burch.) Engl., Bot. Jahrb. **14**: 375 (1891).—Harms in Engl., op. cit. **15**: 573 (1893); in Engl. & Prantl, Pflanzenfam. **3**, 6a, Nachtr. 1: 255 (1897); op. cit. ed. 2, **21**: 490, fig. 224 (1925); in Warb., Kunene-Samb. Exped. Baum: 310 (1903).—R. E. Fr. in Wiss. Ergebn. Schwed. Rhod.-Kongo-Exped. **1**: 159, t. 12 fig. 9 (1914).—Engl., Pflanzenw. Afr. **3**, 2: 600, fig. 266 (1921).—Bremek. in Ann. Transv. Mus. **15**, 2: 248 (1933).—Liebenberg in Bothalia **3**, 4: 525, 532, 534, fig. 12 (1939).— A. & R. Fernandes in Garcia de Orta **6**, 4: 658 (1958); in C.F.A. **4**: 219 (1970).— Watt & Breyer-Brandwijk, Med. Pois. Pl. S. & E. Afr. ed. 2: 828 (1962).—Schreiber in Prodr. Fl. SW. Afr. **89**: 2 (1968).—de Wilde in Meded. Landb Wag. **71**, 18: 237 (1971). Type from S. Africa (Griqueland West).
  Paschanthus repandus Burch., Trav. Int. S. Afr. **1**: 543 (1822).—DC., Prodr. **3**: 336 (1828).—G. Don, Gen. Syst. **3**: 58 (1834).—Schinz in Engl., Bot. Jahrb. **15**, Beibl. 33: 3 (1892).—Harms in Engl. & Prantl, Pflanzenfam. **3**, 6a: 81 (1893).— Marloth, Fl. S. Afr. **2**, 2: 197, fig. 130 (1925). Type as above.
  Modecca paschanthus Harv. in Harv. & Sond., F.C. **2**: 500 (1862) *nom. illegit.*
  Jaeggia repanda Schinz in Verh. Bot. Ver. Prov. Brand. **30**: 254 (1888).—Harms in Engl., op. cit. **24**: 169 (1897). Type from SW. Africa.
  Paschanthus jaeggii Schinz in Mém. Herb. Boiss. no. 20: 23 (1900). Type as for *Jaeggia repanda*.
  Modecca repanda (Burch.) Druce, Report Bot. Exch. Club. Brit. Is. **5**: 1916: 636 (1917). Type as for *Adenia repanda*.

Suberect herb or woody climber 0·2–2 m., arising from a tuberous turnip-like rootstock. Stems annual or perennial, greyish-purple. Leaf-lamina 2–15 × 0·2–2 (6) cm., obovate to linear, subacute to obtuse or retuse, sometimes curved at the apex, margin entire or irregularly repand to lobed, acute to subcordate at the base, membranous to coriaceous, greyish-glaucous, sometimes punctate beneath, penninerved, the nerves 4–10-paired, sometimes brownish-red, prominent beneath and interlocking, the reticulation ± conspicuous; petiole 0·1–1 cm. long. Glands of the lamina-base 2, 1–1·5 mm. in diameter, situated on each side of the base of the midrib; lamina-glands 0–10, 0·5–1 mm. in diameter, submarginal and one apical at the top of the midrib. Stipules 1–1·5 mm. long, narrowly triangular, acute. Sterile tendrils simple, up to 5 cm. long. Cymes sessile or with a peduncle up to 2 cm. long, 1–5-flowered in ♂, 1–2(3)-flowered in ♀ and ♀; bracts and bracteoles 1–2·5(10) mm. long, narrowly triangular; tendrils (0)1, up to 4 cm. long. Flowers polygamous or dioecious. ♂ flower: 15–24 × 2–5 mm. (incl. the 2–3 mm. long stipe), tubular-infundibuliform, greenish-cream; pedicels 1–5(10) mm. long; hypanthium (incl. calyx-tube) 9–14 mm. long, tubiform; calyx-lobes 4–7·5 mm. long, oblong-lanceolate, obtuse, entire; petals 5–8 × 1·5–2 mm., lanceolate, ± obtuse, 1–3-nerved, entire or dentate-fimbriate towards the apex, inserted near the throat of the calyx-tube; corona 0; stamen-filaments 3·5–6 mm. long, free, inserted at c. the middle of the calyx-tube; anthers 4–6 × 0·75–1 mm., obtuse; disk-glands 0; vestigial ovary 0·5–1 mm. (incl. the gynophore) long. ♀ flower: 10–15 × 3–5·5 mm. (incl. the 1–4 mm. long stipe), tubular-campanulate; pedicel 1–3(10) mm. long; hypanthium and calyx-tube 4–5 mm. long; calyx-lobes 4–6·5 mm. long, oblong to lanceolate, obtuse, entire; petals 2–4·5 × 0·33–1·2 mm., lanceolate, obtuse or acute, 1–3-nerved, dentate towards the apex, inserted 3–5 mm. above the base of the hypanthium; corona 0; stamens with the filaments c. 3 mm. long, free, inserted 0·5–2 mm. above the base of the hypanthium and anthers 4–4·5 × 0·75 mm., obtuse; disk-glands 0; pistil 7–9 mm. long; gynophore 3–3·5 mm. long; ovary 3–4 × 2·5 mm., ovate to ellipsoid; styles connate at the base, with the arms 0·5–0·75 mm. long, ending in subglobular papillate stigmas. ♀ flower: 8–11 × 3–4 mm. (incl. the c. 1 mm. long stipe), tubular-campanulate; pedicel 1–2 mm. long; hypanthium and calyx-tube 3·5–5 mm. long; calyx-lobes 3–5 mm. long, oblong to lanceolate, obtuse, entire; petals 1·5–2 × 0·75 mm., lanceolate, obtuse or acute, 1(3)-nerved, entire, inserted 2·5–3·5 mm. above the base of the hypanthium; corona 0; staminodes 2–3 mm. long, free, sometimes with abortive anthers, inserted near the base of the hypanthium; disk-glands 0; pistil 7–8 mm. long; gynophore c. 2 mm. long; ovary 2·5–3 × 2 mm., ellipsoid; styles connate at the base, the arms c. 0·5 mm. long, ending in subglobular, papillate stigmas. Capsule 1·5–2·5(3) × 1·25–2·5 cm., subglobular, with smooth coriaceous pericarp, (2)5–12-seeded. Seeds 7–8 × 7 × 3 mm., suborbicular to broadly ovate, pitted.

Botswana. SE: c. 3 km. S. of Makoro Siding, ♀ fl. & fr., *Leach & Noel* 259 (SRGH*)
Zambia. B: Mongu Distr., Kande L., ♂ fl. 11.xi.1959, *Drummond & Cookson* 6333 (COI; K; SRGH). **Rhodesia.** N: Gokwe, ♂ fl., ♀ fl. & fr. xi.1953, *Bingham* 914 (K; LISC; SRGH). S: c. 16 km. N. of Beitbridge along Fort Victoria road, c. 480 m., st. 21.iii.1967, *Rushworth* 481 (K; SRGH).
Also in Angola, SW. Africa and S. Africa. In woodlands, open places, sandy river beds, among rocks on Kalahari sand.

6. PASSIFLORA L.

Passiflora L., Sp. Pl. **1**: 955 (1753); Gen. Pl. ed. 5: 410 (1754).

Scandent herbs or shrubs usually climbing by tendrils, rarely erect. Leaves alternate, rarely opposite, entire, lobed or partite; petiole often with glands; stipules 2 or absent, sometimes foliaceous. Tendrils usually solitary in the axils of the leaves, sometimes ending the peduncles, rarely absent. Flowers rather large and handsome, axillary, solitary or in racemes, ☿, rarely unisexual; peduncles articulate, often 3-bracteate. Calyx-tube patelliform, campanulate or urceolate to tubular. Sepals 4–5, linear-oblong or linear, often coloured inside, sometimes with horns on the back below the apex. Petals 4–5 or absent, membranous, ± equal to the sepals but more vividly coloured. Corona (faucial and supramedian corona)* of 1 to several series of distinct or ± united filaments, rarely tubular. Operculum (middle or membranous corona) membranous, flat or plicate, entire, lacerate or filamentose, rarely wanting. Nectar-ring (inframedian corona) an annular ridge within or below the operculum, sometimes wanting. Limen (basal corona) close to the base of the gynophore, annular or cupuliform, sometimes wanting. Stamens 4–5, with the filaments adnate to the gynophore, free at the apex; anthers linear, ovate or oblong, 2-celled, dorsifixed. Ovary oblong or subglobose, borne on a gynophore which is usually elongate, rarely absent; styles 3 or 4, subterminal, cylindrical or clavate; stigmas capitate; ovules numerous, rarely few, attached to 3 rarely 4 placentas. Fruit dry or pulpy, sometimes an irregularly 3-valved capsule. Seeds usually numerous, compressed, arillate, with the testa usually scrobiculate and the endosperm fleshy; cotyledons foliaceous.

A genus of c. 370 species in tropical and subtropical America, tropical Asia, Australia and Polynesia. The species occurring in Africa are introduced, some cultivated and naturalized, others only cultivated.

Leaves entire:
  Stem and branches 4-angled, the angles conspicuously winged; petiole also winged,
    without lateral linear glandular appendages, but with 3 pairs of nearly sessile
    glands; seeds more than 5 mm. wide   -   -   -   -   **1.** *quadrangularis*
  Stem terete, not winged; petiole with scattered lateral, linear, glandular appendages;
    seeds not more than 5 mm. wide   -   -   -   -   -   -   **2.** *ligularis*
Leaves 3(5)-lobed or 3(5)-partite:
  Bracts usually 2–4-pinnatifid, with glandular-capitate filiform laciniae; branches and
    petioles softly hirsute, with yellow spreading hairs; leaf-lamina with sparse slender
    hairs and glandular cilia -   -   -   -   -   -   -   **6.** *foetida*
  Bracts not as above; leaf-lamina glabrous, pilosulous or tomentose:
    Leaves glabrous or pilosulous; involucral bracts free; calyx-tube short, campanulate:
      Leaves glabrous, glossy above; petiole with 2 sessile or shortly stipitate glands up
        to 5 mm. below the apex; bracts ovate, 2–2·5 × 1–1·5 cm.; flowers up to 7 cm.
        in diameter; corona-filaments in 4 or 5 rows; fruit 4–5 cm. in diameter
          3. *edulis*
      Leaves hispidulous above and pilosulous beneath, not glossy above; petiole with
        the glands thickly stipitate, more distant from the apex; bracts setaceous,
        2·5–3 mm. long; flowers 2–3 cm. in diameter; corona-filaments in a single
        row; fruit c. 2 cm. in diameter -   -   -   -   -   **4.** *morifolia*
    Leaves tomentose; involucral bracts connate; calyx-tube long and cylindrical
      5. *mollissima*

---

* Like Killip (in Field Mus. Nat. Hist. Bot. Ser. **19**, 1: 16 (1938)) we are using the terminology of Harms (in Engl. & Prantl, Pflanzenfam. **3**, 6a (1893)) giving in brackets the corresponding terminology of Masters (in Mart., Fl. Bras. **13**, 1 (1872)).

1. **Passiflora quadrangularis** L., Syst. Nat. ed. 10, **2**: 1248 (1759).—Sims in Curtis, Bot. Mag. **46**: t. 2041 (1819).—DC., Prodr. **3**: 328 (1828).—Harms in Engl. & Prantl, Pflanzenfam. **3**, 6a: 90 (1893); op. cit. ed. 2, **21**: 503 (1925).—Burtt Davy, F.P.F.T. **1**: 220 (1926).—Killip in Field Mus. Nat. Hist. Bot. Ser. **19**, 2: 335 (1938).—A. & R. Fernandes in Garcia de Orta **6**, 2: 254 (1958); in C.F.A. **4**: 230 (1970).—F. White, F.F.N.R.: 268 (1962).—de Wilde in F.T.E.A., Passifloraceae: 15 (1975). Type from Jamaica.

Strong, herbaceous, glabrous plant climbing by tendrils. Stem stout, 4-angled, the angles conspicuously winged; internodes 12 cm. long or more. Leaf-lamina 10–20 × 8–15 cm., broadly ovate or ovate-oblong, abruptly acuminate at apex, entire-margined, rounded, subtruncate or shallowly cordate at the base, penninerved, the midrib prominent but more so beneath, the principal lateral nerves 10–12 on each side, prominent beneath; petiole 2–5 cm. long, stout, canaliculate above, with 3 pairs of nearly sessile glands; stipules 2–3·5 × 1–2 cm., ovate or ovate-lanceolate, acute at the apex, entire or slightly serrulate at the margin, narrowed at base, thinly membranous. Tendrils stout. Peduncle 1·5–3 cm. long, 3-angled. Bracts 3–5·5 × 1·5–4 cm., cordate-ovate, acute, entire or serrulate towards the base, thin-membranous. Flowers solitary, 7·5–12·5 cm. in diam. Calyx-tube campanulate; sepals 3–4 × 1·5–2·5 cm., ovate or ovate-oblong, concave, greenish or greenish-red outside, white, violet or pinkish inside. Petals 3–4·5 × 1–2 cm., oblong-ovate to oblong-lanceolate, obtuse, white or deeply pink-tinged. Corona in 5 rows, the 2 outer ones consisting of filaments as long as or exceeding the sepals, terete, radiate, banded with reddish-purple and white at base, blue at middle, densely mottled with pinkish-blue in the upper ½, the third row tubercular, the tubercles deep reddish-purple, the fourth row filamentose, the filaments 1–1·5 mm. long, banded with reddish-purple and white, the fifth row 3–7 mm. long, membranous, unequally lacerate. Operculum 4–6 mm. long, membranous, denticulate, white, reddish-purple at the margin. Limen annular, fleshy. Ovary ovoid. Fruit 20–30 × 12–15 cm., oblong-ovoid, cylindrical or longitudinally 3-grooved. Seeds 7–10 × 5–8·5 mm., broadly obcordate or suborbicular, flattened, reticulate at the centre of each face, striate at the margin.

Mozambique. LM: Marracuene, Umbeluzi Experimental Station, 23.xi.1946, *Pedro & Valente* 3094 (LMA).

Native of tropical America, but cultivated and naturalized in several localities in tropical and southern Africa.

2. **Passiflora ligularis** Juss. in Ann. Mus. Hist. Nat. Paris **6**: 113, t. 40 (1805).—Killip in Field Mus. Nat. Hist. Bot. Ser. **19**, 2: 344 (1938).—de Wilde in F.T.E.A., Passifloraceae: 12 (1975) in keys. Type from S. America.

Stout herbaceous glabrous plant, climbing by tendrils. Stems terete, greyish, grooved; internodes up to 8 cm. long. Leaf-lamina 8–17 × 6–13 cm., broadly ovate, abruptly acuminate at apex, entire-margined, deeply cordate at the base, penninerved, membranous, dark green above, paler beneath; petiole 4–10 cm. long, bearing 4–6 scattered, liguliform or filiform glands, 3–10 mm. long; stipules 1–2·5 × 0·8–1·2 cm., ovate-lanceolate or oblong-lanceolate, acute or acuminate, entire or serrulate, narrowed at the base. Tendrils stout and long. Peduncle 2–4 cm. long, solitary or geminate. Bracts 2–3·5 × 1–1·5 cm., connate along ⅕–⅓ of their length, the free parts ovate or ovate-lanceolate, acute, entire, glabrous but tomentose near the margin. Flowers 6–9 cm. in diameter. Calyx-tube short, campanulate. Sepals 2·5–3·5 × 1–1·5 cm., acute, green outside, white inside. Petals c. 3 × 0·8–1 cm., oblong, white or pinkish-white. Corona in 5–7 rows, the filaments of the 2 outer ones as long as the petals, radiate, blue at apex, banded with white and reddish-purple below, the inner rows approximate, the filaments c. 2 mm. long. Operculum membranous, slightly incurved, denticulate, white and reddish-purple at margin. Limen cupuliform, surrounding the base of the gynophore. Ovary ovoid. Fruit 6–8 × 4–5 cm., ovoid, yellowish or purplish. Seeds c. 6 × 4 mm., narrowly obcordate, minutely tridentate at the apex, with the faces irregularly reticulate.

Rhodesia. E: Cashel, Black Mountain, fl. 14.iv.1948, *Chase* 646 (K).

Native from Central and S. America and cultivated in warm regions. In Rhodesia it is probably a garden escape.

3. **Passiflora edulis** Sims in Curtis, Bot. Mag. **45**: t. 1989 (1818).—DC., Prodr. **3**: 329
(1828).—Harms in Engl. & Prantl, Pflanzenfam. **3**, 6a: 91 (1893); op. cit. ed. 2, **21**:
504 (1925).—Engl., Pflanzenw. Afr. **3**, 2: 609 (1921).—Burtt Davy, F.P.F.T. **1**: 220
(1926).—Norlindh in Bot. Notis. **1934**: 106 (1934).—Killip in Field Mus. Nat. Hist.
Bot. Ser. **19**, 2: 393 (1938).—A. & R. Fernandes in Garcia de Orta **6**, 2: 254 (1958);
in C.F.A. **4**: 230 (1970).—F. White, F.F.N.R.: 268 (1962).—de Wilde in F.T.E.A.,
Passifloraceae: 15 (1975). Type a plant cultivated in Europe, obtained from seeds
probably collected in Brazil.

Large herbaceous, glabrous (except ovary) perennial climber, up to 15 m. Stems
± terete, striate; internodes up to 9 cm. long. Leaf-lamina up to 13 × 15 cm.,
3-lobed to below the middle of the lamina, the lobes 2–4 cm. wide, acute or
acuminate, rarely subobtuse, serrate, rounded or shallowly cordate at base,
subcoriaceous, pale green or yellow-green, glossy above, paler and not glossy
beneath; petiole up to 4 cm. long, with 2 glands at the apex, the glands sessile or
shortly stipitate; stipules c. 10 × 1 mm., linear-subulate, entire or minutely
glandular-serrulate. Tendrils c. 10 cm. long, stout, simple. Peduncles up to 6 cm.
long, stout. Bracts 2–2·5 × 1–1·5 cm., ovate, obtuse or acute at the apex, serrate,
pectinate or almost lacerate, often glandular at margin. Flowers solitary, up to
7 cm. in diam. Sepals 3–3·5 × 1 cm., oblong, horned, green outside, white inside.
Petals 2·5–3 × 0·5–0·7 cm., oblong, obtuse, white. Corona with the filaments in 4 or
5 rows, those of the 2 outer ones 1·5–2·5 cm. long, filiform or narrowly liguliform,
crispate towards the apex, white, purple at base, those of the other rows 2–2·5 mm.
long, linear or dentiform. Operculum membranous, incurved, entire or shortly
fimbriate. Limen cupuliform, entire or crenulate. Ovary ovoid or globose,
sericeo-tomentose or glabrous. Fruit 4–5 cm. in diameter, ovoid or globose,
yellow, greenish-yellow or purplish. Seeds 5–6 × 3–4 mm., ovoid, minutely
reticulate.

Zambia. N: Kawambwa Boma, fr. 7.xi.1952, *Angus* 732 (K). W: Ndola West Forest
Reserve, fl. 14.viii.1952, *Angus* 213 (K). **Rhodesia.** C: Salisbury, Mandara, fl. & fr.
1.x.1974, *Biegel* 4638 (SRGH). E: Chirinda, 1216 m., fl. in bud 26.x.1947, *Wild* 2243
(K; SRGH). S: Victoria, Zimbabwe Ruins, *Gouveia & Pedro* 2074 (LMA). **Malawi.**
N: Mzimba, Champoys Forest Nursery, fl. 7.xi.1968, *Salubeni* 1208 (K; SRGH). S:
Mlanje Mt., Lichenya Plateau, 1920 m., fl. 18.xi.1970, *Pawek* 3877 (K). **Mozambique.**
MS: Vila de Manica near Vumba, fl. & fr. 19.vi.1949, *Pedro & Pedrógão* 6777 (LMA).
LM: Goba, near Maiuáua R., fl. 2.xi.1960, *Balsinhas* 163 (COI; K; LMA).
Native of Brazil, but cultivated and naturalized in several regions of tropical and
southern Africa. In hygrophilous and secondary forests, also in bush.

4. **Passiflora morifolia** Mast. in Mart., Fl. Bras. **13**, 1: 555 (1872).—Killip in Field
Mus. Nat. Hist. Bot. Ser. **19**, 1: 107 (1938). Type from Argentina.

Herbaceous perennial climber. Stems yellowish, inconspicuously 4-angled,
grooved, glabrate below, sparingly hispidulous upwards; internodes up to 7 cm.
long. Leaf-lamina 4–11 × 5–15 cm., 3-lobed, the lobes acute, the median ovate or
ovate-lanceolate, usually narrowed at the base, the lateral ones divergent, margin
undulate-dentate or denticulate or subentire, deeply cordate at the base, membranous, 3-nerved, dark green and hispidulous above, paler and pilosulous
beneath; petiole up to 6 cm. long, flattened, hispidulous or pilosulous, with 2
thickly stipitate glands 1·5 × 0·8–1 mm. within 1 cm. of the apex; stipules c.
6 × 3 mm., ovate, acuminate, minutely hispidulous. Peduncles 1–2 cm. long,
solitary or geminate, ± densely hispidulous, not ending in a tendril. Bracts
2·5–3 mm. long, setaceous, at c. 1 cm. below the base of the flower, ± approximate.
Flowers 2–3 cm. in diameter. Calyx-tube patelliform or campanulate. Sepals
10–15 × 3–4 mm., linear-oblong, obtuse, green, densely hispidulous to glabrescent
outside, white mottled with red inside, with a 3·5 mm. long horn-like arista below
the apex. Petals 6–8 × 2–4 mm., linear-lanceolate, obtuse, white. Corona-
filaments in a single row, the filaments 5–6 mm. long, filiform, white, banded with
blue or violet. Operculum membranous, plicate, slightly incurved and crenulate at
the margin. Nectar ring annular, placed midway between the operculum and the
base of the gynophore. Limen membranous, adnate to the floor of the calyx-tube,
the margin free. Ovary subglobose, densely pubescent. Fruit 2 cm. in diameter,
globose, hispidulous, glaucous. Seeds 4 mm. long, very slightly compressed,
obcordate-obovoid, abruptly tapering at the base, coarsely reticulate.

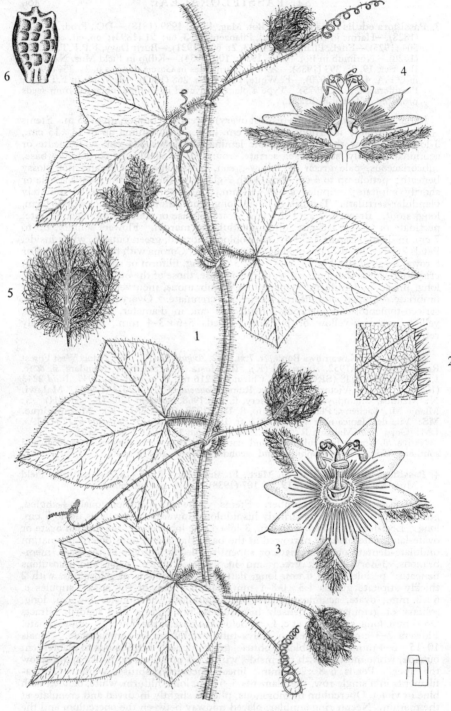

Tab. 98. PASSIFLORA FOETIDA. 1, habit (×½) *Clemens* 3011; 2, marginal section of inferior surface of leaf (×2) *Ross* 35; 3, flower (×1); 4, longitudinal section of flower (×1), 3–4 from *Faulkner* 198; 5, fruit with involucral bracts (×⅔); 6, seed (×5), 5–6 from *Santos Horta* sn.

Rhodesia. C: Gwelo, fr. i.1920, *Walters* (K).
Native of S. America and cultivated in some warm countries.

5. **Passiflora mollissima** (Kunth) Bailey in Rhodora **18**: 156 (1916).—Harms in Engl. &
Prantl, Pflanzenfam. ed. 2, **21**: 506 (1925).—Killip in Field Mus. Nat. Hist. Bot.
Ser. **19**, 1: 291 (1938).—de Wilde in F.T.E.A., Passifloraceae: 14 (1975). Type
from Colombia.
   Tacsonia mollissima Kunth in H.B.K., Nov. Gen. & Sp. **2**: 144 (1817). Type as
above.
   Murucuja mollissima (Kunth) Spreng., Syst. Veg. **3**: 43 (1826). Type as above.

Herbaceous perennial climber reaching to 20 m., with pendulous flowers. Stems
terete, densely and softly yellow-tomentose; internodes 4·5 cm. or more long.
Leaf-lamina 5–10 × 6–12 cm., 3(5)-lobed to c. ⅔ of its length, the lobes 3–4 cm.
broad, ovate or ovate-oblong, acute, the sinuses also acute, sharply serrate-dentate
and glandular at margin, subcordate at the base, membranous, softly pubescent
above, greyish- or yellowish-tomentose beneath; petiole up to 3 cm. long,
canaliculate, with 8–12 small sessile glands near the margin; stipules 7–9 × 3–4 mm.,
subreniform, aristate, denticulate or subentire. Peduncle 2–6 cm. long. Bracts
2·5–3 cm. long, united for ½ to ¾ their length, acute at apex, entire-margined, softly
tomentose. Calyx-tube 6·5–8 cm. long, c. 1 cm. in diameter, glabrous, rarely
pubescent, olive-green, often red outside, white inside. Sepals 2·5–5·5 × 1–1·5 cm.,
oblong, obtuse, shortly horned below the apex. Petals almost as long as the sepals,
obtuse, pink. Corona consisting of a purple band with a few pinkish tubercles or
crenulations. Operculum white, recurved at the margin, subentire. Ovary
9 × 3 mm., oblong, sericeous-tomentose. Fruit 6–7 × 3–3·5 cm., oblong-ovoid,
green when unripe, orange-yellow when ripe, softly pubescent. Seeds up to
6 × 5 mm., broadly obovate, asymmetrical, reticulate.

Rhodesia. C: Salisbury, Lomagundi road, 1600 m., fl. 20.ii.1957, *Ducker* 12 (K;
SRGH).
Native of S. America and cultivated in several warm regions of the world.

6. **Passiflora foetida** L., Sp. Pl. **2**: 959 (1753).—Sims in Curtis, Bot. Mag. **53**: t. 2619
(1826).—DC., Prodr. **3**: 331 (1828).—Engl., Pflanzenw. Afr. **3**, 2: 609 (1921).—
Harms in Engl. & Prantl, Pflanzenfam. ed. 2, **21**: 502 (1925).—Killip in Field Mus.
Nat. Hist. Bot. Ser. **19**, 2: 474 (1938).—A. & R. Fernandes in Garcia de Orta **6**, 2:
254 (1958); in C.F.A. **4**: 231 (1970).—Lawrence in Baileya **8**, 4: 126 (1960).—
J. H. Ross, Fl. Natal: 251 (1972).—de Wilde in F.T.E.A., Passifloraceae: 13 (1975).
TAB. 98. Type of uncertain origin.

Climbing unpleasant-smelling annual or biennial plant up to 4 m. Stems
yellowish, usually hirsute with yellowish ± spreading hairs and capitate-glandular
cilia; internodes 4·5 cm. long or more. Leaf-lamina up to 11 × 11 cm., cordate at
the base, 3(5)-palmatilobed, the lobes acute at the apex, entire or irregularly sinuate
and dentate at the margin, membranous, sparsely hirsute with yellowish spreading
hairs and capitate-glandular cilia; petiole up to 6 cm. long, usually hirsute like the
stems, glandless; stipules deeply cleft into filiform gland-tipped divisions. Sterile
tendrils axillary, 10 cm. long or more, very sparsely hirsute. Peduncle up to 6 cm.
long, solitary. Bracts involucrate, 2–4-pinnatifid or -pinnatisect, rarely 1-
pinnatifid, the segments filiform, gland-tipped. Flowers 2–5 cm. wide, white, pink,
lilac or purplish. Sepals 1·5–2 cm., ovate-oblong or ovate-lanceolate, dorsally horned
below the apex. Petals oblong, oblong-lanceolate or oblong-spathulate, slightly
shorter than the sepals. Corona with the filaments in several rows, those of the 2
outer ones c. 1 cm. long, filiform, the others 1–2 mm. long, capillary. Operculum
membranous, erect, denticulate. Fruit 1·5–3 cm. in diameter, globose or subglobose, yellow to red. Seeds c. 5 × 2·5 mm., ovate-cuneiform, obscurely tridentate
at the apex, coarsely reticulate at the centre of each face.

Mozambique. N: Mogovolas, Posto Agrícola, fl. 20.vii.1934, *Queirós Ribeiro* (LISC).
Z: Mocuba Distr., Namagoa, fl. & fr. 14.ii.1942, *Faulkner* 198 (COI; K). LM: near
Umbeluzi R., c. 30 km. W. of Lourenço Marques, fl. 26.ix.1945, *Gomes e Sousa* 3346
(COI; K; SRGH).
Native of tropical America and naturalized in several regions of tropical and southern
Africa. On the banks of rivers and sometimes as a weed in cultivated fields.

85. CARICACEAE

By F. K. Kupicha

Trees or shrubs with simple or sparingly branched stems, rarely herbs; milky latex present in all parts. Leaves alternate, exstipulate, petiolate, simple and entire or variously lobed or divided. Flowers dioecious, monoecious or ⚥, pentamerous, actinomorphic, in axillary inflorescences or, rarely plants cauliflorous. Calyx small, gamosepalous, subentire or 3–5-lobed, the lobes opposing or alternating with petals. ♂ flower: corolla gamopetalous, with valvate or contorted lobes; stamens 10, biseriate, inserted on the corolla; filaments free or connate into a short staminal ring. ♀ flower: petals ± free; ovary superior, sessile, 5-carpellary, 1-locular or 5-locular, with parietal placentation; style short or absent; stigmas 5, connate at the base, entire or lobed. Ovules ∞, anatropous. Fruit a fleshy many-seeded berry. Seeds with succulent outer sarcotesta and hard endotesca; endosperm present.

A small family of the tropics and subtropics comprising 5 genera, of which *Carica* L., *Leucopremna* Standley, *Jacaratia* Endl. and *Jarilla* Rusby are native to the New World while *Cylicomorpha* Urb. is confined to Africa.

The papaw (or paw-paw or papaya), *Carica papaya* L., is widely grown in the F.Z. area for its fruit, and for the latex which yields papain, a commercially useful proteolytic enzyme. A native of Central America, it was introduced into E. Africa in the 16th or 17th century and is now naturalized in some places, e.g. near the Shire R. in Malawi (fide Greenway in E. Afr. Agric. Journ. **13**: 231 (1948)). The related species *Carica cundinamarcensis* Hook. f. from Ecuador ("mountain pawpaw ") is also grown, though less frequently. It is recorded from Zambia by F. White, F.F.N.R.: 269 (1962).

CYLICOMORPHA Urb.

Cylicomorpha Urb. in Engl., Bot. Jahrb. **30**: 115 (1901).—Harms in Engl. & Prantl, Pflanzenfam. ed. 2, **21**: 519 (1925).

Monoecious trees armed with short conical spines, all parts exuding abundant latex when wounded. Leaves palmately lobed. Inflorescences axillary, the ♂ many-flowered panicles, the ♀ of solitary flowers or short few-flowered racemes. ♂ flowers: calyx cupuliform, obsoletely 3–5-denticulate; corolla tube cylindrical, with 5 lobes shorter than the tube, their aestivation contorted; stamens 10 in 2 whorls in the mouth of the corolla tube, connate into a ring at the base; alterni-petalous anthers borne on short filaments, oppositipetalous ones almost sessile; anthers linear-lanceolate, apiculate, introrse; rudimentary gynoecium present, comprising a small solid hemispherical ovary produced into a slender terete style. ♀ flowers: calyx cupuliform, entire; petals 5, free, oblong-acute, valvate; androecium absent; stigmas 5, free above but connate at the base, linear-oblong, entire, swollen, patent-erect, velvety-papillose all round; ovary ovoid, 5-locular, the numerous ovules borne at the outer angles of the dissepiments. Fruit a subpentagonal, 5-locular, many-seeded berry, the locules not filled with pulp. Seeds with fleshy testa ornamented with a prominent cristate keel and irregular longitudinal interrupted cristate ribs; endosperm copious.

A ditypic tropical African genus.

Cylicomorpha parviflora Urb. in Engl., Bot. Jahrb. **30**: 116 cum fig. (1901).—Hemsley in F.T.E.A., Caricaceae: 2, fig. 1 (1958). TAB. **99**. Syntypes from Tanzania.

    Jacaratia solmsii Urb. in Engl., Pflanzenw. Ost-Afr. **C**: 282 (1895) pro parte quoad specim. usambarensis.

      Cylicomorpha parviflora var. *brachyloba* Urb., op. cit.: 117 (1901). Type from Tanzania.

A fast-growing tree up to 35 m. high with soft white wood and smooth grey bark; trunk simple or sparsely branched at the apex, hollow; upper part of trunk and branches armed with straight conical prickles up to 2·5 cm. long. Leaves clustered

Tab. 99. CYLICOMORPHA PARVIFLORA. 1, apex of shoot showing leaves (× ½) *Drummond &
Hemsley* 3352; 2, leaf, showing maximum number of lobes (× ⅙) *Holst* 8723; 3,
♂ inflorescence (× ⅔); 4, ♂ flower (×2); 5, four stamens with basal tube, internal
view (×4); 6, rudimentary ovary and style of ♂ flower (×4), all from *Drummond &
Hemsley* 3435; 7, ♀ inflorescence (× ⅔); 8, ♀ flower with all petals except one
removed (×2); 9, transversal section of ovary, diagrammatic (×2), all from
Drummond & Hemsley 3448; 10, reconstruction of fruit (× ⅔) *Battiscombe* 1093.
From F.T.E.A.

at the top of the trunk and at branch ends, long-petiolate, glabrous; petioles up to 45 cm., hollow, striate; laminas up to 35 cm. in diameter, ± orbicular in outline, strongly cordate at base, shallowly to deeply palmately 3–5-lobed, the lobes broadly ovate-cuspidate, entire or with subsidiary lobes; young leaves conspicuously mottled with pale yellow or pink veins. Inflorescences axillary, farinose; ♂ up to 40 cm. long, many-flowered, paniculate, with slender branches; ♀ up to 3·5 cm. long, 1–5-flowered, racemose with stout axis. ♂ flowers: calyx ± entire, up to 4·5 mm. long and 4 mm. in diameter; corolla cream, tube to 2·5 cm. long and 4 mm. wide, with lanceolate lobes up to 1·5 cm. long; anthers to c. 5 mm. long; filaments of oppositisepalous stamens to 6 mm. long, of oppositipetalous ones to 3 mm. long, the staminal ring c. 2 mm. in height. ♀ flowers: calyx up to 5 mm. long and 1 cm. in diameter; petals up to 3 × 1 cm., pale greenish abaxially and cream-coloured adaxially, oblong-acute; ovary up to 1·5 × 1 cm.; stigmas to 1·3 cm. long. Fruit to 8 cm. long, smooth, greenish-yellow. Seeds up to 7 mm. long.

Malawi. N: Misuku Hills, Mugesse Forest Reserve, c. 1850 m., 15.ix.1970, *Müller* 1649 (K; SRGH).
Also known from eastern Kenya and Tanzania. Rain forest.

The other species of *Cylicomorpha*, *C. solmsii* (Urb.) Urb., which has larger flowers than *C. parviflora*, is found in Cameroon.

86. CUCURBITACEAE

By C. Jeffrey

Scandent or prostrate tendriliferous annual or perennial herbs or less often woody lianes, rarely erect herbs without tendrils, often with tuberous rootstock. Leaves alternate, palmately veined, simple or pedately compound. Tendrils lateral to the petiole base, simple, distally 2-fid or proximally 2–7-fid, rarely reduced to spines or absent, usually 1 at each node. Flowers unisexual, epigynous, monoecious or dioecious, axillary, variously arranged, the female commonly solitary. Glandular bract-like structures (probracts) sometimes present at base of peduncles. Receptacle-tube shallow to tubular, usually 5-lobed, lobes usually small. Petals usually 5, free or variously united, corolla mostly regular. Stamens basically 5, androecium always variously modified, commonly appearing as 2 double stamens and 1 single stamen, free or variously coherent or united; anther-thecae often convoluted. Staminodes often present in female flowers. Ovary inferior, unilocular or sometimes 3-locular, of usually 3 united carpels; placentation parietal, rarely axillary, placentae often intrusive. Ovules anatropous, 1–many, horizontal, pendulous or ascending; style 1, with 2 or usually 3 stigma-lobes, or styles 3. Fruit a dry or fleshy capsule, berry or hard-shelled pepo, variously dehiscent or indehiscent, rarely a 1-seeded samara. Seeds 1–many, rather large, often compressed, sometimes winged; embryo large; endosperm absent.

A pantropical family of about 600 species, some of which are economically important as cultivated plants. In addition to the genera keyed out below, the genera *Cucurbita* L. and *Sechium* R. Br. are known to be cultivated in the Flora Zambesiaca area. All species and genera known or likely to be cultivated in this area can be determined by reference to the notes and keys given by C. Jeffrey in F.T.E.A., Cucurb.: 2–13 (1967).

KEY TO GENERA (FLOWERING MATERIAL)

```
1.  Tendrils distally 2-fid; styles 3   -   -   -   -   -   -   -   -   2
—   Tendrils simple or proximally 2–5-fid; style 1 -   -   -   -   -   -   3
2.  Stamen 1, central, with thecae ring-like; ovary compressed, containing 1 ovule
                                                         22. Cyclantheropsis
—   Stamens 4–5 with thecae short, straight; ovary trigonous, containing several ovules
                                                         21. Gerrardanthus
3.  Plant erect; tendrils absent   -   -   -   -   -   -   11. Trochomeria
—   Plant scandent or prostrate; tendrils present   -   -   -   -   -   4
```

4. Tendrils represented by weak non-coiling spines - - **4. Acanthosicyos**
—   Tendrils evident, coiling - - - - - - - - - - 5
5. Leaves compound - - - - - - - - - - 6
—   Leaves simple (though sometimes deeply dissected) or absent - - - 7
6. Petals purplish, fringed with filaments - - - - - **1. Telfairia**
—   Petals cream, yellow or orange, not fringed - - - - **2. Momordica**
7. Young stems spotted with darker green - - - - - - - - 8
—   Young stems concolorous (though sometimes calloso-punctate) - - - 9
8. Petals free, at least 1 with an incurved basal scale; ovary spiny - **2. Momordica**
—   Petals shortly united, without basal scales; ovary smooth - - **9. Diplocyclos**
9. Base of leaf-lamina or apex of petiole laterally 2-glandular, glands divergent
                                                            6. Lagenaria
—   Base of leaf-lamina/apex of petiole without a pair of lateral divergent glands 10
10. Subcircular ciliate bracts present at base of some or all petioles - - - 11
—   Subcircular ciliate bracts absent - - - - - - - - 12
11. Petals 1–1·5 mm. long; anther-thecae short, arcuate - - **12. Ctenolepis**
—   Petals 5–60 mm. long; anther-thecae triplicate - - - **11. Trochomeria**
12. Tendrils 3–5-fid - - - - - - - - - - - 13
—   Tendrils simple or 2-fid - - - - - - - - - - 14
13. ♂ flowers solitary; ovary villous; petals 0·7–1·9 cm. long - - **5. Citrullus**
—   ♂ flowers racemose; ovary densely puberulous; petals 2·0–4·5 cm. long
                                                            10. Luffa
14. Petals (or at least 1 of them) with an incurved basal scale, free **2. Momordica**
—   Petals without incurved basal scales, free or more often ± united - - 15
15. Tendrils 2-fid - - - - - - - - - - - 16
—   Tendrils simple - - - - - - - - - - - 20
16. Anther-thecae arcuate; placentae and stigmas 2 - - - **18. Kedrostis**
—   Anther-thecae triplicate or contorted; placentae and stigmas 3 - - - 17
17. ♂ and ♀ flowers in axillary clusters, often co-axillary - - - **9. Diplocyclos**
—   ♂ and ♀ flowers solitary or racemose, not co-axillary - - - - 18
18. Leaf-lobes pinnately lobulate, often deeply so - - - - **5. Citrullus**
—   Leaf-lobes not pinnately lobulate, broadly triangular to elliptic, not lobulate or
    shortly 3-lobulate - - - - - - - - - - 19
19. Receptacle-tube cylindrical, (11)15–32 mm. long in both ♂ and ♀ flowers; anther-
    thecae tightly triplicate; connectives narrow; anther-head oblong **20. Peponium**
—   Receptacle-tube obconic to campanulate and 2–7(15) mm. long in ♂ flowers, shortly
    cylindrical and 2–5 mm. long in ♀ flowers; anther-thecae laxly triplicate; connectives
    broad; anther-head ± spherical - - - - - - **8. Coccinia**
20. Corolla stelliform, the lobes linear, spreading to reflexed, tapering from base to apex,
    5–60 mm. long, united only at the base, greenish-yellow - **11. Trochomeria**
—   Corolla not as above - - - - - - - - - - 21
21. Disk obvious, free from the receptacle-tube laterally, shortly ± cylindrical in ♂
    flowers, annular around base of style in ♀ flowers - - - - 22
—   Disk obscure, indistinct from base of receptacle-tube, or absent - - - 26
22. Stamens 2 or 3, all 2-thecous; ovary smooth, glabrous or almost so **17. Zehneria**
—   Stamens 3, two 2-thecous, one 1-thecous; ovary tuberculate, papillose or spiny or, if
    smooth, then setulose, aculeate, hispid or tomentose - - - - 23
23. ♂ and ♀ flowers subsessile, in axillary clusters - - - - **16. Mukia**
—   ♂ and ♀ flowers distinctly pedicellate, the ♂ solitary or fasciculate, the ♀ solitary
                                                            24
24. Ovary tuberculate, setulose, with the tubercules themselves both laterally and
    apically setulose; receptacle-tube densely setulose; setulae often brownish; anther-
    thecae straight - - - - - - - - - **14. Oreosyce**
—   Ovary smooth, aculeate, hispid or tomentose, or with numerous fleshy spines,
    papillae or tubercules each ending in a hyaline setula, but otherwise glabrous;
    anther-thecae triplicate, or, if straight or merely apically replicate, then receptacle-
    tube shortly hispid - - - - - - - - - - 25
25. Anther-thecae triplicate; ovary softly spiny, papillose or tuberculate, or if smooth
    then stems and petioles coarsely spreading-setulose or stalk of ovary rapidly
    elongating after flowering and burying the developing fruit - **13. Cucumis**
—   Anther-thecae straight or apically replicate; ovary smooth, finely hispid to aculeate-
    setose - - - - - - - - - - **15. Cucumella**
26. Placentae and stigmas 2; petals in ♂ flowers 0·7–5 mm. long, shortly united 27
—   Placentae and stigmas 3; petals in ♂ flowers 7–50 mm. long, united to c. the middle,
    less often shortly united or free - - - - - - - - 28
27. ♂ flowers in lax fascicles or rather lax to rarely dense racemes, with petals 2–5 mm.
    long; ♀ flowers with ovary not developing a cupuliform base - **18. Kedrostis**
—   ♂ flowers in congested racemiform or subcapitate pedunculate or rarely subsessile
    clusters, with petals 0·7–2 mm. long; ♀ flowers with ovary developing a cupuliform
    base - - - - - - - - - - **19. Corallocarpus**

28. Petals united to ½-way or above - - - - - - - - - 29
— Petals free or only shortly united - - - - - - - 30
29. Ovary densely setose; flowers monoecious - - - **7. Raphidiocystis**
— Ovary glabrous or almost so; flowers dioecious - - - **8. Coccinia**
30. Plant annual; leaf-lobes usually pinnately lobulate, often deeply; petals shortly united, 0·7–1·9 cm. long; ovary tomentose - - - - **5. Citrullus**
— Plant perennial; leaf-lobes not lobulate or shortly 3-lobulate; petals free, 1·8–4·7 cm. long; ovary shortly pubescent or ± glabrous - - - - **3. Eureiandra**

KEY TO GENERA (FRUITING MATERIAL)

1. Seeds 3·3–3·5 × 3·2–4 × 1–1·3 cm.; fruits 30–90 × 15–25 cm., bluntly 10-ribbed, ellipsoid with lobed expanded base, tardily dehiscent into 10 longitudinal valves
1. Telfairia
— Seeds much smaller; fruit not as above - - - - - - 2
2. Fruit samaroid, elliptic in outline, compressed, 1-seeded - **22. Cyclantheropsis**
— Fruit not samaroid, not compressed, 1–many-seeded - - - - 3
3. Ripe fruit densely echinate with long, slender, reddish-brown setae, dehiscent at the apex into 10 longitudinal valves and extruding the placentae **7. Raphidiocystis**
— Ripe fruit not as above, if echinate then with somewhat fleshy spines and devoid of reddish-brown setae, indehiscent or dehiscent in some other way - - 4
4. Mature fruit dry, brownish, opening by an apical pore or operculum - - 5
— Mature fruit at least somewhat fleshy, not opening by an apical pore or operculum
6
5. Fruit opening by an apical triradiate pore, trigonous, not fibrous; seeds fusiform, broadly winged at one end - - - - - - **21. Gerrardanthus**
— Fruit opening by an apical operculum, ± terete or 10-angled, fibrous internally; seeds compressed, elliptic in outline, not winged though often with a narrow membranous border - - - - - - - - - **10. Luffa**
6. Surface of fruit with sparse to dense spines, pustules or tubercules or with longitudinal rows of tubercules or with longitudinal fleshy wing-like ridges - - - 7
— Surface of fruit smooth, the fruit in cross-section terete, bluntly angled or ribbed
10
7. Spines, pustules or tubercules of fruit each ending in a setula or in the indurated basal remnants thereof, fruit otherwise not setulose - - - - - 8
— Spines, pustules or tubercules of fruit not ending in a setula, although fruits (including tubercules) sometimes generally hispid or setulose, or fruits with fleshy wings - - - - - - - - - - - - - 9
8. Fruits 4–8 cm. wide, pale yellow when ripe; seeds subcompressed, 2·5–3 mm. thick
4. Acanthosicyos
— Fruits 1·5–3·5 cm. wide, or, if larger, then red when ripe; seeds compressed, 1–2 mm. thick - - - - - - - - - - **13. Cucumis**
9. Ripe fruit setulose, greenish, expelling seeds through pedicel-scar **14. Oreosyce**
— Ripe fruit orange-red, dehiscent by 3 fleshy valves, exposing red-sheathed seeds, or rarely indehiscent - - - - - - - - - **2. Momordica**
10. Fruits baccate, subglobose, 1–1·5 cm. in diameter, solitary, red, 2-seeded; seeds plano-convex or concavo-convex, 7·5–11 × 5–6 × 2–3 mm. - - **12. Ctenolepis**
— Fruits and seeds not as above - - - - - - - - 11
11. Seeds verrucose, tumid, 3·5–5·5 × 2–3 × 1·2–1·7 mm.; fruits subsessile in axillary clusters, bright red, globose, 6–11 mm. in diameter - - - - **16. Mukia**
— Seeds and fruits not as above - - - - - - - - 12
12. Seeds fusiform, c. 12 × 3·5 mm., with deeply sculptured faces, or ellipsoid, tumid, 9–17 × 3–5 × 2–5 mm., with a rugose appendage at each end; fruit fusiform, 1–4-seeded - - - - - - - - - - - **2. Momordica**
— Seeds not as above, neither fusiform nor rugose-appendaged; fruit usually more than 4-seeded - - - - - - - - - - - 13
13. Pericarp at maturity devoid of any reddish coloration, greenish or yellowish or variously mottled in shades of these colours, never bluish - - - - 14
— Pericarp at maturity some shade of red or orange-red, at least partly so, or if devoid of red then bluish, globose, baccate, 8–11 mm. in diameter - - 17
14. Seeds with 2 longitudinal facial subcentral or submarginal flat ridges approaching each other towards the ends, truncate or 2-cornute at one end; fruit green or green mottled with small cream transverse markings - - - - **6. Lagenaria**
— Seeds without facial ridges, tapered or rounded at each end; fruit green, green with paler longitudinal markings, yellowish or brownish - - - - 15
15. Fruit ellipsoid-cylindrical to ovoid-rostrate, usually more than twice as long as broad, 0·5–1·5 cm. wide in cross-section; seeds 0·5–1 mm. thick - - **15. Cucumella**
— Fruit more or less globose, less than twice as long as broad, 1·5–20 cm. wide in cross-section, never rostrate; seeds 1–3 mm. thick - - - - - 16

16. Seeds narrowly elliptic in outline, 5–8 × 2·5–4 × 1·5 mm., if broader or thicker then
    either seeds ovate in outline and plants perennial or fruit subterranean
    13. Cucumis
— Seeds ovate-elliptic in outline, 9–11 × 5–6 × 2·5–2·7 mm.; plant annual; fruit never
    subterranean - - - - - - - - - - **5. Citrullus**
17. Faces of seeds distinctly sculptured or erose; seeds with rather broad grooved
    margins, the groove with a slight median ridge - - - - **2. Momordica**
— Faces of seeds smooth or fibrillose; margins not as above - - - - 18
18. Seeds compressed, thin in comparison with length and breadth, broadly ovate, ovate
    or elliptic in outline, with depressed, flat or slightly convex faces, less than 2 mm.
    thick - - - - - - - - - - - - 19
— Seeds tumid, thick in comparison with length and breadth, subglobose, ovoid or
    pyriform, usually 2·5 mm. or more thick - - - - - - 25
19. Fruit fusiform-rostrate or ovoid-rostrate, acute at apex - - - 20
— Fruit globose, ellipsoid or cylindrical, obtuse to rounded at apex - - 23
20. Mature fruit hispid or aculeate - - - - - - **15. Cucumella**
— Mature fruit glabrous - - - - - - - - - 21
21. Seeds black - - - - - - - - - **20. Peponium**
— Seeds pallid - - - - - - - - - - - 22
22. Seeds 4·5–7 mm. long, 3–4 mm. broad - - - - - **8. Coccinia**
— Seeds 3–4·4 mm. long, 1·7–2·5 mm. broad - - - - **17. Zehneria**
23. Fruit 8–13 mm. long, often clustered - - - - - **17. Zehneria**
— Fruit 15–300 mm. long, usually solitary - - - - - - 24
24. Testa smooth, blackish - - - - - - - - **20. Peponium**
— Testa fibrillose, pallid - - - - - - - - **8. Coccinia**
25. Fruit dehiscent in a calyptriform manner, by circumscissile separation of red upper
    part of fruit from green cupuliform base - - - - **19. Corallocarpus**
— Fruit not so dehiscent - - - - - - - - - 26
26. Seeds in 4 longitudinal rows, developing on 2 placentae, pisiform or asymmetrically
    ovoid, not exceeding 5 mm. long and 3·5 mm. thick; fruit dehiscent by a longitudinal
    slit or indehiscent - - - - - - - - **18. Kedrostis**
— Seeds in 6 longitudinal rows, developing on 3 placentae, not asymmetrically ovoid,
    if pisiform then 5 mm. or more long and 3·7 mm. or more thick; fruit indehiscent
    27
27. Seeds pisiform, 6·5–10 × 6–7·5 × 4·5–6·2 mm.; fruit rostrate - **3. Eureiandra**
— Seeds not pisiform or if so, then not exceeding 7·5 × 5 × 4·5 mm. and fruit not
    rostrate - - - - - - - - - - - - 28
28. Seeds 3–4 mm. broad, 1·4–2·5 mm. thick; testa thin - - **8. Coccinia**
— Seeds 4–6·5 mm. broad, 2·5–5·5 mm. thick; testa thick, hard, whitish
    12. Trochomeria

1. TELFAIRIA Hook.

Telfairia Hook. in Curtis, Bot. Mag. **54**: t. 2751 & 2752 (1827).

Lianes. Leaves pedately 3–7-foliolate. Tendrils proximally bifid. Probracts
present. Flowers large, purplish, dioecious. Male flowers many, racemose.
Receptacle-tube short, broad; lobes dentate. Petals 5, free, fringed with filaments.
Stamens 5 or 3, if 5, then all 2-thecous, if 3, then two 4-thecous, one 2-thecous,
free; filaments inserted on the receptacle-tube; thecae straight or slightly
arcuate. ♀ flowers solitary; perianth similar to that of male flower; ovary ribbed;
ovules many, horizontal; stigma 3-lobed. Fruit very large, fleshy, ribbed, many-
seeded, tardily dehiscent by 10 longitudinal valves. Seeds large, broadly ovate
or suborbicular in outline, compressed, enclosed in a fibrous endocarpic sheath.
A genus of 3 species, confined to tropical Africa.

Telfairia pedata (Sims) Hook. in Curtis, Bot. Mag. **54**: t. 2751 & 2752 (1827).—Hook. f.
    in F.T.A. **2**: 523 (1871).—Hiern, Cat. Afr. Pl. Welw. **1**, 2: 387 (1898).—Brenan,
    T.T.C.L.: 176 (1949).—Keraudren in Fl. Madag., Cucurb.: 56 (1966).—C. Jeffrey
    in F.T.E.A., Cucurb.: 15, t. 1 fig. 1–6 (1967). TAB. **100**. Type: a specimen of a
    cultivated plant (K, holotype).
    Fevillea pedata Sims in Curtis, Bot. Mag. **53**: t. 2681 (1826). Type as above.

Liane up to 30 m. or more in length; young stems herbaceous, glabrous, later
softly ligneous, with thin pallid papyraceous bark. Leaves petiolate, 5–7-foliolate;
median leaflet 5·5–14 × 2–7·5 cm., oblong to broadly elliptic, acuminate, acute,
± sinuate-toothed especially in upper part, subglabrous, pinnately veined, lateral
leaflets similar, the outermost smaller and occasionally lobed near the base;
petiolules 1–6·5 cm. long; petioles 2·5–10 cm. long, glabrous or shortly hairy.

o

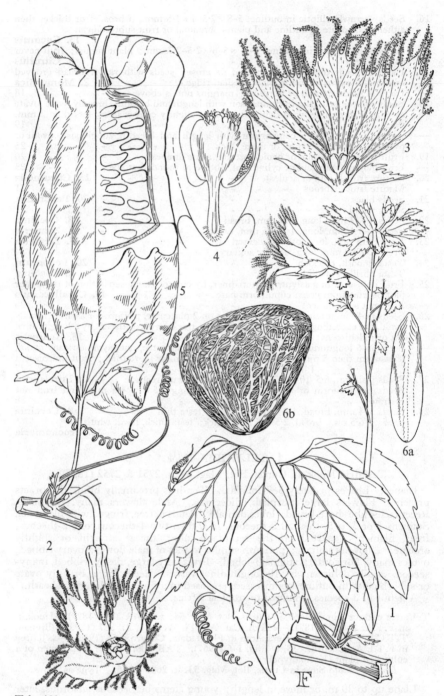

Tab. 100. TELFAIRIA PEDATA. 1, flowering node, ♂ plant ($\times \frac{1}{2}$) *Angus* 2629 & *Borle* 981;
2, flowering node, ♀ plant ($\times \frac{1}{2}$); 3, ♂ flower, dissected ($\times 1\frac{1}{2}$); 4, stamen ($\times 2$);
5, fruit ($\times \frac{1}{4}$); 6a & 6b, seed, lateral and face views ($\times 1$), 2–6 from *Angus* 2629.

Probracts 3–20 mm. long, stalk-like below, expanded and cucullate above. ♂ racemes 6–23·5 cm. long; bracts 4–10 mm. long, broadly ovate, dentate, pubescent, adnate to the pedicels; pedicels 0·5–5 cm. long. Receptacle-tube c. 0·5 mm. long, campanulate, densely shortly hairy; lobes 12–18 mm. long, ovate to lanceolate, acute, sometimes acuminate, shortly laciniate. Petals 2–3·5 cm. long, obovate-cuneate, crinkly, purplish, green-striped near the base. ♀ flowers solitary on 6·5–14 cm. long peduncles, somewhat larger than the male. Fruit 30–90 × 15–25 cm., green, ellipsoid with a lobed expanded base, bluntly 10-ribbed. Seeds numerous, 3·3–3·5 × 3·2–4·0 × 1·0–1·3 cm., compressed; endocarpic sheath persistent, reticulate, obscuring the faintly verrucose testa.

Zambia. N: Mbala (Abercorn), fl. 3.iv.1961, cult., *Angus* 2629 (K; SRGH). **Mozambique.** N: Vila Cabral, serra de Massangulo, st. 12.x.1942, *Mendonça* 798 (LISC). SS: Inhambane, fr. 16.ix.1948, *Myre & Carvalho* 217 (LISC; LMA; SRGH). LM: Lourenço Marques, st. 25.iv.1918, *Borle* 981 (BM; COI; K; PRE).

Probably native in N. Mozambique, also in Tanzania; often cultivated for its edible oily seeds. Coastal rain and riverine forest, 0–1100 m.

2. MOMORDICA L.*

Momordica L., Sp. Pl. **1**: 1009 (1753); Gen. Pl. ed. 5: 440 (1754).
Raphanocarpus Hook. f. in Hook., Ic. Pl. **11**: 67 (1871).
Raphanistrocarpus (Baill.) Pax in Engl. & Prantl, Pflanzenfam. **IV**, 5: 25 (1889).

Prostrate or scandent, usually herbaceous plants. Leaves simple or pedately ternately or 2-ternately 3–21-foliolate. Tendrils simple or 2-fid. Flowers white, cream, yellow or orange, monoecious or dioecious. ♂ flowers solitary, umbellate, shortly racemose or fasciculate, often bracteate; bracts often prominent, more or less cucullate. Receptacle-tube relatively short and broad; lobes entire. Petals 5, free, entire, obtuse to rounded or sometimes retuse, 1–3 with a ventral incurved scale at the base. Stamens 3, two 2-thecous and one 1-thecous, or less often 2, one 3-thecous and the other 2-thecous, free or sometimes with the anthers connate into a head; thecae arcuate, sinuate or triplicate. ♀ flowers solitary, perianth similar to that of ♂ flower but receptacle-tube usually much shorter and lobes often narrower and smaller; ovary ribbed, tuberculate or papillose, rarely smooth; ovules usually many, horizontal, less often few, pendulous and/or erect; stigma 3-lobed. Fruit ellipsoid to fusiform, tuberculate, spiny, winged or ribbed, rarely smooth, indehiscent or dehiscent into 3 valves and exposing seeds enveloped in bright red pulp. Seeds usually compressed, with sculptured faces and grooved margins with a slight erose median ridge.

An Old World genus of c. 40 species, the majority in tropical Africa.

1. Leaves compound, of 3 or more distinct leaflets - - - - - 2
 — Leaves simple, although sometimes deeply lobed - - - - - 7
2. Tendrils simple, unbranched - - - - - - - - 3
 — Tendrils 2-fid - - - - - - - - - - - 4
3. Leaflets 3–5; bract of ♂ inflorescence up to 3 mm. long - - 10. *trifoliolata*
 — Leaflets 9–21; bract of ♂ inflorescence 7–23 mm. long - 11. *cardiospermoides*
4. Leaflets 9–21 - - - - - - - - - 11. *cardiospermoides*
 — Leaflets 3–5 - - - - - - - - - - - 5
5. Ovary and fruit longitudinally ribbed or winged; ♂ receptacle and corolla regular - - - - - - - - - - - - 6
 — Ovary and fruit tuberculate; ♂ receptacle and corolla zygomorphic 9. *anigosantha*
6. ♂ receptacle-tube broadly campanulate, 4–7 mm. long; ♀ receptacle-lobes lanceolate, 4–6 mm. long; seeds oblong in outline, compressed, with sculptured faces, c. 10 × 7 × 3 mm. - - - - - - - - - - 1. *pterocarpa*
 — ♂ receptacle-tube cylindrical below, expanded above, 9–11 mm. long; ♀ receptacle-lobes triangular-acuminate, 6–11 mm. long; seeds subpisiform, tumid, with rugose faces, c. 8 × 7 × 4 mm. - - - - - - - - 2. *friesiorum*
7. Stems and petioles spotted with numerous small darker green marks - 8. *foetida*
 — Stems and petioles concolorous, uniformly green - - - - 8
8. Tendrils simple, unbranched - - - - - - - - 9
 — Tendrils 2-fid - - - - - - - - - - - 13
9. Peduncle of ♂ flowers adnate to petiole of subtending leaf, the flowers thus arising from apex of the petiole, in the basal sinus of the leaf; petals apically 2-lobed,

* By C. Jeffrey and V. Mann.

broadly retuse - - - - - - - - - - - 13. *kirkii*
— Peduncle of ♂ flowers free from petiole of subtending leaf; petals apically obtuse, rounded or only minutely retuse - - - - - - - - 10
10. Leaf-lamina deeply palmately 3–7-lobed usually to the middle or beyond; lobes narrowed towards the base, rhombic, elliptic or obovate in outline, more or less sinuate-lobulate; ♂ flowers solitary - - - - - - - 11
— Leaf-lamina unlobed or usually shallowly 3–5-lobed or -angled; lobes ovate or triangular, usually not narrowed towards the base; ♂ flowers rarely solitary, usually 2–many, if leaves deeply lobed, then ♂ flowers not solitary - - - 12
11. Bract of ♂ flower pallid, with green veins and apex; ♂ pedicel 2–5 mm. long; ovary and fruit with longitudinal rows of small tubercules, otherwise smooth; fruit-stalk 0·6–1·7 cm. long - - - - - - - - - 6. *balsamina*
— Bract of ♂ flower uniformly green; ♂ pedicel 20–95 mm. long; ovary and fruit with longitudinal rows of larger tubercules and with smaller interstitial tubercules; fruit-stalk 3·4–12·2 cm. long - - - - - - - - 5. *charantia*
12. ♂ flowers 1–8, subumbellate; fruit fusiform, 0·4–0·8 cm. in diam.; leaf-lamina not decurrent on to the petiole - - - - - - - - 12. *boivinii*
— ♂ flowers numerous, fasciculate; fruit ovoid rostrate, of larger diam.; leaf-lamina narrowly decurrent on to the petiole - - - - - 4. *henriquesii*
13. Leaf-lamina and bracts deeply palmately 7(or more)-lobate, the lobes subequal and markedly lobulate - - - - - - - - - 7. *repens*
— Leaf-lamina rather shortly palmately 3–5(rarely 7)-lobate, the central lobe clearly largest; lobes unlobed or rarely shortly 3-lobulate - - - - 14
14. Petals with rows of papillae on upper surface; anthers connate into a head, with tightly triplicate thecae and narrow connectives; ovary smooth; fruit rounded at the apex - - - - - - - - - - - - 15
— Petals without rows of papillae on upper surface; anthers more or less free, with rather laxly sinuate thecae and broad connectives, those of the 2 double stamens 2-fid; ovary longitudinally ribbed; fruit longitudinally winged, rostrate - 16
15. Leaf-lamina shortly hispid above; leaf-lobes usually triangular or ovate-triangular, not lobulate; ♂ fascicles subumbelliform - - - - 14. *calantha*
— Leaf-lamina punctate above, almost glabrous; leaf-lobes ovate, shortly 3-lobulate; ♂ fascicles racemiform - - - - - - - 15. *sp. A*
16. Leaf-lamina narrowly decurrent on to the petiole - - - 4. *henriquesii*
— Leaf-lamina not decurrent on to the petiole - - - - 3. *corymbifera*

1. **Momordica pterocarpa** Hochst. ex A. Rich., Tent. Fl. Abyss. **1**: 292 (1848).—Hook. f. in F.T.A. **2**: 536 (1871).—C. Jeffrey in F.T.E.A., Cucurb.: 23, t. 2 fig. 2 (1967).—Binns, H.C.L.M.: 43 (1968). Syntypes from Ethiopia.

Stems prostrate or scandent to 8 m., shortly finely crispate-pubescent, arising from perennial tuberous rootstock. Leaves compound, pedately 3–5(7)-foliolate; median leaflet 4·2–13 × 1·8–8 cm., narrowly to broadly elliptic or ovate, cuneate, rounded or subtruncate at the base, plane or more or less sinuate-denticulate, shortly acuminate, obtuse to acute, apiculate, minutely punctate and shortly setulose on veins beneath, shortly hispid or crispate-setulose above; lateral leaflets smaller, unequal-based; petiolule of median leaflet 0·7–3·7 cm. long. Petiole 0·6–5·5 cm. long, crispate-pubescent. Tendrils 2-fid. Flowers dioecious. ♂ flowers 4–5(18) in bracteate, pedunculate, subumbelliform fascicles; peduncle 5–17 cm. long; bract 10–35 mm. long, closely subtending the flowers, suborbicular, cucullate; pedicels 3–12(22) mm. long. Receptacle-tube 4–7 mm. long, lobes 6–11 mm. long, greenish to black, triangular-lanceolate, obtuse, apiculate. Petals 1·5–3·2 cm. long, white or yellowish, greenish-black at the base, obovate. ♀ flowers on 0·6–3 cm. long peduncles, sometimes apically bracteate; ovary 18–32 × 3–3·5 mm., fusiform, ridged; receptacle-tube 1–3·5 mm. long, lobes 4–6 mm. long, narrowly triangular, obtuse to acute; petals 13–20 mm. long. Fruit 5–6 × 2–2·2 cm., ellipsoid, rostrate, longitudinally winged, bright orange, dehiscent into 3 valves; fruit-stalk 3–8·5 cm. long. Seeds c. 10 × 7 × 3 mm., enveloped by orange pulp, oblong, compressed, with sculptured faces and smooth grooved margins.

Malawi. N: Mzuzu, Marymount dambo, fl. 15.i.1971, *Pawek* 4315 (K; MAL). C: Namitete R. above bridge on Lilongwe to Fort Jameson road, fl. & fr. 5.ii.1959, *Robson* 1478 (BM; K; LISC; PRE; SRGH). S: Limbe, fl. 11.i.1948, *Goodwin* 23 (BM).
E. tropical Africa from Eritrea southwards to Tanzania and Malawi. Woodland and riverine fringes, 1150–1350 m.

A specimen of this species in the Kew herbarium (K) bears a label " Lorenzo Marques, June 1911, *Miss.M.H.Mason* ". It appears doubtful if this upland species could occur at Lourenco Marques and the specimen is possibly wrongly labelled. The occurrence of this species in Mozambique is therefore still uncertain.

2. **Momordica friesiorum** (Harms) C. Jeffrey in Kew Bull. **15**: 356 (1962); in F.T.E.A., Cucurb.: 24, t. 3 fig. 13 (1967). Syntypes from Kenya.
Calpidosicyos friesiorum Harms in Notizbl. Bot. Gart. Berl. **8**: 482 (1923). Syntypes as above.

Stems prostrate or scandent to 4 m. or more, ± crispate-pubescent especially at nodes, arising from perennial tuberous rootstock. Leaves compound, pedately 3–5-foliolate; median leaflet 2·5–13 × 1–5 cm., narrowly ovate to ovate or elliptic, obtuse to subtruncate at the base, remotely denticulate to sinuate-dentate, acutely-acuminate, apiculate, sparsely to distinctly rather finely setulose above and mainly on veins beneath; lateral leaflets smaller, unequal-based; petiolule of median leaflet 0·4–2·7 cm. long; petiole 0·7–5·5 cm. long. Tendrils 2-fid. Flowers dioecious. ♂ flowers 5–6(15), in bracteate pedunculate subumbelliform fascicles; peduncle 4–8 cm. long; bract 10–30 mm. long, closely subtending the flowers, obovate-orbicular, cucullate; pedicels 3–5(10) mm. long. Receptacle-tube 9–11 mm. long, lobes (3)6–8(12) mm. long, purplish, triangular, acute-attenuate. Petals 1–1·5 cm. long, pale yellow, dark at the base, oblong-ovate. ♀ flowers on 1·1–1·3 cm. long peduncles; ovary 30–35 × 4–5 mm., fusiform, ridged. Receptacle-tube 3 mm. long, lobes 6–11 mm. long, triangular-acuminate. Fruit c. 6 × 2 cm., fusiform, rostrate, longitudinally winged or ridged, orange, dehiscent into 3 valves; fruit-stalk 2·3–6 cm. long. Seeds c. 8 × 7 × 4 mm., tumid, with rugose faces and smooth grooved margins.

Malawi. N: Rumpi, Nyika Plateau, fl. 18.v.1970, *Brummitt* 10883 (K; MAL; SRGH).
Eastern tropical Africa from Ethiopia southwards to Tanzania and Malawi. Upland (*Juniperus*) forest, 2135 m.

3. **Momordica corymbifera** Hook. f. in F.T.A. **2**: 539 (1871).—C. Jeffrey in Kew Bull. **15**: 356 (1962) excl. syn. *Momordica henriquesii* Cogn. Type: Mozambique, near Lupata, *Kirk* (K, holotype).

Stems prostrate or scandent to 10 m., sparsely to very densely shortly and finely crispate-pubescent, arising from large woody tuberous rootstock; all parts foetid. Leaves simple. Leaf-lamina 4·7–10·7 × 5·0–10·4 cm., ovate to broadly ovate in outline, deeply cordate, remotely denticulate to sinuate-serrate, finely punctate and shortly and finely pubescent on veins beneath, finely pubescent or asperulous above, unlobed or usually shortly palmately 3–5-lobed, the median lobe largest, broadly triangular to ovate, acuminate, acute, apiculate; lateral lobes much shorter. Petiole 2·1–9·2 cm. long, rather densely shortly finely pubescent. Tendrils 2-fid. Flowers monoecious. ♂ flowers solitary or usually 4–11 in bracteate, pedunculate, subumbelliform or racemiform inflorescences, rarely proliferating into many-flowered fascicles; peduncle 1·1–9·5 cm. long; bracts 2–20 mm. long, usually ovate or broadly ovate, rounded, apiculate, sessile or shortly stalked; pedicels 0·5–2·7 cm. long. Receptacle-tube 3–8 mm. long, lobes 3–6 mm. long, green or blackish, ovate to broadly ovate or obovate-oblong, rounded to acute, apiculate to long-acuminate, apices sometimes recurved. Petals 11–19 mm. long, pale yellow, dark-lined along the veins at the base, recurved. ♀ flowers solitary or paired; peduncle up to 6·2 cm. long, bracteate at apex; bract 7 mm. long; pedicel 3 mm. long; ovary 16–30 × 5–8 mm., ovoid or ellipsoid, rostrate, longitudinally ridged, finely puberulous, the ridges wing-like, toothed; receptacle-tube very short, 1 mm. long, lobes 6–13 mm. long, ovate or ovate-lanceolate, acutely-acuminate; petals 17–26 mm. long. Fruit c. 9 × 5 cm., ellipsoid, shortly rostrate, with 16 longitudinal wing-like ridges; fruit stalk 1 cm. or more long, stout. Seeds 16–16·5 × 12·8–13 × 6·5–7·0 mm., asymmetrically ovate in outline, subcompressed, brown, almost smooth, with grooved margins, enveloped in sheaths of fleshy pulp.

Zambia. B: Sesheke, near Machili, fl. 20.xii.1952, *Angus* 975 (FHO; K). C: Lusaka, fl. 6.xii.1957, *Fanshawe* 4114 (K; SRGH). E: Petauke, E. of Minga, fl. 5.xii.1958, *Robson* 847 (BM; K; LISC; PRE; SRGH). S: Machili, fl. 3.xii.1960,

Fanshawe 5931 (K; SRGH). **Rhodesia.** N: Urungwe, near Rekomitje R., fl. 3.i.1961, *Goodier* 38 (COI; K; SRGH). E: Umtali, road to Macequice, commonage, fl. & fr. immat. 21.xi.1954, *Chase* 5336 (BM; COI; K; LISC; SRGH). **Mozambique.** T: Cabora Bassa, c. 1 km. from Dam, fr. 12.ii.1973, *Torre, Carvalho & Ladeira* 19066 (LISC). MS: Lupata, fl. 1859, *Kirk* s.n. (K).
Not known from elsewhere. Deciduous and semi-deciduous woodland, on sandy soils, 400–1300 m.

4. **Momordica henriquesii** Cogn. in Bol. Soc. Brot. **7**: 228 (1889). Type: Mozambique, Niassa, Mossuril [Mussorile], *Carvalho* 14 (COI, holotype; BR, isotype).
    Momordica pycnantha Harms in Notizbl. Bot. Gart. Berl. **13**: 427 (1936).—C. Jeffrey in F.T.E.A., Cucurb.: 28 (1967). Type from Tanzania.
Stems scandent, very shortly and finely pubescent, becoming woody with grey sulcate bark. Leaf-lamina 4·5–11·3 × 3·5–11·8 cm., broadly ovate, deeply cordate, narrowly decurrent on to petiole, minutely apiculate-denticulate, unlobed or obscurely 3-lobed, shortly acuminate, acute, apiculate, finely pubescent on veins and minutely punctate beneath, minutely hispidulous above. Petiole 1·9–3 cm. long, finely pubescent. Tendrils simple or 2-fid. Flowers dioecious. ♂ flowers numerous in sessile or shortly pedunculate axillary bracteate or ebracteate fascicles; peduncle up to 3·8 cm. long; bracts 2–12 mm. long, elliptic or oblong-elliptic, shortly acuminate; pedicels 15–30 mm. long. Receptacle-tube (1)2–4 mm. long, lobes (3)5–8 mm. long, lanceolate, acutely-acuminate. Petals (0·5)1·2–1·6 cm. long, yellow, dark at the base, obovate, rounded, apiculate. ♀ flowers unknown. Fruit ovoid-rostrate, winged, velutinous. Seeds unknown.

Mozambique. N: Memba, c. 24 km. from Memba towards Mazua, fl. 7.xii.1963, *Torre & Paiva* 9454 (LISC).
Also in SE. Tanzania. Open *Brachystegia* woodland, on red clay soils, 100 m.

5. **Momordica charantia** L., Sp. Pl. **1**: 1009 (1753).—Hook. f. in F.T.A. **2**: 537 (1871).—Ficalho, Pl. Ut. Afr. Port.: 188 (1884).—Hiern, Cat. Afr. Pl. Welw. **1**, 2: 393 (1898).—Warb., Kunene-Samb.-Exped. Baum: 395 (1903).—Gibbs in Journ. Linn. Soc., Bot. **37**: 446 (1906).—Bremek. & Oberm. in Ann. Transv. Mus. **16**: 436 (1935).—Meeuse in Bothalia **8**: 51 (1962).—Keraudren in Fl. Madag., Cucurb.: 30 (1966); in Fl. Cameroun **6**, Cucurb.: 172, t. 3 fig. 3, fig. 12 & t. 31 fig. 1–7 (1967).—C. Jeffrey in F.T.E.A., Cucurb.: 31 (1967).—Binns, H.C.L.M.: 43 (1968).—R. & A. Fernandes in C.F.A. **4**: 263 (1970). Type a cultivated plant.
    Momordica charantia var. *abbreviata* Ser. in DC., Prodr. **3**: 311 (1828).—Hiern, loc. cit. Type from Ceylon.

Stems prostrate or scandent to 5 m., sparsely to densely crispate-pubescent or villous, especially at nodes. Leaf-lamina 1–10 × 1–12·5 cm., broadly ovate to orbicular in outline, cordate, narrowly decurrent on to petiole, punctate and sparsely pubescent to densely villous on veins beneath, sparsely hirsute especially on veins above, deeply palmately 3–7-lobed, lobes variously sinuate-dentate or lobulate, acute to retuse, apiculate. Petiole 0·5–7 cm. long. Tendrils simple. Flowers monoecious, solitary. ♂ flowers: peduncle 0·3–5 cm. long; bract 2–17 mm. long, broadly ovate or reniform, sessile, cordate, amplexicaul, obtuse or retuse, apiculate, green; pedicel 2–9·5 cm. long. Receptacle-tube 1–5 mm. long; lobes 3–7 mm. long, ovate-lanceolate. Petals 1·0–2·5 cm. long, pale to deep yellow, ovate to obovate. ♀ flowers: peduncle 0·2–5 cm. long; bract 1–12 mm. long; pedicel 1–10 cm. long; ovary 8–11(30) × 2–4 mm., ovoid-rostrate to fusiform, ridged, pilose on ridges, tuberculate; receptacle-tube 1–3 mm. long, lobes 2–5 mm. long, lanceolate; petals 0·7–1·2 cm. long. Fruit 2·5–4·8(11) × 1·5–2·3(4) cm., ovoid-rostrate or ellipsoid, longitudinally ribbed and tuberculate, bright orange-red, dehiscent into 3 valves; fruit-stalk 3·4–15 cm. long. Seeds 8–11 × 4·5–8 × 2–3·5 mm., enveloped in sticky red pulp, ovate-elliptic to oblong in outline; faces flattened, sculptured, with sinuate edges; margins grooved.

Zambia. B: Shangombo, fl. 8.viii.1952, *Codd* 7442 (K; PRE). N: Fort Rosebery, Lake Bangweulu, above Samfya beach, fl. & fr. immat. 8.ii.1959, *Watmough* 235 (COI; SRGH). C: Mazabuka, 43 km. E. of Lusaka, by Kafue R., fl. 9.ii.1957, *Noak* 91 (K; SRGH). E: Luangwa valley, 112 km. NW. of Lundazi, Chikwa, fl. & fr. immat. 3.vi.1954, *Robinson* 825 (K; SRGH). S: Gwembe valley, bank of Zongwe R., above Sinazongwe to Mwamba road, fl. & fr. 2.vii.1961, *Angus* 2955 (FHO; K; SRGH). **Rhodesia.** N: Urungwe, Sanyati-Chiroti confluence, fl. & fr. immat. 21.xi.1953, *Wild*

4225 (K; LISC; SRGH). E: Umtali, fl. & fr. immat. 27.xii.1969, *Chase* 8575 (K; SRGH). **Malawi.** N: Karonga Distr., 3 km. N. of Chilumba, St. Anne's, fl. & fr. 23.iv.1969, *Pawek* 2298 (K). C: Salima Bay, fl. 29.ix.1935, *Galpin* 15032 (BOL). S: Ncheu, Kapeni stream, fr. 12.vi.1967, *Jeke* 100 (COI; K). **Mozambique.** N: Johanna, fl. & fr. immat. xi.1862, *Kirk* (K). Z: Maganja da Costa, 50 km. from Vila da Maganja towards the sea, fl. 12.ii.1966, *Torre & Correia* 14520 (LISC). T: Boroma Prov., Sisitso camp, Ulere Station, Messenguese R., fl. 11.vii.1950, *Chase* 2639 (K; SRGH). MS: Cheringomo, Inhamitanga, fr. 22.ix.1944, *Simão* 72 (LISC; LMA). SS: Inhanombe, fl. 25.ii.1955, *E. M. & W.* 569 (BM; LISC; SRGH).
Pantropical, probably an introduction in the New World. Riverine fringes, lake margins, old cultivations, 0–1650 m.

6. **Momordica balsamina** L., Sp. Pl. **2**: 1009 (1753).—Sond. in Harv. & Sond., F.C. **2**: 491 (1862).—Hook. f. in F.T.A. **2**: 537 (1871).—Hiern, Cat. Afr. Pl. Welw. **1**, 2: 394 (1898).—Eyles in Trans. Roy. Soc. S. Afr. **5**: 498 (1916).—Burtt Davy, F.P.F.T. **1**: 227 (1926).—Wild, S. Rhod. Bot. Dict.: 104 (1953); Fl. Vict. Falls: 145 (1953).—Mogg in MacNae & Kalk, Nat. Hist. Inhaca I., Moçamb.: 154 (1958).—Story in Mem. Bot. Survey S. Afr. **30**: 46, t. 55 (1958).—Leistner in Koedoe **2**: 170 (1959).—Meeuse in Bothalia **8**: 49 (1962).—Mitchell in Puku **1**: 170 (1963).—C. Jeffrey in F.T.E.A., Cucurb.: 32 (1967).—Launert & Roessler in Prodr. Fl. SW. Afr. **94**: 20 (1968).—Binns, H.C.L.M.: 43 (1968) " balsaminea ".—R. & A. Fernandes in C.F.A. **4**: 262 (1970). Type a cultivated plant.
Momordica involucrata E. Mey. ex Sond. in Harv. & Sond., loc. cit.—Eyles, loc. cit.—Meeuse, tom. cit.: 50 (1962). Syntypes from S. Africa (Natal).

Stems prostrate or scandent, to 2·7 m., finely rather sparsely crispate-pubescent, especially at nodes. Leaf-lamina 1–9 × 1·2–12 cm., broadly ovate to suborbicular in outline, deeply cordate, minutely punctate and laxly pubescent on veins beneath, very shortly sparsely setulose especially on veins above, deeply palmately 5–7-lobed, lobes sharply to obtusely 3–5-lobulate or sometimes merely sinuate-dentate, acute to rounded or subtruncate, apiculate. Petiole 0·3–6·1 cm. long. Tendrils simple. Flowers monoecious, solitary. ♂ flowers: peduncle 1·6–10·5 cm. long; bract 2–18 mm. long, broadly ovate to suborbicular, sessile, apiculate, pallid with dark green veins and tip; pedicel 2–5 mm. long. Receptacle-tube 2–4·5 mm. long, lobes 4–9 mm. long, ovate, broadly ovate or obovate, acute to rounded, bluntly acuminate or apiculate, blackish or green. Petals 1·0–1·9 cm. long, pale yellow, cream or white, green-veined, dark at the base, obovate-oblong, rounded or slightly retuse, apiculate. ♀ flowers: peduncle 2–5 mm. long; bract 1·5–5 mm. long, green; pedicel 0·4–2·7 cm. long; ovary 4–13 × 2–4 mm., ovoid, rostrate, sparsely tuberculate, puberulous; receptacle-tube 0·5–1 mm. long, lobes narrow, 1–5 mm. long; petals 0·5–1·3 cm. long. Fruit 2·5–6·2 × 1·8–2·8 cm., ovoid, rostrate, tuberculate, bright orange-red or red, dehiscent into 3 valves; fruit-stalk 0·6–2·0 cm. long. Seeds 8·5–11·2 × 5–6·6 × 2–3 mm., enveloped in red foetid pulp, ovate in outline, more or less compressed, faces with slightly depressed centres and elevated sculptured edges; margins grooved.

Caprivi Strip. Katima Mulilo, fl. 24.xii.1958, *Killick & Leistner* 3059 (BM; K; PRE; SRGH). **Botswana.** N: Chobe-Zambezi confluence, fl. 11.iv.1955, *E.M. & W.* 1468 (BM; LISC; SRGH). SW: Chukudu, 410 km. NW. of Molepolole, fl. & fr. immat. 23.vi.1955, *Story* 4970 (K; PRE). **Zambia.** B: Senanga, Kaunga, Mashi R., fl. 6.ix.1961, *Mubita* in *Reynolds* B21 (COI; SRGH). S: Namwala, fl. & fr. 9.i.1957, *Robinson* 2097 (K; SRGH). **Rhodesia.** W: Wankie, fl. & fr. iv.1955, *Davies* 1096 (K; SRGH). C: Marandellas, fl. & fr. 20.iv.1924, *Eyles* 3848 (BOL; PRE; SRGH). E: Umtali, Riverside Drive, fl. 8.i.1956, *Chase* 5943 (BM; COI; K; LISC; LMA; SRGH). S: Ndanga, Lundi R., Chipinda Pools, fl. 24.iv.1961, *Goodier* 1069 (COI; K; SRGH). **Mozambique.** Z: Quelimane, fr. 1908, *Sim* 20779 (PRE). MS: Sofala, 350 km. from Mambone to Nova Lusitânia, fl. & fr. 9.x.1965, *Torre & Pereira* 12364 (LISC). SS: Gaza, Macia, between Macia and S. Martinho do Bilene, fl. & fr. 11.ii.1959, *Barbosa & Lemos* 8366 (COI; K; LISC; LMA). LM: Inhaca I., fl. & fr. 14.xii.1955, *Noel* 179 (PRE; K).
Also in S. and SW. Africa, Angola, Tanzania (coasts only), NE. and W. tropical Africa, tropical Arabia, India and Australia; neotropics probably by introduction. Woodland, wooded grassland and riverine fringes, on sandy soils, 0–1600 m.

It is impossible satisfactorily to distinguish *M. involucrata* from *M. balsamina*; the distinctions given by Meeuse (loc. cit., p. 46) are not clear-cut and the marked ecological disjunction implied by his remarks (loc. cit., p. 51) is not apparent in the F.Z. area.

7. **Momordica repens** Bremek in Ann. Transv. Mus. **15**: 261 (1933).—Meeuse in Bothalia **8**: 51 (1962). Syntypes from S. Africa (Transvaal).
Momordica marlothii Harms in Fedde, Repert. **36**: 269 (1934). Type: Botswana, Palapye, *Marloth* 3329 (B, holotype, †, PRE, isotype).

Stems prostrate or occasionally scandent, to 2 m., rather densely crispate-pubescent, arising from perennial rootstock. Leaf-lamina 1·3–3·1 × 3·4–5·5 cm., suborbicular to transversely elliptic in outline, deeply cordate, densely shortly setulose becoming densely asperous-punctate, slightly succulent, deeply palmately multilobate, the lobes subequal, rather sharply lobulate. Petioles 1·8–6 cm. long, very densely crispate-pubescent. Tendrils 2-fid. Flowers monoecious. ♂ flowers 3–5 in bracteate, pedunculate, subumbelliform inflorescences, only one flower opening at a time; peduncle 1·3–6 cm. long; bracts 5–9 × 7–18 mm., resembling the leaves but smaller, 3–7-lobed; pedicels 2·0–4·4 cm. long. Receptacle-tube 2–5 mm. long, lobes 5–10 mm. long, ovate or oblong-ovate, acute, shortly acuminate or apiculate. Petals 1·4–3·6 cm. long, yellow, dark-veined, dark at the base, broadly obovate, rounded, apiculate. ♀ flowers solitary; peduncle 3–7 cm. long, sometimes bracteate; ovary c. 20 × 5 mm., ellipsoid-rostrate, pubescent, muricate; receptacle-tube 2 mm. long, lobes c. 5 mm. long, ovate-triangular, shortly acuminate, acute, apiculate; petals c. 2·5–3·5 cm. long. Fruit 5–6 × 4·7 cm., ellipsoid or subglobose, rostrate, 10-ribbed, muricate, reddish-brown. Seeds c. 20 × 20 × 7–8 mm., orbicular in outline, more or less compressed, with smooth undulate faces, slightly ridged edges and grooved margins.

Botswana. SE: Mahalapye, fl. & fr. immat. 22.xi.1963, *Yalala* 393 (COI; K; SRGH). **Rhodesia.** W: Bulawayo, i.1898, *Rand* 93 (BM).
Also in Transvaal. Bushland on sandy soil, 1000 m.

8. **Momordica foetida** Schumach. in Schumach. & Thonn., Beskr. Guin. Pl.: 426 (1827).—Swynnerton et al. in Journ. Linn. Soc., Bot. **40**: 74 (1911).—Eyles in Trans. Roy. Soc. S. Afr. **5**: 498 (1916).—Burtt Davy, F.P.F.T.: 227 (1926).—Wiehe, Cat. Fl. Pl. Herb. Div. Pl. Path. Zomba: 17 (1952).—Brenan in Mem. N.Y. Bot. Gard. **8**: 443 (1954).—Meeuse in Bothalia **8**: 47 (1962).—Keraudren in Fl. Cameroun **6**, Cucurb.: 180, t. 2 fig. 7, t. 3 fig. 10 & t. 36 fig. 1–6 (1967).—C. Jeffrey in F.T.E.A., Cucurb.: 29, t. 2 fig. 7 (1967).—Binns, H.C.L.M.: 43 (1968).—R. & A. Fernandes in C.F.A. **4**: 266 (1970). TAB. **101**. Type from Ghana.
Momordica morkorra A. Rich., Tent. Fl. Abyss. **1**: 292, t. 53 (1848).—Eyles, loc. cit. Syntypes from Ethiopia.
Momordica schimperiana Naud. in Ann. Sci. Nat. Bot., Sér. 5, **5**: 23 (1866).—Hiern, Cat. Afr. Pl. Welw. **1**, 2: 394 (1898).—R.E. Fr., Wiss. Ergebn. Schwed. Rhod.-Kongo-Exped. **1**: 312 (1916). Syntypes from Ethiopia.
Momordica foetida var. *villosa* Cogn. in Engl., Bot. Jahrb. **21**: 208 (1895).—Binns, loc. cit. Type from Tanzania.

Stems prostrate or scandent to 7 m., almost glabrous to densely shortly crispate-pubescent, especially at nodes, flecked with darker green, becoming woody with pallid bark when old, arising from perennial tuberous rootstock; all parts foetid. Leaf-lamina 1·3–19·1 × 1·3–18·1 cm., ovate or ovate-triangular to very broadly ovate in outline, broadly and deeply cordate and narrowly decurrent on to petiole, almost plane and minutely denticulate to very strongly sinuate-denticulate, unlobed, acutely-acuminate, apiculate, punctate and almost glabrous to densely shortly crispate-pubescent on veins beneath, shortly scabrid-setulose or punctate above. Petiole 1·3–17 cm. long, green-flecked like the stems. Tendrils simple or 2-fid. Flowers dioecious. ♂ flowers 1–8 (opening singly) in bracteate, pedunculate, subumbelliform inflorescences; peduncle 2–23 cm. long; bract 3–30 mm. long, oblanceolate to broadly rounded, closely subtending the flower, often cucullate; pedicels 0·2–7 cm. long. Receptacle-tube 3–8 mm. long, lobes 5–10 mm. long, ovate or ovate-oblong, obtuse or rounded, dark green to black, bearing a number of stiff emergences on the back. Petals 1·7–3·5 cm. long, pale yellow to orange, deeper orange to black at the base, obovate, rounded. ♀ flowers solitary; peduncle 0·2–6·5 cm. long; bract 3–20 mm. long; pedicel 0·2–18·6 cm. long; ovary 8–26 × 3–11 mm., ellipsoid-rostrate, densely and softly muricate; receptacle-tube 2–2·5 mm. long, lobes 2–11 mm. long, ovate or triangular to oblong-lanceolate, acute to obtuse or rounded; petals 1·5–3·7 cm. long. Fruit 3·5–7·5 × 2·5–5 cm., ovoid or ellipsoid, softly spiny, bright orange, dehiscent into 3 valves; fruit-stalk 5–10(20) cm. long. Seeds 7·2–12 × 5–7·5 × 2–4·5 mm., enveloped in bright

Tab. 101. MOMORDICA FOETIDA. 1, part of shoot, ♂ plant (×½) *Wild* 3900 & *Swynnerton* 94; 2, part of shoot, ♀ plant (×½) *Wild* 3703 & *Eyles* 7951; 3, ♂ flower, basal part, dissected out (×1½); 4, androecium, lateral view and plan (×6); 5, ♀ flower partly dissected (×1); 6, fruit (×1); 7, seed, face and lateral views (×3), 3–7 from *Robson* 248.

red pulp, ovate-oblong or elliptic-oblong in outline, compressed; faces flattened, slightly warted; margin grooved.

Zambia. N: Mbala (Abercorn), Chilongowelo, Tasker's Deviation, fl. i.1955, *Bock* 272(PRE). W: Katenina Hills, fl. 21.xii.1907, *Kässner* 2181 (BM; K). E: Mvuvje R., near Petauke, fl. 5.xii.1958, *Robson* 846 (BM; K; LISC; MO; PRE; SRGH). **Rhodesia.** N: Mazoe, fl. & fr. immat. xi.1906, *Eyles* 473 (BM; BOL; SRGH). C: Salisbury, Enterprise, fl. 28.xi.1952, *Wild* 3900 (K; LISC; SRGH). E: Inyanga, fl. & fr. 24.i.1961, *Wild* 5479 (COI; K; SRGH). **Malawi.** N: Nyika Plateau Valley, c. 4 km. SW. of Rest House, fl. 22.x.1958, *Robson & Angus* 248(BM; K; LISC; MAL; PRE; SRGH). C: Dedza Mt. Forest, fl. 10.xii.1962, *Banda* 465 (COI; LISC; SRGH). S: Zomba, fl. & fr. 2.xi.1949, *Wiehe* N/361 (K). **Mozambique.** N: Vila Cabral, fl. 13.xii.1954, *Bergstrom* (LMA). Z: Milange, Serra Tumbine, fr. 19.i.1966, *Correia* 486 (LISC). MS: Manica, Mavita, Serra Mocuta, fl. & fr. immat. 13.i.1966, *Pereira, Sarmento & Marques* 1155 (LMU). LM: R. Maputo, Salamanga, fl. & fr. 21.xi.1940, *Torre* 2138 (LISC).

Tropical Africa. Rain-forest, woodland, wooded grassland and grassland, often in riverine fringes, 350–2250 m.

9. **Momordica anigosantha** Hook. f. in F.T.A. **2**: 536 (1871).—C. Jeffrey in F.T.E.A., Cucurb.: 24 (1967). Type: Mozambique, Shire R., Shamo, *Kirk* (K, holotype).

Stems scandent or prostrate, to 3 m., glabrous or almost so except at nodes. Leaves compound, pedately 3–5(7)-foliolate; median leaflet 3·5–11 × 1·7–4·7 cm., ovate or elliptic, rounded or subtruncate at the base, almost plane and minutely denticulate, acuminate, acute to obtuse, apiculate, asperulous or minutely punctate and sparsely setulose on veins beneath, sparsely setulose above; lateral leaflets somewhat smaller, unequal-based; petiolule of median leaflets 0·6–1·8 cm. long. Petiole 1·2–5·2 cm. long, shortly pubescent. Tendrils 2-fid. Flowers dioecious. ♂ flowers 4–18(50) in lax racemiform or paniculiform fascicles; peduncle 1·1–7 cm. long, simple or branched, ebracteate or sometimes bearing small leaf-like bracts; pedicels 0·9–3·5 cm. long. Receptacle-tube 5–7 mm. long, asymmetrical, blackish or brownish-green, lobes 4–7 mm. long, lanceolate, acutely-acuminate, unequal, the lower 3 larger than the upper 2. Petals 1·2–2·5 cm. long, cream or yellow, unequal, the lowermost largest, with an orange hippocrepiform mark towards the base. ♀ flowers solitary; peduncle 0·4–1·6 cm. long; bract minute; pedicel 3–6·3 cm. long; ovary 12–17 × 2–4 mm., ovoid, rostrate, tuberculate; receptacle-tube very short, lobes 2–3·5 mm. long, lanceolate, acute; petals c. 2 cm. long, slightly unequal. Fruit 4·5–6·5 × 1·6–3·8 cm., fusiform, rostrate, with 8 longitudinal rows of conical tubercules, bright orange or red, dehiscent into 3 valves; fruit-stalk 5–14·5 cm. long. Seeds 12–14 × 7–8 × 2–3 mm., oblong in outline, compressed; faces flat, sculptured, sinuate at the edges; margin grooved.

Mozambique. N: Cabo Delgado, Macondes, 16 km. from Namacua towards Nangade, fl. 16.iv.1964, *Torre & Paiva* 12053 (LISC). Z: Shamo, fl. & fr. v.1863, *Kirk* (K). MS: Cheringoma, Inhamitanga, fl. 20.ix.1944, *Simão* 49 (LISC; LMA).

Also in E. tropical Africa. Forest and bushland, 350–600 m.

10. **Momordica trifoliolata** Hook. f. in F.T.A. **2**: 537 (1871).—Keraudren in Fl. Madag., Cucurb.: 30 (1966).—C. Jeffrey in F.T.E.A., Cucurb.: 37 (1967). Type from Tanzania.

Stems prostrate or scandent, to 6 m., almost glabrous except at nodes, becoming thickened and pallid-barked when old, arising from woody subterranean tuber. Leaves compound, pedately 3–5(15)-foliolate; median leaflet 1·5–9 × 0·9–6·4 cm., ovate, weakly cordate, obscurely to distinctly sinuate and minutely denticulate, obtuse, apiculate, shortly setulose above; lateral leaflets smaller, somewhat unequal-based; petiolule of central leaflet 0·4–3·2 cm. long; petiole 0·4–2·2 cm. long. Tendrils simple. Flowers dioecious. ♂ flowers solitary, or becoming fasciculate by reduction of leaves and internodes, pedunculate, bracteate; peduncle 0·6–8·0 cm. long; bract 1·5–7(16) mm. long, suborbicular, apiculate; pedicel 0·2–7 mm. long. Receptacle-tube 2–4 mm. long, lobes 3–7·5 mm. long, triangular or broadly ovate, obtuse or rounded. Petals 1·1–2·2 cm. long, yellow, dark at the base, obovate, rounded, apiculate. ♀ flowers solitary, bracteate; bract small; peduncle and pedicel totalling 1·4–2·8 cm. long, the latter longer than the former; ovary 5–8 × 1·5–3·5 mm., cylindrical, glabrous; receptacle-tube obsolescent, lobes

3–4 mm. long, ovate-triangular; petals c. 1·2 cm. long. Fruit 4–6·3 × 2·8–3·5 cm., ovoid, acute but not rostrate, terete or obscurely ribbed, bright red with paler spots, irregularly dehiscent; fruit-stalk 2–3 cm. long. Seeds c. 13 × 7 × 3·5 mm., enveloped in slimy whitish pulp, ovate-rectangular in outline, with sculptured faces and grooved margins.

Mozambique. N: Cabo Delgado, Ingoane, near R. M'Salo, fl. 12.ix.1948, *Pedro &* *Pedrógão* 5176 (LMA).
Also in Madagascar, Ethiopia, Somalia and E. tropical Africa. Ecology in F.Z. area not recorded.

11. **Momordica cardiospermoides** Klotzsch in Peters, Reise Mossamb. Bot.: 150 (1861).—Mitchell in Puku 1: 150, 186 (1963).—C. Jeffrey in F.T.E.A., Cucurb.: 38 (1967).—Binns, H.C.L.M.: 43 (1968). Type: Mozambique, Sena, *Peters* (B†, holotype).
Momordica clematidea Sond. in Harv. & Sond., F.C. 2: 491 (1862).—Eyles in Trans. Roy. Soc. S. Afr. 5: 498 (1916).—Burtt Davy, F.P.F.T. 1: 227 (1926).— Gomes e Sousa in Bol. Soc. Estud. Col. Moçamb. 29–32: 92 (1936).—Wild, Fl. Vict. Falls: 145 (1953).—Meeuse in Fl. Pl. Afr. 33: t. 1295 (1959); in Bothalia 8: 46 (1962). Syntypes from S. Africa (Transvaal).

Stems prostrate or scandent to 10 m., almost glabrous except at nodes, arising from tuberous subterranean rootstock; all parts foetid. Leaves compound, bipedately (9)15–17(21)-foliolate; leaflets commonly 15 in 3 groups of 5, or 17 with median group of 7 and two lateral groups of 5; median leaflet 0·5–6 × 0·4– 4·2 cm., elliptic, ovate or broadly ovate, rounded or cuneate at the base, sinuate-denticulate or sinuate-lobulate or occasionally deeply 3-lobulate, acute to obtuse or rounded, apiculate, minutely punctate, finely setulose on veins above, otherwise glabrous; petiolules of median group of leaflets 0·9–4·5(5) cm. long; petiolules of median leaflet (or leaflets) of median group 0·3–3·5 cm. long. Petiole 0·7–3 cm. long, almost glabrous except at base and apex. Tendrils simple or 2-fid. Flowers monoecious. ♂ flowers solitary, pedunculate, bracteate; peduncles 2·2–15 cm. long; bract 7–23 mm. long, suborbicular, cordate, cucullate; pedicel 6–12 mm. long. Receptacle-tube 3–6 mm. long, lobes 4–8 mm. long, ovate to broadly ovate, shortly acuminate, obtuse to acute. Petals 1·1–3·2 cm. long, pale yellow to orange, dark at the base, broadly obovate, rounded, apiculate. ♀ flowers solitary; peduncle 0·9–6·5 cm. long; bract 2–11 mm. long; pedicel 2–20 mm. long; ovary 7–19 × 2–5 mm., ellipsoid-cylindrical; receptacle-tube 1 mm. long, lobes ovate-oblong to triangular-lanceolate, sometimes acuminate, obtuse to acute, 2–5·5 mm. long; petals 2–2·8 cm. long. Fruit 5–13 × 2·5–7 cm., ellipsoid, smooth, bright red, indehiscent; fruit-stalk 5·6–8·5 cm. long. Seeds 11–14 × 7–9·5 × 2·5–5 mm., ovate-elliptic in outline, compressed, the faces with sinuate markings and edges, margins grooved.

Botswana. N: Chobe-Zambezi confluence, fl. 11.iv.1955, *E.M. & W.* 1479 (BM). SE: Mochudi, fl. 1914, *Harbor* in *Rogers* 6646 (K). **Zambia.** C: Lusaka, Mt. Makulu Research Station, 17 km. S. of Lusaka, fl. 13.v.1956, *Angus* 1288 (K; LISC; SRGH). E: Luangwa R., fl. 25.iii.1955, *E.M. & W.* 1190 (BM; LISC; SRGH). S: Mazabuka, fl. & fr. immat. 28.ii.1963, *van Rensburg* 1510 (K; SRGH). **Rhodesia.** N: Urungwe, Chirundu Bridge, fl. & fr. 26.ii.1953, *Wild* 4049 (K; SRGH). W: Bulalima to Mangwe, Embakwe, fl. 6.v.1942, *Feiertag* (SRGH). C: Hartley, near Cedrela, *Hornby* 3452 (COI; SRGH). E: Umtali, Hillside Golf Club, fl. 16.xii.1951, *Chase* 4281 (BM; COI; K; LISC; SRGH). S: Ndanga, M'tilikwe R., fl. 25.i.1949, *Wild* 2729 (K; SRGH). **Malawi.** S: Port Herald, Makanga, fl. 19.iii.1960, *Phipps* 2560 (COI; SRGH). **Mozambique.** Z: Morrumbala, between Aguas Quentes and Mutarara, fl. & fr. immat. 4.v.1943, *Torre* 5293 (LISC). T: Tete, 3 km. on Changara road, fl. & fr. immat. 13.ii.1968, *Torre & Correia* 17556 (LISC). MS: Amatongas Forest, fl. 19.xi.1911, *Rogers* 10268 (BM; BOL; K). SS: Gaza, between Chibuto and Caniçado, fl. 2.vi.1959, *Barbosa & Lemos* 8561 (COI; LMA). LM: Namaacha, road to Goba, km. 5, *R. & A. Fernandes & Correia* 96 (COI; LMU).
Also in S. Africa (Transvaal, Natal), Swaziland and Tanzania. Woodland and wooded grassland, 0–1400 m.

12. **Momordica boivinii** Baill., Hist. Pl. 8: 407, t. 289–291 (1785); in Bull. Soc. Linn. Paris 1: 309 (1882).—C. Jeffrey in F.T.E.A., Cucurb.: 34, t. 2 fig. 3a–b, t. 3 fig. 1, 2 & 8 (1967).—Launert & Roessler in Prodr. Fl. SW. Afr. 94: 20 (1968).—Binns, H.C.L.M.: 43 (1968). Type from Kenya.

Raphanistrocarpus boivinii (Baill.) Cogn. in Engl., Pflanzenw. Ost-Afr. **C**: 397 (1895).—Burtt Davy, F.P.F.T.: 224 (1926). Type as above.
Raphanocarpus boivinii (Baill.) Chiov., Fl. Somala **1**: 181 (1929).—Meeuse in Bothalia **8**: 44 (1962). Type as above.

Stems prostrate or scandent, to 7·5 m., glabrous or sparsely and finely shortly pubescent, arising from a tuberous napiform rootstock. Leaf-lamina 1·4–7·5 × 0·8–6·7 cm., narrowly to very broadly ovate in outline, deeply cordate, plane or distinctly sinuate and denticulate, finely asperulous beneath, finely more or less densely asperulous or scabrid-punctate above, unlobed or usually shortly palmately 3–5-lobed, the central lobe largest, triangular, shortly acuminate, acute to obtuse, apiculate, the lateral shorter, more rounded. Petiole 0·4–5 cm. long, shortly pubescent. Tendrils simple. Flowers monoecious. ♂ flowers 1–8 in pedunculate bracteate subumbelliform inflorescences, sometimes accompanied by 1–2 ♀ flowers; peduncle 1·1–10 cm. long; bract 3–8 mm. long, narrowly lanceolate to linear; pedicels 5–20 mm. long. Receptacle-tube 2–5 mm. long, lobes 6–17 mm. long, narrowly ovate to lanceolate, acutely-acuminate, dark-coloured. Petals 1·3–2·8 cm. long, yellow to deep orange above, paler beneath, dark at the base, narrowly obovate. Female flowers solitary or 1–2 accompanying male in mixed inflorescence; when solitary, peduncles 1·7–6·4 mm. long and bract subulate, 4–7 mm. long; pedicel 2–6 mm. long; ovary 4·5–12 × 1–1·5 mm., fusiform, ribbed, finely pubescent on ribs; receptacle-tube 0·5 mm. long, lobes 2–5 mm. long, narrowly lanceolate, green; petals 0·8–1·6 cm. long, obovate. Fruit 2·3–10 × 0·4–0·9 cm., fusiform, green or striped with cream, becoming yellow; fruit-stalk 1·8–11 cm. long. Seeds 9–17 × 3–5 × 2–5 mm., ellipsoid, tumid, with terminal verrucose appendices; faces convex, shallowly sculptured, margins slightly flattened.

Botswana. N: Francistown, fl. 31.i.1926, *Rand* 6 (BM). **Zambia.** E: near Katete R., fl. 10.iii.1957, *Wright* 174 (K). **Rhodesia.** N: Shamva, fl. 12.i.1952, *Wild* 3739 (K; SRGH). W: Gwampa Forest Reserve, fl. & fr. 1.xii.1955, *Goldsmith* 54/56 (K; LISC; SRGH). C: Skipton, Marandellas, fl. 5.xii.1946, *Wild* 1615 (K; LISC; SRGH). E: Umtali, Nyamakari R., fl. & fr. 22.ii.1962, *Chase* 7629 (K; SRGH). S: Ndanga, M'tilikwe R., fl. 25.i.1949, *Wild* 2730 (K; LISC; SRGH). **Malawi.** C: Dedza, Mua, fl. & fr. 6.i.1965, *Banda* 613 (COI; K; SRGH). S: Limbe, fl. 8.i.1948, *Goodwin* 3 (BM). **Mozambique.** N: Nampula, fl. & fr. 5.v.1937, *Torre* 1411 (COI; LISC). Z: between Gúruè and Ile, fl. 19.x.1949, *Barbosa & Carvalho* 4534 (LMA). MS: Manica, Dombe, Serra de Machango, fl. & fr. 26.xi.1965, *Pereira & Marques* 932 (LMU). LM: Namaacha, Changalane, fl. & fr. 10.v.1969, *Balsinhas* 1484 (LMA).
Eastern Africa from Ethiopia to Tanzania, Angola, S. Africa (Transvaal, Natal), SW. Africa. Woodland, wooded grassland and riverine fringes, mostly on sandy soils, 150–1400 m.

13. **Momordica kirkii** (Hook. f.) C. Jeffrey in Kew Bull. **15**: 357 (1962); in F.T.E.A., Cucurb. 35, t. 3 fig. 7 (1967).—Binns, H.C.L.M.: 43 (1968). Types: Mozambique, between Sena and Lupata, and near Shigogo, *Kirk* (K, syntypes).
Raphanocarpus kirkii Hook. f. in Hook., Ic. Pl. **11**: 67, t. 1084 (1871); in F.T.A. **2**: 541 (1871).—Gomes e Sousa in Bol. Soc. Estud. Col. Moçamb. **29–32**: 92 (1936). Syntypes as above.

Stems slender, prostrate or scandent to 2m., hispid-setose, arising from tuberous rootstock. Leaf-lamina 1·6–6·3 × 2·1–7·6 cm., broadly ovate or pentagonal in outline, deeply cordate, sinuate-denticulate, sparsely pilose-setose on veins beneath, sparsely pilose and shortly hispid or punctate above, palmately shortly 5-lobed, lobes triangular, shortly acuminate, obtuse, apiculate, the central largest. Petiole 1·2–12·6 cm. long, hispid-setose. Tendrils simple. Flowers monoecious. ♂ flowers (1)3–4, arising from apex of petiole in basal sinus of leaf-lamina, the peduncle adnate to the petiole except for apical 2–3 mm.; bract 4–14 mm. long, ovate-lanceolate or narrowly elliptic, acuminate; pedicels 3–13 mm. long. Receptacle-tube 2–3 mm. long, lobes 5–11 mm. long, ovate-lanceolate or lanceolate, acutely-acuminate, blackish-green. Petals 1·4–2·5 cm. long, orange, obovate-obcuneate, bilobed, retuse, apiculate. ♀ flowers solitary, pedunculate; peduncles 0·3–2·4 cm. long, entirely free from the petioles; ovary 5–21 × 1·2 mm., slender, rostrate, ridged, densely villous or pilose; receptacle-tube very short, lobes 2–4 mm. long, linear-lanceolate, acute; petals 0·6–1·1 cm. long. Fruit 1·8–3·2 × 0·4–0·6 cm., fusiform, ribbed, pubescent especially on ribs, 1–2-seeded; fruit-stalk

1–3·6 cm. long. Seeds c. 12 × 3·5 × 3·5 mm., subfusiform; faces rounded, deeply sculptured, margins grooved.

Caprivi Strip. Singalamwe, fl. 1.i.1959, *Killick & Leistner* 3250 (K; SRGH). **Zambia.** B: Sesheke, fl. 19.xii.1952, *Angus* 964 (K). N: Mpika Distr., Mfuwe, fr. 29.i.1969, *Astle* 5440 (K). S: near Kazangula, fl. 15.iii.1956, *Munro* (K; PRE). **Rhodesia.** N: Kariba, Chirundu, fl. 15.iii.1956, *Simon* 702 (K; LISC; SRGH). W: Wankie, Deka R., fl. 21.vi.1934, *Eyles* 7967 (K; SRGH). E: Melsetter, Odzi R., Hot Springs, fl. & fr. 24.ii.1952, *Chase* 4381 (BM; K; LISC; SRGH). **Malawi.** S: Shire R., 16° S., fl. & fr. 1863, *Kirk* (K). **Mozambique.** T: Chimage, fl. & fr. iii., *P. Guerra* 92 (COI). MS: Chemba, Chiou, fl. & fr. immat. 11.iv.1962, *Balsinhas & Macuácua* 572 (COI; K; LMA).

Also in Tanzania. Semi-deciduous and deciduous woodland and wooded grassland, on sandy soils, 100–1000 m.

14. **Momordica calantha** Gilg in Engl., Bot. Jahrb. **34**: 351 (1904).—C. Jeffrey in F.T.E.A., Cucurb.: 39, t. 2 fig. 1 & 4a–b, t. 3 fig. 6 & 10 (1967). Type from Tanzania.

Stems scandent to 4 m., glabrous or shortly pubescent, arising from tuberous rootstock. Leaf-lamina 5·5–16·5 × 5·5–15 cm., ovate to broadly ovate in outline, cordate, sinuate-denticulate, finely pubescent on veins beneath, shortly hispid above, unlobed or usually palmately 3–5-lobed, central lobe largest, triangular to broadly ovate-triangular or rarely narrowly elliptic, shortly acutely-acuminate, apiculate, lateral lobes shorter. Petiole 1–6·3 cm. long, shortly hispid or scabrid. Tendrils 2-fid. ♂ flowers 5–24 in minutely bracteate umbelliform fascicles; peduncle 0·6–4·5 cm. long; bract up to 2·5 mm. long; pedicels 7–32 mm. long. Receptacle-tube 3–9 mm. long, lobes 3–8 mm. long, oblong, ovate-triangular or triangular, acute to rounded, more or less keeled, green or brownish-green. Petals 2·0–3·0 cm. long, white, black at the base, oblong, rounded, recurved. ♀ flowers solitary, shortly pedicellate; ovary c. 13 × 3–5 mm., ellipsoid, shortly tomentose; receptacle-tube 1 mm. long, lobes c. 8–9 mm. long, linear-lanceolate, acute, recurved; petals 2·1–2·3 cm. long, oblong, rounded. Fruit 5–6·7 × 3·5–4·6 cm., oblong-ellipsoid, smooth, more or less pubescent, bright red, indehiscent; fruit-stalk 3·5 cm. long. Seeds 15–18 × 8–9 × 4–6 mm., oblong, compressed; faces sculptured, margins grooved.

Mozambique. N: Malema, serra Meripa, fl. & fr. 5.ii.1964, *Torre & Paiva* 10473 (LISC).

Also in eastern Zaire, Uganda, Kenya and Tanzania. *Newtonia buchananii* semi-deciduous forest, 1000 m.

SPECIES NOT SUFFICIENTLY KNOWN

15. **Momordica sp. A.**

Stems scandent, sparsely finely pubescent, arising from perennial rootstock, when old becoming woody and grey-barked. Leaf-lamina 13·6 × 15·9 cm., very broadly ovate to subcircular in outline, deeply cordate, remotely sinuate-lobulate, denticulate, shortly pubescent on veins and scabrid-punctate beneath, punctate and almost glabrous above, palmately 7-lobed to below the middle, central lobe ovate in outline, shortly 3-lobulate, acutely-acuminate, apiculate, lateral lobes shorter. Petiole 3·8 cm. long, shortly finely sparsely pubescent. Tendrils 2-fid. ♂ flowers c. 20 in pedunculate racemiform inflorescences; peduncle 2 cm. long, apparently ebracteate at the apex; pedicels 9–18 mm. long. Receptacle-tube 7–8 mm. long, lobes 7–10 mm. long, very broadly ovate, shortly acuminate, rounded. Petals 3 cm. long, yellow, dark at the base, oblong, shortly acuminate, apiculate. ♀ flowers, fruits and seeds unknown.

Mozambique. N: Eráti, 12 km. from Namaqua towards Alua, fl. 8.i.1964, *Torre & Paiva* 9867 (LISC).

Not known from elsewhere. *Brachystegia* woodland, 280 m.

3. EUREIANDRA Hook. f.

Eureiandra Hook. f. in Benth. & Hook., Gen. Pl. **1**: 825 (1867).

Scandent softly ligneous shrubs or herbs. Rootstock perennial, tuberous. Leaves simple, petiolate. Tendrils simple. Flowers yellow or white, dioecious.

♂ flowers solitary or few to many in sessile or pedunculate axillary fascicles. Receptacle-tube short, obconic below, expanded above; lobes lanceolate, entire. Petals 5, free, entire. Stamens 5, 4 in 2 pairs, 1 single, all 1-thecous, or the paired stamens variously to completely united into 2 double 2-thecous stamens, free; filaments inserted in the mouth of the lower part of the receptacle-tube; connectives broad; thecae flexuous. ♀ flowers solitary; perianth similar to that of ♂ flower, receptacle smaller; staminodes 3 or 5, variously developed; ovary ellipsoid, rostrate, terete; ovules many, horizontal; stigma 3-lobed. Fruits ellipsoid, rostrate, fleshy, orange-red, indehiscent. Seeds tumid, margined; testa smooth or fibrillose.

A genus of c. 9 species, confined to tropical Africa.

Aerial stems, except at the very base, annual, not becoming ligneous - 1. *fasciculata**
Aerial stems perennial, becoming softly ligneous and developing a thin pallid papyraceous bark:
  Petals yellow or cream; anther-thecae hairy:
    Anther-thecae conspicuously hairy, hairs long, flexuous; petals yellow; bracts small, filiform - - - - - - - - - - 2. *lasiandra*
    Anther-thecae shortly hairy, hairs straight; petals cream; bracts leaf-like, lobed, stalked, 2–25 mm. long - - - - - - - 4. *eburnea*
  Petals white; anther-thecae glabrous - - - - - - 3. *leucantha*

1. **Eureiandra fasciculata** (Cogn.) C. Jeffrey in Kew Bull. **15**: 353 (1962); in F.T.E.A., Cucurb.: 43, t. 4 (1967).—Binns, H.C.L.M.: 42 (1968). TAB. **102**. Type: Mozambique, Tete, Boroma, *Menyhart* 931 (Z, holotype; K, isotype).
    Momordica fasciculata Cogn. in Bull. Herb. Boiss. **5**: 636 (1897).—Gomes e Sousa in Bol. Soc. Estud. Col. Moçamb. **29–32**: 92 (1936).—Brenan in Mem. N.Y. Bot. Gard. **8**, 5: 443 (1954). Type as above.
    Coccinia polyantha Gilg in Engl., Bot. Jahrb. **34**: 356 (1904). Syntypes from Tanzania.
    Coccinia petersii Gilg, loc. cit. (1904). Type: Mozambique, Manica e Sofala, Sena, *Peters* (B, syntypes †; K, ? isosyntype).

Slender scandent herb. Tuber subterranean, napiform, rough-barked. Stems annual, trailing or climbing to 3 m., finely pubescent; old basal stems arising from tuber becoming thickened, gnarled, with pallid papyraceous bark. Leaf-lamina 2·5–12 × 1·5–12 cm., bright green, membraneous, ovate to broadly ovate in outline, cordate, more or less asperulous or puberulous especially on the veins beneath, unlobed or usually and often deeply palmately 3–5-lobed; lobes broadly triangular or ovate to narrowly elliptic or linear, sometimes lobulate, entire or ± sinuate-dentate, apiculate, the central largest. Petioles 1–3 cm. long, shortly pubescent. ♂ flowers usually 2–6 in short fascicles at the nodes on leafless stems, less often solitary and sometimes axillary to the foliage leaves; pedicels 1–7·5 cm. long, finely pubescent. Receptacle-tube 2·5–7 mm. long, lobes 3–12·5 × 1–2·5 mm., lanceolate, acute, often acuminate, finely pubescent. Petals 1·8–3·5 × 1·4–2·5 cm., deep cream to orange-yellow, rarely white, obovate, rounded, apiculate. Anther-thecae minutely hairy. ♀ flowers solitary; peduncles 4–10 mm. long; ovary 6–22 × 0·5–2·5 mm., shortly pubescent; receptacle-tube 1·5–5 mm. long, campanulate, lobes 3–4·5 mm. long, lanceolate or lanceolate-subulate; petals 2·4–3·2 × 1·1–2·0 cm. Fruit 4·5–8·5 × 2·4–3·5 cm., ellipsoid or cylindrical, rostrate, terete; fruit-stalk stout, 0·5–2·5 cm. long. Seeds 6·5–10·2 × 6–7·5 × 4·5–6·2 mm., subpisiform, smooth.

Zambia. C: Feira road, 12 km. S. from Katondwe Mission, fl. 1.i.1973, *Kornaś* 2906 (K). **Rhodesia.** N: W. of Kanyemba, fl. 27.i.1966, *Müller* 224 (K; SRGH). E: Melsetter, Hot Springs near Odzi R., fl. 29.xii.1948, *Chase* 1409 (K; LISC; SRGH). S: Nuanetsi, Chikadziwa to Litschani's, xi.1956, *Davies* 2147 (K; SRGH). **Malawi.** S: Chikwawa, fl. 5.x.1946, *Brass* 17984 (K; SRGH). **Mozambique.** N: Imala, between Muecate and Imala, fl. 24.x.1948, *Barbosa* 2541 (LISC). Z: between Morrumbala and Derre, fl. 15.x.1963, *Gomes e Sousa* 4805 (K; LMU). T: between Tete and Casula, fl. 12.x.1943, *Torre* 6019 (LISC). MS: Gorongosa, Chitengo, Bela Vista, *Balsinhas* 1043 (COI; LMA). SS: c. 13 km. N. of Cheline, fl. 11.x.1963, *Leach & Bayliss* 11905 (COI; SRGH).
    Also in SE. Tanzania. Woodland and wooded grassland, 50–1050 m.

* Compare also 5 *sp.* **A** (p. 433).

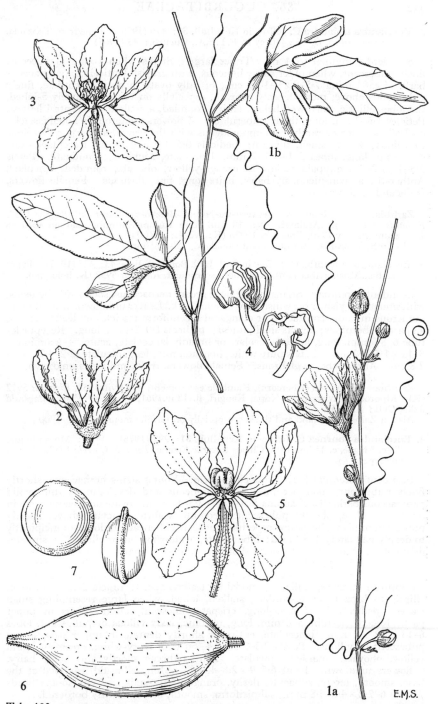

Tab. 102. EUREIANDRA FASCICULATA. 1a & 1b, habit showing variation in leaf-size at time of flowering (×1), 1a from *Milne-Redhead & Taylor* 7546, 1b from *Milne-Redhead & Taylor* 7497; 2, ♂ flower (×1); 3, ♂ flower showing stamens (×1); 4, stamen, dorsal and ventral views (×4), 2–4 from *Milne-Redhead & Taylor* 7497; 5, ♀ flower (×1) *Milne-Redhead & Taylor* 7498; 6, fruit (×1); 7, seed, face and lateral views (×3), both from *Milne-Redhead & Taylor* 7549. From F.T.E.A.

2. **Eureiandra lasiandra** C. Jeffrey in Kew Bull. **30**: 476 (1975). Type from Tanzania.
   Eureiandra sp. B.—C. Jeffrey in F.T.E.A., Cucurb.: 43 (1967).

Scandent perennial to 4 m. Tuber large. Stems when young herbaceous, finely pubescent, when older softly ligneous, with greyish lenticellate papyraceous bark. Leaf-lamina 7·5–8 × 9–10·5 cm., broadly ovate in outline, cordate, finely puberluous above, more densely beneath, especially on veins, palmately 5-lobed, the lobes elliptic, narrowed below, entire, rounded, apiculate, the central largest. Petioles 1·5–2 cm. long, finely puberulous. ♂ flowers numerous in nodal fascicles on leafless stems or rarely contemporaneous with the leaves, all parts puberulous to villous; bracts small, filiform; pedicels 0·5–4·0 cm. long. Receptacle-tube 4–11 mm. long, lobes 5–14 × 1–2 mm., lanceolate, acute, ± acuminate. Petals 2·5–4 × 1·3–3 cm., pale yellow to orange-yellow, obovate, rounded, apiculate. Anther-thecae conspicuously hairy, hairs long, fine, flexuous. Female flowers, fruits and seeds unknown.

Zambia. N: Mporokoso, Mweru-wa-Ntipa, Bulayo-Sumbu road, fl. 5.iv.1957, *Richards* 9044 (K). **Malawi.** N: 10 km. from Rumpi-Katumbi road on road to L. Kazuni, fl. 20.v.1970, *Brummitt* 10936 (K; MAL; SRGH).
Also in S. Tanzania. Bushland and woodland, 750–1350 m.

3. **Eureiandra leucantha** C. Jeffrey in Bull. Jard. Bot. Nat. Belg. **43**: 425 (1973). Type: Zambia, Mbala (Abercorn), Kambole Escarpment, *Richards* 5952 (K, holotype).

Scandent perennial. Stems at first light green, herbaceous, soon softly ligneous, with smooth greyish or brownish papyraceous bark. Leaves unknown. Male flowers numerous in nodal contracted or elongated racemiform fascicles on leafless stems, all parts puberulous; bracts small, linear; pedicels 1·0–3·4 cm. long. Receptacle-tube 6–8·5 mm. long, lobes lanceolate or broadly lanceolate, acute, ± acuminate, 8–13 × 1·5–3·2 mm. Petals pure white, obovate, rounded, apiculate, 3·2–4·7 × 1·8–2·8 cm. Anther-thecae glabrous. Female flowers, fruits and seeds unknown.

Zambia. N: Mbala (Abercorn), Kambole escarpment, fl. 24.viii.1956, *Richards* 5952 (K); Mporokoso, Mweru-wa-Ntipa, Kangiri, fl. 18.iv.1961, *Phipps & Vesey-Fitzgerald* 3283(COI; SRGH).
Also in Zaire. Bushland and thicket, especially on termite mounds, 1050–1200 m.

4. **Eureiandra eburnea** C. Jeffrey in Kew Bull. **31**: 292 (1976). Type: Mozambique, Tete, Mágoè, c. 15 km. from Chicoa towards Estima, *Torre & Correia* 17855 (LISC, holotype).

Scandent perennial herb; tuber napiform. Young stems herbaceous, shortly finely pubescent, soon becoming softly ligneous and developing a thin pallid papyraceous bark. Leaf-lamina 5·5–10·5 × 3·5–11·5 cm., ovate or broadly ovate in outline, cordate, shortly pubescent becoming scabrid-punctate above, shortly pubescent or hispid becoming asperulous or scabrid beneath, unlobed or incipiently to deeply palmately 3–5-lobed; lobes ovate, ovate-rhombic or elliptic, ± sinuate-denticulate, obtuse to acute, often shortly acuminate, apiculate, the central largest, undivided or incipiently to deeply laterally 3–5-lobulate. Petioles 1·5–5·8 cm., shortly and finely spiculate-pubescent. ♂ flowers usually numerous in sessile racemiform fascicles, axillary or nodal on leafless stems; bracts 2–25 mm. long, elliptic to broadly ovate-cordate, stalked, lobed, when large resembling small leaves; pedicels 1–8·5 cm. long, crispate-pubescent, often villous in upper part. Receptacle-tube 5–8 mm. long, obconic, laxly villous or pubescent, lobes 6–13 × 1·2–2·2 mm., lanceolate, entire or shortly dentate in upper part, acute, pubescent or villous. Petals 2·3–3·6 × 1·4–2·6 cm., ivory-white, cream or pallid yellow, obovate, rounded, apiculate. Anther-thecae shortly and finely hairy. ♀ flowers unknown. Fruit 6·5–9 × 2·8–4 cm., oblong-ellipsoid, rounded at the base, smooth, terete, subacute, fleshy, orange; fruit-stalk c. 2 cm. long. Seeds 7·8–9 × 6–7·2 × 4·8–5·8 mm., subpisiform, smooth, black, slightly bordered.

Zambia. E: Luangwa Valley Game Reserve South, fl. 8.iii.1967, *Prince* 343 (K). **Mozambique.** T: 20 km. from Changara towards Cuchumana, Mt. Nhampangué,fr. 22.v.1971, *Torre & Correia* 18560 (LISC).
Not known from elsewhere. Deciduous woodland, including mopane, and associated grasslands, 200–700 m.

SPECIES NOT SUFFICIENTLY KNOWN

5. Eureiandra sp. A.

Resembles *E. fasciculata*, differing in stems glabrous except at nodes and leaves almost glabrous except on main veins above and beneath. Flowers and fruits unknown.

Mozambique. SS: Vilanculos, Mapinhane-Fanhalouro road, 4·5 km. from Mapinhane, st. 21.xi.1968, *R. & A. Fernandes & Pereira* 242 (COI).

Known only from the very imperfect, sterile above-cited gathering, and perhaps identical with *Eureiandra sp. A.* of C. Jeffrey in F.T.E.A., Cucurb.: 41 (1967), an insufficiently-known species from the coastal area of Kenya and Tanzania.

4. ACANTHOSICYOS Hook. f.

Acanthosicyos Hook. f. in Benth. & Hook., Gen. Pl. **1**: 824 (1867).

Prostrate perennial herbs or erect much-branched spiny shrubs. Leaves simple, petiolate, or reduced to minute scales. Tendrils solitary or paired, spiniform. Flowers yellow, dioecious. ♂ flowers solitary or fasciculate. Receptacle-tube campanulate; lobes lanceolate to broadly ovate. Petals 5, almost free, entire. Stamens 3, 2 two-thecous, 1 one-thecous, or 5, all one-thecous; filaments inserted on upper part of receptacle-tube; connectives broad; thecae flexuous. ♀ flowers solitary; perianth similar to that of male flower, receptacle-tube shorter, cylindrical; staminodes 3 or 5, small or prominent; ovary spiny, the spines each terminating in a short seta; ovules many, horizontal; stigma 3- or 5-lobed. Fruit ellipsoid to subglobose, indehiscent, fleshy, many-seeded, covered with stout conical seta-tipped fleshy spines. Seeds elliptic in outline, subcompressed; testa smooth, hard, not bordered.

A genus of 2 species, confined to southern Africa.

Acanthosicyos naudinianus (Sond.) C. Jeffrey in Kew Bull. **15**: 346 (1962).—Launert & Roessler in Prodr. Fl. SW. Afr. **94**: 5 (1968). TAB. **103**. Type from S. Africa (Transvaal).

Cucumis naudinianus Sond. in Harv. & Sond., F.C. **2**: 496 (1862).—Eyles in Trans. Roy. Soc. S. Afr. **5**, 4: 49 (1916). Type as above.

Citrullus naudinianus (Sond.) Hook. f. in F.T.A. **2**: 549 (1871).—Burtt Davy, F.P.F.T. **1**: 230 (1926).—Hutch., Botanist in S. Afr.: 666 (1946).—Story in Mem. Bot. Survey S. Afr. **30**: 47, t. 56 (1958).—Meeuse in Bothalia **8**: 55 (1962).—R. & A. Fernandes in Mem. Junta Invest. Ultramar, Sér. 2, **34**: 99 (1962); in C.F.A. **4**: 277 (1970). Type as above.

Colocynthis naudinianus (Sond.) Kuntze, Rev. Gen. Pl. **1**: 256 (1891).—Leistner in Koedoe **2**: 170 (1949). Type as above.

Perennial herb. Root to 1 m. long, tuberous, fusiform. Stems annual, to 6 m., prostrate, ± hirsute, glabrescent. Tendrils solitary, spiniform. Leaf-lamina 3–18 × 2·5–14 cm., ovate to broadly ovate in outline, cordate, scabrid and ± pilose especially on the main veins above, rather softly asperulous-tomentose to strongly scabrid beneath, usually deeply palmately 5-lobed, the lobes lanceolate to elliptic or sometimes broadly so, usually deeply pinnately lobulate and often ± sinuate at the margins, acute, usually acuminate, apiculate, the central much the largest. Petioles 0·7–7·5 cm. long, hispid, scabrescent. ♂ flowers solitary; pedicels 0·3–2 cm. long, hispid or villous; receptacle-tube 3·5–6 mm. long, pale green, hispid, lobes 3·2–6 × 0·7–1·5 mm., remote, lanceolate, usually slightly expanded above the middle, acuminate, acute. Petals 1·4–2·5 × 0·9–1·3 cm., yellow to white, obovate-elliptic, rounded, apiculate. Stamens 3. ♀ flowers solitary; pedicels 2–8 cm. long; ovary 8–15 × 4–8 mm., ellipsoid; spines short, thick; receptacle-tube 3 mm. long, lobes 3–4 mm. long, subulate; corolla as in ♂ flowers; staminodes 3, small. Fruit 6–12 × 4–8 cm., ellipsoid or subglobose, rather glaucous green, becoming pale yellow when mature, covered with prominent seta-tipped spines; fruit stalk rather slender, c. 6 cm. long. Seeds 7·5–10 × 4–6 × 2·5–3 mm., pallid-stramineous.

Botswana. N: Kwebe Hills, fl. 5.ii.1898, *Lugard* 155 (GRA; K). SW: Chukudu, 435 km. N. of Molepolole, st. 25.vi.1955, *Story* 4983 (K; PRE; SRGH). SE: 6 km. N. of Murumush, fl. 17.ii.1960, *Wild* 5003 (K; SRGH). **Zambia.** B: Sesheke, fl. & fr.

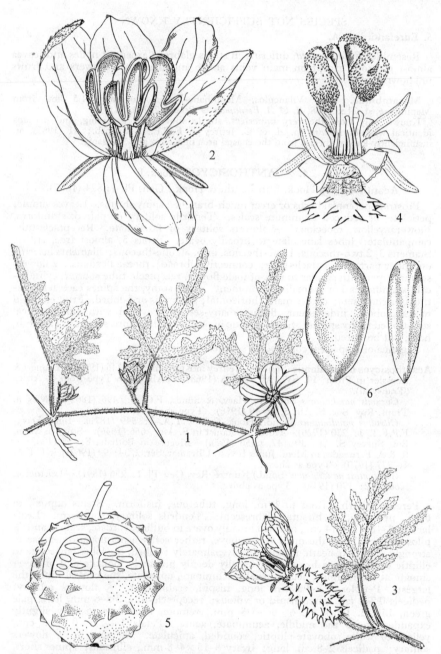

Tab. 103. ACANTHOSICYOS NAUDINIANUS. 1, flowering nodes, ♂ plant (× ½); 2, ♂ flower opened to show stamens (× 2), both from *White* 2046; 3, flowering node, ♀ plant (× ½) *Angus* 10421; 4, ♀ flower, style and staminodes (× 2); 5, young fruit, partially sectioned (× ½); 6, seeds, face and lateral views (× 3½), 4–6 from *Davidson* SRGH 27941.

26.xii.1952, *Angus* 1042 (FHO; K). S: Namwala, fl. 9.i.1957, *Robinson* 2069 (K; SRGH). **Rhodesia.** W: Wankie Game Reserve, Dett Road, fl. & fr. 16.ii.1956, *Wild* 4764 (COI; LISC; SRGH). S: Nuanetsi, between Tswiza and Nyala sidings, fl. & fr. 27.iv.1962, *Drummond* 7766 (K; LISC; SRGH). **Mozambique.** SS: Inhambane, Vilanculos, c. 4 km. from Mapinhane, fl. 19.xi.1948, *R. & A. Fernandes & Pereira* 144 (COI; LMU). LM: Lourenço Marques, fl. 9.xii.1897, *Schlechter* 11672 (BM; BOL; BR; COI; GRA; K; L).

Also in SW. Africa and S. Africa. Woodland, wooded grassland and grassland, on sandy soils, 900–1350 m.

For reference of this species to the genus *Acanthosicyos*, see C. Jeffrey in Kew Bull. **30**: 476 (1975). *Sicyos* being of masculine gender it seems only logical to treat its derivative *Acanthosicyos* also as masculine.

5. CITRULLUS Eckl. & 'Zeyh.

Citrullus Eckl. & Zeyh., Enum. Pl. Afr. Austr. Extratrop.: 279 (1836)
nom. conserv.
Colocynthis Mill., Gard. Dict. abridg. ed. 4 (1754) *nom. rej.*

Prostrate or scandent annual or perennial herbs. Leaves simple, petiolate, usually ± deeply lobed. Tendrils simple, proximally 2–4-fid, or absent. Probracts present. Flowers yellow, monoecious. ♂ flowers solitary. Receptacle-tube short, campanulate; lobes lanceolate, remote. Petals 5, shortly united, entire. Stamens 3, 2 double two-thecous, 1 single one-thecous; filaments inserted on receptacle-tube; connectives broad; thecae flexuous. ♀ flowers solitary; receptacle-tube very short, cylindrical, perianth otherwise similar to that of ♂ flower; staminodes 3; ovary ellipsoid or subglobose, hairy, smooth; ovules many, horizontal; stigma 3-lobed. Fruit subspherical or ellipsoid, greenish or yellow, often mottled, terete, firm-walled, fleshy, indehiscent. Seeds ovate in outline, compressed, smooth or slightly rough.

A genus of 3 species, in Africa and Asia.

Citrullus lanatus (Thunb.) Matsum. & Nakai, Cat. Sem. Spor. Hort. Bot. Univ. Imp. Tokyo: 30 (1916).—Mansfeld in Kulturpfl. Beih. **2**: 421 (1959).—Meeuse in Bothalia **8**: 57 (1962).—R. & A. Fernandes in Mem. Junta Invest. Ultramar., Sér. 2, **34**: 96 (1962); in C.F.A. **4**: 276 (1970).—Keraudren in Fl. Madag., Cucurb.: 155 (1966); in Fl. Cameroun **6**, Cucurb.: 151, t. 27 fig. 7–8 (1967).—C. Jeffrey in F.T.E.A., Cucurb.: 46, t. 5 (1967).—Launert & Roessler in Prodr. Fl. SW. Afr. **94**: 7 (1968). TAB. **104.** Type from S. Africa.
    Cucurbita citrullus L., Sp. Pl. **2**: 1010 (1753). Type a cultivated plant.
    Momordica lanata Thunb., Prodr. Pl. Cap.: 13 (1800). Type as for *Citrullus lanatus.*
    Citrullus vulgaris Eckl. & Zeyh., Enum. Pl. Afr. Austr. Extratrop.: 279 (1836).—Sond. in Harv. & Sond., F.C. **2**: 494 (1862).—Hook. f. in F.T.A. **2**: 549 (1871).—Ficalho, Pl. Ut. Afr. Port.: 190 (1884).— Eyles in Trans. Roy. Soc. S. Afr. **5**, 4: 498 (1916).—R.E.Fr., Wiss. Ergebn. Schwed. Rhod.-Kongo-Exped. **1**: 312 (1916).—Burtt Davy, F.P.F.T. **1**: 229 (1926).—Bremek. & Oberm. in Ann. Transv. Mus. **16**, 2: 436 (1935).—Gomes e Sousa in Bol. Soc. Estud. Col. Moçamb. **29–32**: 92 (1936).—Hutch., Botanist in S. Afr.: 547, 666 (1946).—Suesseng. & Merxm. in Proc. Trans. Rhod. Sci. Ass. **43**: 60 (1951).—Martineau, Rhod. Wild Fl.: 84 (1953).—Williamson, Useful Plants Nyasal.: 35 (1956).—Story in Mem. Bot. Survey S. Afr. **30**: 45, t. 57 (1958).—Nair, Sel. Fam. Zamb. Fl. Pl.: 50 (1967). Type as for *Cucurbita citrullus.*
    Colocynthis citrullus (L.) Kuntze, Rev. Gen. Pl. **1**: 256 (1891).—Leistner in Koedoe **2**: 170 (1959). Type as for *Cucurbita citrullus.*

Annual herb. Stems prostrate or scandent to 10 m., ± villous, glabrescent. Tendrils 2–3-fid. Leaf-lamina 5–20 × 3·5–19 cm., ovate or narrowly ovate in outline, cordate, ± hairy (especially on the veins beneath), becoming scabrid-punctate, usually deeply palmately 3–5-lobed, the lobes elliptic in outline, shallowly to usually deeply ± pinnately lobulate, subentire or obscurely sinuate-denticulate, rounded to subacute, apiculate, the central much the largest. Petioles 2–18·5 cm. long, ± hairy or villous, sometimes scabrescent. Probracts 4–18 mm. long, obovate-spathulate. ♂ flowers on ± villous 12–45 mm. long pedicels; receptacle-tube 2·5–5 mm. long, pale green, lobes 2·5–5 mm. long, lanceolate. Petals 0·7–1·9 × 0·4–1·4 cm., obovate, rounded, apiculate. ♀ flowers on 3–45 mm. long pedicels; ovary 6–15 × 4–8 mm., ellipsoid or subglobose, villous; receptacle-tube

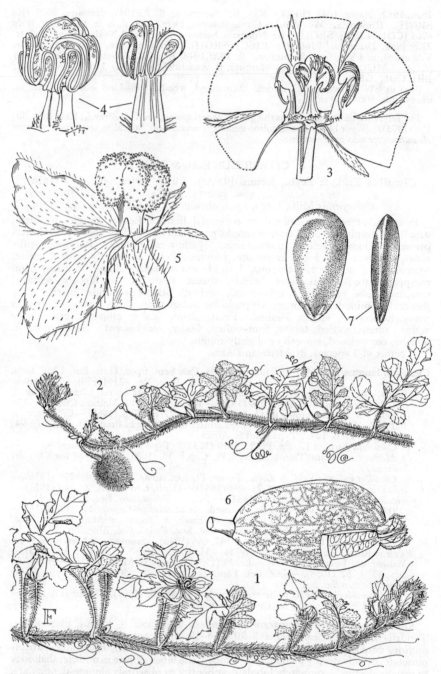

Tab. 104. CITRULLUS LANATUS. 1, habit, with ♂ flower (×½); 2, habit, with ♀ flower (×½), both from *Story* 4999; 3, ♂ flower, androecium (×2); 4, double stamen, dorsal and ventral views (×4); 5, ♀ flower partially dissected (×4); 6, fruit (×1); 7, seed, face and lateral views (×3), 3–7 from *Polhill* 1298.

1·5–2 mm. long, lobes 2·5–5·5 mm. long;   corolla as in ♂ flowers.   Fruit of wild plants 1·5–20 cm. in diameter, subglobose, greenish mottled with darker green, of cultivated plants up to 60 × 30 cm., subglobose or ellipsoid, green or yellowish, concolorous or variously mottled or striped; fruit-stalk 2–5 cm. long.   Seeds c. 9–11 × 5–6 × 2·5–2·7 mm., ovate-elliptic in outline, smooth or slightly verrucose, dark or pale coloured, often mottled, sometimes bordered.

Botswana. N: Kwebe Hills, fl. 5.ii.1898, *Lugard* 154 (K).   SW: Chukudu Pan, fr. 20.vi.1955, *Story* 4937 (K; PRE; SRGH).   SE: Mochudi, Phutodikobo Hill, st. 17.iii.1967, *Mitchison* 77 (K).   **Zambia.** N: Bulaya Pans, fl. 26.iv.1957, *Vesey-FitzGerald* 1187 (SRGH).   E: Lundazi River, fl. 19.xi.1958, *Robson* 680 (BM; K; LE; LISC; P; PRE; SRGH).   S: Zimba, Makoli, fl. vi.1909, *Rogers* 8270 (K; SRGH).   **Rhodesia.** N: Darwin, near R. Nyatundi, *Phipps* 2341 (COI; SRGH).   W: Wankie National Park, W. of Main Camp, on road to Dom, fl. 23.i.1969, *Rushworth* 1441 (COI; K; SRGH).   C: Hartley, Umfuli River, fl. 7.xii.1960, *Rutherford-Smith* 423 (COI; SRGH).   **Malawi.** S: Matikili, fl. 28.i.1938, *Lawrence* 627 (K).   **Mozambique.** N: near Nampula, fl. 18.x.1952, *Barbosa & Balsinhas* 5165 (COI; LMA).   T: Tete, km. 3 on Changara road, fl. 13.ii.1968, *Torre & Correia* 17559 (LISC).   SS: between Chibuto and Chongoene, fl. 6.xi.1952, *Myre* 1288 (LMA).   LM: Lourenço Marques, fl. iii.1944, *Pimenta* 190 (LISC).

Cultivated and adventive in the warmer parts of the world; native of the Kalahari region, including parts of the F.Z. area.   Grassland and bushland, often along water-courses, 50–1350 m.

6. LAGENARIA Ser.

Lagenaria Ser. in Mém. Soc. Phys. Hist. Nat. Genève 3: 25, t. 2 (1825).

Adenopus Benth. in Hook., Niger Fl.: 372 (1849).

Sphaerosicyos Hook. f. in Benth. & Hook., Gen. Pl. 1: 824 (1867).

Vigorous scandent herbs.   Leaves simple; petiole with 2 apical lateral glands. Tendrils proximally 2-fid.   Flowers large, white, monoecious or dioecious, opening in the evenings.   ♂ flowers solitary or racemose.   Receptacle-tube obconic to cylindrical; lobes remote, narrow, usually glandular.   Petals 5, free or nearly so, entire.   Stamens 3, 2 double two-thecous, 1 single one-thecous, free; filaments inserted on the receptacle-tube; connectives broad; thecae flexuous. ♀ flowers solitary; receptacle-tube extremely short, perianth otherwise similar to that of ♂ flower; stigma 3-lobed.   Fruit greenish, terete, hard-shelled, fleshy, indehiscent.   Seeds compressed, oblong or ovate in outline, the faces with two flat submarginal ridges.

A genus of about 6 species, mostly tropical African.

Glands at apex of petiole prominent, patent, inserted just below base of leaf-lamina;
    ♂ flowers usually racemose; flowers usually dioecious; plants perennial:
    Receptacle-tube of ♂ flower 25–50 mm. long, cylindrical; seeds elliptic, rounded at
      the base - - - - - - - - - - 1. *breviflora*
    Receptacle-tube of ♂ flower 7–17 mm. long, obconic; seeds oblong, subtruncate at
      the base - - - - - - - - - - - 2. *sphaerica*
Glands at apex of petiole rather obscure, ± recurved, inserted at the base of the leaf
    lamina; ♂ flowers solitary; flowers monoecious; plants annual - 3. *siceraria*

1. **Lagenaria breviflora** (Benth.) Roberty in Bull. Inst. Fond. Afr. Noire, Sér. A, **16**: 795 (1954).—C. Jeffrey in Journ. W. Afr. Sci. Ass. **9**, 2: 90 (1965); in F.T.E.A., Cucurb.: 49, t. 6 fig. 2 (1967).—Keraudren in Fl. Cameroun 6, Cucurb.: 80, t. 15 fig. 1–4 (1967).—Binns, H.C.L.M.: 42 (1968).—R. & A. Fernandes in C.F.A. **4**: 271 (1970).   TAB. **105** fig. 7–8.   Type from Nigeria.

Adenopus breviflorus Benth. in Hook., Niger Fl.: 372 (1849).—Hook. f. in F.T.A. 2: 528 (1871).—Hiern, Cat. Afr. Pl. Welw. **1**, 2: 389 (1898).—R. & A. Fernandes in Mem. Junta Invest. Ultramar., Sér. 2, **34**: 47 (1962).   Type as above.

Perennial.   Stems annual, to 6 m., sparsely to densely puberulous, sometimes also with longer hairs, scandent.   Leaf-lamina 7·5–20 × 5·5 × 23 cm., broadly ovate in outline, cordate, glossy green and minutely asperulous or scabrid above, paler and shortly and sparsely to moderately puberulous or asperulous beneath, especially on the veins, palmately 3–5-lobed; lobes triangular to elliptic, sinuate-dentate, apiculate, the central largest.   Petioles 1·5–8 cm. long, finely and closely puberulous, sometimes also with sparse to dense longer hairs; glands prominent, patent.   Probracts 6–11 mm. long, narrow.   Flowers usually dioecious.   ♂ flowers racemose; peduncle 1–22 cm. long; bracts small, 2·5–8 mm. long, lobed, dentate;

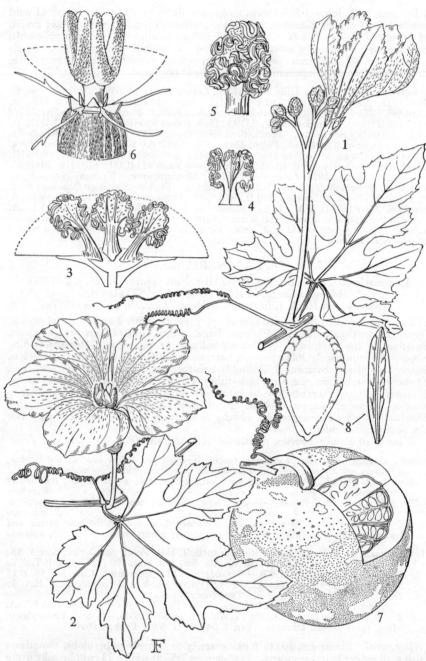

Tab. 105. LAGENARIA SPHAERICA. 1, flowering node, ♂ plant (× ½) *Leach & Rutherford-Smith* 10867 & *Barbosa & Lemos* 8555; 2, flowering node, ♀ plant (× ½) *Courtauld* SRGH 44100; 3 & 4, androecium (× 1); 5, stamen (× 2); 6, ♀ flower, details (× 2), 3–6 from *Drummond & Hemsley* 3439. LAGENARIA BREVIFLORA. 7, fruit, partly sectioned (× ½); 8, seed, face and lateral views (× 3), both from *Richards* 11434.

pedicels 0·5–8·5 cm. long; receptacle-tube 2·5–5 cm. long, cylindrical, finely velutinous, lobes narrow, linear, acute, often glandular, 4–9 mm. long. Petals 3·5–4 × 2·7–3·2 cm., white, obovate. Anthers oblong, included, coherent. ♀ flowers on 2–3 cm. long peduncles; ovary c. 15 × 8 mm., ellipsoid, tomentose; receptacle-lobes and petals similar to those of ♂ flowers. Fruit 8–9·5 × 6·5–8 cm., subglobose, smooth, glossy, dark green with small whitish spots and also larger scattered whitish patches, especially towards the apex; fruit-stalk stout, 2–4 cm. long. Seeds 8·5–11·5 × 5–6 × 2·0–2·4 mm., ovate-elliptic in outline, compressed, bordered.

Zambia. N: Shiwa Ngandu, fl. 10.i.1937, *Ricardo* 163 (BM). W: Fort Rosebery, fr. 28.viii.1952, *Angus* 327 (FHO; K). **Rhodesia.** E: Inyanga, Pungwe Valley, fl. 9.xi.1960, *Wild* 5270 (COI; K; SRGH). **Malawi.** S: Shire R., Elephant Marsh, fl. 1863, *Kirk* (K). **Mozambique.** N: sine loc., fl. immat. 20.ii.1942, *Hornby* 3585 (PRE). MS: Manica, Josi waterfalls, fl. immat. 16.vi.1962, *Chase* 7765 (COI; K; SRGH).
W. & C. tropical Africa from Guinée and Sudan south to Angola; Tanzania; also Brazil. Forest, often riverine, to 1200 m.

2. **Lagenaria sphaerica** (Sond.) Naud. in Ann. Sc. Nat. Bot., Sér. 5, **5**: 9 (1866).— Keraudren, Fl. Madag., Cucurb.: 104, t. 25 fig. 6–9 (1966).—C. Jeffrey in F.T.E.A., Cucurb.: 52, t. 6 fig. 7 (1967). TAB. **105** fig. 1–6. Syntypes from Natal.
    Luffa sphaerica Sond. in Harv. & Sond., F.C. **2**: 490 (1862). Syntypes as above.
    Lagenaria mascarena Naud. in Ann. Sc. Nat. Bot., Sér. 4, **18**: 187 (1863).—Meeuse in Bothalia **8**: 84 (1962). Type a cultivated plant of Comoro Is. origin.
    Sphaerosicyos meyeri Hook. f. in F.T.A. **2**: 532 (1871) *nom. illegit.* Syntypes as for *Lagenaria sphaerica.*
    Sphaerosicyos sphaericus (Sond.) Cogn. in A. & C. DC., Mon. Phan. **3**: 466 (1881).—Gomes e Sousa in Bol. Soc. Estud. Col. Moçamb. **29–32**: 92 (1936).— Mitchell in Puku **1**: 163, 182 (1963). Syntypes as for *Lagenaria sphaerica.*

Perennial. Stems annual, to 10 m. or more, minutely puberulous, prostrate or scandent. Leaf-lamina 5–19 × 4–21·5 cm., broadly ovate in outline, cordate, dark green and minutely asperulous above, paler and finely usually densely puberulous or hispidulous beneath, palmately 5-lobed; lobes shallow to deep, ovate to elliptic, obscurely to usually coarsely sinuate-dentate, often lobulate, obtuse to acute, long-apiculate, the central largest. Petioles 1–12 cm. long, minutely puberulous; glands prominent, patent. Probracts 6–14 mm. long, rather narrow. Flowers dioecious, fragrant. ♂ flowers racemose, rarely solitary; peduncle 1–20 cm. long; bracts small, 2·5–4 mm. long, lobed, dentate; pedicels 0·3–5 cm. long; receptacle-tube 0·9–1·7 cm. long, obconic below, expanded above, minutely puberulous, lobes 3–6 × 1·8–2·5 mm., remote, broadly lanceolate or triangular-lanceolate, acuminate, glandular. Petals 2·5–5·5 × 2–4·5 cm., obovate, rounded, white with green veins. Anthers oblong or ovate, exserted, free. ♀ flowers on 1·5–8·5 cm. long peduncles; ovary 12–20 × 6–15 mm., ellipsoid, densely tomentose; receptacle-lobes and petals similar to those of ♂ flowers. Fruit 7–11 × 6–10 cm., subglobose, smooth, deep green with small paler spots and also larger scattered paler patches; fruit-stalk 2·5–10 cm. long, stout, expanded at apex. Seeds 8·5–11·5 × 5–6 × 2– 2·5 mm., oblong, subtruncate and emarginate at the base, slightly narrowed towards the apex, the faces with 2 flat submarginal ridges.

Zambia. N: Chiengi, st. 12.x.1949, *Bullock* 1229 (K; SRGH). C: Luangwa Valley Game Reserve S., Mfuwe Camp, fl. 2.iv.1966, *Astle* 4774 (K; SRGH). E: Nsadzu Bridge, fl. 27.xi.1958, *Robson* 748 (K). S: Gwembe, 1.iv.1952, *White* 2387 (FHO; K). **Rhodesia.** N: Urungwe, Chewore R., fl. 4.iii.1958, *Goodier & Phipps* 251 (LMA; SRGH). C: Charter Distr., Devuli R. bridge E. end, fl. 15.iv.1963, *Chase* 7985 (COI; SRGH). E: Umtali, N. of Morningside, fl. 26.iii.1954, *Chase* 5209 (BM; COI; K; LISC; SRGH). S: Ndanga, Lundi R., Chipinda Pools, fl. 24.iv.1961, *Goodier* 1068 (COI; K; SRGH). **Malawi.** N: Songue to Karonga, fl. vi.1896, *Whyte* (K). C: Dedza, Ntaha-taha, fl. 11.iv.1969, *Salubeni* 1316 (K; SRGH). S: near Fort Johnston, Farringdon Road, fl. 4.iii.1955, *E.M. & W.* 869 (BM; LISC; SRGH). **Mozambique.** N: between Cuamba and Mutuáli, near R. Lúrio, fl. 24.iv.1961, *Balsinhas & Marrime* 422 (COI; LISC; LMA). MS: Cheringoma, Inhamitanga, fl. 25.ix.1944, *Simão* 106 (LISC; LMA). SS: Gaza, Vila de João Belo, Chipenhe, Régulo Chiconela, fl. 1.iv.1959, *Barbosa & Lemos* 8439 (COI; K; LISC; LMA). LM: between Bela Vista and Salamanga, near Porto Filipe, R. Maputo, fl. 13.viii.1948, *Gomes e Sousa* 3751 (COI).
E. Africa from Somali Republic to S. Africa (Cape Prov.), Madagascar and the Comoro Is. Riverine forest and thicket, 0–1700 m.

3. **Lagenaria siceraria** (Molina) Standley in Publ. Field Mus. Nat. Hist. Chicago, Bot. Ser. **3**: 435 (1930).—Keay in F.W.T.A. ed. 2, **1**: 206 (1954).—Meeuse in Bothalia **8**: 83 (1962).—Keraudren in Fl. Madag., Cucurb.: 103 (1966); in Fl. Cameroun **6**, Cucurb.: 74, t. 15 fig. 6–9 (1967).—C. Jeffrey in F.T.E.A., Cucurb.: 51, t. 6 fig. 9 (1967).—R. & A. Fernandes in C.F.A. **4**: 272 (1970). Type from Chile.

Cucurbita lagenaria L., Sp. Pl. **2**: 1010 (1753). Type a cultivated plant.

Cucurbita siceraria Molina, Sagg. Chil.: 133 (1782). Type as for *Lagenaria siceraria*.

Lagenaria vulgaris Ser. in Mem. Soc. Phys. Hist. Nat. Genève **3**: 25, t. 2 (1825).— Sond. in Harv. & Sond., F.C. **2**: 489 (1862).—Hook. f. in F.T.A. **2**: 529 (1871).— Ficalho, Pl. Ut. Afr. Port.: 186 (1884).—Hiern, Cat. Afr. Pl. Welw. **1**, 2: 391 (1898).—Burtt Davy, F.P.F.T. **1**: 223 (1926).—Gomes e Sousa in Bol. Soc. Estud. Col. Moçamb. **29–32**: 96 (1936).—Williamson, Useful Plants Nyasal.: 74 (1956).— Nair, Sel. Fam. Zamb. Fl. Pl.: 50 (1967). Type as for *Cucurbita lagenaria*.

Vigorous annual. Stems prostrate or scandent, softly villous. Leaf-lamina 3–40 × 4·5–40 cm., broadly ovate or reniform in outline, cordate, shortly and softly puberulous or pubescent, undivided or incipiently palmately 5–7-lobed, the lobes rounded, shallowly sinuate-dentate, apiculate. Petioles 2–30 cm. long, densely pubescent or villous; glands small, somewhat recurved, inserted at the base of the lamina. Probracts undeveloped. Flowers monoecious. ♂ flowers solitary; pedicels 7–31 cm. long, softly hairy or villous; receptacle-tube 1·1– 1·6 cm. long, obconic-cylindrical, villous, lobes remote, 2·5–7 × 1–2 mm., subulate-dentiform or triangular-dentiform. Petals 2–4·5 × 1–3·5 cm., white, obovate. Anthers oblong, included. ♀ flowers on 6–10 cm. long peduncles; ovary 10– 25 × 5–8 mm., subglobose to ellipsoid or cylindrical, densely villous; receptacle-lobes and petals similar to those of ♂ flowers. Fruits large, up to 1 m. long and 20 cm. across, subglobose to cylindrical, often ± biventricose or lageniform, green, becoming yellowish when mature. Seeds 7–20 mm. long, oblong, truncate and emarginate at the base, with 2 flat facial ridges, in some variants rather irregular and rugose.

Zambia. B: near Senanga, fl. 1.viii.1952, *Codd* 7323 (BM; PRE). C: Lusaka, fl. 5.iii.1957, *Noak* 189 (SRGH). **Rhodesia.** N: Mtoko, N. bank of Ruenya R., fl. xii.1965, *Wild* 7483 (COI; K; SRGH). E: Umtali, fr. 14.viii.1953, *Chase* 5066 (SRGH). **Malawi.** N: Mzimba Distr., Katete Mission, Champira, fl. 24.iv.1974, *Pawek* 8500 (K; MAL; MO). **Mozambique.** N: Maniamba, fl. 21.v.1948, *Pedro & Pedrógão* 3782 (LMA). SS: Sul do Save, fl. iii.1944, *Pimenta* 185 (LMA). LM: Lourenço Marques, fl. i.1944, *Pimenta* 187 (LISC).

Pantropical. Cultivated for the fruits used as vegetables and as containers, perhaps nowhere truly wild in the F.Z. area.

7. RAPHIDIOCYSTIS Hook. f.

Raphidiocystis Hook. f. in Benth. & Hook., Gen. Pl. **1**: 828 (1867).—Keraudren & C. Jeffrey in Bull. Jard. Bot. Nat. Belg. **37**: 319–328 (1967).

Scandent herbs. Leaves simple, petiolate. Tendrils simple. Probracts present. Flowers yellow, monoecious. ♂ flowers in sessile axillary fascicles, bracteate. Receptacle-tube short, obconic-campanulate; lobes lanceolate to broadly foliaceous, often pinnately divided into narrow or ± expanded often deflexed lobes. Corolla gamopetalous, 5-lobed, lobes entire. Stamens 3, all two-thecous; filaments inserted on upper part of receptacle-tube; thecae flexuous. ♀ flowers fewer than ♂ and coaxillary with them; perianth similar to that of ♂ flower, the receptacle-lobes usually smaller and less divided; ovary densely setose; ovules many, horizontal; stigma 3-lobed. Fruit subglobose to ellipsoid-cylindrical, fleshy, densely echinate with multicellular brownish setae, dehiscent by 10 apical valves and extruding seeds and placentae. Seeds asymmetrically broadly ovate in outline, compressed; testa smooth, bordered.

A genus of 5 species, 4 in tropical Africa, 1 in Madagascar.

Raphidiocystis chrysocoma (Schumach.) C. Jeffrey in Kew Bull. **15**: 360 (1962); in Journ. W. Afr. Sci. Ass. **9**, 2: 89 (1965); in F.T.E.A., Cucurb.: 53, t. 7 (1967).— Killick & Meeuse in Bothalia **8**: 79 (1962).—R. & A. Fernandes in Mem. Junta Invest. Ultramar., Sér. 2, **34**: 106 (1962); in C.F.A. **4**: 281 (1970).—Keraudren & C. Jeffrey in Bull. Jard. Bot. Nat. Belg. **37**: 325, t. 1 fig. H, t. 2 (1967). TAB. **106**. Type from Ghana.

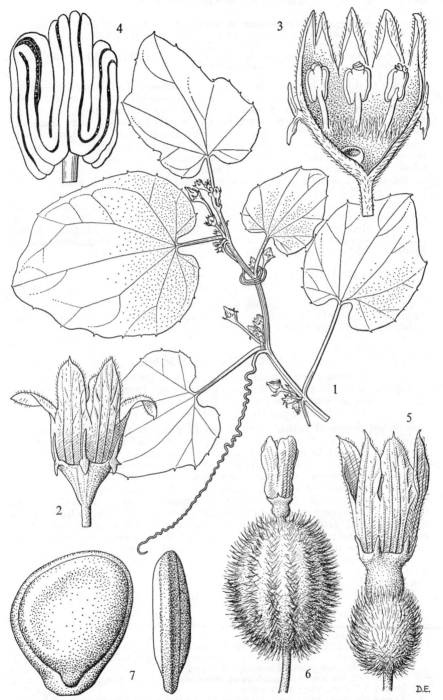

Tab. **106**. RAPHIDIOCYSTIS CHRYSOCOMA. 1, habit (× ⅔); 2, ♂ flower (×2); 3, ♂ flower opened to show stamens (×2); 4, stamen, anther-thecae (×8); 5, ♀ flower (×2); 6, fruit (×1); 7, seed, face and lateral views (×4), all from *Drummond & Hemsley* 3163. From F.T.E.A.

Cucumis chrysocomus Schumach. in Schumach. & Thonn., Beskr. Guin. Pl.: 427 (1827) non sensu Welw. in Ann. Consel. Ultramar **1**: 589 (1859) nec sensu Hiern, Cat. Afr. Pl. Welw. **1**, 2: 396 (1898). Type as above.

Raphidiocystis welwitschii Hook. f. in F.T.A. **2**: 554 (1871).—Hiern, Cat. Afr. Pl. Welw. **1**, 2: 400 (1898). Syntypes from Angola.

Stems to 6 m., pubescent, also ± pilose, scandent. Leaf-lamina 3·5–13 × 3–9 cm., ovate in outline, cordate, remotely apiculate-dentate at the margins, shortly asperulous and rather sparsely pilose, especially on the veins, above, subglabrous to tomentose beneath, unlobed or shortly palmately 3–5-lobed, lobes triangular, the central largest, ± acuminate, apiculate. Petioles 1–6·5 cm. long, densely pilose, hairs ± ascending. Probracts c. 4 mm. long, small, elliptic. ♂ flowers 4–6 in each fascicle; bracts small, to c. 1 cm. long, leaf-like; pedicels 1–3 cm. long, pilose; receptacle-tube 4–6 mm. long, lobes 2–4 mm. long, triangular-lanceolate or lanceolate, undivided or with a few marginal lobes. Corolla c. 1–1·5 cm. long, the lobes united in lower half. ♀ flowers 1–2 in each fascicle; pedicels 0·5–6·5 cm. long; ovary 7–15 × 3–11 mm., ellipsoid, densely setose; receptacle-lobes up to 3 mm. long, triangular-lanceolate or subulate, undivided. Fruits 3–5 × 1·2–3·5 cm., subglobose to ellipsoid, reddish, densely echinate with long, slender, reddish-brown or reddish-yellow setae; fruit-stalk 1–7·5 cm. long, rather slender. Seeds 6–7 × 4·5–6 × 2 mm., broadly and obliquely ovate in outline.

Rhodesia. E: Melsetter, Lusitu R., fr. 24.iii.1962, *Wild* 5734 (COI; K; SRGH).

W. tropical Africa from Guinée to Togo; also Zaire, Uganda, E. Tanzania, N. Angola. Riverine forest, 300 m.

The Rhodesian locality extends the previous known range of this species in eastern Africa by some 1200 km. to the south. This disjunction may well be reduced, however, by intermediate finds in suitable habitats in the F.Z. area.

8. COCCINIA Wight & Arn.

Coccinia Wight & Arn., Prodr. Pl. Ind. Or.: 347 (1834).
Cephalandra Eckl. & Zeyh,. Enum. Pl. Afr. Austr. Extratrop.: 280 (1836).
Physedra Hook. f. in Benth. & Hook., Gen. Pl. **1**: 827 (1867).
Staphylosyce Hook. f., tom. cit.: 828 (1867).

Scandent perennial herbs. Leaves simple, petiolate or sessile, usually very variable in shape. Tendrils simple or proximally 2-fid. Probracts present. Flowers creamy-white to deep orange-yellow, sometimes with a pinkish tinge, dioecious. ♂ flowers solitary, clustered or shortly racemose. Receptacle-tube usually short and broad, obconic-campanulate; lobes entire, often small, remote and dentiform, sometimes larger. Corolla distinctly gamopetalous, 5-lobed, lobes entire. Stamens usually 3, all 2-thecous, less often two 2-thecous, one 1-thecous; filaments inserted on receptacle-tube, usually ± connate or at least coherent into a central column; anthers connate or coherent into a globose head, rarely free; connectives broad; thecae flexuous. ♀ flowers solitary or rarely racemose; ovary smooth; ovules many, horizontal; stigma 3-lobed. Fruit globose to ellipsoid-rostrate or cylindrical, usually terete, thin-walled, fleshy, bright red when mature. Seeds ovate in outline, ± compressed; testa fibrillose or minutely rugulose, pale-coloured.

An Old World genus of c. 30 species, all but 1 confined to Africa.

1. Tendrils all or some 2-fid - - - - - - - - - -	2
— Tendrils all simple, unbranched - - - - - - - -	6
2. Corolla 2·5 cm. or more long - - - - -·- - - -	3
— Corolla not exceeding 2·5 cm. in length - - - - - -	4
3. Leaves densely hispidulous on veins beneath; ♂ receptacle-lobes 12–25 mm. long, ♀ 6·5–19 mm. long; fruit distinctly 10-angled - - - -	3. *schliebenii*
— Leaves subglabrous to hispidulous on veins beneath; ♂ receptacle-lobes 4–10 mm. long, ♀ 3–7 mm. long; fruit terete - - - - - -	4. *grandiflora*
4. Leaves glabrous except for puberulous petioles and veins, ± coriaceous; ♂ racemes rather dense, congested - - - - - - - - - -	5
— Leaves ± setulose, especially on petioles and veins, membranous; ♀ flowers and fruits solitary; ♂ racemes rather lax - - - - -	5. *fernandesiana*
5. Pedicels short and stout, subtended by small thick fleshy bracts; ♀ flowers racemose; fruits 2–4 cm. long - - - - - - - - - -	1. *barteri*

— Pedicels ebracteate; ♀ flowers solitary; fruits 10–19·5 cm. long -   2. *mildbraedii*
6. Leaves sessile, glabrous - - - - - - - -   8. *sessilifolia*
— Leaves evidently petiolate, the petiole if short usually ± hispid or setulose   7
7. Corolla exceeding 2·5 cm. in length, the lobes ovate-lanceolate, acute, apiculate
   4. *grandiflora*
— Corolla not exceeding 2·5 cm. in length, or, if so, then the lobes narrowly ovate to
   ovate, rounded to acute, apiculate - - - - - - - -   8
8. Petioles and veins of lower leaf-surface glabrous, hispid or setulose with recurved,
   straight or ascending hairs - - - - - - - - -   9
— Petioles and veins of lower leaf-surface puberulous with fine crispate hairs
   13. *sp. B*
9. Petioles glabrous or with a very few minute hairs; ♂ receptacle-lobes triangular,
   acute; ♂ flowers racemose; ovary 20–25 mm. long, glabrous -   6. *subglabra*
— Petioles setulose, at least on the upper margins, with stiff straight bristles, or hispid
   with sparse to dense recurved, straight or ascending hairs, or, if ± glabrous, then
   ♂ flowers solitary and ovary 5–13 mm. long - - - - - -   10
10. Leaves glabrous beneath; petiole setulose or pubescent only on upper margins   11
— Leaves ± setulose or hispid beneath, at least on the veins; petioles generally sparsely
   to densely setulose or hispid; ♀ flowers solitary; corolla-lobes ovate, obtuse to acute,
   apiculate - - - - - - - - - - - -   12
11. Corolla 1·2–1·4 cm. long; flowers racemose or clustered - - -   12. *sp. A*
— Corolla 1·5–3·2 cm. long; flowers solitary - - - - -   11. *grandis*
12. Petioles sparsely to densely hispid with rather slender, recurved, straight or usually
   adpressed-ascending hairs; fruits apically ± rostrate, acute - -   7. *adoensis*
— Petioles sparsely to densely setulose or spiculate with rather stout straight spreading
   hairs; fruits not rostrate, or if so, then stems spreading-setulose on ridges   13
13. Seeds lenticular, with thin margins and slightly convex faces; hairs at bases of
   filaments and staminodes eglandular, with tapered apices - -   5. *fernandesiana*
— Seeds flattened, with thick 2-grooved margins and flat faces; hairs at bases of
   filaments and staminodes glandular, with clavate apices - - - -   14
14. Leaves distinctly petiolate, the petiole usually longer than the depth of the basal sinus
   10. *rehmannii*
— Leaves (except the lowermost) subsessile, the petiole usually shorter than the depth
   of the basal sinus - - - - - - - - - -   9. *senensis*

1. **Coccinia barteri** (Hook. f.) Keay in Kew Bull. **8**: 82 (1953); in F.W.T.A. ed. 2, **1**:
215 1954).—C. Jeffrey in F.T.E.A., Cucurb.: 60 (1967).—Keraudren in Fl.
Cameroun **6**, Cucurb.: 128, t. 25 fig. 4–7 (1967).—R. & A. Fernandes in C.F.A.
4: 280 (1970). Type from Nigeria.
   Staphylosyce barteri Hook. f. in Benth. & Hook., Gen. Pl. **1**: 828 (1867); in
F.T.A. **2**: 555 (1871). Type as above.
   Physedra heterophylla Hook. f., tom. cit.: 553 (1871).—Hiern, Cat. Afr. Pl. Welw.
1, 2: 399 (1898).—R. & A. Fernandes in Mem. Junta Invest. Ultramar, Sér. 2, **34**:
104 (1962). Type from Angola.
   Physedra barteri (Hook. f.) Cogn. in A. & C. DC., Mon. Phan. **3**: 525 (1881).—
R. & A. Fernandes, tom. cit.: 105 (1962). Type as for *Coccinia barteri*.

Stems scandent, to 15 m., almost glabrous. Leaf-lamina 7·5–16 × 5–20 cm.,
broadly ovate or ovate in outline, cordate, weakly cordate or subtruncate, dark
green, glossy, rather coriaceous, glabrous except for fine hairs on veins above,
scabrid-punctate, unlobed and 3–5-angled or shallowly to deeply palmately 3–5-
lobed, the lobes triangular, ovate, broadly elliptic, ovate-oblong or oblong-elliptic,
plane and remotely denticulate to rather prominently sinuate-dentate, acute to
obtuse, apiculate, the central largest. Petioles 1–11 cm. long, puberulous.
Tendrils 2-fid. Probracts 3–10 mm. long, broadly spathulate. ♂ flowers in
short, stout, 3–25-flowered bracteate racemes; peduncle 1–1·5 cm. long, puberulous; pedicels very short. Receptacle-tube c. 15 mm. long, cylindrical below,
expanded above, lobes 3–8 mm. long, subulate. Corolla rich yellow or orange-
yellow, the lobes 1–2 × 0·5–1 cm., apiculate, united to above the middle. ♀
flowers in short, dense, 2–12-flowered bracteate racemes, rarely solitary; peduncle
0·3–1 cm. long; pedicels 1·5–7 mm. long; ovary 8–12·5 × 2–3 mm., ellipsoid,
glabrous; receptacle-tube 2–3·5 mm. long, campanulate-cylindrical, lobes small,
1·5–2 mm. long, dentiform; corolla as in ♂ flowers. Fruits 2–4 × 1·5–2·5 cm.
usually in racemose clusters, rarely solitary, ellipsoid, rounded, smooth, red; fruit-
stalk rather stout, 0·3–1 cm. long. Seeds 5 × 2·2–3 × 1–1·2 mm., ovate, compressed,
margined; testa ± fibrillose.

Zambia. W: Kitwe, Ichimpi, fr. 13.vi.1962, *Mutimushi* 179 (SRGH). **Rhodesia.**

E: Melsetter, Chirinda Forest, fr. ii.1962, *Goldsmith* 39/62 (COI; K; SRGH).
Mozambique. MS: Garuso, Bandula Forest, st. 10.ix.1957, *Chase* 6593 (K; SRGH).
W. tropical Africa from Senegal east to Sudan and south to Angola and Mozambique Lowland rain-forest and swamp forest, 700–1050 m.

2. **Coccinia mildbraedii** Harms in Notizbl. Bot. Gart. Berl. **8**: 492 (1923).—C. Jeffrey in F.T.E.A., Cucurb.: 61 (1967). Type from Zaire (Kivu).

Scandent to 10 m. or more. Leaf-lamina 6·5–17·5 × 7·5–19 cm., ovate to suborbicular in outline, cordate, glabrous except on main veins, punctate above and on main veins beneath, palmately 5-lobed, lobes triangular to ovate or elliptic, ± denticulate, acuminate, apiculate, the central largest. Petioles 2·4–9 cm. long, finely pubescent. Tendrils 2-fid. Probracts 2–8 mm. long, spathulate to suborbicular. ♂ flowers in solitary or paired 2–15-flowered racemes; peduncles 1·5–5·5 cm. long, axillary or at nodes on leafless terminal branches and then forming lax compound inflorescences; pedicels 2–9 mm. long. Receptacle-tube 3–6 mm. long, obconic, expanded above, lobes 1–4·5 mm. long, subulate. Corolla pale orange-yellow to brownish, the lobes 8–17 mm. long, united to middle or above. ♀ flowers solitary; peduncles 1·8–3·4 cm. long; ovary 20–32 mm. long, cylindrical; receptacle-tube 4–5 mm. long, lobes 1·5–4 mm. long, triangular to lanceolate; corolla 16–25 mm. long. Fruit 10–19·5 × 2–5 cm., red, cylindrical; peduncle 3–8·5 cm. long. Seeds 6–7·5 × 4–5 × 1·3–1·5 mm., ovoid, compressed; testa becoming ± fibrillose.

Malawi. N: Chitipa, Misuku Hills, Mughesse rain forest, fl. 29.xii.1970, *Pawek* 4216 (K; MAL).
E. tropical Africa from Uganda south to Tanzania and Malawi. Upland rain-forest, 1700 m.

3. **Coccinia schliebenii** Harms in Notizbl. Bot. Gart. Berl. **11**: 685 (1932).—C. Jeffrey in F.T.E.A., Cucurb.: 63 (1967). Type from Tanzania.

Stems vigorous, scandent to 12 m., puberulous. Leaf-lamina 8–24 × 8–28 cm., ovate or broadly ovate in outline, cordate, bullate, densely scabrid-punctate above, densely asperulous on veins beneath, palmately 5-lobed, the lobes triangular, ovate or lanceolate, sinuate-denticulate, rounded to obtuse, apiculate, the central largest. Petioles 3–12 cm. long, densely pubescent. Tendrils 2-fid. Probracts 3–5 mm. long. ♂ flowers solitary or in 2–6 cm. pedunculate bracteate clusters, usually with a co-axillary solitary flower; bracts small, 2–4 mm. long, rounded; pedicels of clustered flowers up to 20 mm. long, of solitary flowers up to 45 mm. long. Receptacle-tube 2–5 mm. long, broad and shallow, lobes 12–25 mm. long, ovate-acuminate. Corolla orange-yellow, green-veined, urceolate, the lobes 4·8–6·2 × 1·4–2·1 cm., united to the middle or above. ♀ flowers solitary; peduncle 20–34 mm. long; ovary 13–24 × 2·5–5 mm., pubescent, 10-ribbed; receptacle-tube 3–4 mm. long, lobes 6·5–19 mm. long; corolla similar to that of ♂ flower. Fruit 6–10 × 2–3 cms., ellipsoid-cylindrical, 10-angled, red; fruit-stalk expanded upwards. Seeds ovate, lenticular, compressed, rather small, unknown in mature condition.

Mozambique. N: Cabo Delgado, Macondes, fl. & fr. immat., 3.i.1964, *Correia* 92 (LISC).
Also in S. Sudan, S. Ethiopia, S. Tanzania. *Parinari* woodland, 800–850 m.

4. **Coccinia grandiflora** Cogn. in Engl., Bot. Jahrb. **21**: 211 (1895).—C. Jeffrey in F.T.E.A., Cucurb.: 64 (1967). Type from Tanzania.

Stems scandent or prostrate, vigorous, shortly pubescent or almost glabrous. Leaf-lamina 7–18 × 7–20 cm., broadly ovate in outline, cordate, deep green, glabrous or shortly pubescent on veins beneath and less so above, punctate, shallowly to deeply palmately 5-lobed, the lobes triangular or broadly ovate to elliptic, subentire to coarsely sinuate-dentate, rounded or obtuse, apiculate, the central largest. Petioles 3·5–9 cm. long, usually pubescent. Tendrils 2-fid, one branch often very weakly developed, rarely simple. Probracts 3–8·5 mm. long, spathulate. ♂ flowers in 1·5–7·5 cm. pedunculate bracteate racemes, usually co-axillary with a solitary flower; bracts small, 2–5 mm. long, glabrous, ± spathulate; pedicels of racemose flowers 2–20 mm. long, of solitary flowers 25–60 mm. long.

Receptacle-tube 5–7 mm. long, obconic below, expanded above, lobes 4–10 mm. long, narrowly triangular-lanceolate, acute. Corolla orange or apricot-yellow, the lobes 2·5–5 × 1–1·5 cm., narrowly ovate, acute, apiculate, united in lower half. ♀ flowers solitary or paired; peduncle 8–22 mm. long; ovary 20–36 × 1·5–2·5 mm., cylindrical, glabrous; receptacle-tube c. 3–5 mm. long, campanulate, lobes 3–7 mm. long, lanceolate, acute; corolla often larger than in ♂ flowers, otherwise similar. Fruit 8–30 × 1·5–2·5 cm., cylindrical, terete, bright red; fruit-stalk stout, 4–6·5 cm. long, expanded upwards. Seeds 4·5–4·7 × 2·5–2·7 × 1·3 mm., ovate, compressed, obscurely bordered.

Rhodesia. E: Umtali, S. slope of Murahwa's Hill, fr. 1.v.1963, *Chase* 8008 (COI; K; LISC; SRGH). **Malawi.** C: Nchisi Mt., fl. 22.ii.1959, *Robson & Steele* 1711 (K). **Mozambique.** MS: Inhamitanga, st. 5.vii.1946, *Simão* 728 (LMA).
Also in SE. Kenya and Tanzania. Rain-forest, 900–1460 m.

5. **Coccinia fernandesiana** C. Jeffrey in Kew Bull. **30**: 478 (1975). Type: Mozambique, Niassa, Eráti, between Namapa and Ocúa, *Lemos & Macuácua* 29 (COI, holotype; K, LISC, LMA, SRGH, isotypes).
    Coccinia senensis sensu C. Jeffrey in F.T.E.A., Cucurb.: 66 (1967) pro parte quoad plantas tanganyikenses, non (Klotzsch) Cogn.

Stems scandent to 2 m., sparsely setulose, becoming white-punctate when older. Leaf-lamina 5·5–17 × 6–12·5 cm., broadly ovate in outline, cordate, scabrid-punctate above, setulose on veins or almost glabrous beneath, shallowly to deeply palmately 3–5-lobed, the lobes triangular, ovate, obovate-elliptic or oblong-lanceolate, denticulate, often sharply so, or coarsely sinuate-dentate, sometimes lobulate, obtuse to acute, apiculate, the central largest. Petioles 0·4–4 cm. long, setulose or hispid, usually short except towards the base of the plant. Tendrils 2–fid, often with one branch very weakly developed, or simple. Probracts 3–7 mm. long, oblanceolate-spathulate. ♂ flowers in pedunculate racemes, co-axillary with solitary flower; peduncles 1·3–5·5 cm. long, shortly setulose; pedicels shortly setulose, those of racemose flowers 4–15 mm. long, of solitary flowers 15–35 mm. long. Receptacle-tube 3·5–5 mm. long, campanulate, shortly setulose, lobes 3–4·5 mm. long, lanceolate, recurved. Corolla yellow or orange-yellow, campanulate, the lobes 1·5–3 × 1–1·5 mm., united to above the middle. ♀ flowers solitary; pedicels 5–13 mm. long, sparsely setulose; ovary 8–11 × 2·5–3 mm., ellipsoid; receptacle-tube 2·5–3 mm. long, narrowly campanulate, lobes 2·3–3·5 mm. long, lanceolate-subulate, narrow, recurved; corolla similar to that of ♂ flower. Fruit 2·5–4 × 1·3–1·8 cm., ellipsoid, terete, obtuse, red; fruit-stalk 1–2 cm. long. Seeds 5·5–7 × 3–5 × 1·2–1·3 mm., broadly ovate or ovate in outline, lenticular, compressed; testa fibrillose.

Mozambique. N: Eráti, between Namapa and Ocúa, near bridge over R. Lúrio, fl. 9.ii.1960, *Lemos & Macuácua* 30 (COI; K; LISC; LMA; SRGH). Z: Milange, 95 km. towards Quelimane, fr. 22.i.1966, *Torre & Correia* 14060 (LISC).
Also in S. Tanzania. Deciduous woodland, 250–500 m.

6. **Coccinia subglabra** C. Jeffrey in Kew Bull. **30**: 479 (1975). Type: Mozambique, Niassa, Nacala, *Torre & Paiva* 9417A (LISC, holotype).

Stems slender, scandent, glabrous, becoming white-punctate when older. Leaf-lamina 5–14 × 6·5–18 cm., broadly ovate in outline, cordate, glabrous or almost so, punctate, deeply palmately 3–5-lobed, the lobes broadly ovate or triangular to elliptic, oblanceolate or oblong-oblanceolate, entire or sinuate-dentate especially towards the apex, acute to obtuse, apiculate, the central largest. Petioles 0·6–4 cm. long, setulose on upper margins or glabrous. Tendrils simple. Probracts 3–5 mm. long, narrow, oblanceolate-spathulate. ♂ flowers in 2–6·5 cm. pedunculate racemes, co-axillary with a 20–65 mm. pedicellate solitary flower; pedicels of racemose flowers 2–7 mm. long. Receptacle-tube 4–6·5 mm. long, obconic, lobes 3–7·5 mm. long, triangular-lanceolate, acute. Corolla-lobes 2·5–3 × 1–2 cm., deep yellow, ovate, acute, united in lower half. ♀ flowers solitary, axillary; pedicels 29–32 mm. long; ovary 20–25 × 1·5–2 mm., cylindrical, glabrous; receptacle-tube c. 3 mm. long, campanulate, lobes 3·5 mm. long, lanceolate, acute. Fruit 4·5–5·5 × 1·2–1·5 cm., cylindrical, terete, obtuse, tapered towards the base, vermilion; fruit-stalk 3·5–4 cm. long. Seeds c. 5 × 3·5 × 1·4 mm., ovate in outline, compressed, lenticular.

Mozambique. N: Mossuril, Cabaceira, *Carvalho* (COI).
Not known elsewhere. Coastal forest, 40–130 m.

7. **Coccinia adoensis** (A. Rich.) Cogn. in A. & C. DC., Mon. Phan. **3**: 538 (1881).—
Burtt Davy, F.P.F.T. **1**: 231 (1926).—Meeuse in Bothalia **8**: 105 (1962).—
C. Jeffrey in F.T.E.A., Cucurb.: 65, t. 8 fig. 1–7 (1967).—Keraudren in Fl.
Cameroun 6, Cucurb.: 134, t. 26 fig. 4–6 (1967).—Launert & Roessler in Prodr. Fl.
SW. Afr. **94**: 8 (1968).—Binns, H.C.L.M.: 42 (1968). Syntypes from Ethiopia.
   Momordica adoensis A. Rich., Tent. Fl. Abyss. **1**: 293 (1847). Syntypes as above.
   Cephalandra pubescens Sond. in Harv. & Sond., F.C. **2**: 493 (1862).—Hook. f. in
F.T.A. **2**: 551 (1871). Syntypes from S. Africa.
   Coccinia hartmanniana Schweinf., Reliq. Kotschy.: 42, t. 27 (1868).—Wiehe, Cat.
Fl. Pl. Herb. Div. Pl. Path. Zomba: 17 (1952).—Binns, H.C.L.M.: 42 (1968).
Types from Sudan.
   Coccinia parvifolia Cogn. in Viert. Nat. Ges. Zürich **52**: 419 (1907).—Burtt Davy,
loc. cit. Type from S. Africa.
   Coccinia pubescens (Sond.) Eyles in Trans. Roy. Soc. S. Afr. **5**, 4: 498 (1916).—
Harms in Notizbl. Bot. Gart. Berl. **8**: 491 (1923).—Hutch., Botanist in S. Afr.: 398
(1946). Syntypes as for *Cephalandra pubescens*.
   Coccinia roseiflora Suesseng. in Proc. Trans. Rhod. Sci. Ass. **43**: 60 (1951). Type:
Rhodesia, Marandellas, *Dehn* 188a (M, drawing).
   Coccinia sp.—Wiehe, loc. cit.—Mitchell in Puku **1**: 121, 148, 183, 190 (1963).
   Cephalandra aff. *diversifolia.*—Wiehe, op. cit.: 16 (1952) & Binns, loc. cit., non
C. *diversifolia* Naud.
   Coccinia palmata sensu Williamson, Useful Plants Nyasal.: 36 (1956) non (Sond.)
Cogn.

Stems annual, arising from subligneous perennial rootstock with tuberiferous
roots, prostrate or scandent, sometimes purple-tinged, sparsely to densely shortly
pubescent. Leaf-lamina 2·5–12·5 × 2·5–15 cm., very variable, narrowly to broadly
ovate or pentagonal in outline, cordate, dark green, almost glabrous except on veins
to shortly pubescent or ± scabrid-punctate above, paler, ± glaucous and almost
glabrous to rather densely pubescent beneath, unlobed or shortly to very deeply
palmately 3–5-lobed, the lobes ovate or triangular to elliptic, narrowly elliptic or
oblanceolate, subentire to strongly sinuate-dentate, sometimes ± lobulate, acute
to obtuse, rounded or sometimes retuse, apiculate, the central largest. Petioles
0·4–5 cm. long, very sparsely to densely rather finely pubescent with short ascending
or spreading hairs. Tendrils simple. Probracts small, 1·5–4 mm. long, obovate-
rotundate to spathulate. ♂ flowers 4–20 in 0·5–9 cm. pedunculate racemes,
usually co-axillary with a 5–70 mm. pedicellate solitary flower; bracts of racemose
flowers small, 1–1·5 mm. long, glandular, often absent; pedicels 1–28 mm. long.
Receptacle-tube 3·5–6·5 mm. long, obconic-campanulate, sparsely shortly hairy or
subglabrous, lobes remote, 1·5–5·5 mm. long, dentiform, subulate or lanceolate.
Corolla pale creamy yellow, salmon-pink or orange, veined with green, brown or
purple, the lobes 1–3 × 0·5–1·5 cm., obovate, apiculate, united to above the middle.
♀ flowers solitary; pedicels 2–20 mm. long; ovary 6–22 × 1–2·5 mm., fusiform,
glabrous, receptacle-tube 1·5–3·5 mm. long, narrowly campanulate; lobes
1–4·5 mm. long, subulate to lanceolate; corolla as in ♂ flowers, but rather narrower.
Fruit 3–8·5 × 1–3 cm., ovoid-ellipsoid to ellipsoid-cylindrical, often shortly
rostrate, smooth, bright red when mature although sometimes remaining greenish
towards the base; fruit-stalk 0·5–2·2 cm. long. Seeds 4·5–6·5 × 3–4 × 1·5–2·5 mm.,
broadly ovate in outline, ± compressed, lenticular.

Caprivi Strip. Mpilila Island, fr. 13.i.1959, *Killick & Leistner* 3361 (K; PRE).
Botswana. N: Chobe, Kazungula, fl. 17.xi.1965, *Dutakels* FH 11/65/4 (K). **Zambia.**
B: 80 km. E. of Mankoya on road to Kafue Hoek, fl. 21.xi.1959, *Drummond & Cookson*
6722 (COI; SRGH). N: Mbala (Abercorn), fl. 31.x.1952, *Robertson* 193 (K; PRE).
W: Kitwe, Ichimpi, fl. 23.xi.1962, *Mutimushi* 227 (NDO; SRGH). C: Mt. Makulu
Research Station, fl. 5.xii.1956, *Simwanda* 83 (COI; SRGH). E: Chadiza, fl. 28.xi.1958,
Robson 760 (BM; K; LISC; MO; PRE; SRGH). S: Choma, Mapanza, fl. 14.xii.1958,
Robinson 2944 (K; PRE; SRGH). **Rhodesia.** N: Gokwe, Cooper Queen, Mgadze
Hunters' Camp, fl. 10.xi.1963, *Bingham* 898 (LISC; SRGH). W: Matopos, Mtsheleli
Valley, 29.xi.1951, *Plowes* 1346 (K; SRGH). C: Rusape, st. 5.ix.1952, *Dehn* in SRGH
40258 (K; LISC; SRGH). E: Umtali, fr. 28.xii.1959, *Chase* 7240 (K; LISC; SRGH).
S: Gwanda, Tuli Experimental Station, fl. 19.xii.1964, *Norris-Rogers* 513 (COI; SRGH).
Malawi. N: Chisenga, foot of Mafinga Mts., fl. 13.xi.1958, *Robson & Fanshawe* 606
(BM; K; LISC; SRGH). C: Lilongwe, Chitedze, fl. 22.iii.1955, *E.M. & W.* 1127

(LISC; SRGH). S: Cholo, Nchima Estate, fr. immat. 20.x.1950, *Wiehe* N/654 (K; SRGH). **Mozambique.** N: Mission Santo Antonio de Unango, fl. i.1934, *Gomes e Sousa* 1634 (COI). Z: Mocuba, Namagoa, fl. 8.iv.1948, *Faulkner* K 248 (K). T: Tete, 63 km. on road to Chicoa, fl. 29.xii.1965, *Torre & Correia* 13874 (LISC). MS: Chupanga, fr. 10.i.1863, *Kirk* (K). SS: Gaza, Massangena, *Magalhães* 163 (LISC).

N. & E. tropical Africa from N. Ghana east to Sudan and south to S. Africa (Transvaal). Woodland, wooded grassland and grassland, 100–1900 m.

8. **Coccinia sessilifolia** (Sond.) Cogn. in A. & C. DC., Mon. Phan. **3**: 534 (1881).—Burtt Davy, F.P.F.T. **1**: 231 (1926).—Hutch., Botanist in S. Afr.: 298, 299, 666 (1946).—Story in Mem. Bot. Survey S. Afr. **30**: 52, t. 62 (1958).—Meeuse in Bothalia **8**: 98 (1962).—Launert & Roessler in Prodr. Fl. SW. Afr. **94**: 9 (1968). Syntypes from S. Africa.

Cephalandra sessilifolia Sond. in Harv. & Sond., F.C. **2**: 493 (1862). Syntypes as above.

Coccinia schinzii Cogn. in Bull. Herb. Boiss. **3**: 419 (1895).—Burtt Davy, loc. cit. Type from S. Africa.

Stems annual, arising from tuberous subterranean rootstock, prostrate or scandent, glabrous. Leaf-lamina 3–9 × 4–13 cm., sessile, ± amplexicaul, broadly ovate in outline, cordate, ± glaucous, glabrous, smooth or sparsely to densely ± strongly scabrid-punctate, deeply palmately 3–5-lobed, the lobes elliptic to lanceolate, entire to coarsely and usually sharply sinuate-dentate or lobulate, sometimes deeply and narrowly so, acute, apiculate. Tendrils simple. Probracts 1·5–3·5 mm. long, small, narrow. ♂ flowers solitary or in sessile or up to 3·5 cm. pedunculate clusters or racemes, when pedunculate usually bracteate and with a co-axillary solitary flower; pedicels of solitary and clustered flowers 0·8–7·5 cm. long, of racemose flowers 1–2 cm. long; bracts small, 1·5–3 mm. long, lanceolate or spathulate. Receptacle-tube 4–5·5 mm. long, campanulate, glabrous; lobes 3·5–5 mm. long, lanceolate, acute. Corolla pale yellow, sometimes with pinkish tinge, green-veined, the lobes 2–3 × 1–1·5 cm., united to above the middle. ♀ flowers solitary; pedicels 0·7–1·5 cm. long, stout; ovary c. 15–18 × 3·5 mm., cylindrical or ellipsoid-rostrate, glabrous; receptacle-tube 3 mm. long, narrowly campanulate, lobes 3–4·5 mm. long, lanceolate; corolla similar to that of ♂ flowers. Fruit 5·5–10·5 × 2–2·5 cm., ellipsoid-fusiform or ellipsoid-cylindrical, often shortly rostrate, bright red; fruit-stalk 0·7–1·5 cm. long, stout, expanded upwards. Seeds 6·5–7 × 3–3·5 × 1·4 mm., asymmetrically ovate in outline, compressed; testa minutely rugulose.

Botswana. N: Aha Hills, fl. 13.iii.1965, *Wild & Drummond* 6953 (COI; K; SRGH). SW: 32 km. W. of Werda, fl. 26.ii.1960, *Wild* 5167 (K; SRGH). SE: Mochudi, 1914, *Harbor* in *Rogers* 6465 (K).

Also in SW. Africa and S. Africa. Wooded grassland.

9. **Coccinia senensis** (Klotzsch) Cogn. in A. & C. DC., Mon. Phan. **3**: 535 (1881).—Gomes e Sousa in Bol. Soc. Estud. Col. Moçamb. **29–32**: 92 (1936).—C. Jeffrey in F.T.E.A., Cucurb.: 66 (1967) pro parte. Type: Mozambique, Sena, *Peters* (B, holotype †).

Cephalandra senensis Klotzsch in Peters, Reise Mossamb. Bot. **1**: 151 (1861).—Hook. f. in F.T.A. **2**: 552 (1871). Type as above.

Stems scandent, sparsely pubescent, setulose on ridges, becoming white-punctate when older, thickened at the base and developing thin, pallid bark. Leaf-lamina 3·5–12 × 4–14 cm., ovate or broadly ovate in outline, deeply cordate, setulose on veins and setulose to strongly scabrid-punctate above, hispid-setulose especially on veins beneath, shallowly to moderately palmately 3–5-lobed, the lobes triangular, ovate-triangular or ovate to ovate-oblong, obovate or elliptic, subentire or remotely and obscurely sinuate-denticulate to coarsely sinuate-dentate, sometimes shortly lobulate, acute to rounded, sometimes shortly retuse, apiculate, the central largest. Petioles 0·4–2·0 cm. long, usually short (less than 1 cm.) except for the lowermost leaves, shortly setulose. Tendrils simple. Probracts small, 3–5 mm. long, oblanceolate-spathulate. ♂ flowers in short sessile or 0·7–3 cm. pedunculate racemes, often with a co-axillary 8–28 mm. pedicellate solitary flower; pedicels of racemose flowers 2–6·5 mm. long. Receptacle-tube 2·8–6·5 mm. long, obconic to broadly campanulate, hispid or setulose, lobes 3·5–7 mm. long, lanceolate, acute. Corolla yellow, lobes 1·2 cm. long, united in lower ⅓. ♀ flowers solitary; pedicels 9–12 mm. long; ovary 8·5–9 × 1·5 mm., cylindrical; receptacle-tube 2 mm. long, campanulate, lobes 3–4 mm. long, lanceolate, acute. Fruits c. 3·2 × 1·7 cm., ellipsoid, smooth, shortly rostrate; fruit-stalk 1–2·1 cm. long, hispid. Seeds 5·8 × 4·2 × 1·5 mm., ovate in outline, compressed, pallid, bordered; testa fibrillose.

Rhodesia. N: near Tete–Salisbury road, fl. 1.iii.1961, *Richards* 14518 (K). **Malawi.** S: Chikwawa, Lengwe Game Reserve, fr. 5.iii.1970, *Brummitt* 8895 (K; LISC; MAL; SRGH). **Mozambique.** T: Tete, 6 km. on road to Changara, fl. 19.iii.1966, *Torre & Correia* 15225 (LISC). MS: near Maronga, ii.1896, *Menyhart* 927 (K).
Not known elsewhere. Grassland, deciduous woodland and thicket, 90–470 m.

10. **Coccinia rehmannii** Cogn. in Bull. Herb. Boiss. **3**: 418 (1895).—Burtt Davy, F.P.F.T. **1**: 231 (1926).—Bremek. & Oberm. in Ann. Transv. Mus. **16**, 2: 437 (1935).—Hutch., Botanist in S. Afr.: 294, 371, 666 (1946).—Story in Mem. Bot. Survey S. Afr. **30**: 51, t. 61 (1958).—Meeuse in Bothalia **8**: 102 (1962).—R. & A. Fernandes in Mem. Junta Invest. Ultramar, Sér. 2, **34**: 112 (1962); in C.F.A. **4**: 281 1970).—Launert & Roessler in Prodr. Fl. SW. Afr. **94**: 8 (1968). TAB. **107**. Type from S. Africa (Transvaal).
Coccinia ovifera Dinter & Gilg in Dinter, Veg. Veldk. Deutsch SW. Afr.: 16 (1912).—Launert & Roessler, loc. cit. Type from SW. Africa.
Coccinia rehmannii var. *littoralis* Meeuse, tom. cit.: 104 (1962). Type from S. Africa (Cape Prov.).
Coccinia palmata sensu Meeuse, tom. cit.: 96 (1962) pro minore parte quoad spec. cit. *Earthy* 1, non (Sond.) Cogn.

Stems annual or perennial, slender, prostrate or scandent, arising from a large tuberous rootstock, almost glabrous to rather densely hispid or setulose, becoming white-punctate when older. Leaf-lamina 2–8 × 2·5–13 cm., broadly ovate in outline, cordate, shortly usually rather sparsely setulose on veins beneath, scabrid-punctate above, pentagonal or shallowly to usually deeply palmately 3–5-lobed, the lobes broadly ovate or obovate to elliptic or oblanceolate, entire or shallowly to quite sharply sinuate-dentate or deeply and sharply lobulate, obtuse, apiculate, the central largest. Petiole 0·5–6 cm. long, sparsely to densely setulose or setose. Tendrils simple. Probracts 1–3 mm. long, small, narrow. ♂ flowers solitary, in sessile few-flowered clusters or in 2–20-flowered 0·8–3 cm. pedunculate racemes, usually with a co-axillary solitary flower; bracts small, 1–5 mm. long, rounded, often absent; pedicels 5–45 mm. long. Receptacle-tube 3–6 mm. long, broadly campanulate, sparsely to usually densely pubescent, setulose or long-pilose, lobes 2–8 mm. long, lanceolate, acute. Corolla whitish-cream to pale yellow, green-veined, the lobes 1–3 × 0·5–1·5 cm., obovate, apiculate, united to above the middle. ♀ flowers solitary; pedicels 2–15 mm. long; ovary 6–13 × 1–2·5 mm., shortly ellipsoid or fusiform, almost glabrous to densely pilose; receptacle-tube 2–3·5 mm. long, narrowly campanulate or shortly cylindrical, lobes 2–5 mm. long, linear-lanceolate or oblanceolate. Fruit 1·5–7·5 × 1–3·5 cm., shortly ellipsoid to ellipsoid-cylindrical, sometimes ± pyriform, smooth, bright red when mature; fruit-stalk 0·4–2 cm. long. Seeds 4·7–7 × 2–3 × 1·1–1·7 mm., asymmetrically ovate-oblong in outline, sometimes slightly arcuate, compressed, with flat faces and 2-grooved margins; testa minutely rugulose.

Botswana. N: Kwebe Hills, fl. 3.iii.1898, *Lugard* 149 (K). SW: Kgalagadi, c. 80 km. NNW. of Tsabong, fr. 1.iii.1963, *Leistner* 3120 (K; PRE). SE: c. 3 km. N. of Shashi, fl. 21.i.1960, *Leach & Noel* 287 (COI; SRGH). **Rhodesia.** W: Bulalima–Mangwe, Embakwe, fl. 1.iv.1942, *Feiertag* in SRGH 45477 (K; SRGH). C: 32 km. W. of Enkeldoorn village, fl. xi.1957, *Miller* 4758A (SRGH). E: Chipinga, E. Sabi, Giriwayo, fl. 19.i.1957, *Phipps* 20 (COI; LISC; SRGH). S: Nuanetsi, 26 km. N. of Nuanetsi R., fl. 12.i.1961, *Leach* 10706 (COI; K; SRGH). **Mozambique.** SS: Gaza, Chibuto, road to Alto Changane, fl. 12.ii.1959, *Barbosa & Lemos* 8377 (K; LMA; SRGH). LM: between Costa do Sol and Marracuene, Muntanhane, fr. 10.xi.1960, *Balsinhas* 230 (K; LMA).
Also in Angola, S. Africa and SW. Africa. Wooded grassland, woodland and coastal dunes, 0–1000 m.

Meeuse (loc. cit., 1962) distinguishes the coastal dune and woodland populations of this species as var. *littoralis*. They are distinguished by their less persistent, less woody stems, thinner leaves with entire or less divided lobes, usually racemose male flowers, larger corollas, and larger, more elongated fruits. However, between the coastal dune variants (which are extreme in these respects) and typical inland var. *rehmannii* there are

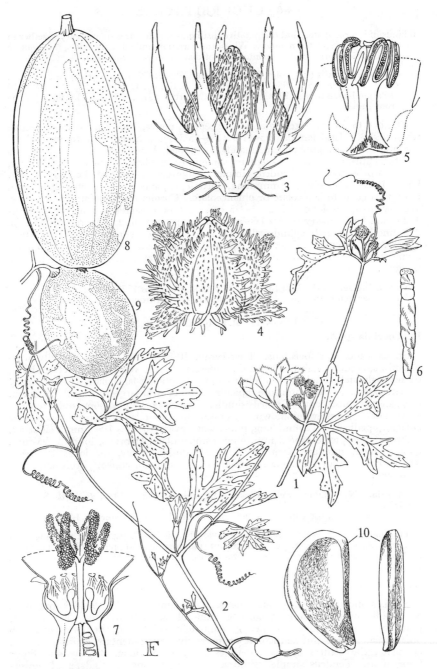

Tab. 107. COCCINIA REHMANNII. 1, part of shoot, ♂ plant (×1) *Wild* 5092; 2, part of shoot, ♀ plant (×½) *Story* 5029; 3, ♂ flower bud, coastal variant (×6) *Lemos & Balsinhas* 133; 4, ♂ flower bud, inland variant (×6) *Wild* 5092; 5, androecium (×3); 6, hair from filament base (×12), both from *Story* 4954; 7, ♀ flower, details, in section (×3) *Lugard* 149; 8, fruit, coastal variant (×1) *Lemos & Balsinhas* 284; 9, fruit, inland variant (×2) *Leistner* 3120; 10, seed, face and lateral views (×6) *Story* 5029.

all kinds of intermediates, and the putative distinguishing features do not vary together in any consistently correlated manner. Whilst the variation undoubtedly exists, it cannot be dealt with satisfactorily by the maintenance of distinct infraspecific taxa.

11. **Coccinia grandis** (L.) Voigt, Hort. Suburb. Calcut.: 59 (1845).—Keay in F.W.T.A. ed. 2, **1**: 215, t. 85 (1954).—C. Jeffrey in F.T.E.A., Cucurb.: 68, t. 8 fig. 8–10 1967).—Keraudren in Fl. Cameroun **6**, Cucurb.: 127, t. 25 fig. 1–3 (1967). Type from India.

Climber to 20 m.; stem glabrous except at nodes, becoming white-punctate when older. Leaf-lamina 3·5–11·5 × 3·5–15·5 cm., broadly ovate in outline, cordate, glabrous, punctate, palmately 3–5-lobed, lobes shallowly triangular to elliptic, entire or ± sinuate-dentate, sometimes lobulate. Petiole 1–5 cm. long, glabrous or almost so. Tendrils simple. ♂ flowers solitary; pedicels 7–70 mm. long. Receptacle-tube 3–7 mm. long, obconic, expanded above, lobes 2·5–6 mm. long, triangular to lanceolate or oblanceolate. Corolla campanulate, pale yellow, green-veined, the lobes 1·5–2 × 1–1·5 cm. ♀ flowers solitary, axillary; pedicels 4–25 mm. long; ovary 5–15 × 1·5–3·5 mm., ellipsoid-cylindrical; receptacle-tube 2–7 mm. long, shortly cylindrical, lobes 2–4 mm. long; corolla-lobes 2–3·2 × 0·7–1·3 cm. Fruit 3–6·5 × 1·5–3·5 cm., ellipsoid, obtuse, red. Seeds c. 6 × 3 × 1·5 mm., asymmetrically ovate in outline, compressed, with flat faces and thick 2-grooved margins.

Mozambique. LM: Vila Luisa, fl. 21.v.1968, *Balsinhas* 1273 (COI; LMA). An escape from cultivation. Native of the Old-World tropics.

SPECIES NOT SUFFICIENTLY KNOWN

12. **Coccinia sp. A.**

Stems scandent, puberulous. Leaf-lamina 10·5–15 × 9–18·5 cm., broadly ovate in outline, cordate, glabrous beneath, pubescent on the veins and punctate above, deeply palmately 5-lobed, the lobes elliptic to oblong-lanceolate, conspicuously sinuate-dentate, acute to obtuse, apiculate, the central largest. Petioles 4–6 cm. long, shortly pubescent. Tendrils simple. Probracts 5–7 mm. long, oblanceolate-spathulate. ♂ flowers unknown. ♀ flowers in rather lax ebracteate racemes or clusters; pedicels 4–12 mm. long, pubescent; ovary 10–15 × 1·5–2 mm., ellipsoid, glabrous; receptacle-tube 3 mm. long, campanulate, glabrous, lobes 2–2·5 mm. long, remote, subulate, dentiform; corolla dusky yellow, the lobes 1·2–1·4 × 0·6–0·8 cm., rounded, apiculate, united to the middle or above. Fruit and seeds unknown.

Zambia. N: Chilongowelo, Tasker's Deviation, fl. 27.ii.1952, *Richards* 883 (K; SRGH). Known only from the gathering cited above. On shrubs by waterfall, 1450 m.

This gathering represents a population obviously allied to *C. barteri*, but differing from that species in its longer, more abundant indumentum, simple tendrils, narrower probracts, laxer ebracteate female inflorescences, and longer ovaries. Its status is still uncertain.

13. **Coccinia sp. B.**

Stems scandent, minutely puberulous. Leaf-lamina 9·5–12 × 9–13·5 cm., ovate or broadly ovate in outline, cordate, puberulous on veins especially beneath, punctate above, unlobed or deeply palmately 3-5-lobed, the lobes ovate to oblanceolate, plane and denticulate or markedly sinuate-dentate, acute to obtuse, apiculate, the central largest. Petioles 3·3–5·5 cm. long, densely and finely puberulous. Tendrils simple. Probracts 3–5 mm. long, spathulate. ♂ flowers unknown. ♀ flowers solitary or up to 4 in shortly pedunculate clusters, usually with a co-axillary solitary flower; pedicels 4–8 mm. long; ovary 16–24 × 1·5–2 mm., cylindrical, glabrous; receptacle-tube 2·5 mm. long, campanulate, lobes 2·3 mm. long, lanceolate, narrow, acute; corolla-lobes 1·6–1·8 cm. long, united to near the middle. Fruits and seeds unknown.

Zambia. S: Mazabuka, Nanga Estate, fl. 7.iii.1963, *van Rensburg* 1620 (K; SRGH). Known only from a very restricted area in the Kafue R. basin. Riverine bush in *Acacia-Hyparrhenia-Setaria* wooded grassland and grassland.

Like *sp. A*, this entity seems allied to *C. barteri*, from which it differs in its less coriaceous leaves, simple tendrils, and much longer ovaries.

EXCLUDED SPECIES

Coccinia aostae Busc. & Muschl. in Engl., Bot. Jahrb. **49**: 499 (1913).

This species, supposed to have been based on *Aosta* 105 from Mozambique, was in fact described from *Schweinfurth* 578, collected in Eritrea; see Gilg in Engl., Bot. Jahrb. **53**: 373 (1915). It is probably a synonym of *C. adoensis*.

Coccinia helenae Busc. & Muschl., tom. cit.: 498 (1913).

This species, supposed to have been based on *Aosta* 87 from Mozambique, was in fact described from *Schweinfurth* 932, collected in the Sudan; see Gilg in Engl., loc. cit. It is a synonym of *C. grandis* (L.) Voigt.

Coccinia palmata (Sond.) Cogn. in A. & C. DC., Mon. Phan. **3**: 540 (1881).

This species is recorded from Mozambique by Meeuse in Bothalia **8**: 96 (1962), on the basis of *Earthy* 1 (BOL). This specimen, which is exceedingly poor, is in fact *C. rehmannii* Cogn., and *C. palmata* must therefore be excluded from the F.Z. area.

Coccinia quinqueloba (Thunb.) Cogn., tom. cit.: 533 (1881).

Williamson, Useful Pl. Nyasal.: 37 (1956) records this species for Malawi, but it is a species of the Cape Province of S. Africa and does not occur in our area. The record probably represents a misidentification of *C. adoensis*.

9. DIPLOCYCLOS (Endl.) Post & Kuntze

Diplocyclos (Endl.) Post & Kuntze, Lex. Phan.: 178 (1903).
Bryonia L. Sect. *Diplocyclos* Endl., Prodr. Fl. Norfolk.: 68 (1833).
Bryonopsis sensu Hook. f. in F.T.A. **2**: 556 (1871) et sensu auct. mult. non Arn.

Scandent perennial herbs; young stems green-spotted. Leaves petiolate, simple, palmately lobed. Tendrils proximally 2-fid. Probracts present, small. Flowers monoecious, in sessile or shortly racemose few- to many-flowered axillary clusters, ♂ and ♀ often co-axillary. ♂ flowers on slender pedicels. Receptacle-tube short and broad; lobes small, remote. Corolla whitish or pale yellow, gamopetalous, 5-lobed, the lobes entire. Stamens 3, 2 double two-thecous, 1 single one-thecous, free; filaments inserted on the receptacle-tube; connectives broad; thecae flexuous. Disk ± undulate, adnate to the receptacle-tube. ♀ flowers subsessile; ovary terete; ovules rather few, horizontal; receptacle-tube smaller and narrower than in ♂ flowers, perianth otherwise similar; stigma 3-lobed. Fruits solitary or clustered, subsessile, baccate, fleshy, globose to ellipsoid, indehiscent, red with white longitudinal stripes or lines of markings. Seeds rather small, pyriform, with highly convex faces and thick 2-grooved, rugulose margins.

A palaeotropical genus of about 4 species, all but one confined to Africa.

Leaves much dissected, the lobes deeply pinnatifid, with spreading lobulate segments
3. *decipiens*
Leaves palmately 5-lobed, the lobes merely sinuate-denticulate or shallowly sinuate-lobulate towards the apex:
  Fruits subglobose, 1·5–2·5 cm. long; ovary 4–5 mm. long - - 1. *palmatus*
  Fruits ellipsoid, 2·5–4·5 cm. long; ovary 7–13 mm. long - - - 2. *tenuis*

1. **Diplocyclos palmatus** (L.) C. Jeffrey in Kew Bull. **15**: 352 (1962); in F.T.E.A., Cucurb.: 73 (1967). Type from Ceylon.
    Bryonia palmata L., Sp. Pl. **2**: 1012 (1753). Type as above.
    Bryonopsis laciniosa sensu Hook. f. in F.T.A. **2**: 556 (1871) pro parte, excl. specim. mossambic et sensu Keay in F.W.T.A. ed. 2, **1**: 214 (1954), non (L.) Naud. sensu stricto.

Stems scandent to 6 m., glabrous. Leaf-lamina 4–14 × 4–15 cm., broadly ovate in outline, cordate, asperulous or ± scabrid-punctate above, setulose on veins beneath, otherwise glabrous, deeply palmately 5-lobed, lobes elliptic or narrowly elliptic, ± sinuate-serrate, denticulate, acuminate, obtuse to acute, apiculate, the central largest. Petiole 2–8 cm. long, coarsely spiculate. Probracts c. 3 mm. long,

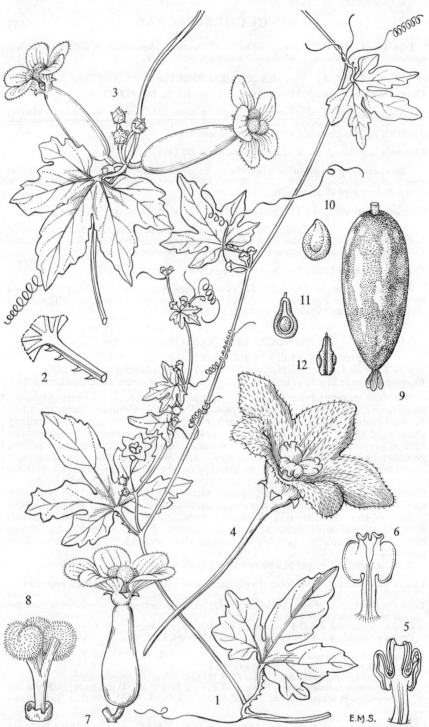

Tab. 108. DIPLOCYCLOS TENUIS. 1, habit ($\times \frac{1}{2}$); 2, aculei on petiole ($\times 1\frac{1}{2}$); 3, flowering node ($\times 2$); 4, ♂ flower ($\times 2$); 5, double stamen, dorsal view ($\times 4$); 6, double stamen, ventral view ($\times 4$); 7, ♀ flower ($\times 2$); 8, style and stigmas ($\times 4$); 9, fruit ($\times 1$), 1–9 from living plant cultivated at Kew; 10, seed with enclosing pocket of flesh ($\times 2$); 11, seed, face view ($\times 2$); 12, seed, lateral view ($\times 2$), 10–12 from *Krauss* 201. From F.T.E.A.

spathulate. ♂ flowers 2–8, usually co-axillary with 1–4 ♀ flowers; pedicels 5–15 mm. long. Receptacle-tube c. 4 mm. long, lobes up to 2 mm. long, triangular-dentiform. Corolla white to greenish-yellow, the lobes 6–9 × 2·5–5 mm. ♀ flowers 1–5, usually co-axillary with up to 8 ♂ flowers; pedicel 1–5 mm. long; ovary 4–5 × 2–3·5 mm. shortly ellipsoid. Fruits 1·5–2·5 × 1·5–2·5 cm., solitary or in clusters of 2–5, subglobose, bright red with silvery white longitudinal stripes or markings; fruit-stalks 1–5 mm. long. Seeds 5–6 × 2·5–3 × 4–4·5 mm.

Rhodesia. E: Chipinga, Gungunyana Forest Reserve, fr. vi.1962, *Goldsmith* 151/62 (COI; SRGH). **Mozambique.** MS: S. of Mt. Bandula, fr. 9.vii.1957, *Chase* 6580 (K; LISC; SRGH).
Old World tropics from S. Tomé & Fernando Po east to Australasia. Rain-forest, 500–1050 m.

2. **Diplocyclos tenuis** (Klotzsch) C. Jeffrey in Kew Bull. **15**: 352 (1962); in F.T.E.A., Cucurb.: 74, t. 9 (1967). TAB. **108**. Type: Mozambique, Quirimba I., *Peters* (B, holotype).
   Bryonia tenuis Klotzsch in Peters, Reise Mossamb. Bot. **1**: 150 (1861). Type as above.
   Bryonopsis laciniosa sensu Hook. f. in F.T.A. **2**: 556 (1871) pro parte quoad specim. mossambic, non (L.) Naud.

Stems to 3 m., glabrous, prostrate or scandent. Leaf-lamina 2–11 × 2·5–13·5 cm., broadly ovate in outline, cordate, scabrid-punctate above, glabrous beneath, palmately 3–5-lobed, lobes narrowly elliptic to obovate, remotely denticulate, plane or sinuate-dentate or sinuate-lobulate, acute to rounded or retuse, apiculate. Petiole 0·5–8 cm. long, remotely spiculate. Probracts 2·5–3 mm. long, spathulate. ♂ flowers 1–8, often co-axillary with 1–2 ♀ flowers; pedicels 6–48 mm. long. Receptacle-tube 2·5–4 mm. long, lobes 0·5–2 mm. long, triangular to lanceolate, dentiform. Corolla whitish, green-veined, the lobes 5–11 × 2·7–5 mm. ♀ flowers 1–3, sometimes co-axillary with 2–3 ♂ flowers; pedicels 1–6 mm. long; ovary 7–13 × 1·5–4·5 mm., elongate-ovoid. Fruits 2·5–4·5 × 1–2·5 cm., solitary or in clusters of 2–3, ovoid to elongate-ovoid, red with white longitudinal markings; fruit-stalks 2–6 mm. long. Seeds 3·5–5 × 2–2·5 × 1·5–3 mm.

Mozambique. N: Larde, fr. 17.vii.1948, *Pedro & Pedrógão* 4586 (LMA). Z: Maganja da Costa, Gobene, 50 km. from Vila da Maganja, fr. 12.ii.1966, *Torre & Correia* 14527 (LISC). MS: Beira, fl. & fr. immat. 7.vii.1913, *Dümmer* 74 (BM). SS: Vilanculos, fr. immat. 27.iii.1952, *Barbosa & Balsinhas* 5023 (COI; LMA). LM: Inhaca I., fl. 6.vi.1958, *Mogg* 27992 (LMU; SRGH).
Also in Tanzania and Kenya. Littoral thicket and forest, 0–200 m.

3. **Diplocyclos decipiens** (Hook. f.) C. Jeffrey in Kew Bull. **15**: 351 (1962); in F.T.E.A., Cucurb.: 74 (1967).—R. & A. Fernandes in Mem. Junta Invest. Ultramar, Sér. 2, **34**: 101 (1962); in C.F.A. **4**: 279 (1970). Type from Angola.
   Cephalandra decipiens Hook. f. in F.T.A. **2**: 552 (1871). Type as above.
   Coccinia decipiens (Hook. f.) Cogn. in A. & C. DC., Mon. Phan. **3**: 539 (1881).—Hiern, Cat. Afr. Pl. Welw. **1**, 2: 400 (1898). Type as above.

Stems scandent, herbaceous, glabrous. Leaf-lamina 3·5–9·5 × 4–12 cm., broadly ovate in outline, cordate, asperulous or scabrid-punctate above, glabrous beneath, much dissected, pedately 5-lobed to the base with the lobes regularly and deeply pinnately lobulate, ultimate segments narrowly to broadly elliptic or ovate-elliptic, sinuate-dentate, obtuse to acute, apiculate. Petiole 1·5–4·5 cm. long, sparsely setulose. Probracts 1–3 mm. long, spathulate. ♂ flowers 5–14, sometimes co-axillary with 1–4 ♀ flowers, pedicels 10–35 mm. long. Receptacle-tube 1·5–3 mm. long, lobes 1–2·7 mm. long, filiform. Corolla pale yellow, lobes 6–7·5 × 2·5–3 mm. ♀ flowers 1–4, sometimes co-axillary with up to 13 ♂ flowers; pedicels c. 2 mm. long; ovary c. 4 × 1·5 mm. Fruits 1·5–2·7 × 1·2–2 cm., solitary or in clusters of 2–4, red with longitudinal rows of white spots, shortly ovate-ellipsoid, slightly rostrate. Seeds c. 5 × 3 × 3·5 mm.

Zambia. N: Mporokoso Distr., L. Mweru, near Kapindi, fl. & fr. immat. 10.iv.1957, *Richards* 9119 (K). W: Kabompo, fl. & fr. immat. 23.iii.1961, *Drummond & Rutherford-Smith* 7252 (COI; K; LISC; SRGH). **Malawi.** N: Rumpi, Njakwa gorge, S. Rukuru R. bridge, fl. & fr. 29.iv.1971, *Pawek* 4757 (K; MAL). C: Dedza, Mua Livulezi Forest, fr. 17.v.1960, *Adlard* 363 (SRGH).
Also in Angola, Zaire (Katanga), S. Tanzania. Deciduous woodland, 1050–1350 m.

10. LUFFA Mill.

Luffa Mill., Gard. Dict. abridg. ed. 4 (1754).

Annual prostrate or scandent herbs. Leaves simple, petiolate. Tendrils proximally 2–6-fid. Flowers yellow or whitish, monoecious. ♂ flowers racemose. Receptacle-tube broadly campanulate; lobes large, enclosing the petals in bud, entire. Petals 5, free, entire. Stamens 5, all one-thecous, or 3, 2 two-thecous & 1 one-thecous; filaments inserted on the receptacle-tube, free; anthers free; connectives broad; thecae much convoluted. ♀ flowers solitary; ovary smooth, ribbed, tuberculate or spiny; ovules many, horizontal; stigma 3-lobed. Fruit globose to cylindrical, rostrate, smooth, ribbed or spiny, dry, brownish, fibrous, dehiscent by an apical operculum. Seeds compressed, oblong-elliptic in outline.

A tropical genus of c. 6 species, two of which are widely cultivated.

Fruits more or less terete; seeds with a narrow membraneous wing-like margin; petals
    deep yellow - - - - - - - - - - - 1. *cylindrica*
Fruits strongly 10-angled; seeds without a narrow membraneous wing-like margin;
    petals pale yellow - - - - - - - - - - 2. *acutangula*

1. **Luffa cylindrica** (L.) M. J. Roem., Syn. Mon. **2**: 63 (1846).—R.E.Fr., Wiss. Ergebn. Schwed. Rhod.-Kongo-Exped. **1**: 312 (1916).—Gomes e Sousa in Bol. Soc. Estud. Col. Moçamb. **29–32**: 92 (1936).—Williamson, Useful Pl. Nyasal.: 77, t. 295 (1956).—R. & A. Fernandes in Mem. Junta Invest. Ultramar, Sér. 2, **34**: 66 (1962).—C. Jeffrey in F.T.E.A., Cucurb.: 76, t. 10 (1967).—Keraudren in Fl. Cameroun **6**, Cucurb.: 84 (1967).—R. & A. Fernandes in C.F.A. **4**: 270 (1970). TAB. **109**. Type a cultivated plant.
    Momordica cylindrica L., Sp. Pl. **2**: 1009 (1753). Type as above.
    Luffa aegyptiaca Mill., Gard. Dict. ed. 8 (1768).—Hook. f. in F.T.A. **2**: 530 (1871).—Ficalho, Pl. Ut. Afr. Port.: 187 (1884).—Hiern, Cat. Afr. Pl. Welw. **1**, 2: 394 (1898).—Keay in F.W.T.A. ed. 2, **1**: 207 (1954).—Mitchell in Puku **1**: 148, 190 (1963). Type a cultivated plant.

Stems prostrate or scandent, to 15 m., finely hairy. Leaf-lamina 6–18 × 6–23 cm., broadly ovate in outline, cordate, dark green, asperulous or scabrid, palmately 5–7-lobed, the lobes triangular to elliptic or narrowly oblong-lanceolate, obscurely sinuate-denticulate to coarsely and deeply sinuate-serrate or rarely lobulate, obtuse to acute, often acuminate, apiculate, the central largest. Petiole 1–15 cm. long, finely ascending-pilose, becoming scabrid-strigose. Tendrils usually 2–4-fid. Probracts 3–7 mm. long, oblong-ovate or obovate. ♂ flowers racemose; peduncles 7–32 cm. long, finely hairy; pedicels 3–12 mm. long, finely hairy; bracts adnate to the pedicels, spathulate, glandular, 2–6 mm. long. Receptacle finely hairy; receptacle-tube 3–7 mm. long, obconic below, expanded above, lobes 9–14 mm. long, triangular, acuminate, sometimes glandular. Petals 2–4·5 × 1–3·5 cm., rich yellow. Stamens 5 or rarely 3. ♀ flowers on 2·5–14·5 cm. peduncles; ovary 20–40 × 2–7 mm., cylindrical, densely puberulous; receptacle-tube 2·5–6 mm. long, lobes 8–16 mm. long, ovate-lanceolate or lanceolate, glandular; corolla as in ♂ flowers. Fruit 6–25 × 2·5–6 cm. or larger, ellipsoid to cylindrical, ± terete; fruit-stalk stout, 1·5–15 cm. long, expanded upwards. Seeds 10–15 × 6–11 × 2–3 mm., broadly elliptic in outline, rounded at the ends, black, smooth, compressed, with a narrow, thin, wing-like margin.

Zambia. N: Mbala, Mpulungu, fl. 22.x.1967, *Simon & Williamson* 1153 (K; SRGH). C: Luangwa bridge, fl. 11.viii.1958, *Best* 147 (K; SRGH). E: Fort Jameson, Lunkwakwa, fl. 23.iii.1955, *E.M. & W.* 1134 (BM; SRGH). S: Lusitu, fl. 25.ix.1959, *Fanshawe* 5214 (SRGH). **Rhodesia.** N: Shashi Mts., NNE. of Bindura, fl. 22.vii.1971, *Mogg* 34290 (K). W: Wankie, Deka R., fl. 21.vi.1934, *Eyles* 8093 (K; SRGH). E: Chipinga, Sabi Valley Experimental Station, fl. xii.1959, *Soane* 216 (COI; SRGH). **Malawi.** N: Kondowe to Kasonga, fl. vi.1896, *Whyte* (K). C: Dedza, Ntaka-taka, fl. 11.iv.1969, *Salubeni* 1318 (K; SRGH). S: Chikwawa, Medalamu village, fl. 5.v.1969, *Banda* 1076 (K; SRGH). **Mozambique.** N: Quissanga, between Mahate and Metuge, fl. 3.x.1948, *Barbosa* 2335 (LISC; LMA). Z: Maganja da Costa, fl. & fr. immat. 6.x.1946, *Pedro* 2132 (LMA). T: Boroma, fl. v–vii.1892, *Menyhart* 638 (K). MS: Maringua, Sabi R., fl. 24.vi.1950, *Chase* 2559 (BM; K; LISC; SRGH). SS: Massengena, fl. vii.1932, *Smuts* P. 365 (K; PRE). LM: Lourenço Marques, fl. & fr. immat. 14.vi.1965, *Marques* 545 (LMU).

Widely distributed in the tropics and subtropics. River margins, also frequently cultivated, 0–1500 m.

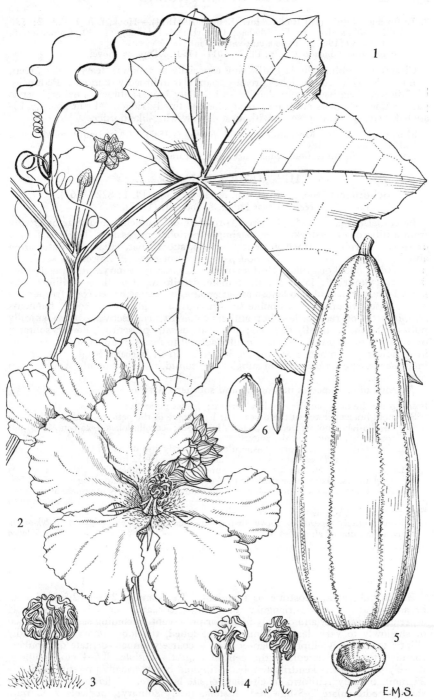

Tab. 109. LUFFA CYLINDRICA. 1, habit (× ⅔) *Tanner* 1474; 2, ♂ flower (× ⅔); 3, stamens (× 2); 4, stamen, ventral and dorsal views (× 2), 2–4 from *Jeffrey* s.n.; 5, fruit with operculum as detached (× ⅔); 6, seed, face and lateral views (× 1½), both from Kew carp. collection. From F.T.E.A.

2. **Luffa acutangula** (L.) Roxb., Hort. Beng.: 70 (1814).—Hook. f. in F.T.A. **2**: 530 (1871).—Keraudren in Fl. Madag., Cucurb.: 101 (1966); in Fl. Cameroun **6**, Cucurb.: 89 (1967). Type a cultivated plant.
     Cucumis acutangulus L., Sp. Pl. **2**: 1011 (1753). Type as above.

Closely resembling the last, differing especially as follows: leaves paler green, more shallowly lobed, the lobes broadly triangular to broadly rounded. Peduncles of ♂ flowers relatively more slender and usually longer; bracts smaller; petals pale yellow; stamens always 3; ovary longitudinally ridged. Fruit sharply 10-angled, usually clavate; seeds without a narrow wing-like border.

Mozambique. N: Moçambique, 1934, *Ribeiro* (LISC). LM: Lourenço Marques, fl. iii.1946, *Pimenta* (LISC).
Known only in cultivation, seldom becoming naturalized.

11. TROCHOMERIA Hook. f.

Trochomeria Hook. f. in Benth. & Hook., Gen. Pl. **1**: 822 (1867).
     Heterosicyos Hook. f., loc. cit. (1867).

Perennial herbs with slender erect, prostrate or scandent annual stems arising from a tuberous rootstock. Leaves simple, petiolate or sessile. Tendrils simple or absent. Probracts when present stipuloid, subcircular, ciliate-laciniate, often absent. Flowers greenish, dioecious, often precocious. ♂ flowers in sessile or pedunculate nodal or axillary clusters or racemes, rarely solitary. Receptacle-tube elongate, cylindrical; lobes minute, remote, dentiform. Corolla ± stelliform, the lobes 5, ovate-triangular to lanceolate-attenuate, spreading or reflexed, free almost to the base. Stamens 3, 2 double two-thecous, 1 single one-thecous; filaments free, inserted on receptacle-tube; anthers coherent; connectives narrow, apically pubescent; thecae tightly flexuous; pistillode usually present. ♀ flowers solitary; ovary ovate to ellipsoid, smooth; ovules several to many, horizontal; perianth as in ♂ flower; staminodes 3; stigma 3-lobed. Fruit baccate, bright red, terete, indehiscent. Seeds ellipsoid to subglobose, tumid, usually smooth, white; testa hard, thick.

A genus of c. 8 species in tropical and southern Africa.

Stems prostrate or scandent; tendrils present; leaves petiolate:
    Stipuloid bracts present at the bases of at least some of the petioles, if absent then the
      corolla-lobes not exceeding 13 mm. in length; corolla-lobes 5–25 mm. long, not
      markedly attenuate at the apex:
      Leaves minutely and sparsely asperulous to shortly and densely hispid or pubescent
        on both surfaces; stems usually hispid; seeds 10–12 mm. long   1. *hookeri*
      Leaves scabrid-setulose or scabrid-punctate above, otherwise glabrous; stems
        glaucous, glabrous or sparsely pubescent; seeds 7·5–9 mm. long   3. *debilis*
    Stipuloid bracts absent; corolla-lobes 20–60 mm. long, markedly attenuate at the apex
                                        2. *macrocarpa*
Stems erect; tendrils absent; leaves sessile:
    Leaves minutely asperulous above, hispidulous beneath   -   -   -   4. *polymorpha*
    Leaves glabrous or almost so -   -   -   -   -   -   -   5. *subglabra*

1. **Trochomeria hookeri** Harv., Gen. S. Afr. Pl. ed. 2: 125 (1868).—Burtt Davy, F.P.F.T. **1**: 224 (1926).—Meeuse in Bothalia **8**: 90 (1962). Type from S. Africa.
     Zehneria garcini sensu Sond. in Harv. & Sond., F.C. **2**: 487 (1862) excl. syn., non (L.) Stocks.

Stems pubescent, prostrate or scandent. Leaf-lamina 2·5–9 × 2·5–10·5 cm., broadly ovate or ovate-triangular in outline, cordate, minutely and sparsely asperulous to shortly and densely hispid or pubescent, becoming scabrid, unlobed or shallowly to very deeply palmately 3–5-lobed, the lobes obovate or broadly elliptic to narrowly elliptic or lanceolate, ± coarsely sinuate-dentate or lobulate, rounded to acute, apiculate, the central largest. Petioles 0·6–4 cm. long, ± densely pubescent. Tendrils simple. Stipuloid bracts usually present, 5–25 × 5–25 mm., broadly reniform to circular, ciliate-laciniate. ♂ flowers in sessile or shortly pedunculate 2–5-flowered clusters, rarely solitary; pedicels 7–45 mm. long, pubescent. Receptacle-tube 10–21 mm. long, sparsely pilose. Lobes 1·5–2·5 mm. long, recurved. Corolla-lobes 1–2·5 mm. long, green or greenish-yellow, reflexed. ♀ flowers on 10 mm. long pedicels; ovary 4·5–12 × 3–7 mm., subglobose to ellipsoid; perianth similar to that of ♂ flower. Fruit 3–4 × 2–3 cm., ovoid-ellipsoid, red; fruit-stalk c. 1·7 cm. long. Seeds c. 10–12 × 5 × 3 mm.

Mozambique. LM: Namaacha, near the frontier post, fl. 25.viii.1967, *Gomes e Sousa & Balsinhas* 4946 (LMA).
Also in Swaziland & S. Africa. Wooded grassland, 50–1200 m.

2. **Trochomeria macrocarpa** (Sond.) Hook. f. in F.T.A. **2**: 524 (1871).—Eyles in Trans. Roy. Soc. S. Afr. **5**, 4: 498 (1916).—Burtt Davy, F.P.F.T. **1**: 225 (1926).—Bremek. & Oberm. in Ann. Transv. Mus. **16**, 2: 437 (1935).—Suesseng. & Merxm. in Proc. Trans. Rhod. Sci. Ass. **43**: 62 (1951).—Meeuse in Fl. Pl. Afr. **30**: t. 1168 (1954); in Bothalia **8**: 88 (1962).—C. Jeffrey in F.T.E.A., Cucurb.: 87, t. 12 fig. 1–7 (1967).—Keraudren in Aubrév., Fl. Cameroun **6**, Cucurb.: 111, t. 23 (1967).—Binns, H.C.L.M.: 43 (1968).—R. & A. Fernandes in C.F.A. **4**: 285 (1970). Syntypes from S. Africa.
   Zehneria macrocarpa Sond. in Harv. & Sond., F.C. **2**: 488 (1862). Type as above.
   Trochomeria macrocarpa var. *kirkii* Cogn. in A. & C. DC., Mon. Phan. **3**: 399 (1881). Type: Malawi, Chiradzura, *Kirk* (K, holotype).
   Trochomeria nudiflora Burtt Davy, tom. cit.: 51, 225 (1926).—Wiehe, Cat. Fl. Pl. Herb. Div. Pl. Path. Zomba: 17 (1952). Type from S. Africa (Transvaal).
   Trochomeria sp.—Wild in Kirkia **5**, 1: 66 (1965).

Stems to 6 m., almost glabrous to shortly pubescent, ± glaucous, prostrate or scandent. Leaf-lamina 1·5–7(9·5) × 2–8(10) cm., broadly ovate in outline, cordate, dark green, glossy and glabrous to asperulous or scabrid-punctate above, sparsely hispid or asperulous beneath, especially on veins, shallowly to very deeply palmately 5-lobed, the lobes ovate or triangular to lanceolate or linear, entire or obscurely to coarsely sinuate-dentate or lobulate, obtuse to acute, apiculate, the central largest. Petioles 0·5–3 cm. long, pilose. Tendrils simple. Stipuloid bracts 3–13 × 2–16 mm. usually absent, when present subcircular, ciliate-laciniate. ♂ flowers in lax sessile or pedunculate 2–13-flowered fascicles or racemes, rarely solitary; pedicels 6–11·5 mm. long, almost glabrous. Receptacle-tube (13)16–25 mm. long, shortly sparsely pilose, lobes 0·4–3 mm. long. Corolla-lobes (1)2–6 cm. long, olive green or dull greenish-yellow, sometimes with a brownish tinge especially towards the apex, strongly reflexed. ♀ flowers on 9–24 mm. pedicels, ovary (6)8–13 × 2–4 mm., glabrous, fusiform; perianth similar to that of ♂ flower. Fruit (3)4–6·5 × 2–3 cm., ellipsoid, usually rostrate, rarely subglobose, bright orange to bright red; fruit-stalk 0·5–4 cm. long, stout. Seeds (8)9–10(11·5) × 5–6·5 × (3·5)4·5–5·5 mm., elliptic in outline, tumid, white, usually smooth, not bordered.

Subsp. **macrocarpa.**

Stipuloid bracts absent, or, if present (not in F.Z. area) then fruit subglobose.

Botswana. SE: 10 km. NW. of Derdepoort, fl., *Codd* 8888 (PRE; SRGH). **Zambia.** B: Mangweshi, fl. 27.vii.1952, *Codd* 7132 (K; PRE). N: Abercorn, fl. 15.ix.1962, *Lawton* 991 (K; NDO). W: Ndola, fl. 29.ix.1954, *Fanshawe* 1579 (K; SRGH). C: Mkushi, fl. 24.ix.1957, *Fanshawe* 3716 (K). E: Petauke, Nyimba-Luembe road, fl. 11.xii.1958, *Robson* 899 (BM; K; LISC; PRE; SRGH). S: Mumbwa, fl. 19.xi.1947, *Brenan & Greenway* 7888 (FHO; K). **Rhodesia.** N: Lomagundi, Umvukwes, vicinity of Imshi Mine, fl. 4.ix.1960, *Leach & Bayliss* 10471. W: Shangani-Bubi, Gwampa Forest Reserve, st. i.1956, *Goldsmith* 82/56 (K; SRGH). C: Salisbury, Ruwa R., fl. 22.ix.1946, *Wild* 1249 (K; SRGH). E: Umtali, Odzani, fl. 23.ix.1958, *Umtali Vet. Dept.* 2 (COI; K; LISC; SRGH). **Malawi.** N: Nyika Plateau, Rufiri Stream, fr. 28.x.1958, *Robson & Angus* 445 (BM; K; LISC; MAL; PRE; SRGH). C: Dedza, fl. 6.x.1949, *Weihe* N/261 (K). S: N. of Chiradzura, fl. 3.x.1859, *Kirk* (K). **Mozambique.** N: Mandimba valley, fl. 9.iv.1942, *Hornby* 4522 (K). Z: Alto Moloqué, Nhanela, fl. x.1945, *Pedro* 1351 (LMA). MS: 16 km. S. of Mt. Espungabera on Maringa road, fl. 22.xi.1960, *Leach & Chase* 10515 (BM; COI; K; LISC; SRGH).
   Throughout N. & E. tropical Africa from Senegal to Sudan and S. Africa (Transvaal). Woodland and wooded grassland, 600–2400 m.

Subsp. *vitifolia* (Hook. f.) R. & A. Fernandes (with stipuloid bracts present and ellipsoid fruit) occurs in Angola and SW. Africa.

3. **Trochomeria debilis** (Sond.) Hook. f. in F.T.A. **2**: 525 (1871).—Burtt Davy, F.P.F.T.: 225 (1926).—Hutchinson, Botanist in S. Afr.: 389, 428 (1946).—Meeuse in Bothalia **8**: 89 (1962).—Launert & Roessler in Prodr. Fl. SW. Afr. **94**, Cucurb.: 21 (1968). TAB. **110.** Type from S. Africa (Cape Prov.).
   Zehneria debilis Sond. in Harv. & Sond., F.C. **2**: 488 (1862). Type as above.

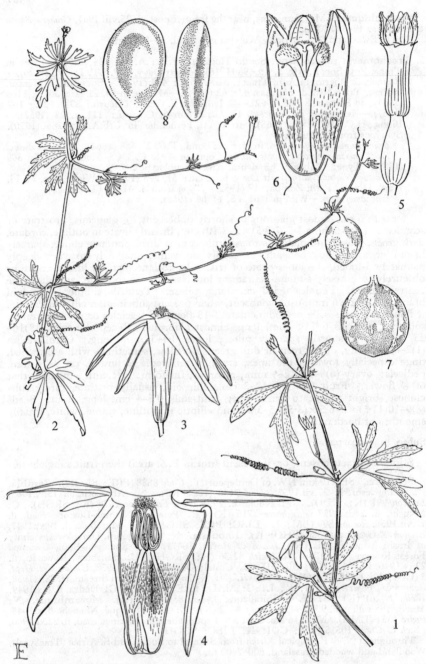

Tab. 110. TROCHOMERIA DEBILIS. 1, part of shoot, ♂ plant (× ½) *Wild & Drummond* 7278; 2, part of shoot, ♀ plant (× ½) *Lugard* 43; 3, ♂ flower (×2); 4, ♂ flower, dissected (×3), both from *Wild & Drummond* 7278; 5, ♀ flower (×3); 6, ♀ flower, receptacle-tube opened (×5); 7, fruit (× ¾), 5–7 from *Lugard* 43; 8, seed, face and lateral views (×3) *McGregor Museum* 2400.

Stems glabrous or shortly and sparsely pubescent, glaucous, prostrate or scandent. Leaf-lamina 1·5–6 × 2–11 cm., broadly ovate in outline, cordate, glaucous, shortly scabrid-setulose or scabrid-punctate above, otherwise glabrous, deeply palmately 5-lobed, the lobes elliptic to narrowly elliptic or linear in outline, rarely entire, usually deeply lobulate with the ultimate segments linear to shortly oblong-lanceolate or ovate-lanceolate, obtuse to acute, apiculate; petioles 0·5–3 cm. long, glabrous or sparsely spiculate. Tendrils simple. Stipuloid bracts 5–20 × 5–23 mm., often absent, when present reniform or suborbicular, laciniate. ♂ flowers solitary or in sessile or shortly pedunculate 2–4-flowered clusters; pedicels 4–20 mm. long, slender. Receptacle-tube 8–16 mm. long, lobes 0·8–3·5 mm. long. Corolla-lobes 0·5–1·3 cm. long, yellowish-green, recurved. ♀ flowers on 0·4–1 cm. long pedicels; ovary 4–7 × 2–3 mm., ellipsoid or ellipsoid-rostrate; perianth similar to that of ♂ flower. Fruit 1·5–3·5 × 1–2·5 cm., sub-globose to ellipsoid, usually ± rostrate, bright red; fruit stalk 0·5–1·8 cm. long. Seeds 7·5–9 × 4·5–5·5 × 2·5–3 mm., elliptic in outline, tumid.
rostrate, perianth similar to that of ♂ flower. Fruit 1·5–3·5 × 1–2·5 cm., subglobose to ellipsoid, usually ± rostrate, bright red; fruit stalk 0·5–1·8 cm. long. Seeds 7·5–9 × 4·5–5·5 × 2·5–3 mm., elliptic in outline, tumid.

Botswana. N: 21 km. W. of Nokaneng, fl. 11.iii.1965, *Wild & Drummond* 6873 (COI; K; SRGH). SW: 32 km. S. of Takatshwane Pan, fl. 21.ii.1960, *Wild* 5102 (BM; K; SRGH). SE: 93 km. NW. of Serowe, fr. 24.iii.1965, *Wild & Drummond* 7278 (COI; SRGH).
Also in S. & SW. Africa. Bushland and wooded grassland, 900–1350 m.

4. **Trochomeria polymorpha** (Welw.) Cogn. in A. & C. DC., Mon. Phan. **3**: 401 (1881).—Hiern, Cat. Afr. Pl. Welw. **1**, 2: 388 (1898).—C. Jeffrey in F.T.E.A., Cucurb.: 88, pro parte, t. 12 fig. 8–10 (1967).—R. & A. Fernandes in C.F.A. **4**: 286 (1970) incl. vars. Type from Angola.
    Heterosicyos polymorphus Welw. in Trans. Linn. Soc., Bot. **27**: 34 (1869).—Hook. f. in F.T.A. **2**: 526 (1871). Type as above.
    Heterosicyos stenolobus Welw., loc. cit. (1869).—Hook. f., loc. cit. (1871). Type from Angola.
    Trochomeria stenoloba (Welw.) Cogn., op. cit.: 402 (1881).—Hiern, tom. cit.: 389 (1898). Type as above.
    Trochomeria bussei Gilg in Engl., Bot. Jahrb. **34**: 343 (1904).—R.E.Fr., Wiss. Ergebn. Schwed. Rhod.-Kongo-Exped. **1**: 313 (1916).—R. & A. Fernandes, tom. cit.: 287 (1970). Type from Tanzania.
    Trochomeria brachypetala R. E. Fr., tom. cit.: 313, t. 37 fig. a–f (1916). Syntypes: Zambia, near Lake Bangweulu, *Fries* 811, 811a (UPS, syntypes).
    Trochomeria multiflora R. Fernandes in Bol. Soc. Brot., Sér. 2, **33**: 189, t. 1 (1959).—R. & A. Fernandes, tom. cit.: 287, t. 34 (1970). Type from Angola.
    Trochomeria teixeirae R. & A. Fernandes in Bol. Soc. Brot., Sér. 2, **43**: 308, t. 2 1969); in tom. cit.: 286, t. 33 (1970). Type from Angola.

Stems (10)20–100 cm. tall, shortly and often densely pubescent, very rarely glabrous, erect or very rarely subscandent, usually unbranched. Leaf-lamina 1·3–10 × 0·2–4·5 cm., sessile, linear to elliptic or broadly elliptic or oblanceolate in outline, finely and usually densely hispidulous or asperulous, entire or obscurely sinuate to bluntly dentate or markedly sinuate-serrate towards the apex, rounded to acute, apiculate, unlobed or shortly to very deeply palmately 3-lobed, the lobes 1·5–15 mm. broad, linear to elliptic, oblong-elliptic, obovate-elliptic or ovate in outline. Tendrils and stipuloid bracts absent. ♂ flowers 2–20 in axillary fascicles or racemes; rhachis of inflorescence up to 40 mm. long; bracts when present 0·5–4 mm. long, subulate to filiform; pedicels 0·1–9 mm. long. Receptacle-tube 7–16 mm. long, finely pilose, lobes 0·3–0·7 mm. long, subulate. Corolla-lobes 2–11 mm. long, deep green to greenish-yellow. ♀ flowers on 2–6 mm. long pedicels; ovary 5–7 × 2–3·5 mm., ovoid to ellipsoid, rostrate; perianth similar to that of ♂ flower. Fruit red, 1–2 cm. in diameter, subglobose; fruit-stalk 0·2–0·6 cm. long. Seeds 5–7·5 × 4–5 × 3·7–4·5 mm., ovoid, tumid, strongly margined, smooth.

Zambia. B: 5 km. W. of Mankoya, fl. 8.xi.1959, *Drummond & Cookson* 6228 (K; SRGH). N: Samfya, fl. 15.xi.1964, *Mutimushi* 1157 (NDO; SRGH). W: Mwinilunga Boma, fr. 2.xii.1937, *Milne-Redhead* 3501 (K). C: 8 km. E. of Lusaka, fl. 26.x.1955,

King 189 (K). E: Katete, fl. 4.xi.1953, *Wright* 39 (K). S: Mazabuka, Choma, Mapanza, fl. & fr. 17.xi.1957, *Robinson* 2494 (K; SRGH). **Rhodesia.** N: Urungwe, Msuku R., fl. 18.xi.1953, *Wild* 4204 (K; SRGH). **Malawi.** N: Karonga Distr., N. of Chilumba, Vintukutu Forest Reserve, 2.i.1973, *Pawek* 6313 (K). **Mozambique.** Z: Moebede road, fl. 15.xi.1948, *Faulkner* 335 (K). MS: Manica, 15 km. from Vila Manica on road to Vila Pery, fl. 25.xi.1965, *Torre & Correia* 13273 (LISC).

Also in SE. Tanzania and Angola. Deciduous woodland on sandy soils, 550–1500 m.

5. **Trochomeria subglabra** C. Jeffrey in Kew Bull. **30**: 480 (1975). Type: Zambia, Mwinilunga, W. of Matonchi Farm, *Milne-Redhead* 2820 (K, holotype).

　　Trochomeria polymorpha sensu C. Jeffrey in F.T.E.A., Cucurb.: 88 (1967) pro minore parte, non (Welw.) Cogn.

Stems 6–40 cm. tall, ± glabrous, erect or ascending, unbranched or sparsely branched from near the base. Leaf-lamina 2–12 × 0·1–3 cm., sessile, linear to broadly elliptic in outline, glabrous or almost so, entire, acute, apiculate, unlobed or very deeply palmately 3–lobed, the lobes 1–3 mm. broad, narrow, linear. Tendrils and stipuloid bracts absent. ♂ flowers solitary or 2–4 in sessile or shortly pedunculate axillary fascicles; bracts usually absent, when present up to 1·5 mm. long; pedicels 4·5–30 mm. long. Receptacle-tube 9·5–16 mm. long, sparsely finely pilose, lobes up to 1 mm. long. Corolla-lobes 5–13 mm. long, greenish-yellow. ♀ flowers on 3–14 mm. long pedicels; ovary 5–7 × 1·5–2·5 mm., ellipsoid-rostrate; perianth similar to that of ♂ flower. Fruits and seeds unknown.

Zambia. N: Mporokoso, Mweru Wantipa, Kabwe Plain, fl. 14.xii.1960, *Richards* 13692 (K). W: Mwinilunga, near Kalenda village, fl. 19.xi.1962, *Richards* 17288 (K). Not known from elsewhere. Ecology uncertain.

12. CTENOLEPIS Hook. f.

Ctenolepis Hook. f. in Benth. & Hook., Gen. Pl. **1**: 832 (1867).
Ctenopsis Naud. in Ann. Sci. Nat. Bot., Sér. 5, **6**: 12 (1866) *nom. illegit.* non De Notar. (1848)
Blastania Kotschy & Peyr., Pl. Tinn.: 15 (1867).

Small scandent herbs with slender stems. Leaves simple, petiolate. Tendrils simple. Probracts stipuloid, suborbicular, fimbriate-ciliate. Flowers very small, yellowish, monoecious. ♂ flowers in short pedunculate racemes. Receptacle-tube shortly campanulate, lobes small, dentiform. Corolla-lobes 5, ovate-lingulate, spreading, united at the base. Stamens 3, 2 double 2-thecous, 1 single 1-thecous; filaments inserted on lower part of receptacle-tube; anthers free, exserted; thecae short, arcuate. Disk absent. ♀ flowers solitary, co-axillary with ♂; ovary subglobose, smooth; ovules few, horizontal; perianth similar to that of ♂ flower; disk absent; stigma usually 2-lobed. Fruit baccate, smooth, subreniform or subglobose, red. Seeds usually 2, rather large, ovate in outline, plano- or concavo-convex, smooth, whitish, sometimes bordered.

2 species in the Old World tropics.

Ctenolepis cerasiformis (Stocks) Hook. f. in F.T.A. **2**: 558 (1871).—Keraudren in Fl. Cameroun **6**, Cucurb.: 49, t. 10 (1967).—C. Jeffrey in F.T.E.A., Cucurb.: 92, t. 14 (1967).—Binns, H.C.L.M.: 42 (1968). TAB. **111**. Syntypes from Sudan and W. Pakistan.

　　Zehneria cerasiformis Stocks in Hook., Journ. Bot. **4**: 149 (1852). Syntypes as above.

　　Blastania fimbristipula Kotschy & Peyr., Pl. Tinn.: 15, t. 7 (1867).—Gomes e Sousa in Bol. Soc. Estud. Col. Moçamb. **29–32**: 91 (1936).—Wiehe, Cat. Fl. Pl. Herb. Div. Pl. Path. Zomba: 16 (1952). Syntypes from Sudan.

　　Blastania cerasiformis (Stocks) Meeuse in Bothalia **8**: 12 (1962). Syntypes as for *Ctenolepis cerasiformis.*

Stems scandent, sparsely shortly setose or almost glabrous, becoming white-punctate, rarely becoming woody and developing a thin pale brown papery bark. Leaf-lamina 2·5–9 × 2·5–13 cm., broadly ovate in outline, cordate, asperulous becoming scabrid-punctate above, asperulous beneath and scabrid-strigose on the main veins, deeply palmately 3–5-lobed, the lobes elliptic or broadly so, coarsely

Tab. 111. CTENOLEPIS CERASIFORMIS. 1, habit ($\times\frac{1}{2}$) *Polhill & Paulo* 2131; 2, ♂ flower opened ($\times9$); 3, double stamen; anther, ventral and dorsal views ($\times36$); 4, flowering node with ♂ inflorescence, young fruit and stipuliform bract ($\times3$); 5, ♀ flower in plan ($\times12$); 6, stigmas ($\times12$), 2–6 from *Polhill* 1308; 7, seed, face and lateral views ($\times3$) *Hornby* s.n.

sinuate or sinuate-serrate, denticulate, sometime lobulate, obtuse to acute, some-times acuminate, apiculate, the central largest. Petioles 0·5–5·5 cm. long, strigose or scabrid-punctate. Stipuloid bracts 6–20 mm. broad and long, finely ciliate. ♂ flowers 3–20, racemose; peduncles 0·5–3·3 cm. long, slender; bracts 1–3 mm. long, small, filiform, or absent; pedicels very slender, shortly ± pilose or almost glabrous. Receptacle-tube 0·5–1 mm. long, lobes 0·5–1 mm. long, narrowly triangular. Petals 1–1·5 × 0·7–1 mm., yellowish, spreading. ♀ flowers on 2–7 mm. pedicels; ovary 2 × 1·5 mm., ellipsoid; receptacle-tube short, 0·5 mm. long, lobes 0·3–1 mm. long, subulate; petals 1·5–2 mm. long. Fruit 1–1·5 cm. in diameter, subglobose, slightly oblate, smooth, red; fruit-stalk 2·5–7 mm. long. Seeds 7·5–11 × 5–6 × 2–3 mm., usually 2, not or scarcely bordered.

Botswana. N: 61 km. N. of Kachikau on road to Kasane, fl. & fr. 11.vii.1937, *Erens* 394 (K; PRE; SRGH). **Zambia.** B: Senanga, Kaunga, near Mashi R. on Angola border, fr. 16.iv.1962, *Mubita* B.89 (SRGH). S: Lusitu, fr. 19.v.1960, *Fanshawe* 5688 (K; SRGH). **Rhodesia.** N: Darwin, Mkumbura R., on road to Tete, fl. & fr. immat. 23.i.1960, *Phipps* 2395 (COI; SRGH). W: Wankie, fr. 1.iii.1963, *Wild* 6068 (K; LISC; SRGH). S: Gwanda, Bubye R., fr. 3.v.1958, *Drummond* 5546 (COI; K; LISC; SRGH). **Malawi.** N: Lupembe, lake shore, 24 km. S. of Karonga, fl. & fr. immat. 11.ii.1953, *Williamson* 155 (BM; SRGH). S: Chikwawa, Ndahwera-Ngabu road, fr. 16.viii.1963, *Salubeni* 85 (SRGH). **Mozambique.** N: Mozambique, fl. & fr. 5.iv.1894, *Kuntze* (K). T: Baroma, Zambesi R., Sisiso, fr. 17.vii.1960, *Chase* 2780 (BM; COI; K; LISC; SRGH). MS: Chemba, Chiou, fl. & fr. 2.iv.1962, *Balsinhas & Macuácua* 540 (COI; K; LISC; LMA). SS: Guija, between Caniçado and Mabalane, fl. & fr. 12.v.1948, *Torre* 7806 (LISC). LM: Magude, fl. 11.iv.1948, *Torre* 7983 (LISC).
Old World tropics from Mauritania & Senegal east to India and in E. Africa south to Transvaal. Lake and river margins in woodland and wooded grassland, 0–1000 m.

13. CUCUMIS L.*

Cucumis L., Sp. Pl. **2**: 1011 (1753); Gen. Pl. ed. 5: 442 (1754).—Deakin, Bohn & Whitaker in Econ. Bot. **25**, 2: 195–211 (1971).

Annual or perennial herbs, the latter with thickened or tuberous rootstock; stems annual, prostrate or scandent. Leaves simple, petiolate. Tendrils simple, rarely more than one at each node. Flowers small to medium or rarely rather large, yellow or pale yellow, monoecious or dioecious, ♂ flowers solitary or in few-flowered axillary fascicles; receptacle-tube campanulate or obconic, ± setulose, lobes usually linear, small. Corolla-lobes 5, united near the base. Stamens 3, two double 2-thecous, one single 1-thecous; filaments short, inserted on receptacle-tube, free; anthers free; connective more or less produced; thecae triplicate. Disk basal, subglobose, free from receptacle-tube. ♀ flowers usually solitary, pedicellate; ovary smooth and densely pubescent or setulose, or covered with soft apically setulose spines or tubercules; ovules numerous, horizontal; perianth as in ♂ flower; staminodes often present, 3, subulate, inserted on receptacle-tube; disk annular, surrounding base of style, free from receptacle-tube; stigma 3-lobed. Fruit usually ellipsoid, oblong-ellipsoid or subglobose, smooth and pubescent to glabrous or usually with dense to scattered apically setiferous spines, tubercules or pustules, firm-walled, fleshy, indehiscent, rarely geocarpic, usually yellowish when ripe, sometimes reddish, often longitudinally striped. Seeds ovate or elliptic in outline, usually smooth, compressed.

Tropics of the Old World, introduced elsewhere; about 25 species, mostly African.

Cucumis sativus L., the cucumber, is known to be cultivated in the F.Z. area. In addition, *Cucumis melo* L., *Cucumis anguria* L. and *Cucumis metuliferus* Naud., which occur wild either native or adventive, are also cultivated.

Tendrils solitary at each node; pedicel of ovary somewhat elongating but not burying the developing fruit; fruit maturing above ground:
  Ovary softly spiny or tuberculate; fruit glabrous, more or less densely softly spiny or tuberculate, spines and tubercles apically setiferous:
    Fruits bright red when ripe, rather sparsely covered with stout fleshy spines; seeds fibrillose - - - - - - - - - - 1. *metuliferus*

* By P. Halliday and C. Jeffrey.

Fruits greenish-yellow to brownish-orange when ripe, rather densely covered with
soft slender spines; seeds smooth:
    Flowers large, corolla-lobes 13–35 mm. long - - - - 5. *quintanilhae*
    Flowers smaller, corolla-lobes usually not exceeding 10 mm. in length:
        Fruit when ripe uniformly yellow or only faintly longitudinally variegated with
        greenish-yellow:
        Plant annual; stems and petioles patent-setulose; fruit-stalk 2·5–21 cm. long,
            usually exceeding 6 cm., expanded upwards - - 2. *anguria*
        Plant perennial; stems and petioles retrorsely scabrid-setose; fruit-stalk
            1·9–5·1 cm. long, not expanded upwards - - - 6. *zeyheri*
Fruit when ripe strongly longitudinally variegated green or white and purplish-
brown, brownish-purple or brownish-orange:
    Plant annual; flowers monoecious; roots fibrous; fruit-stalk rather slender
    or slender, 2–6·8 cm. long:
        Fruit ellipsoid or oblong-ellipsoid, 3·2–9 cm. long, striped pale greenish-
            white and purplish-brown; fruit-stalk 2–4·5 cm. long; spines on fruit
            rather stout, laterally compressed, 3–6 mm. long - 3. *africanus*
        Fruit globose to ellipsoid, 1·8–2·7 cm. long, striped dark and light green or
            white or sometimes purplish-brown, becoming rusty orange with less
            conspicuous striping when ripe; fruit-stalk 4·5–6·8 cm. long, slender;
            spines on fruit weak, terete, slender, 1·5–5 mm. long 4. *myriocarpus*
    Plant perennial; flowers dioecious; roots forming subterranean tubers; fruit-
    stalk stout, about 2 cm. long - - - - - 7. *kalahariensis*
Ovary smooth, uniformly setulose or tomentose, neither spiny nor tuberculate; fruit
without apically setiferous spines or tubercules:
    Plant annual; flowers monoecious; seeds elliptic in outline, about twice as long as
    broad - - - - - - - - - - 8. *melo*
    Plant perennial; flowers dioecious; seeds ovate in outline, up to 1½ times as long
    as broad - - - - - - - - - 10. *hirsutus*
Tendrils 2–8 at each node; pedicel of ovary rapidly elongating and burying the developing
fruit; fruit maturing underground - - - - - - 9. *humifructus*

1. **Cucumis metuliferus** Naud. in Ann. Sci. Nat. Bot., Sér. 4, **11**: 10 (1859).—Hiern,
Cat. Afr. Pl. Welw. **1**, 2: 397 (1898).—Burtt Davy, F.P.F.T. **1**: 228 (1926).—
Gomes e Sousa in Bol. Soc. Estud. Col. Moçamb. **29–32**: 22, 92 (1936) (" meruli-
ferus ").—Suesseng. & Merxm. in Proc. Trans. Rhod. Sci. Ass. **43**: 61 (1951).—
Wild, S. Rhod. Bot. Dict.: 71 (1953).—Williamson, Useful Plants Nyasal.: 43 (1956)
(" metuliferous ").—Story in Mem. Bot. Survey S.!Afr. **30**: 49, t. 59 (1958).— Watt &
Breyer-Brandwijk, Med. & Pois. Pl. S. & E. Afr. ed. 2: 353 (1962).—Meeuse in
Bothalia **8**: 68 (1962).—R. & A. Fernandes in Mem. Junta Invest. Ultramar, Sér. 2,
34: 94 (1962); in C.F.A. **4**: 259 (1970).—Mitchell in Puku **1**: 127 (" metutiferus "),
182, 191 (1963).—Nair, Sel. Fam. Zamb. Fl. Pl.: 51 (1967).—Keraudren in Aubrév.,
Fl. Cameroun **6**, Cucurb.: 142, t. 27 fig. 3–4 (1967).—C. Jeffrey in F.T.E.A.,
Cucurb.: 98 (1967).—Binns, H.C.L.M.: 42 (1968).—Launert & Roessler in Prodr.
Fl. SW. Afr. **94**, Cucurb.: 14 (1968). TAB. **112** fig. H. Type a cultivated
plant.

Annual; stems prostrate or scandent, to 5 m., strongly patent-setose. Leaf-
lamina 3·5–12 × 3·5–13·5 cm., ovate to broadly ovate or subpentagonal in outline,
cordate, hispid-setulose especially on veins beneath, becoming scabrid-punctate,
margins sinuate-denticulate, palmately and usually rather shortly (3)5-lobed,
lobes shallowly triangular to ovate or broadly elliptic, rounded to subacute,
shortly acuminate, apiculate, very rarely lobulate; basal sinus deep, broad,
subrectangular. Petiole 2–11·5 cm. long, strongly patent-setulose. Flowers
monoecious. ♂ flowers solitary or 2–10 in usually sessile fascicles; pedicels 3–
16 mm. long. Receptacle-tube 3·5–6 mm. long, lobes 1·5–5 mm. long. Petals
3–11 mm. long, yellow or pale orange. ♀ flowers solitary, sometimes co-axillary
with ♂; pedicel 5–35 mm. long; ovary 9·5–25 × 7·5–10 mm., ellipsoid, covered
with large soft spines; perianth similar to that of male flower. Fruit 6·5–13 × 3·5–
6·5 cm., oblong-cylindric, rounded at the ends, bluntly trigonous, ornamented with
stout fleshy spines, dark green mottled paler grey-green or yellow, becoming
bright red when ripe; fruit-stalk 2–7 cm. long, strongly spiculate, ridged, not or
only slightly expanded at the apex. Seeds 5·1–7·4 × 3·0–3·7 × 1·0–1·1 mm.,
narrowly ovate in outline, compressed, with rounded margins and depressed or
slightly convex faces, smooth, sericeous-fibrillose.

Botswana. N: 43 km. from Sehitwa on road to Tsau, fr. 4.viii.1955, *Story* 5098 (K;
PRE). SE: 120 km. WNW. of Francistown on Maun road, fr. 2.v.1957, *Drummond* 5295

(K; LISC; SRGH). **Zambia.** B: Masese, fl. 13.iii.1961, *Fanshawe* 6415 (K; SRGH).
N: Mweru-wa-Ntipa, Bulaya Camp, fl. 26.iv.1957, *Vesey-FitzGerald* 1194 (SRGH).
E: Nyamadzi R., fl. & fr. immat. 25.iii.1955, *E. M. & W.* 1174 (BM; LISC; SRGH).
S: Gwembe, fl. 1.iv.1952, *White* 2385 (FHO; K). **Rhodesia.** N: Trelawney, fl.
8.x.1943, *Jack* 215 (K; LISC; PRE; SRGH). W: Wankie, Deka R., fl. 1.iii.1963, *Wild*
6067 (LISC; SRGH). C: Gwelo, Senka lands, fl. & fr. 25.v.1967, *Biegel* 2183 (COI; K;
SRGH). E: Umtali, Burma Valley, fl. 12.i.1959, *Chase* 7039 (K; SRGH). S: Ndanga,
Chipinda Pools, fl. 31.iii.1961, *Goodier* 1050 (COI; K; SRGH). **Malawi.** S: Shire R.,
16° S., fl. 1863, *Kirk* (K). **Mozambique.** N: Mandimba, fl. 1.iv.1942, *Hornby* 3731
(K). Z: Mutarara, R. Zambesi, fl. 6.v.1943, *Torre* 5304 (LISC). T: Mágoè, 17 km. from
Mágoè towards Mágoè Velho, fl. & fr. 3.iii.1970, *Torre & Correia* 18177 (LISC). MS:
between Gondola and Vila Machado, fl. 22.iii.1960, *Wild & Leach* 5223 (K; LMA;
SRGH). SS: Vale do Chinhaquete, between Guijá and Mabalane, 17 km. from
Mabalane, fr. 3.vi.1959, *Barbosa & Lemos* 8592 (COI; K; LISC; LMA). LM: Matola,
on Boane-Chongalane road, fl. 12.v.1968, *Balsinhas* 1266 (COI).
Tropical Africa. Semi-evergreen forest, woodland, wooded grassland and grassland,
also on abandoned cultivated land, often riverine, 300–1100 m.

2. **Cucumis anguria** L., Sp. Pl. **2**: 1011 (1753).—Meeuse in Bothalia **8**: 77 (1962).—
R. & A. Fernandes in Mem. Junta Invest. Ultramar, Sér. 2, **34**: 91 (1962); in C.F.A.
4: 257 (1970).—C. Jeffrey in F.T.E.A., Cucurb.: 104 (1967).—Launert & Roessler
in Prodr. Fl. SW. Afr.: **94**, Cucurb.: 13 (1968). TAB. **112** fig. F. Type a cultivated
plant.
    Cucumis longipes Hook. f. in F.T.A. **2**: 547 (1871).—Hiern, Cat. Afr. Pl. Welw.
1, 2: 395 (1898). Type from Angola.
    Cucumis figarei sensu Eyles in Trans. Roy. Soc. S. Afr. **5**, 4: 498 (1916) non Naud.
    Cucumis africanus var. *myriocarpus* sensu Bremek. & Oberm. in Ann. Transv. Mus.
16, 2: 436 (1935).
    Cucumis sp.—Wiehe, Cat. Fl. Pl. Herb. Div. Pl. Path. Zomba: 17 (1952).
    Cucumis africanus sensu Martineau, Rhod. Wild Fl.: 83, t. 4 (1953); & sensu
Keraudren in Fl. Madag., Cucurb.: 142, t. 34 fig. 7 (1962) non L.f.
    Cucumis ficifolius sensu Wild, Fl. Vict. Falls: 145 (1953) non A. Rich.
    Cucumis anguria L. var. *longipes* (Hook. f.) Meeuse in Blumea, Suppl. **4**: 200
(1958); loc. cit.—Watt & Breyer-Brandwijk, Med. & Pois. Pl. S. & E. Afr. ed. 2:
350 (1962). Type as for *Cucumis longipes*.

Annual; stems prostrate or scandent, to 3 m., patent-setulose. Leaf-lamina
3–10 × 3·5–12·5 cm., broadly ovate in outline, broadly and shallowly cordate,
punctate and setulose becoming scabrid-setose on veins beneath, scabrid-setulose
becoming punctate above, margins ± sinuate-denticulate, palmately (3)5(7)-lobed,
usually deeply, lobes narrowly to broadly elliptic, rhombic, obovate-elliptic or less
often ovate-triangular or triangular, obtuse to rounded, apiculate, the central
largest, often shortly 3-lobulate. Petiole 2–12 cm. long, patent-setulose. Flowers
monoecious. ♂ flowers 2–10 in subsessile fascicles; pedicels 6–30 mm. long.
Receptacle-tube 2–4 mm. long, lobes 1·2–3 mm. long. Petals 3–7·5 mm. long,
yellow. ♀ flowers solitary, sometimes co-axillary with ♂; pedicels 18–105 mm.
long; ovary 7–9 × 6–9 mm., broadly ellipsoid, shortly rostrate, softly spiny;
perianth similar to that of ♂ flower. Fruit 3–4·5 × 2·5–3·5 cm., ellipsoid to subglobose, softly spiny, at least proximally, green or green striped paler green
becoming yellow when ripe; spines 3–11 mm. long; fruit-stalk 2·5–21 cm. long,
expanded upwards. Seeds 5–6 × 2·2–2·7 × 1·0–1·3 mm., elliptic in outline,
compressed, with rounded margins, smooth.

Botswana. N: 6 km. SE. of Tsau, fl. & fr. immat. 18.iii.1965, *Wild & Drummond*
7130 (COI; K; SRGH). **Zambia.** B: Masese, fl. 9.iii.1960, *Fanshawe* 5424 (K). N:
Mwandwisi Valley, fr. immat. iv.1937, *Trapnell* 1767 (K). W: Luanshya, i.1934, *Trapnell*
1397 (K). C: Lusaka, fr. 28.vi.1956, *Angus* 1367 (K; LISC; SRGH). E: Luangwa
Valley Game Reserve S., fl. 17.iv.1967, *Prince* 489 (K). S: Mapanza W., fl. & fr. 3.v.1953,
Robinson 200 (K). **Rhodesia.** N: Miami, fl. iv.1926, *Rand* 30 (BM); Urungwe, 64 km.
N. of Sinoia, fr. 25.x.1959, *Leach* 9495 (COI). C: Salisbury, fr. iv.1932, *Rattray* 513
(SRGH). E: Chipinga, Sabi Valley Experimental Station, fl. ix.1959, *Soane* 75 (COI;
SRGH). S: Gwanda, Bubye Crossing, fl. & fr. 6.v.1958, *Boughey* 2720 (SRGH).
Malawi. N: near Ekwendeni, fl. & fr. 21.v.1970, *Brummitt* 10997 (K; MAL). C:
Lilongwe, Chitedze, fl. 22.iii.1955, *E. M. & W.* 1121 (BM; LISC; SRGH). S: Zomba,
on Blantyre road, fl. & fr. 17.iii.1950, *Wiehe* 447 (K; SRGH). **Mozambique.** T: Tete,
6 km. on Changara road, fl. & fr. 19.iii.1966, *Torre & Correia* 15235 (LISC). MS:
Chemba, Chiou, C.I.C.A. Experimental Station, fl. & fr. 12.iv.1960, *Lemos & Macúdcua* 83
(COI; K; LISC; LMA; SRGH).

Tropics of Africa, Australia and the New World, often cultivated; outside southern Africa probably always as an introduction. Woodland and wooded grassland, often on Kalahari sand, and on abandoned cultivated land, 200–1500 m.

3. **Cucumis africanus** L. f., Suppl.: 423 (1781).—Burtt Davy, F.P.F.T. **1**: 228 (1926) excl. vars.—Meeuse in Bothalia **8**: 72 (1962) pro majore parte.—R. & A. Fernandes in Mem. Junta Invest. Ultramar, Sér. 2, **34**: 93 (1962); in C.F.A. **4**: 258 (1970).— Launert & Roessler in Prodr. Fl. SW. Afr. **94**, Cucurb.: 13 (1968). TAB. **112** fig. D. Type from S. Africa (Cape Prov.).
   Cucumis hookeri Naud. in Gard. Chron. **1870**: 1503 (1870).—Story in Mem. Bot. Survey S. Afr. **30**: 49, t. 58 (1958).—Leistner in Koedoe **2**: 170 (1959).—Watt & Breyer-Brandwijk, Med. Pois. Pl. S. & E. Afr. ed. 2: 353 (1962). Type a cultivated plant.
   Cucumis africanus var. *zeyheri* sensu Bremek. & Oberm. in Ann. Transv. Mus. **16**, 2: 436 (1935).

Annual or perhaps sometimes perennial, or at least sometimes developing a woody rootstock; stems prostrate or scandent, to c. 1 m., patent-setulose or deflexed-setulose, becoming scabrid and developing a thin pallid bark when old. Leaf-lamina 1·6–8·2 × 1·8–7 cm., ovate in outline, cordate, shortly sparsely scabrid-setulose above, densely so beneath, becoming scabrid-punctate, more or less sinuate-denticulate, deeply palmately (3)5-lobed, lobes elliptic, broadly elliptic or ovate-elliptic, obtuse to rounded, shortly acuminate, apiculate, sometimes 3-lobulate, the central largest. Petiole 1·1–6 cm. long, antrorsely or patently scabrid-setulose. Flowers monoecious. ♂ flowers solitary or usually 2–5 in small fascicles; pedicels 2–9 mm. long. Receptacle-tube 3–5 mm. long; lobes 1·5–3 mm. long. Petals 5–11 mm. long, bright yellow. ♀ flowers solitary; pedicel 10–40 mm. long; ovary 8–15 × 3–5 mm., ellipsoid or oblong-ellipsoid, densely softly spiny; perianth similar to that of ♂ flower. Fruit 3·2–6(9) × 1·8–3·3(4·5) cm., ellipsoid to oblong-ellipsoid, rounded at the ends, more or less densely spiny, when ripe strongly striped pale greenish-white and purplish-brown; spines 3–6 mm. long, longitudinally compressed; fruit-stalk 2–4·5 cm. long, slender, not expanded upwards. Seeds 4–7 × 2–3·8 × 1·1–2 mm., elliptic in outline, compressed.

Botswana. N: 16 km. NE. of Toromoja School, fl. & fr. 25.iv.1971, *Pope* 394 (K). SW: 24 km. W. of Ghanzi, fl. 20.iv.1963, *Ballance* 613 (SRGH). SE: Ngami, Mabele-a-Pudi, fl. & fr. 8.v.1930, *van Son* in Herb. Transv. Mus. 28798 (K).
   Also in Angola, S. Africa (Cape Prov.) & SW. Africa. Bushland.

4. **Cucumis myriocarpus** Naud. in Ann. Sci. Nat. Bot., Sér. 4, **11**: 22 (1859).—Exell in Journ. Bot., Lond. **64**: 303 (1926).—Meeuse in Bothalia **8**: 74 (1962).—Watt & Breyer-Brandwijk, Med. Pois. Pl. S. & E. Afr. ed. 2: 353, t. 100 (1962). Syntypes from S. Africa and cultivated plants.
   C. leptodermis Schweickerdt in S. Afr. Journ. Sci. **30**: 460 (1933).—Meeuse, tom. cit.: 76 (1962). Type from S. Africa (Cape Prov.).
   Cucumis merxmuelleri Suesseng. in Proc. Trans. Rhod. Sci. Ass. **43**: 61 (1951). Type: Rhodesia, Marandellas, *Dehn* 746 (M, holotype; K, LISC, SRGH, isotypes).

Subsp. myriocarpus. TAB. **112**, fig. E.

Annual. Stems prostrate or scandent, to 2 m., strongly deflexed-setulose, becoming scabrid, developing a thin pallid bark when older. Leaf-lamina 2·5–10 × 2–7·5 cm., broadly ovate in outline, shallowly cordate, shortly more or less hispid or scabrid-setulose above, hispid or scabrid-punctate and shortly setulose on veins beneath, obscurely sinuate-denticulate, usually very deeply palmately 5-lobed, lobes elliptic, acute to obtuse, shortly acuminate, apiculate, rather deeply 3–5-lobulate, the central longest. Petiole 1·7–10 cm. long, patently or antrorsely aculeate-hispid and also finely deflexed-setulose. Flowers monoecious. ♂ flowers solitary, paired or rarely few-fasciculate; pedicels 5–25 mm. long. Receptacle-tube 2–5 mm. long, lobes 1–3 mm. long. Petals 3–6(10) mm. long, pale yellow. ♀ flowers solitary, co-axillary with ♂; pedicels 15–45 mm. long; ovary 4·5–10 × 2–5 mm., ellipsoid-rostrate, shortly softly spiny; perianth similar to that of ♂ flower. Fruit 1·8–2·7 × 1·5–2·5 mm., globose to ellipsoid, softly weakly spiny, striped dark and light green or white or sometimes purplish or brownish, rusty orange with less conspicuous striping when ripe; spines 1·5–5 mm. long; fruit-

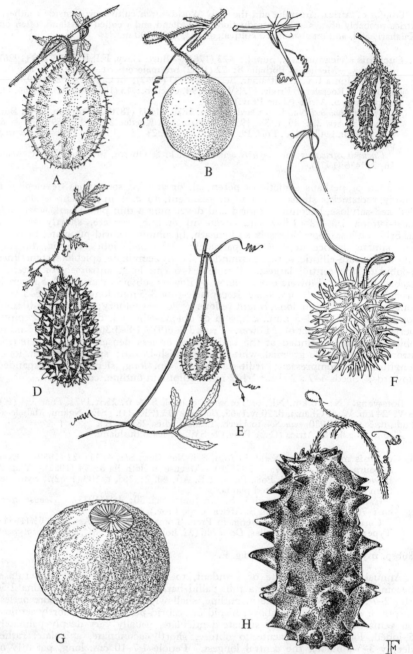

Tab. 112. Fruits of CUCUMIS. A.—CUCUMIS ZEYHERI (× ½) *Chase* 8485; B.—CUCUMIS
HIRSUTUS (× ½) *Milne-Redhead* 3075A. C.—CUCUMIS KALAHARIENSIS (× ½) *Story*
4895. D.—CUCUMIS AFRICANUS (× ½) *Cox* 321. E.—CUCUMIS MYRIOCARPUS subsp.
MYRIOCARPUS (× ½) *Teague* 485. F.—CUCUMIS ANGURIA (× ½) *Trapnell* 1398. G.—
CUCUMIS HUMIFRUCTUS (× ½) *Mitchell* s.n. H.—CUCUMIS METULIFERUS (× ½) *King*
B13:5.

stalk slender, 4·5–6·8 cm. long. Seeds 5·2–6·2 × 2·4–2·9 × 1·2–1·7 mm., elliptic, smooth, with rather rounded margins, pallid.

Botswana. SE: Mochudi, fr. i–iv.1914, *Harbor in Rogers* 6608 (K). **Zambia.** N: Mporokoso, Mweru-wa-Ntipa, fl. 24.ix.1937, *Trapnell* 1791 (K). **Rhodesia.** C: Mlezu School farm, fl. & fr. 4.i.1966, *Biegel* 751 (COI; K; LISC; SRGH). E: Umtali, Odanzi R. valley, fl. & fr. immat. 1915, *Teague* 485 (BOL; K). S: Victoria, Mushandike National Park, fl. immat. 16.ii.1963, *Loveridge* 614 (SRGH). **Mozambique.** SS: Gaza, 6 km. SW. of Malvernia, fl. & fr. 26.iv.1961, *Drummond & Rutherford-Smith* 7531 (COI; K; LISC; SRGH). LM: Maguda, near Chobela, fl. & fr. 20.xii.1955, *Myre & Carvalho* 2298 (LMA).

Also in S. Africa; adventive in Australia and S. Europe. Weed of cultivated land, c. 950 m.

Deakin, Bohn and Whitaker in Econ. Bot. **25**, 2: 195–211 (1971) report on the results of their studies on interspecific hybridization in *Cucumis* L.; they find *C. leptodermis* Schweickerdt and *C. myriocarpus* Naud. cross readily and yield vigorous, fully fertile F_1 hybrids, and support Meeuse's suggestion that " perhaps *C. leptodermis* is no more than a variety or a subspecies of *C. myriocarpus* ". Accordingly, we have treated these taxa as subspecies, see Jeffrey & Halliday in Kew Bull. **30**: 481 (1975). Subsp. *leptodermis* (Schweickerdt) Jeffrey & Halliday is confined to S. Africa.

5. **Cucumis quintanilhae** R. & A. Fernandes in Rev. Bras. Port. Biol. **3**: 269, t. 10 & 11 (1963). Type: Rhodesia, near Beitbridge, *Drummond* 6025 (SRGH, holotype; K, isotype).

Probably annual; stems prostrate or scandent, patent-setulose, later scabrid. Leaf-lamina 6·6–10 × 5·7–8 cm., ovate in outline, weakly cordate, shortly hispid and punctate above, shortly hispid beneath, obscurely and minutely sinuate-denticulate, almost unlobed to deeply palmately 5-lobed, lobes triangular, ovate-triangular or elliptic, acute to rounded, apiculate, sometimes shortly 3-lobulate, the central largest. Flowers monoecious. ♂ flowers solitary or 2–3 in small fascicles; pedicels 17–65 mm. long. Receptacle-tube 5–7 mm. long; lobes 3·5–7 mm. Petals 13–26 mm. long, white or yellow, green-veined. ♀ flowers solitary; pedicel 60–80 mm. long, stout; ovary 8–12 × 4–6 mm., oblong-ellipsoid, softly spiny; receptacle-tube 3–5·5 mm. long, lobes 4–7 mm. long, linear or oblanceolate, entire or shortly lobed; petals 16–37 mm. long. Fruit 4 × 2·5 cm., ellipsoid, green becoming yellow when ripe, spiny; spines 9–12 mm. long. Seeds c. 6·5 × 3 × 1·6 mm.

Rhodesia. S: 3 km. from Beitbridge on West Nicholson road, fl. 15.ii.1955, *E.M. & W.* 398 (BM; LISC; SRGH).

Also in S. Africa (Transvaal). *Colophospermum* bushland.

6. **Cucumis zeyheri** Sond. in Harv. & Sond., F.C. **2**: 496 (1862).—Swynnerton et al. in Journ. Linn. Soc., Bot. **40**: 74 (1911).—Eyles in Trans. Roy. Soc. S. Afr. **5**, 4: 498 (1916).—Suesseng. & Merxm. in Proc. Trans. Rhod. Sci. Ass. **43**: 62 (1951).— Meeuse in Bothalia **8**: 79 (1962). TAB. **112** fig. A. Syntypes from S. Africa (Transvaal).

Cucumis africanus var. *zeyheri* (Sond.) Burtt Davy, F.P.F.T.: 229 (1926).— Bremek. & Oberm. in Ann. Transv. Mus. **16**, 2: 436 (1935).—Hutch., Botanist in S. Afr.: 666 (1946). Syntypes as for *C. zeyheri*.

Cucumis prophetarum subsp. *zeyheri* (Sond.) C. Jeffrey in Kew Bull. **15**: 351 (1962); in F.T.E.A., Cucurb.: 104 (1967). Syntypes as for *C. zeyheri*.

Cucumis africanus sensu Watt & Breyer-Brandwijk, Med. Pois. Pl. S. & E. Afr. ed. 2: 350, t. 99 (1962) non L. f.

Cucumis ficifolius sensu Watt & Breyer-Brandwijk, tom. cit.: 352 (1962) pro parte non A. Rich.

Perennial with fibrous woody tuberous rootstock; stems prostrate or rarely scandent, strongly recurved-setulose or aculeate. Leaf-lamina 2·5–9 × 2–6 cm., ovate or rarely ovate-triangular, weakly cordate, hispid-setulose above and especially on veins beneath, becoming scabrid-punctate above and more densely so beneath, sinuate-serrate or denticulate, rarely unlobed, usually deeply palmately 3–5-lobed, lobes triangular, ovate-triangular or usually elliptic or narrowly elliptic, the central usually much the largest, 3–5-lobulate or lobulate-denticulate, acute to obtuse, apiculate. Petiole 0·5–4 cm. long, strongly and densely recurved-setulose, very scabrid. Flowers monoecious. ♂ flowers solitary; pedicel 3–12(22) mm. long. Receptacle-tube 2·5–4·5 mm. long, lobes 1–4 mm. long. Petals 4·5–9 mm. long, yellow. ♀ flowers solitary; pedicel 14–25 mm. long; ovary 8–11 × 4·5–6 mm., ellipsoid, densely softly setulose; perianth similar to that of ♂ flowers. Fruit 3·4–5·2 × 2·3–3·5 cm., oblong, ellipsoid or obovoid-ellipsoid, softly spiny, con-colorous green or striped dark and light green or green and yellow, concolorous yellow when ripe; spines rather slender, 2–16 mm. long; fruit-stalk 1·9–5·1 cm. long. Seeds 5·2–8·4 × 2·2–4·5 × 1·5–2·0 mm., elliptic in outline, whitish.

Zambia. S: Mazabuka, fl. 25.x.1933, *Herb. Central Res. Station* 452 (PRE). **Rhodesia.** N: Trelawney, fr. 22.ii.1943, *Jack* 119 (SRGH). W: Matopos Research Station, fl. 21.iv.1952, *Plowes* 1420 (K; SRGH). C: Salisbury, fl. 8.iii.1921, *Godman* 41 (BM). E: Umtali, Vumba Mts., fl. & fr. 19.xi.1967, *Chase* 8485 (COI; K; SRGH). S: Victoria, Popoteke Gorge, st. 23.v.1971, *Ngoni* 111 (K). **Mozambique.** LM: near Goba, fl. 27.xii.1953, *Myre & Carvalho* 1743 (K; LISC; LMA); fl. 10.iii.1968, *Balsinhas* 1172 (COI; LMA).

Also in S. Africa, Lesotho and Swaziland. Open woodland and grassland, also a weed of cultivated land, 300–1650 m.

Deakin, Bohn and Whitaker in Econ. Bot. **25**, 2: 195–211 (1971) conclude that their " breeding results failed to furnish strong evidence to support Jeffrey's statement " (in F.T.E.A., Cucurb. 104, 1967) " that *C. zeyheri* is a subspecies of *C. prophetarum* " and believe that " for the present *C. zeyheri* should be retained as a distinct species separate from *C. prophetarum* because they yielded weakly fertile hybrids ". This contrasts with the situation observed with respect to *C. myriocarpus* Naud. and *C. leptodermis* Schweickerdt (see above, p. 467) and we therefore maintain the two taxa distinct at the specific level.

7. **Cucumis kalahariensis** Meeuse in Bothalia **8**: 70 (1962).—Launert & Roessler in Prodr. Fl. SW. Afr. **94**, Cucurb.: 14 (1968). TAB. **112** fig. C. Type cultivated from seed from SW. Africa.

Perennial; stems prostrate, to 3 m., sparsely setulose, rooting at nodes, arising from tuberous roots up to 1 m. below ground. Leaf-lamina 6·5–12·5 × 3·5–10 cm., ovate or ovate-oblong in outline, finely setulose or hispid above and more densely so beneath, becoming scabrid, deeply palmately (3)5(7)-lobed, lobes elliptic, narrowly oblong-elliptic or lanceolate, obtuse to acute, apiculate, sinuate-dentate or sinuate-lobulate, the central much the largest. Petioles 1–5 cm. long, aculeate-setulose, becoming scabrid. Flowers dioecious. ♂ flowers solitary or 2–3 in small fascicles; pedicels 8–10(20) mm. long. Receptacle-tube 5–7 mm. long; lobes 2·5–5 mm. long. Petals 5·5–10 mm. long, yellow. ♀ flowers solitary; pedicels (4)8–10 mm. long; ovary c. 13 × 5 mm., ellipsoid, densely setose; perianth rather larger than in ♂ flowers, otherwise similar. Fruit 3–5·5 × 2–3 cm., oblong or oblong-ellipsoid, rounded at the ends, densely shortly softly spiny, green or greenish-white, longitudinally variegated with brownish-purple; spines 2–5 mm. long, laterally compressed, rather thick; fruit-stalk stout, c. 2 cm. long, slightly expanded upwards. Seeds 6–7 × 3–4 × 1·5–2 mm., elliptic in outline, smooth, pallid, subcompressed.

Botswana. SW: 13 km. N. of Werda, fr. 26.ii.1960, *Wild* 5162 (SRGH). SE: 211 km. NW. of Molepolole, fr. 15.vi.1955, *Story* 4895 (K; PRE; SRGH).

Also in SW. Africa. Bushland and grassland on Kalahari sand.

8. **Cucumis melo** L., Sp. Pl. **2**: 1011 (1753).—Ficalho, Pl. Ut. Afr. Port.: 188 (1884).— Watt & Breyer-Brandwijk, Med. Pois. Pl. S. & E. Afr. ed. 2: 353 (1962).—Meeuse in Bothalia **8**: 61 (1962).—R. & A. Fernandes in Mem. Junta Invest. Ultramar, Sér. 2, **34**: 84 (1962); in C.F.A. **4**: 254 (1970).—Keraudren in Fl. Madag., Cucurb.: 147, t. 34 fig. 5 & 6 (1966); in Fl. Cameroun **6**, Cucurb.: 138, t. 27 fig. 1 & 2 (1967).— C. Jeffrey in F.T.E.A., Cucurb.: 106 (1967).—Nair, Sel. Fam. Zamb. Fl. Pl.: 51 (1967). Type a cultivated plant.
   C. melo var. *agrestis* Naud. in Ann. Sci. Nat. Bot., Sér. 4, **12**: 110 (1859).—Burtt Davy, F.P.F.T.: 228 (1926). Type a cultivated plant.
   C. melo var. *cultus* Kurz in Journ. Asiatic Soc. Bengal **46**: 103 (1877).—Williamson, Useful Plants Nyasal.: 43 (1950). Type a cultivated plant.

Annual; stems prostrate or scandent, to 2 m., patent-setulose or hirsute. Leaf-lamina 2·5–8·5(16) × 3–11(20) cm., broadly ovate to reniform in outline, shallowly cordate, sinuate-denticulate, more or less setulose becoming punctate above and beneath, unlobed and broadly rounded or variously usually faintly palmately (3)5–

lobed, lobes usually broad, shallow and rounded. Petiole 1·5–8(10·5) cm. long, patent-setulose. ♂ flowers solitary or 2–5 in small sessile or shortly pedunculate fascicles; pedicels 3·5–4·5(25) mm. long, often densely setulose. Receptacle-tube 3–8 mm. long; lobes 1–3·5(6) mm. long. Petals 3–10(22) mm. long, yellow or pale apricot yellow. ♀ flowers solitary; pedicel stout, 3–50(80) mm. long; ovary (5)11–16(60) × 2–6 mm., various, often oblong, velutinous; perianth similar to that of ♂ flowers. Fruit 2·5–10 × 2–7·5 cm., various, pyriform, ellipsoid, oblong, globose or obpyriform, green with lighter green or yellow stripes becoming yellow when ripe, in cultivated forms often much larger and very variable in size, shape, ornamentation and colour; fruit-stalk 1–8 cm. long. Seeds 5–8 × 2·5–4 × 1–1·5 mm., elliptic in outline, in cultivated forms often larger.

Zambia. C: Broken Hill, fr. v.1909, *Rogers* 8085 (BOL). E: W. of Nyimba, fl. 12.xii.1958, *Robson* 930 (K). **Rhodesia.** N: Urungwe, Chirundu, fl. 21.ii.1954, *Lovemore* 384 (K; LISC; SRGH). S: Nuanetsi, SW. of Mateke Hills, fl. 12.v.1958, *Guy* SRGH 85925 (SRGH). **Malawi.** S: Port Herald, fl. 1938, *Lawrence* 626 (K). **Mozambique.** MS: Gorongosa, Chitengo, fl. & fr. 21.x.1965, *Balsinhas* 972 (COI; LMA). LM: Gaza, environs of Vila João Belo, fl. 20.iii.1941, *Torre* 2641 (LISC).
Throughout the warmer regions of the Old World, widely cultivated. River margins, also cultivated, 50–1200 m.

9. **Cucumis humifructus** Stent in Bothalia 2: 356 (1927) cum tab. (as " humofructus ").
—Meeuse in Arch. Néerland. Zool. **13**, Suppl.: 314–318 (1958); in Bothalia **8**: 62 (1962).—Watt & Breyer-Brandwijk, Med. Pois. Pl. S. & E. Afr. ed. 2: 353 (1962).—R. & A. Fernandes in Mem. Junta Invest. Ultramar, Sér. 2, **34**: 84 (1962); in C.F.A. **4**: 253 (1970).—C. Jeffrey in F.T.E.A., Cucurb.: 107 (1962).—Launert & Roessler in Prodr. Fl. SW. Afr. **94**, Cucurb.: 14 (1968). TAB. **112** fig. G. Type from S. Africa.
Cucumis sp. nov.—Burtt Davy, F.P.F.T. **1**: 229 (1926).

Annual herb; stems prostrate, to 3 m. or more, crispate-setulose to almost glabrous, rooting at the nodes. Leaf-lamina 3–11(17) × 3–13(22) cm., ovate or broadly ovate to suborbicular in outline, weakly cordate, more or less scabrid-setulose above and on veins beneath, sinuate-denticulate, very shortly palmately 5-lobed or subpentagonal, lobes broadly ovate or broadly triangular, rounded, sometimes shortly acuminate, apiculate. Petiole 1·5–5(11·5) cm. long, crispate-setulose. Tendrils rather weak, 2–8 at each node. Flowers monoecious. ♂ flowers 2–12 in small fascicles; pedicels 5–11(10) mm. long. Receptacle-tube 1·5–3·5 mm. long, lobes 0·7–3 mm. long. Petals 2–7 mm. long, yellow or pale yellow. ♀ flowers solitary, co-axillary with ♂, opening before ♂ towards apex of shoots; pedicel stout, 3·5–11 mm. long, rapidly elongating after flowering and burying the developing fruit; ovary 3–3·5 × 1 mm., ovoid-rostrate, acute, densely retrorsely tomentose; perianth similar to that of ♂ flower, quickly deciduous. Fruit subterranean, c. 4 × 3·5–5·5 cm., subglobose, pale greenish, yellow or whitish, rugose; fruit-stalk 10–25 cm. long. Seeds 15–20 × 7–9 × 2–3 mm., elliptic in outline, compressed, whitish.

Zambia. N: Mbala (Abercorn), Sandpits, fl. 4.i.1961, *Richards* 13778 (K). S: Kalomo, Siantambo, fr. 18.iii.1964, *Mitchell* 24/99 (LISC; SRGH). **Rhodesia.** W: Matobo, Farm Quaringa, fl. ii.1960, *Miller* 7192 (COI; SRGH). C: Hartley, Poole Farm, fl. 14.ii.1953, *Hornby* 3343 (LMA; SRGH).
Eastern and southern tropical Africa from Kenya to SW. Africa and Transvaal. Woodland and grassland, on deep sand, 1350–1500 m.

10. **Cucumis hirsutus** Sond. in Harv. & Sond., F.C. **2**: 497 (1862).—Eyles in Trans. Roy. Soc. S. Afr. **5**, 4: 498 (1916).—R. E. Fr., Wiss. Ergebn. Schwed. Rhod.-Kongo-Exped. **1**: 312 (1916).—Burtt Davy, F.P.F.T. **1**: 228 (1926).—Pole Evans in Mem. Bot. Survey S. Afr. **21**: 32 (1948).—Wild, Fl. Vict. Falls 145 (1953).—Williamson, Useful Plants Nyasal.: 43 (1956).—Wild & Gelfand in Centr. Afr. Journ. Medicine **5**: 295 (1959).—Watt & Breyer-Brandwijk, Med. Pois. Pl. S. & E. Afr. ed. 2: 352 (1962).—Meeuse in Bothalia **8**: 63 (1962).—R. & A. Fernandes in Mem. Junta Invest. Ultramar, Sér. 2, **34**: 85 (1962); in C.F.A. **4**: 254 (1970).—Mitchell in Puku **1**: 127, 180 (1963).—Keraudren in Fl. Madag., Cucurb.: 143, t. 34 fig. 8 & 9 (1966); in Fl. Cameroun **6**, Cucurb.: 140, t. 27 fig. 5 & 6 (1967).—C. Jeffrey in F.T.E.A., Cucurb.: 107 (1967).—Binns, H.C.L.M.: 42 (1968). TAB. **112** fig. B. Syntypes from S. Africa.

Cucumis sonderi Cogn. in A. & C. DC., Mon. Phan. **3**: 489 (1881).—Burtt Davy, loc. cit. Syntypes from S. Africa.

Cucumis welwitschii Cogn., tom. cit.: 490 (1881).—Hiern, Cat. Afr. Pl. Welw. **1**, 2: 395 (1898).—Warb., Kunene-Samb.-Exped. Baum: 395 (1903). Type from Angola.

Cucumis hirsutus Sond. var. *dissectus* Cogn. in Bull. Herb. Boiss. **3**: 418 (1895).— Burtt Davy, loc. cit. Type from S. Africa.

Cucumis spp.—Eyles in Trans. Roy. Soc. S. Afr. **5**, 4: 498 (1916) quoad specim. *Eyles* 50, *Chubb* 322, 328.

Cucumis hirsutus var. *major* Cogn. in Engl., Pflanzenr. **IV**, 275, 2: 134 (1924). Type: Mozambique, Morrumbala, *Kirk* (K, holotype!).

Cucumis seretioides Suesseng. in Proc. Trans. Rhod. Sci. Ass. **43**: 61 (1951). Type: Rhodesia, Rusape, *Dehn* 582 (M, holotype!; SRGH, isotype!).

Cucumis hirsutus var. *welwitschii* (Cogn.) R. & A. Fernandes in Mem. Junta Invest. Ultramar, Sér. 2, **34**: 86 (1962). Type as for *Cucumis welwitschii*.

Perennial; rootstock fibrous, woody; stems prostrate or scandent, to 2·5 m., sparsely to densely patent-setulose, becoming thickened and woody at the base. Leaf-lamina 2–15 × 1–10 cm., broadly ovate or ovate-triangular to narrowly ovate, elliptic or oblong-lanceolate in outline, hastate, weakly cordate, truncate or rounded at the base, obscurely and minutely sinuate-denticulate to coarsely sinuate-serrate, usually rather densely hispid or setulose becoming scabrid-punctate above, densely finely setulose to scabrid-setulose or scabrid-punctate beneath, obtuse to acute, apiculate, unlobed or variously palmately 3–5-lobed, lobes ovate-triangular to linear, the central much the largest. Petiole 0·5–5·5 cm. long, more or less patent-setulose or hispid. Flowers dioecious. ♂ flowers solitary or 2–12 in lax or contracted sessile or pedunculate fascicles; peduncle 2–80 mm. long; pedicels 3–73 mm. long. Receptacle-tube 2·7–9 mm. long; lobes 1–9 mm. long. Petals 3·5–22 mm. long, white, cream, pale yellow or yellow. ♀ flowers solitary or paired; pedicel 7–26 mm. long; ovary 9–13 × 4–12 mm., ellipsoid or ovoid-ellipsoid, densely appressed- or patent-setulose; perianth similar to that of ♂ flower, the petals rather larger, 13–31 mm. long. Fruit 2·5–7 × 1·5–6 cm., globose to oblong-ellipsoid, rounded at the ends, smooth, longitudinally striped light and dark green becoming concolorous yellow and finally brownish-orange when ripe; fruit-stalk 1·8–6·2 cm. long. Seeds 6·5–9 × 4·9–6·3 × 2·1–3·2 mm., ovate in outline, smooth, compressed, not bordered, white.

Botswana. N: Shashe R., fl. 11.iv.1876, *Holub* 1469–1470 (K). SE: Mahalapye aerodrome, fr. 29.i.1963, *Yalala* 352 (SRGH). **Zambia.** N: Kasama, Malole, fl. 5.xii.1960, *Robinson* 4189 (K). W: Kitwe, fl. 5.ii.1954, *Fanshawe* 784 (K; SRGH). C: Broken Hill, fl. x.1904, *Rogers* 8430 (BOL; K; SRGH). E: Chadiza, fl. 28.xi.1958, *Robson* 767 (BM; K; LISC; SRGH). S: Mazabuka, fl. 29.iii.1963, *van Rensburg* 1857 (K; SRGH). **Rhodesia.** N: Conscession, fr. 7.iii.1952, *Wild* 3773 (K; LISC; SRGH). W: Matopos Estate, fl. 8.xii.1949, *West* 3126 (SRGH). C: Marandellas, fl. xii.1926, *Rand* 427 (BM). E: Umtali, Morningside, fl. 18.xi.1954, *Chase* 5331 (BM; K; LISC; SRGH). S: Belingwe, Ngobi Dip, fl. 11.xii.1953, *Wild* 4348 (K; SRGH). **Malawi.** N: Mwanemba, ii–iii.1903, *McClounie* 160 (K). C: Lilongwe, fl. 16.xi.1951, *Jackson* 654 (BM; K). S: Shire Highlands, *Buchanan* 176 (K). **Mozambique.** N: Malema, Mutuali, 29.v.1947, *Pedro* 3295 (LMA). Z: Milange, fl. 13.xi.1942, *Mendonça* 1429 (LISC). MS: Gorongosa, Caça National Park, fl. 4.xi.1963, *Torre & Paiva* 9016 (LISC). LM: Maputo, near Porto Henrique, fr. 27.xi.1958, *Myre & Balsinhas* 3259 (LMA).

Tropical and southern Africa from Cameroon to S. Africa (Cape Prov.). Woodland, wooded grassland and grassland, 150–2500 m.

14. OREOSYCE Hook. f.

Oreosyce Hook. f. in F.T.A. **2**: 548 (1871).
Hymenosicyos Chiov. in Ann. Bot. Roma **9**: 62 (1911).

Small scandent herbs. Leaves simple, petiolate. Tendrils simple. Flowers small, yellow, monoecious. ♂ flowers in few-flowered sessile fascicles, rarely solitary, pedicellate. Receptacle-tube narrowly campanulate, lobes small, filiform. Corolla-lobes 5, united near the base. Stamens 3, two double 2-thecous, one single 1-thecous; filaments short, inserted on the receptacle-tube, free; anthers free; connective shortly produced; thecae straight. Disk basal, subglobose, free from the receptacle-tube. ♀ flowers solitary, pedicellate; ovary setose, obscurely tuberculate; ovules numerous, horizontal; perianth as in ♂ flower; staminodes 3, subulate, inserted on the receptacle-tube; disk annular, surrounding base of

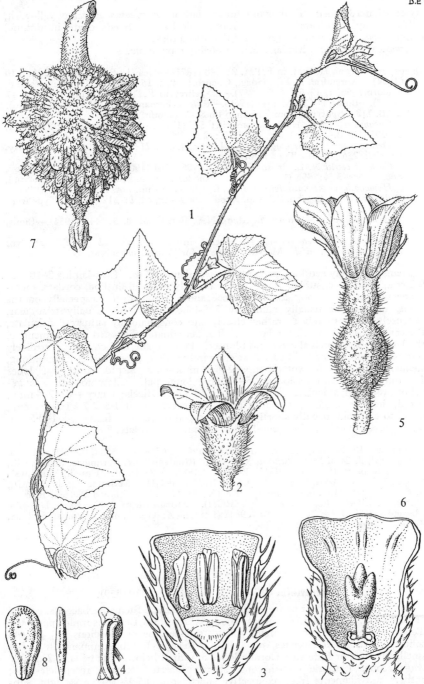

Tab. 113. OREOSYCE AFRICANA. 1, habit (× ⅔) *Fries* 439; 2, ♂ flower (×3); 3, ♂ flower opened to show stamens (×6); 4, double stamen (×8); 5, ♀ flower (×3); 6, ♀ flower opened to show disk (×6); 7, fruit (×1½); 2–7 from *Milne-Redhead & Taylor* 11302; 8, seed, face and lateral views (×3) *Chandler* 1582. From F.T.E.A.

style; stigma 3-lobed. Fruit ovoid, tuberculate or spiny, setose, green or yellowish, dehiscent by expulsion of seeds from pedicel-scar. Seeds ovate in outline, compressed, with broad margins and depressed disk.

Tropical Africa and Madagascar; probably monotypic.

Oreosyce africana Hook. f. in F.T.A. **2**: 548 (1871).—Keay in F.W.T.A. ed. 2, **1**: 210 (1954).—Keraudren in Fl. Madag., Cucurb.: 60, t. 15 (1966); in Fl. Cameroun **6**, Cucurb.: 65, t. 13 fig. 4–8 (1967).—C. Jeffrey in F.T.E.A., Cucurb.: 110, t. 17 (1967).—Binns, H.C.L.M.: 43 (1968).—R. & A. Fernandes in C.F.A. **4**: 252 (1970). TAB. **113**. Syntypes from Fernando Po & Cameroun.

    Cucumis subsericeus Hook. f., tom. cit.: 545 (1871).—Hiern, Cat. Afr. Pl. Welw. **1**, 2: 397 (1898).—Burtt Davy, F.P.F.T.: 228 (1926). Type from Angola.

    Oreosyce triangularis Cogn. in Engl., Bot. Jahrb. **21**: 207 (1895).—Burtt Davy, op. cit.: 226 (1926). Type from Tanzania.

    Cucumis cecilii N. E. Br. in Kew Bull. **1906**: 104 (1906).—Eyles in Trans. Roy. Soc. S. Afr. **5**, 4: 498 (1916). Type: Rhodesia, Inyanga, *Cecil* 225 (K, holotype).

    Hymenosicyos membranifolius (Hook. f.) Chiov. in Ann. Bot. Roma **9**: 63 (1911).—R. E. Fr., Wiss. Ergebn. Schwed. Rhod.-Kongo-Exped. **1**: 312 (1916). Type from Ethiopia.

    Hymenosicyos sp.—Brenan in Mem. N.Y. Bot. Gard. **8**, 5: 443 (1954).—Binns, op. cit.: 42 (1968).

    Oreosyce subsericea (Hook. f.) Meeuse in Bothalia **8**: 22 (1962) excl. syn. *Hymenosicyos bryoniifolio* Merxm. Type as for *Cucumis subsericeus*.

Stems retrorsely setulose, prostrate or scandent to 3 m. Leaf-lamina 2–10 × 2–9·5 cm., triangular, subhastate, ovate or broadly ovate in outline, cordate, membraneous, rather densely hispid-setulose above and beneath, especially on the veins, unlobed or usually palmately 3–5-lobed, the lobes usually triangular, obscurely and remotely to rather closely and conspicuously sinuate-denticulate, the central usually much the largest, often acuminate, obtuse to acute, apiculate, the lateral often much shorter and blunter. Petioles 2·5–11·5 cm. long, retrorsely setulose. ♂ flowers 3–6 in each fascicle, rarely solitary; pedicels 3–15 mm. long, slender, setulose. Receptacle-tube 2–6·5 mm. long, hispid externally; lobes 1–6 mm. long. Corolla-lobes 2·5–11 mm. long, pale yellow to pale orange-yellow, rounded, apiculate. ♀ flowers on 3–10 mm. pedicels; ovary 4–8 × 2–4 mm., ovoid, densely setulose; perianth as in ♂ flower. Fruit 1·5–2·7 × 1·3–2·1 cm., ovoid to ellipsoid, usually shortly rostrate, green or yellow; fruit-stalk 1–2·5 cm. long. Seeds 5–6·5 × 3–3·5 × 0·6–1 mm., obscurely foveolate.

Zambia. B: Kataba, fl. & fr. 24.iv.1961, *Fanshawe* 6518 (K; SRGH). N: Kalambo Farm, Saisi, fl. 24.iii.1955, *Richards* 5106 (K). **Rhodesia.** N: Urungwe, Msukwe R., Dokwa stream, fl. 19.xi.1953, *Wild* 4209 (K; LISC; SRGH). E: Umtali, Murahwa's Hill, fl. & fr. 10.iv.1964, *Chase* 7981 (COI; LISC; SRGH). **Malawi.** C: Dedza Mt., fl. & fr. immat. 24.v.1968, *Salubeni* 1098 (K; SRGH). S: Zomba Plateau, fl. & fr. immat. 7.vi.1946, *Brass* 16315 (K; SRGH). **Mozambique.** N: Cabo Delgado, Macondes, fl. 4.i.1964, *Torre & Paiva* 9842 (LISC). Z: Gúruè, fl. & fr. immat. 5.i.1968, *Torre & Correia* 16942 (LISC).

Also in Fernando Po, Cameroun, E. tropical Africa, Angola, Transvaal & Madagascar. Forest margins, 800–2100 m.

15. CUCUMELLA Chiov.

Cucumella Chiov., Fl. Somala **1**: 183 (1929).

Rather small scandent or prostrate perennial herbs. Stems herbaceous, annual or perennial, becoming somewhat woody when old. Leaves simple, petiolate. Tendrils simple. Flowers small, yellow, monoecious or dioecious. ♂ flowers solitary or in small fascicles, pedicellate. Receptacle-tube campanulate, lobes small, subulate or filiform. Corolla-lobes 5, united below. Stamens 3, 2 double two-thecous, 1 single one-thecous, shortly inserted on the receptacle-tube; anthers and filaments free; connective shortly produced; thecae straight, subparallel, sometimes apically incurved. Disk elevated, convex, free from the receptacle-tube. ♀ flowers solitary, pedicellate; ovary shortly hirsute or hispid; ovules many, horizontal; perianth as in ♂ flower; disk annular, surrounding base of style, free from the receptacle-tube; stigma 3-lobed. Fruit rather small, fleshy, thin-walled or firm-walled, smooth but always ± hirsute or scabrid, ellipsoid or fusiform, sometimes rostrate, indehiscent. Seeds small, ovate or elliptic in outline, compressed, whitish, not or obscurely bordered.

A genus of eight species, seven in Africa (one extending to Madagascar) and one in India.

Fruit ellipsoid, rounded at the apex, green with pale longitudinal stripes, glabrous or sparsely setulose - - - - - - - - - 2. *bryoniifolia*
Fruit ovoid or ellipsoid, rostrate, acute, concolorous, yellow or red, aculeate-spiculate or densely hirsute:
   Pericarp finely and densely shortly hirsute; fruit ellipsoid-cylindrical or fusiform, longitudinally ribbed, orange-red - - - - - - 1. *engleri*
   Pericarp rather coarsely aculeate-spiculate, otherwise glabrous; fruit ellipsoid, abruptly narrowed above into a slender seedless rostrum, terete, yellow - 3. *aëtheocarpa*

1. **Cucumella engleri** (Gilg) C. Jeffrey in Kew Bull. **15**: 350 (1962); in F.T.E.A., Cucurb.: 113, t. 18 fig. 9 (1967).—Binns, H.C.L.M.: 42 (1968). Type from Kenya.
Kedrostis engleri Gilg in Engl., Bot. Jahrb. **34**: 359 (1904). Type as above.

Stems prostrate or scandent, up to 1·2 m. long, shortly hispid. Leaf-lamina 1·2–8 × 1·7–7 cm., broadly ovate or subpentagonal in outline, cordate, when young ± cinereous-tomentose beneath, later hispid or asperulous, especially on veins, shortly hispid above, ± sinuate-denticulate at the margins, unlobed or shortly to moderately palmately 3–5-lobed, lobes ovate-triangular or triangular, obtuse to acute, apiculate, the central largest. Petioles 1–6·7 cm. long, hispid or scabrid. Probably monoecious. ♂ flowers in sessile or sometimes shortly pedunculate 2–12-flowered fascicles, rarely solitary; peduncle up to 5 mm. long; pedicels 5–18 mm. long. Receptacle-tube 2·5–5 mm. long, densely hispid; lobes 1–2 mm. long, filiform. Petals 3–8 mm. long, yellow. ♀ flowers on 5–30 mm. long pedicels; ovary 12–16 × 1·5–2·5 mm., ovoid-cylindrical or fusiform, densely hirsute; receptacle-lobes 2·5–5 mm. long, perianth otherwise as in ♂ flowers. Fruit 1–4 × 0·7–1·5 cm., ellipsoid-cylindrical or fusiform, sometimes ± rostrate, longitudinally ribbed, shortly hispid, orange-red; fruit-stalk 5–30 mm. long. Seeds 3·5–7 × 2–4 × 0·5–1 mm., ovate in outline, lenticular, smooth, not or obscurely bordered.

Zambia. C: Konstrupu Spruit, 24.xii.1907, *Kassner* 2126 (BM). **Malawi.** C: Dedza Mt. Forest, fl. 2.ii.1967, *Jeke* 61 (COI; K; SRGH).
E. tropical Africa. Ecology uncertain.

2. **Cucumella bryoniifolia** (Merxm.) C. Jeffrey in Kew Bull. **15**: 350 (1962); in F.T.E.A., Cucurb.: 114, t. 18 fig. 8 (1967).—Binns, H.C.L.M.: 42 (1968). Type: Rhodesia, Rusape, *Dehn* R25 (M, holotype; SRGH, isotype).
Hymenosicyos bryoniifolius Merxm. in Mitt. Bot. Staatss. München **1**: 205 (1953). Type as above.
Cucumis umbrosus Meeuse & Strey in Bothalia **8**: 67 (1962). Type from S. Africa (Transvaal).
Melothria cinerea sensu Meeuse, tom. cit.: 17 (1962) pro minore parte quoad spec. *Wild* 3045, non (Cogn.) Meeuse.

Stems slender, annual, prostrate or scandent, rather sparsely hispid-spiculate. Leaf-lamina 1–4·5 × 1–6·5 cm., membraneous, ovate or broadly ovate in outline, cordate, shortly asperulous above, hispid or scabrid beneath especially on the veins, plane or remotely and obscurely to coarsely sinuate-denticulate at the margins, shallowly to rather deeply palmately 3–5-lobed, lobes ovate-triangular or ovate to elliptic, oblong-lanceolate or oblanceolate, acute to obtuse or rounded, shortly apiculate, the central largest. Petioles 1–6·5 cm. long, shortly rather finely hispid or scabrid with spreading or recurved hairs. Monoecious. ♂ flowers solitary or rarely 2–3 together; pedicels 11–30 mm. long, slender. Receptacle-tube 3–5 mm. long, obconic; lobes 0·5–1·5(5) mm. long, filiform. Petals 3·5–7 mm. long, yellow. ♀ flowers on 11–24 mm. long pedicels; ovary 4–6 × 1·5–2·5 mm., ovoid or ovoid-rostrate, densely hirsute; perianth similar to that of ♂ flowers. Fruit 2–3 × 1·3–1·5 cm., ellipsoid, terete, obtuse, fleshy, smooth, glabrous or sparsely setulose, dark green with irregular longitudinal whitish bands; fruit-stalk 3–4 cm. long. Seeds c. 3·2 × 2 × 0·5 mm., ovate in outline, compressed, smooth, bordered.

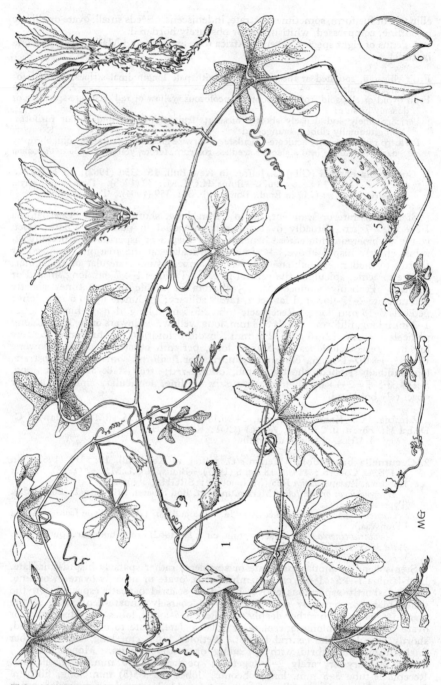

Tab. 114. CUCUMELLA AËTHEOCARPA. 1, habit (×1); 2, ♂ flower (×3); 3, ♂ flower opened to show stamens and disk (×3); 4, ♀ flower (×3); 5, fruit (×2); 6, seed, face view (×5); 7, seed, lateral view (×5), all from *Richards* 17751. From Kew Bulletin.

Zambia. N: 105 km. E. of Kasama, fl. 6.v.1962, *Robinson* 5163 (K; SRGH)·
Rhodesia. N: Mtoko, Mudzi Dam, fl. & fr. 16.ii.1962, *Wild* 5663 (COI; K; SRGH).
W: Matobo, SW. Matopos, Maleme Valley, near Mumongwe, fl. 9.i.1963, *Wild* 5974
(COI; K; SRGH). C: Salisbury, Twentydales, fl. 18.xii.1951, *Wild* 3708 (K; SRGH).
E: Umtali, Zimunya's Reserve, fl. & fr. 11.iii.1956, *Chase* 6008 (BM; COI; LISC;
SRGH). S: Zimbabwe, fl. 4.x.1949, *Wild* 3045 (K; SRGH). **Malawi.** C: Kasungu,
Chamama, Chipala Hill, fl. 16.i.1959, *Robson* 1211 (BM; EA; K; LISC; PRE; SRGH).
S: Fort Lister, fl. 19.xii.1947, *Lupton* 57 (BM).
    Also in Tanzania and S. Africa (Transvaal). Woodland and grassland, usually on
granite outcrops, 750–1450 m. Occurrence in Mozambique given in error by Jeffrey
(l.c.; 1967) for the following species.

3. **Cucumella aëtheocarpa** C. Jeffrey in Kew Bull. **19**: 215, t. 1 (1965); in F.T.E.A.,
    Cucurb.: 114 (1967). TAB. **114.** Type from Tanzania.

    Stems slender, annual, prostrate, rooting at the nodes, setose or aculeate.
Leaf-lamina 1–3 × 2–3·5 cm., broadly ovate in outline, cordate, membraneous,
hispid or scabrid above, scabrid on veins beneath, lobes ovate-oblong or oblanceolate, entire or rather sinuate or sometimes ± lobulate especially in upper part,
obtuse to rounded, apiculate, the central largest. Petioles 0·4–3·5 cm. long,
coarsely antrorsely aculeate or setulose. Monoecious. ♂ flowers solitary; pedicels
6–17 mm. long, slender; receptacle-tube 4–5 mm. long, obconic, lobes filiform,
1–2 mm. long; petals 5–6 mm. long, yellow. ♀ flowers on 4–7 mm. long pedicels;
ovary 6–11 × 2–3·5 mm., ellipsoid, long-rostrate, shortly retrorsely aculeate-
spiculate; perianth similar to that of ♂ flower, petals 6–8 mm. long. Fruits
1·5–2 × 0·5–1 cm., ellipsoid, rostrate, yellow, aculeate; fruit-stalk c. 5 mm. long.
Seeds 3·8 × 2 × 0·6 mm., ovate in outline, compressed, not bordered.

    Mozambique. N: Moçambique, Eráti, 12 km. from Namapa towards Alua, Mount
Geovi, fl. & fr. 8.i.1964, *Torre & Paiva* 9892 (LISC).
    Also in S. Tanzania. Forest with dense shrub layer, usually on granite outcrops,
450–500 m.

16. MUKIA Arn.

Mukia Arn. in Madras Journ. Lit. Sci. **12**: 50 (1840).

    Scandent herbs. Leaves simple, petiolate. Tendrils simple. Flowers very
small, monoecious. ♂ flowers in sessile fascicles, shortly pedicellate.. Receptacle-
tube obconic-campanulate, lobes small, narrow, dentiform or subulate. Corolla-
lobes 5, united below. Stamens 3, two double 2-thecous, one single 1-thecous;
filaments short, inserted on the receptacle-tube, free; connective shortly produced;
thecae straight. Disk basal, subglobose, free from the receptacle-tube. ♀ flowers
in sessile fascicles, rarely solitary, subsessile or shortly pedicellate; ovary ellipsoid
or globose, smooth, ± setose; ovules numerous, horizontal; perianth as in
♂ flower; staminodes 3, subulate, inserted on the receptacle-tube; disk annular,
surrounding base of style; stigma 3-lobed. Fruits subsessile in axillary fascicles,
sometimes solitary, small, baccate, ellipsoid or globose, smooth, red and usually
glabrous when mature. Seeds elliptic in outline, rather tumid, with prominent
margins and flat or convex disk, usually ± verrucose or scrobiculate.
    4 species in the Old World tropics, one extending to Africa.

Mukia maderaspatana (L.) M. J. Roem., Syn. Mon. **2**: 47 (1846).—R. & A. Fernandes
    in Mem. Junta Invest. Ultramar, Sér. 2, **34**: 124 (1962); in C.F.A. **4**: 248 (1970).—
    C. Jeffrey in F.T.E.A., Cucurb.: 115, t. 19 (1967); in Hook., Ic. Pl. **7**: t. 3662
    (1969).—Keraudren in Fl. Cameroun 6, Cucurb.: 69, t. 14 (1967).—Binns,
    H.C.L.M.: 43 (1968). TAB. **115.** Type from India.
        Cucumis maderaspatanus L., Sp. Pl. **2**: 1012 (1753). Type as above.
        Bryonia cordifolia L., loc. cit. Type from Ceylon.
        Bryonia scabrella L. in L. f., Suppl.: 424 (1781). Type from India.
        Mukia scabrella (L.) Arn. in Lond. Journ. Bot. **3**: 276 (1841).—Sond. in Harv. &
    Sond., F.C. **2**: 489 (1862).—Hook. f. in F.T.A. **2**: 561 (1871). Type as above.
        Coccinia cordifolia (L.) Cogn. in A. & C. DC., Mon. Phan. **3**: 529 (1881) sensu
    stricto quoad typum excl. spec. cit. Type as for *Bryonia cordifolia*.
        Melothria maderaspatana (L.) Cogn., op. cit.: 623 (1881).—Hiern, Cat. Afr. Pl.
    Welw. **1**, 2: 403 (1895).—R.E.Fr., Wiss. Ergebn. Schwed. Rhod.-Kongo-Exped. **1**:
    311 (1916).—Burtt Davy, F.P.F.T. **1**: 225 (1916).—Meeuse in Bothalia **8**: 14 (1962).
    Type as for *Mukia maderaspatana*.

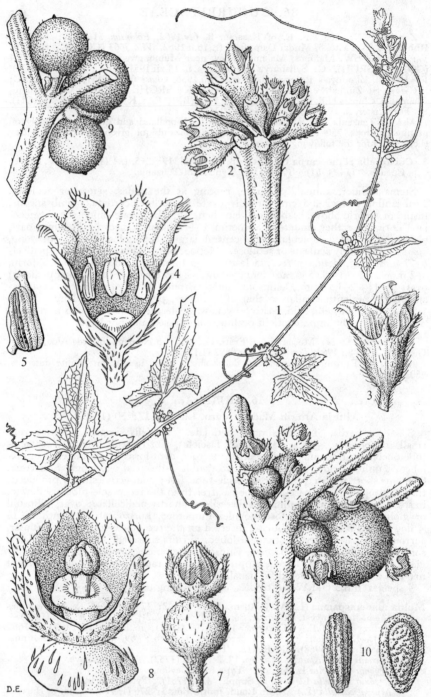

Tab. 115. MUKIA MADERASPATANA. 1, habit (×⅔) *Purseglove* 570; 2, ♂ inflorescence (×4); 3, ♂ flower (×6⅔); 4, ♂ flower opened to show stamens (×10); 5, double stamen (×20), 2–5 from *Drummond & Hemsley* 3041; 6, ♀inflorescence (×4); 7, ♀ flower (×6⅔); 8, ♀ flower opened to show disk (×13½); 9, fruit cluster (×2), 6–9 from *Milne-Redhead & Taylor* 9856; 10, seed, face and lateral views (×4) *Purseglove* 1314. From F.T.E.A.

Stems to 4 m., hispid, prostrate or scandent. Leaf-lamina 1–11 × 1–11 cm., narrowly to broadly ovate or triangular in outline, cordate, hastate or sagittate, hispid or scabrid-setulose, especially on the veins, beneath, more shortly so or scabrid-punctate above, unlobed or usually palmately 3(5)-lobed, the lobes usually triangular, entire or remotely and minutely denticulate to coarsely sinuate-denticulate at the margins, the central much the largest, often acuminate, obtuse to acute, apiculate, the lateral shorter and more rounded. Petioles 0·5–11·5 cm. long, antrorsely hispid or scabrid. ♂ flowers on 1–3 mm. long pedicels. Receptacle-tube 0·7–1·5 mm. long, obconic-campanulate, densely setulose; lobes 0·5–1 mm. long, lanceolate-subulate. Petals 1·2–1·5 mm. long, yellow. ♀ flowers 5–10 in each fascicle; pedicel c. 1 mm. long; ovary 1·5–2 mm. long, globose, setulose; perianth similar to that of ♂ flower. Fruits 2–8 in axillary clusters, rarely solitary, 6–11 mm. in diameter, shortly pedicellate, globose, at first green with paler longitudinal markings, when mature bright red, smooth, glabrous. Seeds 3·5 × 2–3 × 1·5–1·7 mm., ovate in outline, rather tumid, verrucose, bordered.

Zambia. B: Mongu, fl. & fr. 5.i.1960, *Gilges* 829 (COI; SRGH). N: Mbala (Abercorn), fl. i.1955, *Bock* 278 (PRE). W: Solwezi, fl. 21.iii.1961, *Drummond & Rutherford-Smith* 7149 (COI; K; SRGH). C: 100–129 km. E. of Lusaka, Chakwenga headwaters, fl. & fr. 16.xi.1963, *Robinson* 5840 (K; SRGH). E: near Petauke, Mvuvye R., fl. & fr. 5.xii.1958, *Robson* 837 (BM; K; L; LISC; PRE; SRGH). S: Mazabuka, Veterinary Research Station, fr. 27.ii.1963, *van Rensburg* 1479 (K; SRGH). **Rhodesia.** N: Mazoe, Henderson Research Station, fl. & fr. 7.v.1952, *Wild* 3833 (K; LISC; SRGH). C: Salisbury, Atlantica, fl. & fr. 2.v.1963, *Loveridge* 1034 (K; LISC; SRGH). E: Umtali, Imbeza Valley, La Rochelle, fl. & fr. 3.iii.1954, *Chase* 5216 (BM; LISC; LMA; K; SRGH). **Malawi.** N: Kondowe to Karonga, fl. & fr. vii.1896, *Whyte* (K). C: Kasungu, between Chamama and Bua Drift, fl. 15.i.1959, *Robson & Jackson* 1204 (BM; K; LISC; MO; PRE; SRGH). S: Kasupe, Milinje, fl. & fr. 14.ii.1964, *Salubeni* 244 (COI; SRGH). **Mozambique.** N: Amaramba, 40 km. from Nova Freixo towards Mecanhelas, fl. & fr. 17.ii.1964, *Torre & Paiva* 10635 (LISC). Z: Ile region, fr. 21.vi.1943, *Torre* 5537 (LISC). MS: Sena, fl. & fr. i.1859, *Kirk* (K). LM: between Santaca and Catuana, c. 10 km. from Santaca, fl. & fr. 13.iv.1949, *Myre & Balsinhas* 579 (LMA).

Old World tropics from W. Africa to New Guinea and Australia. Flood-plain and valley grasslands, riverine margins and damper places in woodland and wooded grasslands, 300–1250 m.

17. ZEHNERIA Endl.

Zehneria Endl., Prodr. Fl. Norfolk I.: 69 (1833).

Scandent herbs; stems arising from a somewhat thickened perennial rootstock, eventually becoming more or less woody at the base. Leaves simple, petiolate. Tendrils simple. Flowers small to very small, monoecious or dioecious. ♂ flowers solitary or few to many in sessile or pedunculate racemiform or subumbelliform axillary clusters, pedicellate. Receptacle-tube campanulate, lobes small, dentiform. Corolla-lobes 5, united below. Stamens 3 or 2, rarely 4, all 2-thecous; filaments very short, inserted on the middle of the receptacle-tube, or elongated and often hairy, inserted on the base of the receptacle-tube; connective shortly produced; thecae straight or slightly arcuate. Disk basal, elevated, free from the receptacle-tube. ♀ flowers solitary, in monoecious species usually co-axillary with ♂, or few to many in sessile or pedunculate subumbelliform or rarely racemiform axillary clusters, pedicellate; ovary subglobose to fusiform, smooth, glabrous or shortly pubescent; ovules few to many, horizontal; perianth as in ♂ flower; staminodes often present, usually 3; disk annular, surrounding base of style; stigma (2)3-lobed. Fruits solitary or clustered, small, baccate, globose to fusiform, smooth, usually red and glabrous when mature. Seeds ovate or elliptic in outline, compressed, with narrow or sometimes prominent margins and usually smooth, convex, sometimes depressed disk.

A palaeotropical genus of c. 30 species.

Flowers dioecious, the ♂ and ♀ on different plants:
  Fruit globose or ellipsoid, rounded at the apex - - - - 1. *scabra*
  Fruit fusiform, acute - - - - - - - - - 6. *minutiflora*
Flowers monoecious, the male and female on the same plant:
  Stamens subsessile on the receptacle-tube; anther-thecae straight, subparallel; fruit fusiform:

Leaves succulent - - - - - - - - - 7. *pallidinervia*
Leaves not succulent - - - - - - - - - 8. *thwaitesii*
Stamens with rather long, often hairy filaments; anther-thecae shortly arcuate; fruit
ellipsoid or globose:
Flowers very small, the receptacle-tube rather shallowly campanulate, 0·5–1·3 mm.
long; seeds 4·2–4·7 mm. long - - - - - - 5. *marlothii*
Flowers larger, the receptacle-tube deeply campanulate, 2–4 mm. long; seeds up to
3·5 mm. long:
Leaves chartaceous, shortly scabrid-setulose on veins beneath - 3. *sp. A*
Leaves membraneous, glabrous or almost so beneath except for a few setulae on
main veins:
Leaf-lamina ovate or narrowly ovate in outline, not lobed; fruits red
2. *microsperma*
Leaf-lamina ovate to broadly ovate in outline, palmately 5-lobed, usually rather
deeply; fruits bluish - - - - - - - 4. *parvifolia*

1. **Zehneria scabra** (L. f.) Sond. in Harv. & Sond., F.C. **2**: 486 (1862).—Hook. f. in
F.T.A. **2**: 560 (1871).—R. & A. Fernandes in Mem. Junta Invest. Ultramar, Sér. 2,
34: 123 (1962); in C.F.A. **4**: 248 (1970).—C. Jeffrey in F.T.E.A., Cucurb.: 122,
t. 21 fig. 1–8 (1967).—Keraudren in Fl. Cameroun **6**, Cucurb.: 44, t. 9 fig. 6–13
(1967).—Binns, H.C.L.M.: 43 (1968). Type from S. Africa (Cape Prov.).
Bryonia scabra L. f., Suppl.: 423 (1781). Type as above.
Bryonia punctata Thunb., Prodr. Pl. Cap.: 13 (1800) *nom. illegit.* non L. (1753).
Type from S. Africa.
Bryonia cordata Thunb., Fl. Cap.: 149 (1811). Type from S. Africa.
Zehneria longepedunculata A. Rich., Tent. Fl. Abyss. **1**: 287 (1847). Type from
Ethiopia.
Melothria cordata (Thunb.) Cogn. in A. & C. DC., Mon. Phan. **3**: 613 (1881).—
Meeuse in Bothalia **8**: 19 (1962). Type as for *Bryonia cordata.*
Melothria longepedunculata (A. Rich.) Cogn., tom. cit.: 612 (1881).—Swynnerton
in Journ. Linn. Soc., Bot. **40**: 74 (1911).—Eyles in Trans. Roy. Soc. S. Afr. **5**, 4:
498 (1916). Type as for *Zehneria longepedunculata.*
Melothria punctata Cogn., tom. cit.: 615 (1881) *nom. illegit.* non Raf. (1836).—
Burtt Davy, F.P.F.T.: 225 (1916).—Hutch., Botanist in S. Afr.: 250 (1946). Type
as for *Bryonia punctata* Thunb.
Melothria sp. nov. aff. *microsperma.*—Brenan in Mem. N.Y. Bot. Gard. **8**: 443
(1954).
Zehneria oligosperma sensu C. Jeffrey in F.T.E.A., Cucurb.: 124 (1967) pro parte.—
sensu Binns, H.C.L.M.: 43 (1969) non C. Jeffrey (1962).

Stems to 6 m., almost glabrous or sparsely to densely shortly crispate-pubescent
or setulose, prostrate or scandent. Leaf-lamina 1·9–11 × 2–11 cm., membraneous
to subcoriaceous, ovate or broadly ovate in outline, deeply cordate to subtruncate
at the base, deep green and scabrid-punctate above, paler and sparsely setulose to
densely cinereous-velutinous on the veins beneath, obscurely and remotely to
prominently sinuate-denticulate at the margins, unlobed or sometimes palmately
3–5-lobed, the lobes broadly triangular to ovate or elliptic, the central largest,
shortly acuminate, obtuse to acute, apiculate, the lateral shorter. Petioles 0·7–
7·0 cm. long, glabrous to densely antrorsely setulose or crispate-pubescent.
Flowers dioecious. ♂ flowers 3–60 or more in dense subumbelliform or shortly
racemiform sessile or pedunculate axillary clusters, usually co-axillary with a long-
pedicellate solitary flower, rarely solitary; peduncles 0·9–7 cm. long; pedicels
1·5–10 mm. long. Receptacle-tube 2·0–5·5 mm. long, pale green; lobes 0·2–
1·5 mm. long, dentiform. Petals 1·5–3·5 mm. long, white, becoming yellowish
with age; stamens 3. ♀ flowers solitary or 2–10 or more in subumbelliform
sessile or pedunculate axillary clusters; peduncles up to 2·7 cm. long; pedicels
0·4–11(20) mm. long; ovary 2·5–5 mm. long, ellipsoid, rostrate, glabrous or
shortly pubescent. Fruits 1–10 or more in axillary clusters, bright red, usually
glabrous, globose, 8–13 mm. in diameter, or ellipsoid, 10–12 × 7–8 mm.; pedicels
2–68 mm. long. Seeds 2·0–5·6 × 1·1–4·5 × 0·4–1 mm., ovate in outline, smooth,
compressed.

This species is widespread and very variable, and no doubt when more in-
tensively studied will prove divisible into a number of geographical and ecological
subspecies. However, on purely morphological grounds, only one of these,
subsp. *argyrea*, is definitive enough to warrant recognition here; the remaining
variants are therefore treated as forming a polymorphic type subspecies, subsp.
scabra.

Subsp. **scabra.**

Seeds 3·2–5·6 × 2·1–4·5 × 0·5–1 mm.

Zambia. N: 8 km. from Mbala on path to Chipululu Farm, fl. 12.xii.1968, *Sanane* 378 (K). W: Ndola, fr. 26.xii.1954, *Fanshawe* 1754 (K; LISC). C: Broken Hill, fl. xi.1909, *Rogers* 8623 (SRGH). **Rhodesia.** E: Inyanga, Juliasdale, fl. 4.i.1965, *Wild* 6694 (COI; LISC; SRGH). **Malawi.** N: Chitipa Distr., Mafinga Hills near Chisenga, fl. 26.viii.1962, *Tyrer* 583 (BM; SRGH). C: Dedza Mt., fl. 17.v.1967, *Jeke* 91 (COI; K). S: Mlanje, Chambe Basin ridge, Lichulesi side, fr. 14.vi.1962, *Richards* 16691 (K; SRGH). **Mozambique.** N: without locality, fr. 20.ii.1942, *Hornby* 3599 (PRE). MS: Manica, Quinta da Fronteira, near Penhalonga, fl. 16.vi.1962, *Wild* 5831 (COI; K; SRGH).
    S. Africa, tropical Africa, SW. Arabia, peninsular India, Java, Philippine Is. Rain-forest, swamp-forest, river margins and in *Pinus* plantations, 600–2250 m.

Within this subspecies, two main variants can be recognized in our area, one (*scabra* in the strict sense) with leaves hispidulous to densely cinereous-velutinous on the veins beneath, usually drying dark brown, and seeds 3·2–4·0 × 2·1–2·4 × 0·5–0·7 mm., the other with leaves almost glabrous to rather sparsely scabrid-setulose on the veins beneath, usually drying green, and seeds 4·2–5·6 × 2·5–4·5 × 0·8–1·0 mm. The latter tends to be more montane in occurrence, especially to the north of our area in tropical E. Africa, where it also occurs. *Zehneria oligosperma* C. Jeffrey, with which it was formerly confused, differs in its succulent or subsucculent leaves, larger seeds and lower altitudinal distribution. It does not occur in our area.

Subsp. **argyrea** (A. W. P. Zimm.) C. Jeffrey in Kew Bull. **30**: 483 (1975). Type from Tanzania.
    Melothria argyrea A. W. P. Zimmerm., Die Cucurbitaceen **2**: 180 (1922). Type as above.

Seeds 2·0–3·0 × 1·1–1·7 × 0·4 mm.

This subspecies occurs in the F.Z. area in two well-marked variants, which it is convenient to recognize at varietal rank.

Var. **argyrea.**

Fruits c. 10 mm. in diameter, globose.

Mozambique. N: Cabo Delgado, Macondes, 1·6 km. from Mueda towards Nairoto, fr. 18.iv.1964, *Correia* 236 (LISC).
    Also in Angola, Tanzania & SE. Kenya. *Parinari* forest, 800 m.

Var. **chirindensis** C. Jeffrey in Kew Bull. **30**: 484 (1975). Type: Rhodesia, Melsetter, Chirinda Forest, *Goldsmith* 41/61 (COI, holotype; K, SRGH, isotypes).

Fruits 10–12 × 7–8 mm., ellipsoid.

Rhodesia. E: Chirinda Forest, fr. 23.iv.1947, *Wild* 1931 (K; SRGH).
    Not known elsewhere. Clearings and margins of rain-forest, 1040 m.

2. **Zehneria microsperma** Hook. f. in F.T.A. **2**: 559 (1871). Type: Malawi, Chibira, *Meller* (K, holotype!).
    Melothria microsperma (Hook. f.) Cogn. in A. & C. DC., Mon. Phan. **3**: 611 (1881). Type as above.

Stem slender, glabrous or almost so. Leaf-lamina 2·3–7·2 × 1·5–6·2 cm., membraneous, ovate or narrowly ovate, cordate or subtruncate at the base, finely punctate above, glabrous or almost so except for a few short setulae on veins beneath, unlobed, prominently sinuate-denticulate, acuminate, obtuse to acute, apiculate. Petioles 0·7–5·2 cm. long, glabrous or almost so. Flowers monoecious. ♂ flowers solitary or 1–10 in racemiform or subumbelliform male or bisexual (with 1–2 ♀ flowers) pedunculate axillary clusters, usually co-axillary with a long-pedicellate solitary flower; peduncles 0·6–4·8 cm. long; pedicels of racemose flowers 1–5 mm. long, of solitary co-axillary flower 3–12 mm. long. Receptacle-tube 2–4 mm. long; lobes 0·2–0·7 mm. long, dentiform. Petals 1·5–2·5 mm. long. ♀ flowers solitary, sometimes co-axillary with a ♂ inflorescence, or 1–5 in racemi-form or subumbelliform bisexual (with 1–4 ♂ flowers) or rarely wholly ♀ pedunculate axillary clusters; peduncles 0·8–2·5 cm. long; pedicels of racemose flowers 2·5–5 mm. long, of solitary flowers 2·5–12 mm. long; ovary 3–3·5 mm. long, ellipsoid, rostrate; perianth as in ♂ flowers. Fruits 8–10 mm. in diameter, glabrous, globose; pedicels 4–14 mm. long. Seeds 2·5–2·8 × 1·4–1·6 × 0·2 mm., elliptic in outline, smooth, compressed.

Rhodesia. C: Selukwe, Ferny Creek, fl. 8.xii.1953, *Wild* 4289 (K; LISC; SRGH). **Malawi.** S: Chibira, fr. ix.1861, *Meller* (K). **Mozambique.** N: Ribáuè, serra de Ribáuè, fl. & fr. 23.iii.1964, *Torre & Paiva* 11373 (LISC). Z: Massingire, M'bobo, fl. & fr. 4.viii.1962, *Torre* 4498 (LISC). MS: Cheringoma, Inhamitanga, fl. 15.ix.1944, *Simão* 45 (LISC; LMA).
Not known elsewhere. Forest and woodland, 1100–1200 m.

This species is very close to *Z. scabra* subsp. *argyrea* var. *argyrea*, of which it may prove to be no more than a monoecious variant.

3. Zehneria sp. A.

Stems sparsely setulose. Leaf-lamina chartaceous, ovate in outline, cordate, punctate above, shortly rather sparsely scabrid-setulose on veins beneath, sinuate-denticulate at the margins, 3·7–4 × 3–3·5 cm., unlobed or shallowly to moderately deeply palmately 3–5-lobed, the lobes ovate-triangular to ovate, the central largest, acuminate, obtuse, apiculate. Petiole sparsely antrorsely setulose, 1·1–1·7 mm. long. Male flowers 1–5 in sessile male or bisexual axillary clusters; pedicels 2·5–3·5 mm. long. Receptacle-tube 3·5 mm. long, lobes 0·5 mm. long. Petals 1·5 mm. long. Female flowers solitary, co-axillary with 1–2 male flowers; pedicels 2–2·5 mm. long; perianth as in male flower. Fruits (immature) globose, about 8 mm. in diameter; pedicels 3 mm. long.

Zambia. W: Mwinilunga, Kabompo Gorge, fl. & fr. immat. 19.iii.1965, *Robinson* 6652 (K).
Not otherwise known. Shady riverside rocks.

4. **Zehneria parvifolia** (Cogn.) J. H. Ross in Bothalia **10**: 568 (1972). Syntypes from S. Africa (Natal).
　　Melothria parvifolia Cogn. in Bull. Herb. Boiss. **3**: 420 (1895); Meeuse in Bothalia **8**: 18 (1962). Syntypes as above.
　　Melothria cordata sensu Mogg in Macnae & Kalk, Nat. Hist. Inhaca I., Moçamb.: 154 (1958), non (Thunb.) Cogn.

Stems glabrous. Leaf-lamina 1·6–6·5 × 1·8–8·2 cm., membraneous, thin, ovate to broadly ovate in outline, punctate or asperulous above, glabrous or with a few setulae on main veins beneath, moderately to rather deeply or rarely obscurely palmately 5-lobed, the lobes triangular, ovate, broadly elliptic or elliptic, the central largest, rather coarsely and remotely sinuate-denticulate in upper part, sometimes sublobulate, obtuse to acute, apiculate, the lateral shorter. Petioles 0·7–7 cm. long, glabrous or sparsely spiculate. Flowers monoecious. ♂ flowers solitary or 1–9 in shortly racemiform ♂ or bisexual (with 1–2 ♀ flowers) pedunculate axillary clusters, usually co-axillary with a solitary ♀ or sometimes ♂ flower; peduncles 0·3–3·8 cm. long; pedicels 2·5–4(10) mm. long. Receptacle-tube 2·0–3·5 mm. long; lobes 0·2–0·7 mm. long, dentiform. Petals 2–3 mm. long, cream. ♀ flowers solitary and usually co-axillary with ♂ infloresence or 1–2 in shortly racemiform or subumbelli-form bisexual (with 1–7 ♂ flowers) pedunculate axillary clusters; peduncles 0·9–3·0 cm. long; pedicels of racemose flowers 2–7 mm. long, of solitary flowers 2–6(14) mm. long; ovary 2·7–4 mm. long, ellipsoid or broadly ellipsoid, shortly rostrate; perianth as in ♂ flower. Fruits 8–11 mm. in diameter, glabrous, globose, bluish when mature; pedicels 3–14 mm. long. Seeds 2·7–3·5 × 1·6–1·9 × 0·2–0·3 mm., ovate or elliptic, smooth, compressed.

Mozambique. MS: Gaza, Bilene, between Chissano and Licilo, fl. & fr. 5.vii.1958, *Barbosa & Lemos* 8294 (COI; LMA). LM: Marracuene, Babole, fr. 2.x.1957, *Barbosa & Lemos* 7930 (COI; K; LISC; LMA).
Also in Natal. Sublittoral and riverine forest, on sands, 0–30 m.

5. **Zehneria marlothii** (Cogn.) R. & A. Fernandes in Mem. Junta Invest. Ultramar, Sér. 2, **34**: 120 (1962); in C.F.A. **4**: 245 (1970).—Launert & Roessler in Prodr. Fl. SW. Afr. **94**, Cucurb.: 22 (1968). TAB. **116**. Syntypes from S. Africa and SW. Africa.
　　Melothria marlothii Cogn. in Verh. Bot. Ver. Prov. Brand. **30**: 152 (1888).—Meeuse in Bothalia **8**: 15 (1962). Syntypes as above.

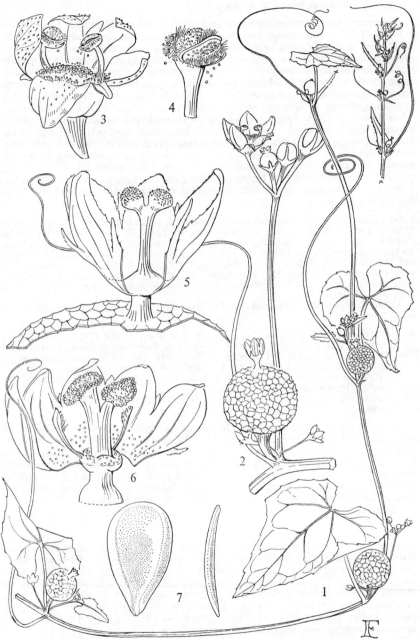

Tab. 116. ZEHNERIA MARLOTHII. 1, part of shoot (×1); 2, flowering node (×6); 3, ♂ flower (×12); 4, stamen (×18); 5, ♀ flower, partly dissected (×12); 6, ♀ flower dissected (×12); 7, seed, face and lateral views (×6), all from *Munro* 17.

Melothria acutifolia Cogn. in Bull. Herb. Boiss. **3**: 419 (1895).—Burtt Davy, F.P.F.T. **1**: 225 (1916).—Bremek. & Oberm. in Ann. Transv. Mus. **16**, 2: 436 (1935). Syntypes from SW. Africa.

Stem scandent to 3 m., glabrous or sparsely setulose at the nodes. Leaf-lamina 1·3–7·0 × 1·0–7·8 cm., membraneous, ovate or broadly ovate in outline, cordate, dark green and rather coarsely scabrid-punctate above, smooth and glabrous or glabrous except for a few setulae on veins beneath, obscurely and remotely to rather prominently sinuate-denticulate at the margins, rather shallowly to moderately deeply palmately 3–5-lobed, the central largest, broadly triangular to ovate, more or less acuminate, acute to obtuse or rounded, apiculate, the lateral shorter. Petioles glabrous or sparsely spiculate, 0·7–6 cm. long. Flowers monoecious. Male flowers 2–15 in shortly racemiform or subumbelliform pedunculate axillary clusters, rarely solitary; peduncle 0·2–2 cm. long; pedicels 1–3 mm. long. Receptacle-tube 0·5–1·3 cm. long, lobes triangular, dentiform, 0·2–0·4 mm. long. Petals white, becoming yellow-green with age, 0·8–1·5 mm. long. Female flowers solitary, co-axillary with male; pedicels 1·5–8 mm. long; ovary 1·5–2·5 mm. long; perianth as in male flower. Fruits globose, glabrous, red, 5–7 mm. in diameter; pedicels 2–7 mm. long. Seeds ovate in outline, smooth, compressed, 4·2–4·7 × 2·7–3·5 × 0·4–0·7 mm.

Botswana. N: Chobi-Zambezi confluence, fl. & fr. 11.iv.1955, *E. M. & W.* 1459 (BM; LISC; SRGH). SW: Chukudu Pan, fr. 20.vi.1955, *Story* 4940 (K; PRE). **Zambia.** B: Sesheke, fl. & fr. iv.1910, *Gairdner* 524 (K). W: 96 km. S. of Mwinilunga on Kabompo road, fr. 1.vi.1903, *Loveridge* 718 (K; SRGH). C: 38 km. W. of Lusaka on old Mumbwa road, fl. & fr. 17.iv.1962, *Angus* 3127 (K; FHO). S: Livingstone, fl. v.1904, *Rogers* 7140 (BOL; K). **Rhodesia.** W: Wankie Game Reserve, fl. 15.ii.1956, *Wild* 4729 (COI; K; LISC; SRGH). S: Beitbridge, Shashi, fl. 3.v.1959, *Drummond* 6061 (K; SRGH).

Also in S. Africa, SW. Africa and Angola. Deciduous woodland, usually on sandy soils, 900–1350 m.

6. **Zehneria minutiflora** (Cogn.) C. Jeffrey in Kew Bull. **15**: 366 (1962); in F.T.E.A., Cucurb.: 126, t. 21 fig. 9 & 10 (1967).—R. & A. Fernandes in Mem. Junta Invest. Ultramar, Sér. 2, **34**: 121 (1962); in C.F.A. **4**: 246 (1970).—Keraudren in Fl. Cameroun **6**, Cucurb.: 37, t. 8 fig. 1–5 (1967). Type from Cameroun.

Zehneria micrantha Hook. f. in F.T.A. **2**: 560 (1871) *nom. illegit*, non F. Muell. (1860). Type as above.

Melothria minutiflora Cogn. in A. & C. DC., Mon. Phan. **3**: 611 (1881). Type as above.

Stems sparsely shortly setulose to densely crispate-pubescent, prostrate or scandent to 3 m. Leaf-lamina 1·5–7·5 × 0·3–7·0 cm., membraneous to subcoriaceous, triangular, broadly ovate, ovate, narrowly triangular, lanceolate or linear in outline, cordate or sagittate or sometimes (when narrower) rounded at the base, scabrid-punctate above, sparsely to densely setulose, hispid or scabrid-setose on veins beneath, plane or ± sinuate and obscurely and remotely to prominently denticulate at the margins, unlobed or usually palmately 3–5-lobed, the lobes ovate or broadly to narrowly triangular, the central largest, usually ± acuminate, obtuse to acute, apiculate, the lateral shorter. Petiole 0·5–5 cm. long, sparsely spiculate or setulose. Flowers dioecious. ♂ flowers 5–25 in usually dense subumbelliform or shortly racemiform pedunculate axillary clusters, usually co-axillary with a long pedicellate solitary flower; peduncles 0·2–4 cm. long; pedicels 2·5–13 mm. long. Receptacle-tube 1·8–3·8 mm. long; lobes 0·2–0·6 mm. long, dentiform. Petals 1·5–2·5 mm. long, white, becoming yellowish with age. ♀ flowers solitary, axillary; pedicels 11–30 mm. long; ovary 4·5–9·5 × 1·5–2 mm., fusiform; perianth as in ♂ flower. Fruit 9–17 × 6–7 mm., fusiform, bright red, thinly fleshy; pedicel 12–32 mm. long. Seed 3–4 × 1·7–2·5 × 1·4–1·6 mm., ovoid, somewhat compressed, smooth, with convex faces.

Zambia. N: Lake Bangweulu, fr. 5.ix.1953, *Fanshawe* 270 (K). W: Fort Rosebery, fl. 30.vii.1952, *White* 3176 (K; FHO). **Rhodesia.** N: Mazoe, Kasipiti, fl. 28.xii.1964, *Loveridge* 1298 (COI; K; SRGH). E: Inyanga, Stapleford-Chinyamuriro, fl. 3.iv.1962, *Wild* 5685 (COI; SRGH). **Malawi.** N: Mzimba Distr., Mzuzu, Lunyanga waterworks, fl. & fr. 21.x.1973, *Pawek* 7433A & B (K; MAL; MO; SRGH; UC). **Mozambique.** MS: Manica, serra Zuira, Tsetsera, fl. 2.iv.1966, *Torre & Correia* 15579 (LISC).

Also in Cameroon, Fernando Po, Ethiopia and E. tropical Africa. Margins of swamps and upland forest, marshes, damp places in grasslands, 900–2100 m.

7. **Zehneria pallidinervia** (Harms) C. Jeffrey in Kew Bull. **15**: 368 (1962); in F.T.E.A., Cucurb.: 127 (1967). Type a cultivated plant from Tanzania.
Melothria pallidinervia Harms in Notizbl. Bot. Gart. Berl. **8**: 614 (1923). Type as above.

Stems glabrous, rooting and forming tubers at the nodes. Leaf-lamina 2–4·8 × 2·9–6·5 cm., succulent, triangular, weakly cordate at the base, slightly punctate above, glabrous beneath, plane or remotely sinuate-denticulate at the margins, unlobed or incipiently palmately 3-lobed. Petioles 0·8–3·3 cm. long, glabrous. Flowers monoecious. ♂ flowers 6–21 in racemiform or subumbelliform pedunculate axillary clusters; peduncles 0·7–4·1 cm. long; pedicels 1–13 mm. long. Receptacle-tube 2–3·5 mm. long; lobes up to 0·5 mm. long, filiform. Petals 2–4 mm. long, white, becoming cream with age. ♀ flowers solitary, co-axillary with ♂; ovary 7–12 mm. long, fusiform, beaked; perianth as in ♂ flower. Fruit 10–13·5 × 5–7 mm., fusiform, red; pedicel 5–7 mm. long. Seeds 3·5–4 × 2·0–2·4 × 0·5–0·7 mm., ovate in outline, compressed, with thick grooved elevated margins and depressed slightly convex disk.

Mozambique. MS: S. of Muda, near Lake Gambue, fr. 20.vi.1961, *Leach & Wild* 11117 (COI; SRGH). LM: Magude, near Chobela, fr. 3.i.1948, *Torre* 7032 (LISC). Also in Uganda, Kenya and Tanzania. Evergreen thicket margins, 0–200 m.

8. **Zehneria thwaitesii** (Schweinf.) C. Jeffrey in Kew Bull. **15**: 371 (1962); in F.T.E.A., Cucurb. 128, t. 21 fig. 11 & 12 (1967).—Keraudren in Humbert, Fl. Madag., Cucurb.: 40, t. 9 fig. 1–9 (1966); in Aubrév., Fl. Cameroun **6**, Cucurb.: 35, t. 8 fig. 6–9 (1967). —R. & A. Fernandes in C.F.A. **4**: 244 (1970). Syntypes from Sudan and Ceylon.
Bryonia deltoidea Arn. in Nova Acta Acad. Caesar. Leop. Carol. **18**: 337 (1836), non Schumach. (1829), *nom. illegit.* Type from Ceylon.
Melothria thwaitesii Schweinf., Reliq. Kotschy.: 44, t. 29 (1868). Syntypes as for *Zehneria thwaitesii*.
Melothria tridactyla Hook. f. in F.T.A. **2**: 452 (1871).—Hiern, Cat. Afr. Pl. Welw. **1**, 2: 402 (1898). Syntypes from Angola, Congo and Mozambique: Manica e Sofala, Shupanga, *Kirk* (K).
Zehneria tridactyla (Hook. f.) R. & A. Fernandes in Mem. Junta Invest. Ultramar, Sér. 2, **34**: 118 (1962). Syntypes as above.

Stems slender, glabrous or sparsely pilose. Leaf-lamina 2–9 × 2·7–9·5 cm., membraneous, broadly ovate, ovate, broadly triangular or triangular in outline, subtruncate, weakly cordate, hastate or sagittate at the base, finely scabrid-punctate above, shortly setulose on veins beneath, plane or usually remotely and minutely sinuate-denticulate at the margins, unlobed or shortly to very deeply palmately 3-lobed, the lobes triangular to linear, ± acuminate, obtuse to acute, apiculate, the central largest. Petioles 0·9–4·0 cm. long, shortly setulose. Flowers monoecious. ♂ flowers solitary, axillary; pedicels 8–24 mm. long, slender. Receptacle-tube 1–2·5 mm. long, lobes 0·3–1·5 mm. long, dentiform. Petals 1·5–6 mm. long, white, becoming cream with age. ♀ flowers solitary, often co-axillary with ♂; pedicels 7–61 mm. long, slender; ovary 6–12 × 1–2 mm., fusiform. Perianth as in ♂ flowers. Fruit 15–45 × 7–14 mm., fusiform, red; pedicel 10–53 mm. long. Seeds 3·2–4·4 × 2·0–2·5 × 0·3–0·5 mm., ovate in outline, smooth bi-convex.

Zambia. N: NE. Mweru-wa-Ntipa, Chisheyla Dambo, fl. 5.viii.1962, *Tyrer* 278 (BM). **Rhodesia.** E: Umtali, fl. 4.xii.1961, *Wild & Chase* 5550 (COI; K; SRGH). **Malawi.** N: Nyika, fl., *Wakefield* (K). S: Mlanje Mt., Lufiri, Great Ruo Gorge, fl. 18.vi.1962, *Richards* 16765 (K; SRGH). **Mozambique.** N: Cabo Delgado, Montepuez, 22 km. from Montepuez towards Nantulo, fl. 8.iv.1964, *Torre & Paiva* 11751 (LISC). Z: between Tacuane and Limbue, 14·2 km. from Tacuane, fl. 25.v.1949, *Barbosa & Carvalho* 2871 (LMA). MS: Lions Creek, fr. 7.iv.1898, *Schlechter* 12199 (K). LM: between Vila Luisa and Manhiça, Boboli Reserve, fr. 17.xi.1950, *Myre & Carvalho* 1037 (LMA).
Tropical Africa, Madagascar, Ceylon, peninsular India. Riverine forest, 400–1200 m.

EXCLUDED SPECIES

Melothria pulchra Busc. & Muschl. in Engl., Bot. Jahrb. **49**: 497 (1913).
This species, supposed to have been based on *Aosta* 515 from Zambia, was in

fact described from *Schweinfurth* 582, collected in Eritrea; see Gilg in Engl., Bot. Jahrb. **53**: 373 (1915). It is a synonym of *Zehneria scabra* (L. f.) Sond. subsp. *scabra*.

18. KEDROSTIS Medic.

Kedrostis Medic., Phil. Bot. 2: 69 (1791).

Prostrate or scandent herbs with tuberous rootstock. Leaves simple, petiolate. Tendrils simple or 2-fid. Flowers small, yellowish, monoecious. ♂ flowers few to many in axillary racemes or fascicles, usually pedunculate. Receptacle-tube campanulate, lobes usually small. Corolla-lobes 5, united below. Stamens 5, all 1-thecous, in 2 pairs with one single, or 3, with the paired stamens ± united, 2 double 2-thecous, 1 single 1-thecous; filaments short, inserted in mouth of receptacle-tube; thecae short, straight or arcuate, lateral. ♀ flowers axillary, solitary; ovary smooth, glabrous or variously pubescent; ovules few to many, horizontal; perianth as in ♂ flowers; staminodes 3–5, small, inserted in mouth of tube; stigma 2-lobed. Fruits solitary, small to medium-sized, succulent, cylindrical-fusiform to globose, often rostrate, red, indehiscent or dehiscent by a longitudinal slit. Seeds asymmetrically ovoid or subglobose, small.

About 25 species in the Old World tropics, the majority in Africa.

♂ peduncles stout, distinct in aspect from the pedicels, which are much more slender; fruit glabrous, 4·5 cm. or more long; seeds subglobose:
  Receptacle-lobes of ♂ flowers 4–9 mm. long, narrowly triangular to lanceolate, of ♀
    flowers 4–4·5 mm. long; fruits rounded at the base, green mottled pale green or
    white when immature - - - - - - - 1. *hirtella*
  Receptacle-lobes of ♂ flowers 0·8–1·3 mm. long, triangular, of ♀ flowers 1 mm. long;
    fruits attenuate at the base, uniformly green when immature - - 2. *leloja*
♂ peduncles slender, similar in aspect to the pedicels; fruit shortly hispid or pilose, up to
  2 cm. long; seeds asymmetrically ovoid:
  Fruits pilose, ovoid-rostrate to subglobose, 9–13 × 5–7 mm.; seeds 4·3–5·1 mm. long;
    plant usually viscid-glandular - - - - - - 3. *foetidissima*
  Fruits shortly and densely hispid, ovoid-rostrate, 16–20 × 7–10 mm.; seeds 5·5–5·8 mm.
    long; plant not viscid-glandular - - - - - 4. *limpompensis*

Note. Meeuse in Bothalia **8**: 35 (1962) records *K. crassirostrata* Bremek. from Botswana, SE. (Machudi) on the basis of *Rogers* 6361 (BOL; PRE) but I have seen no specimens. It is easily recognized by its finely dissected glabrous leaves and sessile conical fruits.

1. **Kedrostis hirtella** (Naud.) Cogn. in A. & C. DC., Mon. Phan. **3**: 644 (1881).—
C. Jeffrey in F.T.E.A., Cucurb.: 133, t. 23 fig. 1–6 (1967).—Keraudren in Fl. Cameroun **6**, Cucurb.: 58, t. 11 fig. 1–8 (1967).—Launert & Roessler in Prodr. Fl. SW. Afr. **94**, Cucurb.: 18 (1968).—R. & A. Fernandes in C.F.A. **4**: 240 (1971). TAB. **117**. Type from Ethiopia.
    Rhynchocarpa hirtella Naud. in Ann. Sci. Nat. Bot., Sér. 4, **16**: 181 (1862).—Hook. f. in F.T.A. **2**: 564 (1871). Type as above.
    Toxanthera natalensis Hook. f. in Hook., Ic. Pl. **15**: 16, t. 1421 (1883). Type from S. Africa (Natal).
    Toxanthera lugardae N.E.Br. in Kew Bull. **1909**: 112 (1909). Type: Botswana, Kwebe, *Lugard* 54 (K, holotype!).
    Toxanthera kwebensis N.E.Br., tom. cit.: 113 (1909). Type: Botswana, Kwebe, *Lugard* 150 (K, holotype!).
    Kedrostis natalensis (Hook. f.) Meeuse in Bothalia **8**: 36 (1962). Type as for *Toxanthera natalensis*.

Prostrate or scandent to 2 m. or more, from tuberous rootstock; stems shortly and finely spreading-hispid or minutely setulose. Leaf-lamina 2·8–7·5 × 4–11 cm., broadly ovate to pentagonal in outline, cordate, ± sinuate-denticulate, densely and finely pubescent to hispid-setulose especially on veins beneath, shortly and finely rather densely pubescent or hispid above, becoming ± scabrid with age, unlobed or incipiently to deeply palmately 3–5(7)-lobed; lobes ovate-triangular to elliptic, obovate or obovate-lanceolate, the central largest, broadly rounded to obtuse, apiculate or long-apiculate. Petioles 1·8–6 cm. long, finely hispid or pubescent, also spreading-setulose on upper margins. Tendrils 2-fid, rarely simple. Flowers monoecious, ♂ and ♀ usually at different nodes. ♂ flowers c. 5–11 in pedunculate axillary racemes; peduncles 0·8–5·5 cm. long; pedicels 1–8 mm. long, distinctly but minutely bracteate. Receptacle-tube 2·2–3 mm. long;

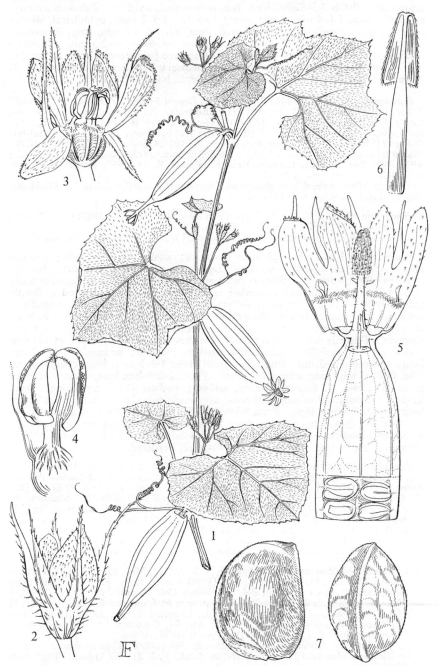

Tab. 117. KEDROSTIS HIRTELLA. 1, part of shoot ($\times\frac{1}{2}$); 2, ♂ flower ($\times4$); 3, ♂ flower, opened to show stamens ($\times4$); 4, double stamen ($\times8$); 5, ♀ flower, dissected out ($\times3$); 6, stigma ($\times5$); 7, seed, face and lateral views ($\times6$) all from *Grosvenor* 469.

lobes 4–9 mm. long, narrowly triangular to lanceolate, acute, attenuate, long-acuminate. Petals 3·5–5 mm. long, yellow or greenish-yellow. ♀ flowers solitary, axillary; pedicel 1–4 mm. long; ovary 10·5–18 × 1·5–2 mm., cylindrical, shortly pubescent; receptacle-tube 1·5–2·5 mm. long, lobes 4–4·5 mm. long, lanceolate-subulate, acute; petals 6–8 mm. long. Fruit 5·8–6·5 × 1·6–2·3 cm., cylindric-fusiform, rounded at the base, rostrate, green spotted pale green or white, becoming red and splitting longitudinally when ripe; pedicel 0·2–1·2 cm. long, stout. Seeds 4·8–5 × 3·8–4 × 3·2–3·5 mm., subglobose, smooth, bordered.

Botswana. N: Kwebe Hills, fl. 16.xii.1897, *Lugard* 54 (K). **Zambia.** S: Machili, fl. 4.xii.1960, *Fanshawe* 5943 (K; SRGH). **Rhodesia.** N: Lomagundi, Oswa road, off Lion's Den-Mangula road, fr. 17.i.1969, *Grosvenor* 469 (COI; K; SRGH). W: Bulawayo, fr. 18.ii.1912, *Rogers* 5660 (BOL). C: Salisbury, Beatrice, fr. 17.ii.1961, *Drewe* 39 (SRGH). E/S: Birchenough Bridge, fl. i.1938, *Obermeyer* 2506 (PRE; SRGH). **Malawi.** N: Rumphi Distr., Njakwa Gorge, fl. 30.xii.1973, *Pawek* 7658 (K; MAL; MO). C: Bua R. drift, Kasungu-Kota Kota road, fl. 13.i.1959, *Robson & Jackson* 1138 (K).
Tropical Africa in the drier regions from Senegal to Transvaal and Angola. Deciduous woodland and bushland, 600–1350 m.

2. **Kedrostis leloja** (J. F. Gmel.) C. Jeffrey in Kew. Bull. **15**: 354 (1962); in F.T.E.A., Cucurb.: 134 (1967). Type from Yemen.
   Turia leloja J. F. Gmel. in L. Syst. Nat. ed. 13, **2**, 1: 403 (1791). Type as above.

Scandent to 1 m., from tuberous rootstock; stems minutely rather sparsely crispate-pubescent or shortly setulose. Leaf-lamina 1·7–5·4 × 3·0–7·8 cm., broadly ovate to suborbicular in outline, cordate, sinuate-denticulate, densely rather finely shortly hispid beneath and somewhat less densely so above, unlobed to deeply palmately 3–5-lobed, lobes ovate to broadly elliptic, obtuse to broadly rounded, apiculate, the central largest, sometimes 3-lobulate. Petioles 0·7–3·1 cm. long, shortly finely pubescent or hispid, also ascending-setulose especially on upper margins. Tendrils simple, rarely 2-fid. Flowers monoecious, ♂ and ♀ at different nodes. ♂ flowers 5–20 in pedunculate axillary racemes; peduncles 1·5–9·2 cm. long; pedicels 1·5–5 mm. long. Receptacle-tube 1·8–3 mm. long, campanulate; lobes 0·8–1·3 mm. long, triangular, acute. Petals 2–2·5 mm. long, cream, yellow or greenish-yellow. ♀ flowers solitary, axillary; pedicel 11–15 mm. long; ovary 14–16 × 1–2 mm., cylindrical, glabrous; receptacle-tube 2 mm. long, lobes 1 mm long; petals 2·5 mm. long. Fruit 4·5–6·5 × 1·5–1·6 cm., fusiform, tapered towards the base and apex, long-rostrate, green, becoming red and splitting longitudinally when ripe; pedicel 0·5–1·5 cm. long, rather slender. Seeds 4·5–5 × 3·4–3·8 × 3·1–3·5 mm., subglobose, smooth or slightly rugose, bordered.

Zambia. C: Katondwe, fl. 4.xii.1964, *Fanshawe* 9026 (K; NDO). **Rhodesia.** N: Gokwe, Sasame R., fr. 20.ii.1963, *Bingham* 962 (COI; SRGH). W: Bulawayo, fl. 1911, *Rogers* 5829 (BOL; K). C: Salisbury, Cranbourne, fl. immat. 1.xii.1948, *Wild* 454 (K). S: Gwanda, Ferguson weir, fl. & fr. 18.xii.1956, *Davies* 2349 (SRGH). **Malawi.** S: Mandala, *Scott Elliot* 8435 (BM). **Mozambique.** N: Monapo, 7 km. from Namialo towards Meserepane, fl. 24.xi.1963, *Torre & Paiva* 9260 (LISC). T: Máguè, Cabora-Bassa, fl. 19.ii.1968, *Torre & Correia* 17718 (LISC).
Yemen, E. Africa from Somalia south to S. Africa (Transvaal). *Colophospermum, Pterocarpus-Acacia* and *Terminalia-Commiphora* forest and woodland, 180–550 m.

3. **Kedrostis foetidissima** (Jacq.) Cogn. in A. & C. DC., Mon. Phan. **3**: 644 (1881).—Hiern, Cat. Afr. Pl. Welw. **1**, 2: 404 (1898).—Burtt Davy, F.P.F.T.: 226 (1926) incl. var.—Gomes e Sousa in Bol. Soc. Estud. Col. Moçamb. **29–32**: 92 (1936).—Meeuse in Bothalia **8**: 26 (1962).—Keraudren in Fl. Cameroun **6**, Cucurb.: 55, t. 11 fig. 9–11 (1967).—C. Jeffrey in F.T.E.A., Cucurb.: 137, t. 23 fig. 11 (1967).—Launert & Roessler in Prodr. Fl. SW. Afr. **94**, Cucurb.: 18 (1968).—R. & A. Fernandes in C.F.A. **4**: 239 (1971). Type from W. Africa.
   Trichosanthes foetidissima Jacq., Collect. **2**: 341 (1789). Type as above.
   Zehneria obtusiloba Sond. in Harv. & Sond., F.C. **2**: 487 (1862). Type from S. Africa.
   Kedrostis foetidissima subsp. *obtusiloba* (Sond.) Meeuse, tom. cit.: 27 (1962). Type as for *Zehneria obtusiloba*.

Rather viscid tuberous-rooted herb, foetid when crushed. Stems to 3 m., prostrate or scandent, rather densely crispate-setulose, glandular, when old developing a thin pallid bark. Tuber napiform, with rugose bark. Leaf-lamina

1·6–9 × 1·6–10·2 cm., narrowly to broadly ovate in outline, cordate or subhastate, obscurely sinuate-denticulate to almost entire, rather thin, bright green, rather sparsely to densely shortly finely hispid above and beneath, unlobed or incipiently to moderately deeply palmately 3–5-lobed, central lobe much the largest, ovate to triangular, obtuse to rounded, shortly acuminate, apiculate; petiole 0·4–5·2 cm. long, hispid, glandular. Tendrils simple. Flowers monoecious, ♂ and ♀ usually co-axillary. ♂ flowers 3–15 in small axillary fascicles or racemes, rarely solitary; peduncles 4·5–30 mm. long, slender; pedicels 3–14 mm. long. Receptacle-tube 1·5–3·5 mm. long; lobes 0·8–2·5 mm. long, lanceolate-subulate. Petals 2–5 mm. long, cream, yellow or greenish-yellow. ♀ flowers solitary; pedicels 1–3·5 mm. long; ovary 2·5–4 × 0·7–1·5 mm., ellipsoid, rostrate, densely finely pilose; receptacle-tube 1·3–2 mm. long; lobes 0·8–1·2 mm. long, triangular-lanceolate to lanceolate-subulate, acute; petals 1·5–3 mm. long. Fruit 9–13 × 5–7 mm., ovoid to subglobose, rostrate or erostrate, finely pilose, pale green with darker longitudinal lines becoming orange then red when ripe, indehiscent; pedicel 1·5–9 mm. long. Seeds 4·3–5·1 × 2·8–3·7 × 2–2·2 mm., asymmetrically ovate in outline, smooth, dark-coloured, bordered, rounded at the base, compressed and shortly 2-lobed at the apex.

Caprivi Strip. Singalamwe, fl. & fr. immat. 1.i.1959, *Killick & Leistner* 3249 (SRGH). **Botswana.** SW: 290 km. NW. of Molepolole, fl. & fr. 17.vii.1955, *Story* 5026 (K). SE: Dikomo Di Kai, fl. & fr. 26.ii.1960, *Wild* 5173 (LISC; SRGH). **Zambia.** B: Sesheke, fl. & fr. 26.xii.1952, *Angus* 1035 (FHO; K). C: Katondwe, fl. & fr. 22.vi.1967, *Fanshawe* 10129 (K; NDO). S: near Kabanga mission, fl. & fr. 28.ii.1963, *Astle* 2207 (SRGH). **Rhodesia.** N: Darwin, Umvukwes, Umsengedzi R., fl. & fr. immat. 23.xii.1952, *Wild* 3975 (K; SRGH). W: Bulawayo, Hillside Dam, fl. & fr. iii.1956, *Miller* 3444 (PRE). C: Salisbury, Prince Edward Dam, fl. & fr. immat. 8.i.1952, *Wild* 3740 (K; SRGH). E: Umtali, fl. & fr. 12.i.1956, *Chase* 5950 (BM; COI; LISC; SRGH). **Mozambique.** T: Boruma (Boroma), fl. & fr. immat. vi.1895, *Menyhart* 929 (K). MS: Lupata, fl. & fr. 19.iv.1860, *Kirk* 276 (K). LM: Matola, fr. 6.v.1967, *Marques* 2031 (LMU).
Tropical Africa, India and Burma, in the drier regions. Woodland and grassland, often on river margins and termite mounds, 50–1350 m.

4. **Kedrostis limpompensis** C. Jeffrey in Kew Bull. **30**: 485 (1975). Type: Mozambique, Caniçado, near Mabalane, *Barbosa & Lemos* 8595 (K, holotype; COI, LMA, LISC, isotypes).

Stems to 6 m., prostrate or scandent, very sparsely shortly crispate-pubescent. Leaf-lamina 2–4·5 × 1·8–5 cm., ovate to broadly ovate in outline, cordate, ± sinuate-denticulate, shortly and finely hispid, sometimes densely, unlobed or incipiently palmately 3–5-lobed, lobes triangular, obtuse to acute, shortly acuminate, apiculate. Petioles 0·8–1·2 cm. long, rather sparsely shortly finely ascending-hispid. Flowers monoecious, ♂ and ♀ sometimes co-axillary. ♂ flowers 3–5 in small pedunculate axillary fascicles; peduncles 3–11 mm. long, slender; pedicels 2–10 mm. long. Receptacle-tube campanulate, 1·5–2 mm. long, lobes 0·7–1 mm. long, ovate-lanceolate, acute. Petals 2·5–3·2 mm. long, yellow-green. ♀ flowers solitary, subsessile; pedicel 1 mm. long; ovary 6–7 × 2–3 mm., ovoid, long-rostrate, densely hirtellous; receptacle-tube 1·5–2 mm. long, campanulate, lobes 0·8–1 mm. long, triangular-lanceolate; petals 3–4 mm. long. Fruit 16–20 × 7–10 mm., subsessile, ovoid, rostrate, shortly and densely hispid, red; pedicel 2–2·5 mm. long. Seeds 5·5–5·8 × 3–3·5 × 2·2–2·4 mm., asymmetrically ovate in outline, slightly verrucose, rounded at the base, compressed and slightly emarginate at the apex.

Rhodesia. S: Beitbridge, Chikwarakwara, fl. 23.ii.1961, *Wild* 5348 (COI; K; SRGH). **Mozambique.** SS: Gaza, Caniçado, near Mabalane, fr. 3.vi.1959, *Barbosa & Lemos* 8595 (COI; K; LISC; LMA).
Also S. Africa (N. Transvaal). *Acacia* bushland, c. 750 m.

19. CORALLOCARPUS Hook. f.

Corallocarpus Hook. f. in Benth. & Hook., Gen. Pl. 1: 831 (1867).

Prostrate or usually scandent herbs; stems arising from a perennial tuberous rootstock, eventually becoming softly more or less woody at the base. Leaves simple, petiolate. Tendrils simple. Flowers small, yellowish, monoecious.

♂ flowers few to many in subsessile or pedunculate subcapitate to racemiform usually congested axillary clusters. Receptacle-tube campanulate; lobes small. Corolla-lobes 5, united below. Stamens 5, all 1-thecous, in 2 pairs with one single, or 3, 2 double 2-thecous, 1 single 1-thecous, the double stamens ± 2-lobed; filaments short, inserted in the mouth of the receptacle-tube; thecae short, straight, lateral. ♀ flowers axillary, solitary or fasciculate; ovary smooth, glabrous or puberulous; ovules few to several, horizontal; perianth as in ♂ flowers; staminodes, when present, 5; stigma 2(3)-lobed. Fruits solitary or fasciculate, small, succulent, ovoid or ellipsoid, often rostrate, red, circumscissile near the persistent greenish cupuliform base. Seeds asymmetrically ovoid or pyriform, rarely subglobose, small.

About 15 species in the Old World tropics, mostly African.

Fruits minutely puberulous or glabrous; petioles and stems spiculate and/or minutely
  puberulous or glabrous:
  Stems glabrous, spiculate on ridges, or minutely puberulous, never both spiculate and
    puberulous; fruits glabrous:
    Fruits not rostrate, rounded at the apex, or if shortly rostrate, then subsessile:
      Stems glabrous or sparsely spiculate; fruits rounded, 6–13 mm. long
                                                                          1. *bainesii*
      Stems minutely puberulous; fruits shortly rostrate, 13–17 mm. long
                                                                          2. *poissonii*
    Fruits rostrate, 14–28 mm. long, distinctly pedicellate; pedicels 5–22 mm. long;
      stems spiculate -   -   -   -   -   -   -   -   -   - 3. *wildii*
  Stems both puberulous and spiculate on ridges; fruits minutely puberulous
                                                                          4. *triangularis*
Fruits shortly pubescent; petioles and stems crispate-pubescent -   -   5. *boehmii*

1. **Corallocarpus bainesii** (Hook. f.) Meeuse in Bothalia **8**: 41 (1962) pro parte.—
   C. Jeffrey in F.T.E.A., Cucurb.: 143 (1967) pro parte excl. syn. *C. poissonii* Cogn. &
   C. bussei Cogn.—Launert & Roessler in Prodr. Fl. SW. Afr. **94**, Cucurb.: 10 (1968)
   pro parte. TAB. **118**. Type: Botswana, Norton Shaw Valley, *Baines* (K, holotype).
      Rhynchocarpa bainesii Hook. f. in F.T.A. **2**: 564 (1871). Type as above.
      Kedrostis bainesii (Hook. f.) Cogn. in A. & C. DC., Mon. Phan. **3**: 644 (1881).
   Type as above.
      Corallocarpus sphaerocarpus Cogn. in Verh. Bot. Ver. Prov. Brand. **30**: 151
   (1888).—Burtt Davy, F.P.F.T. **1**: 227 (1926) pro parte.—Gomes e Sousa in Bol. Soc.
   Estud. Col. Moçamb. **29–32**: 92 (1936). Type from SW. Africa.
      ? *Corallocarpus welwitschii* sensu Bremek. & Oberm. in Ann. Transv. Mus. **16**, 2:
   436 (1935) non (Naud.) Hook. f. ex Welw.

Stems prostrate or scandent, smooth, sparsely spiculate, otherwise glabrous, when older softly woody with smooth brown bark, eventually becoming white-callosed and ridged. Leaf-lamina 2·2–7·5 × 2·7–11·0 cm., broadly ovate in outline, deeply cordate, minutely denticulate, shortly, finely and densely puberulous or scaberulous beneath, less densely so above and with age, deeply palmately 5-lobed, lobes broadly ovate to narrowly elliptic, obtuse to rounded, apiculate, the central largest. Petiole 1·7–5·4 cm. long, sparsely shortly setulose, otherwise glabrous or sparsely moderately densely antrorsely puberulous. ♂ flowers c. 6–25 in congested shortly racemiform pedunculate clusters; peduncle 0·4–6·9 cm. long; pedicels 1·5–4·5 mm. long. Receptacle-tube 1 mm. long; lobes 1 mm. long, lanceolate. Petals c. 1·5 mm. long. ♀ flowers sessile, in many-flowered congested fascicles, rhachis ± incrassate. Fruits 8–9 × 6–6·5 mm., clustered, sessile, shortly ellipsoid, rounded, erostrate, glabrous, bright red. Seeds 3·5–4·5 × 2·4–2·8 × 1·6–2·3 mm., smooth or verruculose, nitid.

Botswana. N: Nata R., near Madsiara drift, fl. & fr. 21.iv.1957, *Drummond & Seagrief* 5157 (K; SRGH). **Zambia.** B: Sesheke, fl. & fr. iv.1910, *Gairdner* 514 (K). **Rhodesia.** N: Darwin, Chimanda Reserve, Mazoe R., near Winda pools, fr. 4.ix.1958, *Phipps* 1298 (K; SRGH). E: Melsetter, Umvumvumvu R. Gorge, fl. & fr. 20.i.1957, *Chase* 6305 (K; SRGH). S: Lower Sabi, Devuli R., fl. & fr. 1.ii.1948, *Wild* 2461 (BR; K; SRGH). **Malawi.** N: L. Nyasa, Likoma Is., fl. & fr., *Johnson* 95 (K). **Mozambique.** T: Boroma, fr. iv.1891, *Menyhart* 928 (BR; K). MS: Cheringoma, between R. Urema a n d Inhaminga, fl. & fr. 6.v.1942, *Torre* 4064 (LISC). LM: Namaacha, Goba, near Fonte dos Libombos, fl. & fr. 10.iii.1968, *Balsinhas* 1170 (COI; LMA).
   Also in SW. Tanzania, SE. Angola, S. Africa (N. Natal, N. Transvaal) & the northern parts of SW. Africa. Deciduous woodland and bushland, 350–1050 m.

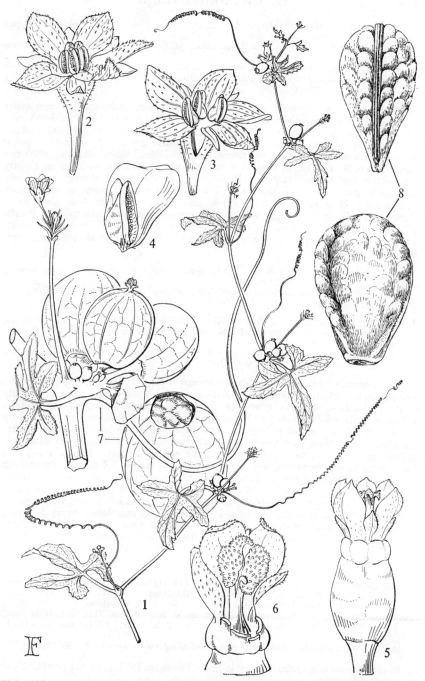

Tab. 118. CORALLOCARPUS BAINESII. 1, part of shoot (× ½) *Johnson & Riddlesdell* s.n. & *Kirk* s.n.; 2, ♂ flower (× 9); 3, ♂ flower partly dissected (× 9); 4, double stamen (× 18), 2–4 from *Lugard* 132; 5, ♀ flower (× 18); 6, ♀ flower, details (× 24), both from *Gairdner* 514 & *Lugard* 132; 7, fruits (× 3) from *Johnson & Riddlesdell* s.n. & *Kirk* s.n.; 8, seeds, lateral and face views (× 9) *Chase* 6305 & *Wild* 2461.

2. **Corallocarpus poissonii** Cogn. in A. & C. DC., Mon. Phan. **3**: 651 (1881). Syntypes from Comoro Is. & Madagascar.
   Corallocarpus bussei Gilg in Engl., Bot. Jahrb. **34**: 363 (1904). Type from Tanzania.
   Corallocarpus bainesii sensu Keraudren in Fl. Madag., Cucurb.: 70, t. 18 fig. 1–7 (1966) & sensu C. Jeffrey in F.T.E.A., Cucurb.: 143 (1967) pro parte, non (Hook. f.) Meeuse.

Stems scandent, minutely and sparsely puberulous, glabrescent, when older softly woody with brownish slightly ridged bark. Leaf-lamina 2·7–8 × 1·5–9·5 cm., broadly ovate in outline, cordate, plane or more or less sinuate, denticulate, rather densely, finely and shortly asperulous beneath, less densely so and becoming scabrid-punctate above, deeply palmately 3–5-lobed, lobes obovate, obovate-elliptic, broadly elliptic or broadly ovate, rounded, apiculate, unlobed or shortly 3-lobulate, the central largest. Petiole 2·1–3·1 cm. long, shortly and finely antrorsely puberulous. ♂ flowers c. 7–9 in shortly racemiform or subcapitate pedunculate clusters; peduncle 1·3–4 cm. long; pedicels 1–3 mm. long. Receptacle-tube 1 mm. long; lobes 0·5–1 mm. long, lanceolate, acute. Petals, 0·7–1·5 mm. long. ♀ flowers 1–6, subsessile; ovary ovoid, rostrate. Fruits 15–18 × 8 mm., subsessile, ovoid, shortly and broadly rostrate, glabrous, red. Seeds not known.

Mozambique. N: Ribáuè, c. 37 km. from Ribáuè towards Lalaua, fr. immat. 22.i.1964, *Torre & Paiva* 10124 (LISC).
Also in SE. Tanzania, Comoro Is. and NW. Madagascar, *Chlorophora-Milletia* and *Brachystegia* forest and woodland, 350–480 m.

3. **Corallocarpus wildii** C. Jeffrey in Kew Bull. **30**: 488 (1975). Type: Rhodesia, Darwin, Umvukwes, Umsengedzi R., *Wild* 3967 (K, holotype; LISC, SRGH, isotypes).
   Corallocarpus sp. C.—C. Jeffrey in F.T.E.A., Cucurb.: 142 (1967) pro majore parte excl. specim. cit. *Koritschoner* 2109.

Stems prostrate or scandent, glabrous except at the very base, sparsely spiculate, arising from a tuberous rootstock. Leaf-lamina 1·8–9 × 2–8 cm., ovate or broadly ovate in outline, cordate, plane or obscurely sinuate, minutely denticulate, minutely scaberulous above and beneath, palmately usually deeply to very deeply 3–5-lobed, lobes triangular, broadly ovate, ovate-elliptic, oblanceolate or linear-lanceolate in outline, the central especially sometimes incipiently to distinctly 3(5)-lobulate, obtuse to acute, apiculate, the central largest. Petiole 1·1–3 cm. long, antrorsely puberulous, antrorsely setulose on ridges. ♂ flowers about 7–30 in congested racemiform or subumbelliform pedunculate clusters; peduncle 1·3–12 cm. long; pedicels 1·5–5·5 mm. long. Receptacle-tube 1·5–2·4 mm. long; lobes 1–1·8 mm. long, ovate-oblong to lanceolate. Petals 1·5–2 mm. long, yellow. ♀ flowers 1–4, usually solitary or paired; pedicels 2–13 mm. long; ovary 5–7 mm. long, narrowly ovoid-rostrate. Fruits 14–28 mm. long, distinctly pedicellate, ovoid, rostrate, glabrous, red except for green apex; pedicels 5–22 mm. long. Seeds 5·1–6 × 2·8–3·4 × 1·9–2·6 mm., asymmetrically ovoid, fibrillose, with ridged 2-grooved margins.

Botswana. N: Dobe, st. 30.xi.1964, *Lee* GN 29 (92) (SRGH). **Zambia.** B: Masese, fl. & fr. 12.i.1961, *Fanshawe* 6122 (SRGH). **Rhodesia.** N: Mazoe, Christon Bank, fr. 13.xii.1964, *Loveridge* 1194 (LISC; SRGH). W: Nyamandhlovu, fl. i.1953, *Plowes* 1541 (K; SRGH). C: Salisbury, Twentydales, fl. & fr. immat. 31.x.1951, *Wild* 3674, 3675 (K; SRGH). E: Umtali, Maranke Reserve, fl. & fr. 11.ii.1953, *Chase* 4785 (BM; SRGH).
Also in SW. Tanzania. *Brachystegia* and *Colophospermum* woodland, 750–1350 m.

4. **Corallocarpus triangularis** Cogn. in Engl., Pflanzenr. **IV**, 275, 1: 171 (1916). Type from SW. Africa.
   Corallocarpus sphaerocarpus var. *scaberrimus* Cogn. in Bull. Herb. Boiss. **3**: 422 (1895).—Burtt Davy, F.P.F.T. **1**: 227 (1926).—Hutch., Botanist in S. Afr.: 666 (1946). Syntypes from S. Africa (Transvaal).
   Corallocarpus sphaerocarpus var. *subhastatus* Cogn., loc. cit. (1895).—Burtt Davy, loc. cit. (1926). Type from S. Africa (Transvaal).
   Corallocarpus sphaerocarpus sensu Story in Mem. Bot. Survey S. Afr. **30**: 45, t. 53 (1958), non Cogn.

Corallocarpus bainesii sensu Meeuse in Bothalia **8**: 41 (1962) pro parte & sensu Launert & Roessler in Prodr. Fl. SW. Afr. **94**, Cucurb.: 10 (1968) pro parte, non (Hook. f.) Meeuse.

Stems to 2 m., prostrate or scandent, shortly puberulous, sulcate, setulose on ridges, when older softly woody, with grey bark, white-callosed on ridges. Leaf-lamina 2·0–5·4 × 2·1–6·6 cm., ovate to subhastate in outline, deeply cordate, finely asperulous or scabrid and more coarsely scabrid-setulose on veins, beneath, finely scabrid above, palmately variously 3(5)-lobed, lobes almost plane to irregularly sinuate-denticulate or lobulate, acute to rounded, apiculate, the central largest. Petiole 1–3·5 cm. long, shortly retrorsely puberulous, setulose on ridges. ♂ flowers about 4–10 in subsessile axillary clusters; penduncles up to 3 mm. long; pedicels 1·5–2·5 mm. long. Receptacle-tube 1·2 mm. long; lobes 1–1·2 mm. long, lanceolate. Petals 1·4–1·5 mm. long, pale yellow to white. ♀ flowers 1–6, subsessile; ovary 3·5 mm. long, ellipsoid, puberulous. Fruits 9–13 × 7–9 mm., 1–6, subsessile, ovoid, not or only shortly rostrate; pedicels up to 2·5 mm. long. Seeds 4–4·5 × 2·8–3 × 2·1–2·6 mm., asymmetrically ovoid, tumid, somewhat compressed at the apex, with slightly rugose faces and prominent margins.

Botswana. N: Kwebe, fr. i.1897, *Lugard & Lugard* 120 (K). SW: 80 km. N. of Kang, fl. & fr. 18.ii.1960, *Wild* 5053 (K; M; SRGH).
Also in S. Africa (NE. Cape, Transvaal) and SW. Africa. *Acacia* bushland, 300–1000 m.

5. **Corallocarpus boehmii** (Cogn.) C. Jeffrey in Kew Bull. **15**: 349 (1962); in F.T.E.A., Cucurb.: 144, t. 4 fig. 8 (1967).—Keraudren in Fl. Cameroun 6, Cucurb.: 53, t. 11 fig. 12–16 (1967). Type from Tanzania.
Kedrostis boehmii Cogn. in Bull. Acad. Belg., Sér. 3, **14**: 357 (1887). Type as above.

Stems to 2 m., prostrate or scandent rather sparsely to densely crispate-pubescent, arising from tuberous rootstock; when old, like the rootstock, with rough greyish bark. Leaf-lamina 1·9–9·7 × 2·2–11 cm., broadly ovate in outline, cordate, plane, minutely and remotely denticulate, densely hirtellous beneath and above, becoming minutely hispid above, rather shallowly to deeply palmately 3–5-lobed; lobes triangular, ovate-triangular, broadly elliptic, obovate-elliptic, elliptic or linear-lanceolate in outline, acute to obtuse, rounded or retuse, apiculate, the central especially sometimes incipiently to distinctly 3–5-lobulate. Petiole 1·2–4·2 cm. long, densely crispate-pubescent. ♂ flowers about 5–10 or more in congested racemiform to subcapitate pedunculate clusters; peduncle 0·4–6 cm. long; pedicels 0·5–2 mm. long. Receptacle-tube 1·5–2 mm. long, lobes 1·5–2 mm. long, lanceolate or ovate-triangular. Petals 2 mm. long, oblong-lanceolate, cream to yellow or greenish-yellow. Flowers 2–7, sessile; ovary 2·5–3 mm. long, ovate-rostrate, densely puberulous. Fruits 1–2(5), 10–15 mm. long, ellipsoid, rounded, not rostrate, densely crispate-pubescent, red, sessile or subsessile; pedicels up to 3·5 mm. long. Seeds 4·0–5·1 × 2·2–3·3 × 1·5–2·0 mm., asymmetrically ovoid-pyriform, slightly compressed, fibrillose or scaly; margins 2 grooved.

Caprivi Strip. C. 32 km. from Singalamwe on road to Katima Mulilo, fl. & fr. immat. 3.i.1959, *Killick & Leistner* 3269 (K; M; SRGH). **Zambia.** N: Chinakila, fl. 10.i.1965, *Richards* 19446 (K). W: Ndola, fl. & fr. 1.iii.1954, *Fanshawe* 893 (K; SRGH). C: Lusaka, fr. 20.iii.1952, *White* 2301 (FHO; K). S: Livingstone, Bombwe Forest, fr. 1934, *Martin* A17/34 (BR; K). **Rhodesia.** W: Matobo, Farm Besna Kobila, fr. ii.1960, *Miller* 7142 (COI; SRGH). C: Salisbury, Hunyani Bridge, fl. & fr. 2.ii.1961, *Rutherford-Smith* 473 (COI; K; LISC; SRGH).
Tropical Africa in the drier regions from Nigeria to the east and south. Woodland and grassland, often in thickets on termite-mounds or rock outcrops, 1050–1450 m.

SPECIES INSUFFICIENTLY KNOWN

Corollocarpus tenuissimus Busc. & Muschl. in Engl., Bot. Jahrb. **49**: 497 (1913).—Wild, Fl. Vict. Falls: 145 (1953). Type: Rhodesia, near Victoria Falls, *Aosta* 146 (B, holotype, †).

Stems slender, terete, succulent, glaucous, glabrous. Leaf-lamina small, ovate-suborbicular or suborbicular in outline, at first densely setulose, soon glabrous, deeply palmately 3–5-lobed, the lobes lanceolate, linear-lanceolate or usually linear, acute or subacute, the central largest, 1–1·5 × 0·3–0·45 cm., the lateral much shorter but a little broader. Petiole slender, glabrous. ♂ flowers 5–8 in subcapitate pedunculate clusters; peduncles short; pedicels 1–1·5 mm. long. Fruit 15–17 × 8–9 mm., glabrous, rostrate; rostrum subfiliform, erect. Seeds turgid.

The above description, which does not fit any plant known to me from the F.Z. area, is adapted from the original. The only known African species to which it could apply would be *C. epigaeus* (Rottl.) Cogn., which does not occur S. of northern Tanzania. In view of the known instances of fabrication by Muschler of specimens attributed to the Aosta expedition—see Engl., Bot. Jahrb. **53**: 366–375 (1915), instances of which are mentioned under *Coccinia* (p. 451) and *Pehneria* (p. 483) above—the possibility cannot be excluded that the type was a specimen collected elsewhere by some other collector. Should it have been based on a genuine collection, *C. tenuissimus* Busc. & Muschl., allowing for some error of description, could well be an older name for *C. wildii* C. Jeffrey. In the absence of the type, however, it must remain a *nomen dubium* which cannot be applied with certainty to any taxon.

20. PEPONIUM Engl.

Peponium Engl. in Engl. & Prantl, Pflanzenfam., Nachtr.: 318 (1897).
Peponia Naud. in Ann. Sci. Nat. Bot., Sér. 5, **5**: 29 (1866) *nom. illegit.*, non Grev. (1863).

Prostrate or scandent perennial herbs. Leaves simple, palmately lobed, with scattered disk glands on lower surface. Tendrils 2-fid. Flowers white or yellow, dioecious. ♂ flowers axillary, racemose, rarely solitary, bracteate, the racemes usually accompanied by a long-pedicellate co-axillary solitary flower. Bracts more or less cucullate, membraneous, glandular. Receptacle-tube elongated, more or less cylindrical; lobes lanceolate, narrow, entire. Petals 5, free or shortly united, entire. Stamens 3, all 2-thecous, or two 2-thecous, one 1-thecous, inserted on receptacle-tube, with free filaments and anthers connate into a narrow cylindrical head; connectives narrow; thecae triplicate. ♀ flowers solitary; perianth similar to that of ♂ flower; ovary terete; ovules numerous, horizontal; stigma 3-lobed. Fruit ovoid or ellipsoid, usually more or less rostrate, terete, smooth, thin-walled, fleshy, indehiscent, red when ripe. Seeds ovoid in outline, compressed, smooth, lenticular, blackish, sometimes obscurely bordered.

A genus of c. 20 species in tropical and southern Africa, Madagascar, Aldabra and the Seychelles.

Leaves glabrous or almost so, even on veins beneath:
 Leaf-lamina thick, coriaceous; leaf-lobes triangular or ovate-triangular, not lobulate;
  tendrils stout - - - - - - - - - - 1. *chirindense*
 Leaf-lamina thin, membranous; leaf-lobes elliptic or broadly elliptic, 3-lobulate;
  tendrils weak - - - - - - - - - - - 2. *sp. A*
Leaves densely hispid, especialy on veins beneath:
 Stems robust, scandent; plant of montane rain-forests - - - - 4. *vogelii*
 Stems rather slender, prostrate; plant of dry coastal forests - - 3. *pageanum*

1. **Peponium chirindense** (Bak. f.) Cogn. in Engl., Pflanzenr. IV, 275, 2: 218 (1924). TAB. **119**. Type: Rhodesia, Chirinda, *Swynnerton* 2102 (BM, holotype).
  Peponia chirindensis Bak. f. in Journ. Linn. Soc., Bot. **40**: 74 (1911).—Eyles in Trans. Roy. Soc. Afr. **5**, 4: 498 (1916). Type as above.

Stems scandent to 8 m. or more, often reaching the canopy of tall trees, rarely prostrate, glabrous or almost so. Leaf-lamina 6·5–15 × 8·5–21 cm., broadly ovate in outline, cordate, usually deeply, coriaceous, dark green, shiny, glabrous except for veins above and occasionally veins beneath, becoming punctate above with age, usually coarsely sinuate-dentate at margins with small apiculate teeth, shallowly to moderately deeply palmately 5-lobed, lobes triangular to ovate-triangular or sometimes rounded, usually shortly acuminate, apiculate, the central largest. Petiole 1·5–7·5 mm. long, glabrous. Tendrils stout. ♂ racemes 7–12-flowered; peduncle 4·5–13·5 cm. long; pedicels of basal solitary flower 2·5–12 cm. long, of racemose flowers 3–32 mm. long; bracts 5–17 mm. long, broadly elliptic, cucullate, rounded to acute. Receptacle-tube 20–30 mm. long, glabrous; lobes 4–8 mm.

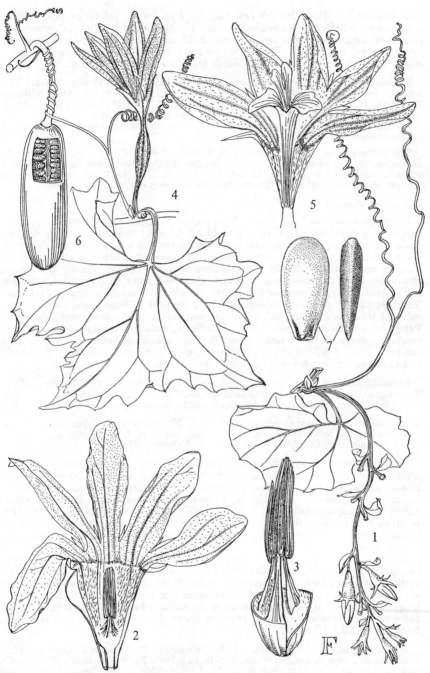

Tab. 119. PEPONIUM CHIRINDENSE. 1, flowering node, ♂ plant (×½) *Kelly* 212; 2, ♂ flower opened (×1); 3, androecium (×2), both from *Dale* SKF 452; 4, flowering node, ♀ plant (×½); 5, ♀ flower opened (×1), both from *Wild* 5269; 6, fruit (×½); 7, seed, face and lateral views (×4), both from *Wild* 3755.

long, lanceolate. Petal 2·6–2·9 cm. long, c. ½ as broad, yellow or pale yellow. ♀ flowers on 2–3·5 cm. long pedicels; ovary 22–30 × 4–6 mm., ellipsoid, glabrous; receptacle-tube 14–21 mm. long, lobes 5–22 mm. long, triangular-lanceolate; petals 2·4–4·2 cm. long. Fruit 8·5–15 × 2·5–8 cm., ovoid-oblong, subrostrate, acute; fruit-stalk 3–5·5 cm. long. Seeds 6·3–9 × 3·3–5 × 1·5–1·7 mm.

Rhodesia. C: Salisbury, Domboshawa, fr. 23.i.1952, *Wild* 3755 (K; SRGH). E: Inyanga, Pungwe Valley, fl. 9.xi.1950, *Wild* 5269 (COI; K; SRGH). **Mozambique.** Z: Gúruè, W. of Picos Namuli, R. Malemo, fr. 5.i.1968, *Torre & Correia* 16941 (LISC). MS: Chimoio, Serra de Garuso, fl. 28.iii.1948, *Barbosa* 1256 (LISC).
Also in Tanzania. Rain-forest, especially on margins and in clearings, sometimes in moist grassland, 600–1800 m.

The pollen studies by Page—see Kew Bull. **30**: 495–502 (1975)—confirm the specific distinctness of this taxon from *P. caledonium* (Sond.) Engl. and show that the putative intermediates cited by C. Jeffrey in F.T.E.A., Cucurb.: 83 (1967) are probably merely variants of the latter species. The two taxa are also very distinct ecologically.

2. Peponium sp. A.

Stems prostrate, to 7 m., arising from napiform rootstock, glabrous or sparsely and weakly setulose. Leaf-lamina 8–11 × 8–11 cm., broadly ovate in outline, shallowly cordate, membraneous, glabrous except for a few setae on the main veins, palmately 5-lobed to about the middle, lobes elliptic to broadly elliptic in outline, more or less deeply 3–5-lobulate, acute to obtuse, shortly acuminate, apiculate; distal margins of lobes coarsely sinuate-dentate, with rounded, apiculate teeth; lateral margins more or less entire. Petiole 1–1·5 cm. long, sparsely setulose. Tendrils weak. ♂ flowers solitary, axillary; pedicel 10·5 cm. long. Receptacle-tube 22 mm. long, glabrous; lobes 10 mm. long, triangular-lanceolate, acute. Petals 3·5 cm. long, yellow. ♀ flowers solitary, axillary; pedicel c. 5 mm. long, setulose; ovary c. 18 × 4 mm., ovoid-rostrate, spreading-setulose; further data lacking. Fruit 3–6·5 × 2–3·5 cm., ellipsoid, shortly rostrate, glabrous; fruit-stalk c. 1 cm. long. Seeds 6·5–7 × 3·2–3·5 × 1·1–1·2 mm.

Mozambique. Z: Île, fl. 26.vi.1943, *Torre* 5578 (LISC). T: Macanga, Monte Furancungo, fr. 15.iii.1966, *Pereira, Sarmento & Marques* 1720 (LMU).
Not known from elsewhere. Moist grassland, 1380–1420 m.

3. Peponium pageanum C. Jeffrey in Kew Bull. **30**: 492 (1975). Type: Mozambique, Cheringoma, Inhamitanga, *Simão* 52 (LISC, holotype; LMA, isotype).
Peponium sp. A.—C. Jeffrey in F.T.E.A., Cucurb.: 82 (1967).

Stems prostrate, much-branched, arising from perennial rootstock, spreading-setulose, sometimes shortly and sparsely so. Leaf-lamina 5–7·5 × 6–9·5 cm., broady ovate in outline, cordate, sinuate-denticulate, setulose or shortly setulose, becoming scabrid-punctate above, hispid-setulose on veins beneath, rarely unlobed, usually palmately 3–5-lobed to half-way or more, lobes ovate-triangular to elliptic in outline, rounded to acute, shortly acuminate, apiculate, often 3-lobulate, sometimes deeply so. Petiole 1–2·2 cm. long, setulose or shortly so. Tendrils weak to rather stout. ♂ racemes 6–8-flowered; peduncle 4–8 cm. long; pedicel of basal solitary flower 1–6 cm. long, of racemose flowers 4–11 mm. long; bracts 4·5–10 mm. long, broadly elliptic, obtuse, apiculate, inconspicuously few-toothed on upper margins. Receptacle-tube 28–32 mm. long, setulose or shortly so; lobes 2–7 mm. long, lanceolate, more or less recurved. Petals 3·5–3·7 cm. long, c. 2 cm. broad, yellow. ♀ flowers unknown. Fruits 7 × 3·5 cm., ellipsoid, more or less rounded except at extreme apex; fruit-stalk stout, 2–3 cm. long. Seeds c. 7 × 4·2 × 1·3 mm.

Mozambique. N: Mozambique, fl. 1863, *Kirk* (K). Z: Maganja da Costa, c. 50 km. from Vila da Maganja, Gobene forest, fl. & fr. 12.ii.1946, *Torre & Correia* 14528 (LISC). MS: Inhamitanga, fl. 20.ix.1944, *Simão* 52 (LISC; LMA). SS: between Nhachengo and Vilanculos, fr. 22.iii.1952, *Barbosa & Bittencourt* 4973 (LMA).
Also in SE. Tanzania. Coastal forests, near sea level.

4. Peponium vogelii (Hook. f.) Engl. in Engl. & Prantl, Pflanzenfam., Nachtr.: 318
    (1897).—C. Jeffrey in F.T.E.A., Cucurb.: 81, t. 11 fig. 1–7 (1967).—R. & A.
    Fernandes in C.F.A. **4**: 283 (1970). Syntypes from Fernando Po and Nigeria.
    Peponia vogelii Hook. f. in F.T.A. **2**: 526 (1871). Syntypes as above.

Stems to 8 m., scandent or prostrate, glabrous or variously hispid or setulose. Leaf-lamina 5–18 × 7–26 cm., broadly ovate in outline, cordate, setulose above and beneath especially on veins, becoming hispid or scabrid-punctate above, sinuate-denticulate at the margins, palmately 5-lobed; lobes triangular to ovate or ovate-elliptic, obtuse to acute, shortly acuminate, apiculate, the central largest. Petiole 2–13 cm. long, setulose or hispid. Tendrils stout. ♂ racemes 4–15-flowered; peduncle 3–21 cm. long; pedicels of basal solitary flower 4–25 cm. long, of racemose flowers 3–25 mm. long; bracts 8–31 mm. long, lanceolate to elliptic or broadly ovate, cucullate, apiculate, usually shortly adnate to the pedicel. Receptacle-tube 15–32 mm. long, setulose; lobes 6–13 mm. long, lanceolate, narrow. Petals 1·9–4·8 cm. long, white to pale yellow, obovate, apiculate. ♀ flowers on 1–5·5 cm. long pedicels; ovary 15–26 × 10–14 mm., ellipsoid, densely pubescent; receptacle-tube 11–22 mm. long; lobes 6–13 mm. long, linear-lanceolate; petals 2·6–3·8 cm. long. Fruit 4–14 × 3–5 cm., ellipsoid, pubescent; fruit-stalk 2–7 cm. long. Seeds 6–10 × 3–5 × 1–1·5 mm.

Malawi. N: Chitipa, Misuku Hills, Mughesse rain forest, fl. 29.xii.1970, *Pawek* 4218 (K; MAL).
Widespread in tropical Africa from Ghana and Ethiopia to Angola, Tanzania and Malawi. Rain-forest, 1700 m.

21. GERRARDANTHUS Hook. f.

Gerrardanthus Hook. f. in Benth. & Hook., Gen. Pl. 1: 840 (1867).

Scandent herbs, stems becoming woody when old, usually arising from perennial tuberous rootstock. Leaves simple, petiolate. Tendrils apically 2–fid. Flowers small, brownish, dioecious, ± zygomorphic. ♂ flowers in paniculoid axillary fascicles. Receptacle-tube broad, shallow, lobes small. Petals 5, free, unequal. Stamens 4, all 1-thecous, in 2 pairs united by their anthers; fifth stamen reduced to a subulate staminode; filaments inserted on receptacle-tube; connective apically ± produced; thecae short, straight. ♀ flowers solitary or few-fasciculate; ovary trigonous; ovules several, pendulous; perianth similar to that of ♂ flowers; styles 3. Fruit an obconic-cylindric, trigonous capsule, dry, brownish, dehiscent by an apical triradiate slit. Seeds elongated, narrow, with an apical membraneous wing.
5 species in tropical and southern Africa.

Fruits 4–6·5 cm. long:
  Buds and ♂ receptacle-lobes obtuse to rounded; body of seeds linear-oblong, compressed; tuber flattened above - - - - - 1. *macrorhizus*
  Buds and ♂ receptacle-lobes acute; body of seeds fusiform, little compressed; tuber markedly convex above - - - - - - - - 2. *lobatus*
Fruits 7–8 cm. long - - - - - - - - - - 3. *sp. A*

1. **Gerrardanthus macrorhizus** Harv. ex Hook. f. in Benth. & Hook., Gen. Pl. 1: 840 (1867).—Meeuse in Bothalia 8: 8 (1962). Type from S. Africa (Natal).

Scandent to 20 m. or more, stems becoming woody when old, arising from flattened tuberous rootstock up to 150 cm. across. Leaf-lamina 2–8 × 2–8 cm., ovate to broadly ovate in outline, cordate, subsucculent, glabrous, unlobed or palmately 3–5–7-lobed; lobes triangular, obtuse, apiculate, the central largest, ± acuminate, sometimes acute. Petiole 1–4 cm. long, glabrous. ♂ flowers 2–7 in corymbiform axillary fascicles; peduncles c. 2 mm. long, obsolescent; pedicels 5–30 mm. long. Receptacle-tube 0·5–1 mm. long; lobes 2–3 mm. long, ovate, obtuse to rounded. Petals 4–7 mm. long, brownish, unequal, 2 rather larger and darker, 3 smaller and paler; anther-connective slightly produced beyond the thecae but not obviously appendaged. ♀ flowers 1–2, axillary; pedicels 1–2 cm. long; ovary c. 15 × 2–2·5 mm. Fruit 4·3–6·5 × 1·4–2·2 cm., cylindric-obconic, smooth, glabrous, obscurely veined. Seeds with compressed verrucose linear-oblong body and broad wing, 29–55 mm. long; body 16–25 × 4–7 mm.; wing 15–30 × 7–14 mm.

Mozambique. LM: Goba, Libombo Mts., fr. 18.iii.1945, *Sousa* 114 (LISC; PRE).
Also in S. Africa (Natal, Cape Prov.) and Swaziland. Lowland forests and hygrophilous bush.

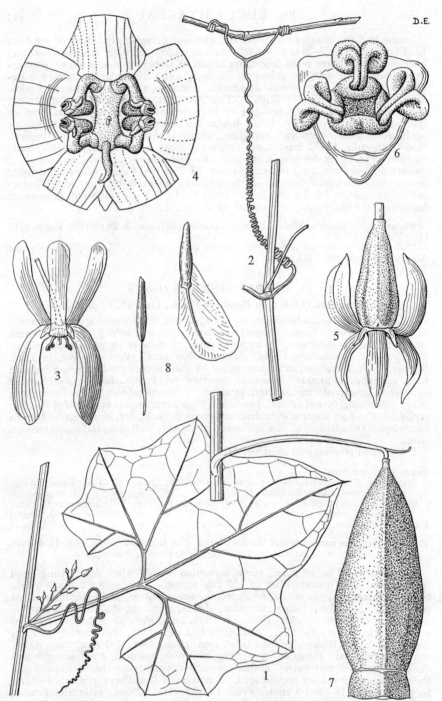

Tab. 120. GERRARDANTHUS LOBATUS. 1, habit (× ⅔) *Schlieben* 2098; 2, apically 2-fid tendril (× ⅔) *Gillett* 14136; 3, ♂ flower (×2); 4, stamens in plan (×6⅔), both from *Milne-Redhead & Taylor* 7243; 5, ♀ flower (×3); 6, styles (×10), both from *Gillett* 14136; 7, fruit (×1⅓) *Drummond & Hemsley* 4158; 8, seed, face and lateral views (×1⅓) *Gillett* 14436. From F.T.E.A.

2. **Gerrardanthus lobatus** (Cogn.) C. Jeffrey in Kew Bull. **15**: 353 (1962); in F.T.E.A.,
Cucurb: 147, t. 25 (1967). TAB. **120**. Type from Kenya.
   Gerrardanthus grandiflorus var. *lobatus* Cogn. in Engl., Pflanzenr. **IV**, 275, 1: 23
(1916). Type as above.

Scandent to 6 m.; stems becoming woody when old, arising from convex
greenish semi-subterranean tuberous rootstock up to 50 cm. or more across.
Leaf-lamina 3–10 × 4·5–12·5 cm., broadly to very broadly ovate in outline,
cordate, subsucculent, almost glabrous, palmately 5–7-lobed or less often unlobed;
lobes triangular or ovate-triangular, obtuse to acute, shortly acuminate, apiculate,
the central largest. Petiole 1·5–8 cm. long, sparsely shortly crispate-pubescent.
♂ flowers 5–50 in lax paniculoid axillary fascicles; peduncles and pedicels 2–
20 mm. long. Receptacle-tube 0·5 mm. long; lobes 2–4 mm. long, ovate-lanceolate
or triangular, acute. Petals brownish yellow, sometimes tinged orange or pinkish,
unequal, the two larger 10–18 mm. long, oblong, erect, the three smaller 9–16 mm.
long, linear, much narrower, reflexed, paler; anther-connective produced
apically into a small appendage. ♀ flowers in fascicles of 2–3 or rarely solitary;
peduncles 1·5–3 cm. long; pedicels 1–8 mm. long; ovary 7–10 × 2–3 mm.,
glabrous; petals unequal, 3 broader, erect, 8–10 mm long, 2 reflexed or spreading,
narrower, 7–8 mm. long. Fruit 4·5–6·3 × 1·8–2·8 cm., cylindric-obconic, smooth,
glabrous, obscurely veined. Seeds 25–44 mm. long, with fusiform little-
compressed body and broad wing; body 12–22 × 1·8–3·5 mm., wing 14–22 × 6–
9 mm.

Malawi. S: Zomba Rock, fl. 1896, *Whyte* (K). **Mozambique.** MS: Chimoio, serra de
Garuzo, fl. 5.iv.1948, *Garcia* 884 (LISC).
Also in Nigeria, Zaire, Uganda, Kenya, Tanzania. Hygrophilous forest.

SPECIES INSUFFICIENTLY KNOWN

3. **Gerrardanthus sp. A.**

Scandent. Fruits c. 7–8 × 2–2·5 cm., dark brown, glabrous, longitudinally
ribbed. Leaves, flowers and seeds unknown.

Malawi. N: Chitipa, Misuku Hills, Mughesse rain forest, fr. 29.xii.1970, *Pawek* 4214
(K; MAL).
Rain forest, 1750 m.

It is impossible to determine the above very imperfect specimen, but it could possibly
represent *G. grandiflorus* Cogn., a species hitherto known only from the coastal forests of
SE. Kenya and NE. Tanzania; see C. Jeffrey in F.T.E.A., Cucurb.: 147 (1967).

22. CYCLANTHEROPSIS Harms

Cyclantheropsis Harms in Engl., Bot. Jahrb. **23**: 169 (1896).

Scandent herbs, stems becoming woody when old, arising from perennial
tuberous rootstock. Leaves simple, petiolate. Tendrils apically 2-fid. Flowers
small, greenish-yellow, dioecious, regular. ♂ flowers in lax, many-flowered,
axillary paniculoid fascicles. Receptacle-tube very shallow, lobes 5, small. Petals
5, free, small. Stamens united into a single central column, inserted on the basal
disk; thecae 2, semi-circular, horizontal, apical on the short central column,
forming a split horizontal ring. ♀ flowers in rather few-flowered axillary fascicles;
ovary compressed, one-locular; ovule solitary, pendulous; perianth similar to
that of ♂ flower; disk 3-angled, basal; styles 3, very short. Fruit compressed,
samaroid, one-seeded, elliptic in outline. Seed elliptic in outline, compressed,
rather large.
Three species, two African, one Madagascan.

Cyclantheropsis parviflora (Cogn.) Harms in Engl., Bot. Jahrb. **23**: 169 (1896).—
Mitchell in Puku **1**: 128 (1963).—C. Jeffrey in F.T.E.A., Cucurb.: 148, t. 26
(1967).—R. & A. Fernandes in C.F.A. **4**: 238 (1971). TAB. **121**. Type from
Zanzibar.
   Gerrardanthus parviflorus Cogn. in A. & C. DC., Mon. Phan. **3**: 936 (1881). Type
as above.

Scandent to 5 m.; stems puberulous, becoming woody and grey-barked when
old, arising from convex semi-subterranean tuberous rootstock. Leaf-lamina

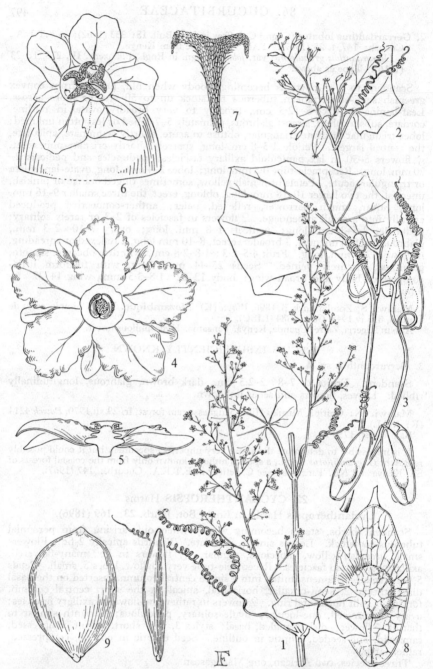

Tab. 121. CYCLANTHEROPSIS PARVIFLORA. 1, part of shoot, ♂ plant (× ½) *Polhill & Paulo* 1060; 2, part of shoot, ♀ plant (× ½) *Polhill & Paulo* 1060 & *Faulkner* 2816; 3, part of shoot, fruiting stage (× ½) *Faulkner* 2816; 4, ♂ flower (× 12); 5, ♂ flower in longitudinal section (× 12), both from *Polhill & Paulo* 1060; 6, ♀ flower (× 12); 7, stigma (× 40); 8, fruit (× 1), 6–8 from *Faulkner* 2816 · 9, seed, face and lateral views (× 3) *Milne-Redhead & Taylor* 7227.

2–12 × 3–11·5 cm., ovate to reniform in outline, cordate, subsucculent, minutely puberulous, unlobed or usually palmately 3–5-lobed, lobes ovate or triangular, obtuse to acute, apiculate, the central largest, usually ± acuminate. Petiole 1–4 cm. long, minutely puberulous. ♂ flowers numerous in lax paniculoid axillary fascicles, pedicellate, peduncles and pedicels slender. Receptacle-tube very shallow; lobes 0·7–1 mm. long, oblong-lanceolate. Petals 1–1·5 mm. long, ovate, greenish-yellow. ♀ flowers in 3–6-flowered axillary racemes; peduncles 5–20 mm. long, pedicels 2–6 mm. long; ovary 3–6 mm. long, 1–1·7 mm. across, compressed; perianth similar to that of ♂ flowers, the receptacle-lobes rather longer; disk obscurely 3-angled; styles very short. Fruits 3·1–5·5 × 1·5–2·8 × 0·5 cm., 1–4 in each raceme, pendulous, brownish, indehiscent, the faces reticulate. Seed slightly rough, 11–21 × 8–16 × 3–5 mm.

Zambia. B: Li Boma, fl. 21.xi.1960, *Fanshawe* 5908 (K). S: Namwala, Kafue National Park, Ngoma, Nakalombwe Hill, fl. 28.xii.1963, *Mitchell* 24/44 (COI; LISC; SRGH). **Rhodesia.** N: Urungwe, Kariba Gorge, fl. 25.ii.1953, *Wild* 4060 (LISC; SRGH). E: Umvumvumvu R. gorge, fr. 24.xii.1947, *Chase* 459 (BM; K; LISC; SRGH). **Malawi.** S: Boadzulu I., fl. 14.iii.1955, *E. M. & W.* 875 (LISC; SRGH). **Mozambique.** T: Cabora Bassa, fl. 20.iv.1972, *Pereira & Correia* 2188 (K; LMU).

Eastern tropical and southern Africa from S. Ethiopia to Angola. Deciduous bushland, wooded grassland and thicket, 180–900 m.

87. BEGONIACEAE
By F. K. Kupicha

Herbs or subshrubs, usually fleshy. Leaves alternate, simple or very rarely pinnate, often asymmetric; stipules free, persistent or deciduous. Flowers monoecious, actinomorphic or zygomorphic, in terminal or axillary cymes, conspicuous; sepals and petals usually not distinguished. ♂ flower: tepals 2–5, free or connate; stamens ∞, filaments free or rarely connate; anthers 2-celled, continuous with the filament, opening lengthwise or rarely by apical pores; rudimentary gynoecium absent. ♀ flowers: tepals ± as in ♂, rarely 6–9; staminodes absent or very small; ovary inferior, (1)2–4(6)-celled, usually angled or winged; styles 2–6, free or connate, usually bifid but occasionally multifid, branches stigmatic and often twisted; placentation axile or rarely (*Hillebrandia*) parietal, placentae entire or lobed; ovules very numerous. Fruit a capsule or berry; seeds minute and very numerous, with reticulate testa, little or no endosperm and straight embryo.

3 genera: *Hillebrandia* Oliv., native to Hawaii, is monotypic; *Symbegonia* Warb., a genus restricted to New Guinea, comprises 10 species; the remaining members of the family belong to the large pantropical genus *Begonia*.

BEGONIA L.

Begonia L., Sp. Pl. 2: 1056 (1753); Gen. Pl. ed. 5: 475 (1754).—Barkley & Golding, The species of the Begoniaceae, ed. 2 (1974).

Herbs and subshrubs, often acaulescent (rarely so in F.Z. area), often with tuberous caudices or creeping rhizomes, or climbing by adventitious roots. Leaves ± asymmetric*, palmate-veined, lobed, toothed or entire, simple or very rarely pinnate (not in F.Z. area), sometimes peltate (not in F.Z. area); stipules free, persistent or deciduous. Flowers monoecious, ± zygomorphic, in terminal and axillary cymes; tepals pink, orange, yellow or white. ♂ flowers: tepals 2–5, free or very rarely connate; stamens ∞, filaments free or very rarely connate; anthers ovoid to oblong, rarely circular or linear, dehiscing by lateral slits or rarely with

---

* In most F.Z. species of *Begonia* the leaves are ± ovate to lanceolate in outline, with the petiole laterally attached. In the species descriptions, the following method of measurement has been adopted: the length of the leaf is the distance from the point of attachment to the leaf apex, not the maximum distance from the base of the lowest leaf-lobe to the apex.

apical pores. ♀ flowers: tepals 2–5(6–9), often persistent; ovary inferior, (2)3(4–6)-locular (always 3-celled in F.Z. area), usually angled or winged; styles 2–6, free or connate, usually bifid but occasionally multifid (not in F.Z. area), branches stigmatic and often twisted; placentation axile, placentae entire or bilamellate (2-lobed in transverse section); ovules ∞. Fruit a capsule or berry, usually with irregular dehiscence; seeds minute and very numerous.

A genus of c. 1400 described species, distributed throughout the tropics with major centres of diversity in northern S. America and from the Himalayas to New Guinea.

Leaf-lamina very sparsely to densely hairy; flowers orange, rarely yellow or pink:
  Flowers pink; fruit a fusiform berry, not winged -   -   -   -   1. *oxyloba*
  Flowers orange, rarely yellow or pink; fruit capsular, prominently 3-winged:
    Plants up to 6 cm. tall; leaves up to 2·2 cm. long, suborbicular -   3. *pygmaea*
    Plants 10–30 cm. tall; leaves 2–11 cm. long, lanceolate to ovate-acute in outline:
      Flowers orange or yellow; tepals in ♀ flowers 5; placentae entire; wings of fruit
        subequal; leaves tapering at apex, acute -   -   -   -   6. *sutherlandii*
      Flowers pink; tepals in ♀ flowers 3; placentae bipartite; wings of fruit unequal;
        leaves ovate in outline, ± obtuse at apex -   -   -   -   5. *rumpiensis*
Leaf-lamina glabrous; flowers pink or white:
  Flowers white or faintly tinged pink, with tepals 1–3 cm. long; bracts often large and
    conspicuous; leaves ovate-acute in outline, 8–16 cm. long, not lobed, margin
    subentire or finely serrate:
    Bracts up to 9 mm. long; petioles 1·5–10 cm. long; leaves distinctly serrate; outer
      tepals of ♂ flowers up to 26 mm. long; placentae entire; fruit prominently and
      unequally 3-winged -   -   -   -   -   -   -   7. *brevibracteata*
    Bracts 15–24 mm. long; petioles 0–4 cm. long; leaves obscurely serrate; outer tepals
      of ♂ flowers 25–34 mm. long; placentae bifid; fruits with 3 subequal narrow
      wings -   -   -   -   -   -   -   -   -   -   8. *nyassensis*
  Flowers pink, with tepals up to 1 cm. long; bracts small and inconspicuous; leaves not
    as above:
    Leaves elliptic to suborbicular in outline, 2–7(10) cm. long, not or very shallowly
      lobed; plants 5–33 cm. tall, often growing on termitaria -   -   2. *princeae*
    Leaves ± ovate in outline, 5–17 cm. long, with 3–5 triangular lobes; plants
      40–100 cm. tall, usually growing in shade among rocks by water:
      Flowers with 4–5 tepals; fruits membranous-walled, prominently winged
                                               4. *sonderana*
      Flowers with 2 tepals; fruits succulent-walled, fusiform, not winged   1. *oxyloba*

Alien *Begonia* species are occasionally found as escapes from cultivation. Specimens of two such species have been seen: *B. coccinea* Hook. and *B. semperflorens* Link & Otto, both native to Brazil. *B. coccinea* is a tall, sturdy plant with entire-margined, linear-lanceolate, glabrous leaves up to 13 cm. long; there are often spiny buds in the leaf-axils; both the flowers and fruit are bright scarlet, the capsules broadly 3-winged. *B. semperflorens*, perhaps the most commonly grown species, has almost symmetrical, ovate, glabrous leaves 4–8 cm. long with crenate margins; the flowers are red, pink or white; the fruits are strongly 3-winged.

1. **Begonia oxyloba** Welw. ex Hook. f. in F.T.A. 2: 573 (1871).—R. Fernandes in C.F.A. 4: 292 (1970). Type from Angola.
    Begonia kummeriae Gilg in Engl., Bot. Jahrb. 34: 87 (1904). Type from Tanzania.

Plants 19–55 cm. tall, annual. Stems glabrous, ascending, unbranched, rooting at lower nodes. Leaves petiolate; lamina 7–14 cm. long, ± ovate in outline, with 5 lobes, cordate at the base; sinuses between lobes ⅕–½ × length of lamina; margins irregularly and very shallowly dentate; superior surface green, glabrous or very sparsely hispid, inferior surface dark green or red when living, becoming green when dry, glabrous; petiole 8–23 cm. long; stipules c. 20 × 8 mm., linear, membranous, entire-margined, those at apex of stem very conspicuous and sheathing the growing point, later caducous; axillary bud sometimes developing into a short conical spine. Flowers in axillary cymes with 2 dichasia; peduncles 15–30 mm. long; primary dichasial branches 2–8 mm. long. Bracts like stipules but very soon caducous. ♂ flowers: tepals 2, c. 9 mm. long, pale pink, suborbicular; filaments c. 1 mm. long. ♀ flowers: tepals 2, c. 7 mm. long, pale pink, sub-orbicular; styles divided to c. ½ their length, the branches stigmatic, almost straight; ovary c. 1 cm. long, fusiform, not winged, tapering into a narrow stipe

c. 1 cm. long; placentae bipartite. Fruit a berry 20–30 × 8–12 mm., ellipsoid, stipe 10–22 mm. long.

Malawi. N: Chitipa Distr., Misuku Hills, Mugesse Forest, Songwe View, 1070 m., fr. 25.iv.1972, *Pawek* 5228 (SRGH). **Mozambique.** Z: Serra do Gúruè, near waterfall of R. Licungo, st. 18.ix.1944, *Mendonça* 2104 (LISC).
Tropical Africa from Sierra Leone eastwards to Kenya, Tanzania and Madagascar; Angola and Zaire. Growing near water, c. 1070 m. Rare in the F.Z. area.

The leaves of *B. oxyloba* are very similar in shape to those of *B. sonderana*, but sterile material of the former can be recognized by the large stipules, the hard spiny buds in the leaf axils and by the occasional presence of scattered hairs on the upper leaf surface.

2. **Begonia princeae** Gilg in Engl., Bot. Jahrb. **30**: 361, cum tab. (1901); in R. E. Fr., Wiss. Ergebn. Rhod.-Kongo-Exped. 1911–1912, **1**: 159 (1914).—Irmsch. in Engl., Bot. Jahrb. **81**: 114 (1961). Type from Tanzania.
    Begonia princeae var. *princeae* forma *vulgata* Irmsch., tom. cit.: 120. Type from Tanzania.
    Begonia princeae var. *princeae* forma *grossidentata* Irmsch., tom. cit.: 119. Type: Zambia, path above Mukoma to Inona Falls, 900 m., 21.xii.1954, *Richards* 3701 (K).
    Begonia princeae var. *rhodesiana* Irmsch. forma *rhodesiana*, tom. cit.: 121. Type: Zambia, Mbala (Abercorn) Distr., Kambole escarpment, 1440 m., 20.ii.1957, *Richards* 8279 (K).
    Begonia princeae var. *rhodesiana* forma *racemigera* Irmsch., tom. cit.: 122. Type: Zambia, Mbala (Abercorn) Distr., Kambole escarpment, 1500 m., 19.ii.1957, *Richards* 8255 (K).
    Begonia princeae var. *racemigera* (Irmsch.) Wilczek in F.C.B., Begoniaceae: 45 (1969). Type as above.

Plants perennial, 5–33 cm. tall. Stems erect, simple or sparsely branched, glabrous; basal tuber small, ellipsoid or globose, pinkish, densely covered with brown hairs. Leaves glabrous, petiolate; laminae 2–7(10) cm. long, ± suborbicular in outline, not or shallowly lobed, truncate to slightly cordate at the base; margins subentire, irregularly crenulate; petioles 0·4–3·5(7) cm. long; stipules 5–22 mm. long, broadly ovate to oblong-acute, brownish, membranous, entire-margined. Flowers in terminal and axillary dichasial cymes with 1–2 dichasia; peduncles 0·3–4·5 cm. long. Bracts 1–2 mm. long, ovate, brownish, membranous. ♂ flowers: tepals 2 + 2, pale to bright pink; outer pair 9–14 × 8–12 mm., suborbicular; inner pair 5–7 × 3–4 mm., obovate; filaments 1–2 mm. long. ♀ flowers: tepals 5, the outer 4 6–12 × 4–6 mm., elliptic to oblong, the innermost smaller; ovary 4–8 × 4–7 mm., ovoid; styles 2·2–3·5 mm. long, divided to ⅓ their length, the branches stigmatic and spirally twisted; placentae bifid. Fruits 6–10 × 5–8 mm. excluding wings, globose, ovoid or ellipsoid; wings very unequal: largest 7–15 mm. wide, triangular, the upper edge at an acute or obtuse angle to the fruit axis; second and third wings much smaller.

Zambia. N: Mbala (Abercorn) Distr., Kawimbe, valley near Nachalanga Hill, 1740 m., fl. & fr. 15.xii.1959, *Richards* 11968 (K; LISC; SRGH). W: Mufulira, 1280 m., fl. & fr. 17.i.1948, *Cruse* 237 (K). C: Kabwe (Broken Hill), fl. & fr. 17.xii.1907, *Kassner* 2013 (BM). E: Lundazi Distr., 120 km. N. of Rumpi, 1400 m., fl. & fr. 26.xii.1970, *Pawek* 4158 (K). S: Mumbwa, *Macaulay* s.n. (K). **Malawi.** N: Nyika, Lohanga Valley, 2130 m., fr. 18.ii.1956, *Chapman* 386 (BM). C: Nkhota Kota Distr., fl. 12.ii.1944, *Benson* 329 (PRE). **Mozambique.** N: between Marrupa and Posto do Maúa, near Messalo R., fr. 4.iv.1961, *Carvalho* 482 (K).
    Tanzania, Angola and Zaire. On termitaria in bush and woodland, 910–2150 m.

Begonia princeae is somewhat variable in leaf shape and also in growth form. Most specimens are leafy throughout, but a few have a pair of large leaves at the base, appressed to the ground, with a more or less leafless scape. There is every intermediate between these extremes, and it is not known whether the different habit types are genetically or environmentally controlled. Irmscher (loc. cit.) has divided *B. princeae* into two varieties and five forms, based on leaf- and habit-variation, but as these infraspecific groups have been made on insufficient evidence it seems preferable to " lump " them in this account.

3. **Begonia pygmaea** Irmsch. in Engl., Bot. Jahrb. **81**: 114 (1961). Type: Zambia, Lunzua (? Lunzuwa), riverine forest, 910 m., fl. & fr. 23.ii.1955, *Richards* 4667 (K, holotype).

Plants perennial, c. 4 cm. tall. Stems glabrous, unbranched, arising from a small globose tuber. Leaves relatively long-petiolate, sparsely white-pilose on superior surface, glabrous below; laminae 1·3–1·5(2·1) cm. long, suborbicular in outline, almost symmetrical, rounded or slightly cordate at the base; margin shallowly crenate; petioles 0·4–5 cm. long; stipules c. 2·5 mm. long, oblong-acuminate. Flowers in axillary, 3-flowered cincinnate cymes; peduncles c. 2·5 cm. long. Bracts 0·8 mm. long, oblong. Flowers unknown. Immature capsule on pedicel 6 mm. long, 4 × 3·8 mm. excluding wings, broadly ovoid, 3-winged; largest wing triangular, the upper margin ± horizontal, 3·5 mm. long, the other two wings with upper margins c. 1·5 mm. long.

Zambia. N: Lunzua (? Lunzuwa), 910 m., fl. & fr. 23,ii.1955, *Richards* 4667 (K). Known so far only from the type. On steep shady bank, c. 910 m.

4. **Begonia sonderana** Irmsch. in Engl., Bot. Jahrb. **81**: 156, t. 7 fig. 1 & 2 (1961).— Letty, Wild Fls. of the Transvaal: t. 101 fig. 2 (1962) sine nom. Type from S. Africa (Transvaal).

Plants perennial, 40–100 cm. tall. Stems glabrous, sparsely branched, arising from a large oblique tuber. Leaves glabrous, petiolate; laminae 5–17 cm. long, ovate in outline, with 3–6 triangular lobes, often cordate at the base; sinuses between lobes ¼–½ × length of lamina; margins irregularly and very shallowly dentate; petioles (1)3·5–14 cm. long; stipules 5–13 mm. long, linear-lanceolate to ovate, reddish-brown when dry, entire. Flowers in terminal and axillary dichasial cymes with 2–3 dichasia; peduncles 6–11 cm. long; primary dichasial branches 1·5–3 cm. long. Bracts 2–5 mm. long, ovate, brownish, membranous, caducous. ♂ flowers: tepals 2+2, pink; outer pair 8–13 × 11–17 mm., suborbicular to transversely elliptic, often cordate at base; inner pair 5–8 × 3–4 mm., obovate; filaments 2–3 mm. long. ♀ flowers: tepals 5, pink; outer ones 6–17 mm. long, elliptic to suborbicular, inner ones shorter and narrower; styles divided to c. ½ their length, the branches stigmatic and spirally twisted; ovary 8–14 × 4–7 mm., oblong, 3-winged; placentae bifid. Capsule 12–16 × 7–9 mm. excluding wings, ellipsoid or cylindrical; wings very variable in shape, elliptic, oblong or triangular with the upper edge ± horizontal; largest wing up to 1 cm. wide, the next almost the same, the third much narrower; wing extending beyond capsule 1–3 mm. distally, 2–4 mm. proximally.

Rhodesia. E: Inyanga, Chipungu Falls, fl. & fr. 30.iii.1949, *Chase* 1635 (BM; K; LISC; SRGH). **Mozambique.** MS: Manica, Serra Zuira, Tsetserra, c. 2000 m., fl. & fr. 5.iv.1966, *Torre & Correia* 15732 (LISC).
Also from S. Africa (Transvaal) and Swaziland. On rocks wetted by spray from waterfalls, among damp mossy boulders in kloof forest, 1650–2000 m. Of restricted distribution, but common within this area.

B. sonderana has often been confused with the two very variable S. African species *B. homonyma* Steud. (syn. *B. caffra* Meisn.) and *B. dregei* Otto & Dietr. (syn. *B. natalensis* Hook.). It can be distinguished from them as follows: *B. sonderana* has bright pink flowers, the ♂ having 4 tepals and the ♀ divided placentae; the other species often have white or pale pink flowers, the ♂ with 2(3–4) tepals and the ♀ with entire placentae.*

5. **Begonia rumpiensis** Kupicha, sp. nov.† Type: Malawi, N. Prov., Rumpi Distr., Livingstonia escarpment, c. 1070 m., fl. & fr. immat. 16.xii.1969, *Pawek* 3088 (K, holotype).

Plants perennial, 10–20 cm. tall. Stems sparsely to densely crispate-hirsute, erect, unbranched, apparently arising from a small tuber. Leaves petiolate; lamina 2–4·5 cm. long, ± ovate in outline, simple or very shallowly and obtusely lobed, the margin subentire to finely serrate; superior surface green, ± glabrous to moderately crispate-hirsute, inferior surface green to dark red, moderately to densely crispate-hirsute especially towards the point of attachment, the indumentum continuing along the petiole; petiole 7–20 mm. long; stipules 5–7 × 2–5 mm.,

* I am most grateful to Dr. O. L. Hilliard for letting me see her account of Begoniaceae for Fl. Southern Afr. (vol. 22, pp. 136–144, 1977) before it went to press. My treatment of *B. homonyma* and *B. dregei* is based on hers.
† Differt a *B. sutherlandii* subsp. *sutherlandii* foliis plus minusve ovatis, 2–4·5 cm. longis, marginis subintegris vel subtiliter serratis; floribus femineis tritepalis roseis; placentis bilobatis; capsulae alis inaequalis.

linear to ovate, membranous, pinkish, with shortly laciniate margin. Flowers in terminal and axillary few-flowered cymes; peduncles 1–2 cm. long; primary dichasial branches c. 1 cm. long; peduncles and branches of inflorescence crispate-hirsute. Bracts similar to stipules. ♂ flowers: tepals 4, pink; outer pair 9–10 × 9–10 mm., ± suborbicular; inner pair 4 × 1·5 mm., narrowly obovate; filaments c. 1 mm. long. ♀ flowers: tepals 3, 6–8 mm. long, pink, suborbicular to broadly elliptic, slightly unequal; styles divided to c. ½ their length, the branches stigmatic and spirally twisted; ovary ellipsoid, 3-winged; placentae bipartite. Capsule 5–6 × 3·5–5 mm. excluding wings, ellipsoid; wings very unequal, triangular, the largest 4–5 mm. wide with the upper edge ± horizontal, the next a little narrower, the third very narrow; wing extending c. 1 mm. beyond capsule at the base but not all at the apex.

Malawi. N: Lake Nyasa, Livingstonia Mission, fl. & fr. 1932, *Sanderson* (BM). Known only from Malawi. On moist banks.

B. rumpiensis resembles *B. sutherlandii* (no. 6) in being hairy and having laciniate-margined stipules, but it is like *B. princeae* in leaf-shape, in having divided placentae and unequal capsule wings. It is the only species of *Begonia* in the F.Z. area having ♀ flowers with 3 tepals.

6. **Begonia sutherlandii** Hook. f. in Curtis, Bot. Mag. **94**: t. 5689 (1868).—Irmsch. in Engl., Bot. Jahrb. **81**: 162 (1961). Type from S. Africa (Natal).

Plants perennial, 10–35 cm. tall. Stems glabrous or sparsely hairy, sparingly branched, arising from a large ellipsoid tuber. Leaves sparsely to fairly densely hairy, petiolate; laminae 2·5–14 cm. long, ovate to lanceolate in outline, the margin often very shallowly lobed, always strongly serrated; petioles 0·7–8 cm. long; stipules 4–10 mm. long, ovate-oblong, fimbriate; bulbils sometimes present in leaf-axils. Flowers in terminal and axillary dichasial cymes with 1–2 dichasia; peduncles 2–6 cm. long; primary dichasial branches 1·3–2·5 cm. long. Bracts 4–12 mm. long, obovate, green, pink or brown, persistent and conspicuous. ♂ flowers: tepals 2 + 2, orange or rarely yellow; outer pair 7–12 × 6–10 mm., suborbicular to ovate or elliptic; inner pair narrower; filaments c. 1·5 mm. long. ♀ flowers: tepals 5, coloured as in the ♂; outer ones 6–10 mm. long, elliptic or ovate, inner ones narrower; styles divided to ½–⅔ their length, the branches spirally twisted; ovary 4·5–9 × 2–5 mm., oblong, 3-winged; placentae entire. Capsule 6–28 × 3–7 mm. excluding wings, ellipsoid or cylindrical; wings 6–14 mm. wide, subequal, triangular, the upper edge horizontal or at an acute or obtuse angle to the fruit axis; wing sometimes extending beyond capsule above and below.

Leaves 2·5–7 cm. long, lanceolate (broadest at the base), the margin relatively coarsely serrated; petiole 0·7–2·5 cm. long - - - - - subsp. *sutherlandii*
Leaves 5–14 cm. long, ± ovate-oblong, broadest at the middle, the margin finely serrated; petiole 1–8 cm. long - - - - - - - - subsp. *latior*

Subsp. **sutherlandii**.
   ? *Begonia sutherlandii* sensu Binns, H.C.L.M.: 23 (1968).
   ? *Begonia flava* sensu Binns, loc. cit.
   Begonia sutherlandii var. *subcuneata* Irmsch., tom. cit.: 164, t. 8 fig. 2 (1961). Type from Tanzania.

Malawi. N: Misuku Hills, Mugesse Forest, c. 1830 m., fl. & fr. ii.1953, *E. G. Chapman* 99 (BM). **Mozambique.** N: Vila Cabral, serra de Massangulo, c. 1450 m., fr. 25.ii.1964, *Torre & Paiva* 10800 (LISC).
   Tanzania, S. Africa (Natal). Shady rock faces, epiphytic on rain forest trees, 1450–1830 m.

Subsp. **latior** (Irmsch.) Kupicha, stat. nov. Type: Zambia, above Mukoma, Inono Falls, 900 m., 21.xii.1954, *Richards* 3699 (B, holotype; K; SRGH).
   Begonia sutherlandii sensu Phillips in Fl. Pl. S. Afr. **8**: t. 283 (1928).—Letty, Wild Fls. of the Transvaal: t. 101 fig. 3 (1962).—Trauseld, Wild Fls. of the Natal Drakensberg: 125, cum 2 photogr. (1969).
   Begonia flava Marais in Fl. Pl. S. Afr. **31**: t. 1233 (1956). Type: Mozambique, Manica Distr., Garuso, 1952, *Schweickerdt* (cultivated at Pretoria Botanic Garden, xi. 1953) in PRE 28540 (PRE, holotype).
   Begonia sutherlandii var. *latior* Irmsch. in Engl., Bot. Jahrb. **81**: 165, t. 9 fig. 1 (1961). Type as for *Begonia sutherlandii* subsp. *latior*.

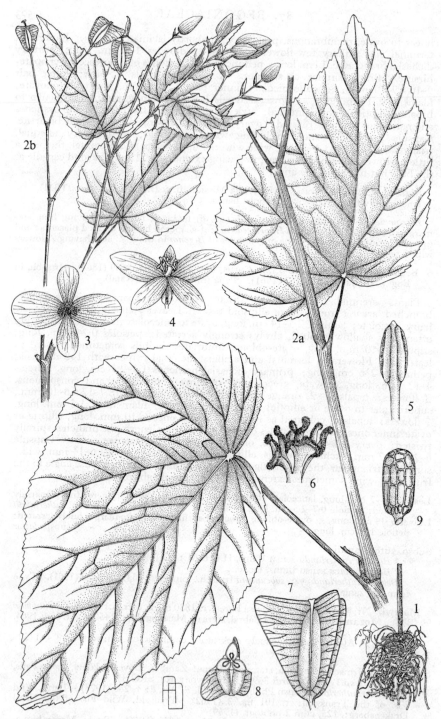

Tab. 122. BEGONIA BREVIBRACTEATA. 1, tuber (×1); 2, habit: (a) basal, (b) apical part of plant (×½); 3, male flower (×1); 4, female flower (×1); 5, dehiscing anther, view of abaxial face (×10); 6, stigmas (×3); 7, fruit (×1½); 8, transverse section through fruit (×1½); 9, seed (×40), all from *Brummitt & Banda* 9195.

Begonia sutherlandii var. *latior* forma *densiserrata* Irmsch., tom. cit.: 169, t. 9 fig. 2 (1961). Type from Tanzania.

Zambia. N: Mbala (Abercorn) Distr., Lunzuwa Falls, fl. 9.i.1952, *Nash* 101 (BM). **Rhodesia.** E: Melsetter Distr., Gweni Mt., fl. 17.xii.1948, *Chase* 1414 (BM; K; SRGH). **Mozambique.** MS: Chimanimani Mts., Musapa Gap, c. 3 km. from Rhodesian border, 910 m., fl. 30.i.1962, *Wild* 5625 (K; PRE; SRGH).

Tanzania, Zaire, S. Africa (from Transvaal to the Cape). Deep ravines, near waterfalls, wet rock crevices, 910–1800 m.

Irmscher (loc. cit.) divided *B. sutherlandii* into four varieties, only one of which (var. *latior*) was represented in the F.Z. area. Var. *subcuneata*, apparently confined to Tanzania, was distinguished from the S. African var. *sutherlandii* by having a slightly different leaf shape and probably by the fact that the two taxa had isolated distributions. Plants belonging to " var. *subcuneata* " have since been found also in N. Malawi and N. Mozambique, and although there is still a disjunction between these populations and those of " var. *sutherlandii* ", they are phenetically so similar that it seems best to treat them as a single taxon—subsp. *sutherlandii*. Subsp. *latior* is very distinct from the type subspecies in the F.Z. area, but in S. Africa these taxa tend to merge.

The Tanzanian var. *rubrifolia* Irmsch., so far known only from the type (from the Njombe Distr.), has small, ovate, reddish leaves. It is thus superficially similar to *B. rumpiensis*, a close geographical neighbour. However, the flower and fruit characters of these two species separate them quite easily.

7. **Begonia brevibracteata** Kupicha, sp. nov.* Type: Malawi, Mlanje Mt., foot of Great Ruo Gorge between hydro-electric station and dam, 870–1060 m., fl. & fr. 18.iii.1970, *Brummitt & Banda* 9195 (BR; EA; K, holotype; LISC; MAL; PRE; SRGH; UPS). TAB. **122**.

Plants perennial, 30–90 cm. tall. Stem glabrous, sparingly branched, arising from a tuber. Leaves glabrous, petiolate; laminae (3)5–12 cm. long, ovate-acute to -acuminate, the margins distinctly serrate; petioles 1·5–10 cm. long; stipules c. 1 cm. long, linear to lanceolate, membranous, deciduous. Flowers in terminal and axillary dichasial cymes with 1–2 dichasia; peduncles 5–9 cm. long; primary dichasial branches 0·5–3 cm. long. Bracts 5–9 × 2–5 mm., ovate, membranous. ♂ flowers: tepals 2 + 2, white or palest pink; outer pair 13–26 × 10–18 mm., elliptic to suborbicular, inner pair 11–20 × 6–10 mm., elliptic; filaments c. 1·5 mm. long. ♀ flowers: tepals 5, coloured as in the ♂; outer pair 17 × 8 mm., oblong-acute; inner pair 12 × 6 mm., elliptic-acute; innermost one 10 × 5 mm., elliptic-acute; styles divided to c. ½ their length, the branches stigmatic and spirally twisted; ovary 7 × 2 mm., linear-oblong, with 3 narrow wings; placentae entire. Capsule 11–18 × 5–9 mm., ellipsoid, the wings triangular, unequal: the largest 7–13 mm. wide, the next 4–10 mm., the third 2–4 mm.

Malawi. S: Zomba Distr., Queen's View, Zomba Mt., fl. & fr. 5.ii.1964, *Salubeni* 224 (SRGH).

Known only from Malawi. Growing in shaded rock clefts and on rocks by water, 850–1680 m.

Specimens of *B. brevibracteata* have been confused with *B. nyassensis* (no. 8), because both species are white-flowered and have almost identical, very restricted geographical distributions. As shown in the key, however, they are well differentiated, and *B. brevibracteata* may be more closely related to *B. sutherlandii* which has undivided placentae.

8. **Begonia nyassensis** Irmsch. in Engl., Bot. Jahrb. **81**: 155 (1961). Type: Malawi, Mlanje Distr., Mlanje Mt., Chambe Plateau, fl. & fr. 24.iii.1958, *Jackson* 2195 (K, holotype; SRGH).

Plants perennial, c. 30 cm. tall. Stems glabrous, reddish, unbranched, arising from a large tuber. Leaves glabrous, the lower ones petiolate, upper ones ± sessile; laminae 9–14 cm. long, ovate-acute to -cuspidate, fleshy, the margin obscurely and irregularly serrate; petioles 0–4 cm. long; stipules 6–20 × 3–10 mm., oblong-acute, entire-margined, deciduous. Flowers scented, in terminal and axillary dichasial cymes with 2–3 dichasia; peduncles 2–11 cm. long; primary dichasial branches 0·5–4 cm. long. Bracts 15–24 × 11–15 mm., ovate, membranous,

* *B. nyassensis* primo adspectu maxime simile, sed bracteis non nisi usque 9 cm. longis, petiolis 1·5–10 cm. longis, foliis valde serratis, floribus minoribus (tepala exteria floris masculi usque 26 mm. longa), placentis simplicis, capsulae alis conspicuis, inaequalis.

conspicuous. ♂ flowers: tepals 2+2, white or palest pink; outer pair 25–34 × 14–22 mm., linear-oblong, inner pair 20–31 × 15–21 mm., broadly oblong; filaments 1·5–3 mm. long. ♀ flowers: tepals 5, coloured as in the ♂; outer two 30–32 × 17–19 mm., subelliptic, inner two 28–30 × 18–23 mm., broadly obovate, innermost one 23–27 × 18–20 mm.; styles divided to c. ½ their length, the branches spirally twisted; ovary 13 × 6–7 mm., linear-oblong, with 3 narrow wings up to 2 mm. wide; placentae bifid. Capsule 16–26 × 12–15 mm., ellipsoid, the wings subequal, up to 5 mm. wide, narrowly triangular, almost parallel-sided and thus making the fruit ± oblong in outline.

Malawi. S: Mt. Mlanje, Chambe basin, c. 1980 m., fl. & fr. 24.i.1967, *Hilliard & Burtt* 4609 (K; SRGH).
Known only from Malawi. Growing in cracks in rock faces, 1830–1980 m.

This splendid *Begonia* would probably be very successful if brought into cultivation.

88. CACTACEAE
By M. L. Gonçalves

Succulent perennials with stems of varied shape and bristles arising from complex axillary structures (areoles). Flowers solitary on areoles, sessile (except in *Pereskia*), bisexual, usually actinomorphic; perianth segments ∞ (5–13 in species of F.Z. area), imbricate in bud, with gradual transition between sepals and petals, fused below to form a tube (hypanthium). Stamens ∞, inserted at base of perianth; anthers 2-thecous, splitting longitudinally. Carpels 3–∞, syncarpous; ovary inferior, unilocular with 3–∞ parietal placentas; ovules ∞; style single with 3–∞ stigmatic lobes. Fruit a berry. Seeds ∞.
A family native to America except perhaps for the genus *Rhipsalis* which is sometimes considered to be indigenous in Africa. Some genera are cultivated as ornamental while *Pereskia aculeata* Mill. and some species of *Opuntia* have become naturalized in Africa.

RHIPSALIS Gaertn.
Rhipsalis Gaertn., Fruct. 1: 137 (1788) *nom. conserv.*
Hariota Adans., Fam. Pl. 2: 249 (1763).

Epiphytes, mostly pendent from trees, sometimes producing adventitious roots on the angled cylindrical stems. Areoles very small, with deciduous setae. Leaves absent or reduced to minute scales. Flowers lateral, small; perianth segments (5)8–13, the outermost 3–5 small, unequal, ± succulent, greenish, the inner 5–8 up to 3 mm. long, translucent. Fruit mucilaginous, white, yellow or pink, naked or with a few scales. Seeds carunculate, with viscid endosperm.
An exclusively American genus except for the F.Z. species and for *R. fasciculata* Haw. and *R. cereuscula* Haw. (Brazil and Madagascar).

Rhipsalis baccifera (J. Mill.) W. T. Stearn in Cact. Succ. Journ. Gt. Brit. 7: 107 (1939) in adnot.—Robyns in F.C.B., Cactaceae: 2 (1968).—Hunt in F.T.E.A., Cactaceae: 5, fig. 2 (1968).—M. L. Gonçalves in C.F.A. 4: 301 (1970). TAB. **123**. Type represented by the description and plate, Class IX, Ord. I (1771) in J. Mill., Ill. Sex. Syst. Linn. (1771–1777).
　　Cassytha baccifera J. Mill., Ill. Sex. Syst. Class IX, Ord. I: t. 29 (1771). Type as above.
　　Rhipsalis cassutha Gaertn., Fruct. 1: 137, t. 28 (1788) (" cassytha " auct. mult.).—Oliv. in F.T.A. 2: 581 (1871).—Swynnerton et al. in Journ. Linn. Soc., Bot. **40**: 75 (1911).—Eyles in Trans. Roy. Soc. S. Afr. 5: 424 (1916).—Pax & Hoffm. in Engl. & Prantl, Pflanzenfam. ed. 2, **21**: 619 (1925).—Burtt Davy, F.P.F.T.: 237 (1926).—Exell in Journ. Bot., Lond. **67**, Suppl. Polypet.: 198 (1929).—Gossw. & Mendonça, Cart. Fitogeogr. Angola: 82 & 94 (1939).—Robyns & Tournay, Fl. Parc Nat. Alb. **1**: 648 (1948).—Keay in F.W.T.A. ed. 2, **1**: 221 (1954).—Brenan in Mem. N.Y. Bot. Gard. **8**: 444 (1954).—Binns, H.C.L.M.: 24 (1968). Type a specimen sent to Gaertner from Kew by Sir Joseph Banks.

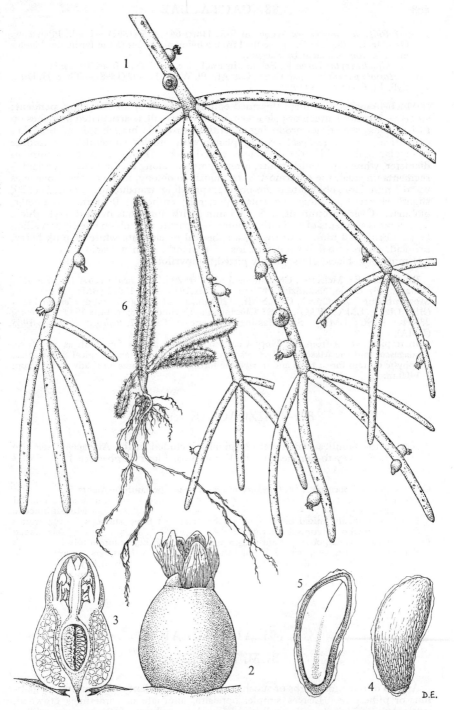

Tab. 123. RHIPSALIS BACCIFERA. 1, habit (× ⅔) *Gardner* in *F.D.*3216 ; 2, old flower (×6);
3, mature bud in longitudinal section (×6), both from *Greenway* 10234 ; 4, seed
(×20) ; 5, seed in longitudinal section (×20), both from *Verdcourt* 49 ; 6, juvenile
plant (×1) *Verdcourt* 156.   From F.T.E.A.

? Rhipsalis zanzibarica Weber in Rev. Hort. **64**: 425 (1892).—Engl., Pflanzenw.
Ost-Afr. **C**: 282 (1895). Described from a living plant sent to the Jardin des Plantes
in Paris from Zanzibar by Sacleux.
Rhipsalis erythrocarpa K. Schum. in Engl., loc. cit. Type from Tanzania.
Hariota parasitica sensu Hiern, Cat. Afr. Pl. Welw. **1**: 407 (1898).—Th. & H. Dur.,
Syll. Fl. Cong.: 235 (1909).

Much-branched succulent unarmed epiphytic shrub up to 3 m. long, pendent;
stems ± furrowed, green or pale green, with ± verticillate articulated branches up
to 1 m. long, sometimes producing adventitious roots at branch apices; branches
cylindrical, green, leafless; young plants with areoles subtended by minute
triangular scales with 1(2) deciduous tufted setae. Flowers up to 10 mm. in
diameter, white or greenish-white, often numerous along the branches. Perianth-
segments unequal, the outermost ± triangular to oblong, unequal, the inner ones
up to 3 mm. long, oblong to ovate-oblong, spreading, translucent. Stamens c. 20,
slightly shorter than inner segments of perianth; anthers ± 0·5 mm. in diameter,
globular. Ovary prominent, c. 3 × 2·5 mm., with numerous ovules; style thick,
± as long as segments of perianth; stigma 3–5-partite with oblong fleshy spreading
lobes. Fruit 4–8 mm. in diameter, a spherical or elongate white or pink berry,
pellucid, surmounted by the remains of perianth, with a viscid pulp. Seeds
c. 1 mm. long, black, glossy, finely pitted, subpyriform.

Rhodesia. E: Melsetter, Chimanimani Mts., fr. 20.i.1965, *Leach* 12678 (K; SRGH).
Malawi. S: Cholo, Nswadzi R., 840 m., fl. 27.ix.1946, *Brass* 17842 (BM; K; SRGH).
Mozambique. Z: Gúruè, Gúruè Mt., 1200 m., fr. 24.ii.1966, *Torre & Correia* 14834
(BR; LISC; LMA; WAG). MS: Cheringoma, Durúndi, fl. & fr. 14.ix.1942, *Mendonça*
239 (K; LISC; LMU). SS: Massinga, Pomene, st. 1.x.1947, *Pedro & Pedrógão* 1985
(LMA).
In tropical Africa from the Ivory Coast to Ethiopia and as far south as S. Africa,
Madagascar and the Mascarene Is.; also tropical Asia and C. and tropical S. America.
Epiphytic on rain forest trees and in riverine forest, also in humus on shady rocks; from
0–1400 m.

89a–d AIZOACEAE sensu lato

Key to the families which were originally included in the Aizoaceae (see also
Dandy, J. E., Key to families and higher groups, Flora Zambesiaca **1**, 1: 55–78
(1960)):

Ovary superior, apocarpous (in *Gisekia*) or syncarpous; perianth segments 5:
  Perianth segments free; stamens 3–∞; fruit of achenes, mericarps or loculicidal
    capsules - - - - - - - - - - **Molluginaceae**
  Perianth segments united below into a tube or almost free; stamens 5–∞; fruit a
    loculicidal or circumscissile capsule - - - - - **Aizoaceae**
Ovary inferior, syncarpous; perianth segments 3–5, united below into a tube:
  Perianth segments 4–5, with 2–3 leaf-like and larger than the others; stamens ∞ in
    several whorls; fruit a loculicidal capsule - - **Mesembryanthemaceae**
  Perianth segments 3–5, united below into a tube; stamens 3–∞, sometimes fasciculate;
    fruit nut-like or drupaceous, winged or horned - - - **Tetragoniaceae**

89a. AIZOACEAE
By M. L. Gonçalves

Annual or perennial herbs or subshrubs, mostly succulent or subsucculent, glabrous or pubescent. Leaves simple, opposite, alternate or sometimes crowded,
exstipulate, sometimes with small stipuliform lobes at base. Inflorescences solitary
or in groups, axillary. Flower ☿, regular. Perianth segments 5, united below into a
tube or almost free, ± herbaceous, imbricate or valvate, persistent. Stamens 5–∞,
hypogynous, sometimes in pairs or in fascicles, when definite alternate with the
perianth segments. Ovary superior, of 1–5 united carpels; loculi as many as

carpels; ovules 1–∞ per loculus; placentation parietal, axile or apical. Fruit capsular, loculicidal or circumscissile. Seeds usually subreniform, not strophiolate; embryo usually curved.

A large family comprising numerous genera. For a discussion on its delimitation see Hoffmann, U. in Engl., Bot. Jahrb. **93**: 247–324 (1973) and Ihlenfeldt, H. D. and Straka, H. in Herre, H., The Genera of the Mesembryanthemaceae, ed. 2, Rotterdam 1973.

Fruit loculicidal:
  Stamens ∞; ovules ∞ - - - - - - - - - **1. Aizoon**
  Stamens 5 or 10; ovules 1 in each locule:
    Stamens 10, in pairs - - - - - - - - - **2. Galenia**
    Stamens 5 - - - - - - - - - - - **3. Plinthus**
Fruit circumscissile:
  Ovary 2–3-locular; styles 2–3; flowers solitary - - - - **4. Sesuvium**
  Ovary 1–2-locular; styles 1–2; flowers solitary or in groups:
    Ovary 1-locular; styles 1; lid of capsule remaining in one piece; flowers solitary
    or in groups - - - - - - - - - **5. Trianthema**
    Ovary 2-locular; styles 2; lid of capsule splitting into 2 valves; flowers in groups
                6. Zaleya

1. AIZOON L.

Aizoon L., Sp. Pl. **1**: 488 (1753); Gen. Pl. ed. 5: 216 (1754).

Many-stemmed annual or perennial herbs. Leaves alternate, exstipulate, hairy. Flowers ⚥, solitary, sessile in leaf-axils or in forks of the branches. Perianth segments 5, united at the base, with a short tube. Stamens ∞, inserted on the tube in fascicles alternate with perianth segments. Ovary superior, of 5 carpels, 5-locular; ovules ∞, pendulous and axillary; styles and stigmas as many as carpels. Fruit a 5-valved loculicidal capsule. Seeds ∞, reniform, concentrically ridged.

A genus of c. 25 species with clearly separated species groups in S. Africa, the Mediterranean and Australia; only the following species is known from S. Africa, tropical Africa and the Mediterranean.

Aizoon canariense L., Sp. Pl. **1**: 488 (1753).—Sond. in Harv. & Sond., F.C. **2**: 469 (1862).—Oliv. in F.T.A. **2**: 584 (1871).—Hutch. & Dalz. in F.W.T.A. **1**: 115 (1927).—Pax & Hoffm. in Engl. & Prantl, Pflanzenfam. ed. 2, **16c**: 223 (1934).—Adamson in Journ. S. Afr. Bot. **25**: 31 (1958).—Jeffrey in F.T.E.A., Aizoaceae: 29, fig. 11 (1961).—Friedr. in Prodr. Fl. SW. Afr. **27**: 15 (1970).—J. H. Ross, Fl. Natal: 162 (1972). TAB. **124**. Type from the Canary Is.
  Aizoon procumbens Crantz, Inst. **1**: 135 (1766). (Type based on illustration: see t. 13 fig. 1 in Hist. Acad. Roy. Sci. Paris **1711**: (1730)).
  Glinus crystallinus Forsk., Fl. Aegypt.-Arab.: 95 (1775). Type from Arabia.
  Veslingia caulifloris Moench, Meth. Pl. Suppl.: 299 (1802). Type from Arabia?

A prostrate often rather thick- and many-stemmed annual or perennial herb. Stem up to c. 40 cm. long, pilose, often also finely papillose. Leaves alternate, leaf-lamina 8–70 × 4–45 mm., suborbicular to oblanceolate-obovate, rounded or bluntly subacuminate at apex, decurrent into the petiole at base, ± pilose on both sides, entire, with petiole 3–18 mm. long. Flowers solitary in leaf-axils or in forks of the branches, sessile, often numerous. Perianth segments 5, c. 3 mm. long, triangular, acute, yellowish inside, greenish or reddish and pilose outside. Stamens c. 12–15, inserted on calyx-tube in fascicles at the base of the sinuses between calyx-lobes. Ovary 5-locular; styles 5, short; stigmas 5. Fruit 5–9 mm. in diameter, usually red or pink, pentagonal, stellate, depressed centrally, splitting into 5 valves, these inflexed and remaining attached to the centre of the ovary. Seeds ∞, reniform, concentrically ridged.

Rhodesia. W: Bulawayo, on Gwanda road, c. 1400 m., fr. 15.v.1958, *Drummond* 5818 (LISC; SRGH). S: 5 km. from Fort Victoria to Umtali, fr. 2.ii.1970, *Figueiredo* 1 (BR; LISC). **Mozambique.** SS: Gaza, Chibuto, Baixo Changana, fr. 23.viii.1963, *Macedo & Macuácua* 1124 (LMA). LM: Maputo, Bela Vista, near Tinonganine, fr. 6.viii.1957, *Barbosa & Lemos* 7769 (COI; LISC).
Tropical and N. Africa, S. and SW. Africa. On roadsides, cultivated ground and rocky soils.

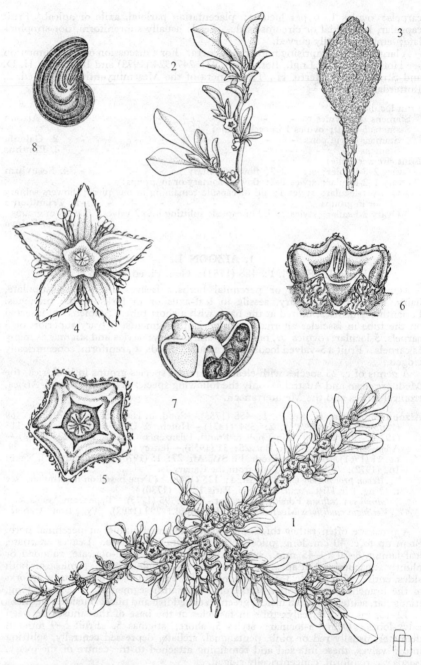

Tab. 124. AIZOON CANARIENSE. 1, habit (× ⅔) *Davis* D48615; 2, immature branch (× 1) *Meinertzhagen* sn; 3, leaf (× 2) *Murbeck* 31/3; 4, flower (× 6) *Jahandiez* 189; 5, developing fruit (× 6); 6, same in lateral view (× 6); 7, mature fruit in partial transverse section showing seeds (× 6), 5–7 from *Jahandiez* 189 & *Meinertzhagen* sn; 8, seed (× 22) from *Jahandiez* 189.

2. GALENIA L.

Galenia L., Sp. Pl. **1**: 359 (1753); Gen. Pl. ed. 5: 169 (1754).
Kolberia Presl, Symb. Bot. **2**: 24, t. 14 (1832).
Sialodes Eckl. & Zeyh., Enum. Pl. Afr. Austr. Extratrop.: 329 (1837).

Shrublets with alternate simple exstipulate leaves. Flowers small, solitary or in
cymes, with bracts and bracteoles. Perianth segments 5, the lobes coloured
inside. Stamens 10, in pairs alternating with the tepals; anthers versatile. Ovary
superior, of (3–4) 5 carpels, (3–4) 5-locular, each locule containing 1 pendulous
ovule; styles free, as many as carpels, their inner surfaces stigmatic. Fruit a
leathery capsule splitting at the top. Seeds solitary, compressed, roundish,
brown, glossy and striate.

A genus of c. 25 species with its main area of distribution in S. Africa extending
into SW. Africa and southern tropical Africa.

Galenia secunda (L.f.) Sond. in Harv. & Sond., F.C. **2**: 474 (1862).—Pax. & Hoffm. in
Engl. & Prantl, Pflanzenfam. ed. 2, **16c**: 224 (1934).—Adamson in Journ. S. Afr.
Bot. **22**: 105 (1956).—Friedr. in Prodr. Fl. SW. Afr. **27**: 55 (1970).—J. H. Ross,
Fl. Natal: 162 (1972). TAB. **125**. Type from S. Africa (Cape Prov.).
Aizoon secundum L.f., Suppl. Pl.: 261 (1781). Type as above.
Aizoon contaminatum Eckl. & Zeyh., Enum. Pl. Afr. Austr. Extratrop. **3**: 326
(1837). Type from S. Africa.
Aizoon elongatum Eckl. & Zeyh., loc. cit. Type from S. Africa.
Aizoon propinquum Eckl. & Zeyh., loc. cit. Type from S. Africa.

A prostrate many-stemmed whitish-grey plant covered with coarse villous
loosely appressed hairs. Stems woody at base and branched from the base.
Leaves alternate, leaf-lamina 5–10 × 3–9 mm., broadly ovate to circular, obtuse,
narrowed at base, coarsely papillose and hairy, often partly or wholly glabrescent
in the later stages, entire, ± deciduous at flowering time. Flowers on short
lateral branches, solitary or in groups of 2–5 on short branchlets, with leaf-like
bracts and very short bracteoles. Perianth segments 5, 2–3 mm. long, coarsely
villous, yellow or white inside. Stamens 10 with yellow, less often red, anthers.
Styles (3) 5. Fruit rounded, depressed on top.

Botswana. N: Toromoja, 21.iv.1971, *Dawson* 2 (K). SW: Mabua Sefhubi Pan,
117 km. S. of Tsane, fl. & fr. 24.ii.1960, *Wild* 5144 (LISC; SRGH).
Also in SW. Africa and S. Africa. On quartzite ridges.

3. PLINTHUS Fenzl

Plinthus Fenzl in Ann. Wien. Mus. **2**: 288 (1841) emend. Verdoorn in Bothalia **4**:
177 (1941).

Erect or sprawling woody undershrubs, the young parts covered with white or
silvery appressed hairs. Leaves small, simple, exstipulate, with axillary fascicles of
smaller leaves. Flowers axillary, solitary. Perianth segments 5, almost free.
Stamens isomerous, inserted at the base of the perianth. Ovary superior, free
from the perianth, 3-chambered, each chamber with 1 ovule; style wanting;
stigmas 3. Fruit loculicidal. Seeds flattened-reniform.

A predominantly S. African genus of 6 species, four of which are also known
from SW. Africa and one from southern tropical Africa.

Plinthus sericeus Pax in Engl., Bot. Jahrb. **48**: 499 (1913).—Pax & Hoffm. in Engl. &
Prantl, Pflanzenfam. ed. 2, **16c**: 225 (1934).—Adamson in Journ. S. Afr. Bot. **27**:
150 (1961).—Friedr. in Prodr. Fl. SW. Afr. **27**: 92 (1970). TAB. **126**. Type from
S. Africa.
Plinthus laxifolius Verdoorn in Bothalia **4**: 178 (1941). Type from S. Africa.
Plinthus karooicus var. *alternifolius* Adamson, tom. cit.: 149 (1961). Type from
S. Africa (Cape Prov.).
Plinthus arenarius Adamson, loc. cit. Type from S. Africa (Cape Prov.).
Plinthus psammophilus Dinter ex Adamson, tom. cit.: 151 (1961) *nom. nud.*

A rounded erect or sprawling white or silvery shrublet c. 50 cm. high with
short internodes. Leaves alternate, leaf-lamina 4–17 × 1–3 mm., usually longer
than the internodes, almost petiolate, those on the main shoots flat, elliptical,
narrowed at both ends and soon deciduous, those in the fascicles smaller, thicker,

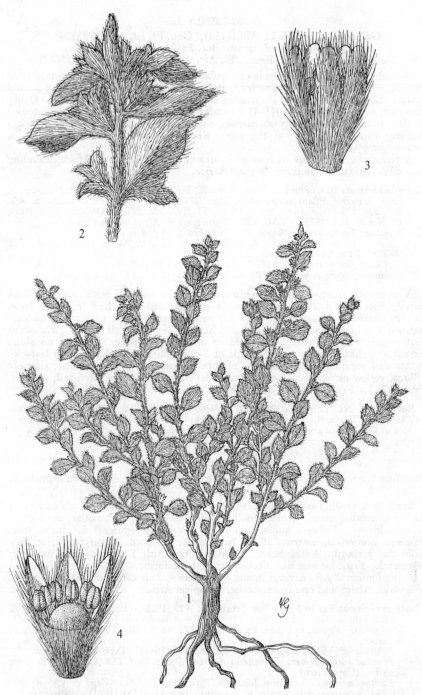

Tab. 125. GALENIA SECUNDA. 1, habit (× ⅔); 2, upper part of branch (×3); 3, flower (×12); 4, flower with facing perianth, segments removed (×12), all from *Zeyher* 2633.

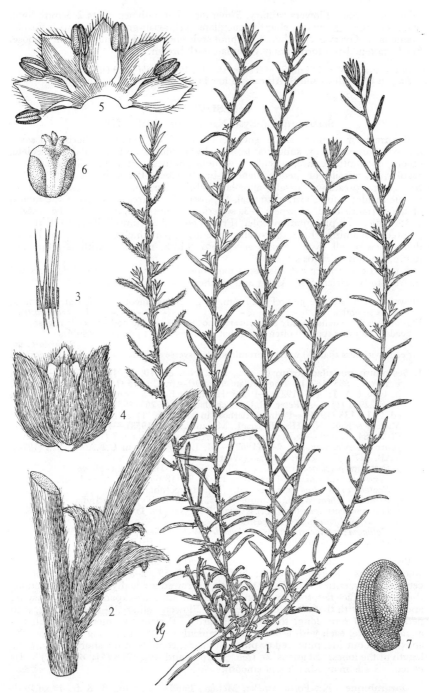

Tab. 126. PLINTHUS SERICEUS. 1, habit (×⅔); 2, part of a branch (×6); 3, hairs (×24);
4, flower (×12); 5, perianth opened up to show insertion of stamen (×12); 6, ovary
(×12); 7, seed, lateral view (×24), all from *Leistner* 2240.

folded upwards.   Flowers solitary, 5-merous.   Perianth segments 2–3 mm. long,
hairy outside, ± acute, slightly cucullate, yellowish-green or whitish inside.
Stamens 5. Ovary 3-chambered, without style and with 3 stigmas. Fruit globose.
Seeds flattened-reniform, finely granular, dark brown.

Botswana. SW: 32 km. S. of Tsane, st. 24.ii.1960, *Wild* 5129 (BM; SRGH).
Also in SW. Africa and S. Africa. In low bush and grassland and on dunes.

4. SESUVIUM L.
Sesuvium L., Syst. Nat. ed. 10: 1058 (1759).

Prostrate or semi-prostrate succulent annual or perennial herbs. Leaves
alternate, opposite or subopposite, usually petiolate and unequal, lanceolate,
oblong or elliptic, entire, succulent, without stipules. Flowers axillary, solitary,
regular, ☿, sessile or pedunculate, 2-bracteolate. Perianth 5-partite; lobes
triangular, spreading, dorsally apiculate behind the tip, green outside, pink or
purple inside; tube short, obconical. Stamens 5–∞, free, inserted at the mouth
of the perianth-tube. Carpels 2–3, superior, united; ovary 2–3-locular; styles
2–3; ovules ∞; placentation axile. Fruit a conical circumscissile capsule, the lid
pointed, remaining whole, thin-walled, the lower part thickened below the line of
dehiscence or all over. Seeds several–∞, black, rounded, with a prominent
hilum, smooth or rugose.
A genus comprising c. 6 species, 1 pantropical, 1 confined to the Galapagos Is.,
the rest to S. Africa, SW. Africa and southern tropical Africa.

Flowers pedicellate, the pedicel 4–20 mm. long; stems smooth, rooting at the nodes;
  seeds smooth - - - - - - - - - 1. *portulacastrum*
Flowers sessile; stems usually rough-warty, not rooting at the nodes; seeds rugose:
  Seeds with more than 20 often somewhat inconspicuous transverse wrinkles
  2. *hydaspicum*
  Seeds with less than 20 usually prominent transverse wrinkles - - 3. *nyasicum*

1. **Sesuvium portulacastrum** (L.) L., Syst. Nat. ed. 10, **2**: 1058 (1759).—Oliv. in
   F.T.A. **2**: 585 (1871).—Engl., Pflanzenw. Ost-Afr. **C**: 175 (1895).—Hutch. & Dalz.
   in F.W.T.A. **1**: 115 (1927).—Pax & Hoffm. in Engl. & Prantl, Pflanzenfam. ed. 2,
   16c: 229 (1934).—A. Peter, Fl. Deutsch Ost-Afr. **2**: 263 (1938).—Keay in F.W.T.A.
   ed. 2, **1**: 135 (1954).—Pedro & Barbosa in Moçamb., Doc. Trim. **81**: 68 (1954).—
   Macnae & Kalk, Nat. Hist. Inhaca I., Moçamb.: 144 (1958).—Jacobsen, Hand. Succ.
   Pl. **2**: 845 (1960).—Jeffrey in F.T.E.A., Aizoaceae: 20, fig. 7 (1961).—Adamson in
   Journ. S. Afr. Bot. **28**: 246 (1962).—M. L. Gonçalves in C.F.A. **4**: 322 (1970).
   TAB. **127**. Type from Curaçao (West Indies).
     Portulaca portulacastrum L., Sp. Pl. **1**: 446 (1753). Type as above.
     Sesuvium pedunculatum Pers., Syn. Pl. **2**: 39 (1807). Type from India.
     Sesuvium revolutum Pers., loc. cit. Type from India.
     Sesuvium repens Willd., Enum. Pl.: 521 (1809). Type from India.
     Halimum portulacastrum (L.) Kuntze, Rev. Gen. Pl. **1**: 263 (1891) (" Halimus ").
   Type as for *Sesuvium portulacastrum*.
     Halimum portulacastrum var. *crithmoides* (Welw.) Hiern, Cat. Afr. Pl. Welw. **1**: 412
   (1898) pro parte quoad specim. *Welwitsch* 2386A (LISU).

Prostrate succulent glabrous perennial herb, rooting at nodes of thick and
smooth stems. Leaves opposite; leaf-lamina 15–55 × 2–8 mm., oblong or oblanceolate, fleshy, entire, rounded at the apex, narrowed gradually below into a petiole,
the upper surface flat, the lower convex, the base scarious-expanded, stem-clasping
and connate with that of the opposing leaf. Flowers solitary, 5–12 mm. long, with
pedicels 4–20 mm. long, thickened upwards. Perianth segments unequal, triangular, acute, each with a fleshy dorsal apiculus c. 1·5 mm. long just below the
apex, green outside, pink, red or purplish inside, connate into a short tube c. ⅓ the
length of the lobes. Stamens ∞, free. Carpels and style (2) 3 (4); ovary (2) 3 (4)-
celled. Seeds many, black and smooth.

Mozambique. N: Porto Amélia, Metuge, Bandar, c. 5 m., fl. & fr. 19.xii.1963,
Torre & Paiva 9619 (COI; EA; LISC; MO; PRE; SRGH). Z: Maganja da Costa,
Rarága Beach, c. 5 m., fl. & fr. 15.ii.1966, *Torre & Correia* 14693 (BR; LISC; LMA; M;
WAG). SS: Bazaruto I., fl. 29.x.1958, *Mogg* 28735 (LISC). LM: Lourenço Marques,
Costa do Sol, fl. & fr. 2.iii.1960, *Balsinhas* 127 (COI; LISC; SRGH).
Pantropical. In mangrove swamps and on sandy dunes and beaches, 0–200 m.

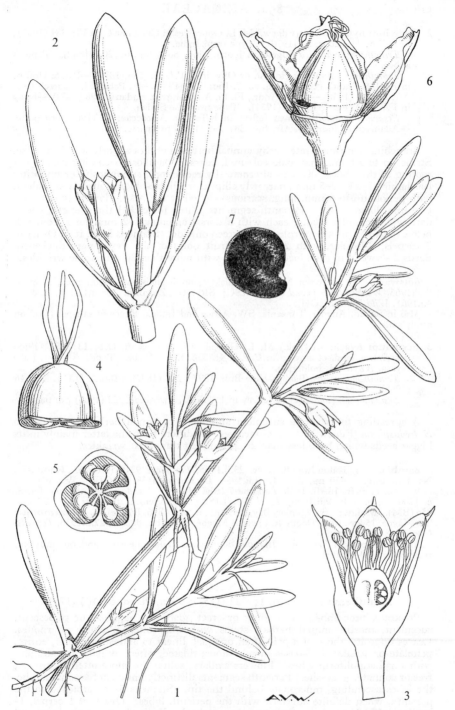

Tab. 127. SESUVIUM PORTULACASTRUM. 1, habit (×1) *Drummond & Hemsley* 3625; 2, flower (×2) *Faulkner* 2108; 3, flower in longitudinal median section (×4); 4, operculum (×8); 5, ovary in cross-section (×8); 6, fruit in process of dehiscing (×4); 7, seed (×12), 3–7 from *Drummond & Hemsley* 3625. From F.T.E.A.

2. **Sesuvium hydaspicum** (Edgew.) M. L. Gonçalves in Garcia de Orta **13**: 381 (1965); in C.F.A. **4**: 326, t. 37 (1970). Type from India.
    Trianthema hydaspica Edgew. in Journ. Linn. Soc., Bot. **6**: 203 (1862). Type as above.
    Trianthema polysperma Hochst. ex Oliv. in F.T.A. **2**: 588 (1871).—Pax & Hoffm. in Engl. & Prantl, Pflanzenfam. ed. 2, **16c**: 230 (1934).—A. Peter, Fl. Deutsch Ost-Afr. **2**: 263 (1938).—F. W. Andr., Fl. Pl. Anglo-Egypt. Sudan **1**: 96 (1950).—Keay in F.W.T.A. ed. 2, **1**: 136 (1954). Type from the Sudan.
    Sesuvium sesuvioides sensu Jeffrey in F.T.E.A., Aizoaceae: 22 (1961) pro parte. —Adamson in Journ. S. Afr. Bot. **28**: 245 (1962) pro parte.

Branching semi-prostrate fleshy annual herb sometimes slightly woody at base. Stems up to 50 cm. long, pale coloured, coarsely papillose becoming rough, not rooting at the nodes. Leaves alternate (by reduction), subopposite or opposite; leaf-lamina 7–23 × 2–5 mm., narrowly elliptic or elliptic-oblong, slightly succulent, with petiole up to 6 mm. long, scarious-winged at the base. Flowers solitary, up to 8 mm. long, sessile. Perianth segments united for c. ⅓–½ their length, equal, narrowly triangular, acute, each with a dorsal apiculus 1–2 mm. long from just below the apex, pink or purple inside, green outside. Stamens 5–∞, free. Ovary of 2 carpels, 2-locular, with 2 styles. Fruit pointed, the lower part thickened. Seeds 2—several in each loculus, rugose, with more than 20 transverse wrinkles.

Botswana. N: between Nata and Odiakwe, on Francistown-Maun road, fl. & fr. 9.iii.1965, *Wild & Drummond* 6821 (LISC; SRGH). **Zambia.** S: Machili, fl. & fr. 4.i.1961, *Wild* 5306 (SRGH).
Also in Sudan, Angola, Tanzania, SW. Africa and India. Edges of salt-pans and on sandy soils.

3. **Sesuvium nyasicum** (Bak.) M. L. Gonçalves in Garcia de Orta **13**: 381 (1965) descript. emend. et ampl.; in C.F.A. **4**: 326, t. 37 (1970). Type: Malawi, Lake Nyasa, *Whyte* (K, holotype).
    Trianthema nyasica Bak. in Kew Bull. **1897**: 268 (1897).—Binns, H.C.L.M.: 16 (1968). Type as above.
    Sesuvium sesuvioides sensu Jeffrey in F.T.E.A., Aizoaceae: 22 (1961) pro parte.

A spreading to prostrate succulent annual or perennial herb very similar to *S. hydaspicum* (Edgew.) M. L. Gonçalves. It differs from the latter mainly in its bigger seeds which have less than 20, prominent transverse wrinkles.

Zambia. C: Katondwe, fl. & fr. 23.ii.1965, *Fanshawe* 9218 (LISC). **Rhodesia.** N: Urungwe, c. 450 m., fl. & fr. 3.ii.1958, *Drummond* 5459 (SRGH). W: Wankie, c. 915 m., fl. & fr. 16.xii. 1956, *Paterson* 7 (SRGH). C: Gwelo, Mlezu School Farm, c. 1280 m., fl. & fr. 9.ii.1966, *Biegel* 890 (LISC). S: Victoria, fl. & fr. 1909, *Monro*, 855 (BM). **Malawi.** S: Monkey Bay, c. 490 m., fr. viii. 1896, *Whyte* (K). **Mozambique** T: Màgoé, 15 km. from Daque R. towards Màgoé, c. 300 m., fl. & fr. 27.ii.1970, *Torre & Correia* 18116 (LISC).
Also in Angola and Tanzania. In moist depressions in mopane woodland, on sandy and rocky soils, 300–1280 m.

5. TRIANTHEMA L.

Trianthema L., Sp. Pl. **1**: 223 (1753); Gen. Pl. ed. 5: 105 (1754).

Prostrate, procumbent, ascending or erect, glabrous, papillose or pubescent, succulent, mostly annual herbs; stems with long often unequal internodes. Leaves opposite, those of a pair often unequal, linear to almost orbicular, entire, petiolate or sessile; leaf-bases membranous, dilated, often connate in pairs and with small stipuliform lobes. Flowers axillary, solitary or more often in groups, free or connate, ± sessile. Perianth segments distinctly united at base into a tube, the lobes ascending, mucronate behind the tip. Stamens 5–∞, attached to the perianth, when definite alternate with the perianth lobes. Ovary of 1 carpel, 1-locular, with 2–∞ ovules; placentation parietal. Fruit circumscissile, enclosed by the perianth, the lid ± truncate, with thickened bands, the apex thin and often depressed. Seeds 2–∞.
A genus of c. 10 species, two in the New World, the others in tropical Africa, Asia and Australia.

Flowers solitary - - - - - - - - - 1. *portulacastrum*
Flowers glomerulate:
  Perianth lobes at least as long as the tube (2. *salsoloides*):
    Stems hispid; leaves oblong-lanceolate:
      Flowers densely crowded; perianth lobes shortly mucronate; leaves deciduous at
      flowering time - - - - - - 2. *salsoloides* var. *salsoloides*
      Flowers in small groups; perianth lobes long mucronate; leaves persistent
          2. *salsoloides* var. *transvaalensis*
    Stems smooth; leaves linear - - - - 2. *salsoloides* var. *stenophylla*
  Perianth lobes shorter than the tube:
    Plant smooth; perianth lobes acute or acuminate - - - - 3. *parvifolia*
    Plant hispid all over; perianth lobes obtuse or subobtuse - - 4. *triquetra*

1. **Trianthema portulacastrum** L., Sp. Pl. **1**: 223 (1753).—Hutch. & Dalz. in
F.W.T.A. **1**: 115 (1927).—Pax & Hoffm. in Engl. & Prantl, Pflanzenfam. ed. 2,
16c: 230 (1934).—Hauman in F.C.B. **2**: 116 (1951).—Keay in F.W.T.A. ed. 2, **1**:
136 (1954).—Jeffrey in F.T.E.A., Aizoaceae: 23 (1961).—Adamson in Journ. S. Afr.
Bot. **28**: 247 (1962).—M. L. Gonçalves in C.F.A. **4**: 327 (1970). TAB. **128**.
Syntypes from Jamaica and Curaçao (W. Indies).
    Trianthema monogyna L., Mant. Pl. **1**: 69 (1767).—DC., Prodr. **3**: 352 (1828).—
Oliv. in F.T.A. **2**: 587 (1871).—Engl., Pflanzenw. Ost-Afr. **C**: 175 (1895).—
R. E. Fr., Wiss. Ergebn. Schwed. Rhod.-Kongo-Exped. **1**: 35 (1914).—A. Peter,
Fl. Deutsch. Ost-Afr. **2**: 261 (1938).—Wild, Fl. Vict. Falls: 144 (1953) *nom. illegit.*
Syntypes as above.

A prostrate somewhat succulent annual herb, subglabrous, the young parts
shortly hairy; stems up to 50 cm. or more, procumbent or ascending, spreading.
Leaves opposite, one of each pair much smaller than the other, leaf-lamina 3–45 ×
3–40 mm., obovate, broadly obovate or almost circular, entire, obtuse, rounded
or retuse, sometimes slightly apiculate at apex, cuneate at base, glabrous or
sparsely hairy on the midrib below, petiolate, stipulate; petiole 2–20 mm. long,
sparsely hairy, expanded into a sheathing membranous base connate with that of
the opposing leaf; stipules up to 3 mm. long, narrowly triangular-acuminate.
Flowers axillary, solitary, partly hidden by the sheathing leaf-bases. Perianth
segments up to 5 mm. long, pinkish or yellowish, the tube longer than the acute
lobes with a dorsal apiculus. Stamens 10–20, inserted on the calyx-tube. Ovary
truncate, 2-lobed; style 1. Seeds 3–12, compressed, rounded, with " ammonite-
like " markings.

Malawi. S: Port Herald, between Muona and Shire R., fr. 20.iii.1960, *Phipps* 2574
(SRGH). **Mozambique.** N: Mossuril, Cabeceira, fr. 1884–85, *Carvalho* (COI). T:
Tete, fr. i.1932, *Pomba Guerra* 43 (COI). LM: Lourenço Marques, fr. 21.iii.1960,
Balsinhas 140A (COI; LISC; LMA; SRGH).
Pantropical weed. In cultivated ground and sandy soil.

2. **Trianthema salsoloides** Fenzl ex Oliv. in F.T.A. **2**: 588 (1781).—Engl., Pflanzenw.
Ost-Afr. **C**: 175 (1895).—Pax & Hoffm. in Engl. & Prantl, Pflanzenfam. ed. 2, **16c**:
230 (1934).—A. Peter, Fl. Deutsch Ost-Afr. **2**: 262 (1938).—F. W. Andr., Fl. Pl.
Anglo-Egypt. Sudan **1**: 96 (1950).—Jeffrey in F.T.E.A., Aizoaceae: 26, fig. 8 no. 8
(1961).—Adamson in Journ. S. Afr. Bot. **28**: 248 (1962).—M. L. Gonçalves in
Garcia de Orta **13**: 382 (1965); in C.F.A. **4**: 328 (1970).—J. H. Ross, Fl. Natal:
162 (1972). Type from the Sudan.

Var. **salsoloides**.—Adamson, loc. cit.—M. L. Gonçalves in C.F.A., loc. cit.
    Trianthema sedifolia sensu Oliv., loc. cit. pro parte quoad specim. *Welwitsch* 1089.
    Trianthema crystallina var. *sedifolia* sensu Hiern, Cat. Afr. Pl. Welw. **1**: 415 (1898).

A succulent, ± papillose annual herb becoming hispid when older; stems up to
c. 50 cm. in length, prostrate, ascending or erect, woody at base, radiating in a
crown from the taproot. Leaves thick, deciduous at flowering time, leaf-lamina
2–15 × 1·5–3 mm., linear to narrowly obovate or oblong-ovate; petiole 0–3 mm.
long; leaf-base expanded, membranous, with 2 stipular lobes. Flowers c. 3 mm.
long, densely crowded especially on short lateral branches, densely hispid, sessile,
often forming a compact compound subspherical mass in the fruiting state.
Perianth lobes 2–3 times as long as the tube with narrow brownish margins and a
short dorsal subapical fleshy mucro; perianth-tube obconical, bearing 5 ±
prominent tubercle-like projections at the base of the sinuses between the lobes.
Stamens 5. Ovary obconical, turbinate, depressed above around the style.
Ovules and seeds 2, sometimes 1 by abortion of the other. Lid of fruit with a
prominent smooth thickened ring.

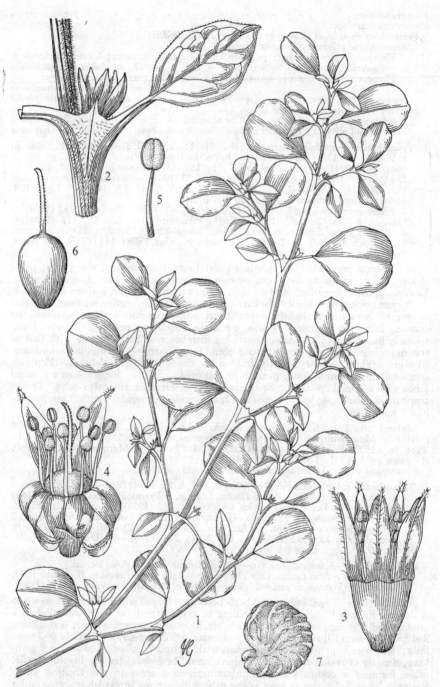

Tab. 128. TRIANTHEMA PORTULACASTRUM. 1, habit (×⅔); 2, part of a branch (×3); 3, flower (×6); 4, flower, facing perianth, segments deflected (×6); 5, stamen (×12); 6, gynoecium (×6); 7, seed, lateral view (×12), all from *Marques* 399.

Mozambique. MS: Gorongosa, fl. & fr. 9.vi.1966, *Macedo* 2132 (LMA). SS: Gaza, Caniçado, adjoining R. Limpopo, 50 m., fr. 13.viii.1938, *Barradas* (K). LM: Maputo, between Umbeluzi and Porto Henrique, fr. 16.iv.1948, *Torre* 7629 (K; LISC; LMU; MO; SRGH). Throughout tropical Africa. On sandy and saline soils.

Var. **transvaalensis** (Schinz) Adamson, loc. cit.—M. L. Gonçalves, loc. cit.; loc. cit.— J. H. Ross, loc. cit. Type from S. Africa (Transvaal).

Trianthema transvaalensis Schinz in Viert. Nat. Ges. Zürich **60**: 396 (1915).—Burtt Davy, F.P.F.T. **1**: 164 (1926). Type as above.

Differs from the type variety in the following characters: stem often reddish; leaves persistent; flowers in small clusters on very short lateral branches; perianth lobes with a long mucro.

Rhodesia. S: Nuanetsi, between Bubye R. and Dumela, fr. 1.v.1961, *Drummond & Rutherford-Smith* 7675 (LISC; SRGH). **Mozambique.** MS: Búzi, c. 19 km. on Nova Lusitania road from Muda, fr. 25.iii.1960, *Wild & Leach* 5212 (COI; SRGH). SS: Gaza, Caniçado, Limpopo R., fr. 13.vii. 1938, *Barradas* (COI). Also in S. Africa (Transvaal). In depressions on basalt and black soils, margins of riverine fringe.

Var. **stenophylla** Adamson, tom. cit.: 249 (1962). Type from S. Africa (Transvaal).

More slender than the other varieties, with smooth stems and leaves; leaves linear, often rolled; flowers crowded on very short lateral branches. Perianth lobes with broad membranous margin.

Botswana. N: Toromoja, 914 m., fr. 21.iv.1971, *Thornton* 1 (K). **Rhodesia.** S: Nuanetsi, between Chikombedzi and Chipinda Pools, fr. 1.v.1962, *Drummond* 7844 (LISC; SRGH). **Mozambique.** LM: Namaacha, 13.4 km. from Changalane to Catuane, fr. 30.iv.1971, *Marques* 2262 (LISC; LMU). Also in S. Africa (Transvaal). In depressions in mopane woodland, on basalt soils.

3. **Trianthema parvifolia** E. Mey. ex Sond. in Harv. & Sond., F.C. **2**: 598 (1862).— Pax & Hoffm. in Engl. & Prantl, Pflanzenfam. ed. 2, **16c**: 230 (1934).—Adamson in Journ. S. Afr. Bot. **28**: 249 (1962). Type from S. Africa.

Trianthema triquetra subsp. *parvifolia* (E. Mey. ex Sond.) Jeffrey in Kew Bull. **14**: 237 (1960). Type as above.

Prostrate glabrous shining herb; stem usually with unequal internodes and often reddish. Leaves small, leaf-lamina 2–10 × 1–2 mm., oblong, obovate to subrotund, smooth, ± succulent, petiole c. as long as the lamina, sheath broad. Inflorescence with c. 10 contiguous, sessile flowers facing upwards; bracts membranous, broad, obtuse or acuminate-apiculate. Perianth lobes triangular, membranous-edged, with a minute mucro behind the tip, c. ½ as long as the smooth tube. Lid of fruit truncate-conical, with a narrow thickened ring ½ way between the edge and the tip.

Botswana. N: Makarikari Pan, 12·8 km. SE. of Natal R. (? Nata R.), 895 m., fl. & fr. 23.iv.1957, *Drummond & Seagrief* 5197 (LISC; SRGH). SW: Tsane Pan, fl. & fr. 24.ii.1960, *Wild* 5125 (SRGH). Also in S. and SW. Africa. In open sandy plains.

4. **Trianthema triquetra** Rottl. ex Willd. in Ges. Naturf. Freunde Berl. **4**: 181 (1803).— Jeffrey in F.T.E.A., Aizoaceae: 25, fig. 8 no. 6–7 (1961).—Adamson in Journ. S. Afr. Bot. **28**: 252 (1962).—M. L. Gonçalves in Garcia de Orta **13**: 382 (1965).—Friedr. in Prodr. Fl. SW. Afr. **27**: 133 (1970).—M. L. Gonçalves in C.F.A. **4**: 329 (1970). Type from India.

Trianthema sedifolia Vis., Pl. Aegypt. & Nub.: 19, t. 3 fig. 1 (1836).—Oliv. in F.T.A. **2**: 588 (1871).—F. W. Andr., Fl. Pl. Anglo-Egypt. Sudan **1**: 96 (1950).— Keay in F.W.T.A. ed. 2, **1**: 136 (1954). Type from the Sudan.

Trianthema glandulosum A. Peter, Fl. Deutsch Ost-Afr. **2**, 2: 30, t. 36 fig. 1 (1932); op. cit. **2**, 3: 262 (1938). Type from Tanzania.

Trianthema crystallina sensu A. Peter, tom. cit: 261 (1938), non (Forsk.) Vahl (1790).

Like *T. parvifolia* but densely covered all over with prominent papillae becoming hard and hispid in the older stages. Stems ridged. Perianth lobes obtuse or subobtuse.

Rhodesia. W: Bulalima Mangwe, Ramaquabane R., fr. 5.vii.1962, *Wild* 5856 (LISC; SRGH). S: Nuanetsi, Malapati Bridge, fr. 24.iv.1961, *Rutherford-Smith* 694 (LISC; SRGH). **Mozambique.** T: Zambeze R., Macombe, fl. & fr. 5.ii.1974, *Macedo* 5542 (LISC; LMA). LM: Namaacha, from Changalane to Catuane, fr. 30.iv.1971, *Marques* 2261 (LISC; LMU).

In tropical Africa. On alluvial flats.

6. ZALEYA Burm. f.

Zaleya Burm. f., Fl. Ind.: 110, t. 31 (1768).
Rocama Forssk., Fl. Aegypt.-Arab.: 71 (1775).

Usually prostrate annual or perennial herbs. Leaves opposite, petiolate, lanceolate or oblanceolate, narrowly or broadly elliptic or ovate, entire, slightly succulent, without stipules. Flowers axillary, in groups, ± sessile, ⚥, regular. Perianth a short tube with 5 membranous-margined lobes, these coloured inside, green outside and with a subapical dorsal mucro. Stamens 5, free, inserted in the upper part of the perianth tube. Ovary superior, 2-carpellate, syncarpous, 2-celled with 2 free stigmas and 2 ovules per cell attached to the interlocular septum. Fruit a 4-seeded capsule, dehiscing by means of a 2-valved operculum.

A genus of c. 6 closely related species in tropical and southern tropical Africa, Asia and Australia.

Zaleya pentandra (L.) Jeffrey in Kew Bull. **14**: 238 (1960); in F.T.E.A., Aizoaceae: 28, fig. 9 (1961).—Adamson in Journ. S. Afr. Bot. **28**: 252 (1962).—Binns, H.C.L.M.: 16 (1968).—Friedr. in Prodr. Fl. SW. Afr. **27**: 135 (1970).—M. L. Gonçalves in C.F.A. **4**: 329 (1970).—J. H. Ross, Fl. Natal: 162 (1972). TAB. **129**. Type from Arabia from seed sent by Forsskal.

    Trianthema pentandra L., Syst. Nat. ed. 12, **2**: 297 (1767); Mant. Pl. **1**: 70 (1767).—Oliv. in F.T.A. **2**: 588 (1871).—Engl., Pflanzenw. Ost-Afr. **C**: 175 (1895).—Burtt Davy, F.P.F.T. **1**: 164 (1926.—Hutch. & Dalz. in F.W.T.A. **1**: 115 (1927).—Pax & Hoffm. in Engl. & Prantl, Pflanzenfam. ed. 2, **16c**: 230 (1934).— Hutch., Botanist in S. Afr.: 672 (1946).—F. W. Andr., Fl. Pl. Anglo-Egypt. Sudan **1**: 96, fig. 61 (1950).—Hauman in F.C.B. **2**: 116 (1951).—Keay in F.W.T.A. ed. 2, **1**: 136 (1954). Type as above.

    Rocama prostrata Forssk., Fl. Aegypt.-Arab.: cviii. et 71 (1775). Type from Arabia.

    Limeum kenyense Suesseng. in Mitt. Bot. Staatss. München **2**: 46 (1950). Type from Kenya.

    Trianthema redimita Melville in Kew Bull. **7**: 268 (1952). Type from Kenya.

A many-stemmed spreading prostrate or procumbent slightly succulent herb, rather coarse and sometimes ± woody at base. Branches up to 60 cm. long, covered with short rough hairs when young, later glabrescent. Leaves opposite, usually unequal; leaf-lamina 8–45 × 2–20 mm., oblanceolate, narrowly to broadly elliptic, obtuse to rounded at the apex, glabrous or pubescent below or on the midrib only, slightly succulent; petiole up to 20 mm. long, membranous-winged and sheathing at base, usually pubescent. Flowers in crowded groups of 5–20, sessile or subsessile, pale green to purple. Perianth lobes 3–5 mm. long, the segments united for c. ½ their length, subacute, shortly mucronate, with narrow membranous edges. Stamens 5. Ovary truncate, with a shallow apical depression. Fruit c. as long as the perianth, splitting c. ⅓ from the base along a smooth thickened rim; valves of the operculum separating. Seeds 4, subspherical-reniform, sulcate.

Zambia. S: Monze, fr. 24.iv.1963, *van Rensburg* 2100 (K; SRGH). **Rhodesia.** W: Wankie, fl. & fr. 21.vi.1934, *Eyles* 8079 (K; LISC). C: Que Que, c. 27 km. from Que Que, c. 1280 m., fl. & fr. 1.v.1966, *Biegel* 1176 (LISC; SRGH). E: Umtali, c. 1100 m., fl. & fr. 25.ii.1956, *Chase* 5978 (BM; SRGH). S: Ndanga, fr. 3.xii.1959, *Poley* 11615 (SRGH). **Malawi.** S: Mlanje, Njobru, fl. & fr. 17.xi.1955, *Jackson* 1766 (K; SRGH). **Mozambique.** N: Nampula, fr. 10.viii.1935, *Torre* 871 (COI; LISC). T: Cabora-Bassa, Estima, fl. & fr. 27.iii.1972, *Macedo* 5106 (LISC; LMA). MS: Chemba, Chiou, fl. & fr. 12.iv.1960, *Lemos & Macudcua* 71 (BM; COI; LISC; SRGH). SS: Gaza, Limpopo, Dumela, fl. & fr. 30.iv.1961, *Drummond & Rutherford-Smith* 7613 (LISC; SRGH). LM: Marracuene, Vila Luisa, fl. & fr. 7.iv.1947, *Barbosa* 131 (COI; LMA; SRGH).

Throughout tropical Africa in the drier areas from Transvaal to Egypt and Senegal; also in Arabia, Palestine and in Madagascar. In open places in woodland, riversides, irrigated land, waste places and cultivated ground, 200–1400 m.

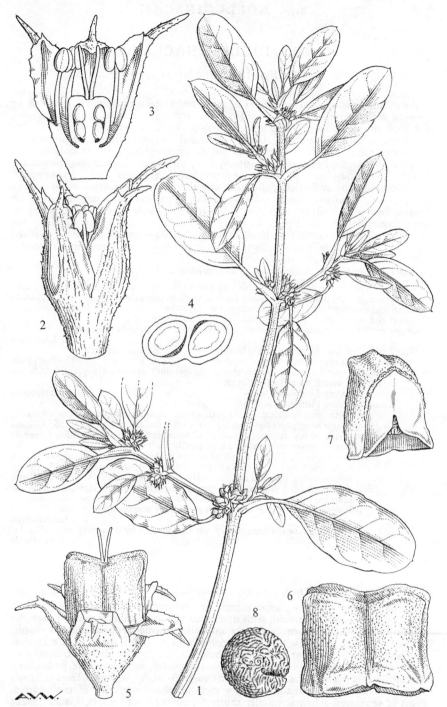

Tab. 129. ZALEYA PENTANDRA. 1, habit (×1); 2, flower (×12); 3, flower, in longitudinal median section (×12); 4, ovary in cross-section (×16); 5, fruit (×8); 6, operculum of fruit (×12); 7, valve of operculum showing commissural face (×12); 8, seed (×10), all from *Drummond & Hemsley* 1046. From F.T.E.A.

89b. MOLLUGINACEAE

By M. L. Gonçalves

Succulent or subsucculent annual or perennial herbs or subshrubs, glabrous or covered with simple or stellate hairs (in *Glinus lotoides*). Leaves simple, alternate, opposite or verticillate, sometimes crowded, with or without stipules. Inflorescences solitary, cymose, loosely dichasial to umbelliform or glomerate, axillary or terminal. Flowers more or less actinomorphic, ☿, rarely unisexual (in *Gisekia africana*). Perianth-segments 5, free, imbricate, herbaceous, persistent. Staminodes often present, sometimes petaloid. Stamens 3–∞, hypogynous, sometimes in pairs or fascicles, alternate with the perianth-segments when definite in number; filaments often enlarged below; anthers 2-locular, opening by longitudinal slits. Ovary superior, 2–5-carpellate, apocarpous (in *Gisekia*) or syncarpous; loculi and stigmas as many as carpels; ovules 1, few or many per loculus; placentation axile or basal. Fruit of achenes, mericarps or loculicidal capsules. Seeds usually subreniform or triangular in outline, sometimes strophiolate; embryo usually curved.

A predominantly S. African family comprising c. 14 genera.

Gynoecium apocarpous; fruit multiple, of achenes; flowers ☿ or unisexual; leaves
    opposite or subopposite, with raphides and without stipules - - **1. Gisekia**
Gynoecium syncarpous; fruits mericarps or capsules; flowers ☿; leaves alternate,
    opposite or whorled, without raphides; stipules absent or present:
Ovary with 2 uniovulate loculi; fruit divided into 2 indehiscent mericarps
                                               2. Limeum
Ovary with 3–5 uni- to pluriovulate loculi; fruit a loculicidal capsule:
    Conspicuous petaloid staminodes present - - - - **3. Corbichonia**
    Conspicuous petaloid staminodes absent (if petaloid staminodes present (*Glinus*)
        then not visible to the naked eye):
        Loculi of ovary 1-ovulate - - - - - - **4. Psammotropha**
        Loculi of ovary pluriovulate:
            Ovules with long funicles; seeds with appendaged strophiole; leaves alternate,
                opposite or apparently whorled, without stipules - - **5. Glinus**
            Ovules with short funicles; seeds without appendaged strophiole; leaves
                alternate, opposite or whorled, generally with stipules:
                Stipules small (c. 2 mm. long), glabrous or hairy, with the margins entire,
                    sometimes caducous or absent; leaves opposite or whorled, linear,
                    oblanceolate, obovate or spathulate - - - - **6. Mollugo**
                Stipules larger (c. 5–6 mm. long), persistent, adnate to the leaf-base, with the
                  margins lobate, with setose hairs; leaves alternate or whorled, linear:
                    Inflorescence cymose; ovary surrounded by a nectariferous disk
                                          7. Pharnaceum
                    Inflorescence umbellate; ovary not surrounded by a nectariferous disk
                                          8. Hypertelis

1. GISEKIA L.

Gisekia L., Mant. Pl. Alt.: 554 (1771).
Miltus Lour., Fl. Cochinch.: 302 (1790).

Herbs, usually prostrate, procumbent or creeping, annual or perennial with annual shoots, many-stemmed, with numerous prominent linear whitish raphides especially on leaves. Leaves mostly opposite, sessile or petiolate, succulent, without stipules, entire, with great variability in size and shape of the lamina. Inflorescences terminal or axillary, umbelliform to loosely dichasial, sessile or pedunculate. Flowers ☿ or sometimes unisexual, small, actinomorphic, pedicellate. Perianth-segments 5, herbaceous, free. Stamens 5–20, hypogynous, free. Carpels 5–15, superior, free, 1-ovulate, with a style on the inner edge of each carpel. Fruit of separated achenes, usually muricate, covered with warts, black. Seeds swollen-reniform, black.

A genus of the Old World tropics comprising c. 4 species.

Stamens (8)10–15(20); carpels 5 or more; flowers ☿ or unisexual; annual or perennial
    herbs - - - - - - - - - - - - **1. *africana***
Stamens 5; carpels 5; flowers ☿; annual herbs - **2. *pharnaceoides* var. *pharnaceoides***

1. **Gisekia africana** (Lour.) Kuntze, Rev. Gen. Pl. **3**: 108 (1898).—Pax & Hoffm. in Engl. & Prantl, Pflanzenfam. ed. 2, **16c**: 192 (1934).—Brenan, T.T.C.L.: 349 (1949).—Hauman in F.C.B. **2**: 101 (1951).—Jeffrey in F.T.E.A., Aizoaceae: 3 (1961).—Adamson in Journ. S. Afr. Bot. **27**: 134 (1961).—Friedr. in Prodr. Fl. SW. Afr. **26**: 5 (1966).—M. L. Gonçalves in C.F.A. **4**: 304, t. 36 (1970).—J. H. Ross, Fl. Natal: 161 (1972). Type: Mozambique, Mozambique I., *Loureiro* (location of type unknown).

Miltus africana Lour., Fl. Cochinch.: 302 (1790). Type as above.

Glinus mozambicensis Spreng., Syst. Veg. **2**: 467 (1825) *nom. illegit.*

Gisekia miltus Fenzl, Nov. Stirp. Dec. Mus. Vindob. **10**: 86 (1839).—Moq. in DC., Prodr. **13**, 2: 28 (1849).—Sond. in Harv. & Sond., F.C. **1**: 156 (1860).—Oliv. in F.T.A. **2**: 594 (1871).—Pax in Engl., Bot. Jahrb. **19**: 132 (1894).—Engl., Pflanzenw. Ost-Afr. **C**: 175 (1895).—Hiern, Cat. Afr. Pl. Welw. **1**: 421 (1898).—A. Peter, Fl. Deutsch Ost-Afr. **2**: 253 (1938) *nom. illegit.*

Gisekia pentadecandra E. Mey. ex Moq. in DC., loc. cit.—Wawra apud Wawra & Peyr. in Sitz.-Ber. Math. Nat. Akad. Wiss. Wien **38**: 563 (1860).—Sond. in Harv. & Sond., loc. cit.—Burtt Davy, F.P.F.T. **1**: 153 (1926). Type from S. Africa.

Gisekia aspera Klotzsch in Peters, Reise Mossamb. Bot.: 136 (1861).—Engl., loc. cit. Type: Mozambique, R. Sena, *Peters* (B, holotype†).

Gisekia miltus var. *pedunculata* Oliv., loc. cit.—Hiern, loc. cit. Type as for *Gisekia aspera*.

Gisekia africana var. *decagyna* Hauman in Bull. Jard. Bot. Brux. **19**: 444 (1949); in F.C.B. **2**: 102 (1951). Type from Angola (Lobito).

Gisekia africana var. *pedunculata* (Oliv.) Brenan in Mem. N.Y. Bot. Gard. **8**, 3: 444 (1954).—Jeffrey in F.T.E.A., Aizoaceae: 3 (1961).—Binns, H.C.L.M.: 16 (1968). Type as for *Gisekia aspera*.

Gisekia africana var. *cymosa* Adamson, tom. cit.: 136 (1961).—J. H. Ross, loc. cit. Type: Mozambique, Lourenço Marques, *Schlechter* 11631 (BM, holotype).

Prostrate to procumbent spreading glabrous herbs, annual or perennial with annual shoots, with several semi-succulent stems often tinged with red up to c. 50 cm. long. Leaf-lamina 5–81 × 1–23 mm., entire, ranging from linear to linear-oblong, oblong, elliptic, lanceolate, oblanceolate, spathulate, obovate or cuneate, the apex retuse, obtuse, rounded or subacute, often apiculate, the margins flat or slightly revolute; petiole 0–5 mm. long. Inflorescences sessile or peduncu-late, often numerous, umbelliform to dichasial, often many-flowered. Flowers ☿ or unisexual, pinkish, scentless; pedicels up to c. 16 mm. long. Perianth-segments 5, 2–4 mm. long in fruit, herbaceous, obtuse or subacute, with red or white membran-ous margins. Stamens (8)10–15(20) with filaments slightly broadened below. Carpels 5 in ☿ flowers, 10–15 in ♀ flowers. Fruit blackish, densely covered with warts.

Caprivi Strip. Katima Mulilo area, c. 915 m., fl. & fr. 24.xii.1958, *Killick & Leistner* 3048 (K). **Botswana.** N: Maun, c. 0·3 km. from river, fl. & fr. ii.1967, *Lambrecht* 25 (LISC). SW: Gomodimo, fl. & fr. 1.iv.1930, *van Son* 28973 (BM; K; SRGH). SE: 8 km. S. of Artesia, fl. & fr. 15.i.1960, *Leach & Noel* 79 (SRGH). **Zambia.** B: near Senanga, 1036 m., fl. & fr. 4.viii.1952, *Codd* 7386 (K; SRGH). C: Chingombe, fl. & fr 26.ix.1957, *Fanshawe* 3733 (K). S: Namwala, fl. & fr. 14.i.1964, *van Rensburg* 2750 (K; SRGH). **Rhodesia.** N: Urungwe, near Mensa, fl. & fr. 22.i.1957, *Phelps* 181 (SRGH). W: Nyamandhlovu, on Victoria Falls line, 1036 m., fl. & fr. xii.1902, *Eyles* 1125 (K; SRGH). E: Chipinga, near junction of Birchenough Bridge and Chipinga roads, 610 m., fl. & fr. 26.iv.1963, *Chase* 8011 (LISC; SRGH). S: 3·2 km. from Beitbridge on West Nicholson road, fl. & fr. 15.ii.1955, *E. M. & W.* 412 (BM; LISC; SRGH). **Malawi.** N: Mwanemba, fl. & fr. iii.1903, *McClounie* 110 (K). C: Chipoka, fl. & fr. 22.iii.1956, *Banda* 226 (BM; LISC; SRGH). S: Chikwawa, lower Mwanza R., 180 m., fl. & fr. 4.x.1946, *Brass* 17975 (K; SRGH). **Mozambique.** N: António Enes, along the coast, fl. & fr. 24.x.1965, *Gomes e Sousa* 4895 (COI; K). Z: Quelimane, on beach, 5 m., fl. & fr. 16.iii.1966, *Torre & Correia* 15177 (LISC). T: near Tete, fl. & fr. ii.1859, *Kirk* (K). MS: Caia, Sena, fl. & fr. 1846, *Peters* (K). SS: Inharrime, Ponta de Zavora, fl. & fr. 4.iv.1959, *Barbosa & Lemos* 8499 (COI; LISC; SRGH). LM: Lourenço Marques, fl. & fr. 6.xii.1897, *Schlechter* 11031 (COI).

Widespread from tropical and southern Africa to Asia. In woodland, open bushland, riverine fringing vegetation, sandy soils and roadsides.

2. **Gisekia pharnaceoides** L., Mant. Pl. Alt.: 562 (1771) (" Gisechia ").—Sond. in Harv. & Sond., F.C. **1**: 155 (1860).—Peters, Reise Mossamb. Bot.: 136 (1861).—Oliv. in F.T.A. **2**: 593 (1871).—Engl., Pflanzenw. Ost-Afr. **C**: 175 (1895).—Hiern, Cat. Afr. Pl. Welw. **1**: 419 (1898) (" pharnacioides ").—Swynnerton et al. in Journ. Linn. Soc., Bot. **40**: 1 (1911).—R. E. Fr., Wiss. Ergebn. Schwed. Rhod.-Kongo-

Tab. 130. GISEKIA PHARNACEOIDES var. PHARNACEOIDES. 1, habit (×1); 2, leaf-base (×3);
3, part of undersurface of leaf showing raphides (×16), 1–3 from *Bruce* 567; 4,
inflorescences (×1); 5, bract (×16); 6, flower (×6); 7, sepal (×10); 8, stamen
(×16); 9, gynoecium (×12); 10, fruit (×6); 11, achene (×10); 12, seed (×10),
4–12 from *Pielou* 58. From F.T.E.A.

Exped. 1: 34 (1914)—Eyles in Trans. Roy. Soc. S. Afr. 5: 350 (1916).—Burtt
Davy, F.P.F.T. 1: 153 (1926).—Pax & Hoffm. in Engl. & Prantl, Pflanzenfam. ed. 2,
16c: 192 (1934).—A. Peter, Fl. Deutsch Ost-Afr. 2: 253 (1938).—Hutch., Botanist
in S. Afr.: 672 (1946).—Hauman in F.C.B. 2: 102 (1951).—Suesseng. & Merxm. in
Proc. Trans. Rhod. Sci. Ass. 43: 183 (1951).—Keay in F.W.T.A. ed. 2, 1: 134
(1954) (" pharnacioides ").—Jeffrey in F.T.E.A., Aizoaceae: 5, fig. 1 (1961).—
Adamson in Journ. S. Afr. Bot. 27: 132 (1961) (" pharnacioides ").—Friedr. in
Prodr. Fl. SW. Afr. 26: 5 (1966).—Binns, H.C.L.M.: 16 (1968).—M. L. Gonçalves
in C.F.A. 4: 305 (1970).—J. H. Ross, Fl. Natal: 161 (1972). TAB. 130. Type from
eastern India, based on a cultivated plant at Uppsala.

Var. **pharnaceoides.**
    Koelreutera molluginoides Murr., Nov. Comm. Götting. 3: 67 (1773). Type from
India.
    Gisekia linearifolia Schumach. & Thonn. apud Schumach. [Beskr. Guin. Pl.:
167 (1827?)] in Kongel. Dansk. Vid. Selsk. Naturvid. Math. Afh. 3: 187 (1828).
Type from Ghana.
    Gisekia molluginoides Wight, Calc. Journ. Nat. Hist. 7: 162 (1847). Type from
India.
    Gisekia rubella Moq. in DC., Prodr. 13, 2: 27 (1849).—Oliv. in F.T.A. 2: 594
(1871).—R. E. Fr., Wiss. Ergebn. Schwed. Rhod.-Kongo-Exped. 1: 35 (1914).—
Hauman in F.C.B. 2: 103 (1951). Type from the Sudan.
    Gisekia congesta Moq. in DC., op. cit.: 28 (1849). Type from Senegambia.
    Gisekia pharnaceoides var. *pedunculata* Oliv., loc. cit.—Hiern, Cat. Afr. Pl. Welw.
1: 420 (1898). Type as for *Gisekia linearifolia.*
    Gisekia pharnaceoides var. *congesta* (Moq.) Oliv., loc. cit.—Hiern, loc. cit. Type
as for *Gisekia congesta.*

Prostrate to procumbent semi-succulent glabrous annual herb with stems up to
c. 60 cm. long, sometimes pink-tinged. Leaf-lamina 5–68 × 1–14 mm., of various
shapes as in *G. africana*; petiole 0–4 mm. long. Inflorescences sessile or pedunculate, umbelliform, congested to very lax, often many-flowered. Flowers ⚥,
pinkish, greenish or yellowish-white, scentless; pedicels up to c. 9 mm. long.
Perianth-segments 5, 1–3 mm. long in fruit, herbaceous, obtuse or subacute.
Stamens 5 with filaments slightly broadened below. Carpels 5. Fruit blackish,
densely covered with warts.
    Botswana. N: Nata R., 900 m., fl. & fr. 21.iv.1957, *Drummond & Seagrief* 5151
(LISC; SRGH). SW: 80 km. N. of Kan, fl. & fr. 18.ii.1960, *Wild* 5067 (K; SRGH).
SE: Mochudi, fl. & fr. i.1915, *Harbor* (SRGH). **Zambia.** B: Mongu, fl. & fr. 23.xii.1959,
Gilges 766 (SRGH). N: Mbala (Abercorn), Lake Tanganyika beyond Kasakalawa,
700 m., fl. & fr. 8.ii.1964, *Richards* 18962 (K). C: Serenje, fl. & fr. 18.ii.1955, *Fanshawe*
2089 (K). S: Livingstone, fl. & fr. 9.ix.1953, *Fanshawe* 282 (K; LISC). **Rhodesia.**
N: Gokwe, fl. & fr. 24.iv.1962, *Bingham* 243 (LISC; SRGH). W: Shangani, Gwampa
Forest Reserve, 915 m., fl. & fr. vii.1955, *Goldsmith* 164 (LISC; SRGH). C: Macheke,
1500 m., fl. & fr. xii.1919, *Eyles* 2012 (K; SRGH). E: Umtali, near Feruka Siding,
fl. & fr. 30.i.1949, *Chase* 1220 (BM; LISC; SRGH). S: Beitbridge, on banks of
Tshilatshokwe R., fl. & fr. 20.iii.1967, *Mavi* 268 (LISC). **Malawi.** N: Mwanemba,
2440 m., fl. & fr. 2.iii.1903, *McClounie* 58 (K). C: road near L. Nyasa Hotel, near Salima,
480 m., fl. & fr. 14.ii.1959, *Robson* 1597 (BM; K; LISC; SRGH). **Mozambique.** N:
Nampula, fl. & fr. 25.ii.1936, *Torre* 931 (BR; COI; LISC). Z: Namacurra, Licungo
riverside, fl. & fr. 28.viii.1949, *Barbosa & Carvalho* 3867 (LMA). MS: Vila Pery,
Chimoio, 700 m., fl. & fr. 24.xi.1965, *Torre & Correia* 13202 (LISC; M; P. WAG).
SS: Bilene, near Maniquenique, fl. & fr. 10.x.1957, *Barbosa & Lemos* 7990 (COI; LISC).
LM: Marracuene road, fl. & fr. 10.viii.1964, *Marques* 8 (COI; LISC; LMU).
    Widespread from tropical Africa, S. Africa and Mascarene Is. to India and Ceylon.
Common weed of open woodland, savanna, riverbanks and sand dunes, waste and ruderal
places, cultivated land especially on sandy soils.

2. LIMEUM L.

Limeum L., Syst. Nat. ed. 10: 995 (1759).
Gaudinia J. Gay in Bull. Sci. Nat. Paris 18: 412 (1829) non Beauv. (1812).
Semonvillea J. Gay, loc. cit.
Dicarpaea K. Presl, Symb. Bot. 1: 37 (1830).
Acanthocarpaea Klotzsch in Peters, Reise Mossamb. Bot. 1: 137, t. 24 (1861).

Glabrous or viscid, glandular-pubescent annual or perennial herbs or subshrubs. Leaves alternate, opposite or fasciculate, linear to orbicular, entire,
sessile or petiolate, sometimes subsucculent, exstipulate. Inflorescences terminal or

lateral, cymose, often umbelliform, sessile or pedunculate, few- or many-flowered, lax or dense. Flowers ☿, whitish or greenish, small, more or less pedicellate. Perianth-segments 5, generally unequal, herbaceous, free. Staminodes 0–5, generally petaloid, arising from the base of the stamens, free. Stamens (5)–7, hypogynous; filaments broadened below, inserted usually on hypogynous disk. Ovary superior, syncarpous, bicarpellate, 2-locular; style 1; stigmas 2. Fruit separating into 2 mericarps. Mericarps indehiscent, 1-seeded, hemispheric-reniform, conical, truncate or compressed-orbicular, winged or wingless, often auricled at base, the outer face reticulate, rugose, spinescent or more or less smooth.

A palaeotropical genus of some 20 species with the centre of distribution in SW. Africa.

Mericarps compressed-orbicular, often winged:
- Inflorescence congested, pseudo-umbellate; mericarps always winged 8. *pterocarpum*
- Inflorescence a lax paniculate cyme, sometimes winged -   -   - 9. *fenestratum*

Mericarps hemispheric-reniform or truncate-turbinate, never winged:
- Plants glabrous, sometimes rough, never glandular-pubescent:
  - Mericarps auriculate; surface of mericarp alveolate, more or less strongly tuberculate-spinescent, especially near the edges -   -   -   -   - 2. *sulcatum*
  - Mericarps not auriculate; surface of mericarp finely rough to reticulate-alveolate, sometimes weakly tuberculate:
    - Plants perennial herbs or subshrubs -   -   -   -   -   - 1. *aethiopicum*
    - Plants annual -   -   -   -   -   - 3. *argute-carinatum* var. *kwebense*
- Plants viscid glandular-pubescent:
  - Plants perennial herbs or subshrubs -   -   -   -   - 4. *dinteri*
  - Plants annual:
    - Mericarps with 2 smooth or furrowed auricles at base:
      - Mericarps truncate-turbinate, with furrowed auricles -   - 7. *arenicola*
      - Mericarps hemispheric-reniform, with smooth auricles -   - 5. *viscosum*
    - Mericarps not auriculate -   -   -   -   -   -   - 6. *myosotis*

1. **Limeum aethiopicum** Burm. f., Fl. Cap. Prodr.: 11 (1768).—Sond. in Harv. & Sond., F.C. **1**: 154 (1860).—Burtt Davy, F.P.F.T. **1**: 154 (1926).—Pax & Hoffm. in Engl. & Prantl, Pflanzenfam. ed 2, **16c**: 194 (1934).—Friedr. in Mitt. Bot. Staatss. München **2**: 141 (1956); in Prodr. Fl. SW. Afr. **26**: 11 (1966).—Guillarmod, Fl. Lesotho: 168 (1971). Type from S. Africa (Cape Prov.).

Glabrous perennial herbs or subshrubs up to 40 cm. tall, woody at base, developing herbaceous shoots; stems erect or ascendent, sparingly branched, striate, minutely scaberulous or smooth. Leaves sessile to shortly petiolate, alternate to subopposite; lamina 5–27 × 1–4 mm., linear to lanceolate or oblanceolate, entire, midrib prominent below; petiole up to 3 mm. long, enlarged at base. Inflorescences mostly terminal, 15–100-flowered, umbelliform, pedunculate. Flowers whitish, small, with pedicels up to 2 mm. long and membranous whitish lanceolate bracteoles. Perianth-segments 5, 2–3 mm. long, slightly unequal, membranous, whitish, with greenish keel, broadly ovate, obtuse, slightly mucronate. Staminodes 0–5, shorter than the perianth-segments, evanescent, with long claw and ovate limb. Stamens (5)7, with filaments much broadened and fimbriate below. Ovary 2-lobed, glabrous. Mericarps 2–3 mm. in diameter, pitted and furrowed, sometimes very shortly tuberculate, not auriculate at base.

Leaves linear-cuneate, lanceolate to elliptic; mericarps with the exterior surface finely marked with a honeycomb-like network of ridges with the angles elevated into small projections, sometimes tuberculate -   -   -   -   - var. *aethiopicum*
Leaves linear; mericarps similar but more coarsely marked all over and the angles of the ridges prominently elevated -   -   -   -   -   - var. *glabrum*

Var. aethiopicum.
Limeum aphyllum L.f., Suppl. Pl.: 214 (1781). Type from S. Africa.
Limeum capense Thunb., Prodr. Pl. Cap.: 68 (1794–1800); Fl. Cap.: 342 (1823). Type from S. Africa.

Botswana. N: Kangwa (Xanwe), 27 km. NE. of Aha Hills, fl. & fr. 12.iii.1965, *Wild & Drummond* 6938 (SRGH). SW: Kaotwe, fl. & fr. 10.iv.1930, *van Son* 28969 (SRGH). SE: 66 km. W. of Lothlakane, fl. & fr. 23.iii.1965, *Wild & Drummond* 7204 (LISC; SRGH). **Rhodesia.** S: Beitbridge, Shashi R., 11 km. downstream from Tuli Police Camp, fl. & fr. 6.v.1959, *Drummond* 6083 (SRGH).

Also in S. Africa. In cracks of limestone in valley bottom near spring, depressions by river and edges of pan.

Var. **glabrum** Moq. in DC., Prodr. **13**, 2: 23 (1849). Syntypes from SW. Africa.
Limeum suffruticosum Schellenb. in Engl., Bot. Jahrb. **48**: 492 (1912). Type from SW. Africa.
Limeum capense sensu Dinter in Fedde, Repert. **18**: 438 (1922) non Thunb. (1794–1800).
Limeum aethiopicum subsp. *namaense* Friedr. in Mitt. Bot. Staatss. München **2**: 142 (1956); in Prodr. Fl. SW. Afr. **26**: 11 (1966). Type from SW. Africa.
Limeum aethiopicum var. *namaense* Friedr., loc. cit. Type as above.

Botswana. SW: 117 km. S. of Tsane, Mabua Sehoba pan, fl. & fr. 22.ii.1960, *de Winter* 7461 (K; SRGH).
Also in SW. Africa. In limestone with embedded fragments of red quartzite forming a small cliff on edge of pan.

2. **Limeum sulcatum** (Klotzsch) Hutch. in Burtt Davy, F.P.F.T. **1**: 46 (1926).—
Friedr. in Mitt. Bot. Staatss. München **2**: 146 (1956); in Prodr. Fl. SW. Afr. **26**: 14 (1966).—M. L. Gonçalves in C.F.A. **4**: 308 (1970).—Guillarmod, Fl. Lesotho: 168 (1971). Type: Mozambique, Sena, *Peters* (B, holotype†).

Annual or perennial herb up to 70 cm. tall, glabrous, sometimes rough, never glandular-pubescent, sometimes a subshrub with several ± erect or trailing herbaceous shoots and woody base; stems not much branched. Leaves 5–50 × 0·5–4 mm., alternate, subopposite or fasciculate, narrowly linear to lanceolate, entire, with prominent midrib below; petiole short, with swollen base. Inflorescences terminal and lateral, 5–50-flowered, pedunculate, umbelliform, laxly branched. Flowers small, greenish-white; pedicels up to 2 mm. long; bracteoles membranous, whitish, lanceolate. Perianth-segments 5, 2–3 mm. long, slightly unequal, membranous, whitish but with greenish keel, broadly ovate, obtuse, mucronate. Staminodes 0(–5), when present very small and caducous. Stamens 7 with filaments enlarged and fimbriate below. Ovary 2-lobed, glabrous. Mericarps sometimes bigger than perianth-segments, with 2 usually large smooth auricles at base, the outer face alveolate-tuberculate to spinescent.

Plants annual; staminodes weakly developed or caducous; mericarps usually rather spinescent with the spines up to 3 mm. long - - - - var. *sulcatum*
Plants usually perennial, herbaceous or suffruticose; staminodes almost always 0; mericarps less distinctly spinescent with spines less than 2 mm. long
var. *scabridum*

Var. **sulcatum.**
Dicarpaea linifolia K. Presl, Symb. Bot. **1**: 38, t. 26 (1790) pro parte excl. typ. Type from S. Africa.
Limeum linifolium (K. Presl) Fenzl in Ann. Wien. Mus. **1**: 342 (1836) pro parte excl. typ. Type as above.
Acanthocarpaea sulcata Klotzsch in Peters, Reise Mossamb. Bot.: 138 (1861). Type as for *Limeum sulcatum.*
Limeum echinatum H. Walt. in Fedde, Repert. **8**: 55 (1910).—Pax & Hoffm. in Engl. & Prantl, Pflanzenfam. ed. 2, **16c**: 194 (1934).—Type from SW. Africa.
Limeum sulcatum var. *gracile* Friedr. in Mitt. Bot. Staatss. München **2**: 148 (1956). Type from SW. Africa.
Limeum sulcatum var. *robustum* Friedr., loc. cit. Type from SW. Africa.

Caprivi Strip. Katima Mulilo area, 914 m., fl. & fr. 26.xii.1958, *Killick & Leistner* 3098 (K). **Botswana.** N: between Odiakwe and Kanyu (Francistown-Maun road), fl. & fr. 9.iii.1965, *Wild & Drummond* 6827 (LISC; SRGH). SW: 33 km. SW. of Takatswane, on road to Lehututu, fl. & fr. 21.ii.1960, *de Winter* 7430 (K; SRGH). SE: Mochudi, fl. & fr. i–iv.1914, *Rogers* 6362 (K). **Rhodesia.** W: Wankie Game Reserve, 914 m., fl. & fr. 15.ii.1956, *Wild* 4735 (SRGH). E: Umtali, 69 km. S. of Umtali on Chipinga road, fl. & fr. 6.i.1969, *Biegel* 2739 (LISC). S: Beitbridge, fl. & fr. 16.ii.1955, *E. M. & W.* 446 (BM; LISC; SRGH). **Mozambique.** Z: Zambeze riverside, fl. & fr. 1884–85, *Carvalho* (COI). MS: Chemba on Tambara road, fl. & fr. 4.iv.1962, *Balsinhas & Macucdua* 548 (COI). LM: Sábiè, Ressano Garcia, Incomati riverside, fl. & fr. iii.1893, *Quintas* 18 (COI).
Also in Angola and S. and SW. Africa. Open sandy ground, woodland and bush, short grassland, roadsides, riverbeds, as weed of cultivated land, especially on sandy soils.

Var. **scabridum** (Klotzsch) Friedr. in Mitt. Bot. Staatss. München **2**: 148 (1956). Type: Mozambique, *Peters* (K, holotype).
Acanthocarpaea scabrida Klotzsch in Peters, Reise Mossamb., Bot.: 139 (1861). Type as above.

Limeum mossambicense Schellenb. in Engl., Bot. Jahrb. **48**: 491 (1912).—Pax & Hoffm. in Engl. & Prantl, Pflanzenfam. ed. 2, **16c**: 194 (1934). Type from Mozambique.

Rhodesia. E: Chipinga, Sabi Experimental Station, fl. & fr. 17.xi.1958, *Whellan* 1566 (SRGH). S: Ndanga, fl. & fr. 22.xii.1951, *Wild* 3717 (LISC; SRGH). **Mozambique.** N: António Enes, fl. & fr. 17.x.1965, *Mogg* 32400 (LISC). Z: Maganja da Costa, Bajone, fl. & fr. 2.x.1949, *Barbosa & Carvalho* 4279 (LISC; LMA). T: Mutarara, opposite Sena, fl. & fr. i.1859, *Kirk* (K). MS: Caia, Sena, fl. & fr. 1846, *Peters* (K). SS: Homoine, fl. & fr. 25.ii.1955, *E. M. & W.* 575 (BM; LISC; SRGH). LM: Maputo, fl. & fr. vi.1914, *Maputoland Expedition* (LISC).
Also in S. Africa (Transvaal). Riverbanks, littoral dunes and other sandy localities.

3. **Limeum argute-carinatum** Wawra apud Wawra & Peyr. in Sitz.-Ber. Math.-Nat. Akad. Wiss. Wien **38**: 563 (1860).—Pax & Hoffm. in Engl. & Prantl, Pflanzenfam. ed. 2, **16c**: 194 (1934).—Friedr. in Mitt. Bot. Staatss. München **2**: 148 (1956); in Prodr. Fl. SW. Afr. **26**: 12 (1966).—M. L. Gonçalves in C.F.A. **4**: 309 (1970). Type from Angola (Lobito).

Var. **kwebense** (N. E. Br.) Friedr., tom. cit.: 149 (1956).—M. L. Gonçalves, loc. cit. TAB.. **131** Type: Botswana, *Lugard* 186 (K, holotype).
Limeum linifolium sensu Oliv. in F.T.A. **2**: 596 (1871) pro parte quoad specim. *Welwitsch* 2425.—Hiern, Cat. Afr. Pl. Welw. **1**: 422 (1898) pro eod. parte.
Limeum kwebense N. E. Br. in Kew Bull. **1909**: 114 (1909). Type as for *Limeum argute-carinatum* var. *kwebense*.

A small prostrate annual herb with widely spreading branches, glabrous in all parts. Leaves alternate, petiolate; lamina 9–40 × 2–9 mm., oblong, oblong-lanceolate or linear-oblong, obtuse at apex, rounded or cuneate at base; petiole up to 4 mm. long. Inflorescence a terminal or subaxillary umbelliform cyme, densely many-flowered; peduncle 6–30 mm. long; bracts up to 2 mm. long, ovate, acuminate, membranous, with green keel. Flowers small, whitish; pedicels up to 2 mm. long; bracteoles membranous, acuminate. Perianth-segments 5, up to 2·5 mm. long, oblong, mucronate-acute, membranous, with a narrowly winged green keel. Staminodes 5, c. 1·5 mm. long, spathulate-elliptic, obtuse, membranous, white. Stamens 7, c. 1·5 mm. long, with filaments dilated at the base. Ovary 2-lobed, with 2 short styles. Mericarps blackish, pitted and furrowed, sometimes very shortly tuberculate, without auricles at base.

Botswana. N: Ngamiland, Tsau-Maun, 930 m., fl. & fr. 25.iii.1961, *Richards* 14850 (K; LISC; SRGH). **Mozambique.** SS: Bazaruto I., fl. & fr. viii. 1937, *Gomes e Sousa* 1972 (COI; K).
Also in Angola, SW. Africa and S. Africa (Transvaal). Sandy and dry soils.

Note. Var. *argute-carinatum*, which is mainly characterized by its broader leaf-laminas, occurs in SW. Africa and S. Africa (Griqueland West and the Transvaal).

4. **Limeum dinteri** Schellenb. in Engl., Bot. Jahrb. **48**: 493 (1912).—Pax & Hoffm. in Engl. & Prantl, Pflanzenfam. ed. 2, **16c**: 194 (1934).—Hutch., Botanist in S. Afr.: 672 (1946).—Friedr. in Mitt. Bot. Staatss. München **2**: 150 (1956); in Prodr. Fl. SW. Afr. **26**: 12 (1966). Type from SW. Africa.
Limeum pseudomyosotis Schellenb., tom. cit.: 494 (1912) pro parte quoad specim. *Lüderitz* 161 et *Dinter* 1180.

Shrublet up to 60 cm. tall, glandular-pubescent in all parts, with decumbent whitish stems. Leaves subsucculent, petiolate, alternate; laminas 6–22 × 2–10mm., elliptic, obovate, usually cuneate at the base, glandular-pubescent on both sides, sometimes rough above; petiole up to 4 mm. long. Inflorescence axillary, many-flowered, loosely corymbose, pedunculate; bracts lanceolate to ovate, acuminate. Flowers small, whitish; pedicels up to 2 mm. long; bracteoles acuminate. Perianth-segments 5, up to 3 mm. long, ovate, margins membranous. Staminodes 5, equalling or shorter than perianth-segments, with very long, narrow claw and flat broad limb. Stamens 7. Ovary 2-lobed, glabrous or very soon glabrescent. Mericarps brownish, sometimes bigger than perianth-segments, surface reticulate-alveolate to tuberculate especially above, ridged longitudinally at base, with 2 basal auricles.

Rhodesia. S: Beitbridge, fl. & fr. 15.ii.1955, *E. M. & W.* 384 (BM; LISC; SRGH). **Mozambique.** SS: Benguerua I., 6 m., fl. & fr. 8.ii.1958, *Mogg* 29105 (LISC).
Also in SW. Africa and S. Africa (Transvaal). Mopane bush and sandy soils.

Tab. 131. LIMEUM ARGUTE-CARINATUM var. KWEBENSE. 1, habit (× ⅔) from *Lugard* 186 and *Richards* 14850; 2, inflorescence (×3); 3, flower (×12); 4, flower, facing perianth-segments removed (×12); 5, androecium (×12); 6, staminode (×12); 7a, outer face, 7b, inner face, of mericarp (×12), 2–7 from *Lugard* 186.

5. **Limeum viscosum** (J. Gay) Fenzl in Nov. Stirp. Dec. Mus. Vindob. **10**: 87 (1839).—
DC., Prodr. **13**, 2: 23 (1849).—Wawra apud Wawra & Peyr. in Sitz.-Ber. Math.-
Nat. Akad. Wiss. Wien **38**: 23 (1860).—Sond. in Harv. & Sond., F.C. **1**: 154
(1860).—Oliv. in F.T.A. **2**: 595 (1871).—Engl., Pflanzenw. Ost-Afr. **C**: 175 (1895).
—Hiern, Cat. Afr. Pl. Welw. **1**: 421 (1898).—Eyles in Trans. Roy. Soc. S. Afr. **5**:
350 (1916).—Burtt Davy, F.P.F.T. **1**: 154 (1926).—Pax & Hoffm. in Engl. & Prantl,
Pflanzenfam. ed. 2, **16c**: 194 (1934).—A. Peter, Fl. Deutsch Ost-Afr. **2**: 252
(1938).—Keay in F.W.T.A. ed. 2, **1**: 134 (1954).—Friedr. in Mitt. Bot. Staatss.
München **2**: 151 (1956).—Jeffrey in F.T.E.A., Aizoaceae: 6, fig. 2 (1961).—Friedr.
in Prodr. Fl. SW. Afr. **26**: 15 (1966).—M. L. Gonçalves in C.F.A. **4**: 310 (1970).—
Guillarmod, Fl. Lesotho: 168 (1971).—J. H. Ross, Fl. Natal: 160 (1972). Type
from Senegal.

A viscid glandular-hairy annual or short-lived perennial herb up to 60 cm. tall,
with diffuse, prostrate, procumbent to ascending stems. Leaves alternate,
petiolate; laminas 5–63 × 2–24 mm., viscid, linear-lanceolate, lanceolate,
oblanceolate-spathulate, ovate, obovate, broadly obovate or almost circular, entire,
retuse, rounded or obtuse, sometimes mucronulate and more or less cuneate at
base; petiole 2–8 mm. long. Inflorescence an extra-axillary, more or less compact,
distinctly pedunculate or subsessile 5–25-flowered cyme. Flowers whitish,
pedicellate, the pedicels up to 7 mm. long; bracteoles small, membranous, acute.
Perianth-segments 5, up to 3 mm. long, unequal, usually broadly ovate with
whitish membranous edges. Staminodes 5, evanescent, with long claw and
expanded broadly ovate limb, cuneate-truncate at base. Stamens 5 or 7; filaments
with the base dilated, inserted on a hypogynous disk. Ovary 2-lobed. Mericarps
2–5 mm. in diameter, the surface with a network of ridges forming a honeycomb
pattern, more or less prominently elevated into small projections at the angles and
with 2 auricles at base.

This species is divided into 3 subspecies and only represented by subsp. *viscosum* in the
Flora Zambesiaca area. Subsp. *nummulifolium* occurs in SW. Africa and S. Africa (Cape
Province); it differs from subsp. *viscosum* by its almost circular leaf-laminas. Subsp.
transvaalense is confined to the Transvaal and Griqueland West and distinguished by its
rather conspicuous staminodes.

Subsp. **viscosum.**

Leaf-laminas obovate-circular to obovate or elliptic:
  Leaf-laminas at most 3 times as long as broad; pedicels up to 3·5 mm. long
                                  var. *viscosum*
  Leaf-laminas at most twice as long as broad; pedicels up to 7 mm. long
                                  var. *dubium*
Leaf-laminas narrowly lanceolate to oblanceolate or oblanceolate-cuneate, 3–7·5 times as
  long as broad; pedicels up to 6 mm. long -   -   -   -   -   var. *kraussii*

Var. **viscosum.**—Hiern, Cat. Afr. Pl. Welw. **1**: 421 (1898) pro parte excl. specim.
Welwitsch 2419.—Friedr. in Mitt. Bot. Staatss. München **2**: 152 (1956).—Jeffrey in
F.T.E.A., Aizoaceae: 6, fig. 2 (1961).—Friedr. in Prodr. Fl. SW. Afr. **26**: 15
(1966).—M. L. Gonçalves in C.F.A. **4**: 310 (1970).—Guillarmod, Fl. Lesotho:
168 (1971). Type from Senegal.
  Gaudinia viscosa J. Gay in Ferussac Bull. Sci. Nat. **18**: 412 (1829). Type as above.
  Limeum viscosum var. *kotschyi* Moq. in DC., Prodr. **13**, 2: 23 (1849).—Burtt Davy,
F.P.F.T. **1**: 154 (1926). Type from the Sudan.
  Limeum kotschyi (Moq.) Schellenb. in Engl., Bot. Jahrb. **48**: 497 (1912).—Pax &
Hoffm. in Engl. & Prantl, Pflanzenfam. ed. 2, **16c**: 194 (1934).—F. W. Andr., Fl. Pl.
Anglo-Egypt. Sudan **1**: 94 (1950). Type as above.

Botswana. N: 16 km. N. of Tsau, fl. & fr. 11.iii.1965, *Wild & Drummond* 6858
(LISC; SRGH). **Zambia.** B: Masese, fl. & fr. 15.v.1962, *Fanshawe* 6818 (K).
Rhodesia. C: Beatrice, 1220 m., fl. & fr. 27.xii.1924, *Eyles* 4424 (K; SRGH).
Mozambique. Z: Pebane, Nabúri, 100 m., fl. & fr. 17.i.1968, *Torre & Correia* 17218
(K; LISC; LMU; MO).
  Also in Senegal, Nigeria, Sudan, Ethiopia, Tanzania and Angola. Sandy flats and
cultivated ground.

Var. **kraussii** Friedr. in Mitt. Bot. Staatss. München **2**: 152 (1956).—Jeffrey in F.T.E.A.,
Aizoaceae: 8 (1961).—J. H. Ross, Fl. Natal: 161 (1972). Type from S. Africa
(Natal).
  Limeum natalense Schellenb. in Engl., Bot. Jahrb. **48**: 495 (1912). Type from
S. Africa (Natal).

Botswana. N: Nata R. near Madsiara drift, 896 m., fl. & fr. 27.v.1957, *Drummond* 5256 (SRGH). SW: Ghanzi, 16 km. W. of Kan, fl. & fr. 18.ii.1960, *Yalala* 55 (SRGH). SE: 33 km. SW. of Takatswane on road to Lehututu, fl. & fr. 21.ii.1960, *de Winter* 7429 (K). **Zambia.** B: Masese, fl. & fr. 9.i.1961, *Fanshawe* 6101 (K; SRGH). S: Namwala on Kalahari sand, 1005 m., fl. & fr. 9.i.1957, *Robinson* 2111 (K; SRGH). **Rhodesia.** W: Shangani, Gwampa Forest Reserve, fl. & fr. 27.i.1955, *Goldsmith* 34/55 (SRGH). C: Gwelo, 1280 m., fl. & fr. 4.i.1966, *Biegel* 750 (LISC). S: Buhera, Makumbe, fl. & fr. i.1967, *Mandaro* 57 (LISC). **Mozambique.** N: Mogincual, fl. & fr. 9.xi.1936, *Torre* 1166 (COI; LISC). SS: Inhambane-Velho, fl. & fr. vi.1938, *Gomes e Sousa* 2142 (K; LISC). LM: between Costa do Sol and Marracuene, fl. & fr. 14.xi.1960, *Balsinhas* 266 (COI; K; LISC; SRGH).
Also in Tanzania, S. Africa (Transvaal and Natal). Open places in mopane woodland, moist areas, roadsides and sandy soils.

Var. **dubium** Friedr. in Mitt. Bot. Staatss. München **2**: 153 (1956).—Type: Botswana, *Rogers* 6461 (Z, holotype).

Rhodesia. S: Gwanda, on road to Tuli Breeding Station, fl. & fr. 19.iii.1959, *Drummond* 5865 (SRGH). **Mozambique.** SS: Chibuto, Alto Changane, fl. & fr. 15.vii.1944, *Torre* 6765 (K; LISC; LMU).
Also in S. Africa (Transvaal). Open deciduous woodland and sandy soils.

6. **Limeum myosotis** H. Walt. in Fedde, Repert. **8**: 56 (1910).—Pax & Hoffm. in Engl. & Prantl, Pflanzenfam. ed. 2, **16c**: 194 (1934).—Friedr. in Mitt. Bot. Staatss. München **2**: 154 (1956); in Prodr. Fl. SW. Afr. **26**: 13 (1966).—M. L. Gonçalves in C.F.A. **4**: 311 (1970). Type from SW. Africa.

Var. **myosotis.**
Limeum viscosum (J. Gay) Fenzl var. *hispidulum* Welw. ex Oliv. in F.T.A. **2**: 596 (1871).—Hiern, Cat. Afr. Pl. Welw. **1**: 421 (1898). Type from Angola.
Limeum viscosum forma *longepedunculatum* Schinz in Verh. Bot. Ver. Prov. Brand. **30**: 256 (1888). Type from SW. Africa.
Limeum pseudomyosotis Schellenb. in Engl., Bot. Jahrb. **48**: 494 (1912) pro parte quoad specim. *Schlechter* 154.
Limeum myosotis var. *confusum* Friedr., tom. cit.: 155 (1956). Type from S. Africa (Cape Prov.).

Hispidulous annual herb, ± glandular-hairy in all parts, with creeping or procumbent to ascending stems. Leaves alternate, petiolate, subsucculent, glaucous, glandular-glabrescent; laminas 11–38 × 3–14 mm., elliptic to broadly elliptic, narrowly spathulate or obovate, cuneate at base, entire; petiole up to 9 mm. long. Inflorescence pseudoaxillary, of rather lax pedunculate corymbose cymes; peduncle up to 3 cm. long. Flowers greenish-white, shortly pedicellate, with small acute membranous bracteoles. Perianth-segments 5, up to 3 mm. long, broadly ovate, with whitish membranous margins. Staminodes 5, generally as long as the perianth-segments, with a narrow claw and rounded limb. Stamens 7. Ovary 2-lobed. Mericarps up to 2 mm. in diameter, not auriculate at the base, the surface marked with a network of ridges forming ± prominently elevated projections at angles.

Botswana. N: Ngamiland, Kwebe, fl. & fr. xii. 1896, *Lugard* 100 (K). SW: 24 km. W. of Ghanzi, fl. & fr. 22.iv.1963, *Ballance* 628 (SRGH).
Also in Angola, SW. Africa and S. Africa. In low scrub.

Note. This very variable species is only represented by the type variety in the F.Z. area. The var. *confusum*, mainly distinguished by its obovate to broadly elliptic leaves, occurs in Angola, SW. Africa and S. Africa (Griqueland West and the Cape Prov.), while var. *rotundifolium*, which has almost circular leaves, is confined to SW. Africa.

7. **Limeum arenicola** Schellenb. in Engl., Bot. Jahrb. **48**: 496 (1912).—Pax & Hoffm. in Engl. & Prantl, Pflanzenfam. ed. 2, **16c**: 194 (1934).—Friedr. in Mitt. Bot. Staatss. München **2**: 156 (1956); in Prodr. Fl. SW. Afr. **26**: 12 (1966). Type from SW. Africa.

Viscid glandular-pubescent annual herb with prostrate or creeping ± sulcate stems. Leaves alternate, subsucculent, petiolate; laminas 5–28 × 4–18 mm., obovate to broadly obovate or almost circular, cuneate at the base and sometimes mucronulate at the apex, entire, rather glandular-pubescent on both sides; petiole up to 12 mm. long. Inflorescence axillary, subsessile, many-flowered, dense. Flowers greenish-white, shortly pedicellate. Perianth-segments 5, up to 2·5 mm. long, unequal, broadly ovate with membranous pink edges. Staminodes nil. Stamens (5)7. Ovary 2-lobed. Mericarps larger than perianth-segments, conical, truncate, ridged radially with 2 furrowed auricles at base.

Botswana. SW: near Mabua Sehoba pan, 117 km. S. of Tsane, fl. & fr. 22.ii.1960, *de Winter* 7453 (K; SRGH).
Also in SW. Africa. In sandy flats.

8. **Limeum pterocarpum** (J. Gay) Heimerl in Engl. & Prantl, Pflanzenfam. **3**, 1b: 9 (1889).—Pax & Hoffm. in Engl. & Prantl, op. cit. ed. 2, **16c**: 194 (1934).—Keay in F.W.T.A. ed. 2, **1**: 134 (1954).—Friedr. in Mitt. Bot. Staatss. München **2**: 158 (1956).—M. L. Gonçalves in Garcia de Orta **13**: 379 (1965).—Friedr. in Prodr. Fl. SW. Afr. **26**: 14 (1966).—M. L. Gonçalves in C.F.A. **4**: 312 (1970). Type from Senegal.

Var. **pterocarpum**.
    Semonvillea pterocarpa J. Gay in Férussac Bull. Sci. Nat. **18**: 412 (1829).—DC., Prodr. **13**, 2: 19 (1849).—Oliv. in F.T.A. **2**: 595 (1871).—Burtt Davy, F.P.F.T. **1**: 153 (1926).—Hutch. & Dalz. in F.W.T.A. **1**: 113 (1927). Type as above.
    Semonvillea punctata J. Gay ex Steud., Nom. Bot. ed. 2, **2**: 556 (1841). Type from Senegal.
    Limeum neglectum Dinter, Deutsch.-SW.-Afr. Fl. For. Landt. Frag.: 61 (1909). Type from SW. Africa.

Erect or rather diffuse nearly glabrous herb up to c. 60 cm. tall, with stems repeatedly branched. Leaves fleshy, alternate, subopposite or fasciculate, petiolate; laminas 8–45 × 2–8 mm., linear, linear-lanceolate or lanceolate, subacute at apex and cuneate at base; petiole up to 5 mm. long. Inflorescence terminal, a pedunculate, pseudoumbellate few–many-flowered cyme with acute whitish membranous bracts. Flowers greenish-white, very shortly pedicellate, with acute whitish membranous bracteoles. Perianth-segments 5, up to 2·5 mm. long, unequal, membranous, ovate or broadly ovate, slightly mucronate, whitish with greenish keel. Staminodes 0–5, evanescent, long-clawed with oblong limb, more or less equalling perianth-segments. Stamens generally 7. Ovary compressed. Mericarps up to 13 mm. in diameter, compressed-orbicular, brownish, opaque, puberulous, with basal sinus and prominently pitted and furrowed centre surrounded by a radiately-nerved membranous wing up to 5 mm. broad.

Botswana. N: Ngamiland, Okovanga, near Tsau, fl. & fr. 15.iii.1961, *Richards* 14779 (K; SRGH). SE: Mahalapye, Morale, 975 m., fl. & fr. x.1957, *de Beer* 502 (SRGH). **Rhodesia.** S: Gwanda, 564 m., fl. & fr. 17.xii.1956, *Davies* 2338 (SRGH).
Also in Senegal, Nigeria, Sudan, Angola, SW. Africa and S. Africa (Transvaal). Bushland, sandy places.

Note. The var. *apterum* occurs in S. Africa (Transvaal) and is characterized by wingless mericarps.

9. **Limeum fenestratum** (Fenzl) Heimerl in Engl. & Prantl, Pflanzenfam. **3**, 1b: 9 (1889).—Pax & Hoffm. in Engl. & Prantl, op. cit. ed. 2, **16c**: 194 (1934).—Friedr. in Mitt. Bot. Staatss. München **2**: 159 (1956); in Prodr. Fl. SW. Afr. **26**: 13 (1966).—M. L. Gonçalves in C.F.A. **4**: 312 (1970).—J. H. Ross, Fl. Natal: 161 (1972). Type from SW. Africa.

Erect or procumbent, sometimes spreading annual (sometimes short-lived perennial) herb up to c. 90 cm. tall, glabrous; stems stiff and freely much-branched. Leaves petiolate, deciduous, alternate, subopposite or fasciculate; laminas 10–70 × 0·5–5 mm., linear to linear-lanceolate or lanceolate, tapering base, mucronate, entire; petiole up to 4 mm. long. Inflorescence a very diffuse terminal panicled cyme. Flowers whitish, subsessile, separated by large internodes, with small whitish membranous bracteoles. Perianth-segments up to 2·5 mm. long, unequal, ovate to oblong, acute, with whitish membranous margins. Staminodes 0–5, evanescent, long-clawed with oblong limb, equalling perianth-segments. Stamens usually 7; filaments ciliate at base. Ovary compressed. Mericarps compressed-orbicular, longer than perianth-segments, puberulous, dorsally tuberculate, the tubercles radiating from the centre, surrounded by a transparent, radiately nerved membranous wing, or wing absent.

Mericarps winged - - - - - - - - - var. *fenestratum*
Mericarps wingless - - - - - - - - - var. *exalatum*

Var. fenestratum.
    Semonvillea fenestrata Fenzl in Nov. Stirp. Dec. Mus. Vindob. **5**: 42 (1839).—
Hook., Ic. Pl. **6**: t. 587 (1843).—Moq. in DC., Prodr. **13**, 2: 20 (1849).—Sond. in
Harv. & Sond., F.C. **1**: 152 (1860).—R. E. Fr., Wiss. Ergebn. Schwed. Rhod.-
Kongo-Exped. **1**: 35 (1914).—Eyles in Trans. Roy. Soc. S. Afr. **5**: 349 (1916).—
Burtt Davy, F.P.F.T. **1**: 153 (1926).—Hutch., Botanist in S. Afr.: 672 (1946).
Type from SW. Africa.
    Limeum glaberrimum Pax in Engl., Bot. Jahrb. **19**: 132 (1894).—Kuntze, Rev. Gen.
Pl. **3**, 2: 108 (1898). Type from SW. Africa.
    Limeum fenestratum var. *perenne* Schinz ex Dinter, Deutsch.-SW.-Afr. Fl. For.
Landt. Frag.: 61 (1909). Type from SW. Africa.
    Semonvillea sol H. Walt. in Fedde, Repert. **8**: 57 (1910). Type from SW. Africa.
    Limeum frutescens Dinter in Fedde, Repert. **18**: 439 (1922). Type from
SW. Africa.
    Limeum fenetratum var. *frutescens* (Dinter) Friedr. in Mitt. Bot. Staatss. München
2: 160 (1956). Type as above.

Botswana. N: 1·6 km. W. of Nata R., 900 m., fl. & fr. 21.iv.1957, *Drummond &*
Seagrief 5149 (LISC; SRGH). SW: 27 km. N. of Kan, fl. & fr. 18.ii.1960, *Wild* 5028
(SRGH). **Zambia.** B: Sioma, along Zambesi R., 914 m., fl. & fr. vii.1921, *Borle* 224
(COI; K; SRGH). S: Namwala, on Kalahari sand, c. 1000 m., fl. & fr. 9.i.1957,
Robinson 2102 (K; LISC; SRGH). **Rhodesia.** W: Nyamandhlovu, Bongolo Farm,
fl. & fr. 11.i.1950, *West* 3063 (SRGH). C: Enkeldoorn, near Umgesi R., fl. & fr. 4.i.1946,
Wild 608 (SRGH). E: Chipinga, fl. & fr. 12.ii.1957, *Goodier* 105 (LISC; SRGH). S:
c. 2 km. Beitbridge on West Nicholson road, fl. & fr. 15.ii.1955, *E. M. & W.* 421 (BM;
LISC; SRGH). **Mozambique.** SS: Chibuto, between Chibuto and Gomes da Costa,
fl. & fr. 14.xi.1957, *Barbosa & Lemos* 8107 (COI; LISC). LM: Lourenço Marques,
Polana, fl. & fr. 7.vii.1965, *Marques* 577 (COI; LISC; LMU).
    Also in Angola, SW. Africa and S. Africa (Transvaal). In mopane woodland, bush,
littoral dunes, on sandy roadsides and as a weed of cultivated land.

Var. exalatum Friedr. in Mitt. Bot. Staatss. München **2**: 160 (1956); Type:
    Mozambique, Marracuene, *Pedro* 17 (K, holotype).

Rhodesia. S: Beitbridge, Chiturupazi, fl. & fr. 22.ii.1961, *Wild* 5331 (SRGH).
Mozambique. N: Mogincual, Quinga beach, c. 5 m., fl. & fr. 28.iii.1964, *Torre &*
Paiva 11431 (LISC). SS: Muchopes, 3 km. from Manjacaze, fl. & fr. 31.iii.1954,
Barbosa & Balsinhas 5509 (BM; LISC). LM: between Lourenço Marques and
Marracuene, fl. & fr. 23.iv.1947, *Barbosa* 154 (COI; LMA).
    Also in S. Africa (Transvaal). In open woodland, on sandy flats and dunes and as a
weed of cultivated land.

3. CORBICHONIA Scop.

Corbichonia Scop., Introd.: 264 (1777).
    Orygia Forssk., Fl. Aegypt.-Arab.: 103 (1775) pro parte excl. typ.

    Glabrous, erect to prostrate, annual or perennial herb. Leaves alternate,
petiolate, subsucculent, from obovate to orbicular, entire, apiculate. Inflorescence
a terminal few–many-flowered cyme. Flowers' ⚥, regular, pedicellate. Perianth-
segments 5, free. Staminodes many, petaloid, slender. Stamens many, hypogynous. Ovary superior, 5-locular, 5-lobed, of 5 united carpels. Ovules many per
loculus; placentation axile. Styles 5, free, sessile, linear. Fruit a 5-valved
loculicidal capsule. Seeds many, subreniform.
    A monotypic genus native to Africa and SW. Asia.

Corbichonia decumbens (Forssk.) Exell in Journ. Bot., Lond. **73**, Suppl. Polypet.,
    Add.: 80 (1935).—Hauman in F.C.B. **2**: 104, t. 9 (1951).—Adamson in Journ.
S. Afr. Bot. **24**: 68 (1958).—Jeffrey in F.T.E.A., Aizoaceae: 9, fig. 3 (1961).—
Friedr. in Prodr. Fl. SW. Afr. **26**: 4 (1966).—M. L. Gonçalves in C.F.A. **4**: 313
(1970).—J. H. Ross, Fl. Natal: 162 (1972). TAB. **132.** Type from Arabia.
    Orygia decumbens Forssk., Fl. Aegypt.-Arab.: 103 (1775).—DC., Prodr. **3**: 455
(1828).—Sond. in Harv. & Sond., F.C. **1**: 136 (1860).—Oliv in F.T.A. **2**: 589
(1871).— Engl., Pflanzenw. Ost-Afr. **C**: 175 (1895).—Hiern, Cat. Afr. Pl. Welw. **1**:
415 (1898).—Swynnerton et al. in Journ. Linn. Soc., Bot. **40**: 1 (1911).—R. E. Fr.,
Wiss. Ergebn. Schwed. Rhod.-Kongo-Exped. **1**: 35 (1914).—Eyles in Trans. Roy.
Soc. S. Afr. **5**: 350 (1916).—Burtt Davy, F.P.F.T. **1**: 154 (1926).—Pax & Hoffm. in

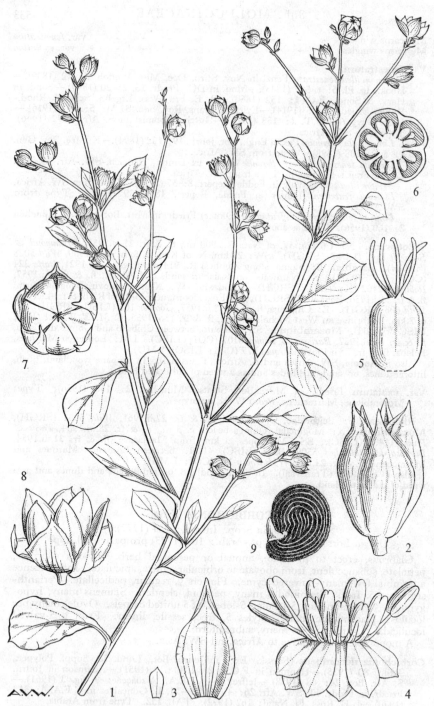

Tab. 132. CORBICHONIA DECUMBENS. 1, habit (×1) *Bogdan* 4321; 2, flower (×8) *Martin* s.n.; 3, sepal (×8); 4, part of rings of staminodes and stamens (×16), both from *Marshall* 15; 5, ovary (×12); 6, ovary in cross-section (×16), both from *Martin* s.n.; 7, capsule, in plan (×4); 8, fruiting calyx (×4); 9, seed (×20), 7–9 from *Purseglove* 1061. From F.T.E.A.

Engl. & Prantl, Pflanzenfam. ed. 2, **16c**: 193, fig. 88 (1934).—A. Peter, Fl. Deutsch Ost-Afr. **2**: 255 (1938).—Gossw.& Mendonça, Cart. Fitogeogr. Angol.: 147 & 192 (1939).—Hutch., Botanist in S. Afr.: 672 (1946). Type as above.
Portulaca decumbens Vahl, Symb. Bot. **1**: 33 (1790). Type from Arabia.
Glinus trianthemoides Heyne in Roth, Nov. Sp. Pl.: 231 (1821). Type from India.
Telephium laxiflorum DC., op. cit.: 366 (1828). Type from Arabia?
Axonotechium trianthemoides (Heyne) Fenzl in Ann. Wien. Mus. **1**: 355 (1836). Type as for *Glinus trianthemoides*.
Orygia mucronata Klotzsch in Peters, Reise Mossamb. Bot.: 140 (1861). Type: Mozambique, Tete, *Peters* (B, holotype†).
Glinus mucronatus (Klotzsch) Klotzsch, op. cit.: 570, t. 25 (1864). Type as above.

Erect, procumbent, prostrate or prostrate-ascending annual or short-lived perennial succulent herb up to 50 cm. high, sometimes with woody rootstock. Stems branched, often sub-woody at base, sometimes very long, the internodes with narrow ridges decurrent from the petiole bases. Leaves alternate, petiolate, fleshy; laminas 4–60 × 3–36 mm., oblanceolate-obovate, obovate-spathulate, obovate or orbicular, entire, apex apiculate, base cuneate, glaucous, sometimes turning purple; petiole 1–13 mm. long, winged. Inflorescences at first terminal but then overtopped and appearing lateral, up to c. 20 cm. long, few–many-flowered, loose, racemose, with lanceolate, membranous bracts up to 3 mm. long. Flowers pink, mauve or magenta, shortly pedicellate. Perianth-segments C. 4 mm. long in fruit, herbaceous, with white membranous margins, broadly ovate. Staminodes many, petaloid, membranous, fugacious, finally longer than perianth-segments. Stamens in 2 whorls. Fruit yellowish-green, shining, c. as long as perianth. Seeds purplish-black, concentrically ridged, with white aril.

Botswana. SW: Mabua Sefhubi Pan, 117 km. S. of Tsane, fl. & fr. 24.ii.1960, *Wild* 5142 (SRGH). **Rhodesia.** N: Gokwe, Bumi estuary, L. Kariba, fl. & fr. 8.i.1962, *Whellan* 2002 (SRGH). W: Bulawayo, fl. & fr. 16.iii.1965, *Best* 769 (LISC; SRGH). E: Melsetter, c. 610 m., fl. & fr. 9.xii.1951, *Chase* 4253 (BM; COI; LISC; SRGH). S: Ndanga, entrance to Lundi Gorge at Chipinda, fl. & fr. 10.i.1960, *Goodier* 769 (LISC; SRGH). **Malawi.** S: Port Herald, just S. of Lilange R., c. 100 m., fl. & fr. 25.iii.1960, *Phipps* 2709 (SRGH). **Mozambique.** Z: Morrumbala (Massigire), Chire R., fl. & fr. 1914, *Mendonça* (LISC). T: Tete, fl. & fr. ii.1932, *Pomba Guerra* 57 (COI). MS: Chemba, on road to Tambara, fl. & fr. 23.iv.1960, *Lemos & Macuácua* 136 (BM; COI; LISC; SRGH). SS: between Guijá and Chibuto, 61·6 km. from Guijá, fl. & fr. 30.iii.1954, *Barbosa & Balsinhas* 5505 (B; LISC; LMA). LM: Sábiè, Ressano Garcia, fl. & fr. 18.ii.1955, *E. M. & W.* 473 (BM; LISC; SRGH).
Widespread in tropical and southern Africa and tropical Asia; also introduced into tropical America. Woodland, savannas with trees, cultivated ground, ruderal places, stony and alluvial soils.

4. PSAMMOTROPHA Eckl. & Zeyh.

Psammotropha Eckl. & Zeyh., Enum. Pl. Afr. Austr. Extratrop.: 286 (1836).

Perennial, often creeping, rosette or cushion herbs. Leaves alternate, opposite or verticillate, densely crowded at base of plant, linear or narrowly lanceolate; stipules scarious. Inflorescence an umbelliform, often glomerate, terminal or axillary cyme; bracts often whorled. Flowers ☿, pedicellate, small. Perianth-segments 5, herbaceous, free. Stamens 5, hypogynous, alternate with the perianth-segments. Ovary of (3–)5 united carpels, (3–)5-lobed, (3–)5-locular; stigmas 3 or 5, more or less free; ovules 1 per loculus; placentation basal. Fruit (3–)5-lobed, (3–)5-celled, a loculicidal capsule. Seeds orbicular-reniform, granular.
A genus of 11 species centred in eastern S. Africa, with two species extending into tropical Africa.

Psammotropha myriantha Sond. in Harv. & Sond., F.C. **1**: 147 (1860).—Oliv. in F.T.A. **2**: 593 (1871).—Hiern, Cat. Afr. Pl. Welw. **1**: 419 (1898).—Burtt Davy, F.P.F.T. **1**: 155 (1926).—Pax & Hoffm. in Engl. & Prantl, Pflanzenfam. ed. 2, **16c**: 195 (1934).—Hauman in F.C.B. **2**: 112 (1951).—Adamson in Journ. S. Afr. Bot. **25**: 57 (1959).—Jeffrey in F.T.E.A., Aizoaceae: 11, fig. 4 (1961).—M. L. Gonçalves in C.F.A. **4**: 314 (1970).—J. H. Ross, Fl. Natal: 161 (1972). TAB. **133**. Type from S. Africa (Transvaal).
Psammotropha myriantha var. *huillensis* Welw. ex Oliv., loc. cit.—Hiern, loc. cit.—Pax & Hoffm. in Engl. & Prantl, loc. cit. Type from Angola.

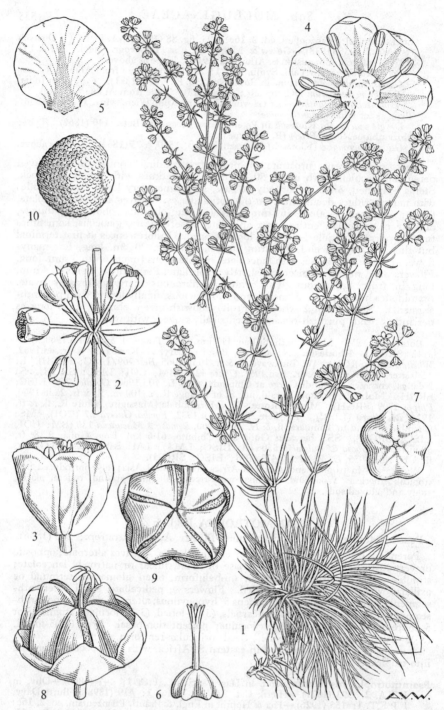

Tab. 133. PSAMMOTROPHA MYRIANTHA. 1, habit and inflorescences (×1) *Eggeling* 6618;
2, part of inflorescence (×4); 3, flower (×12); 4, androecium (×8); 5, sepal (×12);
6, ovary (×12); 7, ovary, in plan (×20), 2–7 from *Paulo* 291; 8, fruiting calyx
(×12); 9, capsule, in plan (×12); 10, seed (×24), 8–10 from *Eggeling* 6618. From
F.T.E.A.

Psammotropha breviscapa Burtt Davy, op. cit.: 49 & 155 (1926). Type from
S. Africa (Transvaal).

Low single- to multiple-tufted herb, or plants cushion-forming. Stems erect,
1–8 cm. long, simple or branched, thick, short; plants sometimes spreading by
means of runners rooting at the nodes. Leaves densely crowded in rosettes,
spreading, ascending or incurved, sessile; laminas 4–45 mm. long, linear or
narrowly lanceolate, 1-nerved, entirely glabrous, often revolute at margin, aristate;
stipules scarious, sometimes caducous. Inflorescences few or many, simple or
branched, with whorled bractiform leaves and flowers in sessile umbelliform
cymes clustered at nodes. Flowers greenish or whitish, pedicellate; pedicels up
to 5 mm. long. Perianth-segments c. 2 mm. long, broadly ovate, green with wide
membranous margins. Stamens on a very small annular disk. Ovary depressed
above, splitting into (3–)5 triangular valves. Seeds blackish-brown.

Zambia. N: Shiwa Ngandu, over hills to Chinkalanga, fl. & fr. 17.i.1937, *Ricardo*
122 (BM). **Rhodesia.** E: Inyangani Mt., 2560 m., fl. & fr. 6.xii.1959, *Wild* 4901
(BM; SRGH). **Mozambique.** MS: Manica, Zuira Mt., Tsetserra, c. 2100 m., fl. &
fr. 5.xi.1965, *Torre & Pereira* 12703 (LISC).
Also in Zaire, Tanzania, Angola and S. Africa. Dry sandy bush, grasslands, in rock
crevices and edge of damp rocks, 460–2600 m.

5. GLINUS L.

Glinus L., Sp. Pl. **1**: 463 (1753); Gen. Pl. ed. 5: 208 (1754).
Paulo-Wilhelmia Hochst. in Flora **27**: 17 (Jan. 1844) non Hochst. in Flora **27**,
Beih.: 4 (after Jan. 1844).

Often more or less prostrate, simply or stellately pubescent or almost glabrous,
sometimes succulent, annuals or perennials with annual shoots. Leaves alternate,
opposite or verticillate, narrowly elliptic to orbicular, entire or serrulate. Perianth
segments 5, herbaceous, free. Staminodes 0–8 or more, usually divided at the
apex, sometimes more or less petaloid. Stamens 3–∞, free, hypogynous. Ovary
superior, of 3 or 5 carpels, 3 or 5-locular, the loculi multi-ovulate; placentation
axile; stigmas sessile, free, as many as carpels. Fruit a 3- or 5-valved loculicidal
capsule. Seeds many, reniform, granulate or smooth, each with a fleshy aril on a
long filiform-appendaged strophiole.
A pantropical genus of c. 10 species.

Indumentum of stellate hairs, or plants glabrescent to glabrous; plants rosette-forming
  1. *lotoides*
Indumentum of straight or crisped hairs or of multicellular prickly pubescence, not
  stellate; plants not rosette-forming:
Indumentum of straight or crisped hairs; stamens 3–10 -   -   2. *oppositifolius*
Indumentum of multicellular prickly hairs; stamens more than 10 -   3. *bainesii*

As the species *G. oppositifolius* and *G. lotoides* hybridize freely and a large number of
intermediates have been recorded, it is sometimes very difficult to distinguish such hybrids
and hybrid derivatives from var. *lanatus* and var. *glomeratus* of *G. oppositifolius*.

1. **Glinus lotoides** L., Sp. Pl. **1**: 463 (1753).—Loefl., Iter Hisp.: 145 (1759).—Spreng.,
    Syst. Veg. **2**: 467 (1825).—Fenzl in Ann. Wien. Mus. **1**: 357 (1836).—Sond. in
    Harv. & Sond., F.C. **1**: 137 (1860).—Peters, Reise Mossamb. Bot.: 141 (1861).—
    Engl., Pflanzenw. Ost-Afr. **C**: 175 (1895).—Warb., Kunene-Samb.-Exped. Baum:
    233 (1903).—R. E. Fr., Wiss. Ergebn. Schwed. Rhod.-Kongo-Exped.: 34 (1914).—
    Burtt Davy, F.P.F.T. **1**: 155 (1926).—Exell in Journ. Bot. Lond., Suppl. Polypet.:
    198 (1929).—Engl. in Engl. & Prantl, Pflanzenfam. ed. 2, **16c**: 222 (1934).—A. Peter,
    Fl. Deutsch. Ost-Afr. **2**: 256 (1938).—Hauman in F.C.B. **2**: 107 (1951).—Keay in
    F.W.T.A. ed. 2, **1**: 135 (1954).—Jeffrey in F.T.E.A., Aizoaceae: 15, fig. 5 (1961).—
    Adamson in Journ. S. Afr. Bot. **27**: 127 (1961).—Friedr. in Prodr. Fl. SW. Afr. **26**:
    6 (1966).—Binns, H.C.L.M.: 16 (1968).—M. L. Gonçalves in C.F.A. **4**: 315
    (1970).—J. H. Ross, Fl. Natal: 161 (1972). Type from Sicily.
      Mollugo dictamnoides Burm. f., Fl. Ind.: 131 (1768).—L., Mant. Pl. Alt. **2**: 243
    (1771). Type from India.
      Mollugo hirta Thunb., Prodr. Pl. Cap.: 120 (1794).—DC., Prodr. **1**: 391 (1824).—
    Hiern, Cat. Afr. Pl. Welw. **1**: 415 (1898).—Eyles in Trans. Roy. Soc. S. Afr. **5**:
    350 (1916).—Wild, Fl. Vict. Falls: 144 (1953). Type from S. Africa.
      Pharnaceum hirtum Spreng., op. cit. **1**: 949 (1825). Type from S. Africa?

Mollugo glinus A. Rich., Tent. Fl. Abyss. **1**: 48 (1847).—Oliv. in F.T.A. **2**: 590 (1871).—Burkill ex Johnston, Brit. Centr. Afr.: 248 (1897). Type from Ethiopia.
Tryphera prostrata Bl. ex DC., op. cit. **13**, 2: 424 (1849). Type from Timor.
Mollugo lotoides (L.) Clarke ex De Wild. & Staner, Contr. Fl. Katanga, Suppl. **4**: 11 (1932). Type as for *Glinus lotoides*.

A decumbent, semi-prostrate or prostrate spreading rosulate herb up to 30 cm. high, diffusely branched, covered with whitish silky stellate hairs. Leaves alternate, opposite or apparently verticillate; laminae 2–37 × 4–23 mm., elliptic, obovate, spathulate or circular, entire; base cuneate, apex subacute to rounded; petiole up to 1·3 cm. long, green when glabrous or glabrescent, whitish when densely covered with persistent stellate silky hairs. Flowers greenish-white, sometimes with a pink tinge, apparently cleistogamous, inconspicuous, clustered at nodes, 2–10 per node; pedicels 1–4 mm. long. Perianth-segments up to 8 mm. long, more or less keeled, acute or mucronate, remaining closed around the fruit. Staminodes 0–8, strap-shaped, 2-fid at apex. Stamens usually numerous. Ovary of (3)5 united carpels; stigmas (3)5. Fruit a capsule, 6 mm. long. Seeds dark brown, usually with ridges, with or without tubercles, the aril c. ½ the length of the seed.

Plants densely whitish-tomentose, the leaf-indumentum persistent - var. *lotoides*
Plants not densely tomentose, greenish; leaves glabrous or glabrescent - var. *virens*

Var. lotoides.

Zambia. N: 26 km. N. of Mwamfuli, 1140 m., 23.xii.1946, *Symoens* 11202 (K). C: Feira, bank of Zambesi R., fl. & fr. 27.ix.1962, *Angus* 3350 (LISC). E: Nsefu Game Camp, 750 m., fl. & fr. 15.x.1958, *Robson* 138 (K; LISC; SRGH). S: Mazabuka, Kabange Village, fl. & fr. 2.x.1963, *van Rensburg* 2515 (K; SRGH). **Rhodesia.** N: Gokwe, bank of Masowa R., fl. & fr. 17.vii.1962, *Bingham* 313 (LISC; SRGH). W: Sebungwe, Sitonka Valley, 1066 m., fl. & fr.x.1955, *Davies* 1592 (COI; SRGH). C: Hartley, fl. & fr. 23.vii.1950, *Hornby* 3205 (SRGH). S: Bikita, 640 m., fl. & fr. 22.vii.1958, *Chase* 6952 (SRGH). **Malawi.** S: Mlanje, fl. & fr. 16.xi.1955, *Jackson* 1761 (K; SRGH). **Mozambique.** Z: between Quelimane and Marral, fl. & fr. 10.x.1941, *Torre* 3617 (COI; EA; J; LISC; LMA). T: Boroma, fl. & fr. 25.vii.1950, *Chase* 2788 (BM; SRGH). MS: Gorongosa, fl. & fr. 12.x.1944, *Mendonça* 2458 (K; LISC; LMU; MO; SRGH). SS: Caniçado, between Guijá and Barragem Village, fl. & fr. 15.xi.1957, *Barbosa & Lemos* 8141 (COI; LISC). LM: without precise locality, fl. & fr. 1917–1918, *Junod* 341 (LISC; LMU).
Widely disseminated in tropical and subtropical places. In woodlands, mixed savanna woodland, grassland, riverine forest, river banks, dry stream beds, sandy soils and roadsides, 600–1440 m.

Var. **virens** Fenzl in Ann. Wien. Mus. **1**: 358 (1836).—Burtt Davy, F.P.F.T. **1**: 155 (1926).—Adamson in Journ. S. Afr. Bot. **27**: 128 (1961).—M. L. Gonçalves in C.F.A. **4**: 316 (1970). Type from SW. Africa.
Mollugo glinus A. Rich. var. *virens* (Fenzl) Oliv. in F.T.A. **2**: 590 (1871). Type as above.
Mollugo hirta Thunb. var. *virens* sensu Hiern, Cat. Afr. Pl. Welw. **1**: 416 (1898).—Engl., Sitz.-Ber. Konigl. Preuss. Akad. Wiss. Berl. **1906**, 2: 878 (1906).—Eyles in Trans. Roy. Soc. S. Afr. **5**: 350 (1916) pro parte.

Botswana. N: Kwebe Hills, 914 m., fl. 28.xii.1897, *Lugard* 73 (K). **Zambia.** B: Shangombo, 1036 m., fl. 9.viii.1952, *Codd* 7463 (BM; COI; K; SRGH). N: Fort Rosebery, fl. & fr. 8.ii.1959, *Watmough* 231 (BM; COI; SRGH). W: Ndola, fl. & fr. 23.vi.1955, *Fanshawe* 2343 (K; LISC). C: 19 km. E. of Lusaka, 1280 m., fl. & fr. 20.xi.1955, *King* 208 (K). E: Lundazi, Tigone Dam, 1200 m., fl. & fr. 19.xi.1958, *Robson* 655 (BM; K; LISC; SRGH). S: Namwala, fl. & fr. 17.xii.1962, *van Rensburg* 1105 (K; SRGH). **Rhodesia.** N: Urungwe, fl. & fr. 30.ii.1957, *Goodier* 425 (SRGH). W: Bulawayo, 1460 m., fl. & fr. xii.1956, *Miller* 4000 (LISC; SRGH). C: Salisbury, 1460 m., fl. & fr. ix.1919, *Eyles* 1806 (K; SRGH). E: Umtali Commonage, 1070 m., fl. & fr. 29.xii.1946, *Fisher* 1141 (SRGH). S: Nuanetsi, Lundi R., 1850 m., fl. & fr. xi.1955, *Davies* 1610 (SRGH). **Malawi.** N: Kondowe to Karonga, 610–1830 m, fl. & fr. vii.1896, *Whyte* (K). S: Zomba, *Scott Elliott* 8499 (K). **Mozambique.** N: Amaramba, Cuamba, fl. & fr. 2.viii.1934, *Torre* 484 (COI; LISC). Z: without locality, fl. & fr. vi.1858, *Kirk* (K). T: Tete, Boroma, fl. & fr. viii.1931, *Pomba Guerra* 52 (COI). MS: Gorongosa, 90 m., fl. & fr. 26.ix.1953, *Chase* 5087 (BM; COI; LISC; SRGH).
The same distribution and habitat as var. *lotoides*.

2. **Glinus oppositifolius** (L.) DC. in Bull. Herb. Boiss., Sér. 2, **1**: 559 (1901).—Burtt Davy, F.P.F.T. **1**: 155 (1926).—Pax & Hoffm. in Engl. & Prantl, Pflanzenfam. ed. 2, **16c**: 222 (1934).—Hauman in F.C.B. **2**: 108 (1951).—Brenan in Mem. N.Y. Bot. Gard. **8**, 3: 444 (1954).—Keay in F.W.T.A. ed. 2, **2**: 135 (1954).—Mogg in Macnae & Kalk, Nat. Hist. Inhaca I., Moçamb.: 144 (1958).—Jeffrey in F.T.E.A., Aizoaceae: 13, fig. 5 (1961).—Adamson in Journ. S. Afr. Bot. **27**: 126 (1961).—Binns, H.C.L.M.: 16 (1968).—M. L. Gonçalves in Garcia de Orta **13**: 379 (1965); in C.F.A. **4**: 316 (1970).—J. H. Ross, Fl. Natal: 161 (1972). TAB. **134** fig. A. Type from Ceylon.

Mollugo oppositifolius L., Sp. Pl. **1**: 89 (1753).—Hiern, Cat. Afr. Pl. Welw. **1**: 416 (1898). Type as above.

Mollugo spergula L., Syst. Nat. ed. 10: 881 (1759).—DC., Prodr. **1**: 391 (1824).— Oliv. in F.T.A. **2**: 590 (1871).—Engl., Pflanzenw. Ost-Afr. **C**: 175 (1895).—Burkill in Johnston, Brit. Centr. Afr.: 248 (1897). Type from India.

Pharnaceum mollugo L., Mant. Pl. Alt. **2**: 561 (1771) *nom. illegit.* Type as for *Mollugo spergula.*

Mollugo parviflora Ser. in DC., loc. cit. Type from India.

Pharnaceum oppositifolium (L.) Spreng., Syst. Veg. **1**: 949 (1825). Type as for *Glinus oppositifolius.*

Mollugo denticulata Guill. & Perr., Fl. Senegamb. Tent. **1**: 45 (1831). Type from Senegal.

Glinus mollugo Fenzl in Ann. Wien. Mus. **1**: 359 (1836).—Sond. in Harv. & Sond., F.C. **1**: 137 (1860).—Peters, Reise Mossamb. Bot.: 142 (1861), *nom. illegit.*

Glinus denticulatus (Guill. & Perr.) Fenzl, tom. cit.: 361 (1836). Type as for *Mollugo denticulata.*

Glinus spergula (L.) Steud., Nom. Bot. ed. 2, **1**: 688 (1840).—Pax in Engl. & Prantl, op. cit. **3**, 1b: 40 (1889).—A. Peter, Fl. Deutsch Ost-Afr. **2**: 257 (1938). Type as for *Mollugo spergula.*

Mollugo glinoides A. Rich., Tent. Fl. Abyss. **1**: 48 (1847). Type from Ethiopia.

Mollugo serrulata Sond. in Linnaea **23**: 15 (1850). Type from S. Africa (Natal).

Mollugo hirta sensu Hiern, Cat. Afr. Pl. Welw. **1**: 415 (1898) pro parte quoad specim. *Welwitsch* 1265.

Decumbent or prostrate creeping herb, glabrous or with appressed hairs especially on youngest parts or in axils, with many stems up to 50 cm. long. Leaves opposite or in whorls of 4–6, unequal; laminas 5–40 × 2–18 mm., lanceolate, obovate, spathulate, elliptic-oblong to almost orbicular, entire or obscurely serrulate in upper ½; apex usually acute, often mucronate; base cuneate, gradually narrowed into a short pedicel; young leaves with a few hairs on midrib, glabrescent, sometimes with many appressed woolly crisped white hairs in axils and on lamina. Flowers small, whitish or cream, in groups of 1–∞ at nodes, on pedicels 2–14 mm. long. Perianth-segments 5, up to 5 mm. long, free, persistently surrounding the fruit. Staminodes 0. Stamens 3–10. Ovary of 3 united carpels; stigmas 3. Fruit up to 5 mm. long. Seeds dark brown with rows of small tubercles; aril less than ⅓ the length of the seed.

Flowers 1–8 at each node:

Pedicels usually more than 5 mm. long; leaves petiolate, without tufts of woolly hairs in the axils, the lamina attenuate at base only - - - var. *oppositifolius*
Pedicels usually not more than 5 mm. long; leaves subsessile, with crisped woolly hairs in axils, lamina attenuate at both ends - - - - - var. *lanatus*
Flowers usually more than 8 at each node; pedicels less than 5 mm. long
var. *glomeratus*

Var. **oppositifolius**.

Botswana. N: Mutsoi, fl. & fr. 9.xi.1967, *Lambrecht* 420 (LISC; SRGH). **Zambia.** B: Mongu, L. Kande, fl. & fr. 11.xi.1959, *Drummond & Cookson* 6335 (LISC; SRGH). N: Mbala (Abercorn), Sumbu Bay L., 780 m., fl. & fr. 29.xii.1963, *Richards* 18710 (K). S: Victoria Falls, fl. & fr. 5.i.1955, *Robinson* 1058 (K). **Rhodesia.** E: Chipinga, S. of Giriwayo Camp, fl. & fr. 13.iii.1957, *Phipps* 616 (SRGH). S: Chibi, Lundi R., 760 m., fl. & fr. 30.x.1955, *Wild* 4692 (LISC; SRGH). **Malawi.** S: Chikwawa, Lower Mwanza R., 180 m., fl. & fr. 3.x.1946, *Brass* 17931 (K; SRGH). **Mozambique.** N: between Mocímboa da Praia and Quiterajo, fl. & fr. 12.ix.1948, *Barbosa* 2097 (LISC). Z: Alto Molócuè, Mamala, 250 m., fl. & fr. 20.xii.1967, *Torre & Correia* 16648 (BR; EA; J; LISC; LMA). MS: Gorongosa, Vanduzi R., fl. & fr. 31.x.1965, *Balsinhas* 1024 (COI; LISC; LMA). SS: Gaza, Chongoene, fl. & fr. 9.vii.1948, *Myre* 63 (LMA). LM: between Manhiça and Palmeira, fl. & fr. 15.i.1965, *Rodrigues & al.* 188 (COI; LISC; LMU).

Widely disseminated in tropical areas. In woodlands, savannas, wet hollows and river banks, usually on sandy soils, 40–780 m.

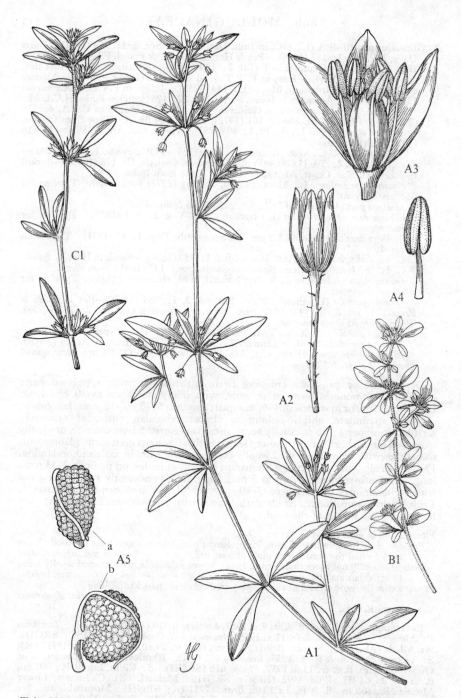

Tab. 134. A.—GLINUS OPPOSITIFOLIUS var. OPPOSITIFOLIUS. A1, habit (× ⅔); A2, flower (×6); A3, flower, facing perianth-segments removed (×12); A4, stamen (×18); A5, seed, (a) lateral view, (b) posterior view (×40), all from *Barbosa* 2097. B.—GLINUS OPPOSITIFOLIUS var. LANATUS. B1, habit (× ⅔), from *Drummond & Rutherford-Smith* 7677. C.—GLINUS OPPOSITIFOLIUS var. GLOMERATUS. C1, habit (× ⅔), from *Fanshawe* 3561.

Var. **lanatus** Hauman in Bull. Jard. Bot. Brux. **19**: 446 (1949); in F.C.B. **2**: 109 (1951).—
M. L. Gonçalves in Garcia de Orta **13**: 379 (1965); in C.F.A. **4**: 317 (1970).
TAB. **134** fig. B. Type from Zaire.

Zambia. N: Mbesuma, fl. & fr. 11.x.1960, *Robinson* 3950 (K: SRGH). E: Luangwa
R., NE. side, fl. & fr. 5.ix.1947, *Greenway & Brenan* 8045 (K). **Rhodesia.** S: Nuanetsi,
near Malipate, fl. & fr. 2.v.1961, *Drummond & Rutherford-Smith* 7677 (LISC; SRGH).
Malawi. S: Mlanje, fl. & fr. 14.xi.1955, *Jackson* 1765 (K; SRGH). **Mozambique.**
MS: Gorongosa, 41 m., fl. & fr. 3.xi.1963, *Torre & Paiva* 9002 (COI; EA; J; LISC).
Also in Zaire and Angola. In savannas, and wet hollows; beside roads and on river
banks.

Var. **glomeratus** M. L. Gonçalves in Garcia de Orta **13**: 379 (1965) (" glomeratum ");
in C.F.A. **4**: 317 (1970). TAB. **134**. fig. C. Type from Angola.
Mollugo hirta var. *virens* sensu Hiern, Cat. Afr. Pl. Welw. **1**: 416 (1898) non
Glinus lotoides var. *virens* Fenzl in Ann. Wien. Mus. **1**: 358 (1836).

Zambia. W: Mwinilunga, Mujileshi R., 1290 m., fl. & fr. 8.xi.1962, *Richards* 16953
(K).
Also in Angola. In swampy ground between ponds, 1290 m.

3. **Glinus bainesii** (Oliv.) Pax in Engl. & Prantl, Pflanzenfam. **3**, 1b: 40 (1889).—Burtt
Davy, F.P.F.T. **1**: 155 (1926).—Pax & Hoffm. in Engl. & Prantl, op. cit. ed. 2, **16c**:
222 (1934).—Friedr. in Prodr. Fl. SW. Afr. **26**: 6 (1966). Type: Rhodesia.
"Koobie to N. Shaw Valley ", *Baines* (GRA, holotype).
Mollugo bainesii Oliv. in F.T.A. **2**: 590 (1871). Type as above.

Prostrate subsucculent shrub up to 40 cm. high, the young parts covered with
spreading hairs and multicellular prickly pubescence, the older parts glabrescent;
stems rather stout, sparsely branched. Leaves verticillate; laminas 2–25 ×
1–6 mm., lanceolate or oblanceolate, entire, acute or mucronate at apex, gradually
narrowed at the base into a petiole up to 4 mm. long; margins slightly revolute
when dry; petiole, midrib (at least on abaxial surface) and occasionally margins
with prickly multicellular pubescence. Flowers white or pale mauve, in groups
of 3–9 per node, most commonly 6, with pedicels lengthening to 2 cm. at the
fruiting stage, usually with some hairs or prickles. Perianth-segments 5, up to
8 mm. long, free, the outer acute, mucronate, with hairs or prickles on the back,
the inner with broad membranous margins. Staminodes 0–7, simple or 3-fid.
Stamens ∞. Ovary of 3 united carpels; stigmas 3. Fruit up to 4 mm. long.
Seeds brown, with smooth or obtusely tuberculate ridges and a small aril.

Botswana. N: Sigara Pan, 48 km. W. of mouth of Nata R., 896 m., fl. & fr. 25.iv.1957,
Drummond & Seagrief 5216 (LISC; SRGH). **Rhodesia.** E: Chiribira Falls, fl. &
fr. 6.vi.1950, *Wild* 28207 (LISC; SRGH). S: Ndanga, Chitsa's Kraal, Sabi R., fl. & fr.
5.vi.1950, *Chase* 2307 (BM; COI; LISC; SRGH). **Mozambique.** MS: Mossurize,
Maringa, Sabi R., fl. & fr. 25.vi.1950, *Chase* 2568 (BM; COI; LISC; SRGH). SS:
Baixo Limpopo, Barragem Village, fl. & fr. 20.xi.1957, *Barbosa & Lemos* 8222 (COI;
LISC). LM: Lourenço Marques, fl. & fr. 7.xii.1897, *Schlechter* 11643 (COI).
Also in SW. Africa and S. Africa (Transvaal). In muddy soil, sandy river banks and dry
river beds, 123–930 m.

6. MOLLUGO L.

Mollugo L., Sp. Pl. **1**: 49 (1753); Gen. Pl. ed. 5: 39 (1754).

Erect or procumbent annual herbs. Leaves basal or in whorls on stem, linear,
oblanceolate, obovate or spathulate, glabrous, entire, the radical ones often rosulate,
the cauline sometimes lacking; stipules small or absent, often caducous. Inflorescences cymose, laxly dichasial or umbelliform. Flowers ⚥, greenish or
whitish, inconspicuous. Perianth-segments 5, free. Stamens 3–5, rarely more,
hypogynous, free. Ovary superior, of 3–5 united carpels, 3–5-locular; ovules
many per locule; placentation axile. Fruit a loculicidal capsule. Seeds many,
compressed, triangular or ovate, estrophiolate.
A pantropical genus of c. 15 species.

Plants with both basal leaves and whorls of persistent linear stem leaves; basal leaves often
    withered away by flowering time - - - - - - - 1. *cerviana*
Plants with basal leaves only, the stems bearing minute bracts; basal leaves persistent
                                 2. *nudicaulis*

1. **Mollugo cerviana** (L.) Ser. in DC., Prodr. **1**: 392 (1824).—Sond. in Harv. & Sond., F.C. **1**: 188 (1860).—Peters, Reise Mossamb. Bot.: 142 (1861).—Oliv. in F.T.A. **2**: 591 (1871).—Engl., Pflanzenw. Ost-Afr. **C**: 175 (1895).—Warb., Kunene-Samb.-Exped. Baum: 232 (1903).—R. E. Fr., Wiss. Ergebn. Schwed. Rhod.-Kongo-Exped. **1**: 34 (1914).—Eyles in Trans. Roy. Soc. S. Afr. **5**: 350 (1916).—Burtt Davy, F.P.F.T. **1**: 155 (1926).—Hutch. & Dalz. in F.W.T.A. **1**: 114 (1927).—Pax & Hoffm. in Engl. & Prantl, Pflanzenfam. ed. 2, **16c**: 227 (1934).—A. Peter, Fl. Deutsch Ost-Afr. **2**: 259 (1938).—Hutch., Botanist in S. Afr.: 672 (1946).—Hauman in F.C.B. **2**: 111 (1951).—Suesseng. & Merxm. in Proc. & Trans. Rhod. Sci. Ass. **43**: 183 (1951).—Wild, Fl. Vict. Falls: 144 (1953).—Keay in F.W.T.A. ed. 2, **1**: 135 (1954).—Adamson in Journ. S. Afr. Bot. **24**: 13 (1957).—Jeffrey in F.T.E.A., Aizoaceae: 16 (1961).—Friedr. in Prodr. Fl. SW. Afr. **26**: 16 (1966).—Binns, H.C.L.M.: 16 (1968).—M. L. Gonçalves in C.F.A. **4**: 319 (1970). Type from Europe.

Pharnaceum cerviana L., Sp. Pl. **1**: 272 (1753). Type as above.

A small glabrous many-stemmed annual herb with slender stems up to 20 cm. high, rather rigid, pale brownish, upright or ascending. Leaves sessile, often glaucous; basal ones 2–25 × 0·3–5 mm., linear, spathulate, obovate or oblanceolate, forming a rosette; cauline ones 3–18 mm. long, in whorls of c. 6 per node, linear. Inflorescences terminal and axillary, sessile or pedunculate, umbelliform, 1–4-flowered; peduncle up to 20 mm. long. Flowers greenish-white, on slender, straight, rather rigid pedicels 5–18 mm. long, usually exceeding the leaves. Perianth-segments 1–3 mm. long, green, sometimes with a brown keel. Stamens (3)5(10). Carpels 3 with 3 short stigmas. Fruit ± globose, as long as the perianth. Seeds brown, compressed, triangular in outline, shining.

Basal leaves linear; peduncles usually absent - - - - - var. *cerviana*
Basal leaves spathulate, obovate or oblanceolate; peduncles usually present
var. *spathulifolia*

Var. **cerviana.**

Mollugo cerviana var. *linearis* Fenzl in Ann.Wien. Mus. **1**: 379 (1836). Type from N. Africa.

Mollugo tenuissima A. Peter in Abh. Königl. Ges. Wiss. Göttingen, N.F. **13**: 254 (1928); Fl. Deutsch, Ost-Afr. **2**, 2: 28, t. 35 fig. 1 (1932); op. cit., **3**: 260 (1938). Type from Tanzania.

Caprivi Strip. Katima Mulilo, on road to Linyanti, 914 m., fl. & fr. 26.xii.1958, *Killick & Leistner* 3099 (K; SRGH). **Botswana.** N: Ngamiland, Kinkogo Tsetse Camp, Okavango, 930 m., fl. & fr. 16.iii.1961, *Richards* 14717 (K; SRGH). SW: Tsabong, fl. & fr. 25.ii.1960, *Wild* 5152 (SRGH). SE: Mahalapye, fl. & fr. 31.i.1912, *Rogers* 6102 (SRGH). **Zambia.** B: Masese, fl. & fr. 10.iii.1960, *Fanshawe* 5450 (SRGH). C: Kapiri Mposhi, fl. & fr. 11.viii.1919, *Burtt Davy* 18088 (BM). S: Choma, fl. & fr. 14.i.1963, *van Rensburg* 1211 (K; SRGH). **Rhodesia.** N: Gokwe, fl. & fr. 24.iv.1962, *Bingham* 239 (SRGH). W: Wankie, Matetsi, 853 m., fl. & fr. iii.1918, *Eyles* 1281 (BM; K). C: Salisbury, fl. & fr. 12.iv.1967, *Mavi* 281 (LISC; SRGH). E: Umtali, fl. & fr. i.1946, *Chase* 177 (BM; SRGH). **Mozambique.** LM: Maputo, Umbeluzi Experimental Station, fl. & fr. 23.xi.1946, *Valente* 3100 (LMA).

Widespread in tropics and subtropics of the Old World and introduced into America. In grassland, on sand dunes, sandy banks, beside rivers, as a weed in gardens and cultivated ground, 853–1000 m.

Var. **spathulifolia** Fenzl in Ann. Wien. Mus. **1**: 379 (1836).—Sond. in Harv. & Sond., F.C. **1**: 188 (1860).—Jeffrey in F.T.E.A., Aizoaceae: 16 (1961).—M. L. Gonçalves in C.F.A. **4**: 320 (1970). Type from India.

Pharnaceum umbellatum Forssk., Fl. Aegypt.-Arab.: 58 (1775). Type from Arabia.

Mollugo umbellata (Forssk.) Ser. in DC., Prodr. **1**: 393 (1824). Type as above.

Mollugo spathulifolia (Fenzl) Dinter in Fedde, Repert. **19**: 236 (1923). Type as for *Mollugo cerviana* var. *spathulifolia*.

Zambia. N: Mbala (Abercorn) Kasaba Game (?Resrve), fl. & fr. 17.ii.1959, *McCallum-Webster* 634 (K). E: near Changwe, between Petauke and Mwape, 650 m., fl. & fr. 16.xii.1958, *Robson* 960 (BM; K; LISC; SRGH). **Malawi.** C: road near L. Nyasa Hotel, near Salima, 480 m., fl. & fr. 14.ii.1959, *Robson* 1605 (K; LISC; SRGH). S: Nsalima, Zomba, fl. & fr. 23.iv.1955, *Banda* 89 (BM; K; SRGH). **Mozambique.** N: Nacala, Nacala Nova, c. 30 m., fl. & fr. 21.xi.1963, *Correia* 42 (LISC). Z: Alto Molócuè, 250 m., fl. & fr. 20.xii.1967, *Torre & Correia* 16647 (LISC). T: Tete, Songo-Tete road, fl. & fr. 1.v.1972, *Macedo* 5252 (LISC; LMA). MS: Marromeu, Chupanga, fl. & fr. xii.1862, *Kirk* (K). LM: Magude, Delagoa Plantation, fl. & fr. 29.xi.1944, *Mendonça* 3113 (LISC).

Widespread in Old World tropics. In wooded savannas, beside rivers on sandy banks, dunes, roadsides and cultivated ground, 30–650 m.

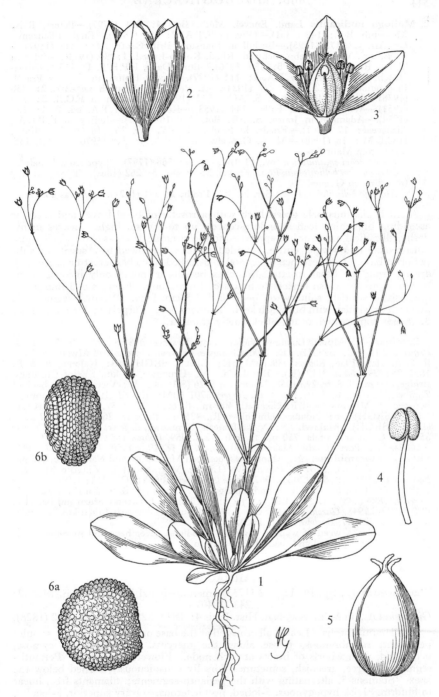

Tab. 135. MOLLUGO NUDICAULIS. 1, habit ($\times\frac{2}{3}$); 2, flower ($\times10$); 3, flower, facing perianth-segments removed ($\times10$); 4, stamen ($\times18$); 5, fruit ($\times18$); 6, seed (a) lateral view, (b) posterior view ($\times40$), 1 & 6 from *Robson* 959, 2–5 from *Pereira & Marques* 834.

2. **Mollugo nudicaulis** Lam., Encycl. Méth. Bot. **4**: 234 (1797).—Peters, Reise Mossamb. Bot.: 142 (1861).—Oliv. in F.T.A. **2**: 591 (1871).—Engl., Pflanzenw. Ost-Afr. **C**: 175 (1895).—Burkill in Johnston, Brit. Centr. Afr.: 248 (1897).— R. E. Fr., Wiss. Ergebn. Schwed. Rhod.-Kongo-Exped. **1**: 34 (1914).—Eyles in Trans. Roy. Soc. S. Afr. **5**: 350 (1916).—Burtt Davy, F.P.F.T. **1**: 155 (1926).— Hutch. & Dalz. in F.W.T.A. **1**: 114 (1927).—Pax & Hoffm. in Engl. & Prantl, Pflanzenfam. ed. 2, **16c**: 227 (1934).—A. Peter, Fl. Deutsch Ost-Afr. **2**: 258 (1938).—Hutch., Botanist in S. Afr.: 672 (1946).—Hauman in F.C.B. **2**: 110 (1951).—Wild, Fl. Vict. Falls: 144 (1953).—Keay in F.W.T.A. ed. 2, **1**: 134 (1954).—Adamson in Journ. S. Afr. Bot. **24**: 16 (1957).—Jeffrey in F.T.E.A., Aizoaceae: 17 (1961).—Friedr. in Prodr. Fl. SW. Afr. **26**: 16 (1966).—Binns, H.C.L.M.: 16 (1968).—M. L. Gonçalves in C.F.A. **4**: 318 (1970). TAB. **135**. Type from Mauritius.

Pharnaceum spathulatum Sw., Fl. Ind. Occ. **1**: 568 (1797). Type from Jamaica.
Pharnaceum bellidifolium Poir., Encycl. Méth. Bot. **5**: 262 (1804). Type probably from French Guiana.
Mollugo bellidifolia (Poir.) Ser. in DC., Prodr. **1**: 391 (1824). Type as above.

A small glabrous pale green herb with a rosette of basal leaves and erect or ascending branching leafless inflorescences, up to 30 cm. high. Leaves many, fleshy, all radical, more or less appressed to the ground; laminas 6–67 × 2–20 mm., oblanceolate or obovate, entire, rounded at apex, narrowed to a sessile or sub-petiolate base. Inflorescence of several 2–3-forked branches, erect or ascending, arising from the basal rosette; branches bearing at nodes only inconspicuous membranous brownish lanceolate bracts c. 2 mm. long. Flowers c. 3 mm. long, greenish or whitish, solitary, on pedicels 1–1·5 cm. long. Perianth-segments 5, greenish or whitish with brownish keel. Stamens 3–5. Carpels, styles and stigmas 3. Seeds compressed, ovate, black, finely granular.

Zambia. N: Mpika, Luangwa Valley Game Reserve, 608 m., fl. & fr. 21.ii.1967, *Prince* 292 (LISC; SRGH). E: near Changwe, between Petauke and Mwape, 650 m., fl. & fr. 16.xii.1958, *Robson* 959 (BM; K; LISC; SRGH). S: Kalomo, fl. & fr. 16.ii.1965, *Fanshawe* 9112 (K). **Rhodesia.** N: Urungwe, 18 km. SE. of Chirundu Bridge, 460 m., fl. & fr. 29.i.1958, *Drummond* 5326 (SRGH). W: Victoria Falls, bank of Zambesi, 914 m., fl. & fr. 12.ii.1912, *Rogers* 5701 (BM; K; SRGH). E: Chipinga, between R. Mwsaowe and R. Cikariati, 400 m., fl. & fr. 22.i.1957, *Phipps* 106 (COI; LISC; SRGH). S: Ndanga, Umtilikwe R., 425 m., fl. & fr. 26 i.1949, *Wild* 2780 (LISC; SRGH). **Malawi.** N: Nyika Plateau, Nymkowa, fl. & fr. xi.1903, *McClounie* 35 (K). C: near Chitala, 750 m., fl. & fr. 12.ii.1959, *Robson* 1552 (BM; K; LISC; SRGH). S: Port Herald, Makanga, 80 m., fl. & fr. 19.iii.1960, *Phipps* 2557 (BM; SRGH). **Mozambique.** N: Montepuez, between Montepuez and Namuno, c. 480 m., fl. & fr. 31.xii.1963, *Torre & Paiva* 9798 (LISC; MO; SRGH). Z: Alto Molócuè, Namala, c. 250 m., fl. & fr. 20.xii.1967, *Torre & Correia* 16646 (COI; EA; LISC; LMA; PRE). T: Tete, between Tete and Changara, c. 350 m., fl. & fr. 5.ii.1970, *Torre & Correia* 17824 (COI; LISC; LMA; P). MS: Chimoio, between Garuzo and Bandula, fl. & fr. 23.ii.1948, *Garcia* 324 (LISC). LM: Maputo, Goba, near Libombos Fountain, fl. & fr. 18.xii.1947, *Barbosa* 760 (LISC).

Pantropical. A weed of wooded savanna, woodland, river beds and margins, waste places, cultivated ground and roadsides, 50–914 m.

7. PHARNACEUM L.

Pharnaceum L., Sp. Pl. **1**: 272 (1753) emend.—Fenzl in Ann. Wien. Mus. **2**: 243 (1840).
Ginginsia DC., in Mém. Soc. Nat. Hist. Paris **4**: 184 (1828); Prodr. **3**: 362 (1828).

Small annual herbs. Leaves all crowded at the base or whorled, linear; stipules persistent, membranous, pilose along the margins. Inflorescences cymose, peduncled, with whorls of bracts at each node. Flowers pedicelled. Perianth-segments 5, free, greenish, sometimes dorsally conspicuously horned below the apex. Stamens 5, alternating with the perianth-segments; filaments free, linear to filiform. Disk hypogynous, 5-lobed, nectariferous. Ovary superior, 3-locular, subglobose, 3-lobed, the locules multiovulate. Stigmas 3, short. Fruit a membranous capsule, 3-locular, loculicidally 3-valved. Seeds several in each locule, globose-lenticular with a thin dorsal ridge, without appendage at the apex of the funicle; testa glossy brown, finely granular.

A genus of c. 25 species endemic to southern Africa.

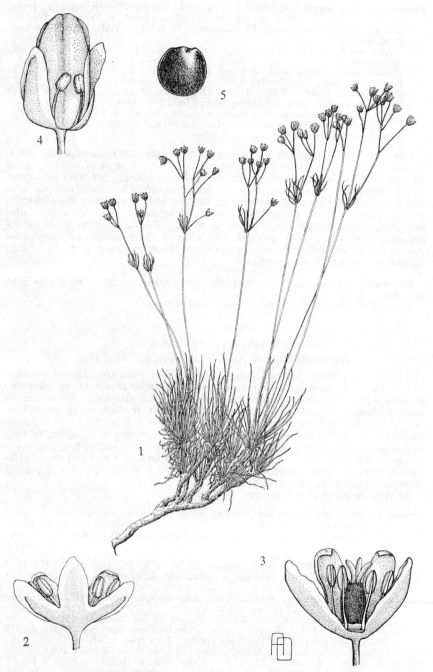

Tab. 136. PHARNACEUM BREVICAULE. 1, habit (×⅔); 2, flower in lateral view (×9);
3, flower partially dissected showing stamens, hypogynous disc and ovary (×10),
1–3 from *Masson* sn; 4, flower with developing capsule (×6) from *Schlechter* 4020;
5, seed (×20) from *Masson* sn.

Pharnaceum brevicaule (DC.) Bartl. in Linnaea **7**: 625 (1832).—Adamson in Journ. S. Afr. Bot. **8**: 273 (1942); op cit. **24**: 39 (1957).—Friedr. in Prodr. Fl. SW. Afr. **26**: 19 (1966).—Guillarmod, Fl. Lesotho: 169 (1971). TAB. **136**. Type from S. Africa.
    Ginginsia brevicaulis DC. in Mém. Soc. Hist. Nat. Paris **4**: 187, t. 17 (1828). Type as above.
    Pharnaceum zeyheri Sond. in Harv. & Sond., F.C. **1**: 141 (1860).—Gibbs in Journ. Linn. Soc., Bot. **37**: 425 (1906).—Eyles in Trans. Roy. Soc. S. Afr. **5**: 350 (1916).—Burtt Davy, F.P.F.T. **1**: 154 (1926).—Pax & Hoffm. in Engl. & Prantl, Pflanzenfam. ed. 2, **16c**: 227 (1934).—Type from SW. Africa.
    Pharnaceum dichotomum sensu Dinter in Fedde, Repert. **22**: 379 (1926).
    Pharnaceum merxmuelleri Friedr. in Mitt. Bot. Staatss. München **2**: 64, t. A (1955). Type from SW. Africa.

Low-growing annual with 1 to several tufts of leaves at or near ground level. Stems woody, often covered by persistent stipules. Leaves in whorls, up to 22 mm. long, linear, sometimes with a thickened margin, set with deciduous setae; stipules 1–4 mm. long, white or yellowish, often undulate, opaque, with coarse hairs at the apex, persistent after leaf-fall and surrounding the leaf-tufts. Inflorescences slender, pedunculate, branched at nodes, the ultimate pseudo-racemose, with whorled linear bracts at nodes. Pedicels capillary. Perianth-segments 5, c. 3 mm. long, greenish with a broad white margin. Nectariferous disk of 5 white or pink acute or subacute lobes. Fruit as long as perianth. Seeds subglobose, brown, finely granular.

Rhodesia. W: Matopos, *Gibbs* 45 (BM). C: Charter, fr. 24.x.1930, *Harvie* 3644 (SRGH).
Also in SW. Africa and S. Africa. In dry sandy soil.

8. HYPERTELIS Fenzl

Hypertelis Fenzl in Ann. Wien. Mus. **2**: 261 (1839).

Annual or perennial glaucous herb, sometimes woody at base. Leaves alternate or whorled, often crowded, linear, entire, succulent, with persistent, membranous stipules adnate to the leaf-base and sometimes stem-clasping. Inflorescences axillary, long-pedunculate, simply umbellate, with few or many pedicellate flowers but no involucral bracts. Perianth-segments 5, free, the outer 2–3 sub-succulent, the inner narrower and membranous. Stamens 5–15, alternate with perianth-segments, singly or (when more than 5) in pairs or fascicles, hypogynous. Nectariferous disk absent. Ovary superior, syncarpous, of 3 or 5 carpels, 3- or 5-locular, with numerous axile ovules; stigmas 3 or 5, short. Fruit a loculicidal capsule. Seeds triangular in outline.
    An African genus of c. 5 species.

Annual or short-lived perennial; stamens 5–8; peduncles, pedicels and perianth-
    segments with wart-like glands - - - - 1. *salsoloides* var. *mossamedensis*
Perennial; stamens 5; peduncles, pedicels and perianth-segments more or less smooth
                                               2. *bowkeriana*

1. **Hypertelis salsoloides** (Burch.) Adamson in Journ. S. Afr. Bot. **24**: 52 (1957).—Friedr. in Prodr. Fl. SW. Afr. **26**: 8 (1966).—M. L. Gonçalves in C.F.A. **4**: 321 (1970). Type from S. Africa.
    Pharnaceum salsoloides Burch., Trav. Int. S. Afr. **1**: 286 (1822).—Hutch., Botanist in S. Afr.: 672 (1946). Type as above.
    Pharnaceum verrucosum Eckl. & Zeyh., Enum. Pl. Afr. Austr. Extratrop. **2**: 286 (1836).—Oliv. in F.T.A. **2**: 592 (1871).—Eyles in Trans. Roy. Soc. S. Afr. **5**: 350 (1916).—Burtt Davy, F.P.F.T. **1**: 154 (1926). Type from S. Africa.
    Hypertelis verrucosa (Eckl. & Zeyh.) Fenzl in Ann. Wien. Mus. **2**: 262 (1839).—Sond. in Harv. & Sond., F.C. **1**: 144 (1860).—Pax & Hoffm. in Engl. & Prantl, Pflanzenfam. ed. 2, **16c**: 227, fig. 100 (1934).—Jacobsen, Handb. Sukk. Pl. **2**: 635, fig. 837 & 838 (1960). Type as above.

Var. **mossamedensis** (Welw. ex Hiern) M. L. Gonçalves in C.F.A. **4**: 321 (1970). Type from Angola.
    Pharnaceum salsoloides var. *mossamedensis* Welw. ex Hiern, Cat. Afr. Pl. Welw. **1**: 418 (1898). Type as above.

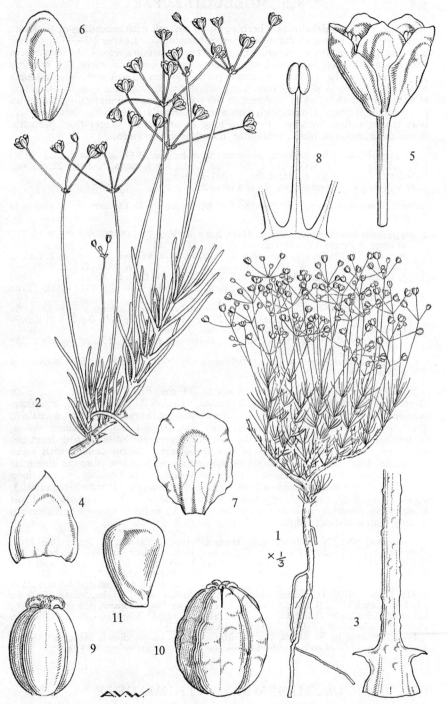

Tab. 137. HYPERTELIS BOWKERIANA. 1, habit (×⅓); 2, inflorescences (×1); 3, leaf-base (×4); 4, bract (×12); 5, flower (×8); 6, outer sepal (×8); 7, inner sepal (×8); 8, stamen (×12); 9, ovary (×12); 10, fruit (×10); 11, seed (×40), all from *Greenway & Rawlins* 8912. From F.T.E.A.

Annual or perennial shrubby herb up to 30 cm. tall with ascending or prostrate stems woody at base and many short erect branchlets. Leaves 1·5–3·5 cm. long, crowded, glaucous, linear, succulent, usually erect; stipules setose. Inflorescences with stout peduncles 3–10 cm. long, smooth or more commonly with few or many projecting glandular warts on the upper part. Flowers up to 4 mm. long, greenish or whitish, on pedicels 7–17 mm. long, usually with wart-like glands, ascending at anthesis but deflexed in bud and at the fruiting stage. Perianth-segments up to 4 mm. long, greenish or whitish with wart-like glands on the back. Stamens 5–8, with bright yellow anthers. Fruit as long as or slightly longer than perianth. Seeds dark brown or black, triangular in outline and smooth.

Mozambique. SS: Chibuto, Maqueze, fl. & fr. 4.xii.1944, *Mendonça* 3248 (COI; EA; LISC; LMU; MO; PRE; SRGH). LM: Marracuene, fl. & fr. 22.vii.1965, *Caldeira & Marques* 598 (COI; LISC; LMU).
Also in Angola (Moçamedes). In salt marshes.

Note. The var. *salsoloides* is distinguished by having up to 15 stamens; it occurs in S. Africa.

2. **Hypertelis bowkeriana** Sond. in Harv. & Sond., F.C. **1**: 145 (1860).—Pax & Hoffm. in Engl. & Prantl, Pflanzenfam. ed. 2, **16c**: 227 (1934).—Adamson in Journ. S. Afr. Bot. **24**: 55 (1957).—Jacobsen, Hand. Sukk. Pl. **2**: 634 (1960).—Jeffrey in F.T.E.A., Aizoaceae: 18, fig. 6 (1961).—Friedr. in Prodr. Fl. SW. Afr. **26**: 8 (1966).— J. H. Ross, Fl. Natal: 162 (1972). TAB. **137**. Type from S. Africa.
    Pharnaceum suffruticosum Bak. in Journ. Linn. Soc., Bot. **20**: 151 (1883). Type from Madagascar.
    Mollugo suffruticosa A. Peter in Abh. Königl. Ges. Wiss. Göttingen, N.F. **13**, 2: 54 (1928); Fl. Deutsch Ost-Afr. **2**, 2: 29, t. 34 (1932); op. cit. **3**: 259 (1938).— Brenan, T.T.C.L.: 349 (1949). Type from Tanzania.
    Mollugo suffruticosa forma *annua* A. Peter, op. cit.: 29 (1932); op. cit.: 259 (1938). Type from Tanzania.
    Hypertelis suffruticosa (Bak.) Adamson, loc. cit. Type as for *Pharnaceum suffruticosum*.

Shrubby perennial glaucous herb up to 30 cm. high, woody at base, with ascending, procumbent or creeping branches. Leaves 23–55 mm. long, alternate, subopposite or crowded in axillary fascicles on short side branches, linear, succulent, glaucous, subcylindrical, often with wart-like glands; stipules membranous, triangular, with somewhat divaricate acute or acuminate lobes, arising from the expanded membranous margins of the leaf-bases. Inflorescences with stout peduncles 4–10 cm. long, smooth or sometimes with a few obscure glandular warts. Flowers up to 4 mm. long, greenish or whitish, on pedicels up to 1 mm. long ascending at first but deflexed in fruit, more or less smooth. Perianth-segments 5, up to 4 mm. long, greenish or whitish, more or less smooth. Stamens 5. Fruit longer than perianth. Seeds brown, triangular in outline, smooth and shining, usually with a distinct ridge.

Botswana. N: Ngamiland, near Tsau, Okavango, 930 m., fl. & fr. 18.iii.1961, *Richards* 14775 (K; SRGH). SE: Gaberones Dam, fl. & fr. ix.1967, *Lambrecht* 328 (LISC). **Rhodesia.** W: Mangwe, Thornville Ranch, fl. & fr. 7.xi.1951, *Plowes* 1317 (SRGH). C: Salisbury, 1470 m., fl. & fr. 17.ix.1911, *Rogers* 5520 (BM; K; SRGH). S: Nuanetsi, near Malipate, fl. & fr. 27.iv.1961, *Drummond & Rutherford-Smith* 7548 (BM; LISC; SRGH). **Mozambique.** MS: Sofala near Divinhe, 7·5 m., fl. & fr. 1.ix.1961, *Leach* 11252 (LISC; SRGH). SS: Chibuto, Capelas Farm, fl. & fr. 11.vi.1960, *Lemos & Balsinhas* 89 (BM; COI; LISC; SRGH). LM: Maputo, Umbeluzi, fl. & fr. 19.iv.1959, *Macuácua* 77 (COI; LISC; SRGH).
From Ethiopia to S. Africa, also in Madagascar. In woodland, mangrove swamp margins, sandy bed of rivers, littoral dunes and salt pan depressions, 0–1470 m.

89c. MESEMBRYANTHEMACEAE

By M. L. Gonçalves

Decumbent or prostrate creeping shrubs or herbs. Leaves opposite, sessile or shortly petiolate, more or less connate below, entire, fleshy, exstipulate. Flowers terminal or axillary, solitary or in few-flowered cymes, bisexual, actinomorphic, haplochlamydeous. Perianth-segments 4 or 5, 2–3 often leafy and larger than others. Staminodes petaloid, ∞, white or pink to purple, persistent. Stamens ∞, in several whorls, with oblong anthers. Ovary inferior, 4- or 5-locular, glandular or eglandular; stigmas as many as loculi; placentation axile or parietal. Fruit a loculicidal capsule with subwoody walls, dehiscing stellately at the apex as valves imbibe moisture. Seeds ∞.

An almost exclusively S. African family of c. 120 genera and c. 2400 species.

Perianth-segments 4; fruit a 4-locular capsule - - - - - **1. Aptenia**
Perianth-segments 5; fruit a 5-locular capsule - - - - **2. Delosperma**

Species of *Lampranthus* and *Carpobrotus* (*C. dimidiatus* (Harv.) L. Bolus and *C. edulis* (L.) N. E. Br.) are often cultivated for ornamental purposes and/or as sand-stabilizers.*

1. APTENIA N. E. Br.

Aptenia N. E. Br. in Gard. Chron. **78**: 412 *in clave* (1925); in Phillips, Gen. S. Afr. Fl. Pl.: 244 (1926).
Mesembryanthemum sensu L. f., Suppl. Pl.: 260 (1781) quoad *M. cordifolium*.
Litocarpus L. Bolus in Fl. Pl. S. Afr. **7**: t. 261 fig. 11 & 12 (1927).

Prostrate succulent short-lived perennial herbs. Leaves opposite, shortly petiolate, flat, cordate to ovate, entire, papillose. Flowers solitary in axils of branches, pedicellate, ebracteate. Perianth-segments 4, free, 2 leafy and larger than others. Staminodes united at the base into a short tube, shorter than perianth. Stamens many, arising from the tube. Ovary inferior, 4-locular; placentation axile; styles absent; stigmas 4. Fruit a 4-locular capsule; valves broader than long, with the apical part thickened abruptly so that the basal termination of the thickening appears vertical; valves with contiguous keels of expansion, without marginal wings or flaps; loculi open when mature; seeds compressed, circular in outline, tuberculate.

A genus of 2 species, known from the coastal districts of the eastern Cape Prov. and from the Pietersburg district of the Transvaal.

Aptenia cordifolia (L. f.) N. E. Br. in Journ. Bot., Lond. **66**: 139 (1928).—Pax & Hoffm. in Engl. & Prantl, Pflanzenfam. ed. 2, **16C**: 211 (1934).—L. Bolus in Journ. S. Afr. Bot. **25**: 371 (1959).—Herre, Gen. Mesembryanthemac.: 78 & 79 (1971).—J. H. Ross, Fl. Natal: 163 (1972). TAB **138**. Type from S. Africa (Cape Prov.).
Mesembryanthemum cordifolium L. f., Suppl. Pl.: 260 (1781). Type as above.
Litocarpus cordifolia (L. f.) L. Bolus in Fl. Pl. S. Afr. **7**: t. 261 fig. 11 & 12 (1927). Type as above.
Aptenia cordifolia (L. f.) Schwant. in Gartenflora **77**: 69 (1928); in Flowering Stones: 5, fig. 1a (1957).—Jacobsen, Hand. Succ. Pl. **3**: 987, fig. 1205 (1960). Type as above.
Tetracoilanthus cordifolius (L. f.) Rappa & Camerone in Lav. Ist. Bot. Giard. Col. Pal. **14**: 64 (1954). Type as above.

Plants freely branching, with prostrate cylindrical branches up to c. 60 cm. long, green later grey, minutely papillose. Leaves distant; lamina 10–40 × 5–20 mm., cordate-ovate, fresh green, fleshy, minutely papillose, with midrib prominent beneath. Flowers solitary, terminal or lateral, short-stalked, purplish-red. Perianth-segments green, the 2 larger ones leafy, ovate or elliptic, the 2 smaller ones conic-subulate, all more or less accrescent in fruit, with a turbinate tube.

---

* Editor's Note. After the volume had gone to press it became known that one species of **Lithops** was found in the F.Z. area. Prof. D. T. Cole of Witwatersrand University kindly supplied the following information: " I have indeed seen and collected *Lithops lesliei* (N. E. Br.) N. E. Br. in Botswana although only just across the border. The colony that I know is situated in the Lobatsi district, roughly 24 km. SSW. of the town. Apart from the live specimens in my collection, a number (5–6) are preserved in spirit and are lodged with the Botanical Research Institute, Pretoria." Further information suggests that this species occurs in several localities in Botswana, growing in isolated patches on kopjies or sides of mountains. We hope to give full treatment to this interesting plant in the addenda of a future volume. Thanks for help are due to Mrs. A. Renew of Mafeking, Dr. A. C. Campbell of Gaberone and Mr. P. Smith, Dept. of Agricultural Research, Botswana.

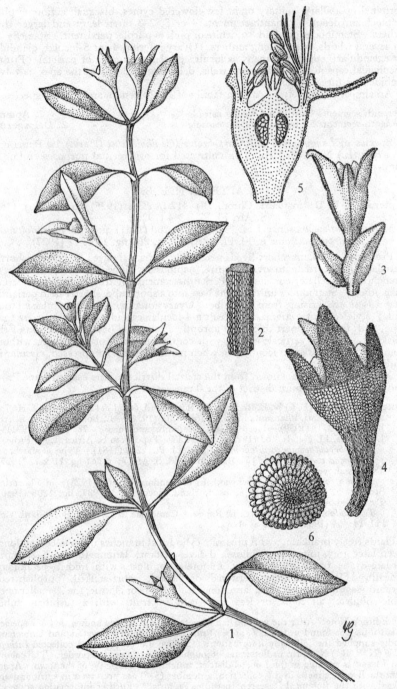

Tab. 138. APTENIA CORDIFOLIA. 1, habit (× ⅔); 2, part of stem showing transverse section (× 2); 3, flower and 2 leaves (× 2); 4, fruit (× 2); 5, flower, longitudinal section (× 4); 6, seed, lateral view (× 16), 1–2, 4, 6 from *Torre* 6531, 3 & 5 from *Leach* 12259.

Staminodes purple or pinkish-purple. Stamens with white filaments. Capsule up to 1 cm. long. Seeds dark brown.

Mozambique. LM: Sábiè, Mangulane, fr. 30.ix.1945, *Gomes e Sousa* 3344 (COI; K; LISC).
Native of S. Africa, naturalized in hot and temperate areas. In open woodland on dry sandy soils.

2. DELOSPERMA N. E. Br.

Delosperma N. E. Br. in Gard. Chron. **78**: 412 & 433 in clave (1925); in Burtt Davy, F.P.F.T. **1**: 157 (1926).
Mesembryanthemum sensu auct. non L., Sp. Pl. **1**: 480 (1753) sensu stricto.

Decumbent or prostrate creeping shrubs or perennial herbs. Leaves opposite, sessile, slightly connate, more or less finely papillose, broadly triangular to cylindrical or ovate, entire. Flowers terminal or axillary, solitary or in cymes, pedicellate, with leaf-like bracts, purple or white. Perianth-segments 5, equal or unequal, the longer ones sometimes horn-shaped or caudate. Staminodes in few series, more or less linear. Stamens whitish, sometimes hairy towards the base. Ovary 5-locular, more or less convex; glands separated, partially crenulated; placentas parietal; stigmas 5, acute, papillose. Fruit 5-locular, keels with membranous marginal wings. Seeds suborbicular, pale brown.
A large genus of c. 140 recorded species mainly from S. Africa (Cape Prov., Natal, Transvaal) and SW. Africa, but also known from tropical E. Africa (Kenya and Tanzania), Ethiopia and Eritrea.

Leaves subcylindrical; flowers white or pinkish-purple, in forked cymes:
  Flowers pinkish-purple - - - - - - - - - 1. *mahonii*
  Flowers white - - - - - - - - - - 2. *steytlerae*
Leaves ovate; flowers white, solitary - - - - - - 3. *tradescantioides*

1. **Delosperma mahonii** (N. E. Br.) N. E. Br. in Gard. Chron. **78**: 412 & 433 (1925); in Burtt Davy, F.P.F.T. **1**: 159 (1926).—L. Bolus in Journ. S. Afr. Bot. **25**: 374 (1959).—Jacobsen, Hand. Succ. Pl. **3**: 1102 (1960).—Herre & Friedr. in Mitt. Bot. Staatss. München **4**: 45 (1961).—Lavis in Journ. S. Afr. Bot. **32**: 209 & 210 (1966).—Guillarmod, Fl. Lesotho: 171 (1971). Type: Rhodesia, Melsetter, *Mahon* (K, holotype).
  Mesembryanthemum mahonii N. E. Br. in Gard. Chron. **32**: 190 (1902).—Eyles in Trans. Roy. Soc. S. Afr. **5**: 351 (1916).—Guillarmod, op. cit.: 170 (1971). Type as above.

Straggling succulent perennial herb up to 25 cm. high, woody at the base, covered with crystalline papillae. Leaves very distant and spreading on the long shoots, crowded and erect on the lateral short shoots, scarcely united at the base; lamina 20–45 × 2–3 mm., thick, subcylindrical, somewhat acute, upper surface furrowed, fresh green. Flowers several together in forked cymes, c. 25 mm. in diameter, pinkish-purple, sweet-smelling, with pedicels up to 12 mm. long. Perianth-segments unequal. Staminodes more or less linear.

Rhodesia. W: Matopos Hills, c. 1520 m., fl. & fr. xi.1902, *Eyles* 1190 (BM; K). C: Salisbury, fl. & fr. 9.xii.1910, *Leach & Baker* 10577 (SRGH). E: Melsetter, c. 1220 m., fl. & fr. 16.v.1902, *Brown* (K). S: near Zimbabwe, fl. 1.vii.1930, *Hutchinson & Gillett* 3339 (K). **Mozambique.** SS: Homoine, fr. 27.v.1947, *Hornby* 2695 (SRGH). Also in S. Africa (Transvaal). In mopane woodland, grassland and on sandy soils.

2. **Delosperma steytlerae** L. Bolus in S. Afr. Gard. **18**: 16 (1928); in Notes Mesem. **1**: 135 (1928); in Journ. S. Afr. Bot. **25**: 374 (1959).—Jacobsen, Handb. Succ. Pl. **3**: 1105 (1960).—Herre & Friedr. in Mitt. Bot. Staatss. München **4**: 45 (1961).—Lavis in Journ. S. Afr. Bot. **32**: 210 (1966). Type: Rhodesia, near Zimbabwe, *Steytler* 670 (K, holotype).

Compact shrub up to c. 15 cm. high, with rough-papillose trailing stems and internodes 5-10 mm. long. Leaves scarcely united at the base; lamina 20–45 × 3–5 mm., thick, subcylindrical, acute at ends, indistinctly keeled, the upper surface grooved. Flowers several together in forked cymes, c. 18–25 mm. in diameter, white, faintly scented, with pedicels 10–20 mm. long. Perianth-segments unequal. Staminodes linear.

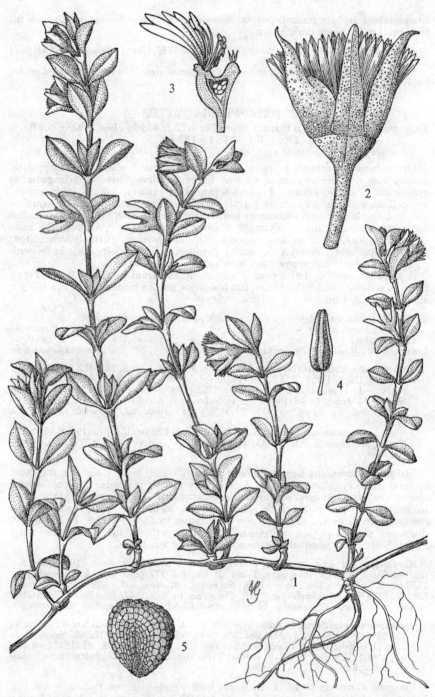

Tab. 139. DELOSPERMA TRADESCANTIOIDES. 1, habit (× ⅔); 2, flower (× 2); 3, flower, partial longitudinal section (× 2); 4, stamen (× 16); 5, seed, lateral view (× 24), 1, 4–5 from *Leach* 12260, 2–3 from *Barbosa* 758.

Rhodesia. S: Zimbabwe Ruins, c. 1220 m., fl. xii.1920, *Eyles* 2788 (SRGH).
Known only from Rhodesia. On granitic rocks.

3. **Delosperma tradescantioides** (Berg.) L. Bolus in Fl. Pl. S. Afr. **7**: t. 261 fig. 1–10
(1927); in Journ. S. Afr. Bot. **25**: 373 (1959).—Jacobsen, Handb. Succ. Pl. **3**: 1107
(1960).—Herre & Friedr. in Mitt. Bot. Staatss. München **4**: 45 (1961).—Herre,
Gen. Mesembryanthemac.: 127, colour illustr. only (1971).—J. H. Ross, Fl. Natal:
164 (1972). TAB. **139**. Type from S. Africa.
Mesembryanthemum tradescantioides Berg. in Engl., Bot. Jahrb. **45**: 224 (1910).
Type as above.

Freely branching low perennial with branches and branchlets creeping and
rooting at the nodes, stem with a light grey-brown membrane when young; young
shoots green, finely papillose, internodes c. 20 mm. long. Leaves united into a
3–4 mm. long sheath; lamina 20–30 × 10–17 mm., ovate, fleshy, tapering to terete
at the base, light green, very finely papillose, upper surface flat, the midrib often
furrowed, the lower surface indistinctly keeled. Flowers c. 15 mm. in diameter,
solitary, lateral, subsessile, white. Perianth-segments unequal. Staminodes linear.

Mozambique. LM: Namaacha, Libombos, Mpondium Mt., 800 m., fl. & fr.
22.ii.1955, *E.M. & W.* 536 (BM; LISC; SRGH).
Also in S. Africa (Transvaal and Natal). Woodland, roadsides and in sandy soils.

89d. TETRAGONIACEAE
By M. L. Gonçalves

Annual or perennial herbs or subshrubs. Leaves alternate, entire, without
stipules. Flowers solitary or fasciculate, axillary, bisexual, actinomorphic.
Perianth-segments 3–5, with the tube adnate to the ovary. Stamens 3–∞, sometimes fasciculate. Ovary inferior or semi-inferior, 2–8(9)-locular with loculus
1-ovulate. Fruit simple, dry, indehiscent, 2–8(9)-locular, pericarp winged or
horned.
A family comprising *Tetragonia* and the monospecific *Tribulocarpus* S. Moore
native to E. tropical and SW. Africa.

TETRAGONIA L.
Tetragonia L., Sp. Pl. **1**: 480 (1753); Gen. Pl. ed. 5: 215 (1754).

Annual or perennial herbs or subshrubs, often minutely papillose. Leaves
subsucculent. Flowers axillary, solitary or fasciculate, subsessile or pedunculate,
greenish or yellowish. Perianth-segments united, the tube produced above the
ovary, 3–5-lobed. Stamens 3–∞, alternate with the perianth-segments, solitary
or fasciculate. Ovary semi-inferior or inferior, (1)2–8(9)-locular, with 1 pendulous
ovule per loculus and as many styles as the loculi, linear and free. Fruit hard,
winged, horned or spiny.
A genus of some 60 species, predominantly S. African but also extending
through Australasia and western S. America.

Stamens more than twice as many as the perianth-segments; ovary wholly inferior;
 fruit (3)4-winged - - - - - - - - - 1. *calycina*
Stamens twice as many as the perianth-segments or fewer; ovary semi-inferior; fruit
 (3)4(5)-horned - - - - - - - - - 2. *tetragonoides*

1. **Tetragonia calycina** Fenzl in Harv. & Sond., F.C. **2**: 466 (1862).—Pax & Hoffm. in
Engl. & Prantl, Pflanzenfam. ed. 2, **16c**: 232 (1934).—Adamson in Journ. S. Afr.
Bot. **21**: 122 (1955).—Friedr. in Prodr. Fl. SW. Afr. **28**: 4 (1967). TAB. **140**. Type
from S. Africa (Cape Prov.).
Tetragonia macroptera Pax in Engl., Bot. Jahrb. **10**: 11 (1888). Type from
SW. Africa.

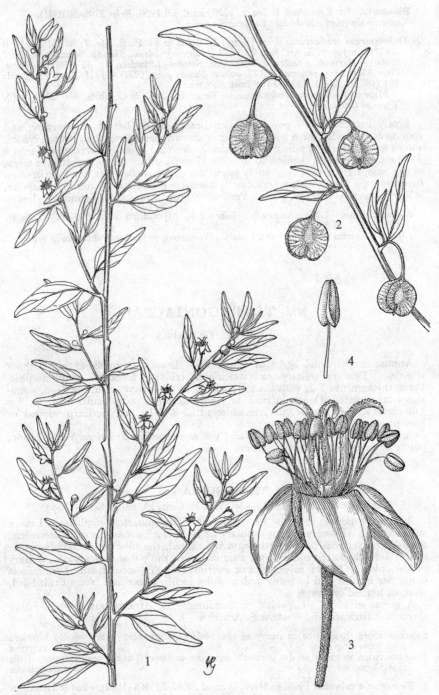

Tab. 140. TETRAGONIA CALYCINA. 1, habit (×⅔); 2, part of fruiting branch (×1½); 3, flower (×6); 4, stamen (×12), all from *Dinter* 379.

Perennial up to 50 cm. high, diffuse, much-branched. Young stems pruinose-papulose, with decurrent ridges, the older terete, pale-coloured, almost smooth. Leaves 2–2·5 × 0·2–0·6 cm., subfleshy, lanceolate, expanded, folded or slightly revolute, papulose especially on the inferior surface, narrowed at base into a petiole up to 4 mm. long. Inflorescences axillary, solitary or geminate, disposed in a loose nearly leafless raceme. Flowers up to 3 mm. long, greenish or yellowish, with slender pedicels 1–2 cm. long, elongating at the fruiting stage. Perianth-segments 4, 2–3 mm. long, oblong, acute and papulose. Stamens c. 15, with linear-oblong anthers more or less as long as the filaments. Ovary wholly inferior, 2–4-locular; styles 2–4, very slightly longer than the stamens. Fruit up to 2·5 cm. long and wide, suborbicular, 2–4-locular, (3)4-winged, notched at top, rounded or rather abruptly narrowed at the base; wings firm, with reticulate veins and distinct intermediate ridges.

Botswana. SW: Minyani Pan, fr. 21.ii.1960, *Wild* 5112 (SRGH).
Also in SW. Africa and S. Africa. In pans.

2. **Tetragonia tetragonoides** (Pall.) Kuntze, Rev. Gen. Pl. **2**: 264 (1891).—Jeffrey in F.T.E.A., Aizoaceae: 34 (1961).—M. L. Gonçalves in C.F.A. **4**: 333 (1970). Type from a plant cultivated in Moscow from seeds of unknown origin.
   Demidovia tetragonoides Pall., Enum. Pl. Hort. Demidof: 150, t. 1 (1781). Type as above.
   Tetragonia expansa Murr. in Comm. Soc. Sci. Göttingen **6**: 13, t. 5 (1783) *nom. illegit.*—Welw. in Ann. Cons. Ultramar, Parte Naõ Off., Sér. 1, **1858**: 557 (1859).—Pax in Engl. & Prantl, Pflanzenfam. **3**, 1: 44, fig. 18 (1889).—Hiern, Cat. Afr. Pl. Welw. **1**: 409 (1898).—Pax & Hoffm. in Engl. & Prantl, Pflanzenfam. ed. 2, **16c**: 232, fig. 103 (1934).—Williamson, Useful Pl. Nyasal.: 117 (1955).—Adamson in Journ. S. Afr. Bot. **21**: 147 (1955). Type as for *Tetragonia tetragonoides*.

A large prostrate or prostrate-ascending somewhat succulent annual with stout trailing stems. Leaves 3–13 × 2–8·5 cm., rhombic, deltate or rhombic-ovate, obtuse or subacuminate at apex, with the base cuneate and more or less decurrent into the petiole, somewhat succulent; petiole 5–25 mm. long. Flowers greenish or yellowish, solitary or paired, axillary, subsessile. Perianth-segments (3)4(5), up to 8 mm. in fruit, acute, green outside, yellowish-green inside, unequal. Stamens 4–many, solitary or fasciculate, with orbicular anthers. Ovary semi-inferior, of 5–8 united carpels and 5–8 styles. Fruit up to 10 × 7 mm., smooth, conical or top-shaped, bearing below each perianth-segment a horn which sometimes gives rise to another flower or branchlet.

Mozambique. SS: Gaza, João Belo, Sepúlveda beach, fl. & fr. 8.x.1968, *Balsinhas* 1356 (LISC). LM: Matola, near the mouth of Umbeluzi R., fl. 27.vii.1963, *Mogg* 32156 (K; SRGH).
Native of New Zealand; widely introduced by cultivation as a vegetable. In salt marsh sandy flats.

90. UMBELLIFERAE
By J. F. M. Cannon

Herbs, frequently with regularly furrowed stems, very rarely shrubs or small trees. Stems usually hollow or with a prominent pith. Leaves alternate, usually compound, often finely divided, occasionally simple and very rarely peltate. Flowers normally ☿, but sometimes some or all of the flowers in an umbel unisexual by reduction. Flowers arranged in simple, or more commonly compound, umbels, rarely verticillate or capitate. Calyx adnate to the ovary and visible only as minute teeth or totally lacking. Petals 5, valvate to slightly imbricate, epigynous, usually white but rarely yellowish, greenish, tinged with pink or very rarely blue. Stamens 5, free, alternating with the petals; anthers 2-celled, opening by longitudinal splits. Styles 2, usually divergent, often connate for part of their length and frequently with a well-developed stylopodium. Ovary inferior, 2-celled with a solitary pendulous ovule in each cell. Fruit dry, usually dividing at maturity into

2 single-seeded mericarps which are supported on a central carpophore derived from the main vascular strand of the fruit. The carpels often have well-developed ribs, and frequently have characteristic resin-canals (vittae) in their walls. The fruit may be laterally or dorsally flattened and have large lateral wings which are associated with wind dispersal. They are sometimes equipped with hooks or spines, an adaptation connected with dispersal in the fur of animals. The seeds are provided with copious endosperm and have minute embryos.

A large family especially characteristic of the warm temperate areas of the world, but well represented in tropical Africa, where its members occur particularly at higher altitudes. Easily recognised by the generally herbaceous habit, the normally characteristic regular disposition of the flowers in umbels and the very distinctive schizocarpic fruit.

Difficulties are often experienced by those who try to identify plants in this very natural family, as many of the characters used are concerned with the fruit, and are rather technical and difficult to observe. This is especially true at the generic level, and in the following key as much emphasis as possible has been placed on non-fruit characters, so that the identification of flowering specimens will not be totally impossible. However, collectors should always select fruiting specimens whenever these are available. The features of the fully ripe fruit can sometimes be deduced from a study of the ovary, or very immature fruit, but sometimes, as in *Peucedanum* and related genera, great changes occur during the course of fruit maturation. The descriptions which follow refer to fully-developed terminal umbels unless otherwise indicated. The presence of a ring of bracts (" involucre ") below the primary umbel, and of a ring of bracteoles (" involucel ") below the secondary (" partial ") umbels is a useful character. The branches of the primary umbel are here called rays, and those of the partial umbels pedicels. The stylopodium (enlarged style bases which rise from the disk) is another valuable feature in identification, and when mentioned in the descriptions refer to ☿ flowers or fruit. ♂ flowers, which may be recognised by their abortive ovaries, occur in some genera, but are not usually specifically mentioned in the accounts. The leaves of members of this family are often complex and difficult to describe. They are most frequently some variant of the compound pinnate or compound ternate pattern, but often the pattern is not completely or regularly expressed. It may sometimes be necessary to cut a rough section of the fruit (see illustrations), so that the number and arrangement of the vittae (oil canals) in the fruit wall can be clearly seen; these are situated in the intervals between the ribs and in the commissural face, i.e. the side of the mericarp which forms the dividing plane of the fruit.

The key to the genera which follows is highly artificial and may not be applicable to species from outside the area covered by this Flora.

In additon to the native and introduced species included in this account, *Ammi majus* has been collected as a " cultivated " garden plant in Salisbury.

I am indebted to L. Constance for help with *Alepidea* and to V. H. Heywood and S. L. Jury for the account of *Agrocharis*.

Low creeping herbs, rooting at the nodes:
    Leaves peltate, or reniform with the margins clearly divided into 5–12 major lobes (see tab. **141**) - - - - - - - - - **1. Hydrocotyle**
    Leaves never peltate; reniform with regular marginal crenulations; linear or obtriangular-reniform with large regular marginal teeth - - **2. Centella**
Erect plants, not generally rooting at the nodes:
    Fruit and ovary with hairs, spines, tubercles or vesicles:
        Fruit and ovary with silky or bristly hairs, or vesicles:
            Bracts and bracteoles 0 - - - - - - - **14. Pimpinella**
            Bracts and bracteoles numerous and well developed:
                Fruit with normal hairs - - - - - **18. Diplolophium**
                Fruit with vesicular hairs - - - - - **19. Physotrichia**
        Fruit and ovary with hooked spines or tubercles:
            Fruit with tubercles:
                Plants almost leafless, " broom-like " - - - - **11. Deverra**
                Plants with obvious leaves - - - - - **4. Alepidea**
            Fruit with hooked spines:
                Leaves ternate or palmate - - - - - - **3. Sanicula**
                Leaves pinnate or compound pinnatisect:
                    Bracts present - - - - - - - - - **6. Agrocharis**
                    Bracts 0 - - - - - - - - - **5. Torilis**

Fruit and ovary glabrous:
  Trees, shrubs and suffrutices:
    Trees or suffrutices flowering before the leaves are produced
                                                            24. **Steganotaenia**
    Shrubs or suffrutices flowering with the leaves -   -   -   9. **Heteromorpha**
  Perennial, biennial or annual herbs:
    Bracts ± connate round the capitulate umbel -   -   -   -   4. **Alepidea**
    Bracts present or absent, not connate, umbels not capitulate:
      Bracts 0:
        Fruit strongly dorsiventrally compressed and often winged:
          Petals bright yellow; plants annual   -   -   -   -   21. **Anethum**
          Petals white, cream or greenish-yellow; plants perennial or biennial:
            Pedicels capillary; leaflets usually linear   -   -   -   23. **Lefebvrea**
            Pedicels relatively stout and stiff; leaflets various:
              Wing of fruit thickened; leaves pinnate -   -   25. **Heracleum**
              Wing of fruit thin; leaves usually compound pinnatisect
                                                            22. **Peucedanum**
        Fruit not strongly dorsiventrally compressed:
          Fruit ± orbicular, annual herbs -   -   -   -   7. **Coriandrum**
          Fruit otherwise:
            Petals yellow:
              Leaves simple -   -   -   -   -   14. **Pimpinella**
              Leaves finely divided:
                Robust, glaucous, spreading weedy plants of up to 2 m ; vittae 1 per
                  interval -   -   -   -   -   -   20. **Foeniculum**
                Slender, single-stemmed plants of up to 1 m. (Nyika Plateau);
                  vittae 3 per interval -   -   -   -   -   13. **Frommia**
            Petals white:
              All umbels pedunculate -   -   -   -   -   14. **Pimpinella**
              Some umbels ± sessile -   -   -   -   -   10. **Apium**
      Bracts present:
        Fruit strongly dorsiventrally compressed:
          Petals bright yellow; plants annual   -   -   -   -   21. **Anethum**
          Petals white, cream or greenish-yellow; plants perennial or biennial:
            Pedicels capillary; leaflets usually linear   -   -   -   23. **Lefebvrea**
            Pedicels relatively stout and stiff; leaflets various:
              Wing of fruit thickened; leaves pinnate -   -   25. **Heracleum**
              Wing of fruit thin; leaves usually compound pinnatisect
                                                            22. **Peucedanum**
        Fruit not strongly dorsiventrally compressed:
          Ultimate leaf-segments linear, entire   -   -   -   -   12. **Aframmi**
          Ultimate leaf-segments variously cut or toothed:
            Leaflet margin deeply cut, teeth without cartilaginous tips
                                                            8. **Conium**
            Leaflet margin ± regularly toothed, teeth often with distinct cartilaginous
              tips:
              Leaflet margins dentate -   -   -   -   -   17. **Berula**
              Leaflets margins very regularly serrate:
                Inflorescence obviously leafy   -   -   -   -   16. **Sium**
                Inflorescence with only much-reduced leaf bases - 15 **Baumiella**

1. HYDROCOTYLE L.

Hydrocotyle L., Sp. Pl. **1**: 234 (1753); Gen. Pl. ed. 5: 109 (1754).

Creeping perennial, rooting at the nodes, glabrous or pubescent. Leaves on
slender petioles, sometimes peltate. Inflorescence a simple umbel (sometimes
almost capitate), a proliferous umbel or an interrupted spike. Peduncle very
short to longer than the subtending leaf. Flowers white to greenish-yellow, calyx
teeth minute; stylopodium depressed (rarely conspicuous). Fruit strongly
laterally compressed, dorsal surface often acute; ribs well developed; carpophore
absent; commissural face of the seed plane to convex. Seed cavity surrounded
by strengthening cells. Vittae absent to well developed.

A genus of c. 75 species with a pantropical distribution with extensions into
temperate regions. There is a strong concentration of species in S. America and
Australia.

Leaves peltate (very rarely with a sinus in *H. bonariensis*):
  Inflorescence an interrupted spike -   -   -   -   -   -   -   1. *verticillata*
  Inflorescence a proliferous umbel -   -   -   -   -   -   -   2. *bonariensis*

Leaves ± reniform with an obvious basal sinus:
Plants completely glabrous; petioles and stems thick, soft and fleshy
3. *ranunculoides*
Plants with hairs on petioles or leaves (very rarely glabrous); stem and petioles fine and almost wiry:
Leaves glabrous on the superior surface - - - - 4. *sibthorpioides*
Leaves with at least a few (usually numerous) bristly hairs on the superior surface
5. *mannii*

1. **Hydrocotyle verticillata** Thunb., Dissert. Hydrocot.: 5 (1798).—Sond. in Harv. & Sond., F.C. 2: 527 (1862).—Eyles in Trans. Roy. Soc. S. Afr. 5: 433 (1916).—Burtt Davy, F.P.F.T. 1: 517 (1926).—R. E. Fr., Wiss. Ergebn. Schwed. Rhod.-Kongo-Exped.: 183 (1914).—Cannon in C.F.A. 4: 336 (1970). Type from S. Africa.
   Hydrocotyle interrupta Elliott, Sketch of the Bot. of S. Carol. & Georgia 1: 345 (1817).—DC., Prodr. 4: 59 (1830). Type from the U.S.A.
   Hydrocotyle vulgaris sensu Thunb., Fl. Cap. 2: 192 (1820); op. cit. ed. 2: 252 (1823) non L. (1753).
   Hydrocotyle vulgaris var. *verticillata* (Thunb.) A. Rich., Mon. Hydrocot.: 27 (1820). Type as for *Hydrocotyle verticillata*.
   Hydrocotyle vulgaris var. *communis* Cham. & Schlecht. in Linnaea 1: 356 (1826). Type as for *Hydrocotyle verticillata*.
Glabrous creeping herb, rooting freely at the nodes; stems delicate, terete, brownish-white. Leaves peltate, on long petioles up to 15 cm. Lamina up to 45 (60) mm. in diameter, circular to broadly elliptic, with 8–13 main veins radiating from the petiole; margin shallowly crenately lobed. Inflorescence an interrupted verticillate spike, approximately equal in length to the subtending leaf (axis occasionally branched to produce 2 or more parallel spikes). Flowers small, inconspicuous, 2–7 in each verticil, initially closely packed together, the axis only expanding to separate the verticils as the fruit begins to mature. Fruit 2 × 4 mm., reddish-brown, laterally flattened, broadly ellipsoid, base cuneate, pedicel 1 mm. to obsolete; apex broadly and shallowly emarginate, stylopodium depressed to almost obsolete, styles short and spreading. Fruit ribs well developed, commissure scarcely depressed.

Caprivi Strip. Lizauli, Mashi R., fl. & fr., *Killick & Leistner* 3254 (K; SRGH). **Botswana.** N: Ngamiland, Thamalakane R., fr. 13.iii.1961, *Richards* 14695 (K). **Zambia.** B: Kalabo, near resthouse, fl. & fr. 13.xi.1959, *Drummond & Cookson* 6402 (K; SRGH). W: L. Ishibu, Ndola, fl. & fr. 18.x.1953, *Fanshawe* 433 (K). C: 9·7 km. SE. of Lusaka, fl. & fr. 25.xi.1957, *King* 390 (K). E: Msoro, 80 km. W. of Fort Jameson, Luangwa Valley, fl. & fr. 9.vi.1954, *Robinson* 850 (K; SRGH). S: Mapanza, fl. & fr. 12.v.1954, *Robinson* 749 (K; SRGH). **Rhodesia.** N: Darwin Distr., Upper Nyatandi R., fr. 27.i.1960, *Phipps* 2450 (BM; K; LMA; SRGH). W: Victoria Falls, fl. ix.1909, *Rogers* 5324 (K; SRGH). C: Salisbury, Highlands, 24.ix.1946, *Wild* 1258 (BM; SRGH). E: Umtali Distr., Golf Course, fl. & fr. 8.ix.1960, *Chase* 7375 (K; SRGH). **Mozambique.** SS: Panda, Inhambane Distr., Domo, Inhassoro Valley, fl. & fr. 27.x.1935, *Lea* 120 (SRGH).
Widespread throughout much of the tropics and subtropics: Australia, New Guinea, S. Africa, Angola, Uganda, U.S.A., Central and S. America. In damp places by rivers and in marshes and ditches.

Very closely related to the European *H. vulgaris* and later work may well show them to be more properly regarded as subspecies. It can be distinguished from *H. vulgaris* by the following characters:

Base of fruit cuneate; leaf veins 8–13; inflorescence c. the same length as the subtending petiole, which is completely glabrous - - - - - - *verticillata*
Base of fruit emarginate; leaf veins 6–10; inflorescence c. ½ the length of the subtending petiole, which always has some long hairs near its apex - - - *vulgaris*

It appears that *H. vulgaris* may be a temperate European derivative of the probably much older tropical *H. verticillata*. Critical experimental work is needed to decide the matter.

2. **Hydrocotyle bonariensis** Lam., Encycl. Méth. Bot. 3: 153 (1789).—Sond. in Harv. & Sond., F.C. 2: 527 (1862).—Hiern in F.T.A. 3: 4 (1877).—Cannon in F.W.T.A. ed. 2, 1: 753 (1958); in C.F.A. 4: 337 (1970). Type from S. America (Uruguay).

Glabrous creeping herbs with slender stems, rooting at the nodes. Leaves peltate (very rarely with a deep narrow sinus extending to the petiole), on long petioles which may exceed 30 cm.; lamina up to 12 cm. in diameter, circular to broadly elliptic, with 12–20 veins radiating from the petiole; margin often shallowly lobed, the lobes with small secondary crenulations. Inflorescence a proliferous umbel on a long slender peduncle which exceeds the subtending petiole. Umbel rays 5–7. When fully expanded, each ray forming a verticillate raceme, with the fruit in interrupted whorls. Bracteoles lanceolate-acute, in whorls under each whorl of flowers. Pedicels slender, 1–20 mm. long; petals white to cream. Fruit ellipsoid, flattened laterally, base cordate, stylopodium depressed to obsolete, styles c. 1 mm. long; commissure not strongly constricted, dorsal and lateral ribs well developed and quite acute in section; pericarp reddish-brown and slightly wrinkled when mature.

Mozambique. T: Tete, fr. ix.1858, *Kirk* (K). SS: Inhambane, Inharrime, near Praia de Zavora, fl. & fr. 22.vi.1960, *Lemos & Balsinhas* 167 (BM; COI; K; LISC; LMA; PRE; SRGH). LM: between Costa do Sol and Marracuene, Muntanhane, fl. & fr. 13.ix.1960, *Balsinhas* 248 (BM; COI; K; LISC; PRE; SRGH).
Widespread in tropical Africa and in tropical and subtropical America. In submaritime habitats, often in sandy soils near brackish lagoons.

3. **Hydrocotyle ranunculoides** L.f., Suppl. Pl.: 177 (1781).—Robyns & Tournay, Fl. Parc Nat. Alb. **1**: 697 (1948).—Cannon in C.F.A. **4**: 337 (1970). Type from Mexico.
Hydrocotyle natans Cyr., Pl. Rar. Regn. Neapol. **1**: 20, t. 6b (1788).—Hiern in F.T.A. **3**: 5 (1877). Type from Italy.
Hydrocotyle adoensis Hochst. in Flora **24**, Intellig. 1: 28 (1841) *nom. nud.*

Glabrous creeping herb, densely rooting at the nodes. Stem prostrate with long or short internodes of 1–9 cm., terete and soft. Leaves arising singly at the nodes on petioles 2–25 cm. long; lamina 0·6–4·5 × 1–6 cm., reniform with a deep basal sinus; margin ± crenately divided into 5–8 lobes, each lobe with small secondary crenulations. Umbels simple with 5–10 flowers, subcapitate on peduncles 1·5–5 cm. long, much shorter than the subtending petioles; rays very short to obsolete. Petals greenish-white. Fruit transversely subrectangular, greenish-brown with red mottlings, laterally compressed, shallowly emarginate at the base and apex, somewhat constricted at the commissure; ribs filiform, only weakly developed; stylopodium depressed, styles short, c. 0·75 mm. long, widely divergent.

Zambia. N: Luitikila R., L. Bwali, veg. 11.xi.1969, *Verboom* 2643 (K). **Malawi.** S: Fort Johnston Bar to Nyenyezi, veg. 17.ix.1929, *Burtt Davy* 21771a (K).
Widely distributed in southern U.S.A., central America and tropical S. America; naturalised in Spain and Italy. In Africa it is known to occur in Ethiopia, Kenya, Uganda, Tanzania, Zaire and Angola in addition to the above localities in Malawi and Zambia. Rooted in marshy ground or free floating in water.

4. **Hydrocotyle sibthorpioides** Lam., Encycl. Méth. Bot. **3**: 153 (1789).—Hiern, Cat. Afr. Pl. Welw. **1**: 423 (1898).—Burtt Davy, F.P.F.T. **1**: 517 (1926).—Cannon in F.W.T.A. ed. 2, **1**: 753 (1958); in C.F.A. **4**: 338 (1970). Type from Mauritius.
Hydrocotyle nitidula A. Rich., Mon. Hydrocot.: 60 (1820).—Hiern in F.T.A. **3**: 4 (1877). Type from Java.
Hydrocotyle monticola Hook. f. in Journ. Linn. Soc., Bot. **7**: 194 (1864). Type from Cameroons.
Hydrocotyle americana var. *monticola* (Hook. f.) Hiern, loc. cit. Type as above.
Hydrocotyle confusa H. Wolff in Notizbl. Bot. Gart. Berl. **9**: 1109 (1927).—Robyns & Tournay, Fl. Parc Nat. Alb. **1**: 696 (1948). Type from Kenya.

Delicate slender creeping herb, rooting at the nodes. Stem filiform, whitish to brownish. Leaves on slender petioles of 3–40 mm. long, glabrous or with scattered flexuous hairs near the tip, solitary or 2–3 together. Lamina reniform with a narrow basal sinus, crenately divided into 5–7 lobes the margins of which have rounded triangular teeth. Superior leaf surface glabrous, the inferior usually with scattered crisped hairs particularly noticeable on young leaves. Inflorescence a simple subcapitate umbel, much shorter than the subtending leaf. Peduncle slender, 2–9 mm. long. Umbel 3–8(10)-flowered, involucre of inconspicuous narrow lanceolate scarious bracts; rays minute to obsolete. Fruit 1·5 × 1·0 mm., greenish to yellowish-brown, broadly ellipsoid, laterally compressed; base broadly and shallowly cordate; apex deeply emarginate; stylopodium depressed to obsolete, styles short, stiff and divergent; ribs well developed.

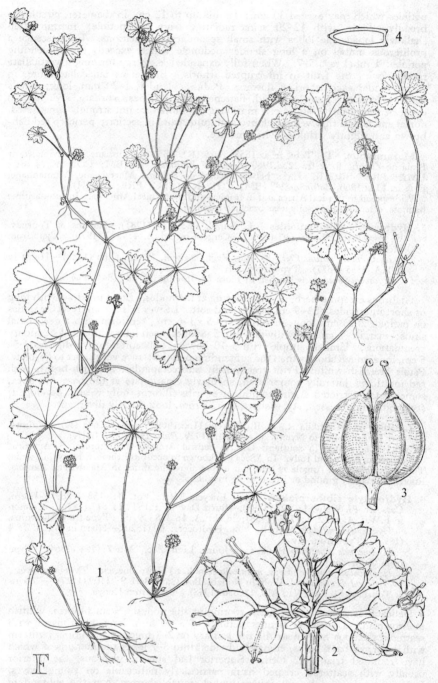

Tab. 141. HYDROCOTYLE MANNII. 1, habit ($\times \frac{1}{2}$); 2, umbel ($\times 9$); 3, fruit ($\times 12$); 4, mericarp in cross-section ($\times 18$), from *Gilliland* B382 and *Taylor* 2399.

Zambia. N: Mbala (Abercorn) Distr., Karambe, Lumi R., fl. & fr. 30.viii.1956, *Richards* 6034 (K). W: Mindolo, fl. & fr. 25.x.1965, *Mutimushi* 1218 (SRGH). **Rhodesia.** E: Melsetter, Glencoe Forest Reserve, fl. & fr. 23.ix.1955, *Drummond* 4968 (BM; LMA; SRGH). **Malawi.** S: Luchenya Plateau, Mlanje Mt., fl. & fr. 10.vi.1962, *Richards* 16622 (K). **Mozambique.** LM: Lourenço Marques, Vasco da Gama, fl. 28.iv.1971, *Balsinhas* 1843 (K).

Widespread in the tropics of the Old World. In Africa in Kenya, Uganda, Tanzania, Angola, Zaire and W. Africa. Introduced and established in both N. and S. America.

Superficially very similar to, and often confused with, *Sibthorpia europaea* (Scrophulariaceae). It can be distinguished in the vegetative state by its glabrous superior leaf surfaces, unlike those of *Sibthorpia* which have numerous appressed bristly hairs.

5. **Hydrocotyle mannii** Hook. f. in Journ. Linn. Soc., Bot. **7**: 194 (1864).—Robyns & Tournay, Fl. Parc Nat. Alb. **1**: 696 (1948).—Cannon in C.F.A. **4**: 338 (1970). TAB. **141**. Type from Fernando Po.

Hydrocotyle hirta R. Br. ex A. Rich., Mon. Hydrocot.: 64 (1820).—Humbert in Bull. Jard. Bot. Brux. **27**: 770 (1957). Type from Australia.

Hydrocotyle moschata sensu Hiern in F.T.A. **3**: 5 (1877).—Bak. f. in Journ. Linn. Soc., Bot. **40**: 75 (1911).—Eyles in Trans. Roy. Soc. S. Afr. **5**: 433 (1916); see note at end of species account.

Hydrocotyle javanica sensu Hiern, tom. cit.: 4 (1877) non Thunb. (1878).

Delicate creeping perennial herb, rooting at the nodes. Stem prostrate, terete, glabrous or with a few scattered hairs. Leaves long-petiolate, 1–2(3) at each node, circular to reniform, basal sinus very narrow. Petioles 1·5–11 cm. long, subglabrous to hirsute, with numerous long, white, bristly hairs especially near the tip. Lamina 1–5 × 0·75–4·5 cm., with 5–12 lobes, the sinuses between the lobes varying from shallow to extending halfway to the centre of the leaf; lobe margins subentire to crenate-dentate; inferior surface with numerous bristly hairs, especially on the veins; superior surface with similar hairs, which may be only very sparsely distributed. Inflorescence a subcapitate umbel; peduncle rather variable in length, but commonly c. ½ the length of the subtending petiole. Umbel of 10–30 flowers on very short to obsolete rays. Petals greenish-white. Fruit c. 1·0 × 1·5 mm., broadly ellipsoid, laterally compressed, brownish when ripe; apex and base distinctly emarginate; stylopodium depressed; styles short, spreading; stigmas slightly clavate; lateral ribs not strongly developed.

Zambia. N: Mbala (Abercorn) Distr., Chilongowelo, fr. 1.iv.1952, *Richards* 1229 (K). **Rhodesia.** E: Umtali Distr., Commonage N. of Palmerston, fl. & fr. 4.xi.1954, *Chase* 5317 (BM). **Malawi.** N: Mzimba Distr., Mzuzu, fl. 11.xii.1970, *Pawek* 4066 (K; UC). C: Dedza Distr., Chongoni Forest Reserve, fl. & fr. 23.x.1967, *Salubeni* 856 (BM; K; SRGH). S: Zomba, *Cameron* 66 (K). **Mozambique.** MS: Vale de Moçambique, Manica, fl. & fr. 25.x.1944, *Mendonça* 2592 (LISC).

From Fernando Po and S. Tomé to the Sudan and thence through E. Africa, Zaire and Angola to the F.Z. area.

The African material is clearly very closely related to *H. moschata* Forst. f. (Fl. Ins. Austr. Prodr.: 22 (1786)) from Australia and New Zealand and to some of the forms included under *H. javanica* Thunb. in Malaysia and India. In the absence of a complete revision of the whole complex, it seems best for the present to retain an African taxon, while drawing attention to the situation. However, if later work, preferably supported by experimental studies, indicates that the whole complex should be united, the correct name will be *H. moschata* Forst. f.

2. CENTELLA L.

Centella L., Sp. Pl., ed. 2, **2**: 1393 (1763); Gen. Pl. ed. 6: 485 (1764).
Solandra L., Syst. Nat. ed. 10: 1269 (1759) non Swartz (1787) *nom. conserv.*
Hydrocotyle Subgen. *Centella* (L.) Benth. & Hook., Gen. Pl. **1**: 873 (1862).

Perennial herbs becoming woody at the base in some species. Stems often procumbent and rooting at the nodes, sometimes decumbent and sprawling over rocks or other plants. Leaves in groups of 2–many, rarely single. Umbels simple and axillary, sessile or pedunculate, with 1–many rays. Flowers ⚥ or ♂, together or in separate umbels (♂ flowers often 3 together on a common pedicel); petals greenish-white, sometimes tinged with red. Involucre of 2 bracts persisting into the fruiting stage. Fruit laterally flattened, somewhat constricted at the commissure; primary ribs usually well developed, sometimes with well-developed secondary ribs also; stylopodium obsolete, styles divergent; carpophore absent;

commissural face of the seed flat. An oil-bearing layer of cells is present beneath the pericarp epidermis and these occasionally differentiate into small vittae. A layer of strengthening cells surrounds the seed cavity.

A genus of 40–50 species of which one (*C. asiatica*) has a pantropical distribution, while all the others are almost completely confined to S. Africa, where they have diversified to form an amazingly rich range of structural types adapted to a wide series of ecological niches.

Leaves reniform to nearly orbicular or very broadly obtriangular with deeply cut teeth:
 Leaves reniform to orbicular with a distinct basal sinus - - - 1. *asiatica*
 Leaves obtriangular with a cuneate base - - - - - -2. *obtriangularis*
Leaves linear:
 Leaves very narrow, c. 1 mm. wide, ± terete - - - - 4. *virgata*
 Leaves broader, c. 2 mm. wide, flat with an obvious midrib - - 3. *glabrata*

1. **Centella asiatica** (L.) Urb. in Mart., Fl. Bras. **11**, 1: 287, t. 78 fig. 1 (1879).—
 Norman in Journ. Bot., Lond. **67**, Suppl. Polypet.: 198 (1929).—Robyns & Tournay,
 Fl. Parc Nat. Alb. **1**: 697 (1948).—Cannon in F.W.T.A. ed. 2, **1**: 753 (1958); in
 C.F.A. **4**: 339 (1970). TAB. **142**. Type from India.
  Hydrocotyle asiatica L., Sp. Pl. **1**: 234 (1753).—Sond. in Harv. & Sond., F.C. **2**:
 527 (1862).—Hiern in F.T.A. **3**: 6 (1877); Cat. Afr. Pl. Welw. **1**: 423 (1898).—
 Eyles in Trans. Roy. Soc. S. Afr. **5**: 433 (1916).—Burtt Davy, F.P.F.T. **1**: 517
 (1926). Type as above.
  Centella coriacea Nannf. in Svensk. Bot. Tidskr. **18**: 416 (1924).—Adamson in
 Journ. S. Afr. Bot. **17**: 4 (1951). Type from S. Africa.

Creeping herb, rooting at the nodes, but sometimes forming a large taproot. Stem terete, with shallow grooves, sometimes purplish. Leaves solitary or in groups of up to 5, pubescent or glabrous; lamina 1–7 cm. wide, reniform to almost orbicular, with a deep basal sinus; margin crenate, glabrous or with scattered hairs on upper part of petiole (petiole sometimes densely hairy when young). Umbels subcapitate; peduncles 1·5–5 cm. long, glabrous or pubescent, usually much shorter than the subtending petiole. Flowers 2–8, ☿, with an involucre of 2 ovate, membranous, persistent bracts. Pedicels slender or obsolete; petals dark crimson to greenish-white, orbicular with a slender inflexed point. Calyx teeth and stylopodium obsolete; styles short, ± divergent. Fruit 3·5 × 3 mm., orbicular to ellipsoid, brown at maturity, deeply constricted at the commissure and flattened laterally, primary ribs prominent when ripe, secondary ribs ± evident; carpophore entire.

Caprivi Strip. Singalamwe, fl. & fr. 31.xii.1958, *Killick & Leistner* 3212 (PRE; SRGH). **Zambia.** B: Kalabo, rest house, fr. 13.xi.1959, *Drummond & Cookson* 6406 (K; SRGH). N: Lukulu, fl. & fr. 28.ix.1962, *Fanshawe* 7061 (K; SRGH). W: Mwinilunga Distr., Matonchi R., fl. & fr. 11.xi.1937, *Milne Redhead* 3191 (BM; K). C: Kafue R., 60 km. S. of Lusaka, veg. 11.xii.1955, *Best* 120 (K; SRGH). E: Lundazi R. below Dam, fl. & fr. 19.xi.1958, *Robson* 678 (BM; K; LISC; SRGH). S: Mapanza, fl. 5.ix.1953, *Robinson* 306 (K). **Rhodesia.** N: Sebungwe, Lusala R., fl. & fr. 31.x.1947, *Whellan* 301 (BM; SRGH). C: Salisbury, Enterprise, fl. 23.xii.1945, *Greatrex* 570 (K; SRGH). E: Melsetter Distr., Mwenzi Farm, fl. & fr. 13.iii.1953, *Chase* 4849 (BM; SRGH). S: Chibi, Razi Dam, fl. & fr. 12.v.1970, *Pope* 298 (K). **Malawi.** N: L. Nyasa near Roangwa, fr. ix.1861, *Kirk* (K). C: Lilongwe, Kamuzu Dam, fr. 6.iii.1968, *Salubeni* 1004 (BM; SRGH). S: Mt. Mlanje, fl. & fr. x.1891, *Whyte* 197 (BM). **Mozambique.** Z: Mocuba, fl. & fr. 31.v.1948, *Faulkner* 276 (K). MS: Mavita, Vale de Moçambique, fr. 25.x.1944, *Mendonça* 2590 (LISC). SS: Gaza, Vila de João Belo, fl. & fr. 7.x.1945, *Pedro* 219 (COI; K; PRE). LM: Lourenço Marques, fl. 10.vii.1957, *Barbosa* 7634 (LM; PRE).

Throughout tropical and some subtropical parts of the world. Has been split into several species on a geographical basis, but these do not seem worthy of recognition, and certainly not at specific level. In damp grassy places or under bushes.

2. **Centella obtriangularis** Cannon in Kirkia **4**: 145 (1964). Type: Mozambique,
 Chimanimani Mts., *Wild* 2893 (BM, holotype).

Glabrous creeping herb, rooting at the nodes; internodes often elongate, up to 19 cm. long. Leaves in groups of 3–10, often on short lateral shoots. Petioles 1–3·5 cm. long; lamina 3–11 × 4–20 mm., very shallowly obtriangular, tending to become transversely elliptic with age, very broadly cuneate at the base, the apex

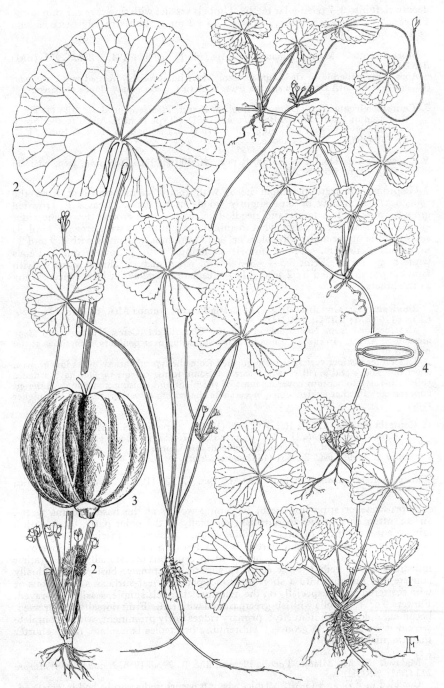

Tab. 142. CENTELLA ASIATICA. 1, habit ($\times\frac{1}{2}$); 2, leaf and base of stem ($\times 1\frac{1}{2}$); 3, fruit ($\times 9$); 4, mericarp in cross-section ($\times 9$), from *Chase* 243 and *Lewalle* 2063.

deeply cleft into 3–7 triangular teeth. Umbels sessile, with fine rays c. 7 mm. long. Flowers ☿ or ♂; petals white. Fruit c. 2·5 × 2 mm., approximately twice as long as the adjacent bract.

Mozambique. MS: Chimanimani Mts., fl. & fr. 8.vii.1949, *Wild* 2955 (BM; SRGH).
Apparently endemic to the Chimanimani Mts., but probably related to *C. calliodus* (Cham. & Schlecht.) Drude, a species from S. Africa. Wet grassy slopes or banks.

3. **Centella virgata** (L.f.) Drude in Engl. & Prantl, Pflanzenfam. **3**, 8: 120 (1898).—
   Adamson in Journ. S. Afr. Bot. **17**: 33 (1951). Type from S. Africa (Cape Prov.).
      Hydrocotyle virgata L.f., Suppl. Pl.: 176 (1781).—Sond. in Harv. & Sond., F.C.
   2: 532 (1862). Type as above.

Var. **gracilescens** Domin in Fedde, Repert. **4**: 300 (1907).—Adamson, tom. cit.: 35 (1951). Type from S. Africa.

Glabrous, prostrate perennial herb, with many stems spreading from the rootstock, which may become slightly woody at the base. Leaves in crowded groups, more or less terete and needle-like, with a fine groove down one side; apex acute, basal sheaths somewhat membranous and tinged with brown. Umbels sessile in the upper axils; rays slender, 4–25 mm. long. Umbels with 1 ☿ and 2– numerous ♂ flowers. Upper umbels sometimes with ♂ flowers only. Petals white to greenish. Fruit 3·5–4 × 3–3·5 mm. when mature, almost smooth, with faint ridges, subtended by 2 triangular-lanceolate bracteoles at least twice as long as the bracts.

Rhodesia. E: Umtali Distr., W. of Castle Beacon, Vumba Mts., fl. & fr. 13.v.1956, *Chase* 6116 (BM; LISC; SRGH).
Also recorded from Namaqualand and the Clanwilliam, Ceres, Tulbagh and Cape areas of S. Africa. In the F.Z. area it occurs sprawling over boulders, sometimes at the margins of streams.
This rather distinct variety is distinguished from the typical variety by its low growth, short internodes and small crowded leaves. Some forms of var. *virgata* have a dense yellowish-rusty tomentum covering most of the plant and causing a strikingly different appearance. Further evidence may necessitate the raising of var. *gracilescens* to a higher rank.

4. **Centella glabrata** L., Pl. Rar. Afr.: 28 (1760). Type from S. Africa.
      Hydrocotyle glabrata (L.) L.f., Suppl. Pl.: 176 (1781). Type as above.
      Hydrocotyle centella Cham. & Schlecht. in Linnaea **1**: 375 (1826).—Sond. in
   Harv. & Sond., F.C. **2**: 531 (1862) *nom. illegit.*

Var. **natalensis** Adamson in Journ. S. Afr. Bot. **17**: 23 (1951). Type from S. Africa (Natal).

A prostrate or sprawling herb, becoming woody at the base. Stems terete, striate, often dark brown to purplish, especially in the older parts, often with a fairly dense covering of crisped white hairs. Leaves in fascicles or pairs or apparently single in younger parts; lamina 1–4 × up to 0·35 cm., very narrowly lanceolate to linear, usually with only 1 main vein, but occasionally with 2 rather indistinct lateral veins; apex acute to slightly mucronate, base very gradually cuneate and merging into a short indistinct petiole; leaf-surfaces subglabrous or with scattered hairs, especially on the margins. Umbels simple, sessile, 1–3-rayed. Flowers ☿ or ♂; corolla whitish-green, tinged with red. Fruit dorsally compressed, becoming oblong with maturity; primary ridges fairly prominent, with incomplete secondary ridges in the grooves; subtending bracteoles lanceolate, much shorter than the mature fruit.

Malawi. S: Mt. Mlanje, Tuchila Plateau, fl. & fr. 29.vii.1956, *Newman & Whitmore* 273 (BM; SRGH).
Confined in F.Z. area to Mt. Mlanje, where it occurs under shrubs and in grassland. Also recorded from Natal, Transkei, East London, Stutterheim and Bathurst areas of S. Africa.
A very striking variety that may merit specific recognition. Var. *glabrata* which is described by Adamson as being " common through the coastal belt on lower slopes and flats of S. Africa " differs from var. *natalensis* chiefly by having distinctly 3-veined

lanceolate-oblong leaves. A very difficult complex which demands a complete re-investigation using experimental methods.

3. SANICULA L.

Sanicula L., Sp. Pl. 1: 235 (1753); Gen. Pl. ed. 5: 109 (1754).

Erect, rarely decumbent herbs, biennial or perennial. Leaves petiolate, sub-sessile, usually somewhat coriaceous, palmately to pinnately divided; petioles with sheathing bases. Stem usually branched in a pseudodichotomous manner. Inflorescence of rather irregular compound umbels with few, unequal rays. Partial umbels with ⚥ and ♂ flowers, the ⚥ one frequently distinctly pedicellate, the ♂ ones sessile to subsessile. Styles short or long and spreading. Stylopodium depressed or obsolete. Fruit ovoid, densely covered with prickles or tubercles. Vittae both large and small, regularly or irregularly arranged. Seed plane or concave on the commissural face.

A genus of 37 species, widely distributed in temperate and subtropical zones, but absent from Australia and New Zealand.

Saniculata elata Buch.-Ham. ex D. Don, Prodr. Fl. Nepal.: 183 (1825).—Shan & Constance in Univ. Cal. Publ. Bot. 25: 47 (1951).—Brenan in Mem. N.Y. Bot. Gard. 8: 445 (1954).—Cannon in F.W.T.A. ed. 2, 1: 753 (1958). TAB. 143. Type from Nepal.

Sanicula europaea var. *capensis* Cham. & Schlecht. in Linnaea 1: 352 (1826). Type from S. Africa.

Sanicula capensis (Cham. & Schlecht.) Eckl. & Zeyh., Enum. Pl. Afr. Austr. Extratrop. 3: 339 (1837). Type as for *Sanicula europaea* var. *capensis*.

Sanicula europaea sensu Sond. in Harv. & Sond., F.C. 2: 533 (1862).—Hiern in F.T.A. 3: 8 (1877).—Bak. f. in Journ. Linn. Soc., Bot. 40: 75 (1911).—R. E. Fr., Wiss. Ergebn. Schwed. Rhod.-Kongo-Exped.: 183 (1914).—Eyles in Trans. Roy. Soc. S. Afr. 5: 434 (1916).

Sanicula europaea var. *elata* (Buch.-Ham.) Boiss. in Bull. Soc. Bot. Fr. 53: 421 (1906).—H. Wolff in Engl., Pflanzenr. 4, 228 Sanic.: 63 (1913).—Burtt Davy, F.P.F.T. 1: 521 (1926).—Robyns & Tournay, Fl. Parc Nat. Alb. 1: 698 (1948). Type as for *Sanicula elata*.

Sanicula natalensis Gandog. in Bull. Soc. Bot. Fr. 65: 32 (1918). Type from S. Africa (Natal).

Erect glabrous herb up to 80 cm. Rootstock with numerous fine wiry roots. Stem dichotomously branched above, with numerous striations. Basal leaves on long petioles up to 30 cm.; lamina up to 14 cm. wide, very broadly ovate or pentagonal in outline, varying from ternate to palmate with up to 5 divisions, these sessile or distinctly petiolulate. Median segment broadly lanceolate to ovate or obovate, acuminate at the apex, 3-lobed, entire or tending towards a pinnatifid state; margins serrate with mucronate tips to the teeth; lateral lobes obliquely and irregularly lanceolate to ovate, with cuneate bases. Cauline leaves similar, becoming gradually reduced above. Bracts 2–3, 3–4 mm. long, narrowly lanceolate. Umbels with 2–3 rays c. 5 mm. long, sometimes with subsessile lateral umbels. Partial umbels 4–7-flowered, with 1–4 flowers ♂. Calyx lobes acute, linear-lanceolate; petals white to yellowish-green; styles c. twice as long as the calyx, spreading. Fruit 2 or 3 in each partial umbel, 2 × 3 mm., shortly ovoid, densely covered with strongly-hooked spines, these yellowish-brown at maturity. Commissural face of seed slightly concave. Commissural vittae 2, with a number of small ones scattered around the fruit wall.

Zambia. W: Mpongwe, 65 km. S. of Luanshya, fl. & fr. 3.ii.1960, *Robinson* 3320 (K; SRGH). E: Lundazi, upper slopes of Kangampande, Nyika, fr. 7.v.1952, *White* 2760 (K). **Rhodesia.** E: Umtali Distr., Engwa, fl. & fr. 2.ii.1955, *E. M. & W.* 96 (BM; LISC; SRGH). **Malawi.** N: Nyika Plateau, km. 56·3 Nyika road, fr. vii.1953, *Chapman* 181 (BM). C: Dedza Distr., Chongoni Forestry School, fr. 8.vii.1961, *Chapman* 1417 (SRGH). S: Zomba Plateau, fr. 7.viii.1946, *Brass* 16310 (K; SRGH). **Mozambique.** MS: Quinta da Fronteira, fl. 15.x.1954, *Chase* 5310 (BM; SRGH).

Widely distributed in tropical and subtropical Asia and Africa. In Africa from Fernando Po to E. Africa and Zaire and thence to S. Africa. Often on mountains, in moist shady situations, often by small streams under forest cover.

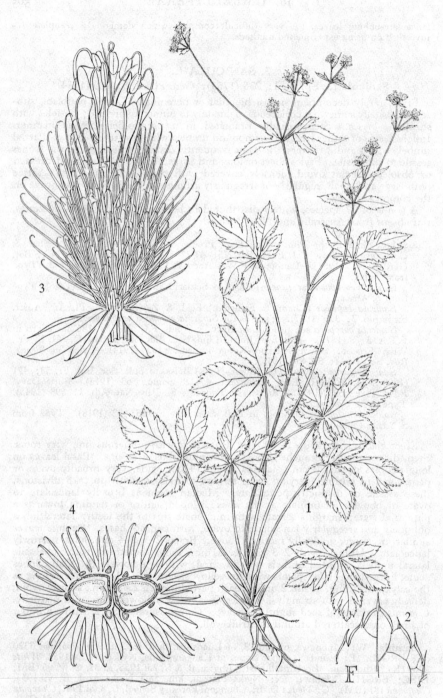

Tab. 143. SANICULA ELATA. 1, habit (×½); 2, leaf margin (×18); 3, fruit (×18), 1–3 from Brenan 9343 and Taylor 3738; 4, mericarps in cross-section (×8), from *Schlieben* 4658.

4. ALEPIDEA De la Roche[*]

Alepidea De la Roche, Eryng. Hist.: 19, t. 1 (1808).—H. Wolff in Engl., Pflanzenr. **4**, 228 Sanic.: 94 (1913).—Weim. in Bot. Notis. **1949**: 217 (1949).

Erect perennial or biennial herbs with wiry stems and simple coriaceous leaves; leaf margins dentate, with the apices of the teeth frequently produced to form long stiff ciliary appendages. Inflorescence comprising a relatively small number of capitulate umbels, each composed of ± sessile ☿ flowers. Bracts very well developed, chartaceous, in ± 2 series, forming a ± connate, conspicuous involucre round the flowers, and probably acting as an insect attractant, comparable to the ray florets of the Compositae. Calyx teeth conspicuous; petals erect, often with a sharply inflexed apex. Stylopodium generally not well developed, depressed and with a thickened margin; styles sometimes with a thickened apical stigmatic region. Fruit ovoid to subglobose; carpophore not strongly developed, undivided. Mericarps usually covered with scaly tubercles; vittae 5, with 1 in each interval and 2 in the commissural face; seed similar in section to the mericarp as a whole.

A very complex group of 20–30 poorly-defined species, widely distributed from S. Africa through E. tropical Africa to Ethiopia. An extremely complex range of variation is presented, especially in the southern part of the area, which has yet to be analysed in a thoroughly satisfactory manner, and most probably awaits a rigorous biosystematic analysis before a stable classification can be achieved. The detailed treatment by Weimarck (1949) does not appear effective in the light of the mass of material now available, particularly for the taxa in the F.Z. area and especially in respect of the subspecies he associated with *A. longifolia* and *A. gracilis*. In the light of this situation, we have adopted a highly conservative view for the present account, which should be regarded as no more than provisional. A definitive treatment must await a complete monographic review of the whole genus, especially with respect to the taxa in S. Africa, which appears to be its centre of diversity.

Mericarps smooth; basal leaves truncate to cordate at base; robust, tall herbs
  1. *amatymbica*
Mericarps tuberculate (rugose in some immature material); basal leaves cuneate to rounded at the base; tall and slender or short herbs:
  Tall herbs with lower cauline leaves little reduced; upper cauline leaves rather sparse:
    Involucral bracts lanceolate to narrowly triangular, acute or acuminate; lower cauline leaves congested; foliar cilia very conspicuous in basal leaves
  2. *gracilis*
    Involucral bracts triangular, obtuse or only abruptly acute; lower cauline leaves not congested; foliar cilia inconspicuous in basal leaves - - 4. *propinqua*
  Low herbs with cauline leaves strongly and progressively reduced, frequently numerous, less commonly rather sparse - - - - - - 3. *swynnertonii*

1. **Alepidea amatymbica** Eckl. & Zeyh., Enum. Pl. Afr. Austr. Extratrop. **3**: 339 (1837).— Bak. f. in Journ. Linn. Soc., Bot. **40**: 75 (1911).—H. Wolff in Engl., Pflanzenr. **4**, 228 Sanic.: 95 (1913).—Burtt Davy, F.P.F.T. **1**: 522 (1926).—Weim. in Bot. Notis. **1949**: 219 (1949). Type from S. Africa.

Robust glabrous herbs up to 1·75 m., with numerous thick lateral roots. Stems terete and strongly grooved. Basal leaves oblong, lamina up to 25 × 9 cm.; apex obtuse, base emarginate to cordate (rarely truncate); margins regularly obtuse-dentate, with the apices of the teeth prolonged into conspicuous cilia. Petioles up to 25 cm. long, with the bases dilated and distinctly stem-clasping. Stem leaves reduced upwards, those of the middle stem sessile, and distinctly auriculate with rounded lobes. Lower leaves on the inflorescence similar to the stem leaves, but with acute apices; upper ones similar but much smaller. Umbels terminal and lateral, dense, with 10–20 subsessile to very shortly stalked flowers. Bracts 6–10 × 2–3 mm., conspicuous, connate at the base, narrowly lanceolate to narrowly elliptic, apices subulate, whitish in colour. Petals greenish-cream. Fruit on thickened pedicels, up to 4 × 3 mm., ovoid to obovoid, ribs inconspicuous; calyx teeth obvious, narrowly triangular; stylopodium low-conical; styles ± persistent, short and recurved. Vittae 5, large and well developed, 1 in each interval and 2

---

[*] With the assistance of L. Constance.

in the commissural face; seed closely following the outline of the pericarp, slightly curved on the inner face.

Rhodesia. E: Umtali, Himalaya, 2275 m., fr. 3.iii.1954, *Wild* 4464 (BM; SRGH). **Mozambique.** N: Vila Cabral, fl. 29.xii.1934, *Torre* 297 (LISC). MS: Tsetserra, 2140 m., fl. 7.ii.1955, *E. M. & W.* 221 (BM; LISC; SRGH).
Also recorded from S. Africa and Lesotho. In open grassland and on stream banks.

One of the most distinct taxa in a genus otherwise noted for poorly-defined species. Placed by Weimarck in a section with only 1 additional species.

2. **Alepidea gracilis** Dümmer in Trans. Roy. Soc. S. Afr. **3**: 11 (1913).—Burtt Davy, F.P.F.T. **1**: 522 (1926).—Brenan in Mem. N.Y. Bot. Gard. **8**: 444 (1954). Type from S. Africa.
  Alepidea gracilis var. *major* Weim. in Bot. Notis. **1949**: 224 (1949). Type from S. Africa.

A fairly robust, glabrous herb up to 120 cm. with numerous fleshy roots. Stem terete with rather fine grooves. Basal leaves spathulate, lamina up to 22 × 6 cm.; apex obtuse, rarely subacute; base usually very long cuneate, rarely more abruptly so; margin dentate to serrate, with conspicuous cilia. Petioles up to 10 cm. long, merging very gradually into the lamina, base slightly dilated. Lower stem leaves similar to the basal ones, but with auriculate bases; upper stem leaves much reduced, usually sparse, with long internodes, but rarely relatively dense. Inflorescence usually a terminal group of long-peduncled umbels arranged in an umbel-like manner, otherwise in a regularly branched arrangement in which a short stalked terminal umbel is flanked by longer stalked laterals, a system which may be repeated on 3 or more levels. Umbels subcapitate, with up to 10 flowers. Bracts conspicuous, connate at the base, usually narrowly triangular to narrowly lanceolate, but sometimes somewhat broader, often with very acute to subulate apices, clear white to pinkish in colour. Petals greenish-white. Fruit 2 × 1·5 mm., obovoid to ovoid with numerous tubercles, but these perhaps not so dense as in some other species of the F.Z. area; ribs relatively conspicuous; calyx teeth obscure; stylopodium depressed; styles short, erect. Vittae 5, 1 in each interval and 1 in the commissural face; seed adnate to the pericarp and slightly concave on the inner face.

Zambia. N: Mbala Distr., fl. 2.i.1968, *Richards* 22848 (K). **Rhodesia.** C: Marandellas, 1000 m., fl. iii.1955, *Davies* 1000 (SRGH). E: Inyanga, Pungwe Falls, 1900 m., fl. 30.iv.1954, *Chase* 5228 (BM; SRGH). S: Victoria, 1200 m., fl. iv.1921, *Mainwaring* 3003 (SRGH). **Malawi.** N: Nyika, Chitipa Distr., 2100 m., *Pawek* 2082 (UC). C: Dedza, fl. 9.iv.1968, *Salubeni* 1044 (BM; SRGH). S: Mt. Mlanje, Luchenya Plateau, 2140 m., fl. 27.vi.1946, *Brass* 16460 (K; SRGH). **Mozambique.** MS: Gorongosa, Mt. Nhandore, fl. 19.x.1965, *Torre & Pereira* 12415 (LISC)
Also recorded as widely distributed in S. Africa. In montane grassland.

3. **Alepidea swynnertonii** Dümmer in Trans. Roy. Soc. S. Afr. **3**: 15 (1913).—Eyles in Trans. Roy. Soc. S. Afr. **5**: 434 (1916). TAB. **144**. Type: Rhodesia, Chimani-mani Mts., *Swynnerton* 6208a (BM; K, holotype).
  Alepidea longifolia subsp. *swynnertonii* (Dümmer) Weim. in Bot. Notis. **1949**: 232 (1949). Type as above.
Relatively slender, glabrous herbs 10–70 cm. tall with thick fleshy roots. Stem usually simple but sometimes branching from the base, terete and clearly grooved. Basal leaves oblong to oblong-spathulate; lamina up to 8 × 3 cm., apex obtuse to subacute, base long cuneate, rarely emarginate, margins dentate-serrate with obvious cilia on the teeth. Petioles up to 10 cm. long, elongating with maturity and very short in some young plants, slightly dilated at the base. Stem leaves generally regularly and progressively reduced upwards, from the base of the stem to the bottom of the inflorescence (a characteristic feature of the species); laminas oblong to narrowly triangular, the larger ones auriculate at the base. Inflorescence usually a group of umbels arranged as an umbel, with additional lower lateral umbels. Umbels long-pedunculate, very dense, subcapitate, up to c. 15-flowered. Bracts conspicuous, connate at the base, very narrowly triangular to lanceolate, white to purplish. Petals greenish-cream. Fruit 2 × 1·5 mm., obovoid, obviously verrucose at maturity; calyx teeth obscure; ribs inconspicuous; stylopodium depressed; styles very short. Vittae 5, 1 per interval and 2 commissural; seed closely adnate to the pericarp, subconvex on the inner face.

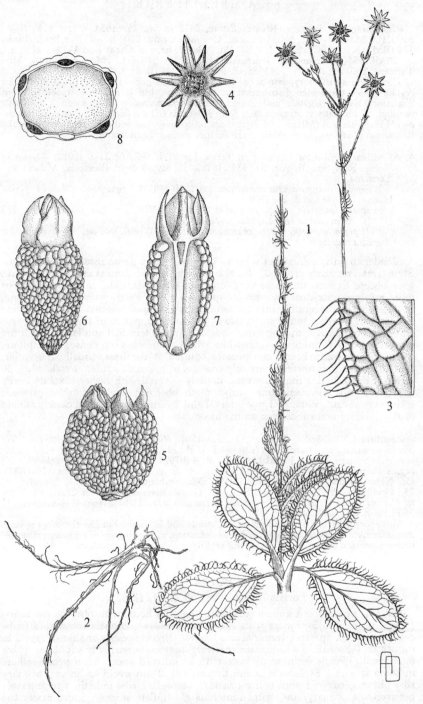

Tab. 144. ALEPIDEA SWYNNERTONII. 1, habit (×½), from *Chase* 7236; 2, rootstock (×½) from *Goodier & Phipps* 207; 3, underside of leaf (×1½); 4, umbel from above (×2), both from *Chase* 7326; 4, fruit (×15); 6, mericarp from exterior (×20); 7, mericarp, commissural view (×20); 8, mericarp in cross-section (×23), 5–8 from *Chase* 5246.

Rhodesia. E: Inyanga, Rhodes Estate, 2438 m., fl. 25.v.1954, *Chase* 5246 (BM; SRGH). **Malawi.** N: Nyika Plateau, Chelinda Camp, 2200 m., fl. 26.x.1958, *Robson* 374 (BM; K; LISC; SRGH). S: Mt. Mlanje, head of Great Ruo Valley, 2100 m., fl. 20.vii.1956, *Newman & Whitmore* 117 (BM; SRGH). **Mozambique.** MS: Tsetsera, 2140 m., fl. & fr. 7.ii.1955, *E. M. & W.* 224 (BM; LISC; SRGH).

The very unsatisfactory state of the taxonomy of this genus makes it impossible to be certain about the wider distribution of the species, but it has been recorded from Tanzania, Kenya, Uganda and Zaire; further confirmation must await a full-scale revision. The many specimens seen from Rhodesia (E) are fairly homogeneous, but the plants from Malawi (Mlanje and Nyika) are more dissimilar and their relationship to the Rhodesian plants would certainly merit further study. In montane grassland.

4. **Alepidea propinqua** Dümmer in Trans. Roy. Soc. S. Afr. **3**: 9 (1913).—Eyles in Trans. Roy. Soc. S. Afr. **5**: 434 (1916). Syntypes from Rhodesia, Malawi and Tanzania.

    ?*Alepidea coarctata* Dümmer, tom. cit.: 10 (1913). Syntypes: Malawi, Nyika Plateau, *Whyte* 156 & 247 (K).

    Alepidea longifolia subsp. *propinqua* (Dümmer) Weim. in Bot. Notis. **1949**: 233 (1949). Syntypes as for *Alepidea propinqua*.

    ? *Alepidea longifolia* subsp. *coarctata* (Dümmer) Weim., loc. cit. Syntypes as for *Alepidea coarctata*.

Slender to fairly robust herbs up to 1·25 m., with a dense mass of fleshy roots. Stem terete, regularly grooved. Basal leaves spathulate; lamina up to 25 × 4·5 cm., apex obtuse to subacute, base very long and evenly cuneate, margins serrate or more rarely serrate-dentate, with conspicuous cilia. Petioles up to 10 cm. long, but merging imperceptibly into the lamina, bases slightly dilated. Lower cauline leaves ovate, cordate to subauriculate at the base; upper stem leaves very sparse and much reduced, with long inter-nodes. Umbels terminal and lateral, with the terminal group forming an umbel-like whorl. Umbels very dense, subcapitate, up to 10-flowered; bracts conspicuous, united at the base, usually triangular, sometimes tending towards narrowly triangular, greenish-white. Petals white to cream. Fruit 2 × 1·5 mm., obovoid, densely covered with rather elongate corky tubercles; ribs inconspicuous; calyx teeth obscure; stylopodium depressed; styles very short. Vittae 5, 1 per interval and 2 commissural; seed closely adnate to the pericarp and subconvex on the inner face.

Zambia. N: Mpika, 1200 m., fr. 5.iv.1961, *Richards* 14996 (K; SRGH). W: Mwinilunga, E. of Lusaro R. (?Lusavo R.), fl. 20.i.1938, *Milne-Redhead* 4263 (BM: K). C: Serenje Distr., Kundalila Falls, fl. 4.ii.1973, *Strid* 2833 (K). **Malawi.** N; Nyika Plateau, Livingstonia, 1200–1500 m., fl. & fr. 4.xi.1953, *Chapman* 176 (BM). C: Nkhota Kota, fl., *Benson* 552 (PRE). **Mozambique.** N: Ribáuè, Mepalué, fl. 25.i.1964, *Torre & Paiva* 10204 (LISC). Z: Gúruè, Serra do Gúruè, fl. 21.ii.1966, *Torre & Correia* 14755 (LISC). MS: Báruè, Serra de Choa, 23 km. from Vila Gouveia, fl. 26.iii.1966, *Torre & Correia* 15433 (LISC).

Also recorded from Tanzania, Kenya, Uganda and Ethiopia, but like the other species this information should not be regarded as definitive, in the absence of a modern comprehensive revision. In montane grassland and open *Brachystegia* woodland.

5. TORILIS Adans.
Torilis Adans., Fam. Pl. **2**: 99 (1763).

Erect or procumbent annual (or rarely biennial) herbs with obvious tap roots. Leaves pinnate to 3-pinnate with sheathing petioles. Umbels compound (subcapitate in some species), terminal and lateral. Bracts present or absent, rays 2 to numerous (obsolete in subcapitate species); bracteoles narrowly subulate; calyx teeth small. Petals white to pinkish, with an inflexed apex; outer petals radiate in some species. Stylopodium shortly conical. Fruit ovoid to linear; primary ribs filiform, covered with spiny bristles; secondary ribs indistinct in intervals between the primary ribs, with numerous glochidiate spines. More rarely the spines of 1 or both mericarps replaced with tubercles. Carpophore 2-cleft at the apex or split to one half of its length; vittae solitary in the intervals and 2 in the commissural face. Seed dorsally flattened, with a concave face towards the commissure, lateral margins of the endosperm not incurved.

A genus of c. 10 species centred in the Mediterranean region.

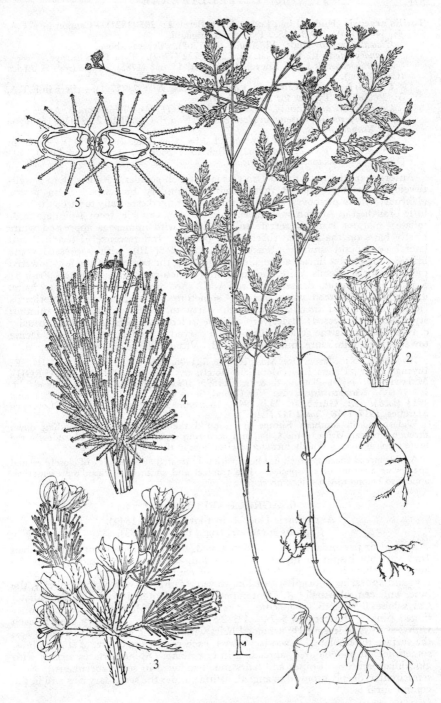

Tab. 145. TORILIS ARVENSIS. 1, habit (×½); 2, leaflet (×6), both from *Taylor* 1254 and
Mackinder sn; 3, umbel (×9); 4, fruit (×15); 5, mericarps in cross-section (×15),
3–5 from *Schlieben* 5054.

Torilis arvensis (Huds.) Link, Enum. Hort. Berol. **1**: 265 (1821).—Cannon in C.F.A. **4**: 340 (1970). TAB. **145**. Type from England.

　　Caucalis arvensis Huds., Fl. Angl.: 98 (1762). Type as above.
　　Scandix infesta L., Syst. Nat. ed. 12, **2**: 732 (1767). Type from Europe.
　　Caucalis capensis Lam., Encycl. Méth. Bot. **1**: 658 (1785). Type from S. Africa (Cape Prov.).
　　Caucalis infesta (L.) Curtis, Fl. Lond. ed. 1, fasc. 6: t. 23 (1791).—Hiern in F.T.A. **3**: 26 (1877). Type as for *Scandix infesta*.
　　Caucalis africana Thunb., Prodr. Pl. Cap.: 49 (1794). Type from S. Africa.
　　Torilis infesta (L.) Spreng. in Neue Schrift. Naturf. Ges. Halle **2**: 24 (1813). Type as for *Scandix infesta*.
　　Torilis africana (Thunb.) Spreng., Pl. Min. Cog. Pug. **2**: 55 (1815).—Sond. in Harv. & Sond., F.C. **2**: 564 (1862).—Robyns & Tournay, Fl. Parc Nat. Alb. **1**: 703 (1948). Type as for *Caucalis africana*.

An erect annual herb up to 1 m., usually much smaller. Root system a well-developed primary taproot, with limited secondary branches. Stem terete, glabrous, with fine grooves, sometimes purplish-tinged especially towards the base, little branched in African specimens. Leaves very variable, from 2-pinnate to 3-foliate; margins coarsely serrate to subentire, with numerous appressed white bristly hairs on the lamina, rhachis and petiole. Inflorescence of terminal and lateral compound umbels; bracts 0(1), rays 2–5(–10), with appressed white bristles; bracteoles linear, acute, almost equalling the pedicels of the few-flowered partial umbels. Petals white, the outer ones sometimes slightly radiate. Fruit 3–5 mm. long, ovoid, densely covered with spreading, spiny glochidiate hairs, usually blackish-green when mature; sometimes 1, or rarely both, mericarps lacking spines and then covered with warty tubercles. Calyx teeth minute; stylopodium depressed; styles very variable in length, in African material usually c. twice as long as the stylopodium. Seed curved in section, with the ends facing towards the commissure, but not recurved. Vittae 5, large and well developed.

Rhodesia. N: Mazoe, Iron Mask Hill, fr. iii.1906, *Eyles* 258 (BM; K; SRGH). E: Inyanga Distr., Danna Kay (Montclair Hotel), fr. 13.iv.1955, *Whellan* 824 (SRGH). **Malawi.** N: Nyika Plateau, fl. & fr. vii.1896, *Whyte* 158 (K). S: Mt. Chiradzulu, *Whyte* (K). **Mozambique.** N: Vila Cabral, Serra Gessi, fl. v.1928, *Pedro & Pedrógão* 4013 (LM). Z: Gúruè, fr. 23.ix.1944, *Mendonça* 2235 (LISC). LM: ?Lourenço Marques, 1917–1918, *Junod* 347 (LISC).

Widespread in southern Europe and around the Mediterranean, extending down through tropical Africa to the Cape, and occurring especially at higher altitudes in the more tropical areas. Widely introduced elsewhere in the world.

A widespread species, which has been variously treated as a complex of closely related species or a group of subspecies. The tropical and southern African representatives appear to belong to the subsp. *arvensis*.

6. AGROCHARIS Hochst.*

Agrocharis Hochst. in Flora **27**: 19 (1844).
　　Caucalis auctt. afr. trop., non L. (1753).

Annual or perennial herbs, weak-stemmed, often semi-scandent to more than 1m., variably hispid to hirsute. Leaves long-petiolate, 2–4 times pinnatisect, ovate to deltate in outline. Inflorescence of compound umbels, with very short rays and appearing subcapitate, or long-rayed. Outer flowers of the umbel ♀, the inner and central usually ♂. Outer petals of the outer flowers slightly radiate. Calyx-lobes 0·5–2 mm. long, acute. Petals greenish or yellowish-white, sometimes tinged pink. Mericarps 4·5–7·0 × 1·5–2·5 mm., oblong. Primary ribs 5, each with 2–3 rows of irregularly arranged hairs with tapering ends and swollen bases, the surface tuberculate. Secondary ribs 4, each with a single row of stout spines, sometimes purple-coloured, spreading in one plane. Spine surfaces striate, with glochidiate apices. (Spines and hairs much reduced in some specimens.) Seed semicircular, deeply grooved to angled. Vittae under the secondary ribs and in the commissural face.

Umbels with very short rays, appearing subcapitate; leaves 2-pinnatisect with ultimate
　　segments deeply and irregularly lobed, lobes ovate to cuneate　　-　　1. *incognita*
Umbels with distinct and often long rays; leaves 3–4-pinnatisect with ultimate segments
　　often subverticillate in appearance, linear to lanceolate -　　-　　- 2. *pedunculata*

* by V. H. Heywood and S. L. Jury

1. **Agrocharis incognita** (Norman) Heywood & Jury, comb. nov. TAB. **146.** Type from Uganda.

    Torilis eminii Engl. in Abh. Königl. Preuss. Akad. Wiss. Berlin **1894**: 58 (1894) *nom. nud.*

    Torilis gracilis Engl., Pflanzenw. Ost-Afr. **C**: 301 (1895) quoad specim. cit., excl. basionym Hook. f. et syn. Vatke. Syntypes from Tanzania and Malawi.

    Torilis gracilis Engl. forma *umbrosa* Engl., loc. cit. Type from Tanzania.

    Caucalis gracilis (Engl.) H. Wolff ex R. E. Fr., Wiss. Ergebn. Schwed. Rhod.-Kongo-Exped.: 183 (1914).—Engl., Pflanzenw. Afr. **3**, 2: 802 (1921) excl. basionym Hook. f. Syntypes as for *Torilis gracilis.*

    Caucalis gracilis subsp. *umbrosa* (Engl.) Engl., loc. cit. (1921). Type as for *Torilis gracilis* forma *umbrosa.*

    Caucalis gracilis forma *umbrosa* (Engl.) H. Wolff in Notizbl. Bot. Gart. Berl. **9**: 1111 (1927). Type as for *Torilis gracilis* forma *umbrosa.*

    Caucalis gracilis forma *typica* H. Wolff, loc. cit. Syntypes as for *Torilis gracilis.*

    Caucalis incognita Norman in Journ. Bot., Lond. **72**: 205 (1934).—Robyns & Tournay, Fl. Parc Nat. Alb. **1**: 701 (1948).—Brenan in Mem. N.Y. Bot. Gard. **8**: 447 (1954). Type as for *Agrocharis incognita.*

An annual to short-lived perennial, straggling to semi-climbing herb up to 2 m., with a slender taproot. Stem terete, striate, often tinged with purple especially in the lower parts, generally rather sparsely branched, with numerous tubercle-based downward-pointing bristles. Leaves 2-pinnatisect, the ultimate segments shallowly to deeply lobed, 2–5 mm. wide, ovate, rather variable but always broader than those of *A. pedunculata*, shortly acuminate at the apex; leaf bases sheathing with a distinct membranous margin; lamina with numerous white bristles on the veins of the inferior surface, these also present on the superior surface but less numerous. Umbels subcapitate, terminal and lateral on long stalks up to 45 cm., with bristly hairs especially towards the apex, becoming scabrid below due to the persistent hair-bases. Bracts numerous, lanceolate to linear, fringed with white hairs; rays of umbels up to 7 mm. long, sparsely hairy. Outer flowers ☿, the inner and central usually ♂. Petals greenish or yellowish-white, sometimes tinged with pink. Fruit 5–7 × 1·5–2·0 mm., narrowly oblong-ellipsoid. Primary ribs with 2–3 rows of irregularly spreading hairs. Secondary ribs with a single row of stout glochidiate spines. Stylopodium shortly conical; styles c. 0·5 mm. long, short and stiff. Vittae well developed, situated under the secondary ribs and in the commissural face.

Rhodesia. E: Umtali, Himalaya Mt., Banti North, fr. 4.xi.1954, *Wild* 4512 (BM; SRGH). **Malawi.** N: Vipya, Kawendama, fr. 15.ii.1962, *Chapman* 1590 (K; SRGH). S: Zomba Mt., Mulunguzi R., fr. 9.iii.1955, *E. M. & W.* 763 (BM; LISC; SRGH). **Mozambique.** MS: Penhalonga Forest, above waterfall, fr. 27.viii.1950, *Chase* 2895 (BM; SRGH).

Also present in the Sudan, Kenya, Uganda, Tanzania and Zaire. In open grassland and under light forest shade.

2. **Agrocharis pedunculata** (Bak. f.) Heywood & Jury, comb. nov.

    Caucalis pedunculata Bak. f. in Trans. Linn. Soc., Lond. Ser. 2, **4**: 15 (1894).—Brenan in Mem. N.Y. Bot. Gard. **8**: 447 (1954). Type: Malawi, Mt. Mlanje, *Whyte* 160 (BM).

    Caucalis longisepala Engl. ex H. Wolff in Engl., Bot. Jahrb. **57**: 223 (1921). Syntypes from Tanzania and Malawi, *Buchanan* 42 & 130 (BM).

    Caucaliopsis stolzii H. Wolff in Engl., tom. cit.: 222 (1921). Type from Tanzania.

    Gynophyge tansaniensis Gilli in Fedde, Repert. **84**: 181 (1973). Type from Tanzania.

An erect to decumbent, pubescent to subglabrous long-lived perennial herb up to 60 cm. with a stout woody taproot. Stem terete, distinctly grooved, sparsely to densely hairy with stiff white backwardly directed subappressed hairs, or sometimes glabrous, sometimes with brownish-purple colouration. Hair-bases tending to become swollen and on old stems remaining as tubercles after the hairs have been eroded. Leaves 3–4-pinnatisect; ultimate segments up to 2 mm. wide, often subverticillate in appearance, linear to linear-lanceolate; petioles with broad sheathing bases. Petiole and rhachis usually with tubercle-based hairs, but sometimes glabrous; lamina with occasional hairs on the veins. Umbels usually on an elongate stalk of 5–50 cm., usually distinct with long rays, rarely subcapitate; bracts variable in shape and number, ranging from entire, linear-

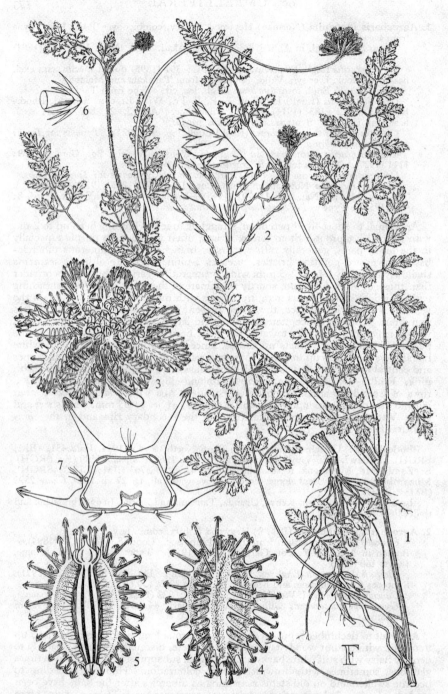

Tab. 146. AGROCHARIS INCOGNITA. 1, habit (×½); 2, underside of leaflet (×6), both from *Chapman* 327 and *Taylor* 1023; 3, fruiting umbel (×3), from *Scheffler* 269; 4, mericarp from exterior (×6); 5, mericarp, commissural view (×6); 6, tip of single glochidiate spine (×24); 7, mericarp in cross-section (×12), 4–7 from *Volkens* 238.

lanceolate to 3-pinnatisect like the leaves, variously hairy. Rays up to 14, up to 3 cm. long, rarely very abbreviated. Outer partial umbels with ☿ outer flowers and ♂ inner and central ones, the innermost with ♂ flowers only; bracteoles similar to the bracts. Petals yellowish-green to occasionally purplish; stylopodium narrowly conical, tapering imperceptibly into erect styles up to 2 mm. long. Fruit 6–7(8) × 2 mm., narrowly oblong-ovoid to cylindrical. Primary ribs with 2–3 rows of short stiff bristly hairs. Secondary ribs with a single row of glochidiate spines. Vittae well developed, situated under the secondary ribs and in the commissural face.

Rhodesia. C: Domboshawa, fr. 7.ii.1938, *Bacon* 6827 (K; SRGH). E: Inyanga Distr., Stapleford, Simla Farm, fr. xi.1934, *Gilliland* 1049 (BM; K; SRGH). **Malawi.** N: Nyika Plateau, L. Kaulime, fl. & fr. 24.x.1958, *Robson* 317a (BM; K; LISC; SRGH). C: Nkhota Kota Distr., Chenga Hill, fl. 9.ix.1948, *Brass* 17592 (K). S: Kirk Range, Gochi Rest House, fl. & fr. 30.i.1959, *Robson* 1347 (BM; K; LISC; SRGH). **Mozambique.** N: Vila Cabral, fl. x–xii.1934, *Torre* 440 (COI; LISC). Z: between Milange and Alto Chindio, 41 km. from Milange, fr. 11.ix.1944, *Barbosa & Carvalho* 4028 (K; LM; LMA). T: Angonia, Vila Mouzinho, fr. 15.x.1943, *Torre* 6043 (LISC).

Southern Sudan, Kenya, Uganda, Tanzania and Zaire. In mountain grassland, especially in frequently burned areas, boggy areas and in open woodland.

7. CORIANDRUM L.

Coriandrum L., Sp. Pl. **1**: 256 (1753); Gen. Pl. ed. 5: 124 (1754).

Slender, glabrous annual herbs with narrow taproots. Leaves thin and membranous, mostly finely dissected (in F.Z. species the basal leaves broadly lobed and strikingly different from the upper ones). Inflorescence of terminal and lateral, few-rayed compound umbels. Petals white to pinkish with a clearly inflexed tip, the outer ones of the umbels noticeably radiate. Calyx teeth obvious and relatively large, often rather unequal. Stylopodium conical; styles long and slender, with a slight terminal stigmatic enlargement. Fruit suborbicular, the mericarps tending to remain joined, even at maturity. Ribs slender, and low; vittae absent or obscure; carpophore divided to the base; seed somewhat concave on the inner face.

A genus of 2 species, probably native to the Mediterranean region.

Coriandrum sativum L., Sp. Pl. **1**: 256 (1753).—Hiern in F.T.A. **3**: 3 (1877). TAB. **147**. Type from Italy.

An erect annual, almost glabrous herb up to 70 cm., with a slender taproot and a strong unpleasant smell. Stems often simple, but in large plants may be profusely branched, with fine regular grooves. Lower leaves 1–2-pinnate, segments 10–20 × 5–10 mm., ovate in overall outline, with a regularly cuneate base, somewhat unevenly divided into pinnate lobes. Upper leaves finely divided into a linear to narrowly linear segments 2–15 × 0·5–1·0 mm., obviously markedly different from the basal leaves but sometimes connected by intermediate types on the lower-middle part of the stem. Umbels terminal and lateral; bracts O; rays 3–7(10), 1–2 cm. long; partial umbels with 2–7 flowers; bracteoles few, linear-lanceolate. Petals white to pinkish-white; the outer ones distinctly radiate. Fruit 2–3 mm. in diameter, suborbicular, slightly longer than broad. Styles long and slender, often eroded from the conical stylopodium before maturity. Calyx teeth narrowly triangular and quite conspicuous. Mericarps with obtuse filiform ribs, with zigzag markings between them. Vittae absent; seed concave on the inner face: carpophore divided to the base.

Rhodesia. N: Nicotina, Darwendale, fl. & fr. 20.viii.1945, *Raynor* in GHS 13480 (BM; SRGH). W: Nyamandhlovu Distr., 84 km. NW. of Bulawayo, fl. & fr. v.1958, *Martin* in GHS 85013 (SRGH). **Mozambique.** SS: Chibuto, Baixo Changa, fr. 23.viii.1963, *Macedo & Macuácua* 1129 (K; LM). LM: Lourenço Marques, fl. & fr. 6.xi.1959, *Pedro* 4975 (LM).

Widely distributed throughout the world as a result of planting and subsequent escape from cultivation. Doubtless introduced into the F.Z. area.

The fruits are widely used for flavouring as the herb Coriander.

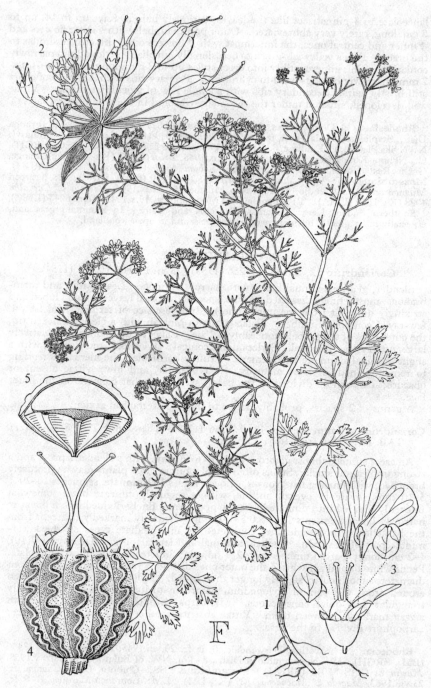

Tab. 147. CORIANDRUM SATIVUM. 1, habit (×½), from *Meinertzhagen* sn; 2, umbel (×4½), from *Hundt* 906; 3, flower (×6), from *Meinertzhagen* sn; 4, fruit (×9); 5, mericarp in cross-section (×9), both from *Hundt* 906.

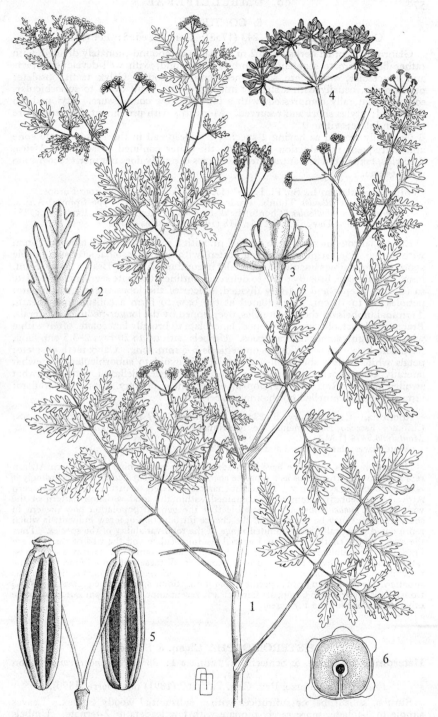

Tab. 148. CONIUM MACULATUM. 1, habit (×½); 2, underside of leaflet (×3); 3, flower
(×10); 4, umbel with mature fruit (×½); 5, fruit (×8); 6, mericarp in cross-section
(×12), all from *Barbosa & Lemos* 7860.

T

8. CONIUM L.

Conium L., Sp. Pl. **1**: 243 (1753); Gen. Pl. ed. 5: 114 (1754).

Glabrous, biennial herbs up to 2 m. Leaves compound-pinnately divided, with rather fine ultimate segments. Umbels compound, with well-developed bracts and bracteoles, which are sometimes semi-caducous. Calyx teeth obsolete; petals white, usually with a slightly inflexed apex. Fruit ovoid to suborbicular, somewhat laterally compressed, with a fairly narrow commissure. Stylopodium depressed; styles short and recurved. Mericarps with prominent ribs, which are often undulate; vittae 0.

Generally treated as having 2 species, 1 widespread in the N. temperate zone and a frequent introduction elsewhere, the other confined to southern Africa. The view taken here is that the African species certainly does not merit recognition at that level.

Conium maculatum L., Sp. Pl. **1**: 243 (1753). TAB. **148**. Type from Europe.
Seseli chaerophylloides Thunb., Prodr. Pl. Cap.: 51 (1794 Type from S. Africa.
Conium chaerophylloides (Thunb.) Eckl. & Zeyh., Enum. Pl. Afr. Extratrop.: 355 (1837).—Burtt Davy, F.P.F.T. **1**: 518 (1926). Type as above.

Erect, glabrous biennial herb up to 2 m. with a somewhat fleshy taproot. Stem with rather fine, regular grooving, often with irregular, characteristic purple spots, but sometimes unspotted especially in southern Africa. Leaves up to 35 cm. long, 2–3-pinnate, broadly ovate to deltate in outline; ultimate segments lanceolate to elliptic, deeply and coarsely divided, apices of the lobes slightly mucronate; petioles up to 10 cm. long, dilated at the base to form a conspicuous sheath. Terminal umbels on short peduncles, overtopped by the longer-peduncled laterals. Bracts and bracteoles well developed, lanceolate to broadly lanceolate, often with a whitish margin, somewhat caducous. Umbels with up to 20 rays 2–3·5 cm. long, partial umbels with 6–15 flowers on pedicels 2–5 mm. long. Calyx teeth obsolete; petals white with a short inflexed tip. Fruit ovoid to suborbicular, somewhat laterally compressed; stylopodium depressed; styles deflexed with somewhat swollen bases. Ribs well developed, either linear or very distinctly undulate; vittae 0; seed channelled on the inner face.

Rhodesia. C: Salisbury, fl. 27.viii.1974, *Biegel* 4559 (K). **Mozambique.** SS: Chibuto, Estacão Experimental de Marriquerique, fl. & fr. 8.ix.1956, *Barbosa &* *Montalvão* 7474 (LMA).
Waste places, damp ground and streamsides. Very poisonous!

No good reasons appear to have been advanced for maintaining the southern African representatives of this genus as a separate species (*C. chaerophylloides*), and accordingly it is here reduced to synonymy. The slight character differences attributed to the southern African representative seem to be contained within the total variation pattern of the widespread *C. maculatum*. It is possible that the general population now present in southern Africa may be the result of the chance introduction of a few individuals which represented genetically a very limited range of the total variability of the species. Thus the stems of the southern African material are immaculate and the ribs of the fruit are linear and not undulate (var. *leiocarpum* in Europe); at the same time the leaves seem to be rather more finely divided than is usual. However, all these states are found within the variability of *C. maculatum* and, while experimental work may throw further light on this situation, it is unlikely to justify specific recognition. Some specimens, e.g. *Biegel* 4559 (K) from a Salisbury rubbish dump, are fairly clearly recent importations from Europe or some other region outside the F.Z. area.

9. HETEROMORPHA Cham. & Schlecht.

Heteromorpha Cham. & Schlecht. in Linnaea **1**: 385 (1826) *nom. conserv.* non Cass. (1817).
Franchetella Kuntze, Rev. Gen. Pl.: 267 (1891) non Pierre (1890).

Shrubs, subshrubs, or suffrutices with a substantial woody caudex. Leaves simple to 3-foliate, more rarely pinnate with few leaflets or 2-ternate. Umbels compound with well-developed bracts and bracteoles, though these tend to be caducous. Flowers with relatively conspicuous triangular calyx teeth; petals broadly ovate-oblong with an inflexed apex. Stylopodium conical, with short, divergent styles. Fruit obovate, slightly laterally compressed, with well-developed lateral wings. One mericarp usually with a winged dorsal rib, the dorsal rib of the other mericarp being filiform. Vittae solitary in the intervals, with 2 in the commissural face; sometimes there is a development of secondary vittae in the wings. Carpophore divided to the base; seed obtusely triangular in section, somewhat convex on the commissural face.

A genus of distinctly woody plants in a family composed characteristically of herbs, a feature it shares in the F.Z. area only with *Steganotaenia*. It can be at once distinguished from shrubs belonging to other families by its typically umbelliferous inflorescences and fruit. A wholly satisfactory classification has yet to be achieved for this genus, which has a remarkable degree of poorly-defined variability. *H. trifoliata*, as here interpreted, has a very wide distribution through S., Central and E. tropical Africa, while the suffrutescent species appear to be confined to Angola, Zaire and Zambia. The relationship of the latter to the truly shrubby species is unclear, and it may be that later work will show that intense ecotypic adaptation, rather than wide genetic divergence, is involved. One gathering from Zambia, Mwinilunga Distr., *Milne-Redhead* 4315 (K), has been referred to the poorly-known *H. angolensis* (Norman) Norman. This is similar to *H. kassneri*, but is a smaller and more slender plant, with ternate or simple, narrowly elliptic basal leaves. These taxa may prove to be conspecific when further material is available.

Robust shrubs (or subshrubs) of 1–3 m., usually much branched from the base
1. *trifoliata*
Suffrutices with ± simple stems rising from a woody caudex   -   -   2. *kassneri*

1. **Heteromorpha trifoliata** (Wendl.) Eckl. & Zeyh., Enum. Pl. Afr. Extratrop.: 342 (1837).—Burtt Davy, F.P.F.T. **2**: 519 (1932).—Miller in Journ. S. Afr. Bot. **18**: 66 (1952).—Martineau, Rhod. Wild Fl.: 57 (1954).—Brenan in Mem. N.Y. Bot. Gard. **8**: 445 (1954).—F. White, F.F.N.R.: 313 (1962).—Cannon in C.F.A. **4**: 341 (1970). TAB. **149**. Type from S. Africa.

Bupleurum trifoliatum Wendl. in Bartl. and Wendl., Beytr. **2**: 13 (1825). Type as above.

Heteromorpha arborescens sensu Hiern in F.T.A. **3**: 10 (1877) pro parte.—Sim, For. Fl. Port. E. Afr.: 69 (1909).—H. Wolff in Engl., Pflanzenr. **4**, 288 Ammin. Het.: 33 (1910) pro parte.—Bak. f. in Journ. Linn. Soc., Bot. **40**: 75 (1911).—R. E. Fr., Wiss. Ergebn. Schwed. Rhod.-Kongo-Exped.: 183 (1914).—Eyles in Trans. Roy. Soc. S. Afr. **5**: 434 (1916).—Steedman, Trees etc. S. Rhod.: 59 (1933).—Burtt Davy & Hoyle, N.C.L.: 74 (1936).—Brenan & Greenway, T.T.C.L. **2**: 626 (1949).—Williamson, Useful Pl. Nyasal.: 65 (1955).

?*Heteromorpha stolzii* H. Wolff in Engl., Bot. Jahrb. **57**: 225 (1921). Type from Tanzania.

Shrubs up to 3 m., much branched from the base, with terete woody stems. Bark of the older parts greyish, that of the younger green to dark brown, with somewhat irregular striations. Generally glabrous but the younger growths sometimes have short, rather irregular, white hairs. Leaves generally 3-foliolate, but sometimes imperfectly pinnate, or even tending towards a 2-pinnate condition; compound ternate leaves are sometimes found on young shoots. Leaflets very variable, up to 10 × 4 cm., lanceolate, obovate, spathulate to linear-lanceolate. Apex acute to subulate, to obtuse with a terminal mucro; base cuneate to rounded, occasionally extending down the petiole or rhachis as an expanded wing; margin ± entire, sometimes slightly undulate. Umbels terminal and lateral on the younger growths. Terminal umbels with up to 25 rays, up to 4·5 cm. long, glabrous or sparsely pubescent; bracts 3–10, narrowly lanceolate, tending to fall before the fruit is fully mature. Partial umbels with up to 25 flowers; bracteoles 0–5, similar to the bracts. Calyx teeth minute, narrowly triangular; petals pale yellow to yellowish-green, with an inturned apex and distinctly darker mid-vein. Fruit up to 7 × 4 mm., oblong to obovoid; stylopodium conical with a somewhat undulate base; styles short and clubbed at the apex. Mericarps variable in wing development, the commonest configuration being one of a pair with lateral wings and a winged dorsal rib, the other with lateral wings but a filiform dorsal rib. Vittae solitary in the intervals and 2 in the commissural face, sometimes with the development of accessory vittae in the wings. Carpophore divided to the base; seed subtriangular in section, with a slightly flattened inner face.

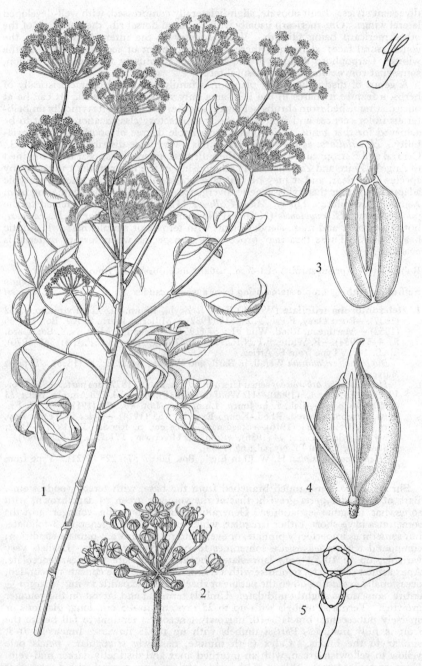

Tab. 149. HETEROMORPHA TRIFOLIATA. 1, habit (×½); 2, partial umbel (×4), both from
Chase 4846; 3, mericarp from exterior (×6); 4, mericarp, commissural view (×6);
5, mericarps in cross-section (×6), 3–5 from *Chase* 701.

Botswana. SE: Serowe, 950 m., fl. & fr. 15.iv.1957, *de Beer* B2 (K; SRGH).
Zambia. N: Nyika, Isoka, fr. 20.vi.1960, *Chapman* 774 (SRGH). W: Sosa Hills,
Ndola, fr. 1.v.1952, *Holmes* (K; SRGH). C: Serenje, fl. 28.iii.1961, *Angus* 2556 (K;
SRGH). S: Mapanza, 1000 m., fl. 12.ii.1955, *Robinson* 1095 (SRGH). **Rhodesia.** N:
Mazoe, 1300 m., fl. & fr. iii.1906, *Eyles* 279 (BM). W: Bulawayo, fr. v.1898, *Rand* 329
(BM). C: Twentydales road, Salisbury, fr. 3.ii.1961, *Rutherford-Smith* 484 (BM). E:
Bideford Hill, Inyanga, fr. 26.iv.1966, *West* 7283 (BM; SRGH). S: Victoria, fl., *Monro*
320 (BM). **Malawi.** N: Nyika Plateau, 2340 m., fr. 20.viii.1946, *Brass* 17342 (BM; K;
PRE; SRGH). C: Namitete R., between Lilongwe and Ft. Jameson, 1150 m., fl.
5.ii.1959, *Robson* 1481 (BM; K; SRGH). S: Luchenya Plateau, Mt. Mlanje, 2100 m.,
fr. 27.vi.1946, *Brass* 16483 (BM; K; PRE; SRGH). **Mozambique.** N: 37 km. E. of
Malema, fl. 16.v.1961, *Leach & Rutherford-Smith* 10877 (K; LISC; SRGH). Z:
Cascata, Gúruè, 1500 m., fr. 18.ix.1944, *Mendonça* 2125 (LISC). T: Zóbuè Mt., fr.
2.x.1942, *Mendonça* 556 (LISC). MS: Gogogo Mt., Gorongoza, 1800 m., fl. 5.vii.1955,
Schelpe 469 (BM). SS: Inharrime, Inhambane, fr. 22.vi.1960, *Lemos & Balsinhas* 162
(BM; K; LISC; LMA; PRE; SRGH). LM: Gaza, between Vila João Belo and
Loumane, fl. 13.viii.1944, *Torre* 6740 (LISC).
Probably very widely distributed from S. Africa, SW. Africa and Angola, through
E. Africa to Ethiopia, but the uncertain state of the systematics of this genus makes
a definitive statement impossible. In montane and riverine woodland, in forest margins
and in secondary regrowth.

Advantage has been taken of the large pith cavity of *Heteromorpha* branches by some
Rhodesian communities, who make use of it in fire making, a hardwood stick being twisted
between the hands in a conical cavity in the *Heteromorpha* branch. Williamson (1955)
states " an infusion of the root mixed with others is drunk to cure colds on the chest and
also for one of the various venereal diseases."

Broad and narrow-leaved variants of this species have been accorded recognition at
various levels, e.g. *H. stenophylla* Welw. ex Schinz in Bull. Herb. Boiss. **2**: 207 (1894) and
Franchetella arborescens var. *platyphylla* Welw. ex Hiern, Cat. Afr. Pl. Welw. **1**: 424 (1898).
Individuals with very narrow leaves do look superficially very different from those with
broad leaves. However, intermediate forms do seem to occur, and until further field and
experimental studies have established the nature of the variation, it is best to note the
variability, but not to accord formal taxonomic recognition.

2. **Heteromorpha kassneri** H. Wolff in Engl., Bot. Jahrb. **57**: 226 (1921). Type from
   Zaire (*Kassner* 2277 (cited as 2211) from Zambia, Bwana Mkubwa (K) is cited as
   " probably the same species ").

Robust suffrutex with a woody caudex and usually simple stems; generally
glabrous, but parts of the younger stems, rays etc. may be somewhat pubescent.
Stems terete, distinctly grooved. Leaves 3-foliolate or pinnate with 2 pairs of
pinnae only. Leaflets up to 9 × 4 cm., variable, ovate, oblong or narrowly elliptic.
Apex acute to obtuse with a terminal mucro; base cuneate, frequently extending
down the rhachis and/or petiole to form a conspicuous wedge-shaped wing.
Umbels terminal and lateral, the terminal ones often characteristically large, with
up to 50 rays up to 7 cm. long; bracts few, linear-lanceolate, semi-caducous.
Calyx teeth minute, narrowly triangular; petals with an inturned apex, probably
pale yellow. Fruit up to 7 × 4 mm., obovoid; stylopodium depressed-conical
with an undulate base; styles short and divergent, clubbed at the apex. Mericarp
wings and other fruit characters as described under *H. trifoliata*.

Zambia. N: 72 km. W. of Kasama, fl. & fr. 4.iv.1961, *Angus* 2706 (SRGH). W:
Mulenga Forest area, 32 km. NW. of Kansanshi, Solwezi Distr., fr. 19.iii.1961, *Drummond
& Rutherford-Smith* 7058 (BM; K; LISC; SRGH). S: Mapanza, fl. & fr. 12.ii.1955,
Robinson 1095 (K).
Known elsewhere from Zaire (Katanga) and southern Tanzania. In open *Brachystegia*
woodland.

The validity of this species will only be established in the field and through cultivation
experiments. Herbarium material gives the impression of a substantial suffrutex with a
woody caudex, with robust, probably simple aerial shoots. *Steganotaenia hockii* q.v.
provides a close parallel in life form in a widely separated genus, but with a similar
restricted distribution. Further ecological studies on the suffrutices in this area will no
doubt illuminate their origin and evolution. The relationship of *H. kassneri* to *H. trifoliata*
is obviously close, even though it presents a rather individual and characteristic general
facies. The other suffrutescent species that have been described from Angola (*H.
gossweileri* and *H. angolensis*) are both reminiscent of *H. kassneri*, but are much smaller
and more slender, and tend towards a further reduction and simplification of the leaves.

10. APIUM L.

Apium L., Sp. Pl. **1**: 264 (1753); Gen. Pl. ed. 5: 128 (1754).

Robust or slender annual, biennial or perennial herbs; stems erect, decumbent or creeping. Flowers white to greenish in compound umbels; bracts and bracteoles 0. Calyx teeth minute or absent; petals ovate to suborbicular, with an inflexed apex. Stylopodium shortly conical or depressed. Fruit ovoid to globose, somewhat laterally compressed, glabrous to slightly setulose, slightly constricted at the commissure. Ribs thick and rounded, vittae solitary in the intervals, with 2 on the commissural face. Carpophore entire or slightly 2-fid; seed round to sub-pentagonal in section, with the commissural face plane.

A genus of 20–30 species, widely distributed through both the temperate and tropical regions, especially in temperate S. America. The cultivated Celery (*A. graveolens* L.) is a member of this genus, and is grown in several very distinct varieties for its leaves, petioles and roots.

Weak annual herbs with fine, linear leaf-segments -   -   -   - 1. *leptophyllum*
Fairly robust biennial to perennial herbs, with broad leaf-segments   -   2. *graveolens*

1. **Apium leptophyllum** (Pers.) F. Muell. ex Benth., Fl. Austr. **3**: 372 (1866).—Sprague in Journ. Bot., Lond. **61**: 129 (1923).—Burtt Davy, F.P.F.T. **1**: 518 (1926).—Cannon in F.W.T.A. ed. 2, **1**: 754 (1958). TAB. **150**. Type from W. Indies.
    Sison ammi Jacq., Hort. Vindob. **2**: t. 200 (1772) non L. (1753) *nom. illegit.*
    Pimpinella leptophylla Pers., Syn. Pl. **1**: 324 (1805). Type as for *Apium leptophyllum.*
    Apium ammi sensu Urb. in Mart., Fl. Bras. **11**, 1: 341 (1879).

Low, weak, glabrous annual herbs up to 50 cm., sometimes branched from the base with a number of stems of equal magnitude, sometimes branched above with a distinct main stem, often semi-decumbent. Stem terete, with relatively coarse grooving. Leaves 3–4-pinnate, with narrowly linear ultimate segments 2–20 × 0·5 mm., those of the basal leaves sometimes relatively shorter and coarser; petioles with distinct sheathing bases with membranous wings. Umbels often subsessile and appearing superficially like a group of simple umbels, mostly lateral in the axils of leaves, but when present the peduncles may reach 2 cm. long. Rays 2–5, 7–15 mm. long; bracts and bracteoles 0; partial umbels with 5–15 flowers on fine, rather irregular pedicels 2–6 mm. long. Calyx teeth minute; petals white. Fruit broadly elliptic to nearly round, quite strongly laterally compressed; stylopodium small, shortly conical; styles minute with almost sessile stigmatic surfaces. Ribs strongly developed and rather obtuse, with only narrow intervals between them at maturity. Carpophore very slightly 2-cleft at the apex; seed subterete to pentagonal in section; vittae minute.

Botswana. N: Maun, fl. & fr. v.1967, *Lambrecht* 207 (BM; SRGH). **Zambia.** W: Mwinilunga Boma, fl. & fr. 3.xii.1937, *Milne Redhead* 3059 (BM; K). **Rhodesia.** W: Bulawayo, Municipal Park, fl. & fr. 30.iv.1958, *Drummond* 5525 (K; SRGH). C: Selukwe, fl. & fr. 8.xii.1953, *Wild* 4310 (BM; SRGH). E: Umtali, La Rochelle, Imbesa, fl. & fr. 14.ix.1956, *Chase* 6201 (BM; K; LISC; LMA; SRGH). **Mozambique.** LM: Lourenço Marques, Vasco da Gama, fl. & fr. 24.x.1971, *Balsinhas* 2248 (K).

Widespread throughout the tropics as a weed, which probably originated in Central America. Recorded from the F.Z. area as a weed of farm lands and lawns.

2. **Apium graveolens** L., Sp. Pl. **1**: 264 (1753). Type from Europe.

An erect biennial or perennial herb, with the characteristic smell of celery, up to 1 m. or sometimes a little more. Stem with prominent rather coarse grooves, rising from a large fleshy taproot. Lower leaves simply pinnate, 10–14 cm. long, with deltate-rhomboid segments up to 4·5 cm. long, often deeply 3-lobed; bases distinctly sheathing; the lower ones obviously petiolate, the uppermost more or less sessile. Stem leaves ternate with rhomboid segments to narrowly lanceolate-elliptic and subentire above. Umbels terminal and lateral, shortly pedunculate or ± sessile in leaf axils. Bracts 0, rays 7–15, up to 2·5 cm. long, rather unequal; bracteoles 0, with flowers on pedicels of 1–5 mm. long. Calyx teeth obsolete; petals greenish-white. Fruit c. 1·5 mm. long, very broadly ovoid; stylopodium depressed; styles c. 0·5 mm. long; carpophore shortly 2-cleft. Vittae solitary in the intervals and 2 on the commissural face.

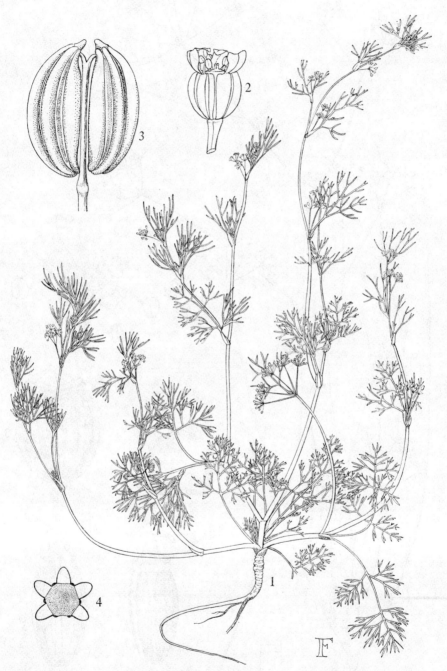

Tab. 150. APIUM LEPTOPHYLLUM. 1, habit (×½); 2, flower (×15); 3, fruit (×15); 4, mericarp in cross-section (×15), from *Welwitsch* 2499a & b and *Best* 482.

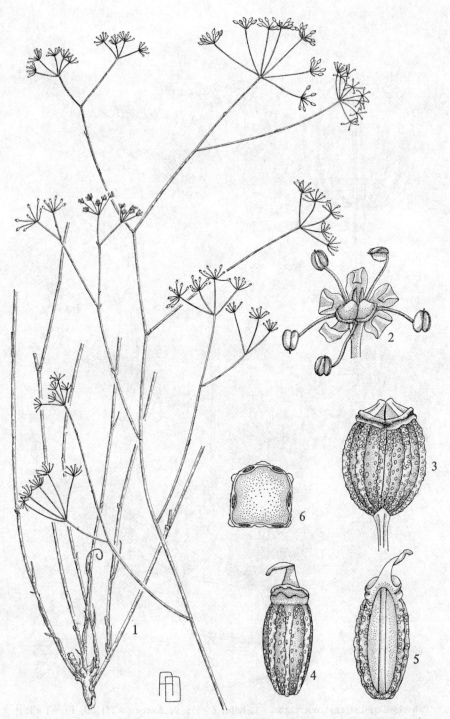

Tab. 151. DEVERRA BURCHELLII. 1, habit (×½), from *West* 2666; 2, flower (×10), from *Codd* 8933; 3, fruit (×15); 4, mericarp from exterior (×15); 5, mericarp, commissural view (×15); 6, mericarp in cross-section (×15), 3–6 from *Dinter* 6200.

Mozambique. LM: Lourenço Marques, Marracuene, Rikatla, *Junod* 246 (LISC).
Widespread in Europe and western Asia and frequently present elsewhere as an
adventive. In tropical Africa occurring as an escape or relic of cultivation. Under
arid conditions the plants may be very dwarf and low-growing, but in other respects
similar to the normal form.

11. DEVERRA DC.

Deverra DC., Coll. Mém. **5**: 49 (1829).—Schreiber in Prodr. Fl. SW. Afr. **103**:
4 (1967).
Pituranthos Viv., Fl. Lybia.: 15, t. 7 (1824) non *Pityranthus* Mart. (1814).
Pityranthus H. Wolff in Engl., Pflanzenr. **4**, 228 Ammin.-Carin.: 97 (1927) =
Pituranthos Viv.

Almost leafless, broom-like subshrubs or shrubs up to 1·25 m., but often very
much shorter. Flowers in terminal and lateral compound umbels. Sepals
obsolete; petals ovate, with a clearly inflexed apex. Fruit ovoid, somewhat
laterally compressed; stylopodium low-conical, merging below into an undulate-
margined disk; styles short, erect, becoming strongly divergent at maturity.
Ribs not very prominent. Vittae 1 in each interval and 2 in the commissural face.
A genus of 10–12 species most of which are distributed from Morocco through
N. Africa to Arabia. The remaining 2 species occur in S. and SW. Africa, one just
extending into the F.Z. area.

Deverra burchellii (DC.) Eckl. & Zeyh., Enum. Pl. Afr. Austr. Extratrop.: 347 (1837).—
Sond. in Harv. & Sond., F.C. **2**: 549 (1862).—Schreiber in Prodr. Fl. SW. Afr. **103**:
5 (1967). TAB. **151**. Type from S. Africa.
Deverra aphylla DC. var. *burchellii* DC., Prodr. **4**: 143 (1830). Type as above.
Pituranthos burchellii (DC.) Benth. & Hook. f. ex Schinz in Bull. Herb. Boiss. **2**:
209 (1894).—H. Wolff in Engl., Pflanzenr. **4**, 228 Ammin.-Carin.: 105 (1927)
("*Pityranthus*"). Type as above.

Almost leafless, subglaucous shrubs or subshrubs up to 1·25 m., arising from a
woody rootstock. Basal leaves with linear leaflets, but soon withering and often
absent in mature plants. Stem leaves reduced to short linear petioles, or frequently
to persistent bases alone, especially on the upper part of the stem. Umbels
terminal and lateral on long peduncles. Rays 6–10, 1·25–2 cm. long; bracts and
bracteoles obsolete or caducous at a very early stage of development. Partial
umbels 5–10-flowered, with pedicels c. 5 mm. long. Calyx teeth obsolete; petals
yellowish-white. Fruit 3–5 × 2–3 mm., ovoid; stylopodium low-conical; styles
short, divergent at maturity; ribs not very obvious, more or less covered with
tubercles that are paler in colour than the main body of the fruit. Carpophore
deeply 2-cleft; vittae 1 in each interval and occupying most of the space between
the ribs, and 2 in the commissural face; seed slightly concave on the inner face.

Botswana. SE: 13 km. NW. of Molepolole, 1000 m., fl. & fr. 2.xii.1954, *Codd* 8933
(BM; SRGH). **Rhodesia.** W: Matobo, fl. & fr. 3.ii.1948, *West* 2666 (BM; SRGH).
Also in S. and SW. Africa. In open woodland on sandy soils in semi-arid situations.

12. AFRAMMI Norman

Aframmi Norman in Journ. Bot., Lond. **67**, Suppl. Polypet.: 199 (1929).

Biennial or perennial herbs with finely divided leaves. Compound umbels
terminal and lateral with long wiry rays. Bracts and bracteoles well developed, the
former entire or slightly pinnatifid, probably frequently semi-caducous. Calyx
teeth broadly triangular; petals yellowish-green, with an inflexed apex; stylopodium low-conical, disk with an undulate margin; styles short, divergent,
stigmatic surfaces slightly clubbed. Fruit ovate to obovate, slightly laterally
compressed. Ribs well developed, obtuse; vittae large, 1 in each interval and 2
in the commissural surface.
An obscure genus of dubious affinities, previously only known from the type
species *A. angolensis* (Norman) Norman. The original author clearly intended to
imply relationship with *Ammi*, but this does not seem very obvious. For the
present it seems best to maintain the genus, with the addition of a further species
that does appear to be closely related to *A. angolensis* with the hope that collectors

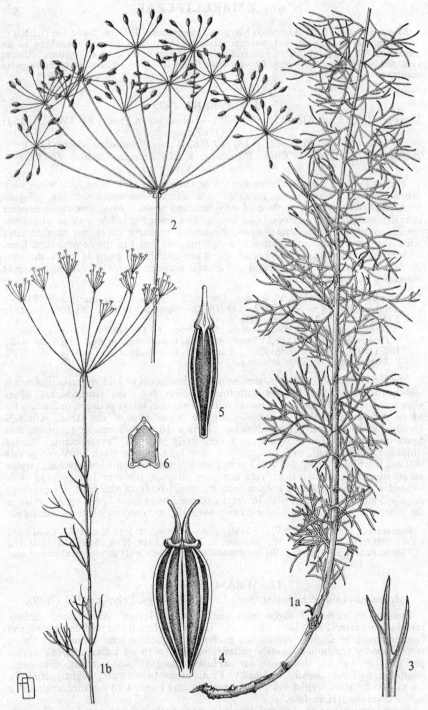

Tab. 152. AFRAMMI LONGIRADIATUM. 1a & 1b, habit (×½) *Drummond & Williamson* 9965; 2, inflorescence (×1) *E. M. & W.* 1281; 3, segment of leaf (×2) *Drummond & Williamson* 9965; 4, fruit (×8); 5, mericarp (×8); 6, transverse section of mericarp (×10), 4–6 from *E. M. & W.* 1281.

will provide more material of these species with fully mature fruit, so that a better assessment of relationships can be made.

Aframmi longiradiatum (H. Wolff) Cannon, comb. nov. TAB. **152**. Type from Zaire.
  Physotrichia longiradiatum H. Wolff in Engl., Bot. Jahrb. **48**: 274 (1913). Type as above.

Erect, glabrous biennial or perennial herbs 1–1·25 m. Rootstock woody, stem terete with fine regular grooves. Stems leafy, the leaves at the base of the stem smaller than those at middle levels. Leaves (2)3–6(10) cm. long, somewhat irregularly 2-ternate to 2-pinnatifid. Leaflets 1–3 cm. long, linear, entire, tapering gradually to the apex; petioles subterete, lamina sometimes extending to form a wing on the rhachis. Umbels terminal and lateral, the terminal ones somewhat irregular with c. 10 wiry wide-spreading rays 3–10 cm. long. Bracts several, irregular, linear or occasionally with traces of pinnatifid lobes; bracteoles c. 2 mm. long, linear, subulate, both probably caducous. Partial umbels with 6–15 flowers, on wiry pedicels 10–22 mm. long (elongating at maturity). Calyx teeth present and more or less persistent, minute, broadly triangular; petals yellowish-green, with an inflexed apex. Fruit c. 5 × 2 mm., obovoid, slightly laterally compressed; ribs rounded and quite conspicuous. Vittae well developed, 1 in each interval and 2 in the commissural face. Seed pentagonal in section, very slightly concave on the commissural face; carpophore not known.

Zambia. N: Kalambo Falls, fl. & fr. 29.iii.1955, *E. M. & W.* 1281 (BM; LISC; SRGH).
  As yet only known from the above locality and from the type locality in Zaire (Katanga).

The Zambian plants have much finer leaves than those of the type, but one specimen, *Drummond & Williamson* 9965 (BM, sheet 2), has relatively broad leaflets and this, together with the highly characteristic, rather irregular umbels, and the spreading wiry rays, suggests that the best course for the present is to equate the Zambian plants with Wolff's *Physotrichia longiradiatum*, and to transfer it to *Aframmi*, where its true affinities seem to lie. The " halbkugeligen papillen " mentioned by Wolff as occurring on the fruit are not apparent on the BM specimen of the type collection; their supposed presence presumably influenced Wolff to place the species in *Physotrichia*. The surface of the fruit of the BM presumed isotype (*Kassner* 2667), is somewhat wrinkled, perhaps as an artifact originating in the drying of the immature tissues, and this may have led Wolff to his interpretation. Since the holotype was presumably destroyed at Berlin, further clarification of this point seems unlikely.

13. FROMMIA H. Wolff

Frommia H. Wolff in Engl., Bot. Jahrb. **48**: 266 (1912); in Engl. & Prantl, Pflanzenfam. **3**, 8, Nachtr. 4: 228 (1915).

Perennial herbs. Leaves 3-pinnate with very finely divided ultimate segments. Umbels compound; bracts and bracteoles 0. Calyx teeth obsolete; petals with strongly inflexed apices. Mature fruit pyriform; stylopodium low-conical; styles short, blunt and deflexed; mericarps slightly dorsally compressed, ribs only slightly developed. Vittae prominent in the mature fruit, 3 in each furrow and 4–6 on the commissure face. Endosperm somewhat curved on the inner face. Carpophore slightly 2-fid at the apex.

A little-known monotypic genus of rather dubious affinities, but possibly rather closely related to *Carum.*

Frommia ceratophylloides H. Wolff in Engl., Bot. Jahrb. **48**: 266 (1912).—Engl., Pflanzenw. Afr. **3**, 2: 809, t. 333 (1921).—H. Wolff in Engl., Pflanzenr. **4**, 228 Ammin.-Carin.: 184, t. 16 (1927).—Cannon in Notes Roy. Bot. Gard. Edin. **32**: 195 (1973). TAB. **153**. Type from Tanzania.

Herb of c. 1 m., with a robust woody rootstock; base of stem with persistent fibrous remains of old leaf-bases. Stem terete, hollow, purplish towards the base, with fine striations, glabrous becoming puberulous above. Leaves 10–12 cm. long, more or less glabrous, mostly grouped near the base of the stem, finely divided with fine, regular, linear segments 2–5 × 0·5 mm.; apices prolonged into a distinct mucro; bases sheathing with membranous margins. (The general appearance of the leaves is strikingly reminiscent of the whole plant of *Ceratophyllum*, a submerged water plant.) Stem leaves few, gradually reduced upwards, and

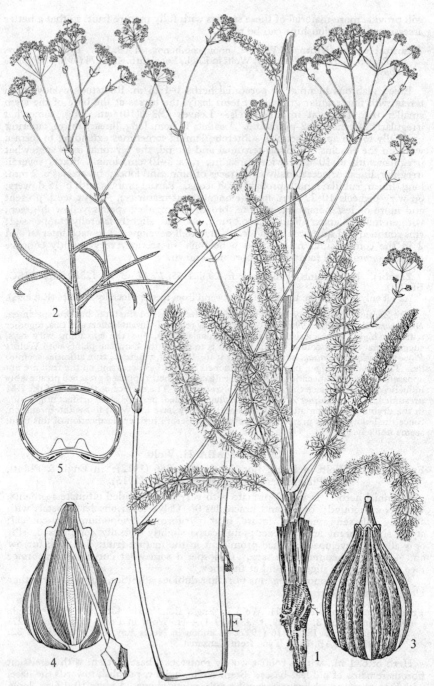

Tab. 153. FROMMIA CERATOPHYLLOIDES. 1, habit ($\times\frac{1}{2}$); 2, leaflet ($\times 3$); 3, mericarp from exterior ($\times 12$); 4, mericarp, commissural view ($\times 12$); 5, mericarp in cross-section ($\times 12$), all from *Whellan* sn.

represented on the inflorescence as membranous bases only. Inflorescence sparsely branched, with glabrous to puberulous peduncles up to 18 cm. long, each bearing a terminal compound umbel and 2 smaller lateral ones, on secondary peduncles. Umbels with c. 7 fine rays c. 2 cm. long; bracts 0. Partial umbels with c. 12 flowers on short pedicels 2–3 mm. long. Petals yellow, with strongly inflexed apices. Fruit 3 × 2 mm., ovoid, dark brown at maturity, slightly constricted at the commissure, scarcely compressed, concave on the commissural face; ribs poorly developed and hardly raised above the general surface of the pericarp. Stylopodium low-conical; styles short, divergent; carpophore very slightly 2-fid at the apex. Endosperm curved and concave on the inner face; vittae very well developed, 3 in each valley and 6 in the commissural face, tending to appear confluent in section in mature fruit.

Zambia. E: Nyika Plateau, Lundazi Distr., Malawi border, fl. & fr. 18.ii.1961, *Richards* 14398 (K). **Malawi.** N: Nyika Plateau, fl. & fr. 14.iii.1961, *Robinson* 4499 (K; SRGH).
Restricted to the Ufipa district of S. Tanzania and the adjoining parts of Malawi and Zambia.

A very unusual umbelliferous plant, easily recognised by the non-specialist by its yellow flowers and highly characteristic *Ceratophyllum*-like leaves.

14. PIMPINELLA L.

Pimpinella L., Sp. Pl. **1**: 263 (1753); Gen. Pl. ed. 5: 128 (1754).

Perennial, biennial or annual herbs. Leaves simple, ternate, simply pinnate to finely divided (occasionally showing a wide range of variation within one population). Umbels compound; bracts and bracteoles 0; calyx teeth obsolete or minute; petals white, cream or sometimes slightly pinkish, apices long-inflexed, the outer ones not noticeably radiate. Fruit ovoid to oblong, somewhat laterally compressed, with a broad commissure. Mericarps with 5 filiform ribs; stylopodium conical to depressed-conical; styles short or long, stigmatic surfaces somewhat clubbed. Carpophore deeply 2-cleft; seed slightly concave on the commissural face.
A large genus of perhaps 200 species, widely distributed through Europe, the Mediterranean and temperate Asia, with extensions into the tropics. Quite well represented in tropical and southern Africa, with some 30–40 species.

In addition to the species recognised in the following account, an obscure plant known only from one very immature gathering (Zambia, Isoka Distr., 18 km. W. of Chitipa, fl. 20. ii. 1970, *Drummond & Williamson* 9833 (BM; SRGH)) appears to be a *Pimpinella* with some affinity to *P. keniensis* Norman. It has white flowers, ternate to pinnate leaves, with regularly denticulate ovate-oblong leaflets. Some remarkable specimens collected by Richards in the Abercorn area (8372, 559 and 8312, all at K) have entire, narrowly ovate, *Plantago*-like leaves, and almost certainly represent a new species. Mature specimens of both these plants are needed so that final conclusions can be reached on their affinities.

Umbel rays pubescent:
  Lower leaves simple, very coriaceous - - - - - - 8. *engleriana*
  Lower leaves ternate, sometimes finely divided (rarely simple), herbaceous
                                1. *stadensis*
Umbel rays glabrous:
  Fruit and ovary pubescent or with vesicular hairs:
    Lower leaves simple (very rarely ternate); fruit with silky or bristly hairs
                                6. *huillensis*
    Lower leaves ternate (sometimes imperfectly, very rarely simple and then broadly trullate); fruit with vesicular hairs - - - - - 7. *kassneri*
  Fruit and ovary glabrous:
    Leaf segments very narrowly linear, grass-like - - - - 10. *sp. B*
    Leaf segments not as above:
      Petals distinctly yellow; leaves simple - - - - - 2. *neglecta*
      Petals white to cream; leaves usually compound and only rarely simple:
        Basal leaves pinnate to finely divided:
          Mature fruit 2–3 mm. long; basal leaves very variable, from simply pinnate to finely divided; (widespread) - - - - - 5. *buchananii*

Mature fruit 3·5–4 mm. long; basal leaves 2-pinnate with very jaggedly
toothed leaflets; (Mt. Mlanje) - - - - - 9. *sp. A*
Basal leaves simple (to ternate):
Several basal leaves present - - - - - - 3. *trifurcata*
One basal leaf only - - - - - - - - 4. *acutidentata*

1. **Pimpinella stadensis** (Eckl. & Zeyh.) D. Dietr., Syn. Pl. 2: 947 (1840).—Sond. in
Harv. & Sond., F.C. 2: 539 (1862).—H. Wolff in Engl., Pflanzenr. 4, 228 Ammin.—
Carin.: 376 (1927). TAB. **154**. Type from S. Africa.
Anisum stadense Eckl. & Zeyh., Enum. Pl. Afr. Austr. Extratrop. 3: 341 (1837).
Type as above.
Pimpinella nyassica Norman in Journ. Bot., Lond. **72**: 205 (1934). Type:
Malawi, Nyika Mts., *Sanderson* 58 (BM).

Relatively slender, probably biennial, herbs up to 60 cm., with narrow taproots;
generally lightly pubescent, rarely becoming glabrescent. Basal leaves (not
always evident in specimens) on long petioles up to 15 cm.; lamina rarely simple
with dentate margins, but usually subternate to ternate, rarely finely divided.
Stem leaves very variable (probably within populations), ternate to 2-ternate,
rarely pinnate; segments dentate to finely divided. Umbels on long peduncles,
bearing much-reduced stem leaves with prominent sheathing bases. Rays 8–15,
10–40 mm. long; bracts and bracteoles 0; partial umbels with up to 15 flowers
on pedicels 3–8 mm. long. Petals white, with inturned apices, the outer scarcely
radiate. Fruit ovoid to very broadly ovoid, slightly laterally compressed, ribs
obvious at maturity. Stylopodium depressed; styles short, spreading at maturity,
clubbed at the apex. Carpophore deeply 2-cleft; vittae 2–3 in the intervals and 2
in the commissural face.

Rhodesia. W: Matobo, Farm Chesterfield, fl. & fr. iv.1959, *Miller* 5910 (SRGH). E:
Melsetter, fl. 25.ii.1956, *Drummond* 5108 (BM; K; SRGH). **Malawi.** N: Nyika Mts.,
fl. xi.1932, *Sanderson* 58 (BM). **Mozambique.** MS: Tsetsera, fl. 7.ii.1955, *E. M. & W.*
253 (BM; LISC; SRGH).
Also apparently widely distributed in S. Africa and possibly also present in Tanzania,
but the relationship of this species with the S. African *P. caffra* remains to be clarified.
In montane grassland and amongst rocks, 1800–2000 m.

Norman (notes on herbarium sheets at BM) thought that *P. stadensis* and *P. caffra* are
conspecific, but the great mass of material accumulated since his time suggests that this is
unlikely. *P. caffra* is here regarded as a robust plant, with regularly cordate almost
coriaceous basal leaves. Plants from the F.Z. area agree well with *Ecklon & Zeyher* 2199
(photo BM), the type collection of *Anisum stadense*. *P. transvaalensis* H. Wolff is another
closely related species and may be conspecific; some specimens from the F.Z. area closely
resemble Wolff's type, but it seems best to retain a conservative view until S. African
Pimpinellas are properly revised.

2. **Pimpinella neglecta** Norman in Journ. Linn. Soc., Bot. **47**: 583 (1927). Type:
Rhodesia, Mazoe Distr., fl. ix.1906, *Eyles* 404 (BM, holotype; SRGH, isotype).

Sinuous, wiry, glabrous perennial or biennial herb up to 45 cm., with a large
elongate or subglobular taproot. Stems terete, with few grooves. Basal leaves
with petioles up to 5 cm., with large membranous sheathing bases. Lamina
emarginate to cordate at the base, apex rounded, margin dentate to denticulate.
Stem leaves much reduced to prominently toothed appendages, the uppermost to
sheathing bases alone. Umbels terminal and lateral; rays 3–7, 1·5–7 cm. long,
relatively robust; bracts and bracteoles 0; partial umbels with 5–7 flowers with
pedicels 5–15 mm. long. Calyx teeth obsolete; petals distinctly yellow, with
incurved apices. Fruit 3 × 1·5 mm., ovoid, somewhat laterally compressed, ribs
filiform. Stylopodium low-conical; styles 1 mm. long, only slightly clubbed at the
apex, slightly divergent and becoming reflexed with maturity. Carpophore
probably 2-cleft (other structural details of fruit not known in mature state).

Zambia. B: Mankoya, Lusna R., fl. & fr. 20.xi.1959, *Drummond & Cookson* 6680
(BM; K; SRGH). W: 67 km. W. of Chingola on Solwezi road, fl. 21.x.1969, *Drummond
& Williamson* 9213 (SRGH). C: between Broken Hill and Bwana Mkubwa, fl. 1916,
Allen 347 (K). **Rhodesia.** N: Mazoe Distr., fl. ix.1906, *Eyles* 404 (BM; **S**RGH)

Apparently confined to the F.Z. area, but very possibly also occurring in Zaire and
Tanzania. In open grassland, c. 1000 m.

A very easily recognised species, with its bright yellow flowers and characteristic habit.

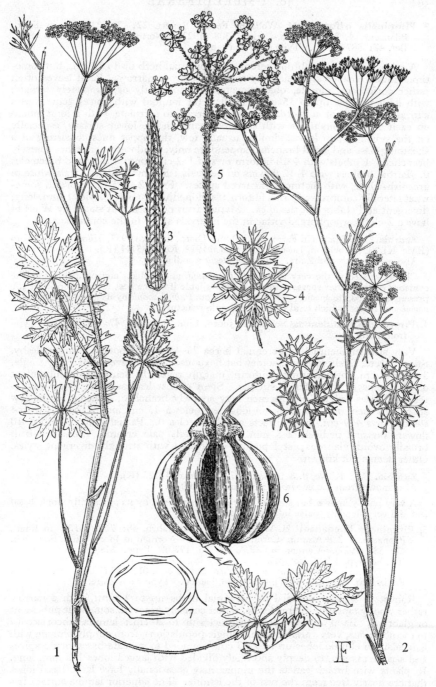

Tab. 154. PIMPINELLA STADENSIS. 1 & 2, habit (×½), from *Drummond* 5108 and *Chase* 4911; 3, underside of leaf-lobe (×6), from *Chase* 4911; 4, another leaf-form variant (×½); 5, flowering umbel (×6), both from *Drummond* 5108; 6, fruit (×18); 7, mericarp in cross-section (×24), both from *Chase* 4911.

3. **Pimpinella trifurcata** H. Wolff in Fedde, Repert. **22**: 348 (1926); in Engl., Pflanzenr. **4**, 228 Ammin.-Carin.: 308 (1927).—Norman in Journ. Linn. Soc., Bot. **47**: 587 (1927). Type from Tanzania.

A robust, wiry, ± glabrous perennial or biennial herb up to 75 cm., but sometimes much smaller and relatively slender; taproot narrow. Basal leaves often rather coriaceous, simple, oblong-ovate, or imperfectly or completely ternate, with long petioles up to 15 cm. Leaflets (or lamina) with apices acute, bases cuneate, rounded or subcordate, margins serrate to serrulate, with a long mucro on each tooth; sometimes with scattered hairs on the lower surface, especially on the veins. Stem leaves similar, usually few, the upper ones consisting of a sheathing base and small branched appendage only. Inflorescence rather sparsely branched. Umbels with 4–10 filiform rays of 1–2 cm. long; bracts and bracteoles 0. Partial umbels with 4–12 flowers on pedicels 1·5–3 mm. long. Petals white to greenish-white, with distinctly incurved apices. Fruit 2 × 1·5 mm., ovoid, somewhat laterally compressed; ribs filiform; stylopodium conical; styles 1 mm. long, divergent and clubbed at the apex. Mature fruit not seen, but stated by Wolff to have a 2-cleft carpophore, 3 vittae in the intervals and 4 in the commissure.

Zambia. W: Matonchi Farm, Mwinilunga Distr., fl. 20.xii.1937, *Milne-Redhead* 3746 (BM; K). C: Chelston, Lusaka, fl. & fr. 1.ii.1973, *Kornaś* 3117 (K). Also in Tanzania and Zaire. In *Brachystegia* woodland.

Closely related to the very variable *P. buchananii* and perhaps conspecific, but on the evidence of the rather sparse evidence now available it seems best to maintain it for the present as a separate species. Distinguished from *P. buchananii* by its generally glabrous, coriaceous leaves, which tend towards a simple structure.

4. **Pimpinella acutidentata** Norman in Journ. Linn. Soc., Bot. **47**: 587 (1927). Type from Zaire.

Very slender biennial or perennial herbs 25–45 cm. Rootstock rather fleshy. Stems terete, finely grooved, somewhat flexuous. Basal leaves apparently single, 3–5 × 5–8 cm., orbicular or reniform, margin regularly denticulate, puberulent on the veins of the inferior surface. Stem almost leafless, with a few much-reduced bases only. Inflorescence very sparsely branched, bearing, in the few specimens seen, only 2 umbels on long peduncles 5–15 cm. long. Umbels with 4 slender rays 4–9 cm. long; bracts and bracteoles 0. Partial umbels with c. 10 flowers on fine pedicels 2–5 mm. long. Petals pale cream. Immature fruit broadly ovoid, glabrous; stylopodium depressed with strongly divergent styles. Other details not known.

Zambia. W: Kitwe, fl. & fr. 4.xii.1973, *Fanshawe* 12125 (K). Also known from Zaire. In open woodland, 1200 m.

A very poorly known but striking species, characterised by its apparently single basal leaf and very slender, sparse inflorescence.

5. **Pimpinella buchananii** H. Wolff in Engl., Bot. Jahrb. **48**: 269 (1912); in Engl., Pflanzenr. **4**, 228 Ammin.-Carin.: 312 (1927).—Norman in Journ. Linn. Soc., Bot. **47**: 588 (1927).—Cannon in C.F.A. **4**: 345 (1970). Type: Malawi, *Buchanan* 709 (B†, holotype; BM, isotype).
? *Pimpinella favifolia* Norman in Journ. Bot., Lond. **61**: 133 (1923).—H. Wolff in Engl., tom. cit.: 313 (1927).—Norman, loc. cit. Type from Zaire.

Robust to slender biennial or perennial herbs up to 1·5 m., with a narrow rather woody taproot. Stems terete, rather coarsely grooved, somewhat pubescent or glabrous. Basal leaves on long peduncles up to 20 cm.; lamina subcoriaceous to membranous, very variable within single populations, from simply pinnate with lanceolate to elliptic lobes up to 6 × 3 cm., rounded to cuneate bases, acute apices and serrate margin, to deeply and finely divided with jagged lobes 1–2 mm. long. In plants with broad leaflets the pinnae may occasionally have large basal lobes that are nearly free from the rest of the lamina. The superior lamina surface has scattered hairs, frequently with bulbous bases, and is usually marked with a pattern of honeycomb-like pits on the epidermis; the inferior surface has scattered hairs or a fairly dense pubescence; in extreme cases both surfaces may be subglabrous. Stem leaves sparse, similar to the basal leaves, reduced on the inflorescence to sheathing bases with minute ternate to pinnate appendages with linear lobes. Inflorescence much branched, with rather small, few-rayed terminal and lateral umbels. Umbels with 3–6 filiform more or less regular rays 0·5–2·5 cm. long. Bracts and bracteoles 0. Partial umbels with 4–11 flowers on pedicels 1–3 mm. Petals very small, white, with incurved apices. Fruit 2 × 1·5 mm., ovoid, somewhat laterally compressed, glabrous, ribs not prominent especially when immature. Stylopodium conical; styles 1 mm. long, clubbed at the apex, divergent and then reflexed in fruit. Carpophore probably 2-cleft at maturity; vittae 2–3 in the intervals and ?4 in the commissural face.

Zambia. 13 km. NW. of Mbala (Abercorn), fl. 11.iv.1961, *Phipps & Vesey-FitzGerald* 3064 (BM; SRGH). W: Solwezi, fl. & fr. 10.iv.1960, *Robinson* 3523 (K; SRGH). C: 10 km. S. of Kapiri Mposhi, fr. 27.iii.1955, *E. M. & W.* 1219 (BM; LISC; SRGH). E: Fort Jameson, fl. 29.iii.1960, *Wright* 250 (K). S: Chishinga Ranch, Gwembe Distr., fl. 1.ii.1962, *Astle* 1346 (SRGH). Malawi. N: Karonga Distr., Mussissi, fl. 30.iv.1947, *Benson* (BM). C: Ntchisi Mt., fl. 14.iii.1967, *Salubeni* 604 (K). S: Namwera Escarpment, fl. 15.iii.1955, *E. M. & W.* 914 (BM; LISC; SRGH). Mozambique. N: Makalongwe Valley near Chiponde, fl. 11.ii.1942, *Hornby* 3573 (PRE). Z: Gúruè, 22 km. from Nintulo near Lioma, fl. 10.ii.1964, *Torre & Paiva* 10514 (LISC).

Also present in Angola, Zaire and Tanzania. In *Brachystegia* woodland and sometimes apparently in moist habitats, 500–1000 m.

A very variable species with an extreme range of leaf morphology, even within individual populations. Final judgement on *P. favifolia* Norman is here reserved. A specimen from Malawi, Chitipa Distr., Nganda Peak, Nyika Plateau, *Pawek* 2081 (UC) compares fairly well with Norman's type from Zaire, in that it has rather coriaceous narrowly lanceolate leaflets with semi-cartilaginous teeth. The flowers are described by the collector as " yellow " and the epidermis has the rather characteristic fine honeycomb pattern to which Norman drew attention. However, the latter does seem to occur in plants that Norman would certainly have determined as *P. buchananii*.

6. **Pimpinella huillensis** Welw. ex Engl. in Abh. Königl. Preuss. Akad. Wiss. Berl. 319 (1892).—H. Wolff in Engl., Pflanzenr. 4, 228 Ammin.-Carin.: 262 (1927).—Norman in Journ. Linn. Soc., Bot. 47: 590 (1927).—Cannon in C.F.A. 4: 347 (1970). Type from Angola.
Pimpinella welwitschii Engl., loc. cit.—Norman, loc. cit.—Brenan in Mem. N.Y. Bot. Gard. 8: 446 (1954). Type from Angola.

Slender to robust biennial or perennial herbs up to 1·5 m., with a rather woody taproot. Stems terete, finely grooved. Basal leaves on long peduncles up to 20 cm.; lamina up to 8 × 6 cm., coriaceous, broadly ovate to broadly lanceolate, very rarely partially 3-lobed or even ternate; bases emarginate to deeply cordate, apices rounded to acute, margins dentate to denticulate, inferior surface often with scattered bristly hairs which are also present on the petioles. Stem leaves similar to the basal ones, gradually reduced to sheathing bracts, with minute ternate-pinnate appendages with linear lobes. Inflorescence usually much branched, with rather small terminal and lateral umbels. Umbels with 5–8(12) ± regular filiform rays 0·5–2·5 cm. long; bracts and bracteoles 0. Partial umbels with c. 10 flowers on rather irregular pedicels 2–5 mm. long. Petals with conspicuously inturned apices, more or less regular. Fruit 1–2 mm. long, ovoid, distinctly laterally compressed, usually more or less densely covered with upwardly directed hairs which obscure the ribs, but may be ± shed as the fruit matures. Stylopodium depressed; styles 1 mm. long, divergent and ± reflexed at maturity, clubbed at the apex. Carpophore deeply 2-cleft; vittae 2–3 in the intervals and ?4 in the commissural face.

Zambia. W: 24 km. S. of Luanshya, fl. 12.iii.1960, *Robinson* 3384 (SRGH). C: Mufulira, fl. 2.iv.1955, *E. M. & W.* 1395 (BM; LISC; SRGH). Rhodesia. N: Goromonzi, Rua R., fl. & fr. 17.iv.1927, *Eyles* 4882 (K; SRGH). C: Salisbury Distr., Ruwani, fl. 22.ii.1952, *Wild* 3764 (BM; SRGH). E: Nyumquarara Valley, fl. ii.1935, *Gilliland* 1393 (BM). Malawi. S: Chipusiri, Ncheu Distr., fl. 17.iii.1955, *E. M. & W.* 963 (BM; LISC; SRGH). Mozambique. N: Vila Cabral, fl. & fr. vi.1934, *Torre* 445 (COI).

Also present in Angola and probably Zaire and Tanzania, but the relationship of this species with the *P. platyphylla* Welw. ex Hiern complex needs further clarification. If the whole group is eventually treated as a single variable species, the correct name will be *P. huillensis*. In open situations in dry bush up to 1500 m.

7. **Pimpinella kassneri** (H. Wolff) Cannon, comb. nov.
Physotrichia kassneri H. Wolff in Engl., Bot. Jahrb. 48: 272 (1913). Type from Zaire.

Biennial herb c. 1 m. tall, with a narrow fleshy taproot. Stems terete and finely grooved. Basal leaves ternate to partially 2-ternate, rarely with simple shortly trullate lamina and regularly dentate margins. Lobes of ternate leaves 2–6 cm. long, linear to irregularly elliptic or oblanceolate; apices obtuse, bases narrowly cuneate, margins ± regularly serrate to irregularly dentate with remote teeth. Petioles 5–15 cm. long, with moderately well-developed sheathing bases. Stem leaves reduced and confined to bases with minute linear appendages on the inflorescence. Inflorescence much branched, with terminal and lateral umbels. Primary umbels with 4–6 ± regular rays 1·5–4 cm. long; bracts and bracteoles 0. Partial umbels with 5–12 flowers on pedicels 3–5 mm. long. Petals white to pale yellow-cream, with distinctly inturned apex. Fruit probably $2 \times 1·5$ mm. when mature, ovoid, densely covered with obtuse vesicular hairs which obscure the ribs and other features of the fruit. Stylopodium depressed to low-conical, characteristically purple-pigmented; styles spreading, with slightly clubbed stigmatic surfaces. Carpophore unknown; vittae single in the intervals and 2 on the commissural face.

Zambia. N: Kali Dambo, fl. & fr. 17.iii.1955, *Richards* 4986 (K). W: Solwezi Distr., Mulenga, fl. & fr. 19.iii.1961, *Drummond & Rutherford-Smith* 7075 (BM; SRGH). **Malawi.** N: Rumpi Distr., Nchena Nchena, fl. 8.iii.1970, *Pawek* 3370 (UC). **Mozambique.** N: Massangulo, fl. & fr. iii.1933, *Gomes e Sousa* 1332 (COI).
An obscure species, known previously only from the type collection (B†, holotype; BM, isotype).

Wolff's allocation of this species to *Physotrichia* seems inappropriate in view of the complete absence of bracts and bracteoles and the small ovoid fruit. Both these features, together with the few-rayed umbels and general habit, suggest *Pimpinella*, especially the general affinity of *P. huillensis*. The combination is here made, with the hope that some attention will be focussed on this little-known taxon.

8. **Pimpinella engleriana** H. Wolff in Engl., Pflanzenr. **4**, 228 Ammin.-Carin.: 261 (1927).—Norman in Journ. Linn. Soc., Bot. **47**: 590 (1927).—Brenan in Mem. N.Y. Bot. Gard. **8**: 446 (1954). Type from Tanzania.
    Pimpinella tomentosa Engl., Bot. Jahrb. **30**: 368 (1901) non Dalz. ex C. B. Cl. (1879). Type from Tanzania.

Robust biennial or perennial herb up to 2 m., with a long, narrow taproot. Stem more or less unbranched, stiff, terete, finely grooved. Basal leaves up to 15×12 cm., on long petioles up to 30 cm., usually more or less coriaceous, cordate to suborbicular, with a deep basal sinus, apex rounded to acute, margins crenate to denticulate, inferior surface with scattered hairs especially on the veins. Stem leaves with long petioles up to 20 cm., but often much shorter, oblong to lanceolate, apex acute to rounded, base deeply cordate to rounded, margins denticulate. Upper leaves much reduced, the uppermost to sheathing bases only, sometimes with a small terminal appendage. Umbels terminal and lateral, often dense and compact, especially when immature. Rays up to 30, 1–3 cm. long, robust and densely hirsute, particularly in flower and young fruiting stages; bracts and bracteoles 0. Partial umbels with 30–40 flowers on robust, densely hirsute pedicels. Petals cream, with long inflexed apices. Fruit ovoid, very densely covered with long hairs which obscure the ribs. Stylopodium depressed; styles 2 mm. long, divergent, stigmatic surfaces clubbed. Carpophore deeply 2-cleft; vittae 2 in the intervals and 4–6 in the commissural face.

Zambia. N: Chitimbwa, Mbala, fl. 30.v.1955, *Nash* (BM). **Malawi.** N: Nyika Plateau, Nganda Hill, fl. 7.ix.1962, *Tyrer* 914 (BM). C: Nkhota Kota Distr., Nchisi Mt., fl. 27.vii.1946, *Brass* 16981 (K; PRE; SRGH). **Mozambique.** N: Vila Cabral, fl. viii.1934, *Torre* 444 (COI).
Also present in Tanzania, Zaire and Burundi. In rough montane grassland and in *Brachystegia* woodland, 1400–1900 m.

A very distinctive species, easily recognised by its dense, flat-topped umbels and very hairy fruit and pedicels.

9. **Pimpinella sp. A.**

This plant, which appears to be closely related to *P. buchananii*, seems to be restricted to Mt. Mlanje (Malawi: S) at an altitude of c. 2200 m. and is represented by the following collections: *Wild* 6143 (BM; K; LISC; SRGH); *Chapman*

(BM; K; SRGH); *Adamson* 370 (K). It is distinguished by its larger fruit and very characteristic jaggedly-cut leaf-lobes. It may well merit specific recognition, but in view of the extreme variability of *P. buchananii*, it seems best for the present to draw attention to this taxon, without according formal recognition, with the hope that field workers will produce further evidence.

10. Pimpinella sp. B.

This species has only been collected on about three occasions and material with fully mature fruit is not available. Its general morphology suggests a *Pimpinella* of the affinity of *P. lineariloba* Cannon, a very poorly known species from Angola. Like that species it has very fine, linear, grass-like ultimate leaf-segments 1–1·5 mm. wide. The umbels are few-rayed, lack bracts and bracteoles and have compact short-pedicelled partial umbels. It is a very distinct plant, but since it is apparently so rare, it seems best to defer formal recognition until its position is further clarified by additional material with mature fruit.

Zambia. N: Kawambwa Distr., Ntumbachushi Falls, fl. 27.ii.1970, *Drummond & Williamson* 10041 (BM; SRGH). C: Serenje Distr., Kanona, fl. & fr. 12.iii.1976, *Hooper & Townsend* 666 (K).

15. BAUMIELLA H. Wolff

Baumiella H. Wolff in Engl., Pflanzenr. **4**, 228 Ammin.-Carin.: 142 (1927).

Perennial herbs with pinnate leaves; leaflets imbricated when young and bearing spinous cartilaginous teeth. Umbels compound; bracts and bracteoles numerous and well developed. Petals suborbicular to broadly obcordate, apex emarginate with an inflexed point; calyx teeth obsolete. Stylopodium well developed, conical with a thickened, undulate margin; styles short, stiff, erect, slightly divergent; stigmatic surfaces slightly thickened. Fruit ovoid, slightly laterally compressed; ribs filiform but quite distinct. Mericarps pentagonal in section; commissural face flat. Vittae 1 in each interval and 2 in the commissural face. Commissural face of the seed flat.

A monotypic genus centred on the F.Z. area and spreading into Angola, Zaire and southern Tanzania. Related to *Carum* but distinguished from it by the strongly-developed conical stylopodium and its unusual leaf structure. Perhaps also related to *Sium* which it resembles in a number of characters.

Baumiella imbricata (Schinz) H. Wolff in Engl., Pflanzenr. **4**, 228 Ammin.-Carin.: 142 (1927). TAB. **155**. Syntypes from Angola.
  Carum imbricatum Schinz in Bull. Herb. Boiss. **2**: 208 (1894). Syntypes as above.
  Bunium imbricatum (Schinz) Drude in Engl., Pflanzenfam. **3**, 8: 194 (1898). Syntypes as above.
  Pimpinella imbricata (Schinz) Engl. in Warb., Kunene-Samb.-Exped. Baum: 324 (1903). Syntypes as above.

Glabrous herb 30–100 cm. tall. Rootstock probably a creeping rhizome with tuberous nodules from which adventitious roots arise at the nodes. Stem terete, hollow with a large central canal, finely grooved, unbranched below the level of the inflorescence. Leaves up to 35 cm. long, simply pinnate, petiole and rhachis terete, fistulose and septate; leaflets rigid, somewhat coriaceous, irregularly ovoid with a cartilaginous margin and strongly developed teeth. Leaf morphology very variable, probably in association with habitat conditions; in extreme cases the leaves may be reduced to little more than fistulose petioles. In young leaves the leaflets are condensed and imbricated. Inflorescence sparsely branched. Umbels compound with up to 25 rays; bracts up to 10 × 2 mm., linear-lanceolate, numerous, becoming subulate at the apex and with distinct membranous margins. Partial umbels with 10–25 flowers on thin, unequal pedicels; bracteoles 8–10, similar to the bracts. Petals greenish-white to yellow. Fruit (not fully mature) 2 × 1 mm., slightly compressed laterally, primary ribs well developed and conspicuous; stylopodium conical, yellowish; styles 0·25 mm. long, brownish; vittae 1 in each interval and 2 in the commissural face.

Zambia. N: 72 km. S. of Mbala on Kasama road, fl. & fr. 30.iii.1955, *E. M. & W.* 1347 (BM; LISC; SRGH). W: Mwinilunga Distr., just SW. of Dobeka bridge, fl. 14.xii.1937, *Milne-Redhead* 3658 (BM; K). C: Mkushi Distr., fl. 4.i.1958, *Robinson*

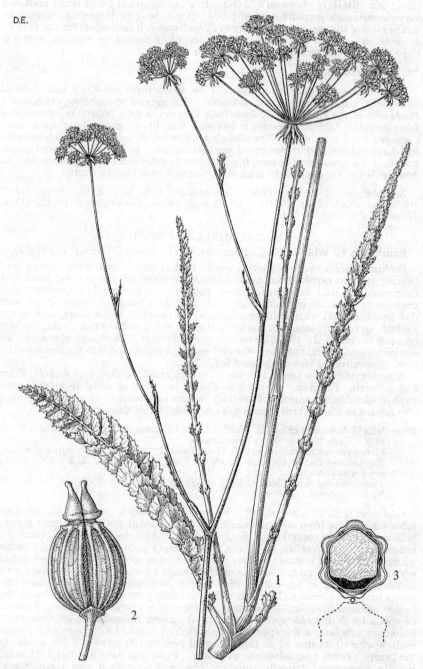

Tab. 155. BAUMIELLA IMBRICATA. 1, habit (× ½), from *Rand* 343; 2, fruit (×10); 3, mericarp in cross-section (×20), both from *Hanham* sn. From C.F.A.

Tab. 156 SIUM REPANDUM. 1, leaves and inflorescence(×½); 2, underside of leaf (×4½); 3, marginal tooth (×12); 4, partial umbel (×2); 5, fruit (×12); 5, mericarp in cross-section (×18), from *Chase* 1408 and *Wild* 3513.

2625 (K; SRGH). **Rhodesia.** C: Marandellas, fl. & fr. 6.vii.1931, *Brain* 5341 (BM; SRGH). E: Inyanga Distr., Troutbeck, sterile 21.i.1948, *Chase* 615 (BM). **Malawi.** N: Nyika Plateau, Rumpi Distr., road to Chelinda, fl. 5.i.1959, *Richards* 10507 (K). **Mozambique.** N: Vila Cabral, plateau near Lichinga, iii.1934, *Torre* 25 (COI; LISC). T: Angónia, fl. 3.x.1942, *Mendonça* 635 (LISC).

Also recorded from eastern Angola, Zaire and southern Tanzania. In bogs, marshes, grassland and woodland.

In drier habitats the leaves are less coriaceous and the cartilaginous margin is reduced. The terminal leaflet may also be much larger than normal in these circumstances.

16. SIUM L.

Sium L., Sp. Pl. **1**: 251 (1753); Gen. Pl. ed. 5: 120 (1754).

Robust glabrous herbs with pinnate leaves. Flowers all ☿, in terminal and lateral compound umbels; bracts and bracteoles numerous and conspicuous. Sepals well developed to minute; petals white to yellowish, more or less equal, relatively broad with an incurved apex, rarely the outer becoming subradiate. Fruit ovoid-cylindrical to globose, slightly compressed laterally and constricted at the commissure. Stylopodium conical or sometimes depressed; styles short and stiff, divergent. Ribs filiform, fairly prominent, lateral ribs marginal; mericarps sub-pentagonal in section, endosperm similar. Carpophore entire or 2-partite at the apex. Vittae superficial, 1–6 in each interval and 2–6 in the commissural face.

A genus of c. 14 species widely distributed through the N. temperate zone.

Sium repandum Welw. ex Hiern, Cat. Afr. Pl. Welw. **1**: 425 (1898).—Burtt Davy, F.P.F.T. **1**: 517 (1926).—H. Wolff in Engl., Pflanzenr. 4, 228 Ammin.-Carin.: 353 (1927).—Cannon in F.W.T.A. ed. 2, **1**: 754 (1958); in C.F.A. **4**: 350 (1970). TAB. **156.** Type from Angola.

A coarse, glabrous perennial herb up to 1·5 m., with a creeping rhizome. Stems terete, hollow, strongly grooved, sometimes tinged with red. Stem leaves up to 30 cm. long, simply pinnate with narrowly oblong-lanceolate leaflets up to 10 × 2 cm.; leaf-bases with well-developed auriculate membranous sheaths; petiole and rhachis terete, hollow except at the point of attachment of the leaflets; apices acute, with a short cartilaginous mucro; margins with regular serrations, cartilaginous, the fine points incurved over the base of the next tooth. Umbels compound, terminal and lateral; rays 7–24; bracts up to 2 cm. long, numerous, well developed, linear, with narrow membranous margins. Rays irregular, 1–4·5 cm. in fruit, bracteoles numerous and similar to the bracts. Flowers all ☿, 20–30 in each partial umbel, on irregular rays 2–8 mm. long. Petals yellowish to greenish-white. Fruit c. 2 × 1 mm., ovate-oblong, truncate below the stylopodium, reddish-brown when ripe, slightly laterally compressed; calyx teeth minute. Stylopodium conical; styles 0·5 mm. long, stiff and divergent, stigmatic surfaces slightly enlarged. Carpophore stiff, undivided or slightly 2-fid. Primary ribs filiform, secondary not developed. Vittae 1 in each interval and 2 in the commissural face; seed conforming to the shape of the fruit.

Zambia. W: Solwezi, veg. 22.vii.1964, *Fanshawe* 8829 (SRGH). S: Mumbwa, *Macaulay* 945 (K). **Rhodesia.** E: Melsetter, Black Mt. Inn, fl. 28.xii.1948, *Chase* 1408 (BM; LISC; SRGH). **Malawi.** S: Kirk Range, Goche, fl. & fr. 31.i.1959, *Robson* 1383 (BM; K; LISC; SRGH). **Mozambique.** T: Angónia, 40 km. from Vila Mousinho towards Zóbuè, 19.vii.1949, *Barbosa & Carvalho* 3682 (COI; LM).

Also known from Angola, Cameroon, Tanzania and S. Africa. At streamsides and often actually in standing water.

17. BERULA Koch

Berula Koch in Röhl., Deutschl. Fl. **2**: 25 (1826).

Robust glabrous herbs with pinnate leaves and hollow stems. Flowers in compound umbels with well-developed bracts and bracteoles; sepals subulate-triangular, quite conspicuous; petals broadly obcordate, with an inflexed point at the apex. Fruit suborbicular, almost didymous, laterally depressed and somewhat constricted at the commissure; ribs poorly developed and not conspicuous, lateral ribs not marginal. Mericarps subpentagonal in section; carpophore adnate to the mericarps. Vittae deeply embedded in the mesocarp.

Tab. 157. BERULA ERECTA. 1, habit (×½), from *Taylor* 2257; 2, leaf margin (×4½);
3, mericarp (×12); 4, mericarp in cross-section (×18), all from *Taylor* 2257.

A monotypic genus related to *Sium*, from which it is separated by the poorly-developed fruit ribs, the non-marginal lateral ribs and the setting of the vittae deep in the mesocarp.

Berula erecta (Huds.) Coville in Contr. U.S. Nat. Herb. **4**: 115 (1893).—Cannon in C.F.A. **4**: 349 (1970). TAB. **157**. Type from England.
  Sium erectum Huds., Fl. Angl. ed. 1: 103 (1762). Type as above.
  Sium angustifolium L., Sp. Pl. ed. 2: 1672 (1763). Type from southern Europe.
  Berula angustifolia (L.) Koch in Röhl., Deutschl. Fl. **2**: 433 (1826). Type as for *Sium angustifolium*.
  Sium thunbergii DC., Prodr. **4**: 125 (1830).—Sond. in Harv. & Sond., F.C. **2**: 539 (1862).—Hiern in F.T.A. **3**: 13 (1877).—Burtt Davy, F.P.F.T. **1**: 519 (1926). Type from S. Africa (Cape Prov.).

Robust glabrous perennial herb up to 2·5 m. high, erect or decumbent, rhizomatous to stoloniferous, rooting at the nodes. Stem hollow, terete, with coarse grooves. Leaves up to 50 cm. long, simply pinnate, with 7–12 pairs of sessile leaflets; petioles up to 12 cm. long, leaf bases narrowly sheathing. Leaflets 1–8 cm. long, coarsely and rather irregularly toothed (some of the lower leaflets sometimes showing a tendency towards the separation of leaflets of the second order). Teeth sometimes with a cartilaginous covering which may extend along the margin towards the next tooth. Upper leaves gradually reduced in size and form upwards. Umbels terminal and lateral on peduncles 2–12 cm. long. Bracts numerous and well developed, narrowly linear with a narrow membranous margin or linear-3-partite with obvious lateral teeth. Rays often rather irregular, 10–20; bracteoles numerous, similar to the bracts but correspondingly smaller. Flowers on fine, sometimes irregular, pedicels. Petals white, calyx teeth triangular-subulate, persistent in fruit. Fruit c. 2 mm. long, brownish at maturity, almost orbicular, slightly broader than long, depressed laterally; stylopodium 0·5 mm. long, small, conical; styles c. 1 mm. long, slightly clubbed at the apex; ribs filiform and inconspicuous at maturity.

Rhodesia. N: Mazoe, Umvukwes, fl. 17.xii.1952, *Wild* 3932 (BM; SRGH). W: Matobo, fl. & fr. 1.i.1949, *West* 2841 (BM; SRGH). C: Salisbury, Windyridge, fl. & fr. 27.i.1946, *Wild* 734 (BM; SRGH).
Widespread through Eurasia and N. America and in Africa from Sudan and Ethiopia through E. Africa to Zaire, Angola and S. Africa. In marshy areas and at the sides of streams.

Sium thunbergii was chiefly distinguished as having cartilaginous teeth on its leaves. This character may be associated with growth in mineral-rich waters and in any case is only shown by a small proportion of African specimens. This, and the other characters mentioned by authors do not appear to justify its separation from the widely distributed *B. erecta*.

18. DIPLOLOPHIUM Turcz.

Diplolophium Turcz. in Bull. Soc. Imp. Nat. Mosc. **20**: 173 (1847).

Robust perennial herbs, stems solid and with relatively fine grooving, nearly glabrous excepting the umbels. Leaves with strongly-developed sheathing bases, ultimate segments filiform, shortly acicular or elliptic to subrotund. Umbels with numerous rays, partial umbels many-flowered; bracts and bracteoles well developed and conspicuous. Petals with a long inflexed point, pubescent on the outer surface. Fruit cylindrical to elliptic, subterete in section, covered with bristly hairs; ridges fairly well developed, obtuse. Calyx teeth obsolete; stylopodium conical; styles exceeding the stylopodium, clubbed at the apex. Vittae conspicuous, 1 in each groove and 2 (or 4) in each commissural face. Carpophore entire or 2-cleft to the base.
A genus of c. 6 species, confined to tropical Africa.

Ultimate leaf segments acicular to filiform - - - - - - 1. *zambesianum*
Ultimate leaf segments elliptic to subrotund - - - - - 2. *buchananii*

1. **Diplolophium zambesianum** Hiern in F.T.A. **3**: 18 (1877).—Eyles in Trans. Roy. Soc. S. Afr. **5**: 434 (1916).—Martineau, Rhod. Wild Fl.: 57 (1954).—Brenan in Mem. N.Y. Bot. Gard. **8**: 446 (1954).—Cannon in C.F.A. **4**: 350 (1970). TAB. **158**. Type: Zambia, Batoka Country, *Kirk* (K).
  Physotrichia arenaria Engl. & Gilg in Warb., Kunene-Samb.-Exped. Baum: 324

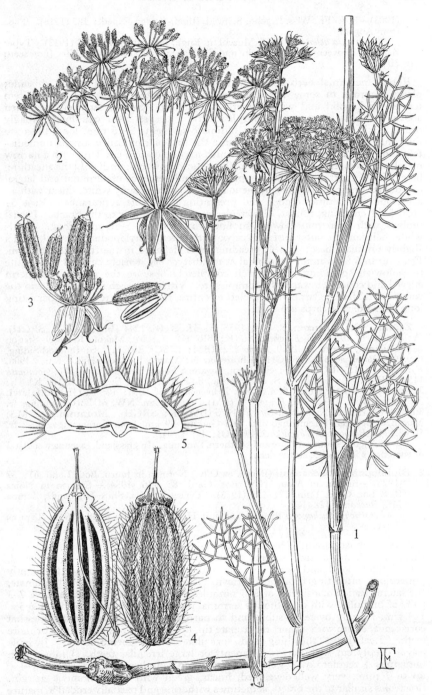

Tab. 158. DIPLOLOPHIUM ZAMBESIANUM. 1, habit (× ½), from *Rutherford-Smith* 7186;
2, umbel (× ½), from *Nash* 103; 3, partial umbel (× 2); 4, external and commissural
views of mericarps (× 6); 5, mericarp in cross-section (× 12), 3–5 from *Gilliland* 6.

(1903)—R. E. Fr., Wiss. Ergebn. Schwed. Rhod.-Kongo-Exped.: 183 (1914). Type from Angola.

Physotrichia helenae Busc. & Muschl. in Engl., Bot. Jahrb. **49**: 482 (1913). Type: Zambia, between L. Bangweulu and L. Tanganyika, *de Aosta* 1150a (not seen; ?B†).

Robust perennial herb 30–200 cm. high; rootstock a hard woody rhizome; plants glabrous to somewhat glaucous, except for the petals and fruit. Stem terete, rigid, solid and with numerous rather fine grooves, sparsely branched above and sometimes tinged purple below. Leaves 5–40 cm. long, broadly triangular in outline, 2–3-pinnate; ultimate segments 2–15 mm. long, linear to filiform. Petioles very short to obsolete; the first pair of leaflets arising immediately above the large sheathing bases, which may be up to 5 cm. long with a narrow membranous margin. Leaves gradually reduced upwards, until only the sheathing base with a short 3-fid lamina is present. Inflorescence of terminal and lateral compound umbels. Bracts conspicuous, greenish to creamy white, linear with an acute apex (sometimes eroded and presenting a truncate appearance). Rays 5–16, 1·5–11 cm. long, robust. Bracteoles numerous, similar to the bracts. Partial umbels with numerous flowers on fine wiry pedicels 5–10 mm. long. Petals white, hairy on the outer surface; calyx teeth absent; stylopodium conical, with a slightly undulate base; styles spreading, later deflexed in the fruiting condition. Fruit up to 6 × 2·5 mm., cylindrical at maturity, very densely covered with white to yellowish spreading hairs which completely obscure the ribs. Endosperm curved and concave towards the commissure. Vittae 1 in each interval and 2 in the commissural face. Carpophore 2-cleft or entire, rather delicate and not persisting long after the mericarps are shed.

Zambia. N: Abercorn, fl. 28.iii.1955, *E. M. & W.* 1244 (BM; LISC; SRGH). W: Ndola, fr. 19.v.1953, *Fanshawe* 16 (K; SRGH). C: Mt. Makulu Research Station near Chilanga, fl. & fr. 1.v.1960, *Angus* 2229 (BM; LISC; SRGH). S: Bowood Siding, fl. v.1909, *Rogers* 8053 (K; SRGH). **Rhodesia.** N: Mazoe, Bernheim Hill, fl. & fr. v.1906, *Eyles* 269 (BM; SRGH). W: Shangani, Gwampa Forest Reserve, fr. x.1956, *Goldsmith* 115/56 (BM). C: Salisbury, S. commonage, fl. & fr. iii.1917, *Eyles* 651 (BM; K). S: 32 km. N. of Fort Victoria, fl. 4.v.1962, *Drummond* 7962 (BM; K; SRGH). **Malawi.** N: Mzimba, fl. 12.vi.1947, *Benson* 94 (BM). C: 13 km. NW. of Fort Manning, fr. 30.v.1961, *Leach & Rutherford-Smith* 11075 (BM; K; SRGH). **Mozambique.** MS: Mucanga, between Furancungo and Vila Gamito, 59 km. from Furancungo, fl. & fr. 10.vii.1949, *Barbosa & Carvalho* 3549 (COI; K; LM).

Also in Angola, southern Zaire and southern Tanzania. In grassland, savanna woodland or sometimes in derelict cultivations.

2. **Diplolophium buchananii** (Benth. ex Oliv.) Norman in Journ. Bot., Lond. **61**: 57 (1923).—Brenan in Mem. N.Y. Bot. Gard. **8**: 446 (1954).—Cannon in Notes Roy. Bot. Gard. Edin. **32**: 198 (1973). Type: Malawi, Shire Highlands, Zomba Mt., *Buchanan* 128 (K).

Physotrichia buchananii Benth. ex Oliv. in Hook., Ic. Pl. **14**: 1358 (1891). Type as above.

A very robust perennial herb 1–2 m. high, often glaucous and sometimes subshrubby. Stems arising several together, relatively simple and unbranched, terete, rigid, solid and with distinct rather fine grooves, glabrous to slightly puberulous, often tinged purplish-brown. Leaves glabrous, 2-ternate to pinnate, the lateral leaflets consisting of 3–5 pinnules, while the central leaflets have 2–3 pairs of pinnules, with or without a terminal pinnule. Ultimate divisions 4–6·5 × 1–4 cm., broadly ovate or subrotund to narrowly elliptic, fleshy and somewhat coriaceous, entire with a short mucronate tip, or with the apex cleft into 3 or more teeth. Leaf bases sheathing and auriculate, rather membranous. Juvenile leaves deeply cut with narrow segments or large irregular teeth. Umbels with numerous ± regular rays 3–9 cm. long, finely and densely pubescent. Bracts 1–4 × up to 7 mm., very well developed, linear, green with a membranous margin. Bracteoles similar to the bracts, sometimes withering and partially eroded in mature fruiting umbels. Flowers numerous in each partial umbel on finely pubescent, regular rays; petals greenish-white to creamy-yellow. Fruit narrowly elliptic with well-developed, obtuse ribs, densely bristly pubescent, especially before maturity. Stylopodium conical; styles clubbed at the apex, reflexed in fruit. Endosperm semi-circular in section, with blunt projections towards the broad ribs, commissural face flat. Vittae 1 in each interval and 2 in the commissural face. Carpophore entire or 2-cleft nearly to the base.

Confined to eastern Rhodesia, southern Malawi and adjacent areas of Mozambique. The plants to the S. of the Zambesi valley have been recognised as a separate species, but since they are distinguished by slight though constant differences only, it seems best to regard them as a subspecies and thus draw attention to the variation in association with geographical separation.

Ultimate leaf segments usually broadly ovate to subrotund, rarely elliptic, often with 2 or
    more large teeth near the tip; rays of fully expanded terminal umbels 5–9 cm. long
                                            subsp. *buchananii*
Ultimate leaf segments narrowly elliptic, always lacking apical teeth; rays of fully
    expanded terminal umbels 3–5 cm. long   -   -   -   -   subsp. *swynnertonii*

Subsp. **buchananii.**

Malawi. S: Zomba Plateau, fl. 6.vi.1946, *Brass* 16289 (BM; K; PRE; SRGH).
Mozambique. N: Ribáuè, serra Mepáluè, fr. 9.xii.1967, *Torre & Correia* 16436 (LISC). Z: Serra do Gúruè, fl. & fr. 20.ix.1944, *Mendonça* 2179 (LISC).
Moist places in rocky montane grassland and in *Brachystegia* woodland.

Subsp. **swynnertonii** (Bak. f.) Cannon in Notes Roy. Bot. Gard. Edin. **32**: 200 (1973).
    Type: Rhodesia, Gazaland, Melsetter, *Swynnerton* 649 (BM).
    Physotrichia swynnertonii Bak. f. in Journ. Linn. Soc., Bot. **40**: 76 (1911).—Eyles in Trans. Roy. Soc. S. Afr. **5**: 434 (1916). Type as above.
    Physotrichia gorungosensis Engl., Pflanzenw. Afr. **3**, 2: 819 (1921). Described from Rhodesia (Melsetter) and Mozambique (Gorongosa), collectors unknown.
    Diplolophium swynnertonii (Bak. f.) Norman in Journ Bot., Lond. **61**: 57 (1923). Type as for *Physotrichia swynnertonii*.

Rhodesia. C: Makoni, summit of mt., fr. vii.1917, *Eyles* 743 (BM; K; SRGH). E: Umtali, edge of Vumba, fl. & fr. 21.ix.1948, *Chase* 1278 (BM; LISC; SRGH). **Mozambique.** MS: Serra de Gorongosa, near Pico Gogôgo, fl. 26.ix.1943, *Torre* 5939 (LISC).
Habitat as for subsp. *buchananii*. The sap of this plant produces blisters on the skin.

19. PHYSOTRICHIA Hiern

Physotrichia Hiern in Journ. Bot., Lond. **11**: 161 (1873).

Biennial or perennial herbs. Leaves mostly basal with nearly bare stems. Flowers in terminal and lateral compound umbels. Sepals well developed, sometimes persisting on the mature fruit. Petals obovate, apex incurved. Fruit elliptic to ovoid, slightly dorsiventrally compressed, covered with vesicular hairs especially on the ribs. Stylopodium conical, styles short and ± divergent. Ribs obtuse; vittae well developed, 1 in each interval and 2 in the commissural face. Carpophore divided to the base; seed concave on the commissural face.

A poorly-known genus of perhaps 5 species confined to central Africa and related to *Diplolophium*, but distinguished by the vesicular hairs on the fruit. Distributed from Angola through southern Zaire and northern Zambia to N. Malawi. Further studies of this genus are needed to clarify its relationships and to further refine its definition.

Basal leaves simple, with broad palmate lobes   -   -   -   -   - 1. *heracleoides*
Basal leaves finely divided, with jagged segments -   -   -   -   - 2. *atropurpurea*

1. **Physotrichia heracleoides** H. Wolff in Engl., Bot. Jahrb. **48**: 273 (1912). TAB. **159.**
    Type from Zaire.

Robust to wiry biennial or perennial herbs up to 80 cm. high, with scattered short bristly hairs which often become dense on the peduncles and rays. Rootstock woody; stem terete with regular rather coarse grooves. Leaves mostly basal, those on the stem much reduced and present on the inflorescence only as sheathing bases with small appendages. Basal leaves simple with 3–5 broadly rounded subpalmate lobes up to 15 × 20 cm., to 2-ternate with leaflets similar to the lobes of the simple leaves; apex broadly to slightly emarginate, base cordate; margins ± regularly denticulate; petioles up to 20 cm. long, with slightly sheathing bases. Umbels terminal and lateral on long peduncles. Terminal umbels with 20–30 rays

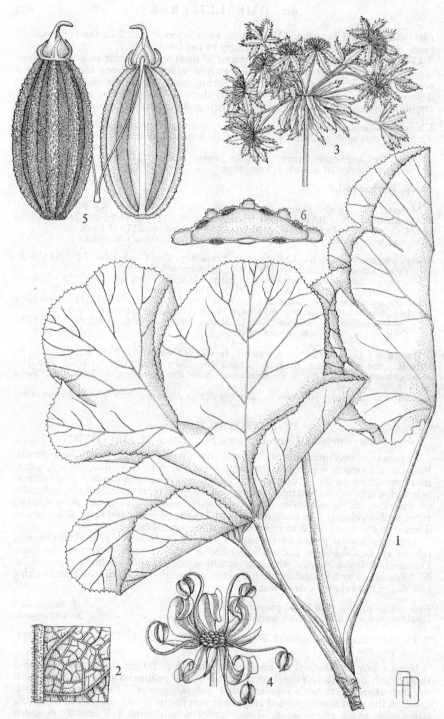

Tab. 159. PHYSOTRICHIA HERACLEOIDES. 1, habit (×½); 2, underside of leaf (×2), both from *Siame* 67a; 3, umbel (×½); 4, flower (×12), both from *E. M. & W.* 1234; 5, fruit (×7); 6, mericarp in cross-section (×10), both from *Siame* 67a.

up to 4 cm. long; bracts and bracteoles very well developed and often conspicuous. Bracts 10–15, 2–2·5 cm. long, linear to narrowly lanceolate, bracteoles similar, up to 1·25 cm. long; both are either entire or may be markedly fimbriate (see notes). Partial umbels with 9–30 flowers on pedicels 5–12 mm. long. Calyx teeth well developed, narrowly triangular to subulate; petals white to pale cream. Fruit 5–3 mm. long, ellipsoid to ovoid, slightly dorsiventrally compressed, covered with vesicular hairs especially on the ribs. Stylopodium conical; styles short and often shed from the maturing fruit. Ribs obtuse; vittae well developed, 1 in each interval, obvious externally and occupying most of the space between the ribs, and 2 in the commissural space. Carpophore divided to the base, seed very slightly concave on the commissural face.

Zambia. N: Danger Hill, 29 km. NNE. of Mpika, fl. & fr. 18.ii.1970, *Drummond & Williamson* 9704 (BM; SRGH). **Malawi.** N: 8 km. N. of Chisenga, fl. & fr. 22.ii.1970, *Drummond & Williamson* 9920 (BM; SRGH).
Also in Zaire and Tanzania. In open *Brachystegia* woodland, 1500–2000 m.

A rather poorly-known species which, as treated here, includes two distinct elements. One has slender stems, with entire bracts and bracteoles, the other is a more robust plant with fimbriate bracts and bracteoles. The former includes the type; the latter, which is illustrated in tab. **159**, has only been collected in Zambia, round Mbala, e.g. *Siame* 67a (BM; SRGH), *E. M. & W.* 1234 (BM; SRGH) and *Richards* (SRGH). Although superficially the plants look strikingly different, some intergradation is apparent and in view of the limited information available it seems best for the present to draw attention to the Abercorn plants without according formal recognition.

2. **Physotrichia atropurpurea** (Norman) Cannon, comb. nov.
  Spuriodaucus atropurpureus Norman in De Wild. & Staner, Contr. Fl. Katanga, Suppl. 3: 120 (1930).—Cannon in C.F.A. **4**: 352 (1970). Type from Angola.

Robust biennial or perennial herb up to 1 m. Rootstock woody, stem terete with regular, rather coarse grooves. Leaves glabrous, mostly basal on long petioles up to 20 cm. long, 3-pinnatisect to finely divided, the ultimate segments up to 4 × 1·5 mm., oblong-lanceolate, with a characteristic shortly mucronate apex. Stem leaves few and much reduced, on the inflorescence only present as sheathing bases with small appendages. Umbels frequently terminal only, occasionally with laterals. Rays 20–30, up to 5 cm. long, robust, pubescent to subglabrous; bracts and bracteoles well developed. Bracts 10–15, c. 15 mm. long, narrowly lanceolate; bracteoles similar, c. 10 mm. long. Partial umbels with 15–25 flowers on pedicels up to 10 mm. long. Calyx teeth well developed, narrowly triangular; petals white to deep purple. Mature fruit not known. Immature fruit elliptic, slightly dorsiventrally compressed, densely covered with vesicular hairs, perhaps ± glabrous in some very immature specimens; stylopodium conical; styles relatively long and divergent.

Zambia. N: Isenga Dambo, Kawambwa, fl. 14.iv.1959, *Mutimushi* 27 (K). W: Mwinilunga Distr., 7 km. N. of Kalene Hill, fl. & fr. 16.iv.1965, *Robinson* 6571 (K; SRGH).
This species, originally described from Angola, is now also known from Zaire and two specimens from the F.Z. area.

Spuriodaucus was erected by Norman to include this and one other species, but no clear indication was given of the diagnostic characters of the genus. Since *S. atropurpureus* shows obvious close affinity with *Physotrichia*, through its strongly-developed bracts and bracteoles, its vesicular hairs and general facies, it is here transferred to that genus. Additional material should be collected by workers in the field, so that the relationships of all the taxa in this general affinity can be properly clarified. *S. asper* Norman from Katanga is clearly closely related and may be conspecific.

20. FOENICULUM Mill.
Foeniculum Mill., Gard. Dict., abridg. ed. 4, 1 (1754).

Biennial or perennial herbs with pinnately divided leaves, with filiform ultimate segments. Umbels compound, bracts and bracteoles 0–few. Sepals absent; petals yellow with strongly inturned apices. Fruit narrowly ovoid, very slightly laterally compressed; ribs fairly robust and obvious at maturity. Vittae solitary in the furrows.

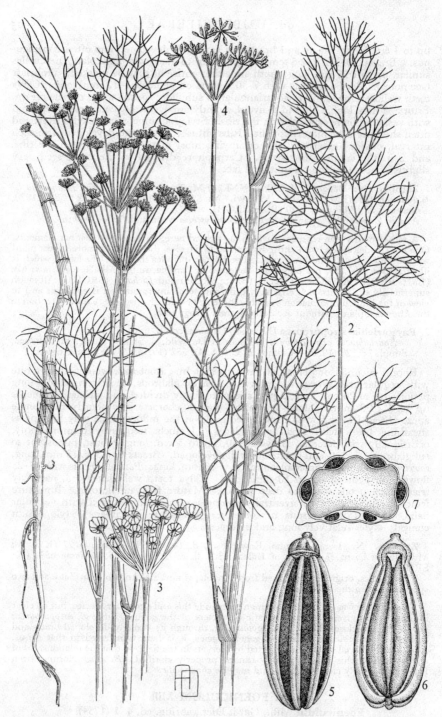

Tab. 160. FOENICULUM VULGARE. 1, habit (×½); 2, leaf (×½); 3, partial umbel with flowers in bud (×5); 4, partial umbel in early fruit (×½); 1–4 from *Lemos & Balsinhas* 137 & 2505; 5, mericarp from exterior (×10); 6, mericarp, commissural view (×10); 7, mericarp in cross-section (×15), 5–7 from *Meinertzhagen* in Carp. Coll. (BM).

A genus of 2–3 species although a number of additional species have been proposed, probably as a result of the very wide introduction of *F. vulgare* around the world, both as a food plant and as a casual.

Foeniculum vulgare Mill., Gard. Dict. ed. 8 (1768).—Cannon in C.F.A. **4**: 353 (1970).
TAB. **160**. Type from Europe.
Anethum foeniculum L., Sp. Pl. **1**: 263 (1753). Type as above.

Robust, glabrous, somewhat glaucous perennial herb up to 2 m. Stems rigid when mature and with many fine distinct ribs. Leaves with well-developed sheathing bases, up to 5 cm. long but frequently much shorter, 3–4-pinnate with finely-divided ultimate segments which are narrowly linear to capillary, becoming reduced upwards towards the inflorescence and finally occurring only as sheathing bases with 0–few linear lobes. Umbels terminal and lateral, rays (4)9–30, 1–11 cm. long, relatively robust and somewhat glaucous. Partial umbels with up to 12 flowers on pedicels 5–10 mm. long; bracts and bracteoles 0 (rarely few). Calyx teeth obsolete; petals bright yellow. Fruit 4–8 mm. long, narrowly ovoid at maturity, very distinctly glaucous-grey especially when immature; stylopodium rather low. Mericarps with ribs that are obvious at maturity, but inconspicuous in young fruit.

Rhodesia. W: Bulawayo, fl. x.1958, *Miller* 5481 (K; SRGH). **Mozambique.**
SS: Chibuto, 4 km. from Maniquenique, fl. 18.vi.1960, *Lemos & Balsinhas* 137 (COI; K; LISC; LMA; PRE; SRGH).
Introduced into the F.Z. area. Very widely distributed in temperate and tropical areas, but probably native only in the Mediterranean.

Superficially similar to *Anethum graveolens* and occurring in similar situations, but readily distinguished when mature fruit are present, as *Anethum* has strongly dorsally compressed fruit similar to those of *Peucedanum*. The whole plant has a strong characteristic smell. Widely used as the culinary herb Fennel and also as selected cultivars, the blanched tuberous stem bases being used like those of celery.

21. ANETHUM L.
Anethum L., Gen. Pl. ed. 5: 127 (1754).

Annual herbs with compound pinnate leaves and filiform ultimate segments. Umbels compound, bracts and bracteoles obsolete. Sepals absent, petals yellow, with strongly incurved apices. Fruit elliptical, strongly dorsiventrally compressed, with some development of lateral wings. Vittae solitary in the furrows.

A monotypic genus, widely distributed in warm temperature and tropical regions, but probably native only in the Mediterranean and parts of western Asia. Its present wide occurrence presumably results from its use as a flavouring herb—Dill.

Anethum graveolens L., Sp. Pl. **1**: 263 (1753).—Cannon in C.F.A. **4**: 353 (1970).
TAB. **161**. Type from Europe.
Peucedanum anethum Baill., Traité Bot. Méd.: 1045 (1884). Type as above.

Robust, rather glabrous annual herb up to 75 cm., with a strong characteristic odour. Stem terete, with numerous fine grooves. Leaves 3–4-pinnate, ultimate segments narrowly linear to filiform. Umbels terminal, equalled or exceeded by long peduncled laterals. Rays numerous, up to 4 cm. long; bracts and bracteoles 0. Partial umbels with up to 35 flowers, but those of small plants may have as few as 6. Sepals obsolete; petals yellow with an obvious inturned apex. Fruit elliptic, strongly dorsiventrally compressed, with a moderately well-developed marginal wing paler in colour than the body of the fruit. Stylopodium low-conical; styles short, clubbed at the apex, divergent and shed before the fruit matures. Dorsal ribs filiform, carpophore 2-cleft to the base; vittae well developed, 1 per interval and 2 in the commissural face.

Mozambique. SS: Chibuto, Baixo Changara, fl. & fr. 22.viii. 1963, *Macedo & Macuácua* 1120 (LMA). LM: Between Quinta da Pedra and Salamanga, Maputo region, fl. & fr. 5.vii.1948, *Gomes e Sousa* 3754 (COI; PRE; SRGH).
Introduced into the F.Z. area.

Superficially similar to *Foeniculum vulgare* Mill. from which it differs in being an annual

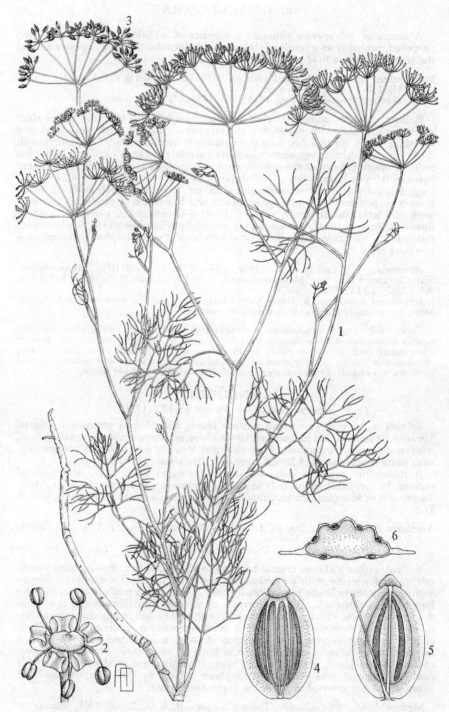

Tab. 161. ANETHUM GRAVEOLENS. 1, habit (×½); 2, flower (×10), both from *Figueredo Gomes & Sousa* 3754; 3, umbel with mature fruit (×½); 4, mericarp from exterior (×8); 5, mericarp, commissural view (×8); 6, mericarp in cross-section (×15), 3–6 from *Schimper* 949.

and in having strongly dorsiventrally compressed fruit. Immature plants are sometimes difficult to identify.

The culinary herb Dill; the foliage and fruit are used in a variety of ways.

22. PEUCEDANUM L.

Peucedanum L., Sp. Pl. **1**: 245 (1753); Gen. Pl. ed. 5: 116 (1754).

Perennial herbs, rarely biennials or annuals. Leaves pinnate or ternate, or compound forms of these, rarely finely divided. Umbels compound, bracts 0 to numerous, probably sometimes caducous; bracteoles several to numerous. Calyx teeth minute or 0. Petals white, cream or yellow, sometimes flushed with pink. Fruit orbicular, oblong or pyriform, very strongly dorsiventrally compressed, with a broad commissure. Lateral wings well developed as expansions of the marginal ribs, sometimes surrounding the stylopodium on both sides, dorsal ribs filiform. Stylopodium well developed, conical; vittae 1–3 in each interval and 2 in the commissural face; carpophore divided to the base.

A large complex and widespread genus of perhaps 200 species, widely distributed in Europe, Asia and Africa, with a closely related genus *Lomatium* in N. America. Well represented in tropical and southern Africa with perhaps 50 species. Its relationships with *Lefebvrea* need clarifying and this is only likely to be achieved through a pan-African review of the whole group.

Ultimate segments narrowly linear, up to 1 mm. wide -   -   -   -   5. *rhodesicum*
Ultimate segments various, not less than 2 mm. wide:
  Petioles and rhachis tomentose, at least at the insertion of the leaflets   1. *claessensii*
  Petioles and rhachis glabrous or with a few scattered hairs:
    Rays and peduncles distinctly muriculate -   -   -   -   -   6. *heracleoides*
    Rays and peduncles ± smooth:
      Fruit subrotund, distinctly emarginate at the base; leaf segments with irregular
        fine serrations -   -   -   -   -   -   -   -   -   2. *linderi*
      Fruit elliptic to pyriform, cuneate to very slightly emarginate at the base; leaf
        segments deeply and jaggedly cut or with regular coarse serrations:
        Stylopodium in fruit enclosed by apical expansions of the wings   7. *angolense*
        Stylopodium ± free:
          Ultimate segments with regularly serrate margin   -   -   4. *eylesii*
          Ultimate segments jaggedly and irregularly cut -   -   -   3. *nyassicum*

1. **Peucedanum claessensii** Norman in De Wild., Pl. Bequaert. **4**, 3: 352 (1928); in Journ. Linn. Soc., Bot. **49**: 510 (1934).—Cannon in Garcia de Orta, Sér. Bot. **1**: 46 (1973). Type from Zaire.

    Peucedanum valerianifolium Bak. in Kew Bull. **1897**: 269 (1897) pro parte, fruiting material only.

Robust herb 1–2·5 m. Stem subterete, grooved, with scattered long spreading hairs, these becoming denser in the upper part of the plant. Leaves up to 50 cm. long in specimens seen but probably much larger, ovate-lanceolate in outline, 2(3)-pinnate. Leaflets lanceolate, sometimes with deep basal lobes, margin rather irregularly serrate. Petioles with large membranous sheathing bases, rhachis and sheath usually fairly densely pubescent, rarely subglabrous. Leaves reduced to sheathing bases on the inflorescence. Umbels terminal and lateral, the laterals exceeding the terminal. Rays 7–16, up to 6·5 cm. long, somewhat irregular. Bracts absent (sometimes the elongate sheath of the leaf under the umbel looks superficially like a bract). Bracteoles few, linear, inconspicuous. Petals pale yellow. Fruit 10–15 × 5–10 mm., pyriform, strongly dorsiventrally compressed; lateral wings up to 2 mm. wide, well developed. Stylopodium depressed-conical, with undulate margin and short, divergent, clubbed styles. Seed-bearing part of the mericarp confined to the upper ⅔, with an obvious sterile lower part. Dorsal ribs filiform; vittae clearly visible externally, 1 per interval and 2 in the commissural face. Carpophore divided to the base.

Zambia. E: Lundazi Distr., Nyika Plateau, fl. & fr. 2.i.1959, *Richards* 10402 (K; SRGH). **Rhodesia.** E: Inyanga, Troutbeck, fl. 21.i.1948, *Chase* 686 (BM; SRGH). **Malawi.** N: Mzuzu Distr., Vipya Range, fr. 24.ii.1961, *Richards* 14456 (K). C: Dedza Distr., fl. 24.x.1956, *Banda* 318 (BM; SRGH). C: Chintembwe Mission near Ntchisi, fr. 22.ii.1959, *Robson* 1715 (K). S: Zomba Mt., fr. 10.i.1957, *Whellan* 1160 (K). **Mozambique.** N: Vila Cabral, fl. & fr. 10.i.1935, *Torre* 439 (COI; LISC). Z:

Milange, fl. & fr. 26.ii.1943, *Torre* 4845 (LISC). MS: Mavita, Manica, Vale do Moçambique, fl. 25.x.1944, *Mendonça* 2801 (LISC).

Also in Zaire and Tanzania. In montane grassland, near streams and occasionally as a weed in cultivated areas, 1066–2133 m.

Peucedanum valerianifolium Bak. is based on a specimen from Malawi: Mt. Zomba, 1200–1800 m., *Whyte* (K). This is a mixed gathering, of which the upper parts are *P. claessensii*, while the lower stem and leaves are *Cephalaria ? retrosetosa* Engl. Since in his diagnosis Baker placed complete reliance on the basal leaves, as is further suggested by his choice of specific epithet, I have no hesitation in typifying his species on the leaves and rejecting the name from consideration in the context of *Peucedanum claessensii*.

2. **Peucedanum linderi** Norman in Journ. Linn. Soc., Bot. **49**: 511 (1934).—Robyns & Tournay, Fl. Parc Nat. Alb. **1**: 713 (1948).—Brenan, Mem. N.Y. Bot. Gard. **8**: 446 (1954).—Cannon in Garcia de Orta, Sér. Bot. **1**: 46 (1973). Type from Zaire.

Robust, almost glabrous herb 1–3 m. high, rootstock fleshy. Stem subterete with fine grooving. Leaves up to 40 cm. long in specimens seen but probably much larger in life, ovate-triangular in outline, 2–3-pinnate. Leaflets ovate-lanceolate, often deeply divided, apex acute, margin rather irregularly serrate, the teeth with short mucronate tips. Leaves progressively reduced upwards, finally to sheaths with minute foliar tips. Umbels terminal and lateral, the laterals usually exceeding the terminals. Rays 7–13, slightly scarious; bracts 0–1, small, linear; bracteoles several, linear, inconspicuous. Petals greenish-white, the outer scarcely radiate. Fruit 7–10 × 6–9 mm., subrotund (rarely to oblong-ovate), strongly dorsiventrally compressed; lateral wings up to 3 mm. wide, well developed, base distinctly emarginate, apex with wings sometimes extended above the level of the stylopodium. Stylopodium low-conical, with a slightly undulate margin, styles clubbed and deflexed after flowering, often shed before the fruit matures. Seed sometimes not fully extended to the base of the mericarp. Dorsal ribs filiform, obtuse. Vittae externally visible, 1 in each interval and 2 in the commissural face. Carpophore divided to the base.

Rhodesia. E: Inyanga Distr., W. face of Inyangani, fr. 9.iv.1957, *Goodier & Phipps* 69 (K; SRGH). **Malawi.** N: Nyika Plateau, fl. & fr. 16.viii.1946, *Brass* 17237 (K; SRGH). **Mozambique.** MS: Tsetsera, fl. 9.xi.1955, *E. M. & W.* 314 (BM; LISC; SRGH).

Mountains of E. Africa from Kenya to Zaire. In the F.Z. area it occurs by mountain streams in *Podocarpus* forest and in rock crevices.

3. **Peucedanum nyassicum** H. Wolff in Engl., Bot. Jahrb. **48**: 282 (1912).—Norman in Journ. Linn. Soc., Bot. **49**: 513 (1912).—Brenan in Mem. N.Y. Bot. Gard. **8**: 446 (1954).—Cannon in Garcia de Orta, Sér. Bot. **1**: 47 (1973). Type: Malawi, Mt. Mlanje, *Whyte* (B†, holotype; BM; K).

Glabrous perennial up to 2 m. high. Stem subterete with relatively fine grooves. Leaves up to 75 × 25 cm., lamina narrowly triangular, 2-pinnate; irregular development of the lowest secondary divisions occurs, giving the appearance of 4 pinnae at each node. Ultimate leaflets up to 45 × 20 mm., lanceolate-elliptic, varying in degree of dissection. Margin irregularly serrate, with prominent mucronate tips to the teeth, becoming almost spinous in some forms with narrow leaflets. Petioles often exceeding the lamina. Leaves progressively reduced upwards, being represented on the inflorescence by sheathing bases with linear appendages. Umbels terminal and lateral. Rays 7–13, 20–50 mm. long, rather irregular. Bracts several, linear, inconspicuous; bracteoles several, similar to the bracts. Flowers on pedicels 8–10 mm. long. Petals yellowish-green, the outer scarcely radiate. Fruit c. 10 × 5 mm. at maturity, oblong to elliptic, strongly dorsiventrally compressed; lateral wings 1–1·5 mm. wide, well developed, sometimes very slightly emarginate at the base. Seed extending through the whole length of the mericarp; stylopodium broadly conical, styles with clubbed apices. Dorsal ribs filiform; vittae visible externally, 1 in each interval and 2 in the commissural face. Carpophore divided to the base.

Rhodesia. E: Inyangani, Pungwe source, 20.x.1946, *Chase* 1425 (K; SRGH). **Malawi.** S: Zomba Forest Nursery, fl. & fr. 9.iii.1955, *E. M. & W.* 738 (BM; LISC; SRGH).

Possibly confined to the F.Z. area, but this can only be finally established after the identity of closely related plants from southern Tanzania has been clarified. In marshy ground and near streams.

Closely related to *P. eylesii* and possibly conspecific.

4. **Peucedanum eylesii** Norman in Journ. Bot., Lond. **70**: 138 (1932); in Journ. Linn. Soc., Bot. **49**: 509 (1934).—Cannon in Garcia de Orta, Sér. Bot. **1**: 47 (1973). Type: Rhodesia, Macheke, *Eyles* 2006 (K, holotype; PRE; SRGH).

Robust glabrous perennial herb 1·5–2·5 m. high. Stem subterete, rather angular, especially in the upper parts, with relatively few, coarse grooves. Lamina triangular in outline, (2) 3-pinnate. Petioles roughly equalling the lamina, with large basal sheaths. Ultimate divisions 15–80 × 6–15 mm., linear-lanceolate to narrowly lanceolate, rather variable but most frequently with a regular serrate margin, with small mucronate tips to the teeth; leaflets sometimes rather jaggedly and irregularly dissected. Uppermost leaves with very narrow linear segments and finally present only as sheathing bases on the inflorescence. Umbels terminal and lateral. Rays 10–15, up to 55 mm. long, rather irregular. Bracts several, small, linear, inconspicuous; bracteoles similar. Flowers with fine pedicels 6–12 mm. long when in fruit. Petals greenish-yellow, the outer scarcely radiate. Fruit c. 10 × 5 mm., oblong to elliptic, strongly dorsiventrally compressed; lateral wings c. 1·5 mm. wide, well developed, slightly emarginate at the base. Stylopodium conical, styles short and clubbed at the apex. Seed occupying only the upper ¾ of the mericarp. Dorsal ribs filiform, well developed; vittae visible externally, 1 in each interval and 2 in the commissural face. Carpophore divided to the base.

Zambia. W: Kitwe, fl. & fr. 15.ii.1955, *Fanshawe* 2654 (K). E: Lundazi Distr., Nyika Rest House, fl. & fr., *Richards* 10402 (K). **Rhodesia.** C: Salisbury, Cleveland, fl. & fr. 24.xi.1945, *Wild* 413 (K; SRGH). E: Umtali Distr., Engwa, fr. 1.ii.1955, *E. M. & W.* 72 (BM; LISC; SRGH). **Malawi.** N: Rumpi Distr., Chelinda Bridge, fl. 9.xii.1966, *Pawek* 512 (SRGH). C: Dedza Distr., Chongoni Forestry School, fl. 18.i.1967, *Salubeni* 509 (SRGH). **Mozambique.** Z: Gúruè, Serra do Gúruè, picada Chá Moçambique, fl. & fr. 3.i.1968, *Torre & Correia* 16856 (LISC). MS: Manica, Rotanda, Tandara, fl. 19.xi.1965, *Torre & Correia* 13173 (LISC).
Probably also present in Tanzania and Zaire. In marshy areas at the side of streams in the shade, 1500–1700 m.

Closely related to *P. nyassicum* and possibly conspecific.

5. **Peucedanum rhodesicum** Cannon in Garcia de Orta, Sér. Bot. **1**: 47 (1973). TAB. **162**. Type: Rhodesia, Inyanga Distr., Mt. Inyangani, *Wild* 4931 (BM, holotype; K; SRGH).

Glabrous perennial herb 1–1·5 m. high. Stem subterete to somewhat angled, with finely prominent grooves, often tinged purple-brown. Leaves with long petioles, lamina up to 50 × 20 cm., ovate to triangular in outline, 3–4-pinnate; ultimate segments 2–6 × 0·5 mm., finely linear with acute apices, sometimes somewhat crisped (resembling those of *Foeniculum*), densely arranged in the lower leaves, more widely spaced in the upper leaves; leaves finally reduced to sheathing bracts on the inflorescence. Umbels terminal and lateral. Rays 10–39, 3–11 cm. long at maturity, irregular. Bracts numerous, up to 15 mm. long, linear, often eroded at maturity; bracteoles similar. Partial umbels with 10–18 flowers on pedicels 6–17 mm. long. Petals greenish-yellow, the outer scarcely radiate. Fruit 10–14 × 5–7 mm., oblong-elliptic, strongly dorsiventrally compressed, base scarcely emarginate; lateral wings 1–2 mm. wide, well developed; seed extending to the base of the mericarp. Stylopodium conical, margin somewhat undulate; dorsal ribs filiform, not very strongly developed. Vittae visible externally, 1 in each interval and 2 in the commissural face. Carpophore divided to the base.

Rhodesia. E: Umtali Distr., Stapleford Forest Reserve, fl. 16.x.1959, *Chase* 7193 (BM; K; LISC; SRGH). **Mozambique.** MS: Manica, Rotanda, Tandara, fl. & fr. 14.xi.1974, *Torre & Correia* 13152 (LISC).
Confined to the mountains between Rhodesia (E) and Mozambique (MS). In mountain grassland, and in scrub, especially in marshy places and near streams.

Tab. 162. PEUCEDANUM RHODESICUM. 1, habit (×½) from *Chase* 7193; 2, flower (×18);
3, mericarp from exterior (×3); 4, commissural view of mericarps (×3); 5, mericarp
(×6), all from Miller 3828a.

6. **Peucedanum heracleoides** Bak. in Kew Bull. **1897**: 268 (1897).—Norman in Journ.
Linn. Soc., Bot. **49**: 514 (1934).—Cannon in Garcia de Orta, Sér. Bot. **1**: 47 (1973).
Type: Malawi, Nyika Plateau, *Whyte* 224 (K).
Peucedanum muriculatum Welw. ex Hiern, Cat. Afr. Pl. Welw. **1**: 429 (1898).—
Norman, loc. cit. Type from Angola.
Peucedanum kingaense Engl., Bot. Jahrb. **30**: 368 (1901). Type from Tanzania.
Peucedanum bequaertii Norman in De Wild., Pl. Bequaert. **4**: 306 (1927). Type
from Zaire.

A very variable plant 0·5–1·5 m. high, with a woody or fleshy taproot. Stem
terete to angled, with fine to coarse grooving, normally bearing very fine muriculate
spines, these most obvious in the younger parts and tending to become eroded in
the older regions. Leaves up to 25 cm. long, pinnate, the leaflets varying from
obtusely ovate through acutely lanceolate to narrowly lanceolate-oblong; leaflets
decurrent on the rhachis and sometimes deeply lobed. Margins shallowly dentate
to deeply serrate, with short mucronate tips. Muriculations present on the petiole,
but becoming more hair-like on the veins of the inferior leaf surface. Umbels
terminal and lateral; rays 8–13, 1–6 cm. long, somewhat irregular. Bracts up to
25 mm. long, several, muriculate, linear to subulate; bracteoles similar. Partial
umbels with flowers on relatively robust pedicels. Petals yellowish. Fruit
10–12 × 5–6 mm., oblong-elliptic to narrowly obovate, strongly dorsiventrally
compressed, with a papillate-muriculate surface; lateral wings 1·5–2 mm. wide,
well developed. Seed confined to the top ¾ of the mericarp; dorsal ribs filiform,
very well developed. Stylopodium shortly conical with an undulate margin;
styles short with clubbed stigmatic surfaces. Vittae 1 per interval and 2 in the
commissural face, visible externally. Carpophore divided to the base, but often
missing after the dispersal of the mericarps.

Zambia. N: Nkali Dambo, fl. 5.i.1952, *Richards* 302 (K). W: Mwinilunga Distr.,
Mwanamitowa R., fr. 15.viii.1930, *Milne-Redhead* 914 (K). **Malawi.** N: Nyika Plateau,
below Rest House, fr. 4.vi.1957, *Boughey* 1630 (K; SRGH).
A very variable complex that seems best treated as one species for the present. Plants
from the F.Z. area are typically robust; those of Zaire (*P. bequaertii*) have narrow leaflets
and particularly well-developed bracts, while the *P. muriculatum* forms from the Huila
area of Angola are shorter variants with small, inconspicuous bracts and bracteoles. In the
F.Z. area this species occurs in montane grassland at c. 2100 m.

7. **Peucedanum angolense** (Welw.) Cannon in C.F.A. **4**: 355 (1970). Type from
Angola.
Lefebvrea angolensis Welw. ex Ficalho in Boll. Soc. Geogr. Lisbon, Sér. 2, **11–12**:
712 (1882).—Hiern, Cat. Afr. Pl. Welw. **1**: 431 (1898). Type as above.
Lefebvrea benguellensis Welw. ex Engl. in Abh. Königl. Preuss. Akad. Wiss. Berl.
1891: 322 (1892).—Hiern, tom. cit.: 430 (1898).—Bak. f. in Journ. Linn. Soc.,
Bot. **40**: 75 (1911). Type from Angola.
Peucedanum buchananii Bak. in Kew Bull. **1897**: 268 (1897). Syntypes: Malawi,
Nyika Plateau, *Whyte* (K); Shire Highlands, 1878, *Buchanan* 167 (K).
Peucedanum benguellensis (Welw.) Eyles in Trans. Roy. Soc. S.Afr. **5**: 434 (1916).
Type as for *Lefebvrea benguellensis*.
?*Lefeburia microcarpa* H. Wolff in Engl., Bot. Jahrb. **57**: 233 (1921). Type:
Malawi, Nyika Plateau, *Whyte* (B†, holotype; BM, photogr.).

A robust glabrous herb up to 3 m. high. Stem subterete with fine grooving,
sometimes purple-spotted. Leaves up to 80 cm. in specimens seen, but certainly
larger in life, 2-ternate to 2-pinnate, usually broadly triangular in outline. Leaflets
rhomboid-elliptic to linear-lanceolate, apex narrowly acute, bases cuneate;
margins irregularly cut with deep teeth. Upper leaves reduced, and represented
on the inflorescence by little more than sheathing bases. Umbels terminal and
lateral; rays 8–14, up to 5 cm. long at maturity. Bracts 0–5, linear, insignificant,
probably caducous; bracteoles up to 6, linear, inconspicuous. Flowers on relatively
robust pedicels, petals yellowish-cream. Fruit 6–13 × 4–10 mm., pyriform to
broadly elliptic, strongly dorsiventrally compressed; lateral wings up to 3 mm.
wide, very strongly developed. Stylopodium broadly conical; styles short,
clubbed at the apices, not deflexed after flowering. Seed confined to the upper ⅔
of the mericarp; dorsal ribs filiform; vittae not so obvious externally as in some
other species, 1 per interval and 2 in the commissural face. Carpophore divided
nearly or completely to the base.

Zambia. B: Masese, fr. 24.v.1962, *Fanshawe* 6844 (SRGH). N: Mbala (Abercorn), fl. & fr. 13.v.1952, *Siame* 31a (BM). W: Mutanda R., 33 km. SSW. of Solwezi, fl. & fr. 21.iii.1961, *Drummond & Rutherford-Smith* 7146 (BM; K). C: Chambesi, Kabwe Distr., fl. & fr. 8.iv.1972, *Kornaś* 1570 (K). S: Mapanza, fl. 10.iv.1955, *Robinson* 1225 (K; SRGH). **Rhodesia.** C: Hunyani, near Hotel, fr. 22.i.1928, *Eyles* 5271 (K; SRGH). E: Inyanga, fl. & fr. 5.iv.1962, *Wild* 5699 (BM; K; SRGH). **Malawi.** N: Marymount, Mzuzu, Mzimba Distr., fr. 5.vi.1969, *Pawek* 2454 (BM; K; UC). **Mozambique.** MS: Mavita, Rotanda, fr. 13.iv.1948, *Barbosa* 1445 (LISC).
Also in Angola, Tanzania and Zaire. In montane forest and in *Brachystegia* woodland.

A very difficult and confusing complex of taxa which has yet to be satisfactorily classified and which biosystematic data, especially those derived from growing many populations under uniform conditions, should do much to clarify. For the present it seems best to take a very conservative view: as understood in this account, *P. angolense* is a very variable complex spread right across S. tropical Africa. There are some indications of geographically based variation, but it seems best to defer formal recognition until a broader view of the group is available. Thus the plants from the border mountains between Rhodesia and Mozambique and around Makoni are exceptionally large, luxuriant and have very large, invariably pyriform fruit. These may be related to *P. uppingtoniae* Schinz, a member of the complex from SW. Africa. Wolff described another species, *Lefeburia microcarpa*, from the Nyika Plateau. This is a small-fruited plant, with diffuse inflorescences and narrow-lanceolate to linear leaf lobes. Specimens with this facies are also found in Zambia, but there seems to be a gradual, more or less continuous pattern of variation from extremes like those mentioned above, connecting with the more central forms of the complex. In this the stature of the plant, the degree of leaf dissection, the number of flowers in the inflorescence, and the form of the fruit are all important variables.

23. LEFEBVREA A. Rich.

Lefebvrea A. Rich. in Ann. Sci. Nat., Sér. 2, **14**: 260, t. 15 fig. 1 (1840).
" *Lefeburea* " Endl., Gen. Suppl. **2**: 69 (1842).
" *Lefeburia* " Lindl., Veg. Kingd. ed. 2: 778 (1847).

Biennial or perennial herbs. Leaves 2-ternate or 2-pinnate (often rather irregularly so). Umbels compound, bracts 0-few, bracteoles few to several but probably caducous. Calyx teeth minute or 0; petals pale yellow to greenish-white, with inturned apices. Fruit broadly elliptic to pyriform, very strongly dorsiventrally compressed with a broad commissure; lateral wings very well developed and surrounding the stylopodium at the apex. Dorsal ribs filiform, stylopodium well developed, conical to elongate-conical. Vittae large and conspicuous, 1 per interval and 2 in the commissural face; carpophore divided to the base.
A rather poorly-defined generic group of 5–10 species widely distributed in tropical Africa. Its relationship to *Peucedanum* is in grave need of clarification. Originally *Lefebvrea* was distinguished by the development of the lateral wings of the fruit apically so as to include the stylopodium, but experience has shown that this feature is present in some species which, in other respects, seem very closely related to typical members of the genus *Peucedanum*. For this account a restricted view has been taken of *Lefebvrea* and only peucedanoid umbels in which the wing character is correlated with long capillary pedicels and rather simple narrow-linear to linear leaflets are included. The species here recognised are clearly closely related to *L. abyssinica* A. Rich., the type species for which the genus was erected.

Leaflets linear, remotely and irregularly serrate to dentate; stylopodium elongate-
    conical, nearly equalling the ovary - - - - - - - 1. *stuhlmannii*
Leaflets narrowly lanceolate, ± regularly serrate; stylopodium conical, ¼–⅓ the length of
    the ovary - - - - - - - - - - - 2. *brevipes*

1. **Lefebvrea stuhlmannii** Engl. in Pflanzenw. Ost-Afr. **C**: 300 (1895).—Cannon in C.F.A. **4**: 358 (1970). TAB. **163**. Type from Uganda.

Robust, glabrous perennial or biennial herbs up to 3 m. Stem terete, finely and regularly grooved. Leaves somewhat irregularly 2-ternate to 2-ternate-pinnate, some of the leaflets having basal lobes on one side which are nearly as long as the principal lobe. Lamina up to 40 cm. long and probably much larger, petioles narrowly sheathing at the base. Leaflets up to 30 × 2 cm., linear, narrowly cuneate at the base, apex gradually tapered, acute; margins entire to remotely and ir-

D.E.

Tab. 163. LEFEBVREA STUHLMANNII. 1, rootstock (×½), from *Carrisso* 1727; 2, leaf(×½), from *Carrisso* 1378; 3, inflorescence (×½); 4, fruit (×3); 5, mericarps in cross-section (×10), 3–5 from *Gossweiler* 12308. From C.F.A.

reguarly serrate to dentate. Upper leaves reduced, finally only represented by sheathing bases with linear appendages on the inflorescence. Umbels terminal and lateral; rays 12–20, up to 18 cm. long, ± regular, rather wiry; partial umbels with 15–30 characteristically fine capillary pedicels 1–3 cm. long. Bracts few, linear and probably caducous; bracteoles several, c. 5 mm. long, subulate, somewhat caducous. Petals pale yellowish-cream. Fruit 10 × 7 mm., elliptic-pyriform, strongly dorsiventrally compressed, with well-developed lateral wings c. 2 mm. wide which surround the stylopodium at the apex. Stylopodium elongate-conical, about equalling the ovary at the flowering stage; styles erect to somewhat divergent, clubbed at the apex. Dorsal ribs filiform; vittae well developed, 1 in each interval and 2 in the commissural face. Carpophore divided to the base.

Zambia. N: L. Chilwa, Mbala (Abercorn), fl. 6.iii.1959, *Richards* 11106 (K). W: Kitwe, fl. 4.iv.1963, *Mutimushi* 265 (SRGH). C: Chakwenga Headwaters, 100 km. E. of Lusaka, fl. & fr. 27.iii.1965, *Robinson* 6521 (K). S: Mumbwa, fl. 12.iii.1962, *Mitchell* 13/40 (BM; K; SRGH). **Rhodesia.** N: Mazoe, fl. & fr. iii.1906, *Eyles* 268 (BM). C: Salisbury, fl. & fr. iii.1919, *Eyles* 1545 (BM; SRGH). E: Umtali, fr. 19.v.1955, *Chase* 5583 (BM; LISC; SRGH). S: Zimbabwe Ruins, fl. 3.vii.1930, *Hutchinson & Gillett* 3358 (BM). **Malawi.** N: Chisenga, fl. & fr. 25.viii.1962, *Tyrer* 573 (BM). C: Dedza, fl. 8.iii.1967, *Salubeni* 578 (K; SRGH). S: Blantyre, fl. & fr. 14.v.1948, *Faulkner* 241 (K; SRGH). **Mozambique.** N: Vila Cabral, fr. 2.v.1934, *Torre* 441 (COI; LISC). T: Angónia, Posto Zootécnico, fl. & fr. 13.v.1948, *Mendonça* 4231 (LISC).

Also in Angola, Zaire and E. Africa. In montane grassland and in open savanna woodland, 1000–2000 m.

One of the easiest species of the Umbelliferae in the F.Z. area to recognise, with its highly characteristic linear leaflets and umbels with slightly drooping fruit on fine capillary pedicels.

2. **Lefebvrea brevipes** Engl. ex H. Wolff in Mildbr., Wiss. Ergebn. Deutsch. Zentr.-Afr. Exped. 1907–1908, **2**: 600 (1913). Type from Tanzania.

?*Lefebvrea naegeleana* H. Wolff in Engl., Bot. Jahrb. **57**: 234 (1921).—Type from Uganda.

Very robust biennial or perennial glabrous herbs up to 3·5 m. high. Stem terete, finely and regularly grooved. Leaves 2-ternate to 2-pinnate, somewhat irregular. Lower leaves 50–60 cm. long and probably much larger. Petioles distinctly sheathing at the base. Leaflets up to 15 × 4 cm., rather variable in outline, from linear-lanceolate to lanceolate, bases rather abruptly cuneate to rounded, apex narrowly acute, rarely extended like a drip-tip; margins ± evenly serrulate to serrate, rarely dentate with projecting teeth. Upper leaves reduced on the inflorescence to sheathing bases with linear-lobed appendages. Umbels terminal and lateral. Rays 10–30, up to 10 cm. long, more or less regular, rather stiff; partial umbels with 15–30 flowers on rather fine pedicels 0·75–1·5 cm. long. Bracts 0 (?caducous); bracteoles very few, subulate, probably caducous. Petals greenish-white. Fruit up to 12 × 9 mm., broadly elliptic to broadly pyriform, strongly dorsiventrally compressed, with well-developed wings c. 2 mm. wide which surround the stylopodium at the apex. Stylopodium conical, c. $\frac{1}{4}$–$\frac{1}{3}$ the length of the ovary at the flowering stage. Dorsal ribs filiform; vittae 1 per interval and 2 in the commissural face. Carpophore divided to the base.

Malawi. C: Lilongwe, fr. 31.iii.1955, *Jackson* 1547 (K; SRGH). S: Zomba Plateau, fl. & fr. 30.v.1946, *Brass* 16083 (K; SRGH). **Mozambique.** N: Maniamba, fl. & fr. 29.ii.1964, *Torre & Paiva* 10910 (LISC). Z: Gúruè, fl. & fr. 8.iv.1943, *Torre* 5118 (LISC). T: Zóbuè, fl. & fr. 17.vi.1941, *Torre* 2862 (LISC).

Also in Zaire, Tanzania and ?Uganda. In open woodland and by stream margins, 1000–1500 m.

Distinguished from the closely related *L. stuhlmannii* by its broader linear-lanceolate to lanceolate leaves, much smaller stylopodium and somewhat larger fruit on rather more robust pedicels.

24. STEGANOTAENIA Hochst.

Steganotaenia Hochst. in Flora 27 Beih.: 4 (1844).—Norman in Journ. Linn. Soc., Bot. **49**: 514 (1934).

Trees, or suffrutices with distinct woody underground rootstocks, deciduous and flowering precociously before the leaves are produced. Leaves simply pinnate. Umbels compound; bracts and bracteoles several, filiform, probably somewhat caducous. Calyx teeth minute; petals greenish, yellowish to creamy-white, rarely purplish. Fruit ovate to obovate, strongly dorsally compressed with a broad commissure; lateral wings well developed, but not thickened; dorsal ribs filiform. Stylopodium strongly depressed and represented only by a low domed structure, with a slightly lobed margin. Vittae absent or rudimentary.

Probably fairly closely related to *Peucedanum*, with which it has been equated by some authors, but at once recognisable by its woody habit, a feature which it shares in the F.Z. area only with *Heteromorpha*. Woody species are very unusual in the family and although a number of shrubs are known *Steganotaenia araliacea*, which is widespread throughout much of tropical Africa, is the only member of the family with a truly arborescent habit.

Tree or large shrub up to 10 m. - - - - - - - 1. *araliacea*
Suffrutex up to 60 cm., with a robust woody caudex - - - - 2. *hockii*

1. **Steganotaenia araliacea** Hochst. in Flora **27**, Beih.: 4 (1844).—Norman in Journ. Linn. Soc., Bot. **49**: 514 (1934).—Burtt Davy & Hoyle, N.C.L.: 74 (1936).—Brenan & Greenway, T.T.C.L. **2**: 626 (1949).—Miller in Journ. S. Afr. Bot. **18**: 66 (1952).—Brenan in Mem. N.Y. Bot. Gard. **8**: 446 (1954).—Cannon in F.W.T.A. ed. 2, **1**: 755 (1958).—F. White, F.F.N.R.: 314 (1962).—Cannon in C F.A. **4**: 358 (1970). TAB. **164**. Type from Ethiopia.

Peucedanum fraxinifolium Hiern ex Oliv in Trans. Linn. Soc., Bot. **29**: 79, t. 42 (1873).—Hiern in F.T.A. **3**: 22 (1877).—Gibbs in Journ. Linn. Soc., Bot. **37**: 446 (1906).—Bak. f. in Journ. Linn. Soc., Bot. **40**: 76 (1911).—Eyles in Trans. Roy. Soc. S. Afr. **5**: 434 (1916).—Burtt Davy, F.P.F.T. **1**: 520 (1926).—Steedman, Trees etc. S. Rhod.: 59 (1933). Type from Ethiopia.

Peucedanum araliaceum (Hochst.) Benth. & Hook. f. ex Vatke in Linnaea **40**: 188 (1876).—Hiern, tom. cit.: 21 (1877).—R. E. Fr., Wiss. Ergebn. Schwed. Rhod.-Kongo-Exped.: 183 (1914). Type as for *Steganotaenia araliacea*.

Peucedanum araliaceum var. *fraxinifolium* (Hiern ex Oliv.) Engl., Pflanzenw. Ost-Afr.: 300 (1895). Type as for *Peucedanum fraxinifolium*.

Peucedanum fraxinifolium var. *haemanthum* Welw. ex Hiern, Cat.Afr.Pl. Welw. **1**: 429 (1898). Type from Angola.

Small glabrous deciduous tree or shrub, from c. 2–10 m. high, with ascending branches; bark grey formed with a small rectangular pattern; upper stem corky, pale grey; leaves normally absent at flowering and fruiting times. Leaves 15–40 cm. long, simply pinnate, densely crowded at the ends of the shoots. Leaflets 3·5–12 × 2–7 cm., in 3–4 pairs plus a terminal one, lanceolate to narrowly lanceolate, rarely nearly circular, borne on short petiolules or sessile. Apex narrowly acute with a prolonged stiff mucro, base rounded to subcuneate; margin regularly serrate, the tips of the teeth produced as long stiff bristles (similar bristles are sometimes also found on the side of the teeth). Inflorescence of 7–10 compound umbels on robust peduncles 6–17 cm. long, densely crowded at the apex of branches, with short membranous basal bracts. Peduncles often with up to 3 smaller lateral umbels in addition to the central one, and these may be composed of ♂ flowers only. Main umbels with 8–15 robust rays of 3–4·5 cm. long. Bracts few, small, almost obsolete. Partial umbels with numerous fine rays of c. 5–7 mm. long; bracteoles few, membranous, very small and inconspicuous. Flowers with greenish-white to yellow or purplish petals; calyx teeth obsolete; stylopodium depressed, forming a disk 2–2·5 mm. in diameter in fruit; styles 1·5 mm. long, divergent, short, slightly clubbed at the apex. Fruit 12–13 × 6–8 mm., usually narrowly obovate, very strongly dorsally compressed, with well-developed lateral wings as wide as the seed-containing part. Mericarps with 3 distinct dorsal ribs which may occasionally be slightly winged. Carpophore entire lying in a groove in the otherwise flat inner surface of the mericarp. Vittae rudimentary to obsolete.

Botswana. N: Kasane, Chobe R., fr. 28.vii.1930, *Van Son* in Transvaal Mus. No. 29032 (BM; SRGH). SE: Mahalapye Village, fl. 2.x.1959, *de Beer* 899 (BM; K; SRGH). **Zambia.** N: Mbala (Abercorn), fl. & fr. 19.ix.1949, *Bullock* 1007 (K; LISC; LMA; SRGH). W: Ndola, fl. & fr. 2.ix.1954, *Fanshawe* 1526 (K; SRGH). C: Mt. Makulu, near Lusaka, sterile 12.vii.1956, *Simwanda* 14 (BM; SRGH). E: near Lupande R. 8 km. S. of Usoro, sterile, *Bush* 61 (K). S: Katambora, Livingstone Distr., fl. & fr. 22.viii.1955, *Gilges* 44 (BM; LISC; SRGH). **Rhodesia.** N: Lomagundi Distr., fr. 3.x.1963, *Jacobsen* 2230 (PRE). W: Wankie, fl. & fr. 20.vi.1934, *Eyles* 8067 (BM; K; SRGH). C: 16 km. from Salisbury on Mazoe road, fl. 2.ix.1960, *Rutherford-Smith* 37

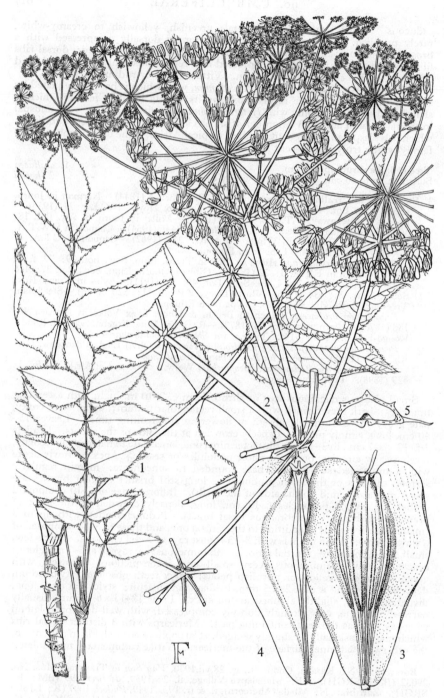

Tab. 164. STEGANOTAENIA ARALIACEA. 1, leaves (×½), from *Chr. Smith* sn and *Swynnerton* 176; 2, inflorescence (×½), from *Büttner* sn; 3, mericarp from exterior (×4½); 4, mericarp, commissural view (×4½); 5, mericarp in cross-section (×4½), 3–5 from *Godman* 203.

(K; LISC; SRGH). E: Umtali commonage, Meneni R., fl. 9.ix.1948, *Chase* 1691 (BM; LISC; SRGH). S: Todd's Hotel, Gwanda Distr., fl. & fr. 10.ix.1947, *West* 2409 (SRGH). **Malawi.** N: Mwanemba, fl. ix.1902, *McClounie* 15 (K). C: Kasungu, fl. & fr. 27.viii.1946, *Brass* 17437 (BM; K; PRE; SRGH). S: Mt. Mlanje, near Nayawani Depot, fl. 26.viii.1956, *Newman & Whitmore* 604 (BM; SRGH). **Mozambique.** N: Mecaloja, Vila Cabral, fl. & fr. 2.ix.1934, *Torre* 443 (COI; LISC). Z: Namagoa Estate, Mocuba, Quelimane Distr., fl. 20.ix.1948, *Faulkner* 304 (COI; K; SRGH). T: 19 km. S. of Tete, fl. 12.viii.1960, *Leach* 10463 (BM; LISC; SRGH). MS: Mt. Maruma, fl. 14.ix.1906, *Swynnerton* 2104 (K; SRGH).

Widespread in tropical Africa from W. Africa and Ethiopia southwards to northern S. Africa. In open woodland and on river banks.

Apparently often planted in villages for its supposed medicinal properties, a factor which may have contributed to its very wide range.

2. **Steganotaenia hockii** (Norman) Norman in Journ. Linn. Soc., Bot. **49**: 515 (1934). Syntypes from Zaire.

    Peucedanum hockii Norman in De Wild. & Staner, Contr. Fl. Katanga, Suppl. 2: 98 (1929). Syntypes as above.

Perennial, glabrous suffrutex with a large underground woody caudex. Stems up to 60 cm. long, nearly simple, more or less terete, with fine striations, leafless at the time of flowering. Leaves elliptic, (sometimes with large lobes on one side to subternate); lamina up to 20 × 10 cm., cuneate, on petioles c. 5 cm. long; margins finely dentate, the teeth with mucronate tips. Umbels terminal and lateral; bracts few, 3–7 mm. long, linear, with long fine tips, inconspicuous and often missing, probably subcaducous; bracteoles 0–2, similar to the bracts. Partial umbels with 4–7 flowers on fine pedicels up to 6 mm. long, but usually shorter. Lateral umbels sometimes with ♂ flowers only. Calyx teeth relatively conspicuous in flowers and young fruit; petals yellowish-green to greenish-brown. Developing fruit, rays and peduncles sometimes tinged with pale purple. Fruit 14–15 × 9–11 mm., pyriform to obovate, strongly dorsally compressed; stylopodium depressed; styles c. 1 mm. long; lateral wings c. 2 mm. wide, well developed, tending to envelop the stylopodium; dorsal ribs filiform. Seeds occupying c. the upper ⅔ of the fruit, with the basal part sterile and narrower. Carpophore unknown; vittae 0.

Zambia. B: Namuzinga Dambo, Sesheke Distr., 16 km. from Machili, fl. 12.viii.1947, *Brenan & Keay* 7695 (K). N: Kasama Distr., Mungwi, young fr. 2.x.1960, *Robinson* 3887 (K). W: Solwezi Distr., R. Jiundu, fl. 14.ix.1930, *Milne-Redhead* 1125 (K). C: Chilanga, 32 km. N. of R. Kafue, fl. 10.x.1909, *Rogers* 8539 (K). S: Mazabuka, Central Research Station, fl. & fr. 27.viii.1931, *Trapnell* SRGH 481 (K).

Also in adjacent areas of Zaire. Occurring in dambos and in dry bush, apparently especially in areas frequently burnt over, and perhaps adapted through its habit to this ecological situation.

Has been suspected of being poisonous to stock.

25. HERACLEUM L.
Heracleum L., Sp. Pl. **1**: 249 (1753); Gen. Pl. ed. 5: 118 (1754).

Perennial, biennial or annual herbs. Leaves pinnately divided, usually with broad segments. Umbels compound; bracts 0; bracteoles several. Calyx teeth minute or obsolete, unequal; petals white (sometimes flushed with pink, especially in bud), often markedly unequal and radiate, those on the outer margin of the umbel being much larger than those on the inside, the whole umbel therefore tending to function like a capitulum in attracting insects, although the individual flowers remain quite distinct. Fruit orbicular to broadly obovate, very strongly dorsally compressed, with a broad commissure. Marginal ridges expanded to form a broad wing, the dorsal ones filiform; wings somewhat thickened and closely appressed to those of the opposing mericarp. Vittae solitary in the intervals, clearly visible from the exterior, and tending to be club-shaped, narrowing towards the base, with 2 in the commissural face.

A genus of c. 70 species, widespread in the N. temperate zone and extending into Africa and Asia under montane conditions.

Tab. 165. HERACLEUM ABYSSINICUM. 1, habit (×½); 2 & 3, umbels with developing fruit (×½); 4, leaflet (×1½); 5, underside of leaflet (×3), 1–5 from *Robson* 389; 6, umbel with mature fruit (×½); 7, mericarps in exterior and commissural views (×3); 8, mericarps in cross-section (×8), 6–8 from *Sebald* 651.

Heracleum abyssinicum (Boiss.) Norman in Journ. Bot., Lond. **74:** 171 (1936).—
Hiern in F.T.A. **3:** 24 (1877). TAB. **165.** Type from Ethiopia.
Malabaila abyssinica Boiss. in Ann. Sci. Nat. Bot., Sér. 3, **1:** 338 (1844).—R. E. Fr.,
Wiss. Ergebn. Schwed. Rhod-Kongo-Exped.: 183 (1914).—Robyns & Tournay, Fl.
Parc Nat. Alb. **1:** 717 (1948). Type as above.

Perennial or biennial herbs with fleshy to woody taproots, up to c. 1 m. tall.
Stem terete, solid, rather coarsely striate with clearly marked ridges, often purplish
especially near the base, covered with numerous short, erect bristly hairs some of
which are glandular; the base covered with fibrous remains of old leaf bases.
Leaves up to 30 × 7 cm., simply pinnate, mostly basal, with 3–7 pairs of ovate-
deltate leaflets and 1 terminal one. Leaves sheathing at the base; petiole
approximately ⅓ the total length of the leaf. Leaflets jaggedly cut and toothed,
sometimes nearly reaching the midrib; teeth ending in a short hyaline mucro.
Petiole and rhachis with hairs similar to those on the stem; superior surface of the
leaflets with scattered, short, fine appressed hairs; inferior surface nearly glabrous
with a few hairs on the veins only; margin with a fringe of short spreading hairs.
Inflorescence of terminal and lateral compound umbels. Rays 7–14, subequal;
bracts 0-several, small, linear and inconspicuous. Partial umbels with 10–20
flowers, mostly ☿ but sometimes with a small group of ♂ flowers at the centre;
bracteoles 0-few, similar to the bracts; pedicels relatively robust. Petals white,
the outer ones more or less radiate. Fruit c. 9 × 5 mm., obovate, strongly dorsally
compressed; calyx teeth obsolete; stylopodium conical with somewhat lobed
base; styles 1·5 times as long as the stylopodium, tending to fall in fruit. Fruit
wings well developed and forming a notch in which the stylopodium is situated at
maturity. Dorsal ridges filiform. Vittae 1 in each interval and 2 in the commissural face. Carpophore deeply 2-cleft.

Malawi. N: Nyika Plateau, Chelinda Stream, Chelinda Camp, fl. & fr. 26.x.1958,
Robson 389 (BM; LISC; SRGH).
From Ethiopia through Zaire, Uganda, Kenya and Tanzania, reaching its southern
limit on the Nyika Plateau. In damp peaty soil in montane habitats.

91. ARALIACEAE

By J. F. M. Cannon

Trees, shrubs, lianes, suffrutices (very rarely herbaceous outside the F.Z. area).
Leaves alternate (rarely opposite), simple, pinnate or digitate; often coriaceous,
glabrous or with a simple or stellate indumentum, the leaves of juvenile shoots
often differing considerably from those of mature foliage; stipules frequently
conspicuous. Flowers small, ☿ (monoecious or dioecious outside the F.Z. area),
actinomorphic; arranged in umbels, racemes or in compound combinations of
these structures. Calyx inconspicuous, with the tube adnate to the ovary. Petals
(4)5(10), valvate or slightly imbricate, usually free but sometimes joined to form a
calyptra. Stamens free, alternating with the petals and usually similar in number,
but occasionally more numerous; anthers opening by longitudinal slits. Ovary
inferior, with 2–8 locules; styles often forming a distinct stylopodium and only free
at the apex, sometimes free throughout. Ovules solitary in each loculus,
anatropous, pendulous. Fruit a berry or drupe. Seeds endospermous, endosperm
smooth or ruminate; embryo very small.
An easily-recognized family of 50–60 genera, occurring throughout the tropics
and especially well represented in the Malaysian region; relatively poorly
represented in temperate areas. Some species are grown in tropical Africa as
decorative plants, including *Schefflera actinophylla*, *Didymopanax morototoni* and
Polyscias guilfoylei var. *laciniata*.

A general review of the family in tropical and southern Africa is given by Bamps
in Bull. Jard. Bot. Nat. Belg. **44:** 101–139 (1974). Tennant, in F.T.E.A., Aral.:
1 (1968), draws attention to several characteristic and well-marked teratological
conditions that may be encountered in species of *Cussonia* and *Polyscias*.

Leaves pinnate - - - - - - - - - - - **2. Polyscias**
Leaves simple, palmate or digitately compound:
    Inflorescence of simple (or compound) spikes or racemes - - - **3. Cussonia**
    Inflorescence compound umbellate or of spikes or racemes of umbels   **1. Schefflera**

1. SCHEFFLERA J. R. & G. Forst. emend. Harms *nom. conserv.*

Schefflera J. R. & G. Forst., Char. Gen. Pl.: 45, t. 23 (1776) emend. Harms in
    Engl. & Prantl, Pflanzenfam. **3**, 8: 35 (1894) *nom. conserv.*
Heptapleurum Gaertn., Fruct. **2**: 472, t. 178, fig. 3 (1791).—Hiern in F.T.A. **3**:
    29 (1877).

Shrubs, trees or lianes, sometimes occurring as epiphytes. Leaves digitate,
long-petiolate; leaflets entire to slightly serrulate or crenulate; stipules often
conspicuous. Inflorescence a panicle of umbellules or of umbellules racemosely
arranged on a main axis (capitulate in some species outside the F.Z. area). Flowers
with 5–10 petals which may be functional or shed as a calyptra as the flower opens.
Stamens equal in number to the petals. Disk flat or slightly raised in the F.Z.
species. Ovary with (2)5–8 locules, the styles prominent and equalling the locules
in number, joined at the base and spreading above. Fruit subglobose to ovoid.
Seeds laterally compressed, ellipsoid, smooth, with non-ruminate endosperm.
    A large genus of some 200 species, occurring in the tropics of both Old and New
Worlds, but with a very strong representation in the Malesian region.

Inflorescence of racemosely arranged umbellules:
  Leaflets with conspicuous, regularly spaced lateral veins, without major branches, or
    only dividing near the margin and with an obvious reticulum of veinules between
    the main laterals - - - - - - - - - 1. *abyssinica*
  Leaflets with rather inconspicuous and irregular lateral veins, which have major
    branches ½ way between the midrib and the margin; reticulum and veinules
    irregular and not very conspicuous - - - - - - 2. *goetzenii*
Inflorescence a paniculate complex of compound umbels:
  Leaves with conspicuous stipules, styles 5–9 - - - - - 3. *myriantha*
  Leaves without stipules, styles 2 - - - - - - - 4. *umbellifera*

1. **Schefflera abyssinica** (A. Rich.) Harms in Engl. & Prantl, Pflanzenfam. **3**, 8: 38
    (1894).—Robyns & Tournay, Fl. Parc Nat. Alb. **1**: 688 (1948).—Brenan & Greenway,
    T.T.C.L.: 61 (1949).—Keay in F.W.T.A. ed. 2, **1**: 751 (1958).—F. White,
    F.F.N.R.: 313 (1962).—Tennant in F.T.E.A., Aral.: 22 (1974).—Bamps in Bull.
    Jard. Bot. Nat. Belg. **44**: 127 (1974); in Distrib. Afr. Pl. **8**: map 234 (1974). Type
    from Ethiopia.
    Aralia abyssinica A. Rich., Tent. Fl. Abyss. **1**: 336 (1847). Type as above.
    Heptapleurum abyssinicum (A. Rich.) Vatke in Linnaea **40**: 191 (1876).—Hiern in
    F.T.A. **3**: 29 (1877). Type as above.
    Schefflera volkensii sensu Burtt Davy & Hoyle, N.C.L.: 33 (1936).

Epiphyte or tree of up to 30 m. Leaves digitate, with long petioles up to 42 cm.;
leaflets 5–7, up to 25(40) × 15(20) cm., broadly ovate or ovate-elliptic to broadly
elliptic, glabrous, coriaceous to subcoriaceous, the margin serrulate to almost
entire, sometimes slightly undulate, the apex usually very long-acuminate, the base
varying from cordate to subtruncate or even cuneate. Petiolules 4–7(12) cm. long,
lateral veins conspicuous and rather widely spaced, with an obvious reticulum of
veins between them. Stipules conspicuous in younger leaves, ovate-acuminate,
probably not persistent. Inflorescence a group of sparsely puberulous racemes of
pedunculate umbellules. Racemes 15–30(40) cm. long, with rusty tomentose
bracts, the umbellules normally confined to the upper ¾ of the axis and subtended
by glabrous to floccose bracteoles. Peduncles of umbellules (0·6)1–1·5(4·5) cm.
long; pedicels usually 5–6 per peduncle (rarely much more numerous), 2–7(11) mm.
long. Styles 5–8, short, 1–1·5 mm. long including the stylopodium, initially
connate but spreading at maturity. Fruit urceolate to subspherical, up to 5 mm.
long, usually glabrous, rarely puberulous.

    Zambia. N: Mporokoso, Lumangwe Falls, Kalungwishi R., veg. 26.ii.1970,
Drummond & Williamson 10024 (BM; SRGH). **Malawi.** N: Misuku, Willindi Forest,
fr. 12.i.1959, *Richards* 10630 (K). C: Dowa Distr., Ntchisi Forest, veg. 3.v.1961,
Chapman 1254 (LISC; SRGH).
    Widely distributed in E. and NE. tropical Africa and the Cameroon; also recorded
from Zaire. In mountain forest at altitudes up to 2800 m.

Tab. 166. SCHEFFLERA GOETZENII.   1, apex of branch with leaves and inflorescences
(× ½); 2, part of inflorescence showing individual flowers (× 6), both from *Chase* 766;
3, infructescence (× ½); 4, fruit (× 6); 5, fruit in cross-section (× 6), 3–5 from
Bruce 240.

2. **Schefflera goetzenii** Harms in von Götzen, Durch Afr. von O. nach W.: 376, 380 (1895).—Robyns & Tournay, Fl. Parc Nat. Alb. **1**: 688 (1948).—Bamps in Fl. Afr. Centr., Aral.: 24 (1974); in Bull. Jard. Bot. Nat. Belg. **44**: 131 (1974); in Distrib. Afr. Pl. **8**: map 237 (1974). TAB. **166**. Type from Zaire.

Schefflera barteri sensu Tennant in Kew Bull. **15**: 331 (1961) pro parte; in F.T.E.A., Aral.: 19 (1968).

Schefflera volkensi ? sensu Steedman, Trees etc. S. Rhod.: 59 (1933).

For a complete synonymy and information on the relationship of this species to *S. barteri* which is widespread in western tropical Africa see Bamps in Bull. Jard. Bot. Nat. Belg. **44**: 131 (1974).

A robust liane, small tree or shrub. Leaves digitate on long petioles up to 57 cm. long, glabrous to slightly puberulous; leaflets 4–11, up to 15(29) × 8(12) cm., subentire with a slightly undulate margin, coriaceous to subcoriaceous; apex mucronate to long-acuminate, base rounded to cuneate, sometimes slightly cordate. Petiolules 0·5–7(10) cm. long; lateral veins rather obscure, branching about midway between the margin and the midrib. Stipules conspicuous and persistent. Inflorescence a group of 3–8(20) racemes of umbellules; racemes up to 20(45) cm. long. Peduncles of umbellules up to 3·2 cm. long, subtended by inconspicuous bracts. Flowers 14(24) in each umbellule; pedicels (2)5(12) mm. long, bracteoles minute or obsolete. Stylopodium short, up to 1·0 mm. long, rather massive. Styles 5–8, free for c. 0·5 mm. at the apex. Fruit urceolate to hemi-ellipsoidal to hemispherical at maturity, c. 4·5 × 4·0 mm., glabrous to puberulous, with longitudinal furrowing.

Rhodesia. E: Umtali Distr., Cloudlands, Vumba, fl. 6.vi.1948, *Chase* 766 (BM; SRGH). **Malawi.** N: Rumphi Distr., Uzumara Forest, 1942 m., fl. 25.v.1975, *Pawek* 9654 (UC; K; M; SRGH; MA). **Mozambique.** N: Mecuburi, Serra Chinga, fr. 2.vi.1968, *Macedo* 3346 (LM). Z: Gúrue, track to Chá Moçambique, fl. 3.i.1968, *Torre & Correia* 16830 (LISC). MS: Gorongosa, near Pico Gogoga, fl. 26.ix.1943, *Torre* 5961 (LISC).

Widely distributed in E. tropical Africa and Zaire. In montane forests extending up to 2500 m.

3. **Schefflera myriantha** (Bak.) Drake in Journ. Bot., Paris **11**: 3 (1897).—Bamps in Fl. Afr. Centr., Aral.: 17, t. 4 (1974); in Bull. Jard. Bot. Nat. Belg. **44**: 135 (1974); in Distrib. Afr. Pl. **8**: map 241 (1974). Type from Madagascar.

Cussonia myriantha Bak. in Journ. Linn. Soc., Bot. **20**: 157 (1883). Type as above.

Schefflera polysciadia Harms in Engl., Pflanzenw. Ost-Afr.: 297 (1895).—Robyns & Tournay, Fl. Parc Nat. Alb. **1**: 690 (1948).—Brenan & Greenway, T.T.C.L.: 61 (1949).—Brenan in Mem. N.Y. Bot. Gard. **8**: 447 (1954).—Tennant in F.T.E.A., Aral.: 21 (1968). Syntypes from Zaire, Uganda (Ruwenzori) and Tanzania.

A robust liane or tree of up to 16 m. or a shrub. Leaves digitate, with long petioles of up to 22 cm. long; leaflets (4)5–7(8), up to 28 × 12·5 cm., narrowly elliptic to subrotund, glabrous, coriaceous, entire or with a slightly undulate margin; apex long-acuminate, more rarely shortly mucronate, the base usually subcordate, but sometimes entire; petiolules up to 8·5 cm. long; lateral veins obvious, numerous and very regular. Stipules conspicuous and persistent. Inflorescence consisting of a group of up to 6 primary branches, each of which bears a terminal triple-branched umbel, and one or more similar but smaller lateral umbels. The primary axis up to 18 cm., often very robust and even woody. The primary umbels bear secondary umbellules, which themselves bear 6–8(12) tertiary rays (the pedicels of the individual flowers). Bracts and bracteoles of the primary and secondary umbels often readily seen in flowering inflorescences, but often less so, or even obsolete in fruit (possibly caducous). Styles 5, at first completely united, but later separating above to reveal the stigmatic surfaces for 1 mm. or a little more. Fruit subspherical, c. 5 mm. in diameter, with a rather truncate apex (the base of the stylopodium).

Malawi. N: Nyika Plateau, fr. 17.viii.1946, *Brass* 17278 (K; SRGH).

Also occurring in E. tropical Africa, Zaire, the Comoro Is., and Madagascar. In riverine and montane forests at altitudes up to 3500 m.

4. **Schefflera umbellifera** (Sond.) Baill. in Adansonia **12**: 147 (1878).—Bernardi in Candollea **24**: 93 (1969).—Bamps in Bull. Jard. Bot. Nat. Belg. **44**: 136 (1974); in Distrib. Afr. Pl. **8**: map 244 (1974). Type from S. Africa.

Cussonia umbellifera Sond. in Linnaea **23**: 49 (1850); in Harv. & Sond., F.C. **2**: 570 (1862).—Bak. f. in Journ. Linn. Soc., Bot. **40**: 76 (1911).—Eyles in Trans. Roy. Soc. S. Afr. **5**: 433 (1916).—Burtt Davy, F.P.F.T. **2**: 514 (1932).—Steedman, Trees etc. S. Rhod.: 58 (1933).—Burtt Davy & Hoyle, N.C.L.: 32 (1936). Type as above.

Tree of up to 15 m. high. Leaves digitately compound, with relatively uniform and regular leaflets. Petiole up to c. 25 cm. long, glabrous. Leaflets elliptic to broadly elliptic (to oblanceolate), up to 12 × 7 cm., usually with distinct petiolules up to 4 cm. long, rarely subsessile, coriaceous, glabrous or with minute scattered hairs, dark glossy green above, paler beneath; margin subentire, slightly undulate; apex obtuse, retuse, emarginate with a terminal mucro or acute; base cuneate (sometimes obliquely so). Inflorescence a paniculate complex of umbels; bracts sometimes present (probably caducous), primary branches up to 15 cm. long, bearing a large terminal compound umbel and much smaller, simple, lateral umbels. Rays of terminal umbels 6–7, up to 4 cm. long (in fruit). Umbellules with 6–14 flowers on pedicels c. 1 cm. long; bracteoles obsolete. Stylopodium indistinct. Fruit globose, sometimes slightly flattened above and below, up to 7 mm. in diameter, glabrous.

Rhodesia. E: Melsetter, fl. 22.ix.1960, *Rutherford-Smith* 170 (BM; SRGH).
Malawi. S: Mt. Mlanje, Chambe Plateau, fl. 23.iv.1957, *Chapman* 361 (BM; K).
Mozambique. Z: Gúruè, Pico Namuli, fl. 9.iv.1943, *Torre* 5127 (LISC). MS: Chimanimani Mts., fl. 6.ii.1958, *Hall* 383 (BM; SRGH).
Also in the Transvaal, Natal and the Cape Province. In mountain forest up to c. 2000 m.

Two varieties of *S. umbellifera* have been recognized by Tennant in Kew Bull. **14**: 220 (1960); they are distinguished as follows:

Leaflets abruptly tapering to the apex, apex generally acute but very rarely emarginate; petiolules 2–12 mm. long - - - - - - - - var. *buchananii*
Leaflets not tapering abruptly at the apex, apex generally obtuse, retuse or emarginate (rarely acute); petiolules 25–30(40) mm. long - - - - var. *umbellifera*

Var. *buchananii* (Harms) Tennant is based on *Cussonia buchananii* Harms in Engl., Bot. Jahrb. **26**: 251 (1899) pro parte quoad specim. a, Nyasaland, *Buchanan* 295. It has been regarded as confined to Mt. Mlanje and Blantyre, and linked by closely similar forms occurring on the border mountains between Rhodesia and Mozambique to the typical variety which has been described as occurring throughout the remainder of the species' range. I consider, however, that these variants are not sufficiently distinct to merit continued recognition.

2. POLYSCIAS J. R. & G. Forst.

Polyscias J. R. & G. Forst., Char. Gen. Pl.: 63, t. 32 (1776) emend. Harms in Engl. & Prantl, Pflanzenfam. **3**, 8: 43 (1894).
Panax sensu Hiern in F.T.A. **3**: 27 (1877).

Trees, often tall and with noticeably regular branching. Leaves petiolate, pinnately compound (bipinnate outside our area), glabrous to densely tomentose with stellate (or simple) hairs. Leaflets simple (or divided in species outside our area), more or less entire. Inflorescence paniculate, regularly compound, racemosely branched in species of our area, the ultimate divisions of racemules or umbellules. Calyx forming a shallow cup, subentire or shallowly 5-toothed. Petals and stamens 5; stylopodium not well developed. Ovary 2-locular (sometimes 5-locular in other areas); styles 2 (or 5), divergent for nearly all their length. Fruit terete (or slightly compressed). Seeds rounded or laterally compressed, ellipsoid, smooth or sometimes ribbed; endosperm non-ruminate.
A genus of some 100 species distributed throughout the tropics of the Old World.

Flowers in racemosely arranged racemules - - - - - - 1. *fulva*
Flowers in racemosely arranged umbellules - - - - - - 2. *albersiana*

1. **Polyscias fulva** (Hiern) Harms in Engl. & Prantl, Pflanzenfam. **3**, 8: 45 (1894).—Steedman, Trees etc. S. Rhod.: 58 (1933).—Burtt Davy & Hoyle, N.C.L.: 32 (1936).—Brenan & Greenway, T.T.C.L.: 60 (1936).—Keay in F.W.T.A. ed. 2, **1**: 750 (1958).—F. White, F.F.N.R.: 313 (1962).—Tennant in F.T.E.A., Aral.: 12

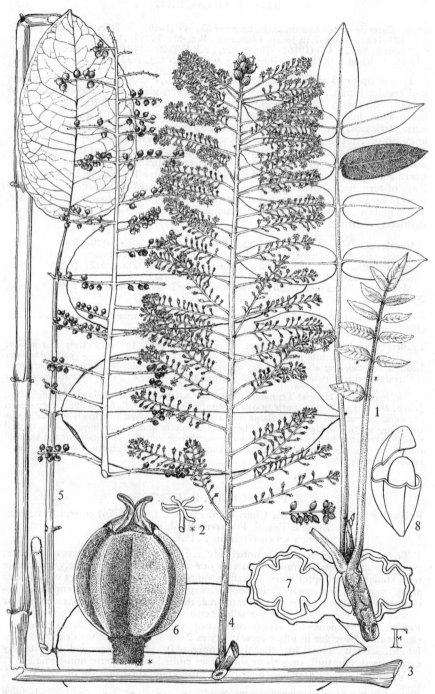

Tab. 167. POLYSCIAS FULVA. 1, young shoot (×½); 2, stellate hair from shoot (×33); 3, mature leaf (×½); 4, branch of inflorescence (×½); 5, branch of infructescence (×½); 6, fruit (×6); 7, fruit in cross-section (×9); 8, seed (×6), 1–5 from *Chase* 1971, 6–8 from *Gossweiler* 6208.

(1968).—Bamps in Fl. Afr. Centr., Aral.: 2 (1974); in Bull. Jard. Bot. Nat. Belg.
44: 122 (1974); in Distrib. Pl. Afr. **8**: map 228 (1974). TAB. **167**. Type from
Fernando Po.
  Panax fulvum Hiern in F.T.A. **3**: 28 (1877). Type as above.
  Polyscias malosana Harms in Notizbl. Bot. Gart. Berl. **3**: 20 (1900). Type:
Malawi, Mt. Malosa, *Whyte* (B, holotype †; K, isotype).
For complete synonymy see Tennant (1968).

A tall tree up to 30 m. high with a long unbranched trunk, eventually dividing
into primary and secondary branches in a very regular manner. Leaves up to
80 cm. long, regularly pinnate, usually imparipinnate though examples lacking the
terminal leaflet are encountered. Leaflets (3)6–7(12)—paired, up to 12(17) × 5(7·5)
cm., sessile or with petioles of up to c. 10 mm. long, coriaceous (often strongly so),
lanceolate-ovate (the terminal leaflet often broader and more ovate than the lateral
paired ones); apices acute, often with a small mucro, rarely obtuse and emarginate;
bases rounded to subcordate (rarely subcuneate); margins entire to slightly
undulate, more or less convolute, becoming leathery and subglabrous, the lower
surface more or less densely tomentose. Inflorescence a complex of compound
racemosely arranged racemes. Primary axes up to 40(70) cm. long, tomentose to
subglabrous, bearing numerous laterals of 8(12) cm. long, densely crowded with
flowers, but elongating somewhat as the fruit matures. Pedicels up to 5 mm. long;
petals greenish to creamy white, flowers honey-scented. Stylopodium somewhat
depressed, styles widely divergent for most of their length. Fruit broadly-ovoid
to subglobose, 3–4(6) × 3–5 mm., glabrous to slightly pubescent, somewhat
ribbed.

Zambia. N: Chinsali, fl. 27.ix.1967, *Fanshawe* 10185 (K). W: Mwinilunga Distr.,
Muzera R., 16·1 km. W. of Kakoma, *White* 3428 (K). **Rhodesia.** E: Umtali Distr.,
Vumba Mts., Thordale, fl. ii.1950, *Chase* 1971 (BM; COI; K; LISC; SRGH).
Malawi. N: S. Vipya, Lwanjati Hill, Champila, fl. 27.i.1956, *Chapman* 274 (BM; COI;
K). C: Ncheu Distr., Dzonze Mt., Kirk Range, fr. 3.vi.1961, *Chapman* 1347 (SRGH).
Mozambique. N: Ribáuè, Serra de Chinga, fl. 12.xii.1967, *Torre & Correia* 16478
(LISC). Z: Gúruè, E. of Pico Namuli, fl. 8.xi.1967, *Torre & Correia* 16007 (LISC).
MS: Manica, Serra Zuira, Planalto Tsetserra, fl. 13.xi.1965, *Torre & Pereira* 12975
(LISC).
  Very widely distributed throughout tropical Africa. In lowland forest, gallery forest
and especially in mountain forest at altitudes up to 2500 m.

2. **Polyscias albersiana** Harms in Engl., Bot. Jahrb. **33**: 182 (1902).—Tennant in
F.T.E.A., Aral.: 15 (1968).—Bamps in Bull. Jard. Bot. Nat. Belg. **44**: 121 (1974);
in Distrib. Pl. Afr. **8**: map 226 (1974). Type from Tanzania.
  Sciadopanax albersiana (Harms) R. Viguier in Bull. Soc. Bot. Fr. **52**: 305 (1905).
Type as above.

A tree up to 20 m., with a regularly branched axis and with pale greyish bark,
especially in the younger parts. Leaves in terminal or subterminal groups,
imparipinnate, up to 60 cm. long with 7–11 pairs of lanceolate to ovate, chartaceous
to coriaceous leaflets. Leaflets up to 13 × 5·5 cm.; margins entire, slightly reflexed;
apices shortly and bluntly acuminate to long and narrowly acuminate; bases
rounded to cordate. Inferior surface glabrous to mealy with a stellate pubescence;
superior surface glabrous, somewhat glossy. Inflorescence of umbellules arranged
rather irregularly in racemes grouped in terminal clusters at the ends of branches.
Racemes up to 30 cm. long, with the peduncles of the umbellules 1–4·5 cm. long.
Flowers pale greenish-yellow, with pedicels 3–15 mm. long. Stylopodium
shallowly conical; styles 2, persistent in fruit. Fruit at first ellipsoid, becoming
broadly ovoid with maturity, up to 7 mm., distinctly ribbed, glabrous.

Mozambique. N: Amaramba, Sierra Mitucué, 20 km. from Nova Freixo, *Torre &
Paiva* 10605 (LISC).
  Only otherwise known from northern and western Tanzania. In upland forest from
1000–2000 m.

Only known in the F.Z. area from the above gathering. The holotype was destroyed in
Berlin and the interpretation here adopted follows that of Tennant in F.T.E.A., which was
based on material from near the type locality.

3. CUSSONIA Thunb.

Cussonia Thunb. in Nov. Act. Soc. Sci. Ups. **3**: 210, t. 12 & 13 (1780).

Trees, shrubs (rarely suffrutices), glabrous to puberulous (more rarely densely pubescent, especially in immature states). Leaves conspicuously petiolate, simple and palmate to digitately compound, often grouped in an " umbrella " arrangement at the ends of branches. Leaflets very variable in shape and marginal definition; apices often long-acuminate. Stipules often quite conspicuous. Inflorescence usually spicate but may be racemose. Floral bracts small or obsolete. Flowers 4–8 mm. in diameter with (4)5 greenish petals. Calyx forming a shallow cup, sometimes with a 4–5-dentate margin. Stamens (4)5. Disk flat, depressed or sometimes fused with the styles to form a well-developed conical stylopodium. Ovary 2-locular (rarely with supernumerary carpels). Styles 2 (3–4 have been rarely reported), connivent for most of their length and often with very short divergent stigmatic surfaces at their apex. Fruit subglobose, urceolate, obovoid or even obconical to wedge-shaped (owing to congestion in densely packed spikes), sometimes fleshy. Seed ovoid or somewhat compressed; endosperm ruminate.

Flowers sessile or subsessile:
 Inflorescence a raceme of small pedunculate spikes - - - - 5. *paniculata*
 Inflorescence a simple spike:
  Spike with a very distinct peduncle:
   Spike very dense; fruit very numerous and so closely packed that they are distinctly
    angular - - - - - - - - - - 6. *spicata*
   Spike more open, with fewer broadly ovoid fruit - - - 7. *arenicola*
  Spike sessile or merging gradually into the peduncle:
   Rhizomatous suffrutex - - - - - - - - 2. *corbisieri*
   Trees and large shrubs:
    Leaves digitately compound (may be palmate in juvenile shoots); spikes long and
     narrow, up to 50 cm. long in flower - - - - 1. *arborea*
    Leaves palmate; spikes short, up to 10 cm. long in flower - 3. *natalensis*
Flowers distinctly pedicellate:
 Leaves digitately compound, leaflets simple; petiolules without obvious wings
                4. *zimmermannii*
 Leaves digitately compound, leaflets tending to pinnate or pinnatifid; petiolules with
  obvious triangular wings:
   Pedicels 5–12 mm. long; trees - - - - - - 8. *zuluensis*
   Pedicels 1–2(4) mm. long; low shrubs - - - - - - 7. *arenicola*

1. **Cussonia arborea** Hochst. ex A. Rich., Tent. Fl. Abyss. **1**: 356, t. 56 (1847).—
F. White, F.F.N.R.: 312 (1962).—Tennant in F.T.E.A., Aral.: 4 (1968).—Bamps
in Fl. Afr. Centr., Aral.: 8, t. 1B (1974); in Bull. Jard. Bot. Nat. Belg. **44**: 104
(1974); in Distrib. Pl. Afr. **8**: map 212 (1974). Type from Ethiopia.
 Cussonia kirkii Seem. in Journ. Bot. Lond. **4**: 299 (1866).—Hiern in F.T.A. **3**: 32
(1877).—Steedman, Trees etc. S. Rhod.: 58 (1933).—Burtt Davy & Hoyle, N.C.L.:
32 (1936).—Tennant in Kew Bull. **14**: 222 (1960).—F. White, loc. cit.—Fanshawe,
Fifty Common Trees of Zambia: 34 (1968).—Tennant, op. cit.: 6 (1968).—Type:
Mozambique, Mt. Morrumbala, *Kirk* (K, holotype).
 A complete synonymy of this widespread species is given by Bamps in Bull. Jard. Bot.
Nat. Belg. (1974).

Robust tree up to 11 m. Leaves digitately compound (palmate in juvenile states). Petiole up to 50(87) cm. long, usually glabrous but sometimes somewhat puberulent in patches. Leaflets 5–9(10), up to 27 × 8 cm., sessile but with long, very narrowly tapering cuneate bases, chartaceous to somewhat coriaceous, glabrous or with occasional hairs; apex acute to acuminate; margins more or less regularly serrate-crenate to more or less entire, rarely deeply and jaggedly cut. Flowering spikes up to 26 together, but frequently of c. 10–12, at first short and congested, but elongating as the flowers mature, and further as the fruit develop, up to 50 cm. in length. Axis of the spike densely puberulous; basal bracts broadly triangular. Flowers sessile with floral bracts that are conspicuous in the un-expanded spikes, but insignificant and more or less caducous before maturity. Stylopodium quite well developed; stigmatic surfaces very small, scarcely freely diverging at the tip. Fruit ovoid-cylindrical to subglobose, 4–5·5 × up to 4 mm., glabrous or more rarely puberulous.

Zambia. N: Mbala (Abercorn) to Kawimbi road, fl. 9.xi.1952, *Robertson* 241 (K-SRGH). W: Mwinilunga Distr., Kalenda Plain, S. of Matonchi, fl. 18.xii.1937, *Milne*;

Redhead 3718 (BM; K; LISC). C: Mt. Makulu, 19·3 km. S. of Lusaka, fr. 1.xi.1957, *Angus* 1474 (BM). **Rhodesia.** N: Gokwe, fl. 18.xii.1963, *Bingham* 997 (BM; K; SRGH). C: Salisbury, fr. 4.xii.1908, *Rand* 1414 (BM). E: Umtali, fl. 1.xii.1953, *Chase* 5151 (BM; K; SRGH). S: Victoria, 1909, *Monro* 718 (BM). **Malawi.** N: 24·1 km. N. of Mzimba, fl. 17.xi.1958, *Robson* 652 (K). C: Dedza Distr., between Chiwao and Kanjoli, fr. 12.xi.1967, *Jeke* 53 (K; SRGH). S: Shire Highlands, fl. no date, *Buchanan* s.n. (K). **Mozambique.** N: Ribáuè near Posto Agricola, fl. 5.xi.1942, *Mendonça* 1265 (LISC). Z: Serra do Sidi, fr. 10.ii.1905, *Le Testu* 656 (BM). T: Chicoa, Serra de Songa, fr. 31.xii.1965, *Torre & Correia* 13940 (LISC). MS: Upper R. Buzi, fl. 24.xii.1905, *Swynnerton* 159 (BM).

Very widely distributed throughout tropical Africa. In open woodland and wooded grassland at altitudes up to 2500 m.

2. **Cussonia corbisieri** De Wild. in Fedde, Repert. **13**: 20 (1914).—Tennant in Kew Bull. **14**: 401 (1960).—F. White, F.F.N.R.: 312 (1962).—Bamps in Fl. Afr. Centr., Aral.: 11, t. 3 (1974); in Bull. Jard. Bot. Nat. Belg. **44**: 109 (1974); in Distrib. Pl. Afr. **8**: map 215 (1974). Type from Zaire.

A robust suffrutex arising from a woody subterranean rhizomatous rootstock, the whole plant less than 1 m. tall. Leaves palmate (rarely subdigitate). Petiole up to 60 cm. long, arising directly from the crown of the rootstock, subglabrous to fairly densely pubescent with crisped hairs. Lamina up to 30 × 30 cm., chartaceous to subcoriaceous, very densely covered with crisped hairs when young, especially on the veins; indumentum appearing much sparser on the mature leaf owing to expansion of the lamina. Lobes (5)6(7), obovate, apex acute to long-acuminate; margin crenate-serrate to rather irregularly serrate-dentate. Inflorescence a group of spikes similar to those of *C. arborea* arising directly from the crown of the rootstock. Spikes at first very crowded, but elongating and becoming less dense at the fruiting stage, up to 60 cm. long. Basal bracts densely puberulous. Flowers sessile, with narrowly linear, pubescent floral bracts. Stylopodium well developed, conical, c. 2 mm. long, stigmatic surfaces free for only a very short distance. Fruit ovoid, slightly ribbed, 8–10 × 6–8 mm.

Zambia. B: 83·7 km. E. of Mankoya on road to Kafue Hook, fl. 21.xi.1974, *Drummond & Cookson* 6724 (SRGH). W: 88 km. W. of Chingola on Solwezi road, veg. 16.iii.1974, *Drummond & Rutherford-Smith* 6945 (K; SRGH). C: Mt. Makulu, fl. 27.ii.1965, *van Rensburg* 3082 (BM; SRGH). E: Sasare to Petauke at 11·3 km., fl. 9.xii.1958, *Robson* 882 (K). S: 40·2 km. from Mumbwa on road to Kafue Hook, fl. 7.xi.1959, *Drummond & Cookson* 6194 (K; SRGH).

Known also, at present, only from Zaire. In open woodland and wooded grassland at elevations up to 2000 m.

This species is very reminiscent of the arborescent *C. arborea* and it seems certain that it is closely related to that species. The nature of this relationship demands further investigation under field conditions. The suffrutescent habit may well be a response to special ecological factors. Collectors have noted that the juvenile leaves of *C. arborea* may be palmate and much more pubescent than those of the mature tree, thus closely resembling those of *C. corbisieri*. Collectors and others with opportunities to make field observations on these species should try to resolve this question.

3. **Cussonia natalensis** Sond. in Harv. & Sond., F.C. **2**: 568 (1862).—Eyles in Trans. Roy. Soc. S. Afr. **5**: 433 (1916).—Burtt Davy, F.P.F.T. **2**: 514 (1932).—Steedman, Trees etc. S. Rhod.: 58 (1933).—Brenan & Greenway, T.T.C.L.: 59 (1949).— Bamps in Bull. Jard. Bot. Nat. Belg. **44**: 112 (1974); in Distrib. Pl. Afr. **8**: map 218 (1974). Type from S. Africa.

Small tree up to 10 m. Leaves palmate, coriaceous, up to 14 × 22 cm.; lobes (3)5, lanceolate to linear-lanceolate, 1–4 cm. wide, with conspicuous midrib, lateral veins and reticulum of veinules. Petiole up to 18 cm. long. Margin regularly serrate (to subentire); apex long-acuminate; base broadly obtuse to deeply cordate. Inflorescence a group of spikes without obvious basal bracts, up to 10 cm. long in flower; peduncle c. ⅓ of the total length of the spike; each flower subtended by a minute deltate floral bract, some of which are also present on the sterile part of the peduncle. Calyx forming a relatively prominent cup. Spikes elongating slightly in fruit. Stylopodium not well developed, narrowly conical; styles free at the apex for c. 5 mm. Fruit subglobose with a flattened top, glabrous.

x

Rhodesia. W: Rhodes Estate, Matopos Hills, fr. ix.1905, *Gibbs* 107 (BM; K). C: Marandellas, veg. 15.ii.1956, *Guy* 1/56 (SRGH). E: Umtali Distr., Lizard Rocks, 20·9 km. S. of Melsetter, fl. 21.xi.1948, *Chase* 1575 (BM; COI; LISC; SRGH).
Also occurring in the Transvaal, Natal and Swaziland. In wooded grassland.

4. **Cussonia zimmermannii** Harms in Engl., Bot. Jahrb. **53**: 361 (1915).—Brenan & Greenway, T.T.C.L.: 59 (1949).—Tennant in F.T.E.A., Aral.: 7 (1968).—Bamps in Bull. Jard. Bot. Nat. Belg. **44**: 119 (1974); in Distrib. Pl. Afr. **8**: map 224 (1974). Syntypes from Tanzania.

Tree up to 20(45) m. high. Leaves digitately compound; petioles glabrous, commonly up to 20 cm. long but lengths of 53 cm. have been recorded. Leaflets 5–7(9), up to 10(25) × 4(8) cm., sessile, chartaceous, glabrous or with a few scattered hairs, oblanceolate (to lanceolate) or ovate (to obovate); apex long-acuminate; base cuneate to narrowly cuneate (sometimes almost creating the impression of a distinct petiole); margins crenate-serrate to subentire. Inflorescence a group of up to 12 spike-like racemes, up to 35 cm. long in fruit. Base of racemes densely clothed with overlapping bracts; floral bracts lanceolate (to ovate). Flowers distinctly pedicellate, pedicels 3–5 mm. Stylopodium well developed, broadly conical; styles short, with shortly recurved stigmatic surfaces. Fruit hemispherical, up to 6 mm. long, glabrous to very slightly puberulous.

Mozambique. N: Nacala, between Itoculo and Netia, fl. 16.x.1948, *Barbosa* 2445 (LISC; LM).
Also recorded from Kenya and Tanzania. In forests at altitudes up to 400 m.

5. **Cussonia paniculata** Eckl. & Zeyh., Enum. Pl. Afr. Austr. Extratrop.: 355 (1837).—Sond. in Harv. & Sond., F.C. **2**: 569 (1862).—Bamps in Bull. Jard. Bot. Nat. Belg. **44**: 114 (1974); in Distrib. Pl. Afr. **8**: map 220 (1974). Type from S. Africa.

Small tree up to 5 m. high. Leaves digitately compound, with very long petioles up to 50 cm. long. Leaflets 7–9(13), up to 27 cm. long, sessile but with long narrow cuneate bases, in some cases becoming almost petiolate, deeply pinnatifid, very variable with the sinus extending sometimes almost to the midrib, but in others only reaching halfway from the margin; lobes equally variable in width. Apex broadly acute with a short terminal mucro; margin sometimes with a few small teeth. Lamina glabrous and with a distinctly glaucous sheen beneath. Inflorescence a group of panicles, each bearing racemosely arranged spikes in subopposite pairs. Main axes up to 35 cm. long, glabrous, finely but distinctly grooved. Spikes subtended by inconspicuous basal bracts and with distinct peduncles up to 5 cm. long. Fertile parts of the spikes at first clearly shorter than the peduncle, but later elongating and often exceeding the peduncle in fruit. Floral bracts minute, more or less lanceolate. Flowers sessile, stylopodium depressed. Fruit subglobose, distinctly fleshy and pea-like (showing clearly the 2 carpels when dry); mature fruit sometimes only setting rather irregularly on the spikes, with many infertile flowers.

Botswana. SE: Bon Accord Farm S. of Lobatsi, veg. 16.i.1960, *Leach & Noel* 108 (K; LISC; SRGH).
Widely distributed in S. Africa. In open wooded grassland at altitudes up to 2100 m.

6. **Cussonia spicata** Thunb. in Nova Acta Soc. Sci. Ups. **3**: 212, t. 13 (1780).—Sond. in Harv. & Sond., F.C. **2**: 568 (1862).—Hiern in F.T.A. **3**: 32 (1877).—Bak. f. in Journ. Linn. Soc., Bot. **40**: 76 (1911).—Eyles in Trans. Roy. Soc. S. Afr. **5**: 433 (1916).—Burtt Davy, F.P.F.T. **2**: 514 (1932).—Steedman, Trees etc. S. Rhod.: 58 (1933).—Burtt Davy & Hoyle, N.C.L.: 32 (1936).—Brenan & Greenway, T.T.C.L.: 59 (1949).—F. White, F.F.N.R.: 312 (1962).—Tennant in F.T.E.A., Aral.: 3 (1968). TAB. **168**. Type from S. Africa.

Tree up to 17 m. high, but commonly 5–10 m., often with an unbranched trunk, but sometimes with sparse branches. Leaves digitately compound (but palmate or even simple and ± entire in juvenile forms); petiole up to 50(67) cm. long, usually glabrous but sometimes slightly pubescent. Leaflets 6–9(12), up to 30(35) × 15(19) cm., sessile or with distinct petiolules, coriaceous, glabrous to subglabrous; lobes very variable in outline. Leaves of mature trees characterized by the presence of prominent wedge-shaped decurrent wings on the petiolules (connate stipules). Spikes very dense, with obvious distinct peduncles and basal bracts. Flowers sessile, with small narrowly lanceolate floral bracts; stylopodium short. Fruit

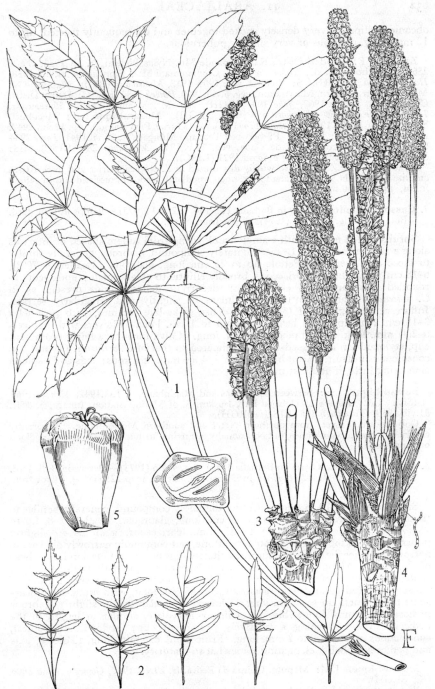

Tab. 168. CUSSONIA SPICATA. 1, mature leaves (×⅓) *Stolz* 1599; 2, individual leaflets showing range of variation (×⅓) *Schlieben* 3859 & *Bruce* 782; 3–4, shoots with inflorescences and infructescences in one (3), partly removed to show arrangment of fruits (×½), 3 from *Hutchinson & Gillett* 3209, 4 from *Hall* 367; 5, fruit (×4); 6, fruit in cross-section (×4), 5–6 from *Hutchinson & Gillett* 3209.

obconical-ellipsoid, very densely packed together and consequently faceted, up to 12 mm. long, glabrous or very slightly puberulous.

Zambia. E: Lundazi Distr., Kangampande Mt., Nyika Plateau, fr. 3.v.1952, *White* 2569 (K). **Rhodesia.** E: Melsetter Distr., Chimanimani Mts., Stonehenge, fr. 5.ii.1958, *Hall* 367 (BM; SRGH). S: Victoria, veg. 1909–1912, *Monro* s.n. (BM). **Malawi.** N: Mzimba Distr., Mzuzu, Marymount, veg. 6.viii.1970, *Pawek* 3680 (K). C: Dedza Distr., Chiwao Hill, fl. 8.ix.1967, *Jeke* 104 (SRGH). S: Mt. Mlanje, Tuchila Plateau, fr. 7.ix.1956, *Newman & Whitmore* 689 (BM; SRGH). **Mozambique.** N: Mecuburi, Serra Chinga, fr. 6.v.1968, *Macedo* 3174 (LM). Z: Gúruè, Pico Namuli, veg. 12.viii.1949, *Andrada* 1855 (COI; LISC). T: between Tete and Múngári, fr. 26.vii.1941, *Torre* 2944 (LISC). MS: Baruè, Serra de Choa, veg. 17.ix.1942, *Mendonça* 286 (LISC). SS: Massinga, veg. xii.1936, *Gomes e Sousa* 1921 (LISC).

In addition to F.Z. area, also widely distributed in E. tropical Africa and in S. Africa and the Comoro Is. In montane forest, open woodland, wooded grassland and in gallery forest, up to 2500 m.

7. **Cussonia arenicola** Strey in Bothalia **11**: 194, t. 2 & 3 (1973).—Bamps in Bull. Jard. Bot. Nat. Belg. **44**: 108 (1974). Type from Natal.

Shrubs of 1–2 m. tall with a single trunk rising from tubers spaced at intervals along a root or rhizome. Leaves digitately compound, glabrous, generally similar to those of *C. spicata.* Petiole up to 25 cm. long, glabrous. Leaflets 4–7(12), 6–18 cm. long, slightly coriaceous, broadly elliptic to oblanceolate; apex acute to rounded and sometimes emarginate; decurrent deltate wings present as in *C. spicata.* Margins entire to undulate to slightly and irregularly serrate. Inflorescence a group of (5)8–15(23) dense spike-like racemes, with peduncles 3–18 cm. long and with basal bracts up to 1 cm. long. Flowers with pedicels 0–2(4) mm. long, floral bracts c. 2 mm. long. Calyx forming a fairly prominent cup, stylopodium depressed-conical compared to that of *C. zuluensis* (8) and less conspicuous in fruit. Fruit broadly ovoid, 4 × 4 mm., glabrous; axis and pedicels elongating little, if at all, at maturity.

Mozambique. SS: between Vilanculos and Mapinhane, fl. 1.ix.1942, *Mendonça* 67 (LISC). LM: Inhaca I., Delagoa Bay, 1·6 km. S. of Marine Station, Inhaca, fl. & fr. 31.viii.1959, *Watmough* 305 (LISC; SRGH).

Only known, as yet, from northern Natal and southern Mozambique. Apparently confined to coastal sand dunes, but occasionally occurring further inland in open woodland on sandy soils.

8. **Cussonia zuluensis** Strey in Bothalia **11**: 195, t. 4 & 5 (1973).—Bamps in Bull. Jard. Bot. Nat. Belg. **44**: 120 (1974); in Distrib. Pl. Afr. **8**: map 225 (1974). Type from Natal.

Small tree up to 10 m. high. Leaves digitately compound, generally similar to those of *C. spicata.* Petiole up to 35 cm. long, glabrous. Leaflets 5–8, up to 15 × 7 cm., with petiolules up to 35 mm., coriaceous, glabrous to slightly puberulent; the lobes very variable in outline, but commonly narrowly obovate to narrowly oblanceolate, with decurrent deltate wings on the petiole similar to those of *C. spicata.* Margins remotely and irregularly serrate, slightly undulate; apex acute, obtuse or even deeply emarginate; base cuneate. Inflorescence a group of densely-flowered pedunculate racemes up to 25 cm. long, with numerous linear-acute basal bracts up to 2 cm. long. Flowers with glabrous or slightly scabrous pedicels 5–12 mm. long; bracteoles linear-lanceolate, similar in length to the pedicels. Calyx forming a relatively prominent cup; stylopodium conical, conspicuous in fruit, up to 2 mm. long. Fruit ovoid, c. 6 mm. long, glabrous; axis and pedicels probably elongating somewhat to maturity.

Mozambique. LM: Maputo, Quinta da Pedra, fr. 29.viii.1947, *Gomes e Sousa* 3606 (K; SRGH).

Widely distributed in Natal, recorded from Swaziland and southern Mozambique. In open forest and scrub at low altitudes on sandy soils; frequently in river valleys.

92. ALANGIACEAE

By J. F. M. Cannon

Trees or shrubs, sometimes spiny (but not in F.Z. area). Leaves alternate, entire to very slightly undulate or lobed; distinctly petiolate, often somewhat asymmetric at the base, stipules 0. Flowers ☿, regular, in few-flowered (in F.Z. area) axillary cymes with articulated pedicels. Calyx truncate or with 4–10 teeth. Petals 4–10, valvate, linear, becoming strongly recurved at maturity, sometimes slightly coherent at the base. Stamens the same number as and alternating with the petals, or up to 2–4 times as many; free or connate at the base, more or less villous adaxially; anthers 2-locular, linear. Disk subglobose, flattened above. Ovary inferior, 1–2-locular, style simple, clavate or 1–3-lobed; ovules solitary, pendulous in the loculi, with 2 integuments. Fruit drupaceous, crowned with the remains of the disk and sepals, 1(2)-seeded. Seeds with the embryo about equalling the endosperm.

A monogeneric family of c. 22 species from the tropics and subtropics of the Old World. It was formally included in the Cornaceae but is distinguished by its articulate pedicels, valvate aestivation and normally uniovulate 1-locular ovary.

ALANGIUM Lam.

Alangium Lam., Encycl. Méth. Bot. **1**: 174 (1783).

Characters as for family.

Alangium chinense (Lour.) Harms in Ber. Deutsch. Bot. Ges. **15**: 24 (1897).—
Bloembergen in Bull. Jard. Bot. Buitenz., Sér. 3, **16**: 169 (1939).—Brenan, T.T.C.L.:
22 (1949).—Verdcourt in F.T.E.A., Alang.: 3 (1958).—F. White, F.F.N.R.: 276
(1962). TAB. **169**. Type from Indo China.
    Stylidium chinense Lour., Fl. Cochinch. **1**: 221 (1790). Type as above.
    Marlea begoniifolia Roxb., Pl. Corom. **3**: 80, t. 283 (1820). Type an illustration of
a plant from India.
    Alangium begoniifolium (Roxb.) Baill., Hist. Pl. **6**: 270 (1876). Type as above.
    Alangium begoniifolium subsp. *eubegoniifolium* Wangerin in Engl., Pflanzenr. **4**,
220b: 21 (1910). Type as above.

A tree of 9–24 m. high. Twigs pubescent when immature and then glabrescent. Leaves (4)10–15(19) × (2·5)4–8(10) cm., varying from narrowly elliptic to broadly ovate, apex acute to mucronate, base rounded or subcuneate to subcordate, often markedly oblique; margins entire to slightly undulate (palmately lobed leaves have been recorded from juvenile plants and from coppiced shoots); petiole 0·5–2·5 cm. long, glabrous or pubescent. Venation typically of 5 main veins ascending from the base of the leaf, some with major laterals and all connected by numerous tertiary veins running approximately parallel and at right angles to the primaries. Flowers with a dense golden pubescence on the exterior, white to yellow, reportedly sweet-scented; 5–8-merous, borne 3–10(23) in sparsely branched axillary cymes less than ½ the length of the subtending leaves. Pedicels 3–3·5 mm. long. Buds flask-shaped. Calyx-tube, 1·5 mm. long ± funnel-shaped, with a narrow spreading limb with 5 minute teeth. Petals strap-shaped, up to 13·5 × 1(1·5) mm., exterior pubescent, interior ± glabrous apart from an area c. ¼ of the way up from the base. Stamens the same number as the petals, 5–9 mm. long, pubescent at the base, anthers narrowly linear. Style pubescent, 7–10 mm. long, with a slightly-lobed stigma. Fruit 8–10 × (4)6–8(9) mm., globose or ellipsoidal, somewhat compressed and occasionally slightly eccentric, distinctly ribbed when dry, puberulous, 1–2-loculate, crowned with the disk and the remains of the sepals.

Zambia. W: Luakera Forest Reserve, source of the Kasombo, Mwinilunga Distr., fl. 5.xi.1952, *Holmes* 964 (K). **Rhodesia.** E: Mtarazi Falls, Inyanga, 1200 m., *Müller* 583 (K; SRGH). **Mozambique.** MS: Quinta da Fronteira, veg. 17.vi.1962, *Chase* 7767 (K; SRGH).
Widespread in E. Africa, Zaire, Angola, Cameroun and Fernando Po, also in tropical Asia, from India to China and Japan and S. to Indo China, Java and the Sunda Is.

Tab. 169. ALANGIUM CHINENSE. 1, flowering branchlet ($\times\frac{1}{2}$); 2, flower ($\times3$); 3, corolla and stamens ($\times3$); 4, gynoecium ($\times3$); 5, longitudinal section through gynoecium ($\times3$), 1–5 from *Gossweiler* 9926; 6, fruit ($\times2$); 7, longitudinal section through fruit ($\times3$); 8, cross-section through fruit ($\times3$), 6–8 from *Gossweiler* 7373.

Generally in upland and lowland areas of high rainfall, but in the F.Z. area so limited in distribution as to make ecological generalizations difficult.

The above description covers material in general, as this species is of very limited distribution within the area of Flora Zambesiaca.

93. CORNACEAE

By J. F. M. Cannon

Trees, shrubs, rarely perennial herbs or woody lianes. Leaves opposite or less commonly alternate, simple, exstipulate. Flowers ☿ or unisexual (then the plants usually monoecious or polygamodioecious), actinomorphic in cymes or panicles, umbels or rarely in capitula and then with large petal-like bracts. Calyx-tube adnate to the ovary, sepals 4–5. Petals 4–5 (rarely lacking), aestivation imbricate or valvate. Stamens the same number as the petals and alternating with them, anthers 2-celled dehiscing laterally or rarely introrsely. Ovary inferior, with (1)2–4 locules. Placentation usually axile, with 1 anatropous pendulous ovule in each loculus. Ovules with 1 integument. Style 1 or several rising from a glandular disk. Fruit typically a drupe, sometimes a berry. Seed with copious endosperm and a small embryo.

A family of c. 10 genera and 90 species, widely distributed in both tropical and temperate regions throughout the world. Includes many species of horticultural interest.

Leaf margins serrate (at least in the apical half), young leaves densely tomentose below, flowers ☿ - - - - - - - - - - - **1. Curtisia**
Leaf margins entire, leaves glabrous below, flowers unisexual - - **2. Afrocrania**

1. CURTISIA Ait.

Curtisia Ait., Hort. Kew. **1**: 162 (1789).—Wangerin in Engl., Pflanzenr. **IV**, 229: 29 (1910).

Small trees or shrubs with a dense tomentum when young, but indumentum becoming much reduced with age. Flowers tetramerous, ☿, arranged in panicles. Petals valvate, ovary 4-locular. Fruit drupaceous with 1 seed in each locule.

A monotypic genus confined to the southern part of Africa.

Curtisia dentata (Burm. f.) C. A. Sm. in Journ. S. Afr. For. Ass. no. **20**: 50 (1951). TAB. **170**. Type from S. Africa.
    Sideroxylon dentatum Burm. f., Rar. Afr. Pl. 235, t. 82 (1738); Fl. Cap. Prodr.: 6 (1768). Type as above.
    Curtisia faginea Aiton, Hort. Kew. **1**: 162 (1789).—Harv. in Harv. & Sond., F.C. **2**: 570 (1862).—Wangerin in Engl., Pflanzenr. **IV**, 229: 30 (1910).—Bak. f. in Journ. Linn. Soc., Bot. **40**: 76 (1911).—Eyles in Trans. Roy. Soc. S. Afr. **5**: 435 (1916).—Burtt Davy, F.P.F.T. **2**: 512 (1932).—Steedman, Trees etc. S. Rhod.: 60 (1933). Type from S. Africa.

A shrub or small tree up to 8(13) m. high. Twigs densely covered with a rusty tomentum when young, becoming dark purplish brown and finally greyish and more or less glabrous. Leaves opposite, up to 10 × 5·5 cm., simple, ovate (elliptic), coriaceous. Margins ± regularly serrate in the apical ½, the serrations becoming reduced in the basal ⅓ to subentire (very rarely the whole margin may be subentire); apex rounded to acute, occasionally with a small mucro; base rounded to cuneate, rarely slightly oblique. Upper surface of the lamina glossy, glabrous but with some hairs along the main vein and some laterals; lower surface densely rusty tomentose when young becoming ± glabrous (apart from the veins) with maturity. Venation penninerved, regular and conspicuous; petioles up to 2·5 cm. long, tomentose when young, becoming less so with age. Inflorescence of terminal panicles, the branches with a dense yellowish-brown tomentum. Bracts obvious, linear-lanceolate, tomentose, subtending the opposite primary branches of the panicle. Flowers sessile to very shortly pedicellate; calyx pubescent, adnate to the

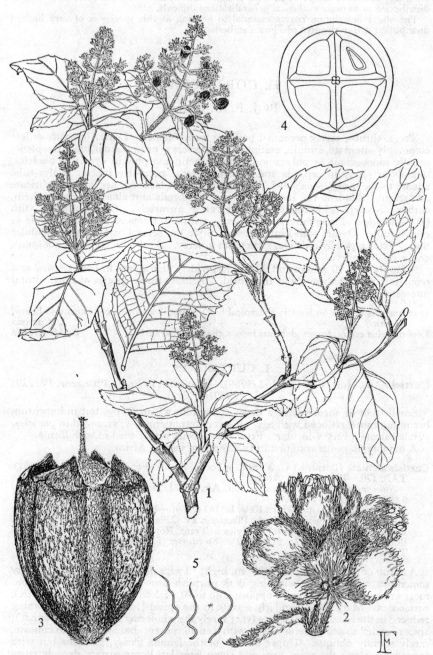

Tab. 170. CURTISIA DENTATA. 1, branches with leaves and inflorescences (×½); 2, group of flowers (×6), both from *Chase* 1463; 3, fruit (×9) *Gilliland* 1398; 4, cross-section of fruit (×6) *Swynnerton* M 37; 5, hair (×18) *Chase* 4168.

ovary but with 4 broadly triangular free teeth; petals 4, 1–1·5 × 0·75 mm., ovate, glabrous within. Stamens 4, alternating with and slightly shorter than the petals. Ovary inferior, turbinate, disk slightly conical, pubescent; style short with stigma with 4 short lobes. Fruit drupaceous, 5–7 × 3–5 mm., subglobose to ovoid, crowned with the remains of the calyx and style, 4-locular with 1 seed in each loculus.

Rhodesia. E: Umtali, Vumba, Leopard Rock Hotel, fl. 5.i.1949, *Chase* 1643 (BM; SRGH). **Mozambique.** MS: Chimanimani Mts., fr. 7.vi.1949, *Wild* 2943 (LISC; SRGH).
Otherwise only known from S. Africa. In montane forest and in stream gullies.

A timber tree valued for its heavy, hard, close-grained wood.

2. AFROCRANIA (Harms) Hutch.

Afrocrania (Harms) Hutch. in Ann. Bot., N.S. **6**: 89 (1942).
Cornus L. Subgen. *Afrocrania* Harms in Engl. & Prantl, Pflanzenfam. **3**, 8: 266 (1898).

Trees with opposite, simple, entire, petiolate leaves. Flowers tetramerous, dioecious. ♂ flowers arranged in cymules forming dense terminal clusters; ♀ flowers in terminal umbels. Inflorescences at first enclosed within 4 deciduous herbaceous bracts which provide protection for the developing flowers. Petals valvate. Ovary 2-locular with 1 ovule in each locule. Fruit drupaceous with the shrivelled remains of the calyx and style at the apex.
A monotypic genus confined to E. and SE. tropical Africa.
Distinguished from *Cornus* and related genera by its dioecious flowers arranged in terminal inflorescences protected by 4 deciduous bracts.

Afrocrania volkensii (Harms) Hutch. in Ann. Bot., N.S. **6**: 90 (1942).—Brenan & Greenway, T.T.C.L.: 173 (1949).—Verdcourt in F.T.E.A., Cornaceae: 1 (1958). TAB. **171**. Type from Tanzania.
Cornus volkensii Harms in Engl., Pflanzenw. Ost-Afr. **C**: 301 (1895).—Wangerin in Engl., Pflanzenr. **IV**, 229: 76, t. 19 (1910).—Burtt Davy & Hoyle, N.C.L.: 40 (1936). Type as above.

Tree up to 18 m. high. Immature stems with sparse appressed hairs, older parts becoming glabrous with age. Leaves 3·5–12(17·5) × 1·5–5(6·3) cm., narrowly elliptic to elliptic to lanceolate; apex acuminate; base regularly broadly to narrowly cuneate. Venation regular with prominent opposite to subopposite lateral veins which curve to run parallel with the midvein; both surfaces with a fine appressed pubescence when young, becoming glabrescent with age. Flowers dioecious in dense terminal inflorescences; the latter at first enclosed within a characteristic globular bud protected by 4 ovate-acuminate herbaceous bracts which are shed as the inflorescence expands. ♂ flowers arranged in cymules, giving the impression of a pseudo-umbel; peduncles and pedicels with appressed pubescence; sepals short, triangular; petals 2 × c. 0·75 mm., with scattered hairs; stamens 4, without filaments; pistillode with abortive narrowly conical style. ♀ flowers in umbels; ovary with adnate calyx-tube cylindrical, c. 3 mm. long, with appressed pubescence and 4 minute free calyx teeth; petals c. 3 × 1 mm., with scattered hairs or glabrous; staminodes 4, alternating with the petals; disk forming a distinct rim round the style which is short, thick and with irregularly clavate stigmatic enlargement at the apex. Fruit drupaceous, ellipsoid, 10–11 × 5 mm. in dried material, with irregularly distributed minute appressed hairs, crowned by the remains of the calyx, disk and style.

Rhodesia. E: Umtali, Banti North, veg. 3.x.1963, *Chase* 7817 (BM; K; LISC; SRGH). **Malawi.** N: Wovwe R., N. of Nganda, Nyika Plateau, fl. 18.ii.1956, *Chapman* 295 (BM; K). S: Chambe Plateau, Mlanje Mt., 20.iv.1957, *Chapman* 360 (BM). **Mozambique.** MS: Vila Pery, fl. 4.iv.1966, *Torre & Correia* 15695 (LISC).
Also in Kenya, Uganda, Tanzania and Zaire. In upland forest often following the course of mountain streams; 1200–3000 m.

Tab. 171. AFROCRANIA VOLKENSII. 1, leaf shoot with flowers of ♂ inflorescence in bud
(×1) *Stolz* 1979; 2, part of ♂ inflorescence (×3); 3, ♂ flower (×6); 4, petal of
♂ flower (×6); 5, ♂ flower with petals removed (×6), 2–5 from *Ross* 924; 6, young ♀
inflorescence enclosed by involucre (× 1·5) *Eggeling* 1050; 7, ♀ inflorescence (×3);
8, ♀ flower (×6); 9, petal of ♀ flower (×6); 10, ♀ flower cut open longitudinally to
show placentation (×6), 7–10 from *Proctor* 170; 11, fruit (×2); 12, diagrammatic
section of fruit (×2), both from *Dyson* s.n. (in spirit). From F.T.E.A.

INDEX TO BOTANICAL NAMES